CONTENTS

Preface **xi**

Acknowledgements **xiii**

PART 1: THE FOUNDATION

CHAPTER 1: THE ATOM

1.1	Matter	**3**
1.2	The atomic theory	**4**
1.3	The size of the atom	**6**
1.4	The electron	**6**
1.5	The atomic nucleus	**6**
1.6	The neutron	**8**
1.7	The fundamental particles	**8**
1.8	Nuclides and isotopes	**9**
1.9	Mass spectrometry	**9**
1.10	Nuclear reactions	**12**
	Questions on Chapter 1	**18**

CHAPTER 2: THE ATOM – THE ARRANGEMENT OF ELECTRONS

2.1	Light from atoms	**21**
2.2	Atomic spectra	**22**
2.3	Electrons in orbits	**23**
2.4	The wave theory of the atom	**27**
2.5	Electronic configurations of atoms	**30**
2.6	The history of the Periodic Table	**32**
2.7	Features of the Periodic Table	**34**
2.8	The electronic configurations of the elements	**38**
2.9	The repeating pattern of the elements	**41**
	Concept map: Atomic structure	**42**
	Questions on Chapter 2	**43**

CHAPTER 3: EQUATIONS AND EQUILIBRIA

3.1	Equations, equilibria and calculations – who needs them?	**46**
3.2	Formulae	**46**
3.3	Equations	**48**
3.4	Relative atomic mass	**51**
3.5	Relative molecular mass	**52**
3.6	The mole	**52**
3.7	Molar mass	**53**
3.8	Empirical formulae	**55**
3.9	Molecular formulae	**55**
3.10	Calculations of percentage composition	**55**
3.11	Equations for reactions of solids	**56**
3.12	Equations for reactions of gases	**57**
3.13	Concentration	**58**
3.14	Volumetric analysis	**62**
3.15	Equations for oxidation–reduction reactions	**65**
3.16	Oxidation number	**67**
3.17	Oxidation numbers and nomenclature	**71**
3.18	Titrimetric analysis, using redox reactions	**73**
3.19	Equilibrium	**74**
3.20	Chemical equilibria	**76**
	Concept map: Quantitative chemistry	**78**
	Questions on Chapter 3	**79**

CHAPTER 4: THE CHEMICAL BOND

4.1 Diamond and graphite – the difference lies in the bonds 83
4.2 Ions 84
4.3 The ionic bond 89
4.4 The covalent bond 97
4.5 Properties of ionic and covalent substances 101
4.6 Covalent compounds 103
4.7 The coordinate bond 106
4.8 Intermolecular forces 108
Concept map: The chemical bond 116
Questions on Chapter 4 117

CHAPTER 5: THE SHAPES OF MOLECULES

5.1 Haemoglobin 120
5.2 The arrangement in space of covalent bonds 120
5.3 Shapes of molecules: a molecular orbital treatment 127
5.4 Delocalised orbitals 130
Concept map: The shapes of molecules 133
Questions on Chapter 5 134

CHAPTER 6: CHEMICAL BONDING AND THE STRUCTURE OF SOLIDS

6.1 Profile: Dorothy Crowfoot Hodgkin (1910–1995) 136
6.2 X ray diffraction 137
6.3 The metallic bond 137
6.4 Ionic structures 138
6.5 Molecular solids 140
6.6 Macromolecular structures 140
6.7 Layer structures 142
6.8 The structure of metals 143
6.9 The structure of ionic crystals 145
6.10 Liquid crystals 147
Concept map: The structure of solids 146
Questions on Chapter 6 147

PART 2: PHYSICAL CHEMISTRY

CHAPTER 7: GASES

7.1 Air bags 151
7.2 States of matter 151
7.3 The gas laws 152
7.4 Avogadro's hypothesis 154
7.5 The ideal gas equation 154
7.6 Dalton's Law of partial pressures 155
7.7 The kinetic theory of gases 156
7.8 Real gases: non-ideal behaviour 157
Concept map: Gases 158
Questions on Chapter 7 159

CHAPTER 8: LIQUIDS

8.1 An incredible liquid 160
8.2 The liquid state 160
8.3 Vaporisation 162
8.4 Molar mass determination 165
8.5 Solutions of liquids in liquids 166
8.6 Partition of a solute between two solvents 172
8.7 Partition chromatography 174
Concept map: Liquids 178
Questions on Chapter 8 179

CHAPTER 9: SOLUTIONS

9.1 An unusual solvent 181
9.2 Solutions of solids in liquids 182
9.3 Recrystallisation 183
Concept map: Solutions 185

CHAPTER 10: THEMOCHEMISTRY

10.1 Sources of energy **186**
10.2 Fossil fuels **186**
10.3 Why do reactions happen?
10.4 Forms of energy **188**
10.5 Exothermic and endothermic reactions **189**
10.6 Enthalpy changes **190**
10.7 Standard enthalpy changes **191**
10.8 Methods for finding the standard enthalpy of reaction **193**
10.9 Hess's Law **196**
10.10 Standard enthalpy change for a chemical reaction **198**
10.11 Average standard bond enthalpy **200**
10.12 The Born–Haber cycle **203**
10.13 Lattice enthalpies **205**
10.14 Enthalpy changes involved when ionic compounds dissolve **207**
10.15 Entropy **209**
10.16 Free energy, the deciding factor **211**
10.17 Link with kinetics **213**
Concept map: Thermodynamics **214**
Questions on Chapter 10 **215**

CHAPTER 11: EQUILIBRIA

11.1 Fritz Haber **219**
11.2 Reversible reactions **219**
11.3 The equilibrium law **222**
11.4 Position of equilibrium **222**
11.5 The effect of conditions on the position of equilibrium **222**
11.6 Examples of reversible reactions **225**
11.7 Oxidation–reduction equilibria **228**
11.8 Phase equilibrium diagrams **228**
Concept map: Equilibrium **230**
Questions on Chapter 11 **230**

CHAPTER 12: ELECTROCHEMISTRY

12.1 Sir Humphry Davy (1778–1867) **233**
12.2 Electrolysis **233**
12.3 Examples of electrolysis **236**
12.4 Explanation of electrolysis **237**
12.5 Applications of electrolysis **239**
Concept map: Electrolysis **241**
12.6 Ionic equilibria **242**
12.7 Indicators **249**
12.8 Buffer solutions **253**
12.9 Salt hydrolysis **256**
12.10 Complex ions **258**
Concept map: Acid-base equilibria **259**
Questions on Chapter 12 **260**

CHAPTER 13: OXIDATION–REDUCTION EQUILIBRIA

13.1 Electrochemical transport **264**
13.2 Electrode potentials **265**
13.3 Redox systems **268**
13.4 Rusting and standard electrode potential **273**
13.5 Electrochemical cells **274**
Concept map: Radox equilibria **277**
Questions on Chapter 13 **278**

CHAPTER 14: REACTION KINETICS

14.1 The speeds of chemical reactions **281**
14.2 Factors which affect the speeds of chemical reactions **282**
14.3 The collision theory **287**
14.4 Catalysis **290**
14.5 Homogeneous catalysis **291**
14.6 Heterogeneous catalysis **292**
14.7 Catalytic converters **293**
14.8 Autocatalysis **294**
14.9 Link with thermodynamics **295**
14.10 The study of reaction kinetics **295**
14.11 Average rate **296**
14.12 Methods of finding the rates of chemical reactions **297**
14.13 The results of measurements of reaction rates **298**
14.14 Order of reaction **300**
14.15 Photochemical reactions **306**
14.16 The effect of temperature on reaction rates **306**
14.17 Theories of reaction rates **307**
14.18 Rate-determining step **311**
Concept map: Chemical kinetics **314**
Questions on Chapter 14 **315**

PART 3: INORGANIC CHEMISTRY

CHAPTER 15: PATTERNS OF CHANGE IN THE PERIODIC TABLE

15.1 Primo Levi's Periodic Table **321**
15.2 Physical properties **322**
15.3 Elements of the s block, d block and p block **326**
15.4 The elements of Period 3 **327**
15.5 Compounds of Period 3 **329**
15.6 Oxides, chlorides and hydrides in the Periodic Table **332**
15.7 Variable oxidation state **335**
Concept map: The Periodic Table **336**
Questions on Chapter 15 **337**

CHAPTER 16: GROUP 0: THE NOBLE GASES

16.1 The 'new gas' **340**
16.2 Members of the group **341**
16.3 Compounds of the noble gases **341**
Questions on Chapter 16 **341**

CHAPTER 17: HYDROGEN

17.1 Occurrence **342**
17.2 Manufacture and uses **343**
17.3 Laboratory preparation and reactions **343**
17.4 Water **343**
17.5 Fluoridation of water **344**
Questions on Chapter 17 **345**

CHAPTER 18: THE s BLOCK METALS: GROUPS 1 AND 2

18.1 Two industries based on salt **346**
18.2 The members of the groups **347**
18.3 Uses **348**
18.4 Occurrence and extraction **349**
18.5 Reactions of Group 1 **349**
18.6 Compounds of Group 1 **350**
18.7 Lithium **355**
18.8 Reactions of Group 2 **356**
18.9 Compounds of Group 2 **357**
Concept map: Group 1 and Group 2 **360**
Concept map: Group 2 **361**
18.10 Extracting metals from their ores **362**
Concept map: Extraction of metals I **363**
Questions on Chapter 18 **364**

CHAPTER 19: GROUP 3

19.1 The aluminium problem **368**
19.2 The members of the group **369**
19.3 Aluminium **369**
19.4 Aluminium compounds **371**
Questions on Chapter 19 **373**

CHAPTER 20: THE HALOGENS

20.1 DDT: a life-saving compound of chlorine **374**
20.2 The members of the group **375**
20.3 Bond formation **376**
20.4 Oxidising reactions **378**
20.5 Commercial extraction **379**
20.6 Reaction with water **382**
20.7 Reaction with alkalis **382**
20.8 Metal halides **384**
20.9 Non-metal halides **385**
20.10 Summary of Group 7 **387**
20.11 Uses of halogens **388**
Concept map: The halogens **389**
Questions on Chapter 20 **390**

CHAPTER 21: GROUP 6

21.1 Oxygen – the breath of life **392**
21.2 The members of the group **393**
21.3 Reactions of oxygen and sulphur **393**
21.4 Allotropes of oxygen **394**
21.5 The ozone layer **394**
21.6 Hydrides of oxygen and sulphur **397**
21.7 Oxides **399**
21.8 Sulphur dioxide **400**
21.9 Sulphuric acid **400**
21.10 Acid rain **405**
Questions on Chapter 21 **408**

CHAPTER 22: GROUP 5

22.1 Two chemical messengers **410**
22.2 The nitrogen cycle **412**
22.3 The members of the group **412**
22.4 The Haber process **413**
22.5 Ammonia **414**
22.6 Nitric acid **416**
22.7 NPK fertilisers **418**
Questions on Chapter 22 **421**

CHAPTER 23: GROUP 4

23.1 Silicon the semiconductor **423**
23.2 A comparative look at Group 4 **425**
23.3 Special features of carbon chemistry **426**
23.4 Sources and uses of Group 4 elements **426**
23.5 The compounds of Group 4 **428**
23.6 The greenhouse effect **435**
Concept map: Group 4 **439**
Questions on Chapter 23 **440**

CHAPTER 24: THE TRANSITION METALS

24.1 Metals and civilisation **442**
24.2 The first transition series **443**
24.3 Physical properties of transition metals **444**
24.4 Chemical properties **445**
24.5 Methods of extraction **445**
24.6 Uses of transition metals **446**
24.7 Oxidation states **446**
24.8 Catalysis by transition metals **447**
24.9 Paramagnetism **447**
24.10 Complex compounds **447**
24.11 Transition metals in gemstones **449**
Concept map: Transition metals I **450**
24.12 Oxides and hydroxides of transition metals **450**
24.13 Oxo-ions of transition metals **452**
24.14 Chlorides **454**
24.15 Sulphides **456**
24.16 Complex ions **456**
24.17 Iron **462**
24.18 Copper **469**
24.19 Zinc **470**
Concept map: Transition metals II **472**
Concept map: Extraction of metals III **473**
Questions on Chapter 24 **474**

PART 4: ORGANIC CHEMISTRY

CHAPTER 25: ORGANIC CHEMISTRY

25.1 Carbon compounds **479**
25.2 Hydrocarbons **480**
25.3 Isomerism among alkanes **481**
25.4 System of naming hydrocarbons **481**
25.5 Alicyclic hydrocarbons **484**
25.6 Aromatic hydrocarbons **484**
25.7 Functional groups **484**
25.8 Reactions of organic compounds **487**
25.9 Isomerism **489**
Questions on Chapter 25 **494**

CHAPTER 26: THE ALKANES

26.1 Methane **495**
26.2 Petroleum oil **495**
26.3 Physical properties **497**
26.4 Reactions of alkanes **498**
26.5 The car **503**
26.6 Catalytic converters **504**
Questions on Chapter 26 **505**

CHAPTER 27: ALKENES AND ALKYNES

27.1 Ethene **507**
27.2 Alkenes **507**
27.3 Source of alkenes **508**
27.4 Physical properties of alkenes **508**
27.5 Structural isomerism **508**
27.6 Reactivity of alkenes **509**
27.7 Combustion **510**
27.8 Addition reactions **510**
27.9 Alkynes **521**
27.10 Ethyne **521**
27.11 Photochemical smog **521**
Questions on Chapter 27 **523**

CHAPTER 28: AROMATIC COMPOUNDS

28.1 Profile: Kathleen Lonsdale (1903–1974) **525**
28.2 Benzene **525**
28.3 Some more names of aromatic compounds **527**
28.4 Physical properties of benzene **527**
28.5 Sources of benzene **528**
28.6 Reactivity of benzene **528**
28.7 Addition reactions of benzene **528**
28.8 Substitution reactions **528**
28.9 Methylbenzene (toluene) **538**
28.10 Reactions of the ring **538**
28.11 Reactions of the side chain **539**
28.12 The effect of substituent groups on the benzene ring **541**
Questions on Chapter 28 **542**

CHAPTER 29: HALOGENOALKANES AND HALOGENOARENES

29.1 Anaesthetics **545**
29.2 Halogenoalkanes **546**
29.3 Physical properties **546**
29.4 Laboratory methods of preparing halogenoalkanes **548**
29.5 Uses of halogenoalkanes **548**
29.6 Chlorofluorocarbons, CFCs **548**
29.7 Reactions **551**
29.8 Reactivity **551**
29.9 The mechanisms of hydrolysis and elimination reactions of halogenoalkanes **555**
29.10 Halogenoarenes **558**
Concept map: Halogenoalkanes and Halogenoarenes **560**
29.11 Grignard reagents **561**
Questions on Chapter 29 **561**

CHAPTER 30: ALCOHOLS AND PHENOLS

30.1 Alcoholic drinks **564**
30.2 Alcohols **565**
30.3 Physical properties **567**
30.4 Industrial sources of alcohols **568**
30.5 Uses of ethanol and methanol **569**
30.6 Reactivity of alcohols **571**
30.7 Reactions of alcohols **572**
30.8 Polyhydric alcohols **578**
30.9 Phenols **579**
30.10 Sources of phenol **580**
30.11 Reactions of phenol **580**
30.12 The explosion at Seveso **585**
Concept map: Alcohols and phenols **586**
Concept map: Relationships between alcohols and other series **587**
Questions on Chapter 30 **588**

CHAPTER 31: ALDEHYDES AND KETONES

31.1 Poisons, flavours and perfumes **590**
31.2 The functional group **591**
31.3 Nomenclature for aldehydes and ketones **591**
31.4 The carbonyl group **592**
31.5 Some members of the series **595**
31.6 Laboratory preparations **595**
31.7 Reactions of carbonyl compounds **597**
31.8 The mechanisms of the reactions of aldehydes and ketones **604**
31.9 Carbohydrates **607**
Concept map: carbonyl compounds **608**
Questions on Chapter 31 **609**

CHAPTER 32: AMINES

32.1 William Perkin and mauve **612**
32.2 Nomenclature for amines **613**
32.3 Natural occurrence **614**
32.4 Physical properties **614**
32.5 Basicity of amines **615**
32.6 Laboratory preparations **617**
32.7 The reactions of amines **619**
32.8 Diazonium compounds **623**
32.9 Quaternary ammonium compounds **626**
Concept map: Amines **627**
Questions on Chapter 32 **628**

CHAPTER 33: ORGANIC ACIDS AND THEIR DERIVATIVES

33.1 The sulphonamide antibiotics **631**
33.2 The carboxyl group **632**
33.3 Nomenclature for organic acids and their derivatives **632**
33.4 Physical properties of acids and their derivatives **633**
33.5 Reactivity of carboxylic acids **635**
33.6 Laboratory preparations of carboxylic acids **637**
33.7 Vitamin C: ascorbic acid **639**
33.8 Reactions of carboxylic acids **639**
33.9 Derivatives of carboxylic acids **643**
33.10 Acid chlorides **643**
33.11 Acid anhydrides **644**
33.12 Aspirin **645**
33.13 Esters **646**
33.14 Fats and oils: soaps and detergents **648**
33.15 Amides **650**
33.16 Nitriles **652**
33.17 Amino acids and proteins **654**
Concept map: Carboxylic acid derivatives **657**
Questions on Chapter 33 **658**

CHAPTER 34: POLYMERS

34.1 Plastics in use **662**
34.2 Long molecules **663**
34.3 Structure and properties **663**
34.4 Addition polymers **665**
34.5 Condensation polymers **669**
34.6 Summary **675**
34.7 Disposal of plastics **676**
Concept map: Polymers **678**
Questions on Chapter 34 **679**

CHAPTER 35: IDENTIFYING ORGANIC COMPOUNDS

35.1 'Bucky balls' **682**
35.2 Methods of purification **683**
35.3 Identifying the elements present **683**
35.4 Empirical, molecular and structural formulae **684**
35.5 Instrumental methods **685**
35.6 The energy levels of a molecule **686**
35.7 Visible–ultraviolet spectra **686**
35.8 Infrared spectrometry **688**
Concept map: Infrared spectrometry **693**
35.9 Mass spectrometry **695**
Concept map: Mass spectrometry **698**
35.10 Nuclear magnetic resonance spectrometry **699**
Concept map: Nuclear magnetic resonance spectrometry **705**
Questions on Chapter 35 **706**

CHAPTER 36: SOME GENERAL TOPICS

36.1 Drugs which alter behaviour **710**
36.2 Synthetic routes **715**
36.3 What are these reagents used for? **719**
36.4 How would you distinguish between the following pairs of compounds? **723**
36.5 Questions on some topics which span chapters **726**
Questions on Chapter 36 **728**

Periodic Table **732**
Basic SI units and derived units **733**
Index of symbols and abbreviations **734**
Answers to numerical problems and selected questions **736**
Appendix: mathematics **750**
Index **761**

ACKNOWLEDGEMENTS

When I wrote the first edition I was fortunate in being able to draw on the counsel of the staff of the Chemistry Department of the University of Hull. I am indebted to Professor R R Baldwin, Professor N B Chapman, Dr P J Francis, Professor G W Gray, FRS, Professor W C E Higginson, and Dr J R Shorter for their advice. I also profited from the guidance of my former supervisor, Professor R P Bell, FRS. My work benefited from the advice on content and presentation which I received from Mr G H Davies, Dr J J Guy and Dr G H Pratt. Parts of the fourth edition have been read by Dr D Lewis, and I thank him for his comments.

The numerical values in the text have been taken largely from *Chemistry Data* by J G Stark and H G Wallace (John Murray 1982).

I thank the following Awarding Bodies for permission to reprint questions from their papers:

AQA for the Associated Examining Board (AEB) and the Northern Examinations and Assessment Board (NEAB)
EDEXCEL for London Examinations (L and L(N))
Hong Kong Examinations Authority (HK)
Northern Ireland Council for the Curriculum Examinations and Assessment (CCEA)
OCR for Oxford and Cambridge Examinations Board (O & C) and the University of Cambridge Local Examinations Syndicate (C)
Welsh Joint Education Committee (W)

The following people and organisations have kindly supplied photographs and given permission for their inclusion.

AEA Technology	Figure 1.10D
BP AMOCO plc	Figures 6.6B, 6.6C
British Steel	Figure 24.17C
Chubb Fire Ltd	Figure 23.5D
David Hoffman Photolibrary	Figure 21.1A
(Sacha Lehrfreund)	Figure 22.7B
Ecoscene (Sally Morgan)	Figure 22.4B
(Joel Creed)	Figure 30.5A
HANNA Instruments	Figure 13.3C
Hydrogas Ltd	Figure 23.5D
ICI	Figures 12.3A, 18.6C, 35.5A
IMI Refiners Ltd	Figure 12.5A
Leslie Garland Picture Library	Figure 14.1B
Martyn Chillmaid	Figures 1.1A, 34.1A, 34.4A
Mary Evans Picture Library	Figure 4.2D
NASA	Figures 21.1B, 21.5B
Perkin Elmer	Figure 35.7B
Pilkington Technology	Figure 23.5F
Rover Group	Figures 7.1A, 24.17E
Science Photolibrary	Figure 1.1B
(Lawrence Livermore Laboratory)	
(TEK IMAGE)	Figure 1.9A
(NOVOSTI)	Figure 1.10A
(Charles D Winters)	Figure 13.3B
Sylvia Cordaiy Photolibrary	Figure 24.17D
Tony Stone Images (Fred Charles)	Figure 14.1A

I acknowledge the sources of the following illustrations:
Figure 4.3H after H Witte and E Wolfel *Reviews of Modern Physics*, 30, 51–5, used by permission of the American Physical Society;
Figure 4.6A after C A Coulson, *Proc. Cam. Phil. Soc.*, 34, 210 (1938), used by permission of the Cambridge Philosophical Society;
Figures 4.6E and 15.2F after Linus Pauling, *The Nature of the Chemical Bond*, Second Edition (1939), used by permission of Cornell University press;

Figure 4.8K adapted from Pauling, Corey and Branson, *Proc. Natl. Acad. Sci.*, US37, 205 (1951).

I thank Stanley Thornes (Publishers) for the commitment which they have shown to the production of this new edition. I thank Adrian Wheaton, Publishing Development Manager, for his advice and John Hepburn, Production Editor, for the meticulous care which he has given to all the editions of this book, Janet Oswell for her careful editing and John Bailey for the picture research. I thank my family for their encouragement and forbearance.

E N Ramsden
1999

PREFACE TO FOURTH EDITION

This text prepares students for AS-level and A-level (A2) examinations in Chemistry. It is assumed that students are approaching AS-level from a study of *GCSE Science: Double Award*. The third edition included additional material to help the student to make the transition from *GCSE Science: Double-Award* to *A-level Chemistry*. The fourth edition takes this approach further. 'Summary boxes' have been added to encourage readers to take stock of what they have learned. 'Concept maps' have been added to help with revision. The introductions to some topics have been rewritten to be more accessible to students with a background of *GCSE Science: Double-Award*.

Each chapter starts with a topic of general interest related to some aspect of the chemistry of that chapter. This may be a famous scientist, a historical anecdote, a contribution of chemistry to medicine or industry or an interesting application. In addition there are topics in many chapters covering environmental concerns (e.g. acid rain, the greenhouse effect and the ozone layer), medicine (e.g. drugs and anaesthetics), agriculture (e.g. fertilisers and pesticides) and industrial accidents.

It is difficult to read a chapter and take in all the information and retain it! To make readers pause for thought, the margin carries a summary of the text. On reaching the end of a section or chapter, a reader can glance back through the summary to see whether he or she has assimilated all the material. If the reader notes any points which need further study, he or she has only to glance at the text alongside the summary to find the relevant passage.

I hope that students like the technique I have devised for integrating descriptive material with diagrams, so that the reader's eye does not have to travel constantly to and fro between a diagram and the text which describes it. I have used this technique largely in the physical chemistry section of the book. The annotated diagrams were consumer-tested and approved by sixth formers in my own school.

At intervals in each chapter 'Checkpoints' are included so that students can pause and test their understanding and, if necessary, revise a section before they pass on to new material. The AS-level and A-level specifications cover so much ground that many teachers find it difficult to take their classes through all the material, while still leaving time for practical work. I hope that teachers will be able to allow students to cover parts of the specification on their own from this text, assisted by the checkpoints, thus reducing the amount of note-taking which needs to be done in class and releasing time for discussion, reinforcement and practical work. Each chapter ends with a searching set of questions, including questions from past examination papers.

Much of the inorganic chemistry is summarised in the form of tables and reaction schemes. Students find these helpful for revision. My preference is to take the s block metals first, follow them with the halogens, and then work through the relevant sections of Groups 6 and 5 to arrive at Group 4 with its interesting gradation from non-metallic to metallic behaviour and end with the transition metals. Teachers whose specifications require a different order will find no difficulty in taking the inorganic chapters in a different sequence.

Some students are preparing for staged assessments, while others are preparing for terminal assessments. Some are preparing for AS-level examinations and others for A2-level examinations. Consequently, changes have been made in the structure of certain chapters to make it easier for students to select text relevant to the modules or units which they are studying. Some topics are present in the core specifications of all Awarding Bodies, and no changes have been made in the structure of these chapters. Some topics are present in optional units, and no changes have been made in the structure of these chapters. Other topics appear in specifications at two levels: a foundation coverage in an AS unit and a more advanced coverage in an A2 unit to be taken in the second year of the course. To meet this development, in the fourth edition some topics are divided into AS-level material and 'Further study' which covers more advanced work for A-level (A2) topics. Options which are not covered in the text but are available separately are *Biochemistry and Food Science*, *Materials Science*, *Detection and Analysis*, *Chemistry of the Environment* and *Chemistry and Society for Hong Kong* by E N Ramsden (Stanley Thornes Publishers).

Some topics are not common to all Awarding Bodies. These topics are marked with asterisks so that students can check their own specification and omit these topics if they wish. I envisage teachers and students selecting the material they require for the unit or module they are studying. Each topic is treated from the beginning without assuming that GCSE work has been remembered.

In the fourth edition, there has been some pruning of topics which are no longer included in specifications and the addition of some new topics. I have divided Chapter 18 into Group 1 and Group 2 so that students who require both groups and those who require Group 2 only will both be able to find their way through it. The coverage of polymers and instrumental analysis is more extensive than in previous editions, in response to the popularity of these topics in the Awarding Bodies' specifications.

In my experience, even students with a fair knowledge of the various series of organic compounds find difficulty in tackling problems which require a knowledge of several series of compounds. The method of converting **A** into **D** by the route

$$\mathbf{A} \longrightarrow \mathbf{B} \longrightarrow \mathbf{C} \longrightarrow \mathbf{D}$$

may well be difficult to formulate. I have tackled this problem in stages by summarising at the end of a chapter the relationship between the series of compounds covered in that chapter and those considered in previous chapters. At the end of the section on organic chemistry all the synthetic routes are summarised in a few reaction schemes. A number of threads which run through the separate chapters on organic chemistry are also drawn together at the end of this section.

New to the fourth edition is the Appendix: Mathematics. It reminds students of the mathematics which they need in order to tackle the calculations in the book. This Appendix will help students with the physical chemistry and inorganic chemistry. Further help and practice is available in *Calculations for A-level Chemistry* by E N Ramsden (Stanley Thornes Publishers).

The fourth edition contains new questions from past examination papers at the ends of the chapters. These are designated (AS/AL) or (AL). These correspond approximately to the standard required for the AS and A2 examinations which will begin in the year 2001. When AS-level and A2-level questions become available, new questions will be substituted.

Answers to numerical questions and selected questions are given. These are not model answers, and the Awarding Bodies take no responsibility whatsoever for the accuracy or method of working in the answers which I give to questions from past papers.

New to this edition is the supporting *Answers Key*. This separate booklet gives outline answers to the questions in the Checkpoints and at the ends of chapters. It also contains a selection of synoptic examination questions with outline answers.

The key skills of communication, application of number, information technology, working with others, improving own learning and performance and problem solving can all be exercised as part of a Chemistry course. During his or her course a student can work towards the Key Skills Qualification. He or she can carry out activities which provide evidence of the key skills of communication, application of number and information technology and build up a portfolio of evidence which can contribute towards the Key Skills Qualification.

There is more to chemistry than the content of any AS- and A-level specification. I assume that all students will be following a course of practical work, but I have not found room for instructions for experiments. I would also like to think that students are reading outside the confines of their specifications. At this level their understanding is sufficiently advanced to open the door to many fascinating topics. Many chapters include examples of the impact of chemistry on medicine, agriculture and the environment and the problems faced by the chemical industry. To keep up to date with developments in chemistry, students will have to read newspapers and periodicals. Their advanced chemistry course equips them to understand many of the scientific issues which they will see reported. I hope they will continue to take an interest in scientific topics long after their examinations are behind them.

E N Ramsden
Oxford 1999

Part 1

The Foundation

1

THE ATOM

1.1 MATTER

You can see a huge variety of forms of matter around you. Rocks and oceans, plants and animals, cars and computers are all varieties of matter. If you want to understand what these forms of matter are composed of and how they work, you need to understand chemistry.

Chemistry is the study of matter and how matter behaves. Matter is anything that has mass and occupies space. Matter seems to be continuous, for example a smooth piece of metal, a rough piece of rock and a glass of water all appear to be continuous. The reality is different: matter is discontinuous; it is made up of particles with spaces in between them.

It was the Greek philosopher Democritus who first considered the idea that matter is made up of particles in about 400 BC. The idea was not accepted because there was no experimental evidence for it. The question remained on the agenda for about 2000 years until a Manchester school teacher, John Dalton, revived the discussion and compiled experimental evidence which convinced people. 'Matter is composed of atoms' said Dalton in his Atomic Theory of 1803. This statement started chemistry on the road from a branch of philosophy to the science which it is today.

Dalton never dreamed that anyone would be able to see an atom. Modern instruments, however, provide direct evidence for the existence of atoms. X ray diffraction patterns are produced by atoms [see Figure 6.2A], and advanced microscopes show the positions of individual atoms [see Figure 1.1B].

Figure 1.1A
Matter

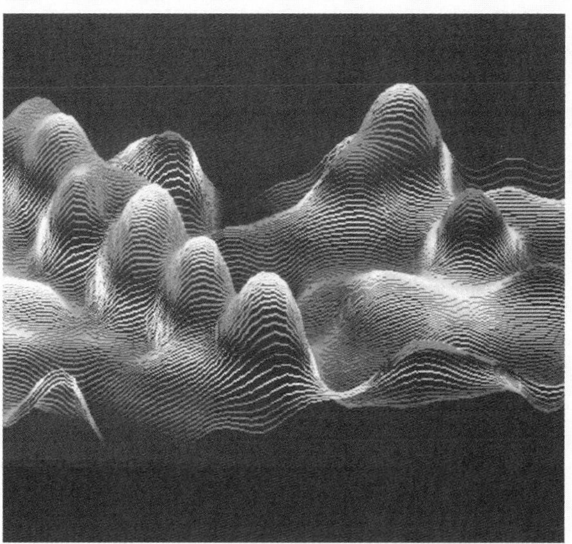

Figure 1.1B
The bumps you can see on the surface of the solid are individual atoms. The surface of DNA is viewed by the technique of scanning tunnelling microscopy. (magnification $\times 1.6 \times 10^6$)

1.2 THE ATOMIC THEORY

The main points in Daltons's theory can be summarised as follows.

1. Matter is composed of tiny particles called atoms, which cannot be created or destroyed or split.
2. All the atoms of any one element are identical: they have the same mass and the same chemical properties. They differ from the atoms of all other elements.
3. A chemical reaction consists of rearranging atoms from one combination into another. The individual atoms remain intact. When elements combine to form compounds, small whole numbers of atoms combine to form compound atoms (as Dalton called them or **molecules** as we call them).

Some points in the atomic theory have been modified since Dalton's time. The atoms of some elements, e.g. uranium, can be split [see § 1.10.8]. Some elements have atoms of more than one kind, which differ slightly in mass; we call these atoms **isotopes** [see § 1.8].

Evidence for the particulate nature of matter

You will be familiar from your earlier work with evidence to support the atomic theory. Checkpoint 1.2 will allow you to check up on how much you remember.

CHECKPOINT 1.2: PARTICLES

1. A purple crystal is dropped into a beaker full of water. After a time, a pink solution has formed. Explain, on the basis of the particle theory of matter, how this happens.

2. Chlorine is denser than air. Yet when air and chlorine are mixed, a homogeneous mixture is obtained [see Figure 1.2A]. Explain, on the basis of the particulate theory of matter, how this happens.

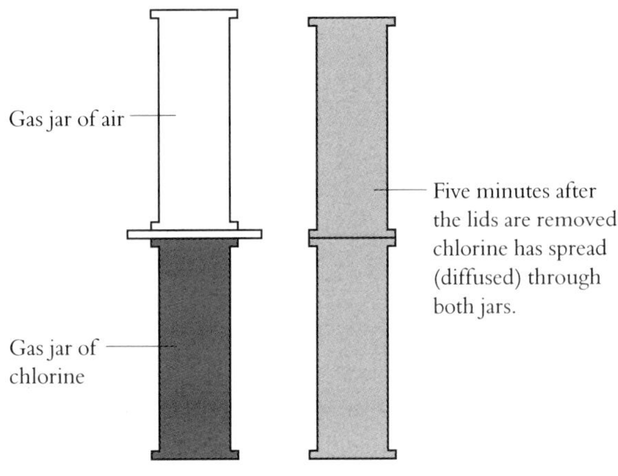

Figure 1.2A
Diffusion of chlorine and air

3. A smoke cell enables one to view the paths of particles of smoke. Figure 1.2B shows the path of a single particle. Explain, on the basis of the particle theory of matter, why the particle moves in this way, constantly changing direction.

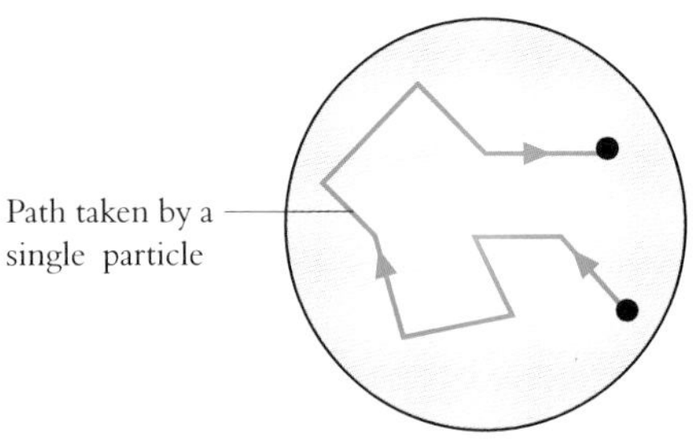

Figure 1.2B
A smoke cell

4. When ice is heated, it melts to form water. When water is heated, is forms water vapour. Explain what happens to particles in the transition (*a*) solid ⟶ liquid and (*b*) liquid ⟶ gas.

5. What has happened to make us disagree with Dalton on the question of whether atoms can be created or destroyed or split?

Evidence from X ray diffraction and electron microscope studies

You know from your previous studies that different kinds of matter consist of different kinds of particles. An element is a substance which cannot be split into simpler substances. Some elements consist of single atoms, e.g. helium consists of helium atoms, He. Other elements consist of molecules, particles which are composed of two or more atoms of the same kind, e.g. nitrogen consists of N_2 molecules, phosphorus consists of P_4 molecules and sulphur consists of S_8 molecules. A compound is a substance which is composed of two or more elements, chemically combined. Some compounds consist of molecules; these particles consist of two or more atoms bonded together. Other compounds consist of charged particles called ions. The numbers of positive and negative charges in a compound are equal. This topic is taken further in Chapters 4 and 6.

1.2.1 PARTICLES IN MOTION

The particulate theory of matter explains the difference in behaviour between solids, liquids and gases.

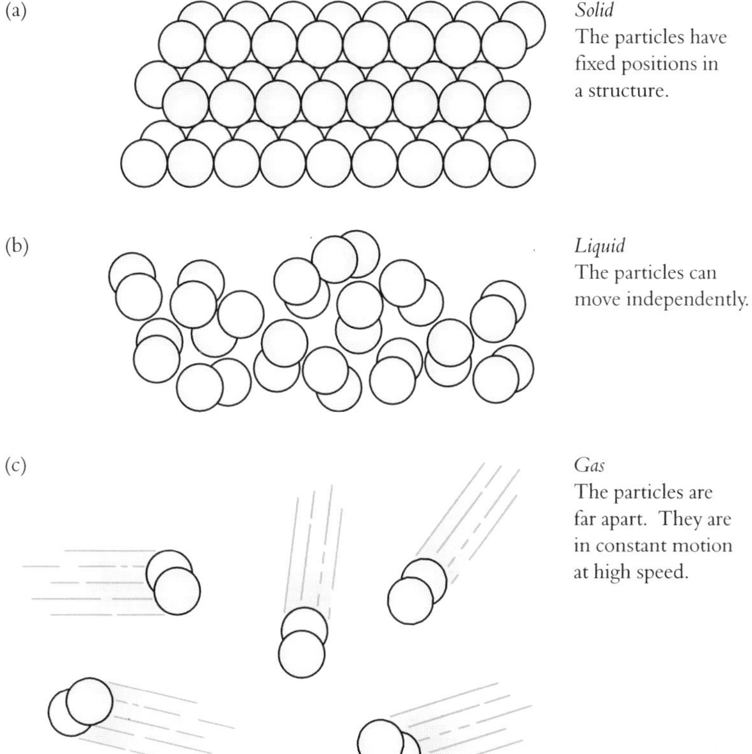
(a)
Solid
The particles have fixed positions in a structure.
(b)
Liquid
The particles can move independently.
(c)
Gas
The particles are far apart. They are in constant motion at high speed.

Figure 1.2C Particles in a solid, a liquid and a gas

Particles in solids, liquids and gases

In a solid, the particles are very close together, and their only motion is vibration about a mean position [Figure 1.2C(a)]. In a liquid the particles are further apart than in a solid. The particles move slowly until they collide with another particle or with the container [Figure 1.2C(b)]. In a gas the particles are far apart. They move in straight lines, changing direction when they collide with another particle or with the walls of the container [see Figure 1.2C(c)]. Particles of a gas move at speeds of the order of 1000 km h^{-1}. The changes in energy that accompany changes of state are defined in § 10.7.2.

1.3 THE SIZE OF THE ATOM

Twentieth-century X ray work [§ 6.2] has shown that the diameters of atoms are of the order of 2×10^{-10} m, which is 0.2 nm (1 nm = 1 nanometre = 10^{-9} m).

The masses of atoms [§ 1.9] range from 10^{-27} to 10^{-25} kg. They are often stated in atomic mass units, u (where 1 u = 1.661×10^{-27} kg).

1.4 THE ELECTRON

Around the year 1900, physicists began to find evidence that atoms are made up of smaller particles. Sir William Crookes was experimenting in 1895 on the discharge of electricity through gases at low pressure. He discovered that a beam of rays was given off by the cathode (the negative electrode). Crookes called the rays **cathode rays**.

Crookes detected cathode rays

The rays also have the properties of particles

Crookes showed that cathode rays also behave like negatively charge particles.

The ratio e/m for cathode ray particles

Sir J J Thomson studied the deflection of cathode rays in electric and magnetic fields. From his measurements he calculated that the ratio of charge/mass, e/m, was -1.76×10^{11} C kg^{-1} (C = coulomb, the SI unit of charge). Since he obtained the same value, regardless of what gas was used or what kind of electrodes were used, he deduced that these negatively charged particles are present in all matter. They were named **electrons**, and were recognised as the particles of which an electric current is composed.

The charge on an electron and its mass

R A Millikan found the value of the electric charge carried by an electron. His 'oil drop' experiments, carried out from 1909 to 1917, are described in many physics books*. From his experiments, he obtained the value of -1.60×10^{-19} C. This amount of charge is called 1 elementary charge unit. Combining this value of charge with Thomson's value of charge/mass gave a value of 9.11×10^{-31} kg for the mass of an electron. This is 5×10^{-4} times the mass of a hydrogen atom.

1.5 THE ATOMIC NUCLEUS

The Thomson model of the atom

In 1898, Thomson surveyed all the evidence that atoms consist of charged particles. He described an atom as a sphere of positive electricity, in which negative electrons are embedded. Other people described this as the 'plum pudding' picture of the atom!

Geiger and Marsden tested the model ...

If this model of the atom is correct, then a metal foil is a film of positive electricity containing electrons. A beam of α particles (helium nuclei, see § 1.8) fired at it should pass straight through. In 1909, Lord Rutherford's colleagues, Geiger and Marsden, tested this prediction [see Figure 1.5A].

... their results ...

They found, as they expected, that α particles penetrated the gold foil. They also found, to their amazement, that a small fraction (about 1 in 8000) of the α particles were deflected through large angles and even turned back on their tracks. Rutherford described this as 'about as incredible as if you fired a 15-inch shell at a piece of tissue paper and it came back and hit you'.

... Rutherford's explanation

Rutherford deduced that the mass and the positive charge must be concentrated in a tiny fraction of the atom, called the **nucleus**. Figure 1.5B shows his interpretation of the results.

* See, e.g., R Muncaster, *A-Level Physics* (Stanley Thornes).

Figure 1.5A
Illustration of Geiger–Marsden experiment

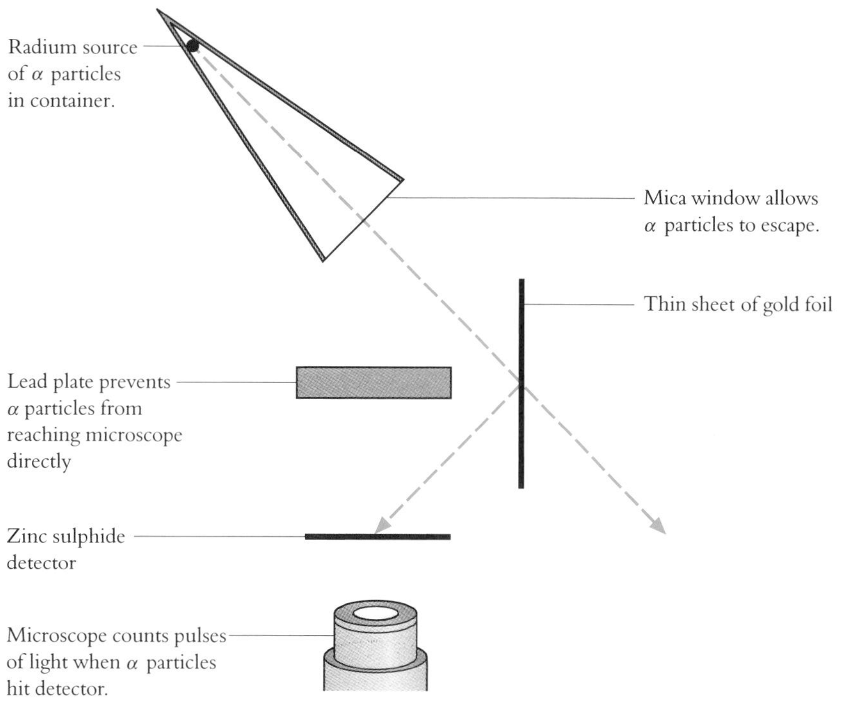

Figure 1.5B
Scattering of α particles by the nuclei of metal atoms

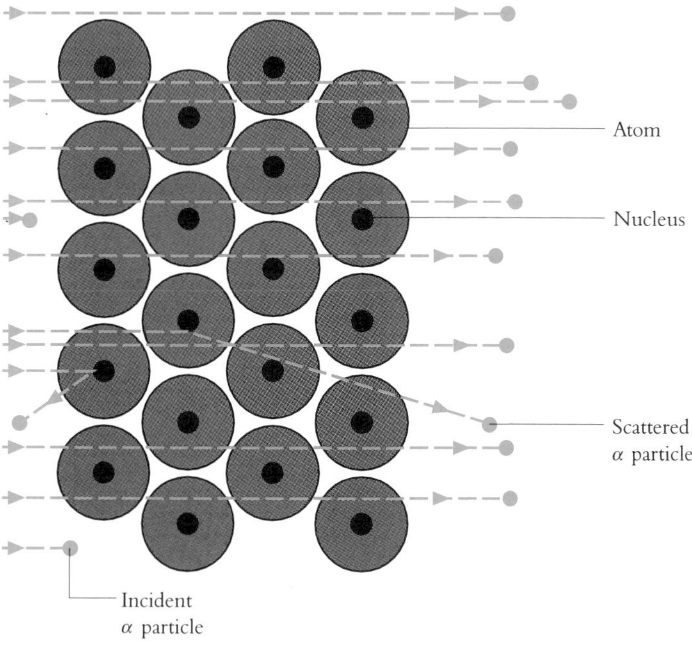

The mass of an atom is concentrated in its nucleus

Only the α particles which collide with the nuclei are deflected; the vast majority pass through the spaces in between the nuclei. The figure is not drawn to scale: a nucleus of the size shown here would belong to an atom the size of your classroom. An atom of diameter 10^{-10} m has a nucleus of diameter 10^{-15} m.

The model of the atom which Rutherford put forward in 1911 was like the solar system [see Figure 1.5C].

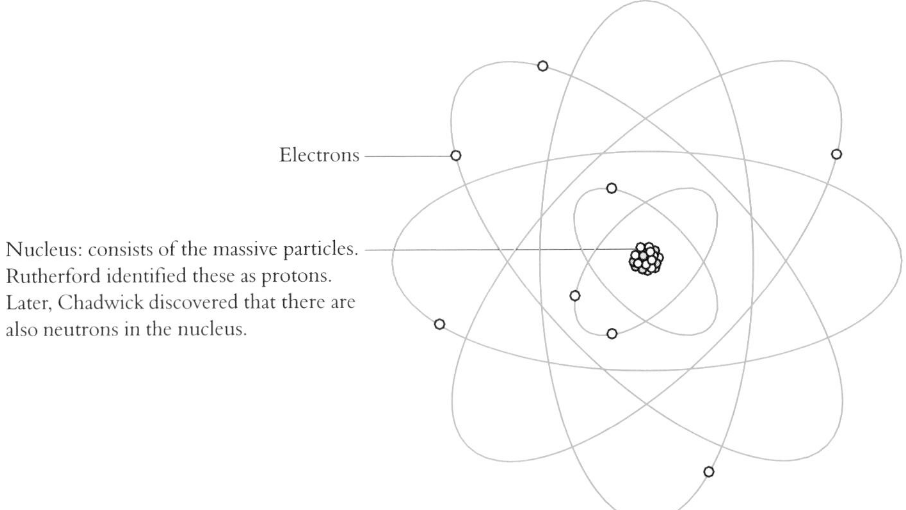

Electrons
Nucleus: consists of the massive particles. Rutherford identified these as protons. Later, Chadwick discovered that there are also neutrons in the nucleus.

Figure 1.5C
The Rutherford atom

Rutherford's model of the atom

The electrons are present in the space surrounding the nucleus. The electrons inhabit a region with a radius one hundred thousand times greater than that of the nucleus. They occupy this space by repelling the electrons of neighbouring atoms. Since electrons are negatively charged, repulsion occurs between the like charges on two electrons if they come close enough together. If another atom approaches too closely, its electrons are repelled by the electrons of the first atom.

1.5.1 THE ARRANGEMENT OF ELECTRONS IN THE ATOM

The electrons are negatively charged; the nucleus is positively charged. What stops the electrons from being pulled into the nucleus by electrostatic attraction? The electrons are in constant motion. They move round and round the nucleus in circular paths called orbits [see Figure 1.5C]. It was suggested that the movement of electrons in orbits round the nucleus would prevent their being pulled in. However, according to the laws of classical physics, an electron moving in a circle round a positive nucleus would gradually lose energy and the electron would spiral into the nucleus. In this way, the Rutherford model of the atom was unsatisfactory. In § 2.3.1, you will see what solution was proposed for this problem.

1.6 THE NEUTRON

The atomic number is the number of protons in the nucleus which equals the number of electrons in the atom

Neutrons have mass but no charge

H G Moseley suggested in 1913 that the multiple charge on the nucleus arose from the presence of **protons**, which contribute the charge. Since atoms are uncharged, the number of electrons must be the same as the number of protons. Atomic masses are greater than the mass of the protons in the atom. To make up the extra mass, the existence of **neutrons** was postulated. These particles should have the same mass as a proton and zero charge. The search for the neutron began. Many years later it was a member of Rutherford's team, J Chadwick, who established the existence of the neutron in 1934.

1.7 THE FUNDAMENTAL PARTICLES

Proton number or atomic number

Nucleon number or mass number

The nucleus was thus shown to consist of protons and neutrons. The number of protons is called the **atomic number** or **proton number**. Protons and neutrons are both **nucleons**. The number of protons and neutrons is called the **nucleon number**, or, alternatively, the **mass number**.

Table 1.1
The mass and charge of sub-atomic particles

Particle	Charge/C	Relative charge	Mass/kg	Mass/u
Proton	$+1.6022 \times 10^{-19}$	+1	1.6726×10^{-27}	1.0073
Neutron	0	0	1.6750×10^{-27}	1.0087
Electron	-1.6022×10^{-19}	−1	9.1095×10^{-31}	5.4858×10^{-4}

1.8 NUCLIDES AND ISOTOPES

Notation for nuclides

The word **nuclide** is used to describe any atomic species of which the proton number and the nucleon number are specified. Nuclides are written as $^{\text{nucleon number}}_{\text{proton number}}\text{Symbol}$ (i.e. $^{\text{mass number}}_{\text{atomic number}}\text{Symbol}$). The species $^{12}_{6}C$ and $^{9}_{4}Be$ are nuclides. Protons are represented as $^{1}_{1}H$, neutrons as $^{1}_{0}n$, α particles as $^{4}_{2}He$ and electrons as $^{0}_{-1}e$.

Isotopes contain the same number of protons and different numbers of neutrons

When an element has a relative atomic mass [§ 3.4] which is not a whole number, it is because it consists of a mixture of **isotopes**. Isotopes are nuclides of the same element. They have the same atomic number but different mass numbers, i.e. they differ in the number of neutrons in the nucleus. Since chemical properties depend upon the nuclear charge and electronic structure of an atom, with mass having little effect, isotopes show the same chemical behaviour. The isotopes of chlorine, $^{35}_{17}Cl$ and $^{37}_{17}Cl$, have the same atomic number, 17. The difference between the mass numbers shows that one isotope has 18 neutrons and the other has 20 neutrons. The chemical reactions of the two isotopes are identical. Their names can be written as chlorine-35 and chlorine-37.

CHECKPOINT 1.8: ATOMS

1. State the number of protons, neutrons and electrons in the following atoms:

   (*a*) $^{39}_{19}K$ (*b*) $^{27}_{13}Al$ (*c*) $^{137}_{56}Ba$ (*d*) $^{226}_{88}Ra$

2. State the number of protons, neutrons and electrons in the following atoms:

   (*a*) $^{12}_{6}C$ (*b*) $^{14}_{6}C$ (*c*) $^{1}_{1}H$ (*d*) $^{2}_{1}H$ (*e*) $^{3}_{1}H$

   (*f*) $^{87}_{38}Sr$ (*g*) $^{90}_{38}Sr$ (*h*) $^{235}_{92}U$ (*i*) $^{238}_{92}U$

   What is the relationship between the different atoms of strontium?

3. 'Dalton was incorrect in saying that all the atoms of a particular element are identical.' Discuss this statement.

4. Why did Chadwick look for the neutron? Why was it hard to find?

1.9 MASS SPECTROMETRY

In mass spectrometry …

… ions are deflected in a magnetic field

Atomic masses are determined by **mass spectrometry**. In the mass spectrometer, atoms and molecules are converted into ions. The ions are separated as a result of the deflection which occurs in a magnetic field. Figure 1.9B shows how a mass spectrometer operates, and Figure 1.9A is a photograph of an instrument. Figure 1.9C shows a mass spectrometer trace for copper(II) nitrate.

Figure 1.9A
A mass spectrometer

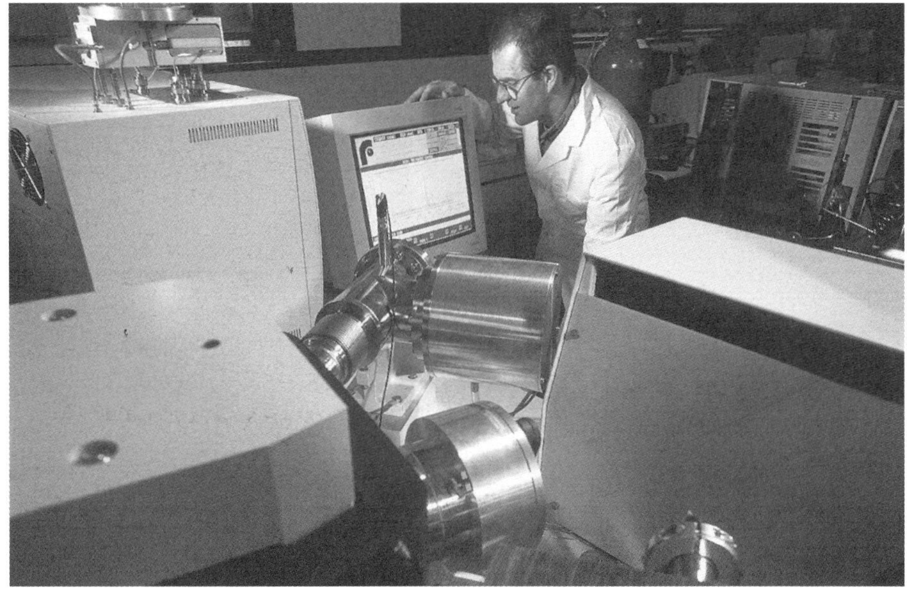

Figure 1.9B
A mass spectrometer: how it works

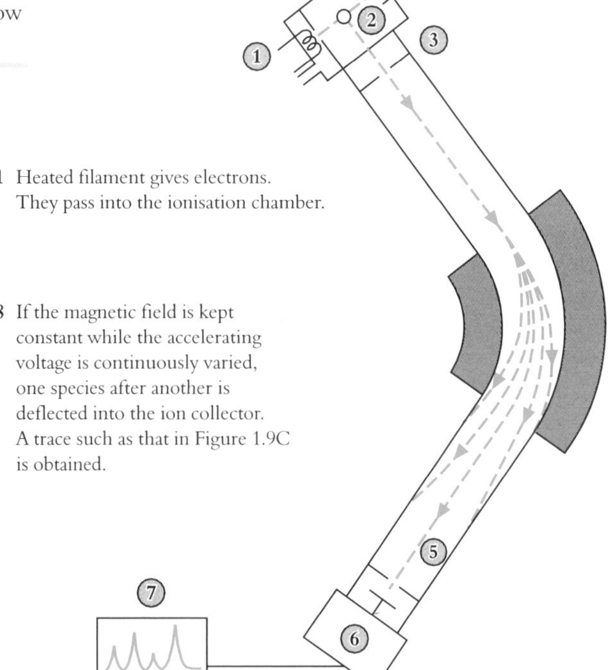

1 Heated filament gives electrons. They pass into the ionisation chamber.

2 The sample is injected as a gas into the ionisation chamber. Electrons collide with molecules of the sample and remove electrons to give positive ions. Some molecules break into fragments. The largest ion is the molecular ion.

3 To this plate, a negative potential is applied (above 8000 V). The electric field accelerates the positive ions.

4 An electromagnet produces a magnetic field. The field deflects the beam of ions into circular paths. Ions with a high ratio of mass/charge are deflected less than those with a low ratio of mass/charge.

5 These ions have the correct ratio of mass/charge to pass through the slit and arrive at the collector.

6 Amplifier. Here the charge received by the collector is turned into a sizeable electric current.

7 Recorder. The electric current operates a pen which traces a peak on a recording.

8 If the magnetic field is kept constant while the accelerating voltage is continuously varied, one species after another is deflected into the ion collector. A trace such as that in Figure 1.9C is obtained.

Notes

(*1*) The height of each peak measures the relative abundance of the ion which gives rise to that peak.

... deflection of ion depends on ratio m/z

(*2*) The ratio of mass/charge for each species is found from the value of the accelerating voltage associated with a particular peak. Many ions have a charge of +1 **elementary charge unit**, and the ratio m/z is numerically equal to m, the mass of the ion (1 elementary charge unit = 1.60×10^{-19} C).

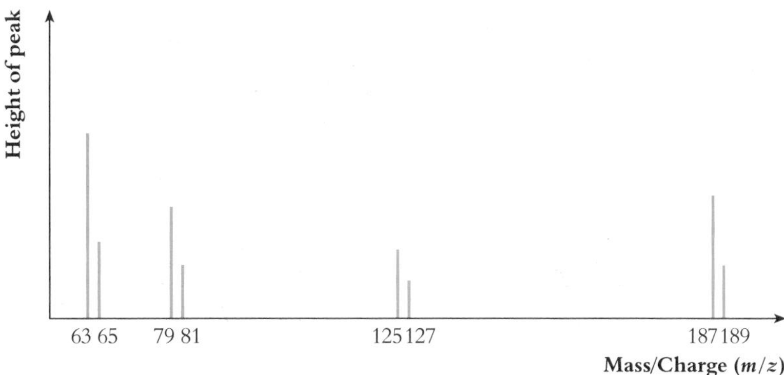

Figure 1.9C
The mass spectrum of copper(II) nitrate

(*3*) The peaks on this trace correspond to the ions

$$63 = {}^{63}Cu^+,\ 65 = {}^{65}Cu^+,\ 79 = {}^{63}CuO^+,\ 81 = {}^{65}CuO^+,$$
$$125 = {}^{63}CuNO_3{}^+,\ 127 = {}^{65}CuNO_3{}^+,\ 187 = {}^{63}Cu(NO_3)_2{}^+,$$
$$189 = {}^{65}Cu(NO_3)_2{}^+$$

1.9.1 USES OF MASS SPECTROMETRY

Determination of the relative atomic mass of an element

Mass spectrometry is used for the determination of relative atomic mass

Figure 1.9D shows the mass spectrum of neon.

The calculation of the average atomic mass of neon ...

The average atomic mass of neon is calculated as follows. Multiply the relative abundance (the height of the peak) by the mass number to find the total mass of each isotope present:

$$\text{Mass of } {}^{22}Ne = 11.2 \times 22.0 = 246.4\ u$$
$$\text{Mass of } {}^{21}Ne = 0.2 \times 21.0 = 4.2\ u$$
$$\text{Mass of } {}^{20}Ne = 114 \times 20.0 = 2280.0\ u$$
$$\text{Total} = 125.4 = 2530.6\ u$$
$$\text{Average mass of Ne} = 2530.6/125.4\ u$$
$$= 20.18\ u$$

... and the relative atomic mass

The average atomic mass of neon is 20.2 u, and the relative atomic mass is 20.2.

Determination of the relative molecular mass of a compound

Mass spectrometry also gives the relative molecular masses of compounds ...

The ion with the highest value of m/z is the molecular ion, and its mass gives the molecular mass of the compound. If isotopes are present, the average molecular mass and the relative molecular mass are found as in the neon example. Some large molecules (e.g. polymers) are fragmented, and do not give molecular ions.

Identification of compounds

... and can be used for the identification of compounds ...

A mass spectrum is obtained, and information about the peak heights and m/z values is fed into a computer. The computer compares the spectrum of the unknown compound with those in its data bank, and thus identifies the compound.

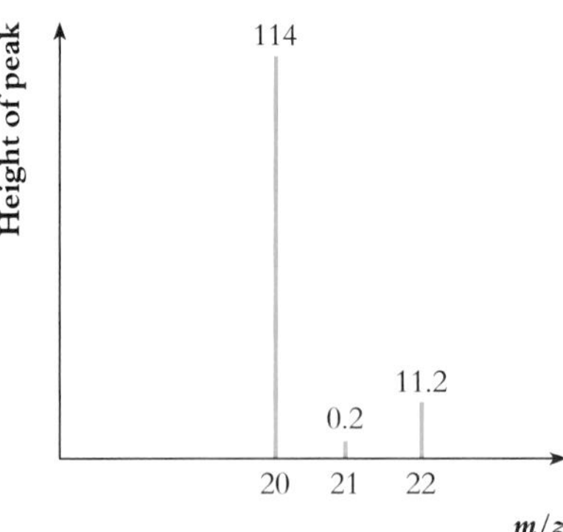

Figure 1.9D
The mass spectrum of neon

Forensic science

... and in forensic science where it is useful as a small sample is enough to give results – see also § 35.9.4

The sensitivity of the mass spectrometer makes it an admirable tool for forensic scientists. The size of sample which they receive for analysis is often very small. A mass spectrum can be obtained on as little as 10^{-12} g. Small amounts of drugs can be identified by mass spectrometry. A fibre left at the scene of a crime can be compared by mass spectrometry with a fibre from a suspect's clothing.

CHECKPOINT 1.9: MASS SPECTROMETRY

1. Describe how, in a mass spectometer, ions are *(a)* formed, *(b)* accelerated, *(c)* separated and *(d)* detected.

2. Define the terms mass number, isotope, relative atomic mass.
   Chlorine has two isotopes of relative atomic masses 34.97 and 36.96 and relative abundance 75.77% and 24.23% respectively.
   Calculate the mean relative atomic mass of naturally occurring chlorine.

3. The mass spectrum of dichloromethane shows peaks at 84, 86 and 88. The intensities of the lines at 84, 86 and 88 u are in the ratio 9 : 6 : 1. What species give rise to these lines? How do you account for the relative intensities of the lines?

4. Figure 1.9E shows the mass spectrum of lead. The heights of the peaks and the mass numbers of the isotopes are shown on the figure. Calculate the average atomic mass of lead.

Figure 1.9E
The mass spectrum of lead

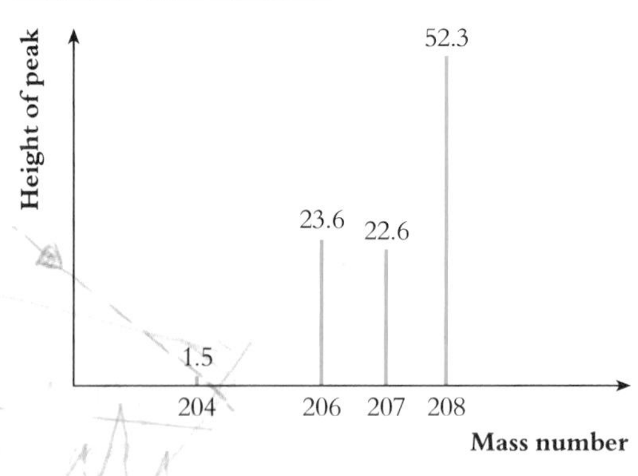

1.10 NUCLEAR REACTIONS

In nuclear reactions new elements are formed

A nuclear reaction is different from a chemical reaction. In a chemical reaction, the atoms which make up the reactants enter into different combinations to form the products, but the nuclei of the atoms are unchanged. In a nuclear reaction, a rearrangement of the protons and neutrons in the nucleus of the atom takes place, and new elements are formed.

1.10.1 RADIOACTIVITY

Discovery

Becquerel discovered a new type of radiation

In 1896, a French physicist called A H Becquerel was experimenting on salts which fluoresced (glowed in the dark). One day he developed a photographic plate which had been left wrapped in a drawer and found to his surprise that the plate had been exposed. He knew that no light could penetrate the wrapping, so Becquerel concluded that the uranium salts in the drawer were to blame. Perhaps the plate had been fogged by some rays coming from the uranium salt. However, there was no known type of radiation that had this effect. Becquerel's instinct told him that it would be worth while investigating this mysterious radiation. He gave the problem to a young research worker called Marie Curie.

Marie Curie found that the radiation was a property of uranium atoms, which she named radioactivity

Marie Curie soon found that the strange effect happened with all uranium salts. It depended on the amount of uranium present but not on the type of compound. She realised that the ability to give off the radiation must be a property of the *atoms* of uranium, and was independent of any chemical bonds which they formed. She realised that this was a completely new type of property, quite different from a chemical reaction. Marie Curie called this property of the uranium atom **radioactivity**.

Marie's husband, Pierre, left his own research work to help her with her exciting new discovery of radioactivity. In 1898 they discovered two new radioactive elements. They called one **polonium** after Marie's native country, Poland. The other element they named **radium**, meaning 'giver of rays'.

Madame Curie is often praised for the arduous work she did in isolating a small quantity of radium from a tonne of the uranium ore **pitchblende**. The work involved back-breaking handling of the ore, crushing it and stirring it with reagents to remove the large mass of unwanted substances. Her real claim to fame, however, is the insight she showed in realising that she was dealing with a new phenomenon which was completely uncharted in either physics or chemistry.

Figure 1.10A Marie Curie in her laboratory

1.10.2 TYPES OF RADIATION

Rutherford suggested that radioactivity is the result of atoms splitting

Why do the atoms of uranium, polonium and radium give off these rays which Marie Curie named radioactivity? The explanation came from the New Zealand-born British physicist Ernest (Lord) Rutherford in 1902. His suggestion was that the atoms split and in the process energy is released. This energy is in the form of radiation. As you can appreciate, this idea was revolutionary. For a century, scientists had accepted Dalton's view that atoms could not be created or destroyed or split. We can now list a large number of elements that have unstable atoms which split to form smaller particles.

Some nuclides are unstable and split up to form smaller atoms

When an atom splits, the nucleus divides and the protons and neutrons in it form two new nuclei. The electrons divide themselves between the two. Sometimes, protons, neutrons and electrons fly out when the original nucleus divides. The process is called **radioactive decay**, and the element is said to be **radioactive**. The particles and energy are called **radioactivity**. Radioactive isotopes have unstable nuclei.

Radioactive substances give three types of radiation ...

... α, β and γ rays

Three types of radiation are given off by radioactive substances. They all cause certain substances, such as zinc sulphide, to luminesce, and they all ionise gases through which they pass. They differ in their response to an electric field, in the manner shown in Figure 1.10B. The uncharged rays, γ (**gamma**) rays, are similar to X rays. They have high penetrating power, being able to pass through 0.1 m of metal. Measurements of m/z identified α (**alpha**) rays as the nuclei of helium atoms and β (**beta**) rays as electrons. β rays can pass through 0.01 m of metal, and α rays can penetrate no more than 0.01 mm of metal.

Figure 1.10B
Effect of an electric field on radiation

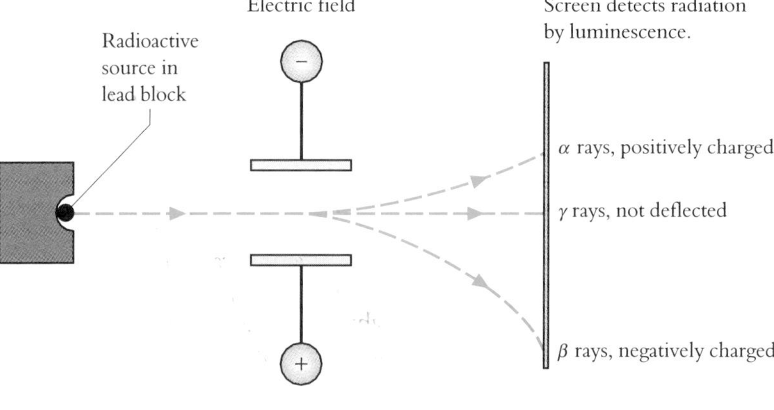

*1.10.3 BALANCING NUCLEAR EQUATIONS

Balancing nucleon (mass) numbers and proton (atomic) numbers

In the equation for a nuclear reaction, the sum of the nucleon numbers (mass numbers) is the same on both sides, and the sum of the proton numbers (atomic numbers) is the same on both sides of the equation. For example, when nitrogen-16 undergoes β decay

$$^{16}_{7}\text{N} \longrightarrow {}^{a}_{b}\text{O} + {}^{0}_{-1}\text{e}$$

Considering mass numbers gives $16 = a + 0 \therefore a = 16$
Considering atomic numbers gives $7 = b + (-1) \therefore b = 8$
The isotope produced is $^{16}_{8}\text{O}$
The equation is $^{16}_{7}\text{N} \longrightarrow {}^{16}_{8}\text{O} + {}^{0}_{-1}\text{e}$

1.10.4 RATE OF RADIOACTIVE DECAY

The rate of radioactive decay is proportional to the number of radioactive atoms present

The half-life is the time taken to decay to half the number of radioactive atoms

The rate at which a radioactive isotope decays cannot be speeded up or slowed down. It depends only on the identity of the isotope and the amount of isotope present. The nature of **nuclear decay** is illustrated in Figure 1.10C. The time taken for a number N_0 of radioactive atoms to decay to $N_0/2$ atoms is called the **half-life**, $t_{1/2}$, of the radioactive isotope. The times taken for $N_0/2$ atoms to decay to $N_0/4$ atoms and for $N_0/4$ atoms to decay to $N_0/8$ atoms are the same and have the same value as $t_{1/2}$. The rate of decay is thus proportional to the number of atoms present. Such reactions are described as **first-order** reactions [§ 14.13.2].

Figure 1.10C Radioactive decay

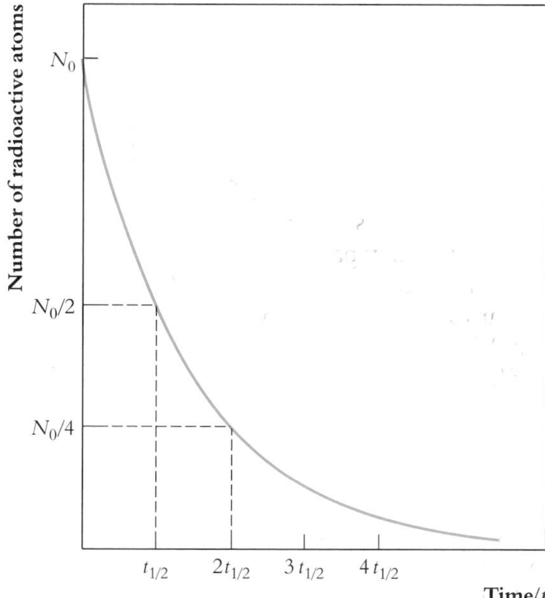

*1.10.5 USES OF RADIOACTIVE ISOTOPES

Radioactivity is used to destroy cancer cells ...

1. Cancerous tissue is destroyed by radioactivity in preference to healthy tissue. A cobalt-60 source (a γ emitter, $t_{1/2} = 5$ years) is used to irradiate cancer patients. The dose which the patient receives must be carefully calculated to destroy only the cancer cells without harming the patient's healthy tissues.

... is used in surgery ...

2. Surgical instruments can be sterilised more effectively by radioactivity than by boiling.

Radioactivity is also used for the detection of leaks ...

3. Underground leaks in water or fuel pipes can be detected by introducing a short-lived radioisotope into the pipe. The level of radioactivity on the surface can be monitored. A sudden increase of surface radioactivity shows where water or fuel is escaping.

... in carbon-14 dating ...

4. **Carbon-14 dating** can be used to calculate the age of plant and animal remains. Living plants and animals take in carbon, which includes a small proportion of the radioactive isotope carbon-14. When a plant or animal dies, it takes in no more carbon-14, and that which is already present decays. The rate of decay decreases over the years, and the activity that remains can be used to calculate the age of the plant or animal material.

... in medicine ...

5. **Tracer studies** use radioactive isotopes to track the path of an element through the body. Radioactive iodine (iodine-131) is administered to patients with defective thyroids to enable doctors to follow the path of iodine through the body. As the half-life is only 8 days, the radioactivity soon falls to a low level. [See Figure 1.10D.]

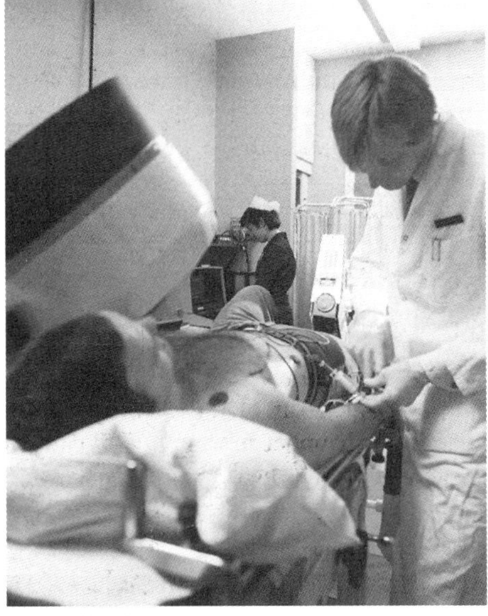

Figure 1.10D Medical uses of radioisotopes. Injection of a short half-life radioactive isotope and a miniature nuclear battery for a heart pacemaker

6. Mechanistic studies sometimes employ radioisotopes. The path of a labelled atom in a molecule can be followed through a sequence of reactions [see esterification, § 33.8.2].

... and in studies of reaction mechanism

7. People who work with radioactive materials take precautions to ensure that they do not receive a high dose of radiation. A radioactive source is surrounded by a wall of lead bricks, except for an outlet through which a beam of radiation can emerge. To handle a powerful source of radiation, people use long-handled tongs. Since radioactivity fogs photographic film, workers exposed to radioactivity wear badges containing film, which is examined periodically so that the dose of radiation they are receiving can be monitored.

CHECKPOINT 1.10: NUCLEAR REACTIONS

1. Write the symbols for the isotopes of chlorine (proton number 17, nucleon numbers 35 and 37).

2. If 8 g of a radioactive isotope decay in a year to 4 g, will 6 g of the same isotope decay to 2 g in the same time? Explain your answer.

3. Supply the missing proton numbers and nucleon numbers:

(*a*) ${}^{14}_{6}C \longrightarrow N + {}^{0}_{-1}e$

(*b*) ${}^{14}_{7}N + {}^{4}_{2}He \longrightarrow O + {}^{1}_{1}H$

(*c*) ${}_{88}Ra \longrightarrow {}^{4}_{2}He + {}^{222}Rn$

(*d*) ${}^{73}As + {}^{0}_{-1}e \longrightarrow {}_{32}Ge$

(*e*) ${}^{24}_{12}Mg + {}^{4}_{2}He \longrightarrow Si + {}^{1}_{0}n$

(*f*) ${}^{19}_{9}F + \longrightarrow {}^{16}_{7}N + {}^{4}_{2}He$

*1.10.6 NUCLEAR FISSION

The mass of the nucleus is less than the sum of the nucleon masses ...

The mass of a nucleus is slightly less than the sum of the masses of the protons and neutrons of which it is composed. The difference in mass is transformed into the **binding energy** of the nucleus. The connection between mass and energy is given in Einstein's equation

$$E = mc^2$$

The mass defect is the source of the binding energy of the nucleus

where E = energy released, m = loss in mass, and c = velocity of light. Since the constant c^2 has a large numerical value, even a very small loss in mass is equivalent to the loss (or release) of a large amount of energy. This is the origin of the substantial quantity of energy released in the **fission** of atomic nuclei.

The first person to obtain energy from **'splitting the atom'** was O Hahn, in 1937. He bombarded uranium-235 with neutrons. Atoms of ^{235}U split into two smaller atoms and two neutrons, with the release of energy:

Splitting the atom of $^{235}_{92}U$

$$^{235}_{92}U + ^{1}_{0}n \longrightarrow ^{144}_{56}Ba + ^{90}_{36}Kr + 2^{1}_{0}n$$

Mass is converted into energy

The sum of the masses of the fission products is less than the mass of the $^{235}_{92}U$ atom

If the sample of uranium-235 is smaller than a certain size, called the **critical mass**, neutrons will escape from the surface. In a large block of uranium-235, neutrons are more likely to meet uranium-235 atoms and produce fission than to escape. Since each nuclear fission produces two neutrons, as shown in Figure 1.10E, a chain reaction is set up. Each time an atom of uranium-235 is split, the mass of the atoms produced is 0.2 u less than the mass of an atom of $^{235}_{92}U$. The lost mass is converted into energy. This is where the energy of the atomic bomb comes from. The atomic bomb consists of two blocks of uranium-235, each smaller than the critical mass. On detonating the bomb, one mass is fired into the other to make a single block larger than the critical mass. The detonation is followed by an **atomic explosion**.

Figure 1.10E
The chain reaction in fission of uranium-235

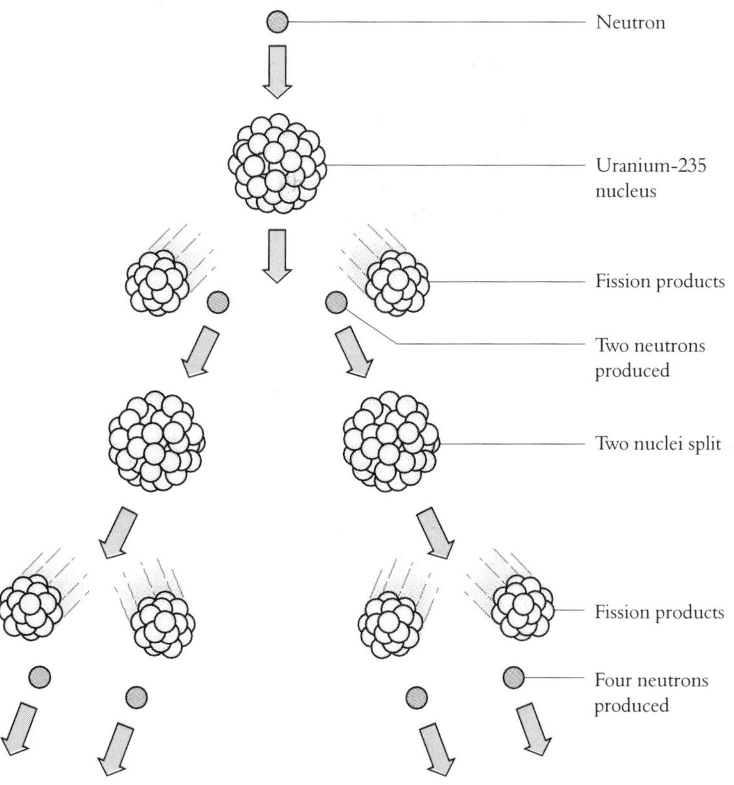

Atomic bombs

The only time atomic bombs have been used in warfare was when two cities in Japan, Hiroshima and Nagasaki, were destroyed in 1945. The death and destruction which followed were on such a terrible scale that nations fighting subsequent wars have avoided using atomic weapons.

In a nuclear reactor ...

... the release of nuclear energy is controlled ...

... boron is used to absorb neutrons

Energy from **nuclear reactors** is obtained by fission of ^{235}U, carried out in a controlled way. The method of controlling the rate of fission to avoid an atomic explosion is to insert rods of boron, an element which is a very good neutron-absorber, into the reactor. If the fission process speeds up, the rods are pushed further into the reactor; if the chain reaction slows down, the rods are pulled out to allow the number of neutrons to increase and speed up the reaction.

QUESTIONS ON CHAPTER 1

1. Explain how a mass spectrometer is used to measure molecular mass.

2. The mass spectrum of C_2H_5Cl shows peaks corresponding to ^1H, ^2H, ^{12}C, ^{13}C, ^{35}Cl and ^{37}Cl. Calculate the mass numbers of the most abundant molecular ion and the heaviest molecular ion. Write the formulae of all the possible ions that contribute to the peak at a mass number of 66.

3. Imagine you have a mixture of hydrogen-1, hydrogen-2 and hydrogen-3, (hydrogen, deuterium and tritium) present as diatomic molecules and that the numbers of atoms of the three species are the same. Sketch the mass spectrum.

4. In 1909, Geiger and Marsden reported the amazing results of their experiments on α particles and thin metal foils.

(a) What is an α particle?
(b) Why did most α particles pass through the foils?
(c) Why were some α particles scattered backwards?
(d) What did they infer from their results about the structure of metal atoms?

5. When chlorine is bubbled through a concentrated aqueous solution of ammonium chloride, a yellow oily liquid, nitrogen trichloride, NCl_3, is formed, together with a solution of hydrochloric acid. Apart from peaks associated with solitary nitrogen atoms (at $m/z = 14$) and chlorine atoms (at $m/z = 35$ and $m/z = 37$), the mass spectrum of nitrogen trichloride contains 9 peaks arranged in 3 groups, ranging from $m/z = 49$ to $m/z = 125$. Predict the m/z values of all 9 peaks, and suggest a formula for the species responsible for each one.

***6.** A series of radioactive decays can be represented

$$^{232}_{90}\text{Th} \xrightarrow{\alpha\ \text{emission}} \mathbf{X} \xrightarrow{\beta\ \text{emission}} \mathbf{Y} \xrightarrow{\beta\ \text{emission}} \mathbf{Z}$$

State the mass number and atomic number of the element **Z**.

***7.** Identify the emitted particles (1) and (2), and state in which groups of the Periodic Table the elements Pb, **X**, **Y** and **Z** occur.

$$^{212}_{82}\text{Pb} \xrightarrow{(1)} {}^{212}_{83}\mathbf{X} \xrightarrow{(2)} {}^{208}_{81}\mathbf{Y} \xrightarrow{\beta\ \text{particle}} \mathbf{Z}$$

***8.** Give values for *a*, *b*, *c* and *d*, and the symbols for **X** and **Y** in the equations

(a) $^{35}_{17}\text{Cl} + {}^{1}_{0}\text{n} \longrightarrow {}^{a}_{b}\mathbf{X} + {}^{1}_{1}\text{H}$

(b) $^{7}_{3}\text{Li} + {}^{2}_{1}\text{H} \longrightarrow 2\,{}^{c}_{d}\mathbf{Y} + {}^{1}_{0}\text{n}$

9. Bromine consists of two isotopes, ^{79}Br (relative abundance 50.5%) and ^{81}Br (relative abundance 49.5%).

(a) Calculate the relative atomic mass to three significant figures.
(b) Copy the figure shown below and sketch on it the peaks you would expect in the mass spectrum of bromine vapour.

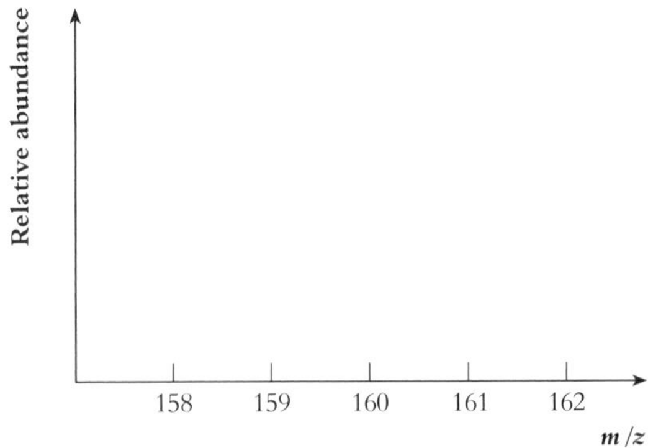

10. The figure below shows a simplified version of the mass spectrum of ethanol, C_2H_5OH. Explain the origin of the six peaks.

Note An asterisk means that the topic is not included in **all** the examination specifications.

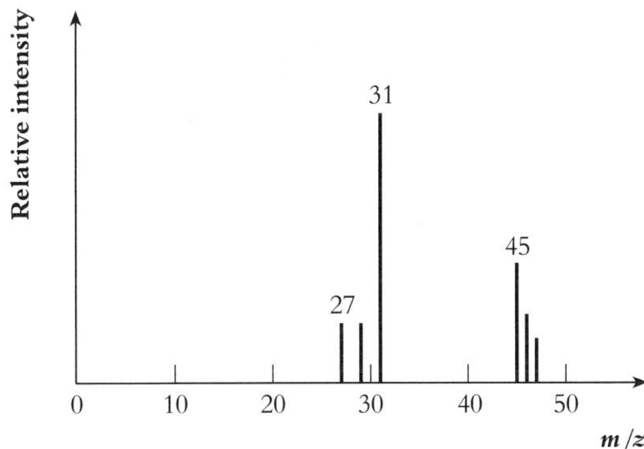

11. A sample of carbon dioxide was prepared from carbon (^{12}C) and oxygen enriched with oxygen-18 and containing $^{16}O_2$ and $^{18}O_2$ in the molar ratio 4 : 1. The mass spectrum of the carbon dioxide contained three peaks, each due to a singly charged molecular ion. What are the relative molecular masses of the three species? Deduce the relative intensities of the three peaks.

12. An organic compound has the composition by mass C 66.7%, H 11.1%, O 22.2%. Its mass spectrum is shown below. The compound is obtained by the oxidation of an alcohol so it is thought to be an aldehyde (with a —CHO group) or a ketone (with a >C=O group).

    (a) Calculate the empirical formula of the compound.
    (b) Interpret the peaks of the mass spectrum to give the molecular formula of the compound.

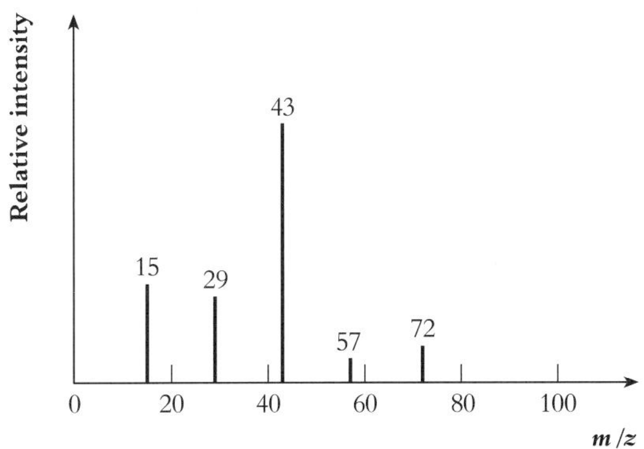

13. *(a)* How is the mass spectrum of an element produced? In your answer explain how the following processes are involved:
    (i) ionisation,
    (ii) acceleration,
    (iii) deflection,
    (iv) detection. **12**

    (b) The mass spectrum of a sample of magnesium contains three peaks with mass/charge ratios and relative intensities shown below.
    (i) Explain why magnesium gives three peaks in its mass spectrum.

m/z	24	25	26
Relative intensity	1	0.127	0.139

    (ii) Use the information in the table to calculate an accurate value for the relative atomic mass of magnesium **6**

18 *NEAB (AS/AL)*

14. *(a)* State the meaning of the term *mass number* of an isotope. **1**
    (b) Define the term *relative atomic mass* of an element. **2**
    (c) A mass spectrometer measures the relative abundance of ions with different values of m/z. Explain the meaning of the symbols m and z. **2**
    (d) A sample of nickel was analysed in a mass spectrometer. Three peaks were observed with the properties shown in the following table

Relative abundance/%	69	27	4
m/z	58	60	62

    (i) Give the symbol, including the mass number and the atomic number, for the ion which was responsible for the peak with $m/z = 58$.
    (ii) Calculate the relative atomic mass of this sample of nickel. **4**

    (e) Complete the electronic configurations for Ni and Ni^{2+}:

    Ni [Ar] ..., Ni^{2+} [Ar] ... **2**

11 *NEAB (AS/AL)*

15. The table below shows some accurate relative atomic masses.

Atom	^{1}H	^{12}C	^{6}Li
Relative atomic mass	1.0078	12.0000	6.0149

    (a) Why is ^{12}C the only atom with a relative atomic mass which is an exact whole number? **1**
    (b) Calculate the mass of 1 mol of $^{1}H^{+}$ ions. The mass of a single electron is 9.1091×10^{-28} g. (Avogadro's number, L, is 6.0225×10^{23}) **2**
    (c) (i) Explain briefly the process by which a sample is ionised in a mass spectrometer.
    (ii) Give **one** reason why it is important to use the minimum possible energy to ionise a sample in a mass spectrometer.
    (iii) After ionisation and before deflection, what happens to the ions in a mass spectrometer; how is this achieved? **5**
    (d) Why is it a good approximation to consider that the relative atomic mass of the $^{6}Li^{+}$ ion, determined in a mass spectrometer, is the same as that of ^{6}Li? **1**

9 *NEAB (AS/AL)*

16. *(a)* A proton, a neutron and an electron all travelling at the same velocity enter a magnetic field. State which particle is deflected the most and explain your answer. **2**

(b) Give two reasons why particles must be ionised before being analysed in a mass spectrometer. **2**

(c) A sample of boron with a relative atomic mass of 10.8 gives a mass spectrum with two peaks, one at $m/z = 10$ and one at $m/z = 11$. Calculate the ratio of the heights of the two peaks. **2**

(d) Compound **X** contains only boron and hydrogen. The percentage by mass of boron in **X** is 81.2%. In the mass spectrum of **X** the peak at the largest value of m/z occurs at 54.

(i) Use the percentage by mass data to calculate the empirical formula of **X**.

(ii) Deduce the molecular formula of **X**. **4**

10 *NEAB (AS/AL)*

17. *(a)* Complete the following equations for nuclear reactions:

(i) $^{226}_{88}Ra \longrightarrow {}^{4}_{2}He +$

(ii) $^{14}_{6}C \longrightarrow {}^{0}_{-1}\beta +$

(iii) $^{60}_{27}Co \longrightarrow {}^{0}_{-1}\beta +$ **3**

(b) (i) Suggest a medical application for any named isotope of your own choice. **2**

(ii) Why is the knowledge of the half-life of this isotope essential to its application? **1**

(c) Write equations for a decay scheme whereby $^{241}_{95}Am$ could emit three particles and become an isotope of americium. **3**

(d) Mass spectrometry can be used to determine isotopic abundances in elements, and fragmentation patterns which are useful in structure determination of organic molecules.

(i) The mass spectrum of magnesium gives peaks at m/e 24 (78.99%), 25 (10.00%) and 26 (11.01%). Calculate the relative atomic mass of naturally-occurring magnesium. Give your answer to 4 significant figures. **2**

(ii) The compounds propanone, CH_3COCH_3, and propanal CH_3CH_2CHO, both have the molecular formula C_3H_6O. The mass spectrum of propanone shows large peaks at m/e 15 and 43, amongst others. Suggest the identity of the species producing these peaks. **2**

(iii) A compound known to have the molecular formula C_3H_6O gives, in its mass spectrum, a large peak at m/e 29. Suggest the identity of a species which could give this peak, and hence identify the compound. **2**

15 *L (AS/AL)*

18. *(a)* State the meaning of the term *atomic number*. **1**

(b) What is the function of the electron gun and the magnet in a mass spectrometer? **2**

(c) The mass spectrum of a pure sample of a noble gas has peaks at the following m/z values.

m/z	10	11	20	22
Relative intensity	2.0	0.2	17.8	1.7

(i) Give the complete symbol, including mass number and atomic number for one isotope of this noble gas.

(ii) Give the species which is responsible for the peak at $m/z = 11$.

(iii) Use appropriate values from the data above to calculate the relative atomic mass of this sample of noble gas. **6**

9 *NEAB (AS/AL)*

2

THE ATOM – THE ARRANGEMENT OF ELECTRONS

2.1 LIGHT FROM ATOMS

What do these events have in common:

- the anniversary of the 'gunpowder plot' by Guy Fawkes in England in 1605
- the anniversary of the liberation of France in 1944 during the Second World War
- the annual celebration of Independence Day on 4th July in the USA?

The thing which they have in common is that all these celebrations employ the emission of light by energetically excited atoms; that is, they all employ fireworks.

A firework contains a fuel, an oxidising agent and one or more metal salts. When the fuse is lit, the oxidising agent oxidises the fuel to gaseous combustion products and heat is given out. How does the heat released in combustion make the metal salts give out light? The heat 'excites' electrons in the metal ions; that is, raises the electrons to higher energy levels. A fraction of a second later, the electrons give out the additional energy as light and return to their normal energy levels. The intense blue, red, green and yellow colours of the fireworks are due to light given out by different elements: copper salts for blue light, strontium salts for red light, barium salts for green light and sodium salts for yellow light.

Light of each colour has a different range of wavelengths. The wavelengths of light (and therefore the colour of light) emitted by the atoms of an element are called the **atomic emission spectrum** of that element. The wavelengths of light which are emitted by an atom depend on how its electrons are arranged. A study of the light that elements emit has given us information about the arrangements of electrons in their atoms.
A knowledge of the arrangements of electrons in the atoms of different elements allows us to understand the classification of elements in the Periodic Table. We can see why the Periodic Table has been successful in systematising and simplifying our study of the huge variety of elements.

Figure 2.1A
A firework display

2.2 ATOMIC SPECTRA

In this chapter we look at atomic spectra. We see how they pose a problem which the Rutherford picture of the atom does not solve.

If sunlight or light from an electric light bulb is formed into a beam by a slit and passed through a prism on to a screen, a rainbow of separated colours is seen. The spectrum of colours is composed of visible light of all wavelengths and is called a **continuous spectrum** [see Figure 2.2A].

Figure 2.2A
A continuous spectrum

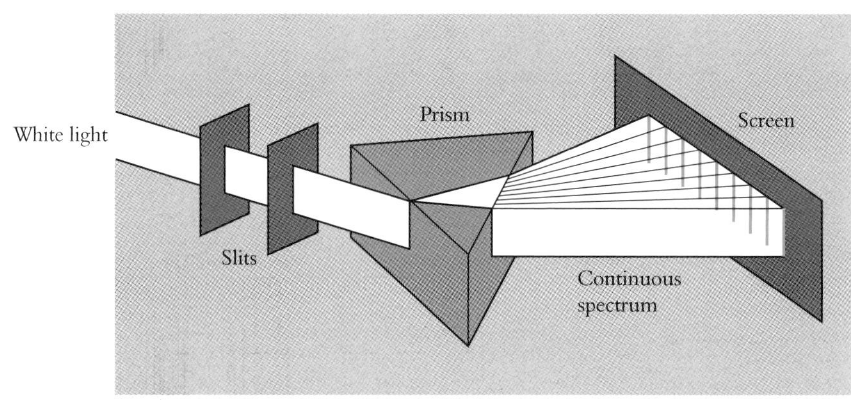

A continuous spectrum is composed of visible light of all wavelengths

Absorption spectra are black lines on a bright background ...

... they are discontinuous spectra

Light of all visible wavelengths is called white light. All atoms and molecules absorb light of certain wavelengths. When white light is passed through a substance, black lines appear in the spectrum where light of some wavelengths has been absorbed by the substance. Spectrometers are instruments used for viewing absorption spectra. The pattern of frequencies absorbed by a substance is called its **absorption spectrum**. It can be used to identify a substance.

If atoms and molecules are heated to sufficiently high temperatures, they emit light of certain wavelengths. Figure 2.2B shows a discharge tube containing a gaseous element. The observed spectrum consists of a number of coloured lines on a black background. The spectrum is called an **atomic emission spectrum** or **line spectrum**.

Figure 2.2B
An emission spectrum (or line spectrum)

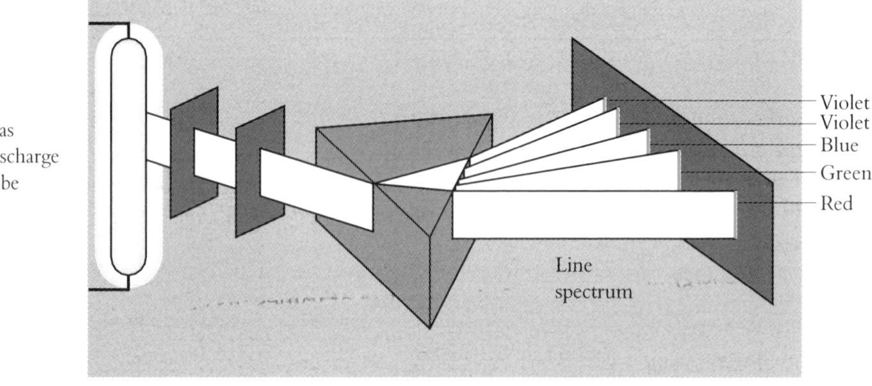

Elements have emission spectra in the visible and ultraviolet region

They are coloured lines on a black background ...

... discontinuous spectra

All substances gives emission spectra when they are excited in some way, by the passage of an electric discharge or by a flame. The atomic emission spectra of elements are in the visible and ultravoilet regions of the spectrum. When sodium or a sodium compound is put into a flame, it colours the flame yellow. A tube of hydrogen gas which has been excited by an electric discharge glows a reddish-pink colour.

Through a spectrometer, the hydrogen emission spectrum is seen to consist of series of lines

Viewed through a spectrometer, the emission spectrum of hydrogen is seen to be a number of separate sets of lines or **series** of lines. These series of lines are named after their discoverers, as shown in Figure 2.2C. The Balmer series, in the visible part of the spectrum, is shown in Figure 2.2D.

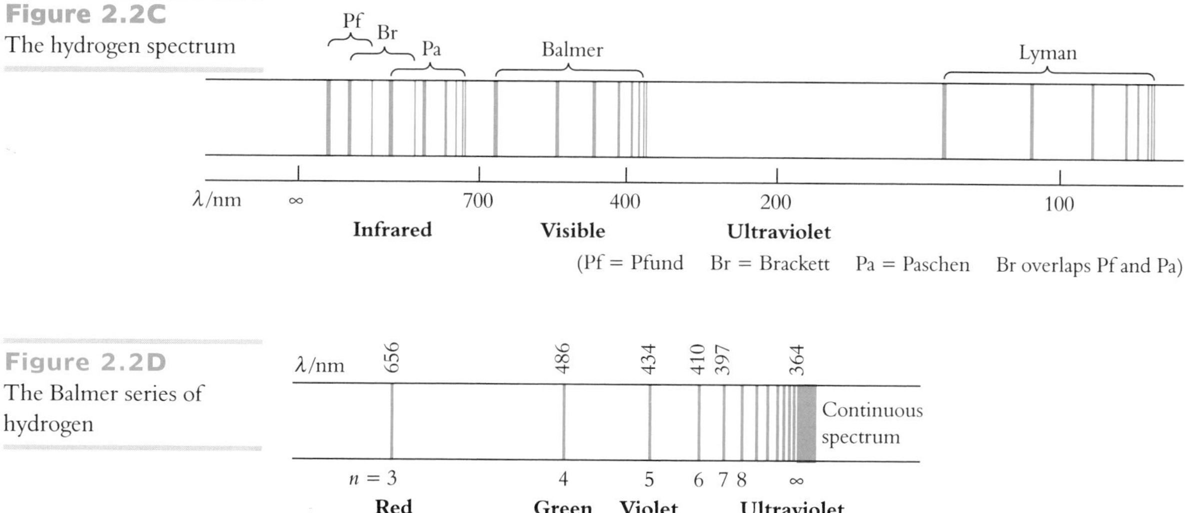

Figure 2.2C The hydrogen spectrum

Figure 2.2D The Balmer series of hydrogen

In each series, the lines become closer together as the frequency increases until at high frequency the lines coalesce

In each series, the intervals between the frequencies of the lines become smaller and smaller towards the high frequency end of the spectrum until the lines run together or **converge** to form a **continuum** of light.

The Rutherford model of the atom does not explain spectral lines

Why do atomic spectra consist of **discrete** (separate) lines? Why do atoms absorb or emit light of certain frequencies? Why do the spectral lines converge to form a continuum? The Rutherford picture of the atom offers no explanation.

2.3 ELECTRONS IN ORBITS

2.3.1 THE BOHR MODEL

Planck theorised that energy is quantised. Bohr suggested that electrons can have only certain amounts of energy ...

In 1913, Niels Bohr (1885–1962) put forward his picture of the atom to answer these questions. Bohr referred to Max Planck's recently developed **quantum theory**, according to which energy can be absorbed or emitted in certain amounts, like separate packets of energy, called **quanta**. Bohr suggested:

1. An electron moving in an orbit can have only certain amounts of energy, not an infinite number of values: its energy is **quantised**.

... and their orbits can have only certain radii

2. The energy that an electron needs in order to move in a particular orbit depends on the radius of the orbit. An electron in an orbit distant from the nucleus requires higher energy than an electron in an orbit near the nucleus.

3. If the energy of the electron is quantised, the radius of the orbit also must be quantised. There is a restricted number of orbits with certain radii, not an infinite number of orbits.

Electrons which absorb photons move to higher orbits

4. An electron moving in one of these orbits does not emit energy. In order to move to an orbit farther away from the nucleus, the electron must absorb energy to do work against the attraction of the nucleus. If an atom absorbs a **photon** (a quantum of light energy), it can promote an electron from an inner orbit to an outer orbit. If sufficient photons are absorbed, a black line appears in the absorption spectrum.

Electrons which fall to lower orbits emit photons of light ...

For an electron to move from an orbit of energy E_1 to one of energy E_2, the light absorbed must have a frequency given by **Planck's equation**:

$$h\nu = E_2 - E_1 \qquad \text{where } \nu = \text{frequency}, \quad h = \text{Planck's constant}$$

The emission spectrum arises when electrons which have been excited (raised to orbits of high energy) drop back to orbits of lower energy. They emit energy as light with a frequency given by Planck's equation [see Figure 2.3A].

Figure 2.3A
The origin of spectral lines

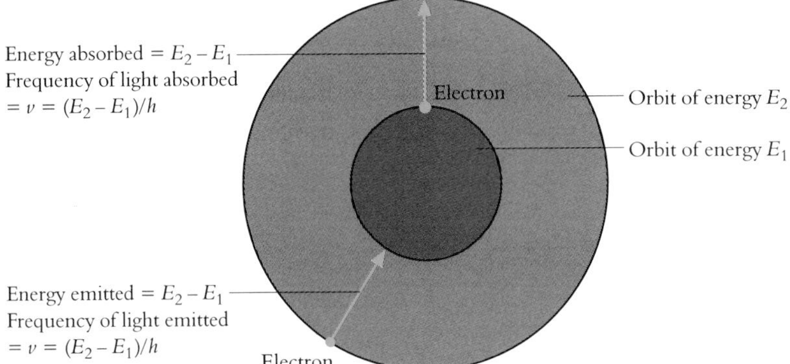

Planck's equation gives the frequencies of light emitted

Bohr gave orbits of different energy different quantum numbers

Bohr assigned **quantum numbers** to the orbits. He gave the orbit of lowest energy (nearest to the nucleus) the quantum number 1. An electron in this orbit is in its **ground state**. The next energy level has quantum number 2 and so on [see Figure 2.3B]. If the electron receives enough energy to remove it from the attraction of the nucleus completely, the atom is **ionised**.

Figure 2.3B
The energy levels at various values of the quantum number, n

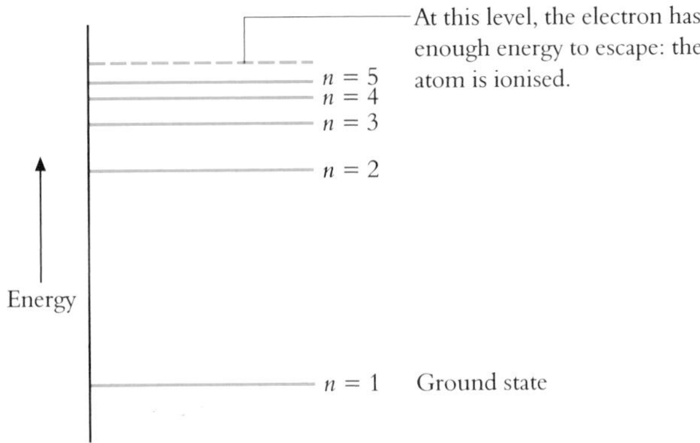

The hydrogen emission spectrum arises as electrons move from orbits of high quantum number to orbits of lower quantum number

Figure 2.3C shows how the lines in the hydrogen emission spectrum arise from transitions between orbits. The Lyman series in the emission spectrum arises when the electron moves to the $n = 1$ orbit (the ground state) from any of the other orbits. The Balmer series arises from transitions to the $n = 2$ orbit from the $n = 3$, $n = 4$ etc. orbits. The Paschen, Brackett and Pfund series arise from transitions to the $n = 3$, $n = 4$ and $n = 5$ orbits from higher orbits.

The frequency of the convergence of spectral lines can be used to give the ionisation energy

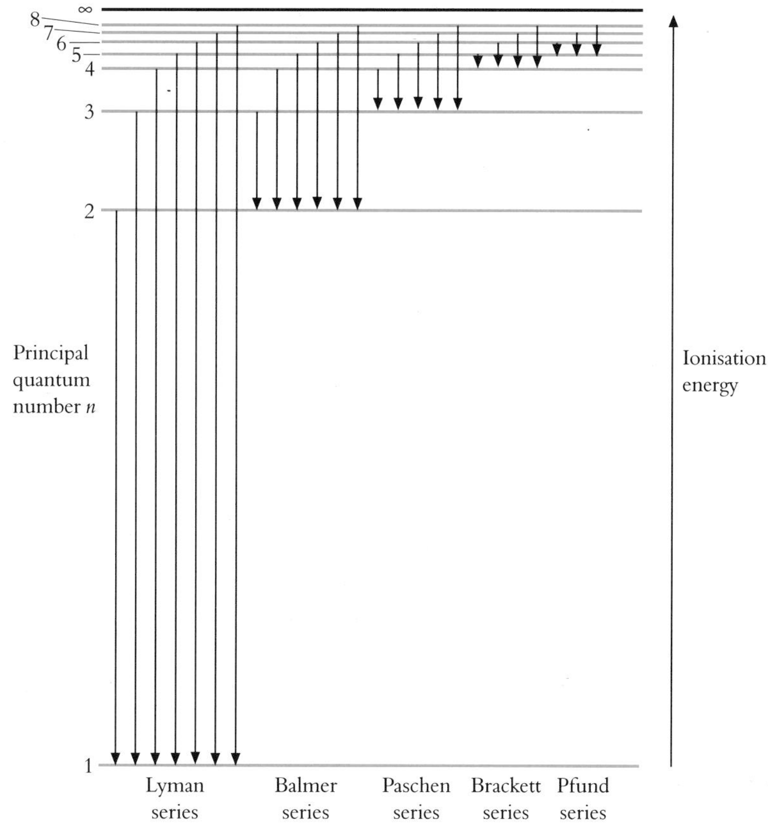

Figure 2.3C
Energy transitions in the hydrogen atom

SUMMARY

Absorption spectra arise when electrons move to higher energy levels. Emission spectra arise when excited electrons move to lower energy levels.

When the lines in the spectrum converge it means that the atom has ionised

In each series of lines, as the frequency increases, each line becomes closer to the previous line until the lines converge, and the spectrum becomes continuous. The Lyman series arises from transitions to the ground state from higher energy levels. The highest frequency lines relate to the highest energy levels. The limit of the Lyman series (the convergence of the lines) corresponds to a transition from the $n = \infty$ orbit (i.e. from an energy level where the electron has escaped from the atom, and the atom has ionised) to the $n = 1$ orbit (the ground state):

$$\mathbf{A}^+ + e^- \longrightarrow \mathbf{A}$$

This transition happens when an electron collides with an ion and returns to the ground state. The convergence frequency can be used to find the **ionisation energy** of the atom.

2.3.2 DETERMINATION OF IONISATION ENERGY

The definition of the first ionisation energy …

The first ionisation energy of an element is the energy required to remove one electron from each of one mole of atoms in the gas phase to form one mole of cations in the gas phase:

$$\mathbf{A}(g) \longrightarrow \mathbf{A}^+(g) + e^-$$

… and using the emission spectrum to find it …

A graphical method can be used to find the value of the ionisation energy from the emission spectrum. The interval between the frequencies of spectral lines becomes smaller and smaller as they approach the continuum.

(*a*) The frequencies of the first lines in the Lyman series are measured. If these are $\nu_1, \nu_2, \nu_3, \nu_4$, etc., the intervals $\Delta\nu = (\nu_2 - \nu_1), (\nu_3 - \nu_2), (\nu_4 - \nu_3)$, etc., can be calculated.

... by a graphical method ...

(*b*) A graph of ν (the lower frequency) against $\Delta\nu$ is shown in Figure 2.3D. It can be extrapolated back to $\Delta\nu = 0$. If there is no interval between lines, this is the beginning of the continuum.

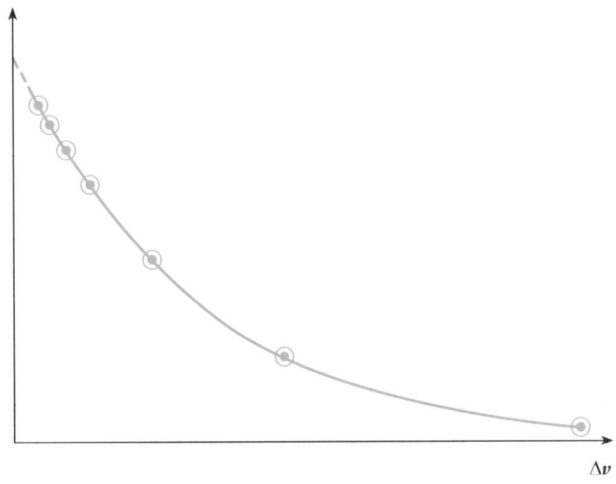

Figure 2.3D
Finding the convergence frequency by a graphical method

SUMMARY

If an electron gains so much energy that it leaves the outermost orbit, the atom has ionised.

(*c*) The value of ν at $\Delta\nu = 0$ is read off and inserted in Planck's equation

$$\Delta E = h\nu$$

(*d*) The value of ΔE is multiplied by Avogadro's constant to give the first ionisation energy for a mole of atoms.

For example, the value of the wavelength at the start of the continuum in the sodium emission spectrum is 242 nm. From this the first ionisation energy of sodium can be calculated. The value is 494 kJ mol^{-1}.

2.3.3 ELECTRON AFFINITY

Energy is given out when an atom gains one electron

When an electron is acquired by an atom, energy is given out, e.g.

$$Cl(g) + e^- \longrightarrow Cl^-(g)$$

The **first electron affinity** is the energy taken in when 1 mole of gaseous atoms accept 1 mole of electrons to become ions. It has a negative value, showing that in fact energy is given out.

Electron affinity is defined ...

The **second electron affinity** of an element is the energy taken in when 1 mole of gaseous ions with a single negative charge accept 1 mole of electrons, e.g.

$$O^-(g) + e^- \longrightarrow O^{2-}(g)$$

It has a positive value because energy is required for the introduction of an electron against the repulsion between e^- and O^-. There is more information about electron affinity in § 10.12.

2.3.4 SOMMERFELD'S QUANTUM NUMBERS

Sommerfeld's second quantum number

Sommerfeld elaborated Bohr's theory in 1916. He proposed that each quantum number governed the energy of a circular orbit and also a set of elliptical orbits of similar energy. He called n the **principal quantum number** and introduced a second quantum number which describes the shape of the elliptical orbits. The **second quantum number**, l, can have values from $(n - 1)$ down to 0. If $n = 4$, $l = 3, 2, 1$ and 0.

2.4 THE WAVE THEORY OF THE ATOM

2.4.1 PARTICLES AND WAVES

The Schrödinger wave equation

According to the wave theory of light, refraction* and diffraction* can be explained by the properties of waves. Other properties of light, such as the origin of line spectra and the photoelectric effect*, need a particle or photon theory for their explanation. The success of the dual theory of light led Louis de Broglie to speculate in 1924 on whether particles might have wave properties. He made the bold suggestion that electrons have wave properties as well as the properties of particles. Erwin Schrödinger used this model to work out a wave theory of the atom. Solutions to the wave equation can only be obtained under certain conditions. An integral number of wavelengths must be fitted into one orbit of the electron round the nucleus.

... Its solution gives the probability of finding the electron at any distance from the nucleus

The solution of the wave equation give the **probability density** of the electron. This is the probability that the electron is present in a given small region of space. The probability that the electron is at a distance, r, from the nucleus is plotted against r for the hydrogen atom in its ground state in Figure 2.4A. The maximum probability of finding the electron is at a distance of 0.053 nm. This is the same as the radius of the orbit occupied by the electron in its ground state according to the Bohr–Sommerfeld model of the atom.

SUMMARY

The wave theory of the electron replaces the idea of finding the electron in a certain position in its orbit with the idea of the probability of finding the electron in a certain volume: the orbital.

The volume of space in which there is a 95% chance of finding the electron is called the atomic orbital

There is a possibility that the electron will be either closer to the nucleus or outside the radius of 0.053 nm. The probability of finding the electron decreases sharply, however, as the distance from the nucleus increases beyond $3r$. The volume of space in which there is a 95% chance of finding the electron is called the **atomic orbital**. There is a 5% probability that the electron will be outside this volume of space at a given instant. On this model, the electron is not described as revolving in an orbit. The electron is said to occupy a three-dimensional space around the nucleus called an atomic orbital. The nucleus is described as being surrounded by a three-dimensional 'cloud of charge' or 'electron cloud'.

The four quantum numbers

Solutions of the wave equation can be obtained if the orbitals are described by four quantum numbers. The first is Bohr's quantum number, n. The second

Figure 2.4A
A probability density diagram for the hydrogen atom in its ground state

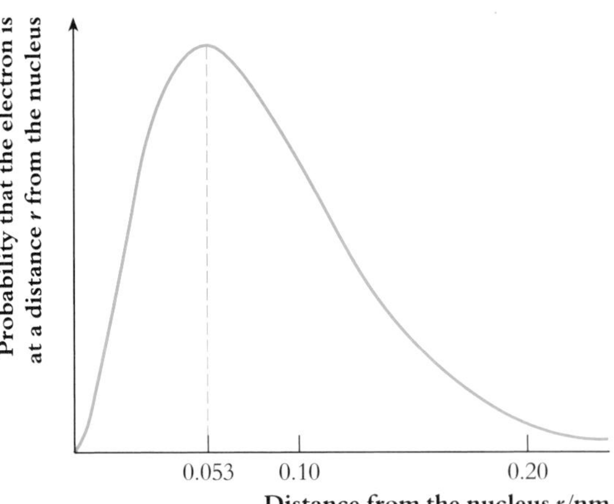

Bohr's quantum number, n. Sommerfeld's quantum number, l

quantum number, l, corresponds to Sommerfeld's quantum number describing the shape of elliptical orbits. The values of l are assigned letters

$l = 0$	1	2	3	4
s	p	d	f	g

If an electron has a principal quantum number $n = 2$ and a second quantum number $l = 0$, it is said to be a 2s electron. For various values of n, the different combinations of the two quantum numbers are

1s
2s 2p
3s 3p 3d
4s 4p 4d 4f
5s 5p 5d 5f 5g

A third quantum number, m_l ...

The wave equation leads to a **third quantum number**, m_l. This gives the maximum number of orbitals for the different values of l.

one	s	orbital
three	p	orbitals
five	d	orbitals
seven	f	orbitals

... and a spin quantum number, m_s

The **fourth quantum number** is called the **spin quantum number**, m_s. It has values of $+\frac{1}{2}$ and $-\frac{1}{2}$. It represents the spin of an electron on its own axis, which can be clockwise or anticlockwise, relative to the orbital of the electron.

The Pauli Exclusion Principle

In his Exclusion Principle, W Pauli stated that *no two electrons in an atom can have the same four quantum numbers*. It follows that, if two electrons in an atom have the same values of n, l and m_l, they must have different values of m_s. Their spins must be opposed. Each orbital can hold two electrons with opposed spins.

2.4.2 SHAPES OF ATOMIC ORBITALS

Figure 2.4B
The shapes of s orbitals

(a) The shape of a 1s orbital

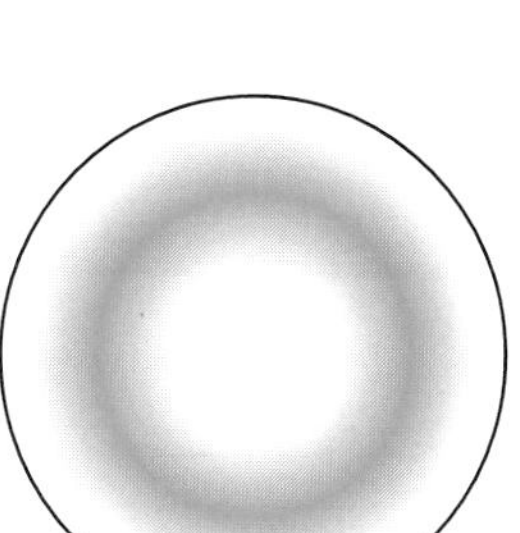

(b) The shape of a 2s orbital

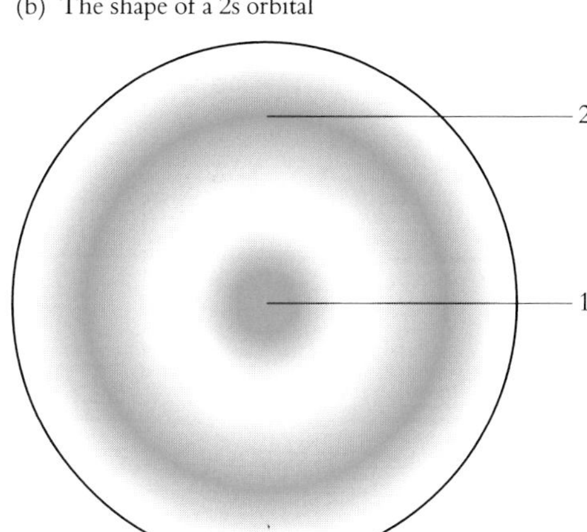

(The density of the shading is a measure of the probability of finding an electron at that distance from the nucleus.)

s orbitals are spherical

p orbitals have an hourglass shape

The shape of an s orbital is spherically symmetrical about the nucleus. The orbital has no preferred direction. The probability of finding an electron at a distance *r* from the nucleus is the same in all directions. [See Figure 2.4B.] A p orbital is not symmetrical: it is concentrated in certain directions. The electron density is shaped like an hourglass. [See Figure 2.4C.] The shapes of d orbitals are shown in Figure 2.4D.

Figure 2.4C
The shape and orientation of p orbitals

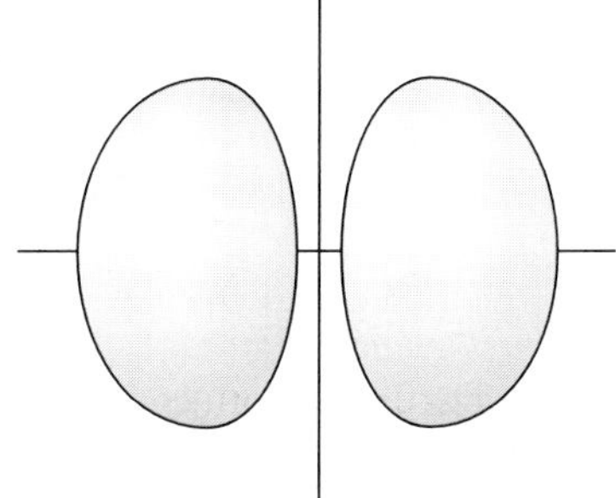

(a) The shape of *one* p orbital

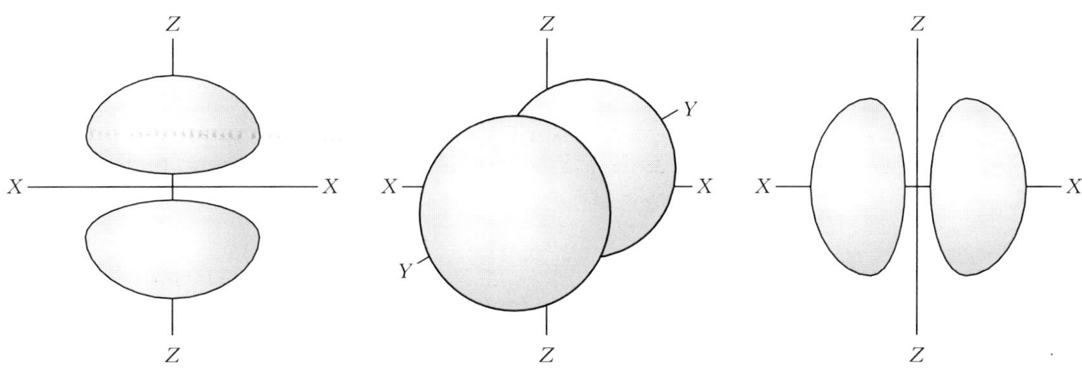

(b) The orientation in space of *three* p orbitals. (Each of the three p orbitals is perpendicular to both of the others. The shape of each orbital is as shown in (a).)

Figure 2.4D
The shape and orientation of d orbitals

There are four d orbitals of shape (a). The lobes lie between the X–Y axes as shown in (a), between the X–Z axes, between the Y–Z axes and, in the fourth case, along the axes X and Y as shown in (b). The fifth orbital has the shape shown in (c).

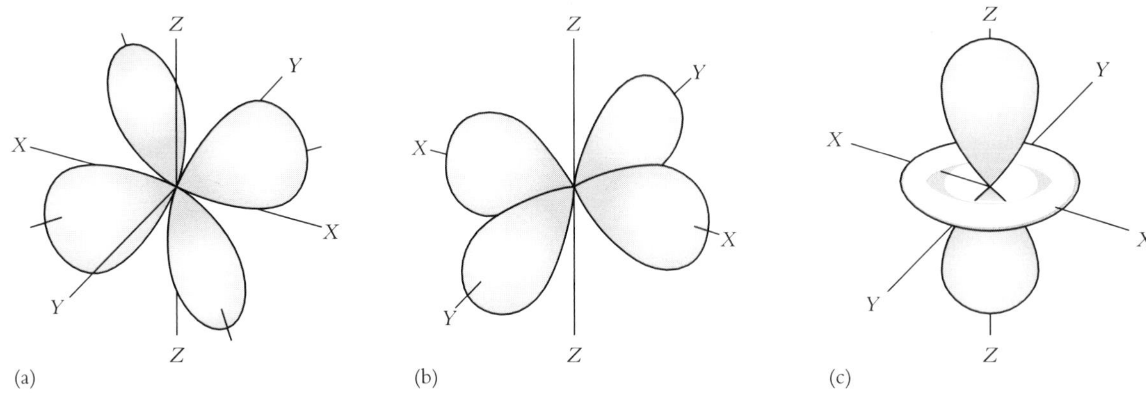

2.5 ELECTRONIC CONFIGURATIONS OF ATOMS

In atoms with more than one electron, there are shells of orbitals with the same principal quantum number

The energy levels of the orbitals of the hydrogen atom are illustrated in Figure 2.3C [§ 2.3.1]. For atoms with more than one electron, the energy levels for each value of n are split between orbitals with different values of l. The relative energy levels of the orbitals are shown in Figure 2.5A. The term **shell** is used for a group of orbitals with the same principal quantum number. A **subshell** is a group of orbitals with the same principal and second quantum numbers, e.g. the 3p subshell.

The arrangement of electrons in atomic orbitals is governed by two factors.

The orbitals of lowest energy are filled first

1. In a normal atom the electrons are arranged so that the energy is at a minimum. Any other arrangement would make the atom an excited atom, which could emit energy and pass to its ground state.

2. The Pauli Exclusion Principle: no two electrons can have the same four quantum numbers.

It is convenient to draw an 'electrons-in-boxes' diagram to show the arrangement of electrons in orbitals.

'Electrons-in-boxes'

electron with $m_s = +\frac{1}{2}$ — [↑↓] — one orbital — electron with $m_s = -\frac{1}{2}$

An s subshell consists of one box, a p subshell of *three* boxes, a d subshell of *five* boxes and an f subshell of *seven* boxes. The boxes are arranged in order of energy in Figure 2.5B.

Electrons occupy lowest energy 'boxes' first

To work out the arrangement of electrons in an atom with 12 electrons, the electrons must be put into the lowest energy boxes first, two to a box, until all the electrons are accommodated [see Figure 2.5B]. It is interesting to do this for the elements in order of atomic number, the order in which they appear in the Periodic Table [p. 732].

Figure 2.5A
The relative energy levels of atomic orbitals (not to scale)

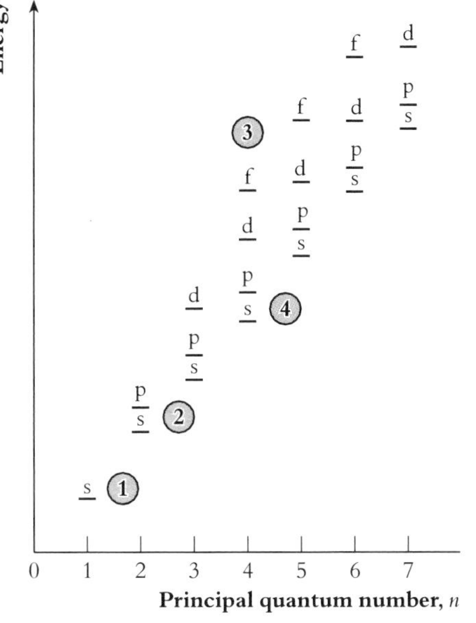

3 Note 4f > 4d > 4p > 4s in energy.

4 Note 4s < 3d in energy. The orbitals of n = 4 overlap those with n = 3.

2 Note 2p electrons have more energy than 2s.

1 Lowest energy: n = 1.

Figure 2.5B
'Electrons-in-boxes', diagram for an atom with 12 electrons

Shells of electrons have the same principal quantum number ...

... and subshells also have the same second quantum number ...

... and are called s, p, d or f subshells

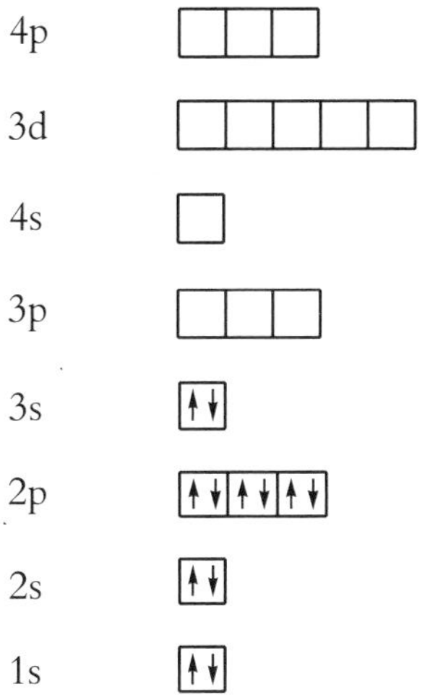

Figure 2.5C
Graph of lg (ionisation energy) against number of electron removed for the potassium atom

The electrons closest to the nucleus require the most energy to remove them

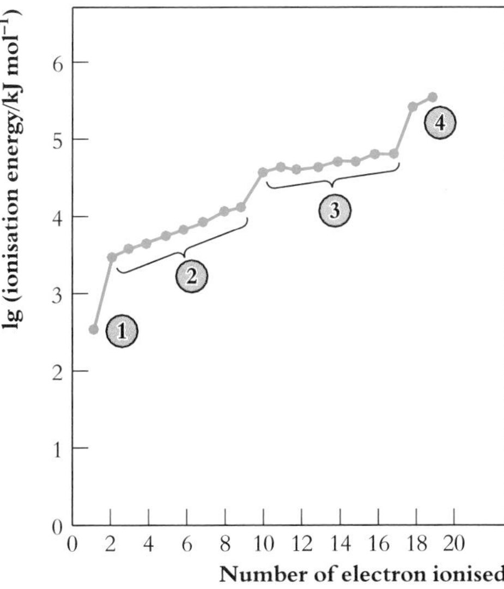

4 The two electrons with the highest ionisation energies are closest to the nucleus, and form the n = 1 shell.

3 These eight electrons are in the next shell, the n = 2 shell.

2 These eight electrons are in the n= 3 shell.

1 This electron has the lowest ionisation energy. It is the easiest to remove. It is in the n = 4 shell.

Measurements of successive ionisation energies support the idea of shells

Evidence for the arrangement of electrons in shells of different energies is provided by values of successive **ionisation energies** for elements. Figure 2.5C shows a graph of the logarithm of the ionisation energy required for the removal of one electron after another from a potassium atom. (A logarithmic plot is used in order to give a condensed graph.) You can see that the electrons fall into four groups. The higher the ionisation energy, the more difficult the electrons are to remove and the nearer they must be to the nucleus.

2.6 THE HISTORY OF THE PERIODIC TABLE

There are over 70 metallic elements, and over 20 non-metallic elements. As you know, elements are classified as metallic and non-metallic elements.

Table 2.6
Properties of metallic and non-metallic elements

Metallic elements	Non-metallic elements
Solids (except mercury, a liquid).	Solids or gases (except bromine, a liquid).
A fresh surface is shiny; corrosion can occur.	Have no one characteristic appearance.
Malleable (can be hammered) and ductile (can be drawn into wire).	Shatter when attempts are made to change the shape.
Conduct heat and electricity.	Are poor thermal conductors and electrical conductors, with exceptions.
The oxides are basic.	The oxides are acidic or neutral.

Classification of elements ...

... as metallic and non-metallic ...

... and metalloid

For a long time chemists look at ways of dividing up the two big groups, metallic and non-metallic elements, into smaller sub-groups. They drew up groups of similar elements, such as the very reactive metals lithium, sodium and potassium. They grouped together a set of slightly less reactive metals, calcium, strontium and barium. Another such group was the very reactive non-metals, chlorine, bromine and iodine. Some elements were discovered which had properties in between metallic and non-metallic. These elements, e.g. silicon, were described as **metalloids**.

Newlands arranged the elements in order of relative atomic mass

In 1866, a British chemist called John Newlands had the idea of arranging the elements in order of their relative atomic masses:

H Li Be B C N O F Na Mg Al Si P S Cl K Ca

[For the symbols of elements see p. 732.]

Newlands noticed that similar elements appeared at regular intervals in the list. He arranged the elements in columns [Figure 2.6A].

Figure 2.6A
Newland's octaves of elements

H	Li	Be	B	C	N	O
F	Na	Mg	Al	Si	P	S
Cl	K	Ca	Cr	Ti	Mn	Fe

He drew up a Law of Octaves ...

In the first column were hydrogen and the very reactive non-metallic elements fluorine and chlorine. In the second column were the very reactive metals sodium, lithium and potassium. In the third column were the metals beryllium, magnesium and calcium. Carbon and silicon both fell into the fifth column, and oxygen and sulphur both fell into the sixth column.

... but his ideas were scorned

Newlands compared his chemical 'octaves' with musical octaves, and called the resemblance the **Law of Octaves**. The comparison was unfortunate: people poured scorn on his ideas.

Mendeleev carried on the work on classification ...

... and drew up his Periodic Table

It was a Russian chemist, Dimitri Mendeleev, who developed Newlands' idea and persuaded chemists to use it. In 1869, Mendeleev summarised his **periodic law** in the statement: *The properties of chemical elements are not arbitrary, but vary with their relative atomic masses in a systematic way*. He arranged the elements in order of increasing relative atomic mass [see Figure 2.6B]. A modern version of his classification, which we call the Periodic Table, is shown on p. 732. You will notice that Mendeleev's Periodic Table lacks the noble gases (helium, neon, argon, etc. in Group 0) because they had not yet been discovered! A vertical row of elements is called a **group** and a horizontal row is called a **period**.

Figure 2.6B
Part of Mendeleev's Periodic Table of 1871

	Gp 1	Gp 2	Gp 3	Gp 4	Gp 5	Gp 6	Gp 7	Gp 8
Row 1	H							
Row 2	Li	Be	B	C	N	O	F	
Row 3	Na	Mg	Al	Si	P	S	Cl	
Row 4	K	Ca	–	–	–	–	–	Ti V Cr Mn Fe Co Ni
Row 5	Cu	Zn	–	–	As	Se	Br	

For the full Periodic Table, see p. 732

Mendeleev's improvements included long periods to accommodate transition metals ...

... spaces ...

... new values of relative atomic mass ...

... predictions about undiscovered elements

Mendeleev made various improvements on Newlands' system.

1. Mendeleev introduced long rows or **periods** for the elements we now call **transition metals**. This meant that the metals Ti, Mn, Fe were no longer placed under the non-metals Si, P, S [see Figure 2.6A].

2. He left spaces. When he saw that arsenic fitted naturally into Group 5 he left two spaces between zinc and arsenic.

3. When elements did not fit comfortably into the slots in the Periodic Table dictated by their relative atomic masses, Mendeleev made new measurements of relative atomic mass. In each case (Cr, In, Pt, Au) the new value justified the arrangement in Mendeleev's Periodic Table.

4. Where he had left gaps in the Periodic Table, Mendeleev predicted that new elements would be discovered to fill the gaps. He had some outstanding successes in predicting the properties of elements. When elements were discovered and found to have the relative atomic mass and the physical and chemical properties Mendeleev had predicted, faith in the Periodic Table soared.

The Periodic Table helped chemists in their search for the elements which were still to be discovered. For example, Mendeleev predicted that an element would be discovered to fill the space under silicon and above tin. His predictions for the element which he called ekasilicon (below silicon) were fulfilled by the properties of the element germanium, which was discovered in 1886 by Winkler.

Similar agreement was found between the predicted properties of eka-aluminium and gallium, which was discovered in 1875 and between ekaboron and scandium, discovered in 1879.

The noble gases were discovered and found to fit into a new group of the Periodic Table ...

The **noble gases** had not been discovered when the Periodic Table was drawn up. As they were discovered one by one, they were found to fit in between the halogens in Group 7 and the alkali metals in Group 1. A separate Group 0 was added to the right-hand side of the table. Argon, however, has a higher relative atomic mass than potassium ($A_r(Ar) = 40$; $A_r(K) = 39$) [§ 3.4]. It made more sense chemically to put potassium with the alkali metals, rather than keep to the order of relative atomic masses. Another example of this kind was the positions of tellurium and iodine. Relative atomic masses placed tellurium under bromine, and iodine under sulphur and selenium; chemical properties placed them in the reverse order.

... Discrepancies were resolved by arranging elements in order of atomic number

Moseley's work on X rays in 1914, solved this problem. He showed that the atomic numbers (proton numbers) of elements are more significant than their relative atomic masses. This discovery was the final step in the validation of the Periodic Table. In the modern Periodic Table elements are arranged in order of proton number (atomic number).

2.7 FEATURES OF THE PERIODIC TABLE

The Periodic Table is divided into the s block, the d block and the p block

What patterns can be seen in the arrangement of the elements in the Periodic Table? First, note the positions occupied by metallic and non-metallic elements. The reactive metals are at the left-hand side of the table, less reactive metallic elements in the middle block and non-metallic elements at the right-hand side.

The metals in Groups 1 and 2 are described as **s block elements** because their outer electron subshells contain s electrons. The metals in the block between Group 2 and Group 3 are called **transition metals**. They are described as **d block elements** because they have incomplete d subshells. Groups 3 to 7 form the **p block** of the Periodic Table [see § 15.5]. As you have seen, their outer electrons are p electrons.

2.7.1 LOOKING AT THE GROUPS

The very reactive alkali metals of Group 1 ...

From the summary in Table 2.7A, you can see how the Periodic Table makes it easier to learn about all the elements. Look at the elements in Group 1: lithium, sodium, potassium, rubidium and caesium. They are all very reactive metals. Their oxides and hydroxides have the general formulae $\mathbf{M}_2O$ and $\mathbf{M}OH$ (where $\mathbf{M}$ is the symbol for the metallic element) and are strongly basic. The oxides and hydroxides dissolve in water to give strongly alkaline solutions. The metals in Group 1 are called the **alkali metals**. Their reactivity increases as you pass down the group. If you know these fact, you do not need to learn the properties of all the metals separately. If you know the properties of sodium, you can predict those of potassium and lithium. Think of having to learn the properties of 106 elements separately! The Periodic Table saves you from this.

The metals in Group 2 are less reactive than those in Group 1. They form basic oxides and hydroxides with the general formulae $\mathbf{M}O$ and $\mathbf{M}(OH)_2$. Their oxides and hydroxides are either sparingly soluble or insoluble. These elements

Table 2.7A
Some reactions of metals

Element	Reaction with air	Reaction with water	Reaction with dilute acids
Group 1 Lithium Sodium Potassium Rubidium Caesium	Burn vigorously to form the strongly basic oxide, $\mathbf{M}_2O$	React vigorously to form hydrogen and a solution of the strong alkali, **M**OH	The reaction is dangerously violent. The vigour of these reactions increases down the group.
Group 2 Beryllium	Burn to form the strongly basic oxide, **M**O	Reacts very slowly.	React readily to give hydrogen and a salt, e.g. $\mathbf{M}Cl_2$. The vigour of these reactions increases down the group.
Magnesium		Burns in steam	
Calcium Strontium Barium		React readily to form hydrogen and the alkali $\mathbf{M}(OH)_2$	
Transition metals Iron	When heated, form oxides, without burning.	Rusts slowly, reacts with steam to form hydrogen and iron oxide.	Reacts to give hydrogen and a salt.
Copper		Does not react.	Does not react.

... the alkaline earth metals of Group 2 ...

are called the **alkaline earths**. Their reactivity increases as you pass down the group. Again, if you know the chemical reactions of one element, you can predict the reactions of other elements in the group.

... and the transition metals

The transition metals are less reactive than those in Groups 1 and 2. Their oxides and hydroxides are less strongly basic and are insoluble.

The noble gases of Group 0

The elements in Group 0 are the **noble gases**, formerly called the **inert gases**. They are present in air. They are the least reactive of the elements. For many years, it seemed as though they took part in no chemical reactions. However, in 1960, two of them, krypton and xenon, were made to combine with the very reactive element, fluorine.

The halogens of Group 7

Group 7 precedes Group 0. It contains a set of very reactive non-metallic elements: fluorine, chlorine, bromine, iodine and astatine. They are called the **halogens** (derived from the Greek: halogen = salt-former) because they react with metals to form salts. Example of their chemical reactions are given in Table 2.7B.

Table 2.7B
Some reactions of the halogens

Halogen	State at room temperature	Reaction with sodium	Reaction with iron	Trend
Fluorine	Gas	Explosive	Explosive	The vigour of these reactions decreases down the group.
Chlorine	Gas	Heated sodium burns in chlorine to form sodium chloride.	Reacts vigorously with hot iron to form iron(III) chloride.	
Bromine	Liquid	Reacts less vigorously to form sodium bromide.	Reacts less vigorously to form iron(III) bromide	
Iodine	Solid	Reacts less vigorously than bromine to form sodium iodide.	Reacts less vigorously than bromine to form iron(II) iodide.	

SUMMARY

The members of a group are similar and show a trend in properties and reactions down the group.

Group 6

At the right-hand side of each period, preceding the halogens, are the non-metallic elements of Group 6: oxygen, sulphur, selenium and tellurium. They are a set of elements with the ability to form two chemical bonds. They show an increase in metallic character from oxygen (a non-metal) to tellurium (a semi-metal or metalloid).

Group 5

The elements nitrogen, phosphorus, arsenic, antimony and bismuth form a group with the ability to form three or five chemical bonds. They show a gradation in properties from nitrogen and phosphorus (non-metals) through metalloid arsenic to antimony and bismuth (metals). These elements are in Group 5.

Group 4 shows the largest gradation in properties from top to bottom of the group

The elements carbon, silicon, germanium, tin and lead have the ability to form four chemical bonds. A marked gradation in properties occurs from carbon (non-metal) through silicon and germanium (metalloids) to tin and lead (metals). In Group 4, there is the maximum gradation in properties down the group.

Group 3

The elements aluminium, gallium, indium and thallium form ions with a charge of +3. Boron is a metalloid, while the rest of the elements in Group 3 are metals.

CHECKPOINT 2.7A: THE PERIODIC TABLE I

Refer to the complete Periodic Table on p. 732.

1. Locate the position in the Periodic Table of francium (Fr, atomic number 87). Make as many predictions as you can about the properties of francium, e.g. whether it is a solid or a liquid or a gas, whether it is a metal or a non-metal, its chemical reactions with air, water and dilute acids.

2. Use the Periodic Table to predict the properties of astatine (At, atomic number 85), e.g. whether it is a solid or a liquid or a gas, whether it is a metal or a non-metal, its chemical reactions.

3. Find the position of beryllium, Be, in the Periodic Table. Say what you think will happen when beryllium is added to cold water.

4. Locate radium (Ra, atomic number 88) in the Periodic Table. Predict *(a)* the formula of radium chloride, *(b)* the nature and formula of radium oxide, and *(c)* the products and speed of the reactions of radium with water.

2.7.2 LOOKING AT THE PERIODS

Period 1 contains hydrogen and helium. Period 2 contains the elements from lithium to neon. Table 2.7C shows some of the patterns which are seen in passing across Period 3. The properties shown are metallic and non-metallic character, the charge on the ions which the element forms, and the structure of the element and its oxide. You will be learning more about structure in Chapters 4 and 6. This table simply tells whether the substance consists of

1. individual atoms, e.g. argon which consists of Ar atoms
2. individual molecules, e.g. sulphur dioxide which consists of SO_2 molecules
3. a giant ionic structure containing millions of ions bonded together, e.g. sodium chloride which consists of Na^+ ions and Cl^- ions in equal numbers
4. a giant covalent structure containing millions of atoms bonded together, e.g. silicon(IV) oxide, SiO_2 which contains silicon atoms and oxygen atoms in the ratio of 2 atoms of oxygen for every silicon atom
5. a giant metallic structure, which consists of millions of atoms of a metallic element bonded together by the metallic bond [see § 6.3].

SUMMARY

Across a period, there is a transition in properties, structure, reactivity and the nature of compounds.

Table 2.7C
Trends across Period 3

Group	1 Na	2 Mg	3 Al	4 Si	5 P	6 S	7 Cl	0 Ar
Character	Metallic			Metalloid	Non-metallic			Noble gas
Reactivity	— Decreases →			–	← Decreases —			–
Structure of element	Giant metallic			Giant covalent	Molecular			Atomic
Ion	Na^+	Mg^{2+}	Al^{3+}	None	P^{3-}	S^{2-}	Cl^-	None
Oxide	Na_2O	MgO	Al_2O_3	SiO_2	P_2O_5 P_2O_3	SO_2 SO_3	Cl_2O Cl_2O_7 etc.	None
Type of oxide	Strongly basic		Some acidic and some basic character		Acidic			None
Structure of oxide	Giant ionic			Giant covalent	Molecular			None
T_m/°C	97.8	650	660	1410	44 (wh) 590 (red)	113 (rh)	−101	−189
T_b/°C	890	1110	2470	2360	280 (wh)	445	−34.7	−186
Electric conductivity/ $\mu\Omega^{-1}\,cm^{-1}$	0.218	0.224	0.382	0.10	10^{-7}	10^{-23}	---	---
First ionisation energy/ $kJ\,mol^{-1}$	494	736	577	786	1060	1000	1260	1520

(Note: wh = white, rh = rhombic, T_m = melting temperature, T_b = boiling temperature)

From left to right across a period, there is a transition

- *from metals to non-metals*
- *from elements which form positive ions to elements which form negative ions*
- *from giant structures to molecular structures*
- *from basic, ionic oxides to acidic, molecular oxides*

CHECKPOINT 2.7B: THE PERIODIC TABLE II

1. Look at the elements in Period 2. Say what you would expect to be the acidic or basic nature of the oxides of lithium, beryllium, boron and carbon.

2. State what types of structure you would expect for the elements lithium, beryllium, nitrogen, oxygen, fluorine and neon.

3. Look at the elements of Period 4. Say what formula or formulae you would expect for the oxides of gallium, germanium, arsenic and selenium.

4. *(a)* What does the change in melting temperature across period 3 tell you about the chemical bonds in the elements?
   (b) Why does the melting temperature increase from Na to Mg to Al?
   (c) Why is the melting temperature of argon so very low?

5. Why is there a big difference in electric conductivity between sodium, magnesium, aluminium and the other elements?

2.8 THE ELECTRONIC CONFIGURATIONS OF THE ELEMENTS

The electronic configurations of elements can be represented by 'electrons-in-boxes'

In the Periodic Table, elements are arranged in order of their proton numbers (atomic numbers). The proton number, Z, is the nuclear charge and also the number of electrons in an atom of the element.

Hydrogen has one electron. It occupies the lowest energy level, 1s. — H — 1s [↑] This can be written as 1s

The orbitals (boxes) of lowest energy are filled first

Helium has two electrons which can occupy the same 1s orbital with opposing spins. — He — 1s [↑↓] written as $1s^2$ The superscript 2 gives the number of electrons in the subshell.

Lithium has three electrons. Two occupy the 1s box, which is then full. The third must go into a box at the next level. — Li — 2s [↑] 1s [↑↓] written as $1s^22s$

Beryllium has four electrons. — Be — 2s [↑↓] 1s [↑↓] $1s^22s^2$

Boron has five electrons. The fifth electron cannot enter the 1s or 2s boxes: it occupies an orbital in the 2p sub-shell. — B — 2p [↑][ ][ ] 2s [↑↓] 1s [↑↓] $1s^22s^22p$

Carbon has six electrons. There are three ways in which the fifth and six electrons can be accommodated in the 2p subshell. The two 2p electrons may be *(a)* in the same box or in different boxes with the spins *(b)* parallel or *(c)* opposed. In fact arrangement *(b)* is favoured.

(a) 2p [↑↓][ ][ ]
(b) 2p [↑][↑][ ]
(c) 2p [↑][↓][ ]

C — 2p [↑][↑][ ] 2s [↑↓] 1s [↑↓] $1s^22s^22p^2$

Electrons prefer to occupy orbitals singly

Only when all the orbitals in a subshell contain an electron do electrons begin to occupy orbitals in pairs

Hund's Multiplicity Rule comes into play here. According to Hund's rule, electrons do not pair in an orbital until all the other orbitals in the subshell have been occupied by a single electron.

Nitrogen ($Z = 7$) obeys Hund's rule by accommodating the three 2p electrons in different boxes with the same spins. — N — 2p [↑][↑][↑] 2s [↑↓] 1s [↑↓] $1s^22s^22p^3$

The 1s shell is filled first ...

... then the 2s subshell, followed by the 2p subshell ...

Oxygen ($Z = 8$). The fourth 2p electron pairs with one of the other three 2p electrons.

O 2p [↑↓|↑|↑]
2s [↑↓]
1s [↑↓] $1s^2 2s^2 2p^4$

... then the 3s subshell, followed by the 3p subshell

Fluorine ($Z = 9$) has the arrangement $1s^2 2s^2 2p^5$

F 2p [↑↓|↑↓|↑]
2s [↑↓]
1s [↑↓] $1s^2 2s^2 2p^5$

The 4s subshell is filled and followed by the 3d subshell ...

Neon ($Z = 10$) $1s^2 2s^2 2p^6$, has a full 2p subshell, thus completing the $n = 2$ shell. The next element, **sodium** ($Z = 11$) has to utilise the $n = 3$ shell, starting with the 3s subshell. The diagrams for sodium and the twelfth element, **magnesium**, are

... before the 4p subshell is filled

Na 3s [↑]
2p [↑↓|↑↓|↑↓]
2s [↑↓]
1s [↑↓] $1s^2 2s^2 2p^6 3s$

Mg 3s [↑↓]
2p [↑↓|↑↓|↑↓]
2s [↑↓]
1s [↑↓] $1s^2 2s^2 2p^6 3s^2$

The following six elements have electrons in the 3p subshell. The configurations of the next six elements are (writing (Ne) for $1s^2 2s^2 2p^6$)

Al	($Z = 13$)	$(Ne)3s^2 3p$
Si	($Z = 14$)	$(Ne)3s^2 3p^2$
P	($Z = 15$)	$(Ne)3s^2 3p^3$
S	($Z = 16$)	$(Ne)3s^2 3p^4$
Cl	($Z = 17$)	$(Ne)3s^2 3p^5$
Ar	($Z = 18$)	$(Ne)3s^2 3p^6$

Electron configuration and the Periodic Table

How do the elements fit into the Periodic Table? So far we have:

Filling the $n = 1$ shell: H and He: First period

Filling the $n = 2$ shell: Li to Ne: Second period

Filling the $n = 3$ shell: Na to Ar: Third period

The next elements start the fourth period. They are

K ($Z = 19$) $(Ar)4s$

Ca ($Z = 20$) $(Ar)4s^2$

Once the 4s subshell is full, the 3d orbitals are filled [see Figure 2.5B, § 2.5]. Over the next 10 elements, electrons enter the 3d subshell. These elements are

The fourth period

Sc ($Z = 21$) $(Ar)4s^2 3d$

to Zn ($Z = 30$) $(Ar)4s^2 3d^{10}$

Filling the d subshell ...

While the d subshell fills, the chemistry of the elements is not greatly affected. The metals scandium to zinc are a very similar set of metals, called **transition metals** [Chapter 24]. The elements gallium ($Z = 31$) to krypton ($Z = 36$) complete the $n = 3$ shell by filling the 4p orbitals. The 18 elements from potassium to krypton comprise the **first long period**.

Electronic configurations can be written as e.g. B(2.3), C(2.4), N(2.5), O(2.6), F(2.7), Mg(2.8.2), where the numbers give the numbers of electrons in the $n = 1$, $n = 2$ and $n = 3$ shells.

Table 2.8
Electronic configurations of the atoms of the first 36 elements

	Z	1s	2s	2p	3s	3p	3d	4s	4p	4d	4f	5s	5p	5d	5f	6s	6p	6d	6f	7s
H	1	1																		
He	2	2																		
Li	3	2	1																	
Be	4	2	2																	
B	5	2	2	1																
C	6	2	2	2																
N	7	2	2	3																
O	8	2	2	4																
F	9	2	2	5																
Ne	10	2	2	6																
Na	11	2	2	6	1															
Mg	12				2															
Al	13				2	1														
Si	14	10 electrons			2	2														
P	15				2	3														
S	16				2	4														
Cl	17				2	5														
Ar	18	2	2	6	2	6														
K	19	2	2	6	2	6		1												
Ca	20							2												
Sc	21						1	2												
Ti	22						2	2												
V	23						3	2												
Cr	24						5	1												
Mn	25						5	2												
Fe	26						6	2												
Co	27	18 electrons					7	2												
Ni	28						8	2												
Cu	29						10	1												
Zn	30						10	2												
Ga	31						10	2	1											
Ge	32						10	2	2											
As	33						10	2	3											
Se	34						10	2	4											
Br	35						10	2	5											
Kr	36	2	2	6	2	6	10	2	6											

CHECKPOINT 2.8: ELECTRONIC CONFIGURATIONS

1. There are six calcium isotopes, of nucleon number 40, 42, 43, 44, 46 and 48. How many protons and neutrons are there in the nuclei?

2. Draw 'electrons-in-boxes' diagrams of the electronic configuration of the following atoms, given the proton number (Z): boron (5), fluorine (9), aluminium (13) and potassium (19).

3. Draw diagrams to show the electronic configurations of the ions: K^+, Cl^-, Ca^{2+}, O^{2-}, Al^{3+}, H^-. (Proton numbers are K = 19, Cl = 17, Ca = 20, O = 8, Al = 13 and H = 1.)

4. Write down the electronic configurations of the atoms with the proton numbers 4, 7, 18, 27, 37. State to which Group of the Periodic Table each element belongs.

5. Write the electronic configurations of the following species (e.g., Li = $1s^2 2s$). Their proton numbers range from Na = 11 to Ar = 18.

   Na^+, Mg^{2+}, Al, Si, P, S, S^{2-}, Cl, Cl^-, Ar

2.9 THE REPEATING PATTERN OF THE ELEMENTS

We have spent time looking at the structure of the atom [Chapter 1] and the electron configurations of the elements [§§ 2.5 and 2.8]. How does this study fit in with the Periodic Table? If you look at the electron configurations of the atoms, some interesting points strike you.

The noble gases have a full outer shell of electrons (helium 2; the other gases 8)

First, notice the elements with a full outer shell of electrons. These are helium (2), neon (2.8), argon (2.8.8), krypton and xenon. These elements are the noble gases (Group 0). Their lack of chemical reactivity has been mentioned. They exist as single atoms. Their atoms do not combine in pairs to form molecules as do the atoms of most gaseous elements (e.g. O_2, H_2). It seems only logical to suppose that it is the full outer shell of electrons that makes the noble gases chemically unreactive.

For members of other groups, the number of electrons in the outermost shell equals the group number

Following each noble gas (that is, with atomic number 1 greater than each noble gas) is an alkali metal [see Table 2.9]. These elements are lithium (2.1), sodium (2.8.1), potassium (2.8.8.1), rubidium and caesium. We can infer that it is because the alkali metals all have a single electron in the outer shell that they all behave in a very similar way.

Following each alkali metal is an alkaline earth metal [see Table 2.9]. The alkaline earths, beryllium (2.2), magnesium (2.8.2), calcium (2.8.8.2), strontium and barium all have two electrons in the outer shell. It seems logical to suppose that it is the similar configuration of electrons that gives the elements similar properties.

Preceding each noble gas (with atomic number 1 less than the noble gas) are the halogens of Group 7: fluorine (2.7), chlorine (2.8.7), bromine and iodine [see Table 2.7B]. We can infer that the halogens all have similar chemical reactions because they all have the same number of electrons in the outer shell.

Table 2.9
A section of the Periodic Table

								Group 0
Period 1	H (1)							He (2)
	Group 1	**Group 2**	**Group 3**	**Group 4**	**Group 5**	**Group 6**	**Group 7**	
Period 2	Li (2.1)	Be (2.2)	B (2.3)	C (2.4)	N (2.5)	O (2.6)	F (2.7)	Ne (2.8)
Period 3	Na (2.8.1)	Mg (2.8.2)	Al (2.8.3)	Si (2.8.4)	P (2.8.5)	S (2.8.6)	Cl (2.8.7)	Ar (2.8.8)
Period 4	K (2.8.8.1)	Ca (2.8.8.2)						

SUMMARY

The position of an element in the Periodic Table is related to the electronic configuration of its atoms.

You can see the following features:

1. The elements are listed in order of increasing atomic number.
2. Elements which have the same number of electrons in the outermost shell fall into the same group of the Periodic Table.
3. The first period contains only hydrogen and helium. The second period contains the elements lithium to neon. The third period contains the elements sodium to argon.

ATOMIC STRUCTURE

Subatomic particles
Proton: mass = 1 u, charge = +1 emu
Neutron: mass = 1u, charge = 0
Electron: mass = 0.005 u,
charge = −1 emu §§ 1.4–7
The subatomic particles can be distinguished by their behaviour in electric and magnetic fields. § 1.10.2

Atomic number = proton number = number of protons in an atom of the element. § 1.7
Mass number = nucleon number = number of protons + neutrons in an atom of the element. § 1.8
Isotopes have the same atomic number but different mass numbers. § 1.8

Mass spectrometry is used for

- determination of M_r
- elucidation of the structure of a compond
- showing the isotopic composition of an element. § 1.9

Radioisotopes give off α, β and γ radiations which differ in penetrating power. They have a characteristic half-life. Isotopes are used in medicine and in industry. § 1.10

THE ATOM
Protons
Neutrons
Electrons
§§1.2–7

The protons and neutrons occupy the **nucleus** of the atom. The electrons move round the nucleus in **orbitals**. § 1.5
In shape, s orbitals are spherical while p orbitals and d orbitals are directed in space. § 2.4.3

The **electron configuration** of an element can be predicted from its **atomic number** and the relative energies of orbitals with the same principal quantum number, e.g. the element of atomic number 15 has the configuration $1s^22s^22p^63s^23p^3$. § 2.5, § 2.8

The **periodic variation** of first ionisation energy is evidence for the energy levels of s, p and d orbitals.
Across a period, ionisation energy increases because the nuclear charge increases.
Down a group, ionisation energy decreases because the valence electrons are shielded from the nucleus by additional shells of electrons. § 15.2

The **first ionisation energy** of an element is the energy absorbed per mole of the reaction:
$X(g) \rightarrow X^+(g) + e^-$ § 2.3.2

First **election affinity**: energy absorbed in $X(g) + e^- \rightarrow X^-(g)$.
Second electron affinity etc. § 2.3.3, § 10.12

The **electron configuration** of an element can be predicted from **successive ionisation energies**. § 2.5

Electron configurations determine the chemical properties of elements and form the basis of the s, p and d blocks of the **Periodic Table**. §§ 2.5–7

4. The chemical properties of elements depend on the electron configurations of their atoms.

5. The electron configuration of its atoms is related to the position of an element in the Periodic Table.

6. The noble gases (Group 0) all have a full outer shell of electrons.

7. The reactive alkali metals (Group 1) have a single electron in the outer shell.

8. The alkaline earth metals (Group 2) have two electrons in the outer shell.

9. The halogens (Group 7) have 7 electrons in the outer shell.

CHECKPOINT 2.9: THE PERIODIC TABLE III

1. *(a)* What are the 'noble gases'?
   (b) In which group of the Periodic Table are they?
   (c) What do the noble gases have in common
   (i) regarding their electron configurations and
   (ii) regarding their chemical reactions?

2. **X** is a metallic element. It reacts slowly with water to give a strongly alkaline solution. In which group of the Periodic Table would you place **X**?

3. **Y** is a non-metallic element. It reacts vigorously with sodium to give a salt of formula Na**Y**. In which group of the Periodic Table would you place **Y**?

4. **Z** is a metallic element which reacts vigorously with water to give a strongly alkaline solution. In which group of the Periodic Table would you place **Z**?

QUESTIONS ON CHAPTER 2

1.

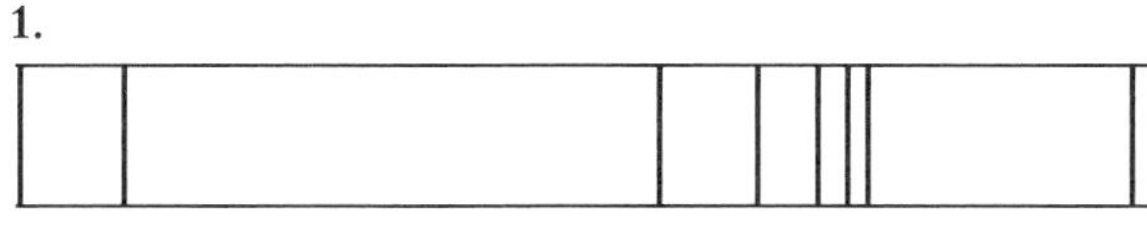

The figure represents the atomic emission spectrum of hydrogen. Explain why it is composed of lines, and say what each line indicates. Why do the lines become closer together as you read from left to right?

2. 'In its ground state, the electron in a hydrogen atom is in a 1s orbital'. Explain this statement.

3. *(a)* Describe how the first ionisation energy of an element can be determined experimentally.
   (b) Results obtained for the ionisation energies of boron are

Electron number	1st	2nd	3rd	4th	5th
Ionisation energy/kJ mol^{-1}	800	2400	3700	25 000	32 800

On graph paper, plot lg (ionisation energy/kJ mol^{-1}) against the number of electron removed. From the graph, deduce the most likely formula of boron chloride.

4. The emission spectrum of hydrogen consists of several series of lines. The series of highest energy is called the Lyman series (see Figure below). Each line in the series is the result of an electronic transition between energy levels.

Figure
Hydrogen emission spectrum, part of the Lyman series

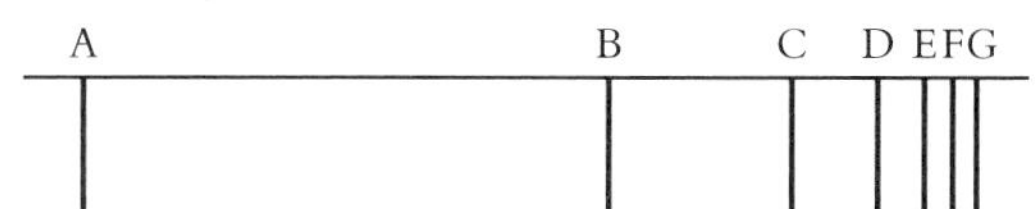

(a) State in which direction the energy increases: A to G or G to A.
(b) State in which direction the frequency increases A to G or G to A.
(c) Explain why the spectrum consists of lines.
(d) What do transitions in the same series all have in common?

5. *(a)* Hydrogen consists of diatomic molecules. Briefly explain how the *atomic* emission spectrum of hydrogen can be obtained.
   (b) How does this spectrum differ from that of light emitted by a tungsten light bulb?
   (c) Define the ionisation energy of hydrogen.
   (d) Explain how the value of the ionisation energy can be obtained from the atomic spectrum.

6. *(a)* In terms of the numbers of sub-atomic particles, state **one** difference and **two** similarities between two isotopes of the same element. **3**

(b) Give the chemical symbol, including its mass number, for an atom which has 3 electrons and 4 neutrons **1**

(c) (i) An element has an atomic number of 23. Its ion has a charge of 3+. Complete the electronic configuration of this ion: [Ar] ...

(ii) To which block in the Periodic Table does this element belong? **2**

(d) (i) Write an equation for the process involved in the first ionisation energy of boron.

(ii) Explain why the second ionisation energy of boron is greater than the first.

(iii) Explain why the fourth ionisation energy of boron is much greater than the third. **6**

12 *NEAB (AS/AL)*

7. *(a)* The graph below shows how the melting points of the elements vary across Period 3.

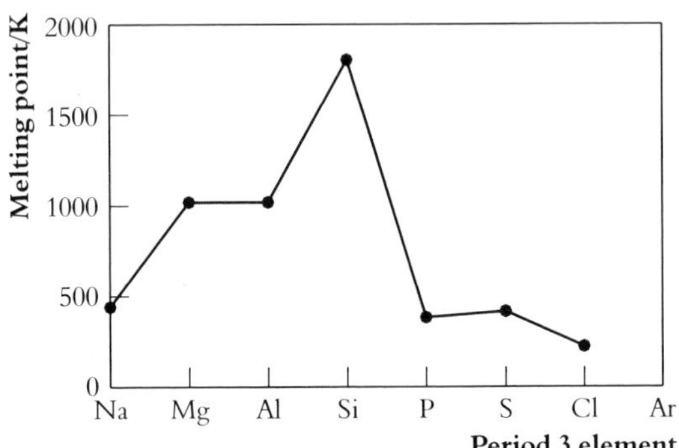

(i) Explain why magnesium has a higher melting point than sodium. **2**

(ii) Copy and complete the graph to show the likely melting point of argon. **1**

(iii) Explain why argon has the melting point which you have shown. **2**

(b) Write an equation for the formation of phosphorus(V) oxide from phosphorus and oxygen. **1**

(c) Write equations to show how sodium oxide and sulphur trioxide react with water and in **each case** predict the approximate pH of the resulting solution. **4**

10 *NEAB (AS/AL)*

8. *(a)* Figure 1 shows the mass spectrum of neon (Ne). Use the isotopic abundances (given in brackets) to calculate the relative atomic mass of neon. $1\frac{1}{2}$

(b) (i) Neon is a monatomic gas. Using your value for the relative atomic mass of neon obtained in part *(a)*, calculate the volume occupied by 3.03 g of neon at 298 K and 1.01×10^5 Pa pressure. $2\frac{1}{2}$

[1 mole of ideal gas occupies 22.4 dm^3 at 273 K and 1.01×10^5 Pa.]

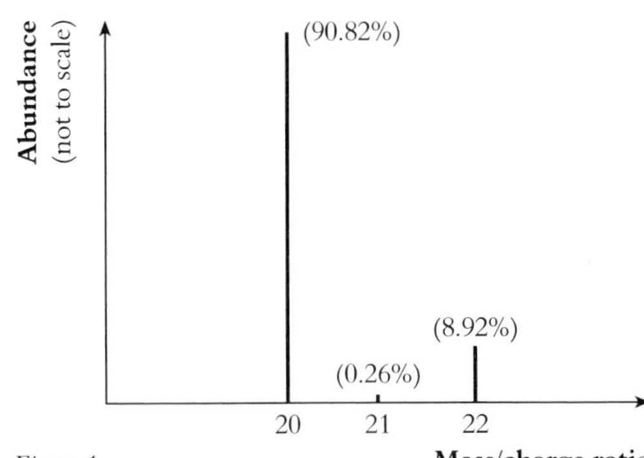

Figure 1

(ii) Calculate the volume which 3.03 g of neon would occupy at 298 K and 5.05×10^5 Pa pressure. **1**

(c) Figure 2 shows a plot of first ionisation energy against atomic number for the first thirteen elements of the Periodic Table.

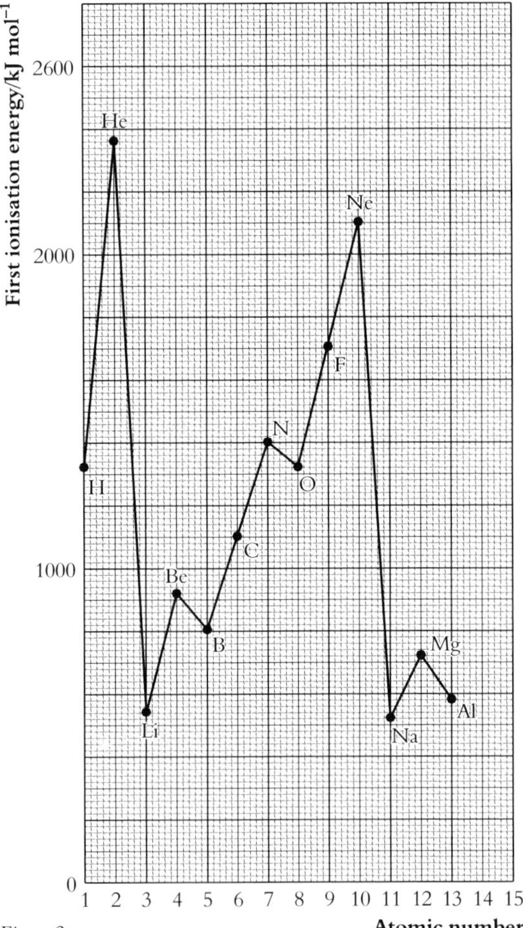

Figure 2

With reference to the *electronic configuration* of the elements explain why:

(i) the first ionisation energy for neon (Ne) is lower than that for helium (He); $1\frac{1}{2}$

(ii) the first ionisation energy for beryllium (Be) is greater than those for lithium (Li) **and** boron (B): **2**

(iii) the first ionisation energy for nitrogen (N) is greater than that for oxygen (O). 1

(d) Give the electronic configuration of an oxygen atom (O) by copying and labelling the boxes below and inserting arrows to represent electrons. $1\frac{1}{2}$

orbital

(e) What is the electronic configuration of the oxide ion (O^{2-}) and which **element** has the same electronic configuration? 1

(For (*b*) see § 3.12)

12 *WJEC (AS/AL)*

9. *(a)* Define the terms **mass number** and **isotope**. 3

(b) *(i)* Write an equation to represent the change associated with the second ionisation energy of sodium. 2

(ii) Examine the following first, second and successive ionisation energies $I_1, I_2 ... I_8$ (kJ mol^{-1}) for the elements **A** and **B**, which are in the *same period* of the Periodic Table.

I. State the Group Number for each of the elements: **A** and **B**.

II. Explain why I_2 is greater than I_1 for element **A**. 2

Element	I_1	I_2	I_3	I_4	I_5	I_6	I_7	I_8
A	496	4563	6913	9544	13352	16611	20115	25491
B	1000	2251	3361	4564	7012	8496	27107	31671

(c) The following is a graph of the logarithm of all the successive ionisation energies against the number of the electron removed, for a particular element.

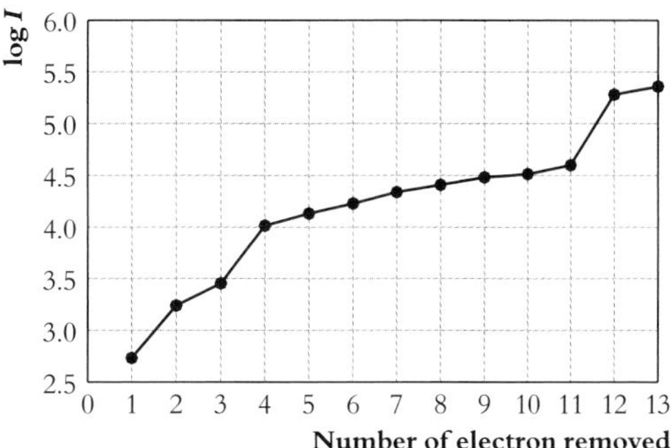

(i) From the graph deduce the electronic configuration of the element. 1

(ii) Explain the shape of the graph. 2

(d) (i) Using outer electrons only, draw a 'dot-and-cross' diagram to show the bonding in the molecule $AlCl_3$.

(ii) Explain why aluminium chloride forms a dimer Al_2Cl_6. 2

(iii) Write a balanced equation for the reaction of Al_2Cl_6 with excess water, given that the products are chloride ions and hexaaquaaluminium(III) ions. 1

(For (*d*) see § 19.4.2) *WJEC (AL)*

10. Consider the incomplete sketch below of the plot of first ionisation energies against atomic (proton) number for the 10 elements sodium to calcium.

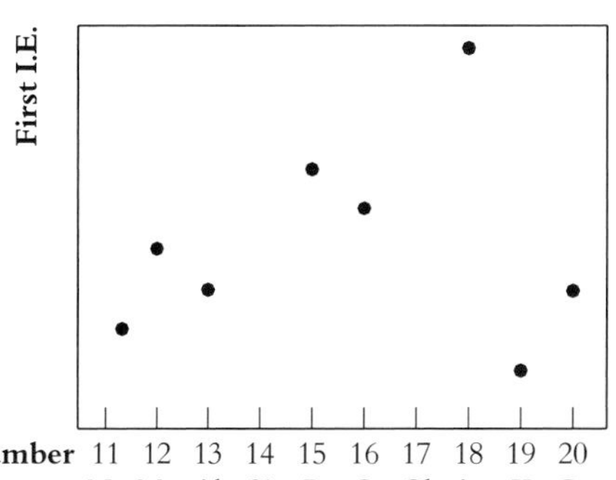

(a) (i) Copy and complete the diagram, by adding the first ionisation energies for silicon and chlorine. 2

(ii) Account for the rise in first ionisation energy between Na and Mg. 2

(iii) Account for the fall in first ionisation energy between Mg and Al. 2

(b) (i) State the overall trend in the atomic radius across the period from sodium to argon. 1

(ii) Account for this change in the atomic radius in terms of the structures of the atoms. 2

(c) The diagram below represents the Periodic Table, with four areas denoted by 1, 2, 3 and 4.

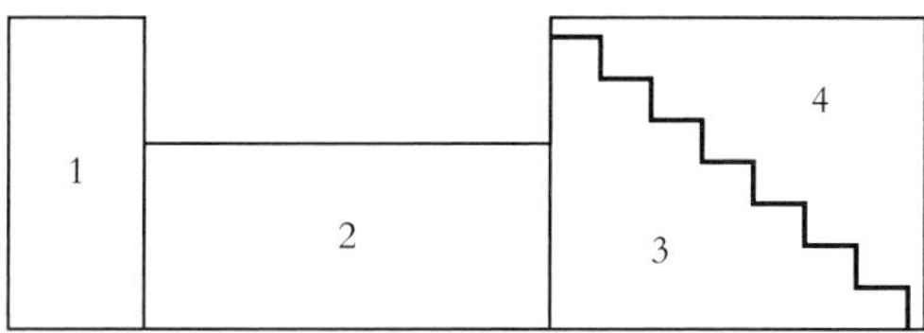

(i) Which area 1, 2, 3 or 4, is most likely to contain the non-metals? 1

(ii) Which area is most likely to contain metals with melting points over 1000 °C? 1

(iii) Which area is most likely to contain elements with oxides which dissolve in water to produce basic solutions? 1

(iv) Which area is most likely to contain elements which form coloured ions? 1

13 *O&C (AS/AL)*

3 EQUATIONS AND EQUILIBRIA

In this chapter we begin to see how equations, a knowledge of equilibria and calculations help chemical industry.

3.1 EQUATIONS, EQUILIBRIA AND CALCULATIONS – WHO NEEDS THEM?

Many cars have air bags which inflate if the car is involved in a collision and cushion the driver. A manufacturer has an order for gas generators for car air bags. How much sodium azide shall the manufacturer put into the gas generator? The bag must generate enough nitrogen to cushion the driver if there is a crash but not enough to push him into the back seat. To find the answer to this question, he has to know the equation for the reaction and he has to be able to relate the mass of solid sodium azide to the volume of nitrogen produced.

How much dynamite shall a miner use for blasting a rock face? He wants enough explosive to bring down enough rock salt to fill fifty trucks but not enough to blast him into the next world. To answer the question, he has to know the chemical composition of dynamite and the equation for the reaction and he has to be able to calculate the force produced from a mass of dynamite detonated.

Smith Brothers have an order for 1000 tonnes of fertiliser. What mass of ammonia do they need to buy so that they can fill the order? It's easy to work out if they know the equation. The manufacturer of ammonia will have to know about systems reaching equilibrium and how to use the best conditions to give the maximum yield of ammonia.

Brown Brothers have an order for 1000 tonnes of concrete. What mass of limestone do they need to buy so that they can fill the order? They can work it out in an instant if they know the equation for the decomposition of limestone. They will also need to know about the best conditions for ensuring that the decomposition of limestone carries on without reaching equilibrium.

3.2 FORMULAE

Every element has a symbol [see Periodic Table, p. 732]. A symbol is a letter or two letters which stand for one atom of the element. Formulae are written for compounds. The formula of a compound consists of the symbols of the elements present and the numbers which show the ratio in which the atoms are present. The compound sulphur dioxide has the formula SO_2. The compound sulphur trioxide has the formula SO_3. The formulae tell you the difference between them. Sulphur dioxide contains 2 oxygen atoms for every sulphur atom: the 2 below the line multiplies the O in front of it. Sulphur trioxide consist of molecules [see Figure 3.2A]. To show three molecules of sulphur dioxide, you write $3SO_2$.

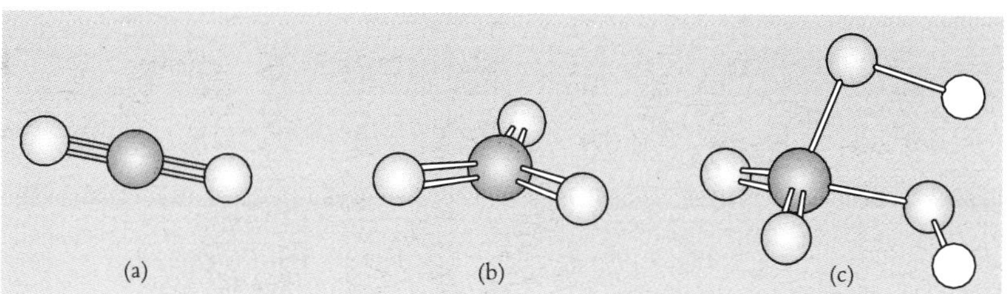

Figure 3.2A Models of molecules of (a) sulphur dioxide, (b) sulphur trioxide, (c) sulphuric acid

The formula of a compound is a set of symbols and numbers. The symbols say what elements are present in the compound. The numbers give the ratio of the numbers of atoms of the different elements in the compound

The formula of sulphuric acid is H_2SO_4. The compound contains two hydrogen atoms and four oxygen atoms for every sulphur atom; to write three molecules, you write $3H_2SO_4$. The 3 in front of the formula multiplies everything after it. In $3H_2SO_4$, there are 6 H, 3 S and 12 O atoms, a total of 21 atoms.

Many compounds do not consist of molecules; they consist of ions. The compound calcium hydroxide is composed of calcium ions, Ca^{2+}, and hydroxide ions, OH^-. There are twice as many hydroxide ions as calcium ions, so the formula for calcium hydroxide is $Ca(OH)_2$. The 2 multiplies the symbols in the brackets. There are 2 oxygen atoms, 2 hydrogen atoms and 1 calcium atom. This is not a molecule of calcium hydroxide; it is a formula unit of calcium hydroxide: one calcium ion and 2 hydroxide ions. A piece of calcium hydroxide contains this formula unit repeated many times. To write $4Ca(OH)_2$ means that the whole of the formula is multiplied by 4. It means 4 Ca, 8 O and 8 H atoms. § 4.3.7 deals with how to work out the formula of a compound. Table 3.2 lists the formulae of some common compounds.

Table 3.2 The formulae of some compounds

Water	H_2O	Aluminium chloride	$AlCl_3$
Sodium hydroxide	NaOH	Aluminium oxide	Al_2O_3
Sodium chloride	NaCl	Carbon monoxide	CO
Sodium sulphate	Na_2SO_4	Carbon dioxide	CO_2
Sodium nitrate	$NaNO_3$	Sulphur dioxide	SO_2
Sodium carbonate	Na_2CO_3	Ammonia	NH_3
Sodium hydrogencarbonate	$NaHCO_3$	Ammonium chloride	NH_4Cl
Calcium oxide	CaO	Hydrogen chloride	HCl
Calcium hydroxide	$Ca(OH)_2$	Hydrochloric acid	HCl(aq)
Calcium chloride	$CaCl_2$	Sulphuric acid	H_2SO_4(aq)
Calcium sulphate	$CaSO_4$	Nitric acid	HNO_3(aq)
Calcium carbonate	$CaCO_3$	Copper(II) oxide	CuO
		Copper(II) sulphate	$CuSO_4$

CHECKPOINT 3.2: FORMULAE

1. How many atoms are present in the following?

(a) C_6H_6 *(b)* P_4O_{10} *(c)* SO_2Cl_2
(d) $C_2H_4Cl_2$ *(e)* $2ZnSO_4$ *(f)* $5CuSO_4$
(g) $Al(NO_3)_3$ *(h)* $2Al(OH)_3$ *(i)* $Fe_2(SO_4)_3$
(j) $3Fe(NO_3)_3$

2. Give the formula of:

(a) sodium hydroxide, *(b)* hydrochloric acid,
(c) ammonia, *(d)* sodium chloride,
(e) calcium oxide, *(f)* calcium hydroxide,
(g) calcium carbonate, *(h)* sulphuric acid,
(i) nitric acid.

3.3 EQUATIONS

You have studied symbols for elements and formulae for compounds [§ 3.2]. These enable you to write equations for chemical reactions.

Example (a) Calcium carbonate decomposes to give calcium oxide and carbon dioxide.

Calcium carbonate ⟶ Calcium oxide + Carbon dioxide

Writing an equation ...

Replacing names with formulae, you can write the chemical equation for the reaction:

$$CaCO_3 \longrightarrow CaO + CO_2$$

On the left-hand side, you have 1 atom of calcium, 1 atom of carbon and 3 atoms of oxygen combined as calcium carbonate. On the right-hand side, you have 1 atom of calcium and 1 atom of oxygen combined as calcium oxide and 1 atom of carbon and 2 atoms of oxygen combined as carbon dioxide. The two sides are equal, and this is why the expression is called an equation.

... adding state symbols

You can give more information if you include state symbols in the equation. These are (s) = solid, (l) = liquid, (g) = gas, and (aq) = in aqueous (water) solution. Putting in the state symbols,

$$CaCO_3(s) \longrightarrow CaO(s) + CO_2(g)$$

tells you that solid calcium carbonate decomposes to form solid calcium oxide and carbon dioxide gas.

Example (b) Magnesium reacts with sulphuric acid to give hydrogen and a solution of magnesium sulphate.

Magnesium + Sulphuric acid ⟶ Hydrogen + Magnesium sulphate

The chemical equation is

$$Mg(s) + H_2SO_4(aq) \longrightarrow H_2(g) + MgSO_4(aq)$$

Hydrogen is written as H_2 because hydrogen gas consists of molecules containing two atoms.

Example (c) Hydrogen and oxygen combine to form water. The word equation is

Hydrogen + Oxygen ⟶ Water

The chemical equation could be

$$H_2(g) + O_2(g) \longrightarrow H_2O(l)$$

This equation is not balanced. There are 2 oxygen atoms on the left-hand side (LHS) and only 1 oxygen atom on the right-hand side (RHS). To balance the O atoms, multiply H_2O on the RHS by 2.

$$H_2(g) + O_2(g) \longrightarrow 2H_2O(l)$$

The O atoms are now balanced, but there are 4H atoms on the RHS and only 2H on the LHS. Multiplying H on the LHS by 2,

$$2H_2(g) + O_2(g) \longrightarrow 2H_2O(l)$$

The equation is now balanced.

Number of atoms on LHS = 4H + 2O
Number of atoms on RHS = 4H + 2O
Number of atoms of each element on LHS = Number of atoms of each element on RHS
The total mass of the reactants = The total mass of the products [see Figure 3.3A]

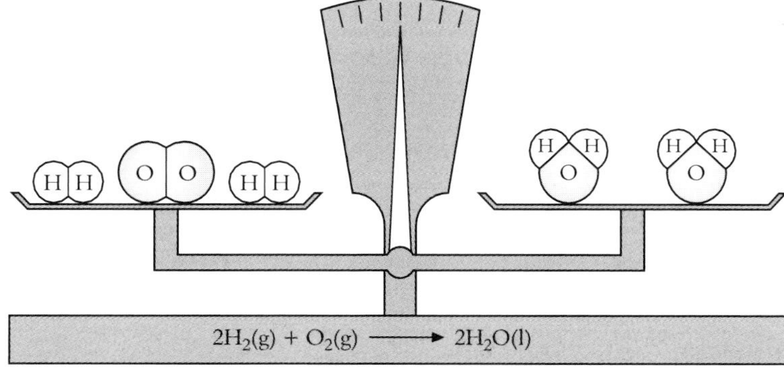

Figure 3.3A
A balanced equation (Example (c))

Example (d) Sulphur dioxide is oxidised by oxygen to sulphur trioxide.

Sulphur dioxide + Oxygen ⟶ Sulphur trioxide

$$SO_2(g) + O_2(g) \longrightarrow SO_3(g)$$

Balancing an equation ...

You can see that the equation is not balanced. There are 4O on the LHS and 3O on the RHS. It is tempting to write O for oxygen on the LHS. You must not do this. Never change a formula. All you can do to balance an equation is to multiply formulae. Instead of changing O_2 to O, multiply SO_2 and SO_3 by 2.

... by multiplying a formula or formulae ...

$$2SO_2(g) + O_2(g) \longrightarrow 2SO_3(g)$$

The equation is now balanced: 2S + 6O on the LHS; 2S + 6O on the RHS [see Figure 3.3B].

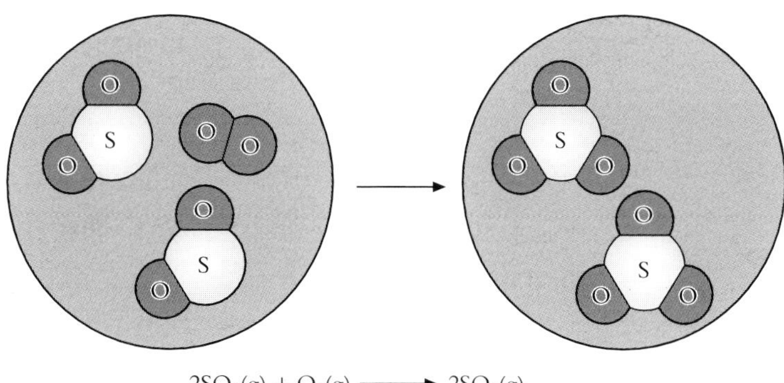

Figure 3.3B
A balanced equation (Example (d))

... never by changing a formula

Example (e) Sodium carbonate reacts with dilute hydrochloric acid to give carbon dioxide and a solution of sodium chloride.

Sodium carbonate + Hydrochloric acid ⟶ Carbon dioxide + Sodium chloride + Water

The chemical equation could be

$$Na_2CO_3(s) + HCl(aq) \longrightarrow CO_2(g) + NaCl(aq) + H_2O(l)$$

When you add up the atoms on the RHS, you find that they are not equal to the atoms on the LHS. The equation is not balanced. Start by balancing Na atoms.

Multiplying NaCl by 2,

$$Na_2CO_3(s) + HCl(aq) \longrightarrow CO_2(g) + 2NaCl(aq) + H_2O(l)$$

Now, there are 2 Na atoms on the RHS and 2Na atoms on the LHS. With 2Cl atoms on the RHS, the HCl on the LHS must be multiplied by 2.

$$Na_2CO_3(s) + 2HCl(aq) \longrightarrow CO_2(g) + 2NaCl(aq) + H_2O(l)$$

The equation is now balanced. Check again:

$$\text{Number of atoms on LHS} = 2Na + C + 3O + 2H + 2Cl$$
$$\text{Number of atoms on RHS} = 2Na + C + 3O + 2H + 2Cl$$

When you are balancing a chemical equation, the only way to do it is to put a number in front of a formula. You never try to alter a formula. In this example, you got 2Cl atoms by multiplying HCl by 2, not by altering the formula to HCl_2, which does not exist. You can multiply a formula, but you cannot change it.

Five steps in writing a balanced chemical equation

The steps in writing a chemical equation are:

1. Write a word equation for the reaction.
2. Write the symbols and formulae for the reactants and products.
3. Add the state symbols.
4. Balance the equation. Multiply the formulae if necessary. Never change a formula.
5. Check again:

   No. of atoms of each element on LHS
   = No. of atoms of each element on RHS

Example (f) Take the reaction between sodium and water to form hydrogen and sodium hydroxide solution. Work through the five steps:

1. Sodium + Water $\longrightarrow$ Hydrogen + Sodium hydroxide solution
2. $Na + H_2O \longrightarrow H_2 + NaOH$
3. $Na(s) + H_2O(l) \longrightarrow H_2(g) + NaOH(aq)$
4. $2Na(s) + 2H_2O(l) \longrightarrow H_2(g) + 2NaOH(aq)$
5. No. of atoms on LHS = 2Na + 4H + 2O

   No. of atoms on RHS = 2Na + 4H + 2O

3.3.1 IONIC EQUATIONS

When we study reactions in solution, we have a choice between two methods of writing the equations. One is to specify completely the reactants and products, e.g.

An ionic equation shows only the ions which react ...

Sodium sulphate + Barium chloride $\longrightarrow$ Barium sulphate + Sodium chloride

$$Na_2SO_4(aq) + BaCl_2(aq) \longrightarrow BaSO_4(s) + 2NaCl(aq)$$

... and omits spectator ions

The other method is to write an equation showing which ions take part in the reaction and ignoring the ions which do not change their bonding during the reaction, e.g.

$$\text{Sulphate ion} + \text{Barium ion} \longrightarrow \text{Barium sulphate}$$
$$SO_4^{2-}(aq) + Ba^{2+}(a) \longrightarrow BaSO_4(s)$$

The other ions present, in this case the sodium ions and chloride ions, pass through the reaction unchanged; they are described as **spectator ions**.

CHECKPOINT 3.3: EQUATIONS

1. Copy these equations, and balance them.

   (a) $Fe_2O_3(s) + C(s) \longrightarrow Fe(s) + CO(g)$
   (b) $Fe_2O_3(s) + CO(g) \longrightarrow Fe(s) + CO_2(g)$
   (c) $NH_3(g) + O_2(g) \longrightarrow NO(g) + H_2O(l)$
   (d) $Cr(s) + HCl(aq) \longrightarrow CrCl_3(aq) + H_2(g)$
   (e) $Fe_3O_4(s) + H_2(g) \longrightarrow Fe(s) + H_2O(l)$
   (f) $C_3H_8(g) + O_2(g) \longrightarrow CO_2(g) + H_2O(l)$

2. Try writing equations for the reactions:

   (a) Hydrogen + Copper(II) oxide ⟶ Copper + Water
   (b) Carbon + Carbon dioxide ⟶ Carbon monoxide
   (c) Magnesium + Sulphuric acid ⟶ Hydrogen + Magnesium sulphate
   (d) Copper + Chlorine ⟶ Copper(II) chloride
   (e) Mercury + Oxygen ⟶ Mercury(II) oxide
   (f) Iron + Sulphur ⟶ Iron(II) sulphide

3. Write balanced chemical equations for the reactions:

   (a) Calcium + Water ⟶ Hydrogen + Calcium hydroxide solution
   (b) Iron + Hydrochloric acid ⟶ Iron(II) chloride solution + Hydrogen
   (c) Iron + Chlorine ⟶ Iron(III) chloride
   (d) Aluminium + Chlorine ⟶ Aluminium chloride
   (e) Zinc + Steam ⟶ Zinc oxide + Hydrogen
   (f) Sodium + Oxygen ⟶ Sodium oxide

4. Write the full equation and the ionic equation for each of the following reactions.

   (a) Iron(II) sulphate + Sodium hydroxide ⟶ Iron(II) hydroxide + Sodium sulphate
   (b) Silver nitrate + Sodium bromide ⟶ Silver bromide + Sodium nitrate
   (c) Lead(II) nitrate + Sodium iodide ⟶ Lead(II) iodide + Sodium nitrate

3.4 RELATIVE ATOMIC MASS

The masses of atoms are very small, from 10^{-24} to 10^{-22} grams. Instead of using the actual masses of atoms, **relative atomic masses** (***A***$_\mathbf{r}$) are used. Originally, they were defined as

Atoms range in mass from 10^{-24} g to 10^{-22} g

$$\text{Original relative atomic mass} = \frac{\text{Mass of one atom of an element}}{\text{Mass of one atom of hydrogen}}$$

Since relative atomic masses are now determined by mass spectrometry, and since volatile carbon compounds are much used in mass spectrometry, the mass of an atom of ${}^{12}_{6}C$ is now taken as the standard of reference:

The definition of relative atomic mass...

$$\text{Relative atomic mass} = \frac{\text{Mass of one atom of an element}}{1/12 \text{ the mass of one atom of carbon-12}}$$

The difference between the two scales is small. On the carbon-12 scale, the relative atomic mass of ${}^{12}_{6}C$ is 12.0000, and the relative atomic mass of ${}^{1}_{1}H$ is 1.0078. The mass of a ${}^{12}_{6}C$ atom is 12.0000 u, and the mass of a ${}^{1}_{1}H$ atom is 1.0078 u [§ 1.3].

3.5 RELATIVE MOLECULAR MASS

The relative molecular mass, M_r, of a compound is the sum of the relative atomic masses of all the atoms in one molecule of a covalent compound

The mass of a molecule is the sum of the masses of all the atoms in it.

> The **relative molecular mass, M_r**, of a compound is the sum of the relative atomic masses of all the atoms in a molecule of the compound.

For example, you find the relative molecular mass of sulphuric acid in this way:

Formula of compound is H_2SO_4

2 atoms of H ($A_r = 1$)	= 2
1 atom of S ($A_r = 32$)	= 32
4 atoms of O ($A_r = 16$)	= 64
Total	= 98

Relative molecular mass, M_r, of $H_2SO_4 = 98$

The relative formula mass of a compound, with the same symbol, M_r, is the sum of the relative atomic masses of all the atoms in one formula unit of an ionic compound

Many compounds consist of ions, not molecules. For ionic compounds, the formula represents a formula unit, rather than a molecule of the compound. A formula unit of sodium sulphate is Na_2SO_4. The term **relative formula mass**, symbol M_r, can be used for ionic compounds. Many people use the term relative molecular mass, M_r, for ionic compounds as well as molecular compounds.

CHECKPOINT 3.5: RELATIVE MOLECULAR MASS

1. Work out the relative molecular masses of these compounds: NaOH, KCl, MgO, $Ca(OH)_2$, HNO_3, $CuCO_3$, NH_4NO_3, $CuSO_4$, $CuSO_4 \cdot 5H_2O$, $Mg(HCO_3)_2$

3.6 THE MOLE

Chemists often want to count out equal numbers of atoms or molecules of different substances. Thanks to the mole concept, it can be done

Very often chemists want to measure out the exact quantities of substances that will react together. What is really useful is to be able to work out these quantities on the basis of the number of atoms (or molecules) of substance **A** that will react with a certain number of atoms of substance **B**. Counting out atoms sounds a tricky business, but, thanks to the mole concept, it can be done! How can we count out numbers of atoms by measuring masses? The key to the calculation is the idea which chemists call the **mole concept**. The origin of the mole concept was the work of a nineteenth century Italian chemist called Amadeo Avogadro. This is how he argued:

We know from their relative atomic masses that one atom of magnesium is twice as heavy as one atom of carbon: $A_r(Mg) = 24$, $A_r(C) = 12$.

Therefore we can say:

If 1 atom of magnesium is twice as heavy as 1 atom of carbon,
then 1 hundred Mg atoms are twice as heavy as 1 hundred C atoms,
and 5 million Mg atoms are twice as heavy as 5 million C atoms,

and it follows that, if we have a piece of magnesium which has twice the mass of a piece of carbon, the two masses must contain equal numbers of atoms.
2 grams of magnesium and 1 gram of carbon contain the same number of atoms; 10 tonnes of magnesium and 5 tonnes of carbon contain the same number of atoms.

The same argument applies to the other elements. Take the relative atomic mass in grams of any element:

12 g Carbon	24 g Magnesium	56 g Iron	40 g Calcium	108 g Silver	238 g Uranium	207 g Lead

All these masses contain the same number of atoms. The number is 6.022×10^{23}.

> The amount of an element that contains 6.022×10^{23} atoms (the same number of atoms as 12 g of carbon-12) is called **one mole** of that element.

The symbol for mole is **mol**. The ratio 6.022×10^{23} mol^{-1} is called the **Avogadro constant**. When you weigh out 12 g of carbon, you are counting out 6×10^{23} atoms of carbon. This amount of carbon is one mole (1 mol) of carbon atoms. Similarly, 48 g of magnesium is two moles (2 mol) of magnesium atoms. You can say that the **amount** of magnesium is two moles (2 mol).

You can have a mole of magnesium atoms, Mg, a mole of magnesium ions, Mg^{2+}, a mole of sulphuric acid molecules, H_2SO_4. One mole of sulphuric acid contains 6×10^{23} molecules of H_2SO_4, that is, 98 g of H_2SO_4 (the molar mass in grams). To write 'one mole of nitrogen' is imprecise: one mole of nitrogen atoms, N, has a mass of 14 grams; one mole of nitrogen molecules, N_2, has a mass of 28 grams.

CHECKPOINT 3.6: THE AVOGADRO CONSTANT

(Take the Avogadro constant to be 6×10^{23} mol^{-1}.)

1. There are 4 billion people in the world. If you had one mole of £1 coins to distribute equally between them, how much would each person receive?

2. What mass of potassium contains *(a)* 6×10^{23} atoms, *(b)* 2×10^{25} atoms?

3. The price of gold is £8.20 per gram; A_r(Au) = 198. Calculate the price of 1 million million atoms of gold.

3.7 MOLAR MASS

Molar mass is defined

The unit is g mol^{-1}

The mass of one mole of a substance is called the **molar mass**, symbol M, unit g mol^{-1}. The molar mass of carbon is 12 g mol^{-1}; that is the relative atomic mass expressed in grams per mole. The term molar mass applies to compounds as well as elements. The molar mass of a compound is the relative molecular mass expressed in grams per mole. Sulphuric acid, H_2SO_4, has a relative molecular mass of 98; its molar mass is 98 g mol^{-1}. Notice the units: relative molecular mass has no unit; molar mass has the unit g mol^{-1}.

> Amount (in moles) of substance = $\frac{\text{Mass of substance}}{\text{Molar mass of substance}}$
>
> Molar mass of element = Relative atomic mass in grams per mole
>
> Molar mass of compound = Relative molecular mass in grams per mole

Sample calculations of amount of substance

Example (a) What is the amount of calcium present in 120 g of calcium?

Method A_r of calcium = 40

Molar mass of calcium = 40 g mol^{-1}

Amount of substance = $\frac{\text{Mass}}{\text{Molar mass}}$

$$\text{Amount of calcium} = \frac{\text{Mass of calcium}}{\text{Molar mass of calcium}} = \frac{120\text{ g}}{40\text{ g mol}^{-1}}$$
$$= 3.0\text{ mol}$$

The amount (number of moles) of calcium is 3.0 mol.

Example (b) If you need 2.50 mol of sodium hydrogencarbonate, what mass of the substance do you have to weigh out?

Method Relative molecular mass of $NaHCO_3$ = 23 + 1 + 12 + (3 × 16) = 84

Molar mass of $NaHCO_3$ = 84 g mol^{-1}

$$\text{Amount of substance} = \frac{\text{Mass of substance}}{\text{Molar mass of substance}}$$

$$2.50\text{ mol} = \frac{\text{Mass}}{84\text{ g mol}^{-1}}$$

$$\text{Mass} = 84\text{ g mol}^{-1} \times 2.50\text{ mol}$$
$$= 210\text{ g}$$

You need to weigh out 210 g of sodium hydrogencarbonate.

CHECKPOINT 3.7: THE MOLE

1. State the mass of:

(a) 3 mol of magnesium ions, Mg^{2+}
(b) 0.50 mol of oxygen atoms, O
(c) 0.50 mol of oxygen molecules, O_2
(d) 0.25 mol of sulphur atoms, S
(e) 0.25 mol of sulphur molecules, S_8

2. Find the amount (moles) of each element present in:

(a) 69 g of lead, Pb
(b) 14 g of iron, Fe
(c) 56 g of nitrogen, N_2
(d) 2.0 g of mercury, Hg
(e) 9.0 g of aluminium, Al

3. State the mass of

(a) 2.0 mol of carbon dioxide molecules, CO_2
(b) 10 mol of sulphuric acid, H_2SO_4
(c) 2.0 mol of sodium chloride, NaCl
(d) 0.50 mol of calcium hydroxide, $Ca(OH)_2$

4. Calculate the molar masses of the following:

(a) $NH_4Fe(SO_4)_2 \cdot 12H_2O$
(b) $Al_2(SO_4)_3$
(c) $K_4Fe(CN)_6$

5. How many moles of substance are present in the following?

(a) 0.250 g of calcium carbonate
(b) 5.30 g of anhydrous sodium carbonate
(c) 5.72 g of sodium carbonate-10-water crystals

6. Use the value of 6.0×10^{23} mol^{-1} for the Avogadro constant to find the number of atoms in

(a) 2.0×10^{-3} g of calcium
(b) 5.0×10^{-6} g of argon
(c) 1.00×10^{-10} g of mercury

3.8 EMPIRICAL FORMULAE

> The **empirical formula** of a compound is the simplest formula which represents its composition.

It shows the elements present and the ratio of the amounts of elements present.

Finding an empirical formula ...

To find an empirical formula, you need to work out the ratio of the amounts of the elements present.

Example A 0.4764 g sample of an oxide of iron was reduced by a stream of carbon monoxide. The mass of iron that remained was 0.3450 g. Find the empirical formula of the oxide.

Method

...A worked example ...

...composition by mass ...

... composition by amount (moles) ...

Elements present	*Iron*		*Oxygen*
Mass/g	0.3450		0.1314
A_r	56		16
Amount/mol	0.3450/56 $= 6.16 \times 10^{-3}$		0.1314/16 $= 8.21 \times 10^{-3}$
Ratio of amounts	1	:	$\frac{8.21 \times 10^{-3}}{6.16 \times 10^{-3}}$
	1	:	1.33
	3	:	4

...empirical formula

Empirical formula is Fe_3O_4.

3.9 MOLECULAR FORMULAE

Finding a molecular formula ...

> The **molecular formula** is a simple multiple of the empirical formula.

... a multiple of the empirical formula

If the empirical formula is CH_2O, the molecular formula may be CH_2O, $C_2H_4O_2$, $C_3H_6O_3$ and so on.

The way to find out which molecular formula is correct is to find out which gives the correct molar mass.

Example A polymer of empirical formula CH_2 has a molar mass of 28 000 g mol^{-1}. What is its molecular formula?

... A worked example

Method

Empirical formula mass = 14 g mol^{-1}

Molar mass = 28 000 g mol^{-1}

The molar mass is 2000 times the empirical formula mass; therefore the molecular formula is $(CH_2)_{2000}$.

3.10 CALCULATION OF PERCENTAGE COMPOSITION

The empirical formula shows percentage by mass composition ...

From the formula of a compound and the relative atomic masses of the elements in it, the percentage of each element in the compound can be calculated. This is called the **percentage composition by mass**.

Example Calculate the percentage mass of water of crystallisation in copper(II) sulphate-5-water.

Method

... A worked example

Formula is $CuSO_4 \cdot 5H_2O$

Relative atomic masses are Cu = 63.5 S = 32 O = 16 H = 1

$$\text{Molar mass} = 63.5 + 32 + (4 \times 16) + (5 \times 18)$$
$$= 249.5 \text{ g mol}^{-1}$$

$$\text{Percentage of water} = \frac{90}{249.5} \times 100$$
$$= 36\%$$

CHECKPOINT 3.10: FORMULAE AND PERCENTAGE COMPOSITION

1. Calculate the percentage by mass of the named element in the compound listed:

(a) Mg in Mg_3N_2, *(b)* Na in NaCl, *(c)* Br in $CaBr_2$

2. Calculate the empirical formulae of the compounds for which the following analytical results were obtained:

(a) 27.3% C, 72.7% O
(b) 53.0% C, 47.0% O
(c) 29.1% Na, 40.5% S, 30.4% O
(d) 32.4% Na, 22.6% S, 45.0% O

3. Find the empirical formulae of the compounds formed in the reactions described below:

(a) 10.800 g magnesium form 18.000 g of an oxide
(b) 3.400 g calcium form 9.435 g of a chloride
(c) 3.528 g iron form 10.237 g of a chloride

4. Weighed samples of the following crystals were heated to drive off the water of crystallisation. When they reached constant mass, the following masses were recorded. Deduce the empirical formulae of the hydrates:

(a) 0.942 g of $MgSO_4 \cdot a\,H_2O$ gave 0.461 g of residue
(b) 1.124 g of $CaSO_4 \cdot b\,H_2O$ gave 0.889 g of residue
(c) 1.203 g of $Hg(NO_3)_2 \cdot c\,H_2O$ gave 1.172 g of residue

3.11 EQUATIONS FOR REACTIONS OF SOLIDS

Equations give us much information ...

Equations tell us not only what substances react together but also what amounts of substances react together.

The equation for the action of heat on sodium hydrogencarbonate

$$2NaHCO_3(s) \longrightarrow Na_2CO_3(s) + CO_2(g) + H_2O(g)$$

tells us that 2 moles of sodium hydrogencarbonate give 1 mole of sodium carbonate. Since the molar masses are $NaHCO_3 = 84 \text{ g mol}^{-1}$ and $Na_2CO_3 = 106 \text{ g mol}^{-1}$, it follows that 168 g of sodium hydrogencarbonate give 106 g of sodium carbonate.

... about the relationship between the amounts of reactants and products

The amounts of substances undergoing reaction, as given by the balanced chemical equation, are called the **stoichiometric** amounts. **Stoichiometry** is the relationship between the amounts of reactants and products in a chemical reaction. If one reactant is present in excess of the stoichiometric amount required to react with another of the reactants, then the excess of one reactant will be left unused at the end of the reaction.

A worked example of a calculation based on an equation

Example What mass of zinc can be obtained from the reduction of 10.00 tonnes of zinc oxide by 10.00 tonnes of charcoal? (1 tonne = 10^3 kg)

Method Write the equation

$$ZnO(s) + C(s) \longrightarrow Zn(s) + CO(g)$$

Amount of ZnO = $10.00 \times 10^6/(65.4 + 16.00) = 1.23 \times 10^5$ mol

Amount of C = $10.00 \times 10^6/12.00 = 8.33 \times 10^5$ mol

Since zinc oxide is present in the smaller amount, the amount of zinc formed is limited by the amount of zinc oxide. From the equation, you can see that 1 mole of ZnO forms 1 mole of Zn.

Amount of Zn = 1.23×10^5 mol

Mass of Zn = $1.23 \times 10^5 \times 65.4 \times 10^{-6}$ tonne

Mass of Zn = 8.04 tonne

CHECKPOINT 3.11: MASSES OF REACTING SOLIDS

1. What mass of pure aluminium oxide must be electrolysed to give 50 tonnes of aluminium?
2. The sulphur present in 0.1000 g of an organic compound is converted into barium sulphate. A precipitate of 0.1852 g of dry $BaSO_4$ is obtained. Calculate the percentage by mass of sulphur in the compound.
3. What is the maximum mass of 2,4,6-trichlorophenol, $C_6H_2Cl_3OH$, that can be obtained from 10.00 g of phenol, C_6H_5OH? A chemist who carried out this conversion obtained 19.54 g of the product. What percentage yield did he obtain?
4. What is the maximum mass of N-benzoylphenylamine, $C_6H_5NHCOC_6H_5$, that can be obtained from 1.00 g of phenylamine, $C_6H_5NH_2$? A chemist who made this derivative obtained 2.04 g. What percentage yield did she obtain?

3.12 EQUATIONS FOR REACTIONS OF GASES

The volume of l mole of any ideal gas is the same

For reactions of gases, it is more usual to consider the volumes of reactants and products, rather than their masses. The volume of 1 mole of any ideal gas is the same; 22.414 dm^3 at 0 °C and 1 atm (standard temperature and pressure) and 24.0 dm^3 at 20 °C and 1 atm (room temperature and pressure).

... the gas molar volume ...

> The gas molar volume is 22.414 dm^3 at stp, 24.0 dm^3 at rtp.

... 22.4 dm^3 at stp ...

... 24.0 dm^3 at rtp

When a reaction involves both solids and gases, the solids are usually measured by mass and the gases by volume.

A worked example of a reaction of solids and gases

Example What mass of potassium chlorate(V) must be decomposed to supply 200 cm^3 of oxygen (measured at stp)? In the presence of a catalyst, decomposition proceeds according to the equation

$$2KClO_3(s) \xrightarrow{MnO_2} 2KCl(s) + 3O_2(g)$$

Method From the equation you can see that

2 mol of $KClO_3$ give 3 mol of O_2

Molar mass of $KClO_3 = 92.5\ g\ mol^{-1}$

Therefore $2 \times 92.5\ g\ KClO_3 \longrightarrow 3 \times 22.4\ dm^3\ O_2$

To supply 200 cm^3 O_2 you need $\dfrac{2 \times 92.5}{3 \times 22.4} \times 200 \times 10^{-3}\ g\ KClO_3$

Mass of $KClO_3$ decomposed = 0.551 g

CHECKPOINT 3.12: REACTING VOLUMES OF GASES

1. What volume of hydrogen is formed when 3.00 g of magnesium react with an excess of dilute sulphuric acid?

2. Carbon dioxide is obtained by the fermentation of glucose:

   $C_6H_{12}O_6(aq) \longrightarrow 6CO_2(g) + 6H_2O(l)$

   If 20.0 dm^3 of carbon dioxide (at stp) are collected, what mass of glucose has reacted?

3. In the preparation of hydrogen chloride by the reaction

   $NaCl(s) + H_2SO_4(l) \longrightarrow HCl(g) + NaHSO_4(s)$

   what masses of sodium chloride and sulphuric acid are required for the production of 10.0 dm^3 of hydrogen chloride (at stp)?

3.13 CONCENTRATION

The concentration of a solution can be stated ...

One way of stating the concentration of a solution is to state the **mass** of solute present in 1 cubic decimetre of solution, e.g. grams per cubic decimetre ($g\ dm^{-3}$). There is another method which is more convenient when it comes to chemical reactions. This is to state the **amount** in moles of a solute present in 1 dm^3 of solution.

... either in grams of solute per cubic decimetre of solution ($g\ dm^{-3}$) ...

If 1 mole of solute is present in 1 dm^3 of solution, the concentration of solute is 1 mole per dm^3 (1 $mol\ dm^{-3}$). The solution is a 1 $mol\ dm^{-3}$ solution or, for short, a 1 M solution [see Figure 3.13A].

Figure 3.13A Solutions of known concentration

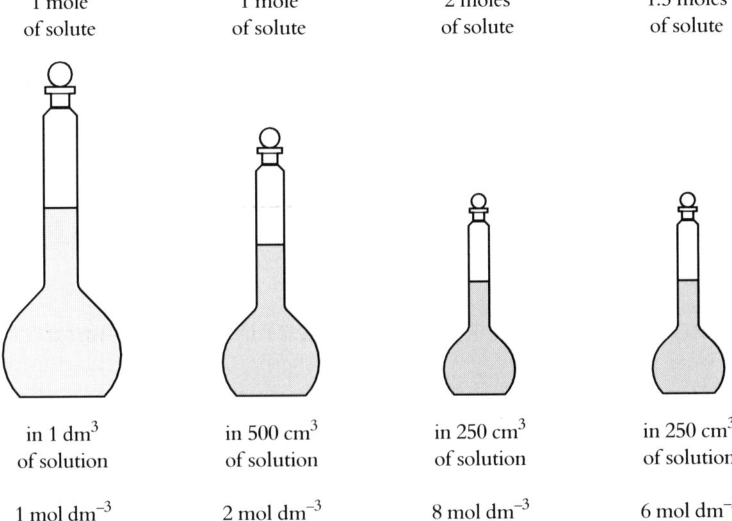

... or in moles of solute per cubic decimetre of solution (mol dm^{-3})

2 moles of solute in 1 dm^3 of solution: concentration = 2 mol dm^{-3} or 2 M.

2 moles of solute in 250 cm^3 of solution: concentration = 8 mol dm^{-3} or 8 M.

1.5 moles of solute in 250 cm^3: concentration = 6 mol dm^{-3} or 6 M.

One cubic decimetre = 1000 cubic centimetres.

One cubic decimetre is also known as one litre, l.

$$1\ dm^3 = 1000\ cm^3 = 1\ litre$$

A standard solution is a solution of known concentration

A solution of known concentration is called a **standard solution**. In strict SI units, concentration is expressed in mol m^{-3} (1 m^3 = 10^3 dm^3).

$$\text{Concentration in moles per litre} = \frac{\text{Amount of solute in moles}}{\text{Volume of solution in dm}^3}$$

Rearranging,

$$\underset{(\text{mol})}{\text{Amount of solute}} = \underset{(\text{dm}^3)}{\text{Volume of solution}} \times \underset{(\text{mol dm}^{-3})}{\text{Concentration}}$$

Be careful with units ...

... amounts in moles ...

... volumes in dm^3 ...

... give concentrations in mol dm^{-3}

Example (a) Calculate the amount of solute present in 250 cm^3 of a solution of hydrochloric acid which has a concentration of 2.0 mol dm^{-3}.

Method

$$\text{Amount (mol)} = \text{Volume (dm}^3) \times \text{Concentration (mol dm}^{-3})$$

$$\text{Amount of solute, HCl} = 250 \times 10^{-3}\ \text{dm}^3 \times 2.0\ \text{mol dm}^{-3} = 0.50\ \text{mol}$$

Note that when you are given the volume in cm^3, you have to change it into dm^3.

Then the units are coherent:

$$\text{Amount (mol)} = \text{Volume (dm}^3) \times \text{Concentration (mol dm}^{-3})$$

Example (b) What mass of sodium carbonate must be dissolved in 1 dm^3 of solution to give a solution of concentration 1.5 M (a 1.5 M solution)?

Method

$$\text{Amount (mol)} = \text{Volume (dm}^3) \times \text{Concentration (mol dm}^{-3}) = 1.00\ \text{dm}^3 \times 1.5\ \text{mol dm}^{-3} = 1.5\ \text{mol}$$

Molar mass of sodium carbonate, $Na_2CO_3 = (2 \times 23) + 12 + (3 \times 16) = 106\ g\ mol^{-1}$

Mass of sodium carbonate = 1.5 mol × 106 g mol^{-1} = 159 g

CHECKPOINT 3.13A: CONCENTRATION

1. Calculate the concentrations of the following solutions:

   (a) 4.0 g of sodium hydroxide in 500 cm^3 of solution
   (b) 7.4 g of calcium hydroxide in 5.0 dm^3 of solution
   (c) 49.0 g of sulphuric acid in 2.5 dm^3 of solution
   (d) 73 g of hydrogen chloride in 250 cm^3 of solution

2. Find the amount of solute present in the following solutions:

   (a) 1.00 dm^3 of a solution of sodium hydroxide of concentration 0.25 mol dm^{-3}
   (b) 500 cm^3 of hydrochloric acid of concentration 0.020 mol dm^{-3}
   (c) 250 cm^3 of 0.20 mol dm^{-3} sulphuric acid
   (d) 10 cm^3 of a 0.25 mol dm^{-3} solution of potassium hydroxide

3.13.1 PREPARING A STANDARD SOLUTION BY WEIGHING

A standard solution is made from a primary standard

Now you know how to calculate the mass of solid which you need to make a standard solution. A standard solution can only be made from a solid which can be obtained 100% (almost) pure. Anhydrous sodium carbonate and sodium hydrogencarbonate can be used to make standard solutions. They are called **primary standards**. Ethanedioic acid, $C_2H_2O_4$, and butanedioic acid, $C_4H_6O_4$, are primary standards which can be used to make standard acid solutions. For other substances, a solution of approximately known concentration is made and then the solution is standardised against a primary standard. You could not make a standard solution of sodium hydroxide. As you were weighing it out, it would absorb water vapour from the air and react with carbon dioxide in the air. You would have to make a solution of approximately known concentration and titrate it against, for example, a standard solution of ethanedioic acid to find its exact concentration.

Method of preparing a standard solution of sodium carbonate

A known mass of the primary standard is dissolved in distilled water ...

1. Calculate the mass of sodium carbonate needed, m_1.

2. Weigh a clean weighing bottle, and record its mass, m_2. [See Figure 3.13B(a).] With a clean spatula, add pure anhydrous sodium carbonate until the combined mass of weighing bottle and sodium carbonate is $m_1 + m_2$.

3. Transfer the sodium carbonate carefully into a clean beaker. [See Figure 3.13B(b).] Use a wash bottle of distilled water so that all the washings run into the beaker. Add about 100 cm^3 of distilled water. Stir with a glass rod until all the solid has dissolved [Figure 3.13B(c)].

4. Pour all the solution carefully through a filter funnel into a graduated flask [Figure 3.13B(d)]. Wash all the solution out of the beaker and off the glass rod [Figure 3.13B(e), (f)].

... and the volume of the solution is made up to a known volume

5. Add distilled water until the level is about 2 cm below the graduation mark on the graduated flask. Add the rest of the distilled water drop by drop from a dropping pipette until the bottom of the meniscus is level with the graduation mark when viewed at eye level [Figure 3.13B(g)]. Insert the stopper of the flask and invert the flask several times to mix the solution [Figure 3.13B(h)].

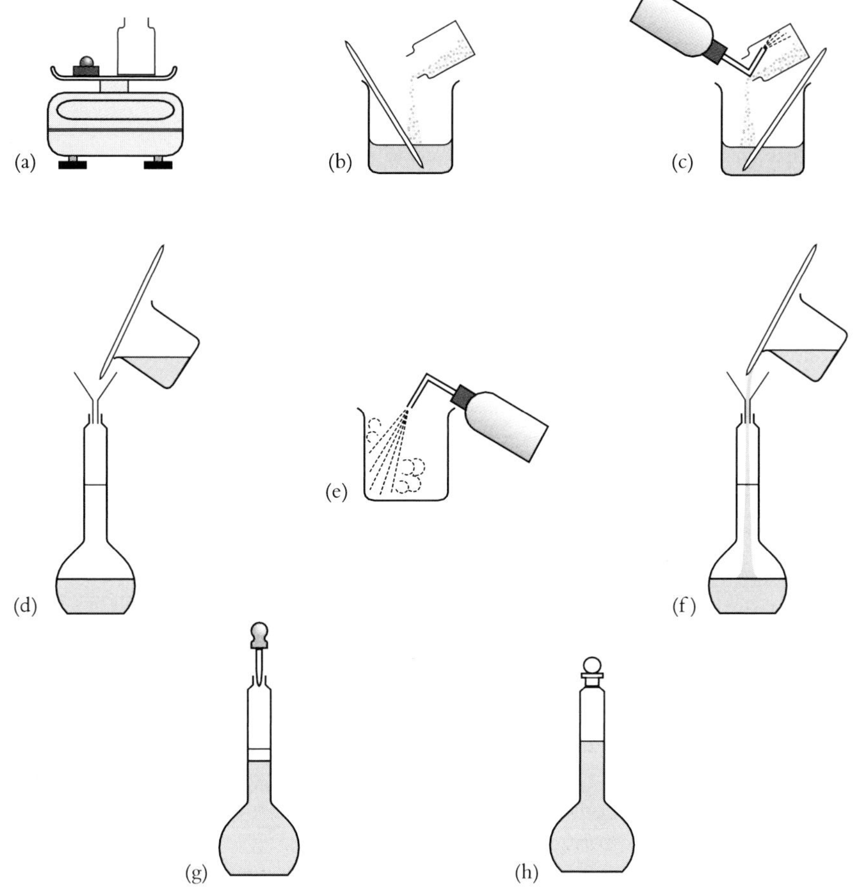

Figure 3.13B Preparing a standard solution

3.13.2 PREPARING A STANDARD SOLUTION BY DILUTION

A standard solution can be made by diluting a more concentrated standard solution

You can prepare a dilute standard solution by diluting a more concentrated standard solution in a measured way.

... either accurately, using a burette and volumetric flask ...

If you want to know the concentration accurately, you use a burette and a volumetric flask. You would not be able to do this with a very concentrated solution, e.g. concentrated sulphuric acid or glacial ethanoic acid. You could use the method to prepare, for instance, a 0.1 mol dm^{-3} solution from a 2 mol dm^{-3} solution. The steps you would follow are:

1. Fill a clean, dry burette with the more concentrated standard solution.
2. Run the calculated volume of the more concentrated solution into a volumetric flask.
3. Make the solution up to the mark with distilled water. Shake.

(*Note* You cannot use very concentrated acids and alkalis in burettes.)

... or approximately using a measuring cylinder and a graduated beaker

If you do not need to know the concentration accurately, the steps to follow are:

1. Use a measuring cylinder to measure the volume of the concentrated solution.
2. Transfer the solution to a graduated beaker.
3. Make up to the mark with distilled water. Stir.

CHECKPOINT 3.13B: SOLUTIONS

1. (a) On Monday, Jerry's teacher gives him some 1.00 mol dm^{-3} acid and instructs him to make a solution which is exactly 0.100 mol dm^{-3}. Say what apparatus Jerry should use and describe what he should do.
   (b) On Tuesday, Jerry is given the same 1.00 mol dm^{-3} acid. This time he is asked to prepare quickly a solution which is between 0.09 mol dm^{-3} and 0.11 mol dm^{-3}. Say what apparatus he should use and what he should do.

2. Explain what dangers you would risk by using a very concentrated acid in a burette.

3. The concentrated hydrochloric acid in the store has a concentration of 12 mol dm^{-3}. The college technician has to fill all the reagent bottles in the lab with approximately 2 mol dm^{-3} hydrochloric acid. There are 24 bottles, each of which holds 250 cm^3.

   Describe how the technician should prepare the solution of dilute hydrochloric acid.

4. Ammonia is bought as '880 ammonia' (a solution of density 0.880 g cm^{-3}), which contains 245 g ammonia per dm^3 of solution. What volume of the concentrated solution would you need to prepare 1.0 dm^3 of 2.0 mol dm^{-3} ammonia solution?

5. (a) Why can sodium hydrogencarbonate be used to prepare standard solutions?
   (b) What mass of sodium hydrogencarbonate would you weigh out to prepare 500 cm^3 of a 0.0100 M solution?
   (c) Describe how you would make up the solution as accurately as possible.

6. You have a large stock bottle of ethanoic acid of concentration 4.00 mol dm^{-3}. You also have a large bottle of 'glacial' ethanoic acid. This is the name given to a concentrated solution of ethanoic acid which freezes at 10 °C. It has a concentration of about 17 mol dm^{-3}.

   (a) How can you make up 1.00 dm^3 of a 0.250 mol dm^{-3} solution of the acid? Say what quantities you would measure and what apparatus you would use.
   (b) Explain how you would prepare 2 dm^3 of 2 mol dm^{-3} ethanoic acid.

3.14 VOLUMETRIC ANALYSIS

The concentration of a solution can be found by volumetric analysis

Volumetric analysis is a means of finding the concentration of a solution. The method is to add a solution of, say, an acid to a solution of, say, a base, in a measured way until there is just enough of the acid to neutralise the base. This method is called **titration** [see Figures 3.14A and 3.14B].

Figure 3.14A
Using a pipette

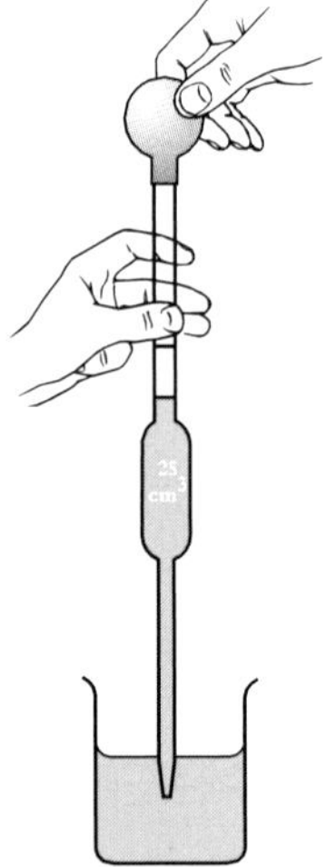

1 Use a pipette filler to suck up the liquid into the pipette.

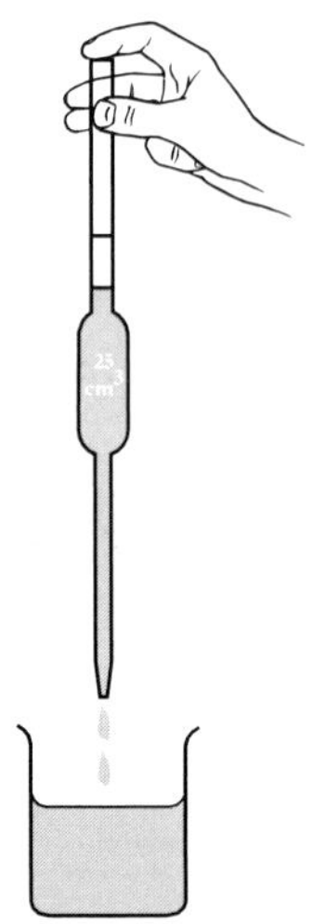

2 Allow liquid to run slowly down to the graduation mark.

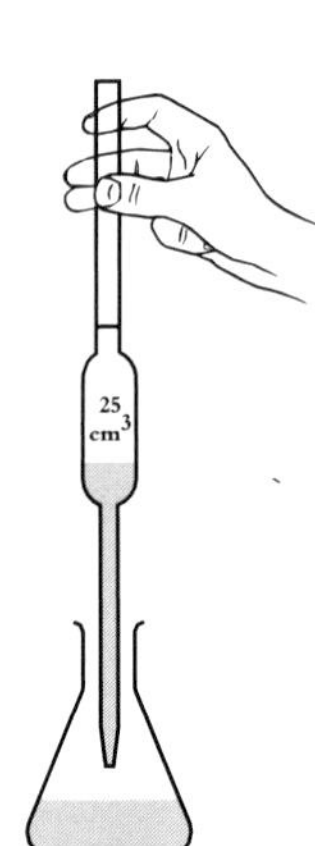

3 Run the liquid into a conical flask.

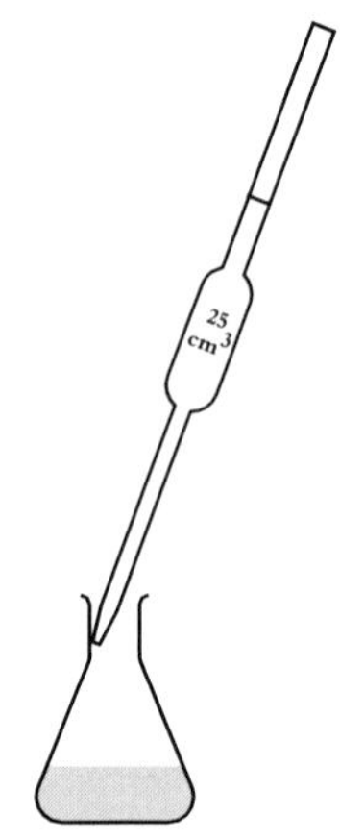

4 Touch the side of the flask with the tip of the pipette.

The method of titration is used ...

... for example an acid of unknown concentration is titrated against a measured volume of a standard solution of a base

The concentration of one of the two solutions must be known, and the volumes of both must be measured. You can use a standard solution of a base to find out the concentration of a solution of an acid. You have to find out what volume of the acid solution of unknown concentration is needed to neutralise a known volume, usually 25.0 cm^3, of the standard solution of a base. An indicator tells when exactly the right volume of solution has been added to achieve neutralisation. You will learn titration in your laboratory periods. Here is a reminder of the practical details:

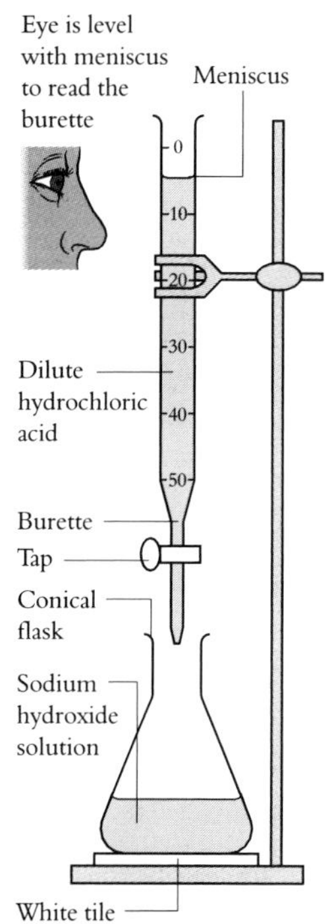

Figure 3.14B
Titration

1. Use a pipette to deliver 25.0 cm^3 of the alkali solution into a clean conical flask [see Figure 3.14A]. Add a few drops of indicator.

2. Wash the burette with a little of the acid solution. Allow the solution to run into the tip of the burette. Read the burette (V_1 cm^3, the bottom of the meniscus) [see Figure 3.14B].

3. Arrange the apparatus as shown in Figure 3.14B. Run the acid solution from the burette dropwise. Use your left hand to open the tap and your right hand to swirl the conical flask (unless you are left-handed). Stop when the indicator just changes colour. This is the 'end-point' of the titration.

4. Read the burette again (V_2 cm^3). Subtract to find the volume of acid used, $(V_2 - V_1)$ cm^3. This 'titre' is the volume of acid needed to neutralise 25.0 cm^3 of alkali.

5. Repeat the titration. Obtain an average titre. From this volume, you can calculate the unknown concentration.

Example (a) By titration, you find that 15.0 cm^3 of hydrochloric acid neutralise 25.0 cm^3 of a 0.100 mol dm^{-3} solution of sodium hydroxide. What is the concentration of hydrochloric acid?

Method

The method of calculating concentration from the results of titration

1. The equation for the reaction,

$$\text{Hydrochloric acid} + \text{Sodium hydroxide} \longrightarrow \text{Sodium chloride} + \text{Water}$$

$$HCl(aq) + NaOH(aq) \longrightarrow NaCl(aq) + H_2O(l)$$

tells you that 1 mole of HCl neutralises 1 mole of NaOH.

Where to start? Start with the substance for which you know both the volume of the solution and the concentration. Work out the amount (moles) of that substance

2. Now work out the amount (mol) of base. You must start with the base because you know the concentration of base, and you do not know the concentration of acid.

$$\text{Amount (mol)} = \text{Volume } (dm^3) \times \text{Concentration } (mol\ dm^{-3})$$

$$\text{Amount (mol) NaOH} = \text{Volume } (25.0\ cm^3) \times \text{Concentration } (0.100\ mol\ dm^{-3})$$

$$= 25.0 \times 10^{-3}\ dm^3 \times 0.100\ mol\ dm^{-3}$$

$$= 2.50 \times 10^{-3}\ mol$$

3. Now work out the concentration of acid.

$$\text{Amount (mol) of HCl} = \text{Amount (mol) of NaOH} = 2.50 \times 10^{-3}\ \text{mol}$$

Also

$$\text{Amount (mol) of HCl} = \text{Volume of HCl(aq)} \times \text{Concentration of HCl(aq)}$$

Therefore, if c mol dm^{-3} is the concentration of HCl,

$$25.0 \times 10^{-3}\ \text{mol} = 15.0 \times 10^{-3}\ \text{dm}^3 \times c\ \text{mol dm}^{-3}$$

$$c\ \text{mol dm}^{-3} = 2.50 \times 10^{-3}\ \text{mol}/15.0 \times 10^{-3}\ \text{dm}^{-3}$$

$$= 0.167\ \text{mol dm}^{-3}$$

The concentration of hydrochloric acid is 0.167 mol dm^{-3}.

Note that, since the volumes are measured to 3 significant figures, e.g. 15.0 cm^3, you quote your answer to 3 significant figures.

A second example

Example (b) 25.0 cm^3 of sulphuric acid of concentration 0.150 mol dm^{-3} neutralised 31.2 cm^3 of potassium hydroxide solution. Find the concentration of the potassium hydroxide solution.

Method

The equation ...

1. The equation,

... the amount of standard reactant ...

$$\text{Sulphuric acid} + \text{Potassium hydroxide} \longrightarrow \text{Potassium sulphate} + \text{Water}$$

$$H_2SO_4(aq) + 2KOH(aq) \longrightarrow K_2SO_4(aq) + 2H_2O(l)$$

tells you that 1 mole of H_2SO_4 neutralises 2 moles of KOH.

... the amount of the second reactant ...

2. Now work out the amount (mol) of acid. You must choose the acid because you do not know the concentration of the base.

$$\text{Amount (mol) acid} = \text{Volume } (25.0\ \text{cm}^3) \times \text{Concentration } (0.150\ \text{mol dm}^{-3})$$

$$= 25.0 \times 10^{-3}\ \text{dm}^3 \times 0.150\ \text{mol dm}^{-3}$$

$$= 3.75 \times 10^{-3}\ \text{mol}$$

... the concentration of the second reactant

3. Now work out the concentration of base.

$$\text{Amount (mol) of KOH} = 2 \times \text{Amount (mol) of } H_2SO_4$$

$$= 7.50 \times 10^{-3}\ \text{mol}$$

Also

$$\text{Amount (mol) of KOH} = \text{Volume of KOH(aq)} \times \text{Concentration of KOH(aq)}$$

Therefore, if c mol dm^{-3} is the concentration of KOH,

$$c\ \text{mol dm}^{-3} = 7.50 \times 10^{-3}\ \text{mol}/31.2 \times 10^{-3}\ \text{dm}^3$$

$$= 0.240\ \text{mol dm}^{-3}$$

The concentration of potassium hydroxide is 0.240 mol dm^{-3}.

CHECKPOINT 3.14: TITRATION

1. 25.0 cm^3 of sodium hydroxide solution are neutralised by 15.0 cm^3 of a solution of hydrochloric acid of concentration 0.25 mol dm^{-3}. Find the concentration of the sodium hydroxide solution.

2. A solution of sodium hydroxide contains 10 g dm^{-3}.
   (a) What is the concentration of the solution in mol dm^{-3}?
   (b) What volume of this solution would be needed to neutralise 25.0 cm^3 of 0.10 mol dm^{-3} hydrochloric acid?

3. 25.0 cm^3 of hydrochloric acid are neutralised by 20.0 cm^3 of a solution of 0.15 mol dm^{-3} sodium carbonate solution.
   (a) How many moles of sodium carbonate are neutralised by 1 mol HCl?
   (b) What is the concentration of the hydrochloric acid?

4. The sixth form decide to test some antacid indigestion tablets. They obtain the results shown in the table by dissolving tablets and titrating the alkali in them against a standard acid.

Brand	**Price (£) of 100 tablets**	**Volume (cm^3) of 0.01 mol dm^{-3} acid required to neutralise 1 tablet**
Stopit	0.91	2.8
Setlit	1.04	3.0
Mendit	1.30	3.3
Basit	1.56	3.6

   (a) Which antacid tablets offer the best value for money?
   (b) What other factors would you consider before choosing a brand?

5. A tanker of acid is emptied into a water supply by mistake. A water company chemist titrates the water and finds that 10.0 dm^3 of water are needed to neutralise 10.0 cm^3 of a 0.010 mol dm^{-3} solution of sodium hydroxide. What is the concentration of hydrogen ions in the water?

6. A solution is made by dissolving 5.00 g of impure sodium hydroxide in water and making it up to 1.00 dm^3 of solution. 25.0 cm^3 of this solution is neutralised by 30.3 cm^3 of hydrochloric acid, of concentration 0.102 mol dm^{-3}. Calculate the percentage purity of the sodium hydroxide.

7. Sodium carbonate crystals (27.8230 g) were dissolved in water and made up to 1.00 dm^3. 25.0 cm^3 of the solution were neutralised by 48.8 cm^3 of hydrochloric acid of concentration 0.100 mol dm^{-3}. Find *n* in the formula $Na_2CO_3 \cdot nH_2O$.

8. A fertiliser contains ammonium sulphate. A sample of 0.500 g of fertiliser was warmed with sodium hydroxide solution. The ammonia evolved neutralised 44.1 cm^3 of 0.100 mol dm^{-3} hydrochloric acid. Calculate the percentage of ammonium sulphate in the sample.

3.15 EQUATIONS FOR OXIDATION–REDUCTION REACTIONS

3.15.1 REDOX REACTIONS

Oxidising agents accept electrons ...

Oxidising agents are substances which can accept electrons from other substances. **Reducing agents** are substances which can give electrons to other substances.

... reducing agents donate electrons

Oxidation and **reduction** occur together. In an **oxidation–reduction** reaction or **redox** reaction, electrons pass from the reducing agent to the oxidising agent.

Iron(II) ions are reducing agents, losing electrons to form iron(III) ions:

$$Fe^{2+}(aq) \longrightarrow Fe^{3+}(aq) + e^- \qquad [1]$$

Write the half-reaction equation for the oxidant

Chlorine is an oxidising agent, accepting electrons to form chloride ions:

$$Cl_2(aq) + 2e^- \longrightarrow 2Cl^-(aq) \qquad [2]$$

Write the half-reaction equation for the reductant. Combine them to give the equation for the redox reaction

Equations [1] and [2] are described as **half-reaction equations**. Free electrons never occur under ordinary laboratory conditions. To represent the real process, the half-reaction equations must be combined. If equation [1] is multiplied by 2 and added to equation [2], the result is

$$2Fe^{2+}(aq) + Cl_2(aq) + 2e^- \longrightarrow 2Fe^{3+}(aq) + 2Cl^-(aq) + 2e^-$$

or, as the electrons cancel out

$$2Fe^{2+}(aq) + Cl_2(aq) \longrightarrow 2Fe^{3+}(aq) + 2Cl^-(aq)$$

The technique of combining the half-reaction equations for the **oxidant** and the **reductant** is useful because often the equations for redox are more complicated than this example.

Reaction between manganate(VII) and iron(II)

Obtaining the equation for the redox reaction between acidic MnO_4^- and Fe^{2+}

Potassium manganate(VII), $KMnO_4$, in acidic solution is a powerful oxidising agent, widely used in titrimetric analysis. The sudden change from purple MnO_4^- ions to pale pink Mn^{2+} ions at the end-point means that no indicator is needed. In balancing the half-reaction equation, $8H^+(aq)$ are needed to combine with 4O in MnO_4^-:

The equation must be balanced for mass ...

$$MnO_4^-(aq) + 8H^+(aq) \longrightarrow Mn^{2+}(aq) + 4H_2O(l)$$

... and for charge

The charge on the left-hand side (LHS) = −1 + 8 = +7 units.
The charge on the RHS = +2 units.
To equalise the charge on both sides of the equation, 5 electrons are needed on the LHS:

$$MnO_4^-(aq) + 8H^+(aq) + 5e^- \longrightarrow Mn^{2+}(aq) + 4H_2O(l) \qquad [3]$$

Potassium manganate(VII) in acidic solution oxidises iron(II) ions. The equation for the reaction is obtained by combining half-reaction equations [1] and [3]. Since Fe^{2+} gives one electron and MnO_4^- needs five, equation [1] is multiplied by 5 and then added to equation [3]:

$$MnO_4^-(aq) + 8H^+(aq) + 5Fe^{2+}(aq) \longrightarrow Mn^{2+}(aq) + 4H_2O(l) + 5Fe^{3+}(aq)$$

Reaction between dichromate(VI) and ethanedioate

Potassium dichromate(VI), $K_2Cr_2O_7$, is an oxidising agent which is most effective in acidic solution. It is reduced to a chromium(III), Cr^{3+}, salt. Balancing the half-reaction equation with respect to mass gives

The redox reaction between acidic $Cr_2O_7^{2-}$ and $C_2O_4^{2-}$

$$Cr_2O_7^{2-}(aq) + 14H^+(aq) \longrightarrow 2Cr^{3+}(aq) + 7H_2O(l)$$

Balancing the equation with respect to charge gives

$$Cr_2O_7^{2-}(aq) + 14H^+(aq) + 6e^- \longrightarrow 2Cr^{3+}(aq) + 7H_2O(l) \qquad [4]$$

The electrons on the RHS cancel the electrons on the LHS of the overall equation

A check shows that the charge on the LHS = −2 + 14 − 6 = +6, and the charge on the RHS = +6 units.

Sodium ethanedioate and ethanedioic acid are oxidised to carbon dioxide. The half-reaction equation is

$$\begin{matrix} CO_2^- \\ | \\ CO_2^- \end{matrix}(aq) \longrightarrow 2CO_2(g) + 2e^- \qquad [5]$$

To obtain the equation for the redox reaction between acidified potassium dichromate and sodium ethanedioate, equation [5] is multiplied by 3 and added to equation [4]:

$$Cr_2O_7^{2-}(aq) + 14H^+(aq) + 3C_2O_4^{2-}(aq) \longrightarrow 2Cr^{3+}(aq) + 7H_2O(l) + 6CO_2(g)$$

Reaction between iodine and sodium thiosulphate

Obtaining the equation for the reaction between I_2 and $S_2O_3^{2-}$

... by combining half-reaction equations ...

... so that the electrons on the RHS cancel the electrons on the LHS of the overall equation

Sodium thiosulphate(VI), $Na_2S_2O_3$, is a reducing agent. It is most often used in titrimetric analysis for reducing iodine to iodide ions, being oxidised in the process to sodium tetrathionate, $Na_2S_4O_6$. When the brown colour of iodine fades as the end-point approaches, a little starch solution is added. This gives an intense blue colour with even a trace of iodine. At the end-point the blue colour vanishes. The two half-reaction equations are

$$2S_2O_3^{2-}(aq) \longrightarrow S_4O_6^{2-}(aq) + 2e^- \quad [6]$$

$$I_2(aq) + 2e^- \longrightarrow 2I^-(aq) \quad [7]$$

Combining the two half-reaction equations gives

$$2S_2O_3^{2-}(aq) + I_2(aq) \longrightarrow S_4O_6^{2-}(aq) + 2I^-(aq)$$

CHECKPOINT 3.15: REDOX REACTIONS

1. Write balanced half-reaction equations for the oxidation of each of the following:

(a) $Sn^{2+}(aq) \longrightarrow Sn^{4+}(aq)$
(b) $Cl^-(aq) \longrightarrow Cl_2(aq)$
(c) $H_2S(aq) \longrightarrow S(s) + H^+(aq)$
(d) $SO_3^{2-}(aq) \longrightarrow SO_4^{2-}(aq) + H^+(aq)$
(e) $H_2O_2(aq) \longrightarrow O_2(g) + H^+(aq)$

Check that the equations are balanced with respect to charge as well as mass. Remember that H_2O is present in all solutions; you will need it to balance some of the equations.

2. Write balanced half-reaction equations for the following reductions:

(a) $Br_2(aq) \longrightarrow Br^-(aq)$
(b) $MnO_2(s) + H^+(aq) \longrightarrow Mn^{2+}(aq)$
(c) $PbO_2(s) + H^+(aq) \longrightarrow Pb^{2+}(aq)$
(d) $IO^-(aq) + H^+(aq) \longrightarrow I_2(aq)$
(e) $ClO_3^-(aq) + H^+(aq) \longrightarrow Cl_2(aq)$

Balance the equations for mass, using H_2O from the solution if needed, and then balance with respect to charge.

3. By combining half-reaction equations, write balanced equations for the following reactions:

(a) $Fe^{3+}(aq) + I^-(aq) \longrightarrow$
(b) $Fe^{3+}(aq) + Sn^{2+}(aq) \longrightarrow$
(c) $MnO_4^-(aq) + Cl^-(aq) + H^+(aq) \longrightarrow$
(d) $MnO_4^-(aq) + H_2O_2(aq) + H^+(aq) \longrightarrow$
(e) $MnO_4^-(aq) + H^+(aq) + I^-(aq) \longrightarrow$
(f) $Cr_2O_7^{2-}(aq) + H^+(aq) + I^-(aq) \longrightarrow$
(g) $Cr_2O_7^{2-}(aq) + H^+(aq) + Fe^{2+}(aq) \longrightarrow$
(h) $MnO_4^-(aq) + H^+(aq) + Sn^{2+}(aq) \longrightarrow$
(i) $Cr_2O_7^{2-}(aq) + H^+(aq) + SO_3^{2-}(aq) \longrightarrow$
(j) $MnO_4^-(aq) + H^+(aq) + SO_3^{2-}(aq) \longrightarrow$
(k) $MnO_4^-(aq) + H^+(aq) + Fe^{2+}(aq) \longrightarrow$
(l) $Cr_2O_7^{2-}(aq) + H^+(aq) + Sn^{2+}(aq) \longrightarrow$
(m) $PbO_2(s) + H^+(aq) + Cl^-(aq) \longrightarrow$
(n) $ClO_3^-(aq) + H^+(aq) + I^-(aq) \longrightarrow$
(o) $Br_2(aq) + I^-(aq) \longrightarrow$
(p) $Cl_2(g) + H^+(aq) + IO^-(aq) \longrightarrow$
(q) $Br_2(g) + H_2S(g) \longrightarrow$

3.16 OXIDATION NUMBER

A method of expressing the combining power of elements is the idea of **oxidation number** or **oxidation state**.

Examples are:

- The oxidation number of sodium in Na^+ is +1.
- The oxidation number of aluminium in Al^{3+} is +3.
- The oxidation number of iodine in I^- is −1.
- The oxidation number of oxygen in O^{2-} is −2.

The use of oxidation numbers is extended to covalent compounds. Some elements are assigned positive oxidation numbers and others are assigned negative oxidation numbers in accordance with certain rules.

3.16.1 RULES FOR ASSIGNING OXIDATION NUMBERS

The oxidation number of an element in the uncombined state is zero ...

1. The oxidation numbers of elements in their uncombined states, such as Na, Ca, Al, are zero. Similarly the oxidation numbers of iodine in I_2, oxygen in O_2 and sulphur in S_8 are zero.

... The oxidation number of an element in an ionic compound is equal to the charge on its ions, e.g. +1, +2, −1, −2 ...

2. In ionic compounds the oxidation number is equal to the charge on the ion. The oxidation number of an element is not always the same. Iron has an oxidation number of +2 in Fe^{2+} and an oxidation number of +3 in Fe^{3+}.

... The oxidation numbers of the elements in a compound add up to zero

3. The sum of the oxidation numbers of all the atoms or ions in a compound is zero.

   In NaCl,

   $$(\text{Ox. No. of Na}) + (\text{Ox. No. of Cl}) = 0$$
   $$(+1) + (-1) = 0$$

   In Na_2O,

   $$2(\text{Ox. No. of Na}) + (\text{Ox. No. of O}) = 0$$
   $$2(+1) + (-2) = 0$$

   In CuS,

   $$(\text{Ox. No. of Cu}) + (\text{Ox. No. of S}) = 0$$
   $$(+2) + (-2) = 0$$

   In $CaBr_2$,

   $$(\text{Ox. No. of Ca}) + 2(\text{Ox. No. of Br}) = 0$$
   $$(+2) + 2(-1) = 0$$

The oxidation numbers of the elements in an ion add up to the charge on the ion

4. The sum of the oxidation numbers of all the atoms in an ion is equal to the charge on the ion. In SO_4^{2-}, the sum of the oxidation numbers (S = +6, O = −2) is

   $$+6 + 4(-2) = -2$$

   which is the charge on the ion.

Reference elements ...

5. Some elements nearly always employ the same oxidation number in their compounds. They are used as reference points in assigning oxidation numbers to other elements. The reference elements are

... which nearly always have the same oxidation number in their compounds

K	Na	+1	H	+1	except in metal hydrides
Mg	Ca	+2	F	−1	
Al		+3	Cl	−1	except in compounds with O and F
			O	−2	except in peroxides, superoxides, fluorides

... Some worked examples

Example (a) What is the oxidation number of thallium in $TlCl_3$?

Method Chlorine always has the oxidation number −1.

Therefore (Ox. No. of Tl) + 3(−1) = 0

and the oxidation number of thallium is +3.

Example (b) What is the oxidation number of Cl in Cl_2O_7?

Method The exceptions to the rule that the oxidation number of Cl equals −1 are compounds with O and F. Oxygen is the reference point with the oxidation number −2.

Therefore 2 (Ox. No. of Cl) + 7(−2) = 0

and the oxidation number of chlorine is +7.

Example (c) What is the oxidation number of Cr in $Cr(CN)_6^{3-}$?

Method The cyanide ion, CN^-, has a charge of −1.

Therefore (Ox. No. of Cr) + 6(−1) = −3

and the oxidation number of chromium is +3.

CHECKPOINT 3.16A: OXIDATION NUMBER

1. State the oxidation numbers of the elements in following atoms or ions:

   Na, Na^+, Ba, Ba^{2+}, Rb^+, Rb, Ga, As, As^{3-}, Br^-, H_2, H^+, F_2, F^-

2. Give the oxidation numbers of the first element in each of the following compounds. Remember the oxidation numbers of the elements in a compound add up to zero. Take oxidation numbers for hydrogen (+1), oxygen (−2), fluorine (−1) and chlorine (−1) as reference points.

   CuO, Cu_2O, H_2S, SO_2, SO_3, PbO, PbO_2, $AlCl_3$, SF_6, SCl_2, $TiCl_4$, V_2O_5

3. What is the oxidation number of the named element in the following species (ions or molecules)?

   (a) N in NO, NO_2, N_2O_4, N_2O, NO_2^-, NO_3^-, N_2O_5

   (b) Mn in $MnSO_4$, Mn_2O_3, MnO_2, MnO_4^-, MnO_4^{2-}

   (c) As in As_2O_3, AsO_2^-, AsO_4^{3-}, AsH_3

   (d) Cr in CrO_4^{2-}, $Cr_2O_7^{2-}$, CrO_3

   (e) I in I^-, IO^-, IO_3^-, I_2, ICl_3, ICl_2^-

3.16.2 CHANGES IN OXIDATION NUMBER

Oxidation–reduction reactions are often discussed in terms of the change in oxidation number of each reactant. In the redox reaction,

$$2Fe^{2+}(aq) + I_2(aq) \longrightarrow 2Fe^{3+}(aq) + 2I^-(aq)$$

When an element is oxidised, its oxidation number increases

When Fe^{2+} is converted into Fe^{3+}, the oxidation number increases from +2 to +3, and we say that Fe^{2+} has been oxidised to Fe^{3+}. When I_2 is converted into I^-, the oxidation number decreases from 0 in I_2 to −1 in I^-, and we say that I_2 has been reduced to I^-.

$$\text{Change in Ox. No. of iron} = \text{No. of atoms} \times \text{Change in Ox. No.} = 2(+1) = +2$$

$$\text{Change in Ox. No. of iodine} = \text{No. of atoms} \times \text{Change in Ox. No.} = 2(-1) = -2$$

$$\text{Sum of changes in Ox. No.} = +2 - 2 = 0$$

When an element is reduced, its oxidation number decreases

In general, when an element is **oxidised**, its oxidation number increases; when an element is **reduced**, its oxidation number decreases. In a redox reaction

$$x\mathbf{A} + y\mathbf{B} \longrightarrow$$

if the oxidation number of **A** changes by $+a$ units, and the oxidation number of **B** changes by $-b$ units

$$\text{then } x(+a) + y(-b) = 0$$

... Some worked examples

Example (a) Consider the reduction of iron(III) ions by a tin(II) salt:

$$Sn^{2+}(aq) + 2Fe^{3+}(aq) \longrightarrow Sn^{4+}(aq) + 2Fe^{2+}(aq)$$

For tin, change in Ox. No. = +2

For iron, change in Ox. No. = −1

And $1(+2) + 2(-1) = 0$

In disproportionation part of a substance is oxidised and part is reduced

Example (b) $3I_2(aq) + 3OH^-(aq) \longrightarrow IO_3^-(aq) + 5I^-(aq) + 3H^+(aq)$

The only element which changes its oxidation number is iodine.

On the LHS, in I_2 Ox. No. of I = 0

On the RHS, in IO_3^- (Ox. No. of I) + 3(−2) = −1

and the Ox. No. of I = +5

In I^-, Ox. No. of I = −1

Iodine has changed from oxidation number zero on the LHS to a combination of Ox. No. +5 and Ox. No. −1 on the RHS. Part of the iodine has been oxidised and part has been reduced. A reaction of this kind is termed a **disproportionation reaction**.

3.16.3 BALANCING EQUATIONS BY THE OXIDATION NUMBER METHOD

Balancing equations ...

The oxidation number method of balancing equations is best explained through an example.

Sum of increases in oxidation number = sum of decreases in oxidation number ...

... a worked example

Example Balance the equation

$$a\text{KIO}_3(aq) + b\text{Na}_2\text{SO}_3(aq) \longrightarrow c\text{KIO}(aq) + d\text{Na}_2\text{SO}_4(aq)$$

Iodine changes from Ox. No. +5 in KIO_3 to +1 in KIO.

Change in Ox. No. of I = −4

Sulphur changes from Ox. No. +4 in Na_2SO_3 to +6 in Na_2SO_4.

Change in Ox. No. of S = +2

Therefore $a(-4) + b(+2) = 0$

If $a = 1$, $b = 2$ and the equation becomes

$$KIO_3(aq) + 2Na_2SO_3(aq) \longrightarrow cKIO(aq) + dNa_2SO_4(aq)$$

By stoichiometry, it follows that $c = 1$ and $d = 2$, giving

$$KIO_3(aq) + 2Na_2SO_3(aq) \longrightarrow KIO(aq) + 2Na_2SO_4(aq)$$

CHECKPOINT 3.16B: EQUATIONS AND OXIDATION NUMBERS

1. Use the oxidation number method to balance the equations:

   (a) $MnO_4^-(aq) + H^+(aq) + Fe^{2+}(aq) \longrightarrow Mn^{2+}(aq) + Fe^{3+}(aq) + H_2O(l)$

   (b) $Sn(s) + HNO_3(aq) \longrightarrow SnO_2(s) + NO_2(g) + H_2O(l)$

   (c) $Cu^{2+}(aq) + I^-(aq) \longrightarrow CuI(s) + I_2(aq)$

   (d) $Cl_2(g) + OH^-(aq) \longrightarrow Cl^-(aq) + ClO^-(aq) + H_2O(l)$

   (e) $Zn(s) + Fe^{3+}(aq) \longrightarrow Zn^{2+}(aq) + Fe^{2+}(aq)$

   (f) $I_2(aq) + S_2O_3^{2-}(aq) \longrightarrow I^-(aq) + S_4O_6^{2-}(aq)$

2. One mole of the compound ICl_x reacts with an excess of potassium iodide solution to give two moles of I_2. Write an equation for the reaction, and state the oxidation number of I in ICl_x.

3. State the oxidation number of the species which is underlined in the following equations. Say whether the species is oxidised or reduced during the reaction. Complete and balance the equations:

   (a) $H_2\underline{O}_2 + \underline{I}^- + H^+(aq) \longrightarrow H_2\underline{O} + \underline{I}_2$

   (b) $\underline{Cu} + \underline{N}O_3^- + H^+(aq) \longrightarrow \underline{Cu}^{2+} + \underline{N}O + H_2O$

   (c) $\underline{Fe}^{2+} + \underline{Cr}_2O_7^{2-} + H^+(aq) \longrightarrow \underline{Fe}^{3+} + \underline{Cr}^{3+} + H_2O$

   (d) $\underline{S}_2O_3^{2-} + \underline{I}_2 \longrightarrow \underline{S}_4O_6^{2-} + \underline{I}^-$

   (e) $K\underline{I}O_3 + K\underline{I} + H\underline{Cl} \longrightarrow K\underline{Cl} + \underline{ICl} + 3H_2O$

   (f) $\underline{Cr}O_4^{2-} + H^+(aq) \longrightarrow \underline{Cr}_2O_7^{2-}$

3.17 OXIDATION NUMBERS AND NOMENCLATURE

3.17.1 SYSTEMATIC NOMENCLATURE

Systems for naming compounds

Oxidation numbers are used in the naming of compounds. Systematic nomenclature is set out by IUPAC (the International Union of Pure and Applied Chemistry) in their *Manual of Symbols and Terminology for Physiochemical Quantities and Units* and also by ASE (the Association for Science Education) in *Chemical Nomenclature, Symbols and Terminology* (3rd edition, 1985). The systems are not quite the same. Although the IUPAC system, coming from an international body, is more widely used, the examination boards follow the ASE system. This book, since it aims to prepare the readers for examinations, also follows the ASE system.

3.17.2 CATIONS

The oxidation state of the element is specified if it is variable

Cations (positive ions) are given the name of the element together with the oxidation number. This system of naming was devised by A Stock:

e.g. Fe^{2+} iron(II) ion Fe^{3+} iron(III) ion

Names of cations

When there is no doubt about the oxidation state because an element assumes one only, then it is omitted:

e.g. Na^+ sodium ion Al^{3+} aluminium ion

3.17.3 ANIONS

Elemental **anions** (negative ions) are named after the element, with the ending *-ide*:

e.g. H^- hydride N^{3-} nitride

Compound anions have names ending in *-ide*, *-ite* or *-ate*:

e.g. OH^- hydroxide NO_2^- nitrite NO_3^- nitrate

Many elements form more than one **oxoanion**, using more than one oxidation state (e.g., NO_2^-, NO_3^-). The names are derived from the name of the element which is combined with oxygen and the ending *-ate*:

e.g. SO_4^{2-} sulphate ion HCO_3^- hydrogencarbonate ion

Both ClO^- and ClO_3^- are chlorate ions. To distinguish between them, the oxidation number of chlorine is added:

Names of anions

e.g.	ClO^-	chlorate(I)	ClO_3^-	chlorate(V)
also	CrO_4^{2-}	chromate(VI)	$Cr_2O_7^{2-}$	dichromate(VI)
	MnO_4^{2-}	manganate(VI)	MnO_4^-	manganate(VII)
	NO_3^-	nitrate(V) or nitrate	NO_2^-	nitrate(III) or nitrite
	SO_4^{2-}	sulphate(VI) or sulphate	SO_3^{2-}	sulphate(IV) or sulphite

The Stock names for the last four examples have not been widely adopted, and people prefer to use nitrate, nitrite, sulphate and sulphite. These names date back to their usage before the changes in nomenclature of 1970.

3.17.4 ACIDS

Acids are named after their anions:

Names of acids

e.g. $HClO$ chloric(I) acid
$HClO_2$ chloric(III) acid
$HClO_3$ chloric(V) acid

Again, the names nitrous acid and sulphurous acid are preferred to the Stock names (nitric(III) and sulphuric(IV)) for the acids HNO_2 and H_2SO_3.

3.17.5 SALTS

Salts are named by combining the names of the cation, with its oxidation number if that is variable, and the anion:

Names of salts

e.g. $FeSO_4$ iron(II) sulphate $NaClO$ sodium chlorate(I)

When a salt is hydrated, the number of water molecules per formula unit is stated:

e.g. $CuSO_4 \cdot 5H_2O$ copper(II) sulphate-5-water

3.17.6 STOICHIOMETRIC FORMULAE

The oxides, sulphides and halides of non-metallic elements are usually named, not by the Stock system but according to their stoichiometry:

Some compounds are named by stoichiometry

e.g.	NO	nitrogen oxide	CS_2	carbon disulphide
	N_2O	dinitrogen oxide	$SiCl_4$	silicon tetrachloride
	NO_2	nitrogen dioxide	$POCl_3$	phosphorus trichloride oxide
	N_2O_4	dinitrogen tetraoxide	$SOCl_2$	sulphur dichloride oxide

Phosphorus compounds are sometimes named by stoichiometry, as above, and sometimes by the Stock system:

e.g. PCl_5 phosphorus pentachloride or phosphorus(V) chloride.

3.18 TITRIMETRIC ANALYSIS, USING REDOX REACTIONS

Titrimetric analysis using redox reactions ...

Redox reactions are used in titrimetric analysis. For example, a solution of unknown concentration of a reductant is titrated against a standard solution of an oxidant. From the volumes of the two solutions and the equation for the reaction, the concentration of the unknown solution can be found.

Example (a) Find the concentration of an iron(II) sulphate solution, given that 25.0 cm^3 of the solution, when acidified, required 19.8 cm^3 of 0.0200 mol dm^{-3} potassium manganate(VII) for oxidation.

Method The equation for the reaction comes first. A combination of the two half-reaction equations, as described in § 3.15.1, gives

... Some worked examples

$$MnO_4^-(aq) + 5Fe^{2+}(aq) + 8H^+(aq) \longrightarrow Mn^{2+}(aq) + 5Fe^{3+}(aq) + 4H_2O(l)$$

The equation indicates that 1 mol of MnO_4^- oxidises 5 mol of Fe^{2+}.

Start with the substance you can find the amount of, in this case potassium manganate(VII). You know the volume and the concentration and can easily find the amount

Amount of MnO_4^- in 19.8 cm^3 = $19.8 \times 10^{-3} \times 0.0200$ mol
= 0.396×10^{-3} mol

Amount of Fe^{2+} in 25.0 cm^3 = 5 × amount of MnO_4^-
= 1.98×10^{-3} mol

Concentration of Fe^{2+} = $(1.98 \times 10^{-3})/(25.0 \times 10^{-3})$ mol dm^{-3}

Concentration of $FeSO_4$ = 7.92×10^{-2} mol dm^{-3}

Example (b) A standard solution is prepared by dissolving 1.185 g of 'AnalaR' potassium dichromate(VI) and making up to 250 cm^3 of solution. This solution is used to find the concentration of a sodium thiosulphate solution. A 25.0 cm^3 portion of the oxidant was acidified and added to an excess of potassium iodide to liberate iodine.

You don't have to remember equations like this. You derive the equation by combining the half-equations

$$Cr_2O_7^{2-}(aq) + 6I^-(aq) + 14H^+(aq) \longrightarrow 3I_2(aq) + 2Cr^{3+}(aq) + 7H_2O(l) \quad [1]$$

When the solution was titrated against sodium thiosulphate solution, 17.5 cm^3 of 'thio' were required. Find the concentration of the thiosulphate solution.

Method Combining the half-reaction equations

$$I_2(aq) + 2e^- \longrightarrow 2I^-(aq)$$
$$2S_2O_3^{2-}(aq) \longrightarrow S_4O_6^{2-}(aq) + 2e^-$$

gives

Start with the dichromate solution. You know the volume and the concentration and can find the amount of dichromate

$$2S_2O_3^{2-}(aq) + I_2(aq) \longrightarrow S_4O_6^{2-}(aq) + 2I^-(aq) \quad [2]$$

Concentration of $K_2Cr_2O_7$ = (1.185/294) × 4 = 0.0161 mol dm^{-3}

Amount of I_2 in 25.0 cm^3 = $0.0161 \times 25.0 \times 10^{-3} \times 3$ mol
(from equation [1]) = 1.208×10^{-3} mol

Amount of thio in 17.5 cm^3 = $1.208 \times 10^{-3} \times 2$ mol
(from equation [2]) = 2.416×10^{-3} mol

$$\text{Concentration of thiosulphate} = (2.416 \times 10^{-3})/(17.5 \times 10^{-3}) \text{ mol dm}^{-3}$$
$$= 0.138 \text{ mol dm}^{-3}$$

CHECKPOINT 3.18: REDOX TITRATIONS

1. A 0.1576 g piece of iron wire was converted into Fe^{2+} ions and then titrated against potassium dichromate solution of concentration 1.64×10^{-2} mol dm^{-3}. From the fact that 27.3 cm^3 of the oxidant were required, calculate the percentage purity of the iron wire.

2. A volume of 27.5 cm^3 of a 0.0200 mol dm^{-3} solution of potassium manganate(VII) was required to oxidise 25.0 cm^3 of a solution of hydrogen peroxide. Calculate the concentration of hydrogen peroxide and the volume of oxygen (at stp) evolved during the titration.

3. Calculate the percentage purity of an impure sample of sodium thiosulphate from the following data. A 0.2368 g sample of the sodium thiosulphate was added to 25.0 cm^3 of 0.0400 mol dm^{-3} iodine solution. The excess of iodine that remained after reaction needed 27.8 cm^3 of 0.0400 mol dm^{-3} thiosulphate solution in a titration.

4. What volume of potassium manganate(VII) solution of concentration 0.0100 mol dm^{-3} will oxidise 50.0 cm^3 of iron(II) ethanedioate solution of concentration 0.0200 mol dm^{-3} in acid conditions?

* See Footnote.

3.19 EQUILIBRIUM

An equilibrium is a state of balance ...

... either static ...

... or dynamic

Imagine that you are looking through the window of a popular restaurant during a busy lunchtime. You can see that all of the restaurant's 200 seats are taken. You come back 30 minutes later, and you see that all the seats are still occupied. However, you can see that people are entering the restaurant and other people are leaving the restaurant. The same situation continues over the next 2 hours. The population remains constant at 200 people while all the time people are entering and leaving the restaurant. There is a balance between the number leaving and the number arriving. This state of balance can be described as a state of **equilibrium**. If the same 200 people sat at the tables all the time, one would say that the situation was **static** (unchanging). However, in the restaurant you are observing, there is motion as some customers arrive and others leave. The thing that remains constant is the balance between the number arriving and the number leaving. This is a **dynamic** (moving) equilibrium.

The restaurant you have been observing can be described as a system. The word **system** is used to describe a part of the universe which one wants to study in isolation from the rest of the universe. There are two kinds of systems: systems in a state of change and systems at equilibrium. A system in which a change in the properties of the system is occurring is described as 'a system in a state of change'. A system in which no change in its properties is occurring is described as 'a system at equilibrium'. An equilibrium may be a **static equilibrium** or a **dynamic equilibrium**.

Consider a system in which a physical change, vaporisation, occurs. Consider what happens when you drop 5 cm^3 of the brown liquid, bromine, into a gas jar and replace the lid [see Figure 3.19A].

* For further practice, see E N Ramsden, *Calculations for A-Level Chemistry* (Stanley Thornes).

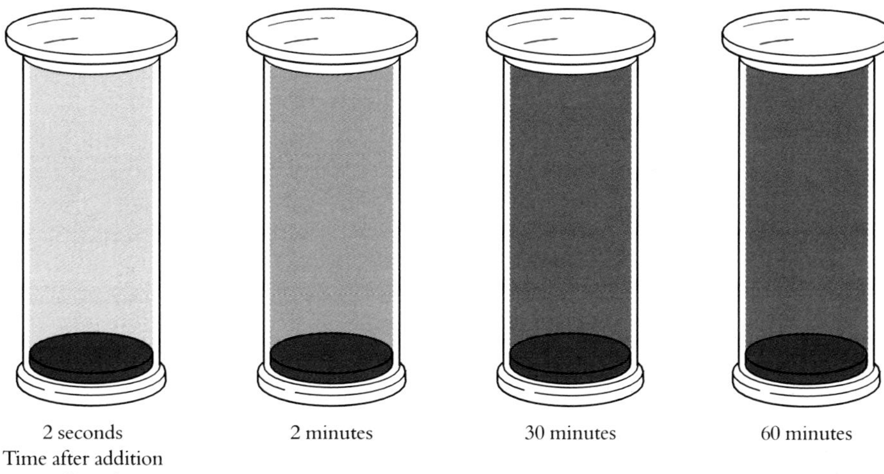

Figure 3.19A
Vaporisation of liquid bromine

A phase is a physically distinct part of a system

As soon as the liquid enters the gas jar, it begins to **vaporise**: some molecules leave the liquid phase and enter the vapour phase. A **phase** is a part of a system which is physically distinct from other parts of the system. The system we are considering is the contents of the closed gas jar. Two phases are present: liquid (bromine) and gas (bromine vapour and air).

A physical change can reach equilibrium ...

After 2 minutes, the gas in the gas jar is brown because it contains bromine molecules as well as air. Vaporisation (or **evaporation**) continues, and after 30 minutes the brown colour of bromine vapour is even more intense. The colour does not continue to deepen for ever. After 60 minutes, it is no more intense than after 30 minutes. It looks as though vaporisation has ceased, and the system is at equilibrium.

... $Br_2(l)$ and $Br_2(g)$ reach equilibrium in a closed system ...

If you could see individual molecules of bromine, however, you would see that the population of bromine molecules in the gas phase is constantly changing. Molecules of bromine are still passing from the liquid to the gas phase but, as fast as they do this, molecules of bromine pass from the gas phase to the liquid phase, that is, they **condense**. The system is at equilibrium because

Rate of vaporisation = Rate of condensation

... and the equilibrium is dynamic

This kind of system is described as being in **dynamic equilibrium**. Dynamic means *moving*, and, at a molecular level, the system is in motion. The properties of the system in bulk are unchanging; the volume of liquid bromine and the concentration of bromine in the gas phase are no longer changing:

$$Br_2(l) \rightleftharpoons Br_2(g)$$

Only a closed system reaches equilibrium

If the system were not closed, it would not come to equilibrium. If the gas jar were open, bromine would continue to vaporise until there was no liquid bromine left.

Another physical change is dissolution (dissolving). Consider what happens when you stir a scoopful of copper(II) sulphate crystals in a beaker of water. As the salt dissolves, the solution becomes a more and more intense blue colour. [See Figure 3.19B.]

A saturated solution is in a state of dynamic equilibrium

After a while, the intensity of the blue colour remains constant, although (provided you have used an excess of crystals) undissolved copper(II) sulphate remains at the bottom of the beaker. The saturated solution is a system at equilibrium. Although nothing more seems to be happening, in fact copper(II)

Figure 3.19B
Dissolution

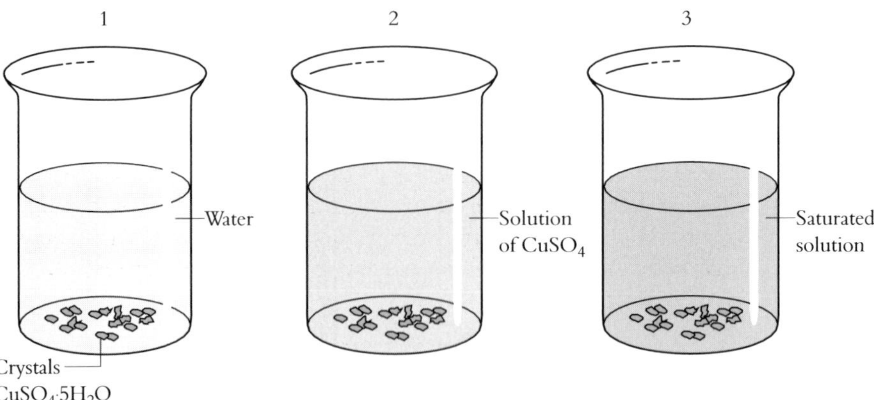
1
2
3
Water
Solution of $CuSO_4$
Saturated solution
Crystals $CuSO_4 \cdot 5H_2O$

sulphate is still dissolving but, as fast as it does so, copper(II) sulphate is crystallising from solution:

$$CuSO_4 \cdot 5H_2O(s) + aq \rightleftharpoons Cu^{2+}(aq) + SO_4^{2-}(aq) + 5H_2O(l)$$

A radioactive tracer can be used to demonstrate the dynamic nature of the equilibrium

There is a way of demonstrating that this system is in dynamic equilibrium. It involves the use of a radioactive **tracer**. If some crystals of $Cu^{35}SO_4 \cdot 5H_2O$, which contain radioactive ^{35}S, are added, you might expect that none would dissolve because the solution is already saturated. After a time, however, it is found that the radioactivity is divided between the solution and the undissolved crystals. The reason is that undissolved solid is constantly dissolving, while solute crystallises from the solution at the same rate. [See Figure 3.19C.]

Figure 3.19C
An experiment using a radioactive tracer

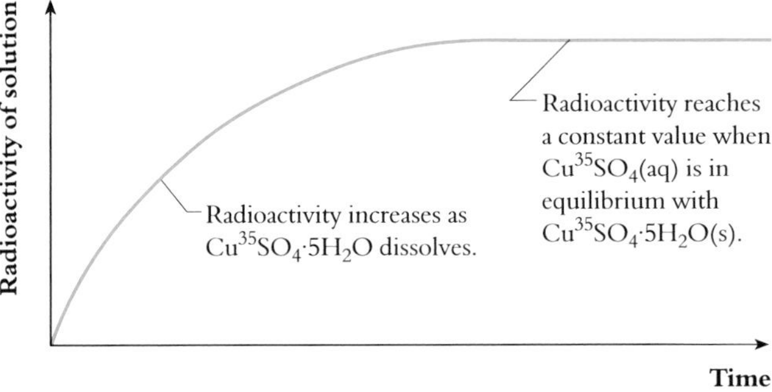
Radioactivity of solution
Time
Radioactivity increases as $Cu^{35}SO_4 \cdot 5H_2O$ dissolves.
Radioactivity reaches a constant value when $Cu^{35}SO_4(aq)$ is in equilibrium with $Cu^{35}SO_4 \cdot 5H_2O(s)$.

3.20 CHEMICAL EQUILIBRIA

The dynamic equilibria described above are physical changes. Chemical reactions can also come to equilibrium.

Chemical reactions, like physical changes, can reach a state of equilibrium

Some chemical reactions take place in one direction almost exclusively. For example, magnesium burns to form magnesium oxide:

$$2Mg(s) + O_2(g) \longrightarrow 2MgO(s)$$

The tendency for magnesium oxide to split up to form magnesium and oxygen is negligible at normal temperatures.

Other chemical reactions take place in both directions at comparable rates. For example, when calcium carbonate is heated strongly, it decomposes:

$$CaCO_3(s) \longrightarrow CaO(s) + CO_2(g)$$

The products formed are the base, calcium oxide, and the acid gas, carbon dioxide. They recombine to form calcium carbonate:

$$CaO(s) + CO_2(g) \longrightarrow CaCO_3(s)$$

In the thermal dissociation of $CaCO_3$...

... $CaO(s) + CO_2(g)$ are in equilibrium with $CaCO_3(s)$ in a closed system ...

When calcium carbonate is heated at a fixed temperature in a closed container, at first calcium carbonate decomposes faster than the products recombine. After a while the amounts of calcium oxide and carbon dioxide build up to a level at which the rate of combination of calcium oxide and carbon dioxide is equal to the rate at which calcium carbonate dissociates. The system has reached a state of dynamic equilibrium:

$$CaCO_3(s) \rightleftharpoons CaO(s) + CO_2(g)$$

... if one of the products is removed the equilibrium is disturbed ...

... because the system is no longer closed

Equilibrium is reached in a closed system. If the container is open, carbon dioxide can escape. The equilibrium is disturbed, and more calcium carbonate dissociates to try to restore the equilibrium. When limestone is heated in a lime kiln, as the aim is to make plenty of quicklime, the carbon dioxide formed is removed by a powerful through draft of air in order to stop the system coming to equilibrium.

The pressure, the temperature and other external factors affect systems in equilibrium. H L Le Chatelier made a study of the way in which systems at equilibrium adjust when external factors are changed. His work is covered in Chapter 11.

3.20.1 CATALYSTS AND EQUILIBRIUM

Catalysts increase the speed with which the equilibrium conditions are reached

Catalysts alter the rates of chemical reactions. In the case of a reaction which reaches a state of dynamic equilibrium, a catalyst increases the rates of both the forward reaction and the reverse reaction by the same ratio. The position of equilibrium is therefore unchanged. What the catalyst does is to decrease the time needed for the system to reach a state of equilibrium. In industrial processes, a catalyst can make a valuable contribution to the economy of the process. In the Haber process, the percentage conversion of nitrogen and hydrogen to ammonia at the temperatures at which plants operate is small. The use of a catalyst to achieve the same percentage conversion in a shorter time increases the productivity of the plant.

CHECKPOINT 3.20: EQUILIBRIUM

1. An aqueous solution of bromine is called 'bromine water'. Some bromine molecules react with water molecules:

$$Br_2(aq) + H_2O(l) \rightleftharpoons HBr(aq) + HBrO(aq)$$

Hydrobromic acid   Bromic(I) acid

The products, hydrobromic acid and bromic(I) acid are both strong acids. The reaction is reversible, and a solution of bromine in water reaches an equilibrium state in which the concentrations of all the species are constant.

Predict what change in the equilibrium will happen as the result of the addition of a small amount of sodium hydroxide. In which direction will the equilibrium be displaced, from left to right or from right to left? Predict what colour change you will see. How could you reverse the colour change?

Do an experiment to check your predictions.

2. A solution of bismuth trichloride in concentrated hydrochloric acid contains four substances: bismuth trichloride, $BiCl_3$, bismuth chloride oxide, BiOCl, hydrochloric acid and water. All four substances are in equilibrium:

$$BiCl_3(aq) + H_2O(l) \rightleftharpoons BiOCl(s) + 2HCl(aq)$$

Bismuth chloride oxide is a white solid which is insoluble in water.

(a) Explain why adding water makes the solution change from clear to cloudy.

(b) Suggest how you could make the solution clear again.

QUANTITATIVE CHEMISTRY

The physical quantity **amount of substance** has the unit the **mole**. One mole of any element contains 6.02×10^{23} atoms. One mole of a compound contains 6.02×10^{23} molecules or formula units. The ratio $6.02 \times 10^{23}\ mol^{-1}$ is called the **Avogadro constant**. § 3.6, § 3.7

The percentage by mass composition of a compound is found by analysis. Conversion to molar composition gives the **empirical formula** of the compound. The molar mass converts the empirical formula into the **molecular formula**. § 3.8, § 3.9

The masses of reactants and products in a chemical reaction can be found by **analysis**. Converted into amounts (in moles) of reactants and products they give the **chemical equation** for the reaction. And vice versa: the chemical equation can be used to calculate the amount (and mass) of product from a known mass of reactant. § 3.11

Chemical reactions can be represented by

- full chemical equations
- ionic equations. § 3.3

The **molar volume of a gas** is

- $24.0\ dm^3$ at rtp
- $22.4\ dm^3$ at stp. § 3.12

From the volumes of gaseous reactants and products can be calculated the **equation** for the reaction. And vice versa: the equation tells the volumes of gaseous reactants and products. § 3.12

The **concentration** of a solution is expressed in $g\ dm^{-3}$ or in $mol\ dm^{-3}$. § 3.13

Titration can be used to give the reacting volumes of solutions and from the volumes and concentrations the equation follows. Titration of a solution of unknown concentration against a standard solution, with the equation for the reaction, gives the unknown concentration. § 3.13, § 3.14

QUESTIONS ON CHAPTER 3

1. Explain what is meant by

(a) stoichiometric equation
(b) oxidation
(c) reduction
(d) disproportionation

2. State what has been oxidised and what has been reduced in the following reactions:

(a) $Zn(s) + 2HCl(aq) \longrightarrow ZnCl_2(aq) + H_2(g)$
(b) $CH_4(g) + 4Cl_2(g) \longrightarrow CCl_4(l) + 4HCl(g)$
(c) $NH_4^+NO_3^-(s) \longrightarrow N_2O(g) + 2H_2O(l)$
(d) $IO_3^-(aq) + 5I^-(aq) + 6H^+(aq) \longrightarrow 3I_2(aq) + 3H_2O(l)$
(e) $2CrO_4^{2-}(aq) + 2H^+(aq) \longrightarrow Cr_2O_7^{2-}(aq) + H_2O(l)$
(f) $2CuCl(aq) \longrightarrow Cu(s) + CuCl_2(aq)$

3. What volume of 0.250 mol dm^{-3} sodium hydroxide solution is required to neutralise 25.0 cm^3 of 0.150 mol dm^{-3} sulphuric acid?

4. Explain what is meant by the terms

(a) equilibrium *(b)* dynamic equilibrium

5. A 25.0 g measure of household ammonia was dissolved in water and made up to 500 cm^3. A 25.0 cm^3 portion of this solution required 29.4 cm^3 of 0.250 mol dm^{-3} sulphuric acid for neutralisation. What is the percentage by mass of ammonia in the cleaning fluid?

6. Arsenic can be oxidised to arsenic(V) acid, H_3AsO_4. This acid oxidises I^- ions to I_2, which can be estimated by titration against a standard thiosulphate solution:

$$As + 5HNO_3 \longrightarrow H_3AsO_4 + 5NO_2 + H_2O$$
$$H_3AsO_4 + 2HI \longrightarrow H_3AsO_3 + I_2 + H_2O$$

If 0.1058 g of a sample containing arsenic required 28.7 cm^3 of a 0.0198 mol dm^{-3} solution of sodium thiosulphate in the final titration, what is the percentage of arsenic in the sample?

7. Complete and balance the following half-equations by adding e.g. electrons, hydrogen ions, hydroxide ions and water molecules. Say whether each is an oxidation or a reduction.

(a) $SO_3^{2-}(aq) \longrightarrow SO_4^{2-}(aq)$
(b) $Cr_2O_7^{2-}(aq) \longrightarrow Cr^{3+}(aq)$ in acidic solution
(c) $I_2(aq) \longrightarrow 2I^-(aq)$
(d) $S_2O_3^{2-}(aq) \longrightarrow S_4O_6^{2-}(aq)$
(e) $Fe^{2+}(aq) \longrightarrow Fe^{3+}(aq)$
(f) $MnO_4^-(aq) \longrightarrow Mn^{2+}(aq)$

8. Balance the following equations using the half-equations in Question 7. The reactions occur in acidic conditions. Say which is the oxidising agent and which is the reducing agent.

(a) the reaction of thiosulphate and iodine
$S_2O_3^{2-}(aq) + I_2(aq) \longrightarrow S_4O_6^{2-}(aq) + I^-(aq)$
(b) the reaction of manganate(VII) and sulphurous acid
$MnO_4^-(aq) + SO_3^{2-}(aq) \longrightarrow Mn^{2+}(aq) + SO_4^{2-}(aq)$
(c) the reaction between iron(II) ions and dichromate(VI) ions
$Fe^{2+}(aq) + Cr_2O_7^{2-}(aq) \longrightarrow Fe^{3+}(aq) + Cr^{3+}(aq)$
(d) $SO_3^{2-}(aq) \longrightarrow SO_4^{2-}(aq)$
(e) $Cr_2O_7^{2-}(aq) \longrightarrow Cr^{3+}(aq)$ in acidic solution
(f) $I_2(aq) \longrightarrow 2I^-(aq)$
(g) $S_2O_3^{2-}(aq) \longrightarrow S_4O_6^{2-}(aq)$
(h) $Fe^{2+}(aq) \longrightarrow Fe^{3+}(aq)$
(i) $MnO_4^-(aq) \longrightarrow Mn^{2+}(aq)$

9. Power stations emit exhaust gases containing the pollutant sulphur dioxide. One way of tackling the pollution is to pass the exhaust gases through an aqueous suspension of powdered limestone. The reaction can be represented:

$$2SO_2(g) + 2CaCO_3(s) + O_2(g) \longrightarrow 2CaSO_4(aq) + 2CO_2(aq)$$

Gypsum is allowed to crystallise out as $CaSO_4 \cdot H_2O(s)$.

(a) A power station produces 55 000 tonnes of gypsum per year (1 tonne = 1000 kg). How many moles of gypsum is this?
(b) How many moles of sulphur dioxide were used in the formation of this mass of gypsum?
(c) What volume of sulphur dioxide is this (assuming that 1 mole of gas occupies 24 dm^3 at the operating temperature)?
(d) What use can be made of the gypsum produced?

10. A copper coin of mass 2.00 g consists of copper alloyed with small quantitites of tin and zinc. The problem is to find the percentage of copper in the coin. The coin was dissolved in moderately concentrated nitric acid to form a solution of copper(II) nitrate. The solution was made up to 250 cm^3. A 25.0 cm^3 portion was neutralised and added to an excess of potassium iodide solution. Iodine was liberated:

$$2Cu^{2+}(aq) + 4I^-(aq) \longrightarrow 2CuI(s) + I_2(aq)$$

The liberated iodine required 30.0 cm^3 of 0.100 mol dm^{-3} sodium thiosulphate solution in a titration.

$$I_2(aq) + 2Na_2S_2O_3(aq) \longrightarrow 2I^-(aq) + S_4O_6^{2-}(aq)$$

(a) Which of the two reactions for which equations are given above are redox reactions?
(b) Which indicator could be used in the titration?
(c) (i) What amount (in moles) of sodium thiosulphate was used in the titration?
(ii) What amount of iodine was titrated?
(iii) What mass of copper was present in the coin?
(iv) What is the percentage of copper in the coin?

11. Bronze is an alloy of copper and tin. The problem is to find the percentage by mass of tin in the alloy. Potassium manganate(VII) oxidises tin(II) ions to tin(IV) ions.

A 9.40 g sample of powdered bronze was warmed with an excess of dilute sulphuric acid to convert the tin into tin(II) sulphate. After filtration, the solution was made up to 250 cm^3.

In a titration, 25.0 cm^3 of the solution of tin(II) sulphate required 19.0 cm^3 of 0.0200 mol dm^{-3} potassium manganate(VII) solution for oxidation.

(a) Why was the solution obtained from bronze filtered?
(b) What conditions were employed for the titration?
(c) How was the end-point of the titration spotted?
(d) Write the half-equations for (i) the oxidation of tin(II) to tin(IV) (ii) the reduction of manganate(VII) to manganese(II).
(e) Combine the half-equations to give the equation for the reaction.
(f) Calculate the percentage by mass of tin in the alloy.

12. *(a)* The first ionisation energy of sodium is +500 kJ mol^{-1}. Write an equation to show this change. **2**
(b) (i) What type of bonding is present in sodium oxide, Na_2O? **1**
(ii) What type of oxide is Na_2O? **1**
(iii) Write an equation for the reaction of Na_2O with water. **1**
(c) Sodium also forms sodium peroxide, Na_2O_2. Treatment of this with dilute sulphuric acid gives hydrogen peroxide, H_2O_2, and sodium sulphate.
(i) What is the oxidation number of oxygen in sodium peroxide? **1**
(ii) Write an equation for the conversion of sodium peroxide to hydrogen peroxide. **1**
(d) Hydrogen peroxide reacts with acidified potassium dichromate(VI) as follows:

$$Cr_2O_7^{2-}(aq) + 3H_2O_2(aq) + 8H^+(aq) \longrightarrow 2Cr^{3+}(aq) + 7H_2O(l) + 3O_2(g)$$

A solution containing 14.7 g dm^{-3} of potassium dichromate(VI), $K_2Cr_2O_7$, is reacted with 20.0 cm^3 of a 0.100 mol dm^{-3} solution of hydrogen peroxide. Calculate the volume of potassium dichromate(VI) solution required. **4**
(e) When dilute aqueous ammonia is added to a solution of nickel(II) sulphate, a green precipitate **B** is formed which dissolves in excess of the same ammonia solution to give a blue solution **C**.
(i) Give the formula of the green precipitate **B**. **1**
(ii) Suggest a formula for the ion responsible for the blue colour in **C**. **1**
(iii) Write an equation for the conversion of **B** to **C**. **1**
(iv) What type of reaction is this? **1**

15 *L (AS/AL)*

13. Silver used to be alloyed with copper to make coins. Before 1921 **silver** coinage in the UK was mostly silver (composition by mass: Ag 92.5%, Cu 7.5%). After 1927, until World War II, coins were **half-silver** (composition by mass: Ag 50%, Cu 40%, Ni 5%, Zn 5%). Today, silver is not used in UK coins.

(a) Silver is one of the best metallic conductors. Explain electronic conduction in silver. **2**
(b) Calculate the percentage molar composition of a UK **silver** coin before 1921. **2**
(c) The percentage of silver in an alloy may be found by dissolving the alloy in nitric acid and reacting the silver nitrate produced with a solution of sodium chloride to give a precipitate of silver chloride. After purification the precipitate is dried and weighed.
(i) Write a balanced equation for the reaction of silver nitrate with sodium chloride.
(ii) Find the percentage by mass of silver in an alloy from the following information:

A sample of 1.245 g of the alloy was dissolved in nitric acid and reacted with an excess of sodium chloride. The purified silver chloride had a mass of 1.519 g. **4**

(Relative atomic masses: Ag = 107.868; Cl = 35.453)

CCEA (AS/AL)

14. When burnt in oxygen, 24.00 g of magnesium produces 40.00 g of magnesium oxide, MgO. When burnt in nitrogen, 24.00 g of magnesium would produce 33.33 g of magnesium nitride, Mg_3N_2. When magnesium is burnt in air it produces both magnesium oxide and magnesium nitride.

The following experimental results were obtained when a sample of magnesium ribbon was burnt in air. [A_r (Mg) 24.0; A_r (O) 16.0; A_r (N) 14.0]

Mass of test-tube = 12.20 g
Mass of test-tube + magnesium ribbon = 18.20 g
Mass of test-tube after burning = 21.70 g

(a) What do you understand by the chemical term *mole*? **3**

(b) Use the numerical data to confirm that one mole of magnesium nitride is obtained from three moles of magnesium. **2**

(c) Write balanced chemical equations for the formation of:
(i) *magnesium oxide* (ii) *magnesium nitride* **2**

(d) (i) Calculate the total mass of product formed when 1.0 mol of magnesium is burnt in air. **2**
(ii) Calculate the mass of magnesium oxide formed when 1.0 mol of magnesium is burnt in air.
[Hint: let x = mass of magnesium oxide formed] **4**
(iii) What is the molar ratio of MgO to Mg_3N_2 formed in the reaction? **1**
(iv) Suggest a reason for the amount of Mg_3N_2 formed being small. **1**

15 *O&C (AS/AL)*

15. *(a)* The concentration of sulphur dioxide in air can be measured using an acidified solution of potassium manganate(VII) which oxidises sulphur dioxide to sulphate ions, SO_4^{2-}.
(i) State, in terms of electrons, what happens to a species when it is oxidised.
(ii) State the oxidation states of sulphur in SO_2 and in SO_4^{2-}
(iii) Deduce the half equation for the oxidation of sulphur dioxide to sulphate ions in the presence of water.
(iv) Write the half equation for the reduction of manganate(VII) ions in acid solution. **6**

(b) Air, contaminated with sulphur dioxide, was bubbled through an acidified solution of potassium manganate(VII). It was found that 0.5 m^3 of air decolorised 100 cm^3 of 0.01 M $KMnO_4$.
(i) Calculate the number of moles of $KMnO_4$ which were decolorised.
(ii) In this analysis, $KMnO_4$ reacts with SO_2 in the ratio 2 : 5.

Use this information to calculate the number of moles of SO_2 in 0.5 m^3 of air and hence the mass of SO_2 per m^3 of this sample of air. **4**

10 *NEAB (AS/AL)*

16. *(a)* Use the concept of oxidation states to deduce whether either of the reactions given below involves redox processes. Explain your answers and, where appropriate, identify the element which is being oxidised.
(i) $2NH_3 + 3Cl_2 \longrightarrow N_2 + 6HCl$
(ii) $CuO + 2HCl \longrightarrow CuCl_2 + H_2O$ **5**

(b) (i) Potassium manganate(VII), $KMnO_4$, can be used in the quantitative estimation of ethanedioate ions, $C_2O_4^{2-}$, in an acidified aqueous solution. In this reaction, ethanedioate ions are converted into carbon dioxide. Deduce half equations for the redox processes involved and hence derive an equation for the overall reaction. **3**
(ii) A 1.93 g sample of a crystalline ethanedioate salt was dissolved in water and made up to 250 cm^3. 25.0 cm^3 of this solution, after acidification, was found to react with 30.4 cm^3 of 0.0200 M $KMnO_4$. Calculate the percentage by mass of ethanedioate ions in the original salt.

NB If you are unable to deduce the overall ratio $C_2O_4^{2-} : MnO_4^-$ for this reaction you may assume the ratio 5 : 3 (This is not the correct ratio). **6**

14 *NEAB (AS/AL)*

17. Antimony, symbol Sb, proton (atomic) number 51, has been known since about 4000 BC. Nowadays, its main use is to harden and to strengthen lead alloys.

(a) A typical sample of antimony consists of two isotopes and has the following composition, by mass: ^{121}Sb, 57.25%; ^{123}Sb, 42.75%.
(i) How could this information be obtained experimentally? **1**
(ii) What can be deduced about the atomic structure of these two isotopes? **2**
(iii) Calculate the relative atomic mass of the antimony sample. **2**

(b) Antimony is produced in a two-stage process from the sulphide ore, Sb_2S_3.

The ore is first roasted in oxygen to form the oxide.

$$2Sb_2S_3(s) + 9O_2(g) \longrightarrow Sb_4O_6(s) + 6SO_2(g)$$

The oxide is then reduced with carbon.

$$Sb_4O_6(s) + 3C(s) \longrightarrow 4Sb(s) + 3CO_2(g)$$

(i) State the oxidation state of antimony in its oxide. **1**
(ii) Showing each stage of your working clearly, calculate the volume of carbon dioxide that would be produced by the processing of 10 moles of Sb_2S_3. [Assume that 1 mole of a gas occupies 24 dm^3 under the experimental conditions.] **3**

9 *C (AS/AL)*

18. When potassium nitrate, KNO_3, is heated, it decomposes according to the equation given below.

$$2KNO_3(s) \longrightarrow 2KNO_2(s) + O_2(g)$$

(a) Deduce the oxidation state of nitrogen in (i) KNO_3 (ii) KNO_2 **2**

(b) Calculate the maximum mass of KNO_2 which could be obtained if 1.55 g KNO_3 were fully decomposed by heat. **4**

(c) Another 1.55 g sample of KNO_3 was partly decomposed by heating. The residue was dissolved in water and the volume of the solution made up to 250 cm^3. A 25.0 cm^3 portion of this solution was found to react, in acid solution, with 28.9 cm^3 of 0.0150 M $KMnO_4$. The reaction can be represented by the equation

$$2MnO_4^- + 5NO_2^- + 6H^+ \longrightarrow 2Mn^{2+} + 5NO_3^- + 3H_2O$$

(i) Calculate the number of moles of $KMnO_4$ which reacted with 25.0 cm^3 of the solution.
(ii) Calculate the total number of moles of KNO_2 present in the residue.
(iii) Calculate the mass of KNO_2 present in the residue.
(iv) Use the result obtained in parts *(b)* and *(c)*(iii) to calculate the percentage conversion of KNO_3 to KNO_2 in this experiment. **7**

13 *NEAB (AS/AL)*

19. The concentration of hydrochloric acid in the human stomach is approximately 0.1 mol dm^{-3}. Excess of this acid causes discomfort referred to as 'heartburn' or 'acid indigestion'. Remedies designed to neutralize some of this excess acid often contain compounds such as magnesium hydroxide, $Mg(OH)_2$, and sodium hydrogencarbonate, $NaHCO_3$.

(a) The hydrogencarbonate ion, HCO_3^-, is capable of behaving as either an acid or a base. Write equations, with state symbols, to show the hydrogencarbonate ion acting as
(i) an acid by reacting with water **2**
(ii) a base by reacting with hydroxonium ions, $H_3O^+(aq)$, and giving off a gas. **2**

(b) Magnesium hydroxide reacts with hydrochloric acid according to the equation

$$Mg(OH)_2(s) + 2HCl(aq) \longrightarrow MgCl_2(aq) + 2H_2O(l)$$

(i) The molar mass of magnesium hydroxide is 58.0 g mol^{-1}. Calculate the number of moles of hydrochloric acid which can be neutralized by 1.00 g of magnesium hydroxide. **2**
(ii) Calculate the volume of 0.100 M HCl which can be neutralized by 1.00 g of magnesium hydroxide, giving your answer to 3 significant figures. **2**

8 *L(N) (AS/AL)*

20. The concentration of hydrogen peroxide in a solution can be determined by titrating an acidified solution against aqueous potassium manganate(VII) added from a burette. The potassium manganate(VII) reacts with the colourless aqueous solution of hydrogen peroxide as shown in the equation given below.

$$5H_2O_2 + 2MnO_4^- + 6H^+ \longrightarrow 5O_2 + 8H_2O + 2Mn^{2+}$$

(a) State the role of hydrogen peroxide in this reaction. **1**
(b) Identify a suitable acid for use in this titration. **1**
(c) State the colour change at the end-point of this reaction. **1**
(d) After acidification with a suitable acid, 25.0 cm^3 of a dilute aqueous solution of hydrogen peroxide were found to react with 18.1 cm^3 of 0.0200 M $KMnO_4$. Calculate the molar concentration of hydrogen peroxide in the solution. **4**

7 *NEAB (AS/AL)*

21. Standard solutions containing thiosulphate ions are used to measure the amount of iodine in a solution. The result can be used to find the reacting quantities in some redox reactions.

20.0 cm^3 of 0.100 M copper sulphate solution was reacted with an excess of potassium iodide. The iodine formed reacted with exactly 40.0 cm^3 of 0.0500 M sodium thiosulphate solution.

(a) Calculate the number of moles of thiosulphate ions, $S_2O_3^{2-}$, used in the reaction. **1**
(b) Complete the ionic equation for the reaction between iodine and thiosulphate ions.

$$I_2(aq) + 2S_2O_3^{2-}(aq) \longrightarrow$$ **2**

(c) Calculate the number of moles of iodine molecules, I_2, in the reaction mixture. **1**
(d) Calculate the number of moles of copper(II) ions, $Cu^{2+}(aq)$, in 20.0 cm^3 of 0.100 M copper sulphate solution which reacted with the potassium iodide. **1**
(e) Use oxidation numbers and your answers to *(c)* and *(d)* to complete the equation for the reaction between copper(II) ions and iodide ions. Justify your balancing of the equation.

$$Cu^{2+}(aq) + I^-(aq) \longrightarrow I_2(aq) +$$ **2**

6 *L (AS/AL)*

4 THE CHEMICAL BOND

4.1 DIAMOND AND GRAPHITE – THE DIFFERENCE LIES IN THE BONDS

One of the most famous of diamonds is the Hope diamond. This huge blazing diamond with a bluish tint was mined in India, and was once the eye of a statue of Sita, a Hindu goddess. The diamond was stolen and was said to have been cursed by the goddess Sita. Over the centuries a string of owners met with bad luck. Marie Antoinette was guillotined during the French Revolution in 1792. Henry Hope bought the diamond in 1830, but his unlucky family lost all their father's wealth and had to sell the diamond. Another owner, a European prince, gave the diamond to an actress in the Folies Bergères but afterwards shot her in a fit of jealous rage. A Greek owner drove his car over a precipice, killing himself and his family. The Sultan of Turkey acquired the diamond and shortly afterwards was overthrown in 1909 by a revolution. The history of bad luck did not deter Mrs Evalyn Walsh from buying the Hope diamond, but bad luck dogged her all the same, with her children dying in accidents and her husband becoming mentally ill. When she died a dealer bought the stone and gave it to the Smithsonian Institute in Washington, USA, so that it could be enjoyed by people in all walks of life.

Why have people always been fascinated by diamonds? The brilliance of diamonds is due to their ability to reflect light. The fire of diamonds is due to their ability to disperse white light into flashes of light of all the colours of the spectrum. There is a big difference between the refractive index of diamond for red light and that for violet light. Diamond has another outstanding characteristic. It is the hardest of natural materials. This is why diamond is used for grinding, cutting, etching, polishing, drilling and many other industrial applications. Diamond is a simple substance; it is a form of the element carbon. The reason for its amazing properties – high refractive index and hardness – is the way in which atoms of carbon are bonded together in diamond.

Graphite is a shiny dark grey solid. It is soft and, when you rub it, it leaves traces on your fingers. Although graphite appears to be so different from diamond, it is another form of the element carbon. We expect non-metallic elements to be non-conductors of electricity, but graphite is the exception. It is widely used as an electrical conductor. The striking differences between diamond and graphite are due to the different ways in which carbon atoms are bonded.

In this chapter we shall be looking at the different kinds of chemical bonds and the ways in which substances with different bonds have different characteristics.

4.2 IONS

4.2.1 SOME QUESTIONS

The background to the ionic theory

- Why are all copper(II) compounds blue, regardless of the non-metallic part of the compound?
- Why are all dichromates orange, regardless of the metallic part of the compound?
- Why do aqueous solutions of some substances conduct electricity?
- Why does a layer of copper appear on the negative electrode when a direct electric current is passed through a solution of a copper salt [see Figure 4.2A, § 4.2.3]?
- Why is chlorine evolved at the positive electrode when a direct electric current passes through a concentrated aqueous solution of a chloride?
- Why do X ray diffraction patterns of salts show a regular patter of dots [see Figure 6.2A, § 6.2]?

To answer these questions and others, the **ionic theory** was developed. According to the ionic theory, some compounds consist of tiny particles which carry an electric charge, either positive or negative, and are called **ions**. The ionic theory was developed to explain the behaviour of compounds when a direct electric current is passed through the molten compound or an aqueous solution of the compound.

4.2.2 WHICH SUBSTANCES CONDUCT ELECTRICITY?

Substances can be divided into four groups according to their ability to conduct a direct current of electricity [see Table 4.2].

Table 4.2
Electrical conductors and non-conductors

Solids: Metals and alloys and graphite conduct electricity

Liquids: Solutions of acids, alkalis and salts conduct electricity

Solids	
Electrical conductors	Non-conductors, i.e. insulators
All metallic elements All alloys One non-metallic element, the graphite allotrope of carbon	Non-metalic elements, e.g. sulphur Many compounds, e.g. polyethene Crystalline salts, e.g. sodium chloride, copper(II) sulphate
Liquids	
Electrolytes	Non-electrolytes
Solutions of acids and alkalis and salts; such liquids are called **electrolytes**. Chemical changes occur at the electrodes. For example, copper(II) chloride solution changes into copper and chlorine and water.	The liquids which do not conduct electricity are water and organic compounds, such as ethanol. They are called **non-electrolytes**.

4.2.3 WHAT HAPPENS WHEN COMPOUNDS CONDUCT ELECTRICITY?

Important terms: cell, electrode, anode, cathode, electrolyte

Crystals of copper(II) chloride do not conduct electricity, and distilled water does not conduct electricity. Yet when the two substances are mixed to make a solution, the solution conducts. Figure 4.2A shows what happens. The objects which conduct electricity into and out of the cell are called **electrodes**. The electrodes are usually made of elements such as platinum and graphite, which do not react with electrolytes. The electrode connected to the positive terminal of the battery is called the **anode**. The electrode connected to the negative terminal is called the **cathode**.

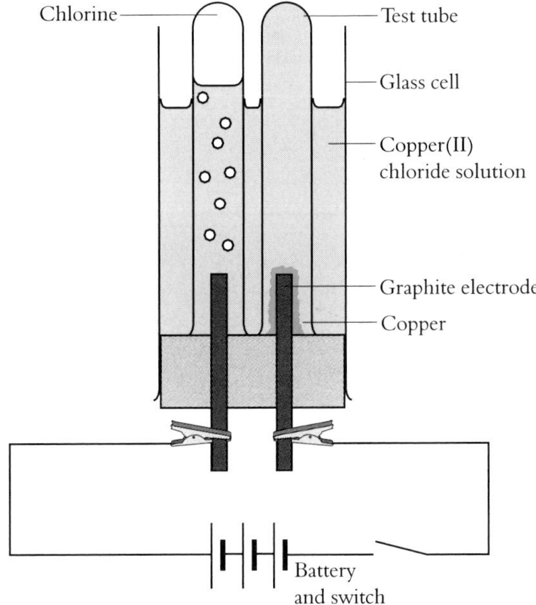

Figure 4.2A
Electrolysis of copper(II) chloride

Solutions of some compounds are electrolysed ...

... split up by an electric current ...

... to form new substances

In the process, the solute, copper(II) chloride, is split up. Copper appears as a layer on the cathode. Chlorine is evolved at the anode. Copper(II) chloride has been **electrolysed**, that is, split up by an electric current. The process is called **electrolysis**. Copper(II) chloride is an **electrolyte**. An electrolyte is a compound which conducts electricity when molten or in solution. The compound is split up in the process. Any container in which electricity produces a chemical change (or in which a chemical change produces electricity) is called a **cell**.

Such compounds are called electrolytes

In explaining electrolysis, we must take into account that, when copper salts are electrolysed, copper always appears at the negative electrode, never at the positive electrode. This must show that there is a positive charge associated with copper in all its salts. We accept that copper, the element, is composed of copper atoms. Does the combined copper in copper(II) chloride and other salts consist of a different kind of copper atoms with positive charges? Chlorine appears always at the positive electrode, never at the negative electrode. This indicates that there is a negative charge associated with the chlorine combined with copper in copper(II) chloride. Chlorine in its salts must consist of a type of chlorine atom with negative charge.

They consist of ions ...

... positive ions (cations) ...

... and negative ions (anions)

In 1834, Michael Faraday suggested a theory to explain electrolysis. He suggested the existence of tiny particles of matter carrying positive or negative electric charges. We now call them **ions**. Positively charged ions are called **cations**. Negatively charged ions are called **anions**.

... and include all salts

According to the ionic theory, the compound copper(II) chloride consists of positively charged copper ions and negatively charged chloride ions. During electrolysis, positively charged copper ions travel to the negative electrode. When they reach the negative electrode, the ions lose their charge: they are discharged to form copper atoms. Negatively charged chloride ions travel in the positive electrode. When they meet the positive electrode, they lose their charge: they are discharged, and molecules of chlorine are formed [see Figure 4.2B].

In electrolysis ions are discharged at the electrodes

Figure 4.2B
What happens at the electrodes

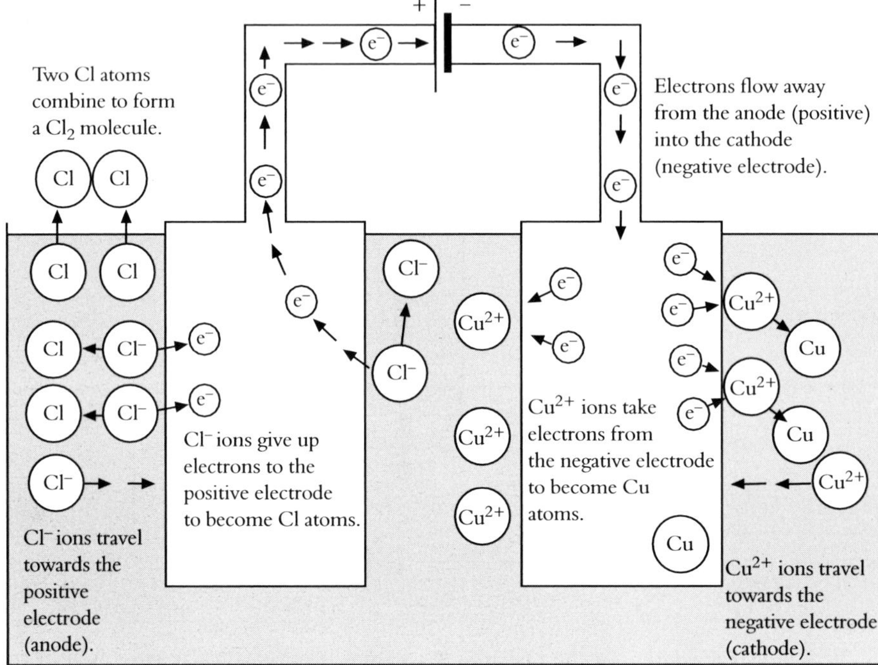

Solid copper(II) chloride does not conduct electricity, and water is a very poor electrical conductor. However, an aqueous solution of the salt is a good conductor and is an electrolyte. To explain this difference, scientists suggest that, in a solid, the ions are fixed in position, held together by strong attractive forces between positive and negative ions. In a solution of a salt, the ions are free to move [see Figure 4.2C].

Figure 4.2C
The ions in a solution are free to move

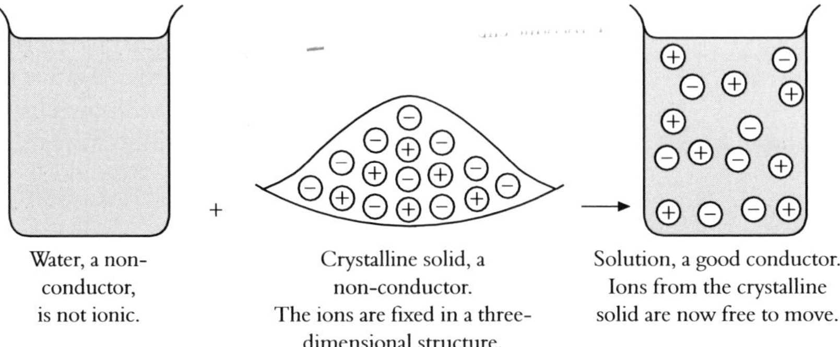

PROFILE: MICHAEL FARADAY

The scientist who did the first work on electrolysis was Michael Faraday (1791–1867). He was the son of a poor blacksmith, and at the age of 14 he became an apprentice bookbinder. He educated himself by reading the books he was asked to bind and by joining various self-improvement groups. Faraday became especially interested in chemistry and electricity. When he attended a course of lectures by Humphry Davy at the Royal Institute, Faraday made detailed notes, illustrated them with excellent diagrams, and bound them in a book. Davy was impressed with the young man's enthusiasm and offered him a job as a laboratory assistant.

Faraday was soon contributing actively to the research. The variety of his achievements is amazing. He made the first electric motor, the first transformer and the first dynamo. He formulated the First and Second Laws of Electrolysis. He discovered benzene and did research on steel, optical glass and the liquefaction of gases. Faraday became a superb scientific lecturer. He gave weekly evening lectures at the Royal Institution to popularise science and started a tradition of Christmas lectures for children which is still continued.

Figure 4.2D Michael Faraday lecturing to children

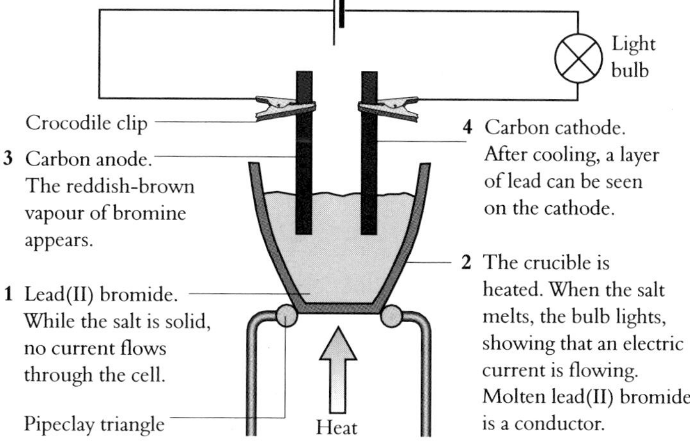
Light bulb
Crocodile clip
3 Carbon anode. The reddish-brown vapour of bromine appears.
4 Carbon cathode. After cooling, a layer of lead can be seen on the cathode.
2 The crucible is heated. When the salt melts, the bulb lights, showing that an electric current is flowing. Molten lead(II) bromide is a conductor.
1 Lead(II) bromide. While the salt is solid, no current flows through the cell.
Pipeclay triangle
Heat

Figure 4.2E Electrolysis of molten lead(II) bromide

4.2.4 MOLTEN SALTS

Some compounds are electrolysed when molten. These compounds are composed of ions

In a solution of a salt, the ions are free to move, and the solution can be electrolysed. Another way of enabling the ions to move is to melt the salt. You have probably seen the experiment shown in Figure 4.2E. Lead(II) bromide is a convenient salt to use because it has a fairly low melting point. The experiment shows that molten lead(II) bromide is electrolysed. In the molten salt, ions are free to move.

4.2.5 HOW DOES AN ATOM BECOME AN ION?

A metal atom may lose one or more electrons to become a cation

Atoms are uncharged. The number of protons in an atom is the same as the number of electrons [see § 1.6]. If an atom either gains or loses an electron, it will become electrically charged. Metal atoms and hydrogen atoms form positive ions. For example, a sodium atom loses one electron to form a positively charged sodium ion.

Sodium atom ⟶ Electron + Sodium ion
Na(11 protons, 11 electrons) ⟶ e^- + Na^+ (11 protons, 10 electrons)
uncharged charge = + 1

A magnesium atom loses 2 electrons to become a magnesium ion, Mg^{2+}.

An aluminium atom loses 3 electrons to become an aluminium ion, Al^{3+}. The charge on a cation may be +1, +2 or +3.

An atom of a non-metal may gain one or more electrons to become an anion

Non-metallic elements form negative ions (anions). They do this by gaining electrons. A chlorine atom gains one electron to become a chloride ion, Cl^-.

Chlorine atom + Electron ⟶ Chloride ion
Cl(17 proton, 17 electrons) + e^- ⟶ Cl^- (17 protons, 18 electrons)
uncharged charge = −1

An oxygen atom gains 2 electrons to become an oxide ion, O^{2-}. Some anions contain oxygen combined with another element. Examples are: hydroxide ion, OH^-; nitrate ion, NO_3^-; sulphate ion, SO_4^{2-}. An anion may have a charge of −1, −2 or −3. Table 4.3, § 4.3.7, lists the symbols and formulae of some common ions.

4.2.6 NON-ELECTROLYTES

Non-electrolytes consist entirely of molecules

Some liquids do not conduct electricity. It follows that these substances do not contain ions. They consist of uncharged particles called molecules [see § 4.4]. Compounds formed between metallic and non-metallic elements are usually electrolytes, and compounds formed between non-metallic elements are generally non-electrolytes, e.g. ethanol (alcohol), or weak electrolytes [see § 4.2.7].

4.2.7 WEAK ELECTROLYTES

Weak electrolytes consist mainly of molecules

Some substances conduct electricity to a very slight extent. For example, ethanoic acid is a poor conductor; it is a weak electrolyte. A solution of ethanoic acid contains a small concentration of ions, which make it conduct. The compound exists mainly in the form of molecules, which do not conduct [see § 4.4].

CHECKPOINT 4.2: ELECTRICAL CONDUCTORS

1. *(a)* Divide the following list into (i) electrical conductors and (ii) non-conductors:

   solid wax, molten wax, ethanol (alcohol), distilled water, aqueous ethanol, copper, wood, steel, sodium chloride crystals, sugar crystals, sugar solution, tetrachloromethane (CCl_4), brass, polythene, molten magnesium chloride, solid sodium hydroxide, molten sodium hydroxide, sodium hydroxide solution, PVC, petrol, silver

   (b) Say which of the conductors in the list are electrolytes.

2. Briefly explain the terms: electrolysis, electrolyte, cell, anode, cathode, ion, anion, cation.

3. Explain *(a)* why ions move toward electrodes and *(b)* why solid copper(II) sulphate is not an electrical conductor.

4. Answer the questions at the beginning of § 4.2.1.

4.3 THE IONIC BOND

Let us look in more detail at how atoms are able to form ions. How can one atom be able to give up an electron to form a cation? How can another atom be able to accept an electron to form an anion?

4.3.1 SODIUM CHLORIDE

Look at the electron configuration in a sodium atom [Figure 4.3A]. There is just one electron more than there is in an atom of the noble gas, neon.

Figure 4.3A
Atoms of sodium and neon

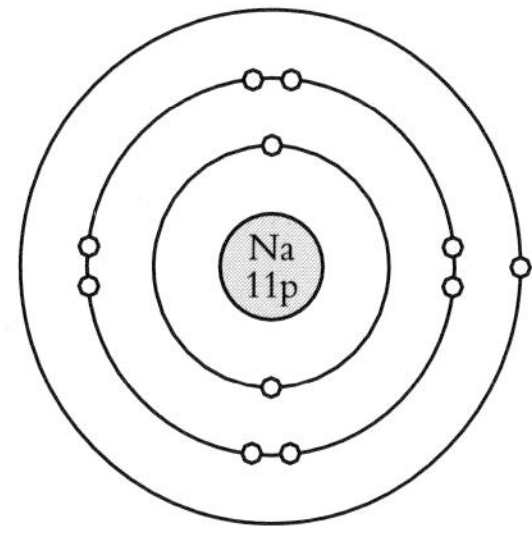

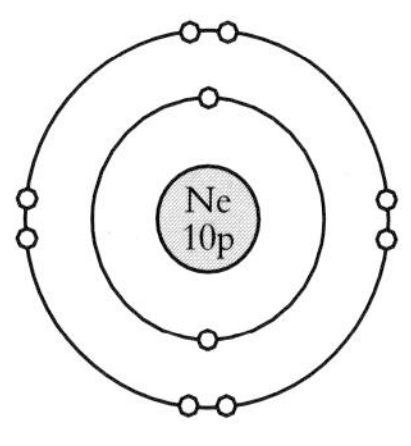

The noble gases have a full outer shell of electrons. This electron configuration is very stable

The noble gases [see § 2.6] are helium, neon, argon, krypton, xenon and radon. They take part in hardly any chemical reactions. With the electron configuration ns^2np^6 ($1s^2$ for He) the noble gases have a full outer shell of electrons [Figure 4.3B] and it seems probable that it is this electron configuration that makes them stable, that is, chemically unreactive.

Figure 4.3B
The electron configurations in helium neon and argon

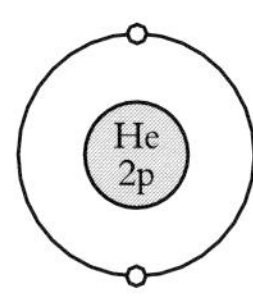

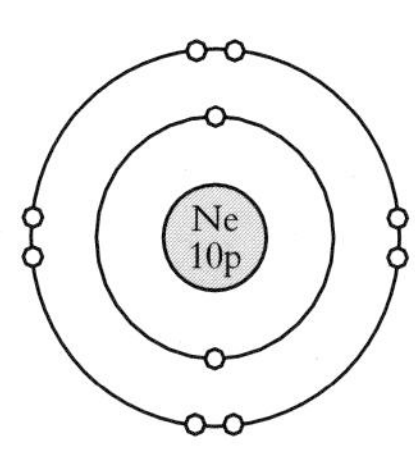

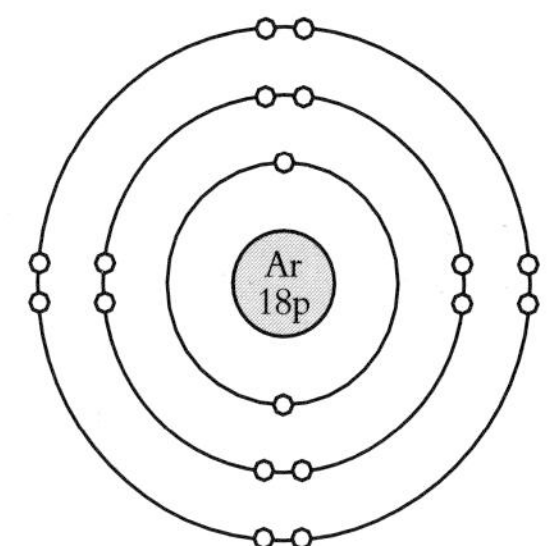

To achieve a full outer shell of electrons, a sodium atom must lose one electron ...

When a sodium atom Na(2.8.1) loses the lone electron from its outermost shell, the outer shell that remains contains 8 electrons, the same as the noble gas neon Ne(2.8). This electron configuration is associated with stability. A sodium atom cannot lose an electron unless another atom will accept it. It can, however, give an electron to a chlorine atom. In fact, sodium burns vigorously in chlorine to form sodium chloride.

... and a chlorine atom must gain one electron

How can a chlorine atom accept an electron? With the electron configuration Cl(2.8.7), one more electron gives chlorine the same electron arrangement as the noble gas argon Ar(2.8.8). A full outer shell brings with it stability [Figure 4.3C].

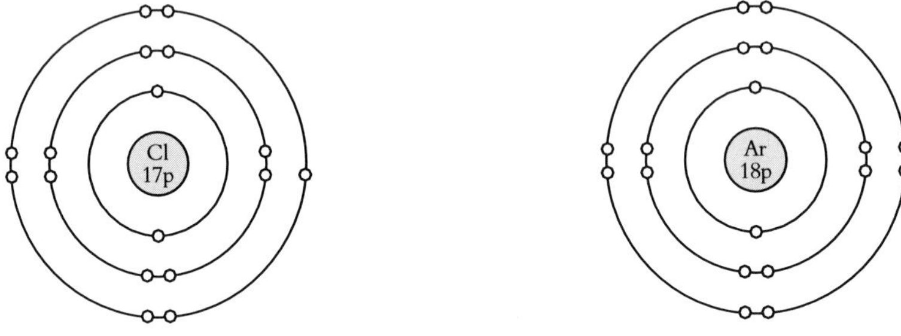

Figure 4.3C
The configuration of electrons in chlorine and argon

Figure 4.3D shows what happens when an atom of sodium donates an electron to an atom of chlorine. A full outer shell is left behind in sodium, and a full outer shell is created in chlorine.

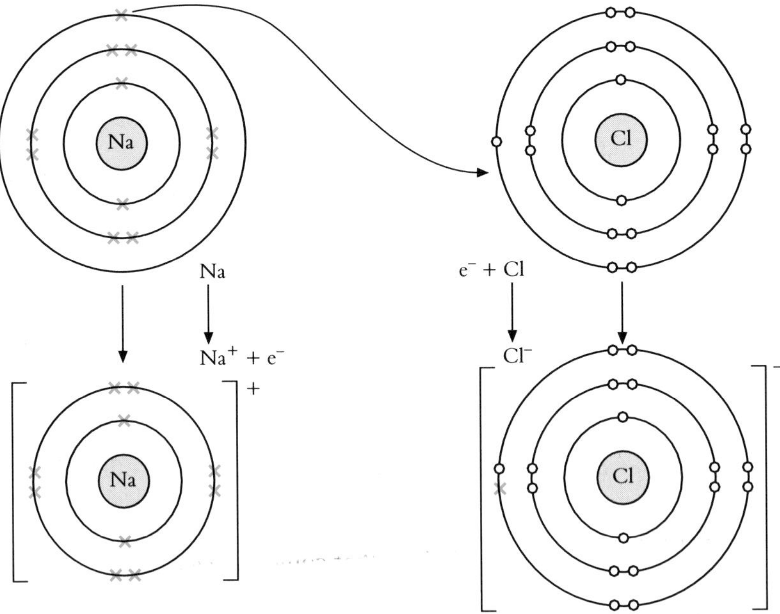

Figure 4.3D
The formation of sodium chloride. (The sodium electrons have been shown as × and the chlorine electrons as O.)

The oppositely charged ions are held together by an electrostatic attraction called an ionic bond or electrovalent bond

The electrostatic attraction between oppositely charged ions holds the ions Na^{+} and Cl^{-} together. This electrostatic attraction is the chemical bond in the compound, sodium chloride. It is called an **ionic bond** or **electrovalent bond**. Sodium chloride is an ionic or electrovalent compound. The compounds which conduct electricity when they are melted or dissolved are electrovalent compounds.

Figure 4.3E
A sodium atom and a sodium ion

A sodium ion is positively charged

A chloride ion is negatively charged

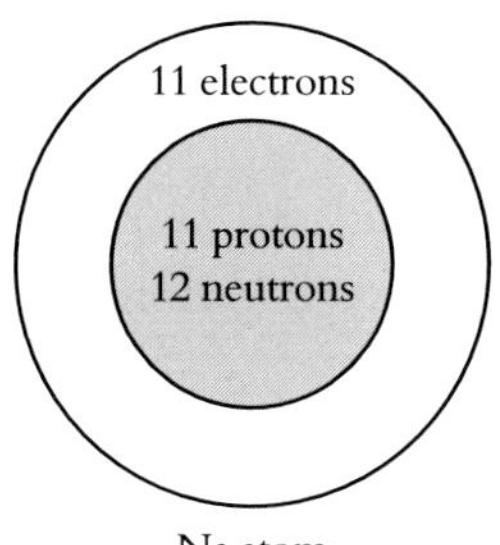

Na atom

A sodium atom has 11 protons, 11 electrons and 12 neutrons. Charge = +11 – 11 = 0 charge unit.

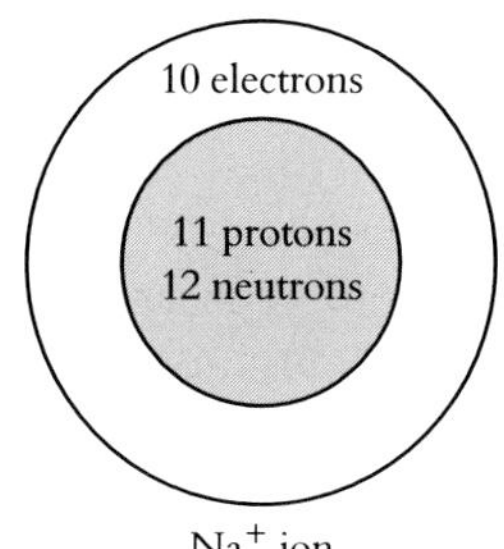

Na^+ ion

After it loses one electron, it has 11 protons and 10 electrons. Charge = +11 – 10 = +1 charge unit. The sodium atom, Na, has become a sodium ion, Na^+.

Figure 4.3F
A chlorine atom and a chloride ion

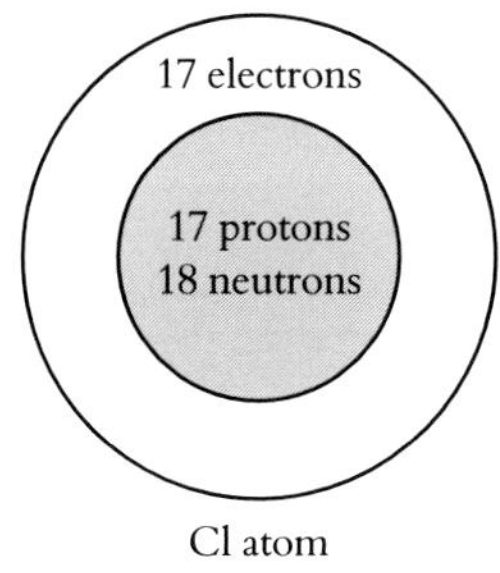

Cl atom

A chlorine atom has 17 protons, 17 electrons and 18 neutrons. Charge = +17 – 17 = 0 charge unit.

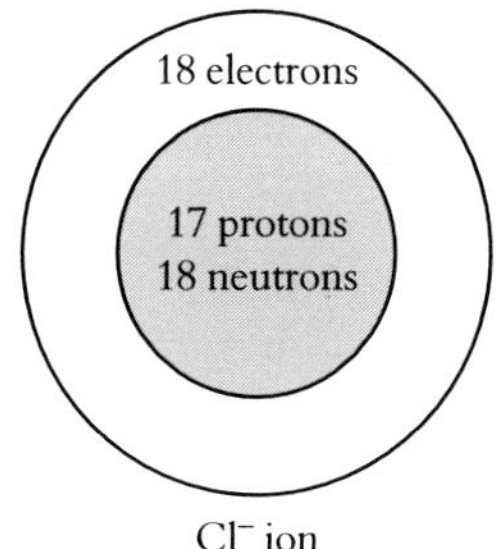

Cl^- ion

After it gains an electron, making 17 protons and 18 electrons, the charge is +17 – 18 = –1 charge unit. The chlorine atom, Cl, has become a chloride ion, Cl^-.

SUMMARY

An ionic bond is formed when an atom of a metallic element gives an electron or electrons to an atom of a non-metallic element. The ions formed are held together by an electrostatic attraction.

The theory of the chemical bond is due to the work of W Kossel and G N Lewis. In 1916, working independently, they both put forward the theory that the formation of chemical bonds can be explained by the tendency of atoms to give or receive electrons in order to attain a noble gas type of electron configuration.

4.3.2 CRYSTALS

Ionic compounds from crystals

A feature of ionic compounds is that they form **crystals**. The crystals of sodium chloride are perfect cubes. In a dilute solution of sodium chloride, sodium ions and chloride ions are moving about independently of other ions. When the solution is evaporated to the point of crystallisation, the ions are much closer together. A sodium ion attracts chloride ions, as shown in Figure 4.3G(a). Each chloride ion attracts other sodium ions, and a three-dimensional arrangement of ions called a **crystal structure** is built up [see Figure 4.3G(b)]. There is no pair of Na^+ and Cl^- ions that could be regarded as a molecule of sodium chloride. The formula NaCl represents the ratio in which ions are present in the crystal structure. A pair of ions $Na^+ Cl^-$ is called a **formula unit** of sodium chloride.

A crystal of sodium chloride is a three-dimensional structure of sodium ions and chloride ions

The crystal is uncharged because the number of sodium ions is equal to the number of chloride ions. The bonds between positive and negative ions are strong. This is why solid sodium chloride does not conduct electricity and is not electrolysed. In the solid, the ions cannot move out of their positions in the three-dimensional structure. When the salt is melted or dissolved, the ions are free to move and can travel towards the electrodes [see Figure 4.2C].

Figure 4.3G
The arrangement of ions in a sodium chloride crystal

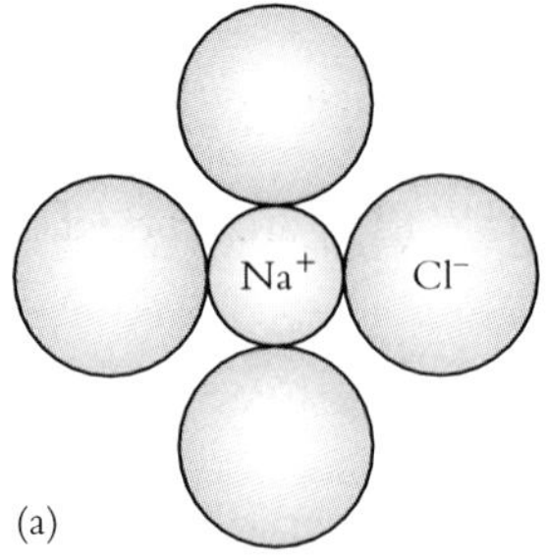

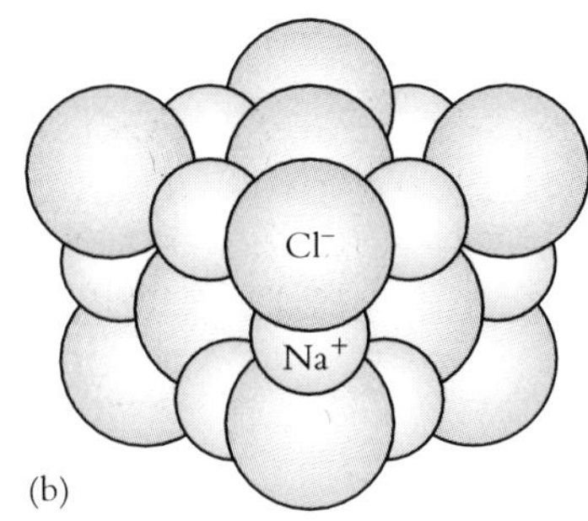

X ray analysis demonstrates the existence of ions

An impressive demonstration of the existence of ions is the use of X ray analysis [§ 6.2] to obtain electron density maps. The one for sodium chloride is shown in Figure 4.3H. It consists of regions of charge which are isolated from other regions of charge. This is the picture one would expect for a structure consisting of separate Na^+ and Cl^- ions. The technique is now sufficiently accurate to show that there are ten electrons, not eleven, associated with each sodium nucleus: the species present in Na^+, not Na.

Figure 4.3H
Electron density map for sodium chloride

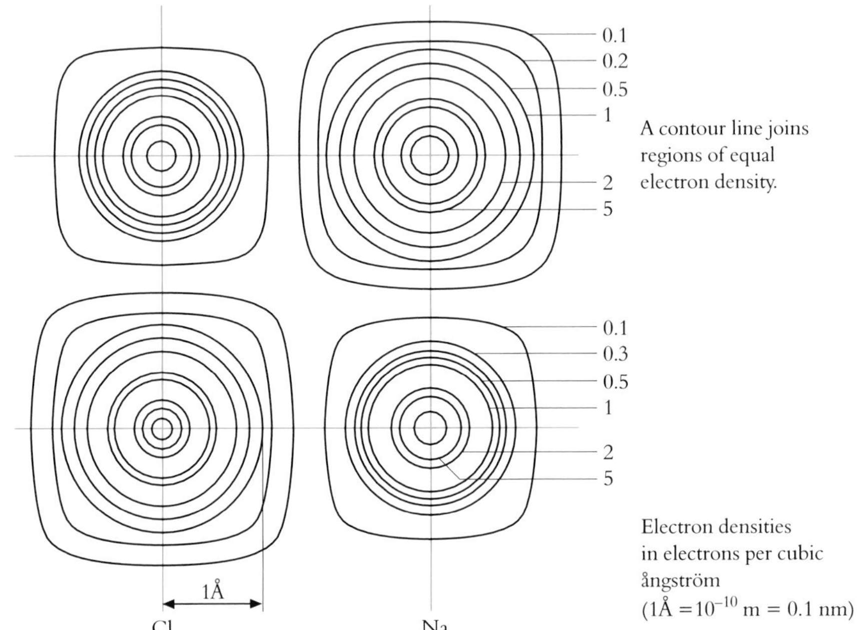

4.3.3 MAGNESIUM FLUORIDE

Magnesium fluoride contains two F^- ions for every Mg^{2+} ion

Magnesium (2.8.2) has two electrons in the outermost shell. It has to lose these to attain a stable, full outer shell of electrons. Fluorine (2.7) needs to gain an electron. For a magnesium atom to lose two electrons and a fluorine atom to gain only one electron, one magnesium atom has to combine with two fluorine atoms.

$$Mg\ (2.8.2) \longrightarrow Mg^{2+}\ (2.8) + 2e^-$$

$$F(2.7) + e^- \longrightarrow F^-(2.8)$$

$$Mg + F_2 \longrightarrow Mg^{2+}\ 2F^-$$

4.3.4 MAGNESIUM OXIDE

Magnesium oxide consists of Mg^{2+} ions and O^{2-} ions in equal numbers

The product formed when magnesium burns in air is magnesium oxide. One atom of magnesium, Mg (2.8.2), gives two electrons to one atom of oxygen, O (2.6). The ions Mg^{2+} (2.8) and O^{2-} (2.8) are formed. The electrostatic attraction between them is an ionic bond.

Figure 4.31
The formation of an ionic bond between atoms of magnesium and oxygen

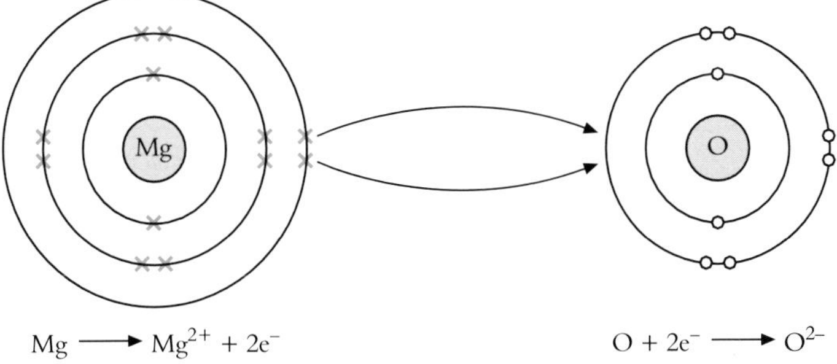

$Mg \longrightarrow Mg^{2+} + 2e^-$ $\quad$ $O + 2e^- \longrightarrow O^{2-}$

4.3.5 THE CHARGE ON THE IONS OF AN ELEMENT

Metals form ions with charge +1, +2 or +3

Non-metals form ions with charge −1, −2 or −3

Electrovalent compounds (ionic compounds) are formed when a metallic element combines with a non-metallic element. Metallic elements form positive ions, e.g. sodium ions, Na^+, magnesium ions, Mg^{2+} and aluminium ions, Al^{3+}. Non-metallic elements form negative ions, e.g. chloride ions, Cl^-, and oxide ions, O^{2-}.

Refer to Table 2.9, which lists the electron configurations of the atoms of some elements.

- Can you see from the electron configuration why potassium and sodium form ions K^+ and Na^+, while magnesium and calcium form ions Mg^{2+} and Ca^{2+}? What do the electron configurations of potassium and sodium have in common? What do the electron configurations of magnesium and calcium have in common?

- Can you see from the electron configurations why fluorine and chlorine form ions F^- and Cl^-, while oxygen and sulphur form ions O^{2-} and S^{2-} and nitrogen forms ions N^{3-}?

- In general,

The charge on the ions of a metallic element	=	Number of electrons in the outermost shell of an atom of that element
The charge on the ions of a non-metallic element	=	8 − (Number of electrons in the outermost shell of an atom of that element)

Non-metallic elements often form oxo-ions (which contain oxygen), e.g. sulphate, SO_4^{2-}, and nitrate, NO_3^-.

4.3.6 IONS AND THE PERIODIC TABLE

The charge depends on the number of electrons in the outermost shell ...

- Compare the charges on the ions of the following elements with the group in which you find them in the Periodic Table [p. 732].

  Ions: Na^+, K^+, Ca^{2+}, Mg^{2+}, Ba^{2+}, Sr^{2+}, Al^{3+}, N^{3-}, O^{2-}, S^{2-}, F^-, Cl^-, Br^-, I^-

- Can you see the relationship?

... and therefore on the position of the element in the Periodic Table

Charge on ions of metallic element = Group number of element

Charge on ions of non-metallic element = 8 – Group number of element

CHECKPOINT 4.3A: BONDING IN IONIC COMPOUNDS

1. Lithium is an alkali metal with the electron configuration Li(2.1). Fluorine is a halogen with the electron configuration F(2.7). Draw the configuration of electrons *(a)* in the atoms of Li and F and *(b)* in a pair of ions in the compound lithium fluoride.

2. Sodium is a silvery-grey metal. It has to be kept under oil because it reacts rapidly with oxygen and water vapour in the air. Chlorine is a poisonous green gas. Sodium chloride is a white, crystalline solid, which we eat as 'table salt'. Explain how the sodium in sodium chloride differs from sodium metal. Explain how the chlorine in sodium chloride differs from chlorine gas.

3. Refer to Figure 4.3D for the formation of sodium chloride. Draw a similar diagram to illustrate the formation of magnesium fluoride, MgF_2.

4. Use 'dot and cross' diagrams to show how the following atoms combine:

   (a) Li(2.1) and O(2.6), *(b)* Na(2.8.1) and F(2.7), *(c)* Be(2.2) and F(2.7), *(d)* Al(2.8.3) and F(2.7), *(e)* Ca(2.8.8.2) and O(2.6).

5. There are two differences between

   (a) a sodium ion, Na^+(2.8) and a neon atom, Ne(2.8)
   (b) a chloride ion, Cl^-(2.8.8) and an argon atom, Ar(2.8.8).

   What are the differences?

6. Say in which group of the Periodic Table you would expect to find the elements which form the following ions:

   Rb^{2+}, Ga^{3+}, Sr^{2+}, At^-, P^{3-}, Se^{2-}, Cs^+, Ra^{2+}

4.3.7 FORMULAE OF IONIC COMPOUNDS

Electrovalent compounds consist of positive and negative ions. A compound is neutral because the charge on the positive ion (or ions) is equal to the charge on the negative ion (or ions). In zinc chloride, $ZnCl_2$, one zinc ion, Zn^{2+}, is balanced in charge by two chloride ions, $2Cl^-$.

You can work out the formula of an ionic compound by balancing the charges on the ions

Let us use this principle of balancing the charges to work out the formulae of electrovalent compounds.

Compound:	*Zinc chloride*
Which ions are present?	Zn^{2+}, Cl^-
How can you balance the charges?	One Zn^{2+} ion needs two Cl^- ions.
How many ions are needed?	Zn^{2+} and $2Cl^-$ ions
What is the formula?	$ZnCl_2$

Compound:	*Sodium carbonate*
Which ions are present?	Na^+, CO_3^{2-}
How can you balance the charges?	Two Na^+ are needed to balance one CO_3^{2-}.
How many ions are needed?	$2Na^+$ and CO_3^{2-}
What is the formula?	Na_2CO_3

Compound:	*Calcium hydroxide*
Which ions are present?	Ca^{2+}, OH^-
How can you balance the charges?	Two OH^- ions balance one Ca^{2+} ion.
How many ions are needed?	Ca^{2+} and $2OH^-$
What is the formula?	$Ca(OH)_2$ The brackets tell you that the 2 multiplies everything inside them. There are 2 O atoms and 2 H atoms, in addition to the 1 Ca ion.

Compound:	*Iron(II) sulphate*
Which ions are present	Fe^{2+}, SO_4^{2-}
How can you balance the charges?	One Fe^{2+} ion balances one SO_4^{2-} ion.
How many ions are needed?	Fe^{2+} and SO_4^{2-}
What is the formula?	$FeSO_4$

Compound:	*Iron(III) sulphate*
Which ions are present?	Fe^{3+}, SO_4^{2-}
How can you balance the charges?	Two Fe^{3+} would balance three SO_4^{2-} ions.
How many ions are needed?	$2Fe^{3+}$ and $3SO_4^{2-}$
What is the formula?	$Fe_2(SO_4)_3$ The brackets tell you that all the atoms inside them must be multiplied by the 3 that follows them. The formula contains 2 Fe, 3 S and 12 O.

You need to learn the symbols and charges of the ions in Table 4.3. Then you can work out the formula of any electrovalent compound containing these ions.

You will notice that the sulphates of iron are named iron(II) sulphate and iron(III) sulphate. The Roman numerals, II and III, show which type of ion, Fe^{2+} or Fe^{3+}, is present. This is always done with the compounds of elements of variable oxidation number.

Table 4.3
The symbols and formulae of some common ions

Ion	Symbol	Ion	Symbol
Aluminium	Al^{3+}	Bromide	Br^-
Ammonium	NH_4^+	Carbonate	CO_3^{2-}
Barium	Ba^{2+}	Chloride	Cl^-
Calcium	Ca^{2+}	Dichromate(VI)	$Cr_2O_7^{2-}$
Copper(I)	Cu^+	Hydrogencarbonate	HCO_3^-
Copper(II)	Cu^{2+}	Hydroxide	OH^-
Hydrogen	H^+	Iodide	I^-
Iron(II)	Fe^{2+}	Manganate(VII)	MnO_4^-
Iron(III)	Fe^{3+}	Nitrate	NO_3^-
Lead(II)	Pb^{2+}	Nitrite	NO_2^-
Magnesium	Mg^{2+}	Oxide	O^{2-}
Mercury(II)	Hg^{2+}	Phosphate	PO_4^{3-}
Potassium	K^+	Silicate	SiO_3^{2-}
Silver	Ag^+	Sulphate	SO_4^{2-}
Sodium	Na^+	Sulphide	S^{2-}
Zinc	Zn^{2+}	Sulphite	SO_3^{2-}

CHECKPOINT 4.3B: FORMULAE OF IONIC COMPOUNDS

1. Write the formulae of the following ionic compounds:

   (a) silver chloride, *(b)* potassium nitrate, *(c)* silver nitrate, *(d)* zinc bromide, *(e)* magnesium iodide, *(f)* copper(II) bromide, *(g)* ammonium chloride, *(h)* ammonium sulphate, *(i)* calcium hydroxide, *(j)* aluminium chloride, *(k)* sodium hydrogencarbonate, *(l)* sodium sulphite, *(m)* ion(II) hydroxide, *(n)* iron(III) hydroxide, *(o)* aluminium oxide.

2. Name the following compounds:

   (a) AlI_3, *(b)* $CuCO_3$, *(c)* $Zn(OH)_2$, *(d)* AgBr, *(e)* $Cu(NO_3)_2$, *(f)* $FeBr_2$, *(g)* $FeBr_3$, *(h)* Al_2O_3, *(i)* $KMnO_4$, *(j)* Na_2SiO_3, *(k)* Na_3PO_4, *(l)* KNO_2, *(m)* $K_2Cr_2O_7$, *(n)* $Ca_3(PO_4)_2$, *(o)* Na_2SO_3, *(p)* $BaSO_3$, *(q)* $Ca(HCO_3)_2$.

4.3.8 IONIC RADII

The sum of the cationic and anionic radii is equal to the interionic distance in a crystal

The distance between the centres of the ions is the sum of the **cationic radius** and the **anionic radius**. Some values of ionic radii are shown in Figure 15.2D [§ 15.2].

4.3.9 ENERGY CHANGES IN COMPOUND FORMATION

Electrovalent bonds are formed if the reaction between elements to give an ionic solid is exothermic

A detailed picture of the energy changes involved in the formation of an ionic compound was drawn by the theoreticians M Born and F Haber. They considered, for example, the formation of sodium chloride from its elements:

$$Na(s) + \tfrac{1}{2}Cl_2(g) \longrightarrow NaCl(s)$$

They analysed the reaction as the sum of five steps. These are

1. Vaporisation of sodium $Na(s) \longrightarrow Na(g)$ *Endothermic*
2. Ionisation of sodium $Na(g) \longrightarrow Na^+(g) + e^-$ *Endothermic*
3. Dissociation of chlorine $\tfrac{1}{2}Cl_2(g) \longrightarrow Cl(g)$ *Endothermic*
4. Ionisation of chlorine $Cl(g) + e^- \longrightarrow Cl^-(g)$ *Exothermic*
5. Combination of ions to form a crystalline solid $Na^+(g) + Cl^-(g) \longrightarrow NaCl(s)$ *Exothermic*

The sum of the five energy changes is exothermic. [See § 10.12 for a fuller treatment.]

CHECKPOINT 4.3C: IONS

1. Write the formulae of the electrovalent compounds which contain the following pairs of ions:

   (a) Mg^{2+} and N^{3-}, *(b)* Al^{3+} and F^-, *(c)* Al^{3+} and S^{2-}, *(d)* Fe^{2+} and O^{2-}, *(e)* Fe^{3+} and O^{2-}, *(f)* Co^{3+} and SO_4^{2-}, *(g)* Ni^{2+} and NO_3^-

2. Write the electron configuration of each of the following ions, and give the name of the *isoelectronic* noble gas (which has the same electron configuration):

   Li^+, N^{3-}, Be^{2+}, K^+, S^{2-}

   (Atomic numbers are Li = 3, N = 7, Be = 4, K = 19, S = 16.)

3. Write the electron configurations for the following cations:

   Mn^{2+}, Cu^+, Cu^{2+}, Zn^{2+}

   (Atomic numbers are Mn = 25, Cu = 29, Zn = 30.)

The driving force behind the reaction is the fact that sodium metal and chlorine molecules can pass to a lower energy level by forming ionic bonds. The formation of sodium chloride is exothermic. This picture of electrovalency proves to be more fruitful than the simple picture of attaining a noble gas electron configuration. Elements will not form an ionic compound if it is at a higher energy level than the elements. They may combine by the formation of covalent bonds.

4.4 THE COVALENT BOND

Some compounds are non-electrolytes. Since these compounds do not conduct electricity, they cannot consist of ions. Non-electrolytes contain a type of chemical bond which differs from the ionic bond.

In the single covalent bond, two atoms share one pair of electrons. By sharing, the bonded atoms both gain a full outer shell of electrons

Two atoms of chlorine combine to form a molecule, Cl_2. Both chlorine atoms have the electron configuration Cl(2.8.7). G N Lewis suggested that the bond involves each of the two chlorine atoms sharing one of its outermost electrons – **valence electrons** as they are termed – with the other chlorine atom. The two atoms have to approach sufficiently closely for their atomic orbitals to overlap. The shared pair of electrons is called a **covalent bond**. They occupy the same orbital with opposing spins [§ 2.5]. The Cl_2 molecule can be represented as shown in Figure 4.4A.

Figure 4.4A
Ways of representing the chlorine molecule

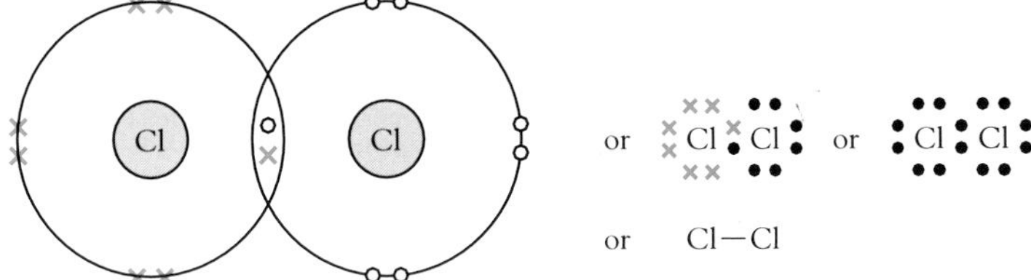

There are not really two types of electron, but it makes it easier to count the electrons if those from one atom are represented as crosses and those from the other as dots. By sharing a pair of electrons, each of the chlorine atoms has obtained eight electrons in its outer shell: it has 'completed its octet'. Electrons are shared when half-filled atomic orbitals of adjacent atoms overlap in space.

The H_2 molecule can be shown as in Figure 4.4B. Each hydrogen atom shares its electron with another hydrogen atom to gain a full outer s shell of 2 electrons.

Figure 4.4B
Ways of representing the hydrogen molecule

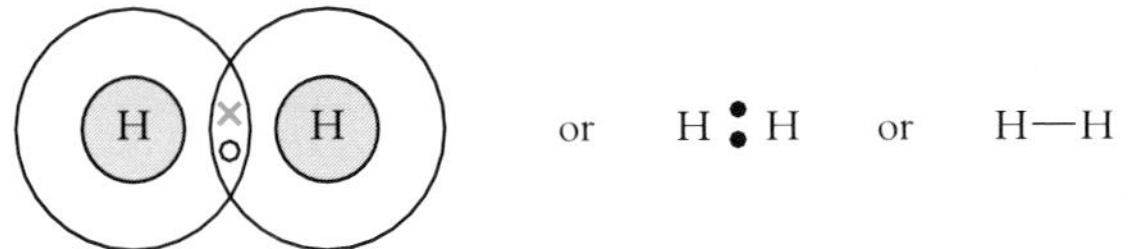

The HCl molecule can be shown as in Figure 4.4C.

Figure 4.4C
Ways of representing the hydrogen chloride molecule

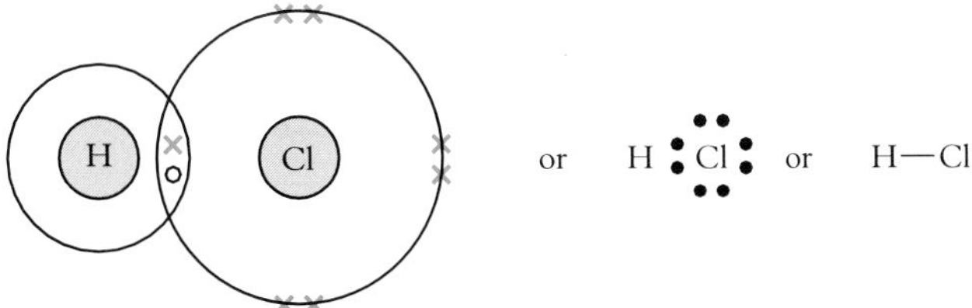

Each hydrogen atom has a full s shell with two electrons and each chlorine atom has a full octet.

Covalent bonding is important in carbon compounds. The carbon atom, with four valence electrons, can attain a full octet by sharing one electron with each of four hydrogen atoms. The bonding in methane, CH_4, can be shown by a 'dot-and-cross' diagram [see Figure 4.4D]. The hydrogen electrons are shown as dots and the carbon electrons as crosses. Carbon has completed its octet, and hydrogen has attained the noble gas configuration of helium by completing its 1s shell.

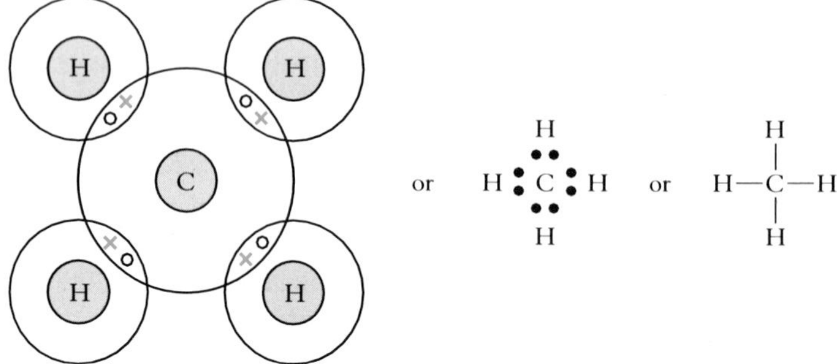

Figure 4.4D
Ways of representing the bonding in methane

In the ethene molecule [Figure 4.4E] each carbon atom shares one pair of electrons with a hydrogen atom and two pairs of electrons with the other carbon atom. When atoms share two pairs of electrons there is a double bond between them. Each carbon atom forms single bonds to two hydrogen atoms and a double bond to the other carbon atom.

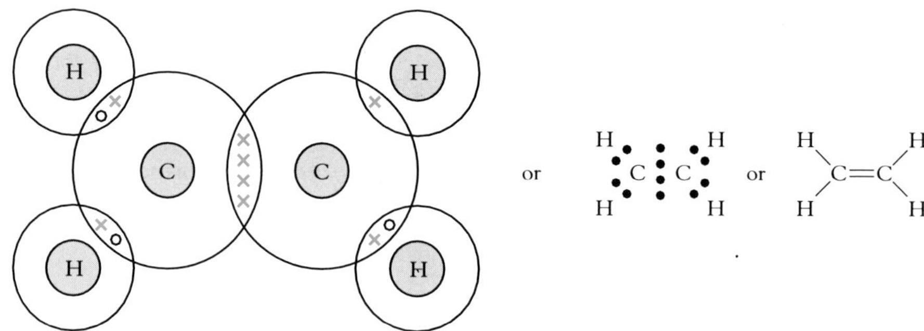

Figure 4.4E
Ways of representing the ethene molecule

A double bond is formed when two atoms share two pairs of electrons

In carbon dioxide, the carbon atom shares two electrons with each of two oxygen atoms, in order to give all three atoms a full octet of valence electrons. [See Figure 4.4F.]

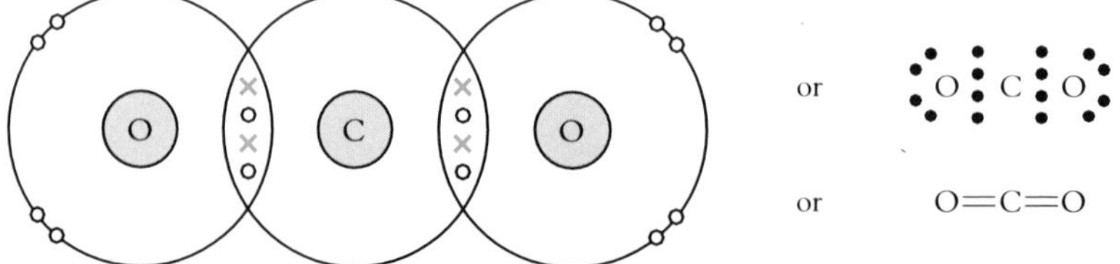

Figure 4.4F
Ways of representing the carbon dioxide molecule

As each shared pair of electrons is a covalent bond, the two pairs of shared electrons between carbon and oxygen constitute a double bond. The pairs of electrons on the oxygen atoms which are not shared are described as **'lone pairs' of electrons**.

A triple bond is formed when two atoms share three pairs of electrons

In a molecule of nitrogen, each nitrogen atom, with five electrons, needs to share three of its electrons with the other atom of nitrogen in order to complete its octet. The bonding can be written as shown in Figure 4.4G.

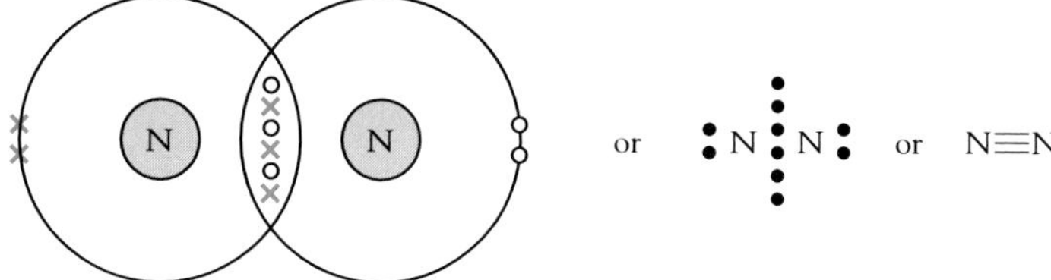

Figure 4.4G
Ways of representing the N_2 molecule

The valence bond method considers the overlapping of atomic orbitals in isolation from the rest of the molecule

This method of describing molecules is called the **valence bond method**. It considers the atoms in a molecule in isolation from the rest of the molecule, except that one or more electrons in the outer shell of one atom are accommodated in the outer shell of another atom by an overlapping of atomic orbitals.

A different approach is the **molecular orbital treatment** [§ 4.6.1].

CHECKPOINT 4.4: BONDING IN COVALENT COMPOUNDS

1. The electron configurations of hydrogen and fluorine are H(1) and F(2.7). Show by means of a sketch (like those in Figures 4.4A to 4.4G) what happens when H and F combine to form a covalent bond in HF.
2. Sketch the configuration of electrons in a molecule of the covalent compound CH_3Cl. (Electron configurations are C(2.4), H(1) and Cl(2.8.7).)
3. Sketch the configuration of electrons in a molecule of the covalent compound, NH_3. (N has the electron configuration (2.5).)
4. By means of a 'dot-and-cross' diagram, such those in Figures 4.4A to 4.4G, sketch the arrangement of electrons in the molecule O_2.
5. Sketch the arrangement of electrons in a molecule of ethene, $H_2C{=}CH_2$.

4.4.1 FORMULAE OF COVALENT COMPOUNDS

The formula of a covalent compound depends on the number of pairs of electrons shared by the bonded atoms

The formulae of covalent compounds are decided by the number of pairs of electrons shared between atoms, as shown in the 'dot-and-cross' diagrams in Figures 4.4D to 4.4G.

4.4.2 STRUCTURES OF COVALENT SUBSTANCES

Individual molecules

Many covalent substances are composed of individual molecules

Covalent substances can be solids, liquids or gases. Gases consist of individual molecules. The molecules move independently, with negligible forces of attraction between them. In liquids there are significant forces of attraction between the molecules. However, these are weaker forces than those in solids, and allow molecules to change their relative positions easily, so the substance can flow.

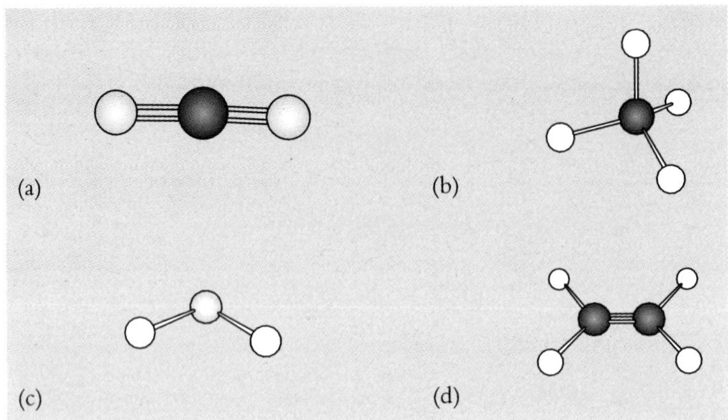

Figure 4.4H
Models of molecules of (a) CO_2, (b) CH_4, (c) H_2O, (d) $H_2C{=}CH_2$

4.4.3 MOLECULAR STRUCTURES

In some covalent elements and compounds molecular structures form as weak forces act between molecules

In some covalent substances, the attractive forces between molecules are strong enough to make the substances solids. For example, iodine is a shiny black crystalline solid. Forces of attraction operate between the molecules of iodine to hold them in a three-dimensional structure [see Figure 4.4I]. There are two kinds of chemical bonds in it: the **intramolecular** bonds (inside the molecules), which are strong covalent I—I bonds, and the **intermolecular forces** (between individual molecules), which are weak forces of attraction between I_2 molecules. The origin of these weak forces of attraction is discussed in § 4.6.2. A structure of this kind, which consists of individual molecules bonded together in a regular arrangement by weak intermolecular bonds, is called a **molecular structure**. Other examples are described in § 6.5.

In a molecular structure, the intermolecular bonds are easily broken to allow molecules to move independently, that is, to enter the liquid phase. Such covalent substances have melting temperatures lower than those of ionic substances. Many are liquid or gaseous at room temperature. The boiling temperatures of many covalent substances are low for the same reason.

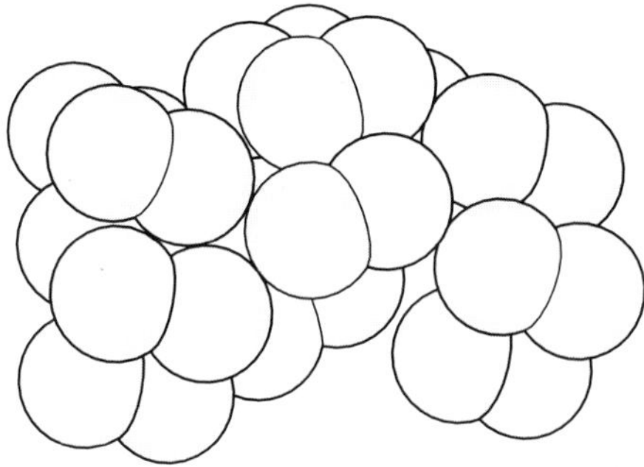

Figure 4.4I
The structure of an iodine crystal

Molecules can associate through the operation of van der Waals forces and other forces

Polymers have molecules which consist of a large number of repeating units. Poly(ethene) has the formula $(CH_2)_n$ where $n = 2000–4000$. The molecules are long chains with van der Waals forces of attraction [see § 4.8.2] between them which are strong enough to hold the molecules together but weak enough to make the material plastic. In polyamides, e.g. nylon, there are hydrogen bonds between molecules which make the material very strong [see §§ 4.8.3, 34.3]. In some polymers, e.g. epoxy resins, there are cross-links between chains, resulting in an inflexible material [see § 34.5.2]. [See Figure 4.4J.]

Figure 4.4J
Polymer molecules

(a) Disorganised strands of the polymer low-density poly(ethene)

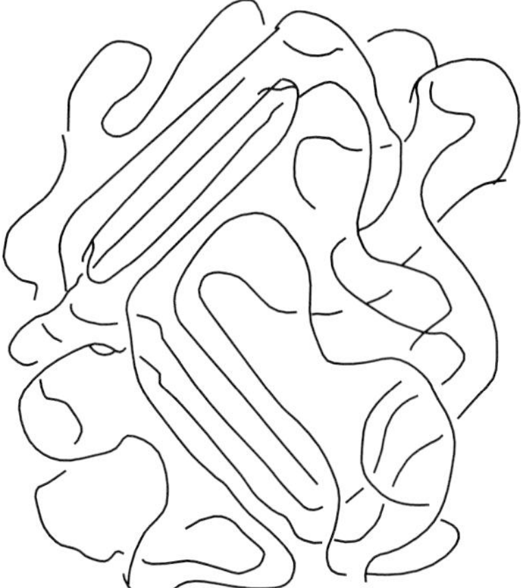

(b) Strands of high-density poly(ethene) showing regions of crystallinity

Macromolecular structures

Other covalent substances are composed of macromolecules ...

... e.g. diamond and graphite

Some covalent substances do not consist of individual molecules. They are **macromolecular structures**, or **giant molecular structures**. One example is diamond. Diamond is the hardest naturally occurring substance. The extraordinary properties of diamond arise from its structure. A crystal of diamond contains millions of atoms. Figure 4.4K shows how every carbon atom is joined by covalent bonds to four other carbon atoms. It is very difficult to break this macromolecular structure, and this is why diamond is so hard.

Other macromolecular structures, e.g. graphite [Figure 6.7A], boron nitride (BN), silicon(IV) oxide [Figure 6.6D] and silicon carbide (SiC) are described in § 6.6. The strong covalent bonds holding these giant molecules together give the substances high melting temperatures (higher than most ionic solids) and make them **involatile** (i.e. they have high boiling temperatures).

Figure 4.4K
The bonding in diamond

Macromolecules involve strong covalent bonds

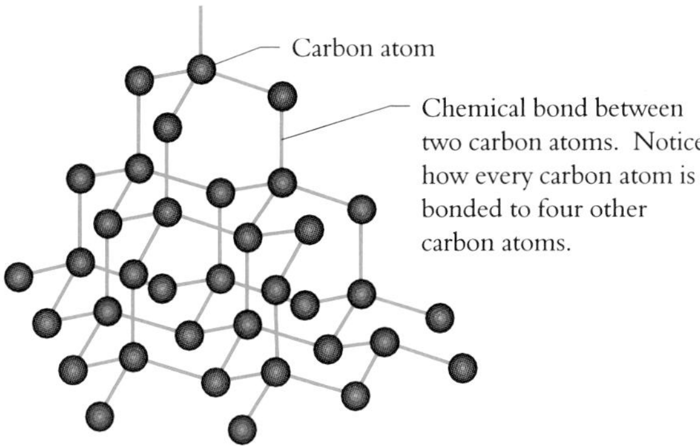
Carbon atom
Chemical bond between two carbon atoms. Notice how every carbon atom is bonded to four other carbon atoms.

4.5 PROPERTIES OF IONIC AND COVALENT SUBSTANCES

The physical characteristics of substances, for example, whether they are solids, liquids or gases, depend on the types of chemical bond in the substances. Chemical behaviour also depends on the type of bond present.

Table 4.5A Characteristics of ionic and covalent substances

The properties of an element or a compound depend on the type of chemical bond present

The physical state (s, l, g), the melting and boiling temperatures, the solubility and the electrolytic conductivity depend on whether ionic bonds or covalent bonds or weak intermolecular bonds are present

Ionic bonding	Covalent bonding
Ionic compounds are formed when metallic elements combine with non-metallic elements	Atoms of non-metallic elements combine by forming covalent bonds.
(a) Atoms of metallic elements form positive ions. Elements in Groups 1, 2 and 3 of the Periodic Table form ions with charges +1, +2 and +3, e.g. Na^+, Ca^{2+}, Al^{3+}. (b) Elements of non-metallic elements form negative ions. Elements of Groups 6 and 7 of the Periodic Table form negative ions with charges −2 and −1, e.g. O^{2-} and Br^-.	The maximum number of covalent bonds that an atom can form is equal to the number of electrons in the outer shell. Often, an atom does not use all its electrons in covalent bond formation.
The ionic bond is a strong electrostatic attraction between ions of opposite charge. An ionic compound is composed of a giant regular structure consisting of millions of ions. Ionic compounds are crystalline because of this structure. The strong attraction between ions of opposite charge and the giant structure make it difficult to separate the ions and break the crystal structures. This is why ionic compounds have high melting temperatures and boiling temperatures.	There are three types of covalent substances: (a) Individual molecules. The bonds which hold the atoms together are strong, but there are negligible forces of attraction between molecules. Such covalent substances are gases, e.g. HCl, SO_2, CO_2, CH_4. (b) Molecular structures. Between individual molecules there are weak forces of attraction. Such covalent substances are low boiling temperature liquids, e.g. ethanol, and low melting temperature solids, e.g. iodine [Figure 4.4I] and solid carbon dioxide, which are molecular structures. (c) Macromolecular structures or giant molecular structures. Large numbers of atoms link in chains or sheets, e.g. graphite [Figure 6.7A] or in 3-dimensional structures, e.g. diamond [Figures 4.4K and 6.6A] and quartz [Figure 6.6D]. Substances with giant molecular structures have high melting and boiling temperatures.
Ionic compounds are electrolytes.	Covalent compounds are non-electrolytes.
Many ionic compounds dissolve in water. They are insoluble in organic solvents.	Covalent compounds are often insoluble in water. Many dissolve in organic solvents such as ethanol or propanone.

CHECKPOINT 4.5: CHEMICAL BONDS

1. Say what kind of bonding you would expect between the following pairs of elements:

   (a) Li and Br, *(b)* Sr and Cl, *(c)* Ba and S, *(d)* Ca and O, *(e)* S and O, *(f)* F and Cl, *(g)* Cl and O.
   [See the Periodic Table on p. 732 for help.]

2. From the information in Table 4.5B, say what you can about the chemical bonds in **A**, **B**, **C**, **D** and **E**.

Substance	State	Melting temperature (°C)	Does it conduct electricity?
A	Liquid	−60	Does not conduct electricity.
B	Solid	890	Conducts electricity when molten.
C	Solid	720	Does not conduct electricity when molten.
D	Solid	85	Does not conduct electricity when molten.
E	Gas	−100	Does not conduct electricity

Table 4.5B

4.6 COVALENT COMPOUNDS

4.6.1 THE MOLECULAR ORBITAL TREATMENT

The molecular orbital method considers the molecule as a whole ...

... and calculates the electron density over the whole molecule

We studied the valence bond method of describing molecules in § 4.4. Another approach is to consider the *entire* molecule as a unit. Each electron is under the influence of all the nuclei and the electrons in the molecule. The atomic orbitals are replaced by molecular orbitals. The molecular orbital method of viewing the covalent bond uses **quantum mechanics** to calculate the distribution of electron density over the molecule. Figure 4.6A shows the results of the calculation for the hydrogen molecule. The contour lines join regions of the same electron density.

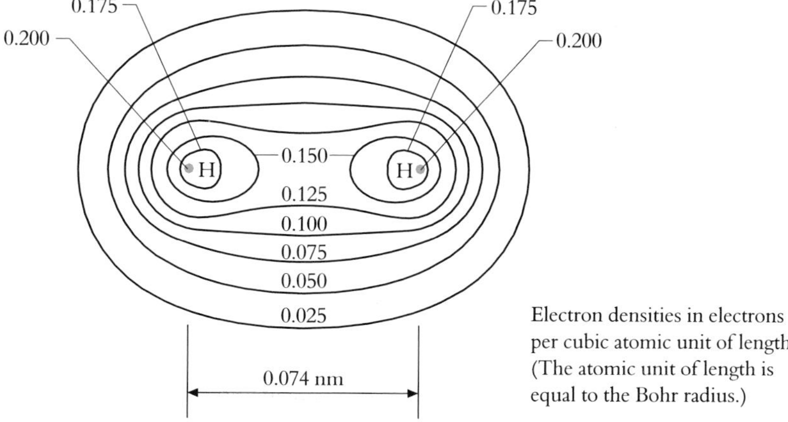

Figure 4.6A Electron density map for the hydrogen molecule

In the molecular orbital the 2H in H_2 are held together by the attraction of the nuclei for the electron cloud between the two nuclei

The highest electron density is near each nucleus. There is also a region of high electron density between the two nuclei called the **electron cloud**. The electron density between the nuclei screens the nuclei from one another and prevents repulsion between the two positive nuclei from driving the atoms apart. Although there is a force of repulsion between the positively charged nuclei, the force of attraction between each nucleus and the electron cloud between the two nuclei is greater. It is this attraction that holds the atoms together in the molecule. [See Figure 4.6B].

Figure 4.6B
Attraction and repulsion in the H_2 molecule

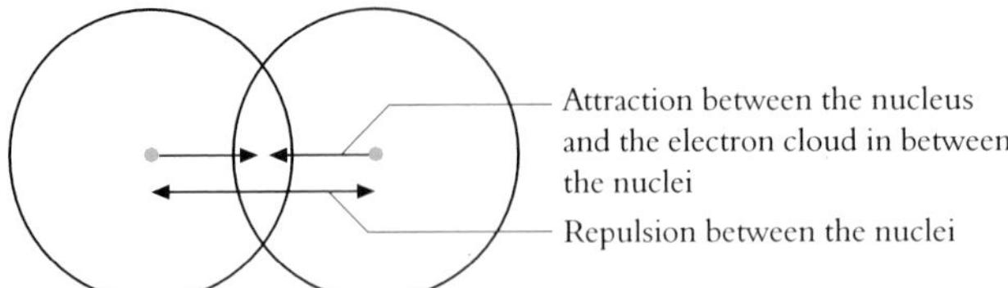

4.6.2 COVALENT RADII

Covalent radii (atomic radii)

The distance between the nuclei of covalently bonded atoms is the sum of their **covalent radii**. Covalent radii, which are also called **atomic radii**, are *additive*. The sum of the covalent radii of chlorine and hydrogen gives the length of the covalent bond in hydrogen chloride [see Figure 4.6C]. This figure also shows the **van der Waals radii**, which will be covered in § 4.8.2.

Figure 4.6C
Covalent radii and van der Waals radii for hydrogen and chlorine

In a covalent bond, distance between nuclei = sum of the covalent radii of the bonded atoms

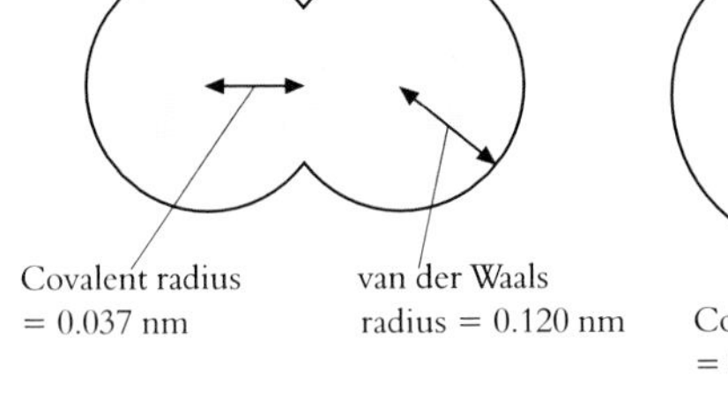

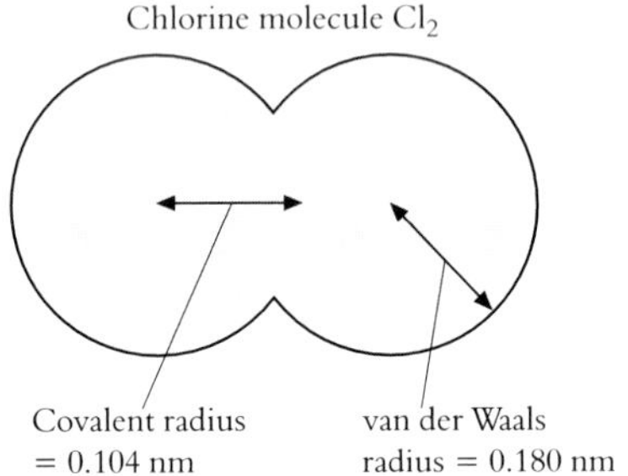

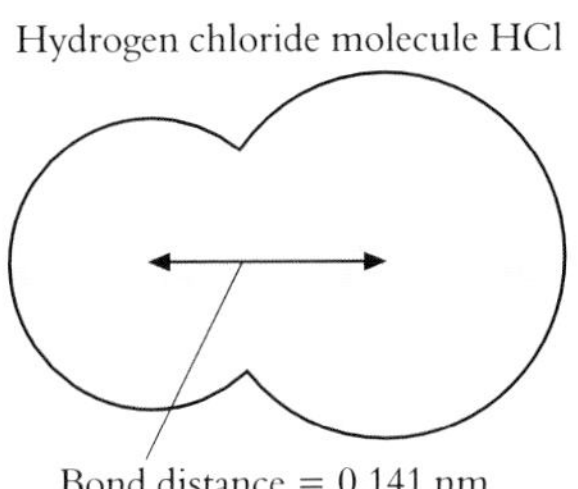

4.6.3 ELECTRONEGATIVITY AND THE COVALENT BOND

Electronegativity is the ability of an atom in a covalent bond to attract electrons

Fluorine is the most electronegative of elements

In the bond between identical atoms, e.g. H—H, the electron density of the bonding orbital is distributed symmetrically between the bonded atoms. In a bond between different atoms, the bonding electrons may be more attracted to one of the bonded atoms than to the other. In the molecule HF, for example, the electron density of the bonding electrons lies more towards the fluorine atom than towards the hydrogen atom. The ability of an atom in a covalent bond to attract the bonding electrons is called **electronegativity**. Thus, fluorine is more **electronegative** than hydrogen. Electronegativity is not a quantity which can be measured or to which a unit can be assigned. Pauling derived a scale of relative electronegativity values. He assigned a value of 4.00 to fluorine, the most electronegative of elements. Some of his values are:

Various electronegativity values

H	Li	Be	B	C	N	O	F
2.1	1.0	1.5	2.0	2.5	3.0	3.5	4.0

If the bonded atoms differ in electronegativity a covalent bond is polar ...

... it has a dipole

The HF molecule is described as **polar**: the centre of the negative charge (due to the electrons) does not coincide with the centre of positive charge (due to the nuclei). The molecule has a **dipole**: electric charges of equal magnitude and opposite sign separated by a small distance. There are many other covalent molecules which, like HF, are polar. There is no sharp distinction between an ionic bond and a covalent bond [see Figure 4.6D].

Figure 4.6D
Bond types

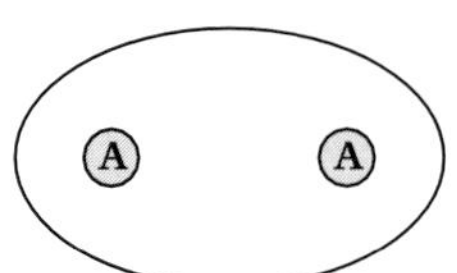

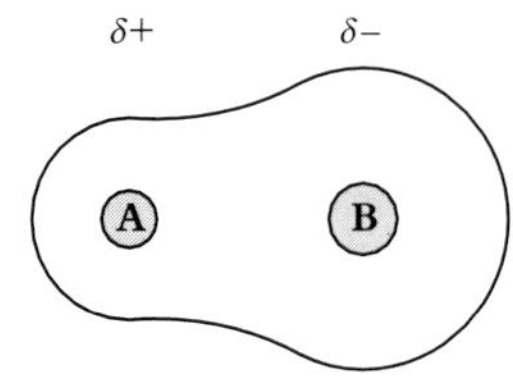

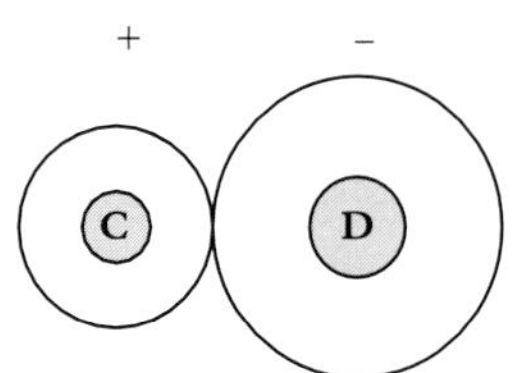

There is no sharp distinction between an ionic bond and a covalent bond

(a) Covalent bond. The electron density is symmetrically distributed.

(b) Polar covalent bond. The bonding electron density is greater in the region of **B**. **A** has a small positive (δ^+) charge, while **B** is slightly negative (δ^-).

(c) Ionic bond. The electron cloud of $\mathbf{C}^+$ does not come under the influence of $\mathbf{D}^-$. The electron cloud of $\mathbf{D}^-$ is not distorted by $\mathbf{C}^+$.

Some covalent bonds have a degree of ionic character

The term **ionic bond** is used for bonds which are predominantly ionic. The term **covalent bond** is used for non-polar bonds, such as C—I, and also bonds in which there is a considerable degree of polarity. Bonds such as $C^{\delta+}—Cl^{\delta-}$ and $C^{\delta+}—O^{\delta-}$, are termed **polar covalent bonds**. The curve which Pauling drew to relate the percentage of ionic character in a bond to the difference in electronegativity between the bonded atoms is shown in Figure 4.6E. Approximate values for some covalent bonds are given:

Bond	C—I	C—H	C—Cl	C—F
Ionic character/%	0	4	6	40

Figure 4.6E
Curve relating percentage of ionic character in a bond to the difference in electronegativity between the bonded atoms

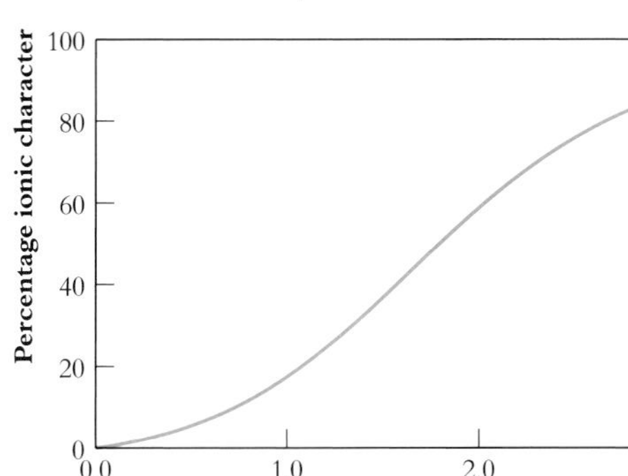

In some ionic compounds there is a degree of sharing of electron clouds between cations and anions

Pauling's electronegativity values can be applied to a bond between two atoms, either in a diatomic molecule or between two atoms considered in isolation from other atoms and bonds. They should *not* be applied to a formula unit in a crystal structure as they will give a wrong impression. If applied to $CaCl_2$, NaCl or LiF, electronegativity values will predict the nature of the bonding in an ion pair considered in isolation, e.g. $Na^+(g)\ Cl^-(g)$. In reality, we are interested in crystalline sodium chloride, in which each Cl^- ion is part of a crystal structure and its electron cloud is influenced symmetrically by six Na^+ ions around it. The net result is that the electron cloud of the Na^+ ion becomes rather cubical in shape, being drawn into the spaces between the ions in the structure, but sodium chloride is still overwhelmingly ionic in nature.

The deformation of the spherical electron clouds is more pronounced in lithium fluoride. In calcium chloride it does not occur. The reasons for these differences were explained by K Fajans.

4.6.4 FAJANS' RULES OF BOND TYPE

The distortion of an electron cloud by a neighbouring charged ion or **dipole** is called **polarisation**. Comparing ions with similar nuclear charges, large ions are more easily polarised than small ions. The electrons in small ions are more closely controlled by the positively charged nucleus.

K Fajans formulated two rules to predict the proportions of ionic and covalent character in the bond formed between two atoms.

Fajans' rules predict the ionic and covalent character of bonds

1. Bonds will tend to be ionic if the ions formed are small in charge. For example, sodium chloride is likely to be ionic because Na^+ and Cl^- bear unit charges, whereas aluminium chloride is likely to be covalently bonded because Al^{3+} ions are highly charged.
2. Bonds will tend to be ionic if the radius of the possible cation is large (e.g. the alkali metals) and the radius of the possible anion is small (e.g. the smaller halogens).

A cation with a small radius and a high charge will polarise anions ...

... with an increase in the covalent character of the bonds

The sizes of some ions are shown in Figure 15.2D, § 15.2. Compare the radii of the ions Na^+, Mg^{2+} and Al^{3+}. The high charge on the Al^{3+} ion results in a small radius because the remaining electrons are drawn in close to the nucleus. The combination of a high charge and a small radius gives Al^{3+} a high **charge density** (i.e. charge/volume ratio), and this enables the cation to attract the electron cloud of a neighbouring anion (or of a molecule). The electron cloud of the anion will be distorted in such a way as to increase the electron density near the cation: the anion has become **polarised**.

A large anion is easily polarised ...

... and its bonds will have some covalent character

An anion is larger than the atom from which it was formed, and has one or two more electrons. The nucleus is less able to attract the electrons closely than it is in the parent atom. In a large anion, the electrons are further from the nucleus and less under its control than in a small anion, making the larger anion easier to polarise. If the cation is small and the anion is large, the cation will be able to polarise the anion, and there will be some sharing of the electron cloud of the anion: i.e. the bond will have some covalent character.

CHECKPOINT 4.6: THE COVALENT BOND

1. Draw dot-and-cross diagrams to show the bonding in
   (a) H—C≡N
   (b) CCl_4
   (c) O=S(Cl)Cl
   (d) SO_2Cl_2 (O=S(=O)(Cl)Cl)
2. Distinguish between *(a)* the intramolecular bonds, and *(b)* the intermolecular forces in chloroform, $CHCl_3$. What makes you think that the bonds in *(a)* are strong and the forces in *(b)* are weak?
3. Sodium chloride melts at 800 °C. Tetrachloromethane, CCl_4, is a liquid at room temperature. Explain how this difference arises.

4.7 THE COORDINATE BOND

A coordinate bond is a covalent bond ...

... in which both electrons in the bond come from one of the bonded atoms

A donor atom shares a lone pair of electrons with an acceptor atom

Once formed, a coordinate bond is like a covalent bond

A **coordinate bond** is a covalent bond in which the shared pair of electrons is provided by only *one* of the bonded atoms. One atom is the **donor**, the other is the **acceptor**, and the bond is sometimes called the **dative covalent bond**. Once formed, a coordinate bond has the same characteristics as a covalent bond. For an atom to act as a donor, it must have at least one pair of unshared electrons (a lone pair) in its outermost shell (the valence shell). The acceptor has at least one vacant orbital in its outer shell. It may be a metal cation or a transition metal atom or an atom in a molecule.

In water, H—$\overline{\text{O}}$—H, the oxygen atom has two lone pairs of electrons. They can be shared with an atom which needs them to complete it valence shell. A proton H^+ has an empty 1s orbital. By accepting a pair of electrons from oxygen, it

The oxonium ion, H_3O^+, is formed by coordination

achieves a full s shell:

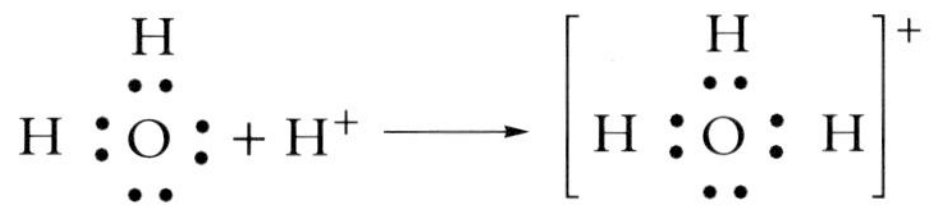

$$H:\ddot{\underset{\cdot\cdot}{O}}: + H^+ \longrightarrow \left[H:\ddot{\underset{\cdot\cdot}{O}}:H\right]^+$$

(with H above O in each structure)

Water coordinates to metal ions ...

... an example is $[Ca(H_2O)_6]^{2+}$

The species formed is an **oxonium** ion, H_3O^+. The positive charge contributed by the proton is spread over the whole ion. The proton has one unit of positive charge spread over a surface area which is minute compared with other ions. The high charge density makes it extremely reactive: the proton cannot exist by itself. It is stabilised by the coordination of water molecules.

Water also coordinates to metal ions. The fact that bonds are formed between metal ions and water is responsible for the solubility of many salts. Energy is required to break the bond holding ions together in a crystal structure. If energy is given out when coordinate bonds form between metal ions and water, this may swing the balance in favour of solution. Metal ions are hydrated, e.g. $[Ca(H_2O)_6]^{2+}$:

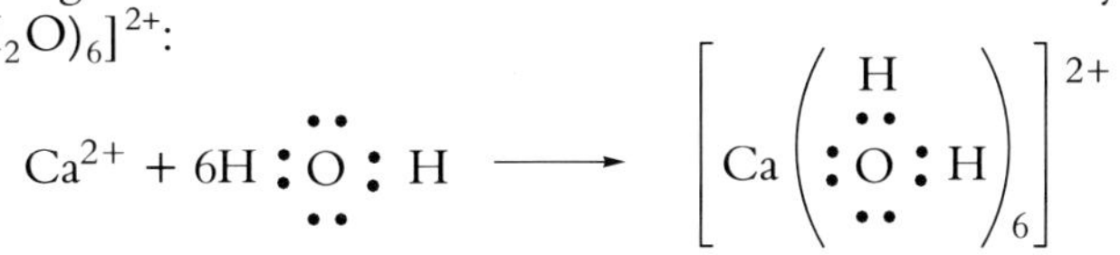

$$Ca^{2+} + 6H:\ddot{\underset{\cdot\cdot}{O}}:H \longrightarrow \left[Ca\left(:\ddot{\underset{\cdot\cdot}{O}}:H\right)_6\right]^{2+}$$

(with H above O in the product)

The lone pair on N in NH_3 is responsible for the formation of coordination compounds by NH_3

The nitrogen atom in ammonia, H—N: (with two further H atoms on N), has an unshared pair of electrons which it is able to share with an atom that needs two electrons to complete its octet. A molecule of ammonia forms a coordinate bond to a hydrogen ion in the formation of an ammonium ion. The lone pair of the nitrogen atom coordinates into the valence shell of the hydrogen ion:

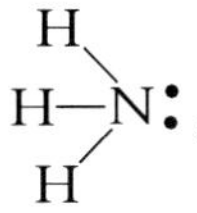

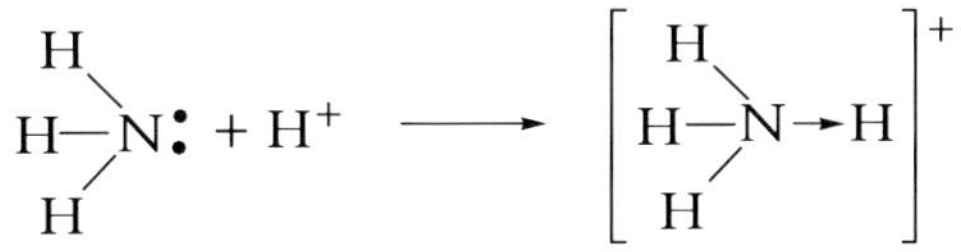

$$H_3N: + H^+ \longrightarrow [H_3N \rightarrow H]^+$$

The symbol → is used for a coordinate bond, an arrow pointing from the donor towards the acceptor.

Ions such as $CuCl_4^{2-}$ are formed by coordination ...

... they are called complex ions ...

... formed by combination of an ion with an oppositely charged ion or with a molecule

Copper(II) chloride is blue in solution. In the presence of a high concentration of chloride ions, the solution turns a very deep green. The colour is due to the formation of $CuCl_4^{2-}$ ions. Coordinate bonds form between Cl^- ions and Cu^{2+} ions:

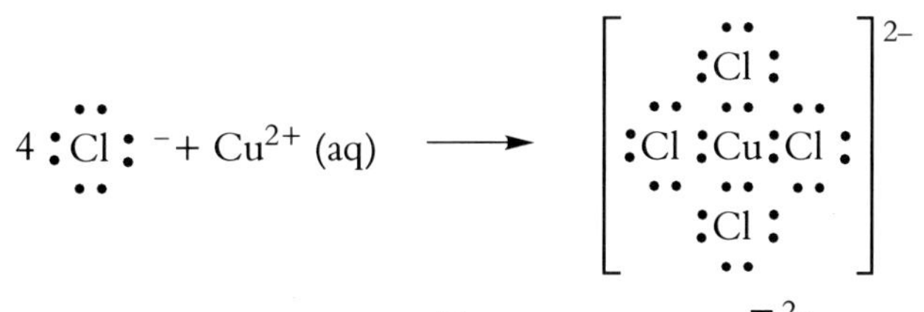

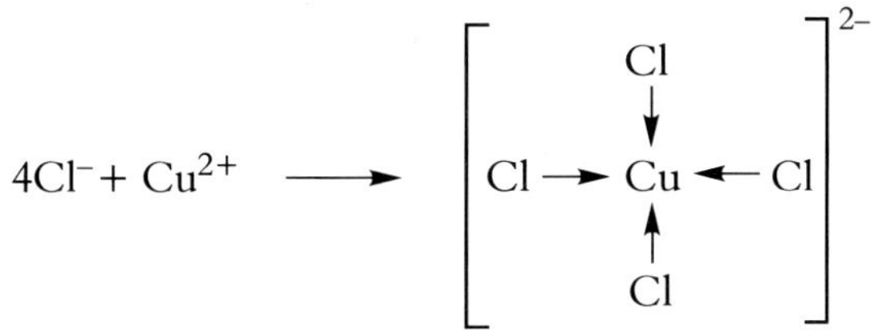

$$4\,:\ddot{\underset{\cdot\cdot}{Cl}}:^- + Cu^{2+}(aq) \longrightarrow [CuCl_4]^{2-}$$

(product drawn as Cu surrounded by four :Cl: groups, each sharing an electron pair with Cu)

$$4Cl^- + Cu^{2+} \longrightarrow \left[\begin{array}{ccc} & Cl & \\ & \downarrow & \\ Cl \rightarrow & Cu & \leftarrow Cl \\ & \uparrow & \\ & Cl & \end{array}\right]^{2-}$$

The copper atom now has eight electrons in its valence shell. Ions such as $CuCl_4^{2-}$ and $Cu(NH_3)_4^{2+}$, which are formed by the combination of an ion with an oppositely charged ion or a molecule are called **complex ions**.

CHECKPOINT 4.7: THE COORDINATE BOND

1. Draw a dot-and-cross diagram to show the arrangement of valence electrons in the complex ion $Cu(H_2O)_4^{2+}$. It is formed by coordination of H_2O molecules on to a Cu^{2+} ion.

2. Draw a dot-and-cross diagram to show the bonding in the complex ion $Fe(H_2O)_6^{3+}$. How many electrons has the iron atom in its valence shell? How can it accommodate this number?

3. Explain the bonding in the compound NH_3BF_3.

4. The formula for carbon monoxide can be written $|C \equiv O|$. Draw a dot-and-cross diagram for this molecule. The lone pair on the oxygen atom can be used to form a coordinate bond to a nickel atom. By means of a diagram, show the bonding in nickel carbonyl, $Ni(CO)_4$.

5. Explain the bonding in the tetraammine copper ion, $Cu(NH_3)_4^{2+}$.

6. The slightly soluble compound lead(II) chloride dissolves in concentrated hydrochloric acid to form a soluble complex ion. What do you think the formula of this ion might be? Explain the bonding in the ion.

4.8 INTERMOLECULAR FORCES

Dipole–dipole interactions, van der Waals forces and hydrogen bonds are all intermolecular forces

Intermolecular forces and bonds are of a number of types: dipole–dipole interactions, van der Waals forces and the hydrogen bond. Polar molecules (§ 4.6.3) have a dipole. A dipole consists of two electric charges of equal magnitude and opposite signs separated by a small distance.

4.8.1 DIPOLE–DIPOLE INTERACTIONS

Attractions between dipoles result in an ordered arrangement of molecules

In the solid state, polar molecules interact to form an ordered arrangement. Dipole–dipole interactions between the molecules lead the molecules to pack in such a way that partial positive charges will be adjacent to partial negative charges [see Figure 4.8A].

Solvation helps solids to dissolve

Polar solids dissolve in polar solvents. The energy required to break up the crystal is recouped by the energy released when polar solute molecules interact with polar solvent molecules. [See Figure 4.8A.] This interaction is called **solvation**; if the solvent is water, it is called **hydration** [§ 17.4].

Figure 4.8A
The process of dissolution

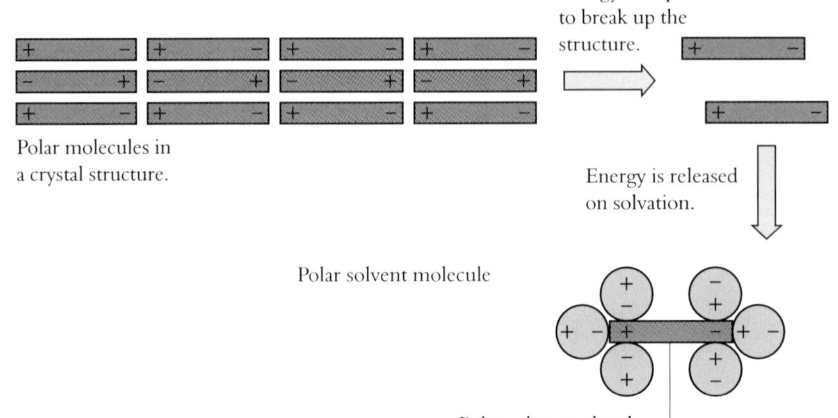

Energy is required to break up the structure.
Polar molecules in a crystal structure.
Energy is released on solvation.
Polar solvent molecule
Polar solute molecule

4.8.2 VAN DER WAALS FORCES

There are different types of van der Waals forces ...

When molecules pack together in the liquid or solid state, there must be forces of attraction between them. J D van der Waals postulated the existence of forces of attraction and repulsion that are neither ionic nor covalent. Such forces arise in a number of ways and are collectively called **van der Waals forces**. Dipole–dipole interactions [§ 4.8.1] between polar molecules are one type of van der Waals force.

... Attractive forces exist between non-polar molecules ...

... These forces are thought to be due to the momentary polarisation of molecules ...

Attractive forces exist also between non-polar molecules. Even atoms of the noble gases are attracted to one another to a slight degree; this is why the noble gases can be liquefied. Consider two non-polar molecules which are very close together. Since they are non-polar, the arrangement of electrons is *on the average* symmetrical. Yet, *at any given instant*, the electron distribution in one molecule may be unsymmetrical. There may be a dipole in the molecule *for an instant*. Figure 4.8B shows how a temporary dipole in one molecule **A** can attract the electron cloud of a neighbouring molecule **B**. This means that both molecules will have dipoles, and the direction of the dipoles will be such that they attract one another. Since the electrons are moving about at high speed, the attraction has a fleeting existence. In the next instant, the dipole in **A** may be in the opposite direction.

Figure 4.8B
Attraction between momentary dipoles

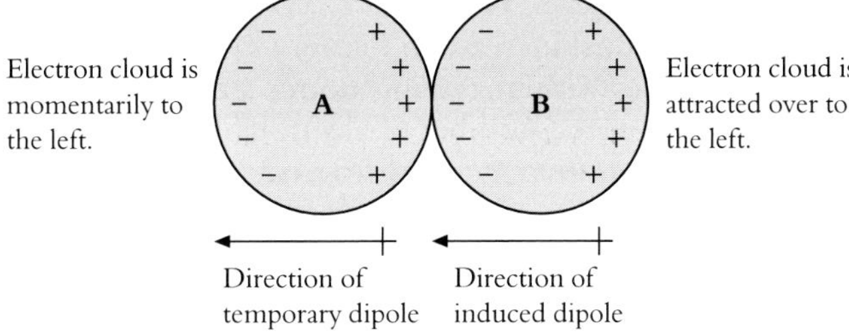
Electron cloud is momentarily to the left.
A
B
Electron cloud is attracted over to the left.
Direction of temporary dipole
Direction of induced dipole

... leading to a permanent attraction

Again, the dipole which it **induces** in **B** will result in an attraction. The dipoles are temporary, but the net attraction which they produce is permanent.

The ease with which an electron cloud is distorted is called polarisability

The ease with which an electron cloud is distorted, and therefore the ease with which a dipole is induced, is called **polarisability**. Polarisability increases with the number of electrons in a molecule, and the strength of the forces of attraction therefore increases with molar mass. The shape of molecules is another factor: elongated molecules are more easily polarised than compact, symmetrical molecules.

Figure 4.8C
Van der Waals forces determine the distance between atoms in the liquid state

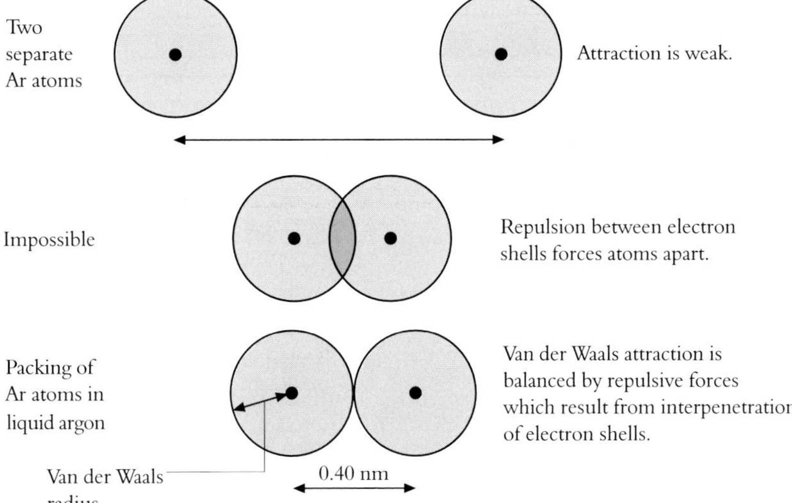
Two separate Ar atoms
Attraction is weak.
Impossible
Repulsion between electron shells forces atoms apart.
Packing of Ar atoms in liquid argon
Van der Waals attraction is balanced by repulsive forces which result from interpenetration of electron shells.
Van der Waals radius
0.40 nm

Van der Waals forces determine distances between atoms in liquids and solids

If two molecules are far apart, there will be no induction of dipoles and no attractive forces between them. Should the molecules move too close together, repulsion between electron shells will predominate over the induction effect and drive the molecules apart. Figure 4.8C shows how closely argon atoms can approach in the liquid state. Half the distance between argon atoms at their closest distance of approach is the **van der Waals radius** of the atom.

Van der Waals forces are strong between large linear molecules ...

... e.g. polymer molecules ...

Van der Waals forces between small molecules are weak. Between molecules with long chains of atoms, giving many points of contact, multiple van der Waals forces operate. This is why in the series of alkanes [§ 26.3] ethane, C_2H_6, is a gas at stp, hexane, C_6H_{14}, is a liquid, and octadecane, $C_{18}H_{38}$, is a solid. Branched-chain hydrocarbons are more **volatile** (have lower boiling temperatures) than unbranched-chain hydrocarbons. Hydrocarbon polymers have molecules which are continuous chains, containing thousands of repeating units. The long, strand-like molecules of poly(ethene), $\text{+}CH_2{-}CH_2\text{+}_n$ [§ 27.8.10] can align themselves to give thousands of contacts between atoms and set up very strong attractions through the operation of multiple van der Waals forces. Poly(ethene) is an extremely tough material, which is used for the manufacture of laboratory and kitchen ware.

Van der Waals forces in the noble gases

... and also exist between atoms of the noble gases

There is a gradual increase in the very low melting and boiling temperatures of the noble gases with increasing atomic size (see Table 4.8). As the size of the atoms increases the number of electrons increases and the magnitude of the van der Waals forces increases. Hydrogen and oxygen have diatomic molecules between which the only intermolecular forces are van der Waals forces. Their boiling and melting temperatures are similar to those of noble gases of comparable molar mass.

Table 4.8
The noble gases and other gases

Element		Boiling temperature/°C	Melting temperature/°C	Atomic number	M_r
Helium	He	−269	−272 at 26 atm	2	4
Neon	Ne	−246	−249	10	20
Argon	Ar	−186	−189	18	40
Krypton	Kr	−152	−157	36	84
Xenon	Xe	−107	−112	54	131
Radon	Rn	− 62	− 71	86	222
Hydrogen	H_2	−253	−259	1	2
Oxygen	O_2	−183	−218	8	32
Nitrogen	N_2	−196	−210	7	28

You will notice that the gases have a very narrow band of temperature in which the liquid state exists. The reason is that the intermolecular forces in the liquid are very similar to those in the solid. This is a feature of van der Waals forces.

CHECKPOINT 4.8A: INTERMOLECULAR FORCES I

1. What is the origin of van der Waals forces?

2. Why are van der Waals forces stronger in xenon than in neon?

3. *(a)* State the number of electrons in the molecule of (i) helium (ii) hydrogen and the boiling temperature of each gas.
   (b) State the number of electrons in the molecular of (i) argon (ii) oxygen and the boiling temperature of each gas.
   (c) Describe the relationship between the number of electrons and the boiling temperature. What can you infer from the relationship?

4.8.3 THE HYDROGEN BOND

The nature of the bond

When hydrogen is combined with an electronegative atom intermolecular hydrogen bonds can be formed

The bond in hydrogen fluoride is a polar covalent bond [§ 4.6.3] which can be written as $\overset{\delta+}{H}—\overset{\delta-}{F}$. If two polar HF molecules are close enough, there is an attraction between the positive end of one molecule and the negative end of the other molecule:

$$\overset{\delta+}{H}—\overset{\delta-}{F}\cdots\cdots\overset{\delta+}{H}—\overset{\delta-}{F}$$

Since the attraction between the hydrogen atom in one molecule and the fluorine atom in the next molecule is stronger than the repulsion between the two hydrogen atoms and between the two fluorine atoms, the two molecules are bonded together.

The attraction between a hydrogen atom in one molecule and an electronegative atom, such as fluorine, in another molecule is called a **hydrogen bond**. Hydrogen bonding holds a number of molecules of hydrogen fluoride together in liquid hydrogen fluoride [see Figure 4.8D]. A molecule of HF forms an even stronger hydrogen bond to a fluoride ion, F^-. The ion $[F\cdots\cdots H\cdots\cdots F]^-$ is formed. Hydrogen fluoride forms acid salts, e.g. KHF_2, containing this anion.

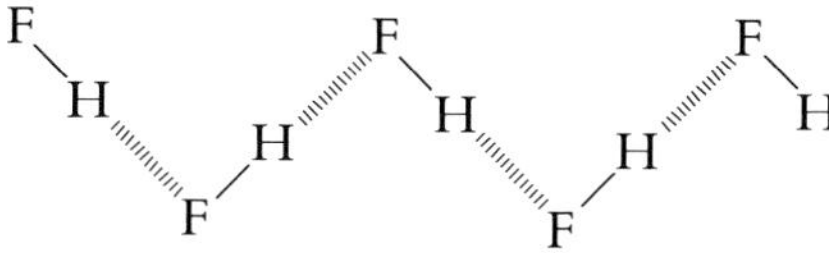

Figure 4.8D Hydrogen bonding in HF(l)

Hydrogen is unique

The strength of H bonds

Hydrogen bonds form when a hydrogen atom is covalently bonded to one of the electronegative atoms, fluorine, chlorine, oxygen or nitrogen. Hydrogen bonding is different from the dipole–dipole interactions in other polar molecules. A hydrogen atom has no inner, non-bonding electrons to set up forces of repulsion with the non-bonding electrons of the other atom. The strength of a hydrogen bond is $\frac{1}{10}$ to $\frac{1}{20}$ of that of a covalent bond.

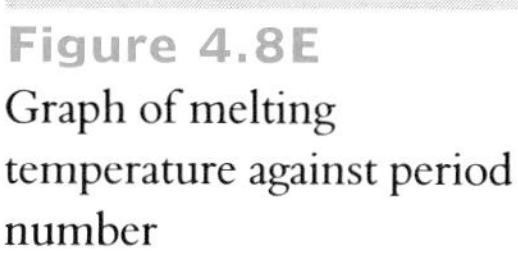

Figure 4.8E Graph of melting temperature against period number

Figure 4.8F Graph of boiling temperature against period number

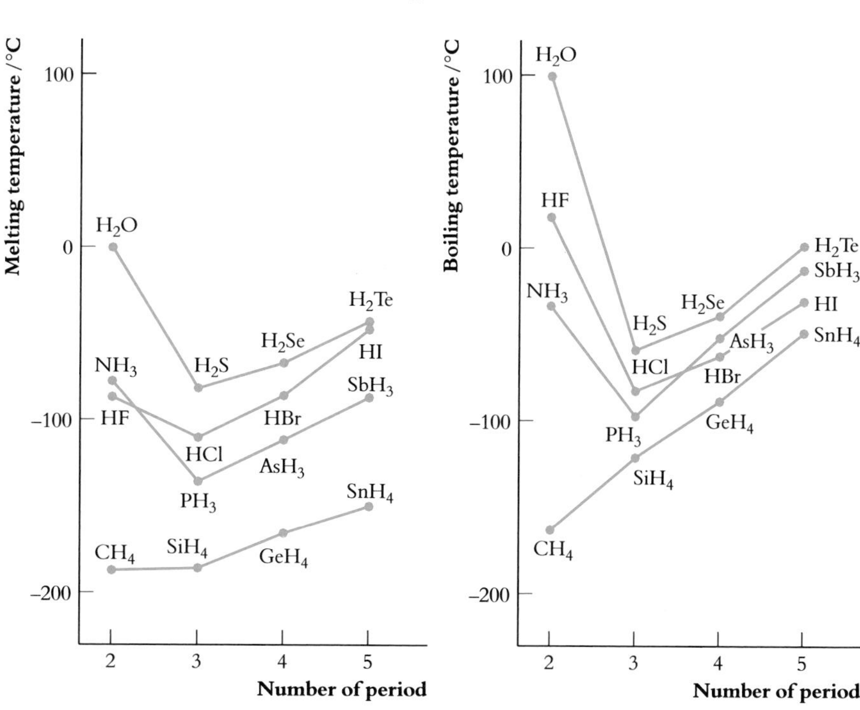

The attraction between a hydrogen atom in one molecule and an electronegative atom, e.g. oxygen or fluorine, in another molecule is called a hydrogen bond.

Evidence for H bonds is given by the melting and boiling temperatures of HF, H_2O, NH_3

Some evidence for the existence of the hydrogen bond is shown in Figures 4.8E and 4.8F. A comparison is made of the melting temperatures and boiling temperatures of hydrogen fluoride, water and ammonia with those of other hydrides in the same groups of the Periodic Table. The melting temperatures and boiling temperatures of these compounds are much higher than those of other hydrides in the same groups. The molecules of HF, H_2O and NH_3 must be held together by intermolecular bonds stronger than those between molecules of the other hydrides. Since fluorine, oxygen and nitrogen are the most electronegative of elements, the intermolecular forces are thought to be hydrogen bonds. [See Figure 4.8G.]

Figure 4.8G
Hydrogen bonding in $H_2O(l)$ and $NH_3(l)$

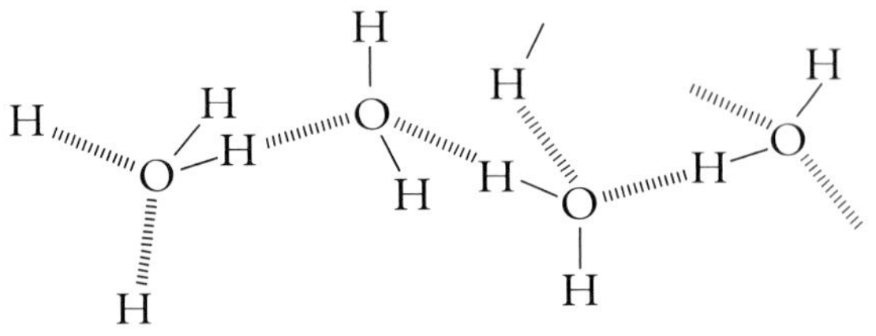

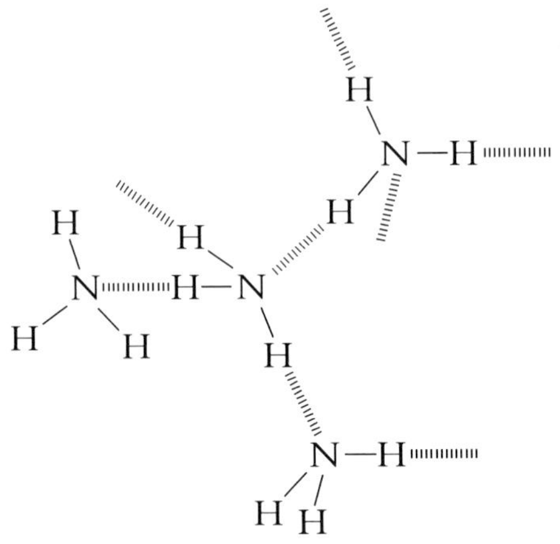

When the molar masses of carboxylic acids are found from measurements in the vapour phase and from solutions in organic solvents, the values are often up to twice the values calculated from the formulae. The molecules are thought to dimerise through the formation of hydrogen bonds [see Figure 4.8H].

Figure 4.8H
H bonding in RCO_2H

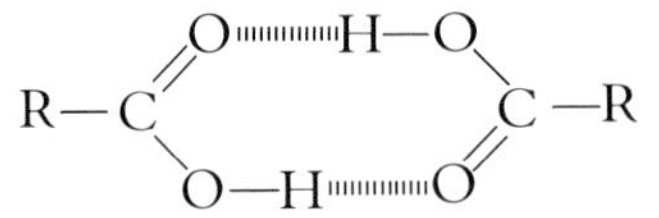

Some compounds can form **intramolecular hydrogen bonds** between two groups in the same molecule. [See Figures 4.8K to 4.8N and Question 5, Checkpoint 4.8B.]

Solubility considerations

Substances dissolve in water if they can form H bonds with it

Water is a hydrogen-bonded association of water molecules. A substance such as ethanol, C_2H_5OH, will dissolve in water as molecules of ethanol can displace water molecules in the association. New hydrogen bonds form between molecules of ethanol and water [see Figure 4.8I]. Halogenoalkanes such as chloroethane, C_2H_5Cl, do not form hydrogen bonds with water and are only slightly soluble. There are more references to solubility and hydrogen bonding in Part 4: Organic Chemistry.

Figure 4.8I
Hydrogen bonds between alcohols and water

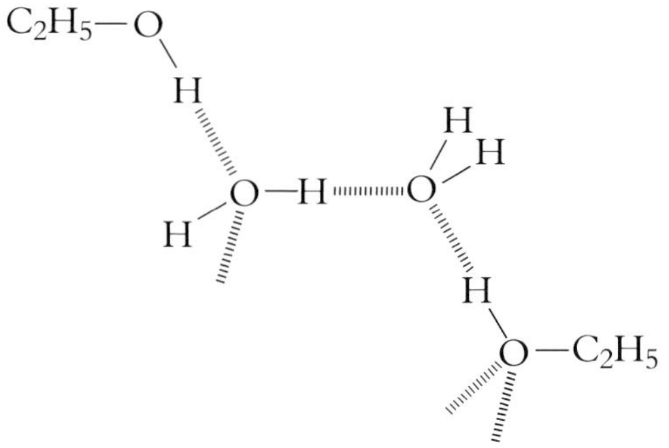

C_2H_5—O
H
H
H
O—H·····O
H
H
O—C_2H_5

Volatility

Hydrogen bonding occurs in alcohols and amines

In the liquid state, the molecules of alcohols are associated by hydrogen bonding. Energy must be supplied to break these bonds when the liquid is vaporised, making the boiling temperatures of alcohols higher than those of non-associated liquids, e.g. alkanes, of comparable molar mass. Amines also are hydrogen-bonded in the liquid state [§ 32.4].

The structure of ice

Liquid water is associated by H bonding

The bonds in H_2O are inclined at approximately the tetrahedral angle of 109.5°. The lone pairs occupy the other apices of the tetrahedron [see Figure 5.2F, § 5.2.3]. Liquid water contains associations of water molecules [see Figure 4.8G]. In ice, the arrangement of molecules is similar, but the regularity extends throughout the whole structure [see Figure 4.8J].

Figure 4.8J
Hydrogen bonding in ice
Notes Each H_2O molecule uses both its H atoms to form hydrogen bonds and is also bonded to two other H_2O molecules by means of their H atoms. The arrangement of bonds about the O atoms is tetrahedral

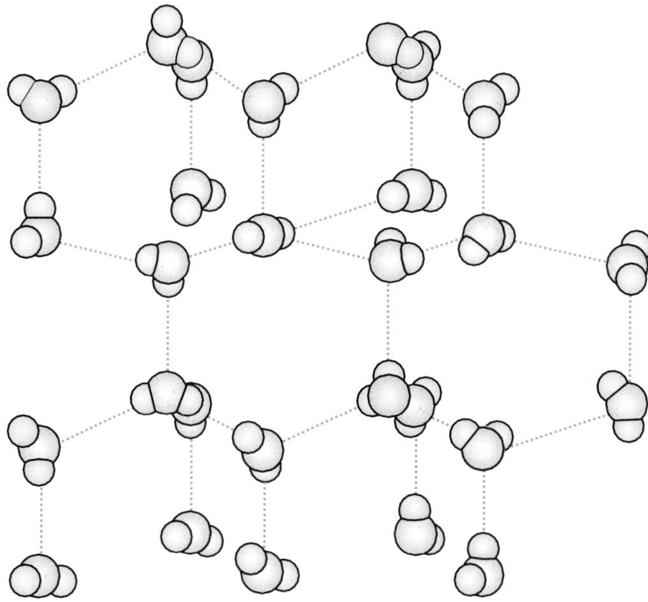

Hydrogen bonding extends throughout the whole structure in ice ...

The structure spaces the molecules further apart than they are in liquid water. This is why, when water freezes, it expands by 9%, and why ice is less dense than water at 0 °C. The underlying structure of ice resembles that of diamond [see Figure 6.6A, § 6.6].

... This explains why ponds freeze from the surface downwards

The fact that ice is less dense than water at 0 °C explains why ponds and lakes freeze from the surface downwards. Water reaches its maximum density at 4 °C. As it cools further, the water at the surface becomes less dense and therefore stays on top of the slightly warmer water until it freezes. The layer of ice on the surface helps to insulate the water underneath from further heat loss. Fish and plants survive under the ice in Canadian lakes and rivers for months.

The structure of proteins

Protein molecules have H bonds ...

Hydrogen bonding is important in protein molecules. Proteins consist of long chains of formula

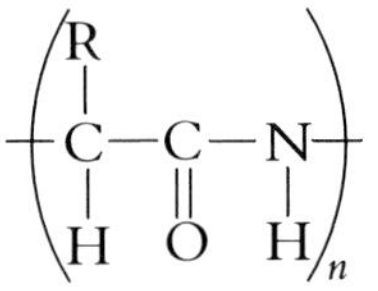

$$\left(\begin{array}{ccccc} R & & & & \\ | & & & & \\ -C & - & C & - & N- \\ | & & \| & & | \\ H & & O & & H \end{array}\right)_n$$

... between $\rangle C{=}O$ and $\rangle N{-}H$ groups

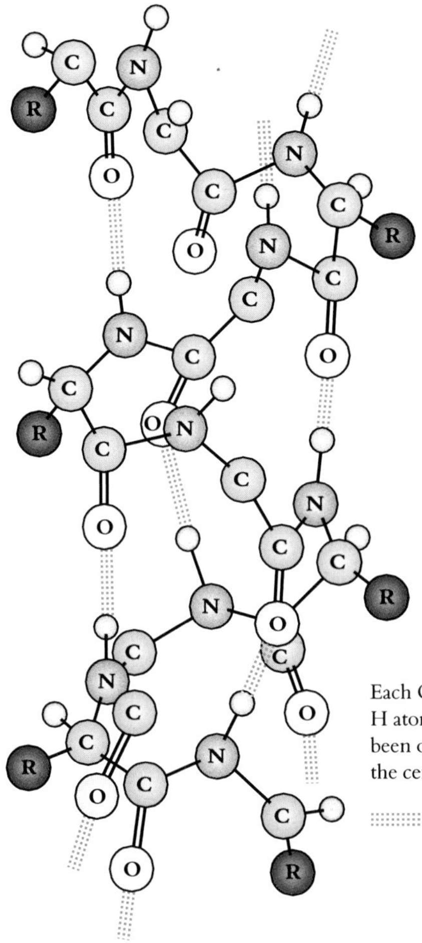

R can be a number of groups. Since both the $\rangle \overset{\delta+}{C}{=}\overset{\delta-}{O}$ group and the $\rangle \overset{\delta-}{N}{-}\overset{\delta+}{H}$

Figure 4.8K
A protein chain with the α helical structure

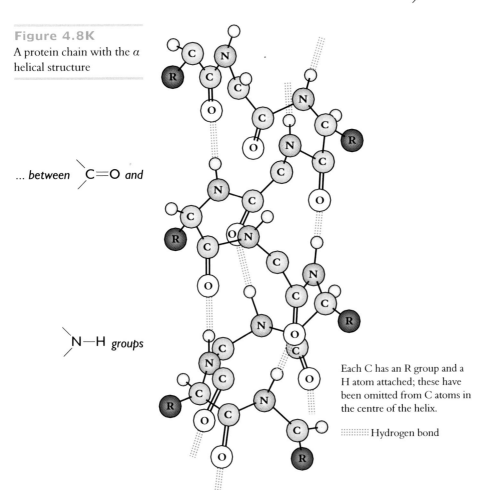

Each C has an R group and a H atom attached; these have been omitted from C atoms in the centre of the helix.

Hydrogen bond

groups are polar, hydrogen bonding can occur between them:

$$\begin{matrix} \diagdown^{\delta+} & \delta- & \delta+ & \delta-\diagup \\ C{=}O \cdots\cdots H{-}N \\ \diagup & & & \diagdown \end{matrix}$$

The helical structure of proteins is sustained by H bonds

A single protein molecule contains many hydrogen bonds. They are one of the forms of intramolecular attraction which hold the protein in a three-dimensional arrangement described as the *secondary structure* of the protein. Figure 4.8K shows the α helical structure proposed by Pauling and co-workers as a result of their X ray diffraction studies on protein molecules. An ***α* helix** is a spiral, which, looking away from you, is spiralling in a clockwise direction.

The double helix

Nucleic acids contain phosphate groups, sugar groups and bases

Hydrogen bonding is important in the famous **double helix** of DNA. **Chromosomes** are the bodies in the nuclei of the cells of living organisms which carry genetic information. They contain macromolecular substances called **nucleic acids**. These are of two types: ribonucleic acid, RNA, and deoxyribonucleic acid, DNA. The macromolecular chains in DNA are of the type

```
—P—S—P—S—P—S—P—
   |     |     |
   B     B     B
```

where P is a phosphate group and S is the sugar deoxyribose.

The double helix of DNA is held in this configuration by H bonds

B is one of the four bases: adenine, thymine, cytosine and guanine. DNA consists of two macromolecular strands spiralling round each other in a double helix, as shown in Figure 4.8L. The strands are held together by hydrogen bonding between the bases, as shown in Figure 4.8M.

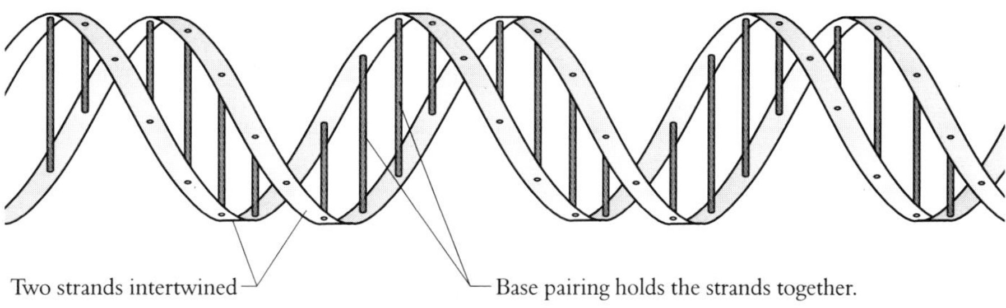
Two strands intertwined
Base pairing holds the strands together.

Figure 4.8L The double helix

Figure 4.8M Hydrogen bonding between strands in DNA

```
—P—S—P—S—P—S—P—S—P—
   |     |     |     |
   B     B     B     B
   ⋮     ⋮     ⋮     ⋮
   B     B     B     B
   |     |     |     |
—P—S—P—S—P—S—P—S—P—
```

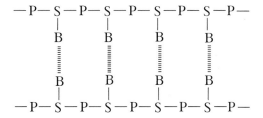

Of the four bases, *thymine* can pair up with *adenine* by hydrogen bonding and *cytosine* can form hydrogen bonds with *guanine*. The double helix brings these base pairs into contact so that they can form the bonds that keep the structure intact. Figure 4.8N shows the details of the hydrogen bonding.

THE CHEMICAL BOND

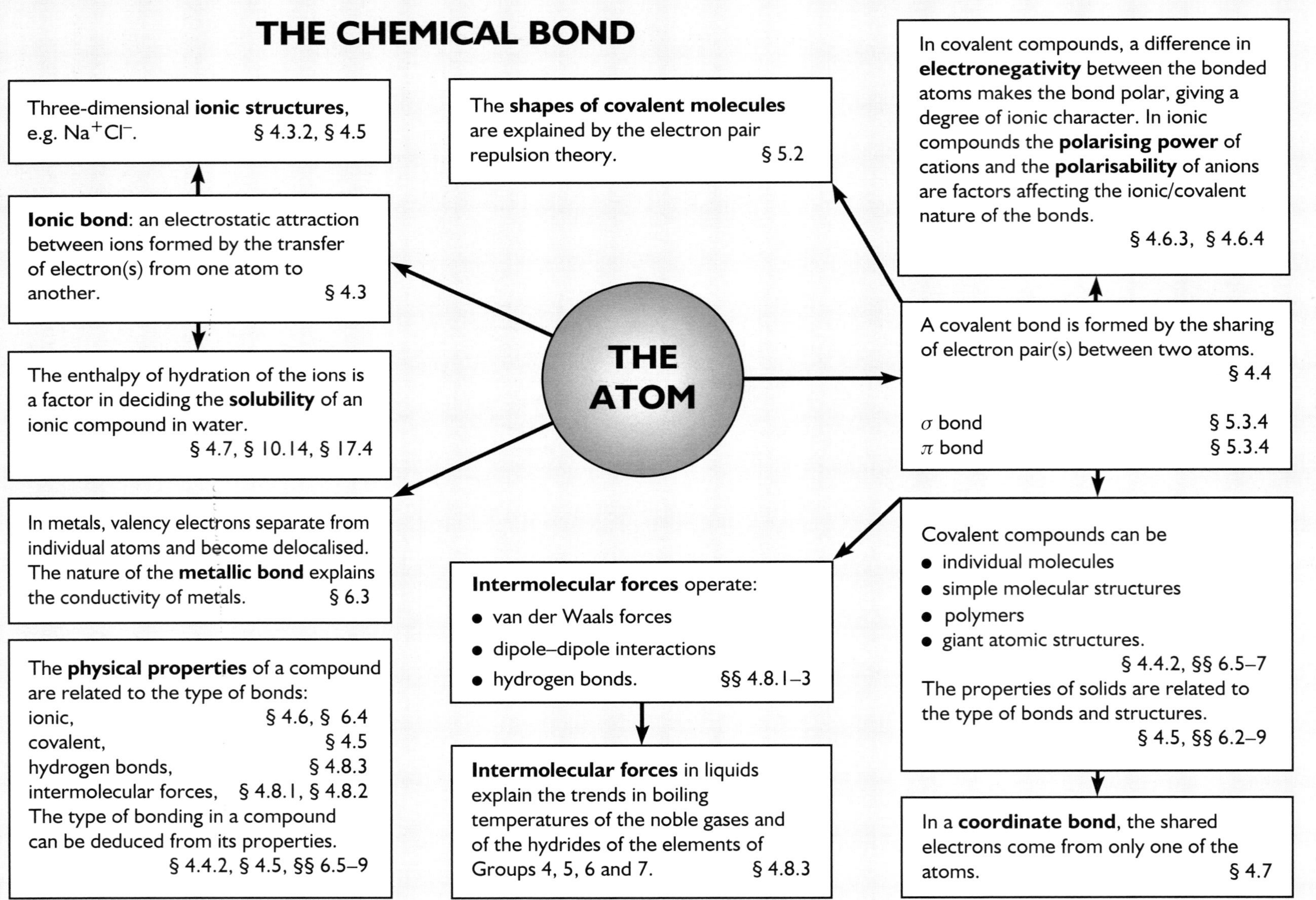

Three-dimensional **ionic structures**, e.g. Na^+Cl^-. § 4.3.2, § 4.5
Ionic bond: an electrostatic attraction between ions formed by the transfer of electron(s) from one atom to another. § 4.3
The enthalpy of hydration of the ions is a factor in deciding the **solubility** of an ionic compound in water. § 4.7, § 10.14, § 17.4
In metals, valency electrons separate from individual atoms and become delocalised. The nature of the **metallic bond** explains the conductivity of metals. § 6.3
The **physical properties** of a compound are related to the type of bonds:
ionic, § 4.6, § 6.4
covalent, § 4.5
hydrogen bonds, § 4.8.3
intermolecular forces, § 4.8.1, § 4.8.2
The type of bonding in a compound can be deduced from its properties. § 4.4.2, § 4.5, §§ 6.5–9
The **shapes of covalent molecules** are explained by the electron pair repulsion theory. § 5.2
THE ATOM
Intermolecular forces operate:
• van der Waals forces
• dipole–dipole interactions
• hydrogen bonds. §§ 4.8.1–3
Intermolecular forces in liquids explain the trends in boiling temperatures of the noble gases and of the hydrides of the elements of Groups 4, 5, 6 and 7. § 4.8.3
In covalent compounds, a difference in **electronegativity** between the bonded atoms makes the bond polar, giving a degree of ionic character. In ionic compounds the **polarising power** of cations and the **polarisability** of anions are factors affecting the ionic/covalent nature of the bonds. § 4.6.3, § 4.6.4
A covalent bond is formed by the sharing of electron pair(s) between two atoms. § 4.4
σ bond § 5.3.4
π bond § 5.3.4
Covalent compounds can be
• individual molecules
• simple molecular structures
• polymers
• giant atomic structures. § 4.4.2, §§ 6.5–7
The properties of solids are related to the type of bonds and structures. § 4.5, §§ 6.2–9
In a **coordinate bond**, the shared electrons come from only one of the atoms. § 4.7

CHECKPOINT 4.8B: INTERMOLECULAR FORCES II

1. What conditions are necessary for the formation of a hydrogen bond?

2. How do hydrogen bonds contribute to the structure of ice? How do we know that some hydrogen bonding is present in water?

3. What reason do we have for postulating the existence of van der Waals forces? What is responsible for the van der Waals attractive forces, and what gives rise to repulsion?

4. What types of intermolecular forces operate in *(a)* HBr(g), *(b)* Br_2(g), *(c)* ICl(g) and *(d)* HF(l)?

5. Sketch hydrogen bonding in *(a)* liquid ethanol, *(b)* aqueous ethanol, *(c)* liquid ethanoic acid, *(d)* aqueous ethanoic acid and *(e)* liquid ammonia. How many hydrogen bonds can be formed by one molecule of (i) H_2O and (ii) NH_3?

6. If water were not hydrogen-bonded, what would you expect for its boiling temperature and melting temperature and the relative densities of the liquid and solid states?

QUESTIONS ON CHAPTER 4

1. Explain the 'octet theory' of valency. Point out examples of the success of the theory in explaining the formation of chemical bonds. Discuss cases in which the octet theory is inadequate.

2. Discuss the bonding in *(a)* NaH, *(b)* NH_4^+, *(c)* $BeCl_2$, *(d)* HF(l), *(e)* CCl_4 and *(f)* Cu. [For *(f)*, see § 6.3.]

†3. Put the following substances in order of increasing boiling temperatures, giving reasons for your choice:

C_4H_9OH, $CH_3CH_2CH_2CH_2CH_3$, $(CH_3)_4C$, N_2

†4. Which members of the following pairs would you expect to have the higher boiling temperature?

(a) C_3H_8 and CH_3OCH_3
(b) $CH_3CH_2NH_2$ and CH_3CH_2OH
(c) CH_3CH_2OH and C_2H_6
(d) C_3H_8 and $(CH_3)_2C{=}O$

Give reasons for your choice.

5. Briefly explain the importance of the hydrogen bond to living creatures.

6. *(a)* The following diagram shows part of a radioactive decay series.

State which **one** of the following best describes the emissions involved in the decay from ^{238}Pa to ^{234}U.

A 1 α-emission and 3 β-emissions
B 2 α-emissions and 2 β-emissions
C 3 α-emissions and 1 β-emission
D 4 α-emissions **1**

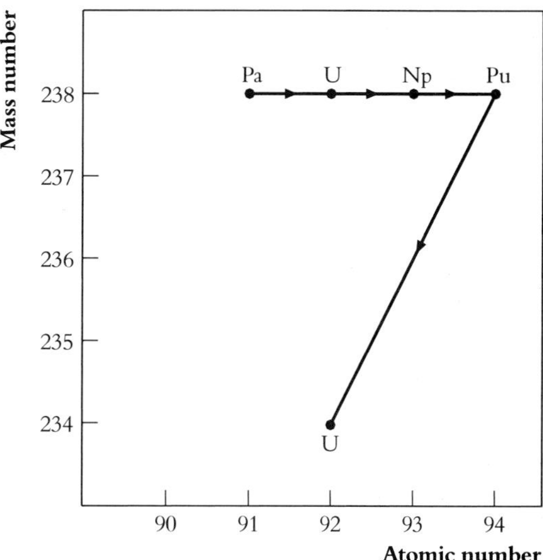

(b) The plot overleaf is of the logarithm of successive ionisation energies (I.E.) for all the electrons in a gaseous potassium atom.
(i) Explain why
I. the first ionisation energy has such a low value, **1**
II. the plot shows a general increase in the values of these ionisation energies. **1**
(ii) I. Give a reason for the change in ionisation energies between the 1st and 2nd ionisations. **1**
II. Name the orbital from which the 10th electron is removed. **1**
III. State why the 18th and 19th ionisations are so high. **1**

† *Note* A dagger indicates a question which should be tackled on re-reading as it involves material which is covered in later chapters.

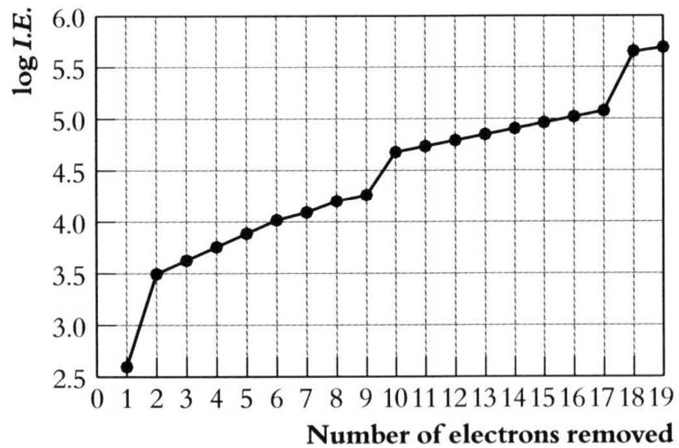

(c) From the hydrides with formulae H_2O, NH_3 and CH_4 select the **one** with

(i) the highest boiling temperature, **1**

(ii) the least polar bonds, **1**

(iii) two lone pairs of electrons. **1**

(A hydride may be used once, more than once, or not at all.)

(d) Write down the electronic structure of the vanadium atom in terms of the appropriate s, p and d orbitals by copying and labelling the boxes below and inserting arrows to represent electrons.

Orbital

....

1

10 *WJEC (AS/AL)*

7. *(a)* The figure shows a plot of first ionisation energy against atomic number for the elements of atomic number 10 to 18. (The letters are not the chemical symbols for the elements concerned.)

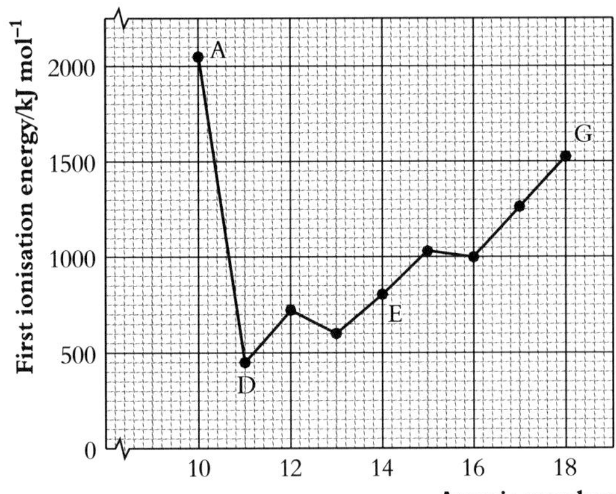

(i) I. Explain **briefly** why the ionisation energy of the element labelled **G** is less than that of the element labelled **A**. $\frac{1}{2}$

II. Explain **briefly** why the ionisation energy of the element labelled **D** is considerably less than that of the element labelled **A**. $\frac{1}{2}$

(ii) Draw the shape of the orbital from which the electron is lost when,

I. element **A** is ionised,

II. element **D** is ionised. **1**

(iii) Give the electronic configuration of the element labelled **E** on the figure by copying and labelling **each** of the following boxes and inserting arrows, in appropriate boxes, to represent electrons.

Orbital

....

$1\frac{1}{2}$

(iv) **State** and **explain** which of the elements **D** or **E** will form a cation more readily. **1**

(b) (i) Explain the principle of the mass spectrometer. **2**

(ii) The mass spectrum of the chlorine molecule (Cl_2) is shown below.

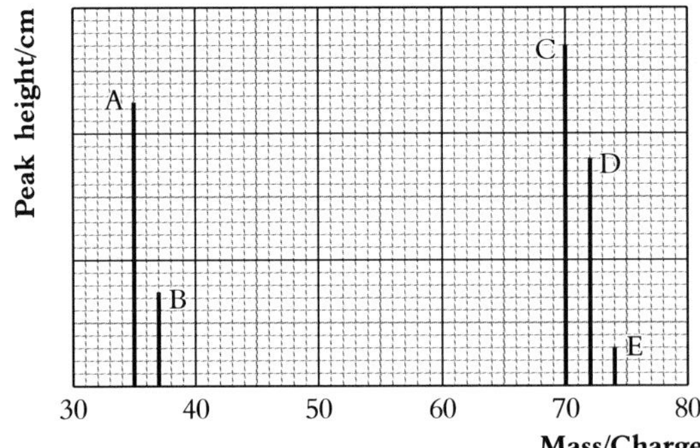

Give the formulae of the ions that give rise to peaks **D** and **E**. **1**

(c) (i) State what is meant by,

I. a **dipole**, and II. a **polar** bond. **1**

(ii) Explain what is meant by the term **electronegativity**. **1**

(iii) The electronegativity values, derived by Pauling, for a series of atoms are listed below.

Atom	H	Li	Be	B	C	N	O	F
Electronegativity	2.1	1.0	1.5	2.0	2.5	3.0	3.5	4.0

Use these values to predict the polarities, e.g. $X^{\delta+}—Y^{\delta-}$ of the following bonds: Li—H, C—O, B—C **$1\frac{1}{2}$**

11 *WJEC (AS/AL)*

8. The table below shows the melting points of the Period 3 elements except for silicon.

Element	Na	Mg	Al	Si	P	S	Cl	Ar
mp/K	371	923	933		317	392	172	84

(a) Explain in terms of bonding why the melting point of magnesium is higher than that of sodium. **3**

(b) State the type of bonding between atoms in the element silicon and name the type of structure which silicon forms. **2**

(c) Predict the approximate melting point of silicon. **1**

(d) Explain why chlorine has a lower melting point than sulphur. **2**

(e) Predict the approximate melting point of potassium and give one reason why it is different from that of sodium. **2**

10 *NEAB (AS/AL)*

9. *(a)* Naturally occurring boron contains two isotopes with mass numbers 10 and 11. The relative atomic mass of boron is 10.81.

(i) Calculate the percentage abundance of ^{10}B in naturally occurring boron. **2**

(ii) Draw the mass spectrum of naturally occurring boron. **1**

(b) Boron is used to absorb neutrons in a nuclear reactor.

Give the chemical symbol of the element **X** and state the value of a and b in the equation below.

$$^{11}\mathbf{B} + {}^{1}_{0}\mathrm{n} \longrightarrow {}^{b}_{a}\mathbf{X} + \gamma$$ **2**

(c) (i) Draw dot-and-cross diagrams to show the bonding in boron trifluoride, $\mathbf{BF_3}$, and ammonia, NH_3. **2**

(ii) By reference to your diagrams, explain how the two molecules bond together to form a compound. **2**

(d) (i) Graphite is a black slippery substance which conducts electricity. Its structure is made up of planes of hexagonal rings of carbon atoms. Explain why graphite is
I. slippery; II. conducts electricity. **2**

(ii) Boron nitride, BN, is a white slippery substance with the graphite type structure. Suggest, by means of a sketch or otherwise, the arrangement of boron and nitrogen atoms in one of the planes of atoms in boron nitride. **1**

12 *WJEC (AS/AL)*

5 THE SHAPES OF MOLECULES

5.1 HAEMOGLOBIN

The function of haemoglobin is to combine with oxygen, transport it in the blood stream to parts of the body where it is needed and then release the oxygen. The haemoglobin molecule has a complicated shape, as you can see from Figure 5.1A. The molecule consists of four protein chains, two each of different types, and four haem groups. The protein chains are held in a complicated three-dimensional arrangement by the different types of bonds that operate between them, including dipole–dipole interactions, van der Waals forces and hydrogen bonds (see Chapter 4). The four haem groups contain iron in the form of Fe^{2+}. Each haem group is able to combine with oxygen through the formation of a coordinate bond from a lone pair on an oxygen molecule to the Fe^{2+} ion. The haem groups are situated on the outside of the molecule where oxygen can easily reach them. The shape of the molecule enables it to perform its biological function.

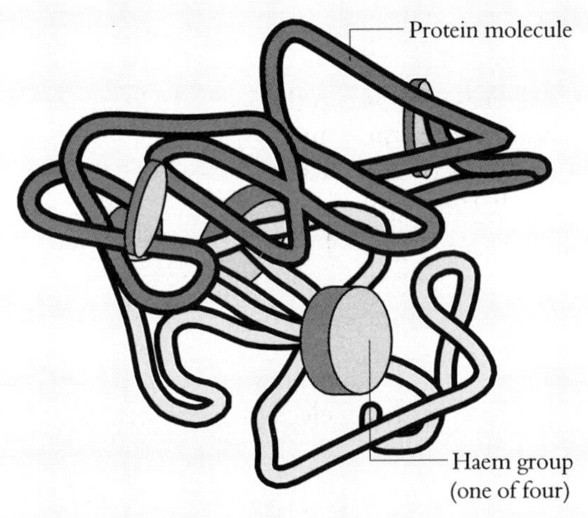

Figure 5.1A
The haemoglobin molecule

The molecules that begin our study in this chapter are less complicated than haemoglobin, but they illustrate the need for thinking about molecules in space and not just as they appear when written in two dimensions on a sheet of paper.

5.2 THE ARRANGEMENT IN SPACE OF COVALENT BONDS

Ionic bonds are not directed in space whereas ...

Electrovalent bonds are the electrostatic attractions that exist between oppositely charged ions. Since ions radiate a spherically symmetrical positive or negative field, ionic bonds are *non-directional*.

... covalent bonds have a preferred direction in space

When atoms approach one another closely, their atomic orbitals overlap and molecular orbitals are formed. Covalent bonds are formed when a bonding pair of electrons enters a molecular orbital of low energy. The bonding electrons must have opposing spins, in accordance with the Pauli Exclusion Principle [§ 2.4.1].

The more the atomic orbitals overlap, the more stable will be the molecular orbital formed. The strongest bonds will be formed if the atoms approach in such a way that there is maximum overlap between atomic orbitals. It follows that a covalent bond will have a *preferred direction*. A covalent molecule will have a shape which is determined by the angles between the bonds joining the atoms together.

$BeCl_2$ and $SnCl_2$ differ in shape ...

... BCl_3 and NH_3 differ in shape

There must be a reason why beryllium chloride, $BeCl_2$, is a linear molecule without a dipole moment, while tin(II) chloride, $SnCl_2$, is a bent molecule with a dipole moment. There must also be a reason why the four atoms in boron(III) chloride, BCl_3, are coplanar, whereas in ammonia the nitrogen atom lies above three coplanar hydrogen atoms. [See Figure 5.2A.]

Figure 5.2A
$BeCl_2$ compared with $SnCl_2$ and BCl_3 compared with NH_3

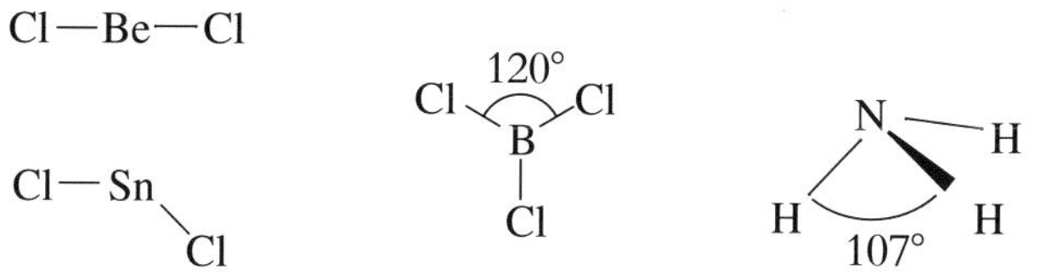

Cl—Be—Cl

Cl—Sn
\\
Cl

120°
Cl Cl
B
|
Cl

N—H
H H
107°

The differences can be explained by the Sidgwick–Powell theory, which is the Valence Shell Electron Pair Repulsion Theory

A theory to account for the shapes of molecules was put forward by Sidgwick and Powell in 1940. It is known as the **Valence Shell Electron Pair Repulsion Theory**. Sidgwick and Powell considered the shapes of small molecules and molecular ions, such as $BeCl_2$, BCl_3, NH_3, NH_4^+ and CH_4. They pointed out that the arrangement of electron pairs around the central atom in a molecule depends on the number of electron pairs. Between each electron pair and any other electron pair there is a force of electrostatic repulsion, which forces the orbitals as far apart as possible. Any lone pairs of electrons on the central atom occupy atomic orbitals, and they too repel the bonding pairs of electrons and affect the geometry of the molecule.

SUMMARY

The number of electron pairs in the valence shell of an atom in a molecule decides the angle between its bonds and the shape of the molecule.

5.2.1 LINEAR MOLECULES

SUMMARY

Two bonds : bond angle 180°; linear molecule.

The molecules of gaseous beryllium chloride, $BeCl_2$, are linear. Beryllium, in Group 2 of the Periodic Table, has two electrons in its valence shell, and forms two covalent bonds. A **linear** arrangement of the atoms (a bond angle of 180 °) puts the two electron clouds as far apart as possible:

Cl—Be—Cl

Other linear molecules are

Examples of linear molecules

H—C≡C—H   H—C≡N   O=C=O

The electron pairs in a multiple bond are assumed on the Sidgwick–Powell theory to occupy the position of one electron pair in a single bond.

5.2.2 TRIGONAL PLANAR MOLECULES

The arrangement of 3 pairs of valence electrons is trigonal planar

When there are three pairs of electrons around the central atom, the bonds lie in the same plane at an angle of 120° to one another. Three atoms form a triangle about the central atom, and the arrangement is described as **trigonal planar**. An example is boron trichloride, BCl_3. Boron, in Group 3 of the Periodic Table, has

Lone pairs of electrons can determine the shape of the molecule

three valence electrons and forms three covalent bonds. Gaseous tin(II) chloride, $SnCl_2$, has a dipole moment, proving that the molecule is not linear. The reason is that tin, in Group 4, is using only two of its four electrons for bond formation. The lone pair of electrons repel the bonding pairs and a trigonal planar arrangement of orbitals results. [See Figure 5.2B.] This arrangement maximises the angle between the electron pairs and minimises the repulsion between them.

Figure 5.2B
The trigonal planar arrangement of electron pairs in BCl_3 and $SnCl_2$

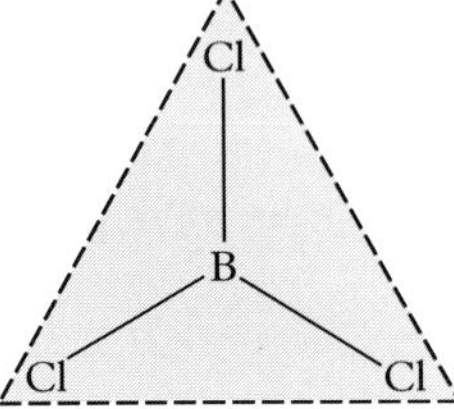

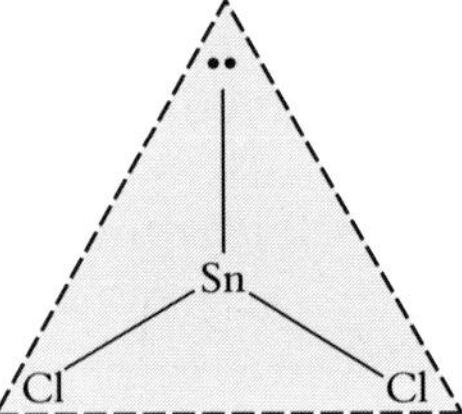

Other structures based on a trigonal planar arrangement are ethene, the nitrate ion and sulphur dioxide [see Figure 5.2C].

Figure 5.2C
The arrangement of electron pairs in $CH_2{=}CH_2$, NO_3^- and SO_2

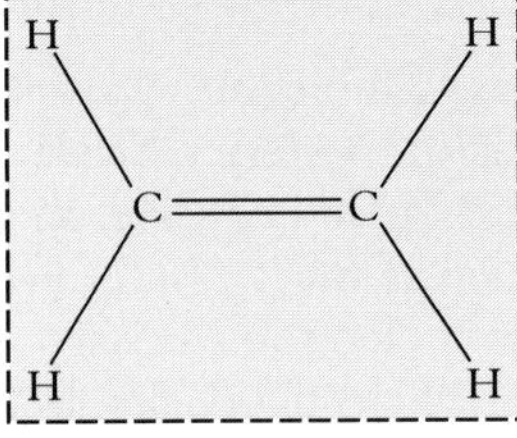

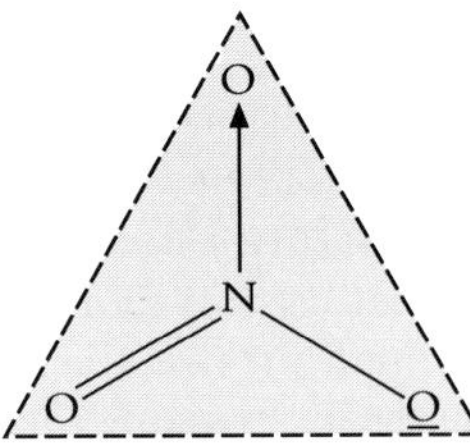

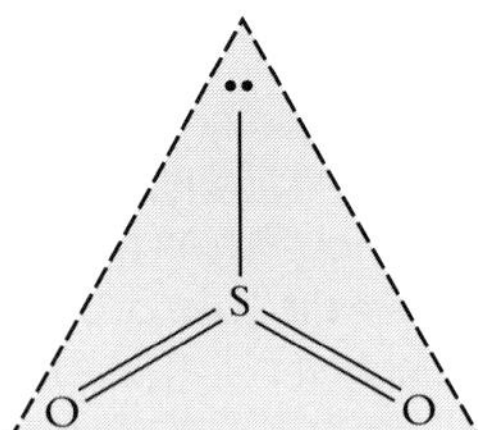

SUMMARY

When an atom forms 3 bonds or 2 bonds +1 lone pair the bond angle = 120° and the arrangement of electron pairs is trigonal planar.

There is an interesting result of the coplanar arrangement. The formula of but-2-ene is $CH_3CH{=}CHCH_3$. There are two structures with this formula:

(a) $CH_3—C—H$ / $||$ / $CH_3—C—H$ (b) $CH_3—C—H$ / $||$ / $H—C—CH_3$

Structure (a), in which the hydrogen atoms are on the same side of the double bond, is called *cis*-but-2-ene, and structure (b), with the hydrogen atoms on opposite sides of the double bond, is called *trans*-but-2-ene. The existence of *cis*- and *trans*- forms of compounds is covered in § 25.9.2. [See Figure 5.2D.]

Figure 5.2D
cis- and *trans*- but-2-ene

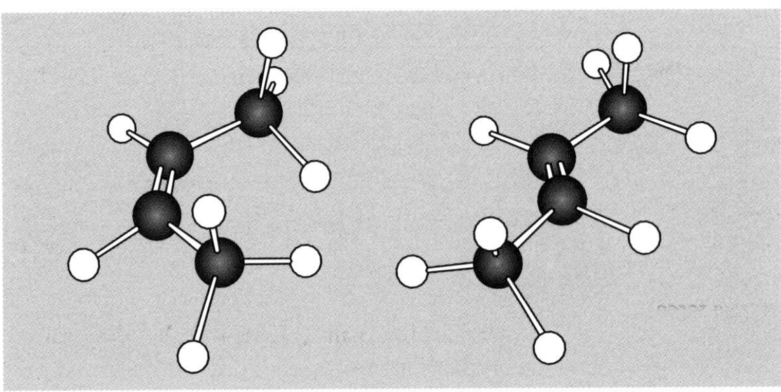

5.2.3 TETRAHEDRAL MOLECULES

4 electron pairs adopt a tetrahedral configuration

The molecules CH_4, NH_3, NH_4^+ and H_2O all have four pairs of electrons around the central atom. Whether they are bonding pairs or lone pairs of electrons, they experience mutual repulsion. To minimise this repulsion, the four electron orbitals take up the spatial arrangement that maximises the angle between the orbitals. This is the **tetrahedral** arrangement. [See Figures 5.2E to 5.2G.]

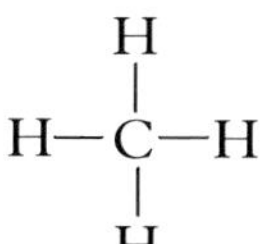

Figure 5.2E
The bonding in CH_4, NH_3, NH_4^+, H_2O

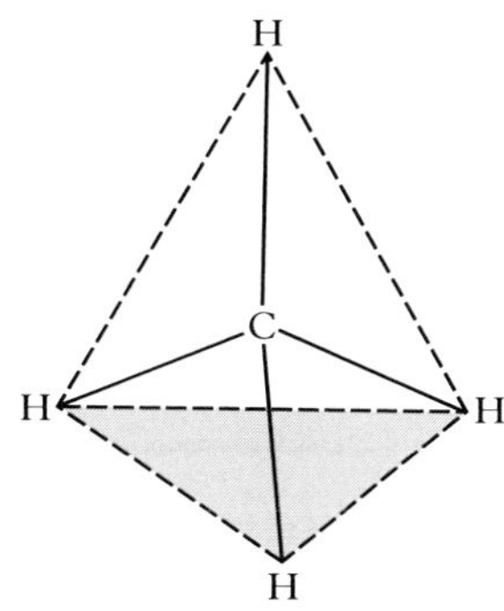

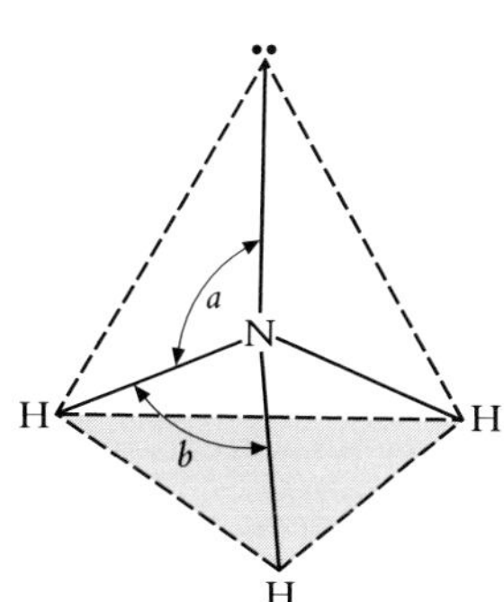

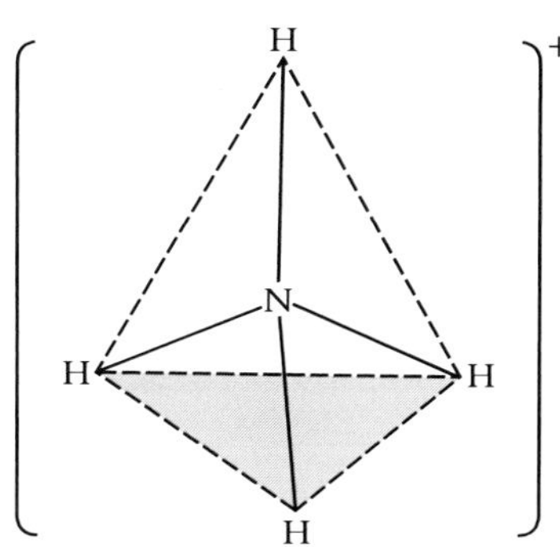

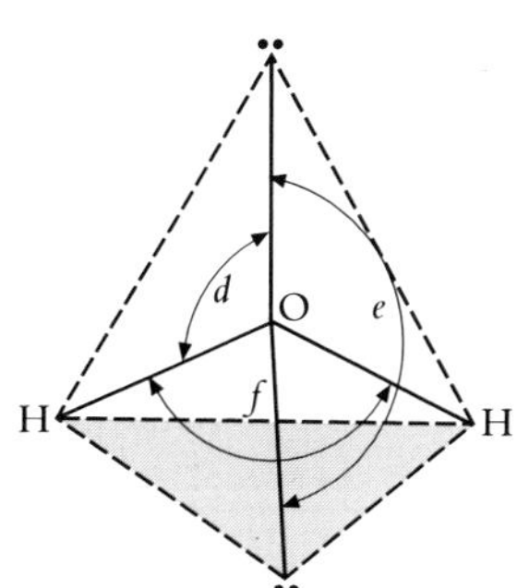

Figure 5.2F
The tetrahedral arrangement of valence electron pairs in CH_4, NH_3, NH_4^+ and H_2O

The arrangements of atoms are of course *not* all the same. In CH_4 and NH_4^+, the atoms form a tetrahedron, in NH_3 they form a trigonal pyramid, and in H_2O they form a bent line.

In CH_4 and NH_4^+ all the bonds are the same. Once formed, a coordinate bond is the same as a covalent bond. The structures are perfect tetrahedra with the tetrahedral angle of 109.5° between each pair of bonds. In NH_3 the bond angle is 107°, and in H_2O it is 104.5°.

Lone pairs are closer to the nucleus than bonding pairs and exert a greater repulsive force

To account for these differences from the expected bond angle, it was suggested that, since lone pairs are closer to the nucleus than bonding pairs, they will exercise a greater force of repulsion. Repulsion between electron pairs decreases in the order

Lone pair : lone pair repulsion > Lone pair : bonding pair repulsion > Bonding pair : bonding pair repulsion

Repulsion between the lone pair and the bonding pairs in NH_3 makes the angle *a* in Figure 5.2F greater than the tetrahedral angle (109.5°) and consequently the angle *b* less than 109.5°. Similarly in H_2O, angles *d* and *e* are greater than 109.5°, and the angle *f* between the H—O—H bonds is 104.5°. Other structures based on the tetrahedron are the sulphate and sulphite ions [see Figure 5.2H].

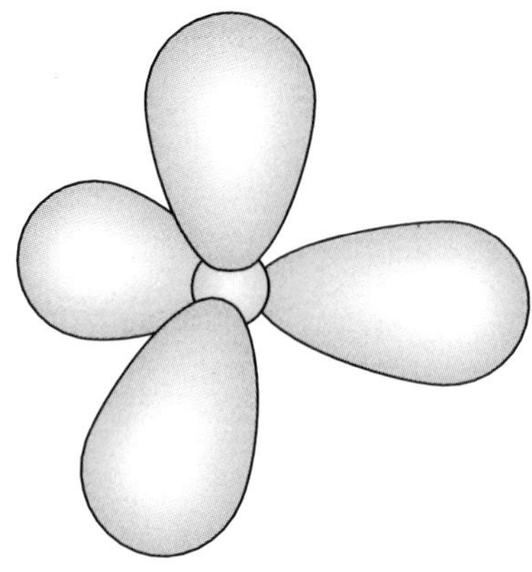

Figure 5.2G
The bonding orbitals in methane

SUMMARY

When an atom forms 4 bonds or 3 bonds +1 lone pair or 2 bonds +2 lone pairs, the bond angle = 109.5°, and the arrangement of electron pairs is tetrahedral.

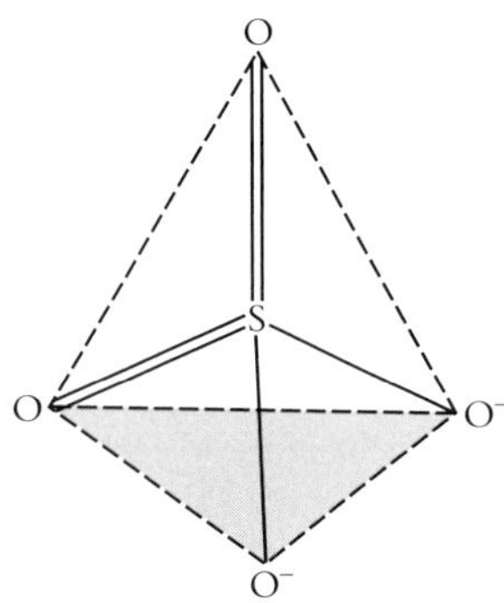

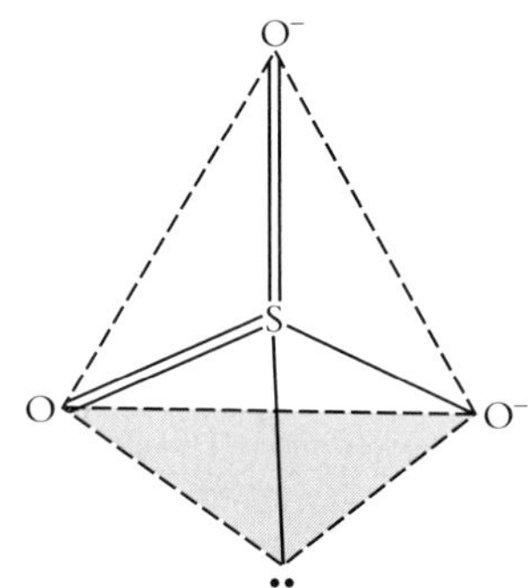

Figure 5.2H
The shapes of SO_4^{2-} and SO_3^{2-}

5.2.4 CHIRALITY

Six planes of symmetry run through the tetrahedron of hydrogen atoms in CH_4, and the centres of positive charge and negative charge coincide in the carbon atom at the centre of the tetrahedron. [See Figure 5.2I.]

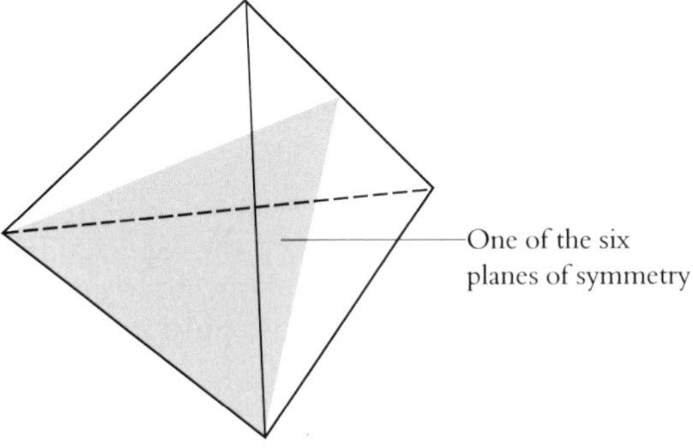

Figure 5.2I
The symmetry of the tetrahedron

Four different groups bonded to carbon destroy the symmetry of the tetrahedron ...

... so that the C atom is chiral ...

If four different atoms or groups are attached to a carbon atom, there is no longer a plane of symmetry in the molecule. Nor is there a centre of symmetry or an axis of symmetry. The carbon atom in CHClBrF is an **asymmetric** carbon atom. There are two ways of drawing a tetrahedral arrangement for this formula. You will see from Figure 5.2J that one molecule is the mirror image of the other; they are called **enantiomers**. The molecules cannot be superimposed, being related in the same way as a left hand and a right hand [see Figure 5.2K]. They show the geometric property of **chirality**.

Figure 5.2J
Enantiomers of bromochlorofluoromethane

... and CHCl BrF exists in two forms ...

... enantiomers ...

... which are mirror images ...

... optical isomers

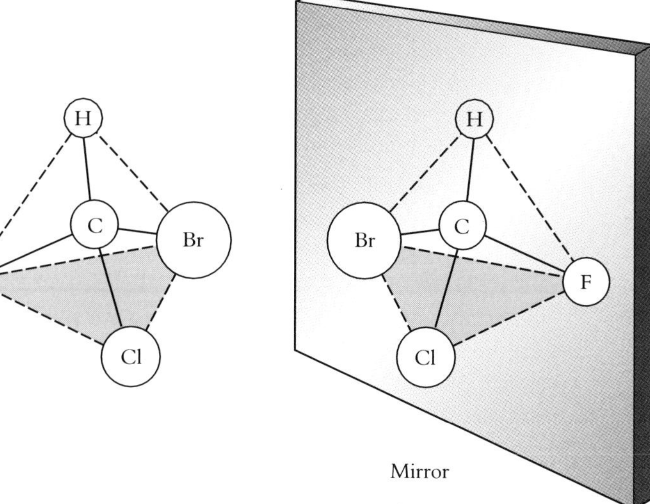

Figure 5.2K
Left hand and right hand

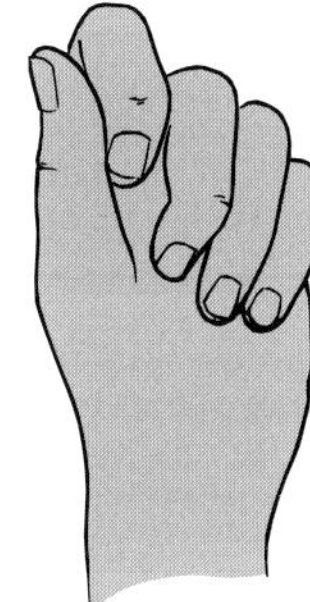

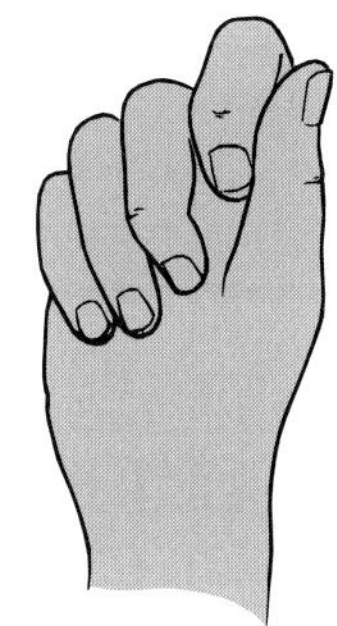

Optical isomerism is one type of stereoisomerism

There are two different compounds with the two different kinds of molecules shown in Figure 5.2J. They are **isomers** (different compounds with the same formula). Since they differ in the spatial arrangement of atoms, they are called **stereoisomers**. This type of stereoisomerism, involving enantiomers, is called **optical isomerism** [§ 24.16.4 and § 25.9.3].

Figure 5.2L
The octahedral shape of the electron pairs in SF_6, IF_5, ICl_4^-

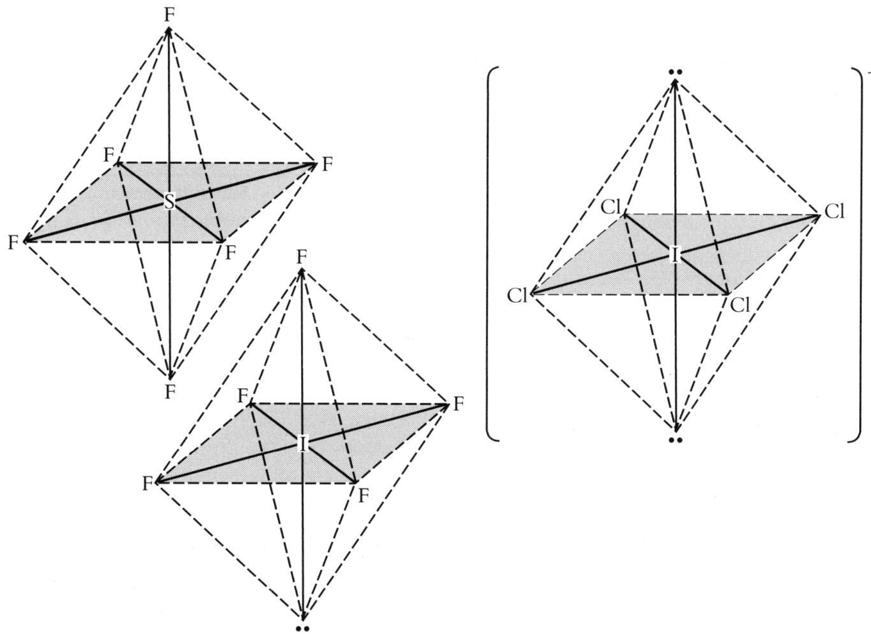

5.2.5 STRUCTURES WITH 6 PAIRS OF VALENCE ELECTRONS

Some atoms have more than 8 electrons in the valence shell

Structures with more than four pairs of electrons about the central atom may occur if the element is in the third period or a later period. This is known as **expansion of the octet**.

SUMMARY

When an atom forms 6 bonds or 5 bonds + 1 lone pair or 4 bonds + 2 lone pairs, the arrangement of electron pairs is octahedral.

Structures with *six* pairs of electrons around the central atom are sulphur(VI) fluoride, SF_6, iodine(V) fluoride, IF_5, and the ICl_4^- ion. The **octahedral arrangement** of electron pairs is shown in Figure 5.2L.

The arrangement of *atoms* in IF_5 is **square pyramidal**, a lone pair occupying the sixth position in the octahedron. In ICl_4^- the four chlorine atoms are in a **square planar** configuration, with lone pairs occupying the axial positions of the octahedron.

5.2.6 SUMMARY

Table 5.2
A summary of the shapes of molecules

No. of valence electrons	No. of bond pairs	No. of lone pairs	Total electron pairs	Arrangement of orbitals	Arrangement of atoms	Example
4	2	0	2	Linear	Linear	$BeCl_2$
6	3	0	3	Trigonal planar	Trigonal planar	BF_3
8	4	0	4		Tetrahedral	CH_4
8	3	1	4	Tetrahedral	Trigonal pyramidal	NH_3
8	2	2	4		Bent line	H_2O
12	6	0	6		Octahedral	SF_6
12	5	1	6	Octahedral	Square pyramidal	IF_5
12	4	2	6		Square planar	ICl_4^-

CHECKPOINT 5.2: SHAPES OF MOLECULES

1. *(a)* Take three long balloons, blow them up and tie the ends. Hold the three tied ends between your finger and thumb. What positions do the three balloons adopt?
   (b) Add a fourth balloon and notice the positions which the balloons take up.

2. In BF_3, how many electron pairs are there around the B atom? Sketch the spatial distribution of bonds.

3. In BrF_3, how many electrons does Br use for bond formation? How many lone pairs does Br possess? What is the total number of electron pairs around the Br atom? Sketch the spatial distribution of bonds. How is this arrangement described?

4. Sketch the spatial distribution of bonds in HOBr. How would you describe the shape of *(a)* the electron orbitals, *(b)* the molecule?

5. Sketch the arrangement of bonds in CCl_4. If the $C^{\delta+}$—$Cl^{\delta-}$ bond has a dipole moment of x debyes, what is the dipole moment of the CCl_4 molecule?

6. In the compound XeF_4, how many electrons is the noble gas xenon using for bond formation? How many lone pairs does xenon have? What is the total number of electron pairs around the central Xe atom? Sketch the arrangement of bonds. What shape is the molecule?

7. Explain how the Sidgwick–Powell theory predicts the shape of the following molecules: *(a)* $SnCl_4$, *(b)* PH_3, *(c)* PF_5, *(d)* BH_3 and *(e)* BeH_2.

8. Sketch the spatial arrangement of bonds in the following: *(a)* F_2O, *(b)* $SeCl_4$, *(c)* SO_3, *(d)* PF_6^- and *(e)* $COCl_2$.

*5.3 SHAPES OF MOLECULES: A MOLECULAR ORBITAL TREATMENT

An alternative to the VSEPR treatment is the molecular orbital approach

The valence shell electron pair repulsion theory provides a simple treatment of the shapes of covalent molecules. A more precise treatment of the spatial distribution of covalent bonds about a central atom involves a consideration of the atomic orbitals used in bond formation. The shapes of atomic orbitals are described in § 2.4.2 and shown in Figures 2.4B to 2.4D. When **atomic orbitals** overlap, **molecular orbitals** are formed. The **molecular orbital approach** is illustrated in this section.

5.3.1 METHANE: sp^3 HYBRID ORBITALS

The electronic configuration of carbon in its normal, C, and excited, C*, states and hydrogen, H, are shown below:

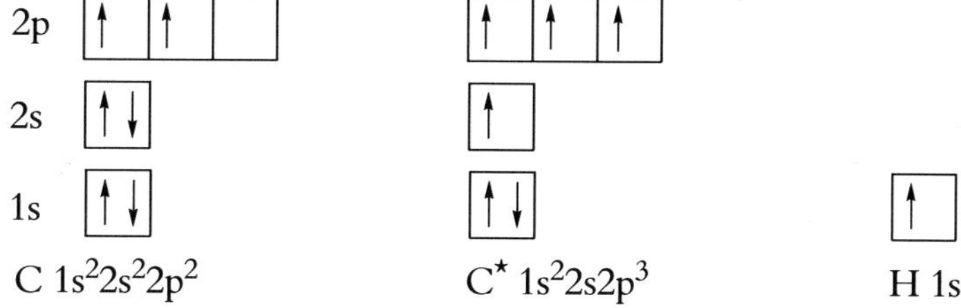

In carbon, an s electron is promoted to a p orbital

Each carbon atom, C, has two unpaired electrons, and one might expect carbon to form two bonds. It would not then attain a neon-like structure: it needs to share four electrons to do this. A sharing of four electrons can be achieved by promoting one of the 2s electrons into the 2p level. The excited carbon atom, C*, might be expected to form two different kinds of bond, using one s orbital and three p orbitals. Actually, the electron density distributes itself evenly through four bonding orbitals, which are called sp^3 hybrid orbitals (Figure 5.3A).

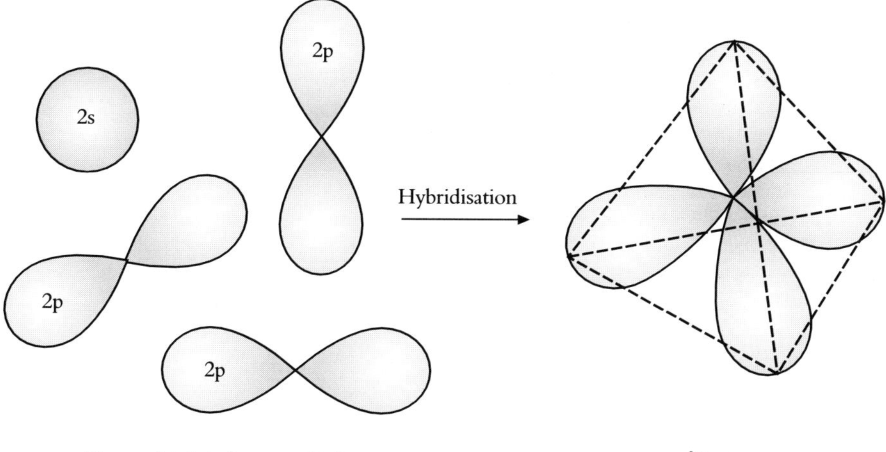

Figure 5.3A The sp^3 hybrid orbitals of carbon

One s orbital and three p orbitals form four sp^3 hybrid orbitals

The sp^3 atomic orbital is more concentrated in direction than a p orbital [see Figure 5.3B]. An sp^3 orbital is therefore able to overlap more extensively and form stronger bonds than a p orbital.

Figure 5.3B
Comparison of atomic orbitals

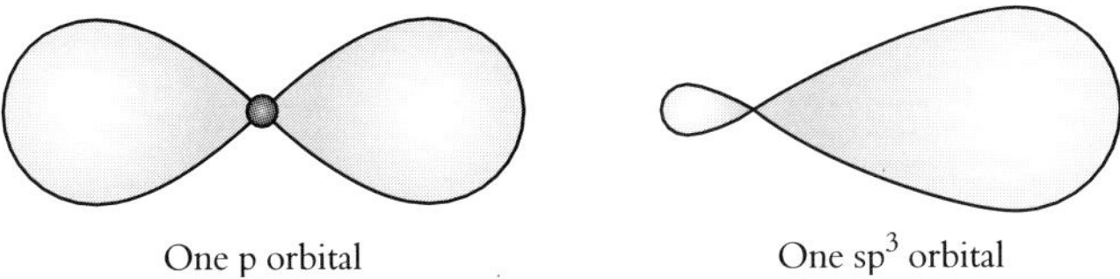

The overlapping of the four sp^3 orbitals of carbon and four s orbitals of hydrogen in methane is shown in Figure 5.3C. Experimental evidence for the tetrahedral arrangement is gained from X ray diffraction studies of diamond. The angle between the bonds is shown to be 109.5° [see Figure 6.6A, § 6.6].

Figure 5.3C
Overlapping of atomic orbitals in methane

In methane, four sp^3 hybrid orbitals of C overlap with four s orbitals of four H atoms

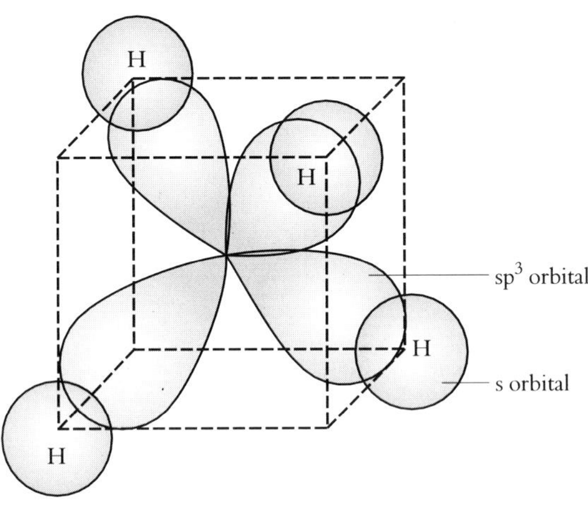

5.3.2 DOUBLE BONDS

A double bond is less than twice as strong as a single bond ...

... and a triple bond has less than three times the strength of a single bond

Carbon forms double bonds in compounds such as carbon dioxide, O═C═O, and ethene, $H_2C{=}CH_2$. The double bond is not simply two single bonds. The amount of energy required to break a certain bond in a mole of molecules is called the **standard bond enthalpy** [§ 10.11]. Standard bond enthalpies of carbon–carbon bonds are

C—C	346 kJ mol^{-1}
C═C	610 kJ mol^{-1}
C≡C	837 kJ mol^{-1}

The C═C bond is less than twice as strong as a C—C bond, and the C≡C bond is less than three times as strong as a C—C bond.

Two p orbitals and one s orbital can hybridise to form three coplanar sp^2 hybrid orbitals

In a molecule of ethene, each carbon atom uses a 2s orbital and two of the three 2p orbitals to form three sp^2 hybrid bonds [see Figure 5.3F]. The electronic configurations of carbon are shown below:

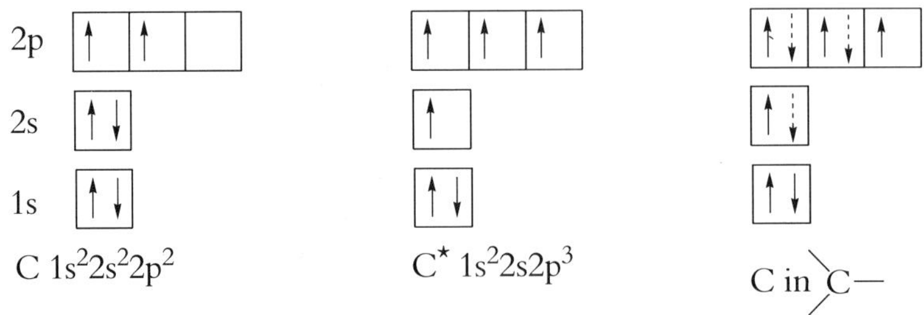

The three sp^2 hybrid orbitals are coplanar with an angle of 120° between them.

Figure 5.3D
The sp^2 hybrid orbitals of carbon

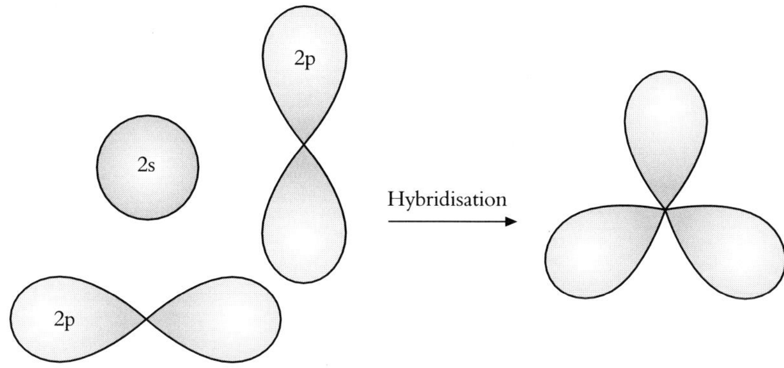

Each C has an unhybridised p orbital ...

The carbon-carbon bond formed when the sp^2 orbitals of neighbouring carbon atoms overlap is called a σ, **sigma**, bond. In σ bonds, e.g. any single bond, overlap of atomic orbitals occurs along the line joining the two bonded atoms. There is an unhybridised p orbital at right angles to the plane of the three sp^2 orbitals, and the p orbitals on adjacent carbon atoms are close enough to overlap. The overlapping occurs at the sides of the orbitals [see Figure 5.3E].

Figure 5.3E
The ethene molecule

Overlapping of atomic orbitals along the line joining the two bonded atoms is a σ (sigma) bond

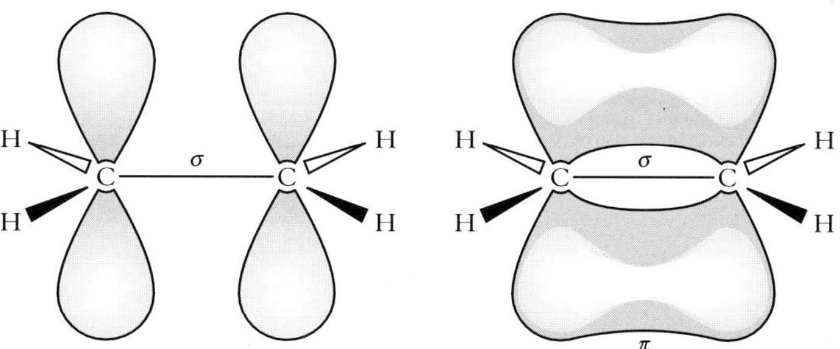

(a) The C atoms have unhybridised p orbitals.

(b) Sideways overlap between the two parallel p orbitals produces one π orbital.

... Sideways overlap between p orbitals is called a π (pi) bond ...

This type of bond, produced by sideways overlapping of p orbitals above and below the plane of the sp^2 bonds, is called a π, **pi**, bond. It is not as strong as a σ bond since there is less overlapping of orbitals [see Figure 5.3F]. This is why the C=C bond is less than twice as strong as a C—C bond.

... π bonds are less strong than σ bonds

Since overlapping of p orbitals on adjacent carbon atoms can occur only when the p orbitals are parallel, the two H_2C— structures must be coplanar, i.e., lie in the same plane. If one CH_2 group twists with respect to the other, the amount of overlapping of p orbitals will decrease and the π bond will be partially broken. Since it requires energy to break a bond, the most stable arrangement of the molecules is the one in which all six atoms lie in the same plane [see Figure 27.5B(a), § 27.5.3].

For π bonds to be formed the atoms in $H_2C{=}CH_2$ must be coplanar

Figure 5.3F
The difference between σ and π bonds

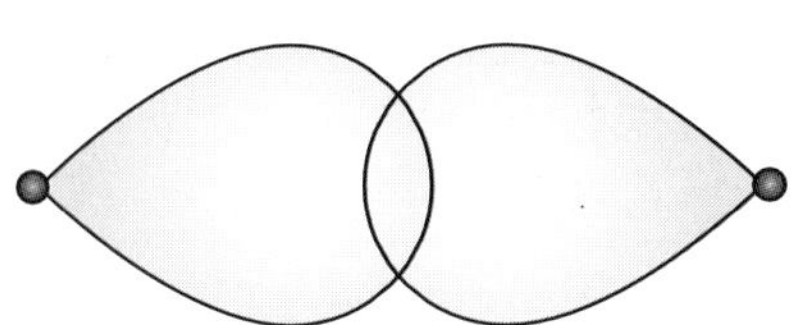

(a) σ bonding.
The orbitals point towards each other.

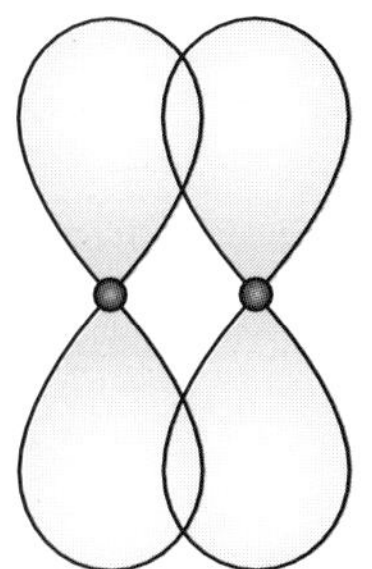

(b) π bonding.
The orbitals are parallel and overlap sideways.

π bond formation is restricted to small atoms

The formation of strong π bonds is restricted to members of the second period: carbon, nitrogen and oxygen. In larger atoms, strong π bonds are not formed because, being removed from the line between the centres of the atoms, the π bond becomes rapidly weaker as the size of the atom increases.

CHECKPOINT 5.3: BONDING

1. Deduce the shapes of the following species:
   AsH_3, PH_4^+, H_3O^+, CS_2, $CH_2{=}C{=}CH_2$, $HC{\equiv}N$.

2. What is the arrangement of bonds around the central atom in each of the following species:
   CH_4, BF_3, NF_3, ICl_4^-, BrO_3^-, ClO_4^-, $CHCl_3$?

3. Write structures which show the arrangement of electrons in the bonding orbitals of:
   O_2, CO_2, NO_3^- and CN^-
   (e.g., $\ddot{\underset{\cdot\cdot}{O}}::\ddot{\underset{\cdot\cdot}{O}}$)

4. What can you deduce from the fact that, whereas water has a dipole moment, carbon dioxide has none?

5. Write electron structures for: PH_3, NH_3, NH_4Cl, H_2O, H_2O_2, SiH_4, HOCl
   (e.g., $\begin{matrix} H:\ddot{\underset{\cdot\cdot}{P}}:H \\ H \end{matrix}$)

6. The ammonium ion and methane are said to be *isoelectronic*. What does this mean? Why do the compounds have different chemical properties?

7. *(a)* Sketch the arrangement of bonds in $CF_2{=}CF_2$.
   (b) Explain why there are two isomers with the formula $CFCl{=}CFCl$.

5.4 DELOCALISED ORBITALS

Localised electrons are to be found between the nuclei of two bonded atoms ...

In the compounds discussed so far, the electrons in the σ and π bonds have been located in the region *between* the nuclei of the bonded atoms. They are **localised** electrons. In some molecules, some of the electrons are **delocalised**: they do not remain between a pair of atoms.

5.4.1 BENZENE

The benzene molecule has sp[2] hybrid bonds between adjacent carbon atoms ...

... σ bonds ...

... and each carbon atom has an unhybridised p orbital perpendicular to the plane of the ring

Benzene, C_6H_6, is an aromatic hydrocarbon [§ 25.6]. Kèkulé [§ 28.2] proposed the structural formula for benzene which is shown here.

```
        H
        |
  H     C     H
    \ ⁄   ⫽ ⁄
     C     C
     ‖     |
     C     C
    ⁄ \   ⫽ \
  H     C     H
        |
        H
```

The p orbitals overlap sideways ...

... to form π bonds

An alternating system of single and double bonds is called a **conjugated double bond system**. Between each pair of adjacent carbon atoms is a σ bond, formed by overlapping of sp^2 hybrid orbitals. Since sp^2 bonds are coplanar, all the carbon atoms lie in the same plane and form a regular hexagon. The unhybridised p orbitals of the carbon atoms are perpendicular to the plane of the hexagonal benzene 'ring'.

Delocalised electrons do not remain between a pair of bonded atoms

As in ethene [see Figure 5.3G], overlapping of p orbitals on adjacent carbon atoms gives rise to π bonds. In benzene, the p orbitals are able to overlap all round the ring (see Figure 5.4A). The electrons in the p orbitals cannot be regarded as located between any two carbon atoms: they are free to move between all the carbon atoms in the ring. They are described as **delocalised** and are represented as an annular cloud of electron density above and below the plane of the molecule.

Electrons in the π bonds in benzene are delocalised

The formula of benzene is often written as shown in Figure 5.4A(c) to represent the delocalisation of π electrons.

Figure 5.4A
The benzene ring

Bonding electrons are usually localised: they remain between the nuclei of bonded atoms

Some molecules have electrons which are delocalised

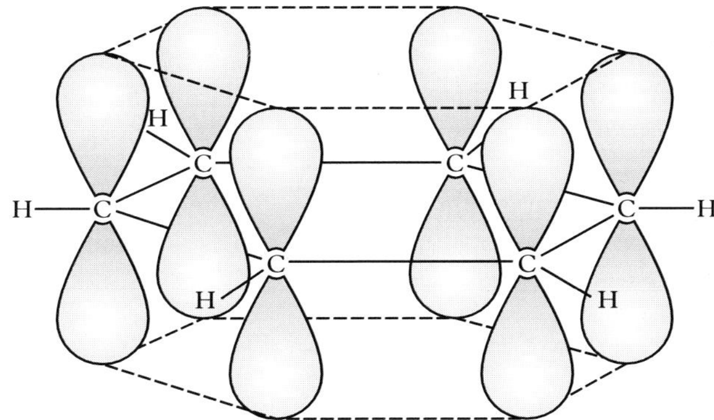

(a) Each of the C atoms has an unhybridised p orbital. These overlap sideways.

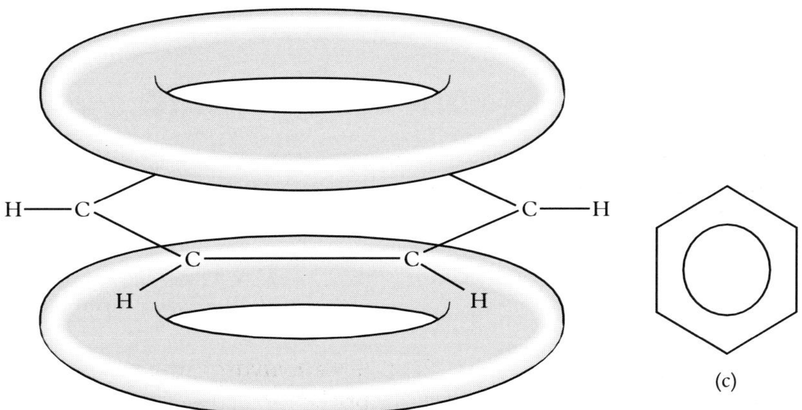

(c)

(b) The resulting distribution of π electron charge can be represented by two doughnut-shaped regions, one above and one below the hexagon of carbon atoms.

Figure 5.4B
The bonding orbitals in ethene and benzene

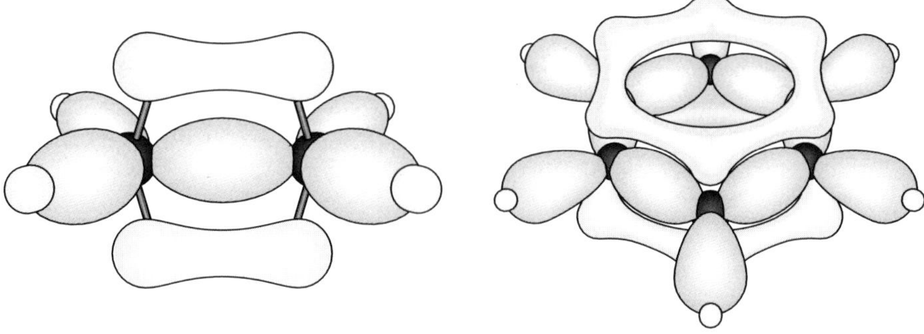

Is the standard enthalpy of benzene equal to the sum of the standard bond enthalpies?

The delocalisation of π electrons confers stability on benzene. One can calculate the standard enthalpy content of benzene on the basis of the structure

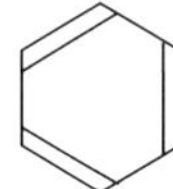

by adding the average standard bond enthalpies for the bonds [§ 5.3.4 and § 10.11]. The sum of 3(C—C) bonds plus 3(C═C) bonds plus 6(C—H) bonds is -5350 kJ mol^{-1}. The standard enthalpy content can also be found from measurement of the standard enthalpy of combustion [§§ 10.7.1, 10.8.2].

No: delocalisation of electrons confers stability on benzene

The experimental value of -5550 kJ mol^{-1} is more negative than the theoretical value. This means that benzene is *more* stable than one would expect it to be on the basis of the formula

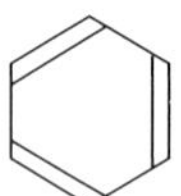

The *difference*, -200 kJ mol^{-1} is called the **delocalisation energy** of benzene.

CHECKPOINT 5.4: HYBRID BONDS

1. How is it known that the four C—H bonds in methane are equivalent?

*__2.__ What is meant by the terms: *(a)* localised molecular orbital and *(b)* delocalised molecular orbital?

*__3.__ Describe the formation of the second bond between the two carbon atoms in ethene. Explain why the ethene molecule, $H_2C{=}CH_2$, is planar.

†__4.__ Why was the Kekulé structure for benzene adopted? Why was it superseded? [Refer to Chapter 28.]

THE SHAPES OF MOLECULES

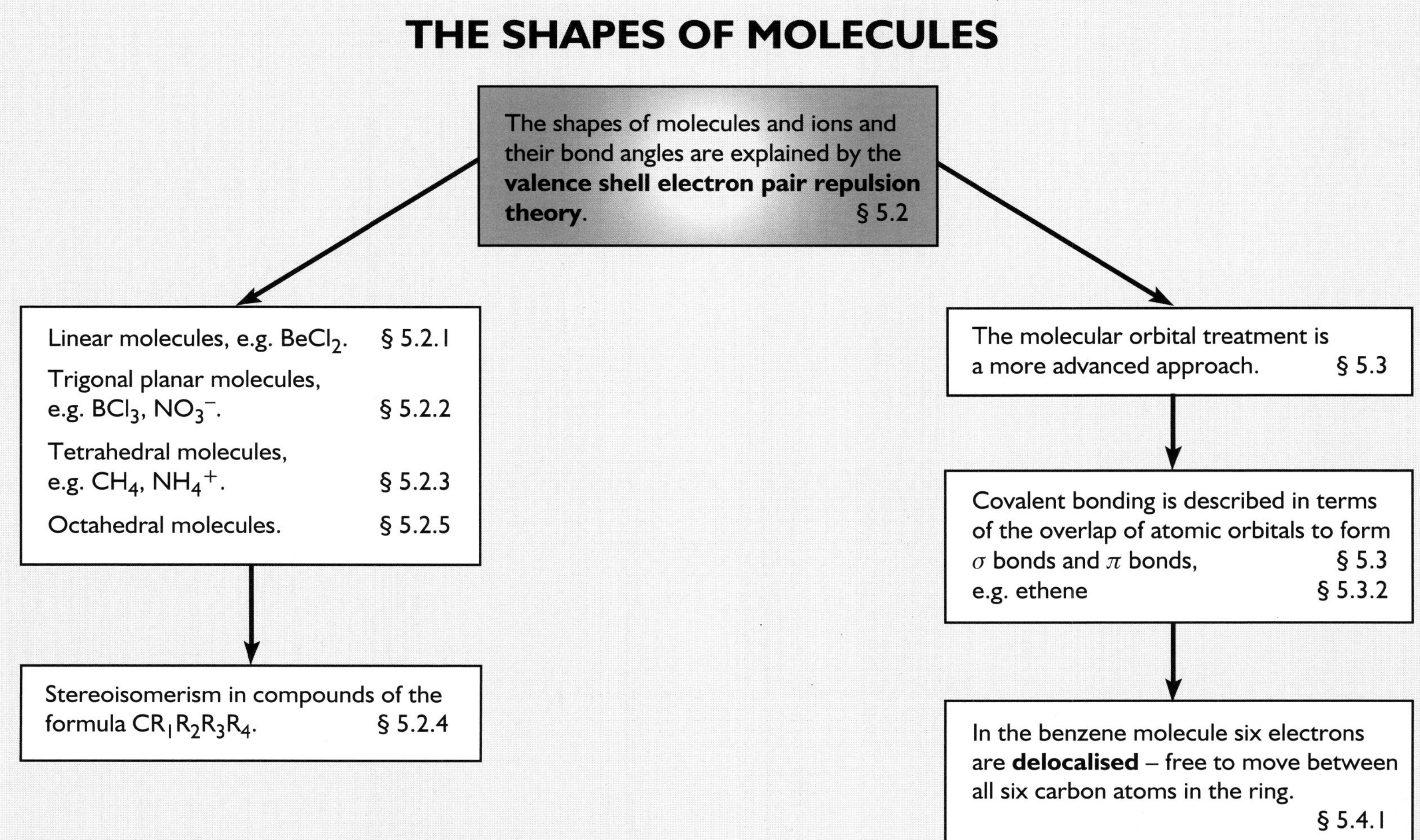
The shapes of molecules and ions and their bond angles are explained by the **valence shell electron pair repulsion theory**. § 5.2
Linear molecules, e.g. $BeCl_2$. § 5.2.1
Trigonal planar molecules, e.g. BCl_3, NO_3^-. § 5.2.2
Tetrahedral molecules, e.g. CH_4, NH_4^+. § 5.2.3
Octahedral molecules. § 5.2.5
Stereoisomerism in compounds of the formula $CR_1R_2R_3R_4$. § 5.2.4
The molecular orbital treatment is a more advanced approach. § 5.3
Covalent bonding is described in terms of the overlap of atomic orbitals to form σ bonds and π bonds, § 5.3
e.g. ethene § 5.3.2
In the benzene molecule six electrons are **delocalised** – free to move between all six carbon atoms in the ring. § 5.4.1

QUESTIONS ON CHAPTER 5

1. State the valence shell electron pair repulsion theory. Sketch the arrangement of atoms in *(a)* $BeCl_2$, *(b)* BCl_3 *(c)* CCl_4, *(d)* NH_3, *(e)* H_2O and *(f)* IF_5. Explain how the arrangement of bonds is predicted on the VSEPR theory.

2. Give the formula of a molecule whose atoms occupy each of the following shapes: *(a)* linear, *(b)* planar trigonal, *(c)* tetrahedral and *(d)* octahedral. State the angle between the bonds in each structure.

Some molecules which are based on a tetrahedral structure have bond angles different from the regular tetrahedral angle. Give an example, and explain the difference.

3. 'The shapes of simple molecules can be deduced from a consideration of the bonding electrons employed'. Discuss this statement. Apply the principle to the shapes of *(a)* NH_3, *(b)* H_2O, *(c)* $H_2C{=}CH_2$, *(d)* I_3^- and *(e)* SF_6.

4. *(a)* What are the main features of the electron pair repulsion model for accounting for the shapes of molecules?

(b) By considering the numbers of lone and bonding pairs of electrons, predict the general shapes of the following molecules or ions: F_2O, H_3O^+, ClF_4^-.

5. Explain the following statements:

(a) Two different compounds have the molecular formula

$$\begin{array}{ccc} H & & OH \\ & C & \\ CH_3 & & Cl \end{array}$$

(b) The spatial arrangement of atoms in BF_3 is different from that in NH_3.

(c) XeF_4 is planar, whereas CCl_4 is tetrahedral.

(d) The bond angle in NH_3 is greater than that in H_2O.

6. *(a)* (i) State the shapes of the following molecules. Using the VSEPR (Valence-Shell Electron Pair Repulsion) principle explain how the shapes of these simple molecules arise, (diagrams are not required): BF_3, CH_4, SF_6. **3**

(ii) The bond angle $\widehat{HNH}$ in NH_4^+ is 109°. The bond angle in $\widehat{HNH}$ in NH_3 is 107°. The bond angle $\widehat{HOH}$ in H_2O is 104.5°.

Explain these differences in bond angles in terms of repulsion between the different types of electron pairs. **3**

(iii) Use the VSEPR principle to predict the shapes of the following molecules: $BeCl_2$, BrF_5. **2**

(b) The figure shows a graph of melting temperature against relative molecular mass for the hydrides shown in line **A** and line **B**.

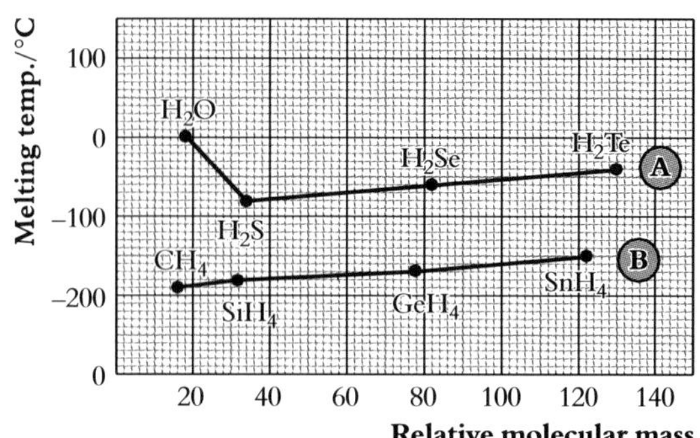

(i) Explain the trend in the melting temperatures of the hydrides represented by line **B**. **1**

(ii) Explain why the first hydride in the group represented by line **A** has a higher melting temperature than the others. **1**

(c) Bonding between peptide chains is very important biologically. The structure of a peptide chain is shown below.

$$H_3\overset{+}{N}{-}\underset{\substack{|\\CH_2SH}}{CH}{-}\overset{\substack{O\\||}}{C}{-}\overset{\substack{H\\|}}{N}{-}\underset{\substack{|\\(CH_2)_4\overset{+}{N}H_3}}{CH}{-}\overset{\substack{O\\||}}{C}{-}\overset{\substack{H\\|}}{N}{-}\underset{\substack{|\\CH_2CO_2^-}}{CH}{-}\overset{\substack{O\\||}}{C}{-}O^-$$

Give **two** features of this peptide chain which can result in bonding **between** chains. **1**

11 *WJEC (AS/AL)*

7. *(a)* In a copy of the table, give the formulae of the chlorides of the elements of Period 3, other than silicon. **2**

Element	Na	Mg	Al	Si	P	S
Formula of chloride				$SiCl_4$		

(b) Calculate the percentage by mass of silicon in silicon tetrachloride. **2**

(c) (i) Draw a dot and cross diagram to show the bonding in silicon tetrachloride.

(ii) Draw the shape of this molecule. Explain your answer in terms of the Electron Pair Repulsion Theory.

(iii) State the shape of a molecule of $AlCl_3$ and explain why it is different from that of $SiCl_4$. **6**

(d) (i) Give an equation for the reaction of $SiCl_4$ with cold water.

(ii) How does the behaviour of carbon tetrachloride with cold water compare with this? Explain any differences. **4**

14 *L (AS/AL)*

8. *(a)* Copy and complete the table by writing the numbers of protons, neutrons and electrons for each of the species shown. **2**

Species	Number of protons	Number of neutrons	Number of electrons
$^{20}_{10}Ne$			
$^{23}_{11}Na^{+}$			
$^{32}_{16}S^{2-}$			
an α particle			

(b) The figure shows a plot of first ionisation energy against atomic number for the elements of atomic number 10 to 18. (The letters are not the chemical symbols for the elements concerned.)

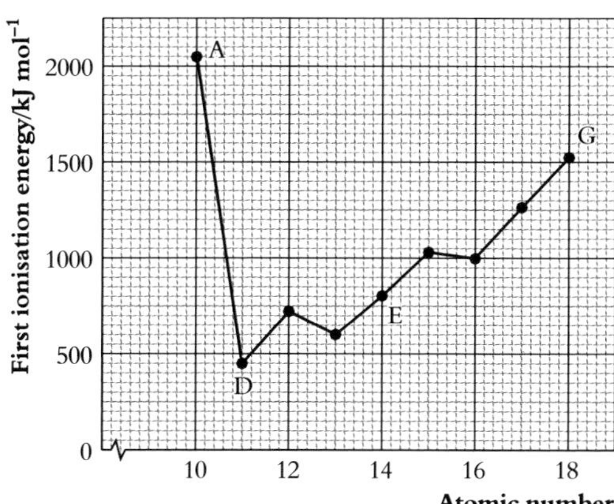

(i) I. Explain **briefly** why the ionisation energy of the element labelled **G** is less than that of the element labelled **A**.
II. Explain **briefly** why the ionisation energy of the element labelled **D** is considerably less than that of the element labelled **A**. **1**

(ii) Draw the shape of the orbital from which the electron is lost when,
I. element **A** is ionised,
II. element **D** is ionised. **1**

(iii) Give the electronic configuration of the element labelled **E** on the figure by copying and labelling **each** of the following boxes and inserting arrows, in appropriate boxes, to represent electrons.

Orbital

....

□ □ □□□ □ □□□ □□□□□ □

1½

(iv) **State** and **explain** which of the elements **D** or **E** will form a cation more readily. **1**

(c) State whether each of the following statements, concerning the covalent bond is true or false. (More than one statement may be correct.)
(i) The covalent bond has a pair of electrons with opposed spins.
(ii) Atoms which bond together covalently **usually** achieve an inert gas structure. **1**

(d) (i) Arrange in order of **increasing** repulsion (smallest first) the repulsion between: two bonding pairs of electrons, two lone pairs of electrons, and a lone pair and a bonding pair of electrons, arranged around a central atom. **½**
(ii) Explain how this sequence arises. **1**

(e) (i) State what is meant by,
I. a *dipole*, and II. a *polar* bond. **1**
(ii) I. The C — H bonds within the methane molecule are strong.
Liquid methane boils at a very low temperature.

Explain why there is no conflict between these two statements.

II. **Explain** why ammonia (NH_3) boils at −33 °C, whereas the heavier phosphine molecule (PH_3) boils at −86 °C. **2**

12 *WJEC (AS/AL)*

9. *(a)* What is a polar covalent bond? **1**
(b) In what circumstances will a covalent bond be polar? **1**
(c) In what circumstances will an anion be polarised? **1**
(d) How does a polarised anion differ from an unpolarised anion? **1**
(e) (i) Draw diagrams to show the shapes of the following molecules and in each case show the value of the bond angle on the diagram:

$BeCl_2$, NCl_3, SF_6.

(ii) State which of the above molecules is most likely to form a co-ordinate bond with a hydrogen ion. Give a reason for your answer. **8**

12 *NEAB (AS/AL)*

6

CHEMICAL BONDING AND THE STRUCTURE OF SOLIDS

6.1 PROFILE: DOROTHY CROWFOOT HODGKIN (1910–1995)

X ray crystallography has unravelled the molecular structures of three life-saving compounds: penicillin, vitamin B_{12} and insulin. This feat was accomplished by Dorothy Crowfoot Hodgkin. Dorothy Crowfoot was the daughter of archaeologists who worked in Egypt and the Sudan. Dorothy studied at the Universities of Oxford and Cambridge before becoming a fellow of Somerville College, Oxford. She decided to continue the work in X ray crystallography which she had started in Cambridge. Early in her scientific career, she married Thomas Hodgkin who gave his wife unfailing support and encouragement in all the challenges she undertook.

One of her first problems was penicillin. Although penicillin had been discovered by Sir Alexander Fleming in 1929, it was not isolated as a pure compound until 1941. At that time, Howard Florey and Ernst Chain were working to develop a method of culturing penicillin so that it could be used as a bactericide to treat casualties in the Second World War. They gave Dorothy Hodgkin 10 mg of the substance. By 1945 she had worked out its exact structure, allocating a position in space to each of the 41 atoms. The technique she used was X ray diffraction.

The next problem was even more taxing. In 1948 Glaxo laboratories gave Dorothy Hodgkin some beautiful red crystals of a substance which was effective against pernicious anaemia. The discoverers of this compound called it vitamin B_{12}. At first sight, Dorothy estimated that, with over 90 atoms, the structure of vitamin B_{12} would take many years to work out. However, she was able to use an advanced computer in the USA to perform the calculations, and the structure was completed in 1956. As a result of this work, vitamin B_{12} was synthesised.

The structure of insulin, with over 800 atoms, was a formidable task. With the new computing power at her disposal, Dorothy Hodgkin completed the work in 1969. In 1964 she received the Nobel Prize for Chemistry. Two other women have won this prize: Marie Curie in 1911 and her daughter Irene Joliot-Curie in 1935.

6.2 X RAY DIFFRACTION

A crystal is a regular three-dimensional arrangement of particles

A solid in which the arrangement of atoms or ions or molecules follows a regular three-dimensional design has a crystalline form. The surfaces of the crystal are planes, called **faces**. They intersect at angles that are characteristic for the substance. The ordered arrangement of particles in the crystal is called the **crystal structure**.

A crystal acts as a three-dimensional diffraction grating for X rays

When a beam of X rays meets a crystal, X rays interact with electrons, and the beam is scattered. The scattered X rays must be made to produce a visible pattern, e.g., on a photographic film. From the pattern of scattering [see Figure 6.2A] crystallographers can work out the structure of the crystal.

Figure 6.2A shows a developed X ray film. The central spot has been produced by undeflected X rays, and the circles of spots are produced by X rays which have been diffracted through various angles by the planes of atoms or ions in the crystal.

Figure 6.2A An X ray diffraction pattern

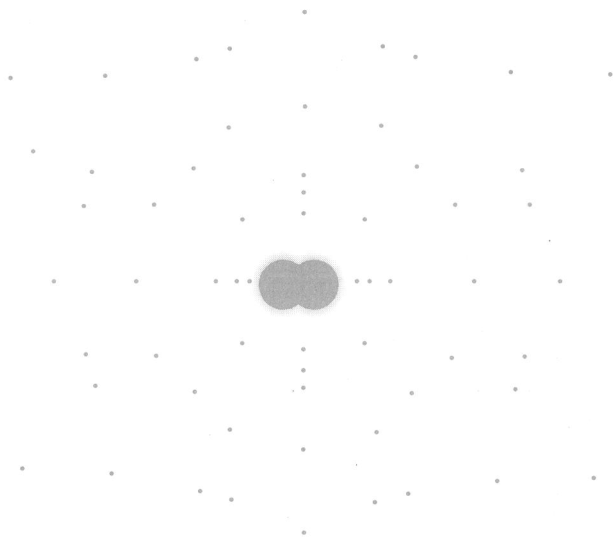

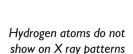

Hydrogen atoms do not show on X ray patterns

Since it is the *electrons* of atoms that scatter X rays, the small atoms in a compound, which have few electrons, especially hydrogen atoms, are difficult to detect. The structures of metals, ionic compounds and macromolecular substances (such as diamond and graphite) have been worked out from X ray diffraction measurements.

6.3 THE METALLIC BOND

The nature of the metallic bond must be responsible for the properties of metals

Modern technology is based on the use of metals. Most of our machines and most of our forms of transport are made of metal. There must be some feature of the bond between metal atoms that gives metals their special properties. Many metals are strong and can be deformed without breaking; many are **malleable** (can be hammered) and **ductile** (can be drawn out under tension). They are shiny when freshly cut and good conductors of heat and electricity. Any theory of the metallic bond must account for all these physical properties.

In a metal ...

... atomic orbitals overlap to form molecular orbitals ...

... the valence electrons become delocalised ...

... metal cations are formed ...

... these are attracted to the electron cloud

The outer shell electrons of a metal (the valence electrons) are relatively easily removed, with the formation of metal cations. When two metal atoms approach closely, as in a metal structure, their outer shell orbitals overlap to form molecular orbitals. If a third atom approaches, its atomic orbitals can overlap with those of the first two atoms to form another molecular orbital. For a large number of atoms, a larger number of molecular orbitals are formed, extending over three dimensions. As a consequence of the multiple overlapping of atomic orbitals, the outer electrons from each atom come under the influence of a very large number of atoms. They are free to move through the structure and are no longer located in the outer shell of any one atom: they are **delocalised**. The removal of the electrons leaves behind metal cations. The reason why the cations are not pushed apart by the repulsion between them is that, in a pair of cations, each cation is attracted to the delocalised electron cloud between them [see Figure 6.3A].

Figure 6.3A
Deformation of metal structure

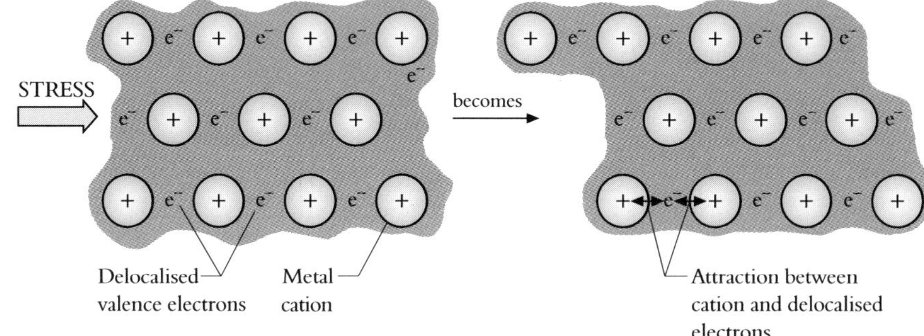

The metallic bond explains the strength of metals ...

This theory of the metallic bond explains the physical properties of metals. The metallic bond is not directed in space. If a stress is applied to the metal, the structure can change in shape without fracturing [see Figure 6.3A]. This contrasts with the effect of stress on an ionic structure [see Figure 6.4C, § 6.4].

... also their thermal conductivity ...

The high thermal conductivity of metals is accounted for. When heat is supplied to one end of a piece of metal, the kinetic energy of the electrons is increased. The increase is transmitted through the system of delocalised electrons to other parts of the metal.

... and electrical conductivity ...

Electrical conductivity can also be explained. If a potential difference is applied between the ends of a metal, the delocalised electron cloud will flow towards the positive potential.

... and the shiny appearance of metals

The shiny appearance of metals fits in which the theory of the nature of the metallic bond. The metal contains a large number of molecular orbitals at a large number of different energy levels. When light falls on to the metal, electrons are excited. A large number of transitions between energy levels is possible, with a whole range of frequencies being absorbed. As electrons return to lower energy levels, light is emitted and makes the metal shine.

6.4 IONIC STRUCTURES

Ionic structures are regular three-dimensional arrangements of ions

The alkali metal halides are ionic compounds. The ions are arranged in a regular three-dimensional structure. The melting and boiling temperatures of ionic compounds are high, owing to the strong forces of electrostatic attraction between the ions in the crystal. When the salts are melted or dissolved, the ions become free to move, and the salts conduct electricity. The crystals of alkali metal halides are *cubic* in shape, and X ray analysis shows two kinds of structures. The sodium chloride structure is illustrated in Figure 6.4A and that of caesium chloride in Figure 6.4B.

Figure 6.4A
The sodium chloride structure

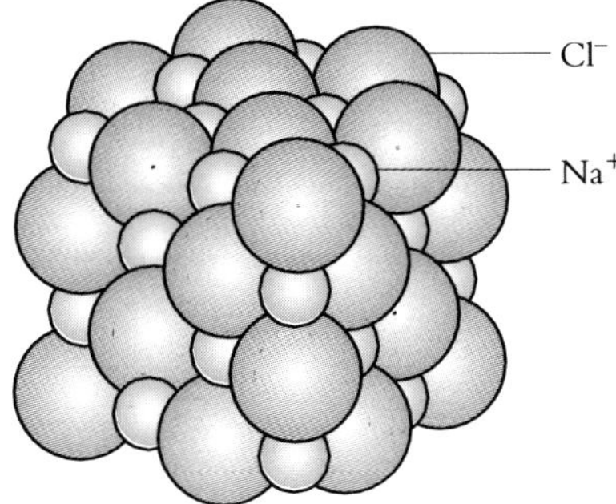

Figure 6.4B
The caesium chloride structure

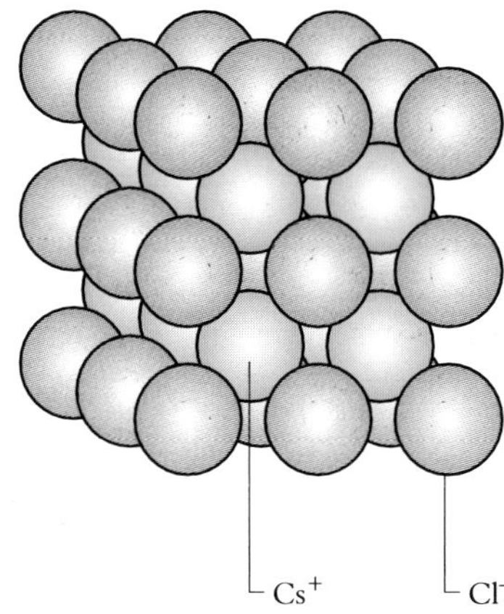

The ratio: (radius of cation/radius of anion) decides the type of crystal structure

The number of anions that can surround one cation is called the ***coordination number*** *of the cation*

Sodium chloride shows 6 : 6 coordination

Caesium chloride shows 8 : 8 coordination

Unlike metals, ionic crystals are brittle

Why do different salts adopt different crystalline structures? The best arrangement of ions in a structure is the one with the lowest energy. This is the structure that allows the greatest number of contacts between oppositely charged ions without pushing together ions with the same charge. The sodium ion has a radius of 0.181 nm. It is large enough to form contacts with six chloride ions: the coordination number of sodium is 6. Six sodium ions surround each chloride ion: the coordination number of chlorine is 6. The structure shows 6 : 6 coordination. Since the caesium ion, with radius 0.168 nm, is larger than the sodium ion, a larger number of chloride ions can surround it. The structure shows 8 : 8 coordination.

Ionic crystals are brittle. Figure 6.4C shows what happens when an ionic crystal is subjected to stress. A slight dislocation in the crystal structure brings similarly charged ions together. Repulsion between the like charges fractures the crystal. This is a different picture from the effect of stress on a metallic crystal, where deformation of the structure does not result in fracture [see Figure 6.3A].

Figure 6.4C
An ionic structure is easily fractured

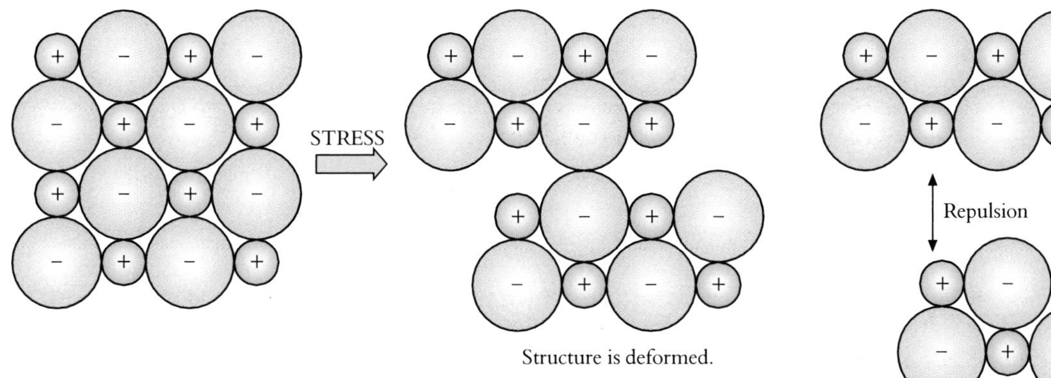

6.5 MOLECULAR SOLIDS

Van der Waals forces hold the molecules together in solid argon and in solid iodine ...

Some solids are held together by weak attractions between individual molecules. They are described as **molecular solids** and said to have a **molecular structure**. At very low temperatures, even the noble gases can be solidified. Figure 6.5A shows the structure of atoms in solid argon. The van der Waals forces between the atoms are very weak, and if the temperature rises above −170 °C the solid melts. Liquid argon consists of separate atoms. Figure 6.5B shows the structure of solid iodine.

Figure 6.5A Solid argon

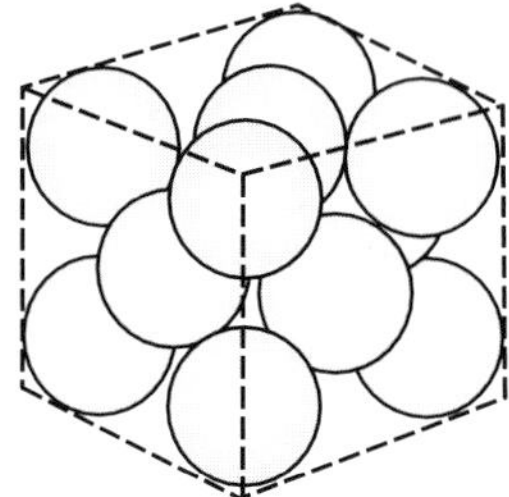

Figure 6.5B Solid iodine

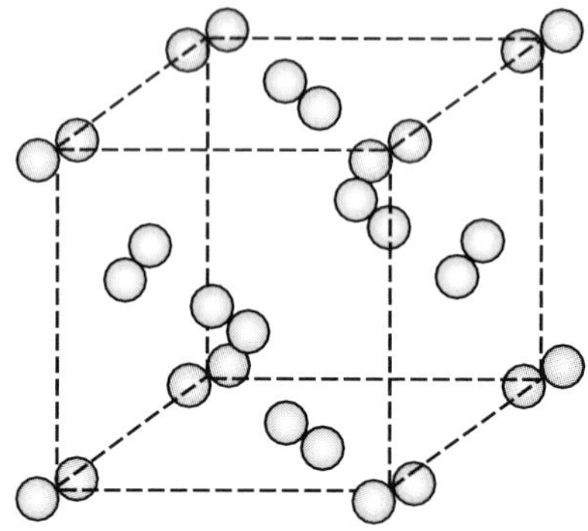

The I_2 molecules in the centre of each face are shown in blue.

solids with molecular structure are volatile ...

... e.g. iodine ...

Iodine is a molecular solid up to a temperature of 30 °C. The atoms are covalently bonded in pairs as I_2 molecules. Operating between the molecules are the much weaker van der Waals forces. As a result of the regular arrangement of molecules, iodine is a crystalline solid with regular faces, which give a shiny appearance. When solid iodine is heated, the van der Waals forces are broken and individual molecules are set free. The vapour phase, which is purple, consists of individual I_2 molecules. Bromine and chlorine adopt similar structures at lower temperatures.

... e.g. solid CO_2

Carbon dioxide is well known in its solid form as 'dry ice' or 'drikold'. Above −78 °C is sublimes, absorbing heat from its surroundings to do so. From this property arises the widespread use of solid carbon dioxide as a refrigerant, both in laboratory work and in the food industry. Pop singers sometimes like to enhance their performance by having lumps of dry ice on stage. As it sublimes, it cools the moist air, and swirling clouds of water droplets form. Solid carbon dioxide has a structure resembling that of iodine, which is shown in Figure 6.5B.

6.6 MACROMOLECULAR STRUCTURES

A macromolecular structure is held together by covalent bonds, e.g. diamond ...

A number of solids have the kind of structure described as **macromolecular** or **giant molecular**. Covalent bonds between atoms bind all the atoms into a giant molecule. Diamond, an **allotrope** [§ 21.4] of carbon, is one of the hardest substances known and has a macromolecular structure [see Figure 6.6A]. Each carbon atom forms four bonds (sp^3 hybrid bonds) to four other carbon atoms. The giant molecular structure which results is very strong. It is different from a molecular structure, in which, although the bonds between the atoms in a molecule are strong covalent bonds, the intermolecular forces of attraction are weak. Diamond remains a solid up to a temperature of 3500 °C, at which it sublimes.

Figure 6.6A
The structure of diamond

Such substances are hard and involatile, e.g. diamond and silica

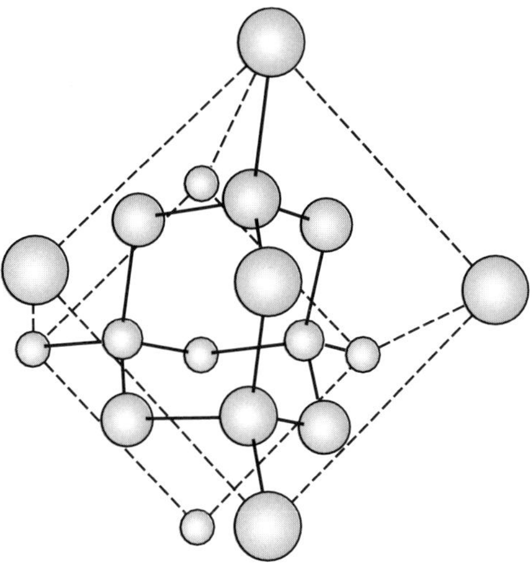

Each C atom is surrounded by 4 others: the coordination number is 4.

... in which the strong covalent bonds result in a hard, abrasive character

The hard, abrasive character of diamond finds it many uses. Diamond-tipped tools are used for cutting and engraving, and diamond-tipped tools are used by oil prospectors for boring through rock [see Figure 6.6B]. The high **refractive index*** of diamond gives it the sparkle that makes it the most prized of jewels [see Figure 6.6C].

Figure 6.6B
Diamond-studded drill bits

Figure 6.6C
A selection of diamonds

Other macromolecular structures are SiC, BN and SiO_2

Other solids with a diamond-like structure are silicon carbide $(SiC)_n$, and boron nitride $(BN)_n$. The formula unit, BN, is isoelectronic with the unit CC. Silicon(IV) oxide, SiO_2 (**silica**), also forms a three-dimensional structure. The Si—O bonds about each silicon atom are tetrahedrally distributed and each oxygen atom is bonded to two other silicon atoms [see Figure 6.6D]. The structure occurs in **quartz** and other crystalline forms of silica. Quartz remains solid up to a temperature of 1700 °C.

* See R Muncaster, *A-Level Physics* (Stanley Thornes).

Figure 6.6D
Silicon(IV) oxide structure

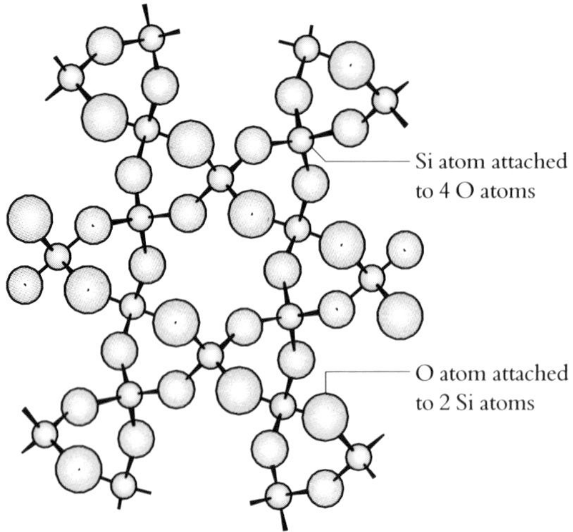

6.7 LAYER STRUCTURES

Layer structures have covalent bonds within each layer and weak van der Waals forces between layers e.g. graphite

Graphite, the other allotrope of carbon, has a **layered** structure. Within each layer, every carbon atom uses three coplanar sp^2 hybrid orbitals to bond to three other carbon atoms. A network of coplanar hexagons is formed, with a C—C bond distance of 0.142 nm. Between layers, the distance is 0.335 nm. The weak van der Waals forces of attraction between the layers allow one layer of bonded atoms to slide over another layer. The structure, which is shown in Figure 6.7A, accounts for the properties of graphite. It is a lubricant, whereas diamond is abrasive. The unhybridised p electrons form a delocalised cloud of electrons similar to the metallic bond. They enable graphite to conduct electricity and are responsible for its shiny appearance.

Figure 6.7A
The structure of graphite

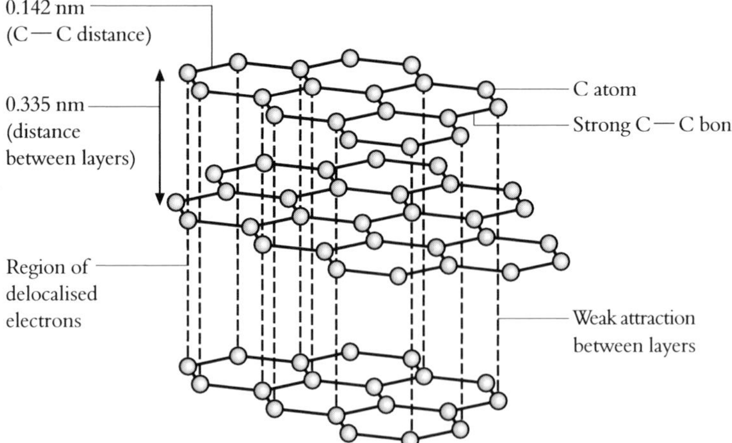

CHECKPOINT 6.7: COVALENT STRUCTURES

1. *(a)* Why is it easy to rub away carbon atoms from graphite?
   (b) Why is graphite used as a lubricant?
   (c) Why is it impossible to rub away carbon atoms from a diamond?
   (d) What characteristics of (i) diamond and (ii) graphite are useful in industry?

2. Solid iodine consists of shiny black crystals. Iodine vapour is purple. What is the difference in chemical bonding between solid and gaseous iodine?

3. 'Dry ice' is a name given to solid carbon dioxide.
   (a) In what way does solid carbon dioxide resemble ice?
   (b) In what way is solid carbon dioxide (i) better than ice as a refrigerant and (ii) more convenient than ice?
   (c) Pop singers sometimes use dry ice to produce swirling clouds on stage. These clouds are clouds of condensed water vapour. Explain (i) where the water vapour comes from and (ii) why it is cool enough on stage to condense water vapour.

4. From Figures 6.6A, § 6.6, and 6.7A, § 6.7, which would you expect to have the higher density, diamond or graphite? Why is diamond used in cutting tools but not graphite?

5. What is meant by the statement that the electrons in diamond are *localised*, whereas graphite has *delocalised* electrons. Which electrons in graphite are delocalised? How do they affect the properties of graphite?

6. Sketch the structure of silicon carbide, SiC, which resembles diamond. Why do you think that carborundum (SiC) is used as an abrasive? Why does silicon carbide not exist in a graphite-type of structure?

7. Boron nitride, BN, has a structure like that of graphite. Explain the bonding in BN.

FURTHER STUDY ▼

*6.8 THE STRUCTURE OF METALS

Types of metal structures

When identical spheres pack together so as to minimise the space between them, a close-packed structure is formed ...

Metal atoms pack closely together in a regular structure. There is no way of packing spheres to fill a space completely without leaving gaps between them. Arrangements in which the gaps are kept to a minimum are called **close-packed** arrangements. X ray studies have revealed three main types of metallic structures. In the **hexagonal close-packed structure**, and the **face-centred-cubic close-packed structure**, the metal atoms pack to occupy 74% of the space. In the **body-centred-cubic structure**, the atoms occupy 68% of the total volume.

Figure 6.8A shows a face-centred-cubic close-packed structure. Since every atom is in contact with 12 others (6 in the same layer, 3 in the layer above and 3 in the layer below) it is said to have a **coordination number** of 12. The high coordination numbers in these structures arise from the non-directed nature of the metallic bond.

A unit cell

Also shown in Figure 6.8A is the **unit cell**. A unit cell is the smallest part of the crystal that contains all the characteristics of the structure. The whole structure can be generated by repeating the unit cell in three directions.

A body-centred-cubic structure is less close-packed

The less closely packed body-centred-cubic structure is shown in Figure 6.8B(a). With one atom at each of the eight corners of a cube and one in the centre touching these eight, the coordination number is 8. Figure 6.8B(b) shows an expanded view, and (c) shows the unit cell with tie-lines to show that the coordination number is 8.

Figure 6.8A
Face-centred-cubic close-packed structures

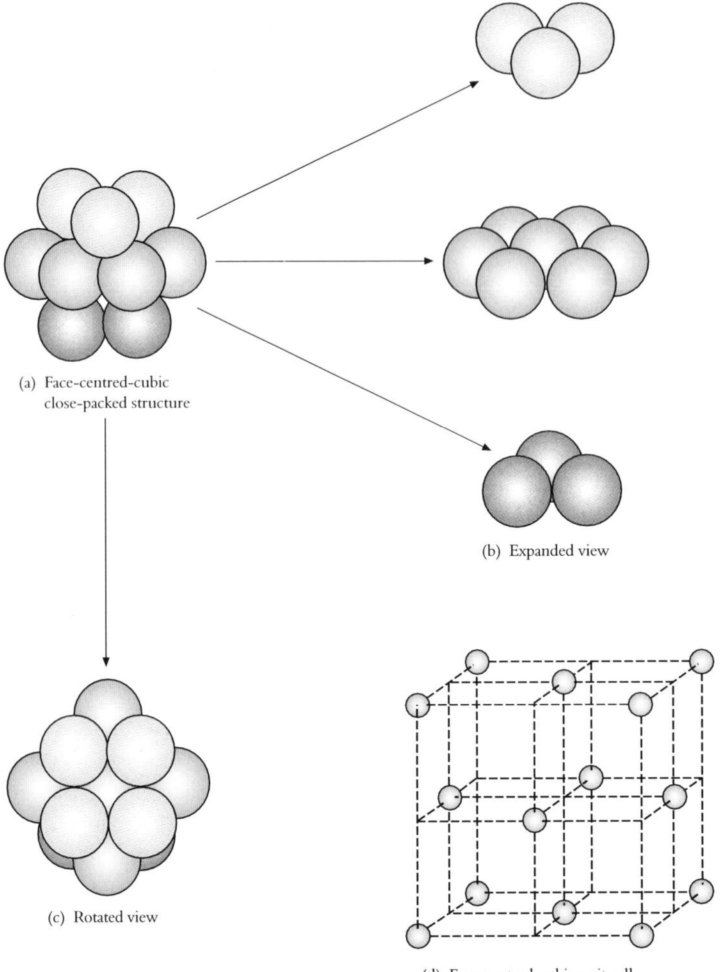

Figure 6.8B
Body-centred-cubic structure

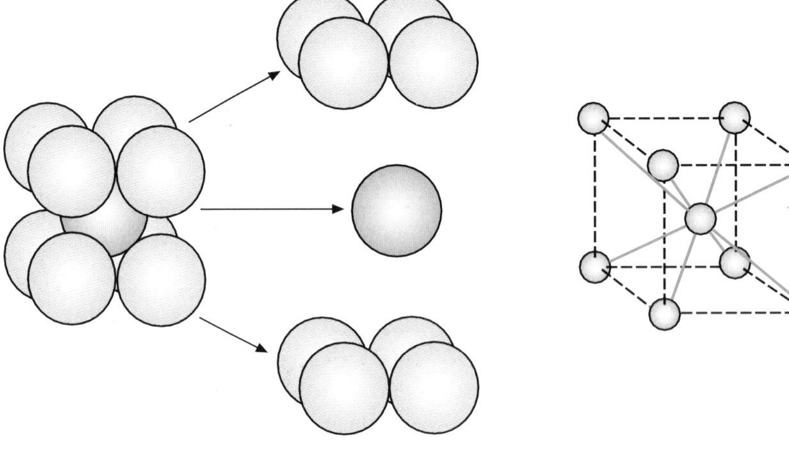

(a) Body-centred-cubic structure

(b) Expanded view

(c) Unit cell with tie-lines

SUMMARY

Metals are crystalline. The atoms pack together in close-packed arrangements: the hexagonal and face-centred structures and in the body-centred cubic structure which is less close-packed. The coordination number is the number of atoms with which each atom is in contact.

CHECKPOINT 6.8: METALLIC STRUCTURES

1. Why might someone describe an entire piece of metal as a large molecule?
2. How does the nature of the metallic bond account for the properties of metals?
3. Explain what is meant by *(a)* a close-packed structure, *(b)* a coordination number of 8.

FURTHER STUDY ▲

6.9 THE STRUCTURE OF IONIC CRYSTALS

In § 6.4, we studied the crystalline structure of salts such as sodium chloride [Figure 6.4A] and caesium chloride [Figure 6.4B]. In this section we look at the reason why different salts adopt different crystalline structures.

The number of anions that can surround one cation is called the ***coordination number*** *of the cation. It depends on the ratio: (radius of cation/radius of anion). The type of crystal structure adopted by an ionic compound depends on the coordination numbers of the cation and anion. Sodium chloride shows 6 : 6 coordination. Caesium chloride shows 8 : 8 coordination*

The best arrangement of ions in a structure is the one with the lowest energy. This is the structure that allows the greatest number of contacts between oppositely charged ions without pushing together ions with the same charge.

The sodium ion has a radius of 0.181 nm. It can be surrounded by six chloride ions. Figure 6.9A shows the unit cell of the sodium chloride structure. The unit cell is the smallest part of the crystal that shows all the features of the crystal structure. The whole crystal can be formed by multiplying the unit cell many times in each dimension. You can see that six chloride ions surround each sodium ion: the coordination number of sodium is 6. Six sodium ions surround each chloride ion: the coordination number of chlorine is 6. The structure shows 6 : 6 coördination.

The caesium chloride structure [Figures 6.4B and 6.9B] is different. Since the caesium ion, with radius 0.168 nm, is larger than the sodium ion, a larger number of chloride ions can surround it. The structure shows 8 : 8 coordination.

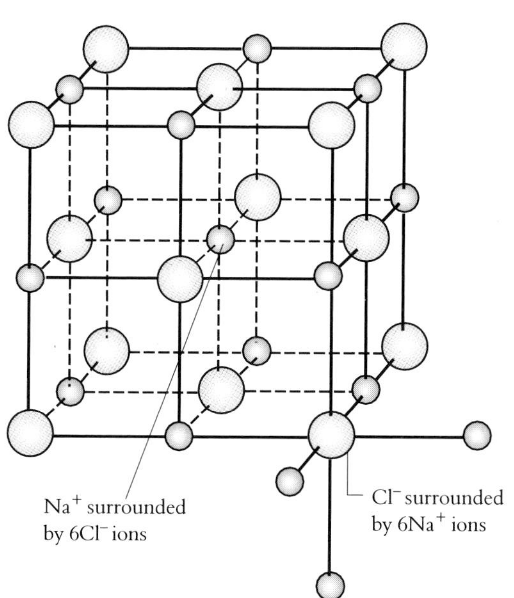
Na^+ surrounded by $6Cl^-$ ions
Cl^- surrounded by $6Na^+$ ions

Figure 6.9A Sodium chloride structure: the unit cell (showing only the centres of the ions)

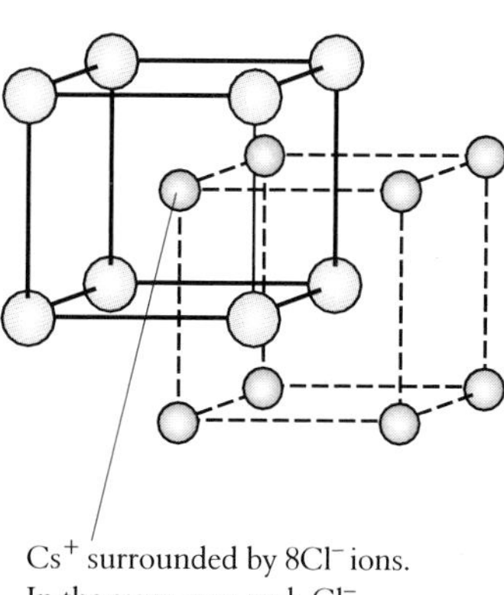
Cs^+ surrounded by $8Cl^-$ ions. In the same way, each Cl^- is surrounded by 8 Cs^+ ions.

Figure 6.9B Caesium chloride structure: the unit cell (showing only the centres of the ions)

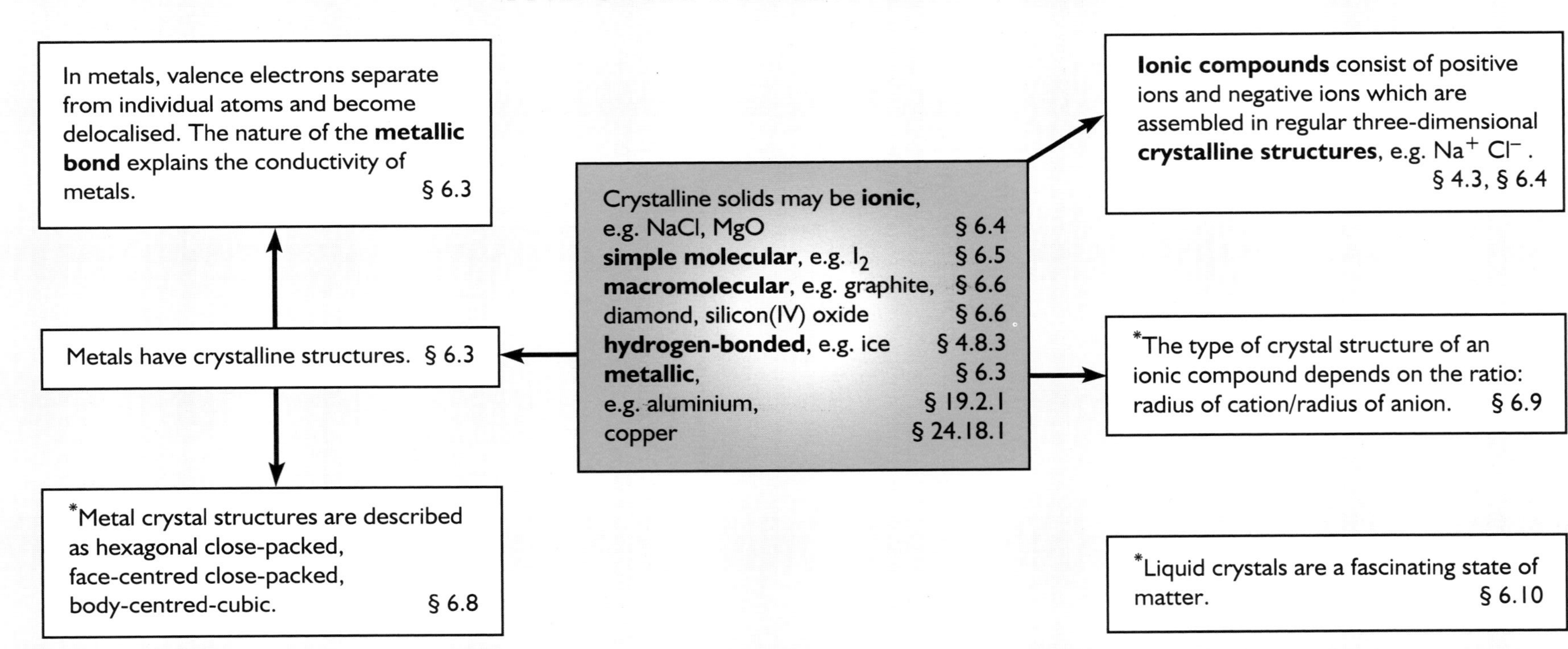
THE STRUCTURES OF SOLIDS
Crystalline solids may be ionic, e.g. NaCl, MgO § 6.4
simple molecular, e.g. I2 § 6.5
macromolecular, e.g. graphite, § 6.6
diamond, silicon(IV) oxide § 6.6
hydrogen-bonded, e.g. ice § 4.8.3
metallic, § 6.3
e.g. aluminium, § 19.2.1
copper § 24.18.1
Ionic compounds consist of positive ions and negative ions which are assembled in regular three-dimensional crystalline structures, e.g. Na+ Cl−. § 4.3, § 6.4
*The type of crystal structure of an ionic compound depends on the ratio: radius of cation/radius of anion. § 6.9
*Liquid crystals are a fascinating state of matter. § 6.10
Metals have crystalline structures. § 6.3
In metals, valence electrons separate from individual atoms and become delocalised. The nature of the metallic bond explains the conductivity of metals. § 6.3
*Metal crystal structures are described as hexagonal close-packed, face-centred close-packed, body-centred-cubic. § 6.8

6.10 LIQUID CRYSTALS

Pure solids melt sharply, the temperature remaining constant at the melting temperature until all the solid has melted. There are, however, many crystalline solids which pass at a sharp **transition temperature** to a turbid liquid phase before finally melting to form a clear liquid. The turbid liquids phases can flow as liquids do and possess some degree of order, with the result that these turbid liquids resemble crystals in certain optical properties. They are known as **liquid crystals**.

Viewed in plane-polarised light under a microscope, liquid crystals show characteristic coloured patterns. The structures and therefore the colours change with temperature so liquid crystals can be used as thermometers.

A mixture of liquid crystals which changes colour over about 3 °C in the range of body temperature is used for **skin thermography**. Skin overlying veins and arteries is slightly warmer than in other areas, and the difference in temperature can be detected by liquid crystals. Specialists can use the technique of skin thermography to detect blockages in veins and arteries. The technique has been successful in the early diagnosis of breast cancer. When a layer of cholesteric material is painted or sprayed on to the surface of the breast, a tumour shows up as a 'hot area', which is coloured blue.

Room thermometers contain liquid crystals with a suitable temperature range. Figures show up in different colours as the temperature changes.

The digital displays you see in watches and calculators contain liquid crystals. The structure of the liquid crystal can be changed by the use of a very small electric field. If the change in orientation results in a change in optical properties, the liquid crystal can be used to display information, e.g., the time, or date. The timing of a watch display is controlled by a quartz crystal. A small electric current induces quartz to resonate at 32 768 oscillations per second. A quartz watch has no mechanical moving parts, giving it a big advantage over traditional watches.

QUESTIONS ON CHAPTER 6

1. What types of intramolecular and intermolecular bonds exist in *(a)* solid argon, *(b)* solid bromine, *(c)* diamond, *(d)* graphite and *(e)* silica?

2. Crystals of salts fracture easily, but metals are deformed under stress without fracturing. Explain the difference.

***3.** What is the coordination number of an ion? What is the coordination number of the cation in *(a)* the NaCl structure and *(b)* the CsCl structure? What is the reason for the difference?

4. Say what properties you would expect of substances which are *(a)* metals, *(b)* ionic compounds, *(c)* composed of individual covalent molecules, and *(d)* macromolecular covalent compounds.

5. Explain the bonding present in the solids sodium chloride, sodium, phosphorus(V) chloride (PCl_5), graphite and ice. Point out how the type of bonding determines the physical properties of the solids.

6. Give two examples of *(a)* ionic solids, *(b)* molecular solids and *(c)* covalent macromolecular solids. What are the factors that determine whether each of these types of solid will dissolve in water?

7. Describe the structure of each of the following solids. Explain how the type of chemical bonds in each solid determines the structure: sodium chloride, ice, poly(ethene), aluminium, iodine.

8. *(a)* Describe the nature of the chemical bonds present in the following solids:
(i) diamond (ii) graphite (iii) copper.
(b) Explain the following observations.
(i) Diamond is hard and an electrical insulator. Graphite is soft and a good electrical conductor.
(ii) Sodium is softer than copper.
(iii) Both sodium and copper are good electrical conductors.

***9.** *(a)* What is meant by the coordination number of a cation?
(b) What is the unit cell of a crystal structure?
(c) Explain why sodium chloride and caesium chloride have different crystalline structures.

10. *(a)* The diagram below represents part of a sodium chloride crystal with the position of one sodium ion shown by a plus (+) sign in a circle.

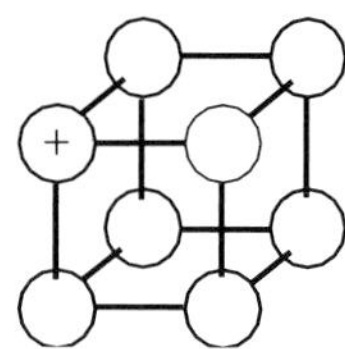

(i) Mark with minus (–) signs all the circles in a copy of the diagram which show the positions of chloride ions.

(ii) How many nearest sodium ions surround each chloride ion in a sodium chloride crystal? **2**

(b) Describe a simple test to show that sodium chloride is ionic. **2**

(c) A crystal of aluminium chloride vaporises when heated to a relatively low temperature. In the gas phase, aluminium chloride exists as a mixture of $AlCl_3$ and Al_2Cl_6 molecules. The Al_2Cl_6 molecule is formed when $AlCl_3$ molecules are linked by two co-ordinate (dative covalent bonds). The structure of Al_2Cl_6 is shown below with one of the co-ordinate bonds labelled.

(i) Explain the meaning of the term *co-ordinate bond*.

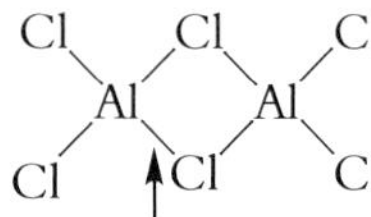

(ii) Using an arrow, indicate on a copy of the diagram the other co-ordinate bond.

(iii) Explain briefly why solid aluminium chloride vaporises at a relatively low temperature. **5**

9 *NEAB (AS/AL)*

11. *(a)* Describe the nature of the attractive forces which hold the particles together in magnesium metal and in magnesium chloride. **4**

(b) Name the type of bond between aluminium and chlorine in aluminium chloride and explain why the bonding in aluminium chloride differs from that in magnesium chloride. **3**

(c) Write an equation, including state symbols, to show what happens when magnesium chloride dissolves in water. Explain, in terms of bonding, the nature of the interaction between water and magnesium in this solution. **3**

10 *NEAB (AS/AL)*

12. *(a)* Real gases do not behave ideally.

(i) State the **two** major assumptions which are made for ideal gas behaviour. **2**

(ii) I. State the *physical conditions* under which real gases approach ideal gas behaviour, and
II. explain these conditions in terms of the two assumptions in *(a)* (i). **2**

(b) The structures of diamond and graphite are shown in the diagram below.

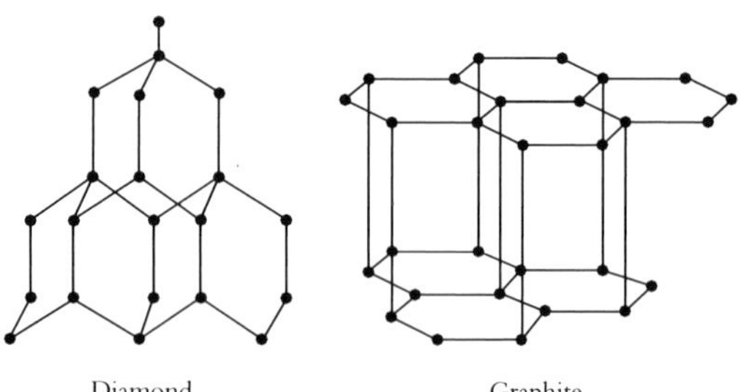

Diamond　　　Graphite

Explain the relationship between the structures of diamond and graphite and their I. hardness and II. electrical conductance. **3**

(c) (i) State the meaning of the term **coordination number** of an ion in a crystal structure. **1**

(ii) The structures of sodium chloride and caesium chloride are shown below.
What is the coordination number of the **chloride** ions in **each**? **1**

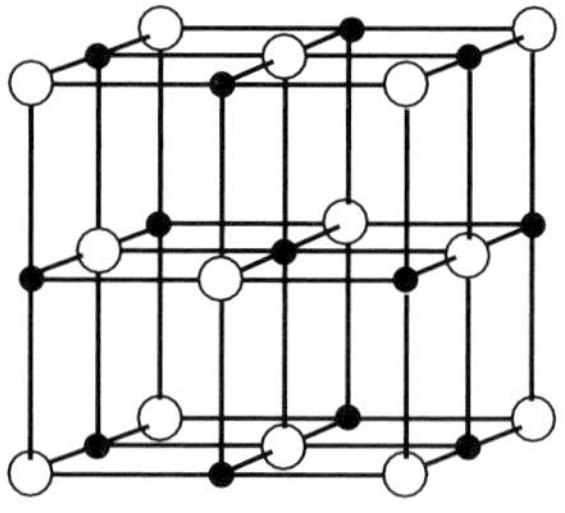

Sodium chloride

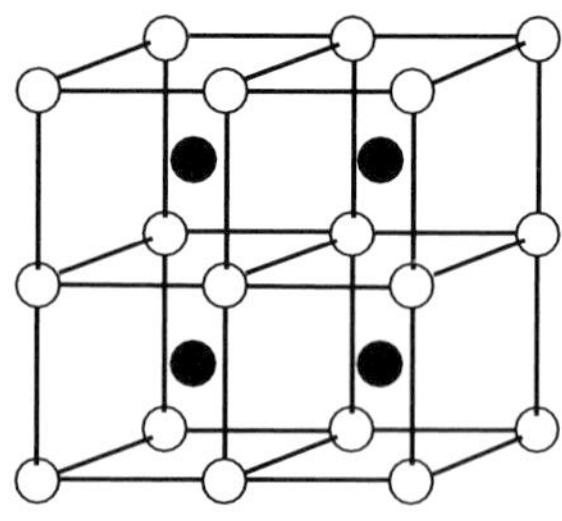

Caesium chloride

(iii) Explain why
I. Sodium chloride is soluble in water, and
II. ethanol is soluble in water while hydrocarbons are not. **3**

12 *(WJEC)*

For *(a)* see Chapter 7

Part 2

Physical Chemistry

7 GASES

7.1 AIR BAGS

If a car is in a collision, the driver often suffers injury or death by being thrown forward on to the dashboard. Many cars are now fitted with an air bag embedded in the steering wheel. If the car is in a collision and comes rapidly to a stop, the air bag inflates between the driver and the dashboard and cushions the impact. The air bag inflates within 50 microseconds. After the driver hits the bag, gas diffuses out through small holes in the bag. Air bags are becoming more widespread now that their effectiveness in saving lives has been proved. The bags are made of nylon, sometimes with a rubber coating. They do not contain air; they contain nitrogen. On impact, electronic sensors at the front of the car set off a chemical reaction in a gas generator. The generator contains sodium azide, NaN_3, potassium nitrate and silica. The electrical impulse from the sensor detonates the sodium azide:

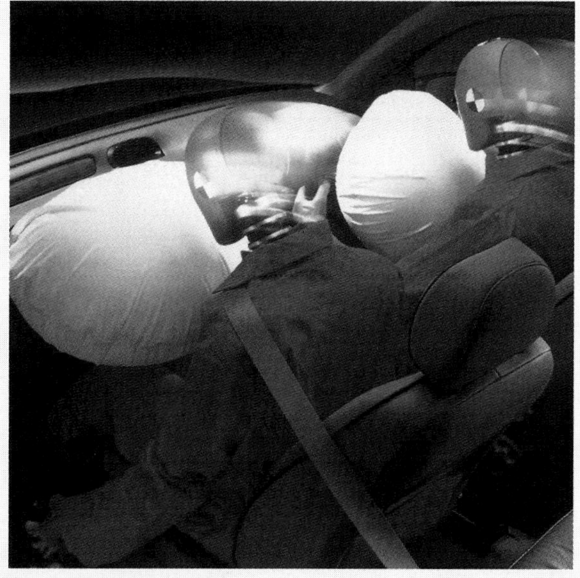

Figure 7.1A
Air bag

$$2NaN_3(s) \longrightarrow 3N_2(g) + 2Na(s)$$

The sodium formed reacts with potassium nitrate:

$$10Na(s) + 2KNO_3(s) \longrightarrow K_2O(s) + 5Na_2(s) + N_2(g)$$

The third component of the mixture, silicon(IV) oxide, reacts with potassium oxide and sodium oxide to form a silicate glass:

$$K_2O(s) + Na_2O(s) + SiO_2(s) \longrightarrow \text{silicate glass}$$

The reason why the air bag works is that a small mass of sodium azide (126 g) can generate a large volume of nitrogen, about 70 dm^3. Why does a gas occupy such a large volume compared with a comparable mass of solid? The kinetic theory of gases, which is discussed in this chapter, will answer the question.

7.2 STATES OF MATTER

Gases differ from liquids and solids ...

Many pure substances can exist in all of the three states of matter: solid, liquid or gas, depending on conditions of temperature and pressure. There are characteristics in which gases differ from solids and liquids.

... in compressibility ...

1. Gases are highly compressible. Increasing the pressure on a gas can decrease its volume to a small fraction of the original value. Decreasing the pressure will allow the gas to expand to its original volume.

... in the degree of expansion on heating ...

2. Gases expand on heating. The increase in volume with temperature is typically about one hundred times greater than for a liquid.

... in ease of flow ...

3. Gases flow much more freely than solids or liquids. They diffuse, that is, spread from a region of higher concentration to a region of lower concentration. This allows gases to mix with other gases in any proportions to form a homogeneous mixture.

... in density

4. Gases have low densities. Compare for example iron 7.9×10^3 g dm^{-3}, water 1.0×10^3 g dm^{-3}, oxygen 1.4 g dm^{-3}. When a gas is cooled its density increases. When oxygen is cooled to −183 °C it liquefies with a density of 1.15 g cm^{-3}. This shows that the same mass of oxygen occupies as a gas 800 times the volume which it takes up in the liquid state. It follows that the molecules are much further apart in the gas than in the liquid state. It also shows that the molecules of a gas are not bonded in a structure as they are in solids and are more free to move than in a liquid.

7.3 THE GAS LAWS

7.3.1 BOYLE'S LAW

The gas laws state the results of experiments on gases

Boyle's Law relates the pressure and volume of a gas: PV = constant

A gas exerts pressure on the walls of its container. Robert Boyle measured the volume of gas at different pressures. The results of his work are known as **Boyle's Law**. The law can be expressed as: *for a fixed mass of gas at constant temperature the product of pressure and volume is constant*:

$$PV = \text{Constant}$$

7.3.2 CHARLES' LAW

Charles' Law deals with the effect of temperature on volume: V/T = constant ...

The French scientists J L Gay-Lussac and J A C Charles investigated the effect of temperature on the volume of a gas. They found that the volume increased in a linear manner with temperature [see Figure 7.3A].

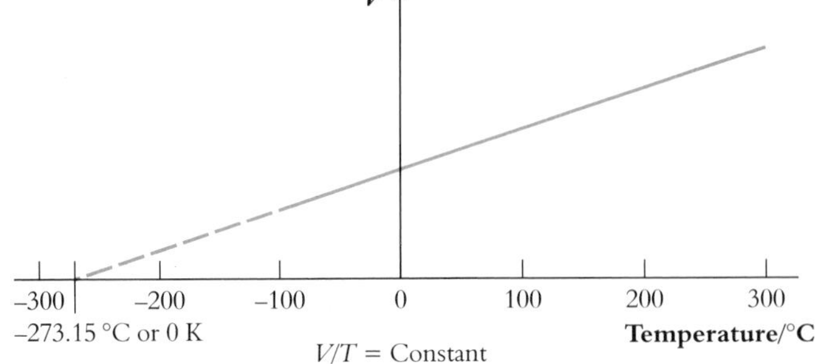

Figure 7.3A Graph illustrating Charles' Law (Plot of volume against temperature (°C) for a gas at constant pressure)

When the graph is extrapolated (the broken line beyond the experimental points) it cuts the temperature axis at −273.15 °C. The same value is obtained for all gases. It looks as though all gas volumes would become zero at −273.15 °C, but in fact gases liquefy or solidify long before they reach this temperature.

... when T is the absolute temperature ...

... measured in kelvins on the Kelvin scale

Lord Kelvin, half a century later, used Charles' linear relationship between gas volume and temperature (as shown in Figure 7.3A) to define a new temperature scale. The temperature −273.15 °C is the zero of the **absolute temperature scale** or **Kelvin scale**. Temperatures on this scale are measured in kelvins and obtained by adding 273.15 to the temperature on the Celsius scale.

273.15 K = 0 °C

$$\text{Temperature/K} = \text{Temperature/°C} + 273.15$$
$$273.15\,\text{K} = 0\,°\text{C}$$

You can see that in Figure 7.3A the volume of gas is directly proportional to the temperature in kelvins: $V \propto T$. Charles's law can be expressed as

$$V/T = \text{Constant where } T \text{ is the temperature in kelvins}$$

7.3.3 THE EQUATION OF STATE FOR AN IDEAL GAS

Gases which obey Boyle's law and Charles' Law under all conditions are called ideal gases

Gases do not obey Boyle's Law and Charles' Law closely under all conditions. A gas which obeys both laws is said to behave ideally, to be an **ideal gas**. By combining Boyle's Law (PV = Constant) and Charles' Law (V/T = Constant) one obtains the relationship

Combining both laws gives the equation of state for an ideal gas:

$P_1V_1/T_1 = P_2V_2/T_2$

$$PV/T = \text{Constant.}$$

This is often written as

$$\frac{P_1V_1}{T_1} = \frac{P_2V_2}{T_2}$$

Standard temperature and pressure, stp, are 0 °C and 1 atm

Room temperature and pressure, rtp, are 20 °C and 1 atm

This is the **equation of state for an ideal gas**. It enables one to calculate the effect of a change in temperature and pressure on the volume of a gas. One cannot compare gas volumes unless they are stated at the same temperature and pressure. Gas volumes are usually compared at 0 °C and 1 atmosphere (atm). These conditions are referred to as **standard temperature and pressure** (stp). Sometimes volumes are quoted at **room temperature and pressure** (rtp): 20 °C and 1 atm.

Units

The SI unit of force is the newton. When a force of one newton moves through one metre, one joule of work is done.

$$1 \text{ newton metre (N m)} = \text{one joule (J)}$$

Pressure units

The SI unit of pressure is the newton per square metre (N m^{-2}), called the pascal (Pa):

$$1 \text{ atmosphere} = 1.0132 \times 10^5 \text{ N m}^{-2} = 1.0132 \times 10^5 \text{ Pa}$$
$$= 760 \text{ mm mercury}$$

Volume units

The SI unit of volume is the cubic metre, m^3, but cubic decimetres, dm^3, cubic centimetres, cm^3, and litres, l, are also used.

$$1\text{ m}^3 = 10^3\text{ dm}^3 = 10^6\text{ cm}^3$$
$$1\text{ dm}^3 = 1\text{ litre}$$

Temperatures must be in kelvins in the ideal gas equation of state.

Example If the volume of a gas collected at 60 °C and 1.05×10^5 N m^{-2} is 60 cm^3, what would be the volume of gas at stp?

A sample calculation of the volume of a gas at stp ...

Method The experimental conditions are

$$P_1 = 1.05 \times 10^5 \text{ N m}^{-2}$$
$$T_1 = 273 + 60 = 333 \text{ K}$$
$$V_1 = 60 \text{ cm}^3$$

Standard conditions are

$$P_2 = 1.01 \times 10^5 \text{ Nm}^{-2}$$
$$T_2 = 273 \text{ K}$$

Since

$$\frac{P_1V_1}{T_1} = \frac{P_2V_2}{T_2}$$

The volume of gas at stp

$$V_2 = \frac{1.05 \times 10^5 \times 60 \times 273}{1.01 \times 10^5 \times 333} \text{ cm}^3$$
$$= 51 \text{ cm}^3$$

SUMMARY

Ideal gases obey the equation of state for an ideal gas: PV/T = constant.

7.4 AVOGADRO'S HYPOTHESIS

Gay-Lussac's experiments ...

Gay-Lussac studied chemical reactions between gases. He noticed that there is always a very simple ratio between the volumes of gases that react together. For instance,

... showed that gas volumes combine in simple ratios

1 volume of hydrogen + an equal volume of chlorine ⟶ 2 volumes hydrogen chloride

1 volume of oxygen + 2 volumes of hydrogen ⟶ 2 volumes of water vapour

Avogadro's hypothesis offered an explanation

To explain Gay-Lussac's results, Avagadro in 1811 suggested that equal volumes of gases measured at the same conditions of temperature and pressure, contain the same number of molecules. This theory is called **Avogadro's hypothesis**. On the basis of this theory one can interpret the observation

1 volume of hydrogen + 1 volume of chlorine ⟶ 2 volumes of hydrogen chloride

as 1 molecule of hydrogen + 1 molecule of chlorine ⟶ 2 molecules of hydrogen chloride

The volume occupied by one mole of gas is the same for all gases: the gas molar volume = 22.414 dm³ at stp, 24.056 dm³ at rtp

It follows from Avogadro's hypothesis that if equal volumes of gases contain equal numbers of molecules then the volume occupied by one mole of molecules must be the same for all gases. It is called the **gas molar volume**. The value is 22.414 dm^3 at stp (0 °C and 1 atm) or 24.056 dm^3 at rtp (20 °C and 1 atm).

7.5 THE IDEAL GAS EQUATION

The ideal gas equation:

PV = nRT

For gases which obey the equation of state for an ideal gas,

$$\frac{P \times V}{T} = \text{Constant (for a given mass of gas)}$$

R is the universal gas constant, 8.314 J K⁻¹ mol⁻¹

Applying the equation to one mole of gas, $V = V_m$ the gas molar volume, and the constant is the **universal gas constant**, symbol R. Then

$$PV_m = RT$$

For n moles of gas,

$$PV = nRT$$

Since $R = PV_m/T$, R can be found by inserting values of P in N m^{-2}, V_m in dm^3 and T in kelvins into the equation. Then $R = 8.314$ J K^{-1} mol^{-1} (joules per kelvin per mole).

Using the ideal gas equation

1. The gas molar volume is used in calculations on the volumes of gases taking part in chemical reactions [see § 3.12].
2. The molar mass of a volatile liquid can be found by weighing a measured volume of its vapour [see § 8.4].

7.6 DALTON'S LAW OF PARTIAL PRESSURES

The partial pressure of a gas is the contribution which it makes to the total pressure. It is equal to the pressure which the gas would exert if it alone occupied the container

In a mixture of gases each gas behaves as if it were the only gas present, assuming there are no chemical interactions between the gases. In air, which is a mixture of approximately $\frac{4}{5}$ nitrogen and $\frac{1}{5}$ oxygen by volume, $\frac{4}{5}$ of the air pressure is due to nitrogen and $\frac{1}{5}$ is due to oxygen. The contribution which each gas makes to the total pressure is called the **partial pressure**. The partial pressure of a gas is the pressure which it would exert if it alone occupied the container. Dalton's Law of Partial Pressures states that in a mixture of gases which do not react chemically the total pressure is the sum of the partial pressures of the components.

Mole fraction of substance = moles of substance/total moles

The partial pressure of each gas depends on the total pressure and on the **mole fraction** of the gas. Mole fractions are one method of expressing the composition of a mixture. In air one fifth of the molecules are oxygen molecules so the mole fraction of oxygen $= \frac{1}{5}$. Four fifths of the molecules are nitrogen molecules so the mole fraction of nitrogen $= \frac{4}{5}$. In general, the mole fraction of **A** in a mixture of **A** and **B** is

$$\text{Mole fraction of } \mathbf{A} = \frac{\text{Moles of } \mathbf{A}}{\text{Moles of } \mathbf{A} + \text{Moles of } \mathbf{B}} = \frac{n_\mathbf{A}}{n_\mathbf{A} + n_\mathbf{B}}$$

From Avogadro's Hypothesis, it follows that

$$\frac{n_\mathbf{A}}{n_\mathbf{A} + n_\mathbf{B}} = \frac{\text{Volume of } \mathbf{A}}{\text{Total volume}}$$

A calculation of partial pressure

Example 4.00 dm^3 of oxygen at a pressure of 400 kPa and 1.00 dm^3 of nitrogen at a pressure of 200 kPa are introduced into a 2.00 dm^3 vessel. What is the total pressure in the vessel?

Method When oxygen contracts from 4.00 dm^3 to 2.00 dm^3, the pressure increases from 400 kPa to

$$400 \times \frac{4.00}{2.00} = 800 \text{ kPa}$$

The partial pressure of oxygen in the vessel is 800 kPa.

When nitrogen expands from 1.00 dm^3 to 2.00 dm^3, the pressure decreases from 200 kPa to

$$200 \times \frac{1.00}{2.00} = 100 \text{ kPa}.$$

The partial pressure of nitrogen is 100 kPa.

$$\text{The total pressure} = p(O_2) + p(N_2) = 800 + 100 = 900 \text{ kPa}$$

The total pressure in the vessel is 900 kPa.

CHECKPOINT 7.6: PARTIAL PRESSURES

1. 50.0 cm^3 of carbon dioxide at 10^5 N m^{-2} are mixed with 150 cm^3 of hydrogen at the same pressure. If the pressure of the mixture is 1.00×10^5 N m^{-2}, what is the partial pressure of carbon dioxide?

2. A mixture of gases at a pressure 1.01×10^5 N m^{-2} has the volume composition of 30% CO, 50% O_2, 20% CO_2.
   (a) What is the partial pressure of each gas?
   (b) If the carbon dioxide is removed by the addition of some pellets of sodium hydroxide, what will be the partial pressures of O_2 and CO?

3. Into a 10.0 dm^3 vessel are introduced 4.00 dm^3 of methane at a pressure of 2.02×10^5 N m^{-2}, 12.5 dm^3 of ethane at a pressure of 3.50×10^5 N m^{-2} and 1.50 dm^3 of propane at a pressure of 1.01×10^5 N m^{-2}. What is the pressure of the resulting mixture?

4. A mixture of 20% NH_3, 55% H_2 and 25% N_2 by volume has a pressure of 9.80×10^4 N m^{-2}.
   (a) What is the partial pressure of each gas?
   (b) What changes will take place in the partial pressures of hydrogen and nitrogen if the ammonia is removed by the addition of solid phosphorus(V) oxide?

7.7 THE KINETIC THEORY OF GASES

According to the kinetic theory ...

... most of a gas is empty space ...

... in which molecules are scattered with large distances between them

The gas molecules are constantly moving ...

... in straight lines ...

... until they collide with one another or with the walls of the container ...

... exerting no forces between molecules ... except when they collide

The average molecular speed is proportional to the absolute temperature

There is a need for a theory to explain the properties of gases described by the gas laws [§§ 7.1–7.6].

The kinetic theory of gases was put forward by R J Clausius (1857) and J C Maxwell (1859). The theory tackles these questions by making three fundamental assumptions. These are:

1. The gas molecules themselves occupy only a tiny fraction of the volume in which a gas is contained. The distance between molecules is many times the diameter of a molecule. The kinetic theory model views a sample of gas as a nearly empty space. The molecules are tiny masses scattered through this space.

2. The gas molecules are in constant motion. The name of the theory comes from the Greek word *kinein* to move. The molecules move in straight lines until they collide with each other or with the walls of the container. Except during collisions, the molecules exert no force upon one another. The collisions are **perfectly elastic**; this means that the molecules bounce apart with no loss of energy.

3. The molecules have a range of speeds. The speed of a molecule depends on its kinetic energy. As the temperature increases the kinetic energy increases and, as a result, the average speed increases. The average kinetic energy is proportional to the absolute temperature; therefore at a certain temperature all gases have the same average kinetic energy.

How do these postulates of the kinetic theory explain the behaviour of gases?

The theory explains how gases exert pressure ...

Boyle's law and Charles' Law ...

how gases diffuse ...

Dalton's Law of partial pressure

1. Pressure. When a moving object collides with a surface it exerts a force. The collisions of gas molecules with the walls of the container exert a force which results in a pressure (pressure = force/area). The greater the number of molecules and the more frequently they collide with the container walls, the greater is the pressure.
2. Boyle's Law ($V \propto 1/P$). When the volume of a gas decreases, the distances between the molecules and the walls are shorter so collisions are more frequent and the pressure exerted *by* the gas therefore increases. When the pressure exerted *on* a gas increases the molecules move closer together and the volume decreases.
3. Charles' Law ($V \propto T$). As the temperature increases the kinetic energy increases and the average molecular speed increases. Molecules hit the walls more frequently and with more force, increasing the pressure exerted by the gas. As a result the walls move outwards: the volume increases until constant pressure is restored.
4. Diffusion. The reason that gases diffuse rapidly is that their particles are in constant motion.
5. Dalton's Law of Partial Pressures ($P_{total} = P_\mathbf{A} + P_\mathbf{B}$). Imagine 0.4 mol of gas **A** in a container. The number of collisions per second gives rise to a pressure $P_\mathbf{A}$. Adding 0.2 mol of gas **B** increases the number of molecules and therefore the frequency of collisions by 50%. Each gas exerts pressure in proportion to the number of molecules of that gas present (the mole fraction).

Calculations based on the ideal gas equation and the kinetic energy equation provide some interesting information about molecules. At rtp, a molecule of nitrogen travels at an average speed of 0.5 km s^{-1}. The average distance it travels between collisions is 180 times its diameter (diameter 3.7×10^{-10} m, distance 6.6×10^{-8} m). The collision frequency is 8×10^9 collisions s^{-1}.

The discussion has considered *average* values of molecular energy and speed. The kinetic theory allows one to calculate how the *individual* values of molecular energy are distributed between different energy levels. This is important in the study of reaction kinetics; see § 14.3.

7.8 REAL GASES: NON-IDEAL BEHAVIOUR

Real gases do not behave as the ideal gas model at low temperatures and high pressures

The behaviour of an ideal gas is shown in Figure 7.3A. Real gases depart from this behaviour at high pressures and low temperatures. Two of the assumptions of the kinetic theory have to be modified under these conditions.

The molecules are closer together and forces operate between them. At sufficiently low temperature and high pressure, a gas liquefies

1. There are no forces between molecules. When gases are compressed (at e.g. 100–200 atm), the molecules come close enough for van der Waals forces to operate between them. These forces make the gas more compressible than it is at low pressure, and $PV < RT$. Some gases liquefy. For every gas there is a temperature called the **critical temperature** above which the gas cannot be liquefied by pressure without further cooling. Gases with high critical temperatures (e.g. carbon dioxide, 31 °C), are easily liquefied, and these gases show the biggest deviations from ideal behaviour. At still higher pressures (e.g. 200–1000 atm), the molecules are pushed so close together that repulsion operates between them, and $PV > RT$.
2. The volume of the molecules is negligible compared with the volume occupied by the gas. This is no longer true in a highly compressed gas.

GASES

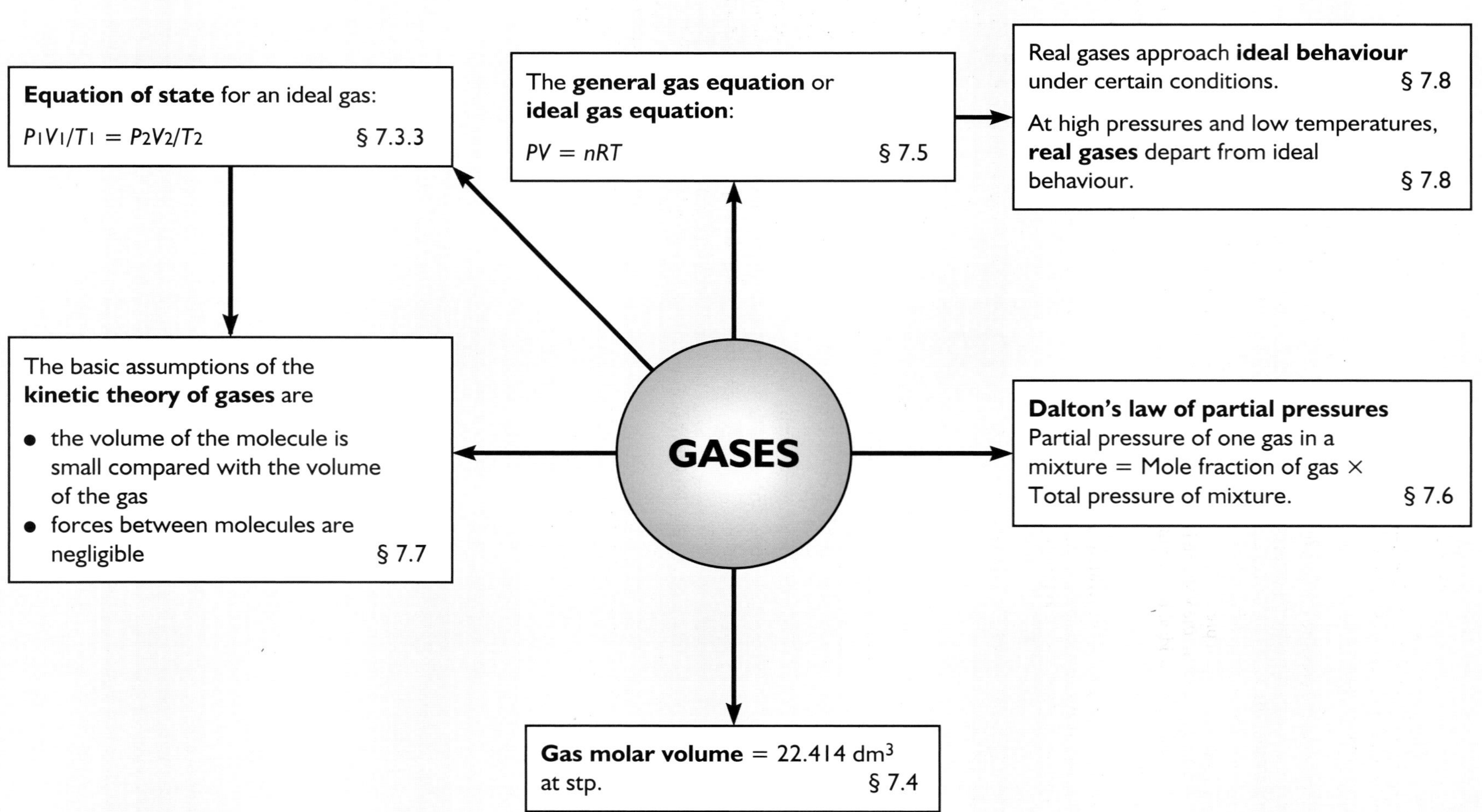
Equation of state for an ideal gas:
$P_1V_1/T_1 = P_2V_2/T_2$ § 7.3.3
The general gas equation or ideal gas equation:
$PV = nRT$ § 7.5
Real gases approach ideal behaviour under certain conditions. § 7.8
At high pressures and low temperatures, real gases depart from ideal behaviour. § 7.8
The basic assumptions of the kinetic theory of gases are
• the volume of the molecule is small compared with the volume of the gas
• forces between molecules are negligible § 7.7
GASES
Dalton's law of partial pressures
Partial pressure of one gas in a mixture = Mole fraction of gas × Total pressure of mixture. § 7.6
Gas molar volume = 22.414 dm^3 at stp. § 7.4

QUESTIONS ON CHAPTER 7

1. How does the kinetic theory of gases explain *(a)* the diffusion of gases *(b)* gas pressure and *(c)* the compressibility of gases?

2. *(a)* How does the pressure of a given mass of gas in a fixed volume change as the temperature rises? How does the kinetic theory explain this change?
(b) At stp, a certain mass of gas has a volume of 1.00 dm^3. At 30 atm pressure, the volume is 31.2 cm^3. At 60 atm, the volume is 14.9 cm^3. Does the gas show ideal behaviour? Justify your answer.

3. The ideal gas equation can be written as $PV = nRT$. Use this equation to calculate the volume of one mole of an ideal gas at 300 K and 100 kPa pressure.

4. Why does 3 mol of N_2 take up more space than 2 mol of NaN_3 [see § 7.1]?

5. What is the effect on the volume of one mole of an ideal gas when each of the following occurs?

(a) The pressure is doubled at constant temperature.
(b) The pressure is reduced by a factor of three at constant T.
(c) The absolute temperature is increased by a factor of 2.5 at constant pressure.
(d) Three moles of the gas are added at constant T and P.

6. What is the effect on the pressure of one mole of an ideal gas when each of the following happens?

(a) The temperature changes from 600 K to 300 K at constant volume.
(b) The temperature changes from 300°C to 600°C at constant volume.
(c) The volume is decreased from 8 dm^3 to 2 dm^3.
(d) Half the gas escapes through a leaky valve.

7. A weather balloon has a volume of 60.0 dm^3. It is released at sea level at 101 kPa and 23 °C. The balloon can expand to a maximum volume of 860 dm^3. It rises to an altitude at which the temperature is −5 °C and the pressure is 6.70 kPa. Does it reach its maximum volume?

8. Haemoglobin transports oxygen from the lungs to the rest of the body. One molecule of haemoglobin combines with four molecules of oxygen. If 1.00 g haemoglobin combines with 1.53 cm^3 of oxygen at 37 °C and 0.987 kPa, what is the molar mass of haemoglobin?

9. *(a)* Analysis of an organic compound **X**, which contains carbon, hydrogen and oxygen only, gave 54.5% by mass of carbon and 9.1% by mass of hydrogen.
(i) Calculate the empirical formula of **X**.
(ii) In a separate experiment, 0.539 g of a vaporised sample of **X** occupied 200 cm^3 at a temperature of 100 °C and a pressure of 95.0 kPa. Use these results to calculate the relative molecular mass of **X**.
(iii) Deduce the molecular formula for **X**. **8**
(b) A compound with molecular formula $C_3H_6O_2$ burns completely in oxygen to form carbon dioxide and water only.
(i) Write an equation for this reaction.
(ii) Calculate the volume of oxygen gas (at 298 K and 100 kPa) which is required to burn completely one mole of $C_3H_6O_2$ **3**

11 *NEAB (AS/AL)*

8 LIQUIDS

8.1 AN INCREDIBLE LIQUID

The Italian chemist Primo Levi in his book* *The Mirror Maker* marvels at the mechanism by which a spider spins a continuous solid thread. Inside the spider's body is a reservoir of the viscous liquid of which the thread is made. When this liquid is extruded through holes in the spider's spreader, it solidifies as soon as it reaches the air. It is essential that the thread solidify instantaneously because the spider is hanging from the thread while it spins. How does the change from liquid to solid take place in a fraction of a second?

We know of a number of methods by which liquids solidify. Does the liquid thread simply freeze as it is extruded? No, this is not the case because the spider's body is at the same temperature as the air. Does liquid evaporate and leave behind a solid solute? Water is the only solvent in the spider's body, and one can see that a spider's web is not water-soluble because it can withstand droplets of dew. Does polymerisation take place to form molecules which are large enough to have strong forces of attraction between them and are therefore solid? Chemists know of no polymerisation that takes place in a fraction of a second. The clue to the mechanism is that the spider is hanging from the thread; the thread is under traction. The viscous liquid is composed of long molecules which are rolled up and can slide over one another. When the liquid is extruded under traction the molecules become unravelled and extend to form parallel threads. Forces of attraction are then able to operate along the length of the molecules [see § 4.8] and the viscous liquid becomes a solid.

* *The Mirror Maker* by Primo Levi (Methuen)

Figure 8.1A
A spider's web

8.2 THE LIQUID STATE

The kinetic theory of gases, which was covered in Chapter 7, has applications to the liquid state. Whether a substance is a gas or liquid or solid depends on the balance between two factors. One is the kinetic energy of the particles, which tends to disperse them. The other is the strength of the attractions between the particles, which tends to draw them together. The kinetic energy of the particles is proportional to the absolute temperature. The strength of the attractions is greater for polar molecules than for non-polar molecules and is greatest for the ions in an ionic solid. The strength of an attraction between particles decreases as the distance between them increases.

Kinetic energy of particles ...

... tends to disperse them

In a gas the strength of the attractions is smaller than the energy of movement so the particles are far apart. In a liquid the attractions are stronger because the particles are closer together, but their kinetic energy allows the particles to move about. In a solid the attractions dominate, and the particles are fixed in three-dimensional structures [see Chapter 6].

Attractive forces between particles ...

... tend to draw them together

In a gas, kinetic energy predominates

Since the particles are close together, a liquid, unlike a gas, has a definite volume. Having little space between particles, liquids are compressed only slightly by the application of pressure. The distance between the particles depends on a balance between forces of attraction between particles and forces of repulsion between neighbouring electron clouds [see Figure 8.2A]. The forces of attraction, being weaker than those in solids, are not strong enough to hold the liquid in a definite shape. A liquid flows to fit the shape of its container. The attractions between particles make liquids flow and diffuse much more slowly than gases. The particles in a solid are not much closer than those in a liquid but they are bonded together. This is why solids have a definite shape, are even less compressible than liquids and do not flow (although some solids flow extremely slowly).

Figure 8.2A Distance between molecules of a liquid

Distance between molecules in a liquid.

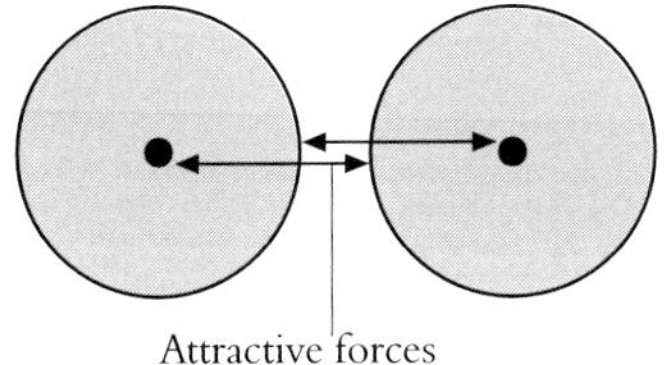

Attractive forces between molecules.

In a solid, attractive forces predominate

In a liquid, the attractive forces between molecules are weak enough to allow the molecules to move

Imaginary compression of liquid forces molecules closer together.

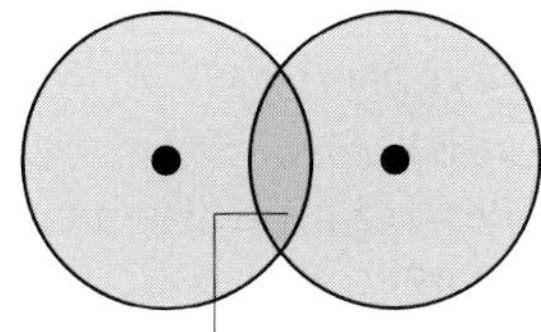

Interpenetration of electron clouds causes repulsion, and molecules move further apart.

Changes of state

The changes of state between liquid, gas and solid are:

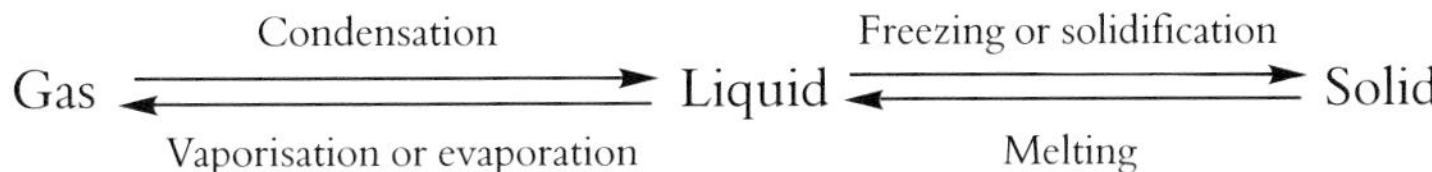

Changes of state are governed by the balance between the kinetic energy of moving particles and intermolecular forces. As the temperature of a liquid increases, the kinetic energy increases and the most energetic particles can overcome the attractive forces, break free from the body of the liquid and move independently: they vaporise. When a gas is cooled, the kinetic energy of the particles decreases and forces of attraction draw the slower moving particles together: the gas condenses. Similarly, when a liquid is cooled sufficiently, attractive forces overcome kinetic energy, and the liquid solidifies (freezes).

SUMMARY

Solid, liquid or gas: the physical state of a substance depends on the balance between kinetic energy, which tends to disperse particles, and intermolecular forces, which tend to draw particles together. The balance changes with temperature.

When solids melt there is usually an expansion of about 10%. When liquids vaporise the expansion is much greater, for example one volume of water forms 1300 volumes of water vapour. Heat must be taken in to overcome forces of attraction between molecules of liquids and solids: melting and vaporisation are endothermic changes. When forces of attraction are set up between the molecules of a gas and they change into a liquid, and again when the molecules of liquid become ordered and solidify, heat is given out: condensation and freezing are exothermic changes.

The liquid state resembles the solid state more than the gaseous state. The quantity of heat required to melt one mole of a substance in its solid state is much smaller than the quantity of heat required to vaporise one mole of the liquid form of the same substance. This indicates that the change from solid to liquid involves less disruption of intermolecular attractions than the change from liquid to gas. For standard enthalpy of vaporisation and standard enthalpy of melting see § 10.7.2.

CHECKPOINT 8.2 LIQUIDS

1. *(a)* Why are liquids more compressible than solids and less compressible than gases?
   (b) Why do liquids flow more easily than solids and less easily than gases?

2. *(a)* Heat must be supplied to boil a liquid. What happens to this heat?
   (b) Why is less heat required to melt 1 mole of solid **X** than to vaporise 1 mole of liquid **X**?

3. What factors decide the distance between molecules in a liquid?

8.3 VAPORISATION

Molecules with high energy escape from the liquid to the vapour phase

Since molecules of liquid are in constant motion, some of them will have enough energy to escape from the liquid into the vapour state. (The term **vapour** is applied to a gas below its critical temperature. The critical temperature is the temperature below which a gas can be liquefied by an increase in pressure, without further cooling. Those that escape will be molecules with energy considerably above average, sufficient to remove them against the attraction of other molecules of liquid. The molecules that remain in the liquid will be of lower energy than those that escape, and the temperature of the liquid will fall. Since the fraction of molecules with high energy increases as the temperature rises, the rate of **vaporisation** or **evaporation** increases with temperature.

At equilibrium, the rate of evaporation equals the rate of condensation

Evaporation will continue until no liquid remains. If the liquid is in a closed container, however, the molecules in the vapour state will collide with the walls of the container, and some will be directed back towards the liquid. Some of these will re-enter the liquid, i.e. **condense**. [See Figure 8.3A.] **Equilibrium** will be reached when the rate at which molecules of liquid evaporate is equal to the rate at which molecules of vapour condense.

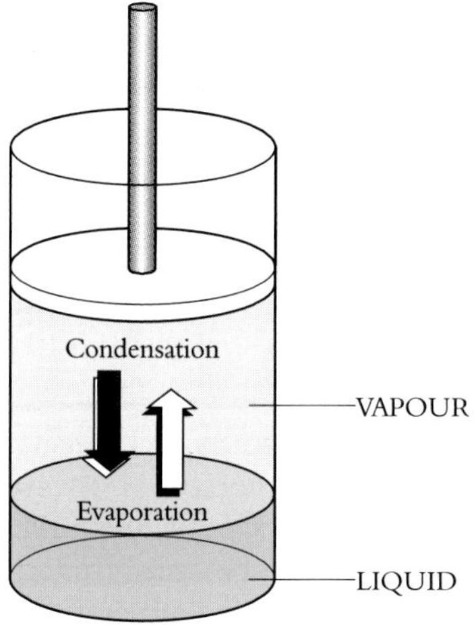

Figure 8.3A
A liquid and its vapour reach equilibrium in a closed container. The rate of evaporation is equal to the rate of condensation

FURTHER STUDY

8.3.1 SATURATED VAPOUR PRESSURE

The maximum vapour pressure developed by a liquid is its saturated vapour pressure at that temperature

As the molecules of vapour collide with the walls of the container they exert a pressure. The maximum vapour pressure that can be developed by a liquid is called the **saturated vapour pressure** of that liquid. Since the fraction of molecules with high energy increases with temperature, evaporation increases as the temperature rises and the saturated vapour pressure increases with temperature. The magnitude of the saturated vapour pressure depends on the *identity* of the liquid and the *temperature*; it does not depend on the amount of liquid present. Solids have vapour pressures too, but, at room temperature, the vapour pressures of most solids are low.

A plot of saturated vapour pressure against temperature is shown in Figure 8.3B.

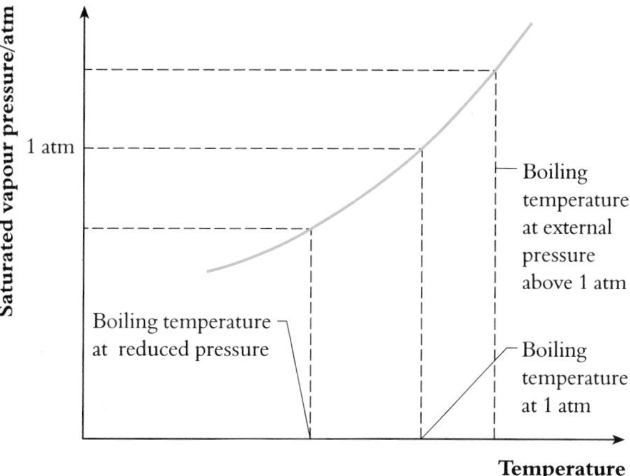

Figure 8.3B
Variation of saturated vapour pressure with temperature

8.3.2 DISTILLATION

Saturated vapour pressure, svp, increases with temperature

If a liquid is heated in an open container, its saturated vapour pressure increases until it becomes equal to atmospheric pressure. When this happens, bubbles of vapour form in the interior of the liquid and escape into the atmosphere because the vapour pressure is high enough to push the air aside: the liquid boils. The **boiling temperature** of a liquid is defined as the temperature at which its saturated vapour pressure, svp, is equal to the external pressure, normally 1 atm.

When the svp equals the external pressure, the liquid boils

While a liquid is boiling, the heat taken in is used to produce molecules with enough energy to escape into the vapour phase. The average kinetic energy of the molecules remaining in the liquid does not increase, and its temperature remains constant. Figure 8.3B shows how the boiling temperature increases at pressures greater than 1 atm and decreases at pressures less than 1 atm.

Distillation under reduced pressure is used for the purification of some substances

Distillation under reduced pressure (sometimes called **vacuum distillation**) is used in the purification of substances which decompose at their boiling temperatures. The apparatus shown in Figure 8.3C could be used. By means of vacuum pumps, pressures down to 10^3 or 10^2 N m^{-2} can be obtained.

The measurement of boiling temperature

A pure liquid can be identified by its boiling temperature at a certain pressure (normally 1 atm). When a boiling temperature is measured, the thermometer must measure the temperature of the vapour in equilibrium with the boiling liquid [see Figure 8.3C]. It should not be immersed in the liquid because **superheating** (heating above the boiling temperature) can occur if a liquid is heated rapidly.

Figure 8.3C Distillation apparatus

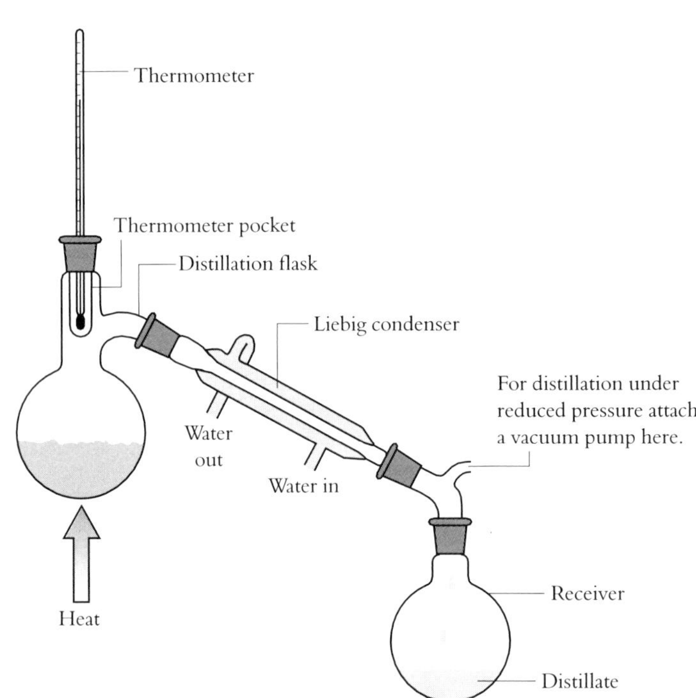

SUMMARY

The saturated vapour pressure (svp) of a liquid is the maximum vapour pressure that can be exerted at the stated temperature. It does not depend on the amount of liquid present. The svp increases with a rise in temperature. When the svp equals the external pressure, normally 1 atm, the liquid boils.

CHECKPOINT 8.3 VAPOUR PRESSURE

1. *(a)* Explain the term *saturated vapour pressure*.
   (b) Boiling temperatures are water 100 °C, ethanol 78 °C, ethoxyethane 35 °C. Place the liquids in order of increasing saturated vapour pressure (svp) at 20 °C, and justify your order.
   (c) Explain how the svp of a liquid depends on (i) the temperature (ii) intermolecular forces in the liquid (iii) the volume of liquid.

2. While a volatile liquid standing in a beaker evaporates its temperature remains the same as that of the room. If the same liquid is contained in an insulated flask while it vaporises into the air, its temperature falls below that of the surrounding air. Explain the difference in behaviour.

3. Why is distillation under reduced pressure often used in the purification of chemicals?

*8.4 MOLAR MASS DETERMINATION

The volume of vapour formed by a known mass of liquid is measured ...

The gas syringe method of determining the molar mass of a volatile liquid is shown in Figure 8.4A. A weighed quantity of liquid is injected into a gas syringe, where it vaporises, and the volume occupied by the vapour is measured. The method does not give a very accurate result. It is used in conjunction with accurately known empirical formulae to establish the molecular formulae of compounds.

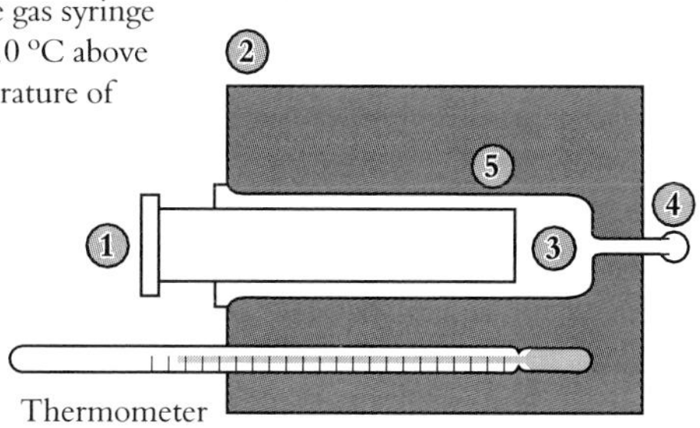

2 Furnace keeps the gas syringe at a temperature 10 °C above the boiling temperature of the liquid.

1 Gas syringe

Thermometer

3 The gas syringe contains air. The volume v_1 cm^3 is noted.

4 Self-sealing rubber cap. Liquid is injected through a hypodermic needle from a weighed syringe (m_1 g). The syringe is reweighed (m_2 g). The mass of liquid injected = $(m_1 - m_2)$ g.

5 When the liquid is injected, the plunger moves backwards. The volume of gas in the barrel is read (v_2 cm^3). Volume of vapour = $(v_2 - v_1)$ cm^3.

Figure 8.4A
Gas syringe method of molar mass determination

... and its molar mass is calculated ...

... A sample calculation

Method of calculation

$$PV = nRT = \frac{m}{M}RT$$

where $V = v_2 - v_1$, and $m = m_1 - m_2$
T = temperature recorded by thermometer,
P = the measured atmospheric pressure
R = the gas constant = 8.314 J K^{-1} mol^{-1}
$\therefore$ M, the molar mass of the gas, can be found.

Example $m_1 = 20.255$ g, $m_2 = 20.120$ g, $v_1 = 9.80$ cm^3, $v_2 = 65.8$ cm^3, $T = 363$ K, $P = 1.01 \times 10^5$ N m^{-2}

When these experimental results are inserted into the equation

$$PV = \frac{m}{M}RT$$

$$M = \frac{0.135 \times 8.314 \times 363}{1.01 \times 10^5 \times 56.0 \times 10^{-6}}$$

The molar mass, $M = 72$ g mol^{-1}

* Check your specification.

CHECKPOINT 8.4: MOLAR MASS OF VOLATILE LIQUIDS

($R = 8.314 \text{ J K}^{-1} \text{ mol}^{-1}$, $1 \text{ atm} = 1.01 \times 10^5 \text{ N m}^{-2}$)

1. When 0.184 g of a liquid was injected into a gas syringe at 45 °C, the volume of gas formed was 55.8 cm^3 at 1 atm. What value does this indicate for the molar mass of the liquid? By referring to the empirical formula CH_2 obtain an accurate value for the molar mass of the liquid.

2. When 0.125 g of a liquid was injected into a gas syringe at 50 °C, it formed 36.8 cm^3 of vapour at atmospheric pressure. The empirical formula of the compound is CH_2Cl. What value of molar mass do the experimental data give? What is the accurate value?

3. A sample of 0.108 g of a liquid of empirical formula CH_2 was vaporised. The volume of vapour was 50.8 cm^3, measured at 22 °C and $9.8 \times 10^4 \text{ N m}^{-2}$. Calculate the molar mass of the liquid.

4. By vaporising 0.100 g of a liquid at 100 °C and $1.01 \times 10^5 \text{ N m}^{-2}$, 20.0 cm^3 of vapour are obtained. Find the molar mass of the liquid.

8.5 SOLUTIONS OF LIQUIDS IN LIQUIDS

8.5.1 RAOULT'S LAW

A **solution** is a homogeneous (i.e., the same all through) mixture of two substances. When one liquid dissolves in another, the saturated vapour pressure of the solution depends on the saturated vapour pressures of the components and on the composition of the mixture. One way of expressing the composition of a mixture of liquids is to state the **mole fraction** of each constituent. By definition

Definition of mole fraction

$$\text{Mole fraction of } \mathbf{A} \text{ in a mixture of } \mathbf{A} \text{ and } \mathbf{B} = \frac{\text{Number of moles of } \mathbf{A}}{\text{Total number of moles}}$$

Using the symbol x for mole fraction

$$x_\mathbf{A} = \frac{n_\mathbf{A}}{n_\mathbf{A} + n_\mathbf{B}}$$

Figure 8.5A shows the saturated vapour pressures of mixtures of liquids **A** and **B**. **A** and **B** are completely miscible: they dissolve in each other in all proportions. The mole fraction of **A**, $x_\mathbf{A}$ is 1.0 at the left hand side of the graph and the saturated vapour pressure is that of pure **A**, $p^0{}_\mathbf{A}$. As the proportion of **B** increases, the mole fraction of **A** decreases and the saturated vapour pressure of **A** decreases along the line $p_\mathbf{A}$ until it reaches zero at $x_\mathbf{A} = 0$. At the left hand side of the graph, the mole fraction of **B**, $x_\mathbf{B} = 0$ and the vapour pressure of **B** = 0. At the right hand side, $x_\mathbf{B} = 1$, and $p_\mathbf{B} = p^0{}_\mathbf{B}$. The total vapour pressure is given by $p_\mathbf{A} + p_\mathbf{B}$.

It is possible to calculate the saturated vapour pressure of each liquid, **A** and **B**, in the mixture. **Raoult's Law** (after F Raoult) states that <u>the saturated vapour pressure of each component is equal to the product of its saturated vapour pressure when pure and its mole fraction</u>. Thus for **A**, the svp ($p_\mathbf{A}$) is given by the svp of pure **A** ($p^0{}_\mathbf{A}$) multiplied by its mole fraction $x_\mathbf{A}$.

$$p_\mathbf{A} = p^0{}_\mathbf{A} \times x_\mathbf{A}$$

Figure 8.5A
Vapour pressure – composition graph for a solution of two liquids

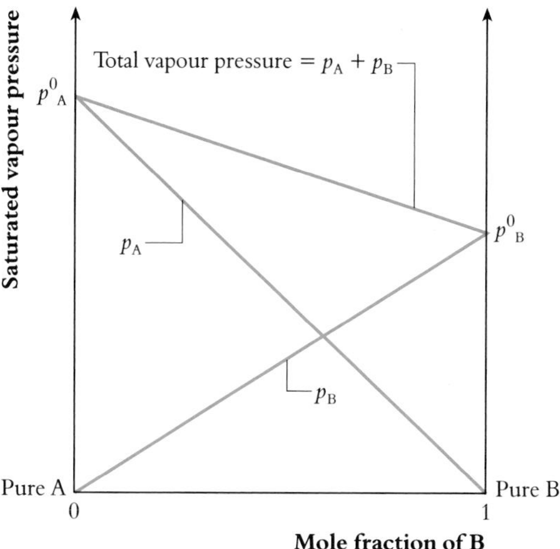

Raoult's Law is obeyed by ideal mixtures. These are mixtures in which the intermolecular forces **A—A**, **A—B** and **B—B** are all equal and, as a result, there is neither a heat change nor a volume change on mixing.

It follows from Raoult's law that the vapour above a mixture is richer than the liquid in the more volatile component, that with the higher saturated vapour pressure. The composition of the vapour in equilibrium with an ideal mixture of **A** and **B** at a constant temperature is shown in Figure 8.5B.

Figure 8.5B
Composition of liquid and vapour for an ideal mixture of liquids

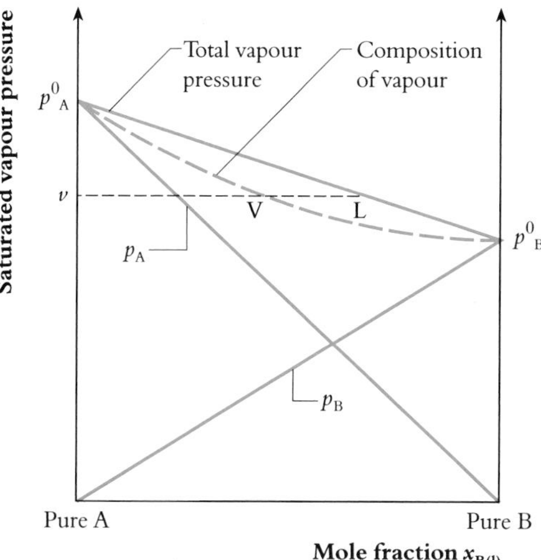

When the total vapour pressure is v, the composition of the liquid is shown by the point L and the composition of the vapour by the point V. The vapour is richer than the liquid in the more volatile component, that with the higher saturated vapour pressure. The difference in composition between the liquid and the vapour is exaggerated in the figure.

A plot of boiling temperature against composition is shown in Figure 8.5C. Liquid **B** with the lower saturated vapour pressure has the higher boiling temperature. Neither the liquid curve nor the vapour curve is linear, even for an ideal mixture. The same shape is obtained whether boiling point is plotted against the mole fraction or the percentage composition by mass.

Figure 8.5C
Boiling temperature – composition curve for an ideal mixture of liquids

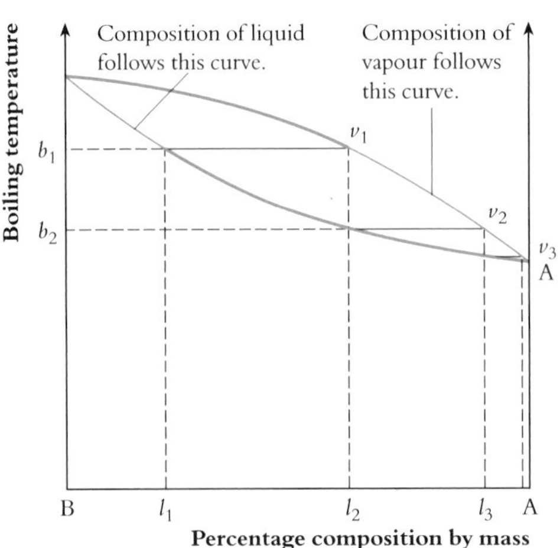

8.5.2 FRACTIONAL DISTILLATION

The process of fractional distillation is used to separate the components of a mixture of liquids. It depends on the difference in their boiling temperatures and on the fact that the vapour above a mixture of liquids is richer in the component with the lower boiling temperature. An apparatus for fractional distillation is shown in Figure 8.5D.

Fractional distillation separates the components of a mixture according to their boiling temperatures

The fractionating column has a large surface area on which ascending vapour and descending liquid come into contact. A mixture rich in the most volatile component distils over at the top of the column, where the thermometer registers its boiling temperature. As distillation continues, the temperature rises towards the boiling temperature of the next most volatile component. The receiver is changed to collect the second component. In this way, the components are distilled over at their boiling temperatures.

Each time that equilibrium is established between liquid and vapour, the vapour becomes richer in the more volatile component

The principles underlying fractional distillation are illustrated in Figure 8.5C. Imagine that a liquid of composition l_1 is heated until it begins to boil at a temperature b_1. Then the vapour in equilibrium with the liquid has composition v_1. If the vapour is condensed by meeting the cold surface of a distillation column, it condenses to form a liquid of composition l_2. This liquid starts to trickle down the column towards the distillation flask. If it is heated, it begins to boil at b_2, to form a vapour of composition v_2. If this vapour is condensed, it forms a liquid of composition l_3. By repeated vaporisation and condensation, the composition of the vapour is made to follow the curve $v_1v_2v_3$, becoming richer and richer in **A**, the more volatile component. The liquid is becoming richer in the less volatile component, **B**, and its composition follows the curve from l_1, towards **B**. The longer the column, the more vaporisation followed by condensation steps will be achieved, and the closer to pure **A** and pure **B** will the **distillate** and **residue** become.

8.5.3 CONTINUOUS FRACTIONAL DISTILLATION

Fuels are obtained from crude oil by fractional distillation

The oil industry uses fractional distillation. Crude petroleum oil is vaporised and fed into a massive fractionating column, which may be 30 to 60 m high and 3 to 6 m in diameter. Different fractions, such as gasoline and kerosene, are drawn off continuously from the column at different levels. Figure 8.5E shows how liquid and vapour attain equilibrium at each level, so that low boiling temperature fractions rise to the top of the column and high boiling temperature fractions are drawn off from the bottom of the column.

Figure 8.5D
Fractional distillation

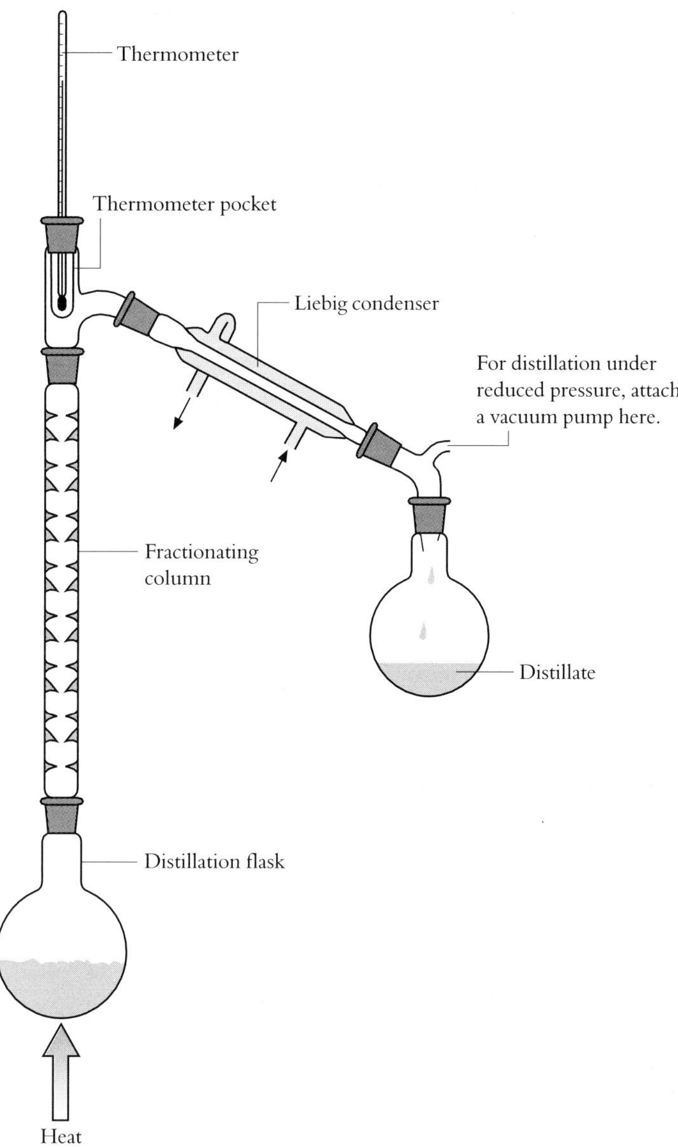

8.5.4 NON-IDEAL SOLUTIONS

Raoult's Law is obeyed by ideal solutions ...

... non-ideal solutions have vapour pressures greater or less than those predicted

Solutions which have a vapour pressure greater than that predicted from Raoult's Law are said to show a **positive deviation** from the law. Those with a vapour pressure lower than the calculated value are said to show a **negative deviation**. [See Figure 8.5F.] Typical of a pair of liquids showing a slight positive deviation are hexane and ethanol. The molecules are very different, and molecules of ethanol can more easily escape from a mixture of ethanol and hexane than from pure ethanol, where hydrogen bonding holds molecules together. A slight negative deviation from Raoult's Law is shown by a mixture of trichloromethane, $CHCl_3$, and ethoxyethane, $C_2H_5OC_2H_5$. Forces of attraction between the two kinds of molecules tend to prevent the molecules escaping into the vapour phase. When a pair of liquids shows a very large positive deviation from Raoult's Law, the vapour pressure–composition curve has a maximum. A system with a very large negative deviation shows a minimum [see Figure 8.5F].

Figure 8.5E
Continuous fractional distillation of petroleum oil

3 Ascending vapour must pass through *bubble caps*, which bring it into contact with the liquid in the plates.

2 Baffle plates, trays in which condensate collects.

1 Vaporised petroleum oil is fed in.

4 Overflow pipes carry liquid from baffle plates down to a lower level.

Petroleum gases
Aviation fuel
Gasoline
Kerosene
Gas oil
Diesel oil
Lubricating oils and waxes
Fuel oil
Bitumen

Figure 8.5F
Saturated vapour pressure–composition curves for non-ideal solutions

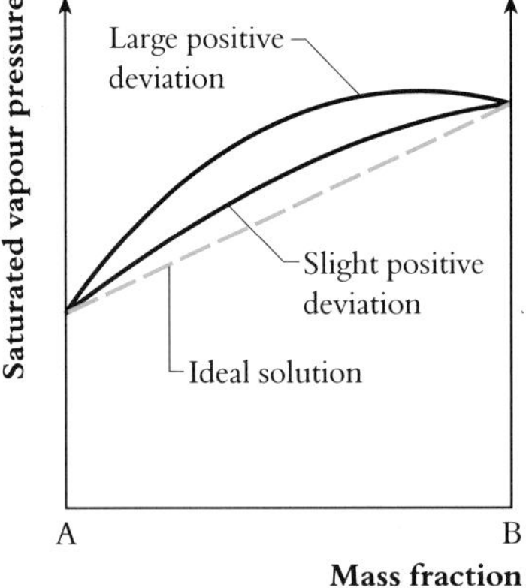

(a) Positive deviation

Saturated vapour pressure
Ideal solution
Slight negative deviation
Large negative deviation
C
D
Mass fraction

(b) Negative deviation

Figure 8.5G
Boiling temperature–composition curves for non-ideal solutions

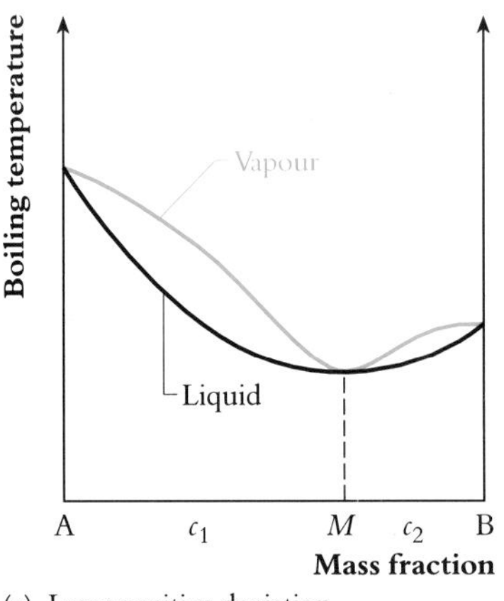

(a) Large positive deviation

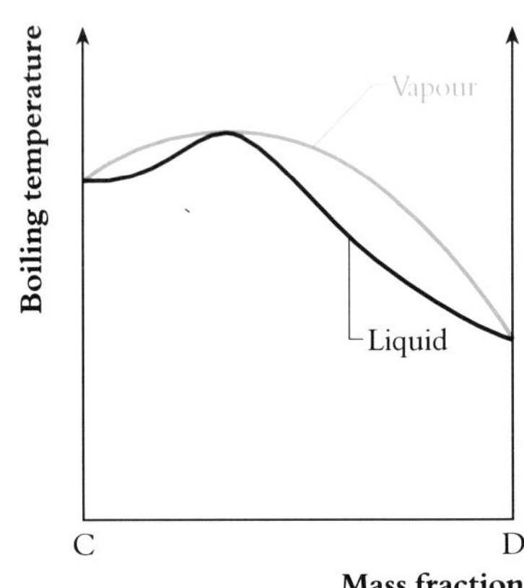

(b) Large negative deviation

Boiling temperature–composition curves for systems which show gross deviations from Raoult's Law are shown in Figure 8.5G.

If there is a maximum or minimum in the svp–composition curve, an azeotrope distils

A maximum in the saturated vapour pressure curve results in a minimum in the boiling temperature–composition curve and vice versa. An application of the principles of fractional distillation [§ 8.5.2] shows that, in Figure 8.5G(a), a liquid of composition c_1 can be separated by distillation into almost pure **A** and a mixture of composition **M**. A liquid of composition c_2 can be separated by distillation into almost pure **B** and the mixture **M**. If a liquid of composition **M** is boiled, the vapour has the same composition as the liquid. This means that the composition, and therefore the boiling temperature, T_b, do not change as distillation is continued. The mixture is therefore called an **azeotropic** or **constant-boiling mixture**. Examples are nitric acid/water (T_b is a maximum) and benzene/ethanol (T_b is a minimum). Azeotropes are *not* classified as compounds because their compositions vary with pressure.

8.5.5 STEAM DISTILLATION

A substance which decomposes at its boiling temperature can be distilled safely in steam

A number of organic compounds decompose on heating and cannot be distilled at atmospheric pressure. **Steam distillation** is one solution to the problem. Steam is passed into a heated liquid, and a mixture of water and the liquid distils over at a temperature between 100 °C and the boiling temperature of the liquid [Figure 8.5H]. The method depends on Raoult's law. The saturated vapour pressure of the mixture of water and the organic liquid is equal to the sum of the saturated vapour pressures of the pure components. When the total saturated vapour pressure is equal to 1 atm, the mixture distils. The mixture therefore distils below the temperature at which the saturated vapour pressure of the pure liquid would reach 1 atm. For example, in the preparation of phenylamine [§ 32.6.2] a mixture of phenylamine and water distils at 98 °C at 1 atm, well below the boiling temperature of phenylamine, 190 °C. Phenylamine is extracted from the distillate is with ethoxyethane [see § 8.6].

Figure 8.5H
Steam distillation

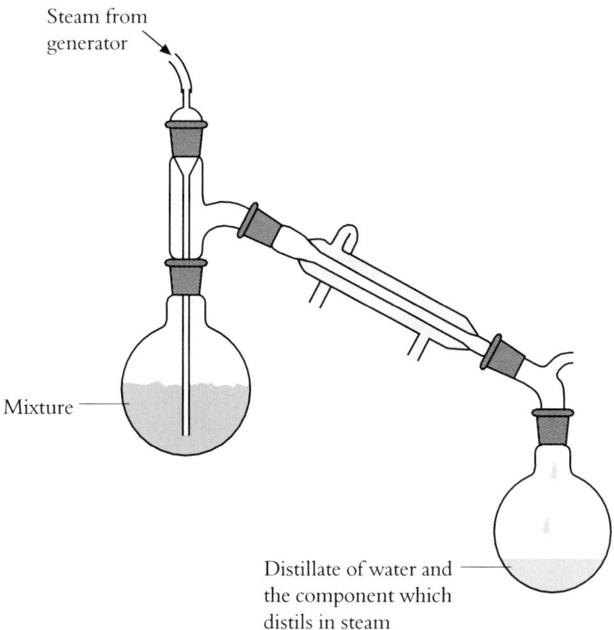

SUMMARY

Raoult's Law

svp of component of a mixture = svp of pure component × mole fraction of component

(svp = saturated vapour pressure)

A high svp means a low boiling temperature. Fractional distillation separates liquids on the basis of their boiling temperatures which are determined by their saturated vapour pressures.

CHECKPOINT 8.5: RAOULT'S LAW

1. **X** and **Y** are two immiscible liquids which form an ideal mixture. At 20 °C, their saturated vapour pressures are $p^0_X = 25$ kPa, $p^0_Y = 45$ kPa. The mixture contains 1 mole of **X** and 4 moles of **Y** at 20 °C. Calculate the partial pressures of **X** and **Y** and the total pressure.

2. Liquids **A** and **B** have saturated vapour pressures of 15 kPa and 40 kPa at 25 °C. Calculate the saturated vapour pressure of an ideal solution of 2 moles of **A** in 3 moles of **B** at 25 °C.

3. Explain how fractional distillation separates crude oil into different products.

4. When a mixture of two liquids shows a positive deviation from Raoult's Law, is the total saturated vapour pressure greater or less than calculated? Is the boiling temperature higher or lower than expected?

8.6 PARTITION OF A SOLUTE BETWEEN TWO SOLVENTS

A pair of immiscible liquids form two layers. When a solute is added it may dissolve in both liquids. In this case the solute will divide itself between the two layers. It may well be more soluble in one solvent than the other. For example iodine dissolves both in water (with a low solubility) and in organic solvents such as 1,1,1-trichloroethane. The way in which iodine is distributed – or partitioned – between two solvents can be studied by shaking the mixture in a separating funnel [see Figure 8.6A].

Figure 8.6A
Partition of iodine between two solvents

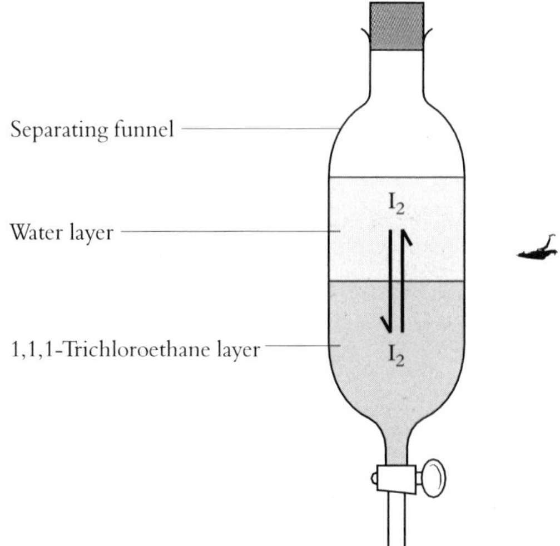

The mixture is shaken well and allowed to stand while the layers separate. Iodine can be seen in the water layer as a pale brown solution and in the organic layer as a pink solution. Iodine molecules pass from one layer to another until equilibrium is reached.

$$I_2(aq) \rightleftharpoons I_2(\text{organic layer})$$

Then the two layers can be run off separately and analysed. Titration against sodium thiosulphate solution gives the concentration of iodine in each layer. The results of experiments with different amounts of iodine and different volumes of solvents show that the ratio of iodine concentrations in the two solvents is always the same.

$$\frac{[I_2(aq)]}{[I_2(\text{trichloroethane})]} = \text{Constant}$$

The constant is called the **partition coefficient** or **distribution coefficient**. It is constant for a particular temperature. The general form of this equation is expressed as:

$$\frac{[\text{Solute in upper layer}]}{[\text{Solute in lower layer}]} = \text{Partition coefficient}$$

It holds provided that there is insufficient solute to saturate either solvent.

8.6.1 SOLVENT EXTRACTION

Often in the preparation of an organic compound the product is obtained as an aqueous solution. The chemist needs to separate the product from water. Solvent extraction is used. Organic compounds are more soluble in organic solvents than in water.

Ethoxyethane (ether) is a good solvent for many organic compounds, it is immiscible with water and it has a low boiling temperature (35 °C). An aqueous solution of the product is shaken with ethoxyethane in a separating funnel. The product distributes itself between the two layers in accordance with its partition coefficient between ethoxyethane and water. If the partition coefficient is high, a large fraction of the product will pass into the ethoxyethane layer. The ether layer is separated from the aqueous layer and dried by the addition of an anhydrous salt such as anhydrous calcium chloride. With its low boiling temperature, ether is easily distilled off to leave the product. The distillation is done in a fume cupboard because ether is very flammable and the vapour is denser than air and tends to accumulate at bench level where it can catch fire.

SUMMARY

When a solute dissolves in both of a pair of immiscible liquids the distribution law states:
[solute in upper layer]/[solute in lower layer] = Constant, k
k = distribution coefficient or partition coefficient.
Solvent extraction is used to extract an organic product from an aqueous solution.

It is more efficient to use a certain volume of ether in portions for repeated extractions than to use it all in one operation. Consider an aqueous solution of 5.00 g of a product **X** in 1.00 dm^3 of water. The partition coefficient [**X** in ether]/[**X** in water] = 80.

(a) Let the mass of **X** extracted by 100 cm^3 of ether = a g

$$\frac{a/100}{(5.00 - a)/1000} = 80.0 \quad \text{and } a = 4.44 \text{ g}$$

(b) Let the masses of **X** extracted by two successive 50 cm^3 portions of ether be (i) b g and (ii) c g.
then

(i) $$\frac{b/50}{(5.00 - b)/1000} = 80.0 \quad \text{and } b = 4.00 \text{ g}$$

(ii) $$\frac{c/50}{(1.00 - c)/1000} = 80.0 \quad \text{and } c = 0.80 \text{ g}$$

The total is 4.80 g of **X** when two portions of ether are used compared with 4.44 g of **X** when the ether is used as one portion.

CHECKPOINT 8.6: PARTITION

1. The partition coefficient of **S** between ethoxyethane and water is 5.0. A solution containing 10.0 g of **S** in 5000 cm^3 of water is extracted with 100 cm^3 of ethoxyethane. Which of the following gives the mass of **S** extracted from the water?

   A 2.5 g **B** 5.0 g **C** 7.5 g **D** 10.0 g **E** 12.5 g

2. An aqueous solution contains 5.0 g of **X** in 100 cm^3 of solution. The partition coefficient of **X** between water and an organic solvent is 0.200. Calculate the mass of **X** extracted by shaking 100 cm^3 of the aqueous solution with

   (a) 50 cm^3 of the solvent
   (b) two successive 25 cm^3 portions of the solvent.

8.7 PARTITION CHROMATOGRAPHY

8.7.1 THE THEORY

Chromatographic separations use repeated partition of solutes between two solvents

It has been shown in an example in § 8.6.1 and in Question 2, Checkpoint 8.6, that repeated extractions with organic solvents are effective in removing a considerable quantity of solute from an aqueous solution. The technique of repeated extractions can be applied to the separation of a number of solutes in a solution, provided that the solutes differ in their solubility in a second solvent.

In column chromatography ...

... the stationary phase is adsorbed on a solid which packs the column ...

In partition chromatography, many extractions are performed in succession in one operation. The solutes are partitioned between a **stationary phase** and a **mobile phase**. The stationary phase is a solvent (often water) **adsorbed** (bonded to the surface) on a solid. This may be paper or a solid such as alumina or silica gel, which has been packed into a column or spread on a glass plate. The mobile phase is a second solvent which trickles through the stationary phase.

8.7.2 COLUMN CHROMATOGRAPHY

... The moving phase is a second solvent ...

The *column* is a glass tube packed with an **adsorbent** (e.g., alumina, silica, starch, magnesium silicate). The solid adsorbent is made into a slurry with water, poured in and allowed to settle. [See Figure 8.7A.] A solution of the mixture to be analysed is poured on to the top of the column so that the components can be adsorbed on the column. The second solvent, called the **eluant**, is allowed to trickle slowly through the column.

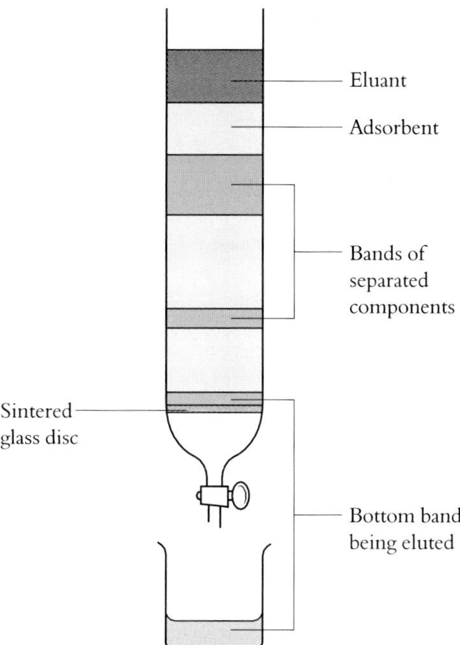

Figure 8.7A
Column chromatography

... Solutes partition between the solvents ...

... The rate at which a solute travels down the column depends on its partition coefficient

Partition of the solutes takes place between the moving solvent and the stationary adsorbed water, and the eluant carries the solutes with it, to be readsorbed further down the column. The rates at which the solutes travel down the column depend on their partition coefficients. If these are sufficiently different a complete separation is achieved, and the components become spread out along the column in order of their partition coefficients. The separated bands are called a **chromatogram**. The components can be recovered by dissolving them separately out of the column – **eluting** them – with more solvent, and then evaporating off the solvent.

8.7.3 PAPER CHROMATOGRAPHY

In paper chromatography ...

... the stationary phase is water adsorbed on paper ...

... The solvent ascends ...

In paper chromatography, the principle is the same as for column chromatography, but the adsorbent is a thin film of water held on paper. A solution of the mixture to be separated is applied to a strip of chromatography paper. This is hung in a glass tank so that the end dips into solvent in the bottom of the tank [see Figure 8.7B]. The solvent may be water, ethanol, glacial ethanoic acid, butanol, etc., or a mixture of any of these. As the solvent rises through the paper, it meets the sample, and the solutes partition between the mobile phase and the stationary phase. The separation is stopped when the solvent has travelled nearly to the top of the paper. The distance travelled by the **solvent front** is measured. Then for each component, the $\mathbf{R_F}$ **value** is calculated, by

$$R_F = \frac{\text{Distance travelled by the component}}{\text{Distance travelled by the solvent front}}$$

The R_F value can be used to assist in identifying the components of a mixture. The separated components can be obtained by cutting the paper into strips and dissolving out each compound.

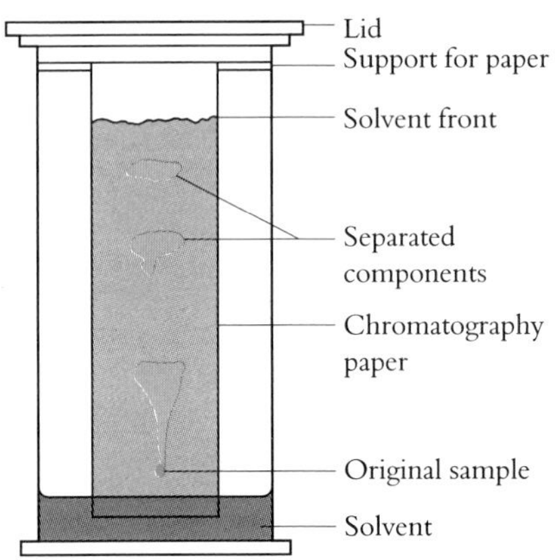

Figure 8.7B Paper chromatography

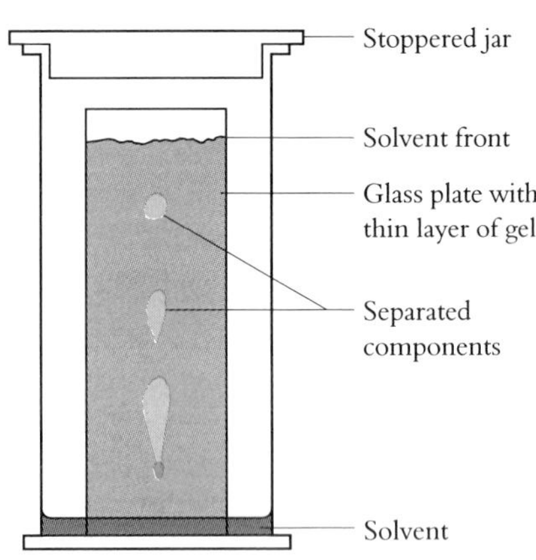

Figure 8.7C Thin layer chromatography

... or descends through the paper

There is a different design of chromatography tank, with a trough at the top of the tank to hold the solvent. The paper hangs down from this trough, and solvent descends through the paper. The two techniques are naturally described as **ascending** and **descending paper chromatography**.

8.7.4 THIN LAYER CHROMATOGRAPHY

A thin layer of adsorbent solid can be used for chromatography

The solid adsorbent may be in the form of a *thin layer* on the surface of a glass plate. The adsorbent (e.g., silica gel or calcium sulphate) is made into a thick paste and spread evenly over a glass plate. The thin layer of paste is allowed to dry and baked in an oven. Spots of the mixture are applied and a chromatogram is developed in the same manner as for paper chromatography. [See Figure 8.7C.] Thin layer chromatography has the advantage that a variety of different adsorbents can be used. Since the thin layer is more compact than paper, more equilibrations take place while the solvent front travels over the same distance. Often separations can be achieved in a few centimetres, and coated microscope slides are frequently used for thin layer chromatography.

8.7.5 GAS–LIQUID CHROMATOGRAPHY

In gas–liquid chromatography, the stationary phase is a liquid adsorbed on a solid ...

... The mobile phase is a stream of carrier gas

In gas–liquid chromatography, a gas (or volatile liquid or solid) is separated into its components by equilibration with a liquid. The liquid is the stationary phase. It is spread on the surface of inert solid particles which pack the column. Long (5 to 10 m) and narrow (2 to 10 mm bore), the column is often wound into a coil. The mobile phase is a stream of 'carrier gas', nitrogen or a noble gas. When a sample of volatile mixture is injected into the carrier gas, each component sets up a partition equilibrium between the vapour phase and the liquid phase. Some components are less soluble in the liquid phase than others and emerge from the column ahead of the others. A detector records each component as it leaves the column. [See Figure 8.7D.] The emerging gases can also be analysed by a mass spectrometer [§ 1.9.1].

SUMMARY

Partition of a solute between two immiscible solvents is the basis of chromatography. Many partitions take place in succession. The solutes partition between a mobile phase and a stationary phase. The stationary phase is a solvent adsorbed on a solid, e.g. paper or alumina. The mobile phase is a second solvent. Column chromatography, paper chromatography, thin layer chromatography and gas–liquid chromatography all employ partition.

Figure 8.7D
Gas–liquid chromatography

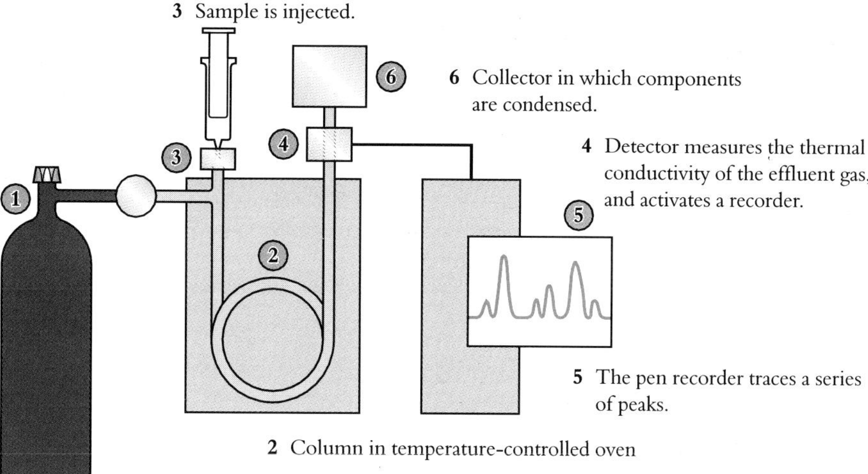

3 Sample is injected.
6 Collector in which components are condensed.
4 Detector measures the thermal conductivity of the effluent gas, and activates a recorder.
5 The pen recorder traces a series of peaks.
2 Column in temperature-controlled oven
1 Cylinder of carrier gas with valve to control the flow rate.
1
2
3
4
5
6

8.7.6 ION EXCHANGE

Ion exchange resins replace cations by $H^+(aq)$ ...

... and anions by $OH^-(aq)$

SUMMARY

Ion exchange is a type of partition. The method can be used to purify water. A resin removes cations and anions and replaces them with hydrogen ions and hydroxide ions to give deionised water.

Ion exchange is a type of partition of ionic compounds. The **ion exchange resin** is a polymer which contains at intervals polar groups which can remove undesirable cations (or anions) and replace them with other cations (or anions). The water softener Permutit®, is sodium aluminium silicate, which replaces calcium and magnesium ions in hard water by sodium ions. To purify water, a resin must replace all the cations and anions present by hydrogen ions and hydroxide ions. A combination of a **cation exchanger** and an **anion exchanger** is needed. Cation exchange resins often contain sulphonic acid groups, $—SO_3^-H^+$, and anion exchangers often contain quaternary ammonium groups, e.g., $—\overset{+}{N}(CH_3)_3OH^-$. If water passes slowly through a **deioniser**, equilibrium is set up at each level between hydrogen ions attached to the resin and hydrogen ions in solution:

$$—SO_3^-H^+(\text{resin}) + Na^+(aq) \rightleftharpoons —SO_3^-Na^+(\text{resin}) + H^+(aq)$$

As the water moves on, two of the components in the equilibrium ($H^+(aq)$ and $Na^+(aq)$) are removed and a fresh equilibrium is established. Each successive equilibration increases the replacement of metal cations by hydrogen ions. At the same time, hydroxide ions are replacing other anions. If tap water is run slowly through an ion exchange resin, 'deionised' water of high purity can be obtained.

FURTHER STUDY ▲

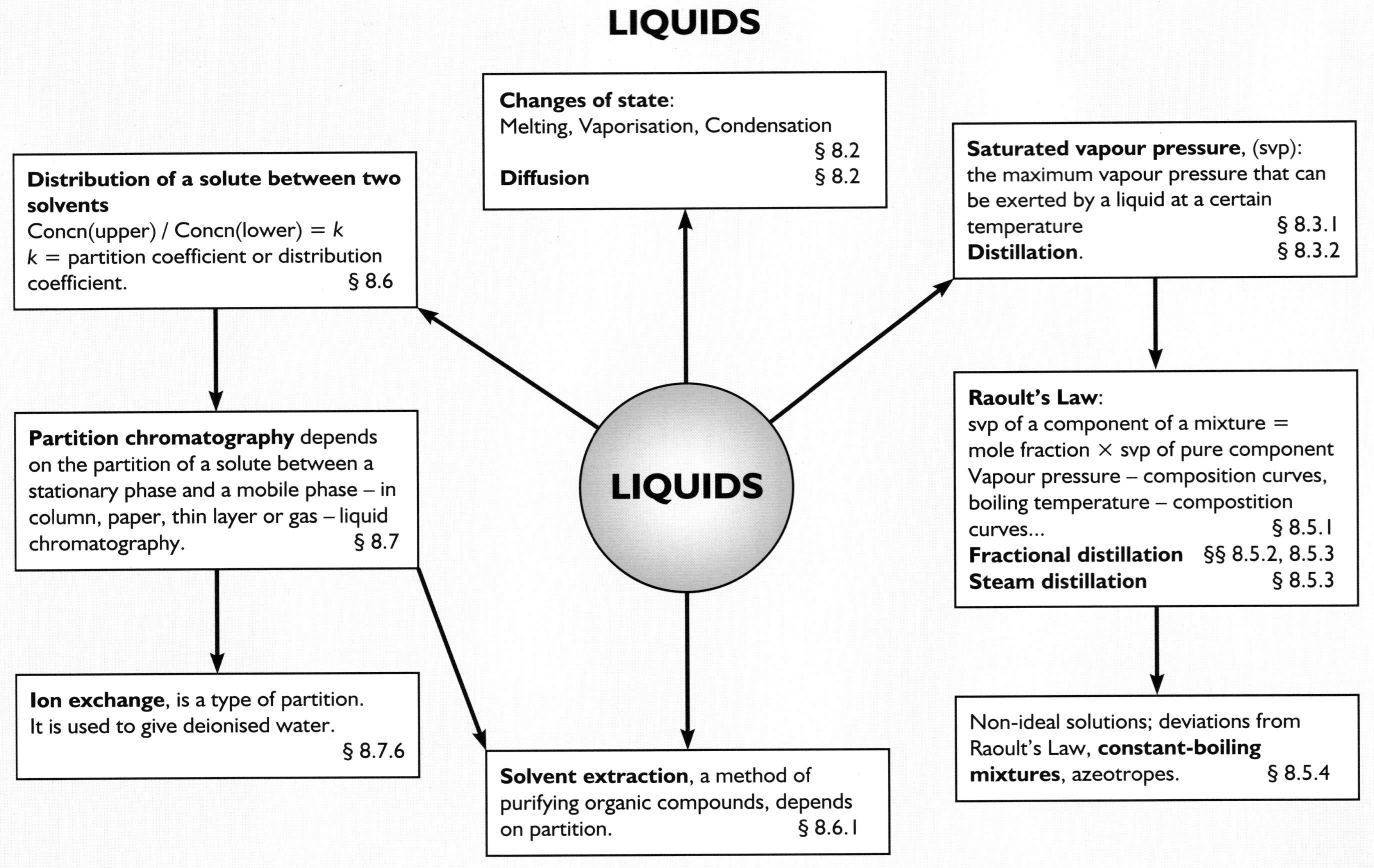
LIQUIDS
Changes of state: Melting, Vaporisation, Condensation § 8.2
Diffusion § 8.2
Distribution of a solute between two solvents
Concn(upper) / Concn(lower) = k
k = partition coefficient or distribution coefficient. § 8.6
Saturated vapour pressure, (svp): the maximum vapour pressure that can be exerted by a liquid at a certain temperature § 8.3.1
Distillation. § 8.3.2
LIQUIDS
Partition chromatography depends on the partition of a solute between a stationary phase and a mobile phase – in column, paper, thin layer or gas – liquid chromatography. § 8.7
Raoult's Law: svp of a component of a mixture = mole fraction × svp of pure component
Vapour pressure – composition curves, boiling temperature – compostition curves... § 8.5.1
Fractional distillation §§ 8.5.2, 8.5.3
Steam distillation § 8.5.3
Ion exchange, is a type of partition. It is used to give deionised water. § 8.7.6
Solvent extraction, a method of purifying organic compounds, depends on partition. § 8.6.1
Non-ideal solutions; deviations from Raoult's Law, constant-boiling mixtures, azeotropes. § 8.5.4

QUESTIONS ON CHAPTER 8

1. Explain
(a) how liquids mix
(b) why some of a liquid evaporates below its boiling temperature.

2. Describe practical methods which you have used in the laboratory for column chromatography, paper chromatography and thin-layer chromatography. Give examples of mixtures which can be separated by these methods.

3. Discuss the physical principles involved in
(a) the ethoxyethane extraction of a product from aqueous solution
(b) column chromatography.

4. Compare the fractional distillation of an ideal mixture, such as hexane ($T_b = 69$ °C) and heptane ($T_b = 98.5$ °C) with that of a non-ideal mixture, such as ethanol ($T_b = 78$ °C) and water. Draw boiling temperature–composition curves for the mixtures.

5. Define the term *partition coefficient*. A compound **Z** has a partition coefficient of 4.00 between ethoxyethane and water. Calculate the mass of **Z** extracted from 100 cm^3 of an aqueous solution of 4.00 g of **Z** by two successive extractions with 50 cm^3 of ethoxyethane.

*__6.__ *(a)* State Raoult's Law as applied to mixtures of miscible liquids.
(b) At 50 °C the vapour pressure of hexane is 54.0 kPa and that of heptane 22.0 kPa. Draw graphs on the same axes showing how the partial vapour pressures of hexane and heptane and the total vapour pressure vary with the mole fraction for mixtures of hexane and heptane. Assume the mixture obeys Raoult's Law. Label the graphs.
(c) Draw a temperature–composition diagram for the liquid and vapour phases of mixtures of hexane and heptane. Label the diagram.

*__7.__ *(a)* Draw diagrams to show how boiling temperature varies with composition for mixtures of
(i) methanol ($T_b = 65$ °C) and ethanol ($T_b = 78$ °C), a mixture which obeys Raoult's Law almost ideally
(ii) cyclohexane ($T_b = 81$ °C) and ethanol ($T_b = 78$ °C), a mixture which shows a positive deviation from Raoult's Law.
Label the diagrams. Show the compositions of the liquid and the vapour.
(b) Explain why methanol and ethanol form an ideal mixture while a mixture of cyclohexane and ethanol shows a positive deviation.

Note: * means 'check your specification'.

*__8.__ State Raoult's Law.
(a) Some liquid mixtures which show a positive deviation from the law can form a constant–boiling (azeotropic) mixture. Explain how this happens with the aid of a boiling temperature–composition diagram.
(b) (i) What is meant by a negative deviation from Raoult's law?
(ii) A mixture of **A** and **B** shows a negative deviation from Raoult's Law. Why is it that when **A** and **B** are mixed the temperature rises?

9. *(a)* Briefly explain the term 'solvent extraction'.
(b) Without going into calculations, explain why it is more effective to use successive small volumes of solvent rather than one large volume.

10. You are given the problem of finding the partition coefficient of ethanoic acid between water and 2-methylpropan-1-ol. Describe what measurements you would make.

11. *(a)* (i) Sketch a graph of **V** (the volume of an ideal gas) against 1/**P** (the reciprocal of its pressure) at constant temperature. **1**
(ii) State **two** factors which cause real gases to deviate from ideal behaviour. **2**
(b) (i) Sketch a graph to show the variation of the saturated vapour pressure of a liquid with temperature. **1**
(ii) Explain why the boiling temperature of a liquid varies with the external pressure. **1**
(iii) Give **one** reason why some liquids are purified by distillation under reduced pressure. **1**
(c) Calculate the molar mass of a volatile organic liquid, given that 0.597 g of the liquid when vaporised gives an ideal gas of volume 153 cm^3 at 100 °C and 1.01×10^5 Pa pressure (1 atmosphere). **3**
[1 mole of ideal gas occupies 2.24×10^4 cm^3 at 0 °C and 1.01×10^5 Pa pressure (1 atmosphere).]
9 *WJEC (AS/AL)*

12. *(a)* (i) Iodine, (I_2), is a **molecular solid** at temperatures up to 30 °C. At temperatures in excess of 30 °C iodine sublimes (changes state from a solid to a vapour).
I. Explain the term **molecular solid**, by reference to the bonding present within solid iodine. $\mathbf{1\frac{1}{2}}$
II. Explain the structural changes which occur when iodine passes directly from the solid to the vapour phase. $\mathbf{1\frac{1}{2}}$
(ii) Describe a simple model for bonding in metals and use this to explain the electrical conductivity of metals. **2**

(b) Using a simple kinetic–molecular model describe
(i) the nature of the liquid state, and $1\frac{1}{2}$
(ii) vaporisation and the variation of vapour pressure with temperature. $2\frac{1}{2}$

(c) A sample of helium was compressed at **273 K** from a volume of 183 cm^3 to a volume of 0.3 cm^3.
(i) Calculate the final pressure of the helium sample if the initial pressure was 4.78 Pa. $1\frac{1}{2}$
(ii) Calculate the number of moles of helium present in the above sample. $1\frac{1}{2}$
[1 mole of ideal gas occupies 2.24×10^4 cm^3 at 273 K and 1.01×10^5 Pa.]

12 *WJEC (AS/AL)*

13. (a) (i) The graph below shows the variation of pressure with volume at 298 K for 1 mol each of an ideal gas and a real gas.

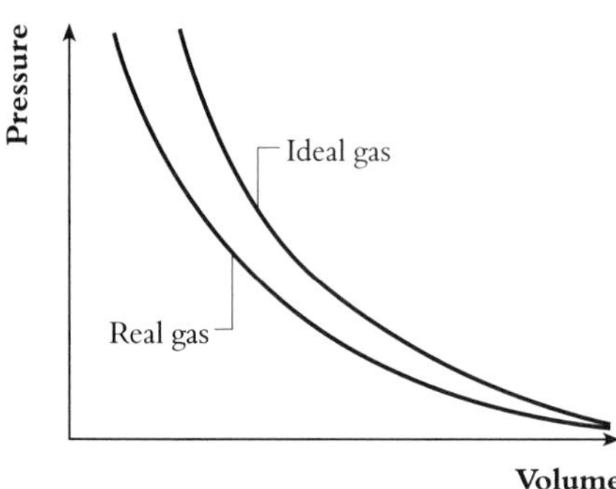

Explain why
(1) at low pressures, the curve for the real gas approaches that for the ideal gas
(2) at 298 K and when contained in vessels of identical volume, 1 mol of N_2 exerts a greater pressure than 1 mol of HCl.

(ii) At 298 K, 1.0 dm^3 of N_2 at 0.20 Pa pressure was mixed with 2.0 dm^3 of O_2 at 0.40 Pa in a 4.0 dm^3 container. Assuming that both N_2 and O_2 behave ideally, calculate the pressure of the gaseous mixture at 298 K. **5**

(b) Briefly describe each of the following intermolecular forces and illustrate your answer with an appropriate example in each case: dipole–dipole interactions, hydrogen bonding, van der Waals' forces **4**

(c) (i) Two miscible liquids *A* and *B* form ideal solutions when mixed. At 298 K, the vapour pressures of pure *A* and pure *B* are 32 kPa and 16 kPa respectively.

For a mixture of 1 mol of *A* and 3 mol of *B* at 298 K, calculate:
(1) the vapour pressure of the mixture
(2) the mole fraction of *A* in the vapour which is in equilibrium with the mixture.

(ii) Predict whether the following pairs of liquids when mixed, would give solutions showing positive deviation or negative deviation from Raoult's Law. Explain your prediction.

(1) CH_3CH_2OH and 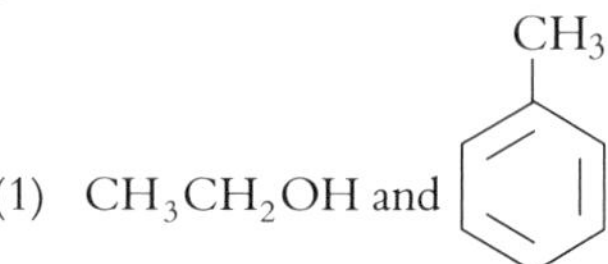

(2) $C_2H_5OC_2H_5$ and $CHCl_3$

8 *HKALE (AL)*

14 (a) Explain the meaning of the terms
(i) *empirical formula* **2**
(ii) *molecular formula*. **2**

(b) An organic compound, **X**, has the following composition by mass:

C = 38.7% H = 9.7% O = 51.6%

[A_r(C) 12.0; A_r(H) 1.0; A_r(O) 16.0]

What is the *empirical* formula of the compound **X**? **2**

(c) In an experiment to determine the relative molecular mass of **X**, 0.500 g of **X** occupied 2.34×10^{-4} m^3 at a pressure of 1.00×10^5 Pa and a temperature of 350 K. [Gas constant, $R = 8.31\ J\ K^{-1}\ mol^{-1}$; $1\ J = 1\ Pa\ m^3$]
(i) State the ideal gas equation. **1**
(ii) Use the ideal gas equation to determine the number of moles of **X** used in this experiment. **2**
(iii) From your answer to (ii) determine the relative molecular mass of **X**. **1**
(iv) From your answers to parts *(b)* and *(c)*(iii) deduce the molecular formula of **X**. **1**

11 *O&C (AS/AL)*

9 SOLUTIONS

9.1 AN UNUSUAL SOLVENT

There are many people who prefer to drink decaffeinated coffee. They like the taste of coffee but they do not want the stimulation that caffeine gives, which keeps some people awake at night. Organic solvents were originally used to dissolve the caffeine out of coffee beans. Benzene, which is flammable, was the first used and was replaced by trichloroethene, which is now found to be a carcinogen. Dichloromethane is still used, but a better solvent is taking over. This is supercritical fluid carbon dioxide.

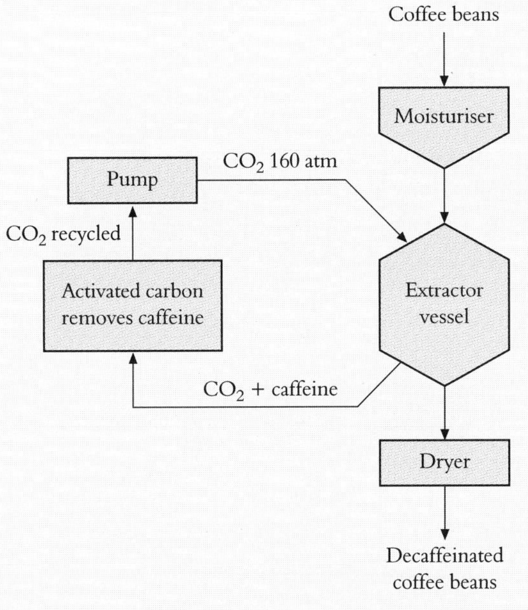

Figure 9.1A
Dissolving the caffeine out of coffee beans

Carbon dioxide has a critical temperature (above which it cannot be liquefied) of 31 °C and a critical pressure (below which it cannot be liquefied) of 74 atm. Above its critical temperature and at a pressure above its critical pressure, carbon dioxide is a **supercritical fluid**, that is, a fluid in which the distinction between liquid and gas disappears. The properties of supercritical fluids, being in between those of liquids and gases, find them many uses. Carbon dioxide as a supercritical fluid is a valuable solvent. It has the solvent properties of a liquid and flows like a gas. The first success of supercritical fluid carbon dioxide was in the decaffeination of coffee beans.

There are many advantages of supercritical fluid carbon dioxide over organic solvents. It does not damage the ozone layer. The use of carbon dioxide does not add to the greenhouse effect because the gas is extracted from the air during the production of oxygen. The carbon dioxide solvent can be recycled. Carbon dioxide is preferred over other supercritical fluids because it is readily available, completely safe to use in food processing, non-flammable and supercritical at lower pressure than some alternatives.

Hop flavours for use in beer making are also extracted by supercritical fluids on a large scale. On a smaller scale oils, flavours and essences are extracted for the pharmaceutical industry from plants and micro-organisms.

9.2 SOLUTIONS OF SOLIDS IN LIQUIDS

A saturated solution is defined

A solution which contains the maximum amount of solute that will dissolve at the temperature is **saturated**. If it contains less solute than this it is **unsaturated**, and when more solute is added more will dissolve until the solution is saturated. In some cases when a hot saturated solution is prepared and then cooled, more than the usual maximum amount of solute remains. The solution is then **supersaturated**. If a small crystal is added to 'seed' it or the solution is stirred, the excess of solute crystallises, leaving a saturated solution.

A saturated solution is in equilibrium with excess solute [see Figure 9.2A]. The rate at which solute particles dissolve is the same as the rate at which solute particles recrystallise

$$\text{Solute(undissolved)} \underset{\text{Recrystallises}}{\overset{\text{Dissolves}}{\rightleftharpoons}} \text{Solute(dissolved)}$$

Figure 9.2A
Equilibrium in a saturated solution

A saturated system and undissolved solid are a system in equilibrium

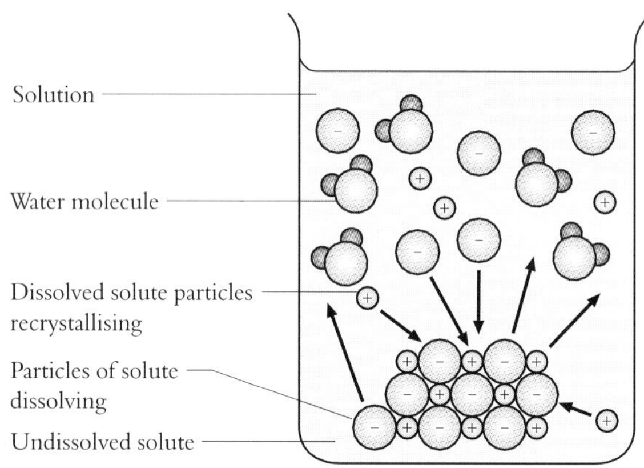

Solubility is defined ...

... its units given ...

... and methods of measurement

The maximum quantity of solid that dissolves in a certain quantity of liquid at a stated temperature is the **solubility** of the solid. It can be stated in g solute dm^{-3} of solution or mol dm^{-3} solution or g solute kg^{-1} solvent. The solubility of a solute can be found by (a) titration against a suitable reagent (b) evaporating known mass of solution to dryness and weighing the solute that remains and (c) instrumental methods such as visible-ultraviolet spectrometry [§ 35.7].

Figure 9.2B
A solubility curve

In this solubility curve, solubility = mass of solute per kg of solvent

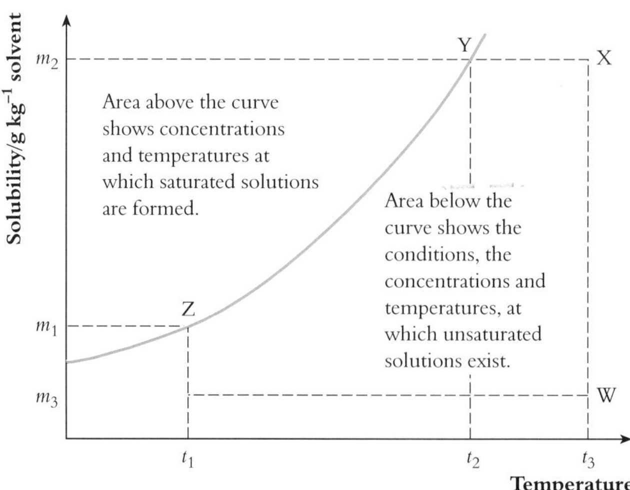

Dissolution may be endothermic or exothermic

Solubility is affected by temperature. Most solids increase in solubility with temperature. For these solids, dissolution is endothermic:

$$\text{Solute} + \text{Solvent} + \text{Heat} \rightleftharpoons \text{Solution}$$

Some solids decrease in solubility with temperature. These are solids for which dissolution is exothermic:

$$\text{Solute} + \text{Solvent} \rightleftharpoons \text{Solution} + \text{Heat}$$

This is discussed in § 10.15. A graph of solubility against temperature is called a **solubility curve** [see Figure 9.2B].

A solubility curve shows the effect of temperature on solubility ...

... and enables one to predict the quantity of solute that will crystallise when a solution is cooled

A solution at point **X** contains m_2 g of solute in 1 kg of solvent at a temperature t_3. It is an unsaturated solution, and when it is cooled it does not deposit solute until it reaches a temperature t_2. At t_2, solute is deposited, and the concentration of the solution decreases. On further cooling, more solute is deposited, and the concentration of the solution again decreases. As the temperature falls, the concentration follows the curve **YZ**. At the temperature t_1, the mass of solute that can remain in solution is m_1 g. The mass of solute that has crystallised between t_2 and t_1 is $(m_2 - m_1)$ g. A solution **W**, containing m_3 g of solute in 1 kg of solvent at t_3, can be cooled from t_3 to t_1 without depositing any solute as the solubility is never exceeded.

9.3 RECRYSTALLISATION

Solids are purified by **recrystallisation** from a suitable solvent, e.g., water, ethanol, propanone. The impure material is dissolved in the minimum quantity of hot solvent. The hot solution is filtered to remove insoluble impurities. A heated Buchner funnel and flask [see Figure 9.3A] are used to prevent cooling of the solution and crystallisation of the solute. The filtered solution is allowed to cool so that the solute crystallises out. The impurities are present in smaller quantities and remain in solution at the lower temperature. The crystals are filtered, washed in the funnel with a little cold solvent, and dried. The purity of a solid is assessed by finding its melting temperature. [See § 35.2.]

Figure 9.3A
Filtration under reduced pressure

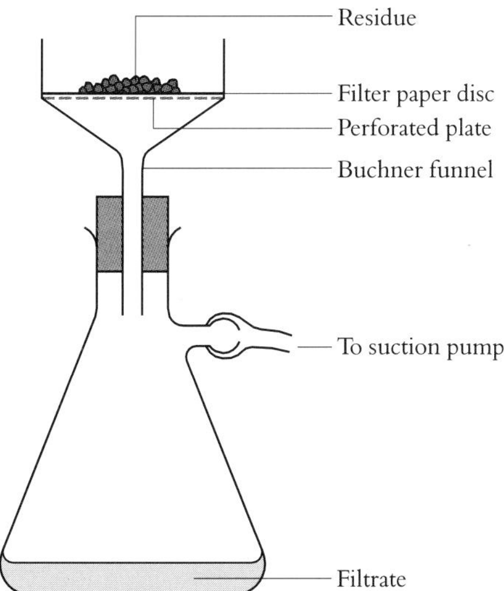

Recrystallisation of a solute is used as a method of purification. Impurities remain in solution while the main solute crystallises

Example A chemist has prepared 10 g of compound **A**. About 0.5 g of impurity **B** are present. For both **A** and **B** the solubilities in water are 100 g kg^{-1} at 80 °C, 30 g kg^{-1} at 20 °C. Will the chemist be able to obtain pure **A** by recrystallisation?

Method Prepare a saturated solution of the mixture of **A** and **B** in 100 g of water at 90 °C. Filter the hot solution to remove insoluble impurities. Warm the filtrate to redissolve any solid that has crystallised. Cool the solution from 90 °C to 20 °C. Since at 90 °C 10 g of **A** will dissolve but at 20 °C only 3.0 g of **A** will dissolve, 7.0 g of **A** will crystallise. Since at 20 °C, the mass of **B** that can remain in solution is 3.0 g, and since only about 0.5 g is present, no **B** will crystallise. The crystals will be pure **A**. They can be filtered and dried.

SUMMARY

A saturated solution contains as much solute as can be dissolved at a stated temperature.

Solubility = mass of solid dissolved in a stated mass or volume of solvent at a stated temperature.

A solubility curve is a graph of solubility against temperature.

The technique of recrystallisation depends on the solubilities of a substance and the impurities present with it.

CHECKPOINT 9.3: SOLUTIONS

1. You are given a bottle of solid **A** and three aqueous solutions of **A**. One is saturated, one unsaturated and one supersaturated. How would you find out which was which?

2. Solubility curves for a number of ionic solids are given below.

Figure 9.3B
Some solubility curves

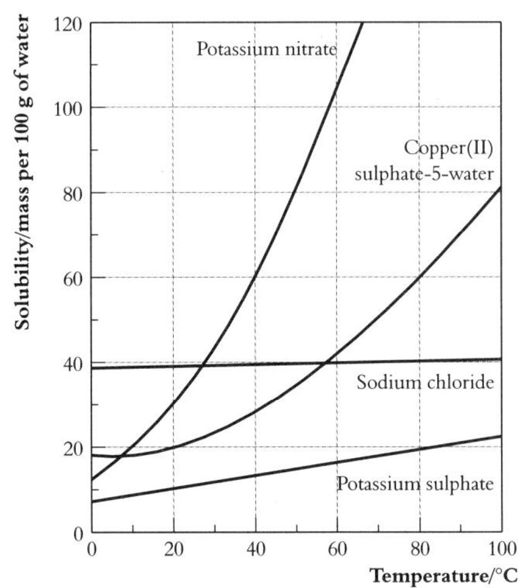

(a) (i) What mass of copper(II) sulphate-5-water will dissolve in 100 g of water at 100 °C?
(ii) What happens when the solution is cooled to 50 °C?

(b) What is the solubility at 60 °C of
(i) potassium nitrate
(ii) sodium chloride?

(c) A solution is saturated at 60 °C with potassium nitrate and with sodium chloride. A 100 g sample of solution is cooled from 60 °C to 20 °C. What mass of crystals forms?

3. A saturated solution of $CaSO_4$(aq) has some undissolved $CaSO_4$(s) lying at the bottom of the container. A little $Ca^{35}SO_4$ is mixed with the undissolved $Ca^{32}SO_4$. The isotope ^{35}S is radioactive, and, after a while, both the solution of $CaSO_4$(aq) and the undissolved $CaSO_4$(s) are found to be radioactive. Explain how the solution has been able to take up $CaSO_4$ in spite of being saturated.

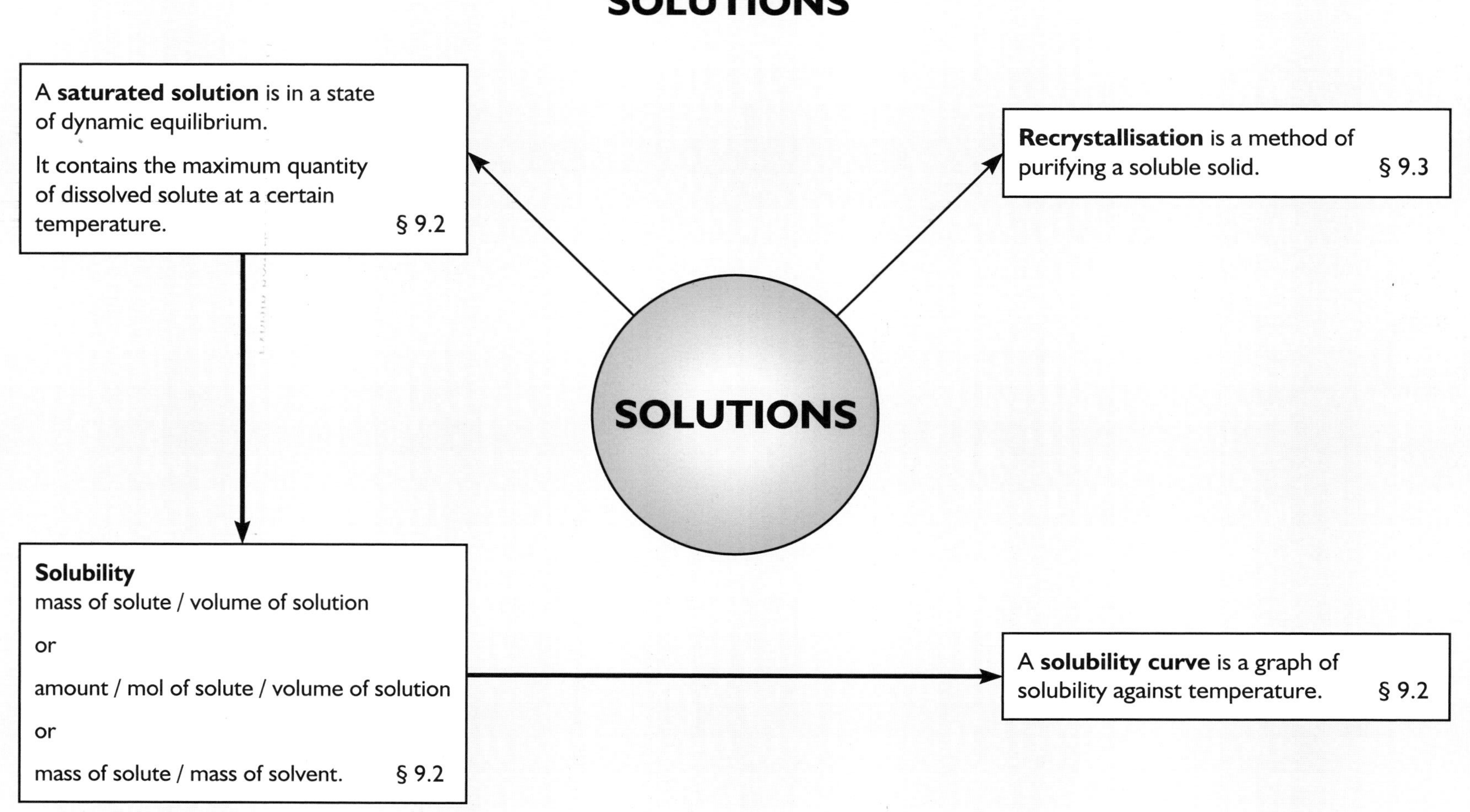
SOLUTIONS
A saturated solution is in a state of dynamic equilibrium.
It contains the maximum quantity of dissolved solute at a certain temperature. § 9.2
Recrystallisation is a method of purifying a soluble solid. § 9.3
SOLUTIONS
Solubility
mass of solute / volume of solution
or
amount / mol of solute / volume of solution
or
mass of solute / mass of solvent. § 9.2
A solubility curve is a graph of solubility against temperature. § 9.2

10

THERMOCHEMISTRY

10.1 SOURCES OF ENERGY

As the twenty first century begins, the human race needs urgently to consider its use of energy. The world's supply of fuel is diminishing. For some nations this threatens a drop in the standard of living. For less developed countries, it may be a question of survival. There are many chemical aspects of energy production and utilisation. There is no greater scientific challenge offered to chemists and engineers today than the problem of energy.

Reliance on wood and coal as sources of energy at the start of the twentieth century has been overtaken by the use of petroleum. The fossil fuels [§ 10.2] are still our major source of energy, but the pollution they cause and the limited reserves mean that fossil fuels cannot provide for our energy needs for ever. It is estimated that 90% of our known oil reserves could be used up by 2020, and oil companies are actively exploring to find new reserves. Deposits of coal are much greater than known reserves of oil.

Coal, biomass and hydrogen are the most important fuels for the future. Coal is a polluting fuel [see § 10.2], but there are ways of making clean fuels from it. In the process of coal gasification, coal is heated to form methane and carbon.

Biomass (animal and plant matter) can be used as a source of energy. Sugar, cellulose and other plant materials are converted into fuels. Ethanol is one such fuel, which is mixed with petrol and sold as gasohol [see § 30.5.3]. Methane is produced in biogas generators through the microbial breakdown of plant and animal waste [see § 26.1].

Hydrogen produces about one-third as much energy per mole as methane on combustion. The product of combustion, water vapour, is non-polluting. Hydrogen is a renewable fuel because the water produced returns as rain. However, energy must be spent in making hydrogen from water by electrolysis.

To conserve our resources and to develop new fuels we need to understand how energy is released in chemical reactions, such as combustion. This is one of the areas we study in thermochemistry.

10.2 FOSSIL FUELS

Fossil fuels were formed from decayed plant and animal matter

They derive their energy indirectly from the Sun

Most of the energy used in the world comes from the combustion of organic compounds These include the complex organic compounds in wood and the hydrocarbons of natural gas, coal, petroleum oil and other **fossil fuels**. These are so-called because they were formed from the decayed animal and vegetable matter of living things that inhabited the Earth millions of years ago. The ultimate source of the energy in fossil fuels is the Sun. Solar radiation enables green plants to carry out photosynthesis, the process during which they synthesise carbohydrates. Fossil fuels are **finite resources**: when we have used up Earth's reserves no more will be formed.

Fossil fuels are of enormous value to society

Wood and fossil fuels have many advantages. The appliances needed to burn them and harness their energy are easy to construct and simple to use. The products of combustion, provided that there is an ample supply of oxygen, are the harmless substances carbon dioxide and water. There are some drawbacks, however. Some fossil fuels contain sulphur compounds, and measures must be taken to remove the combustion product sulphur dioxide [see Acid rain, § 21.10]. There is also the problem of the enhanced greenhouse effect [see § 23.6]. As the demand for energy and the combustion of fossil fuels have increased over the last century, the level of carbon dioxide in the atmosphere has increased and with it the threat of global warming.

The combustion of fossil fuels to release energy is an important area of study

Obtaining energy from fossil fuels

We obtain energy from fuels by combustion, e.g. the combustion of the hydrocarbon methane in natural gas:

$$CH_4(g) + 2O_2(g) \longrightarrow CO_2(g) + 2H_2O(l)$$

and octane in petrol:

$$2C_8H_{18}(l) + 25O_2(g) \longrightarrow 16CO_2(g) + 18H_2O(l)$$

The products are carbon dioxide and water, provided that sufficient oxygen is present. If the supply of oxygen is insufficient carbon monoxide and carbon are produced also. In the combustion of octane in a plentiful supply of air, 40 MJ of energy are obtained for every litre of octane used, enough to drive an average car 7–9 miles.

Combustion is **oxidation** with the release of energy. When combustion is accompanied by a **flame** we call it **burning**. Substances which release a lot of energy on combustion are used as **fuels**.

10.3 WHY DO REACTIONS HAPPEN?

This chapter will tackle the question of why chemical reactions and physical changes happen

It is important for us to understand combustion. We can ask the question, 'Why do fuels burn?' 'Why does methane react rapidly with oxygen to form carbon dioxide and water vapour?' 'Why do carbon dioxide and water vapour *not* react to form methane, however long you leave them together? There are other puzzling questions apart from combustion. 'Why does a garden fork rust if it is left outside?' 'Why does a rusty fork never become shiny again?' 'Why does a sugar lump dissolve in a cup of coffee?' 'Why does it never reappear out of the solution?' Such questions lead to the more fundamental questions, 'Why do chemical reactions happen?' and 'Why do physical changes happen?'

Some changes are spontaneous ...

... happening without external help

Our experience of such changes tells us that there are **spontaneous changes**, which occur by themselves and there are changes that do not occur unless we make them happen. Spontaneous changes seem to have a driving force that sends them in a certain direction. To send such a process in the reverse direction, we have to expend energy. For example, with a little heating, magnesium burns rapidly to form magnesium oxide. To reverse the process is difficult: to split magnesium oxide into magnesium and oxygen, energy must be supplied in the electrolysis of molten magnesium oxide. Some spontaneous processes release energy, for example when water vapour condenses to form liquid water, heat is given out. Other spontaneous processes take in energy, for example when ammonium nitrate dissolves in water the temperature of the solution falls.

This chapter investigates the importance of heat changes in chemical and physical changes ...

... and also changes in the degree of disorder in a system ...

... and reaches the conclusion that the change in free energy is the factor that decides whether a change will happen

In this chapter we work towards an understanding of why chemical reactions and physical changes happen. We study methods of measuring the heat changes that take place in chemical and physical changes. Since laboratory reactions usually take place at constant pressure, we explore the convenience of studying heat change at constant pressure, a quantity called the **change in enthalpy**. We refine this to **standard enthalpy change** because we must compare enthalpy changes under identical conditions.

It turns out that the standard enthalpy change is not always successful in predicting whether a reaction will happen. We have to consider another quantity, called **entropy**. This is the degree of disorder in a system. For example, in ice the molecules are ordered in a crystal structure, while in liquid water they are free to move so a certain mass of liquid water contains more entropy than the same mass of ice. A positive change in entropy means that the degree of disorder in the system increases.

The two factors, enthalpy change and entropy change are included in the **free energy change**. Finally we have an answer to the question, 'Will a change happen?' A change happens if the free energy decreases, that is if the free energy of the products is less than the free energy of the reactants.

So, a change happens if it is accompanied by a decrease in free energy. This is not the end of the story, however. We still want to know how fast a change will happen. The answer to that question will be tackled in Chapter 14: Reaction kinetics.

10.4 FORMS OF ENERGY

There are many forms of energy ...

... All are either kinetic or potential energy ...

Heat is kinetic energy

The energy of chemical bonds is potential energy

There are many forms of energy: heat, light, chemical energy, nuclear energy, etc., but basically there are only two kinds of energy, **kinetic energy** and **potential energy**. The energy which an object possesses because it is moving is called kinetic energy. The amount of kinetic energy that an object possesses is the amount of work that it can do before it comes to rest, having used up all its kinetic energy. The energy which an object possesses because of its position or because of the arrangement of its component parts is called potential energy. **Heat** is a form of kinetic energy: it is the kinetic energy associated with the motion of atoms and molecules. The energy of chemical bonds is a form of potential energy, arising from the positions of atoms and molecules with respect to one another. Any piece of matter therefore possesses energy, potential energy, due to the arrangement of its particles. When the matter changes into a different type of matter, in a change of state or a chemical reaction, the positions of the particles relative to one another change: the potential energy of the matter has changed. Energy is taken in or given out by the piece of matter to bring about the change.

The study of the energy changes that accompany chemical reactions is called **thermochemistry** or **chemical thermodynamics**.

The First Law of Thermodynamics states that energy is conserved

Energy can be converted from one form into another. Electrical energy can be converted into heat energy. Our bodies can convert the energy of the chemical bonds in food into other kinds of energy. Calculations on energy conversions show that energy is never created and never destroyed. Observations on physical changes and chemical reactions are summarised in the **First Law of Thermodynamics**. This law states that *energy can be changed from one form into another, but it can neither be created nor destroyed*.

Heat is the transfer of energy from regions of high temperature to regions of low temperature

The transfer of energy is **heat**. When a fuel burns the energy released may be used to do work, as in driving a vehicle, or to heat the surroundings. Heat energy leaves a system if the system is at a higher temperature than its surroundings. (Energy flows out of us when we touch a cold object; energy flows into us when we step into a hot bath.) What we describe as a **system** is any substance or reaction mixture in which we are interested. The **surroundings** are the container and everything else outside the system. In practice we normally consider the immediate surroundings, not the whole of the universe.

SUMMARY

Matter possesses potential energy due to the positions of particles relative to one another. These positions change in a chemical reaction and in a change of state. This is why such changes must gain energy from the surroundings or lose energy to the surroundings.

10.5 EXOTHERMIC AND ENDOTHERMIC REACTIONS

Chemical reactions are classified as exothermic – in which heat is given out – and endothermic – in which heat is taken in

A chemical reaction that releases heat is described as an **exothermic reaction**. A chemical reaction that takes in heat is described as an **endothermic reaction**.

- Combustion is an exothermic reaction.
- Living organisms obtain energy from the combustion of carbohydrates, such as glucose:

  $$C_6H_{12}O_6(aq) + 6O_2(g) \longrightarrow 6CO_2(g) + 6H_2O(l)$$

  This combustion, cellular respiration, takes place in many stages, each of which is governed by enzymes. As a result energy is released gradually as the organism needs it and the cells in which respiration takes place do not become overheated.
- The reduction of a metal oxide, e.g. iron(III) oxide by aluminium is so strongly exothermic that it is called the thermit reaction

  $$2Al(s) + Fe_2O_3(s) \longrightarrow 2Fe(s) + Al_2O_3(s)$$

  The iron produced is molten and can be used to weld pieces of iron together.
- The dissolution of sulphuric acid in water is a strongly exothermic process. If you pour water into sulphuric acid, enough heat can be generated to boil the solution. This is why you always dilute sulphuric acid by pouring the concentrated acid into a large volume of water. The amount of heat generated is the same but if the solution does boil it will be a dilute solution that splashes out, not concentrated sulphuric acid.
- The process of photosynthesis is endothermic. Plants which contain the enzyme chlorophyll convert carbon dioxide and water into sugars. The process takes place only in sunlight as the energy of sunlight is converted into the energy of chemical bonds in the sugars.

$$6CO_2(g) + 6H_2O(l) \xrightarrow[\text{chlorophyll}]{\text{sunlight}} C_6H_{12}O_6(aq) + 6O_2(g)$$

- The dissolution of ammonium nitrate in water is endothermic. The energy needed to separate ammonium ions and nitrate ions in the solid is greater than the energy given out when the ions are hydrated [see § 10.15].

Changes of state are also classified as exothermic and endothermic

Changes of state can be exothermic or endothermic [see Figure 10.7B].

- Vaporisation is endothermic; for example heat must be supplied to convert liquid water into water vapour. Condensation is exothermic.
- Melting is endothermic; freezing is exothermic.

10.6 ENTHALPY CHANGES

Matter posseses energy in the form of ...

... kinetic energy and ...

... potential energy ...

... the sum of which = internal energy

Matter contains energy. This is in the form of kinetic energy and potential energy. The kinetic energy of matter is the energy of motion at a molecular level. The atoms or ions or molecules in a solid are vibrating and rotating and translating (moving from one place to another). The potential energy of matter arises from the positions of the atoms relative to one another. Bond-breaking and bond-making involve changes in potential energy. When two solids react, as in the reaction

$$Ca(s) + I_2(s) \longrightarrow CaI_2(s)$$

there is little change in kinetic energy, but there is a big change in potential energy as the bonding in the product is different from the bonding in the reactants. The kinetic energy and potential energy together make up the **internal energy** of matter.

Internal energy of matter =	Kinetic energy (changes little during chemical reaction)	+	Potential energy (changes during chemical reaction)

Figure 10.6A
Heat changes in chemical reactions

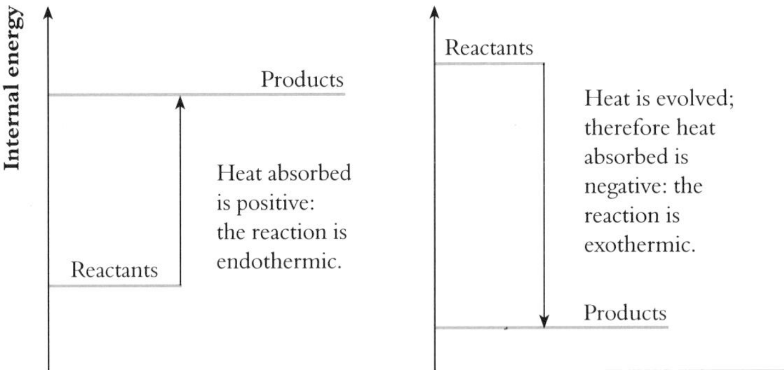

Heat may be given out or taken in during a chemical reaction

Frequently during the course of a chemical reaction, heat is either given out or taken in from the surroundings. Provided no work is done by the system on the surroundings, the heat absorbed is given by:

$$\begin{pmatrix}\text{Heat absorbed}\\ \text{in reaction}\end{pmatrix} = \begin{pmatrix}\text{Internal energy}\\ \text{of products}\end{pmatrix} - \begin{pmatrix}\text{Internal energy}\\ \text{of reactants}\end{pmatrix}$$

If the reaction takes place at constant pressure, any gases formed are allowed to escape and they do work in forcing back the atmosphere. The heat absorbed during the reaction is equal to the increase in internal energy plus the work done in expansion.

$$\begin{pmatrix}\text{Heat absorbed at}\\ \text{constant pressure}\end{pmatrix} = \begin{pmatrix}\text{Change in}\\ \text{internal energy}\end{pmatrix} + \begin{pmatrix}\text{Work done on}\\ \text{surroundings}\end{pmatrix}$$

The quantity of heat absorbed at constant pressure is called the change in enthalpy, ΔH

Since most laboratory work is carried out at constant pressure, it is usual to deal with the heat absorbed at constant pressure. This quantity is called the change in enthalpy, ΔH. The change in enthalpy is equal to the heat absorbed at constant pressure. Reactions of solids and liquids do not involve large changes in volume, the work done on the surroundings is small, and the change in enthalpy ΔH is close to the change in internal energy.

10.6.1 STANDARD CONDITIONS

Standard conditions: gases at 1 atm, solutions 1 M, solids in their standard states

The value of an enthalpy change depends on the temperature, on the physical states of the reactants (s, l, g), the pressures of gaseous reactants and products and the concentrations of solutions. Enthalpy changes are therefore stated under standard conditions and are denoted by the symbol $\Delta H^{\ominus}$. **Standard conditions** are: gases at a pressure of 1 atm, solutions at unit concentration and substances in their standard states. The **standard state** of a substance is the pure substance in a specified state (solid, liquid or gas) at 1 atmosphere pressure. A standard enthalpy change may be written $\Delta H^{\ominus}_T$, meaning the standard enthalpy change at temperature T or $\Delta H^{\ominus}$, in which case the temperature is 298 K (25 °C). Thus ΔH refers to heat absorbed at constant pressure and $\Delta H^{\ominus}$ refers to heat absorbed at constant pressure under standard conditions.

10.7 STANDARD ENTHALPY CHANGES

The standard enthalpy changes for some physical and chemical changes are defined below.

10.7.1 CHEMICAL REACTIONS

Standard molar enthalpy change ΔH⊖ ...

... of formation ...

Standard enthalpy of formation, $\Delta H^{\ominus}_F$, is the heat absorbed per mole when a substance is formed from its elements in their standard states. For example, the standard enthalpy of formation of sodium chloride is calculated for the reaction between solid sodium and gaseous chlorine molecules, Na(s) and Cl_2(g). It follows from this definition that all elements in their standard states have a value of zero for their standard enthalpies of formation. Figure 10.7A illustrates how the value of $\Delta H^{\ominus}$ may be negative (as for sodium chloride) or positive (as for ethyne). If the enthalpy absorbed is negative, enthalpy is released and the reaction is exothermic.

... of combustion ...

Standard enthalpy of combustion, nation $\Delta H^{\ominus}_C$, is the heat absorbed per mole when a substance is completely burned in oxygen at 1 atm. Since heat is usually evolved in such a reaction, $\Delta H^{\ominus}_C$ will be negative.

... of neutralisation ...

Standard enthalpy of neutralisation is the heat absorbed per mole when an acid and a base react to form water under standard conditions.

... of reaction ...

Standard enthalpy of reaction, $\Delta H^{\ominus}_R$, is the heat absorbed in a reaction at 1 atm between the number of moles of reactants shown in the equation for the reaction.

Figure 10.7A Standard enthalpy of formation

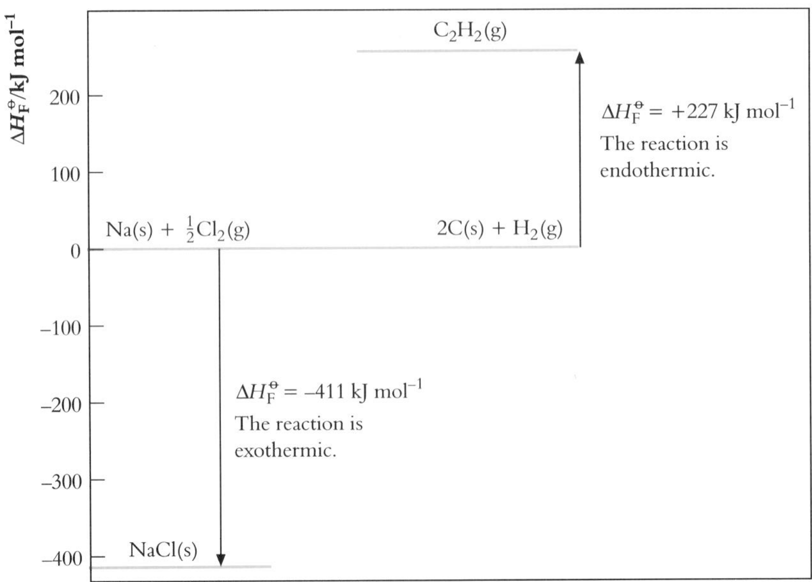

In the reaction

$$4H_2O(l) + 3Fe(s) \longrightarrow Fe_3O_4(s) + 4H_2(g)$$

there is no reason why the standard enthalpy of reaction should be related to 1 mole of iron or 1 mole of steam or 1 mole of iron oxide. Instead, it is related to the whole reaction, as written, between 4 moles of water and 3 moles of iron.

... of dissolution ...

Standard enthalpy of dissolution is the heat absorbed per mole when a substance is dissolved at 1 atm in a stated amount of solvent. This may be 100 g or 1000 g of solvent or it may be an 'infinite' amount of solvent, i.e., a volume so large that on further dilution there is no further heat exchange.

of atomisation

Standard enthalpy of atomisation is the enthalpy absorbed per mole when a substance decomposes to form gaseous atoms.

10.7.2 CHANGES OF STATE

Melting

A change of state ...

... melting, freezing, vaporisation, condensation ...

... involves a change in enthalpy

The particles of a solid are arranged in a regular three-dimensional structure. When a solid is heated the particles gain energy and vibrate more vigorously. Eventually they are able to leave their positions in the structure and move past one another; they flow as a liquid. Melting is endothermic.

- The **enthalpy of melting** or **enthalpy of fusion** of a substance is the enthalpy change per mole when a substance melts.
- The **enthalpy of freezing** of a substance is the enthalpy change per mole when a substance freezes.

Definitions of ...

... ΔH melting ...

... ΔH freezing ...

These changes do not take place under standard conditions. Enthalpy changes of melting and freezing are quoted at 1 atm and at the temperature at which the change of state takes place, that is, the melting temperature. The enthalpy of freezing is the negative of the enthalpy of melting:

$$\Delta H_{freezing} = -\Delta H_{melting}$$

Vaporisation

... ΔH *vaporisation* ...

The particles of a liquid are in motion. Forces of attraction keep them in the liquid phase. When a liquid is heated the average energy of the molecules increases. Some molecules, with energy above average, are able to break away from the attraction of other molecules and enter the vapour phase. Vaporisation is endothermic. The **enthalpy of vaporisation** is the difference in enthalpy between the vapour and the liquid state of a substance per mole of substance:

$$\Delta H_{\text{vaporisation}} = \text{Enthalpy of vapour} - \text{Enthalpy of liquid}$$

Enthalpy changes of vaporisation are quoted at 1 atm and at the temperature at which the change of state takes place.

Condensation

... ΔH *condensation* ...

Condensation is the reverse of vaporisation, and is exothermic.

$$\Delta H_{\text{condensation}} = -\Delta H_{\text{vaporisation}}$$

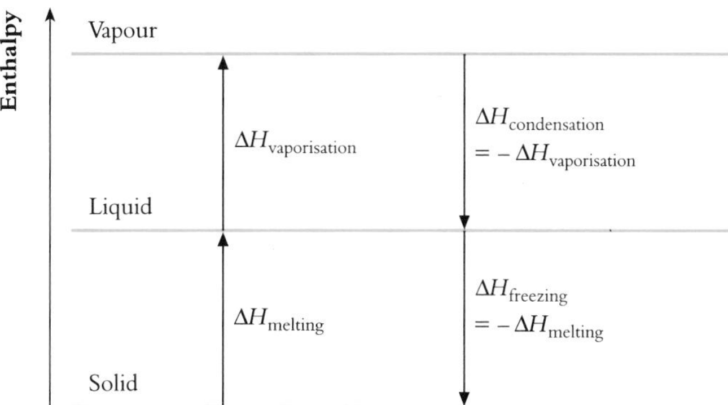

Figure 10.7B Enthalpy changes of changes of state

SUMMARY

The internal energy of matter is the sum of the kinetic energy and potential energy which it possesses.

Change in enthalpy = Change in internal energy + Work done on surroundings
= Heat absorbed at constant pressure

Standard enthalpy change, $\Delta H^{\ominus}$, = molar enthalpy change under standard conditions. These are: substances in their normal physical states, gases at 1 atm, solutions at 1 mol dm^{-3}. The temperature must be stated.

Definitions of $\Delta H^{\ominus}$ for some chemical reactions and ΔH for some changes of state are given.

10.8 METHODS FOR FINDING THE STANDARD ENTHALPY OF REACTION

10.8.1 THE EXPERIMENTAL METHODS

The methods of finding the standard enthalpy changes of chemical reactions depend on measuring the change in temperature that results as heat is given out

Measurements in a calorimeter are used to find enthalpy changes.

or taken in. The reaction is carried out in a container called a **calorimeter** (after the old unit of heat the calorie; 1 calorie = 4.18 J). The calorimeter is insulated from the surroundings to minimise the loss of heat to the surroundings. In advanced work, the heat which the calorimeter loses to or gains from the surroundings is measured and compensated for.

10.8.2 RELATIONSHIP BETWEEN TEMPERATURE AND HEAT

The basis of the method is to deduce the heat exchange from the temperature change. When an object receives heat, its temperature rises. The rise in temperature depends on the **heat capacity** of the object. Heat capacity is defined by the equation

Heat supplied = Mass × Specific heat capacity × Rise in temperature = m × c × ΔT

$$\text{Heat capacity} = \frac{\text{Heat supplied}}{\text{Rise in temperature}}$$

The greater the mass of the object, the more heat is required to raise its temperature. The quantity **specific heat capacity**, symbol **c**, is defined by.

$$\text{Specific heat capacity} = \frac{\text{Heat capacity}}{\text{Mass}}$$

It has the unit J K^{-1} g^{-1} or kJ K^{-1} kg^{-1}.

The change in temperature is given by

$$\text{Rise in temperature} = \frac{\text{Heat supplied}}{\text{Specific heat capacity} \times \text{Mass}}$$

Rearranging the equation,

$$\text{Heat supplied} = \text{Mass} \times \text{Specific heat capacity} \times \text{Rise in temperature}$$
$$\text{Heat} = m \times c \times \Delta T$$

SUMMARY

Heat supplied = Rise in temperature × Specific heat capacity × Mass
Specific heat capacity = $\frac{\text{Heat capacity}}{\text{Mass}}$

Example Water has specific heat capacity 4.18 J K^{-1} g^{-1}. 100 g of water rise in by 2.0 K. How much heat has the water received?

Method $\text{Heat} = m \times c \times \Delta T$
$= 100\ \text{g} \times 4.18\ \text{J K}^{-1}\ \text{g}^{-1} \times 2.0\ \text{K} = 936\ \text{J}$

10.8.3 STANDARD ENTHALPY OF NEUTRALISATION

Neutralisation is the formation of water from hydrogen ions and hydroxide ions:

$$H_3O^+(aq) + OH^-(aq) \longrightarrow 2H_2O(l)$$

The heat released when a known amount of water is formed can be found by measuring the temperature rise produced in a calorimeter. A very simple type of calorimeter consists of an expanded polystyrene cup with a lid. Expanded polystyrene is a good thermal insulator, and the temperature rise that occurs in the cup can be measured before the loss of heat to the surroundings is serious.

Example 25.0 cm^3 of 1.00 mol dm^{-3} hydrochloric acid at 21.5 °C were placed in a polystyrene cup. 25.0 cm^3 of 1.00 mol dm^{-3} sodium hydroxide at 21.5 °C were added. The mixture was stirred, and the temperature rose to 28.2 °C. The density of each solution = 1.00 g cm^{-3}, and the specific heat capacity of each solution = 4.18 J K^{-1} g^{-1}. Calculate the standard molar enthalpy of neutralisation.

Method Mass of solution = 50.0 g

$\Delta T = 28.2\ °C - 21.5\ °C = 6.7\ °C = 6.7\ K$

$$\text{Heat} = m \times c \times \Delta T$$
$$= 50.0\ g \times 4.18\ J\ K^{-1}\ g^{-1} \times 6.7\ K = 1400\ J$$

$$\text{Amount of water formed} = 25.0 \times 10^{-3}\ dm^3 \times 1.00\ mol\ dm^{-3}$$
$$= 2.50 \times 10^{-2}\ mol$$

$$\text{Standard molar enthalpy of neutralisation} = 1400\ J/2.50 \times 10^{-2}\ mol$$
$$= -56.0\ kJ\ mol^{-1}$$

The value you will find in tables is -57.9 kJ mol^{-1}. The low experimental value shows that some heat has been lost to the surroundings.

In this example, the heat capacity of the calorimeter is taken to be that of the aqueous solutions in it. The heat capacity of the polystyrene cup is ignored. In more accurate measurements, the heat capacity of the calorimeter is measured and this quantity is used in the calculation.

10.8.4 EXPERIMENTAL METHOD FOR FINDING ENTHALPY OF COMBUSTION

Figure 10.8A shows a simple method for obtaining an approximate value for the enthalpy of combustion of a fuel.

$$\text{Fuel(l)} + O_2(g) \longrightarrow CO_2(g) + H_2O(g)$$

Note that the enthalpy of combustion is not $\Delta H^{\ominus}_{combustion}$ because conditions are not standard. The water formed is not produced in the standard state – liquid; it is water vapour.

Figure 10.8A Apparatus for finding enthalpy of combustion

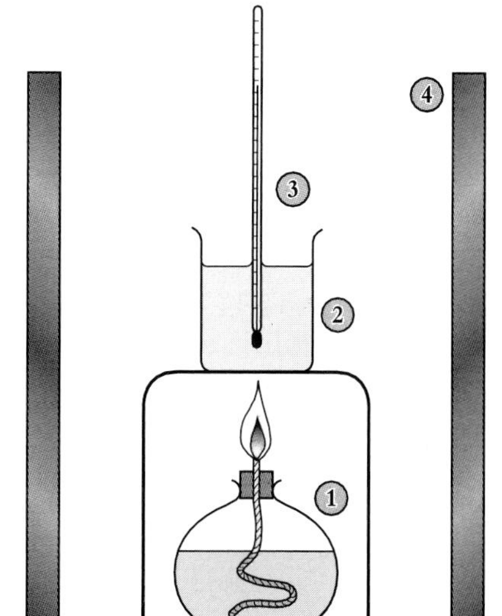
4
3
2
1
4 Shield reduces heat loss to surroundings.
3 Thermometer records rise in temperature, t °C.
2 Metal calorimeter contains a known mass of water, m_2 g.
1 Spirit burner contains fuel. Weighing before and after burning gives mass of fuel burnt, m_1 g.

A worked example of enthalpy of combustion

Example When ethanol was burnt in the apparatus shown in Figure 10.8A, the results were: $m_1 = 1.50$ g, $m_2 = 500$ g, $t = 19.5$ °C. Find the enthalpy of combustion of ethanol. Compare the experimental value with the listed value of −1368 kJ mol^{-1}.

Method Heat evolved = 500 × 4.18 × 19.5 = 40.8 kJ

Amount of ethanol burnt = 1.50/46 = 0.0326 mol

Molar enthalpy of combustion = 40.8/0.0326 = 1250 kJ mol^{-1}

The enthalpy of combustion is −1250 kJ mol^{-1}. This is lower than the listed value because some of the heat evolved is lost to the surroundings.

CHECKPOINT 10.8: COMBUSTION

1. A cocktail contains 33 g of ethanol. If $\Delta H^\ominus_C$ (ethanol) is −1370 kJ mol^{-1}, calculate the standard enthalpy released when this amount of ethanol is combusted inside the tissues of the drinker. Your result will be slightly in error because the value of $\Delta H^\ominus_C$ which you have been given is not exactly appropriate. What value of $\Delta H^\ominus_C$ should be used in the calculation?

2. You want to boil a kettle. The kettle contains 2.00 kg of water at 20 °C. What mass of natural gas (methane) must be burned to raise this quantity of water to 100 °C? (Assume no heat is lost.) (Specific heat capacity of water = 4.18 J g^{-1} K^{-1}; $\Delta H^\ominus_C$ (CH_4) = −890 kJ mol^{-1}.)

3. *(a)* What does the term 'enthalpy' mean?
   (b) How does the enthalpy change indicate whether a process is exothermic or endothermic?
   (c) 20.0 g of water cools from 20.0 °C to 10.0 °C. Is the process exothermic or endothermic? Calculate the enthalpy change. (Specific heat capacity of water = 4.18 J K^{-1} g^{-1}.)
   (d) 50 g of copper is heated from 10 °C to 100 °C. Calculate the enthalpy change. (Specific heat capacity of copper = 0.38 J K^{-1} g^{-1}.)

4. A calorimeter has a heat capacity of 6.00 kJ K^{-1}. The combustion of 2.00 g of magnesium raised the temperature from 20.20 °C to 22.32 °C. Calculate the enthalpy change of combustion.

5. 50.0 cm^3 of 0.500 mol dm^{-3} sodium hydroxide solution and 50.0 cm^3 of 0.500 mol dm^{-3} nitric acid, both at 20.0 °C, were mixed and stirred in a calorimeter with negligible heat capacity. The temperature of the mixture rose to 23.2 °C.
   (a) Calculate the change in enthalpy for the neutralisation.
   (b) Calculate the change in enthalpy per mole of water formed.

10.9 HESS'S LAW

Imagine that you are a climber at point A on a mountain. You climb to camp B. You have reached an altitude of 1000 m. A companion decides to do it the hard way; he climbs from A to peak C, at an altitude of 1500 m, and then descends to camp B. When he arrives at camp B he is at the same altitude as you: 1000 m. Going from A to B directly or from A to B via C, the total gain in altitude is the same. The altitude of a camp on a mountainside depends on the position of the camp on the mountain and not on how you get there.

Enthalpy can be compared with altitude. The enthalpy of a system depends on the substances present and their physical states and temperatures and not on how the system was prepared. Now look at a physical change.

Figure 10.9A
Climbing a mountain

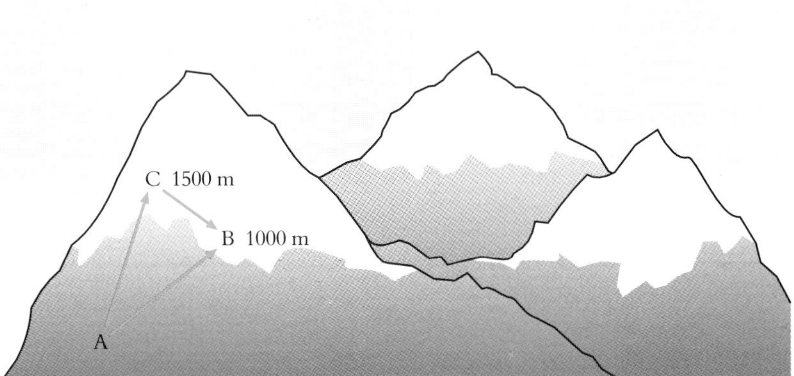

Figure 10.9B
Vaporisation

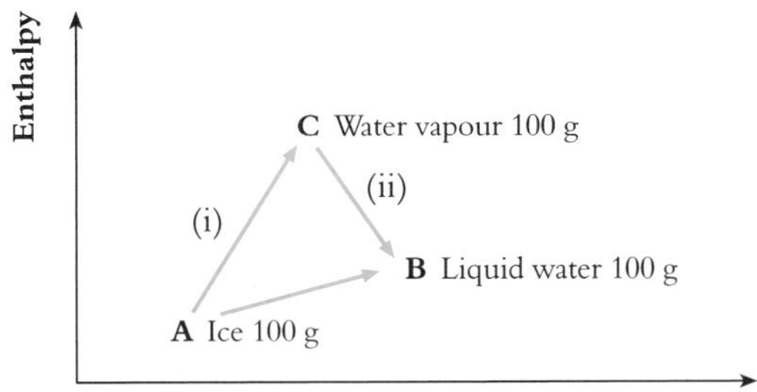

You may find it helps to compare enthalpy with altitude. The difference in altitude between two places on a mountainside does not depend on the route you take from one to the other. The change in enthalpy when system ***A*** *changes to system* ***B*** *does not depend on the reaction route from* ***A*** *to* ***B***

The system **B** consists of 100 g of liquid water at temperature T. To arrive at **B** from **A** by route **A** ⟶ **B**, 100 g of ice is melted.

$$\Delta H = H_{\text{liquid}} - H_{\text{ice}}$$

To arrive at **B** from **A** by route **A** ⟶ **C** ⟶ **B**, 100 g of ice is converted into vapour (by sublimation) and the water vapour is condensed to liquid. The enthalpy changes in (i) **A** ⟶ **C** and (ii) **C** ⟶ **B** are given by

$$\Delta H(\text{i}) = H_{\text{vapour}} - H_{\text{ice}}$$

$$\Delta H(\text{ii}) = H_{\text{liquid}} - H_{\text{vapour}}$$

For route **A** ⟶ **C** ⟶ **B**

$$\begin{aligned}\Delta H &= \Delta H(\text{i}) + \Delta H(\text{ii})\\ &= H_{\text{vapour}} - H_{\text{ice}} + H_{\text{liquid}} - H_{\text{vapour}}\\ &= H_{\text{liquid}} - H_{\text{ice}}\end{aligned}$$

SUMMARY

Hess's Law: If a change can take place by more than one route, the overall change in enthalpy is the same, whichever route is followed.

This ΔH is the same as for route **A** ⟶ **B** so to calculate the change in enthalpy from **A** to **B** you need to know the enthalpies of the initial state **A** and the final state **B**, but you do not need to know what happened on the route from **A** to **B**. J Hess summarised this in **Hess's Law**, which states

> If a change can take place by more than one route, the overall change in enthalpy is the same, whichever route is followed.

Now look at a chemical reaction.

10.10 STANDARD ENTHALPY CHANGE FOR A CHEMICAL REACTION

The standard enthalpy change for a chemical reaction [§ 10.7.1] can be calculated from the standard enthalpies of formation of all the products and reactants involved. For example, in the reaction of hydrogen chloride with ammonia

Finding $\Delta H^\ominus$ of reaction from $\Delta H_F^\ominus$ of the reactants and the products ...

$$\underset{(-46)}{NH_3(g)} + \underset{(-92.3)}{HCl(g)} \longrightarrow \underset{(-315)}{NH_4Cl(s)}$$

Written underneath each species is its standard enthalpy of formation in kJ mol^{-1}. The standard enthalpy of this reaction $\Delta H_R^\ominus$ is given by

$$\begin{aligned}\Delta H_R^\ominus &= (-315) - (-46 + (-92.3)) \\ &= -177 \text{ kJ mol}^{-1}\end{aligned}$$

The negative sign means that the reaction is exothermic.

... using Hess's Law

The standard enthalpy of reaction depends only on the difference between the standard enthalpy of the reactants and the standard enthalpy of the products and not on the route by which the reaction occurs. This is an example of Hess's Law, which states that, if a reaction can take place by more than one route, the overall change in enthalpy is the same, whichever route is followed.

In the set of reactions shown in the **enthalpy diagram** in Figure 10.10A, the standard enthalpy change in going from **A** to **B** by the direct route 1 is $\Delta H_1^\ominus$. The standard enthalpy change in going from **A** to **B** by the indirect route 2 is $\Delta H_2^\ominus + \Delta H_3^\ominus$.

$$\Delta H_1^\ominus = \Delta H_2^\ominus + \Delta H_3^\ominus$$

and the standard enthalpy change is the same for the different reaction routes. It follows that the standard enthalpy change for the reaction **B** $\longrightarrow$ **A** is $-\Delta H_1^\ominus$.

Figure 10.10A
An enthalpy diagram

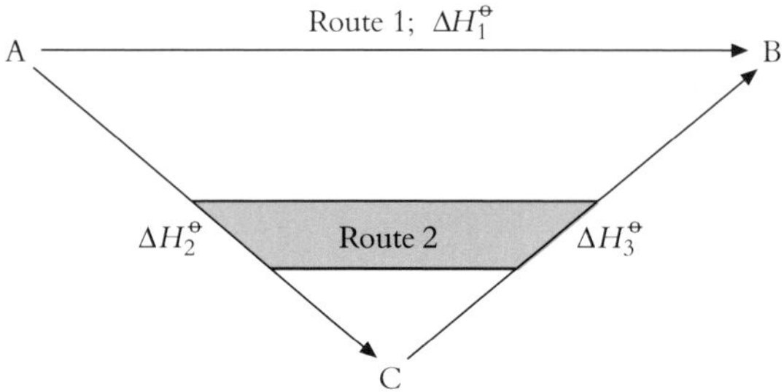

$\Delta H^\ominus$ is the same for different reaction routes from reactants to products

10.10.1 FINDING THE STANDARD ENTHALPY OF FORMATION OF A COMPOUND INDIRECTLY

$\Delta H_F^\ominus$ of a compound may be found directly by experiment or indirectly, by applying Hess's law ...

Sometimes, the standard enthalpy of formation of a compound [§ 10.7.1] can be measured directly by allowing known amounts of elements to combine and measuring the heat evolved. Other reactions are difficult to study, and the standard enthalpy of reaction must be found indirectly.

To find the standard enthalpy of formation of ethyne from practical measurements is impossible as attempts to make ethyne from carbon and hydrogen

$$2C(s) + H_2(g) \longrightarrow C_2H_2(g)$$

will result in the formation of a mixture of hydrocarbons. However, standard enthalpies of combustion can be measured accurately for all the substances in the equation. They are

[a] $C(s) + O_2(g) \longrightarrow CO_2(g);\ \Delta H^{\ominus}_C = -394\ \text{kJ mol}^{-1}$

[b] $H_2(g) + \frac{1}{2}O_2(g) \longrightarrow H_2O(l);\ \Delta H^{\ominus}_C = -286\ \text{kJ mol}^{-1}$

[c] $C_2H_2(g) + 2\frac{1}{2}O_2(g) \longrightarrow 2CO_2(g) + H_2O(l);\ \Delta H^{\ominus}_C = -1300\ \text{kJ mol}^{-1}$

The method of calculating $H^{\ominus}_F\ (C_2H_2)$ from these values of $\Delta H^{\ominus}_C$ involves three steps:

1. Write the equation for the combustion of ethyne, since this is the reaction for which $\Delta H^{\ominus}$ can be measured.

2. Under each species, write $\Delta H^{\ominus}_F$. You can see from equations [a] and [b] that the standard enthalpies of formation of carbon dioxide and water are −394 and −286 kJ mol^{-1} respectively.

   $C_2H_2(g) + 2\frac{1}{2}O_2(g) \longrightarrow 2CO_2(g) + H_2O(l);\ \Delta H^{\ominus}_3 = -1300\ \text{kJ mol}^{-1}$

   $\Delta H^{\ominus}_F(C_2H_2)\ \ 0 \qquad 2(-394) \qquad (-286)$

3. Since

   $\Delta H^{\ominus} = \text{Total } H^{\ominus} \text{ of products} - \text{Total } H^{\ominus} \text{ of reactants}$

   $\Delta H^{\ominus}_3 = -1300 = 2(-394) + (-286) - \Delta H^{\ominus}_F(C_2H_2)$

   $\Delta H^{\ominus}_F(C_2H_2) = +226\ \text{kJ mol}^{-1}$

   The standard enthalpy of formation of ethyne is 226 kJ mol^{-1}.

Ethyne is described as an **endothermic compound** since $\Delta H^{\ominus}_F$ is positive.

$$\Delta H^{\ominus}_{\text{Formation}} = \Delta H^{\ominus}_{\text{Combustion}} \text{ of elements } - \Delta H^{\ominus}_{\text{Combustion}} \text{ of compounds}$$

10.10.2 FINDING THE STANDARD ENTHALPY OF A REACTION

When a physical change or a chemical reaction takes place,

- Standard enthalpy change = Standard enthalpy of products − Standard enthalpy of reactants.
- The standard enthalpy of a substance = its standard enthalpy of formation.

The equation for the hydration of ethene to form ethanol is

Calculating the standard enthalpy of reaction ...

$$CH_2{=}CH_2(g) + H_2O(l) \longrightarrow C_2H_5OH(l)$$
$$(+52) \qquad (-286) \qquad (-278)$$

Written beneath each species is the value of ΔH^{θ}_F in kJ mol^{-1}. The **standard enthalpy of reaction** is given by

$$\Delta H^{\ominus}_R = -278 - (52 - 286)$$
$$= -44\ \text{kJ mol}^{-1}$$

The reaction is exothermic by 44 kJ mol^{-1}. In short

... The method

$$\Delta H^{\ominus}_{\text{Reaction}} = \Sigma\Delta H^{\ominus}_{\text{Formation}} \text{ of products } - \Sigma\Delta H^{\ominus}_{\text{Formation}} \text{ of reactants}$$

CHECKPOINT 10.10: ENTHALPY CHANGES

1. State the sign of the standard enthalpy change in the following:

   (a) the combustion of octane
   (b) the condensation of steam
   (c) the freezing of water
   (d) the electrolysis of water
   (e) the combustion of sodium in chlorine.

2. Why is it necessary to compare enthalpy changes under standard conditions?

10.11 AVERAGE STANDARD BOND ENTHALPY

Each type of bond has an average standard bond enthalpy ...

When bonds are broken, energy is required, When bonds are made, energy is given out. There is a certain amount of energy associated with every type of chemical bond, such as the C — C bond, the C = C bond and the C — H bond. When methane is atomised in the gaseous state, the energy required per mole is the energy needed to break four moles of C — H bonds. The standard enthalpy change is given by

$$CH_4(g) \longrightarrow C(g) + 4H(g);\ \Delta H^{\ominus}_{at} = 1662\ \text{kJ mol}^{-1}$$

The standard enthalpy of one mole of C — H bonds in methane is therefore 415.5 kJ mol^{-1}. The standard enthalpy of atomisation is different from the standard enthalpy of formation:

$$C(s) + 2H_2(g) \longrightarrow CH_4(g);\ \Delta H^{\ominus} = -75\ \text{kJ mol}^{-1}$$

This includes the standard enthalpies of atomisation of carbon(graphite) and hydrogen, as shown in Figure 10.11A.

Figure 10.11A
Standard enthalpy terms in the formation of methane

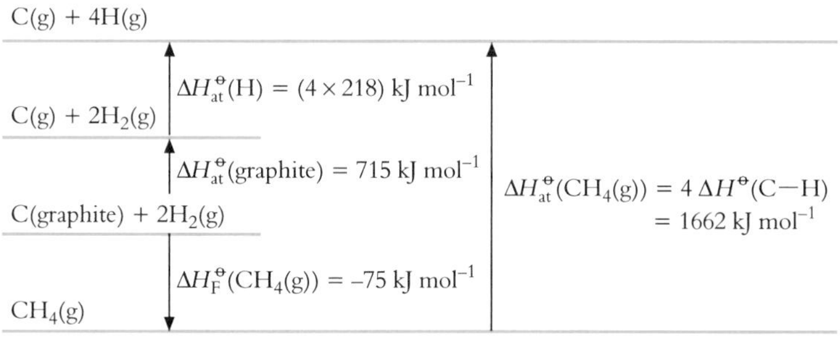

C(g) + 4H(g)
$\Delta H^{\ominus}_{at}(H) = (4 \times 218)$ kJ mol^{-1}
C(g) + 2H$_2$(g)
$\Delta H^{\ominus}_{at}$(graphite) = 715 kJ mol^{-1}
C(graphite) + 2H$_2$(g)
$\Delta H^{\ominus}_{F}(CH_4(g)) = -75$ kJ mol^{-1}
CH$_4$(g)
$\Delta H^{\ominus}_{at}(CH_4(g)) = 4\,\Delta H^{\ominus}$(C—H) = 1662 kJ mol^{-1}

... or bond energy term

The amount of energy associated with a bond, such as the C—H bond, is not exactly the same in all compounds. An average value of 413 kJ mol^{-1} can be given. It takes an average of 413 kJ to break one mole of C—H bonds in different compounds. This value is called the **average standard bond enthalpy** of the C—H bond or the **bond energy term**. Values of average standard bond enthalpies are tabulated, and examples of their use are given below. Since true standard bond enthalpies vary from compound to compound, calculations based on average standard bond enthalpies give approximate results.

Method of calculation of $\Delta H^{\ominus}_F$ from bond energy terms

Example Calculate the standard enthalpy of formation of ethane, using the following information. The average standard bond enthalpies are: C—C = 348, C—H = 413 kJ mol^{-1}. The standard enthalpies of atomisation are: C(s) = 718 kJ mol^{-1}, and $\frac{1}{2}H_2(g) = 218$ kJ per mole of H atoms formed.

Method In the C_2H_6 molecule are

one C — C bond of standard enthalpy = 348 kJ mol^{-1}

six C — H bonds of standard enthalpy = 2478 kJ mol^{-1}

The total standard enthalpy of the bonds = 2826 kJ mol^{-1}

This is the energy given out when the atoms combine:

$$2C(g) + 6H(g) \longrightarrow C_2H_6(g);\ \Delta H^{\ominus} = -2826 \text{ kJ mol}^{-1}$$

The standard enthalpy of formation of C(g) from C(s) is 718 kJ mol^{-1}. The standard enthalpy of formation of H(g) from $\frac{1}{2}H_2(g)$ is 218 kJ mol^{-1}. These values of the standard enthalpy content of each species can be put into the equation:

$$2C(g) + 6H(g) \longrightarrow C_2H_6(g);\ \Delta H^{\ominus} = -2826 \text{ kJ mol}^{-1}$$

$$2(718) \quad 6(218) \qquad \Delta H_F^{\ominus}$$

$$\Delta H^{\ominus} = \Delta H_F^{\ominus}(\text{product}) - \Delta H_F^{\ominus}(\text{reactants})$$

$$-2826 = \Delta H_F^{\ominus} - 2(718) - 6(218)$$

$$\Delta H_F^{\ominus} = -82 \text{ kJ mol}^{-1}$$

The standard enthalpy of formation of ethane is −82 kJ mol^{-1}.

10.11.1 STANDARD ENTHALPY OF REACTION FROM AVERAGE STANDARD BOND ENTHALPIES

When the standard enthalpy change for a reaction cannot be measured, an approximate value can be obtained by using average standard bond enthalpies. During a reaction, energy must be supplied to break bonds in the reactants, and energy is given out when the bonds in the products form. The standard enthalpy of reaction is the difference between the sum of the average standard bond enthalpies of the products and the sum of the average standard bond enthalpies of the reactants.

Method for calculation of ΔH_R from bond energy terms

Example Calculate the standard enthalpy change of the reaction

$$(CH_3)_2C{=}O(g) + HCN(g) \longrightarrow (CH_3)_2C(OH)(CN)\ (g)$$

Mean standard bond enthalpies/kJ mol^{-1} are C=O, 743; C — H, 412; C — O 360; C — C, 348; H — O, 463.

Method Bonds broken are one C=O of $\Delta H^{\ominus}$ = 743 kJ mol^{-1}
one C — H of $\Delta H^{\ominus}$ = 413 kJ mol^{-1}

Total standard enthalpy absorbed = 1156 kJ mol^{-1}

Bonds created are one C — O of $\Delta H^{\ominus}$ = 360 kJ mol^{-1}
one O — H of $\Delta H^{\ominus}$ = 463 kJ mol^{-1}
one C — C of $\Delta H^{\ominus}$ = 348 kJ mol^{-1}

Total standard enthalpy absorbed = −1171 kJ mol^{-1}

Standard enthalpy change of reaction = −1171 + 1156
= −15 kJ mol^{-1}

10.11.2 STANDARD ENTHALPY OF DELOCALISATION OF BENZENE

In § 5.4.1 the Kekulé structure for benzene, with alternate single and double carbon-carbon bonds, and the delocalised structure [see also § 28.6] were described. The standard enthalpy of formation of benzene can be predicted by adding the average standard bond enthalpies for the Kekulé formula:

Benzene is more stable than predicted by calculation

The difference is the standard delocalisation enthalpy

The answer is +209 kJ mol^{-1}. The experimental value is +49 kJ mol^{-1}; that is, benzene is more stable than predicted by 160 kJ mol^{-1}. This is powerful evidence for regarding the benzene molecules not as a ring of alternate single C—C and double C=C bonds but as a molecule with identical C—C bonds intermediate, in character between single and double bonds and stabilised by delocalisation:

The standard enthalpy of delocalisation of benzene is 160 kJ mol^{-1}.

SUMMARY

Standard bond dissociation enthalpies are found by experiment and are known accurately.

Average standard bond enthalpies are known approximately. They are used to give approximate values for $\Delta H_F^\ominus$, $\Delta H_R^\ominus$ and $\Delta H_{Delocalisation}^\ominus$

CHECKPOINT 10.11: STANDARD ENTHALPY CHANGE

1. *(a)* What is meant by the 'standard state' of a substance?
   (b) Define the term 'standard enthalpy of reaction'. Mention two methods by which it can be determined without actually carrying out the reaction.

2. Standard enthalpies of combustion are graphite −393.5 kJ mol^{-1} and diamond −395.4 kJ mol^{-1}. Calculate the change in enthalpy for the change graphite ⟶ diamond.

3. Standard enthalpies of combustion are: rhombic sulphur −296.83 kJ mol^{-1}; monoclinic sulphur −297.16 kJ mol^{-1}. Calculate the change in standard enthalpy for the change rhombic sulphur ⟶ monoclinic sulphur.

4. Calculate the standard enthalpy for the reduction of hydrazine, N_2H_4, to ammonia:

   $$N_2H_4(l) + H_2(g) \longrightarrow 2NH_3(g)$$

   Use the values:

   $$N_2(g) + 2H_2(g) \longrightarrow N_2H_4(l);\ \Delta H^\ominus = +50.63\ \text{kJ mol}^{-1}$$

   $$N_2(g) + 3H_2(g) \longrightarrow 2NH_3(g);\ \Delta H^\ominus = -92.22\ \text{kJ mol}^{-1}$$

5. *(a)* Explain why it is difficult to obtain an experimental value for the standard enthalpy of reaction for the combustion of methane to carbon monoxide:

   $$2CH_4(g) + 3O_2(g) \longrightarrow 2CO(g) + 4H_2O(l)$$

   (b) Calculate a value, given the following data:

   $$CH_4(g) + 2O_2(g) \longrightarrow CO_2(g) + 2H_2O(l);\quad \Delta H^\ominus = -890\ \text{kJ mol}^{-1}$$

   $$2CO(g) + O_2(g) \longrightarrow 2CO_2(g);\quad \Delta H^\ominus = -566.0\ \text{kJ mol}^{-1}$$

6. Write thermochemical equations (chemical equations showing the standard enthalpy of reaction) for the standard enthalpies of formation of *(a)* $CO_2(g)$, *(b)* $KClO_3$, *(c)* C_2H_5OH, *(d)* Al_2O_3, *(e)* CH_3CO_2H

7. When 2.00 g of magnesium reacts with nitrogen to form magnesium nitride, Mg_3N_2, the heat evolved is 12.7 kJ. Calculate the standard enthalpy of formation of magnesium nitride.

8. Values of average standard bond enthalpies/kJ mol^{-1} are: C—C 348, C=C 612, C—H 413, H—H 436. Standard enthalpy of atomisation of carbon/kJ mol^{-1} = 715. Use these values to calculate the standard enthalpy changes for the following reactions:

(*a*) $H_2C{=}CH_2(g) + H_2(g) \longrightarrow H_3C{-}CH_3(g)$

(*b*) $2H_2C{=}CH_2(g) \longrightarrow$ cyclobutane ($CH_2{-}CH_2$ / $CH_2{-}CH_2$, joined in a ring)

(*c*) $4C(s) + 4H_2(g) \longrightarrow$ cyclobutane ($CH_2{-}CH_2$ / $CH_2{-}CH_2$, joined in a ring)

(*d*) Comment on whether you would choose (ii) or (iii) as a method of making cyclobutane.

(*e*) The average standard C—C bond enthalpy in cyclobutane is in fact 320 kJ mol^{-1}. Suggest why it is smaller than the usual value for C—C of 348 kJ mol^{-1}.

FURTHER STUDY

10.12 THE BORN–HABER CYCLE

The **Born–Haber cycle** is a technique for applying Hess's Law to the standard enthalpy changes which occur when an ionic compound is formed. Think of the reaction between sodium and chlorine to form sodium chloride. The reaction can be considered to occur by means of the following steps, even though the reaction itself may not follow this route.

The formation of an ionic compound can be treated as a sequence of separate steps

Vaporisation of sodium
$Na(s) \longrightarrow Na(g)$; $\Delta H^\ominus_S$ = Standard enthalpy of sublimation or vaporisation of sodium [1]

Ionisation of sodium
$Na(g) \longrightarrow Na^+(g) + e^-$; $\Delta H^\ominus_I$ = Standard ionisation enthalpy of sodium [2]

Dissociation of chlorine molecules
$\frac{1}{2}Cl_2(g) \longrightarrow Cl(g)$; $\frac{1}{2}\Delta H^\ominus_D = \frac{1}{2}$ Standard bond dissociation enthalpy of chlorine molecules [3]

Ionisation of chlorine atoms
$Cl(g) + e^- \longrightarrow Cl^-(g)$; $\Delta H^\ominus_E$ = Electron affinity of chlorine [4]

Reaction between ions
$Na^+(g) + Cl^-(g) \longrightarrow NaCl(s)$; $\Delta H^\ominus_L$ = Standard lattice enthalpy [5]

The standard enthalpy changes can be defined as follows:

Definitions of $\Delta H^\ominus$ terms involved in the steps of the Born–Haber cycle

1. The **standard enthalpy of sublimation or vaporisation** is the enthalpy absorbed when 1 mole of sodium atoms is vaporised.
2. The **standard enthalpy of ionisation** of sodium is the enthalpy required to remove 1 mole of electrons from 1 mole of gaseous sodium atoms.
3. The **standard bond dissociation enthalpy** of chlorine is the enthalpy required to dissociate 1 mole of chlorine molecules into atoms.
4. The **electron affinity** of chlorine is the enthalpy absorbed when 1 mole of chlorine atoms accept 1 mole of electrons to become chloride ions. It has a negative value, showing that this reaction is exothermic.
5. The **standard lattice enthalpy** is the enthalpy absorbed when 1 mole of sodium chloride is formed from its gaseous ions; its value is negative. (The standard lattice dissociation enthalpy is the enthalpy absorbed when 1 mole of sodium chloride is separated into its gaseous ions; it has a positive value.)

Figure 10.12A
Born–Haber cycle for sodium chloride

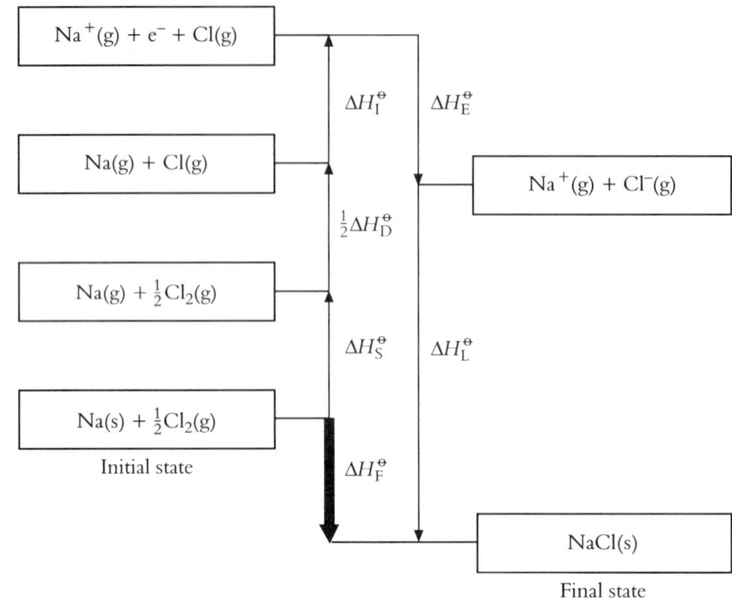

The standard enthalpy changes in Steps **1** to **5** are represented in Figure 10.12A. The steps in a Born–Haber cycle are represented as going upwards if they absorb energy and downwards if they give out energy.

Application of Hess's Law in the Born–Haber cycle ...

... to give ΔH_F ...

... or to give electron affinity

Applying Hess's Law to this cycle, it follows that the sum of the standard enthalpy terms [1] to [5] is equal to the difference in standard enthalpy between the product, sodium chloride, and the reactants, solid sodium and gaseous chlorine molecules, that is the standard enthalpy of formation of sodium chloride. Inserting numerical values gives

$$\Delta H_F^\ominus = \Delta H_S^\ominus + \tfrac{1}{2}\Delta H_D^\ominus + \Delta H_I^\ominus + \Delta H_E^\ominus + \Delta H_L^\ominus$$
$$= +109 + 121 + 494 - 380 - 755 = -411 \text{ kJ mol}^{-1}$$

In practice, it is easier to measure standard enthalpies of formation than to measure some of the other steps. The electron affinity is the hardest term to measure experimentally: Born–Haber cycles are often used to calculate electron affinities.

The Born–Haber cycle for magnesium oxide is shown in Figure 10.12B.

You will be able to identify $\Delta H_1^\ominus$, $\Delta H_2^\ominus$, $\Delta H_3^\ominus$ and $\Delta H_5^\ominus$ by comparison with the sodium chloride cycle. In the case of oxygen, $\Delta H_4^\ominus$ is an endothermic term. The reason is that, although energy is given out in the process

$$O(g) + e^- \longrightarrow O^-(g)$$

If the lattice enthalpy is strongly exothermic, this may be the term that decides whether an ionic compound is formed

energy is required for the introduction of a second electron against the repulsion of O^-, in the process

$$O^-(g) + e^- \longrightarrow O^{2-}(g)$$

making $\Delta H_4^\ominus$ endothermic. The only exothermic term is the lattice enthalpy. The reason why the formation of the ionic lattice is strongly exothermic is that both ions Mg^{2+} and O^{2-}, are small, and both carry two units of charge. When these opposite charges are brought close together, there is a big release of energy. This example shows the importance of lattice enthalpy in determining whether ionic compounds are formed [see Question 3, Checkpoint 10.14].

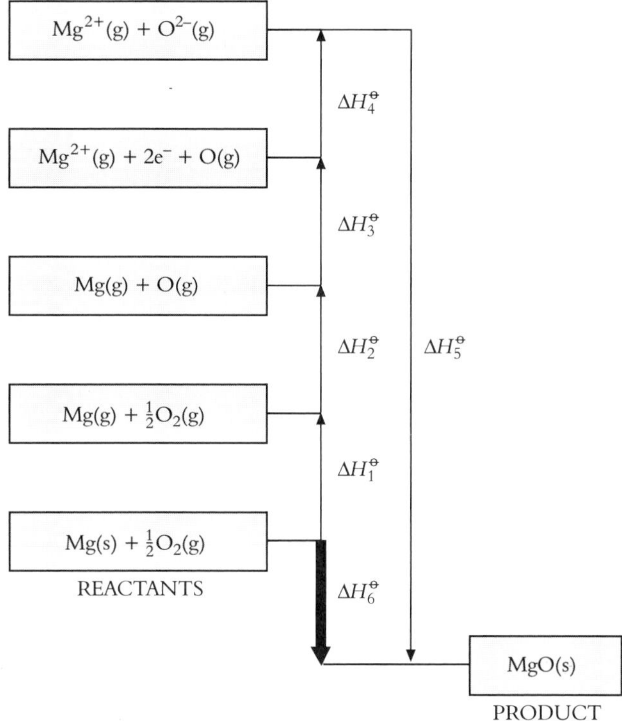

Figure 10.12B Born–Haber cycle for magnesium oxide

SUMMARY

The **Born–Haber cycle** shows the standard enthalpy terms which are involved when an ionic solid is formed. The cycle is usually employed to calculate electron affinity.

10.13 LATTICE ENTHALPIES

Small ions can approach closely

Some values of lattice enthalpies/kJ mol^{-1} are NaCl −771, KCl −707, NaF −918, CsF −747, NaI −699, MgO −3791, BaO −3054. Where do the differences in lattice enthalpies arise? The question can be answered with the help of ionic radii. Ionic radii/nm are Na^+ 0.102, K^+ 0.138, Cs^+ 0.170, Mg^{2+} 0.072, Ba^{2+} 0.136, F^- 0.133, Cl^- 0.180, I^- 0.219, O^{2-} 0.140.

Multiple charged ions have strong forces of attraction between them

- Comparing NaCl and KCl, the ion Na^+ has a smaller radius than K^+ and can approach more closely to the Cl^- ion. When the ions are closer together the forces of attraction between them are stronger.
- Comparing NaCl and NaF, F^- has a smaller radius than Cl^- and can approach more closely to Na^+.
- Comparing MgO and BaO, Mg^{2+} is much smaller than Ba^{2+}. In MgO the ions are much closer together than in BaO and the forces of attraction between them are stronger.
- Comparing NaCl and MgO, in MgO the ions Mg^{2+} and O^{2-} have two charges each and the forces of attraction are greater than between singly charged ions.

Lattice enthalpy is proportional to product of charges on ions/sum of ionic radii

In general,

$$\text{Lattice enthalpy} \propto \frac{(\text{Charge on cation}) \times (\text{Charge on anion})}{\text{Sum of ionic radii}}$$

10.13.1 CALCULATED AND EXPERIMENTAL VALUES

The lattice enthalpy of an ionic compound can be found experimentally ...

... and also calculated from a theoretical model

The lattice enthalpy of a compound can be found experimentally by means of a Born–Haber cycle. It is also possible to calculate the standard lattice enthalpy of a compound from a theoretical model. A simple model of an ionic bond is two point charges separated by the interionic distance. The potential energy required to separate two ions can be calculated from their charges and the distance between them. The model can be extended to a three-dimensional ionic structure to give a value for the standard lattice enthalpy. Table 10.13 compares such theoretical values with experimental values.

Table 10.13 Calculated values and experimental values (kJ mol^{-1}) of lattice enthalpies

Compound	Calculated value (from model)	Experimental value (from Born–Haber cycle)
NaCl	766	788
NaBr	731	719
NaI	686	670
KCl	692	718
KBr	667	656
KI	631	615
AgCl	769	921
AgBr	759	876
AgI	736	862
ZnS	3430	3739

The values agree well for some ionic compounds ...

... but not for all

Where the difference in electronegativity between the combined elements is small ...

... agreement is poor

This indicates some covalent character in the bonds

For the alkali metal halides, there is good agreement between the values calculated from the model and the experimental values. The simple ionic model is thus justified for the alkali metal halides. For the silver halides and zinc sulphide, the agreement is less good. The simple ionic model needs modification for these compounds. Where the differences in electronegativity are great, e.g. in sodium iodide, the ionic model gives good agreement with experimental values. Where the differences in electronegativity are smaller, e.g. in zinc sulphide, the bonding is actually stronger than predicted by the ionic model. The explanation put forward is that the bonding is not purely ionic. There is an incomplete transfer of valence electrons from cation to anion. As a result, the electron density in between the cation and the anion is higher than calculated; that is, the bond has partial covalent character; see § 4.6.4.

SUMMARY

For an ionic compound,

Lattice enthalpy $\propto$ product of charges on ions/sum of ionic radii

Values of lattice enthalpy calculated from a theoretical model in many cases agree with experimental values. Those which do not agree are for ionic compounds with some degree of covalent character.

CHECKPOINT 10.13: LATTICE ENTHALPY

1. Refer to the ionic radii in § 10.13. Explain why

(a) the lattice enthalpy of sodium fluoride is much greater than that of caesium fluoride

(b) the lattice enthalpy of sodium chloride is greater than that of sodium iodide

(c) the lattice enthalpy of magnesium oxide is greater than that of sodium chloride.

2. The theoretical model of an ionic bond as two point charges allows a calculation of the lattice energy of an ionic compound. The calculated value can be compared with an experimental value for the lattice energy. Explain why there is good agreement between the two values for sodium iodide and less good agreement for zinc sulphide.

10.14 ENTHALPY CHANGES INVOLVED WHEN IONIC COMPOUNDS DISSOLVE

Figure 10.14A shows what happens when a salt dissolves in water. The polar nature of water molecules

$$H^{\delta+} - O^{\delta-} - H^{\delta+}$$

explains why water is a good solvent for ionic compounds.

Figure 10.14A Dissolution of a salt in water

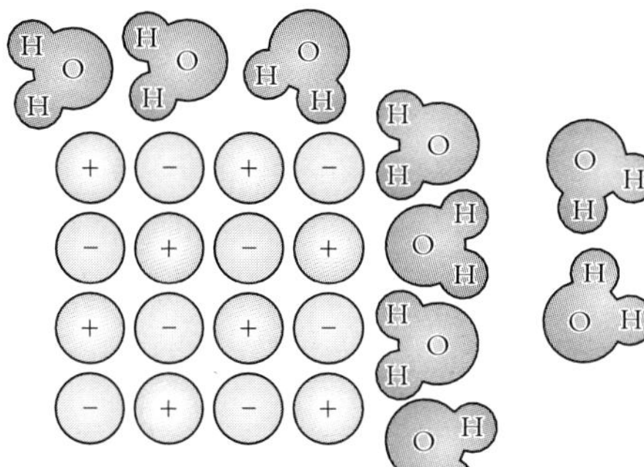

(a) Water molecules are attracted to ions in the crystal. The $\delta-$ oxygen atoms are attracted to the cations, and the $\delta+$ hydrogen atoms are attracted to the anions.

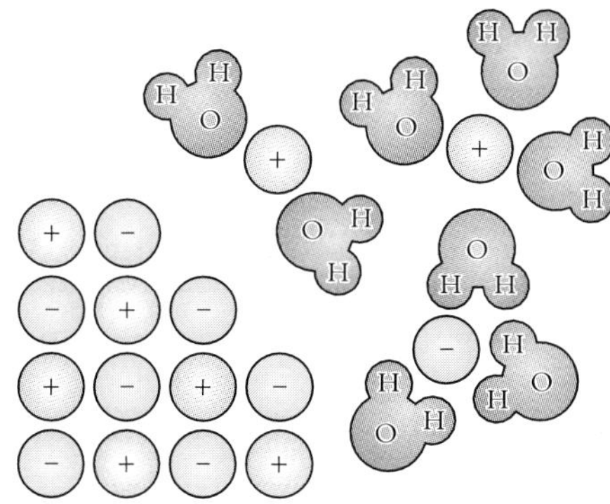

(b) As forces of attraction come into play, energy is given out. This compensates for the energy required to break up the crystal structure. Ions leave the structure.

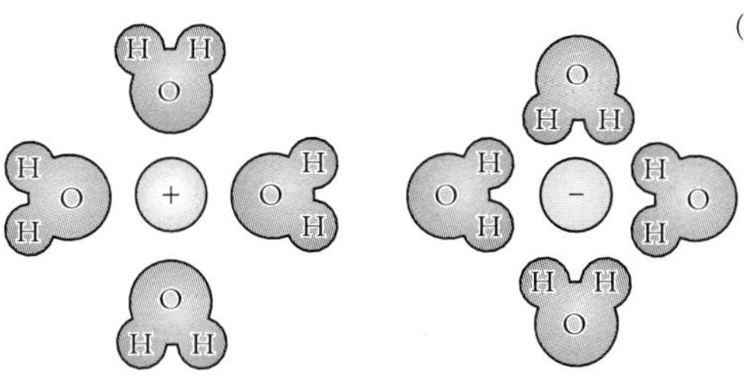

(c) These hydrated ions are called *aqua* ions. Many are surrounded by six water molecules.

When an ionic solid dissolves in a solvent, two enthalpy terms are involved.

When salts dissolve, energy is required to separate the ions ...

1. The ions must be separated from the ionic lattice. The energy required is the lattice dissociation enthalpy.

... Solvation of the ions releases energy ...

2. The separate ions interact with the molecules of solvent. If the solvent is polar, a charged ion can be attracted to one end of a polar solvent molecule. The energy released as these attractive forces come into play is compensation for the energy required to dissociate the lattice:

$\Delta H^{\ominus}$ solvation is defined

$$\Delta H^{\ominus}_{\text{Dissolution}} = \Delta H^{\ominus}_{\text{Lattice dissociation}} + \Delta H^{\ominus}_{\text{Solvation}}$$

(a positive quantity) (a negative quantity)

Standard enthalpy of solvation $\Delta H^{\ominus}_{\text{Solvation}}$ is the enthalpy absorbed at 1 atm per mole when gaseous ions are converted into solvated ions:

$$X^+(g) \longrightarrow X^+(\text{solvated})$$

The way in which $\Delta H^{\ominus}_{\text{Lattice dissociation}}$ and the sum of $\Delta H^{\ominus}_{\text{Solvation}}$ for the two ions contribute to $\Delta H^{\ominus}_{\text{Dissolution}}$ is shown below. In this example, enthalpy considerations favour dissolution, and $\Delta H^{\ominus}_{\text{Dissolution}}$ is negative. The salt dissolves exothermically.

Figure 10.14B

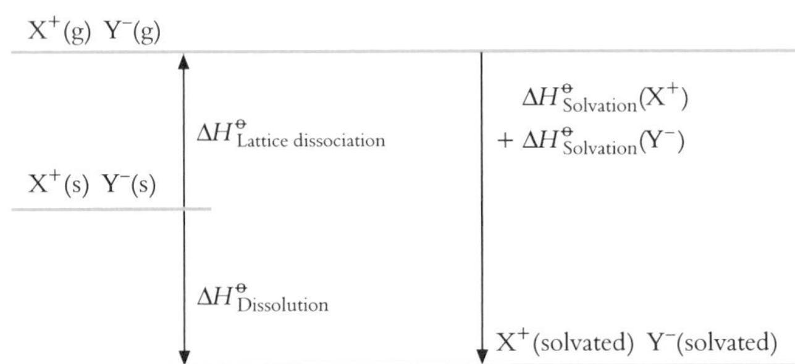

There are, however, salts which dissolve endothermically. This is because the value of $\Delta H^{\ominus}_{\text{Dissolution}}$ does not give the whole picture. Entropy considerations, which are discussed in § 10.15 may outweigh a positive value of $\Delta H^{\ominus}_{\text{Dissolution}}$.

As water molecules have large dipoles, they can interact strongly with solute ions. These powerful interactions result in a high negative enthalpy of solvation, which makes water a good solvent. Non-polar solvents, such as hydrocarbons, do not dissolve ionic substances because there is no negative enthalpy of solvation to compensate for the positive lattice dissociation enthalpy.

SUMMARY

For an ionic solid.

$$\Delta H^{\ominus}_{\text{Dissolution}} = \Delta H^{\ominus}_{\text{Lattice dissociation}} + \Delta H^{\ominus}_{\text{Solvation}}$$

On balance $\Delta H^{\ominus}_{\text{Dissolution}}$ may be positive or negative.

A positive value favours dissolution.

CHECKPOINT 10.14: STANDARD ENTHALPY OF REACTION

1. Refer to Figure 10.12B. To what do the symbols $\Delta H_1^\ominus$ to $\Delta H_5^\ominus$ refer? What is the significance of an arrow pointing upwards? Given that the values, in kJ mol^{-1} are: $\Delta H_1^\ominus = +153$, $\Delta H_2^\ominus = +248$, $\Delta H_3^\ominus = +2180$, $\Delta H_4^\ominus = +745$, and $\Delta H_5^\ominus = -3930$, calculate $\Delta H_6^\ominus$, the standard enthalpy of formation of magnesium oxide.

2. The following values are $\Delta H_F^\ominus$/kJ mol^{-1} at 298K:

   $CH_4(g) - 76$; $CO_2(g) - 394$; $H_2O(l) - 286$;
   $H_2O(g) - 242$; $NH_3(g) - 46.2$;
   $C_2H_5OH(l) - 278$; $C_8H_{18}(l) - 210$;
   $C_2H_6(g) - 85$

   Calculate the standard enthalpy changes at 298 K for the reactions:

   (a) $C_2H_6(g) + 3\frac{1}{2}O_2(g) \longrightarrow 2CO_2(g) + 3H_2O(l)$
   (b) $C_2H_5OH(l) + 3O_2(g) \longrightarrow 2CO_2(g) + 3H_2O(l)$
   (c) $H_2(g) + \frac{1}{2}O_2(g) \longrightarrow H_2O(l)$
   (d) $C_8H_{18}(l) + 12\frac{1}{2}O_2(g) \longrightarrow 8CO_2(g) + 9H_2O(l)$

3. The Born–Haber cycle for rubidium chloride is shown in Figure 10.14C: The letters **A** to **F** represent standard enthalpy changes. Give the names of these quantities. Their values in kJ mol^{-1} are: **A** = −431, **B** = +86, **C** = +122, **D** = +408, **F** = −675. Calculate the value of **E**.

Figure 10.14C
Born–Haber cycle for rubidium chloride (not to scale)

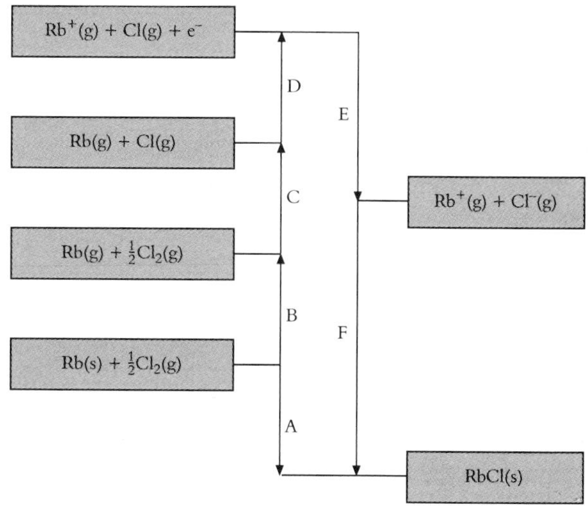

10.15 ENTROPY

A spontaneous change happens without external help

A **spontaneous change** is a change that happens naturally; it tends to occur without the need to be driven by an external influence. One can accept that exothermic reactions can happen spontaneously because they involve a decrease in the enthalpy of the system. It is rather analagous to a ball spontaneously rolling downhill. It is more difficult to explain why an endothermic reaction can happen spontaneously with an increase in the enthalpy of the system. One does not expect a ball to roll uphill spontaneously: some force must be applied to it. There must be a factor in addition to the enthalpy change that enables some endothermic changes to occur spontaneously. What are the conditions that decide whether a change can occur spontaneously?

Consider some spontaneous changes.

Spontaneous changes involve an increase in disorder ...

- When gases come into contact they diffuse to form a homogeneous mixture. Instead of a tidy arrangement of different molecules in different containers, the molecules become mixed up in a random arrangement. The pattern is that mass tends to become more disordered.

... e.g. diffusion ...

... dissolution ...

... melting ...

- Potassium chloride dissolves endothermically. The solid is a highly ordered arrangement of potassium ions and chloride ions. When it dissolves, this regular arrangement is replaced by a random distribution of ions in solution.

$$KCl(aq) + aq \longrightarrow KCl(aq);\ \Delta H^\ominus = +19 \text{ kJ mol}^{-1}$$

... dissociation ...

The extent of disorder is measured by the quantity entropy

- When ice melts the regular hydrogen-bonded structure of ice (see Figure 4.8J) is replaced by the fluid association of water molecules in liquid water. When water vaporises the association of water molecules is replaced by individual molecules moving independently in the gas phase. Both of these changes of state occur spontaneously and endothermically.

$$H_2O(s) \longrightarrow H_2O(l); \Delta H^\ominus = +6.0\ \text{kJ mol}^{-1}$$
$$H_2O(l) \longrightarrow H_2O(g); \Delta H^\ominus = +44\ \text{kJ mol}^{-1}$$

- Ammonium carbonate dissociates spontaneously

$$(NH_4)_2CO_3(s) \longrightarrow 2NH_3(g) + CO_2(g) + H_2O(g); \Delta H^\ominus = +68\ \text{kJ mol}^{-1}$$

One mole of a solid with an ordered crystalline structure is replaced by four moles of gas. The system has become less ordered.

In all these changes a relatively ordered system has been replaced by a more disordered system.

The entropy of a substance increases when it is heated ...

... when its particles are allowed to move more freely

The extent of disorder of a system is measured by a quantity called **entropy** (symbol S). Low entropy means little disorder; high entropy means a high degree of disorder. Natural changes tend to increase the degree of disorder of matter and of energy. This observation is embodied in the **Second Law of Thermodynamics**, which states, Entropy tends to increase. The entropy of a substance can be increased by heating. This increases the motion of particles and therefore the degree of disorder. Entropy can be increased by providing more space into which particles can spread.

... when it melts ...

... when it vaporises

The standard entropies of gases are higher than those of similar solids and liquids at the same temperature

A solid is an ordered three-dimensional structure with low entropy. When a solid melts, the particles have more freedom to move: the liquid is more disordered than the solid and has higher entropy. When a liquid freezes, its particles adopt the ordered structure of the solid with a decrease in entropy. When a substance vaporises the particles have almost complete freedom of movement and they occupy a much higher volume than in the liquid or solid state. Gases therefore have much higher entropies than liquids and solids at the same temperature. A perfect crystal at temperature zero K has perfect order and zero entropy. As the substance is heated the entropy increases. All substances are more disordered at room temperature than at zero K.

Change in entropy

... Estimating whether a change involves an increase or a decrease in entropy, S ...

Entropy is given the symbol S, standard entropy, $S^\ominus$, and change in entropy, ΔS. An increase in the degree of disorder of a system shows in a positive value of ΔS. It is often possible to tell whether a reaction has a positive or negative value of ΔS by an inspection of the equation. The equation

$$NH_4NO_3(s) \longrightarrow N_2O(g) + 2H_2O(g)$$

... and predicting the sign of $\Delta S^\ominus$

shows that 1 mole of the crystalline solid, ammonium nitrate, forms 1 mole of a gas, dinitrogen oxide, and 2 moles of water vapour, a total of 3 moles of gas. The value of ΔS is positive. Under standard conditions, the water formed is a liquid, and the increase in entropy is less: $\Delta S^\ominus$ has a smaller positive value. Other reactions for which it is easy to predict the sign of $\Delta S^\ominus$ are

$$CaO(s) + H_2O(l) \longrightarrow Ca(OH)_2(s); \Delta S^\ominus \text{ negative}$$
$$CaCO_3(s) \longrightarrow CaO(s) + CO_2(g); \Delta S^\ominus \text{ positive}$$

SUMMARY

Spontaneous exothermic reactions are common.

Endothermic reactions are only spontaneous if there is a large increase in entropy in the reaction mixture, that is, a large increase in the degree of disorder.

Solid state ⟶ Liquid state ⟶ Gaseous state
—— increase in entropy ⟶

CHECKPOINT 10.15: ENTROPY

1. Which substance has the higher molar entropy?
   (a) **A**: O^2 at 20 °C and 1 atm or **B**: O_2 at 20 °C and 4 atm
   (b) **A**: N_2 at 20 °C and 1 atm or **B**: N_2 at 100 °C and 1 atm
   (c) **A**: $I_2(g)$ or **B**: $I_2(s)$
   (d) **A**: $Br_2(l)$ or **B**: $Br_2(g)$
   (e) **A**: graphite or **B**: diamond
   (f) Explain why the value $\Delta H^{\ominus}_{\text{Melting}} = +6.0\ \text{kJ mol}^{-1}$ for water is lower than the value $\Delta H^{\ominus}_{\text{Vaporisation}} = +44\ \text{kJ mol}^{-1}$.

2. Does each of the following reactions involve an increase or a decrease in standard entropy?
   (a) $N_2(g) + 3H_2(g) \longrightarrow 2NH_3(g)$
   (b) $N_2O_4(g) \longrightarrow 2NO_2(g)$
   (c) $HCl(g) + NH_3(g) \longrightarrow NH_4Cl(s)$
   (d) $C_2H_4(g) + H_2(g) \longrightarrow C_2H_6(g)$

3. Arrange the following in order of increasing entropy:
   (a) 1 mol $H_2O(l)$ 100 °C, 1 atm
   (b) 1 mol $H_2O(s)$ 0 °C, 1 atm
   (c) 1 mol $H_2O(l)$ 0 °C, 1 atm
   (d) 1 mol $H_2O(g)$ 100 °C, 1 atm
   (e) 1 mol $H_2O(l)$ 25 °C, 1 atm
   (f) 1 mol $H_2O(g)$ 100 °C, $\frac{1}{2}$ atm

4. State the sign of the entropy change in the following reactions:
   (a) $NH_3(g) + HCl(g) \longrightarrow NH_4Cl(s)$
   (b) $COCl_2(g) \longrightarrow CO(g) + Cl_2(g)$
   (c) $PCl_3(g) + Cl_2(g) \longrightarrow PCl_5(g)$
   (d) $N_2(g) + 3H_2(g) \longrightarrow 2NH_3(g)$
   (e) $C_6H_{12}(l) + 9O_2(g) \longrightarrow 6CO_2(g) + 6H_2O(g)$

5. Explain why the entropy of a solid is less than that of the gaseous form of the same substance.

10.16 FREE ENERGY, THE DECIDING FACTOR

Both factors, the change in enthalpy and the change in entropy, are important in deciding whether a physical or chemical change will occur. They are combined in the equation

The free energy G is a measure of the enthalpy of a system that is free to cause change: G = H – TS

Free energy, G = enthalpy, H – (Temperature/K × Entropy, S)

$$G = H - TS$$

from which it follows that, at constant temperature,

$$\Delta G = \Delta H - T\Delta S$$

A process occurs spontaneously if it is accompanied by a decrease in free energy, that is, if ΔG is negative

The value of ΔG decides whether a reaction can happen. All spontaneous physical and chemical changes involve a decrease in free energy, a negative value of ΔG. What factors make ΔG negative?

$$\Delta G = \Delta H - T\Delta S$$

This is the case with strongly exothermic changes ...

... ΔH is large and negative ...

... and with changes that involve a big increase in entropy ...

... ΔS is positive

- In an exothermic reaction,

  ΔH is negative. If ΔS has a positive value, then
  ΔG = (negative value) – (positive value) = negative value

  If the reaction is strongly exothermic but ΔS has a small negative value,

  ΔG = (large negative value) – (small negative value) = negative value

- In an endothermic reaction

  ΔH is positive, therefore ΔG is negative only if $T\Delta S$ has a large positive value: ΔG = (positive value) – (large positive value) = negative value

  A change is assisted by a decrease in enthalpy (ΔH negative) and by an increase in entropy (ΔS positive).

SUMMARY

The change in free energy in a process is given by

$\Delta G = \Delta H - T\Delta S$

ΔG must be negative for a change to occur spontaneously; therefore if ΔH is positive $T\Delta S$ must have a large positive value.

To give $T\Delta S$ a large positive value, ΔS must be positive: entropy must increase. A large value of ΔS helps, and a high temperature helps to make the $T\Delta S$ term outweigh a positive value of ΔH. The value of the entropy change is multiplied by the temperature; therefore the role of entropy change in deciding whether or not a reaction is spontaneous becomes more important at higher temperatures.

A reaction with a negative value of ΔG is said to be **feasible**. This means that, if the reaction takes place, it will go in the direction of the reactants forming the products and not in the reverse direction. To say that a reaction is feasible does not tell anything about the rate of the reaction. The reactants may have to surmount an energy barrier, the **activation energy**, before they can react [§ 14.9].

What happens if the value of ΔG for a change is positive? The reaction can be made to occur by driving it with energy from an external source. The reaction

$$2H_2O(l) \longrightarrow 2H_2(g) + O_2(g);\ \Delta G^\ominus = +494\ \text{kJ mol}^{-1}$$

can be made to occur by sending an electric current through water.

CHECKPOINT 10.16: FREE ENERGY

1. Give examples of spontaneous processes which are *(a)* exothermic, *(b)* endothermic, *(c)* accompanied by an increase in entropy, *(d)* accompanied by a decrease in entropy.

2. Is the change in standard entropy for each of the following processes positive or negative?

   (a) 1 mol solid ethanoic acid $\longrightarrow$ 1 mol liquid ethanoic acid
   (b) 1 mol liquid ethanol $\longrightarrow$ 1 mol gaseous ethanol
   (c) 1 mol $N_2O_4(g) \longrightarrow$ 2 mol $NO_2(g)$
   (d) 1 mol $O_2(g)$ + 1 mol $N_2(g) \longrightarrow$ 2 mol $NO(g)$
   (e) $\frac{1}{2}$ mol $O_2(g)$ + 1 mol $Cu(s) \longrightarrow$ 1 mol $CuO(s)$

3. Predict whether ΔG will be positive, negative, probably positive or probably negative for a reaction that is:

   (a) exothermic and accompanied by an increase in entropy of the reactants
   (b) endothermic and accompanied by an increase in entropy of the reactants
   (c) exothermic and accompanied by a decrease in entropy of the reactants
   (d) endothermic and accompanied by a decrease in entropy of the reactants
   (e) Can a change in temperature affect the sign of ΔG in *(a)*, *(b)*, *(c)* or *(d)*? Explain your answer.

4. Calculate the change in standard free energy and determine whether the reaction

   $$Fe_2O_3(s) + 3H_2(g) \longrightarrow 2Fe(s) + 3H_2O(g)$$

   will take place at 300 K and at 800 K. Use the data:

	$Fe_2O_3(s)$	$H_2(g)$	$Fe(s)$	$H_2O(g)$
Standard enthalpy/ kJ mol^{-1}	–822	0	0	–242
Standard entropy/ J K^{-1} mol^{-1}	90.0	131	27.0	189

10.17 LINK WITH KINETICS

Which decides whether reactions happen, thermodynamics or kinetics?

Thermodynamic studies tell us $\Delta G^{\ominus}$ and therefore whether a change is feasible ...

If $\Delta G^{\ominus}$ for a change is negative, we can state that the reactants are thermodynamically unstable relative to that change. Methane is thermodynamically stable with respect to dissociation into its elements:

$$CH_4(g) \longrightarrow C(g) + 2H_2(g);\ \Delta G^{\ominus} = +51\ \text{kJ mol}^{-1}$$

Methane is thermodynamically unstable relative to oxidation:

$$CH_4(g) + 2O_2(g) \longrightarrow CO_2(g) + 2H_2O(l);\ \Delta G^{\ominus} = -580\ \text{kJ mol}^{-1}$$

... Kinetic studies tell us the energy of activation and therefore how fast a change will happen

Thermodynamic instability ($\Delta G^{\ominus}$ is negative) relative to a certain change is a necessary condition for that change to occur, but it is not the only necessary condition. Methane does not react with air at a measurable rate at room temperature (in the absence of UV light). The reason is that a large amount of energy, called the **activation energy**, must be supplied to enable the molecules to react. The concept of activation energy is covered in Chapter 14. Briefly, the idea is that, although the product molecules have a lower free energy content than the reactant molecules, there is an energy barrier in between the reactant and the product molecules. The reactant molecules must be given enough energy to surmount this barrier in order to react. Very, very few molecules of reactant acquire this activation energy at room temperature. In consequence, the overall rate of reaction is so slow as to be unobservable. The reaction is described as being **under kinetic control**. If the temperature is raised sufficiently, the proportion of energetic molecules increases, and the reaction takes place at a measurable rate. Substances which react slowly because the activation energy of the reaction is high and the temperature is too low are described as **unreactive** or **non-labile** or **inert**.

SUMMARY

Will a change happen? Thermodynamics does not tell the whole story. Even when a reaction is favoured by thermodynamics, it may take place so slowly as to be unobservable. The reason is the kinetic factor. The activation energy must be supplied before reaction can happen; see Chapter 14.

FURTHER STUDY ▲

THERMOCHEMISTRY

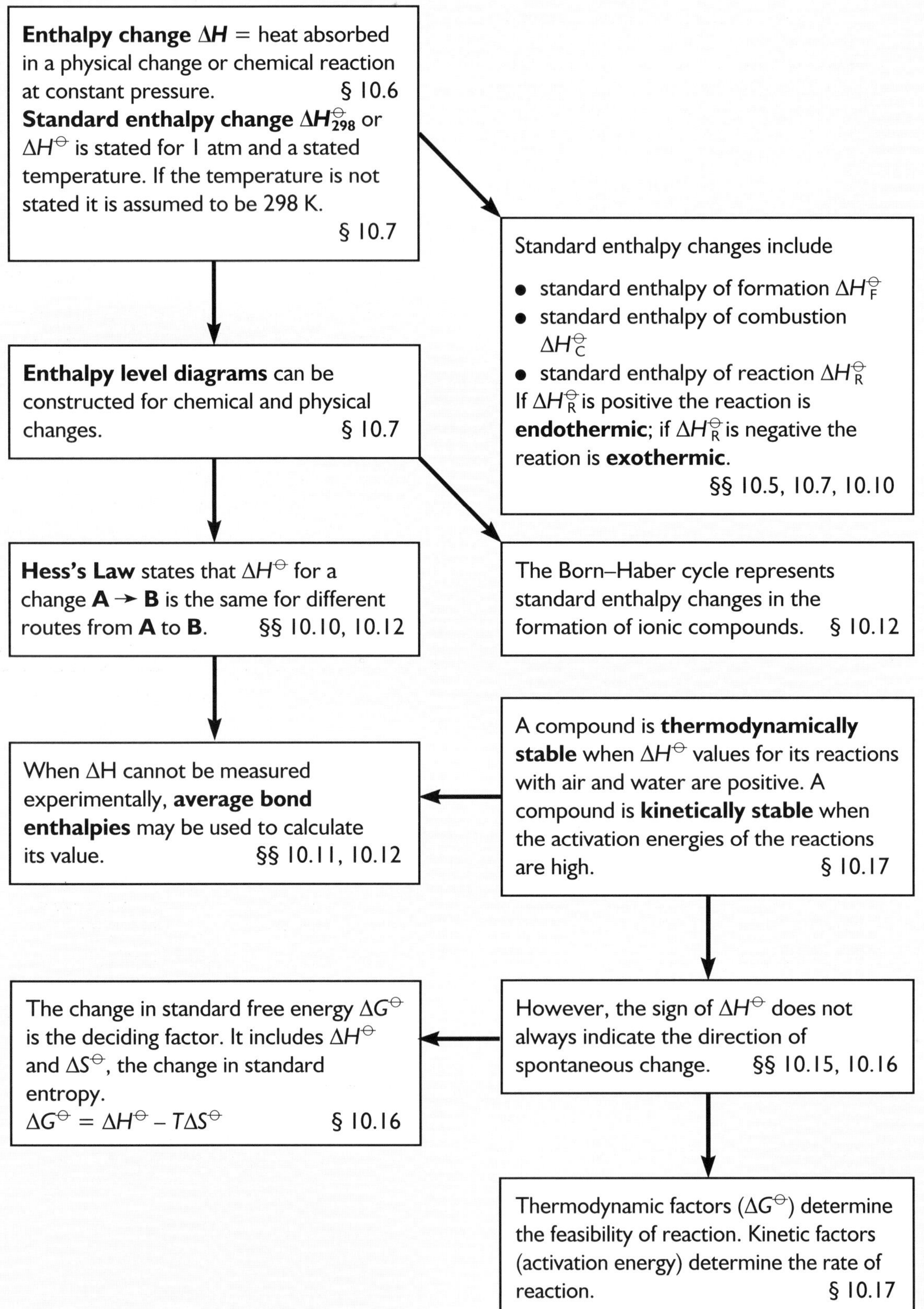
Enthalpy change ΔH = heat absorbed in a physical change or chemical reaction at constant pressure. § 10.6
Standard enthalpy change $\Delta H^{\ominus}_{298}$ or $\Delta H^{\ominus}$ is stated for 1 atm and a stated temperature. If the temperature is not stated it is assumed to be 298 K. § 10.7
Standard enthalpy changes include
• standard enthalpy of formation $\Delta H^{\ominus}_{F}$
• standard enthalpy of combustion $\Delta H^{\ominus}_{C}$
• standard enthalpy of reaction $\Delta H^{\ominus}_{R}$
If $\Delta H^{\ominus}_{R}$ is positive the reaction is **endothermic**; if $\Delta H^{\ominus}_{R}$ is negative the reation is **exothermic**. §§ 10.5, 10.7, 10.10
Enthalpy level diagrams can be constructed for chemical and physical changes. § 10.7
Hess's Law states that $\Delta H^{\ominus}$ for a change **A** → **B** is the same for different routes from **A** to **B**. §§ 10.10, 10.12
The Born–Haber cycle represents standard enthalpy changes in the formation of ionic compounds. § 10.12
When ΔH cannot be measured experimentally, **average bond enthalpies** may be used to calculate its value. §§ 10.11, 10.12
A compound is **thermodynamically stable** when $\Delta H^{\ominus}$ values for its reactions with air and water are positive. A compound is **kinetically stable** when the activation energies of the reactions are high. § 10.17
The change in standard free energy $\Delta G^{\ominus}$ is the deciding factor. It includes $\Delta H^{\ominus}$ and $\Delta S^{\ominus}$, the change in standard entropy.
$\Delta G^{\ominus} = \Delta H^{\ominus} - T\Delta S^{\ominus}$ § 10.16
However, the sign of $\Delta H^{\ominus}$ does not always indicate the direction of spontaneous change. §§ 10.15, 10.16
Thermodynamic factors ($\Delta G^{\ominus}$) determine the feasibility of reaction. Kinetic factors (activation energy) determine the rate of reaction. § 10.17

QUESTIONS ON CHAPTER 10

1. *(a)* Describe how you would find the standard enthalpy of neutralisation of a strong acid by a strong base.

(b) When you have studied § 12.7, explain why the same value is obtained for the neutralisation by sodium hydroxide of nitric, hydrochloric and sulphuric acids but a different value is obtained with ethanoic acid [§ 12.9.1].

2. Construct a Born–Haber cycle for the formation of solid potassium chloride from its elements in their standard states. Use the data below, and calculate the standard enthalpy of formation of KCl(s).

$K(s) \longrightarrow K(g)$; $\Delta H^{\ominus} = 90\ \text{kJ mol}^{-1}$

$K(g) \longrightarrow K^+(g) + e^-$; $\Delta H^{\ominus} = 418\ \text{kJ mol}^{-1}$

$\frac{1}{2}Cl_2(g) \longrightarrow Cl(g)$; $\Delta H^{\ominus} = 122\ \text{kJ mol}^{-1}$

$Cl(g) + e^- \longrightarrow Cl^-(g)$; $\Delta H^{\ominus} = -348\ \text{kJ mol}^{-1}$

$Cl^-(g) + K^+(g) \longrightarrow KCl(s)$; $\Delta H^{\ominus} = -718\ \text{kJ mol}^{-1}$

3. Use the data below to find

(a) the C — C bond enthalpy in ethane and

(b) the value of $\Delta H^{\ominus}$ for $3C(g) + 8H(g) \longrightarrow C_3H_8(g)$

$C(g) + 4H(g) \longrightarrow CH_4(g)$; $\Delta H^{\ominus} = -1664\ \text{kJ mol}^{-1}$

$2C(g) + 6H(g) \longrightarrow C_2H_6(g)$; $\Delta H^{\ominus} = -2827\ \text{kJ mol}^{-1}$

4. Construct a Born–Haber cycle, and use it to find the standard lattice enthalpy of cadmium(II) iodide.

$Cd(s) \longrightarrow Cd(g)$; $\Delta H^{\ominus} = +113\ \text{kJ mol}^{-1}$

$Cd(g) \longrightarrow Cd^{2+}(g) + 2e^-$; $\Delta H^{\ominus} = +2490\ \text{kJ mol}^{-1}$

$I_2(s) \longrightarrow I_2(g)$; $\Delta H^{\ominus} = +19.4\ \text{kJ mol}^{-1}$

$I_2(g) \longrightarrow 2I(g)$; $\Delta H^{\ominus} = +151\ \text{kJ mol}^{-1}$

$I(g) + e^- \longrightarrow I^-(g)$; $\Delta H^{\ominus} = -314\ \text{kJ mol}^{-1}$

$Cd(s) + I_2(s) \longrightarrow CdI_2(s)$; $\Delta H^{\ominus} = -201\ \text{kJ mol}^{-1}$

5. *(a)* State Hess's Law.

(b) Find the standard enthalpy change for the reaction

$$CO(g) + 2H_2(g) \longrightarrow CH_3OH(l)$$

Use the data

$CO(g) + \frac{1}{2}O_2(g) \longrightarrow CO_2(g)$; $\Delta H^{\ominus} = -283\ \text{kJ mol}^{-1}$

$H_2(g) + \frac{1}{2}O_2(g) \longrightarrow H_2O(l)$; $\Delta H^{\ominus} = -286\ \text{kJ mol}^{-1}$

$CH_3OH(l) + 1\frac{1}{2}O_2(g) \longrightarrow CO_2(g) + 2H_2O(l)$; $\Delta H^{\ominus} = -715\ \text{kJ mol}^{-1}$

(c) Explain how Hess's Law underwrites the method of calculation.

6. The standard enthalpies of hydrogenation of cyclohexene and benzene are −120 and −208 kJ mol^{-1} respectively.

(a) Explain why the value for benzene is not three times that for cyclohexene.

(b) Estimate the standard enthalpies of hydrogenation of cyclohexa-1,3-diene and cyclohexa-1,4-diene.

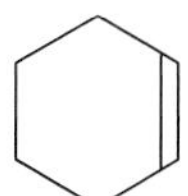

Cyclohexene

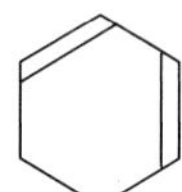

Cyclohexa-1,3-diene

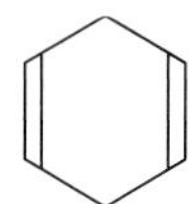

Cyclohexa-1,4-diene

7. A pellet of potassium hydroxide weighing 0.166 g is added to 50.0 g of water in a styrofoam cup. The temperature of the water rises from 19.4 to 20.2 °C. Find the standard enthalpy of solution of KOH.

8. *(a)* Describe the dissolution of an ionic solid in water discussing the energetics of the process.

(b) Suggest why, in general, $\mathbf{A}^{2+}\mathbf{B}^{2-}$ compounds are less soluble than $\mathbf{A}^{+}\mathbf{B}^{-}$ compounds.

(c) Suggest why

(i) LiF is less soluble than NaF

(ii) LiCl is more soluble than NaCl

(iii) NaF is less soluble than NaCl.

	LiCl	**LiF**	**NaCl**	**NaF**
Lattice enthalpies/ kJ mol^{-1}	−843	−1029	−775	−968
Hydration enthalpies/ kJ mol^{-1}	−883	−1023	−778	−965

9. *(a)* What do you understand by the term *entropy*?

(b) For each of the following reactions, say, with reasons, whether the entropy is likely to increase or decrease or stay approximately the same:

(i) $N_2(g) + 3H_2(g) \longrightarrow 2NH_3(g)$

(ii) $SO_2(g) + Cl_2(g) \longrightarrow SO_2Cl_2(g)$

(iii) $H_2NCO_2NH_4(s) \longrightarrow CO_2(g) + 2NH_3(g)$

(iv) $CO(g) + H_2O(g) \longrightarrow CO_2(g) + H_2(g)$

10. Describe how you could measure the molar enthalpy of combustion of ethanol in the laboratory. Say what precautions you would take to minimise error. Explain how you would calculate ΔH_C from the results.

11. *(a)* Use the data in the table to calculate the lattice enthalpy of calcium oxide.

Change in standard enthalpy	kJ mol^{-1}
Formation of calcium oxide	−635
Atomisation of calcium	+193
Sum of first two ionisation energies of calcium	+1740
Dissociation of oxygen, per mole of atoms	+250
Sum of first two electron affinities of oxygen	+702

(b) The standard enthalpy of formation of magnesium oxide is similar to that of calcium. The lattice enthalpy of magnesium oxide is about 300 kJ mol^{-1} greater than that of calcium oxide. Suggest which enthalpy terms in the Born–Haber cycle are responsible for the difference.

12. Calcium chloride $CaCl_2$ is a stable compound. Neither CaCl nor $CaCl_3$ exists. According to calculations,

(a) $CaCl_3$ would have a large positive standard enthalpy of formation, and
(b) CaCl would have a small negative value.

Explain *(a)* and *(b)* in terms of the Born–Haber cycle.

13. The table gives information about the fuels methane and carbon

Substance	$CH_4(g)$	$CO_2(g)$	$H_2O(l)$	C(s)	$O_2(g)$
Standard enthalpy of formation/ kJ mol^{-1}	−75.0	−394.0	−286.0	0	0

(a) Why are zero values of $\Delta H_F^\ominus$ assigned to carbon and oxygen?
(b) Calculate the standard enthalpy of combustion of coal (assuming that it is carbon) and natural gas (methane).
(c) Say which fuel produces more energy (i) per gram of fuel (ii) per mole of carbon dioxide formed.
(d) Which of the two fuels is likely to contribute less to the greenhouse effect per kilojoule of energy produced? Explain your answer.

14. *(a)* Explain; with an example, the term *lattice enthalpy*.

What is a Born–Haber cycle? Using the data below, construct a Born–Haber cycle based on crystalline potassium chloride and use this to calculate the lattice enthalpy of potassium chloride.

What does the magnitude of lattice enthalpy tell us about bonding in an ionic crystal? **9**

Process	$\Delta H^\ominus$/kJ mol^{-1}
Dissociation enthalpy of chlorine: $Cl_2(g) \longrightarrow 2Cl(g)$	+242
Electron affinity of chlorine: $Cl(g) + e^- \longrightarrow Cl^-(g)$	−349
Sublimation enthalpy of potassium: $K(s) \longrightarrow K(g)$	+89
Enthalpy of formation of potassium chloride	−437
First ionisation enthalpy of potassium	+419

(b) Using an example of your own choice in each case, explain the terms *enthalpy of hydration* and *enthalpy of solution* when applied to ionic species.

The lattice enthalpy of potassium iodide is +649 kJ mol^{-1}. The enthalpy of hydration of potassium ions is −321 kJ mol^{-1} and that of iodide ions is −296 kJ mol^{-1}. Use these data to determine the enthalpy of solution of potassium iodide. Explain why potassium iodide dissolves in water spontaneously, even though the process is endothermic. **10**

(c) Explain, with an example, the term *mean bond enthalpy*.

Use values of mean hond enthalpies given in the table below to calculate the enthalpy of hydrogenation of ethene.

The enthalpy of hydrogenation of benzene is −208 kJ mol^{-1}. Explain why this is not simply three times the enthalpy of hydrogenation of ethene. **11**

Bond	Mean bond enthalpy/kJ mol^{-1}
H—H	436
H—C	413
C—C	348
C=C	612

30 *NEAB (AL)*

15. *(a)* In $\Delta H^\ominus$, what does the symbol $^\ominus$ indicate? **1**

(b) Some mean bond enthalpies are given below.

Bond	C—C	H—H	Cl—Cl	C—H
Mean bond enthalpy/ kJ mol^{-1}	348	436	242	412

(i) Write the equation for the reaction used to define the bond enthalpy of a chlorine–chlorine bond. Include state symbols.

(ii) Why is the term *mean bond enthalpy* used in the table instead of just *bond enthalpy*?
(iii) Use the data above to predict what happens first when a sample of propane, C_3H_8, is cracked in the absence of air and explain your prediction. **5**

(*c*) Use the following data to calculate the standard enthalpy of formation of propane. **4**

$$C_3H_8(g) + 5O_2(g) \longrightarrow 3CO_2(g) + 4H_2O(l) \quad \Delta H^\ominus = -2220 \text{ kJ mol}^{-1}$$

$$H_2(g) + \tfrac{1}{2}O_2(g) \longrightarrow H_2O(l) \quad \Delta H^\ominus = -286 \text{ kJ mol}^{-1}$$

$$C(s) + O_2(g) \longrightarrow CO_2(g) \quad \Delta H^\ominus = -394 \text{ kJ mol}^{-1}$$

10 *NEAB (AS/AL)*

16. When methane is burned in air, the following reactions, **I**, **II**, and **III**, are all feasible.

I $CH_4(g) + 2O_2(g) \rightleftharpoons CO_2(g) + 2H_2O(g)$
$\Delta H^\ominus_{298} = -802 \text{ kJ mol}^{-1}$
$\Delta S^\ominus_{298} = -5.10 \text{ J K}^{-1} \text{mol}^{-1}$

II $CH_4(g) + \tfrac{3}{2}O_2(g) \rightleftharpoons CO(g) + 2H_2O(g)$
$\Delta H^\ominus_{298} = -519 \text{ kJ mol}^{-1}$
$\Delta S^\ominus_{298} = +81.7 \text{ J K}^{-1} \text{mol}^{-1}$

III $CH_4(g) + O_2(g) \rightleftharpoons C(s) + 2H_2O(g)$
$\Delta H^\ominus_{298} = -409 \text{ kJ mol}^{-1}$
$\Delta S^\ominus_{298} = -8.00 \text{ J K}^{-1} \text{mol}^{-1}$

Some further thermodynamic data for reaction **II** are shown in the table below. Similar data for reactions **I** and **III** are shown in graphical form below.

Temperature/K	$\Delta G^\ominus$**/kJ mol**$^{-1}$
1500	−642
3000	−764
4000	−846

(*a*) State the meaning of the term *mean bond enthalpy*. Show how mean bond enthalpy data could be used to estimate the enthalpy change in reaction **I**. **5**

(*b*) (i) Calculate the value of $\Delta G^\ominus_{298}$ for reaction **II** and, on a copy of the graph below, plot the value you calculate, together with values from the table above.
(ii) Explain how the gradient of the graph for reaction **I** is related to the entropy change, ΔS, for that reaction.
(iii) Explain why $\Delta S^\ominus_{298}$ for reaction **I** is similar to that for reaction **III**, whereas $\Delta S^\ominus_{298}$ for reaction **II** is very different from the other two. **10**

(*c*) (i) Explain the term *feasible reaction* and explain how feasibility is related to the change in free energy, $\Delta G^\ominus$, and to the value of the equilibrium constant, K_c.
(ii) All three reactions **I**, **II** and **III**, are said to be feasible at temperatures between 0 K and 4000 K. Justify this assertion and explain why, despite this, methane is not oxidised appreciably by air at room temperature. Deduce which of the three reactions is the most feasible at 3000 K. Hence explain why a smoky Bunsen flame (which results from reaction **III**) must be the consequence of a reaction which does not reach equilibrium. **5**

(*d*) (i) Write expressions for the equilibrium constants K_c(I) (which refers to reaction **I**) and K_c(II) (which refers to reaction **II**), and also for K_c(IV) (the equilibrium constant for reaction **IV** below).

IV $CO(g) + \tfrac{1}{2}O_2(g) \rightleftharpoons CO_2(g)$

Show that $K_c(\text{IV}) = \dfrac{K_c(\text{I})}{K_c(\text{II})}$

(ii) There exists a temperature T_x, at which reactions **I** and **II** are equally feasible. Deduce the value of T_x from your graph and using the expression in part (*d*) (i), deduce also the value of K_c(IV) at this temperature. Derive the units of K_c(IV). **10**

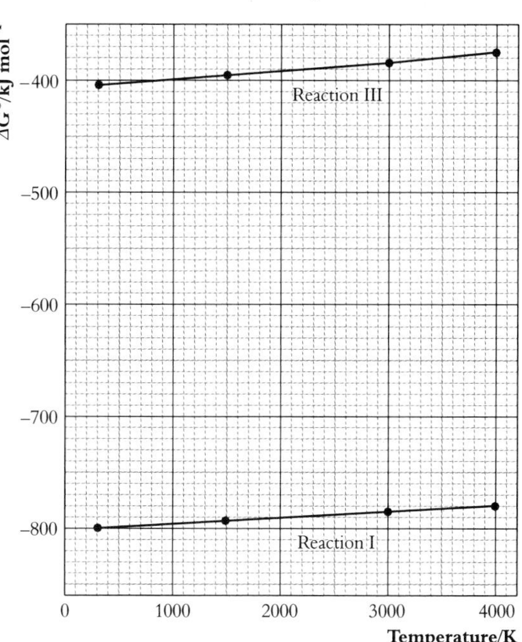

For part (*d*) see Chapter 11.

30 *NEAB (AL)*

17. Use the data in the table below to answer the following questions. Give chemical equations and calculate numerical values of ΔS wherever possible.

Species	$S^\ominus$/ J K^{-1} mol^{-1}	Species	$S^\ominus$/ J K^{-1} mol^{-1}
C(graphite)	6	$H_2O(g)$	189
C(diamond)	3	$H_2O(l)$	70
$H_2(g)$	131	$CH_4(g)$	186
CO(g)	198	CaO(s)	40
$CO_2(g)$	214	$CaCO_3(s)$	90

(a) Entropy can be linked to disorder. When H_2O goes from ice to water and then to steam, the value of its standard entropy, $S^\ominus$, alters. Illustrate this alteration by sketching a labelled graph of entropy against temperature from 200 K to 450 K, paying particular attention to the (approximate) relative sizes of all the entropy changes in your graph. **4**

(b) At all temperatures below 100 °C, steam at atmospheric pressure condenses spontaneously to form water. Explain this observation in terms of ΔG and calculate the enthalpy of vaporisation of water at 100 °C. **4**

(c) Explain why the reaction of 1 mol of methane with steam to form carbon monoxide and hydrogen ($\Delta H^\ominus = +210$ kJ mol^{-1}) is spontaneous only at high temperatures. **6**

(d) Explain why the change of 1 mol of diamond to graphite ($\Delta H^\ominus = -2$ kJ mol^{-1}) is feasible at all temperatures yet does not occur at room temperature. **3**

(e) The reaction between 1 mol of calcium oxide and carbon dioxide to form calcium carbonate ($\Delta H^\ominus = -178$ kJ mol^{-1}) ceases to be feasible above a certain temperature, T_s. Determine the value of T_s. **2**

19 *NEAB (AL)*

18. Use the following data to answer parts (*b*) and (*c*) of this question.

$F_2(g) \longrightarrow 2F(g)$	$\Delta H_1^\ominus = +158$ kJ mol^{-1}
$Ag^+(g) + F^-(g) \longrightarrow AgF(s)$	$\Delta H_2^\ominus = -969$ kJ mol^{-1}
$Ag(s) \longrightarrow Ag(g)$	$\Delta H_3^\ominus = +278$ kJ mol^{-1}
$Ag^+(g) + F^-(g) \longrightarrow Ag^+(aq) + F^-(aq)$	$\Delta H_4^\ominus = -991$ kJ mol^{-1}
$Ag(s) + \frac{1}{2}F_2(g) \longrightarrow AgF(s)$	$\Delta H_5^\ominus = -203$ kJ mol^{-1}
$Ag(g) \longrightarrow Ag^+(g) + e^-$	$\Delta H_6^\ominus = +731$ kJ mol^{-1}

(a) Explain what is meant by the term *standard enthalpy change* and write equations to illustrate the meaning of the terms *electron affinity* and *second ionisation enthalpy*. **5**

(b) Calculate the enthalpy of solution of silver fluoride. **3**

(c) (i) Construct a Born–Haber cycle for the formation of silver fluoride.
(ii) Use your cycle and the data given above to calculate the value of the electron affinity of fluorine. **7**

15 *NEAB (AL)*

19. Sodium bromide is formed from its elements at 298 K according to the equation

$$Na(s) + \tfrac{1}{2}Br_2(l) \longrightarrow NaBr(s)$$

The lattice dissociation enthalpy of solid sodium bromide refers to the enthalpy change for the process

$$NaBr(s) \longrightarrow Na^+(g) + Br^-(g)$$

The electron addition enthalpy refers to the process

$$Br(g) + e^- \longrightarrow Br^-(g)$$

Use this information and the data in the table below to answer the questions which follow.

Standard enthalpies		$\Delta H^\ominus$/kJ mol^{-1}
$\Delta H_f^\ominus$	formation of NaBr(s)	−361
$\Delta H_{ea}^\ominus$	electron addition to Br(g)	−325
$\Delta H_{sub}^\ominus$	sublimation of Na(s)	+107
$\Delta H_{diss}^\ominus$	bond dissociation of $Br_2(g)$	+194
$\Delta H_i^\ominus$	first ionisation of Na(g)	+498
$\Delta H_L^\ominus$	lattice dissociation of NaBr(s)	+753

(a) Construct a Born–Haber cycle for sodium bromide. Label the steps in the cycle with symbols like those used above rather than numerical values. **6**

(b) Use the data above and the Born–Haber cycle in part (*a*) to calculate the enthalpy of vaporisation, $\Delta H_{vap}^\ominus$, of liquid bromine. **3**

9 *NEAB (AL)*

11 EQUILIBRIA

11.1 FRITZ HABER

Equilibrium was a nuisance to Fritz Haber. At the beginning of the twentieth century European farmers imported their fertiliser, sodium nitrate, from Chile. With an abundant supply of nitrogen in the air, it was evident that a clever chemist could fix this nitrogen to make nitrogen compounds, and European farmers would no longer have to ship fertiliser across the Atlantic. Haber was one of the chemists who took up the challenge. He tried the simplest way of fixing atmospheric nitrogen: combining it with hydrogen to make ammonia

$$N_2(g) + 3H_2(g) \rightleftharpoons 2NH_3(g)$$

The way would then be clear for the manufacture of ammonium salts for use as fertilisers. There was a problem. The reaction came to equilibrium after only a small percentage of the mixture had reacted. Haber understood Le Chatelier's Principle (§ 3.19, 3.20), and saw that the yield should be higher at higher pressure. A young British chemist, Robert le Rossignol, joined Haber's team and constructed high-pressure equipment, in which he used a wide variety of catalysts. At a pressure of 200 atm and 600 °C, Haber and le Rossignol achieved the grand percentage of 5% conversion. With clever engineering this was enough. The ammonia was liquefied, the unreacted gases from the equilibrium mixture were recycled, the expensive catalyst osmium was replaced with an iron catalyst, and the process became economically viable. The scaling up of the process was done by Carl Bosch. His work involved overcoming the problems of working at high pressure, for example the reaction of hydrogen with the carbon content of the cast iron apparatus.

The Haber process was working in time for the start of the First World War. The ammonia produced was used to make nitric acid and explosives. Haber contributed to the German war effort also by organising the use of chlorine against the British and French forces in France. After the war he received the Nobel Prize in chemistry, although many people resented this award because of his gas warfare work. His wife was so upset by Haber's contribution to the war that she committed suicide in 1916.

In 1933, Germany passed laws to prevent Jews from being employed as civil servants or university personnel. Although a Jew, Haber was exempt on account of his war service, but he nevertheless resigned in protest. He sought refuge in England, and was given a post at Cambridge University. On his way from England to Italy in 1934 he suffered a heart attack and died in Switzerland.

An introduction to this topic has been made in Chapter 3.

11.2 REVERSIBLE REACTIONS

Some reactions take place in both directions ...

Chemical reactions which take place in both directions are called reversible reactions. An example of a reversible reaction between gases is the reaction between hydrogen and iodine to form hydrogen iodide:

... e.g., the reaction between hydrogen and iodine

$$H_2(g) + I_2(g) \rightleftharpoons 2HI(g)$$

M Bodenstein made a detailed study of this reaction from 1890 to 1900. His method was to seal in glass bulbs either a mixture of hydrogen and iodine or pure hydrogen iodide, and place the bulbs in a thermostat bath. After a time interval, he cooled the bulbs rapidly so that chemical reaction stopped. Then he analysed the contents of each bulb to find the amounts of hydrogen, iodine and hydrogen iodide present.

Figure 11.2A
Bodenstein's results for a set of experiments at 448 °C

From dissociation of 1 mol HI and from combination of 0.5 mol H_2 + 0.5 mol I_2 the mixtures of HI, H_2 and I_2 obtained have the same composition

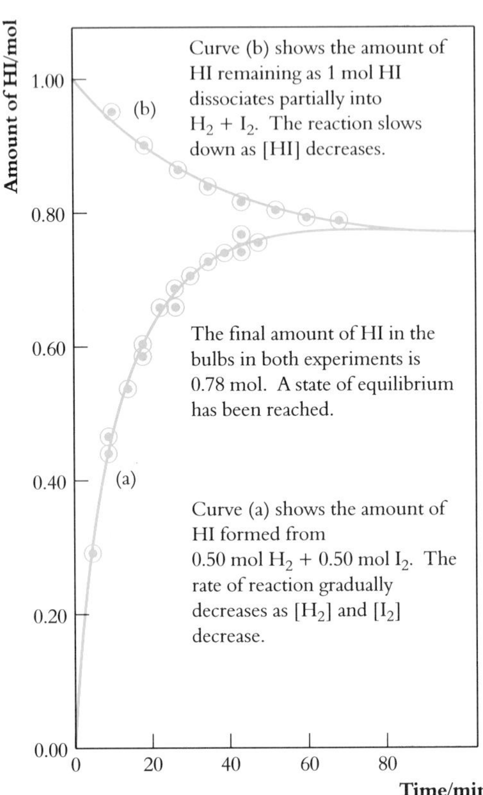

Amount of HI/mol
1.00
0.80
0.60
0.40
0.20
0.00
0
20
40
60
80
Time/min
(b)
(a)
Curve (b) shows the amount of HI remaining as 1 mol HI dissociates partially into $H_2 + I_2$. The reaction slows down as [HI] decreases.
The final amount of HI in the bulbs in both experiments is 0.78 mol. A state of equilibrium has been reached.
Curve (a) shows the amount of HI formed from 0.50 mol H_2 + 0.50 mol I_2. The rate of reaction gradually decreases as $[H_2]$ and $[I_2]$ decrease.

Figure 11.2A shows a set of results at a certain temperature and pressure. If the reaction between 0.50 mol H_2 and 0.50 mol I_2 went to completion, 1.00 mol HI would be formed. It appears that, with 0.78 mol HI formed, no further reaction takes place. The system has reached equilibrium. The same amount of HI is present in the bulbs when equilibrium is reached from the other direction, by the dissociation of hydrogen iodide.

A state of dynamic equilibrium is reached ...

... when the forward and reverse reactions occur at the same rate

In an equilibrium mixture of hydrogen, iodine and hydrogen iodide, it is not obvious that chemical reactions are still occurring. In fact, both the forward and the reverse reactions are still taking place. Since the rates of the forward and reverse reactions are equal, the concentration of each species remains constant. The system is said to be in **dynamic equilibrium**. It is possible to prove this by injecting iodine containing a small quantity of radioactive iodine-131 into an equilibrium mixture. Radioactive iodine appears in the hydrogen iodide, showing that the synthesis of hydrogen iodide and the decomposition of hydrogen iodide are still occurring.

The equilibrium constant K_c

The equilibrium concentrations of HI, I_2 and H_2 fit the simple law

$$\frac{[HI]^2}{[H_2][I_2]} = K_c$$

Square brackets represent concentrations in mol dm^{-3}. The ratio, K_c, has a constant value at a particular temperature, no matter what amounts of HI, I_2 and H_2 are taken initially. K_c is the **equilibrium constant** for the reaction in terms of concentration.

The esterification reaction reaches equilibrium ...

An example of a reversible reaction in solution is the esterification of ethanoic acid by ethanol to make ethyl ethanoate. This ester is hydrolysed by water to ethanoic acid and ethanol:

$$CH_3CO_2H(l) + C_2H_5OH(l) \rightleftharpoons CH_3CO_2C_2H_5(l) + H_2O(l)$$

Figure 11.2B shows what happens when ethanoic acid and ethanol are mixed and allowed to come to equilibrium.

Figure 11.2B
An esterification reaches equilibrium

... because the ester is partially hydrolysed to form acid + alcohol

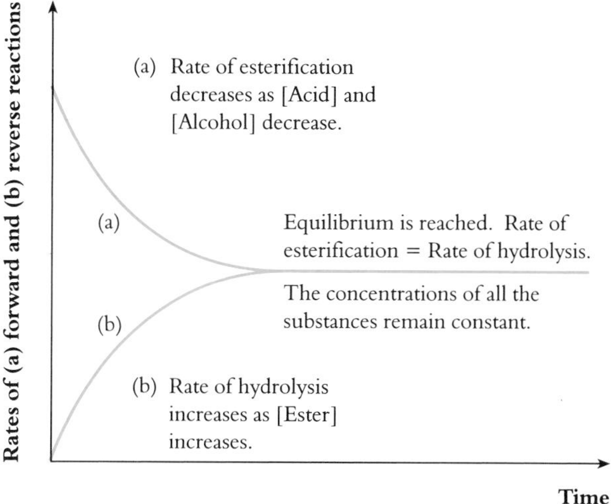
Rates of (a) forward and (b) reverse reactions
(a) Rate of esterification decreases as [Acid] and [Alcohol] decrease.
(a)
Equilibrium is reached. Rate of esterification = Rate of hydrolysis.
The concentrations of all the substances remain constant.
(b)
(b) Rate of hydrolysis increases as [Ester] increases.
Time

The concentrations of the reactants and products at equilibrium fit the equilibrium law

The equilibrium concentrations of acid, alcohol, ester and water obey the law

$$\frac{[CH_3CO_2C_2H_5][H_2O]}{[CH_3CO_2H][C_2H_5OH]} = K_c$$

K_c is the equilibrium constant for the esterification.

Esterification is catalysed by inorganic acids. The presence of a catalyst does not alter the equilibrium constant. Its effect is to decrease the time needed for the system to reach a state of equilibrium.

These two reactions are examples of **homogeneous** equilibria, the first in the gas phase and the second in the liquid phase. If the reactants are in different phases, the equilibrium is described as **heterogeneous**. An example of a heterogeneous equilibrium is the reaction between steam and heated iron to form iron(II) iron(III) oxide Fe_3O_4, and hydrogen:

The reaction between iron and steam reaches equilibrium provided that it takes place in a closed container

$$3Fe(s) + 4H_2O(g) \rightleftharpoons Fe_3O_4(s) + 4H_2(g)$$

If the reaction takes place in a closed container, the system reaches equilibrium. In an open container, hydrogen escapes. In an effort to restore equilibrium, more iron reacts with steam to make more hydrogen. When this reaction was used industrially to make hydrogen, the yield was increased by passing a stream of steam over heated iron, and not allowing the system to come to equilibrium [see also § 11.6.3].

11.3 THE EQUILIBRIUM LAW

The equilibrium law summarises the results of a vast amount of research on reactions such as the three mentioned above. For a reaction

$$a\mathbf{P} + b\mathbf{Q} \rightleftharpoons c\mathbf{R} + d\mathbf{S}$$

the equilibrium constant, expressed in terms of concentrations, K_c, is given by the **Equilibrium Law**:

$$K_c = \frac{[\mathbf{R}]^c[\mathbf{S}]^d}{[\mathbf{P}]^a[\mathbf{Q}]^b}$$

A statement of the equilibrium law

The law can be stated as follows. If a reversible reaction is allowed to reach equilibrium, then the product of the concentrations of the products (raised to the appropriate powers) divided by the product of the concentrations of the reactants (raised to the appropriate powers) has a constant value at a particular temperature. The 'appropriate power' is the coefficient of that substance in the stoichiometric equation for the reaction. The dimensions of K_c are concentration$^{(c+d-a-b)}$, and the units vary from one equilibrium to another.

11.4 POSITION OF EQUILIBRIUM

The position of equilibrium is not the same as the equilibrium constant

The proportion of products to reactants in the equilibrium mixture is described as the **position of equilibrium**:

$$a\mathbf{P} + b\mathbf{Q} \rightleftharpoons c\mathbf{R} + d\mathbf{S}$$

If the conversion of **P** and **Q** into **R** and **S** is small, the position of equilibrium lies to the left.
K_c is small.

If the equilibrium mixture is largely composed of **R** and **S**, the position of equilibrium lies to the right.
K_c is large.

A change in conditions, e.g., pressure, will affect the position of equilibrium

The equilibrium constant K_c is not the same as the position of equilibrium. While K_c is constant at a particular temperature, a change in external conditions can alter the position of equilibrium. Chemists are often interested in finding the best conditions for operating a manufacturing process. They want to shift equilibrium reactions in the direction of forming the products. The study of the factors which alter the position of equilibrium is commercially important.

11.5 THE EFFECT OF CONDITIONS ON THE POSITION OF EQUILIBRIUM

11.5.1 LE CHATELIER'S PRINCIPLE

Le Chatelier's Principle describes the effects of external factors on equilibria

Le Chatelier studied the influence of pressure, temperature and concentration on equilibria. His views are known as **Le Chatelier's Principle**. This states that in any equilibrium, when a change is made to some external factor (such as temperature or pressure), the change in the position of equilibrium is such as to tend to change the external factor in the opposite direction. The system cannot completely cancel the change in the external factor, but it moves in the direction that will minimise the change.

11.5.2 CHANGES IN CONCENTRATION

When there is a change in concentration ...

An aqueous solution of bismuth(III) chloride is cloudy because of the hydrolysis

$$BiCl_3(aq) + H_2O(l) \rightleftharpoons BiOCl(s) + 2HCl(aq)$$

... the position of equilibrium changes; the equilibrium constant is unchanged

If a little concentrated hydrochloric acid is added, the position of equilibrium shifts in the direction that will absorb acid, i.e., from right to left. The solution cannot absorb *all* the acid added. The hydrolysis of bismuth(III) chloride is much reduced, and a clear solution results.

11.5.3 CHANGES IN PRESSURE

When there is a change in pressure ...

It is mainly with gaseous reactions that changes in pressure are important (except for geological reactions). Le Chatelier's Principle can be applied to the formation of ammonia:

... the position of equilibrium changes; the equilibrium constant is unchanged

$$N_2(g) + 3H_2(g) \rightleftharpoons 2NH_3(g);\ \Delta H^{\ominus} = -92\ \text{kJ mol}^{-1}$$

If the pressure of an equilibrium mixture of nitrogen, hydrogen and ammonia is increased, the equilibrium shifts in the direction that tends to decrease the pressure. It does this by decreasing the total number of molecules present, i.e., by moving from left to right of the equation. Although the position of equilibrium has moved towards the ammonia side of the equation, the equilibrium constant has not changed.

11.5.4 CHANGES IN TEMPERATURE

The equilibrium constant of an exothermic reaction decreases with temperature ...

The formation of ammonia is exothermic. If the temperature is raised, the system can absorb heat by the dissociation of ammonia into nitrogen and hydrogen. The equilibrium constant for the formation of ammonia is decreased. This is a different kind of effect from the influence of pressure, where the position of equilibrium is altered but the equilibrium constant remains the same.

In the case of an endothermic reaction, such as

$$N_2(g) + O_2(g) \rightleftharpoons 2NO(g);\ \Delta H^{\ominus} = 180\ \text{kJ mol}^{-1}$$

... and that of an endothermic reaction increases with temperature

increasing the temperature increases the equilibrium constant.

For an exothermic reaction, e.g. the formation of ammonia,

$$N_2(g) + 3H_2(g) \rightleftharpoons 2NH_3(g);\ \Delta H^{\ominus} = -92\ \text{kJ mol}^{-1}$$

the equilibrium constant decreases with a rise in temperature.

For $\Delta G^{\ominus}$ see § 10.16

The relationship between the equilibrium constant, the temperature and the standard enthalpy of reaction is:

$$\ln K = \text{Constant} - \Delta H^{\ominus}/RT$$

A relationship exists between the standard free energy change for the reaction $\Delta G^{\ominus}$ and the equilibrium constant K:

$$\Delta G^{\ominus} = -RT \ln K \quad \text{or} \quad \Delta G^{\ominus} = -2.303\ RT \log K$$

The equilibrium constant K for a system is related to the standard free energy change $\Delta G^{\ominus}$ ...

where R is the gas constant, 8.3 J K^{-1} mol^{-1} and T is the temperature in kelvins.

When $K = 1$, $\Delta G^{\ominus} = 0$.

When $K > 1$, $\Delta G^{\ominus}$ is negative.

... $\Delta G^{\ominus} = -RT \ln K$ For a reaction to be feasible, $\Delta G^{\ominus} \leqslant 0$ and $K \geqslant 1$

For a reaction to be feasible, that is, for the reactants to be converted into the products, there must be a decrease in standard free energy, that is

$$\Delta G^{\ominus} \leqslant 0 \text{ and } K \geqslant 1$$

There is further discussion of the importance of choosing the most favourable conditions for carrying out reversible reactions in the Haber process [§ 22.5] and the Contact process [§ 21.9]. Catalysts do not alter the position of equilibrium [§ 14.4]. They change only the rate at which equilibrium is attained.

11.5.5 PHYSICAL CHANGES

Le Chatelier's Principle applies to physical changes as well as chemical changes. In the reversible change of state

$$\text{Ice} \rightleftharpoons \text{Water}; \Delta H^{\ominus} = 6.00 \text{ kJ mol}^{-1}$$

Physical changes also obey Le Chatelier's Principle

When pressure is exerted on a mixture of ice and water in equilibrium, the system adjusts in such a manner as to tend to decrease its volume, i.e., ice melts. (The density of water increases from 0 °C to 4 °C.) When the temperature rises, the system adjusts itself in such a manner as to tend to resist an increase in temperature. This can be accomplished by the melting of ice, which is an endothermic process. As long as there is any ice left, the temperature will remain constant at 0 °C (at 1 atm).

SUMMARY

Conditions affect the position of equilibrium. There are two kinds of effects here.

1. Change in the position of equilibrium.
The effect of a change in pressure, for example in the formation of ammonia in the Haber process, is to alter the position of equilibrium while the equilibrium constant remains unchanged.

2. Change in the equilibrium constant.
Equilibrium constants vary with temperature.
For an exothermic reaction, the equilibrium constant decreases with a rise in temperature.
For an endothermic reaction, the equilibrium constant increases with a rise in temperature.

Catalysts
Catalysts do not change the equilibrium constant or the position of equilibrium; they change only the rate at which equilibrium is attained.

CHECKPOINT 11.5: LE CHATELIER'S PRINCIPLE

1. Consider the equilibrium

$$N_2(g) + 3H_2(g) \rightleftharpoons 2NH_3(g); \Delta H^{\ominus} = -92 \text{ kJ mol}^{-1}$$

At 25 °C and 10 atm, the percentage conversion to ammonia in the equilibrium mixture is 5%. If the pressure is increased to 40 atm, while the temperature remains the same, will the percentage conversion to ammonia be greater or less? If the temperature is increased to 250 °C, while the pressure remains at 10 atm, will the equilibrium amount of ammonia be greater or less? What will happen if some iron and molybdenum are added to the equilibrium mixture?

2. Consider the equilibrium

$$2NO_2(g) \rightleftharpoons N_2O_4(g); \Delta H^{\ominus} = -54 \text{ kJ mol}^{-1}$$

How are (*a*) the extent of conversion of NO_2 to N_2O_4 and (*b*) the equilibrium constant affected by (i) an increase in temperature, (ii) an increase in pressure and (iii) an increase in volume at constant temperature? Explain your answers.

3. In the equilibrium

$$PCl_5(g) \rightleftharpoons PCl_3(g) + Cl_2(g);\ \Delta H^{\ominus} = 90\ \text{kJ mol}^{-1}$$

what is the effect on (*a*) the position of equilibrium and (*b*) the equilibrium constant of (i) increasing the temperature, (ii) decreasing the volume of the container (iii) adding a catalyst, (iv) adding $Cl_2(g)$ and (v) adding a noble gas? Explain your answers.

4. The freezing of water at 0 °C is represented

$$H_2O(l, \rho = 1.00\ \text{kg dm}^{-3}) \rightleftharpoons H_2O(s, \rho = 0.92\ \text{kg dm}^{-3})$$

Explain why the application of pressure to ice at 0°C makes it melt. Why do most other solids not show the same behaviour?

11.6 EXAMPLES OF REVERSIBLE REACTIONS

11.6.1 REACTION 1: ESTERIFICATION

Ethanoic acid and ethanol react to form ethyl ethanoate and water

$$CH_3CO_2H(l) + C_2H_5OH(l) \rightleftharpoons CH_3CO_2C_2H_5(l) + H_2O(l)$$

The mixture reaches equilibrium:

$$\frac{[CH_3CO_2C_2H_5][H_2O]}{[CH_3CO_2H][C_2H_5OH]} = K_c$$

The reaction is catalysed by acid. The value of K_c can be found by the following method.

A method of finding K_c for the esterification reaction between ethanol and ethanoic acid

1. A number of mixtures are made up, containing ethanol, ethanoic acid, ethyl ethanoate and water. Each mixture is different. Every mixture contains a small amount of hydrochloric acid to catalyse the reaction.
2. The mixtures are put into stoppered bottles and left for a week in a thermostat bath.
3. At the end of the week, the contents of each flask are titrated against standard sodium hydroxide solution, using phenolphthalein as indicator. The titration gives the amount of CH_3CO_2H + the amount of HCl. The amount of HCl is still the same as that present initially. The amount of CH_3CO_2H is found by subtraction.
4. The amounts of the other substances present are calculated.

A sample calculation A 10.0 cm^3 mixture contains the initial amounts/mol ethanol 0.0515; ethanoic acid 0.0525; water 0.0167; ester 0.0314; $H^+(aq)$ 1.00×10^{-3}.

The equilibrium amount of ethanoic acid = 0.0255 mol.
Since the amount of ethanoic acid has decreased by 0.0270 mol, ethanol has decreased by the same amount, and ester and water have both increased by this amount.

Species	CH_3CO_2H +	$C_2H_5OH \rightleftharpoons$	$CH_3CO_2C_2H_5$ +	H_2O
Initial amount/mol	0.0525	0.0515	0.0314	0.0167
Equilibrium amount/mol	0.0255	0.0245	0.0584	0.0437

Since

$$K_c = \frac{[CH_3CO_2C_2H_5][H_2O]}{[CH_3CO_2H][C_2H_5OH]}$$

All the substances are present in the same volume of solution, therefore

$$K_c = \frac{0.0584 \times 0.0437}{0.0255 \times 0.0245} = 4.1$$

From calculations on all the mixtures, an average value of K_c at the chosen temperature can be found. For this equilibrium K_c is dimensionless because

$$K_c = \frac{[\text{acid/mol dm}^{-3}][\text{water/mol dm}^{-3}]}{[\text{ester/mol dm}^{-3}][\text{alcohol/mol dm}^{-3}]}$$

therefore the concentration units cancel out, and K is a number.

11.6.2 REACTION 2: THE REACTION BETWEEN HYDROGEN AND IODINE

In the reaction between hydrogen and iodine

$$H_2(g) + I_2(g) \rightleftharpoons 2HI(g)$$

For gaseous reactions, it is usual to employ K_p, the equilibrium constant in terms of partial pressures

$$K_c = \frac{[HI]^2}{[H_2][I_2]}$$

Since this is a reaction between gases, the concentration of each gas can be expressed as a partial pressure [§ 7.6]. Then

$$K_p = \frac{p_{HI}^2}{p_{H_2} \times p_{I_2}}$$

K_p is the equilibrium constant in terms of partial pressures.

A detailed treatment of the H_2 and I_2 reaction and ...

You can see that, since the pressure units cancel out in the expression, K_p is dimensionless. K_p predicts the same equilibrium composition, no matter what the pressure may be. This is not the case for all gaseous equilibria [contrast Reaction 3, below].

... a worked example

Example When 1.00 mol hydrogen and 1.00 mol iodine are allowed to reach equilibrium in a 1.00 dm^3 flask at 450 °C and 1.01 × 10^5 N m^{-2}, the amount of hydrogen iodide at equilibrium is 1.56 mol. Calculate K_p at 450 °C.

Method

P = total pressure	$H_2(g)$	$+ I_2(g) \rightleftharpoons$	$2HI(g)$	*Total*
Initial amount/mol	1.00	1.00	0	2.00
Equilibrium amount/mol	$1 - a$	$1 - a$	$2a$	2
Since $2a = 1.56$ $a = 0.78$ mol				
Equilibrium amount/mol	0.22	0.22	1.56	2.00
Equilibrium partial pressure	$\left(\frac{0.22P}{2.00}\right)$	$\left(\frac{0.22P}{2.00}\right)$	$\left(\frac{1.56P}{2.00}\right)$	P

$$K_p = \frac{p_{HI}^2}{p_{H_2}p_{I_2}} = \left(\frac{1.56P}{2.00}\right)^2 \bigg/ \left(\frac{0.22P}{2.00}\right)^2 = 50$$

11.6.3 REACTION 3: THE REACTION BETWEEN IRON AND STEAM

The heterogeneous reaction between iron and steam ...

In the reaction between steam and heated iron

$$3Fe(s) + 4H_2O(g) \rightleftharpoons Fe_3O_4(s) + 4H_2(g)$$

$$K_p = \frac{p_{H_2}{}^4}{p_{H_2O}{}^4}$$

Solid reactants do not appear in the expression for K_p

The solids do not appear in the expression. Their vapour pressures remain constant (at a constant temperature) as long as there is some of each solid present. These constant vapour pressures are incorporated into the value of the constant K_p.

Example A mixture of iron and steam was allowed to reach equilibrium at 600 °C. The equilibrium pressures of hydrogen and steam were 3.2 kPa and 2.4 kPa respectively. Calculate the value of the equilibrium constant in terms of partial pressures.

Method

$$K_p = \frac{p_{H_2}{}^4}{p_{H_2O}{}^4} = \left(\frac{3.2}{2.4}\right)^4 = 3.16$$

The equilibrium constant, K_p is 3.2. It is dimensionless.

CHECKPOINT 11.6: EQUILIBRIUM CONSTANTS

1. Write an expression for the equilibrium constant, K_p, for each of the following reactions:

   (a) $CS_2(g) + 4H_2(g) \rightleftharpoons CH_4(g) + 2H_2S(g)$

   (b) $4NH_3(g) + 5O_2(g) \rightleftharpoons 4NO(g) + 6H_2O(g)$

   (c) $2NO_2(g) + 7H_2(g) \rightleftharpoons 2NH_3(g) + 4H_2O(g)$

2. Write an expression for the equilibrium constant, K_c, for each of the following reactions:

   (a) $Sn^{2+}(aq) + 2Fe^{3+}(aq) \rightleftharpoons Sn^{4+}(aq) + 2Fe^{2+}(aq)$

   (b) $Ag^+(aq) + Fe^{2+}(aq) \rightleftharpoons Fe^{3+}(aq) + Ag(s)$

   (c) $2Cr^{3+}(aq) + Fe(s) \rightleftharpoons 2Cr^{2+}(aq) + Fe^{2+}(aq)$

3. Equilibrium is established in the reaction

   $\mathbf{A}(aq) + \mathbf{B}(aq) \rightleftharpoons 2\mathbf{C}(aq)$

   If equilibrium concentrations are $[\mathbf{A}] = 0.25$, $[\mathbf{B}] = 0.40$, $[\mathbf{C}] = 0.50$ mol dm^{-3}, what is the value of K_c?

4. The following reaction was allowed to reach equilibrium:

   $2\mathbf{D}(aq) + \mathbf{E}(aq) \rightleftharpoons \mathbf{F}(aq)$

   The initial amounts of the reactants present in 1.00 dm^3 of solution were 1.00 mol **D** and 0.75 mol **E**. At equilibrium, the amounts were 0.70 mol **D** and 0.60 mol **E**. Calculate the equilibrium constant, K_c.

5. The gases SO_2, O_2 and SO_3 are allowed to reach equilibrium. The partial pressures of the gases are $p_{SO_2} = 0.050$ atm, $p_{O_2} = 0.025$ atm, $p_{SO_3} = 1.00$ atm. Find the values of K_p for the equilibria

   (a) $SO_2(g) + \frac{1}{2}O_2(g) \rightleftharpoons SO_3(g)$

   (b) $2SO_2(g) + O_2(g) \rightleftharpoons 2SO_3(g)$

6. A mixture contained 1.00 mol of ethanoic acid and 5.00 mol of ethanol. After the system had come to equilibrium, a portion of the mixture was titrated against 0.200 mol dm^{-3} sodium hydroxide solution. The titration showed that the whole of the equilibrium mixture would require 289 cm^3 of the standard alkali for neutralisation. Find the value of K_c for the esterification reaction.

11.7 OXIDATION–REDUCTION EQUILIBRIA

Redox reactions reach a position of equilibrium ...

Like other reactions, oxidation–reduction reactions do not always go nearly to completion. In many cases, a state of equilibrium is reached. The reactants and products may be present at equilibrium in comparable amounts. If solutions of the reducing agent, iron(II) sulphate, and the oxidising agent, iodine, are mixed, the resulting solution will contain Fe^{3+}, Fe^{2+}, I_2 and I^-:

... for example, $Fe^{2+} + I_2$ ...

$$I_2(aq) + 2Fe^{2+}(aq) \rightleftharpoons 2I^-(aq) + 2Fe^{3+}(aq) \qquad [1]$$

When solutions of the oxidising agent, Fe^{3+}, and the reducing agent, I^-, are mixed, the resulting solution will again contain all four species:

... $Fe^{3+} + I^-$

$$2Fe^{3+}(aq) + 2I^-(aq) \rightleftharpoons 2Fe^{2+}(aq) + I_2(aq) \qquad [2]$$

The equilibrium constant

If the same amounts of iron and iodine have been used to make the two solutions, the compositions of solutions [1] and [2] will be the same. The reason is that equilibrium is established, and the position of equilibrium is the same, whether it is reached by route [1] or route [2]. The concentrations of the four species at equilibrium are related by the equilibrium constant. For reaction [2], the equilibrium constant is

$$K_c = \frac{[Fe^{2+}]^2[I_2]}{[Fe^{3+}]^2[I^-]^2} = 6 \times 10^7 \text{ mol}^{-1}\text{ dm}^3$$

Redox reactions are used in volumetric analysis; [§ 3.18]

The concentration of each reactant and product is raised to the appropriate power, i.e., the coefficient before that reactant in the stoichiometric equation. The value of K_c is high because the position of equilibrium lies far over to the right hand side of equation [2]. In 1 dm^3 of solution made by adding 0.1 mol of Fe^{3+} and 0.05 mol of I_2, the concentration of Fe^{3+} remaining at equilibrium is only 5×10^{-5} mol dm^{-3}.

With other redox reactions, the **position of equilibrium** (the extent to which the reaction reaches completion) lies closer to the reactants side.

*11.8 PHASE EQUILIBRIUM DIAGRAMS

Definition of a phase

A **phase** is a homogeneous part of a system which is physically distinct from other parts of the system. A mixture of gases is one phase. A solution is one phase. A saturated solution in the presence of an excess of solute is a two-phase system.

In a phase diagram ...

When a substance exists in different physical states, the conditions under which each state exists can be represented by a **phase diagram**. An **area** on a phase diagram represents one phase. A **line** represents the conditions under which two phases can exist in equilibrium. A **triple point** describes the conditions under which three phases can coexist.

11.8.1 WATER

Figure 11.8A shows the phase diagram for water.

Water is one of the few substances which have a solid–liquid line that slopes to the left with increasing pressure. The solid is less dense than the liquid at the freezing temperature.

*Check your specification.

Figure 11.8A
Phase equilibrium diagram for water (not to scale)

... an area = one phase ...

... a line = two phases ...

... a point = three phases, and is named a triple point

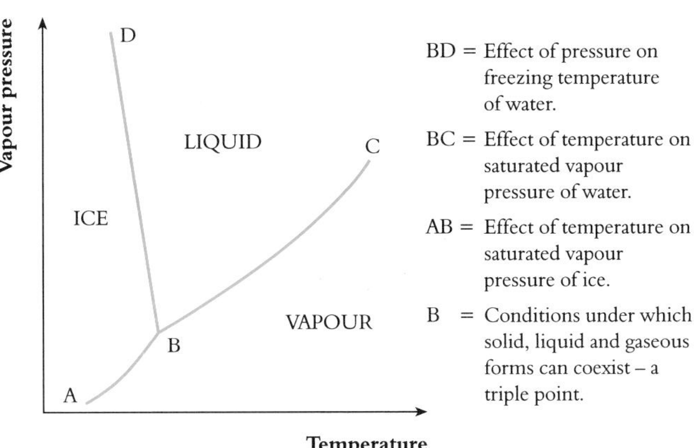

11.8.2 CARBON DIOXIDE

Figure 11.8B shows the phase diagram for carbon dioxide.

Figure 11.8B
Phase equilibrium diagram for carbon dioxide (not to scale)

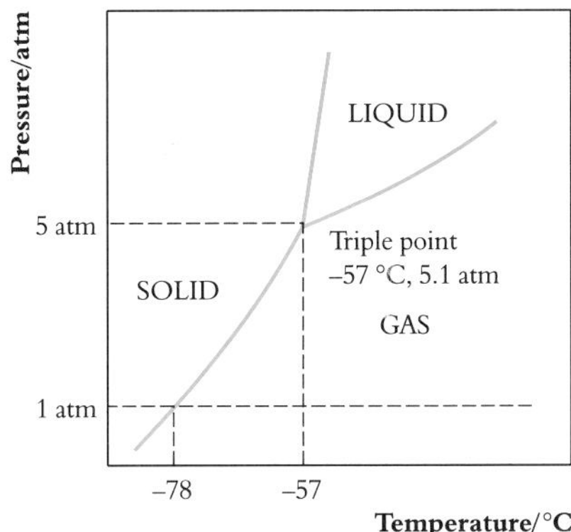

Many other chapters deal with systems which come to equilibrium; they include chromatography § 8.7, acids and bases Chapter 12 and electrochemical cells Chapter 13.

CHECKPOINT 11.8: PHASE EQUILIBRIA

1. Figure 11.8C shows the pressure–temperature phase diagram for substance **X**.
   (a) Trace the figure. On your copy, mark the areas S (solid) L (liquid) and G (gas).
   (b) Name the point A, and explain its significance.
   (c) Say, with an explanation, how the freezing temperature of pure X varies with an increase in pressure.

2 *(a)* What is sublimation?
   (b) Which line in Figure 11.8B represents sublimation?
   (c) State one respect in which the phase diagram of carbon dioxide differs from that of water.

Figure 11.8C
Pressure-temperature phase diagram

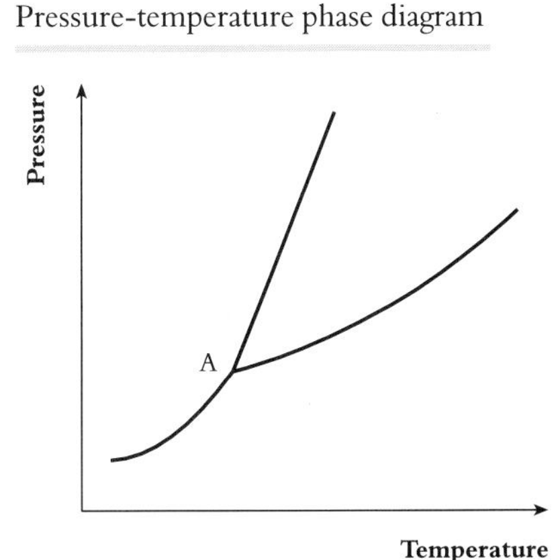

EQUILIBRIUM

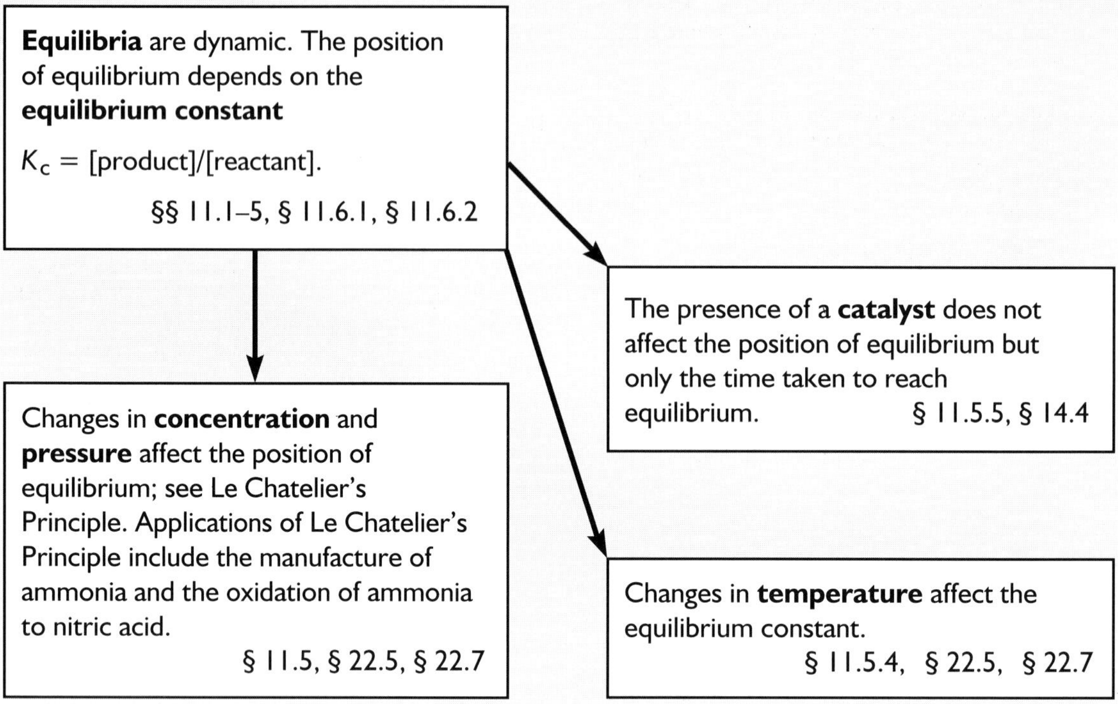

Equilibria are dynamic. The position of equilibrium depends on the **equilibrium constant**

K_c = [product]/[reactant].

§§ 11.1–5, § 11.6.1, § 11.6.2

The presence of a **catalyst** does not affect the position of equilibrium but only the time taken to reach equilibrium. § 11.5.5, § 14.4

Changes in **concentration** and **pressure** affect the position of equilibrium; see Le Chatelier's Principle. Applications of Le Chatelier's Principle include the manufacture of ammonia and the oxidation of ammonia to nitric acid.

§ 11.5, § 22.5, § 22.7

Changes in **temperature** affect the equilibrium constant.

§ 11.5.4, § 22.5, § 22.7

QUESTIONS ON CHAPTER 11

1. Study the gaseous equilibrium

$N_2O_4(g) \rightleftharpoons 2NO_2(g)$;

$\Delta H^{\ominus} = 54 \text{ kJ mol}^{-1}$ *Endothermic*

(a) Write an expression for the equilibrium constant, K_p, in terms of partial pressures.
(b) State, with reasons, the effect on the equilibrium position of (i) an increase in total pressure and (ii) an increase in temperature.

2. The **Contact process** involves the equilibrium

$2SO_2(g) + O_2(g) \rightleftharpoons 2SO_3(g)$

(a) Write an expression for K_p.
(b) At 1000 K and 1.00 atm, the equilibrium mixture contains 27.0% SO_2 and 40.0% O_2 by volume. Find the value of K_p.
(c) Use the expression from *(a)* to show how the ratio of SO_3/SO_2 in equilibrium mixtures is related to the total pressure in the presence of a fixed excess of oxygen.
(d) Explain why industrial plants operate at pressures of around 2 atm.

3. Study the equilibrium

$H_2O(g) + CO(g) \rightleftharpoons H_2(g) + CO_2(g)$

(a) Write an expression for K_p.
(b) When 1.00 mol of steam and 1.00 mol of carbon monoxide are allowed to reach equilibrium, 33.3% of the equilibrium mixture is hydrogen. Calculate the value of K_p. State the units of K_p.
(c) Would you expect an increase in total pressure to affect the yield of hydrogen? Explain your answer.

4. An aqueous solution is made by dissolving 1.00 mol of $AgNO_3$ and 1.00 mol of $FeSO_4$ in water and making up to 1.00 dm^3. When the equilibrium

$Ag^+(aq) + Fe^{2+}(aq) \rightleftharpoons Fe^{3+}(aq) + Ag(s)$

is established, $[Ag^+] = [Fe^{2+}] = 0.44 \text{ mol dm}^{-2}$, and $[Fe^{3+}] = 0.56 \text{ mol dm}^{-3}$. Find K_c for the equilibrium.

5. Iron is added to a solution containing the ions, Cr^{3+}, Cr^{2+} and Fe^{2+}. The equilibrium concentrations of the ions present in solution are $[Cr^{3+}] = 0.030 \text{ mol dm}^{-3}$, $[Cr^{2+}] = 0.27 \text{ mol dm}^{-3}$, $[Fe^{2+}] = 0.11 \text{ mol dm}^{-3}$. Find K_c for the equilibrium

$2Cr^{3+}(aq) + Fe(s) \rightleftharpoons 2Cr^{2+} + (aq) + Fe^{2+}(aq)$

6. For the equilibrium

$2\mathbf{P}(g) + \mathbf{Q}(g) \rightleftharpoons 2\mathbf{R}(g)$

K_c is numerically equal to 6.0. Into a 1.00 dm^3 flask are introduced 3.0 mol of **P**, 3.0 mol of **Q** and 3.0 mol of **R**.

(a) State the unit in which K_c is expressed.
(b) Is the mixture at equilibrium?
(c) If not, what must the volume of the flask be in order for such a mixture to exist in equilibrium at the temperature for which K_c is given?

7. Into a 1.00 dm^3 vessel at 1000 K are introduced simultaneously 0.500 mol of SO_2, 0.100 mol of O_2 and 0.700 mol of SO_3.

For the equilibrium

$$2SO_2(g) + O_2(g) \rightleftharpoons 2SO_3(g)$$

the value of K_c at 1000 K is 2.8×10^2 mol dm^{-3}.

(a) Is the mixture at equilibrium?
(b) If it is not at equilibrium, in which direction will the reaction proceed?

8. Ethanol and ethanoic acid react according to the equation

$$C_2H_5OH(l) + CH_3CO_2H(l) \rightleftharpoons CH_3CO_2C_2H_5(l) + H_2O(l)$$

3.0 g of ethanoic acid and 2.3 g of ethanol were equilibrated at 100 °C for an hour and then quickly cooled in an ice bath. 50 cm^3 of 1.0 mol dm^{-3} aqueous sodium hydroxide were added. When the mixture was titrated with 1.0 mol dm^{-3} hydrochloric acid, 33.3 cm^3 of acid were required.

(a) Why was the mixture cooled rapidly in an ice bath?
(b) State *Le Chatelier's Principle* and predict the effect of adding ethanol to the reaction mixture.
(c) Write an expression for K_c and calculate its value.

9. This question refers to the Haber process for the synthesis of ammonia. The equation which represents the reaction is given below.

$$N_2(g) + 3H_2(g) \rightleftharpoons 2NH_3(g); \quad \Delta H^{\ominus}(298\text{ K}) = -92\text{ kJ mol}^{-1}$$

(a) Give **one** source each of nitrogen and hydrogen for this process. 1
(b) Write the expression for the equilibrium constant, K_p, for the above process. If the pressure is measured in atmospheres what will be the units of K_p? $1\frac{1}{2}$
(c) State *Le Chatelier's principle.* 1
(d) State and explain the effect on the above equilibrium of
I. increasing the pressure, and 1
II. removing ammonia from the mixture of gases. 1
(e) I. Describe the effect on the equilibrium yield of increasing the operating temperature of an exothermic reaction. 1
II. Considering your response to (*e*) I. above explain why the ammonia synthesis is operated at a temperature of 400 °C. $2\frac{1}{2}$
(f) Name the catalyst used in the Haber process. $\frac{1}{2}$
(g) I. Describe the function of a catalyst. $\frac{1}{2}$
II. Describe the effect of a catalyst on the position of equilibrium. 2

12 *WJEC (AS/AL)*

10. Sulphuric acid is manufactured by the Contact process. One stage in the Contact process involves the reaction between sulphur dioxide and oxygen.

$$2SO_2(g) + O_2(g) \rightleftharpoons 2SO_3(g) \quad \Delta H = -197\text{ kJ mol}^{-1}$$

(a) On an industrial scale, the Contact process is typically carried out at 450 °C and 200 kPa with a vanadium(V) oxide catalyst.
(i) Comment on this choice of temperature and pressure.
(ii) Sulphur trioxide reacts with water to form sulphuric acid.
1 Write the equation for this reaction.
2 Describe one hazard of this reaction and how it is avoided in the Contact Process. 8
(b) When 10.0 mol of sulphur dioxide were reacted with 5.0 mol of oxygen at 450 °C, 90% of the sulphur dioxide was converted into sulphur trioxide.
(i) Calculate how many moles of sulphur dioxide, of oxygen and of sulphur trioxide were present in the equilibrium mixture.
(ii) Write down expressions for the mole fraction of each gas in the equilibrium mixture.
(iii) Assuming that the total pressure of the equilibrium mixture was 200 kPa, calculate the partial pressure of each gas in the equilibrium mixture.
(iv) Write an expression for the equilibrium constant, K_p, for this reaction.
(v) Calculate the value of K_p and state its units. **9**
(c) State **two** uses of sulphuric acid. 2

19 *C (AS/AL)*

11. Each of the equations **A**, **B**, **C** and **D** represents a dynamic equilibrium.

A $N_2(g) + O_2(g) \rightleftharpoons 2NO(g) \quad \Delta H^{\ominus} = +180\text{ kJ mol}^{-1}$

B $N_2O_4(g) \rightleftharpoons 2NO_2(g) \quad \Delta H^{\ominus} = +58\text{ kJ mol}^{-1}$

C $3H_2(g) + N_2(g) \rightleftharpoons 2NH_3(g) \quad \Delta H^{\ominus} = -92\text{ kJ mol}^{-1}$

D $H_2(g) + I_2(g) \rightleftharpoons 2HI(g) \quad \Delta H^{\ominus} = -10\text{ kJ mol}^{-1}$

(a) Explain what is meant by the term *dynamic equilibrium*. 1
(b) Explain why a catalyst does not alter the position of any equilibrium reaction. 2
(c) The units of the equilibrium constant, K_c for one of the above reactions are mol dm^{-3}. Identify the reaction **A**, **B**, **C** or **D** which has these units for K_c and write the expression for K_c for this reaction. 2

(d) The graphs below show how the yield of product varies with pressure for three of the reactions **A**, **B**, **C** and **D** given above.

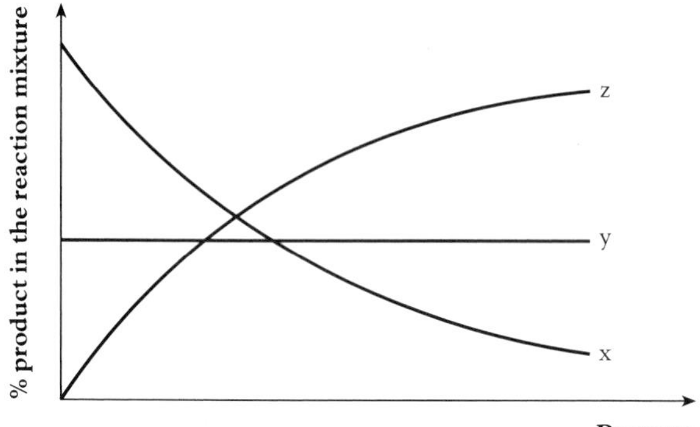

(i) Identify a reaction from **A**, **B**, **C** and **D** which would have the relationship between yield and pressure shown in graphs x, y and z.
(ii) Explain why an industrial chemist would not use a very low pressure for the reaction represented in graph x.
(iii) Explain why an industrial chemist may not use a very high pressure for the reaction represented in graph z.
(iv) Add to a copy of the above graphs a line to show how the product yield would vary with pressure if the reaction which follows curve z was carried out at a temperature higher than that of the original graph. 7

12 *NEAB (AS/AL)*

12. Dinitrogen tetroxide and nitrogen dioxide exist in the following equilibrium.

$N_2O_4(g) \rightleftharpoons 2NO_2(g)$

When 11.04 g of dinitrogen tetroxide were placed in a vessel of volume 4.80 dm^3 at fixed temperature, 5.52 g of nitrogen dioxide were produced at equilibrium under a pressure of 100 kPa. Use these data, where relevant, to answer the questions that follow.

(a) Calculate the equilibrium number of moles of each of the gases in the vessel. 2
(b) (i) Write an expression for the equilibrium constant, K_c, for the above equilibrium.
(ii) Calculate the value of the equilibrium constant, K_c, and state its units. 4
(c) (i) Write an expression for the equilibrium constant, K_p, for the above equilibrium.
(ii) Calculate the mole fraction of NO_2 present in the equilibrium mixture.
(iii) Calculate the value of the equilibrium constant, K_p, and state its units. 5
(d) State and explain the effect on the mole fraction of NO_2 when the pressure is increased at constant temperature. 3

14 *NEAB (AL)*

13. Ammonia is manufactured by the Haber–Bosch process by mixing hydrogen and nitrogen at 700 K and 200 atmospheres pressure and using an iron catalyst with a potassium hydroxide promoter.

$N_2(g) + 3H_2(g) \rightleftharpoons 2NH_3(g)\ \Delta H = -92\text{ kJ}$

(a) Explain the term **dynamic equilibrium**. 2
(b) Explain why the use of high pressure favours ammonia formation. 2
(c) Analysis of an equilibrium mixture, obtained by mixing nitrogen and hydrogen, showed 24.0 g of NH_3, 13.5 g of H_2 and 60.3 g of N_2 to be present. The total pressure of the system was 10 atmospheres.
(i) Calculate the mole fraction of each of these substances present at equilibrium: N_2, H_2, NH_3.
(ii) Calculate K_p and state its units. 4
(iii) Explain the effect of increasing temperature on the value of K_p. 2
(d) The reaction of ammonia with oxygen is exothermic

$4NH_3 + 5O_2 \rightleftharpoons 4NO + 6H_2O$

However, mixtures of ammonia and oxygen do not catch fire spontaneously. Use this information to explain the difference between **kinetic** and **thermodynamic stability**. 8

18 *CCEA (AL)*

14. Phosphorus(V) chloride dissociates at high temperatures according to the equation

$PCl_5(g) \rightleftharpoons PCl_3(g) + Cl_2(g)$

83.4 g of phosphorus(V) chloride are placed in a vessel of volume 9.23 dm^3. At equilibrium at a certain temperature, 11.1 g of chlorine are produced at a total pressure of 250 kPa. Use these data, where relevant, to answer the questions that follow.

(a) Calculate the number of moles of Cl_2 and PCl_5 in the vessel at equilibrium. 2
(b) (i) Write an expression for the equilibrium constant, K_c, for the above equilibrium.
(ii) At equilibrium, the number of moles of PCl_3 and Cl_2 are the same. Calculate the value of the equilibrium constant, K_c, and state its units. 4
(c) (i) Write an expression for the equilibrium constant, K_p, for the above equilibrium.
(ii) Calculate the mole faction of chlorine present in the equilibrium mixture.
(iii) Calculate the partial pressure of PCl_5 present in the equilibrium mixture.
(iv) Calculate the value of the equilibrium constant, K_p, and state its units. 7

13 *NEAB (AL)*

12 ELECTROCHEMISTRY

12.1 SIR HUMPHRY DAVY (1778–1829) AND MICHAEL FARADAY (1791–1867)

The science of electrochemistry was developed by Davy and Faraday.

Sir Humphry Davy is famous for his invention of the miner's safety lamp. He refused to patent his invention, saying, 'My sole object was to serve the cause of humanity'. A group of mine owners who appreciated the benefit to their workers presented Davy with a silver dinner service. The silver dinner service was sold and the funds used to endow the Royal Society Davy medal, which is awarded annually for the most important chemical discovery in Europe and North America. The first person to receive it, in 1877, was Robert Bunsen.

Davy was born in Penzance, Cornwall, in 1778, the eldest son of a woodcarver. On leaving school at 16, he was employed by a surgeon, who encouraged Davy to educate himself. At the age of 19, Davy went to work in an institute for medical research in Bristol. Within a few months, he had discovered the anaesthetic properties of laughing gas, N_2O. Davy was quick to take advantage of Volta's invention of the galvanic pile, which gave a reliable source of electricity that could be used to investigate electrolysis. It enabled Davy to isolate by electrolysis six new elements: sodium, potassium, magnesium, calcium, strontium and barium. He demonstrated that chlorine is an element and identified iodine. Davy was appointed to a lectureship in the Royal Institution. He gave spectacular lectures which were so popular that traffic jams formed in the street outside whenever he lectured. At the age of 34, Davy was knighted.

Another of Davy's discoveries was Michael Faraday (see § 4.2). Faraday put the new science of electrochemistry on a quantitative basis. (see § 12.2). While Davy had a social life and a number of hobbies, including travel, hunting and fishing, Faraday's time was fully occupied with his work and his religion. He had no students and did all his experiments himself. He turned down offers of a knighthood and the presidency of the Royal Society. He made time to be active in church work. Like Davy, Faraday was an inspiring lecturer. He followed Davy as professor of chemistry in the Royal Institution and continued the popular science lectures for the public that Davy had started.

12.2 ELECTROLYSIS

Some compounds conduct electricity when in the molten state or in solution [§§ 4.2.3, 4.2.4]. When compounds conduct electricity, chemical changes occur and new substances are formed. The compounds which conduct electricity and are decomposed by it are called **electrolytes**, and the chemical changes **electrolysis**. The container in which electrolysis takes place is called a **cell**, and the conductors which carry electricity into and out of the cell are called **electrodes**. In an electrolysis cell, an electric current brings about a chemical change. There is another type of cell, a chemical cell in which a chemical change gives rise to an electric current, which you will meet in Chapter 13. The reactions in both types of cell are called **electrochemical reactions**.

12.2.1 AMOUNT OF SUBSTANCE LIBERATED

The electrolysis of a solution of a silver salt to deposit a layer of silver on the cathode can be carried out as shown in Figure 12.2A

Figure 12.2A
Electrolysis of silver nitrate solution

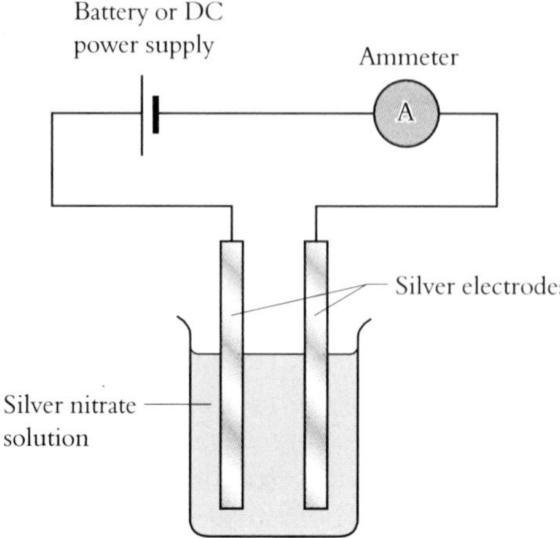

With this apparatus, it is possible to find out how much electric charge is needed to deposit a certain mass of silver. If the cathode is weighed before and after the passage of the current, the difference gives the mass of silver deposited. If the current is measured with a milliammeter and the electrolysis is timed, one can work out the quantity of electric charge which has passed. Electric charge is measured in coulombs (C).

One coulomb of charge = One ampere of current flowing for one second
Charge in coulombs (C) = Current in amperes (A) × Time in seconds (s)

The cathode process is

$$Ag^+(aq) + e^- \longrightarrow Ag(s)$$

According to this equation

1 mol silver ions + 1 mol electrons ⟶ 1 mol silver atoms

One can measure the mass of silver discharged by a certain quantity of electric charge. For example, in an experiment in which 40.0 mA flowed for 2.00 hours,

Mass of silver deposited on the cathode = 0.332 g

Quantity of electricity = Current × Time
= 0.040 A × 2.00 × 60 × 60 s
= 288 C

Quantity of electricity needed to discharge 1 mol silver = 288 C × 108/0.322
= 96.5×10^3 C

This quantity is 96 500 coulombs. This quantity of electric charge must be the charge on one mole of electrons. The ratio 96 500 coulombs per mole ($C\ mol^{-1}$) is called the **Faraday constant**.

Similar electrolysis experiments show the quantities of electricity required to discharge one mole of each element are:

Quantity of electricity required for 1 mol Ag = 96.5×10^3 C
Quantity of electricity required for 1 mol Cu = 193×10^3 C
Quantity of electricity required for 1 mol Cr = 289×10^3 C
Quantity of electricity required for 1 mol H_2 = 193×10^3 C
Quantity of electricity required for 1 mol O_2 = 386×10^3 C

When 96.5×10^3 C of electricity pass through the cell, the amounts of metals deposited are Ag 1 mol; Cu $\frac{1}{2}$ mol; Cr $\frac{1}{3}$ mol

The cathode processes are:

$$Ag^+(aq) + e^- \longrightarrow Ag(s)$$
$$Cu^{2+}(aq) + 2e^- \longrightarrow Cu(s)$$
$$Cr^{3+}(aq) + 3e^- \longrightarrow Cr(s)$$

You can see from these equations that 1 mole of electrons will discharge 1 mole of Ag^+ ions, $\frac{1}{2}$ mole of Cu^{2+} ions or $\frac{1}{3}$ mole of Cr^{3+} ions.

In general

$$\text{Amount of element discharged/mole} = \frac{\text{Number of moles of electrons}}{\text{Number of charges on one ion of the element}}$$

[For the formation of hydrogen and oxygen see § 12.4.2.]

A worked example

Example Find how long it will take to deposit 1.00 g of chromium when a current of 0.120 A flows through a solution of chromium(III) sulphate solution.

Method

Since

$$Cr^{3+}(aq) + 3e^- \longrightarrow Cr(s)$$

3 mol electrons are needed to deposit 1 mol chromium (i.e., 52 g chromium)
3/52 mol electrons deposit 1 g chromium
Number of coulombs required = 96 500 × 3/52 C

Since

Number of coulombs = Amperes × Seconds = $0.120 \times t$ (t = time/s)
$0.120 \times t = 96\,500 \times 3/52$
$t = 46\,400$ s (approximately)

The current passes for 46 400 s (12.9 h).

12.2.2 DETERMINATION OF THE AVOGADRO CONSTANT

Electrolysis experiments give a means of determining the Avogadro constant. The basis of the method is as follows.

- One mole of ions, electrons or atoms contains the Avogadro constant of particles [§ 3.6].

- The charge on a singly charged ion, e.g. Ag^+, is the same as the charge on an electron [§ 12.2.1].
- One mole of singly charged ions, e.g. Ag^+, require one mole of electrons for discharge.

Faraday constant F = electric charge per mole of substance ...

- The quantity of electricity required per mole to discharge singly charged ions in electrolysis is the Faraday constant, 9.649×10^4 C mol^{-1}.

Since

... Avogadro constant L = number of particles per mole of substance ...

Faraday constant = Electric charge per mole of substance, and

Avogadro constant = Number of particles per mole of substance

... e = charge on one electron ...

Faraday constant/Avogadro constant = Electric charge/Number of particles
= Electric charge on one particle
= Charge on one electron

... Therefore F = Le ...

That is, $F = Le$

where F = Faraday constant, L = Avogadro constant, e = electronic charge

... since e is known and F can be found by experiment ...

The charge on one electron is known to be 1.602×10^{-19} C.

Substituting values of F and e into the equation: $L = F/e$

... L can be calculated

$$L = 9.649 \times 10^4 \text{ C mol}^{-1}/1.602 \times 10^{-19} \text{ C} = 6.022 \times 10^{23} \text{ mol}^{-1}$$

The value of the Avogadro constant is 6.022×10^{23} mol^{-1}.

12.3 EXAMPLES OF ELECTROLYSIS

When a molten salt is electrolysed, the products are predictable. When an aqueous solution is electrolysed, hydrogen and oxygen sometimes appear at the cathode and anode. The products formed from a few electrolytes are shown in Table 12.3.

Table 12.3
Products of electrolysis

(a) Using inert electrodes (platinum or graphite)		
Electrolyte	Cathode	Anode
$PbBr_2$(l)	Pb(s)	Br_2(g)
NaCl(l)	Na(s)	Cl_2(g)
$CuCl_2$(aq)	Cu(s)	Cl_2(g)
NaCl(aq)	H_2(g)	Cl_2(g)
KNO_3(aq)	H_2(g)	O_2(g)
$CuSO_4$(aq)	Cu(s)	O_2(g)
H_2SO_4(aq)	H_2(g)	O_2(g)
NaOH(aq)	H_2(g)	O_2(g)
(b) When the electrodes take part in the reactions		
Electrolyte	Copper cathode	Copper anode
$CuSO_4$(aq)	Cu(s) deposited	Cu(s) dissolves to form Cu^{2+} ions

Figure 12.3A Manufacture of sodium hydroxide and chlorine by electrolysis of brine in diaphragm cells at ICI, Lostock

12.4 EXPLANATION OF ELECTROLYSIS

12.4.1 ELECTROLYSIS OF AN AQUEOUS SALT

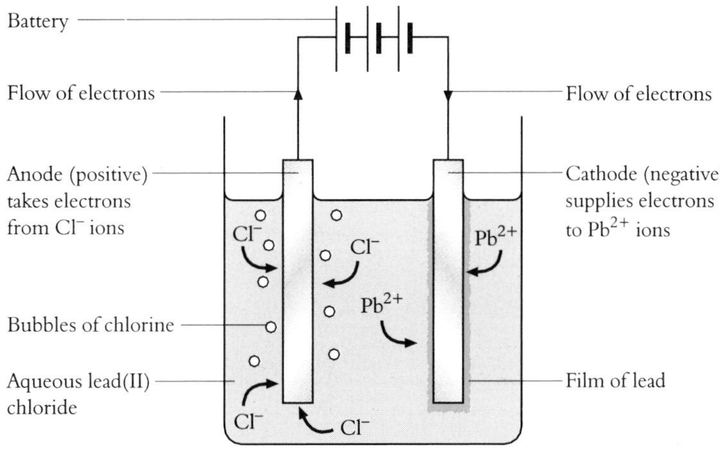

Figure 12.4A Electrolysis of aqueous lead(II) chloride

Figure 12.4A shows what happens when aqueous lead(II) chloride is electrolysed. Lead cations are attracted to the cathode. This negative electrode supplies electrons which discharge the cations to form atoms of lead.

Equations for electrode processes

$$Pb^{2+}(aq) + 2e^- \longrightarrow Pb(s)$$

The chloride anions are attracted to the anode. Being positively charged, the anode takes electrons from the anions, thus discharging them:

$$Cl^-(aq) \longrightarrow Cl(g) + e^-$$

The electrode process is followed by the formation of chlorine molecules:

$$2Cl(g) \longrightarrow Cl_2(g)$$

The flow of current through the external circuit

As the anode takes electrons from Cl^- ions and the cathode supplies electrons to Pb^{2+} ions, electrons flow through the external circuit from anode to cathode. The electric current is conducted through the cell by ions and through the external circuit by electrons.

12.4.2 THE PART PLAYED BY WATER

The products of electrolysis are not always as predictable as in the previous example. When aqueous sodium nitrate is electrolysed the products are hydrogen at the cathode and oxygen at the anode. Sodium ions are present at the cathode but are not discharged. A small concentration of hydrogen ions is present due to the dissociation of water:

$$2H_2O(l) \rightleftharpoons H_3O^+(aq) + OH^-(aq) \text{ water}$$

The hydrogen ions are discharged:

$$H_3O^+(aq) + e^- \longrightarrow H(g) + H_2O(l)$$

followed by $2H(g) \longrightarrow H_2(g)$

Although the concentration of hydrogen ions is only 10^{-7} mol dm^{-3} in water, after these hydrogen ions are discharged more are formed by the dissociation of more water molecules. This gives a vast supply of hydrogen ions, and sodium ions remain in solution while hydrogen is evolved.

At the anode, both nitrate ions and hydroxide ions (from the dissociation of water) are present. Hydroxide ions are easier to discharge than nitrate ions so nitrate ions remain in solution. The electrode process is:

$$OH^-(aq) \longrightarrow OH(aq) + e^-$$

followed by

$$4OH(aq) \longrightarrow O_2(g) + 2H_2O(l)$$

How much hydrogen and oxygen are evolved?

The equations for the electrode processes allow us to calculate the mass and volume of hydrogen and oxygen evolved in electrolysis [see also § 12.2.1].

The evolution of 1 mole of H_2 requires 2 moles of electrons.

Therefore 2 g of H_2 or 22.4 dm^3 at stp of H_2 are liberated by 193 000 C.

The evolution of 1 mole of O_2 requires 4 moles of electrons.

Therefore 32 g of O_2 or 22.4 dm^3 at stp of O_2 are liberated by 386 000 C.

12.4.3 THE ELECTROCHEMICAL SERIES

An order can be drawn up for cations to show the relative ease of discharge ...

... The same can be done for anions ...

If a solution contains copper(II) ions and zinc ions, during electrolysis zinc ions remain in solution, while copper(II) ions are discharged. If a solution contains bromide ions and iodide ions, iodide ions are discharged, while bromide ions remain in solution. Cations can be arranged in order, according to their relative ease of discharge at a cathode. Anions can be arranged in order of their relative ease of discharge at an anode. The list of ions in order of ease of discharge is called the **electrochemical series**. A shortened version of it is shown in Table 12.4 [see also Table 13.3, § 13.3.2].

In the electrolysis of a solution containing a mixture of ions, to find out which of, the ions will be discharged, you read from the bottom of the electrochemical series. When all the ions of that species have been discharged, then the ions next higher up in the series are discharged.

Table 12.4
The electrochemical series

... The result is the electrochemical series

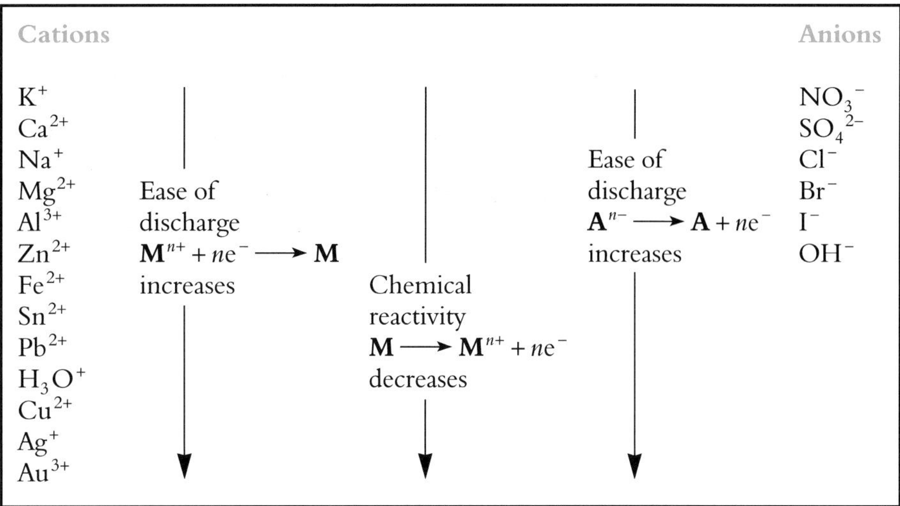

Cations				Anions
K^+, Ca^{2+}, Na^+, Mg^{2+}, Al^{3+}, Zn^{2+}, Fe^{2+}, Sn^{2+}, Pb^{2+}, H_3O^+, Cu^{2+}, Ag^+, Au^{3+}	Ease of discharge $\mathbf{M}^{n+} + ne^- \longrightarrow \mathbf{M}$ increases ↓	Chemical reactivity $\mathbf{M} \longrightarrow \mathbf{M}^{n+} + ne^-$ decreases ↓	Ease of discharge $\mathbf{A}^{n-} \longrightarrow \mathbf{A} + ne^-$ increases ↓	NO_3^-, SO_4^{2-}, Cl^-, Br^-, I^-, OH^-

The effect of concentration

If the concentration of one ion is very much greater than the concentration of another, this factor may interfere with the expected order of discharge ...

Sometimes this rule does not predict the observed result. The electrolysis of sodium chloride solution is used in the manufacture of chlorine (§ 18.6.5), although the electrochemical series predicts that oxygen should be formed at the anode, not chlorine. In solutions of halides, the relative concentrations of halide ions and hydroxide ions affect the result. The concentration of hydroxide ions in water is only 10^{-7} mol dm^{-3}. In an aqueous solution of a halide of concentration 0.1 mol dm^{-3}, the concentration of halide ions is 10^6 times greater than the concentration of hydroxide ions. Although, according to the electrochemical series, hydroxide ions should be discharged, the concentration of halide ions is so much greater than the concentration of hydroxide ions that halide ions are discharged.

Similarly, inspection of the electrochemical series would lead you to expect that hydrogen would be evolved in the electrolysis of a lead(II) salt in aqueous solution. In fact, lead is deposited at the cathode. Since the concentration of lead ions is many powers of ten greater than the concentration of hydrogen ions in an aqueous solution of a lead(II) salt, the effect of concentration overrides the positions of the ions in the electrochemical series.

SUMMARY

The electrochemical series lists

- cations in order of ease of discharge at a cathode. This is the reverse of the order of reactivity of the metals
- anions in order of ease of discharge at an anode.

Sometimes exceptions arise as a result of large differences in concentration.

The electrochemical series gives the order in which metal ions are discharged.

$$M^{n+} + ne^- \longrightarrow M(s)$$

This reaction is the opposite of that which happens when metals react:

$$M(s) \longrightarrow M^{n+} + ne^-$$

For this reason the order of discharge of ions is the reverse of the order of reactivity of the metals.

12.5 APPLICATIONS OF ELECTROLYSIS

There are many applications of electrolysis

1. Extraction of sodium by the electrolysis of molten sodium chloride in the Downs process [§ 18.4].
2. Manufacture of sodium hydroxide by the electrolysis of aqueous sodium chloride in a diaphragm cell [§ 18.6.5].
3. Manufacture of sodium chlorate(I) and sodium chlorate(V) [§ 18.6.5] by the electrolysis of aqueous sodium chloride.

4. Manufacture of chlorine by the electrolysis of sodium chloride as in 1 and 2 above.
5. Manufacture of hydrogen by the electrolysis of brine (aqueous sodium chloride) as in 2 above.
6. Extraction of magnesium and calcium by electrolysis of the molten chlorides [§ 18.4].
7. Extraction of aluminium by the electrolysis of molten aluminium oxide [§ 19.3.2].
8. Anodisation and dyeing of aluminium [§ 19.3.1].
9. Purification of copper by electrolysis, using a lump of impure copper as the anode [§ 24.18].
10. Electroplating, e.g., chromium plating [§ 24.17.6].

Figure 12.5A Purification of copper by electrolysis

12.5. THE WRECK OF THE *TITANIC*

The liner *Titanic* was thought to be unsinkable, yet she sank on her maiden voyage in 1912 after colliding with an iceberg. The wreck was discovered by divers in 1985, and 1800 objects have been recovered from the sea bed. Electricité de France, EDF, has a laboratory which specialises in the restoration of objects recovered from shipwrecks and archaeological digs. Jewellery, watches, coins and many other objects have been restored.

EDF used an electrochemical process to remove the crust-like deposit of salts which had formed on the objects. This avoided the use of abrasive cleaning products. EDA wired the metal objects as cathodes in electrolysis cells. The chloride ions and other anions in the crust travelled towards the anode. Non-conducting objects, e.g. porcelain and wood, were placed between two electrodes so that ions passed through the object and flushed out the deposit. The electrochemical treatment of each object took between 200 and 1000 hours. The objects were rinsed with de-ionised water. To avoid cracking, they were dried slowly at low temperature and pressure.

CHECKPOINT 12.5: ELECTROLYSIS

1. The same current was passed through molten sodium chloride and through molten cryolite containing aluminium oxide. If 4.60 g of sodium were liberated in one cell, what mass of aluminium was liberated in the other?

2. What mass of each of the following substances will be liberated by the passage of 0.200 mol of electrons?

   (a) Mg(s) from $MgCl_2$(l), *(b)* Cl_2(g) from NaCl(aq), *(c)* Cu(s) from $CuSO_4$(aq), *(d)* Pb(s) from $Pb(NO_3)_2$(aq), *(e)* H_2(g) from NaCl(aq).

3. During the electrolysis of a 1.00 mol dm^{-3} solution of copper(II) sulphate at 20 °C, a current of 0.100 A was passed for 1.00 h. The mass of the copper cathode increased by 0.118 g. State how the change in mass would be affected if the experiment were repeated

   (a) at a current of 0.200 A, *(b)* at 30 °C, *(c)* with a 2.00 mol dm^{-3} solution of copper(II) sulphate, *(d)* for a time of 2.00 h.

4. When a metal of relative atomic mass 207 is deposited by electrolysis, a current of 0.0600 A flowing for 66 min increases the mass of the cathode by 0.254 g.

   Find the number of moles of metal deposited and the number of moles of electrons that have passed. Deduce the number of units of charge on the cations of this metal.

ELECTROLYSIS

The **electrochemical series** lists

- cations in order of ease of discharge at a cathode
- anions in order of ease of discharge at an anode.

Exceptions arise as a result of large differences in concentration. § 12.4.3

The order of ease of discharge of metals is the reverse of their order of **reactivity**.

Electrolysis is the passage of a current through a molten or aqueous compound. Substances are liberated at the electrodes. The amounts of substances liberated depend on the quantity of electric charge passed and on the charge on the ions. § 12.2.1
The masses of solids and the masses and volumes of gases liberated can be calculated. § 12.2.1

Industrial uses of electrolysis include

- the electrolysis of brine § 12.5, §18.6.5
- the extraction and anodising of aluminium § 12.5, § 19.3.1, § 19.3.2
- the purification of copper § 12.5, § 24.18

The **Faraday constant** is the charge on one mole of electrons = 96 500 C mol^{-1}.

$F = Le$

where L = Avogadro constant and e = charge on one electron § 12.2.2
The **Avogadro constant** can be found by an electrolytic method. § 12.2.2

The **rusting of iron** is an electrochemical reaction § 13.4, § 24.17.6

12.6 IONIC EQUILIBRIA

12.6.1 ACIDS AND BASES

There have been many attempts to define acids and bases. The sour taste and the effect on vegetable colourings, such as litmus, characterised acids. The soapy feel and detergent power characterised alkalis. Acids were seen to react with alkalis and also with some other compounds to give salts. The term **base** replaced the term **alkali** as meaning the opposite of an acid. A **base** was defined as a substance which would react with an acid to form a salt and water. An acid was defined by Liebig in 1838 as a compound that contains hydrogen that can be displaced by a metal. A big advance was the Ostwald–Arrhenius theory of electrolytic dissociation in 1880. They defined acids as substances that produce hydrogen ions in solution. Bases, they said, produce hydroxide ions in solution and neutralise acids by the reaction

$$H^+ + OH^- \longrightarrow H_2O$$

Before long this theory ran into difficulties. It was noted that, while hydrochloric acid conducts electricity, pure hydrogen chloride does not. Should hydrogen chloride be classified as an acid or does it only become an acid on contact with water? Bases such as ammonia neutralise acids by picking up hydrogen ions, rather than by providing hydroxide ions

$$NH_3 + H^+ \longrightarrow NH_4^+$$

H^+ ions cannot exist in solution: they are so small that the charge density is high, and they must be solvated ...

... In water, H_3O^+ ions are formed

As work proceeded, doubt was cast on the existence of the hydrogen ion, H^+, in solution. The proton, H^+, is very small (10^{-15} m diameter) compared with other cations (around 10^{-10} m diameter). The electric field in its neighbourhood is so intense that it attracts any molecule with unshared electrons, such as H_2O. The reaction

$$H^+ + H_2O \longrightarrow H_3O^+$$

was shown (by spectroscopic measurements) to liberate 1300 kJ mol^{-1}. As the reaction is so exothermic, unhydrated protons do not exist in solution. The hydrated proton, H_3O^+, is called the **oxonium ion**, and is also referred to as the hydrogen ion. In this text, H_3O^+ will be called a hydrogen ion. Its structure is shown in Figure 12.6A with that of the ammonium ion for comparison.

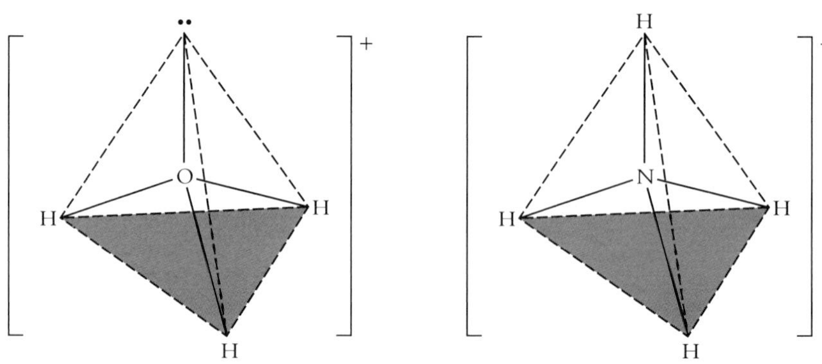

Figure 12.6A
The oxonium and ammonium ions (H_3O^+ and NH_4^+)

The Brönsted–Lowry definition of acids and bases

Brönsted and Lowry described acids as proton-donors and bases as proton-acceptors

Acknowledging that the proton does not exist in solution made it necessary to review the definition of acids. The best definition of acids and bases is that proposed T M Lowry and also, independently, by J N Brönsted in 1923. They said that an acid is a substance which can donate a proton to another substance.

A base is a substance which can accept a proton from another substance. The relationship is

$$\text{Acid} \rightleftharpoons \text{Base} + \text{H}^+$$

This equation does not represent an actual reaction in solution since the proton, H^+, cannot exist in solution. The acid and base which are related in this way, by the exchange of a proton, are called a **conjugate acid–base pair**. Since

$$\text{Acid 1} \rightleftharpoons \text{Conjugate base 1} + \text{H}^+$$

and $$\text{Base 2} + \text{H}^+ \rightleftharpoons \text{Conjugate acid 2}$$

a reaction between an acid and a base is

Conjugate acid–base pairs

$$\text{Acid 1} + \text{Base 2} \rightleftharpoons \text{Conjugate base 1} + \text{Conjugate acid 2}$$

Reactions between acids and bases

Acid 1 is transformed into its conjugate base, and Base 2 is transformed into its conjugate acid. No substance can act as an acid in solution unless a base is present to accept a proton: the reactions of acids are reactions between acids and bases. Similarly, all reactions of bases in solution are acid–base reactions.

Examples of Brönsted–Lowry acids and bases

Examples of **Brönsted–Lowry acids** are

$$HCl + H_2O \longrightarrow H_3O^+ + Cl^- \qquad [1]$$

$$HSO_4^- + H_2O \rightleftharpoons H_3O^+ + SO_4^{2-} \qquad [2]$$

$$CH_3CO_2H + H_2O \rightleftharpoons H_3O^+ + CH_3CO_2^- \qquad [3]$$

Water is amphoteric or amphoprotic ...

... can act as an acid or as a base

Examples of **Brönsted–Lowry bases** are:

$$NH_3 + H_2O \rightleftharpoons NH_4^+ + OH^- \qquad [4]$$

$$RNH_2 + H_2O \rightleftharpoons RNH_3^+ + OH^- \qquad [5]$$

$$HSO_4^- + H_3O^+ \rightleftharpoons H_2SO_4 + H_2O \qquad [6]$$

$$CH_3CO_2^- + H_2O \rightleftharpoons CH_3CO_2H + OH^- \qquad [7]$$

In [1], [2] and [3], water is acting as a proton-acceptor, a base; in [4], [5], [6] and [7], water is acting as a proton-donor, an acid. Water is described as an **amphoteric** or **amphiprotic** solvent. The HSO_4^- ion is an amphoteric or amphiprotic species.

The Lewis definition of acids and bases

There are reactions which appear to us, on common-sense grounds, to be acid–base reactions, and which do not come within the scope of the Brönsted–Lowry definition. Such reactions are

$$CaO + SO_3 \longrightarrow CaSO_4$$

$$NH_3 + BF_3 \longrightarrow NH_3BF_3$$

A Lewis base gives a lone pair of electrons to a Lewis acid ...

... with the formation of a covalent bond

To accommodate reactions of this type, G N Lewis (during 1930 to 1940) proposed a fresh definition of acids and bases. He described as acid–base reactions those reactions in which an unshared electron pair in the base molecule is accepted by the acid molecule, with the formation of a covalent bond. For example, in the reaction

$$H_3N\!: + BF_3 \longrightarrow H_3N \longrightarrow BF_3$$

A comparison of the two definitions

ammonia is the base, and boron trifluoride is the acid. The Lewis definition of a base includes the Brönsted–Lowry bases because a species with a lone pair of electrons will accept a proton from a Brönsted–Lowry acid. Lewis acids, such as BF_3 and SO_3, are not acids in the Brönsted–Lowry sense, and acids such as HCl, H_2SO_4 and CH_3CO_2H are not acids according to the Lewis definition.

The Brönsted–Lowry description of acids and bases lends itself readily to a quantitative treatment of the strengths of acids and bases [§ 12.6.3 and § 12.6.5]. No such quantitative treatment is possible for Lewis acids and bases.

> The Brönsted–Lowry definition:
> An acid is a substance which can donate a proton to another substance.
> A base is a substance which can accept a proton from another substance.

12.6.2 THE IONIC PRODUCT FOR WATER

The hydrogen ion concentration in a solution can be denoted by means of the **pH** of the solution. The pH of a solution is the negative logarithm to the base ten of the hydrogen ion concentration in mol dm^{-3}. **pOH** is related in the same way to the hydroxide ion concentration:

pH and pOH ...

$$pH = -\lg[H_3O^+/\text{mol dm}^{-3}]$$
$$pOH = -\lg[OH^-/\text{mol dm}^{-3}]$$

Water is only slightly ionised:

$$2H_2O \rightleftharpoons H_3O^+ + OH^-$$

The product of the concentrations of hydrogen ions and hydroxide ions is equal to 1.00×10^{-14} mol^2 dm^{-6} at 25 °C. This product is called the **ionic product** for water, K_w.

... The ionic product for water K_w and pK_w

$$[H_3O^+][OH^-] = K_w = 1.00 \times 10^{-14}\ \text{mol}^2\ \text{dm}^{-6}$$
$$pH + pOH = pK_w = 14 \text{ at } 25\,°C$$

This relationship is true of aqueous solutions as well as water.

12.6.3 CALCULATION OF pH AND pOH FOR STRONG ACIDS AND BASES

The pH of a solution of a strong acid or strong base is simply calculated.

Example (a) What is the pH of a solution of hydrochloric acid of concentration 0.1 mol dm^{-3}?

Method

Since

Calculation of pH ...

$$[H_3O^+] = 0.1\ \text{mol dm}^{-3} = 10^{-1}\ \text{mol dm}^{-3}$$
$$\lg[H_3O^+] = -1 \text{ and } pH = 1$$

Example (b) What is the pH of a solution of sodium hydroxide of concentration 0.01 mol dm^{-3} at 25 °C?

Method

Since

... *Calculation of pOH*

$$[OH^-] = 0.01 = 10^{-2}\ mol\ dm^{-3}$$

$$pOH = 2$$

$$pH = 14 - pOH = 12 \text{ at } 25\ °C$$

At 25 °C, solutions with a pH of 7 are neutral, solutions with a pH less than 7 are acidic, and solutions with a pH greater than 7 are alkaline. The relationship between pH and $[H_3O^+]$ and $[OH^-]$ is illustrated in Table 12.6.

Table 12.6
pH values of acidic and alkaline solutions

$[H_3O^+]$/ mol dm^{-3}	1	10^{-1}	10^{-2}	10^{-3}	10^{-4}	10^{-5}	10^{-6}	10^{-7}	10^{-8}	10^{-9}	10^{-10}	10^{-11}	10^{-12}	10^{-13}	10^{-14}
$[OH^-]$/ mol dm^{-3}	10^{-14}	10^{-13}	10^{-12}	10^{-11}	10^{-10}	10^{-9}	10^{-8}	10^{-7}	10^{-6}	10^{-5}	10^{-4}	10^{-3}	10^{-2}	10^{-1}	1
pH	0	1	2	3	4	5	6	7	8	9	10	11	12	13	14
	Strongly acidic			Weakly acidic			Neutral			Weakly alkaline			Strongly alkaline		

CHECKPOINT 12.6A: ACIDS AND BASES

1. Define the terms *Brönsted acid* and *Brönsted base*. Give two examples of each, explaining how they fit the definitions. Define the terms *Lewis acid* and *Lewis base*, and give one example of each.

†2. Explain the following statements:

(*a*) $[Al(H_2O)_6]^{3+}$ is classified as an acid [§ 19.4.1].
(*b*) $ClCH_2CO_2H$ is a stronger acid than CH_3CO_2H [§ 12.6.6].
(*c*) The Friedel–Crafts catalyst, $AlCl_3$, is classified as a Lewis acid [§ 28.8.7].

3. How are Lewis acids and bases different from Brönsted–Lowry acids and bases and from Arrhenius acids and bases?
In the equilibrium

$$NH_3 + H_2O \rightleftharpoons NH_4^+ + OH^-$$

which species is the acid and which the base, according to the definitions of (*a*) Arrhenius and (*b*) Lowry and Brönsted?

4. According to the Brönsted–Lowry theory, water can function as an acid and as a base. Quote one reaction in which water acts as an acid and one in which it acts as a base.

5. At 0 °C, $K_w = 1.14 \times 10^{-15}\ mol^2\ dm^{-6}$. Find:

(*a*) pK_w, at 0 °C
(*b*) the pH at 0 °C of a 0.01 mol dm^{-3} solution of sodium hydroxide

6. Name the Lewis acid and the Lewis base in the following reactions:

(*a*) $RCOBr + FeBr_3 \rightleftharpoons RCO^+FeBr_4^-$
(*b*) $Ag^+ + 2NH_3 \longrightarrow Ag(NH_3)_2^+$
(*c*) $CH_3CO_2^- + 2HF \longrightarrow CH_3CO_2H + HF_2^-$

12.6.4 WEAK ELECTROLYTES

The degree of ionisation of a weak electrolyte

Weak electrolytes consist of molecules, some of which dissociate to form ions. The fraction of molecules which dissociate is called the **degree of ionisation** or **degree of a dissociation**. For a weak acid H**A**. An equilibrium is set up between undissociated molecules H**A** and the ions $H_3O^+$and $\mathbf{A}^-$:

$$HA + H_2O \rightleftharpoons H_3O^+ + A^-$$

The equilibrium constant [§ 11.2] is called the **acid dissociation constant**, K_a, and is given by

The dissociation constant of a weak acid

$$K_a = \frac{[H_3O^+][A^-]}{[HA]}$$

The square brackets show concentrations in mol dm^{-3}, e.g. $[H_3O^+]$ is the concentration of hydrogen ions. The expression for K_a does not include $[H_2O]$ because the concentration of water remains constant in dilute solutions. The acid dissociation constant can be calculated from the pH of a solution of the acid of known concentration.

Example (a) A solution of ethanoic acid of concentration 0.100 mol dm^{-3} has a pH of 2.88. Calculate the value of K_a.

Method

$$K_a = \frac{[H_3O^+][CH_3CO_2^-]}{[CH_3CO_2H]}$$

The concentrations $[H_3O^+]$ and $[CH_3CO_2^-]$ are equal.

$[CH_3CO_2H] = 0.100$ mol dm^{-3}. In fact $[CH_3CO_2H]$ is slightly less than 0.100 mol dm^{-3} because some molecules have dissociated. For most weak electrolytes, the difference can be neglected.

Since pH = 2.88, $[H_3O^+] = \text{antilg}\,(-2.88) = 1.32 \times 10^{-3}$ mol dm^3.

Then $K_a = (1.32 \times 10^{-3}\text{ mol dm}^{-3})^2/0.100\text{ mol dm}^{-3}$
$= 1.74 \times 10^{-5}$ mol dm^{-3}.

$$pK_a = -\lg K_a = 4.76$$

A weak base which is partially ionised in solution has a basic dissociation constant, K_b.

$$B + H_2O \rightleftharpoons BH^+ + OH^-$$

$$K_b = \frac{[BH^+][OH^-]}{[B]}$$

The concentration $[H_2O]$ remains constant and is not included in the equation. The basic dissociation constant can be calculated from the pH of a solution of the base of known concentration.

Example (b) If the pH of a 0.100 mol dm^{-3} solution of ethylamine is 11.85, what is its basic dissociation constant, K_b?

Method When ethylamine dissociates,

$$C_2H_5NH_2 + H_2O \rightleftharpoons C_2H_5NH_3^+ + OH^-$$

$$K_b = \frac{[C_2H_5NH_3^+][OH^-]}{[C_2H_5NH_2]}$$

Since pH = 11.85, pOH = 14.0 – 11.85 = 2.15
and $[OH^-]$ = antilg(–2.15) = 7.08×10^{-3} mol dm^{-3}

$$K_b = [OH^-]^2/[C_2H_5NH_2] = (7.08 \times 10^{-3}\ \text{mol dm}^{-3})^2/0.100\ \text{mol dm}^{-3}$$
$$= 5.01 \times 10^{-4}\ \text{mol dm}^{-3}$$

$pK_b = -\lg (5.01 \times 10^{-4}) = 3.30$

The value of K_a for an acid is a quantitative measure of the strength of the acid in all its typical reactions ...

... K_b for a base is equally important

The value of its dissociation constant tells you how strong an acid is and how vigorously it will take part in the reactions which are typical of acids. It enables you to calculate the pH of a solution of the acid [§ 12.6.3]. It is a measure of the effectiveness of an acid in acid-catalysed reactions. All these aspects of acid behaviour are covered by one physical constant. From the values of K_a, you can say that chloroethanoic acid is 80 times stronger than ethanoic acid. The value of the dissociation constant of a base is equally important. From the values of K_b, you can say that methylamine is 23 times as strong a base as ammonia. The quantitative measure of the strengths of acids and bases provided by the K_a and K_b values is a splendid feature of the Brönsted–Lowry treatment of acids and bases. No such quantitative treatment can be made of Lewis acids and bases.

Tables often list pK values of acids and bases. These are defined as

$$pK_a = -\lg(K_a/\text{mol dm}^{-3})$$
$$pK_b = -\lg(K_b/\text{mol dm}^{-3})$$

The higher the value of K_a (or K_b), the lower the value of pK_a (or pK_b), and the stronger is the acid (or base).

12.6.5 HOW TO CALCULATE THE pH OF A SOLUTION OF A WEAK ACID OR A WEAK BASE

In order to calculate the pH of a solution of a weak acid or base, you must know the concentration of the solution and the dissociation constant of the acid or base. The converse is also true: if the pH of a solution is measured, you can use it to find the dissociation constant of the weak electrolyte.

Example (a) Calculate the pH of a 1.00×10^{-2} mol dm^{-3} solution of butanoic acid, for which $K_a = 1.51 \times 10^{-5}$ mol dm^{-3}.

Method Use the equation

A sample calculation of the pH of a solution of a weak acid ...

$$K_a = \frac{[H_3O^+][C_3H_7CO_2^-]}{[C_3H_7CO_2H]}$$

The concentrations, $[H_3O^+]$ and $[C_3H_7CO_2^-]$ are equal.

The concentration $[C_2H_7CO_2H]$ is very little less than 1.00×10^{-2} mol dm^{-3}. Since the degree of ionisation is small, the approximation $[C_3H_7CO_2H] = 1.00 \times 10^{-2}$ mol dm^{-3} is made to simplify the calculation. Then

$$[H_3O^+]^2 = 1.51 \times 10^{-5} \times 1.00 \times 10^{-2}\ \text{mol}^2\ \text{dm}^{-6}$$
$$[H_3O^+] = 3.89 \times 10^{-4}\ \text{mol dm}^{-3}$$
$$\text{pH} = 3.42$$

... and a calculation of a dissociation constant from the pH of a solution of a weak base

Example (b) A solution of dimethylamine of concentration 1.00×10^{-2} mol dm^{-3} has a pH of 7.64 at 25 °C. Calculate the dissociation constant of the base

Method The dissociation of the base can be represented by

$$(CH_3)_2NH + H_2O \rightleftharpoons (CH_3)_2NH_2^+ + OH^-$$

Thus

$$K_b = \frac{[(CH_3)_2NH_2^+][OH^-]}{[(CH_3)_2NH]}$$

$$= [OH^-]^2/(1.00 \times 10^{-2}\ \text{mol dm}^{-3})$$

Since

$$pH = 7.64,\ pOH = 14.0 - 7.64 = 6.36 \text{ at } 25\ °C$$

$$[OH^-] = \text{antilg}(-6.36) = 4.37 \times 10^{-7}\ \text{mol dm}^{-3}$$

$$K_b = (4.37 \times 10^{-7})^2/(1.00 \times 10^{-2}) = 1.91 \times 10^{-11}\ \text{mol dm}^{-3}$$

12.6.6 HOW SUBSTITUENTS AFFECT THE STRENGTH OF ACIDS AND BASES

If **X** is more electronegative [§ 4.6.3] than carbon, the acid $\mathbf{X}CH_2CO_2H$ is a stronger acid than CH_3CO_2H. An example is chloroethanoic acid, $ClCH_2CO_2H$, which is 80 times stronger than ethanoic acid. The chlorine nucleus in the anion $ClCH_2CO_2^-$ attracts the electrons in the Cl—C bond, enabling the charge to be spread through the anion more than it is in $CH_3CO_2^-$:

$$\left\{ Cl \leftarrow CH_2-C\langle^{O}_{O} \right\}^- \qquad \left\{ H-CH_2-C\langle^{O}_{O} \right\}^-$$

Electron withdrawing substituents make acids stronger, bases weaker ...

The reduction of the charge located on the oxygen atoms makes $ClCH_2CO_2^-$ a weaker proton acceptor (a weaker base) than $CH_3CO_2^-$, and therefore makes $ClCH_2CO_2H$ a stronger acid than CH_3CO_2H. Dichloroethanoic acid is stronger still, and trichloroethanoic acid is as strong as some mineral acids. Values of K_a, are given below:

CH_3CO_2H	1.8×10^{-5} mol dm^{-3}	Cl_3CCO_2H	2.2×10^{-1} mol dm^{-3}
$ClCH_2CO_2H$	1.4×10^{-3} mol dm^{-3}	HCO_2H	1.7×10^{-4} mol dm^{-3}

... Electron donating substituents make acids weaker, bases stronger

Methanoic acid HCO_2H, is a stronger acid than ethanoic acid. It is inferred from this that the group $—CH_3$ is less electronegative than hydrogen:

$$\left\{ H-C\langle^{O}_{O} \right\}^- \qquad \left\{ H_3C-C\langle^{O}_{O} \right\}^-$$

By donating electrons to the $—CO_2^-$ group, $—CH_3$ makes it a stronger base (a better proton acceptor) and makes CH_3CO_2H a weaker acid than HCO_2H.

Bases are proton acceptors. If a group **X** is substituted for hydrogen in methylamine, then $\mathbf{X}CH_2NH_2$ will have a different value of K_b from CH_3NH_2. Groups such as $—CH_3$, which increase the availability of electrons at the nitrogen atom, increase its power to attract protons, i.e., its basicity. Thus dimethylamine, $(CH_3)_2NH$ is a stronger base than methylamine. Values of basic dissociation constants are

$$CH_3NH_2\ 4.4 \times 10^{-4}\ \text{mol dm}^{-3} \qquad (CH_3)_2NH\ 5.9 \times 10^{-4}\ \text{mol dm}^{-3}$$

CHECKPOINT 12.6B: pH AND DISSOCIATION CONSTANTS

1. The value of K_w, the ionic product for water, increases with temperature. Will the pH of pure water be greater or less than 7 at 100 °C?

2. The pH values of 0.100 mol dm^{-3} solutions of hydrochloric acid and ethanoic acid are 1 and approximately 3 respectively. How does this difference arise?

3. Give an expression for the dissociation constant, K_a, of a weak acid. How is pK_a related to K_a? If two acids, H**A** and H**B**, have pK_a values of 3.4 and 4.4, what can you say about the relative strengths of the two acids?

4. Explain what is meant by a conjugate acid–base pair. Give two examples.

5. Find the pH of the following solutions at 25 °C:

   (a) 0.001 00 mol dm^{-3} HCl(aq)
   (b) 2.50×10^{-2} mol dm^{-3} $HClO_4$(aq) (a strong acid)
   (c) 3.60×10^{-5} mol dm^{-3} HNO_3(aq)
   (d) 2.50×10^{-2} mol dm^{-3} NaOH(aq)
   (e) 3.00×10^{-3} mol dm^{-3} $Ca(OH)_2$(aq)

6. Find the pH of the following solutions:

   (a) 1.00×10^{-2} mol dm^{-3} ethanoic acid (pK_a = 4.76)
   (b) 0.100 mol dm^{-3} methanoic acid (pK_a = 3.75)

7. Find the dissociation constants of the following acids from the data:

   (a) A solution of 2.00 mol dm^{-3} hydrogen cyanide has a pH of 4.55.
   (b) A solution of 1.00 mol dm^{-3} iodic(I) acid has a pH of 5.26.

12.7 INDICATORS

Indicators are weak acids or weak bases ...

The indicators used in acid–base titrations are weak acids or weak bases. The ions are of a different colour from the undissociated molecules. Litmus is a weak acid, which can be represented by the formula HL. In solution

$$HL + H_2O \rightleftharpoons H_3O^+ + L^-$$

... Litmus

The molecules HL are red, and the anions L^- are blue. The dissociation constant of the indicator is

$$K_a = \frac{[H_3O^+][L^-]}{[HL]}$$

In acid solution, $[H_3O^+]$ is high, and H_3O^+ ions combine with L^- ions to form HL molecules, which are red. In alkaline solution, H_3O^+ ions are removed to form molecules of water, and HL molecules react with OH^- ions to form L^- ions, which are blue:

$$HL + OH^- \rightleftharpoons L^- + H_2O$$

Their molecules and ions differ in colour ...

If [HL] = [L^-], the indicator appears purple. Since

$$[H_3O^+] = K_a \frac{[HL]}{[L^-]}$$

this happens when

$$[H_3O^+] = K_a$$

and $\quad pH = pK_a$

... They change colour over 2 units of pH ...

If the ratio $[HL]/[L^-] \geqslant 10/1$, the solution appears red. If the ratio $[HL]/[L^-] \leqslant 1/10$, the solution appears blue. Thus litmus changes from red to blue over a range of 2 pH units.

The indicator methyl orange is a weak base, which can be represented as B [Figure 12.7C]:

$$B + H_3O^+ \rightleftharpoons BH^+ + H_2O$$

... Methyl orange

The molecules B are yellow, and the cations BH^+ are red. If the ratio $[B]/[BH^+] \geqslant 10/1$, the indicator appears yellow; if the ratio $\leqslant 1/10$, the indicator appears red. The indicator changes from yellow to red over 2 pH units. At a pH equal to pK_b, the ratio $[B]/[BH^+] = 1$, and the indicator is orange.

12.7.1 ACID–BASE TITRATIONS

To understand how indicators show the end-point in acid–base titrations, it is necessary to calculate the way in which pH changes during the course of a titration.

Titration of a strong base into a strong acid

Consider the titration of 25.0 cm^3 of 0.100 mol dm^{-3} hydrochloric acid with 0.100 mol dm^{-3} sodium hydroxide solution. At the beginning of the titration

$$[H_3O^+] = 1.00 \times 10^{-1} \text{ mol dm}^{-3};\ pH = 1.00$$

When 24.0 cm^3 of sodium hydroxide have been added

$$[H_3O^+] = (1.0 \text{ cm}^3 \text{ of } 0.100 \text{ mol dm}^{-3} \text{ acid})/(49.0 \text{ cm}^3 \text{ of solution})$$

$$[H_3O^+] = (1.0 \times 10^{-3} \times 0.100)/(49.0 \times 10^{-3}) = 2.04 \times 10^{-3} \text{ mol dm}^{-3}$$

$$pH = 2.69$$

Figure 12.7A Changes in pH during titration of base (0.100 mol dm^{-3}) into 25.0 cm^3 of acid (0.100 mol dm^{-3})

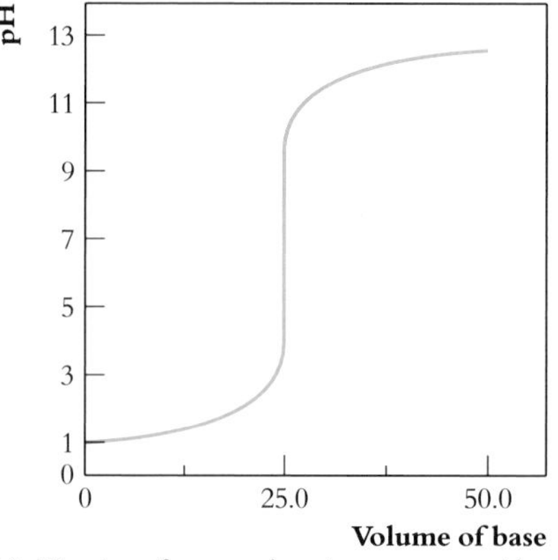

(a) Titration of a strong base into a strong acid

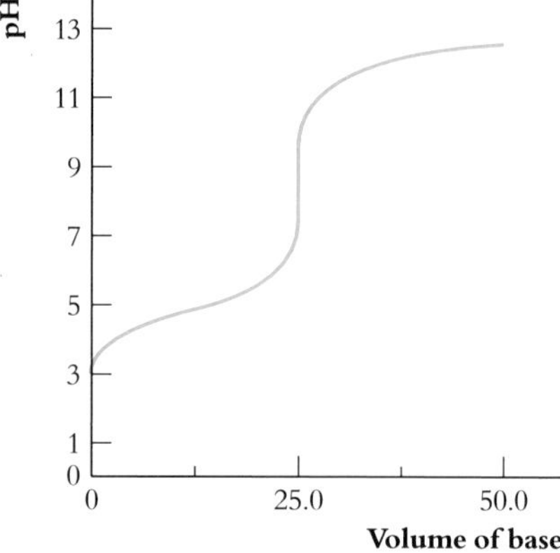

(b) Titration of a strong base into a weak acid

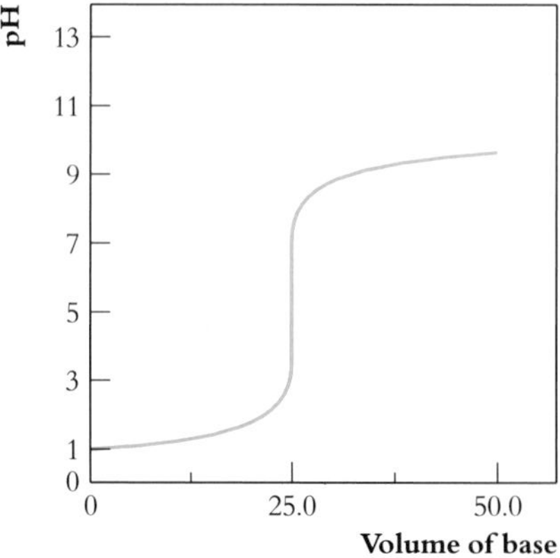

(c) Titration of a weak base into a strong acid

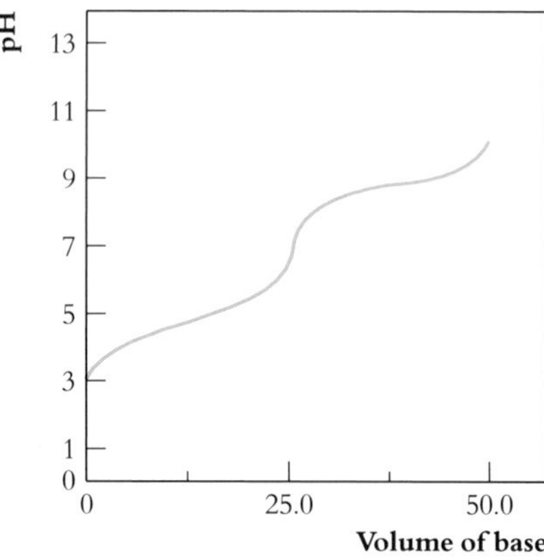

(d) Titration of a weak base into a weak acid

Calculation of pH changes during titration of a strong acid by a strong base ...

Similar calculations have been employed to give all the pH values plotted in Figure 12.7A(a). The figure shows how pH varies during the course of the titration. [See Question 4, Checkpoint 12.7.]

... pH changes rapidly at the end-point ...

... and many indicators can be used

The end-point of the titration occurs when 25.0 cm^3 of alkali have been added. The pH changes rapidly from 3.5 at 24.9 cm^3 of alkali to 10.5 at 25.1 cm^3 of alkali. Any indicator which changes colour over the range of pH 3.5 to 10.5 can be used to show the end-point of the titration.

Titration of a strong base into a weak acid and a weak base into a strong acid

The weak acid ethanoic acid can be titrated with sodium hydroxide solution. At the beginning of the titration, 25.0 cm^3 of 0.100 mol dm^{-3} ethanoic acid have a pH given by

$$[H_3O^+]^2 = K_a[CH_3CO_2H]$$

$$\therefore \quad pH = 2.88$$

Calculation of pH changes during the titration of a weak acid by a strong base ...

... in which litmus and phenol phthalein can be used

When 20.0 cm^3 of alkali have been added

$$[CH_3CO_2H] = 5.0\ cm^3 \times 0.100\ mol\ dm^{-3}/45.0\ cm^3$$

$$[CH_3CO_2^-] = 20.0\ cm^3 \times 0.100\ mol\ dm^{-3}/45.0\ cm^3$$

$$[H_3O^+] = \frac{K_a[CH_3CO_2H]}{[CH_3CO_2^-]} = K_a \times \frac{5.0}{20.0}$$

$$pH = pK_a - \lg(5.0/20.0) = 5.35$$

and of a weak base by a strong acid ...

... in which litmus and methyl orange can be used

The results of calculations of pH over the range of the titration are shown in Figure 12.7A(b). Similar calculations on the course of the titration of the weak base ammonia with a strong acid are shown in Figure 12.7A(c). You can see in Figure 12.7A(b) that the pH changes from 5 to 10.5 rapidly at the end-point. Indicators which change in this region are litmus and phenolphthalein. In Figure 12.7A(c), you can see that the pH changes rapidly from 3 to 7 at the end-point. Indicators which can be used are methyl orange and litmus. Phenolphthalein cannot be used in the titration of a weak base as it changes at pH 9.

Titration of a weak acid and a weak base

... of a weak acid with a weak base

In the titration of a weak acid and a weak base, a titration curve like that in Figure 12.7A(d) is obtained. The change in pH at the end-point is gradual, and indicators will change colour gradually. No indicator will give a sharp end-point. The way out of this difficulty is to titrate the weak acid against a strong base and the weak base against a strong acid.

Titration of a carbonate

... and of a carbonate ...

Figure 12.7B shows the pH changes during the titration of sodium carbonate with hydrochloric acid. The two stages in the titration are

$$Na_2CO_3(aq) + HCl(aq) \longrightarrow NaHCO_3(aq) + NaCl(aq)$$

$$NaHCO_3(aq) + HCl(aq) \longrightarrow NaCl(aq) + H_2O(l) + CO_2(g)$$

Figure 12.7C shows indicators which will change at the two equivalence-points. Use can be made of this two-stage titration to estimate sodium carbonate in a mixture of sodium carbonate and sodium hydroxide or sodium carbonate and sodium hydrogencarbonate.

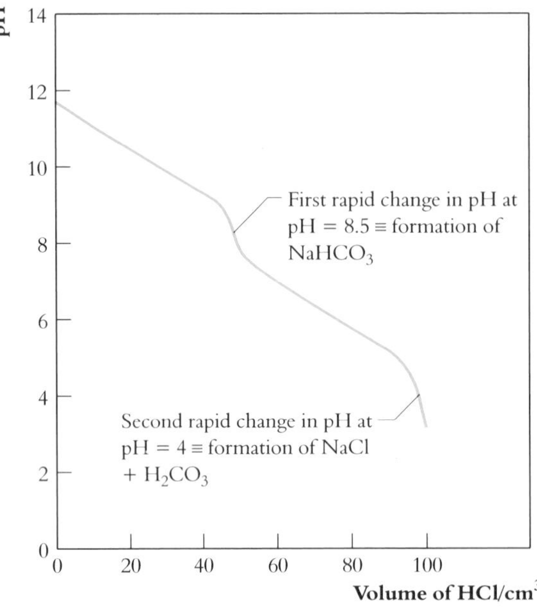

Figure 12.7B Changes in pH during titration of 50 cm^3 of sodium carbonate solution (0.100 mol dm^{-3}) with hydrochloric acid (0.100 mol dm^{-3})

The pH range of indicators

The range over which different indicators can be used is shown in Figure 12.7C. Universal Indicator is a mixture of several indicators.

Figure 12.7C Changes in pH for some common indicators

<table>
<tr><th>pH
Indicator</th><th>1</th><th>2</th><th>3</th><th>4</th><th>5</th><th>6</th><th>7</th><th>8</th><th>9</th><th>10</th><th>11</th></tr>
<tr><td>Thymol blue</td><td>Red</td><td>Change</td><td colspan="6">Yellow</td><td>Change</td><td colspan="2">Blue</td></tr>
<tr><td>Methyl orange</td><td colspan="3">Red</td><td>Change</td><td colspan="7">Yellow</td></tr>
<tr><td>Methyl red</td><td colspan="4">Red</td><td colspan="2">Change</td><td colspan="5">Yellow</td></tr>
<tr><td>Litmus</td><td colspan="5">Red</td><td colspan="3">Change</td><td colspan="3">Blue</td></tr>
<tr><td>Bromothymol blue</td><td colspan="5">Yellow</td><td colspan="2">Change</td><td colspan="4">Blue</td></tr>
<tr><td>Phenophthalein</td><td colspan="8">Colourless</td><td colspan="2">Change</td><td>Red</td></tr>
<tr><td>Universal Indicator</td><td colspan="3">Red</td><td>Orange</td><td colspan="2">Yellow</td><td colspan="2">Green</td><td>Blue</td><td colspan="2">Violet</td></tr>
</table>

CHECKPOINT 12.7: TITRATION

1. A vegetable extract contains a dye which is a weak acid with $K_a = 10^{-5}$ mol dm^{-3}. At pH 2 the dye is yellow, and at pH 8 it is blue. Explain how this dye could be used for acid–base titrations. Give an example of a titration for which it could be used, and one for which the indicator would be unsuitable.

2. Explain why it is possible to use one indicator for a number of different titrations which reach equivalence-point at different values of pH.

   By means of a diagram, explain why phenolphthalein is not used in the titration of a weak base into a strong acid. What kind of indicator should be used for such a purpose?

3. Sketch the change in pH that occurs during the titration of aqueous ammonia into hydrochloric acid of the same concentration.

   Suggest an indicator which could be used for the titration, and name one that should be avoided.

4. This question refers to Figure 12.7A(a), § 12.7.1. Refer to the worked examples in § 12.7.1 if necessary.

   Calculate the pH of a solution produced by adding to 25.0 cm^3 of 0.100 mol dm^{-3} hydrochloric acid the following volumes of 0.100 mol dm^{-3} sodium hydroxide solution:

   (a) 24.5 cm^3, *(b)* 24.7 cm^3, *(c)* 24.9 cm^3, *(d)* 25.1 cm^3, *(e)* 25.5 cm^3.

   Check to see whether your results fit the shape of Figure 12.7A(a).

12.8 BUFFER SOLUTIONS

A buffer solution absorbs small amounts of $H^+(aq)$ and $OH^-(aq)$ without a change in pH

A **buffer solution** is one which will resist changes in pH due to the addition of small amounts of acid and alkali. An effective buffer can be made by preparing a solution containing both a weak acid and also one of its salts with a strong base, e.g., ethanoic acid and sodium ethanoate. This will absorb small amounts of hydrogen ions because they react with ethanoate ions to form molecules of ethanoic acid:

The combination of a weak acid and its salt with a strong base acts as a buffer...

$$CH_3CO_2^- + H_3O^+ \rightleftharpoons CH_3CO_2H + H_2O$$

Hydroxide ions are absorbed (in small amounts) by combining with ethanoic acid molecules to form ethanoate ions and water:

$$OH^- + CH_3CO_2H \rightleftharpoons CH_3CO_2^- + H_2O$$

... A weak base and its salt with a strong acid act as a buffer

A solution of a weak base and one of its salts formed with a strong acid, e.g., ammonia solution and ammonium chloride, will act as a buffer. If hydrogen ions are added, they largely combine with ammonia, and, if hydroxide ions are added, they largely combine with ammonium ions:

$$NH_3 + H_3O^+ \rightleftharpoons NH_4^+ + H_2O$$
$$OH^- + NH_4^+ \rightleftharpoons NH_3 + H_2O$$

12.8.1 CALCULATION OF THE pH OF A BUFFER SOLUTION

The pH of a buffer solution consisting of a weak acid H**A** and its salt with a strong base is calculated from the equation

$$K_a = \frac{[H_3O^+][\mathbf{A}^-]}{[H\mathbf{A}]}$$

$$[H_3O^+] = K_a \frac{[H\mathbf{A}]}{[\mathbf{A}^-]}$$

$$pH = pK_a + \lg \frac{[\mathbf{A}^-]}{[H\mathbf{A}]}$$

Since the salt is completely ionised and the acid only slightly ionised, we can assume that all the anions come from the salt, and put

$$[\mathbf{A}^-] = [\text{Salt}]$$
$$[\text{H}\mathbf{A}] = [\text{Acid}]$$

Calculation of the pH value of a buffer ...

$$\therefore \quad \text{pH} = \text{p}K_a + \lg\frac{[\text{Salt}]}{[\text{Acid}]}$$

An effective buffering action is obtained at pH values fairly close to pK_a. For a buffer made from a weak base **B** and its salt with a strong acid $\mathbf{B}H^+X^-$, a similar treatment shows that

$$\text{pH} = \text{p}K_w - \text{p}K_b + \lg\frac{[\text{Base}]}{[\text{Salt}]}$$

Example Three solutions contain ethanoic acid ($K_a = 1.80 \times 10^{-5}$ mol dm^{-3}) at a concentration 0.10 mol dm^{-3} and sodium ethanoate at a concentration (*a*) 0.10 mol dm^{-3}, (*b*) 0.20 mol dm^{-3}, (*c*) 0.50 mol dm^{-3}. Calculate the pH values of the three solutions.

Method

... a sample calculation

$$\text{pH} = \text{p}K_a + \lg\frac{[\text{Salt}]}{[\text{Acid}]}$$

In solution (*a*) pH = 4.75 + lg(0.10/0.10)
= 4.75 + lg1.0
= 4.75

In solution (*b*) pH = 4.75 + lg(0.20/0.10)
= 4.75 + lg2.0
= 5.05

In solution (*c*) pH = 4.75 + lg(0.50/0.10)
= 5.45

The pH values are (*a*) 4.75, (*b*) 5.05, (*c*) 5.45

12.8.2 THE BIOLOGICAL IMPORTANCE OF BUFFER SOLUTIONS

Protein molecules are held in a three-dimensional configuration by interactions between protein chains...

Protein molecules are long chains of amino acid groups (see § 33.17). Forces of attraction between polar groups such as $—CO_2^-$ and $—NH_3^+$ hold the protein chain in a complicated three-dimensional structure. Should the medium become too acidic, $—CO_2^-$ groups would be converted into $—CO_2H$ groups.

$$—CO_2^-(aq) + H^+(aq) \rightleftharpoons —CO_2H(aq)$$

...including interactions between polar groups, e.g. $—CO_2^-$ and $—NH_3^+$

This would interfere with the attraction between $—CO_2^-$ groups and $—NH_3^+$ groups. Should the medium become too alkaline, $—NH_3^+$ groups would be converted into $—NH_2$ groups.

$$—NH_3^+(aq) + OH^-(aq) \rightleftharpoons —NH_2(aq) + H_2O(l)$$

These interactions depend on the pH

Again this would interfere with the attraction between $—NH_3^+$ groups and $—CO_2^-$ groups.

The attractions between acidic and basic groups therefore exist only within a limited range of pH values. Outside this range, the three-dimensional structure of the protein is destroyed. Blood proteins would coagulate if the pH of blood were allowed to depart far from neutral pH.

Figure 12.8A
Some interactions between groups in a protein molecule

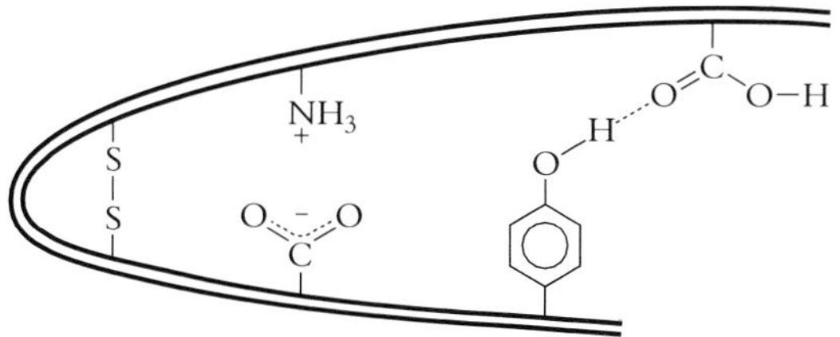

Enzymes are proteins and their activity depends on their three-dimensional structure ...

... which depends on pH

Enzymes can work only within a narrow range of pH

The reactions that take place in living things are catalysed by enzymes. Enzymes are proteins, and the ability of enzymes to act as catalysts depends on their three-dimensional structure. Outside the range of pH which maintains the structure of the protein, the enzyme cannot do its job. There is an optimum pH for each enzyme, and the enzyme will only function over a range of 2 or 3 pH units. Most cells and tissues require a pH value close to 7.

How cell contents are buffered

Cell contents are buffered to provide a suitable pH ...

... by phosphates ...

Buffers keep the cell pH as constant as possible. Many tissues contain hydrogenphosphate ions and, since phosphoric acid is a weak acid, the equibrium is set up:

$$H_2PO_4^-(aq) \rightleftharpoons H^+(aq) + HPO_4^{2-}(aq)$$

When the hydrogen ion concentration increases, the equilibrium is displaced from right to left, thus removing hydrogen ions from solution. If the hydrogen ion concentration decreases, $H_2PO_4^-$ ions dissociate to release H^+ ions. Thus the buffer tends to maintain a constant hydrogen ion concentration.

... by hydrogencarbonates ...

Blood contains carbonate ions and hydrogencarbonate ions in equilibrium:

$$HCO_3^-(aq) \rightleftharpoons H^+(aq) + CO_3^{2-}(aq)$$

The system acts as a buffer, absorbing small amounts of hydrogen ion by moving from right to left and releasing small amounts of hydrogen ion by moving from left to right.

If hydroxide ions accumulate in the tissue, hydrogencarbonate ions can absorb them by forming carbonate ions and water:

$$HCO_3^-(aq) + OH^-(aq) \rightleftharpoons CO_3^{2-}(aq) + H_2O(l)$$

... by amino acids, peptides and proteins

Amino acids have the formula $H_2NCH\mathbf{R}CO_2H$, where **R** can be a number of different groups. They can ionise as weak acids and as weak bases. Both the weakly acidic carboxyl group and the weakly basic amino group can act as buffers.

$$H_2NCH\mathbf{R}CO_2H(aq) + H_2O(l) \rightleftharpoons H_3N^+CH\mathbf{R}CO_2H(aq) + OH^-(aq)$$
$$H_2NCH\mathbf{R}CO_2H(aq) + H_2O(l) \rightleftharpoons H_2NCH\mathbf{R}CO_2^-(aq) + H_3O^+(aq)$$

Peptides and proteins are condensation polymers of amino acids [§ 33.17.2] and share this buffering action. Amino acids, peptides and proteins are important buffering agents in living tissues.

SUMMARY

Buffer solutions are important in controlling the pH of biological systems, e.g. blood and medicines and cosmetics. Biological systems depend on enzymes, and enzymes work only within limited ranges of pH.

CHECKPOINT 12.8: BUFFERS

1. How does a buffer maintain an almost constant pH, even when small amounts of acid or alkali are added to a solution? What two components are present in a buffer?

2. Solution (*a*) contains 0.100 mol dm^{-3} potassium chloride solution. Solution (*b*) contains 0.100 mol dm^{-3} ammonium ethanoate. Both solutions are neutral. Explain why the addition of 1.00 cm^3 of 1.00 mol dm^{-3} hydrochloric acid to 1.00 dm^3 of solution (*a*) changes its pH to 3, but the same treatment has very little effect on the pH of solution (*b*).

3. Given a 0.100 mol dm^{-3} solution of dimethylamine, $(CH_3)_2NH$, what would you need in order to prepare a buffer solution? Explain how the buffer would react to the addition of (*a*) hydrogen ions and (*b*) hydroxide ions.

4. What is the pH of a solution that has been prepared by the addition of 50.0 cm^3 of 0.200 mol dm^{-3} sodium hydroxide to 50.0 cm^3 of 0.400 mol dm^{-3} ethanoic acid ($K_a = 1.75 \times 10^{-5}$ mol dm^{-3})?

5. Write the equilibrium for the ionisation of the weak acid RCO_2H. Explain how a solution of the acid can absorb small amounts of (*a*) hydrogen ions (*b*) hydroxide ions without a change in pH.

6. Write the equilibrium for the ionisation of the weak base RNH_2. Explain how a solution of the base can absorb small amounts of (*a*) hydrogen ions (*b*) hydroxide ions without a change in pH.

7. Glycine, $H_2NCH_2CO_2H$, is the simplest amino acid.
   (*a*) Write an equation to show how glycine ionises (i) as an acid (ii) as a base.
   (*b*) Write an equation to show how glycine can combines with small amounts of hydrogen ions without a significant change in pH.
   (*c*) Write an equation to show how glycine can combine with small amounts of hydroxide ions without a significant change in pH.

8. Briefly explain why a fairly constant pH is important for living tissues.

12.9 SALT HYDROLYSIS

Many salts dissolve in water to give neutral solutions. Some salts, however, react with water to form acidic or alkaline solutions. These reactions are described as **salt hydrolysis**.

Some salts react with water ...

Sodium carbonate is the salt of a strong base and a weak acid. The carbonate ion, CO_3^{2-}, is a base, a proton acceptor. The reaction

... Their solutions are acidic or alkaline ...

$$CO_3^{2-}(aq) + H_2O(l) \rightleftharpoons OH^-(aq) + HCO_3^-(aq)$$

makes a solution of sodium carbonate alkaline. Solutions of sodium hydrogencarbonate are less strongly alkaline because HCO_3^- is a weaker base than CO_3^{2-}:

$$HCO_3^-(aq) + H_2O(l) \rightleftharpoons H_2CO_3(aq) + OH^-(aq)$$

... Such reactions are called salt hydrolysis

Other basic anions are S^{2-}, HS^-, CN^-, $CH_3CO_2^-$ and the anions of many other organic acids.

The hydrolysis of a salt of a weak base and a strong acid, e.g., ammonium chloride, results in an increase in the concentration of hydrogen ions:

$$NH_4^+(aq) + H_2O(l) \rightleftharpoons NH_3(aq) + H_3O^+(aq)$$

The hydrolysis of aluminium salts is covered in § 19.4.1, and transition metal salts in § 24.14.2.

12.9.1 ENTHALPY OF NEUTRALISATION

The reaction that happens in neutralisation is the combination of hydrogen ions and hydroxide ions to form water

The value of $\Delta H^{\ominus}_{Neutralisation}$ relates to the neutralisation of one mole of $H^+(aq)$ ions

$$H^+(aq) + OH^-(aq) \longrightarrow H_2O(l);\ \Delta H^{\ominus}_{Neutralisation}$$

Values of $\Delta H^{\ominus}_{Neutralisation}$ are:

hydrochloric acid + sodium hydroxide -57.9 kJ mol^{-1}
ethanoic acid + sodium hydroxide -56.1 kJ mol^{-1}
hydrogen sulphide + sodium hydroxide -32.2 kJ mol^{-1}

The dissociation constants of the acids at 25 °C are:

hydrochloric acid: completely dissociated
ethanoic acid: 1.7×10^{-5} mol dm^{-3}
hydrogen sulphide: 8.9×10^{-8} mol dm^{-3}

When hydrochloric acid neutralises sodium hydroxide, the products are sodium chloride and water. All the hydrogen ions and all the hydroxide ions react to form water molecules.

When ethanoic acid reacts with sodium hydroxide, the products are sodium ethanoate and water. Since sodium ethanoate is the salt of a weak acid, it is partially hydrolysed:

$$CH_3CO_2^-(aq) + H_2O(l) \rightleftharpoons CH_3CO_2H(aq) + OH^-(aq)$$

In the neutralisation of a weak acid the salt formed is alkaline due to salt hydrolysis therefore the value of $\Delta H^{\ominus}$ is lower

The neutralisation of a weak base yields an acidic salt, and $\Delta H^{\ominus}$ is lower

Thus hydroxide ions are produced in the hydrolysis and the solution is alkaline. Neutralisation has not been complete, and the value of the standard enthalpy of neutralisation is lower than that of hydrochloric acid and sodium hydroxide.

When hydrogen sulphide is neutralised by sodium hydroxide, the products are sodium sulphide and water. Hydrogen sulphide is a weaker acid than ethanoic acid, and sodium sulphide is therefore hydrolysed to a greater extent than sodium ethanoate.

$$S^{2-}(aq) + H_2O(l) \rightleftharpoons HS^-(aq) + OH^-(aq)$$

The value of $\Delta H^{\ominus}$ is therefore lower than for the neutralisation of ethanoic acid.

CHECKPOINT 12.9: SALT HYDROLYSIS

1. Predict whether the pH of the following solutions will be 7 or > 7 or < 7. Give your reasons.

(a) 0.10 mol dm^{-3} ammonium chloride
(b) 0.010 mol dm^{-3} methylammonium chloride
(c) 0.10 mol dm^{-3} potassium cyanide
(d) 0.10 mol dm^{-3} sodium methanoate.

2. The pH readings below refer to the titration of sodium hydroxide solution into 25.0 cm^3 of 0.100 mol dm^{-3} ethanoic acid.

V/cm^3	0	4.0	6.0	8.0	10.0	12.0	14.0	14.4	14.6	14.8	15.0	15.2	15.4	16.0
pH	2.8	3.8	4.2	4.6	5.1	5.5	6.2	6.5	6.8	7.6	9.0	9.8	10.5	11.4

Plot *V*, the volume of sodium hydroxide solution, against pH. Use your graph to find the following quantities:

(a) the pH at the equivalence point
(b) the concentration in mol dm^{-3} of sodium hydroxide.

Name an indicator which could be used for this titration.

3. Explain why the values of $\Delta H^{\ominus}_{Neutralisation}$ for these two reactions are the same:

$$NaOH(aq) + HCl(aq) \longrightarrow NaCl(aq) + H_2O(l)$$

$$2KOH(aq) + H_2SO_4(aq) \longrightarrow K_2SO_4(aq) + 2H_2O(l)$$

4. Ammonia solution is titrated with hydrochloric acid. From the heat evolved, the value of $\Delta H^{\ominus}_{Neutralisation}$ is calculated as -56.2 kJ mol^{-1}. Explain why the value is lower than that for hydrochloric acid and a strong base.

12.10 COMPLEX IONS

Complex ions are formed by the combination of a cation with a neutral molecule (or molecules) or an oppositely charged ion (or ions). Coordinate bonding is involved [§ 4.7]. Examples are $[Al(H_2O)_6]^{3+}$, $[Ag(NH_3)_2]^+$, $[CuCl_4]^{2-}$ and $[Al(OH)_4]^-$.

Complex ions have dissociation constants ...

The formation of a complex ion is an equilibrium reaction. In the formation of tetracyanozincate ions, $[Zn(CN)_4]^{2-}$, there is set up an equilibrium:

$$Zn^{2+}(aq) + 4CN^-(aq) \rightleftharpoons Zn(CN)_4{}^{2-}(aq)$$

The dissociation constant of the complex ion is

$$K_d = \frac{[Zn^{2+}][CN^-]^4}{[Zn(CN)_4{}^{2-}]} = 2.0 \times 10^{-15}\ \text{mol}^4\ \text{dm}^{-12}$$

The low value of the dissociation constant shows that the complex ion is very stable. The units are (concentration)^5/(concentration), i.e. mol^4 dm^{-12}.

... The reciprocal 1/dissociation constant is the stability constant

The reciprocal of the dissociation constant is called the **stability constant** of the complex. A high stability constant shows a stable complex.

Edta can be used in complexometric titration ...

A number of metal ions form complexes with edta. These complexes are so stable that they can be used for measuring the concentration of the metal ions by **complexometric titration**. The disodium salt of edta is a primary standard. It is used in alkaline solution so that all the carboxyl groups are ionised as

$$(^-O_2CCH_2)_2NCH_2CH_2N(CH_2CO_2{}^-)_2$$

... with an indicator

The end-point in the titration is shown by an indicator which forms a coloured complex with the metal ion which is being titrated. If Eriochrome Black T is used as indicator, the metal-indicator complex colour of red is seen at the beginning of the titration. As the edta solution is added, metal ions are removed from the indicator and complex with edta. At the end-point, the blue colour of the free indicator is seen:

$$\text{Metal-indicator (red)} + \text{edta} \longrightarrow \text{Metal-edta} + \text{Indicator (blue)}$$

A worked example of complexometric titration

Example Hardness in water is caused by calcium and magnesium ions. Both these ions complex strongly with edta. To a 200 cm^3 sample of tap water were added an alkaline buffer and a few drops of Eriochrome Black T. A volume 3.50 cm^3 of 0.100 mol dm^{-3} edta was used in titration. Find the concentration of calcium and magnesium ions in the water. The complexes have the formulae Ca(edta) and Mg(edta).

Method Amount of edta in titration = $3.50 \times 0.100 \times 10^{-3} = 3.50 \times 10^{-4}$ mol

Amount of $Ca^{2+} + Mg^{2+} = 3.5 \times 10^{-4}$ mol

$$[Ca^{2+}] + [Mg^{2+}] = 3.50 \times 10^{-4}/(200 \times 10^{-3})$$
$$= 1.75 \times 10^{-3}\ \text{mol dm}^{-3}$$

ACID–BASE EQUILIBRIA

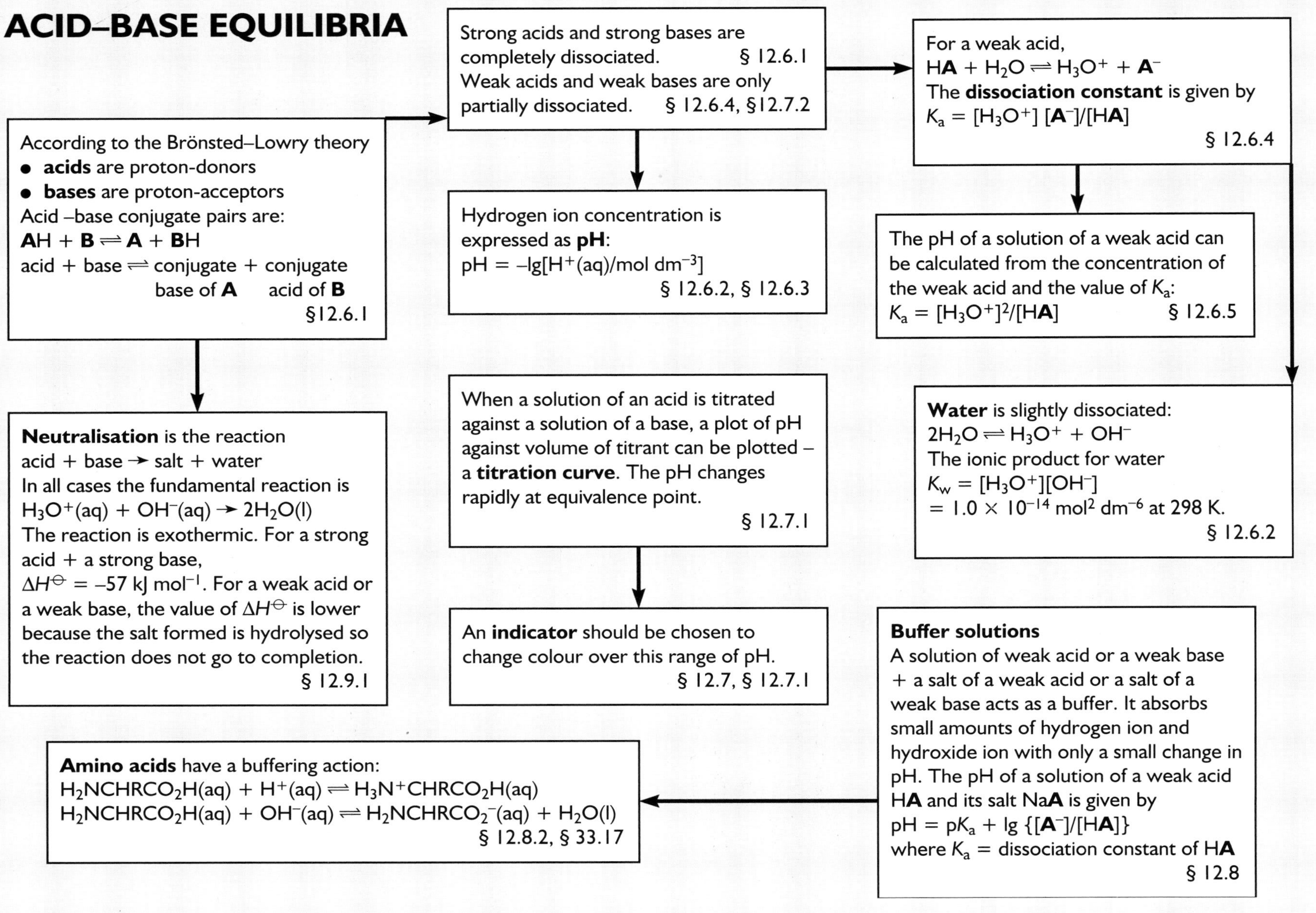

According to the Brönsted–Lowry theory
• **acids** are proton-donors
• **bases** are proton-acceptors
Acid –base conjugate pairs are:
AH + **B** ⇌ **A** + **B**H
acid + base ⇌ conjugate base of **A** + conjugate acid of **B**
§12.6.1
Strong acids and strong bases are completely dissociated. § 12.6.1
Weak acids and weak bases are only partially dissociated. § 12.6.4, §12.7.2
For a weak acid,
H**A** + $H_2O \rightleftharpoons H_3O^+$ + **A**$^-$
The **dissociation constant** is given by
$K_a = [H_3O^+][A^-]/[HA]$
§ 12.6.4
Hydrogen ion concentration is expressed as **pH**:
pH = –lg[H^+(aq)/mol dm^{-3}]
§ 12.6.2, § 12.6.3
The pH of a solution of a weak acid can be calculated from the concentration of the weak acid and the value of K_a:
$K_a = [H_3O^+]^2/[HA]$ § 12.6.5
Neutralisation is the reaction
acid + base → salt + water
In all cases the fundamental reaction is
$H_3O^+(aq) + OH^-(aq) \rightarrow 2H_2O(l)$
The reaction is exothermic. For a strong acid + a strong base,
$\Delta H^\ominus = -57\ kJ\ mol^{-1}$. For a weak acid or a weak base, the value of $\Delta H^\ominus$ is lower because the salt formed is hydrolysed so the reaction does not go to completion.
§ 12.9.1
When a solution of an acid is titrated against a solution of a base, a plot of pH against volume of titrant can be plotted – a **titration curve**. The pH changes rapidly at equivalence point.
§ 12.7.1
Water is slightly dissociated:
$2H_2O \rightleftharpoons H_3O^+ + OH^-$
The ionic product for water
$K_w = [H_3O^+][OH^-]$
$= 1.0 \times 10^{-14}\ mol^2\ dm^{-6}$ at 298 K.
§ 12.6.2
An **indicator** should be chosen to change colour over this range of pH.
§ 12.7, § 12.7.1
Buffer solutions
A solution of weak acid or a weak base + a salt of a weak acid or a salt of a weak base acts as a buffer. It absorbs small amounts of hydrogen ion and hydroxide ion with only a small change in pH. The pH of a solution of a weak acid H**A** and its salt Na**A** is given by
$pH = pK_a + lg\{[A^-]/[HA]\}$
where K_a = dissociation constant of H**A**
§ 12.8
Amino acids have a buffering action:
$H_2NCHRCO_2H(aq) + H^+(aq) \rightleftharpoons H_3N^+CHRCO_2H(aq)$
$H_2NCHRCO_2H(aq) + OH^-(aq) \rightleftharpoons H_2NCHRCO_2^-(aq) + H_2O(l)$
§ 12.8.2, § 33.17

CHECKPOINT 12.10: COMPLEX IONS

1. Explain these observations, with the help of your knowledge of inorganic chemistry:

   (a) The solubility of $PbCl_2$ in water increases on the addition of concentrated hydrochloric acid.

   (b) The addition of aqueous ammonia to aqueous magnesium sulphate gives a white precipitate which is soluble in aqueous ammonium chloride.

   (c) Nickel hydroxide is precipitated when aqueous ammonia is added to an aqueous solution of nickel sulphate. The precipitate dissolves on addition of (i) an excess of aqueous ammonia to form a deep blue solution (ii) aqueous potassium cyanide to form an orange solution (iii) concentrated hydrochloric acid to form a blue solution.

QUESTIONS ON CHAPTER 12

1. Write equations for the reactions that occur when each of the following is separately dissolved in water:

   (a) $HCl(g)$, *(b)* $NH_3(g)$, *(c)* $CH_3CO_2H(l)$, *(d)* $C_2H_5NH_2(g)$, *(e)* $H_2NCH_2CO_2H(s)$.

2. Classify the following species as oxidising agent/reducing agent/Brönsted acid/Brönsted base. Quote one reaction (with its equation) for each species to illustrate its typical behaviour. Some species fall into more than one category:

   (a) $NH_3(aq)$, *(b)* $NH_4^+(aq)$, *(c)* $HSO_3^-(aq)$, *(d)* $(CO_2H)_2$, *(e)* $Fe(H_2O)_6^{3+}(aq)$.

3. Explain the significance of the Faraday constant, 96 478 C mol^{-1}. Write an equation for the electrode process that results in the evolution of oxygen at the anode during the electrolysis of an aqueous solution. Calculate the time in minutes for which a current of 0.500 A would need to be passed in order to yield 500 cm^3 (at stp) of oxygen.

4. Sketch titration curves showing the change in pH which occurs during the addition of an excess of sodium hydroxide of concentration 1.0 mol dm^{-3} to 25 cm^3 of *(a)* 1.0 mol dm^{-3} hydrochloric acid, *(b)* 1.0 mol dm^{-3} sulphuric acid, *(c)* 1.0 mol dm^{-3} ethanoic acid ($pK_a = 4.75$).

   Explain why the pH at the end-point in *(c)* is different from that in *(a)*.

5. *(a)* Explain why an aqueous solution of ammonium chloride is acidic.

   (b) Explain why a mixture of ammonium chloride and ammonia in solution has a buffering action.

6. Describe an experiment which you could carry out to find the number of units of charge on a metal cation.

7. A current of 0.200 A is passed for 2.00 h through 200 cm^3 of 0.0500 mol dm^{-3} aqueous silver nitrate. Calculate the volume (at stp) of hydrogen evolved.

8. A 50 cm^3 portion of aqueous ammonia (1.00 mol dm^{-3}) is titrated with 1.00 mol dm^{-3} hydrochloric acid. Draw a graph to show roughly (without doing detailed calculations) the change in pH that occurs as titration proceeds, until 60 cm^3 of the acid have been added.

   What kind of indicator would be suitable for use in this titration ($K_b(NH_3) = 1.8 \times 10^{-5}$ mol dm^{-3})?

9. *(a)* Explain the Brönsted–Lowry theory of acids and bases.

   (b) Explain the meanings of the terms 'concentration' and 'strength' as applied to a base.

   (c) Explain how the dissociation constant of a base measures the strength of the base.

   (d) Calculate the pH of
   (i) 0.010 mol dm^{-3} hydrochloric acid
   (ii) 0.010 mol dm^{-3} ethanoic acid
   (iii) a buffer solution containing 0.010 mol dm^{-3} of ethanoic acid and 0.010 mol dm^{-3} of sodium ethanoate.

   (For ethanoic acid, $K_a = 1.8 \times 10^{-5}$ mol dm^{-3}.)

10. *(a)* Write an expression for the acid dissociation constant, K_a, of ethanoic acid.

    (b) The ionisation of ethanoic acid is endothermic. State how you would expect the value of K_a to change with a rise in temperature.

    (c) Describe how you would measure the pH of an aqueous solution of ethanoic acid.

    (d) 50 cm^3 of aqueous ethanoic acid (0.1 mol dm^{-3}) are added slowly to 25 cm^3 of aqueous sodium hydroxide (0.1 mol dm^{-3}). Sketch a graph to show how the pH of the solution changes as ethanoic acid is added.

    (e) 10.0 cm^3 of aqueous ethanoic acid (0.100 mol dm^{-3}) are mixed with 10.0 cm^3 of aqueous sodium ethanoate (0.200 mol dm^{-3}) at 298 K. Calculate the pH of the resulting solution. (For ethanoic acid, $K_a = 1.8 \times 10^{-5}$ mol dm^{-3} at 298 K.)

    (f) Explain why this solution acts as a buffer.

11. *(a)* Solution A is 0.15 M lactic acid, and solution B is a mixture of equal volumes of 0.30 M lactic acid and 0.10 M aqueous sodium hydroxide solution.

(i) Calculate the pH of A and of B at 298 K.

(ii) A few drops of dilute hydrochloric acid are added to 50.0 cm^3 of A and of B.

Compare the effect of such action on the pH of the two solutions. Explain your answer.

(iii) 50.0 cm^3 of A and of B are titrated with a strong base.

Compare the pH of the two solutions at the respective equivalence points of the titrations. Explain your answer. **9**

Notes:

(1) Lactic acid is a monoprotic acid and its K_a at 298 K is 1.38×10^{-4} mol dm^{-3}.

(2) For (ii) and (iii), you may assume that the volume changes are negligible.

(b) A gaseous compound D contains carbon and hydrogen only, and has a density of 1.15 g dm^{-3} at 95.3 kPa pressure and 298 K. Assuming that D behaves ideally, calculate its molar mass and deduce its molecular formula. **4**

(1 kPa = 1×10^3 N m^{-2})

(c) (i) Draw the three-dimensional structure of BF_3.

(ii) BF_3 reacts with NH_3 to form an adduct, $BF_3 \cdot NH_3$. Account for the formation of the adduct and draw its three-dimensional structure. **4**

HKALE (AL)

12. *(a)* For the reaction below state, giving a reason, which of the species on the **left-hand side** of the equation is behaving as a Lowry–Brönsted acid. **2**

$$C_6H_5COO^- + H_2O \longrightarrow C_6H_5COOH + OH^-$$

(b) Explain why an aqueous solution of sodium chloride should have a pH of 7 whereas an aqueous solution of sodium ethanoate should have a pH significantly greater than 7. **2**

(c) Calculate the pH of hydrochloric acid of concentration 0.1 mol dm^{-3}. **1**

(d) The approximate value of $[H^+(aq)]$ in an aqueous solution of a weak acid can be calculated from

$$[H^+(aq)]^2 = K_a c$$

where K_a is the acid dissociation constant and c is the concentration of the weak acid in mol dm^{-3}.

Using this expression, or otherwise, explain how an increase in the concentration, c, of a weak acid will affect the concentration of the aqueous hydrogen ions and the pH of the solution. **2**

WJEC (AS/AL)

13. *(a)* When ethanoic acid is dissolved in water, the following equilibrium is established:

$$CH_3CO_2H + H_2O \rightleftharpoons CH_3CO_2^- + H_3O^+ \quad \text{I}$$

When hydrogen chloride dissolves in ethanoic acid, the equilibrium established is:

$$CH_3CO_2H + HCl \rightleftharpoons CH_3CO_2H_2^+ + Cl^- \quad \text{II}$$

Comment on the role of the ethanoic acid in:

(i) equilibrium I; **1**

(ii) equilibrium II. **1**

(b) What is the relationship between the species $CH_3CO_2H_2^+$ and CH_3CO_2H? **1**

(c) The value of K_a for ethanoic acid at 298 K is 1.74×10^{-5} mol dm^{-3} and for methanoic acid, HCO_2H, it is 1.60×10^{-4} mol dm^{-3} at the same temperature.

(i) Write an expression for K_a for CH_3CO_2H. **1**

(ii) Hence calculate the pH of a 0.100 mol dm^{-3} solution of CH_3CO_2H at 298 K.

(d) The pH of a 0.050 mol dm^{-3} solution of HCO_2H is 2.55. Using this, together with the data in *(c)* and your answer to *(c)*(ii):

(i) state which of the two acids is the stronger; **1**

(ii) comment on the relative pH values of the two acids. **3**

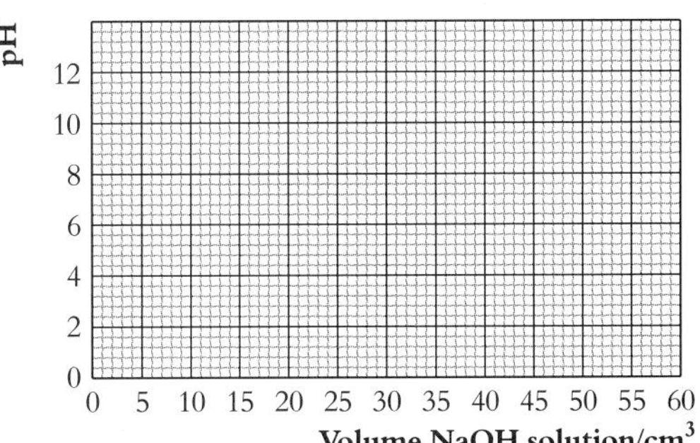

(e) (i) Sketch with reasonable accuracy, on a copy of the axes above, how the pH changes during the titration of 20.0 cm^3 of a 0.100 mol dm^{-3} solution of methanoic acid with 0.050 mol dm^{-3} sodium hydroxide solution. **3**

(ii) Select using the data below a suitable indicator for this titration. Give a brief reason for your choice based on the curve drawn in *(e)*(i). **2**

Indicator	pH Range
Bromocresol green	3.5–5.4
Bromothymol blue	6.0–7.6
Phenol red	6.8–8.4

15 *L (AS/AL)*

14. Propanoic acid occurs naturally in Swiss cheese in a concentration that can be as high as 1%. Its sodium and calcium salts are food additives used in processed cheeses to retard the formation of moulds.

(a) Propanoic acid is a weak acid with an acid dissociation constant of 1.22×10^{-5} mol dm^{-3}.
(i) Write an equation for the ionisation of propanoic acid. **2**
(ii) Calculate pK_a for propanoic acid. **2**

(b) Sodium propanoate is made by the reaction of sodium hydroxide with propanoic acid.
(i) Write an equation for the reaction. **1**
(ii) If the reaction were carried out by titration using 0.1 M solutions of the hydroxide and the acid, state the name of a suitable indicator. **1**

(c) A mixture of sodium propanoate and propanoic acid acts as a buffer solution.
(i) What is meant by the term **buffer solution**? **2**
(ii) Explain how the mixture of sodium propanoate and propanoic acid acts as a buffer solution. **5**
(Up to 2 marks may be obtained for the quality of language in this part)
(iii) Calculate the pH of the solution formed by adding 15.0 cm^3 of 0.1 M sodium hydroxide to 30.0 cm^3 of 0.1 M propanoic acid. **3**
(iv) Name a buffer solution found in a biological system and explain its importance. **2**

13 *CCEA (AS/AL)*

15. *(a)* (i) With the aid of an example, explain the meaning of the terms *strong acid*, *weak acid*, *strong base* and *weak base*.
(ii) Describe and explain the use of buffer solutions. **10**

(b) Assuming the temperature to be 25 °C, what is the pH of
(i) 0.05 mol dm^{-3} sulphuric acid, H_2SO_4
(ii) 0.01 mol dm^{-3} sodium hydroxide? **3**

(c) Benzoic acid, $C_6H_5CO_2H$, is a weak acid with an acid dissociation constant, K_a, of 6.3×10^{-5} mol dm^{-3} at 25 °C.
(i) Calculate the pH of 0.020 mol dm^{-3} benzoic acid at this temperature.
(ii) Draw a sketch graph of the change in pH which occurs when 0.020 mol dm^{-3} potassium hydroxide is added to 25 cm^3 of 0.020 mol dm^{-3} benzoic acid until in excess.
(iii) The pK_a values of some indicators are

thymol blue	1.7
congo red	4.0
thymolphthalein	9.7

Which of these indicators would be most suitable for determining the end point of the titration between the benzoic acid and potassium hydroxide in (ii)? Explain your answer. **12**

25 *C (AS/AL)*

16. *(a)* (i) State what is meant by the term **Lowry–Brönsted base**. **1**
(ii) For the reaction represented by the equation below, write the formulae of the **two** Lowry–Brönsted bases. **1**

$$C_6H_5COOH + H_2O \rightleftharpoons C_6H_5COO^- + H_3O^+$$

(b) (i) Write an expression for the acid dissociation constant. K_a, for ethanoic acid, CH_3COOH. **1**
(ii) The hydrogen ion concentration in aqueous ethanoic acid of concentration 0.10 mol dm^{-3} at 298 K is 1.34×10^{-3} mol dm^{-3}. Calculate the pH of the ethanoic acid **1**
(iii) The hydrogen ion concentration in hydrochloric acid of the same concentration, as the acid in *(b)*(ii), is 1.0×10^{-1} mol dm^{-3}. Calculate the pH of the hydrochloric acid. **1**
(iv) Explain why the two calculated pH values are different although the concentrations in *(b)*(ii) and *(b)*(iii) are the same. **2**

(c) Explain why
(i) the pH value of aqueous sodium ethanoate is greater than that of pure water at the same temperature, **1**
(ii) the pH value of aqueous sodium chloride is the same as that of pure water at the same temperature. **1**

(d) 20.00 cm^3 of aqueous sodium carbonate is neutralised, using methyl orange as indicator, by 25.00 cm^3 of hydrochloric acid of concentration 0.1000 mol dm^{-3}.

$$Na_2CO_3 + 2HCl \longrightarrow 2NaCl + H_2O + CO_2$$

Calculate the concentration of the aqueous sodium carbonate in mol dm^{-3}. **2**

11 *WJEC (AS/AL)*

17. *(a)* State **two** identical features of and **two** differences between the titration curves for the separate addition of 0.75 M sodium hydroxide to 25 cm^3 samples of
(i) 1.5 M hydrochloric acid,
(ii) 1.5 M ethanoic acid. **4**

(b) What characteristic features must all acid–base indicators possess? Name an indicator that could be used for **both** of the titrations in part *(a)* above and explain the reasons for your choice. **4**

(c) An aqueous solution of the weak acid **HA**, which has an initial pH of 2.50, is titrated against sodium hydroxide. When the pH reaches 4.30, [HA] = [A$^-$]. Calculate the value of the acid dissociation constant **HA** and hence the original concentration of the acid in solution. **7**

15 *NEAB (AS/AL)*

18. *(a)* Explain how an aqueous solution containing ethanoic acid and sodium ethanoate resists changes in pH when contaminated with small amounts of acid or alkali. **2**

(b) It can be shown that for solutions described in *(a)*

$$pH = pK_a + \log \frac{[\text{sodium ethanoate}]}{[\text{ethanoic acid}]}$$

where $pK_a = -\log K_a$.

(i) State the significance of the value of the pH when

[sodium ethanoate] = [ethanoic acid]. **1**

(ii) Calculate the mass of sodium ethanoate which must be dissolved in 1 dm^3 of ethanoic acid of concentration 0.10 mol dm^{-3} to produce a solution with a pH value of 5.5 at 298 K. It can be assumed that there is no volume change on dissolving the salt.

(At 298 K the value of K_a for ethanoic acid is 1.8×10^{-5} mol dm^{-3}.) **3**

(c) (i) Name an indicator which would be suitable for the titration of aqueous ethanoic acid with aqueous sodium hydroxide, giving a reason for your choice. **1**

(ii) State whether the pH of an aqueous solution of pure ammonium chloride would be greater or less than 7 at 298 K, giving a reason for your answer. **1**

(d) The pH of human blood plasma is maintained at a value between 7.39 and 7.41 by the buffering action of dissolved carbon dioxide, hydrogencarbonate ions and carbonic acid, H_2CO_3. Write equations to show what reactions may occur when the blood absorbs small amounts of acid or small amounts of alkali. **2**

10 *WJEC (AS/AL)*

19. *(a)* When the molar concentration of hydrogen ions and the molar concentration of hydroxide ions in a sample of water are multiplied together a value is obtained which is constant at a fixed temperature.

(i) Name this constant.

(ii) State the value of this constant at 298 K.

(iii) State qualitatively how the value of this constant changes when the temperature of the water is increased. Explain your answer. **5**

(b) A 4.00 dm^3 sample of an aqueous solution of a strong acid was neutralised exactly by the addition of 13.4 g of anhydrous sodium carbonate. The neutralisation reaction is given by the equation

$$2H^+ + CO_3^{2-} \longrightarrow H_2O + CO_2$$

(i) Calculate the number of moles of Na_2CO_3 in the 13.4 g sample.

(ii) Calculate the number of moles of H^+ ions present in 4.00 dm^3 of the acid solution. Use this result to calculate the hydrogen ion concentration in the solution and hence its pH.

(iii) Calculate the pH of the resulting solution formed if 13.4 g of NaOH, rather than 13.4 g of Na_2CO_3, had been added to 4.00 dm^3 of the original acid solution. **11**

16 *NEAB (AS/AL)*

20. *(a)* Water partially dissociates into ions.

(i) Write an equation for this dissociation.

(ii) Write an expression for the ionic product, K_w, of water.

(iii) At 303 K the ionic product of water, K_w, has a value of 1.47×10^{-14} mol^2 dm^{-6}. Use this value, together with your knowledge of the value of K_w at 298 K, to deduce the sign of the enthalpy change when water dissociates into ions. **5**

(b) (i) Calculate the pH of a 0.300 M solution of NaOH at 298 K.

(ii) Calculate the pH of the solution formed when 25.0 cm^3 of 0.300 M NaOH are added to 225 cm^3 of water at 298 K.

(iii) Calculate the pH of the solution formed when 25.0 cm^3 of 0.300 M NaOH are added to 75.0 cm^3 of 0.200 M HCl at 298 K. **9**

14 *NEAB (AS/AL)*

13 OXIDATION–REDUCTION EQUILIBRIA

13.1 ELECTROCHEMICAL TRANSPORT

There is intense interest in the aim of replacing the petrol-driven vehicle. The incentive comes from the damage done by the pollution produced by the internal combustion engines (§ 26.5) and the fact that the Earth's reserves of crude oil are limited.

Already thousands of fork-lift trucks and milk delivery vans are powered by lead–acid storage batteries. The range of these vehicles limits their use. For a milk delivery van with a short delivery round, a battery is better than a petrol engine because the electric motor can be switched on and off with no trouble whereas a petrol engine must be left to idle. Improvements have been made in recharging: new battery chargers can recharge the lead–acid battery in 3 hours. A small car weighing 1 tonne requires 20 kWh of energy to travel 55 km (35 miles). It can do this on 4 kg (1 gallon) of petrol. An electrically operated car would require 1.5 tonne of lead–acid batteries, occupying 0.5 m^3 of space. The prototype electric cars are two-seaters with a range of 65 km (40 miles) and maximum speed 100 km h^{-1}.

Research into battery technology continues. Characteristics of a battery are:

- storage capacity, the quantity of charge the battery can supply before recharging
- energy density, watt hours per kg (Wh kg^{-1})
- power density, watts per kg (W kg^{-1})

Characteristics of a number of cells are tabulated.

Table 13.1 Performance of some galvanic cells and fuel cells

Cell	Potential/ V	Energy density/ Wh kg^{-1}	Power density/ W kg^{-1}	Comments
Lead–acid	2.04	15	150	Massive
Nickel–cadmium	1.48	40	150	Costly
Silver–zinc	1.70	110		Very costly
Sodium–sulphur		200		Operates > 250 °C
Hydrogen–oxygen fuel cell	1.20	1000–2000	100	

Fuel cells

A fuel cell converts chemical energy into electrical energy in a galvanic cell with a continuous supply of reactants from outside the cell. The hydrogen–oxygen fuel cell is described in § 13.5.4. Any ionic redox reaction can in theory be used in a fuel cell. In practice the reaction must take place at a reasonable rate and the voltage developed must be adequate.

The main attraction of fuel cells is the high efficiency of energy conversion. The energy output of a fuel cell ΔG can be compared with the maximum amount of energy that can theoretically be obtained from the fuel, that is the enthalpy of combustion of the fuel ΔH. For the hydrogen–oxygen fuel cell the thermal efficiency of the cell is theoretically 95%. In practice, side effects reduce the efficiency to 80%.

Which source of power will the car of the future use? Storage batteries have good power densities ($W\ kg^{-1}$) but poor energy densities ($Wh\ kg^{-1}$). Fuel cells have poor power densities but good energy densities. A power source consisting of a combination of battery and fuel cell may be the answer. During steady cruising the fuel cell would supply power to the electric motor and also recharge the battery. On steep gradients and at times when rapid acceleration is needed, as in starting up and for overtaking, the battery could be used to provide bursts of power. In this chapter we shall look at the chemistry that made such developments possible.

13.2 ELECTRODE POTENTIALS

A piece of metal consists of metal cations and a cloud of delocalised valence electrons [§ 6.3]. If a strip of metal is placed in a solution of its ions, some of the cations in the metal may dissolve leaving a build-up of electrons on the metal

$$\mathbf{M}(s) \longrightarrow \mathbf{M}^{2+}(aq) + 2e^-$$

The metal will become negatively charged. Alternatively, metal ions may take electrons from the strip of metal and be discharged as metal atoms

$$\mathbf{M}^{2+}(aq) + 2e^- \longrightarrow \mathbf{M}(s)$$

The electrode potential of a metal may be positive or negative

In this case, the metal will become positively charged. The potential difference between the strip of metal and the solution depends on the nature of the metal and on the concentration of the ions involved in the equilibrium at the metal surface. Zinc acquires a more negative potential than copper, since it has a greater tendency to dissolve as ions and a smaller tendency to be deposited as metal.

The two metals, zinc and copper in solutions of their ions, may be combined as shown in Figure 13.2A to make an electrochemical cell. The solutions of zinc sulphate and copper(II) sulphate are separated by a porous partition. The metals form the electrodes of the cell, and are connected through a voltmeter. Since the electrode reactions are

$$Zn(s) \longrightarrow Zn^{2+}(aq) + 2e^-$$
$$Cu^{2+}(aq) + 2e^- \longrightarrow Cu(s)$$

Two metals and solutions of their ions combine to make a cell which can produce a current ...

zinc is negatively charged, and copper is positively charged, and electrons flow through the external circuit from zinc to copper. Zinc dissolves from the zinc electrode, and copper is deposited on the copper electrode. Ions flow through the porous partition. The overall cell reaction is

$$Zn(s) + Cu^{2+}(aq) \longrightarrow Zn^{2+}(aq) + Cu(s)$$

... and is called an electrochemical cell ...

... or sometimes simply a chemical cell

The type of cell in which a chemical reaction results in the production of an electric current is called an **electrochemical cell** (or a **galvanic** or **voltaic** cell). The electromotive force (emf) of the cell is a measure of the tendency of electrons to flow through the external circuit. Under **reversible conditions**, the emf is equal to the difference between the potentials of the two electrodes. To achieve reversible conditions, an electronic voltmeter is used to measure emf. This has a resistance so high that the current which it takes from the cell is negligible. When no current flows, the cell is operating reversibly.

Cells are always shown with the more negative electrode at the left-hand side. In this case, zinc is more negative than copper so the zinc electrode is shown on the left.

Figure 13.2A A zinc–copper electrochemical cell

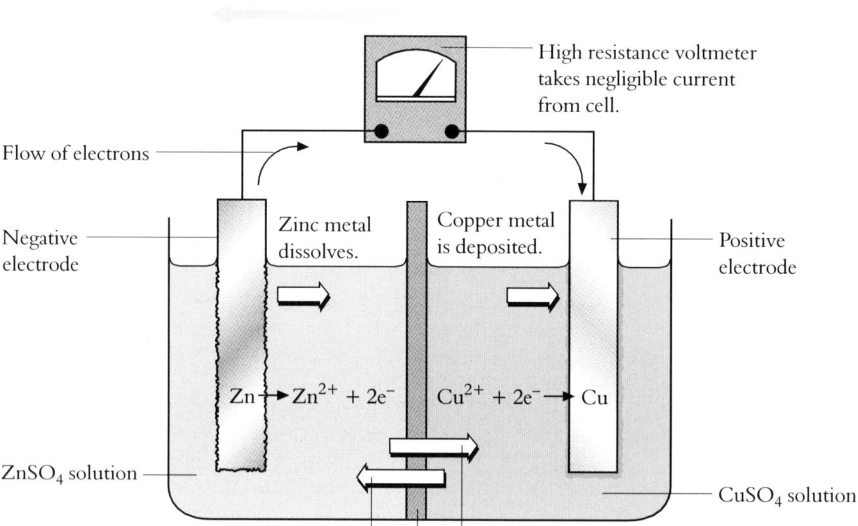

Cells are shown with the more negative electrode at the left-hand side

Convention for the sign of the emf of a cell ...

... $E_{cell} = E_{RHS\ electrode} - E_{LHS\ electrode}$

By convention, the emf, E, is taken to act from left to right through the cell. Thus

$$E_{\text{cell}} = E_{\text{RHS electrode}} - E_{\text{LHS electrode}}$$

Then a positive emf indicates that electrons are flowing from left to right through the external circuit [in Figure 13.2A, from Zn to Cu]. The flow of conventional electricity (positive electricity) is therefore from right to left through the external circuit [from Cu to Zn in Figure 13.2A].

The cell in Figure 13.2A can be represented as

$$\text{Zn(s)}\,|\,\text{ZnSO}_4\text{(aq)}\,\vdots\vdots\,\text{CuSO}_4\text{(aq)}\,|\,\text{Cu(s)}$$

Its emf is given by

$$E = E_{\text{Cu}} - E_{\text{Zn}}$$

13.2.1 STANDARD ELECTRODE POTENTIAL

Definition of the standard electrode potential, $E^{\ominus}$, of a metal

If the metal is immersed in a solution of its ions of concentration 1 mol dm^{-3} at 25 °C, then the potential acquired under these standard conditions is the **standard electrode potential** of that metal, $E^{\ominus}$. In the cell shown in Figure 13.2A, if each electrode is immersed in a 1 mol dm^{-3} solution of its ions at 25 °C, then its potential will be its standard electrode potential. The values are $E^{\ominus}_{Cu} = +0.34$ V, $E^{\ominus}_{Zn} = -0.76$ V. The voltmeter will then register an emf of 1.1 V. This emf is the difference, $E^{\ominus}_{Cu} - E^{\ominus}_{Zn}$.

13.2.2 FINDING ELECTRODE POTENTIALS

If a cell is constructed from a standard electrode (i.e., one of known potential) and an electrode of unknown potential, the emf of the cell can be measured and used to find the unknown electrode potential. For a cell with the standard electrode on the left-hand side

$$E_{\text{cell}} = E_{\text{electrode of unknown potential}} - E_{\text{standard electrode}}$$

The potential of an electrode is found by combining it with a standard electrode to make a cell, and finding the emf of the cell ...

The **standard hydrogen electrode** is the reference electrode with which other electrodes are compared. It consists of a platinised platinum electrode immersed in a solution of 1 mol dm^{-3} hydrogen ions. Hydrogen gas at a pressure of 1 atm is bubbled over the platinum electrode [Figure 13.2B]. On the surface of the platinum, equilibrium is established between hydrogen gas and hydrogen ions.

$$H_2(g) + 2H_2O(l) \rightleftharpoons 2H_3O^+(aq) + 2e^-$$

A potential develops on the surface of the platinum. It is assigned a value of zero volts.

The standard electrode potentials of other systems can be found by combining them with a standard hydrogen electrode and measuring the emf of the cell formed.

... $E^{\ominus}$ for another electrode can be found by combination with a standard hydrogen electrode ...

A cell with a standard hydrogen electrode is shown with the standard hydrogen electrode on the left-hand side.

A cell which combines a standard zinc electrode and a standard hydrogen electrode is shown in Figure 13.2B.

The two compartments in the figure are connected by a **salt bridge**. This contains an electrolyte such as potassium chloride, which conducts electricity but does not allow mixing of the two solutions in the half-cells. The emf of this cell is −0.76 V. The voltmeter shows that electrons flow through the external circuit from zinc to the hydrogen electrode, showing that zinc has a standard electrode potential of −0.76 V.

...but the hydrogen electrode is difficult to operate

The hydrogen electrode is not a convenient reference electrode to use in measurements: maintaining a stream of hydrogen at 1 atm takes careful management. It is usual to employ a secondary standard such as the saturated calomel electrode. This electrode contains mercury and solid mercury(I) chloride (calomel) in contact with a saturated solution of potassium chloride [see Figure 13.3A]. An equilibrium is set up between Hg^+ ions and Hg(l). The emf of a cell composed of a saturated calomel electrode and a standard hydrogen electrode is 0.244 V at 25 °C. Since

$$E_{\text{cell}} = E_{\text{saturated calomel electrode}} - E_{\text{standard hydrogen electrode}} = +0.244 \text{ V at } 25\ ^\circ\text{C}$$

Figure 13.2B
Apparatus for finding the standard electrode potential of the $Zn(s) \longrightarrow Zn^{2+} + 2e^-$ system

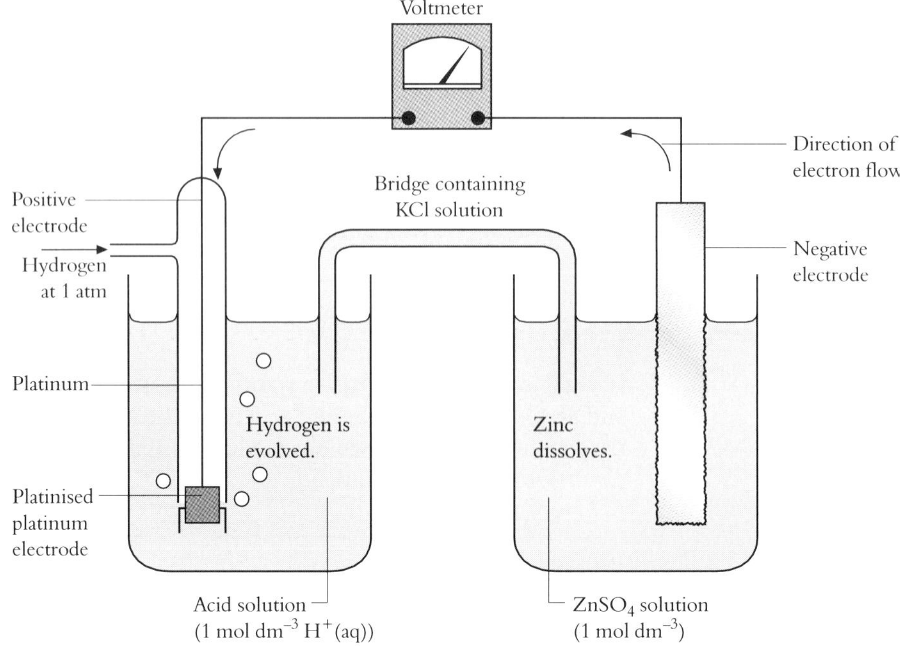

More convenient is the calomel electrode

and since $E_{\text{standard hydrogen electrode}}$ = zero, $E_{\text{saturated calomel electrode}}$ = +0.244 V at 25 °C. The saturated calomel electrode is easier to use than the standard hydrogen electrode. It can be combined with electrodes of unknown potential, and from the emf of the cell the unknown electrode potential can be found.

SUMMARY

To find an electrode potential, combine the electrode with a standard electrode and measure the emf of the cell.
Standard electrodes include
- **the standard hydrogen electrode**
- **the saturated calomel electrode.**

13.3 REDOX SYSTEMS

In the electrode process

$$Zn(s) \longrightarrow Zn^{2+}(aq) + 2e^-$$

Redox systems have electrode potentials

zinc is acting as a reducing agent, supplying electrons. Oxidation–reduction systems other than metals in equilibrium with their ions also have electrode potentials. If a piece of platinum wire is immersed in an acidic solution of manganate(VII) ions, the equilibrium

$$MnO_4^-(aq) + 8H^+(aq) + 5e^- \rightleftharpoons Mn^{2+}(aq) + 4H_2O(l)$$

Definition of standard electrode potential for a redox half-cell

takes electrons from the piece of platinum, which therefore becomes positively charged. If each of the species involved in the equilibrium has a concentration of 1 mol dm^{-3}, the potential acquired by the platinum is the **standard electrode potential** or **standard reduction potential** of the redox system.

The convention is to write each half-cell reaction as a reduction process:

$$\text{Oxidant} + ne^- \rightleftharpoons \text{Reductant}$$

The standard electrode potentials quoted in tables refer to the reaction written in this way, i.e., as a reduction reaction. They are therefore reduction potentials. Reactions which proceed from left to right more readily than the reduction of hydrogen ions

$$H^+(aq) + e^- \rightleftharpoons \tfrac{1}{2}H_2(g)$$

are given a positive standard electrode potential. Oxidising agents have high positive standard electrode potentials. Reducing agents, such as metals, have highly negative standard electrode potentials.

13.3.1 IONS OF THE SAME ELEMENT IN DIFFERENT OXIDATION STATES

An element can have a standard electrode potential for the equilibrium between different oxidation states. The standard electrode potential of the system Fe^{3+}/Fe^{2+} can be measured as in Figure 13.3A. In this cell,

$$E_{cell} = E_{Fe^{3+}/Fe^{2+}} - E_{calomel(sat)}$$

$$0.526\,V = E_{Fe^{3+}/Fe^{2+}} - 0.245\,V$$

$$E_{Fe^{3+}/Fe^{2+}} = 0.771\,V$$

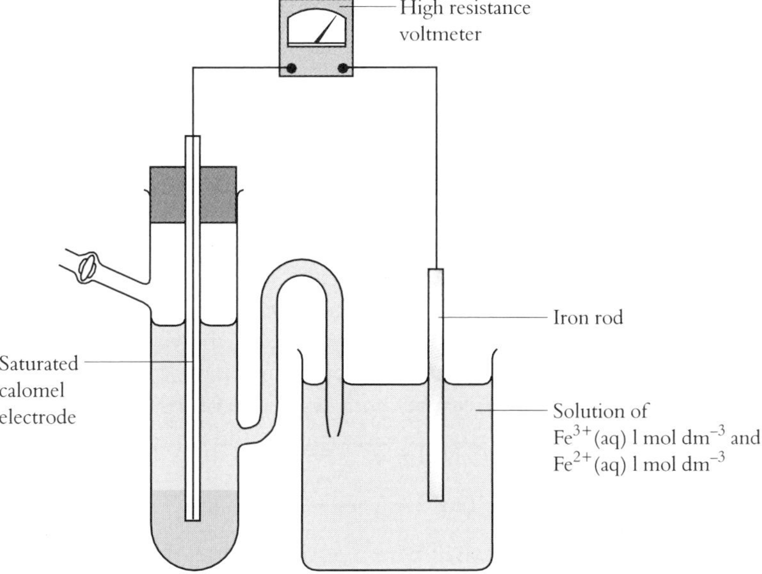

Figure 13.3A Apparatus for measuring the standard electrode potential for an element in different oxidation states

13.3.2 THE ELECTROCHEMICAL SERIES

When all the redox systems are arranged in order of their standard electrode potentials, the electrochemical series is obtained. Table 13.3 shows some common redox systems:

> The electrochemical series is the arrangement of the standard electrode potentials of redox systems in order of magnitude.

Table 13.3
Standard electrode potentials

Reaction	$E^{\ominus}$/V (at 298 K)
$Li^+(aq) + e^- \rightleftharpoons Li(s)$	−3.04
$K^+(aq) + e^- \rightleftharpoons K(s)$	−2.92
$Ca^{2+}(aq) + 2e^- \rightleftharpoons Ca(s)$	−2.87
$Na^+(aq) + e^- \rightleftharpoons Na(s)$	−2.71
$Mg^{2+}(aq) + 2e^- \rightleftharpoons Mg(s)$	−2.38
$Al^{3+}(aq) + 3e^- \rightleftharpoons Al(s)$	−1.66
$Zn^{2+}(aq) + 2e^- \rightleftharpoons Zn(s)$	−0.76
$Fe^{2+}(aq) + 2e^- \rightleftharpoons Fe(s)$	−0.44
$Cr^{3+}(aq) + e^- \rightleftharpoons Cr^{2+}(aq)$	−0.41
$Ni^{2+}(aq) + 2e^- \rightleftharpoons Ni(s)$	−0.25
$Sn^{2+}(aq) + 2e^- \rightleftharpoons Sn(s)$	−0.14
$Pb^{2+}(aq) + 2e^- \rightleftharpoons Pb(s)$	−0.13
$2H^+(aq) + 2e^- \rightleftharpoons H_2(g)$	0.00
$Sn^{4+}(aq) + 2e^- \rightleftharpoons Sn^{2+}(aq)$	0.15
$Cu^{2+}(aq) + 2e^- \rightleftharpoons Cu(s)$	0.34
$I_2(s) + 2e^- \rightleftharpoons 2I^-(aq)$	0.54
$Fe^{3+}(aq) + e^- \rightleftharpoons Fe^{2+}(aq)$	0.77
$Ag^+(aq) + e^- \rightleftharpoons Ag(s)$	0.80
$Br_2(l) + 2e^- \rightleftharpoons 2Br^-(aq)$	1.07
$MnO_2(s) + 4H^+(aq) + 2e^- \rightleftharpoons Mn^{2+}(aq) + 2H_2O(l)$	1.23
$Cr_2O_7^{2-}(aq) + 14H^+(aq) + 6e^- \rightleftharpoons 2Cr^{3+}(aq) + 7H_2O(l)$	1.33
$Cl_2(g) + 2e^- \rightleftharpoons 2Cl^-(aq)$	1.36
$Ce^{4+}(aq) + e^- \rightleftharpoons Ce^{3+}(aq)$ (in $H_2SO_4(aq)$)	1.45
$PbO_2(s) + 4H^+(aq) + 2e^- \rightleftharpoons Pb^{2+}(aq) + 2H_2O(l)$	1.47
$MnO_4^-(aq) + 8H^+(aq) + 5e^- \rightleftharpoons Mn^{2+}(aq) + 4H_2O(l)$	1.52
$Ce^{4+}(aq) + e^- \rightleftharpoons Ce^{3+}(aq)$ (in $HNO_3(aq)$)	1.61
$H_2O_2(aq) + 2H^+(aq) + 2e^- \rightleftharpoons 2H_2O(l)$	1.77
$F_2(g) + 2e^- \rightleftharpoons 2F^-(aq)$	2.87

13.3.3 MAKING PREDICTIONS

The criterion for a spontaneous cell reaction is that $E^{\ominus}_{cell}$ is positive

What reactions occur when two electrodes combine to form a cell? When two electrodes combine to form a cell, the value of $E^{\ominus}$ for the cell must be positive if the cell reaction is to happen spontaneously. For example, when copper and silver are in contact with solutions of their ions, two equilibria are set up:

$$Cu^{2+}(aq) + 2e^- \rightleftharpoons Cu(s);\ E^{\ominus} = +0.34\ V$$

$$Ag^+(aq) + e^- \rightleftharpoons Ag(s);\ E^{\ominus} = +0.80\ V$$

When two redox systems combine to form a cell …

… the system with the more positive value of $E^{\ominus}$ acts as the oxidant …

So that $E^{\ominus}$ shall have a positive value, the reactions that take place are

$$Ag^+(aq) + e^- \longrightarrow Ag(s);\ E^{\ominus} = +0.80\ V$$

$$Cu(s) \longrightarrow Cu^{2+}(aq) + 2e^-;\ E^{\ominus} = -0.34\ V$$

$$\text{Total: } Cu(s) + 2Ag^+(aq) \longrightarrow Cu^{2+}(aq) + 2Ag(s);\ E^{\ominus} = +0.46\ V$$

If solutions containing Ce^{4+}, Ce^{3+}, Fe^{3+} and Fe^{2+} are mixed, the redox equilibria in the solution are

$$Fe^{3+}(aq) + e^- \rightleftharpoons Fe^{2+}(aq);\ E^{\ominus} = +0.77\ V$$

$$Ce^{4+}(aq) + e^- \rightleftharpoons Ce^{3+}(aq);\ E^{\ominus} = +1.45\ V$$

The redox reaction that takes place is that for which $E^{\ominus}$ is positive, i.e.

$$Ce^{4+}(aq) + Fe^{2+}(aq) \longrightarrow Ce^{3+}(aq) + Fe^{3+}(aq);\ E^{\ominus} = +0.68\ V$$

A redox reaction will go almost to completion between two redox systems which differ by 0.3 V or more in their electrode potentials.

Reactions which are shown in this way to be feasible do not always occur. The reason is that kinetic factors may make the reaction so slow that no change can be observed; see §§ 10.17, 14.3.

... and the value of $E^{\ominus}$ for the cell is positive

Will a cell reaction happen?
Yes, if $E^{\ominus}$ for the cell is positive.
Will the cell reaction go to completion?
Yes, if $E^{\ominus} > 0.3$ V.
How fast will the reaction occur?
This depends on kinetic factors.

*13.3.4 STANDARD ELECTRODE POTENTIAL AND STANDARD FREE ENERGY CHANGE

$E^{\ominus}$ is related to $\Delta G^{\ominus}$ ...

... for a feasible reaction, $\Delta G^{\ominus}$ is negative ...

... and $E^{\ominus}$ is positive

There is a relationship between electrochemistry and thermochemistry. In the electrode reaction,

$$\mathbf{M}^{n+}(aq) + ne^- \rightleftharpoons \mathbf{M}(s)$$

if the standard free energy change is $\Delta G^{\ominus}$ then the value of the standard electrode potential, $E^{\ominus}$ at the same temperature is given by

$$\Delta G^{\ominus} = -nFE^{\ominus}$$

where n is the number of electrons transferred in the electrode reaction and F is the Faraday constant.

For example, $E^{\ominus} = +0.34$ V at 298 K for the electrode reaction

$$Cu^{2+}(aq) + 2e^- \rightleftharpoons Cu(s)$$

The value of $\Delta G^{\ominus}$ is given by

$$\Delta G^{\ominus} = -2 \times 96\,500 \times (-0.34)$$
$$= 65.6\ \text{kJ mol}^{-1}$$

13.3.5 THE pH METER

The potential of a glass electrode depends on the pH of the solution in which it is immersed

A pH meter is an instrument with a probe that can be immersed in a solution and a dial that displays the pH of the solution [see Figure 13.3B]. The probe contains a glass electrode and a saturated calomel electrode which acts as the reference electrode. The glass electrode has a platinum wire sealed into a thin-walled glass tube containing a buffer solution (see Figure 13.3C). The potential of the glass electrode depends on the pH of the solution in which it is placed. It is given, at 25 °C, by

$$E = \text{Constant} - 0.0592\ \text{pH}$$

Since

$$\text{emf of cell} = E_{\text{glass electrode}} - E_{\text{saturated calomel electrode}}$$

and since the potential of the saturated calomel electrode is constant, the emf depends on the pH of the solution. The dial of the pH meter gives a direct reading of pH.

Figure 13.3B
a pH meter

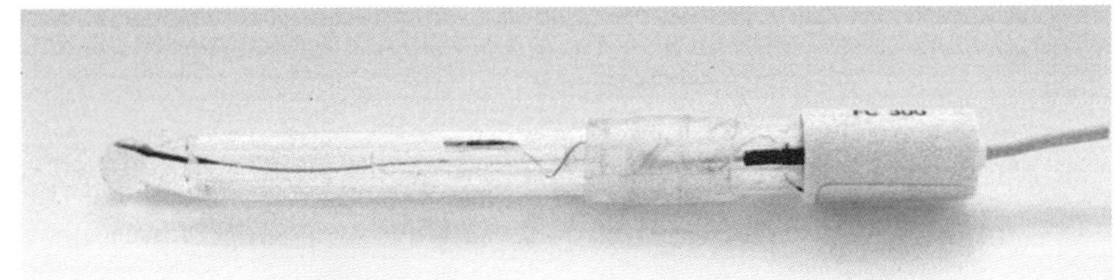

Figure 13.3C
A glass electrode (Hanna instruments)

*13.3.6 THE EFFECT OF CONCENTRATION ON ELECTRODE POTENTIAL

The value of the electrode potential E of a metal depends ...

The value of the electrode potential of a metal depends on the standard electrode potential of the metal and also on the concentration of the metal ions in solution. Consider a silver electrode in equilibrium with a solution of silver ions,

$$Ag^+(aq) + e^- \rightleftharpoons Ag(s)$$

... on the standard electrode potential E$^\ominus$ ...

... on the temperature ...

... and on the concentration of metal ions in solution

Le Chatelier's Principle predicts what will happen to the electrode potential if the concentration of silver ions is changed. In a more dilute solution, more silver atoms will ionise in an attempt to restore the equilibrium concentration of silver ions, more electrons will accumulate on the metal, and the electrode potential will become more negative (will have a larger negative value or a smaller positive value). On the other hand, if the concentration of metal ions is increased, some ions will take electrons from the electrode to form silver atoms, and the electrode potential will become less negative (have a smaller negative value or a larger positive value). Experiment shows that the relationship is:

$$E = E^\ominus + 0.0592 \text{ lg } [Ag^+(aq)]$$

The electrode potential increases with the concentration of metal ions

For the zinc electrode, the relationship is:

$$E = E^\ominus + (0.0592/2) \text{ lg } [Zn^{2+}(aq)]$$

In general,

$$E = E^\ominus + (0.0592/z) \text{ lg [ion]}$$ where z is the charge on the metal ions.

For the glass electrode, described above,

$$E = E^{\ominus} + (0.0592/2)\lg[H^+(aq)] = E^{\ominus} - 0.0592\,pH$$

In the dichromate redox system

$$Cr_2O_7^{2-}(aq) + 14H^+(aq) + 6e^- \rightleftharpoons 2Cr^{3+}(aq) + 7H_2O(l)$$

E for the dichromate redox system depends on the pH

Under standard conditions, all concentrations equal 1 mol dm^{-3}. Increasing the hydrogen ion concentration above 1 mol dm^{-3} drives the equilibrium towards the right-hand side and increases the electrode potential, making dichromate a more powerful oxidising agent. This is why potassium dichromate solution is acidified for use as an oxidising agent.

CHECKPOINT 13.3: ELECTRODE POTENTIALS

1. Does a high positive standard electrode potential for a redox system indicate that the system acts as an oxidising agent or a reducing agent?

2. Explain the terms *(a)* electrode potential, *(b)* standard electrode potential.

Explain how you could find the standard electrode potential for the system

$Fe^{3+}(aq)|Fe^{2+}(aq)|Pt$

3. This cell is set up:

$Ag(s)|Ag^+(aq, 1\ mol\,dm^{-3}) ⁞⁞ Cu^{2+}(aq, 1\ mol\,dm^{-3})|Cu(s)$

(a) State the emf of the cell. Refer to Table 13.3.
(b) Write the equation for the chemical reaction that takes place in the cell when the copper and silver electrodes are connected by an external circuit.
(c) State the direction in which electrons flow through the external circuit.

4. Which electrode, anode or cathode, is associated in an electrochemical cell with oxidation?

5. Refer to Table 13.3. Which of the following reactions will occur spontaneously? (Assume all concentrations are 1 mol dm^{-3}.)

(a) $Fe(s) + Zn^{2+}(aq) \longrightarrow Fe^{2+}(aq) + Zn(s)$
(b) $Fe(s) + Sn^{2+}(aq) \longrightarrow Fe^{2+}(aq) + Sn(s)$
(c) $Sn^{4+}(aq) + 2I^-(aq) \longrightarrow Sn^{2+}(aq) + I_2(s)$
(d) $Zn(s) + Mg^{2+}(aq) \longrightarrow Zn^{2+}(aq) + Mg(s)$
(e) $Zn(s) + Sn^{2+}(aq) \longrightarrow Zn^{2+}(aq) + Sn(s)$
(f) $Sn^{4+}(aq) + 2Fe^{2+}(aq) \longrightarrow Sn^{2+}(aq) + 2Fe^{3+}(aq)$
(g) $Cr_2O_7^{2-}(aq) + 14H^+(aq) + 6Cl^-(aq) \longrightarrow 2Cr^{3+}(aq) + 7H_2O(l) + 3Cl_2(g)$
(h) $2Ce^{4+}(aq) + 2Br^-(aq) \longrightarrow 2Ce^{3+}(aq) + Br_2(l)$

6. Write the equation for the manganate(VII)–manganese(II) equilibrium. Deduce which will be the stronger oxidising agent, acidified manganate(VII) or neutral manganate(VII). Explain your reasoning.

13.4 RUSTING AND STANDARD ELECTRODE POTENTIAL

Rusting is a serious problem. A large fraction ($\frac{1}{8}$) of the annual UK production of steel (20 million tonnes) is needed simply to replace iron lost through rusting. Rust is hydrated iron(III) oxide, $Fe_2O_3 \cdot xH_2O$. Both water and air are needed for rusting to occur. It is an electrochemical process, with different parts of an iron structure acting as cathodes and anodes. At an anodic region, the process which occurs is

Rusting is an electrochemical process

The mechanism of rusting

$$Fe(s) \longrightarrow Fe^{2+}(aq) + 2e^-$$

At a cathodic region, the process is

$$O_2(aq) + 2H_2O(l) + 4e^- \longrightarrow 4OH^-(aq)$$

The presence of dissolved acids and salts in water increases its conductivity and speeds up the process of rusting. If cathodic and anodic areas are close together, precipitation of iron(II) hydroxide, $Fe(OH)_2(s)$, occurs. Air oxidises this to rust, hydrated iron(III) oxide:

Prevention of rusting ...

$$2Fe(OH)_2(s) + \tfrac{1}{2}O_2(aq) + H_2O(l) \longrightarrow Fe_2O_3 \cdot xH_2O(s)$$

Cathodic protection

... by cathodic protection

If a block of a metal higher in the electrochemical series is connected to iron, then that metal acts as the anode, and it is corroded while iron remains intact. Zinc and magnesium are often used. Underground pipes are protected by attaching bags of magnesium scraps at intervals and replacing these from time to time when they have been corroded. The hulls of ships are protected by attaching blocks of zinc, which are sacrificed to protect the iron. This technique is called **sacrificial protection** [see Figure 13.4A].

Sacrificial protection ...

Figure 13.4A
Sacrificial protection of iron by zinc

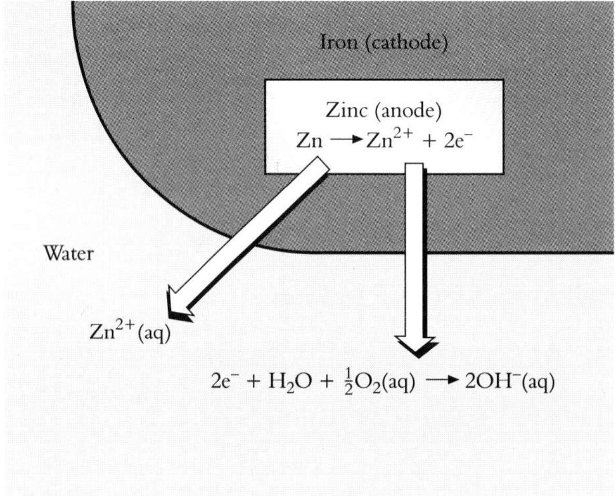

... by a more reactive metal ...

... or the application of a negative potential ...

... are the techniques used in cathodic protection

Another method of cathodic protection is to make iron the cathode by connecting it to the negative side of a battery while a conductor such as graphite is connected to the positive end. The negative potential on the iron structure inhibits the formation of $Fe^{2+}(aq)$ ions.

CHECKPOINT 13.4: RUSTING

1. The rusting of iron is an electrolytic process. Different regions of the iron act as cathodes and anodes.

(a) Write the equation for the reaction of iron at an anodic region.

(b) Write the equation for the reaction of oxygen at a cathodic region.

(c) Which regions on the surface of the iron are most likely to behave as anodes?

(d) Why does the presence of sodium chloride accelerate rusting?

(e) Which of the following metals could be used for sacrificial protection of iron: zinc, nickel, tin, magnesium, lead? Explain your answer.

13.5 ELECTROCHEMICAL CELLS

In electrochemical cells, often referred to as 'chemical cells', a chemical reaction produces an electric current.

13.5.1 DANIELL CELL

The Daniell cell utilises zinc and copper in solutions of their ions as shown in Figure 13.2A, § 13.2. The emf of the cell is 1.1 V.

13.5.2 DRY CELLS

Dry cells were invented to overcome the difficulty of electrolyte solution leaking out of cells such as the Daniell cell. In dry cells, the electrolyte is made into a paste. An example is shown in Figure 13.5A.

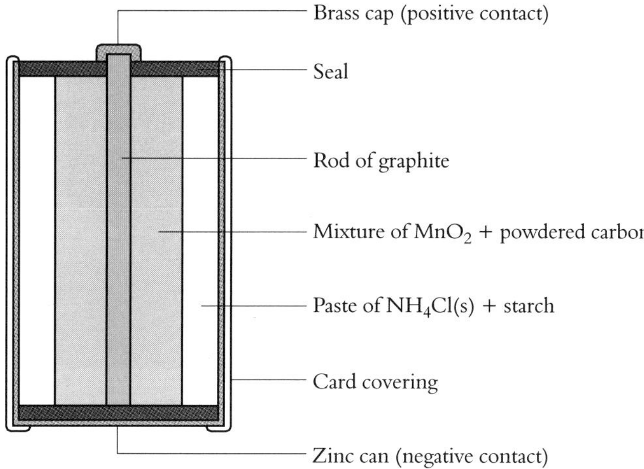

Figure 13.5A
A dry cell

Portable dry cells ...

This type of cell is used in radios, flashlights and clocks as it is portable. The emf of the cell shown in Figure 13.5A is 1.5 V. The initial electrode processes are

... batteries ...

Anode: $Zn(s) \longrightarrow Zn^{2+}(aq) + 2e^-$

Cathode: $2NH_4^+(aq) + 2e^- \longrightarrow 2NH_3(g) + 2H_2(g)$

13.5.3 THE LEAD–ACID BATTERY

The lead–acid battery is used in vehicles to power the starter motor. It consists of six 2 V cells connected in series. Each cell consists of two lead plates dipping into a 30% solution of sulphuric acid. One plate is coated with lead(IV) oxide, PbO_2. The lead plate and the lead(IV) oxide-coated plate have different electrode potentials so when they are connected a current flows between them. The electrode processes are:

Positive plate: $2PbO_2(s) + 4H^+(aq) + SO_4^{2-}(aq) + 2e^- \longrightarrow PbSO_4(s) + 2H_2O(l)$

Negative plate: $Pb(s) + SO_4^{2-}(aq) \longrightarrow PbSO_4(s) + 2e^-$

When PbO_2 and Pb have been converted into $PbSO_4$, there is no difference between the plates, and the cell can no longer give a current. The battery must be charged by passing a direct electric current through it. This reverses the electrode processes in the cells to produce lead and lead(IV) oxide. The battery can again give a current. When the vehicle is in motion, it drives a generator which charges the battery. If there is too much stopping and starting, the battery loses its charge and becomes flat until it is recharged.

During discharge sulphuric acid is formed; during charge it is used up. Since sulphuric acid is much denser than water, the state of the battery can be assessed by measuring the density of the battery acid with a hydrometer.

Figure 13.5B
Part of a lead–acid battery

Positive electrode connects the positive plates of many cells.

Positive and negative electrodes of cells are linked to give greater voltage.

Cell divider

Separator

Negative electrode connects the negative plates of many cells.

13.5.4 FUEL CELLS

Fuel cells are a promising source of energy for the future

A fuel cell is a cell which converts the chemical energy of a continuous supply of reactants into electrical energy. Fuel is supplied to one electrode and an oxidant, usually oxygen, to the other [see Figure 13.5C]. A great deal of research is being done on fuel cells as they are a promising source of energy for the future. The American Gemini space probes and Apollo moon probes used hydrogen–oxygen fuel cells. The astronauts used the product of the reaction to supplement their drinking water.

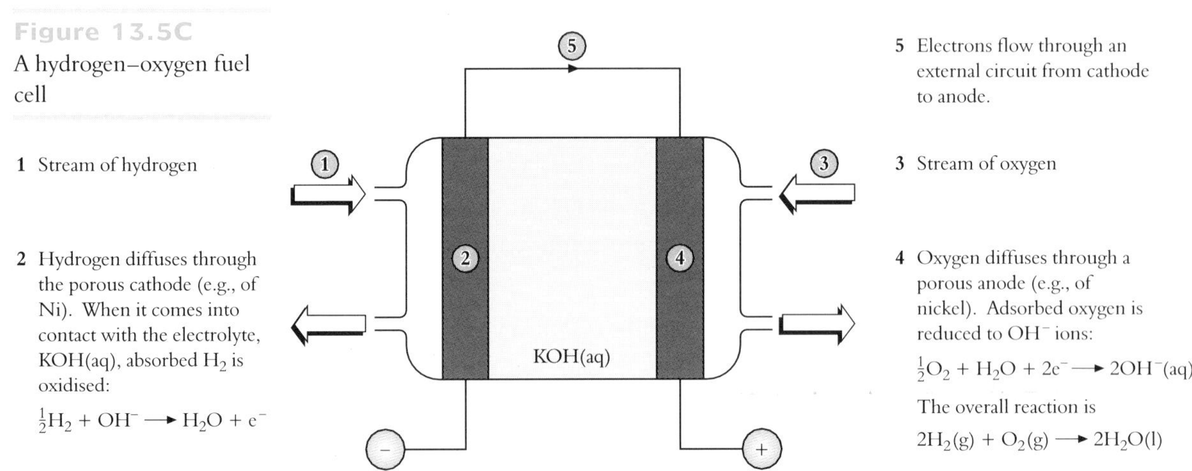

5 Electrons flow through an external circuit from cathode to anode.
1 Stream of hydrogen
3 Stream of oxygen
2 Hydrogen diffuses through the porous cathode (e.g., of Ni). When it comes into contact with the electrolyte, KOH(aq), absorbed H_2 is oxidised:
$\frac{1}{2}H_2 + OH^- \longrightarrow H_2O + e^-$
4 Oxygen diffuses through a porous anode (e.g., of nickel). Adsorbed oxygen is reduced to OH^- ions:
$\frac{1}{2}O_2 + H_2O + 2e^- \longrightarrow 2OH^-(aq)$
The overall reaction is
$2H_2(g) + O_2(g) \longrightarrow 2H_2O(l)$
KOH(aq)
−
+

Figure 13.5C
A hydrogen–oxygen fuel cell

REDOX EQUILIBRIA

For **a redox system**,
oxidant + $ne^- \rightleftharpoons$ reductant
An electrode immersed in the system acquires a potential that depends on the position of the equilibrium between the oxidant and reductant. § 13.2

The **standard electrode potential** $E^\ominus$ refers to 298 K and a 1 mol dm^{-3} solution of ions. § 13.2.1

An **electrode potential** is measured by finding the emf of a cell composed of the electrode and a standard electrode, e.g. the hydrogen electrode, which has $E^\ominus = 0$, or the calomel electrode. § 13.2.2

Listing oxidants and reductants in order of $E^\ominus$ values gives an **electrochemical series**. § 13.3.2

Corrosion
Metals high in the electrochemical series have a high negative value of $E^\ominus$ and corrode more rapidly than metals lower in the electrochemical series. § 13.4 § 24.17.6

Cells
Electrodes of different $E^\ominus$ values combine to form a voltaic cell. emf of cell = $E^\ominus$ RHS electrode – $E^\ominus$ of LHS electrode. Storage cells and batteries are discussed. § 13.5

Concept of **oxidation number**, e.g.
Ox. No. of I in $I_2 = 0$,
Ox. No. of Fe in $Fe^{3+} = +3$,
Ox. No. of S in $SO_4^{2-} = +6$,
Ox. No. of O in $O^{2-} = -2$.
§ 3.16.1

Oxidation = increase in Ox. No.
Reduction = decrease in Ox. No.
§ 3.16.2

In a **redox reaction**, sum of increases in Ox. No. of oxidised species = sum of decreases in Ox. No. of reduced species, e.g.

$2Fe^{3+} + 2I^- \rightarrow 2Fe^{3+} + I_2$

§ 3.16.3

Half-equations
For the oxidising agent manganate(VII),
$MnO_4^- + 8H^+ + 5e^- \rightarrow Mn^{2+} + 4H_2O$
For the oxidising agent iodine,
$I_2 + 2e^- \rightarrow 2I^-$
For the reducing agent thiosulphate,
$2S_2O_3^{2-} \rightarrow S_4O_6^{2-} + 2e^-$
Balance the number of electrons given and accepted, and add the half-equations to give the equation for the redox reaction, e.g.
$2S_2O_3^{2-} + I_2 \rightarrow S_4O_6^{2-} + 2I^-$
§ 3.15

Redox titrations
§ 3.15, § 3.18

$E^\ominus$ values can be used to predict the direction in which a redox reaction will happen. For a redox reaction between **A** and **B**
$\mathbf{A} \rightarrow \mathbf{A}^+ + e^-$; $E^\ominus_\mathbf{A}$
$\mathbf{B} + e^- \rightarrow \mathbf{B}^-$; $E^\ominus_\mathbf{B}$
If $E^\ominus_\mathbf{A} + E^\ominus_\mathbf{B}$ is positive, the reaction will proceed
$\mathbf{A} + \mathbf{B} \rightarrow \mathbf{A}^+ + \mathbf{B}^-$
If $E^\ominus_\mathbf{A} + E^\ominus_\mathbf{B}$ is negative, the reverse reaction will proceed
$\mathbf{A}^+ + \mathbf{B}^- \rightarrow \mathbf{A} + \mathbf{B}$
§ 13.3.2

Disproportionation
$Cu^{2+} + e^- \rightleftharpoons Cu^+$; $E^\ominus = +0.15$ V
$Cu^+ + e^- \rightleftharpoons Cu$; $E^\ominus = +0.52$ V
To give a positive $E^\ominus$, the reaction that takes place is disproportionation:
$2Cu^+ \rightleftharpoons Cu + Cu^{2+}$; $E^\ominus = +0.37$ V
§ 3.16.2

QUESTIONS ON CHAPTER 13

1. Refer to Table 13.3, § 13.3.2.

Which of the following species are oxidised by manganese(IV) oxide: Br^-, Ag, I^-, Cl^-?

2. Which of the following species are reduced by Sn^{2+}: I_2, Ni^{2+}, Cu^{2+}, Fe^{2+}?

3. The rusting of iron is prevented by *(a)* a coating of paint or *(b)* a layer of zinc or *(c)* a layer of tin. How do these different methods work? What is meant by 'sacrificial protection'? Give examples of its use.

4. Refer to Table 13.3. § 13.3.2. Relate the differences between the chemical properties of the elements mentioned below to the differences between the standard electrode potentials of the following systems:

(a) $K^+(aq) \mid K(s)$
(b) $Mg^{2+}(aq) \mid Mg(s)$
(c) $Ca^{2+}(aq) \mid Ca(s)$
(d) $Fe^{2+}(aq) \mid Fe(s)$
(e) $Cu^{2+}(aq) \mid Cu(s)$
(f) $Cl_2(g) \mid Cl^-(aq)$

5. Two half-cells are

(a) $Co^{2+}(aq, 1\ mol\ dm^{-3}) \mid Co(s)$
(b) $Cl_2(g, 1\ atm) \mid Cl^-(aq, 1\ mol\ dm^{-3}) \mid Pt$

State which will be the positive and which the negative electrode when the two half-cells are connected. Write the equation for the cell reaction. If the emf of the cell is 1.63 V, what is $E^\ominus$ for the cobalt half-cell? [See Table 13.3, § 13.3.2. for $E^\ominus(Cl_2/Cl^-)$.]

6. Calculate the standard emfs of the following cells at 298 K:

(a) $Ni(s) \mid Ni^{2+}(aq) \,\vdots\vdots\, Sn^{2+}(aq), Sn^{4+}(aq) \mid Pt(s)$
(b) $Pt(s) \mid I_2(s), I^-(aq) \,\vdots\vdots\, Ag^+(aq) \mid Ag(s)$
(c) $Pt(s) \mid Cl_2(g), Cl^-(aq) \,\vdots\vdots\, Br_2(l), Br^-(aq) \mid Pt(s)$
(d) $Sn(s) \mid Sn^{2+}(aq) \,\vdots\vdots\, Ag^+(aq) \mid Ag(s)$
(e) $Ag(s) \mid Ag^+(aq) \,\vdots\vdots\, Cu^{2+}(aq) \mid Cu(s)$
(f) $Fe(s) \mid Fe^{2+}(aq) \,\vdots\vdots\, Cu^{2+}(aq) \mid Cu(s)$
(g) $Zn(s) \mid Zn^{2+}(aq) \,\vdots\vdots\, Pb^{2+}(aq) \mid Pb(s)$

7. Refer to Table 13.3, § 13.3.2 and to the standard reduction potentials listed here:

$E^\ominus(VO_2^+(aq) \mid VO^{2+}(aq)) = +1.00\ V$
$E^\ominus(Cd^{2+}(aq) \mid Cd(s)) = -0.40\ V$
$E^\ominus(BrO_3^-(aq) \mid Br_2(g)) = +1.52\ V$
$E^\ominus(S_4O_6^{2-}(aq) \mid S_2O_3^{2-}(aq)) = 0.090\ V$

(a) State which of the species MnO_4^-, Ce^{4+}, $Cr_2O_7^{2-}$, VO_2^+, Fe^{3+} are able to liberate chlorine from an acidic solution of sodium chloride.
(b) Write a balanced equation for the reaction between MnO_4^- and VO^{2+} in acid solution.
(c) Find $E^\ominus$ for the cell

$Zn(s) \mid ZnSO_4(1\ mol\ dm^{-3}) \,\vdots\vdots\, CdSO_4(1\ mol\ dm^{-3}) \mid Cd(s)$

(d) Put the following into order of their power as oxidising agents in acid solution: $Cr_2O_7^{2-}$, Cl_2, MnO_4^-, I_2, BrO_3^-, $S_4O_6^{2-}$

8. *(a)* Define the term **oxidation**. 1

(b) Determine the oxidation number of the metal in each of the following species: 3
(i) MnO_4^{2-}, (ii) CuI, (iii) $Fe(CN)_6^{4-}$

(c) A sample of rusty iron wire was completely dissolved in 1.00 dm^3 aqueous sulfuric acid producing a solution containing both $FeSO_4$ and $Fe_2(SO_4)_3$.

A 25.0 cm^3 sample of this solution was titrated with aqueous potassium manganate(VII), $KMnO_4$, of concentration 0.0200 mol dm^{-3}. It required 27.2 cm^{-3} for complete reaction, according to the following equations:

$$MnO_4^- + 8H^+ + 5e^- = 4H_2O + Mn^{2+}$$
$$Fe^{2+} = Fe^{3+} + e^-$$

(i) Give the overall equation for this reaction. 2
(ii) Calculate the concentration of Fe^{2+}(aq) in the acidic solution. 3
(iii) Sufficient zinc metal was added to the remaining acidic solution so that all the iron(III) was reduced to iron(II). A 25.0 cm^3 sample of this solution now required 29.0 cm^3 of the same potassium manganate(VII) solution for titration.

Calculate the percentage of the iron which had rusted. 3

(d) A different wire had the composition 74% iron, 18% chromium and 8% nickel by mass. In this case. when subjected to a similar procedure the two titration values (without and with addition of zinc) were identical.
(i) Account for this observation. 2
(ii) Suggest a use for such a metal. 1

(e) An examination of standard redox potentials led one student to suggest that, in the experiment outlined in part (*c*),

'I would have expected the zinc to reduce iron(III) all the way to iron metal.'

Using the data below suggest why the reduction to iron did not happen. 2

			$E^\ominus$/V
$Zn^{2+} + 2e^-$	=	Zn	−0.76
$Fe^{2+} + 2e^-$	=	Fe	−0.44
$Fe^{3+} + 3e^-$	=	Fe	−0.04
$2H^+ + 2e^-$	=	H_2	0.00
$Fe^{3+} + e^-$	=	Fe^{2+}	+0.77

17 *O&C (AS/AL)*

9. A student measuring standard electrode potentials set up the following cell, using a standard hydrogen electrode

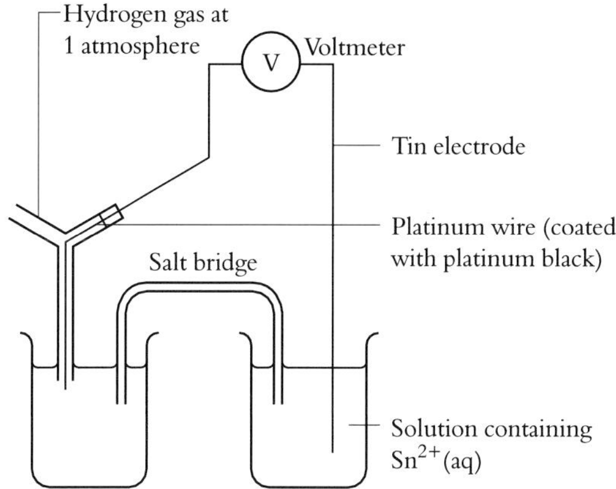

(a) Name a solution which could be used in the left hand beaker. What should its concentration be? **2**

(b) What solution is suitable for use in the salt bridge? **1**

(c) Write the cell diagram for the arrangement above. **2**

(d) The *Book of data* gives the following standard electrode potentials:

$Fe^{2+}(aq)|Fe(s)$ −0.44 V

$Sn^{2+}(aq)|Sn(s)$ −0.14 V

(i) Calculate the standard electrode potential of the following cell. Include a sign and unit in your answer.

$Sn(s)|Sn^{2+}(aq) \vdots\vdots Fe^{2+}(aq)|Fe(s)$ **2**

(ii) Which is the better reducing agent, iron or tin? Explain your answer in terms of the way electrons would flow if the cell delivered a current. **2**

(e) Cans which are used to hold acidic fruit juice are made from iron coated with tin. The tin surface inside the can is often coated with a layer of lacquer.

Use standard electrode potential data to explain what would happen if both the tin and lacquer coating are scratched, exposing the iron. **2**

11 *L(N) (AL)*

10. *(a)* The diagram below shows an electrochemical cell connected to a digital voltmeter. An electromotive force of 0.83 V was recorded at 298 K.

(i) Write half-equations for the reaction at the anode and at the cathode, and give their corresponding standard electrode potentials.

(ii) Write the overall equation for the electrochemical reaction.

(iii) Write the cell diagram for the electrochemical cell, using the IUPAC convention. **6**

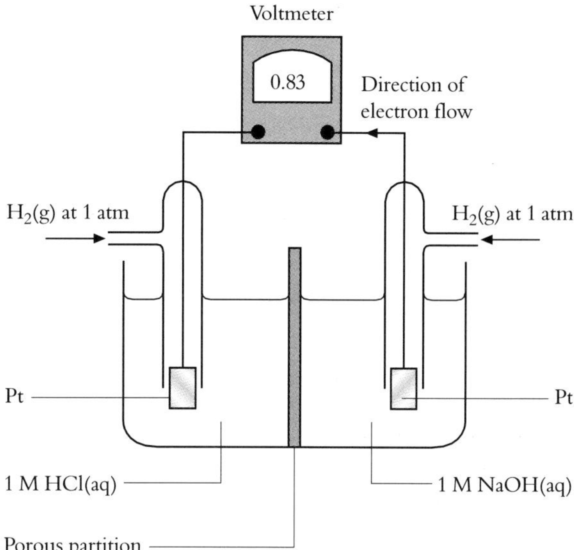

(b) Discuss and write relevant equations for the electrochemical processes involved when broken surfaces of the following are exposed to moist air.
(i) galvanized iron (ii) tin-plated iron **4**

(c) A solution is formed by mixing equal volumes of 0.20 M $CH_3CO_2H(aq)$ and 0.20 M $CH_3CO_2Na(aq)$.

(i) Identify *all* Brönsted acids and *all* Brönsted bases in the solution.

(ii) Calculate the concentration of *each* chemical species, excluding H_2O, present in the solution at 298 K.

(K_a of CH_3CO_2H $= 1.76 \times 10^{-5}$ mol dm^{-3} at 298 K) **7**

HKALE (AL)

11. This question concerns redox behaviour, and the following data will be found useful.

	$E^{\ominus}$/V
$Fe^{3+}(aq) + e^- \rightleftharpoons Fe^{2+}(aq)$	+0.77
$\frac{1}{2}Cl_2(g) + e^- \rightleftharpoons Cl^-(aq)$	+1.36
$MnO_4^-(aq) + 8H^+(aq) + 5e^- \rightleftharpoons Mn^{2+}(aq) + 4H_2O(l)$	+1.52

(a) In potassium manganate(VII) titrations, the solutions are acidified with dilute sulphuric acid.

(i) Using the data above, explain why dilute hydrochloric acid is not used for this purpose. **3**

(ii) Why is potassium manganate(VII) usually placed in the burette, despite the difficulties it presents in reading the burette? **2**

(b) A test for Mn^{2+} ions in solution is to react them with sodium bismuthate(V), $NaBiO_3$, in the presence of nitric acid. A purple colour will develop owing to the formation of MnO_4^- ions in the solution. The ionic half equation for the reduction of BiO_3^- ion is:

$$BiO_3^- + 6H^+ + 2e^- \rightleftharpoons Bi^{3+} + 3H_2O$$

(i) Use this half equation and the data above to write an ionic equation for the oxidation of the Mn^{2+} ion. **2**

(ii) Suggest, qualitatively, how the $E^{\ominus}$ value for $BiO_3^- \mid Bi^{3+}$ compares with that for $MnO_4^- \mid Mn^{2+}$. **1**

(c) An alloy contains iron and manganese only. On warming with dilute nitric acid 2.30 g of this alloy gave a solution containing iron(III) ions and manganese(II) ions. Treatment of this solution with excess sodium bismuthate(V) completely oxidised all the Mn^{2+} ions present to MnO_4^- ions.

The excess bismuthate(V) ions were then completely destroyed and the solution made up to 250 cm^3 with distilled water and thoroughly shaken.

Titration of 25.0 cm^3 portions of this solution required 25.0 cm^3 of standard 0.100 mol dm^{-3} iron(II) sulphate solution.

(i) Write the equation for the reaction occurring during the titration. **2**

(ii) Calculate the percentage of manganese present in the alloy. **5**

15 *L (AL)*

12. The data given below are taken from the electrochemical series.

Reaction at 298 K	$E^{\ominus}$/V
$MnO_4^{2-}(aq) + 4H^+(aq) + 2e^- \longrightarrow MnO_2(s) + 2H_2O(l)$	+1.55
$MnO_4^-(aq) + 8H^+(aq) + 5e^- \longrightarrow Mn^{2+}(aq) + 4H_2O(l)$	+1.51
$MnO_4^-(aq) + e^- \longrightarrow MnO_4^{2-}(aq)$	+0.60
$Cu^+(aq) + e^- \longrightarrow Cu(s)$	+0.52
$Cu^{2+}(aq) + e^- \longrightarrow Cu^+(aq)$	+0.15

Disproportionation is the term used for a reaction in which an element changes from a single oxidation state to two different oxidation states, one being higher and the other lower than the original. This can be illustrated by the reaction

$$Cl_2(aq) + H_2O(l) \longrightarrow 2H^+(aq) + Cl^-(aq) + OCl^-(aq)$$

in which Cl(0) becomes Cl(−1) and Cl(1).

Use the information given above, wherever relevant, to answer the questions that follow.

(a) What is meant by the term *electrochemical series*? **1**

(b) On warming, OCl^- ions change into Cl^- and ClO_3^-. Write an equation for this disproportionation reaction and determine the oxidation state of chlorine in the ClO_3^- ion. **2**

(c) How does the overall redox potential show that a reaction is spontaneous? **1**

(d) The copper(I) ion is unstable in aqueous solution and undergoes spontaneous disproportionation to form $Cu^{2+}(aq)$ ions and a solid precipitate. Identify the solid precipitate, construct an overall equation for this disproportionation and use values from the electrochemical series to calculate the overall $E^{\ominus}$ value for the reaction. **4**

(e) The manganate(VI) ion, MnO_4^{2-}, is also unstable and undergoes spontaneous disproportionation to form manganate(VII) ions and solid manganese(IV) oxide. Construct an overall equation for this disproportionation and use values from the electrochemical series to calculate the overall $E^{\ominus}$ value for the reaction. **4**

12 *NEAB (AL)*

13. *(a)* Standard electrode potentials, $E^{\ominus}$, are measured relative to a standard reference electrode. What is the standard reference electrode and what is its potential? **2**

(b) State **three** conditions which must apply when values of $E^{\ominus}$ are being determined. **3**

(c) What is the function of a *salt bridge*, and what might it contain? **2**

(d) What is meant by the *electrochemical series*? **2**

(e) Consider the following standard electrode potentials.

$Fe^{2+}(aq) + 2e^- \longrightarrow Fe(s)$ $E^{\ominus}/V = -0.44$

$Zn^{2+}(aq) + 2e^- \longrightarrow Zn(s)$ $E^{\ominus}/V = -0.76$

State which species is reduced if these two half-cells are joined together in an electrochemical cell. Explain your answer. **1**

12 *NEAB*

14 REACTION KINETICS

14.1 THE SPEEDS OF CHEMICAL REACTIONS

Chemical reactions occur within a wide range of speeds. Some reactions, for example the explosion shown in Figure 14.1A, seem to happen as soon as the reactants come into contact – in a fraction of a second. Other reactions, such as rusting, take a moderate length of time, up to several months. And still others take much longer, for example the ageing process which turns a human baby into an adult over a period of decades. At the extreme end of the time scale are the reactions which made coal out of dead plants over hundreds of millions of years.

Why do people want to know how fast a chemical reaction will occur? How fast does a medicine act? How fast does blood clot? The answers to these questions could mean the difference between life and death. How long does it take for cement to harden, for a fabric to take up a dye, for a monomer to polymerise to form a plastic? The answers to these questions could mean the difference between profit and loss, between the success and failure of an industry. The same variables, such as concentration, temperature and pressure, affect the rates of these very different processes. Most of the variables can be adjusted to speed up a reaction to maximise the yield of product within a certain time or to slow down an unwanted reaction.

Figure 14.1A
An explosion

Figure 14.1B
Rusting, a slow process

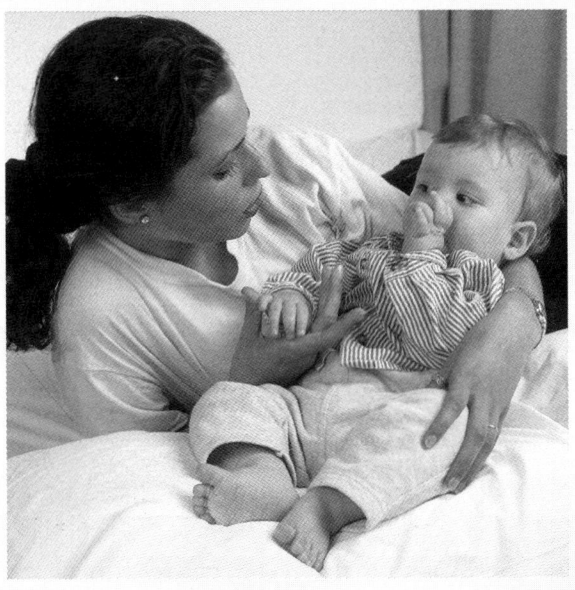

Figure 14.1C
Ageing is a slow process

The study of the factors which affect the speeds of chemical reactions is called **reaction kinetics**. In this topic we shall begin by examining the factors that affect reaction rates and go on from this to study how the rate of a reaction can be expressed quantitatively in the form of a **rate law**. We shall examine the stages which a reaction may pass through and model the species that exists for a fraction of a second at the instant when the chemical bonds in the reactants are breaking and the new bonds in the products are forming.

CHECKPOINT 14.1: SOME REVISION QUESTIONS

- Why do flour mills consider there is a danger that finely divided flour can explode, yet you probably do not think of flour as a dangerous substance?
- The fireworks called 'sparklers' contain iron dust glued to an iron rod. Why do they burn rapidly in air whereas a solid iron rod can be heated in a flame without catching fire?
- Why does a soluble aspirin work faster than a tablet that has to be swallowed whole?
- Why do medical doctors use a higher concentration of penicillin to treat meningitis than to treat tonsillitis?
- Why do surgeons lower a patient's body temperature during heart surgery? The answer can mean the difference between life and death to a patient having open-heart surgery.
- Why do we raise the temperature of foods to cook them – accelerating reactions which lead to the breakdown of cell walls and decomposition of proteins?
- Fritz Haber wanted to make the very unreactive gas nitrogen combine with hydrogen to form ammonia – which could then be used to make fertilisers. How did he do it?

Before you reach the end of this chapter, you will be able to answer these questions.

14.2 FACTORS WHICH AFFECT THE SPEEDS OF CHEMICAL REACTIONS

A number of factors can be changed to alter the speed of a chemical reaction. These are:

1. the size of the particles of a solid reactant
2. the concentrations of reactants in solution
3. the pressures of gaseous reactants
4. the temperature

5. the presence of light
6. the addition of a catalyst.

In § 14.2, we will take a look at these factors in a qualitative manner before going on to a fuller, quantitative treatment in later sections of this chapter.

14.2.1 PARTICLE SIZE

The reaction between calcium carbonate and dilute hydrochloric acid is used to prepare carbon dioxide

$$CaCO_3(s) + 2HCl(aq) \longrightarrow CO_2(g) + CaCl_2(aq) + H_2O(l)$$

... the size of particles of a solid reactant ...

The calcium carbonate used is in the form of marble chips. The reaction can be used to find out whether large lumps of a solid react at the same speed as small lumps of the same solid. Figure 14.2A shows an apparatus which could be used.

The mass of the flask and contents is noted at various times after the start of the reaction. When the mass is plotted against time, results such as those in Figure 14.2B are obtained. The results show that the smaller the size of the particles of calcium carbonate, the faster the reaction takes place [see Question 3, Checkpoint 14.2].

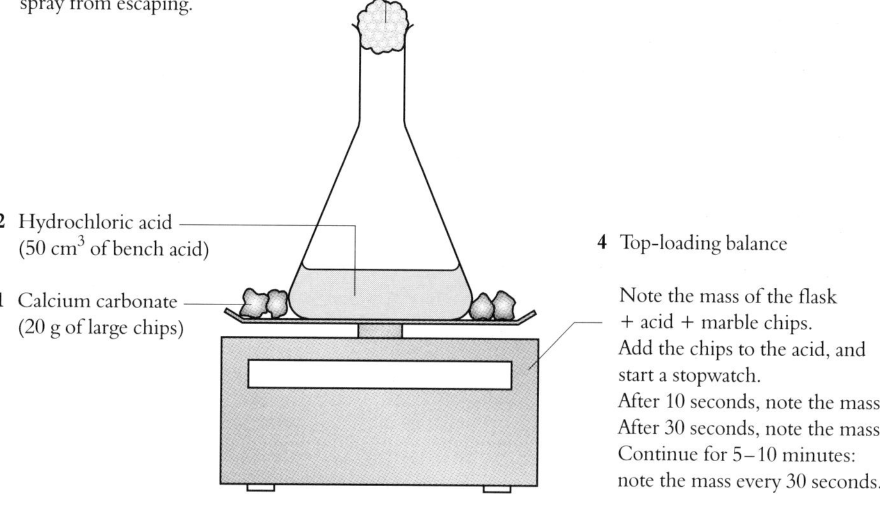
3 Cotton wool prevents spray from escaping.
2 Hydrochloric acid (50 cm³ of bench acid)
1 Calcium carbonate (20 g of large chips)
4 Top-loading balance
Note the mass of the flask + acid + marble chips.
Add the chips to the acid, and start a stopwatch.
After 10 seconds, note the mass.
After 30 seconds, note the mass.
Continue for 5–10 minutes: note the mass every 30 seconds.

Figure 14.2A An apparatus for following the loss in mass during a reaction

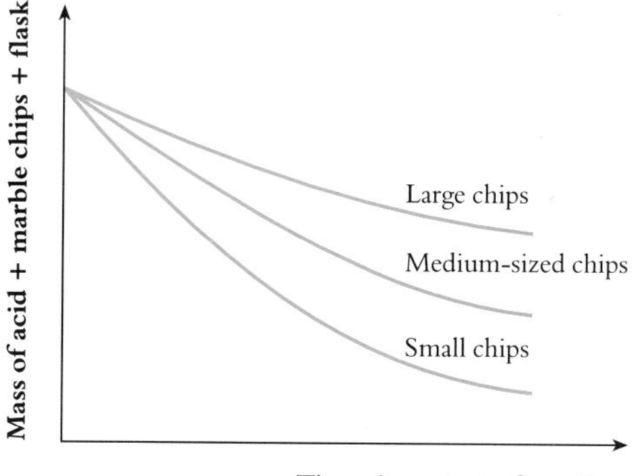
Mass of acid + marble chips + flask
Large chips
Medium-sized chips
Small chips
Time since start of reaction

Figure 14.2B Results obtained with different sizes of marble chips

14.2.2 CONCENTRATION

... the concentration of a reactant in solution ...

Many chemical reactions take place in solution. The concentration of a reactant may affect the speed of the reaction. In the reaction between sodium thiosulphate and acid,

$$\underset{\text{Sodium thiosulphate}}{Na_2S_2O_3(aq)} + 2HCl(aq) \longrightarrow \underset{\text{Sulphur}}{S(s)} + SO_2(g) + 2NaCl(aq) + H_2O(l)$$

Sulphur appears as very small particles of solid suspended in the solution. A method of studying the speed at which sulphur is formed is shown in Figure 14.2C. Graphs of readings of time and 1/time against the concentration of thiosulphate are shown in Figures 14.2D and 14.2E.

The faster the reaction takes place, the shorter the time before the deposit of sulphur is dense enough to hide the cross from view. That is, the speed of the reaction is inversely proportional to the time taken for the reaction to finish.

Speed of reaction $\propto$ 1/Time

Figure 14.2C Measuring the time needed for reaction

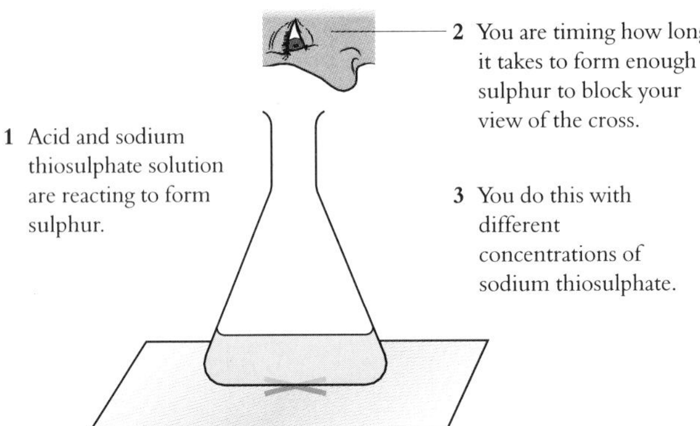

1 Acid and sodium thiosulphate solution are reacting to form sulphur.
2 You are timing how long it takes to form enough sulphur to block your view of the cross.
3 You do this with different concentrations of sodium thiosulphate.

You can see from the graph in Figure 14.2E that

Speed of reaction $\propto$ Concentration of sodium thiosulphate

If you repeat the experiment, keeping the concentration of thiosulphate constant and varying the concentration of acid, you will find that

Speed of reaction $\propto$ Concentration of acid

The reason why the speed increases with concentration is that the ions are closer together in a concentrated solution. The closer together they are, the more frequently do the ions collide. The more often they collide, the greater is their chance of reacting.

Figure 14.2D A graph of time against concentration

The results show that the cross disappears soonest when the solution is most concentrated.

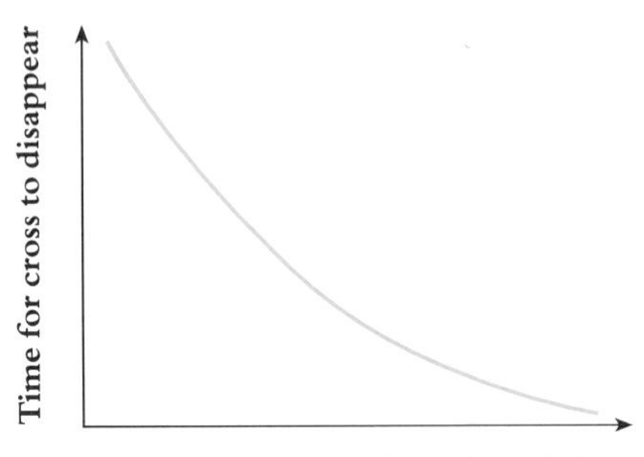

Figure 14.2E
A graph of 1/time against concentration

When 1/(time for cross to disappear) is plotted against concentration, a straight line is obtained. This shows that for this reaction

1/Time ∝ Concentration

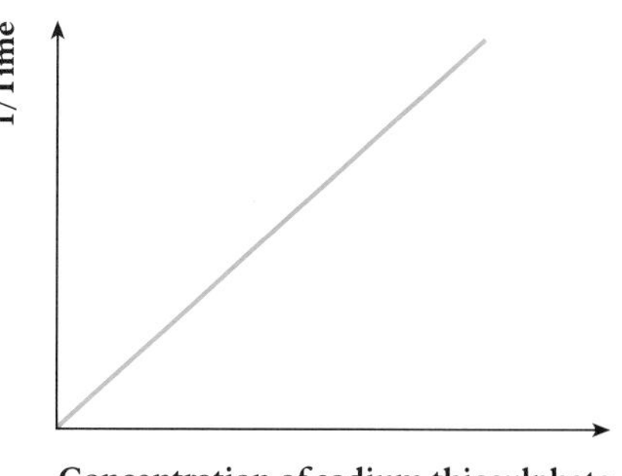

The pressures of gaseous reactants

... the pressure of a gaseous reactant ...

When gases react, molecules have first to collide before they can react. If the pressure is increased, molecules of gas are pushed closer together. As a result, they collide more frequently and react more rapidly.

14.2.3 TEMPERATURE

... the temperature ...

To study the effect of altering the temperature on the rate of a chemical reaction, one reaction that could be chosen is the thiosulphate–acid reaction [see Figure 14.2C]. Figure 14.2F shows a typical plot of values of the time taken for the cross to be obscured by sulphur against the temperature. A plot of 1/time against temperature shows a steep increase in the speed of the reaction as the temperature rises [see Figure 14.2G]. This reaction goes approximately twice as fast at 30 °C as it does at 20 °C.

At the higher temperature, the ions or molecules of solute have more energy. They move with a higher velocity and collide more frequently and with more force. Once the ions have collided, there is a chance that they will react. The increased collision frequency results in a higher rate of reaction.

14.2.4 LIGHT

... the presence of light ...

Light is another form of energy that will speed up chemical reactions. One example is photosynthesis, the reaction by which green plants synthesise sugars. The reaction takes place in sunlight and in the presence of chlorophyll, the green pigment in plants.

$$6CO_2(g) + 6H_2O(l) \longrightarrow \underset{\text{Sucrose}}{C_6H_{12}O_6(aq)} + 6O_2(g)$$

Another example is the formation of silver from silver salts that takes place when a photographic film is exposed to light.

Figure 14.2F
A graph of time for cross to disappear against temperature

Figure 14.2G
A graph of 1/time against temperature for the reaction

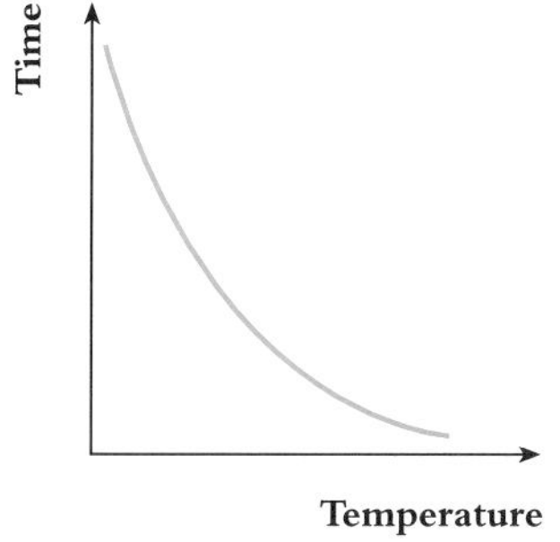

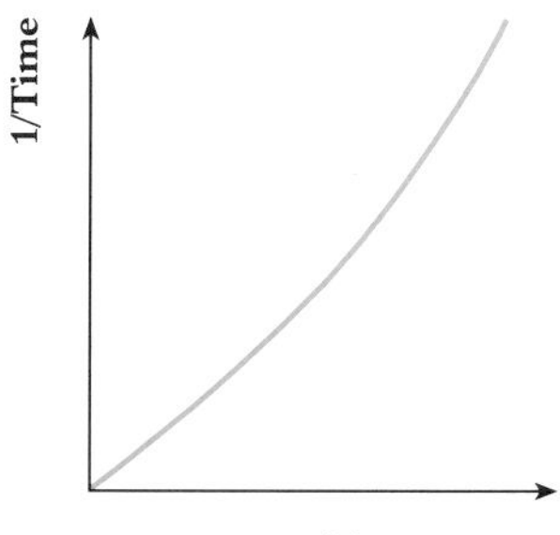

14.2.5 CATALYSTS

... and the presence of a catalyst

Photosynthesis takes place only in green plants. This is because chlorophyll, the pigment in green plants, must be present. Chlorophyll is a catalyst for this reaction. A **catalyst** is a substance which increases the speed of a reaction without being used up in the reaction.

Catalysts are important in industry. A manufacturer tries to find a catalyst which will enable a reaction to give a good yield at a low temperature. Then fuel bills will be lower and profits will be higher. When plastics are manufactured, often the monomer is polymerised under high pressure. If a catalyst can be found to enable the reaction to give a good yield at low pressure, then the industrial plant will not have to withstand high pressure; it can be constructed of less robust materials at lower cost. Table 14.2 lists some important industrial catalysts.

All the processes which take place in plants and animals need catalysts. The catalysts which occur in living things are called **enzymes**. They catalyse a specific reaction, such as the digestion of a certain protein.

Table 14.2
Some important industrial catalysts

Catalyst	Reaction
Platinum	Oxidation of ammonia to give nitrogen monoxide, a step in the manufacture of nitric acid [see § 22.7]
Vanadium(V) oxide	Oxidation of sulphur dioxide to sulphur trioxide, a step in the manufacture of sulphuric acid [see § 21.9]
Nickel	Hydrogenation of unsaturated compounds to form saturated compounds; used in the manufacture of margarine [see § 17.3]
Iron	The combination of nitrogen and hydrogen to form ammonia in the Haber process [see § 22.6]

14.2.6 THE IMPORTANCE OF REACTION RATES

The rates of chemical reactions vary greatly

Chemical reactions vary enormously in speed. Some proceed slowly over a period of months (e.g., the rusting of iron); others take weeks to reach completion (e.g., the fermentation of ethanol). Some reactions are fast (e.g., the precipitation of insoluble salts); others are so fast as to be explosive (e.g., the reaction between hydrogen and oxygen).

Many factors affect the rate of a reaction ...

Industrial chemists are interested in knowing how fast a reaction takes place as the speed is a factor in deciding whether a manufacturing process can be carried out profitably. Many factors influence the rate of a chemical reaction, and it is important to discover the conditions under which a reaction will proceed most economically. The rate at which the product is formed is only one factor. The cost of the energy consumed if a high temperature is needed must be computed. If the process requires high pressure, the cost of a plant which is robust enough to withstand the conditions will be high.

CHECKPOINT 14.2: REACTION RATES

1. There is a danger in coal mines that coal dust may catch fire. Explain why coal dust is more dangerous than coal.

2. 'Relief' is a remedy for acid indigestion. Which will work faster to relieve pain, Relief indigestion tablets or Relief indigestion powder?

3. In the reaction between marble chips and acid, assume that the particles are cubes and the length of the side is *a*.

   (a) What is (i) the surface area of a cube of side *a*, (ii) the volume of a cube of side *a*, (iii) the ratio: Surface area/Volume?

   (b) As *a* decreases, does the ratio Surface area/Volume increase or decrease? How does your answer explain the change in the rate of reaction when large chips are replaced with small chips?

4. Some manganese(IV) oxide (a catalyst) was added to a solution of hydrogen peroxide in the apparatus shown in the figure. Oxygen was liberated in the reaction

$$\underset{\text{Hydrogen peroxide}}{2H_2O_2(aq)} \longrightarrow \underset{\text{Oxygen}}{O_2(g)} + \underset{\text{Water}}{2H_2O(l)}$$

The table shows readings of the volume of oxygen (measured at rtp) collected at various times after the start of the reaction.

   (a) Plot a graph of the volume of oxygen against the time after the start of the experiment.

   (b) What amount of oxygen (in mol) was formed in the reaction?

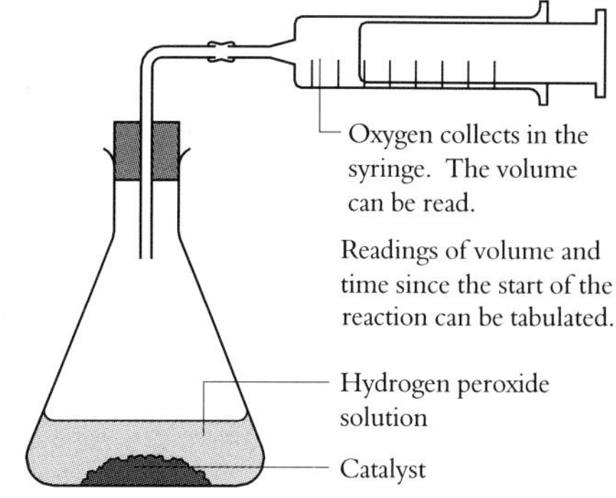
Oxygen collects in the syringe. The volume can be read.
Readings of volume and time since the start of the reaction can be tabulated.
Hydrogen peroxide solution
Catalyst

Volume of oxygen/cm^3	Time/min
37	1
67	2
87	3
107	4
114	5
120	6
120	7
120	8

   (c) What mass of hydrogen peroxide was present?

   (d) Find the time taken for the decomposition of half the mass of hydrogen peroxide.

   (e) On your graph, sketch the curve that you would predict if the reaction were carried out at 30 °C.

   (f) Suggest another way of altering the speed of the reaction, and sketch the curve that would result from the change that you suggest.

14.3 THE COLLISION THEORY

The **collision theory** was developed to describe reactions in the gas phase. You will know from experience that most reactions go faster when the temperature is raised. The sort of reactions you have been studying in § 14.2 take place about twice as fast at 30 °C than at 20 °C. As the temperature rises the average energy of the molecules increases and the rate of reaction increases. It seems logical to deduce that the two observations are connected: molecules need energy in order to react. The collision theory proposes that molecules must first collide in order to react. If they collide at low speed – and low energy – they bounce apart again without reacting. If they collide at higher speed, with more than a certain energy, then bonds can be broken and new bonds made. The minimum energy which colliding molecules need in order to react is called the **activation energy** of the reaction, E_A.

Figure 14.3A shows activation energy as a barrier between the reactants and the products. A graph of this kind, showing the progress of the reaction is called a

reaction profile. The activation energy E_A is the difference between the potential energy of the reactants and the potential energy at the height of the barrier. As the reactant molecules approach, their potential energy (energy due to position) increases. It reaches a maximum as the molecules collide and bonds are distorted. It decreases as the atoms rearrange to form molecules of products and the product molecules separate. If the reactant molecules have energy less than E_A, they cannot pass over the barrier; they roll down the left-hand side again and separate. If they have energy E_A or greater, they pass over the top of the barrier and roll down the right-hand side to form the products.

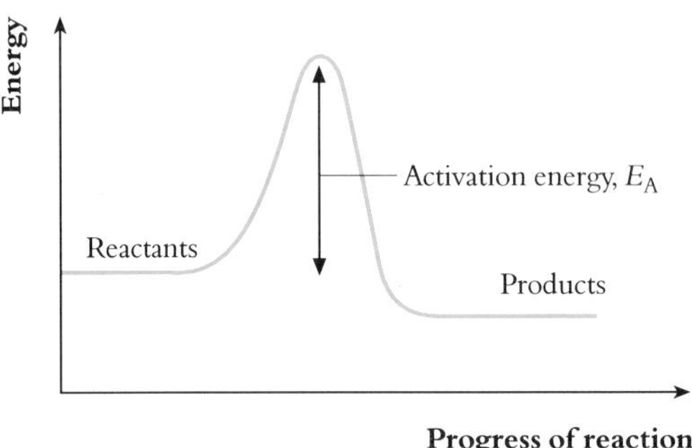

Figure 14.3A A reaction profile for an exothermic reaction, showing the activation energy E_A

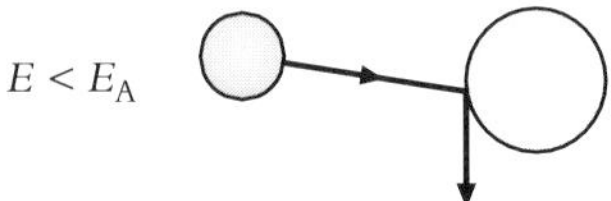

(a) The colliding molecules have kinetic energy less than E_A. They bounce apart without reacting.

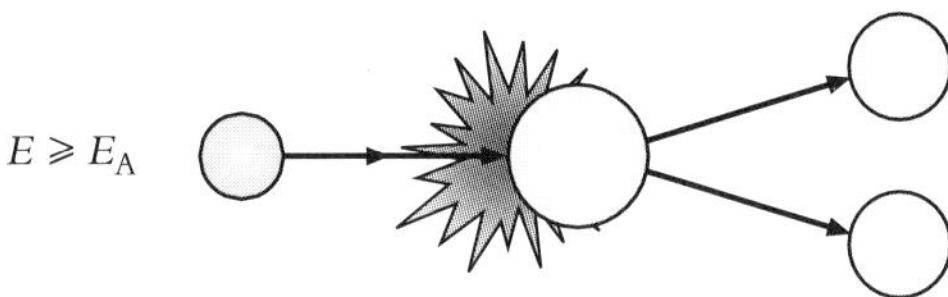

(b) The colliding molecules have kinetic energy equal to (or greater than) E_A. They react to form the products.

Figure 14.3B The energy of colliding molecules

The effect of a rise in temperature on reaction rate is more complicated than simply increasing the average kinetic energy of the molecules. The energy of the molecules has a spread of values, and this changes with temperature. From the kinetic theory equation [§ 7.8] J C Maxwell and L Boltzmann calculated the manner in which the energies of the molecules are distributed between below average energy, average energy and above average energy. They plotted the fraction of the total number of molecules which have energy x against the value of molecular energy x, as in Figure 14.3C. As the temperature rises, the shape of the Maxwell–Boltzmann distribution curve changes. At low temperature (T_1), a very large fraction of the molecules have energy close to the average energy, with only a very small fraction of the molecules having very low energies or very high energies. At a higher temperature (T_2), there is a wider spread of energies. At a still higher temperature (T_3), the spread increases further. It follows that the proportion of molecules with energy greater than or equal to average is greater at high temperatures. In all cases the area under the curve is the same as the fractions add up to unity.

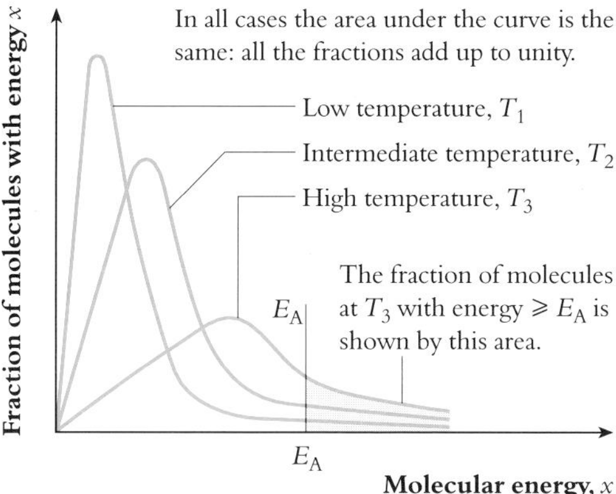

Figure 14.3C
The Maxwell–Boltzmann distribution of molecular energies (not to scale: E_A is typically 30–40 times the value of x at the top of the curve.)

Now look at the value of the activation energy, E_A. The fraction of molecules with energy $\geqslant E_A$ at each temperature is represented by the shaded area below the curve. Maxwell and Boltzmann calculated that

$$\text{Fraction of molecules with energy} \geqslant E = e^{-E/RT}$$

You can see that the shaded area under the curve, the fraction of the total number of molecules with energy $\geqslant E_A$, is greatest at the highest temperature, T_3. So, with a rise in temperature,

- the average kinetic energy of the molecules increases
- the fraction of molecules with higher than average kinetic energy increases.

Molecules must collide, they must have enough kinetic energy – and what else? They must also collide in an orientation which favours reaction; see Figure 14.3D which shows the reaction,

$$Cl(g) + HI(g) \longrightarrow HCl(g) + I(g)$$

Figure 14.3D
Approach of Cl to HI
(a) without reaction
(b) followed by reaction

(a) Cl atom approaches HI molecule from the iodine side.

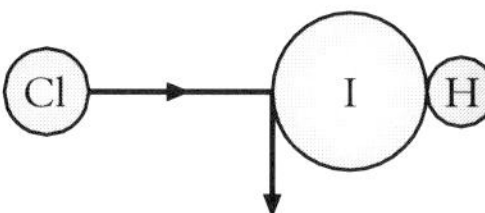

Repulsion between the chlorine atom and the iodine atom in the hydrogen iodide molecule prevents reaction from occurring.

(b) Cl atom approaches HI molecule from the hydrogen side.

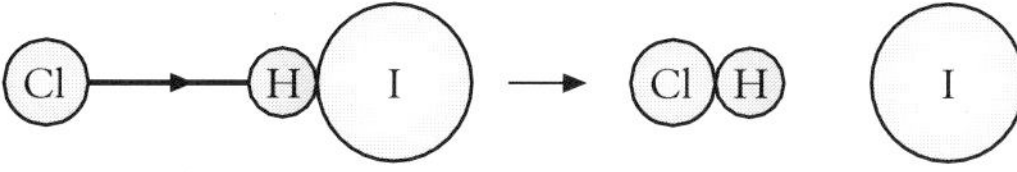

Reaction occurs as the chlorine atom bonds to the hydrogen atom and displaces the iodine atom.

The collision theory was developed for gaseous reactions, but it has been found to apply to reactions in solution also. Although solvent molecules prevent molecules of reactant coming together as frequently as they would in the gas phase, once molecules have come together they are less able to escape and will collide repeatedly. The repeated collisions may well lead to reaction.

As the temperature rises

- the average energy of the molecules increases
- the spread of values of molecular energies changes
- the fraction of the total number of molecules with high energy increases.

The collision theory states that in order to react, molecules must (a) collide (b) collide with enough energy to react (c) collide in a favourable orientation.

CHECKPOINT 14.3: COLLISION THEORY

1. State what is meant by
   (a) the standard enthalpy change of reaction
   (b) the activation energy of reaction.

   Illustrate your answers with diagrams.

2. *(a)* What does the shaded area under the curve represent?
   (b) Which is the higher temperature, T_1 or T_2?
   (c) There are two factors which are responsible for increasing the rate of a chemical reaction as the temperature rises? What are these factors?

3. In a reaction **A** + **B** ⟶ **C**, reaction does not take place every time that **A** and **B** collide. Why is this?

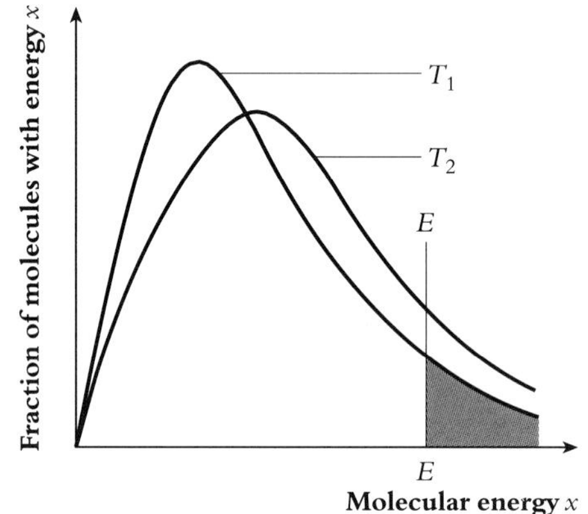

14.4 CATALYSIS

A catalyst speeds up a reaction without being used up in the reaction. The collision theory of reaction rates provides a model to show how this can happen. It is found experimentally that a catalyst lowers the activation energy for a reaction. According to the theory, it does this by providing an alternative route from reactants to products; see Figure 14.4A.

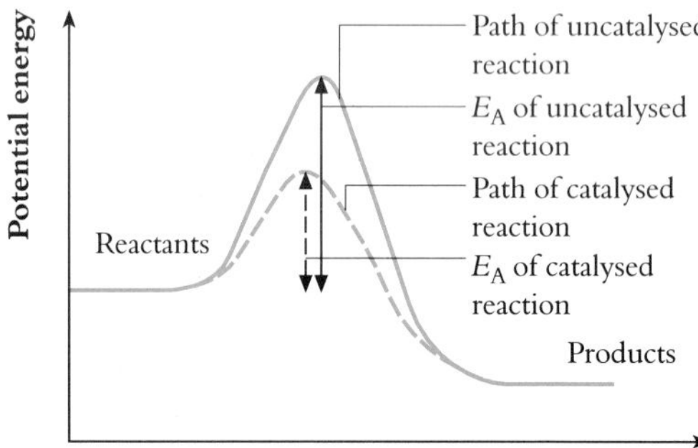

Figure 14.4A Energy profiles for a catalysed and an uncatalysed reaction

A catalyst increases the rate constant for the reaction. If the catalyst and the reactants are in the same phase, the process is described as **homogeneous catalysis**. **Heterogeneous catalysis** takes place at the surface of a catalyst which is in a different phase from the reactants (e.g., a solid catalysing a reaction between gases). The reactants are adsorbed on to the surface of the catalyst, where bonds are broken and new bonds are formed. The products are then desorbed from the surface. Catalysis is an extremely widespread phenomenon, and only a few examples are given here.

Catalysis may be homogeneous or heterogeneous

14.5 HOMOGENEOUS CATALYSIS

14.5.1 ACID–BASE CATALYSIS

Ester hydrolysis is an example of homogeneous catalysis

Many reactions are catalysed by acids and bases. An example is the acid-catalysed hydrolysis of esters to give a carboxylic acid and an alcohol or a phenol:

$$RCO_2R' + H_2O \xrightleftharpoons{H^+(aq)} RCO_2H + R'OH$$

14.5.2 ENZYMES AS CATALYSTS

Biological reactions are catalysed by enzymes

Enzymes are the substances that catalyse biological reactions. They are proteins. Enzymes can bring about reactions in aqueous solution, at the pH and temperature of living organisms. They bring about reactions under these mild conditions when many of the non-enzymic reactions we are familiar with require high temperature or high pressure or high or low pH to give a good yield.

They are proteins, and their catalytic activity depends on their tertiary structure. Enzymes are specific ...

... catalysing the reaction of one substrate only or a few related substrates

Compared with other catalysts you have met, enzymes are very **specific**; that is, an enzyme catalyses the reactions of only one substance or a very limited range of substances. The substance which an enzyme enables to react is called the **substrate**. Most enzymes act within cells, and cells contain many substances. It is important that an enzyme targets its own substrate and leaves other substances untouched.

Like other proteins, enzymes have a three-dimensional configuration, the tertiary structure of the protein. This structure is dependent on pH and temperature. Since the catalytic power of the enzyme depends on its tertiary structure, enzyme activity is sensitive to changes in pH and temperature.

The active site; the lock and key theory

The reason for the specificity ...

... is that the substrate fits the active site of the enzyme ...

... as a key fits a lock. The active site binds the substrate ...

... and catalyses its reaction

The reason for the specificity of enzymes for substrates is that the substrate must fit into the **active site** of the enzyme. This site is only a small region, perhaps 5% of the enzyme's surface. It is a crevice in the enzyme molecule into which the substrate molecule fits. The fit between an enzyme and its substrate has been compared with that of a lock and a key. The **lock and key theory** of the active site was originated by Emil Fischer in 1894. Just as it takes the correct key to open a lock, it takes the correct enzyme to bond to the substrate and catalyse its reaction. The very precise fit is the reason why enzymes are so specific. The bonding between enzyme and substrate may involve electrostatic attraction, hydrogen bonding and van der Waals forces. The bonds formed between the enzyme and the substrate weaken the bond in the substrate which is to be broken and therefore catalyse the reaction.

Figure 14.5A
The substrate binds to the active site of the enzyme and reacts

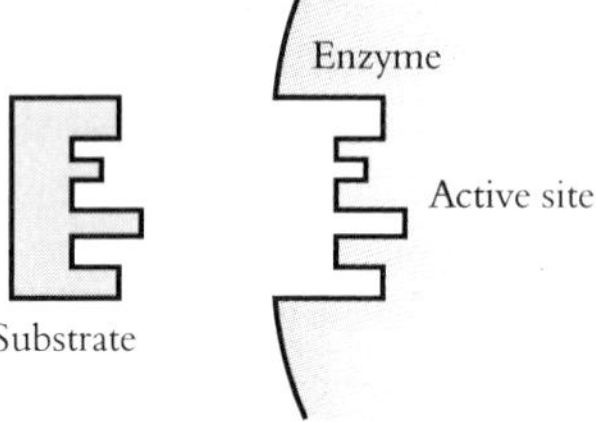
Enzyme
Active site
Substrate

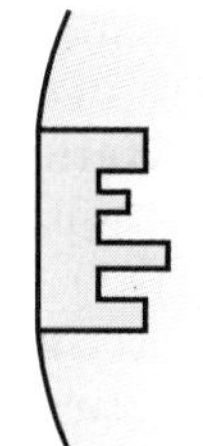

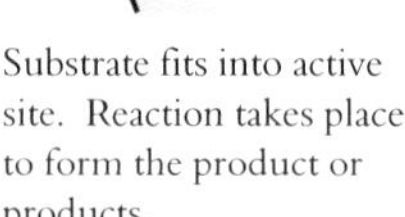
Substrate fits into active site. Reaction takes place to form the product or products.

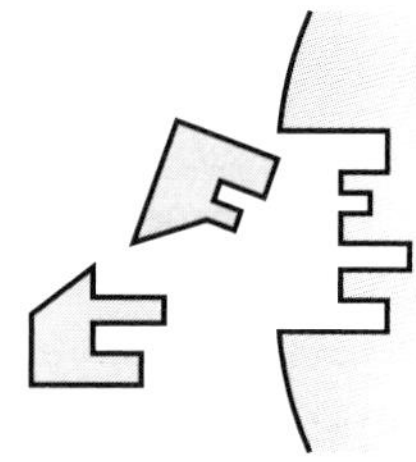
The products leave the active site.

Immobilised enzymes

Many industrial processes are **batch processes** which use soluble enzymes. The enzyme and substrate are mixed in a reaction vessel and left until a good yield of product has been formed. Usually the enzyme is discarded because recovery would be too costly. The cost of the process can be reduced if the enzyme can be recovered and used again. Making the enzyme insoluble by attaching it to an inert support, e.g. a resin such as Sephadex®, achieves this aim. The technique is called **immobilising** the enzyme. The enzyme and its support can be packed into columns and substrate can be passed over the enzyme in a **continuous process** over long periods. Alternatively, the enzyme and its support may be in the form of insoluble granules which can be used in a batch reactor and then separated by centrifugation and re-used.

The technique of immobilising enzymes allows enzymes to be used in continuous processes

Some examples of the use of immobilised enzymes

The use of immobilised enzymes has halved the cost of production of **high-fructose syrup** by the hydrolysis of starch to glucose followed by the conversion of glucose to fructose, which is a sweeter sugar.

An industrial process converts penicillin into **semi-synthetic penicillin** which will attack bacteria that are resistant to penicillin. The process uses an immobilised enzyme.

14.6 HETEROGENEOUS CATALYSIS

Transition metals

Many examples of heterogeneous catalysis involve transition metals. Their empty d orbitals allow them to bond with many substances to form reactive intermediates.

In the Haber process for the manufacture of ammonia, the catalyst is a mixture of iron and vanadium:

Many transition metals act as heterogeneous catalysts ...

$$N_2(g) + 3H_2(g) \xrightarrow{Fe/V} 2NH_3(g)$$

This reaction is of enormous industrial importance because ammonium salts are used as fertilisers.

In the hydrogenation of alkenes, a process which is used in the conversion of liquid oils into solid fats which can be sold for a higher price, nickel is used. It is finely divided to increase the area of surface over which the reactants can come into contact.

... Reaction takes place at the surface

$$R_2C{=}CR_2(g) + H_2(g) \xrightarrow{Ni} R_2CH{-}CHR_2(g)$$

The oxidation of sulphur dioxide in the Contact process is catalysed by platinum or by vanadium(V) oxide. If platinum is used, it readily absorbs impurities present in the sulphur dioxide and loses its activity: it is easily 'poisoned'.

'Cracking'

Catalysis is important in the petroleum industry

The 'cracking' of hydrocarbons in the petroleum industry is catalysed by a mixture of silica and alumina:

$$\underset{\text{Octane}}{C_8H_{18}(g)} \xrightarrow{Al_2O_3/SiO_2} \underset{\text{Butane}}{C_4H_{10}(g)} + \underset{\text{Butene}}{C_4H_8(g)}$$

The importance of 'cracking' is that it converts high boiling point fractions obtained from the fractional distillation of crude oil into more valuable low boiling point fractions which are used as fuel for motor vehicles and aircraft [see § 26.4.2].

14.7 CATALYTIC CONVERTERS

A catalytic converter is used to convert pollutants in vehicle exhaust gases into harmless products. The choice of catalyst is a transition metal ...

... with the right ability to absorb the reactants and to release the products

Carbon monoxide and hydrocarbons are present in vehicle exhaust gases as a result of the incomplete combustion of hydrocarbons in the engine. Oxides of nitrogen are present as a result of the combination of oxygen and nitrogen in the air at the temperature of the car engine. These pollutants can be removed from the exhaust gases by fitting the vehicle with a **three-way catalytic system**. This can both oxidise hydrocarbons and carbon monoxide and also reduce oxides of nitrogen. The exhaust gas is passed over an oxidation catalyst of e.g. platinum and palladium and then, with an intake of air, over a reduction catalyst of e.g. platinum, palladium, cobalt or nickel. The pollutants are removed by the reactions

$$\text{hydrocarbons + oxygen} \longrightarrow \text{carbon dioxide and water}$$
$$\text{carbon monoxide + oxygen} \longrightarrow \text{carbon dioxide and water}$$
$$\text{nitrogen monoxide + carbon monoxide} \longrightarrow \text{nitrogen + carbon dioxide}$$

For more detail, see § 26.6.

Choice of catalyst

The choice of catalyst in catalytic converters is a transition metal [§§ 15.2, 15.3]. These metals are able to act as catalysts because they have a variable oxidation state. The reactants must be adsorbed on the catalyst and, after the reaction, the products must be desorbed. Some transition metals, e.g. tungsten, adsorb too strongly and are therefore not effective. Others, e.g. silver, adsorb too weakly to catalyse the reaction. Platinum and nickel and other transition metals have the right ability to adsorb the reactants and release the products.

Surface area of catalyst

The catalyst is more effective when spread over a large surface area

A catalyst is more effective when it has a large surface area over which it can adsorb the reactants. This is achieved by applying it to a framework such as that shown in the catalytic converter shown in Figure 26.6A.

Poisoning of catalysts

Catalyst poisons must be avoided ...

... in vehicles and in the Haber process

Catalysts can be 'poisoned' by contact with a substance which is adsorbed strongly and prevents the catalyst from adsorbing the reactants. Lead poisons the catalytic converters in vehicles. This is why unleaded petrol must be used in vehicles with catalytic converters.

Haber process

... and in the Haber process

The Haber process manufactures ammonia from nitrogen and hydrogen. The source of hydrogen is natural gas, which is mainly methane, and which reacts with steam in a process called steam reforming:

$$CH_4(g) + H_2O(g) \rightleftharpoons CO(g) + 3H_2(g)$$

The reaction is catalysed by nickel, which is poisoned by sulphur compounds. It is therefore necessary to remove sulphur compounds from the supply of natural gas. A reaction with hydrogen converts them into hydrogen sulphide. This is removed from the stream of gas by reacting with zinc oxide to form zinc sulphide.

14.8 AUTOCATALYSIS

Potassium manganate(VII) in acid solution is a strong oxidising agent. For example ethanedicarboxylic acid is oxidised to carbon dioxide and water.

In autocatalysis the product of the reaction is a catalyst for the reaction

$$2MnO_4^-(aq) + 6H^+(aq) + 5(CO_2H)_2(aq) \longrightarrow 2Mn^{2+}(aq) + 10CO_2(g) + 8H_2O(l)$$

The reaction rate increases as the reaction proceeds and then slows down. The increase in rate happens because the reaction is catalysed by Mn^{2+}, and Mn^{2+} ions are formed by the oxidation. The later decrease in rate is due to the reactants being used up. The catalysis of a reaction by a product of the reaction is called **autocatalysis**.

CHECKPOINT 14.8: CATALYSTS

1. Why is it important to industrial chemists to find catalysts which allow reactions to proceed *(a)* at lower temperatures *(b)* at lower pressures?

2. Give an example of the importance of a catalysed reaction in *(a)* the food industry *(b)* the fertiliser industry *(c)* the petrochemical industry.

3. Explain the following terms: catalyst, enzyme, substrate, active site.

4. What is meant by the specificity of enzyme action? Why are enzymes more specific than other catalysts?

5. What is an immobilised enzyme? Explain the advantages of using immobilised enzymes over free enzymes.

6. Many cars are fitted with catalytic converters.
   (a) What is a catalyst?
   (b) What is the catalyst in a catalytic converter, and what reaction does it catalyse?
   (c) What is the advantage of a catalytic converter in a car?

7. *(a)* What problem arises when a too-powerful adsorbent is employed as a catalyst in a reaction?
   (b) Why is a honeycomb ceramic structure coated with platinum and other metals employed as a catalytic converter instead of a lump of platinum?
   (c) What is meant by the 'poisoning' of a catalyst?

14.9 LINK WITH THERMODYNAMICS

For a reaction to happen spontaneously ...

... there must be a decrease in free energy ...

... and the activation energy must not be too high

In Chapter 10 you saw that there is a condition for changes to proceed spontaneously – to happen without external help. The condition is that the change involves a decrease in free energy, ΔG. The value of the standard free energy, $\Delta G^{\ominus}$ is given by

$$\Delta G^{\ominus} = \Delta H^{\ominus} - T\Delta S^{\ominus}$$

where $\Delta G^{\ominus}$ = Change in standard free energy
$\Delta H^{\ominus}$ = Change in standard enthalpy
$\Delta S^{\ominus}$ = Change in standard entropy

If $\Delta G^{\ominus}$ for a reaction is negative, we say that the reactants are thermodynamically unstable with respect to the reaction: the reaction can occur. If $\Delta G^{\ominus}$ for a reaction is positive, we say that the reactants are thermodynamically stable with respect to the reaction: the reaction cannot occur. We say the reaction is **under thermodynamic control**. Thermodynamic instability ($\Delta G^{\ominus}$ negative) is not the only condition necessary for a change to occur. The molecules must possess enough energy to climb the energy barrier between reactants and products (Figure 14.3A). For example methane is a fuel; the oxidation of methane is thermodynamically unstable with respect to the products of oxidation:

$$CH_4(g) + 2O_2(g) \longrightarrow CO_2(g) + 2H_2O(l); \Delta G^{\ominus} = -580 \text{ kJ mol}^{-1}$$

Yet oxidation does not proceed at room temperature because at room temperature very, very few molecules of methane and oxygen collide with energy equal to the activation energy. As a result, the reaction takes place so slowly as to be unobservable, and we say the reaction is **under kinetic control**. At higher temperatures the fraction of molecules which collide with enough energy to react increases and the reaction takes place more rapidly.

SUMMARY

Reaction under thermodynamic control: $\Delta G^{\ominus}$ for the reaction is positive,
Reaction under kinetic control: $\Delta G^{\ominus}$ for the reaction is negative, but the activation energy is high.

CHECKPOINT 14.9: THERMODYNAMICS

1. What is a spontaneous reaction? Do spontaneous reactions always take place rapidly? Give examples. In what way does a catalyst affect the spontaneous reaction?

2. According to thermodynamics, an exothermic reaction is *feasible*. Why do some exothermic reactions proceed very slowly?

FURTHER STUDY ▼

14.10 THE STUDY OF REACTION KINETICS

We shall now take the study of reaction kinetics further. We need to go into more detail of measurements of the rates of reactions and interpreting the results of those measurements.

... Reaction kinetics is the study of the factors that affect the rates of chemical reactions

The study of the factors that affect the rates of chemical reactions is called **reaction kinetics**. Such studies throw light on the **mechanisms** of reactions. All reactions take place in one or more simple steps, and the sequence of steps is called the mechanism. The number of reacting species (molecules, atoms, ions or free radicals) that take part in a reaction step is the **molecularity** of that step. Reaction steps are described as **unimolecular**, **bimolecular** or **trimolecular** [§ 14.17.2]. The rate of the overall reaction the rate at which product is formed – is determined by the rate of the slowest step. This is called the **rate-determining** step; see § 14.18. Examples of reaction mechanisms are covered in §§ 27.8.5, 27.8.6, 28.8.2, 28.8.7, 29.9, 31.8 and 33.13.1.

14.11 AVERAGE RATE

The rate of a reaction is the rate of change of concentration of a reactant or a product

The average rate of a chemical reaction over a certain interval of time is equal to the change in the concentration of a reactant or product that occurs during that time divided by the time. When an ester is hydrolysed, if the concentration of the ester decreases from 1.00 mol dm^{-3} to 0.50 mol dm^{-3} in 1.00 hour, then the average rate of reaction over this time interval can be given:

$$\text{Average rate} = (1.00 - 0.50)\ \text{mol dm}^{-3}/(1.00 \times 60 \times 60)\ \text{s}$$
$$= 1.39 \times 10^{-4}\ \text{mol dm}^{-3}\ \text{s}^{-1}$$

In the example in Figure 14.11A, the rate of reaction slows down as the concentration of reactant decreases

The unit of reaction rate is concentration time^{-1}. Consider a reaction of the type

$$\mathbf{A} \longrightarrow \mathbf{B}$$

where 1 mole of the reactant produces 1 mole of the product. Figure 14.11A shows how the concentration of product increases and the concentration of reactant decreases as the time which has passed since the start of the reaction increases.

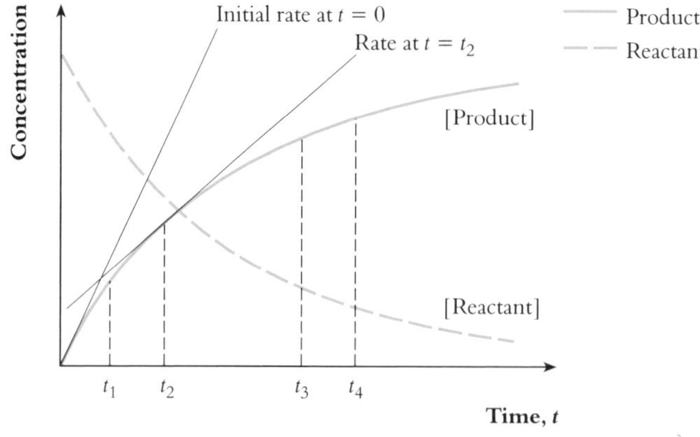

Figure 14.11A Variation of concentrations of reactant and product with time

The rate at the very beginning of the reaction is called the initial rate

You can see that less product has been formed between t_3 and t_4 than in an equal time interval between t_1 and t_2. The rate of reaction decreases as the reaction proceeds and the reactant is gradually used up. One can only state the rate of reaction at a certain time. At time t_2 the rate of reaction is the gradient of the tangent to the curve at this point [see Figure 14.11A]. The rate at the start of the reaction, when an infinitesimally small amount of the reactant has been used up, is called the **initial rate** of the reaction. In Figure 14.11A the gradient of the tangent to the curve at $t = 0$ gives the initial rate.

The rate of the reaction

$$\mathbf{A} \longrightarrow \mathbf{B}$$

is the rate of decrease in concentration of **A** or the rate of increase in concentration of **B**:

$$\text{Rate} = -\frac{d[\mathbf{A}]}{dt} = \frac{d[\mathbf{B}]}{dt}$$

where [**A**] = concentration/mol dm^{-3} of **A**. In the reaction

$$BrO_3^-(aq) + 5Br^-(aq) + 6H^+(aq) \longrightarrow 3Br_2(aq) + 3H_2O(l)$$

$$\text{Rate} = -\frac{d[BrO_3^-]}{dt} = -\frac{1}{5}\frac{d[Br^-]}{dt} = \frac{1}{3}\frac{d[Br_2]}{dt} \text{ and so on}$$

The rate of a reaction is found by measuring some property of a reactant or a product at various times after the start of the reaction. Some of the methods of 'following' the reaction in this way will now be described.

14.12 METHODS OF FINDING THE RATES OF CHEMICAL REACTIONS

14.12.1 CHEMICAL METHODS

Titration can be used to follow the change in the concentration of the reactant or the product

The progress of a reaction can often be followed by chemical analysis. The reaction is carried out in a thermostatically controlled water bath. Solutions of the reactants of known concentrations are mixed, and a stop clock is started. A sample of the reacting mixture is withdrawn with a pipette, and the reaction is stopped. This may be done by removing one of the reactants by a chemical reaction. Alternatively, the reaction may be suddenly slowed down by cooling or by dilution. This is done by pipetting a sample into a freezing mixture or into an excess of the solvent. A titration is then performed to find the concentration of one of the reactants or one of the products.

Example The alkaline hydrolysis of an ester

$$CH_3CO_2C_2H_5(aq) + NaOH(aq) \longrightarrow CH_3CO_2Na(aq) + C_2H_5OH(aq)$$

The course of an alkaline hydrolysis of an ester is followed by measuring the concentration of alkali at various times after the start of the reaction

Solutions of ester and alkali of known concentrations are allowed to reach the temperature of a thermostat bath. The solutions are mixed, and the time of mixing is noted. A sample of the reaction solution is withdrawn by pipette, and run into about four times its volume of ice-cold water. The dilution and cooling reduce the rate of reaction almost to zero. The alkali that remains is titrated against standard acid, using phenolphthalein as indicator. Ethanoate ions do not affect the colour change of this indicator. The analysis is repeated at various intervals of time after the start of the reaction.

14.12.2 PHYSICAL METHODS

A change in gas volume

The volume of gas evolved can be measured after various time intervals

In a reaction in which a gas is formed, the volume of gas can be recorded at various times [see Figure 14.12A]. Examples are the reaction of a metal with an acid and the decomposition of hydrogen peroxide:

$$Mg(s) + 2HCl(aq) \longrightarrow H_2(g) + MgCl_2(aq)$$

$$2H_2O_2(aq) \longrightarrow 2H_2O(l) + O_2(g)$$

Figure 14.12A Measuring the evolution of a gas in a reaction

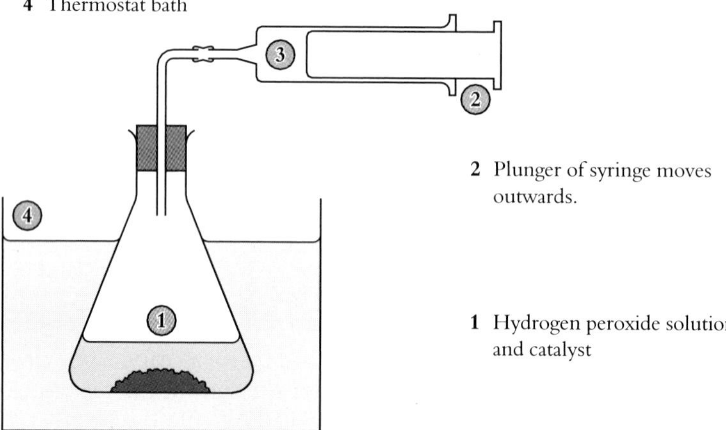
3 Oxygen. Volume is recorded at certain times after the start of the reaction.
4 Thermostat bath
2 Plunger of syringe moves outwards.
1 Hydrogen peroxide solution and catalyst
3
2
4
1

A change in gas pressure

An increase or decrease in gaseous pressure can be used to follow many gaseous reactions

Some reactions between gases involve an increase in the number of moles of gas, e.g.

$$2N_2O_5(g) \longrightarrow 2N_2O_4(g) + O_2(g)$$

If the reaction takes place at constant volume, the resulting increase in pressure can be followed. The reaction

$$2H_2(g) + O_2(g) \longrightarrow 2H_2O(g)$$

is accompanied by a decrease in the number of moles of gas and by a decrease in pressure at constant volume.

Changes in the absorption spectrum

The use of a spectrophotometer enables light absorption to be measured

Many substances absorb light, either in the visible region or, more frequently, in the ultraviolet region. A **spectrophotometer** is an instrument for measuring the absorption of light at various wavelengths. If a reaction is carried out in a cell inside a spectrophotometer, the change in the absorption spectrum can be used to follow the course of the reaction [see § 35.7].

Electrical properties

A change in conductance indicates a change in the concentration of ions ...

A change in the conductance of a solution occurs if ions are used up or created during a reaction.

... as does a change in electrode potential

The potential of an electrode in contact with a solution of its ions changes if the concentration of ions changes. Potentiometric methods [see § 13.2] can often be used to follow the course of a reaction which involves ions.

Thermal conductivity

Thermal conductivity

A gaseous reaction can be followed by measuring the thermal conductivity of the mixture of reacting gases.

14.13 THE RESULTS OF MEASUREMENTS OF REACTION RATES

Methods such as those described enable us to measure the rates of chemical reactions. It is interesting to use such methods to find out how the rate of reaction depends on the concentrations of the reactants, the temperature and other factors. The results of such studies can be interpreted to give a detailed picture of what happens during a chemical reaction.

14.13.1 THE EFFECT OF PARTICLE SIZE ON REACTION RATE

Solids react faster in finely divided form

Reactions of solids take place faster when the solids are in a finely divided state. This is because the ratio of surface area to mass is greater in small particles than in large particles, and the area over which the solid can come into contact with liquid or gaseous reactants is greater. Examples are the reaction of powdered zinc and granulated zinc with acids and the reaction of powdered calcium carbonate and marble chips with acid.

14.13.2 THE EFFECT OF CONCENTRATION ON REACTION RATE

Consider a reaction between **A** and **B**:

$$\mathbf{A} + \mathbf{B} \longrightarrow \text{Products}$$

The way in which rate is related to concentration is governed by the order of the reaction

The rate of reaction depends on the concentrations of **A** and **B**, but one cannot simply say that the rate of reaction is proportional to the concentration of **A** and proportional to the concentration of **B**. The relationship is

$$\text{Reaction rate} \propto [\mathbf{A}]^m[\mathbf{B}]^n = k[\mathbf{A}]^m[\mathbf{B}]^n$$

An expression of this kind is called a **rate equation**. The powers m and n are usually integers, often 0, 1 or 2, and are characteristic of the reaction. The reaction is of **order** m with respect to **A** and of order n with respect to **B**. The overall order of reaction is $(m + n)$. The proportionality constant k is called the **rate constant** or **velocity constant** for the reaction.

The rate constant for a reaction relates the rate to the concentrations of the reactants

Example (a) A solution of **Q**, of concentration 0.20 mol dm^{-3} undergoes a **first-order** reaction at an initial rate of 3.0×10^{-4} mol dm^{-3} s^{-1}. Calculate the rate constant.

Method Since, for a first-order reaction

$$\text{Initial rate} = k[\mathbf{Q}]_0$$

$$3.0 \times 10^{-4}\ \text{mol dm}^{-3}\,\text{s}^{-1} = k \times 0.20\ \text{mol dm}^{-3}$$

$$\text{The rate constant } k = 1.5 \times 10^{-3}\,\text{s}^{-1}$$

The unit of a first-order rate constant is time^{-1}.

First-order and second-order rate constants have different dimensions

Example (b) A **second-order** reaction takes place between the reactants **P** and **Q**, which are both initially present at concentration 0.20 mol dm^{-3}. If the initial rate of reaction is 1.6×10^{-4} mol dm^{-3} s^{-1}, what is the rate constant?

Method

$$\text{Initial rate} = k[\mathbf{P}]_0[\mathbf{Q}]_0$$

$$\therefore 1.6 \times 10^{-4}\ \text{mol dm}^{-3}\,\text{s}^{-1} = k \times (0.20\ \text{mol dm}^{-3})^2$$

$$\text{The rate constant, } k = 4.0 \times 10^{-3}\,\text{dm}^3\,\text{mol}^{-1}\,\text{s}^{-1}$$

A second-order rate constant has the unit concentration^{-1} time^{-1}.

14.14 ORDER OF REACTION

The stoichiometric equation for a reaction does not reveal the order of the reaction

The order of a reaction does not follow from its stoichiometric equation. The reaction between bromate(V) ions, bromide ions and hydrogen ions to give bromine is represented by the equation

$$BrO_3^-(aq) + 5Br^-(aq) + 6H^+(aq) \longrightarrow 3Br_2(aq) + 3H_2O(l)$$

The results of kinetic measurements give the following rate equation:

$$-\frac{d[BrO_3^-]}{dt} \propto [BrO_3^-][Br^-][H^+(aq)]^2$$

The reaction is first order with respect to bromate(V), first order with respect to bromide, second order with respect to hydrogen ion and fourth order overall. The negative sign means that $[BrO_3^-]$ decreases with time.

14.14.1 ORDER OF REACTION FROM INITIAL RATE

In a reaction

$$\mathbf{A} \longrightarrow \mathbf{X}$$

The order of a reaction can be found by comparing the initial rates of two reactions at known initial concentrations

the rate of reaction $= k[\mathbf{A}]^n$, where n = the order of reaction. Two experiments to find the rate of reaction, two 'runs', are done at different concentrations of **A**. $[\mathbf{A}_0]_1$ = initial concentration of **A** in Run 1, and $(v_0)_1$ = the initial rate in Run 1.

$$(v_0)_1 = k[\mathbf{A}_0]_1{}^n$$

$$(v_0)_2 = k[\mathbf{A}_0]_2{}^n$$

The ratio

$$\frac{(v_0)_1}{(v_0)_2} = \left(\frac{[\mathbf{A}_0]_1}{[\mathbf{A}_0]_2}\right)^n$$

The order of reaction can be found by comparing the initial rates of reactions at different concentrations, as in the example below.

Example The following results were obtained for a reaction between **A** and **B**:

A worked example: how to find the order by comparing initial rates

Run	Concentrations/mol dm^{-3}		Initial rate/mol dm^{-3} s^{-1}
	[**A**]	[**B**]	
(a)	0.50	1.0	2.0
(b)	0.50	2.0	8.0
(c)	0.50	3.0	18
(d)	1.0	3.0	36
(e)	2.0	3.0	72

What is the order of reaction (*a*) with respect to **A** (*b*) with respect to **B**? (*c*) What is the rate equation for the reaction? (*d*) Calculate the rate constant.

Method Let the rate equation be

$$\text{Rate} = k[\mathbf{A}]^m[\mathbf{B}]^n$$

(a) Compare runs (d) and (e) in which [**B**] is constant; therefore Rate $\propto$ $[\mathbf{A}]^m$.
When [**A**] is doubled the reaction rate doubles; therefore Rate $\propto$ $[\mathbf{A}]^1$
The reaction is first order with respect to **A**, and

$$\text{Rate} = k[\mathbf{A}][\mathbf{B}]^n$$

(b) Compare runs (b) and (a) in which [**A**] is constant; therefore Rate $\propto$ $[\mathbf{B}]^n$
When [**B**] is doubled the rate is quadrupled; therefore Rate $\propto$ $[\mathbf{B}]^2$
The rate is second order with respect to **B**.

$$\text{Rate} = k[\mathbf{A}]^m[\mathbf{B}]^2$$

(c) Combining the results of *(a)* and *(b)*

$$\text{Rate} = k[\mathbf{A}][\mathbf{B}]^2$$

(d) The rate constant can now be calculated from the results of any run, for example *(c)*

$$18 = k \times 0.5 \times (3.0)^2$$

$$k = 4.0\ \text{dm}^6\,\text{mol}^{-2}\,\text{s}^{-1}$$

Once the order has been found, the rate constant can be found

This third-order rate constant has the unit concentration^{-2} time^{-1}.

CHECKPOINT 14.14A: INITIAL RATES

1. Explain the terms: *rate of reaction*, *order of reaction*, *stoichiometry of reaction* and *rate constant*.

2. Mention three physical methods which can be used to follow the course of a chemical reaction. What are the advantages of physical methods over chemical methods?

3. For the reaction

$$\mathbf{A} + \mathbf{B} \longrightarrow \mathbf{C}$$

the following results were obtained for kinetic 'runs' at the same temperature:

$[\mathbf{A}]_0$/mol dm^{-3}	$[\mathbf{B}]_0$/mol dm^{-3}	Initial rate/ mol dm^{-3} s^{-1}
0.20	0.10	0.20
0.40	0.10	0.80
0.40	0.20	0.80

Find

(a) the rate equation for the reaction
(b) the rate constant
(c) the initial rate of a reaction, when $[\mathbf{A}]_0 = 0.60$ mol dm^{-3} and $[\mathbf{B}]_0 = 0.30$ mol dm^{-3}

4. Tabulated are values of initial rates measured for the reaction

$$2\text{A} + \text{B} \longrightarrow \text{C} + \text{D}$$

Experiment	[A]/ mol dm^{-3}	[B]/ mol dm^{-3}	Initial rate/ mol dm^{-3} min^{-1}
1	0.150	0.25	1.4×10^{-5}
2	0.150	0.50	5.6×10^{-5}
3	0.075	0.50	2.8×10^{-5}
4	0.075	0.25	7.0×10^{-6}

(a) Find the order with respect to **A**, the order with respect to **B** and the overall order of the reaction.
(b) Find the value of the rate constant.
(c) Find the initial rate of reaction when $[\mathbf{A}]_0 = 0.120$ mol dm^{-3} and $[\mathbf{B}]_0 = 0.220$ mol dm^{-3}.

14.14.2 ZERO-ORDER REACTIONS

If the rate is independent of concentration, the reaction is zero order ...

In a **zero-order** reaction, the rate is independent of the concentration of the reactant. A plot of the concentration of the reactant, [**A**], against time has the form shown in Figure 14.14A.

Figure 14.14A
A plot of concentration against time for a zero-order reaction

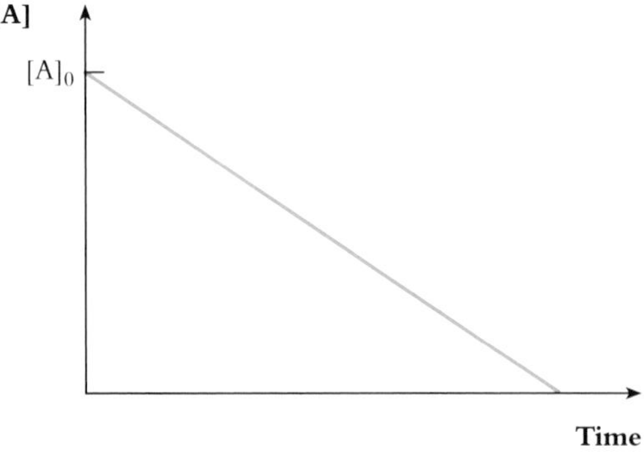

The rate equation for a zero-order reaction is

$$\text{Rate} = k[\mathbf{A}]^0$$

and the rate constant has the unit concentration time^{-1}.

... One example is the reaction of iodine and propanone ...

In the iodination of propanone [§ 14.18.1], the reaction rate does not change if the concentration of iodine is changed:

$$CH_3COCH_3(aq) + I_2(aq) \xrightarrow[\text{buffer}]{\text{acid}} CH_3COCH_2I(aq) + HI(aq)$$

The reaction is said to be zero order with respect to iodine.

... Another example is the adsorption of reactants in gaseous reactions

Sometimes, reactions between gases are zero order with respect to one of the reactants. This often indicates that this reactant has been adsorbed on the surface of the vessel. The rate of reaction then depends on the frequency with which molecules of the non-adsorbed gas collide with the inside of the vessel. This frequency is proportional to the concentration of the non-adsorbed reactant.

14.14.3 FIRST-ORDER REACTIONS

If the reaction

$$\mathbf{A} \longrightarrow \text{Products}$$

is a first-order reaction, the rate equation will be

$$\text{Rate} = k[\mathbf{A}]$$

Since the rate of reaction is proportional to the concentration of **A**, when half of **A** has reacted the rate will be half the initial rate. When three-quarters of **A** have reacted the rate will be one quarter of the initial rate. The graph of [**A**] against time is shown in Figure 14.14B.

This will remind you of the shape of the graph of radioactive decay [Figure 1.10C] which is a first-order reaction.

Half-life

The time taken to go to half-completion is called the **half-life** of the reaction, $t_{1/2}$.

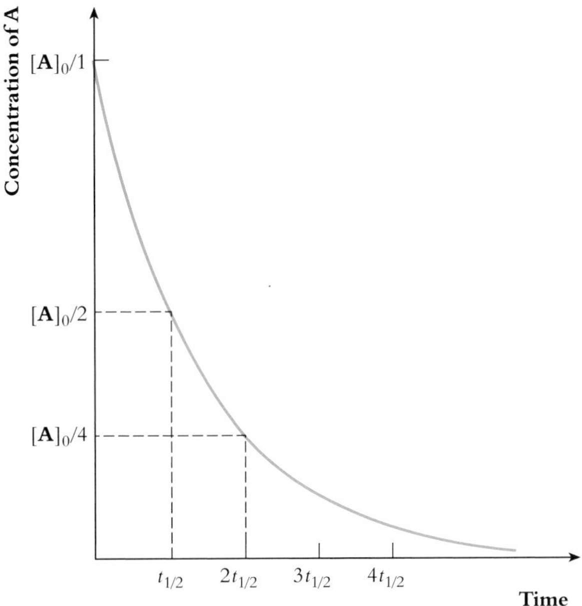

Figure 14.14B A plot of [Reactant] against time for a first-order reaction; $t_{1/2}$ = half-life

Example (a) The half-life of radium is 1590 years. How long will it take for a sample of radium to decay to 25% of its original radioactivity?

Method To decay to 50% of its original activity takes one half-life = 1590 years. To decay from 50% to 25% of its original activity takes another half-life of 1590 years.
Total = 3180 years

Example (b) A radioactive sample of caesium-136 decays from 480 cpm to 60 cpm in 42 days. What is its half-life?

Method Decrease in amount of caesium-136 = (60 cpm/480 cpm) × 100% = 12.5%

The isotope has decayed from 100% ⟶ 50% ⟶ 25% ⟶ 12.5% that is 3 half-lives.
Therefore 42 days = 3 half-lives and $t_{1/2}$ = 14 days.

14.14.4 PSEUDO-FIRST-ORDER REACTIONS

The acid-catalysed hydrolysis of an ester, e.g., ethyl ethanoate

$$CH_3CO_2C_2H_5(l) + H_2O(l) \xrightarrow{\text{acid}} CH_3CO_2H(aq) + C_2H_5OH(aq)$$

If the concentration of one reactant is very large, the reaction appears to be zero order with respect to that reactant

is first order with respect to ester and first order with respect to water. If water is present in large excess, only a small fraction of the water will be used up in the reaction. The concentration of water is practically constant, and the rate depends on the concentration of ester alone:

$$-d[CH_3CO_2C_2H_5]/dt = k'[CH_3CO_2C_2H_5]$$

k' = a first-order rate constant. The reaction appears to be zero order with respect to water.

CHECKPOINT 14.14B: FIRST-ORDER REACTIONS

1. A radioactive element has a count rate of 120 cpm. After 3 hours this has dropped to 15 cpm. What is the half-life of the isotope?

2. If the half-life of a radioactive element is 150 s, what percentage of the isotope will remain after 600 seconds?

3. The results listed were obtained for a 'run' on the reaction

$$\mathbf{A} \longrightarrow \mathbf{B} + \mathbf{C}$$

Plot [**A**] against *t*.

Time/s	[A]/mol dm^{-3}
0	0.800
400	0.580
800	0.400
1200	0.280
1600	0.200
2000	0.140
2400	0.100

(a) From the graph, find the order of the reaction with respect to **A**.
(b) By drawing a tangent, find the initial rate of reaction.
(c) Calculate the rate constant for the reaction.

4. The decomposition of benzene diazonium chloride is first order:

$$C_6H_5N_2Cl(aq) \longrightarrow C_6H_5Cl(aq) + N_2(g)$$

The following results give the volume of nitrogen at 50 °C, measured at various time intervals after the start of the reaction, obtained in the decomposition of 500 cm^3 of a 1.10×10^{-3} mol dm^{-3} solution at 50 °C:

Time/min	2.0	4.0	6.0	9.0	12.0	16.0	22.0	28.0
Volume of N_2/cm^3	1.7	3.4	4.9	6.6	8.1	9.5	11.2	12.2

(a) Plot the volume of nitrogen evolved against the time interval.
(b) Calculate the volume of nitrogen at 50 °C that will be formed at $t = \infty$ from the amount of benzene diazonium chloride specified. Enter this on the graph.
(c) From the graph, estimate the half-life of the reaction, $t_{1/2}$.
(d) What evidence have you that the reaction is first order?
(e) Obtain the initial rate of reaction.
(f) From it, calculate the rate constant.

14.14.5 SECOND-ORDER REACTIONS

In the second order reaction

$$\mathbf{A} + \mathbf{B} \longrightarrow \text{Products}$$

the rate equation is

$$\text{Rate} = -\mathrm{d}[\mathbf{A}]/\mathrm{d}t = -\mathrm{d}[\mathbf{B}]/\mathrm{d}t = k[\mathbf{A}]\,[\mathbf{B}]$$

14.14.6 SUMMARY OF THE DEPENDENCE OF THE CONCENTRATIONS OF REACTANT AND PRODUCT ON TIME

The manner in which the rate of reaction depends on the concentration of the reactant is shown in Figure 14.14C. For a zero-order reaction, the rate remains constant as the concentration of reactant changes. For a first-order reaction, the rate of reaction is directly proportional to the concentration of the reactant. For a second-order reaction, the rate of reaction increases with the concentration as shown in the figure. It is therefore sometimes possible to tell the order of a reaction from an inspection of a graph of rate against concentration over a wide range of concentration.

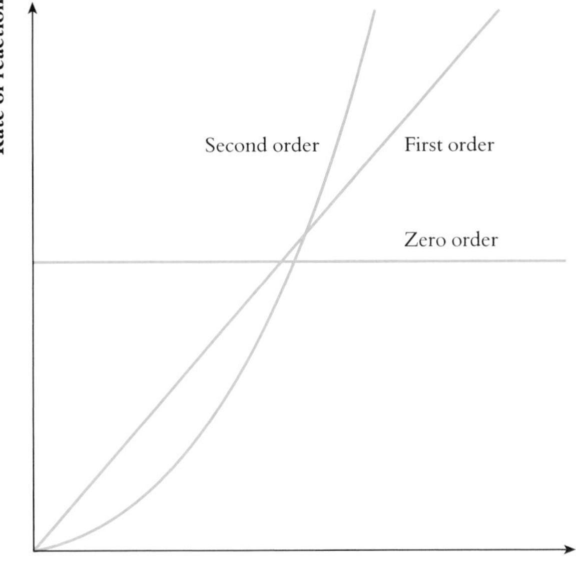

Figure 14.14C
Graphs of rate against concentration

CHECKPOINT 14.14C: ORDER OF REACTION

1. Dinitrogen oxide decomposes according to the equation,

$2N_2O(g) \longrightarrow 2N_2(g) + O_2(g)$

When the concentration of dinitrogen oxide decreases from 5×10^{-3} mol dm^{-3} to 2×10^{-3} mol dm^{-3}, the rate of decomposition falls by a factor of 2.5. What is the order of reaction?

2. In the reaction between nitrogen monoxide and bromine,

$2NO(g) + Br_2(g) \longrightarrow 2NOBr(g)$,

When the concentration of nitrogen monoxide is reduced from 4×10^{-2} mol dm^{-3} to 2×10^{-2} mol dm^{-3} the rate falls by a factor of 4. What is the order of reaction with respect to nitrogen monoxide?

3. Bromine is formed by the reaction between bromate(V) ions and bromide ions and acid:

$BrO_3^-(aq) + 5Br^-(aq) + 6H^+(aq) \longrightarrow 3Br_2(aq) + 3H_2O(l)$

The results of some experiments on the reaction are shown below.

Experiment	Concentration/ mol dm^{-3}			Initial rate/ mol BrO_3^- dm^{-3} s^{-1}
	BrO_3^-	Br^-	H^+	
1	0.10	0.10	0.10	1.2×10^{-3}
2	0.10	0.30	0.10	3.6×10^{-3}
3	0.20	0.10	0.10	2.4×10^{-3}
4	0.20	0.10	0.20	9.6×10^{-3}

Write the rate law for the reaction, state the overall order of reaction, and find the value of the rate constant.

4. The results of experiments on the reaction of persulphate ions with iodide ions,

$S_2O_8^{2-}(aq) + 3I^-(aq) \longrightarrow 2SO_4^{2-}(aq) + I_3^-(aq)$

are shown below. Write the rate equation for the reaction and calculate the rate constant.

Experiment	Concentration/ mol dm^{-3}		Initial rate/ (mol $S_2O_8^{2-}$) dm^{-3} s^{-1}
	$S_2O_8^{2-}$	I^-	
1	0.15	0.20	1.08
2	0.25	0.20	1.80
3	0.25	0.10	0.90

5. The ester ethyl ethanoate is hydrolysed by warming with dilute sulphuric acid to ethanol and ethanoic acid. Describe the experiments you would do to find the order of this reaction with respect to the ester. Say what measurements you would make and explain how you would calculate the order from them.

6. Thermal decomposition occurs when 1-chloropropane $CH_3CH_2CH_2Cl$, is heated in the absence of air. The following results were obtained at 400 °C.

Plot a graph of concentration of 1-chloropropane against time. Use your graph to deduce the order of reaction. Explain your reasoning.

Time/minutes	Concentration of 1-chloropropane/10^{-4} mol dm^{-3}
0	40
10	34
20	28
30	24
40	19
60	14
90	8.7
120	5.0

7. Ethyl ethanoate is hydrolysed by aqueous alkali.

$$CH_3CO_2C_2H_5(l) + OH^-(aq) \longrightarrow CH_3CO_2^-(aq) + C_2H_5OH(aq)$$

In a study of this reaction the following results were obtained.

Initial rate of reaction/ mol dm^{-3} s^{-1}	[$CH_3CO_2C_2H_5$]/ mol dm^{-3}	[OH^-]/ mol dm^{-3}
1.13×10^{-3}	0.076	0.076
1.13×10^{-3}	0.038	0.152
5.65×10^{-4}	0.019	0.152

(a) Calculate the order of reaction with respect to (i) OH^-, (ii) $CH_3CO_2C_2H_5$
(b) Write the rate equation for the reaction.
(c) Calculate the value of the rate constant.

8. In a reaction

$$\mathbf{R} \longrightarrow \mathbf{S}$$

the initial rate is 1.7×10^{-4} mol dm^{-3} s^{-1} when $[\mathbf{R}]_0 = 0.25$ mol dm^{-3}. Predict the initial rate when $[\mathbf{R}]_0 = 0.75$ mol dm^{-3} if the reaction is *(a)* zero order, *(b)* first order and *(c)* second order.

9. The results tabulated refer to the isomerisation

trans-CHCl=CHCl ⟶ *cis*-CHCl=CHCl

Time/s	0	600	900	1200	1500	1800
Trans-isomer/mol	1.00	0.90	0.85	0.81	0.77	0.73

From a suitable plot, find the order of reaction and the rate constant.

14.15 PHOTOCHEMICAL REACTIONS

Some reactions take place faster in the presence of light

Reactions with very high rates often involve free radicals. When a covalent bond splits **homolytically**, each of the bonded atoms or groups takes one of the bonding pair of electrons, and each product is termed a **free radical**:

$$\mathbf{A:B \longrightarrow A\cdot + \cdot B}$$

Light energy may split bonds to form free radicals

Energy must be supplied to break the bonds and produce free radicals. In thermal reactions, this energy comes from collisions with other molecules. Those reactions which are started by the absorption of light energy are called **photochemical reactions**.

The chlorination of methane is a photochemical reaction

An important photochemical reaction is the chlorination of methane in sunlight to chloromethane and other derivatives [§ 26.4.7]. It proceeds by a chain reaction.

14.16 THE EFFECT OF TEMPERATURE ON REACTION RATES

Reaction rates increase with temperature ...

An increase in temperature increases the rate of a reaction by increasing the rate constant. Figure 14.16A shows plots of rate constant k against temperature T and of ln k against $1/T$.

The variation of rate constant with temperature was studied by Arrhenius and found to fit the equation

$$k = Ae^{-E/RT}$$

Figure 14.16A
(a) Plot of k against T,
(b) Plot of $\ln k$ against $1/T$

A high activation energy means a low rate constant

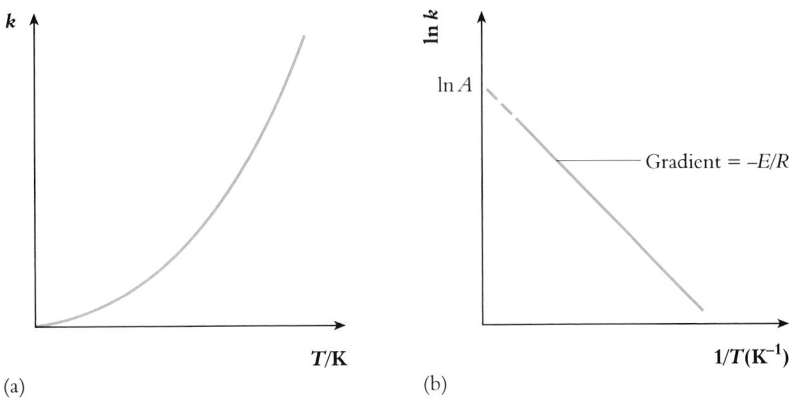

... The dependence of rate on temperature fits the Arrhenius equation

This equation is known as the **Arrhenius equation**. R is the gas constant [§ 7.5] and A and E are constant for a given reaction. The constants can be found by using the equation in logarithmic form:

$$\ln k = \ln A - E/RT$$

ln k against 1/T gives a linear plot

A plot of $\ln k$ against $1/T$ is a straight line of gradient $-E/R$ and an intercept on the y-axis of $\ln A$. If $\lg k$ is plotted against $1/T$, the gradient of the line is $-E/2.303R$, and the intercept is $\lg A$.

The significance of the temperature dependence is discussed in § 14.17.

14.17 THEORIES OF REACTION RATES

14.17.1 THE COLLISION THEORY; TAKING IT FURTHER

Gas molecules must collide in order to react ...

... The rate of reaction is much less than the rate of collision

From the Kinetic Theory of Gases, was developed the **Collision Theory of bimolecular reactions in the gas phase**. In a reaction between two gaseous substances **A** and **B**, a molecule of **A** must collide with a molecule of **B** before reaction can occur. It has been shown that the collision frequency Z is proportional to the product [**A**] [**B**]. If every collision results in reaction, the rate of reaction will equal the collision frequency. Theoreticians have calculated the collision frequency, and have found that the rate constant for a bimolecular gaseous reaction should be of the order of 10^{11} dm^3 mol^{-1} s^{-1}. Although there are reactions which have rate constants of this magnitude, e.g.

$$H\cdot + Br_2 \longrightarrow Br\cdot + HBr$$

such reactions are few. The reason why most reactions are slower than this is that only a small fraction of collisions results in reaction.

In the Arrhenius equation

$$k = Ae^{-E/RT}$$

Molecules must not only collide: they must have enough energy to react

E is termed the **energy of activation**, A is the rate of collision between molecules, and E is the energy which the colliding molecules must possess before a collision will result in reaction.

The Arrhenius equation ties in with the Maxwell–Boltzmann equation

The fraction $e^{-E/RT}$ appears in the Maxwell–Boltzmann calculations [§ 14.3]. They calculated that for a value of molecular energy E/J mol^{-1}, the fraction of molecules at temperature T that possess energy $\geqslant E$ is given approximately by the value of $e^{-E/RT}$. Calculation of k from the Arrhenius equation shows that a reaction with an activation energy of 50 kJ mol^{-1} should approximately double in rate over 10 K. This behaviour is observed for many such reactions.

The term A in the Arrhenius equation

In the Arrhenius equation, it was suggested that the term A might be the collision frequency. Calculated values of collision frequencies, however, are often higher than values obtained from plots such as Figure 14.16A(b), § 14.16. The explanation must be that molecules must collide not only with sufficient energy for reaction to occur, but also in a favourable orientation in space. The term A is therefore thought to be the rate at which molecules collide in an orientation which is favourable to reaction.

The Collision Theory applies to reactions in solution as well as to reactions in the gas phase

The Collision Theory was developed for bimolecular reactions in the gas phase, but it has been found to apply to reactions in solution. Although solvent molecules prevent molecules of reactant coming together as frequently as they would in the gas phase, once they have come together, reactant molecules are less able to escape, and will collide repeatedly. These repeated collisions between a pair of molecules may well lead to reaction. Some reactions which have been studied both in the gas phase and in solution are found to have similar rate constants in the two phases.

> The Collision Theory of bimolecular reactions states that, in order to react, molecules must *(a)* collide, *(b)* collide in a favourable orientation and *(c)* collide with enough energy to react.

14.17.2 THE TRANSITION STATE THEORY

The Transition State Theory looks at the course of a collision in detail

The **Transition State Theory** is concerned with what actually happens during a collision. It follows the energy and orientation of the reactant molecules as they collide and seeks an explanation of why such a small fraction of collisions results in reaction.

When two molecules approach each other in a collision, the electron clouds experience a gradual increase in their mutual repulsions, and the molecules begin to slow down. While this is happening, the kinetic energy of the molecules is being converted into potential energy. If the molecules had little kinetic

Figure 14.17A Collisions between molecules

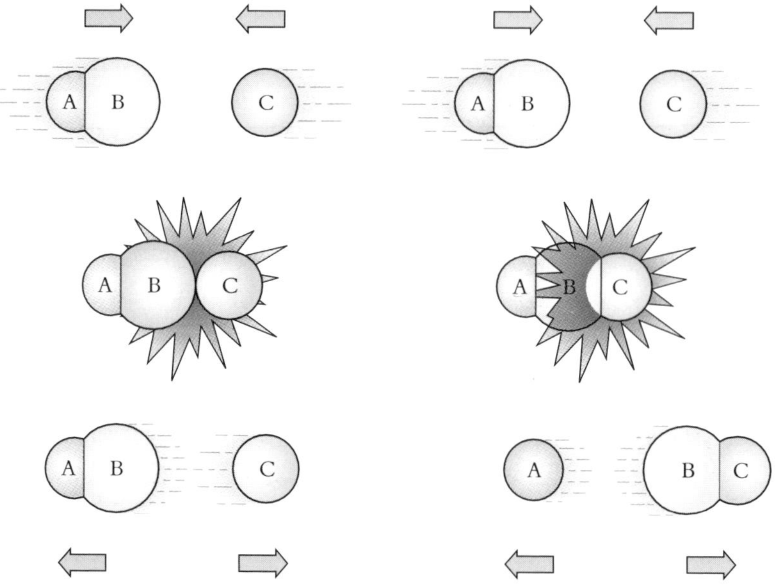

(a) Two slow-moving molecules collide. The electron clouds do not interpenetrate.

(b) Two fast-moving molecules collide. Atoms approach closely, and electron clouds interpenetrate. This leads to reaction.

energy to begin with, i.e., if they were not moving very fast, they will come to a stop before their electron clouds have interpenetrated very much, and then fly apart again without reacting [see Figure 14.17A(a)].

Kinetic energy is converted into potential energy as reacting molecules collide

When two fast-moving molecules collide, they have a lot of kinetic energy that can be converted into potential energy. They are able to overcome the forces of repulsion between their electron clouds and to approach each other closely. The interpenetration of the electron clouds that occurs permits a rearrangement of valence electrons, with the breaking of old bonds and the formation of new bonds, i.e., a chemical reaction [see Figure 14.17A(b)].

A methyl radical and a molecule of hydrogen chloride may react if they collide:

$$CH_3\cdot + HCl \longrightarrow CH_4 + Cl\cdot$$

As the reactant molecules move along the reaction coordinate towards becoming the products, the potential energy passes through a peak ...

The change in potential energy that takes place during the course of one reactive collision (i.e., one that changes the reactant molecules into the product molecules) is shown in Figure 14.17B. The horizontal axis is called the **reaction coordinate**. Positions along the reaction coordinate represent the distance that the reacting species have moved towards forming the products. As $CH_3\cdot$ and HCl approach one another, their potential energy increases to a maximum. The arrangement of atomic nuclei and bonding electrons at the potential energy maximum is called the **transition state**.

... The arrangement of atomic nuclei and bonding electrons at this peak is called the transition state

It can be represented as

$$H_3C\text{-----}H\text{-----}Cl$$

As the new bond, $H_3C—H$, forms, it assists the breaking of the old bond, H—Cl. Once formed, the transition state is transformed into the products. The difference in potential energy between the transition state and the reactants is called the activation energy E_A. The number of reacting species that take part in the formation of the transition state is the **molecularity** of the reaction step [§ 14.2.6]. This reaction step is **bimolecular**.

Figure 14.17B
Potential energy diagram for a single reactive collision

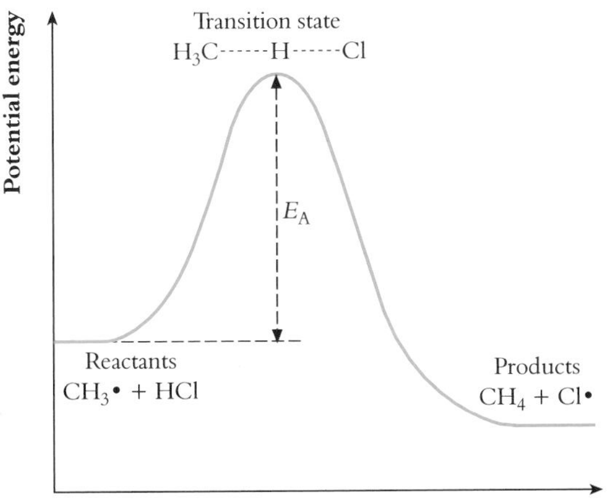

Now consider 1 mole of reactants. Figure 14.17C shows the relationship between the standard enthalpy $H^{\ominus}$ of one mole of the reactants, one mole of the products and one mole of the activated complex. Such a diagram is described as an **enthalpy profile** (or **energy profile**) of the reaction.

Figure 14.17C
An enthalpy profile for a reaction

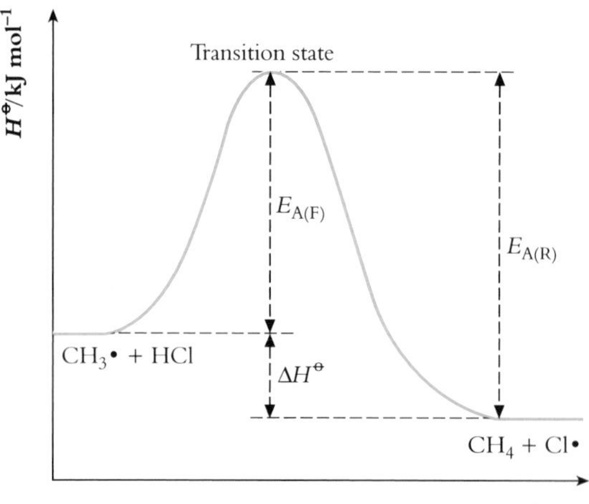

$E_{A(F)}$ and $E_{A(R)}$ in Figure 14.17C represent the activation energies of the forward and reverse reactions. $\Delta H^\ominus$ is the standard enthalpy of reaction:

$$\Delta H^\ominus = H^\ominus_{\text{Products}} - H^\ominus_{\text{Reactants}}$$

As can be seen from Figure 14.17C

$$\Delta H^\ominus = E_{A(F)} - E_{A(R)}$$

The Transition State Theory applies to reactions in solution as well as to gaseous reactions. The alkaline hydrolysis of primary halogenoalkanes (e.g., C_2H_5Br) is discussed in § 29.9.1.

Some reactions take place via a reactive intermediate

Some reactions tale place via a **reactive intermediate**. One example is the nitration of benzene [§ 28.8.2]. Another is the hydrolysis of tertiary halogenoalkanes [§ 29.9.2]:

$$(CH_3)_3CBr \longrightarrow (CH_3)_3C^+ + Br^-$$
$$(CH_3)_3C^+ + H_2O \longrightarrow (CH_3)_3COH + H^+(aq)$$

The reactive intermediate is the carbocation, $(CH_3)_3C^+$. It is preceded by and followed by a transition state. The enthalpy profile for a reaction of this kind is shown in Figure 14.17D.

Figure 14.17D
Enthalpy profile of a reaction which involves a reactive intermediate

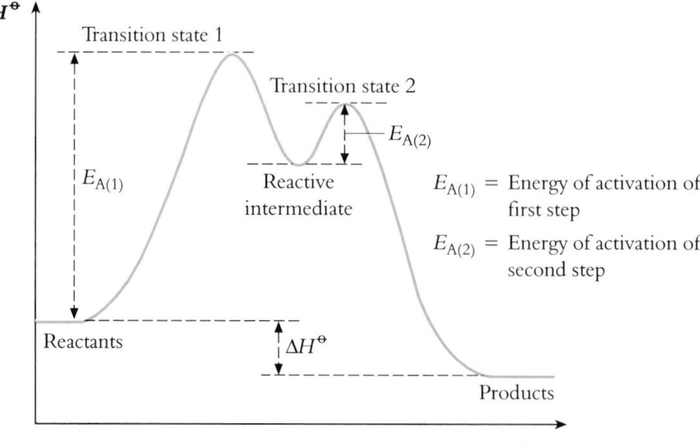

CHECKPOINT 14.17: REACTION KINETICS

1. Explain the following statements:
   (a) The rate of a bimolecular reaction cannot be calculated from the collision frequency alone.
   (b) The increase in the rate of a chemical reaction with an increase in temperature is much greater than the corresponding increase in the collision frequency.
   (c) The presence of a catalyst can make a big change in the rate of a chemical reaction.

2. The Collision Theory postulates that collisions between molecules are necessary before reaction can occur. The theory was developed for bimolecular reactions in the gas phase. How does the Collision Theory apply to reactions in solution?

3. Describe what is meant by a *chain reaction*. By referring to a chosen example, explain what are chain-initiating steps, chain-propagating steps and chain-terminating steps.

4. Why does the probability that a collision will result in reaction depend on the orientation of the colliding molecules? Illustrate your answer with reference to the reactions

$$2HI(g) \longrightarrow H_2(g) + I_2(g)$$

$$CH_3Br + I^- \xrightarrow{\text{in propanone}} CH_3I + Br^-$$

$$(CH_3)_3N + C_2H_5I \xrightarrow[\text{solvent}]{\text{in an organic}} (CH_3)_3\overset{+}{N}C_2H_5 + I^-$$

14.18 RATE-DETERMINING STEP

When a reaction takes place in a number of steps, the rate of the reaction is determined by the rate of the slowest step: the **rate-determining step**. It is like an assembly line in a factory. Imagine Mr Murphy, Mr Shah, Mr Evans and Mr O'Connor all working at different stages on the line. Mr Murphy can do his adjustments on 40 units an hour, Mr Shah and Mr Evans can both process 30 units an hour, while Mr O'Connor, with a more demanding task, can only manage 20 units an hour. The rate at which units leave the assembly line is 20 units an hour – the rate of the slowest step.

Two reactions which take place in a number of steps will be described.

14.18.1 THE IODINATION OF PROPANONE

In a kinetic study of the iodination of propanone ...

The iodination of propanone

$$CH_3COCH_3(aq) + I_2(aq) \longrightarrow CH_3COCH_2I(aq) + HI(aq)$$

is a reaction with interesting kinetics. The reaction is acid-catalysed. From the equation, one might postulate that the rate equation would be

$$-\frac{d[I_2]}{dt} = k[CH_3COCH_3]^a[I_2]^b$$

A study of the reaction will involve finding a and b.

First: find the order with respect to iodine

... the concentration of propanone is kept approximately constant

It is arranged that the propanone concentration is much greater than the iodine concentration, e.g., $[CH_3COCH_3] = 1.00\ \text{mol dm}^{-3}$ and $[I_2] = 0.00500\ \text{mol dm}^{-3}$ so that, at the end of a run, $[CH_3COCH_3] = 0.995\ \text{mol dm}^{-3}$, a decrease of 0.5%. One can say that $[CH_3COCH_3]$ is effectively constant, and

$$-\frac{d[I_2]}{dt} = k_1[I_2]^b \text{ so that } b \text{ can be found.}$$

Solutions of known concentration of (a) propanone, (b) iodine in potassium iodide, and (c) an acid buffer of known pH are prepared and brought to the required temperature in a thermostat bath. The reaction is started by pipetting volumes of the three solutions into a flask, and a stop watch is started. After a few

The reaction is started by mixing the reagents ...

... A sample of the solution is pipetted ...

... The reaction is stopped ...

... Titration gives the iodine concentration

minutes, a sample of the reacting mixture is pipetted from the solution into a sodium hydrogencarbonate solution. This stops the reaction instantly by neutralising the acid. The time at which the reaction stops is recorded. The iodine that remains is determined by titration against a standard solution of sodium thiosulphate. The analysis is repeated at intervals of a few minutes. The volume of thiosulphate required is plotted against the time elapsed since the start of the reaction. Figure 14.18A shows the plot obtained. It is a straight line: the gradient of the graph does not change as the concentration of iodine decreases. This shows that the rate of reaction remains constant as the iodine concentration decreases:

$$-\frac{d[I_2]}{dt} = \text{Constant}$$

It is found that the rate of the reaction remains constant as the iodine concentration decreases

The reaction is zero order with respect to iodine, and the rate equation becomes

$$-\frac{d[I_2]}{dt} = k[CH_3COCH_3]^a$$

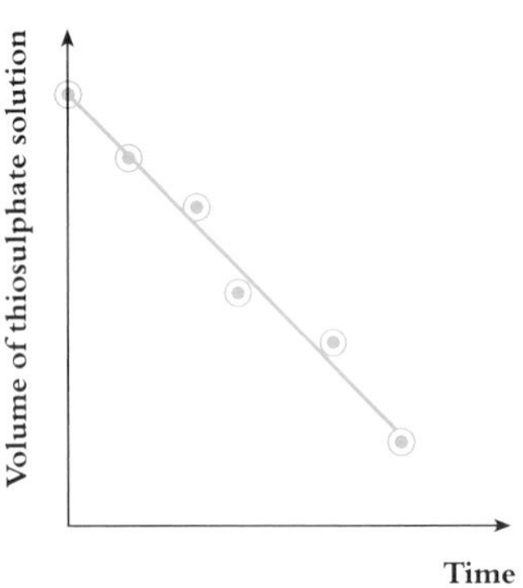

Figure 14.18A Graph of results of iodine titration

Second: find the order with respect to propanone

Runs at different propanone concentrations

The procedure is repeated with different concentrations of propanone. It is found that doubling the concentration of propanone doubles the rate of the reaction: the reaction is first order with respect to propanone:

$$-\frac{d[I_2]}{dt} = k[CH_3COCH_3]$$

Third: find the effect of acid concentration

Runs at different acid concentrations

A set of runs in buffers with different values of pH and fixed concentrations of propanone and iodine shows that the rate of reaction is proportional to the hydrogen ion concentration. The rate equation is

$$-\frac{d[I_2]}{dt} = k[CH_3COCH_3][H^+(aq)]$$

Interpretation of the rate equation

Iodine must be present for iodination to occur, but its concentration does not appear in the rate equation. To explain this, it is suggested that the reaction takes place in steps. It is the rate of the slowest step that determines the rate of the overall reaction. If iodine is involved in a step which is too fast to be rate-determining, it will not appear in the rate equation. The suggested mechanism for the reaction is

From the results of kinetic studies, a mechanism for the reaction can be worked out

$$CH_3-\underset{\underset{O}{\|}}{C}-CH_3 + H^+(aq) \underset{}{\overset{\text{slow}}{\rightleftharpoons}} CH_3-\underset{\underset{\overset{+}{O}-H}{\|}}{C}-CH_3 \overset{\text{fast}}{\rightleftharpoons} CH_3-\underset{\underset{O-H}{|}}{C}=CH_2 + H^+(aq) \quad [1]$$

This is the slow, rate-determining step. An enol is formed. Since halogens are electrophiles [§ 25.8.3], iodine reacts rapidly with the $\rangle C{=}C\langle$ bond through its π-electrons:

$$CH_3{-}\underset{\underset{O-H}{|}}{C}{=}CH_2 + I{-}I \xrightarrow{\text{fast}} CH_3{-}\underset{\underset{O-H}{|}}{\overset{+}{C}}{-}CH_2{-}I + I^- \qquad [2]$$

The intermediate formed has a positive charge on a carbon atom, and fast loses a proton to form the iodoketone:

$$CH_3{-}\underset{\underset{O-H}{|}}{\overset{+}{C}}{-}CH_2{-}I \xrightarrow{\text{fast}} CH_3{-}\underset{\underset{O}{\|}}{C}{-}CH_2{-}I + H^+(aq) \qquad [3]$$

A curved arrow shows the movement of a pair of electrons [see § 27.8.5].

14.18.2 THE OXIDATION OF IODIDE TO IODINE BY HYDROGEN PEROXIDE

The great team of Augustus Harcourt and William Esson devoted their working lives to research on reaction kinetics. One of the reactions they studied in the early part of the twentieth century was the oxidation of iodide by hydrogen peroxide. This reaction, which is still known as the Harcourt–Esson reaction, is an interesting investigation for you to make during the course of a practical session.

The overall reaction is

$$H_2O_2(aq) + 2H_3O^+(aq) + 2I^-(aq) \longrightarrow I_2(aq) + 4H_2O(l)$$

Kinetic measurements showed that the reaction is second order. The rate is proportional to the iodide ion concentration and to the hydrogen ion concentration. To explain the order of reaction, Harcourt and Esson suggested that the reaction takes place in stages.

1. The first step is a slow reaction between H_2O_2 and I^-:

$$H_2O_2(aq) + I^-(aq) \xrightarrow{\text{slow}} IO^-(aq) + H_2O(l) \qquad [1]$$

2. The second step is the establishment of equilibrium by the weak acid, iodic(I) acid, HIO:

$$IO^-(aq) + H_3O^+(aq) \overset{\text{fast}}{\rightleftharpoons} HIO(aq) + H_2O(l) \qquad [2]$$

   This happens rapidly.

3. Lastly, iodine is formed in a fast reaction between iodic(I) acid, hydrogen ions and iodide ions:

$$HIO(aq) + H_3O^+(aq) + I^-(aq) \xrightarrow{\text{fast}} I_2(aq) + 2H_2O(l) \qquad [3]$$

This mechanism agrees with the stoichiometry of the overall reaction. You can check up by adding [1] + [2] + [3]. The kinetics and overall order of reaction are those of the slowest step, [1], which is the rate-determining step for the whole reaction. This research work of Harcourt and Esson illustrates one of the important applications of reaction kinetics, which is to throw light on the mechanisms of chemical reactions.

FURTHER STUDY ▲

CHEMICAL KINETICS

For a chemical reaction to take place, particles of the reactants must collide. They must have enough energy to surmount the energy barrier between reactants and products. §§ 14.2, 14.17

The rate of reaction is affected by

- the temperature
- the concentrations of reactants in solution
- the pressures of gaseous reactants
- the surface areas of solid reactants.

These factors affect the **frequency of collisions** between reacting paricles. §§14.2, 14.3, 14.13, 14.16

The **rate equation** for a reaction is derived from measurements of the concentrations of reactants or products after measured intervals of time. Rate = k[**A**]^m [**B**]^n where k = **rate constant**, m = **order of reaction** with respect to **A**, n = order of reaction with respect to **B**. § 14.14

Link with thermochemistry § 14.9

The rate of a reaction increases with rising temperature. This increases the **energy** with which the reacting particles collide. §§ 14.2, 14.16, 14.17

The **Maxwell–Boltzmann distribution** is a curve which plots the fraction of a set of molecules with a certain energy against the energy. Curves at different temperatures show that, as well as an increase in average energy, the fraction of molecules with high energy increases at high temperature. § 14.3, 14.17.1

For a reaction **A** → **X**, **initial rate** of reaction = k[**A**]$_0{}^m$ where [**A**]$_0$ = initial concentration of **A** and m = order with respect to **A**. Comparison of initial rates at different [**A**]$_0$ gives m. §§ 14.11, 14.14.1

First-order reactions have a **half-life** which is independent of [reactant]. § 14.14.3

There is an energy barrier between reactants and products. In order to react, reactants must possess enough energy to surmount this barrier. This is the **activation energy**. § 14.3, 14.17

Catalysts provide a different reaction route with a lower activation energy. §§ 14.2.5, 14.4–8

Experimental techniques for following the course of reaction include titrimetric analysis, spectrophotometric analysis, measurement of the pressure of a gaseous reactant, measurement of the electric potential or conductance. Measurements are treated graphically. The concentration of reactant or product is plotted against time elapsed since the start of the reaction. § 14.2.1–3, 14.12

The results of kinetic experiments give evidence for the mechanism of a chemical reaction. § 14.18.1–2 Many reactions take place in a number of steps. In many cases only one step is the **rate-determining step**. § 14.18 Many reactions proceed via a **transition state**. § 14.17.2

QUESTIONS ON CHAPTER 14

1. Explain what is meant by the terms *rate constant* and *order of reaction*.

2. For the hydrolysis of ethyl ethanoate

$$CH_3CO_2C_2H_5(l) + H_2O(l) \longrightarrow CH_3CO_2H(aq) + C_2H_5OH(aq)$$

the rate equation is

Rate = $k[CH_3CO_2C_2H_5]$

(a) What is the order of reaction?
(b) Why does $[H_2O]$ not come into the rate equation?

3. In the reaction

$$A \longrightarrow B \longrightarrow C$$

what are the units of the rate constant for

(a) a zero-order reaction, (b) a first-order reaction and (c) a second-order reaction?

4. The radioactive isotope $^{24}Na^+(aq)$ is injected into an animal. If the half-life is 15 hours, how long will it take before the radioactivity has fallen to 25% of the original value?

5. A radioactive isotope decays from an initial count of 160 cpm to 20 cpm in 27 days. What is its half-life?

6. The following results were obtained in a study of the reaction

$$S_2O_8^{2-}(aq) + 2I^-(aq) \longrightarrow 2SO_4^{2-}(aq) + I_2(aq)$$

between peroxodisulphate and iodide ions:

Experiment	$[S_2O_8^{2-}]$/ mol dm^{-3}	$[I^-]$/ mol dm^{-3}	Initial rate $\frac{d[S_2O_8^{2-}]}{dt}$ /mol dm^{-3} s^{-1}
1	0.040	0.040	9.6×10^{-6}
2	0.080	0.040	1.92×10^{-5}
3	0.080	0.020	9.6×10^{-6}

Find the order of reaction with respect to (a) $S_2O_8^{2-}$, (b) I^- and (c) find the rate constant. What is the initial rate of the reaction when $[S_2O_8^{2-}]_0 = 0.12$ mol dm^{-3}, and $[I^-]_0 = 0.015$ mol dm^{-3}?

7. What is meant by the *mechanism* of a chemical reaction? Explain why, when chemical reactions take place in more than one step, the overall rate of the reaction is determined by the rate of the slowest step. Give an example of a reaction of this kind.

8. (a) Write an equation for the decomposition of hydrogen peroxide.
(b) Explain how you could find the rate of decomposition of hydrogen peroxide in aqueous solution, in the presence of a catalyst, using (i) a gas syringe, (ii) a standard solution of potassium manganate(VII).
(c) Tabulated below are values of the initial rate of decomposition at different concentrations of hydrogen peroxide. Plot a graph of initial rate against concentration.

$[H_2O_2]$/mol dm^{-3}	0.100	0.175	0.250	0.300
Rate/10^{-4} mol dm^{-3} s^{-1}	0.593	1.04	1.48	1.82

(d) From the graph, find the order of reaction and the rate constant.

9. (a) (i) Sketch on a copy of the given axes the distributions of molecular energies for a given mass of a gas at two temperatures, $\mathbf{T}_1$, and $\mathbf{T}_2$, where $\mathbf{T}_1$ is higher than $\mathbf{T}_2$. **2**
(ii) State how the areas under your curves are related to each other. **1**
(iii) Using the curves explain why the rates of reaction increase with temperature. **1**

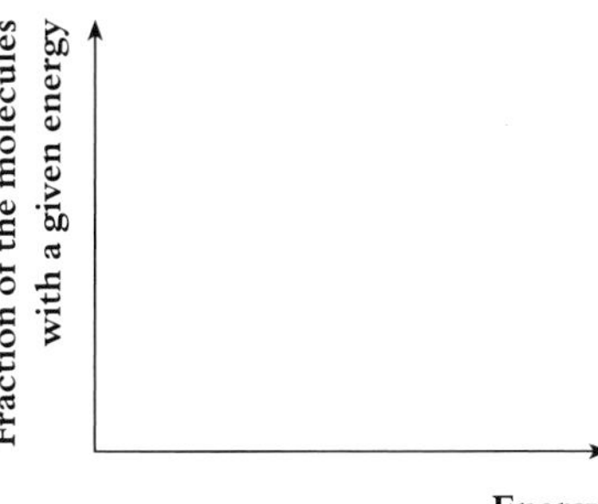

(b) The decomposition of gaseous dinitrogen pentoxide is represented by

$$2N_2O_5(g) \longrightarrow 4NO_2(g) + O_2(g)$$

and is first order with respect to dinitrogen pentoxide.
(i) Write down the rate equation for the decomposition.
(ii) A proposed mechanism for the reaction is made up of the following steps:

$$N_2O_5 \longrightarrow N_2O_3 + O_2 \qquad \text{I}$$

$$N_2O_3 \longrightarrow NO + NO_2 \qquad \text{II}$$

$$NO + 2N_2O_5 \longrightarrow 3NO_2 + N_2O_5 \qquad \text{III}$$

From your rate equation in (b)(i), write down which of the steps would be expected to be the rate-determining step. **1**

(iii) Data for the decomposition of dinitrogen pentoxide at 333 K are shown below.

Pressure of dinitrogen pentoxide remaining/Pa	Rate of reaction/Pa per second
1200	1.04
1000	0.88
800	0.69
400	0.34

Plot these data on graph paper and hence calculate the value of the rate constant at 333 K, giving the appropriate unit. **4**

10 *WJEC (AS/AL)*

10. *(a)* The rate of reaction between propanone, CH_3COCH_3, and iodine to give iodopropanone, CH_3COCH_2I, is found to be independent of $[I_2]$, but directly proportional to $[H^+]$ and directly proportional to [propanone].

(i) Construct the balanced stoichiometric equation for the overall reaction.

(ii) Write the *rate equation* for this reaction, and state the overall order and the units of the rate constant.

(iii) Suggest with reasons which of the following two possible mechanisms, **A** or **B**, fits the observed kinetic data (**X** and **Y** are intermediates):

$$\left.\begin{array}{lll} CH_3COCH_3 + H^+ & \longrightarrow \mathbf{X} & \text{(slow)} \\ \mathbf{X} + I_2 & \longrightarrow \text{products} & \text{(fast)} \end{array}\right\} \mathbf{A}$$

$$\left.\begin{array}{lll} CH_3COCH_3 + I_2 & \longrightarrow \mathbf{Y} & \text{(slow)} \\ \mathbf{Y} + H^+ & \longrightarrow \text{products} & \text{(fast)} \end{array}\right\} \mathbf{B}$$

(iv) Describe the roles of I_2 and H^+ in this reaction.

(v) The reaction between propanone and bromine proceeds by a similar mechanism. How would you expect the rate of this reaction to compare with that of the above reaction? Explain your answer. **10**

(b) Oxides of nitrogen in the atmosphere contribute to the formation of acid rain by catalysing the oxidation of SO_2 to SO_3. Write equations to show how this occurs, and describe the type of catalysis observed here. **2**

C (AL)

11. The following information refers to a procedure to determine the *order of reaction* with respect to iodide ions for the reaction represented by the equation

$$2I^-(aq) + H_2O_2(aq) + 2H^+(aq) \longrightarrow 2H_2O(l) + I_2(aq)$$

Rate is measured by the time taken for the iodine produced to react with a small *fixed amount* of sodium thiosulphate added to the constant volume system. The faster the iodine is produced, the shorter the time taken for the sodium thiosulphate to be used up. The reciprocal of this time can be used as a measure of the initial rate of reaction. The results are given below.

Experiment	[KI(aq)]/ mol dm^{-3}	Time (t)/s	Reciprocal of time ($1/t$)/s^{-1}
I	0.004	74	0.0135
II	0.006	49.4	0.0202
III	0.008	37	0.0270
IV	0.010	30	0.0333
V	0.012	25	0.0400

(a) (i) In the experiments the concentrations of acid and hydrogen peroxide were far more concentrated than that of potassium iodide. Explain why this was necessary. **1**

(ii) In each of the experiments the aqueous hydrogen peroxide was always added last. State why this was necessary. **1**

(iii) Explain why the volume of the system was kept constant for all the experiments. **1**

(b) (i) Plot a graph of the initial rate ($1/t$) on the vertical axis against the concentration of potassium iodide used on the horizontal axis. **2**

(ii) Use your graph to determine the order, **n**, of the reaction with respect to iodide ions. Carefully state your reasoning. **3**

(c) Further studies show that the rate equation for the reaction is

$$\text{Rate}(\text{mol dm}^{-3}\,\text{s}^{-1}) = k[I^-(aq)]^n[H_2O_2(aq)]$$

(i) From this overall rate equation, state what can be deduced about the role of aqueous hydrogen ions. **1**

(ii) State the units of k in the above rate equation. **1**

10 *WJEC (AS/AL)*

12. For the reaction:

$$2NO(g) + O_2(g) \rightleftharpoons 2NO_2(g)$$

the rate equation for the forward reaction is:

$$\text{Rate} = k[NO]^2[O_2]$$

(a) (i) Deduce the units for the rate constant k. **1**

(ii) State and explain how the rate of reaction would change if the concentration of O_2 was doubled, all other factors remaining constant. **2**

(iii) State and explain how the rate of reaction would change if the concentration of NO was halved, all other factors remaining constant. **2**

(b) (i) What is the overall order of the reaction? **1**

(ii) Explain on the basis of collision theory why this reaction is unlikely to occur in a single step. **2**

(c) The first step in a possible mechanism for the reaction above is:

$$2NO \rightleftharpoons N_2O_2$$

(i) Draw dot-and-cross diagrams to show the electronic structures of the two molecules NO and N_2O_2. What feature of the electronic structure of NO would suggest that this is a likely first step in the reaction? **4**

(ii) Explain why the enthalpy change for this step is -163 kJ mol^{-1}, given that the average bond energy for the N—N bond in compounds of nitrogen is $+163$ kJ mol^{-1}. **2**

(iii) Explain why this step does not control the rate of the reaction. Assuming there is one further step in the reaction write an equation for this. **2**

(d) (i) Deduce the effect of increasing the temperature on the position of equilibrium in the first step of the mechanism in *(c)*. **2**

(ii) Discuss the economic implications of increasing the temperature on the overall process, if NO_2 was made industrially by this process. **3**

(e) NO_2 reacts with aqueous sodium hydroxide according to the following equation:

$$2OH^- + 2NO_2 \longrightarrow NO_2^- + NO_3^- + H_2O$$

(i) What type of reaction is this? Justify your answer. **2**

(ii) Deduce the ionic half equations for this reaction. **2**

25 *L (AL)*

13. This question concerns the reaction between aqueous potassium manganate(VII), $KMnO_4$, and an aqueous solution containing both ethanedioic acid, $H_2C_2O_4$, and sulphuric acid.

In the reaction the purple manganate(VII) ion is reduced slowly, in the presence of hydrogen ions, to manganese(II) ions. The resulting solution is colourless.

(a) Design an experiment to determine how the rate of this reaction varies with temperature. **5**

(b) During this reaction the concentration of the Mn^{2+} ions varies with time as shown on the sketch below. The temperature of the mixture remains constant throughout.

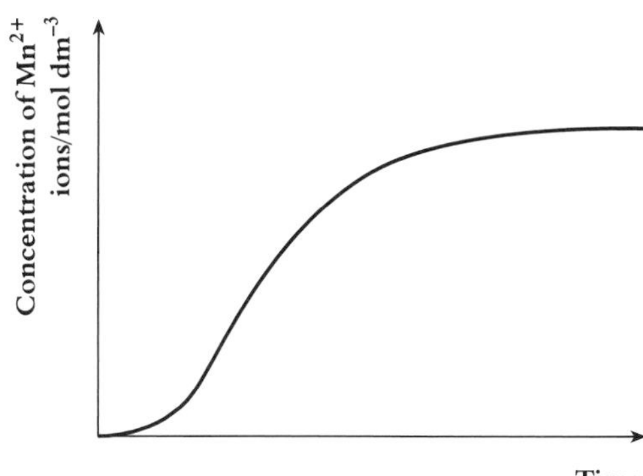

Comment on and explain the shape of the curve. **4**

(c) In the reaction the ethanedioate ion from ethanedioic acid is oxidised to carbon dioxide. Construct ionic half equations and use them to write an overall ionic equation for the reaction of manganate(VII) ions and ethanedioate ions in the presence of hydrogen ions. **5**

(d) 12.5 cm^3 of 0.0800 mol dm^{-3} sodium hydroxide solution was required to completely neutralise both the sulphuric acid and ethanedioic acid in 10.0 cm^3 of the acid solution. Calculate the total acid concentration. **4**

18 *L (AL)*

14. *(a)* Explain why, at a given temperature, hydrochloric acid reacts faster with powdered calcium carbonate than it does with lumps of calcium carbonate. **2**

(b) The collision theory states that reactions can only occur when particles collide. Give a reason why collisions between particles do not always lead to a reaction. **1**

(c) The initial rate of the reaction between substances **A** and **B** was measured in a series of experiments and the following rate equation was deduced.

Rate = k[**A**]^2[**B**]

(i) Copy and complete the table of data below for the reaction between **A** and **B**.

Expt	Initial [A]/ mol dm^{-3}	Initial [B]/ mol dm^{-3}	Initial rate of reaction/ mol dm^{-3} s^{-1}
1	3.0×10^{-2}	4.0×10^{-2}	1.6×10^{-5}
2	6.0×10^{-2}	4.0×10^{-2}	
3	3.0×10^{-2}		6.4×10^{-5}
4		16.0×10^{-2}	1.6×10^{-5}

(ii) Using the data for experiment 1, calculate a value for the rate constant, k, and state its units.

(iii) State the effect on the rate constant, k, of increasing the temperature.

(iv) State the effect on the rate constant, k, of increasing the concentration of **B** at a fixed temperature. **8**

11 *NEAB*

15. Hydrogen peroxide is a covalent compound which, in solution, can act as an oxidising or reducing agent.

(a) It oxidises bromide ions according to the following ionic equation:

$$H_2O_2 + 2Br^- + 2H^+ \longrightarrow Br_2 + 2H_2O$$

The rate law for the reaction is:

$$\text{Rate} = k[H_2O_2][H^+][Br^-]$$

(i) What is the overall order of the reaction?
(ii) What are the units of k?
(iii) Suggest, without experimental detail, how the rate of the reaction may be determined.
(iv) If the concentration of bromide was doubled, assuming all other factors were kept constant, what would be the effect on the rate?
(v) Explain how raising the pH of the reaction mixture affects the rate. **9**

(b) Acidified permanganate is rapidly reduced by hydrogen peroxide.

$$5H_2O_2 + 2KMnO_4 + 3H_2SO_4 \longrightarrow K_2SO_4 + 2MnSO_4 + 8H_2O + 5O_2$$

(i) Describe the colour change which takes place.
(ii) Rewrite the equation as a balanced ionic equation. **3**

(c) Hydrogen peroxide is used in the test for manganese(II) ions. Describe this test giving experimental details and observations. **2**

(d) (i) Draw the structure of hydrogen peroxide using dots and crosses for the outer electrons.
(ii) Suggest and explain the shape of a hydrogen peroxide molecule. **4**

NI

16. *(a)* What distinguishes *heterogeneous* from *homogeneous* catalysis?
Discuss the important features of **each** of the following types of catalysis by stating the type of catalysis involved and explaining the mode of action of the catalyst, giving suitable examples where appropriate.
(i) acid catalysis
(ii) catalysis by transition metal ions in solution
(iii) the removal of atmospheric pollutants from car exhausts
(iv) enzyme catalysis. **15**

(b) Giving suitable examples where appropriate, explain the main features of heterogeneous catalysis. Pay particular attention to:
(i) the way in which heterogeneous catalysts work
(ii) the fact that a given heterogeneous catalyst can be specific to a given reaction
(iii) the relative activity of different metal catalysts
(iv) ways in which catalytic activity might be lowered. **15**

30 *NEAB*

17. *(a)* 2-Bromo-2-methylpropane, $(CH_3)_3CBr$, and sodium hydroxide react together according to the following equation

$$(CH_3)_3CBr + OH^- \longrightarrow (CH_3)_3COH + Br^-$$

The following data give the results of three experiments used to determine the rate equation for the reaction at 25 °C.

Expt	Initial $[(CH_3)_3CBr]$/ mol dm^{-3}	Initial $[OH^-]$/ mol dm^{-3}	Initial rate of reaction/ mol dm^{-3} s^{-1}
1	1.0×10^{-3}	2.0×10^{-1}	3.0×10^{-3}
2	2.0×10^{-3}	2.0×10^{-1}	6.0×10^{-3}
3	2.0×10^{-3}	4.0×10^{-1}	6.0×10^{-3}

From these results it can be deduced that the rate equation is:

$$\text{Rate} = k[(CH_3)_3CBr]$$

(i) Show how the data can be used to deduce that the reaction is first order with respect to $(CH_3)_3CBr$ and zero order with respect to OH^-.
(ii) Calculate a value for the rate constant at this temperature and state its units.
(iii) Calculate the initial rate of reaction when the initial concentration of $(CH_3)_3CBr$ is 4.0×10^{-3} mol dm^{-3} and the initial concentration of OH^- is 1.0×10^{-1} mol dm^{-3}. **6**

(b) Name and outline a mechanism for the reaction of the primary haloalkane, CH_3CH_2Br, with aqueous sodium hydroxide. **4**

(c) Give a simple test to distinguish between the organic products of the reactions in part (*a*) and part (*b*) by stating the reagent used and what you would see with each compound. **3**

13 *NEAB*

Part 3

Inorganic Chemistry

15

PATTERNS OF CHANGE IN THE PERIODIC TABLE

15.1 PRIMO LEVI'S PERIODIC TABLE

Primo Levi was an Italian chemist who was born in 1919. In the 1930s Italy was governed by the fascist dictator Mussolini, and Jews, including Primo Levi, suffered from racial discrimination. In 1939 the Second World War began, and Italy later joined forces with Germany. In 1943 the Allies invaded Italy, the fascist government was overthrown, and Italy changed sides to fight against Germany for the rest of the war. Primo Levi joined a group of Italian partisans to fight against the Germans who still occupied northern Italy. His group was betrayed by a spy who infiltrated them, and he was arrested and sent to Auschwitz concentration camp in Germany. Primo Levi survived the camp in large measure because of his knowledge of chemistry. He was assigned, with a few other chemists, to work in the IG Farben Buna plant, making synthetic rubber for the German war effort. At one time he avoided starvation by drinking glycerol, which he found in the lab, to provide him with precious calories. He also oxidised kerosene to fatty acids and ate them to supplement the wretched diet that the camp provided.

Primo Levi told the story of his imprisonment in a book called 'The Periodic Table' (published in English by Michael Joseph Ltd, 1985, and Abacus 1986). Each chapter is devoted to an element.

In the chapter called cerium Levi tells the story of how he found rods of cerium in the laboratory where he was working as a prisoner. He and a co-prisoner, Alberto, used the rods to make flints for cigarette lighters which they sold to civilian workers who came into the laboratory daily. The money they made bought bread to help them to stay alive on concentration camp rations. Sadly, Alberto did not survive the war; he died on a forced march from Auschwitz to another camp.

In the chapter called vanadium, Primo Levi describes how after the war he was working in Italy in a factory making varnish. The firm imported a shipment of resin from Germany. It proved unsatisfactory because the varnish made from it did not dry properly. When Levi took up the matter with the German supplier he found that he was dealing with Dr Lothar Muller, who had supervised his work as a prisoner in Auschwitz.

The chromium chapter is set in Italy just after the war. Levi was given the problem of what to do with a large batch of anti-rust lead chromate paint which had 'livered', that is, turned into a gelatinous mass that looked like a lump of orange liver. On studying the instructions for making the paint, he found that 'add 2 or 3 drops of base' had been changed to 'add 23 drops of base'. He decided to try mixing and stirring the paint with ammonium chloride, which is sufficiently acidic to neutralise the excess of base. To his delight, the paint came out of the mixer fluid and smooth. The management were so pleased that they gave him a rise and two bicycle tyres – a great treasure in post-war Italy.

The chapter called arsenic refers to an earlier time when Levi was working as an analytical chemist. A client came in to ask him to analyse a packet of sugar. This was in the days before spectroscopic methods of analysis [see Chapter 35]. The first test Levi did was to dissolve a sample. He was suspicious when he obtained a cloudy solution. When he burnt a sample he obtained the smell of burnt sugar and also another smell which he suspected was arsenic. He confirmed this by passing hydrogen sulphide through the solution and obtaining a yellow precipitate of arsenic sulphide. When the client came back to learn the results of the analysis he took the news with amazing calm, explaining that he suspected that the packet came from a business rival. He did not want to sue, but he would have a quiet word with his rival.

There are 21 chapters in Primo Levi's book, all of them fascinating. You should read it! In this book we are going to study the Periodic Table in a systematic way, focussing on relationships between elements.

15.2 PHYSICAL PROPERTIES

The Periodic Table

The Periodic Table was introduced in § 2.6. The reason why this arrangement of the elements was proposed is that the elements show **periodicity** in their physical and chemical properties. **Periodic** means repeating after a regular interval. The way in which alkali metals, halogens and noble gases occur at regular intervals was described in § 2.9. The regular intervals are a consequence of the way in which electron shells are filled [§ 2.8].

The physical properties of elements show periodicity, i.e., they repeat after an interval of 8 or 18 elements

Some of the physical properties which show periodicity are illustrated in Figures 15.2A to 15.2F. The melting temperature [see Figure 15.2A] of an element depends on the strength of the bonds and also on the structure of the element. When metals melt, some metallic bonding remains in the liquid phase. When macromolecular substances, such as carbon, melt, nearly all the bonds have to be broken to melt the solid. The standard enthalpy of melting varies in a similar way with the proton number (atomic number) of the element.

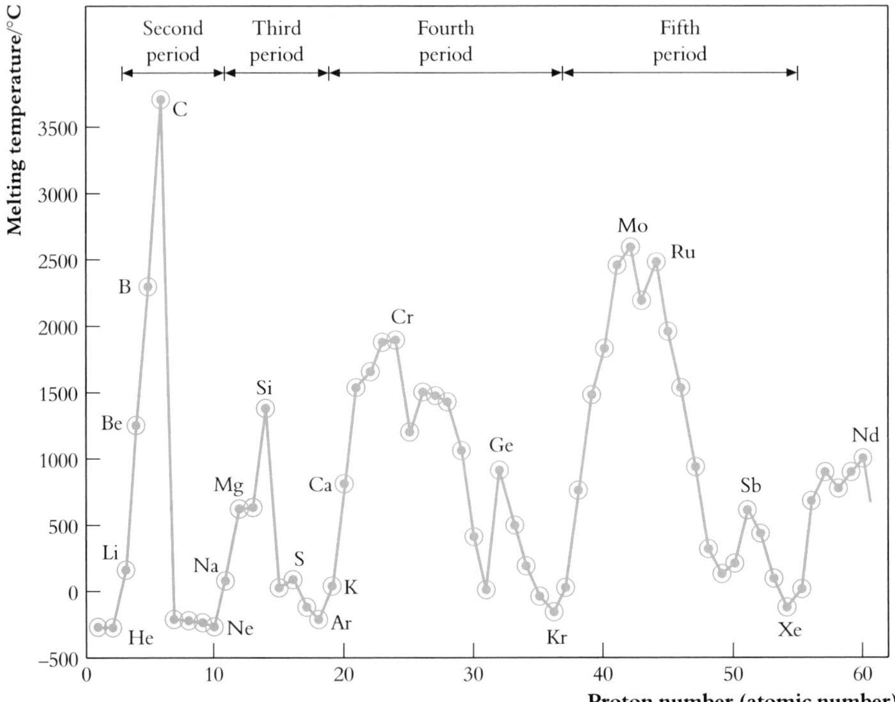

Figure 15.2A Periodicity of melting temperatures of the elements

The variation of boiling temperature with proton number is shown in Figure 15.2B. The standard enthalpies of vaporisation [§ 10.7.2] vary in a similar manner.

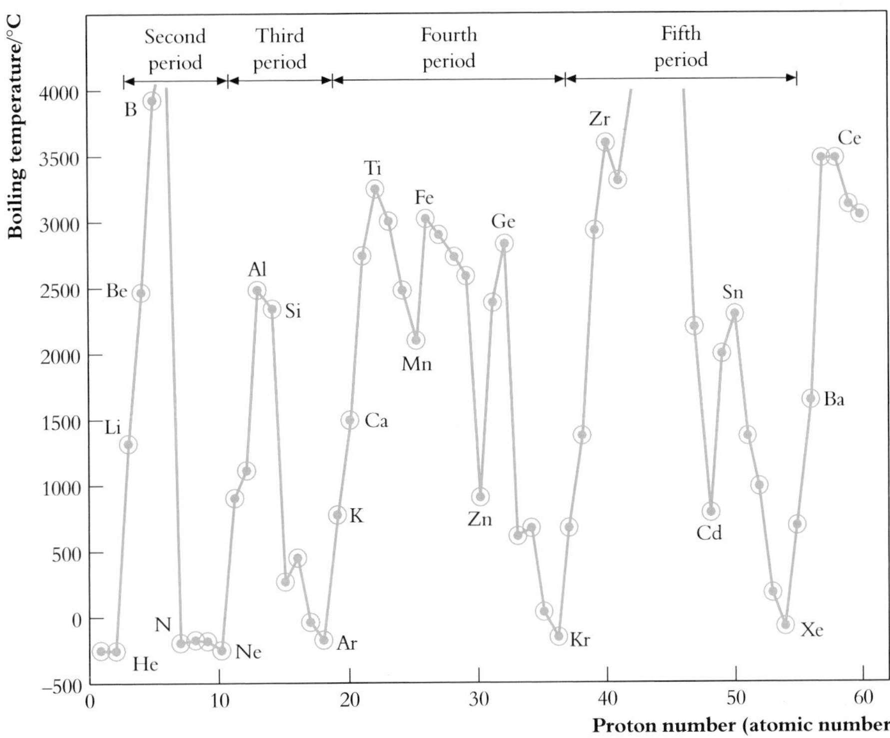

Figure 15.2B Periodicity of the boiling temperatures of the elements

Atomic radius is plotted against proton (atomic) number in Figure 15.2C. The sizes of some atoms and ions are compared in Figure 15.2D.

The radii of atoms and ions increase from top to bottom of a group ...

From Figure 15.2D, you can see that ionic and atomic radii increase from top to bottom of any group. Both the nuclear charge and the number of electron shells increase down a group. Although the increasing nuclear charge decreases the sizes of individual shells, the effect of adding a shell is the dominating effect.

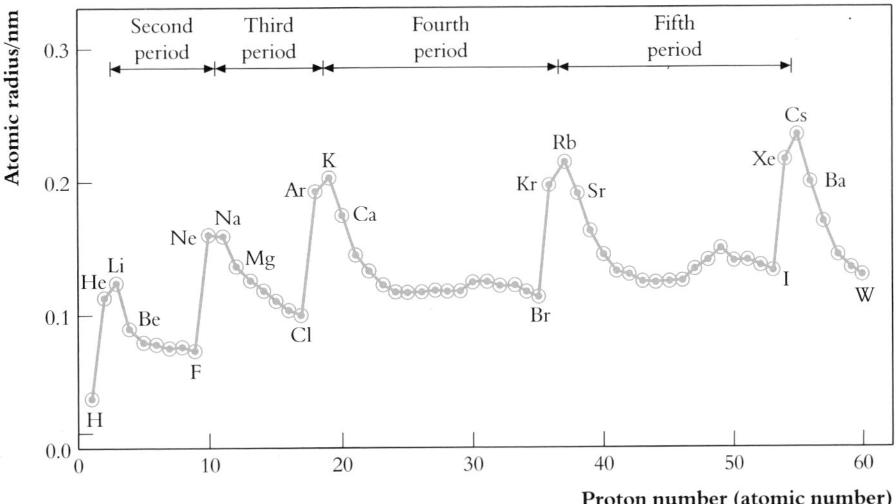

Figure 15.2C Variation of atomic radius with proton number. (The values for noble gases are van der Waals radii. The values for the other elements are single covalent bond radii)

Figure 15.2D
Atomic and ionic radii
(○ = 0.10 nm = 100 pm)

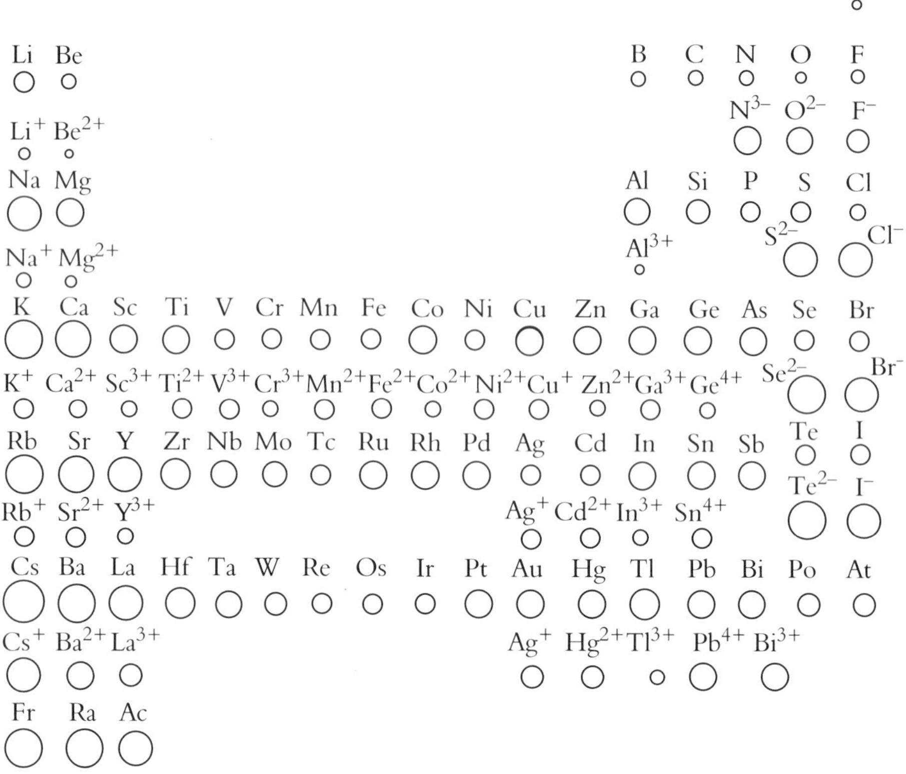

... as shells are added ...

... and decrease from left to right across a period as the effective nuclear charge increases

Covalent and ionic radii both decrease from left to right across any period of the Periodic Table. In the second period, (Li–F), the nuclear charge increases from 3 to 9. As the nuclear charge increases, it pulls the $n = 1$ electrons closer to the nucleus, and the radius of the $n = 1$ shell decreases. The effect on the $n = 2$ electrons is complicated by the fact that they are *screened* or shielded from the nucleus by the $n = 1$ shell, so that the effective nuclear charge is less than the actual nuclear charge. For example, in lithium the outermost electron is attracted by a nucleus with a charge of +3 screened by two electrons. The net nuclear charge is closer to +1 than +3. In beryllium, the $n = 2$ electrons are attracted by a nucleus which has a charge of +4, and is screened by 2 electrons, which make its effective charge close to +2. Nevertheless, reading from left to right across a period, the effective nuclear charge increases and causes a steady decrease in atomic radius across a period [see Figure 15.2D]. A comparison of ions with the same numerical charge, e.g., $\mathbf{M}^{2+}$, shows that ionic radii follow the same pattern.

A series of transition metals differ little in atomic radius

Transition metals show a very small change in atomic radius across a series. In the first transition series, beginning with scandium, the size of the atoms is governed by the $n = 4$ shell while the additional electrons are entering the $n = 3$ shell.

Ionisation energy varies in a periodic manner ...

... reaching a peak at each noble gas

The first ionisation energy [see Figure 15.2E] shows periodic variation. It reaches a peak at each noble gas. From helium to lithium and from neon to sodium, the ionisation energy decreases sharply. The additional electrons (a 2s electron in Li and a 3s electron in Na) are much more easily removed than the electrons in the noble gases. Across the periods from lithium to neon and from sodium to argon, the increasing nuclear charge makes it more difficult to remove an electron, and the ionisation energy increases. The increasing ionisation energies for the removal of successive electrons are shown in Figure 2.5C, § 2.5.

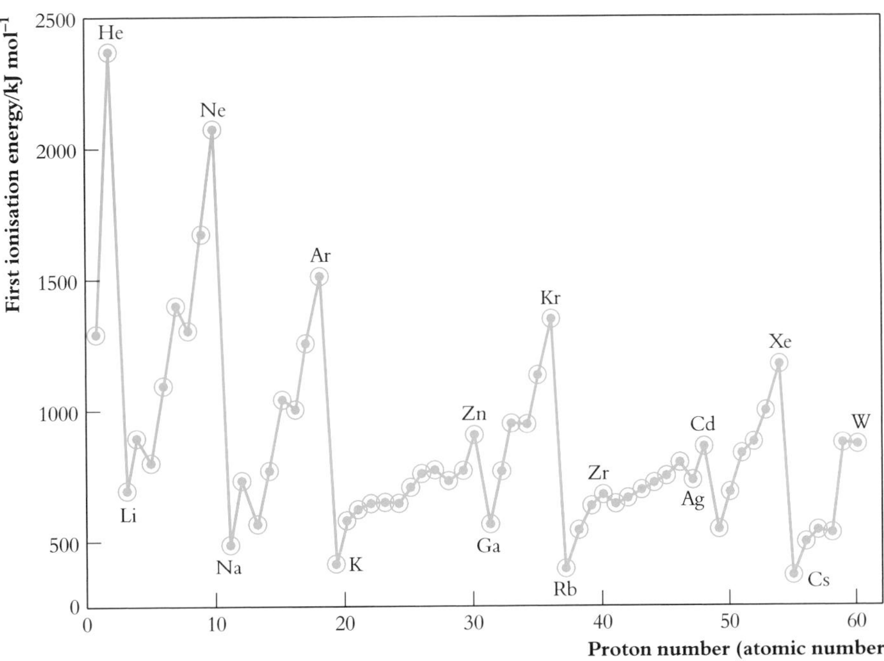

Figure 15.2E
Variation of first ionisation energy with proton number

Electron affinity varies in a regular manner

... becoming more negative across a period ...

... less negative down a group

Electron affinity [§§ 2.3.3, 10.12] is an important quantity in determining whether a non-metallic element will form ionic compounds. Across a period, as nuclear charge increases, electron affinity increases (becomes more negative, more exothermic). If an electron enters a shell close to the nucleus it becomes more tightly bound, with the release of more energy, than if it enters a shell distant from the nucleus. Electron affinity therefore decreases (becomes less negative, less exothermic) from top to bottom of a group.

Electronegativity increases across a period from left to right and decreases down a group

The property of elements called electronegativity was covered in § 4.6.3. The difference in electronegativity between bonded atoms determines the percentage of ionic character in a covalent bond [see Figure 4.6E, § 4.6.3]. Across a period, as nuclear charge increases, electronegativity increases. It decreases down a group as the number of electron shells increases. An element with a very low electronegativity is described as **electropositive**. Figure 15.2F shows how the electronegativity of an element is related to its position in the Periodic Table.

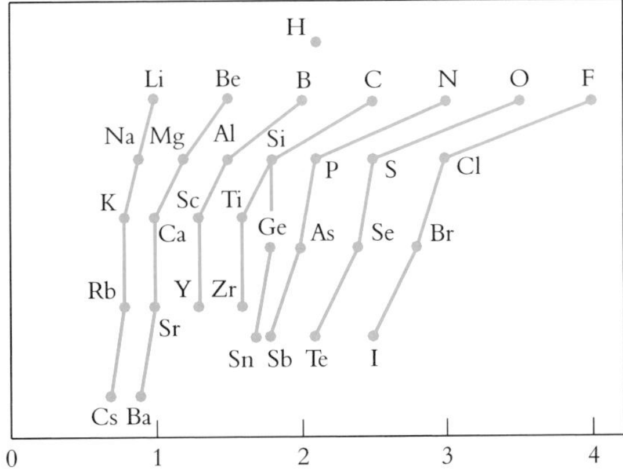

Figure 15.2F
The electronegativity scale of the elements, showing relation to the Periodic Table (after L Pauling)

Other physical properties which illustrate periodicity are the standard enthalpy of atomisation, the standard enthalpy of hydration of ions and atomic volume.

15.3 ELEMENTS OF THE S BLOCK, D BLOCK AND P BLOCK

The Periodic Table is divided into the s block, the d block and the p block; see Figure 15.3A. Elements whose valence electrons are in an s subshell are in the s block, elements whose valence electrons are in a d subshell are in the d block, and elements with valence electrons in a p subshell occupy the p block. The properties of an element are related to its electron configuration, and therefore to its position in the Periodic Table; see Table 15.3.

Figure 15.3A The Periodic Table in outline

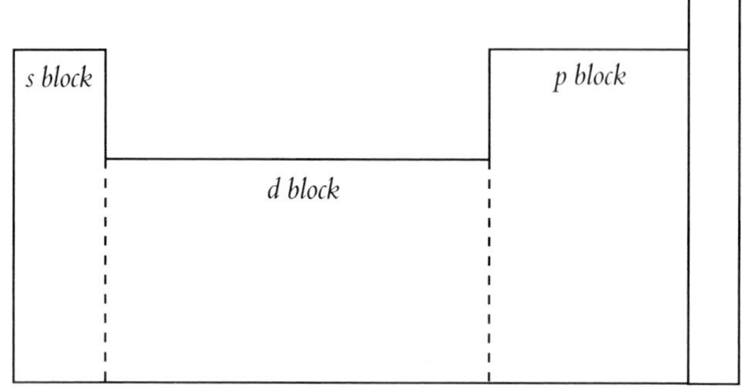

Table 15.3 Properties of s block, d block and p block elements

The properties of s block, d block and p block elements

	s block	d block	p block
Structure and bonding	Form simple ionic compounds.	Oxides and chlorides are either ionic with a high degree of covalent character or covalent. The structure may be layered or macromolecular.	The oxides, hydroxides and chlorides are covalently bonded molecular compounds.
Acid–base properties	The oxides and hydroxides are basic; the chlorides are neutral.	The oxides and hydroxides are basic and insoluble with some amphotericity. The chlorides are hydrolysed to give acidic solutions.	The oxides, chlorides and hydroxides are acidic.
Redox properties	Strong reductants, e.g. the action of sodium on water.	Weak reductants, e.g. the action of iron on water.	Trend from weak reductants through weak oxidants to very strong oxidants in Group 7.
Solubility and complexing properties	Do not form complexes.	Form many complexes with a wide range of ligands. Some chlorides are insoluble.	The anions act as ligands in complex formation.

15.4 THE ELEMENTS OF PERIOD 3

The changing properties across a period are related to the change in the nuclear charge and size of the atoms and the increasing number of outer-shell electrons. The major properties which change across a period are:

- structure and bonding,
- acid–base properties,
- redox properties,
- solubility and complexing properties.

The third period starts with sodium in Group 1 and continues through magnesium, aluminium, silicon, phosphorus, sulphur, chlorine and argon.

The third period ... ... Na, Mg, Al, Si, P, S, Cl, Ar ... ... with the exception of the noble gas argon ... ... shows gradation in properties ...

Standard electrode potential/V

Na	Mg	Al	Si	P	S	Cl
−2.71	−2.37	−1.66	–	–	−0.48	+1.36
Powerful reducing agents			Very weak reducing agent	Weak reducing agents and weak oxidising agents		Powerful oxidising agent

Examples of redox reactions

The first element in the equation is acting as a reducing agent:

$$2Na(s) + 2H_2O(l) \longrightarrow H_2(g) + 2NaOH(aq)$$
$$Mg(s) + 2H^+(aq) \longrightarrow Mg^{2+}(aq) + H_2(g)$$
$$2Al(s) + 3Cl_2(g) \longrightarrow 2AlCl_3(s)$$
$$Si(s) + 2Cl_2(g) \longrightarrow SiCl_4(l)$$
$$2P(s) + 5O_2(g) \longrightarrow 2P_2O_5(s)$$
$$S(s) + Cl_2(g) \longrightarrow SCl_2(l)$$

... from reducing agents to oxidising agents ...

The first element in the equation is acting as an oxidising agent:

$$S(s) + Cu(s) \longrightarrow CuS(s)$$
$$Cl_2(g) + H_2(g) \longrightarrow 2HCl(g)$$
$$S(l) + H_2(g) \longrightarrow H_2S(g)$$

Vigour of reaction with oxygen

... in reaction with oxygen ...

Na Mg Al	Si	P S	Cl Ar
React vigorously	Reacts slowly	React vigorously	Do not react

Electronegativity

... in electronegativity ...

Na	Mg	Al	Si	P	S	Cl
0.9	1.2	1.5	1.8	2.1	2.5	3.0

Compound formation

... in the type of bonding ...

Na	Mg	Al	Si	P	S	Cl	Ar
Form cations and therefore ionic compounds			Form covalent compounds		Form ionic compounds and covalent compounds		Does not form stable compounds

... in oxidation states ...

Oxidation states in hydrides, oxides and chlorides

Na	Mg	Al	Si	P	S	Cl	Ar
+1	+2	+3	+4 in oxides and chlorides	+3, +5 in oxides and chlorides	+6 in oxides, +4 in oxides and chlorides +2, +1 in chlorides	+7, +5, +3, +1 in oxides −1 in HCl	0
in hydrides oxides and chlorides							

... and reaction with hydrogen

Reaction with hydrogen

Na	Mg	Al	Si	P	S	Cl	Ar
React vigorously when heated to form ionic, basic hydrides		Do not react			React when heated to form covalent, acidic hydrides		Does not react

Table 15.4
Summary of reactions of the elements of Period 3

The reactions of the elements with chlorine, oxygen, hydrogen and cold dilute sulphuric acid

Element	Heat in dry chlorine	Heat in dry oxygen	Heat in dry hydrogen	Add cold dilute sulphuric acid
Na	Very vigorous reaction gives Na^+Cl^-	Very vigorous reaction gives $(Na^+)_2O^{2-}$ + $(Na^+)_2O_2^{2-}$	Very vigorous reaction gives Na^+H^-	Dangerously violent reaction gives $H_2(g)$ + $(Na^+)_2SO_4^{2-}$
Mg	Vigorous reaction gives $Mg^{2+}(Cl^-)_2$	Very vigorous reaction gives $Mg^{2+}O^{2-}$	Vigorous reaction gives $Mg^{2+}(H^-)_2$	Very vigorous reaction gives $H_2(g)$ + $Mg^{2+}SO_4^{2-}$
Al	Vigorous reaction gives Al_2Cl_6	Vigorous reaction gives $(Al^{3+})_2(O^{2-})_3$	No reaction	Vigorous reaction, after oxide layer is removed, gives $H_2(g)$ + $(Al^{3+})_2(SO_4^{2-})_3$
Si	Slow reaction gives $SiCl_4$	Slow reaction gives SiO_2	No reaction	No reaction with dilute acid
P	Slow reaction gives PCl_3, PCl_5	Vigorous reaction gives P_2O_3, P_2O_5	No reaction	No reaction with dilute acid
S	Slow reaction gives SCl_2, S_2Cl_2	Slow reaction gives SO_2	Very slow reaction gives H_2S	No reaction
Cl	No reaction	No reaction	Vigorous reaction gives HCl	No reaction
Ar	No reaction	No reaction	No reaction	No reaction

CHECKPOINT 15.4: THE ELEMENTS OF PERIOD 3

1. *(a)* Write the electronic configurations of the following elements, given their atomic numbers: Na(11), Ca(20), V(23), N(7).
   (b) How would you expect the oxides of Na, V and N to differ in bonding, structure, formulae and acid–base character?

2. *(a)* Describe the changes across the elements of Period 3 in
   (i) electronegativity
   (ii) standard electrode potential
   (iii) the character of the oxides as oxidants or reductants.
   (b) Give one example of a Period 3 element acting as (i) an oxidant (ii) a reductant.

3. *(a)* Name the type of bonding present in gaseous hydrogen chloride. By means of a dot-and-cross diagram, showing the outer electrons only, show how the bond is formed from the two atoms.
   (b) Explain why the bond is polar.
   (c) Write an equation for the reaction of hydrogen chloride with water.
   (d) Name the type of bonding present in sodium chloride. By means of a dot-and-cross diagram, showing the outer electrons only, show how the bond is formed from the two atoms.

15.5 COMPOUNDS OF PERIOD 3

The chemical behaviour of compounds is determined by the character of the bonds in the compound.

15.5.1 OXIDES AND HYDROXIDES

Ionic oxides are basic

Electrovalent oxides are basic. The high negative charge/volume ratio of the O^{2-} ion makes it act as a proton-acceptor, e.g.

$$2Na^+O^{2-}(s) + H_2O(l) \longrightarrow 2Na^+(aq) + 2OH^-(aq)$$

Covalent oxides are acidic or neutral

Covalent oxides are either acidic or neutral. An oxide like SO_3 cannot act as a proton-donor, but it can give rise to oxonium ions by acting as an electron-acceptor (a Lewis acid, see § 12.6.1).

$$SO_3(s) + 2H_2O(l) \longrightarrow H_3O^+(aq) + HSO_4^-(aq)$$

Oxides of intermediate bond character are amphoteric.

Oxides

No oxide of argon exists. The oxides of Period 3 are shown in Table 15.5A.
Reaction with water

The compounds of Period 3, Na to Cl, are reviewed. The chlorides of Period 3 show gradation ...
... from solid through liquid to gas ...
... from ionic to covalent bonding ...
... from dissolving in water to being hydrolysed by water

e.g.

$$Na_2O(s) + H_2O(l) \longrightarrow 2NaOH(aq)$$
sodium oxide → sodium hydroxide

$$SO_3(g) + H_2O(l) \longrightarrow H_2SO_4(aq)$$
sulphur(VI) oxide (sulphur trioxide) → sulphuric acid

$$SO_2(g) + H_2O(l) \longrightarrow H_2SO_3(aq)$$
sulphur(IV) oxide (sulphur dioxide) → sulphurous acid

$$Cl_2O_5(g) + H_2O(l) \longrightarrow 2HClO_3(aq)$$
chlorine(V) oxide → chloric(V) acid

Table 15.5A

Oxides	Na_2O	MgO	Al_2O_3	SiO_2	P_2O_5	SO_2 SO_3	Cl_2O_7
Structure	Ionic	Ionic	Ionic with covalent character	Macromolecular covalent		Molecular covalent	Molecular covalent
Acid–base character	Basic	Basic	Amphoteric	←		Acidic	→

Hydroxides

The oxides show gradation ...

... from ionic and basic to covalent and acidic

Na Mg	Al	Si P	S Cl	Ar
Ionic	Amphoteric	Acidic	Strongly acidic	None
Strongly basic				

The hydroxides show gradation ...
... from basic to acidic

The alkali metals in Group 1 are named for their hydroxides which are strongly basic and are alkalis (soluble bases). The alkaline earths in Group 2 are named for their strongly basic hydroxides. In the p block the corresponding compounds are strongly acidic, e.g. sulphurous acid, H_2SO_3, sulphuric acid, H_2SO_4, chloric(I) acid, HClO, chloric(V) acid, $HClO_3$.

15.5.2 HALIDES

Many ionic chlorides dissolve in water

Electrovalent chlorides simply dissolve in water, with the exception of a few sparingly soluble chlorides, e.g.

$$Na^+Cl^-(s) + aq \longrightarrow Na^+(aq) + Cl^-(aq)$$

Many covalent chlorides are hydrolysed

Covalent chlorides may, like CCl_4, neither dissolve in nor react with water or they may, like PCl_3, be hydrolysed to an acid and hydrogen chloride:

$$PCl_3(l) + 3H_2O(l) \longrightarrow H_3PO_3(aq) + 3HCl(aq)$$

Chlorides of polarising cations are acidic in solution

The phosphorus atom has empty d orbitals, into which electrons from the oxygen atoms in a water molecule can coordinate. Coordination is followed by hydrolysis. Chlorides of polarising cations may be partially hydrolysed:

$$BiCl_3(aq) + H_2O(l) \rightleftharpoons BiOCl(s) + 2HCl(aq)$$

Table 15.5B

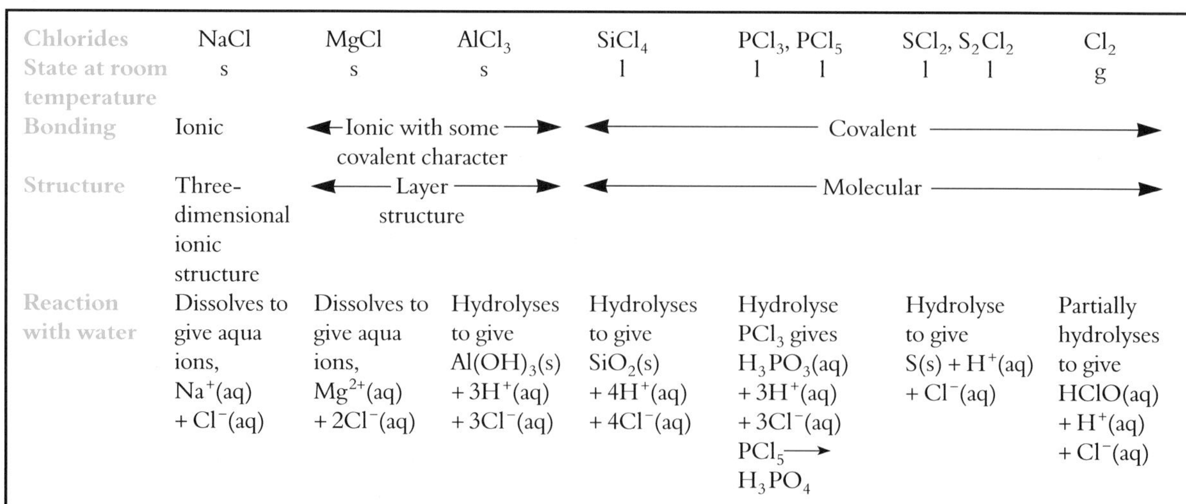

Chlorides	NaCl	MgCl	$AlCl_3$	$SiCl_4$	PCl_3, PCl_5	SCl_2, S_2Cl_2	Cl_2
State at room temperature	s	s	s	l	l l	l l	g
Bonding	Ionic	← Ionic with some covalent character →		←	Covalent		→
Structure	Three-dimensional ionic structure	← Layer structure →		←	Molecular		→
Reaction with water	Dissolves to give aqua ions, $Na^+(aq) + Cl^-(aq)$	Dissolves to give aqua ions, $Mg^{2+}(aq) + 2Cl^-(aq)$	Hydrolyses to give $Al(OH)_3(s) + 3H^+(aq) + 3Cl^-(aq)$	Hydrolyses to give $SiO_2(s) + 4H^+(aq) + 4Cl^-(aq)$	Hydrolyse PCl_3 gives $H_3PO_3(aq) + 3H^+(aq) + 3Cl^-(aq)$ $PCl_5 \longrightarrow H_3PO_4$	Hydrolyse to give $S(s) + H^+(aq) + Cl^-(aq)$	Partially hydrolyses to give $HClO(aq) + H^+(aq) + Cl^-(aq)$

The solutions of chlorides and other salts of highly polarising cations are acidic on account of salt hydrolysis [§§ 12.9, 19.4.1].

$$[Al(H_2O)_6]^{3+}(aq) + H_2O(l) \rightleftharpoons [Al(OH)(H_2O)_5]^{2+}(aq) + H_3O^+(aq)$$

Chlorides

No chloride of argon exists. The chlorides of Period 3 are shown in Table 15.5B.

15.5.3 HYDRIDES

Ionic hydrides

Ionic hydrides are formed by Group 1 and 2 metals

The hydrides react with water and burn in air

Ionic hydrides are formed by the metals of Groups 1 and 2. They are made by passing hydrogen over the heated metal. These hydrides are crystalline solids with structures similar to the corresponding halides. They react vigorously with water to give hydrogen, and burn vigorously in air:

$$H^-(s) + H_2O(l) \longrightarrow H_2(g) + OH^-(aq)$$

$$2H^-(s) + O_2(g) \longrightarrow O^{2-}(s) + H_2O(g)$$

Figure 15.6F shows where in the Periodic Table ionic hydrides are formed.

Covalent hydrides

Covalent hydrides are molecular and often gaseous

Covalent hydrides are molecular. Except for water and hydrogen fluoride, which are associated by hydrogen bonding, covalent hydrides are gaseous.

Interstitial hydrides

In a metallic structure (e.g. Figures 6.8A and B) there are holes between the metal atoms. Some of these holes are small enough to retain hydrogen atoms with the formation of an **interstitial hydride**. The composition of an interstitial hydride depends on the hydrogen pressure. The formula may be between $\mathbf{M}H_2$ and $\mathbf{M}H_0$. (where **M** is the symbol of the metal).

Trends across period 3 hydrides

	Na Mg	Al	Si P	S Cl	Ar
Bonding	Ionic	Unstable >150 °C	Unstable	Covalent	None formed
Reaction with water	Basic	Reacts $\longrightarrow H_2$	Insoluble	Acidic	

The hydrides show gradation ...
... from ionic and basic to covalent and acidic

The hydrides of Groups 1 and 2 are ionic and basic, e.g.

$$Na^+(s)H^-(s) + H_2O(l) \longrightarrow Na^+(aq)OH^-(aq) + H_2(g)$$

The hydrides of Groups 6 and 7 are covalent and acidic, e.g.

$$HCl(g) + H_2O(l) \longrightarrow H_3O^+(aq) + Cl^-(aq)$$

15.5.4 SALTS

Thermal stability is related to bond character

Nitrates, sulphates and carbonates in which the electrovalent character of the bonds is high are stable to heat. Salts with more polarising cations are more easily decomposed by heat: calcium carbonate is decomposed, whereas sodium carbonate is stable. Hydrogencarbonates do not exist after Group 2, and solid hydrogencarbonates are obtained only in Group 1.

Some compounds are tabulated

	Na	Mg	Al	Si	P	S	Cl	Ar
Carbonate	✓	✓	← None formed →					
Hydrogencarbonate	✓	✓	← None formed →					
Sulphate	✓	✓	✓	← None formed →				
Nitrate	✓	✓	✓	← None formed →				

CHECKPOINT 15.5

1. Choose from the elements: Na, Mg, Al, Si, P, S, Cl, Ar.
   (a) Name an element which exists as molecules containing (i) 1 atom, (ii) 2 atoms, (iii) 4 atoms, (iv) 8 atoms.
   (b) Say which elements are (i) s block, (ii) d block, (iii) p block.

2. Magnesium chloride is a solid of high melting temperature, aluminium chloride is a solid which sublimes readily at 180 °C, and silicon tetrachloride is a volatile liquid. Explain how the differences in chemical bonding account for these differences in volatility.

3. Draw dot-and-cross diagrams of the outer electron shells to show the bonding in NaCl, MgS, PH_3 and $SiCl_4$.

4. Choose from the elements: Na, Mg, Al, Si, P, S, Cl, Ar.
   (a) List the elements that react readily with cold water to form alkaline solutions.
   (b) List the elements that have hydrides with low boiling temperatures.
   (c) Give the formulae of the oxides of the elements in their highest oxidation states and the formulae of the corresponding acids or hydroxides.
   (d) List the elements that form nitrates.
   (e) What is the most ionic compound that can be formed by the combination of two of these elements?
   (f) Which element has both metallic and non-metallic properties?
   (g) List the elements that normally exist as molecules.

5. Describe the pattern in the chemistry of the chlorides of the elements from Na to S in Period 3 of the Periodic Table.

15.6 OXIDES, CHLORIDES AND HYDRIDES IN THE PERIODIC TABLE

In § 15.5 we have looked at the gradation in properties across Period 3. In this section are summarised for the Periodic Table as a whole:

- the bonding in oxides, chlorides and hydrides [Figure 15.6A].

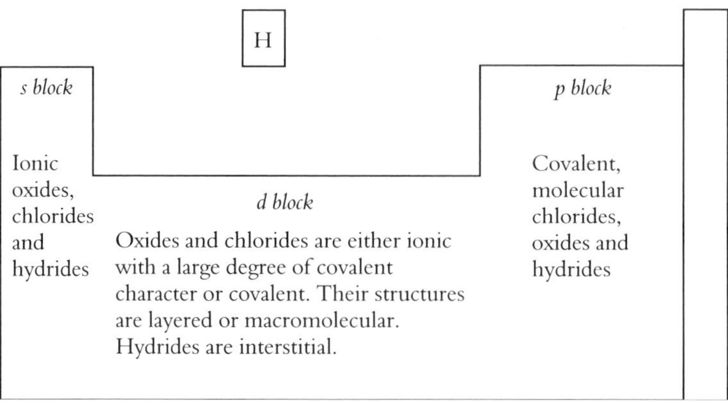

Figure 15.6A Bonding in oxides, chlorides and hydrides

- redox properties [Figure 15.6B]
- the structure of oxides and their acid–base character [Figures 15.6C and D]
- the nature of chlorides [Figure 15.6E]
- the nature of hydrides [Figure 15.6F]
- the solubility of salts and complex ion formation [Figure 15.6G].

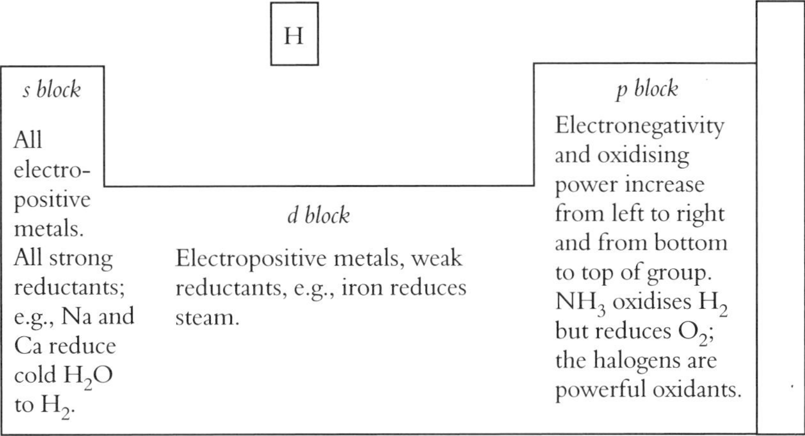

Figure 15.6B
Redox properties and the Periodic Table

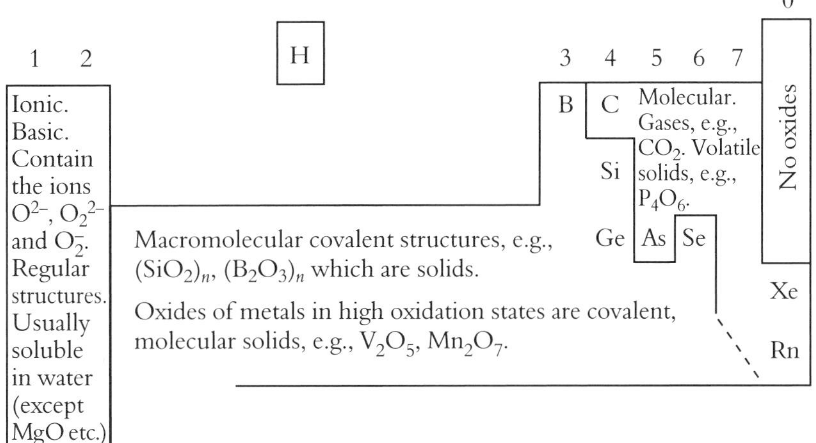

Figure 15.6C
Classification of oxides according to structure

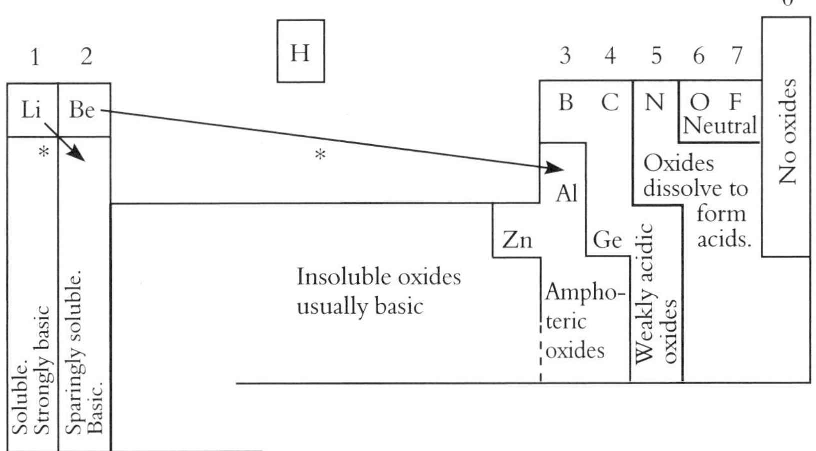

Figure 15.6D
The acid–base character of oxides

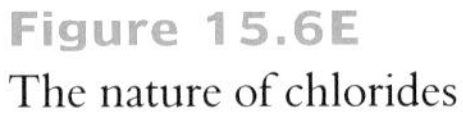

Figure 15.6E
The nature of chlorides

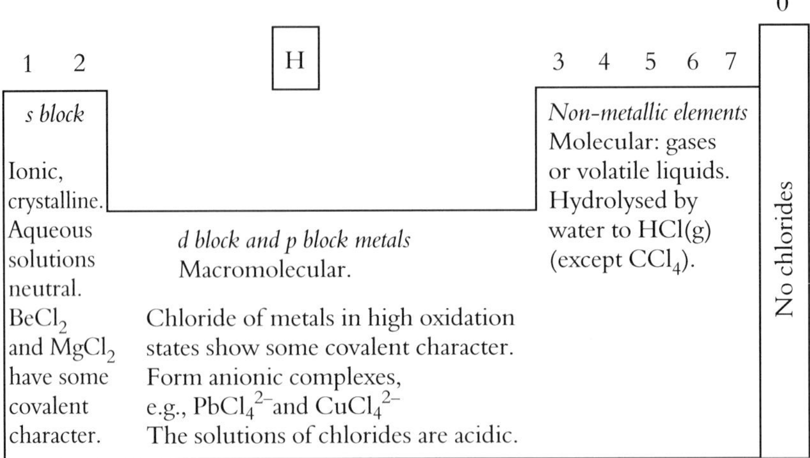
0
1 2
H
3 4 5 6 7
s block
Ionic, crystalline. Aqueous solutions neutral. $BeCl_2$ and $MgCl_2$ have some covalent character.
d block and p block metals
Macromolecular.
Chloride of metals in high oxidation states show some covalent character. Form anionic complexes, e.g., $PbCl_4^{2-}$ and $CuCl_4^{2-}$
The solutions of chlorides are acidic.
Non-metallic elements
Molecular: gases or volatile liquids. Hydrolysed by water to HCl(g) (except CCl_4).
No chlorides

Figure 15.6F
Hydrides

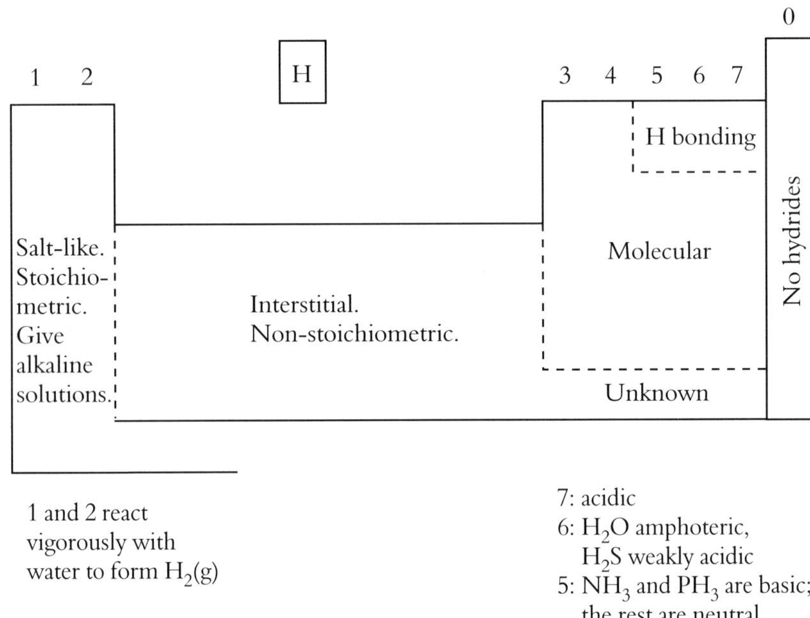
0
1 2
H
3 4 5 6 7
H bonding
Salt-like. Stoichio-metric. Give alkaline solutions.
Interstitial. Non-stoichiometric.
Molecular
Unknown
No hydrides
1 and 2 react vigorously with water to form H_2(g)
7: acidic
6: H_2O amphoteric, H_2S weakly acidic
5: NH_3 and PH_3 are basic; the rest are neutral.

Figure 15.6G
Solubility of salts and complex ion formation

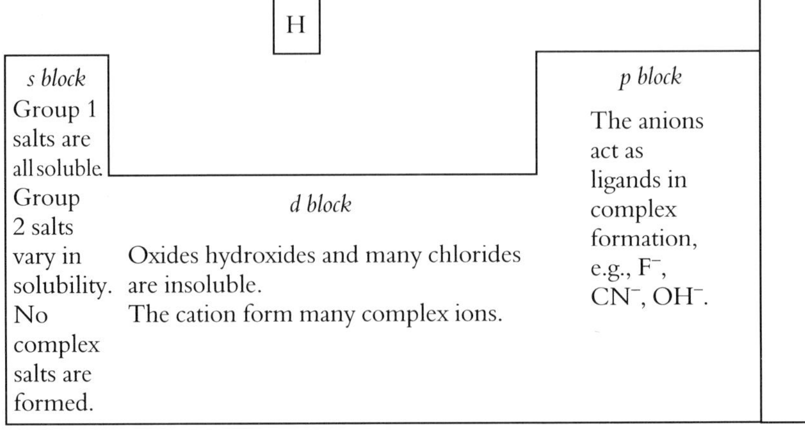
H
s block
Group 1 salts are all soluble. Group 2 salts vary in solubility. No complex salts are formed.
d block
Oxides hydroxides and many chlorides are insoluble.
The cation form many complex ions.
p block
The anions act as ligands in complex formation, e.g., F^-, CN^-, OH^-.

15.7 VARIABLE OXIDATION STATE

The compounds of a metal in a high oxidation state show more covalent character than compounds of the metal in a lower oxidation state

In this section an attempt has been made to relate the properties of the compounds of an element to the position of the element in the Periodic Table. The oxidation state of the element also influences the nature of its compounds. Since the polarising power of a cation $\mathbf{M}^{4+}$ is greater than that of $\mathbf{M}^{2+}$, **M**(IV) compounds will show more covalent character than **M**(II) compounds. For example, lead(II) chloride is a crystalline solid which dissolves sparingly in water to form a solution of Pb^{2+}(aq) and Cl^-(aq) ions. Lead(IV) chloride is a liquid, which is rapidly hydrolysed by water to lead(IV) dichloride oxide and hydrogen chloride:

$$PbCl_4(l) + H_2O(l) \longrightarrow PbOCl_2(aq) + 2HCl(aq)$$

Lead(II) oxide, PbO, is basic, while lead(IV) oxide is amphoteric, reacting with alkalis to form plumbates(IV), containing the anion, $Pb(OH)_6{}^{2-}$.

CHECKPOINT 15.7: BLOCKS OF THE PERIODIC TABLE

1. Classify the following (i) according to structure and (ii) according to acid–base character:

(a) Na_2O
(b) CaO
(c) SiO_2
(d) Fe_2O_3
(e) I_2O_5
(f) Cl_2O_7

2. Which member of each of the following pairs is the more acidic oxide?

(a) BaO and CO
(b) MnO and Mn_2O_7
(c) NO and N_2O_5
(d) Cr_2O_3 and CrO_3
(e) CaO and Al_2O_3
(f) SO_2 and SeO_2

3. Explain why oxygen and sulphur differ in

(a) the ionic character of **MO** and **MS** (where **M** is a metal)
(b) the boiling temperatures of H_2O and H_2S
(c) the formulae of their fluorides, F_2O and SF_6.

4. Consider the oxides Na_2O, MgO, SiO_2, and SO_3.

(a) Draw a dot-and-cross diagram (outer elec-trons only) to show the bonding in MgO.
(b) Choose an oxide from the list which
  (i) is insoluble in water
  (ii) reacts with water to form a strongly alkaline solution
  (iii) reacts with water to form a weakly alkaline solution
  (iv) reacts with water to form a strongly acidic solution
  (v) consists of simple molecules at room temperature.
(c) Give equations for the reactions mentioned in *(b)* (ii) and (iv).

5. *(a)* Predict the differences in chemical properties between (i) $SnCl_2$ and $SnCl_4$ (ii) SnO and SnO_2.
(b) Predict the difference in acid–base behaviour between (i) SO_2 and SO_3 (ii) P_2O_3 and P_2O_5.

6. From the first 20 elements in the Periodic Table give the symbol of an element which fits each of the following descriptions:

(a) exists as free atoms
(b) has the highest first ionisation energy
(c) forms a chloride XCl, which dissolves readily in water to form a neutral solution
(d) forms a liquid chloride which reacts with water and which has a tetrahedral molecule
(e) forms two chlorides in which its oxidation states are +3 and +5
(f) forms an oxide with a giant molecular structure
(g) forms two acidic oxides of formulae XO_2 and XO_3.

7. *(a)* Describe and explain the trend in the size of the atoms of the Group 2 elements (Be–Ba) passing down the group.
(b) State and explain which Group 2 element has the highest first ionisation energy.
(c) (i) Explain why beryllium chloride has more covalent character than barium chloride.
  (ii) How would you expect the properties of beryllium chloride to illustrate its covalent character?

THE PERIODIC TABLE

PERIODIC TABLE

Elements are arranged in order of atomic number

s block elements

Strong reductants.

Compounds are ionic.

Oxides and hydroxides are basic.

Chlorides are neutral.

Do not form complexes. § 15.3

d block elements

Weak reductants.

Oxides and chlorides have a high degree of covalent character.

Oxides are insoluble and are basic or amphoteric.

Chlorides are hydrolysed.

Form many complexes. § 15.3

p block elements

Oxidising power increases L → R across a period.

Compounds are covalent.

Oxides and hydroxides are acidic.

Chlorides are acidic.

Form complex ions. § 15.3

Physical properties of the elements that vary in a periodic manner are: § 15.2

melting temperature	Figure 15.2A
boiling temperature	Figure 15.2B
atomic radius	Figure 15.2C
ionic radius	Figure 15.2D
first ionisation energy	Figure 15.2E
electronegativity	Figure 15.2F

Trends across Period 3 from L → R

Elements

- Standard electrode potential changes from a high negative value to a high positive value.
- The elements change from powerful reductants to powerful oxidants.
- Electronegativity increases. § 15.4

Compounds

- Cations: Electrovalent character of compounds decreases L → R
- Anions: Electrovalent character of compounds increases L → R

§ 15.5

Compounds in relation to the periodic table

Across Period 3 from L → R

Oxides and hydroxides

- ionic and basic → amphoteric → acidic → strongly acidic → none (Ar) § 15.5.1 and Figures 15.6A, B, C, D

Halides

- ionic → covalent,
- crystalline → layer structure → molecular
- dissolve → hydrolysed § 15.5.2 and Figures 15.6A, E

Hydrides

- ionic and basic (Groups 1 & 2) → unstable → covalent and acidic (Groups 6, 7) § 15.5.3 and Figures 15.6A, F

Salts

Carbonates and hydrogencarbonates in Groups 1 & 2

Sulphates and nitrates in Groups 1, 2, 3 § 15.5.4

Complexes Figure 15.6G

Variable oxidation state and bond character § 15.7

QUESTIONS ON CHAPTER 15

1. *(a)* Which of the cations, Ca^{2+} or Al^{3+}, has the greater polarising power?
(b) Which of the chlorides, $CaCl_2$ or $AlCl_3$, is the more covalent in character?

One of these chlorides sublimes when it is heated. Which one do you suppose it is?

2. *(a)* Compare the polarising power of Be^{2+} and Ba^{2+}.
(b) What would you predict about the degree of covalent character in $BeCl_2$ and $BaCl_2$?

One of the chlorides is hydrolysed in aqueous solution. Which one do you suppose it is?

3. Of the compounds $SnCl_2$ and $SnCl_4$, which one has the more covalent character? One of the chlorides is a crystalline solid, and one is a fuming liquid. Which is which?

4. *(a)* Define the term: first ionisation energy of an element.
(b) State two factors that determine the first ionisation energies in a given group of the Periodic Table.
(c) Write the electronic configurations of the elements with atomic numbers 8, 10 and 11.
(d) Write the elements in order of increasing first ionisation energy. Explain your answer.
(e) Refer to the table of first ionisation energies.

	Atomic number	**First ionisation energy/kJ mol^{-1}**
Group 2		
Be	4	900
Mg	12	736
Ca	20	590
Group 3		
B	5	799

(i) Explain why the first ionisation energy of beryllium is greater than that of boron.
(ii) Explain the change in magnitude of first ionisation energies of Group 2 elements from beryllium to calcium.

5. Refer to the section of the Periodic Table shown below.

Group	1	2	3	4	5	6	7
	Li	Be	B	C	N	O	F
	Na	Mg	Al	Si	P	S	Cl

Give the symbol of the element which is

(i) the most reactive non-metal
(ii) the most reactive metal
(iii) the element with the smallest atom
(iv) the element which forms the largest anion
(v) the most electronegative element
(vi) a p block metal
(vii) forms an amphoteric oxide X_2O_3
(viii) has the highest boiling temperature
(ix) can conduct electricity and form a dioxide.

6. Refer to the table of atomic radii and melting temperatures.

Element	Na	Mg	Al	Si	P	S	Cl
Atomic radius/nm	0.157	0.136	0.125	0.117	0.110	0.104	0.099
Melting temperature/°C	98	651	660	1410	44	114	−101

(a) Explain the trend shown by atomic radii across Period 3.
(b) Explain how the melting temperatures of the elements are related to structure and bonding.
(c) Discuss the variation in acid–base character of the oxides across the period.
(d) Describe the reaction with water of the chlorides of the elements sodium to phosphorus.

7. *(a)* What is meant by periodicity in the Periodic Table?
(b) Mention five physical properties which show periodicity.
(c) Mention five chemical properties which show periodicity.
(d) Summarise how the chemistry of the elements and compounds changes across a period by considering the elements sodium to chlorine and their hydrides.

8. *(a)* Consider the following oxides:

CaO, Al_2O_3, Na_2O, MgO, P_2O_5, SO_2, SiO_2

State which of these oxides are
(i) basic (ii) acidic (iii) neutral (iv) amphoteric
(b) Describe briefly how you could test to see whether a solid oxide was basic, acidic, neutral or amphoteric.
(c) Consider the following chlorides:

CCl_4, HCl, $MgCl_2$, $NaCl$, PCl_5, $PbCl_2$, $PbCl_4$, $SiCl_4$

State which of the chlorides
(i) have predominantly ionic bonds
(ii) have predominantly covalent bonds
(iii) do not clearly belong to either (i) or (ii).
(d) Discuss the differences in behaviour between ionic and covalent chlorides.

9. *(a)* Define the term: first electron affinity.
(b) Refer to the table of first electron affinities of Period 2.

Element	Li	B	C	N	O	F	Ne
Electron affinity/ kJ mol^{-1}	−60	−28	−122	+7	−142	−328	+29

Discuss the difference between the values for carbon and nitrogen with reference to their electronic configurations.
(c) Predict whether the electron affinity of beryllium will be more or less negative than that of lithium, giving your reasons.
(d) The second electron affinity of oxygen is +844 kJ mol^{-1}. Comment on this value in comparison with the values of first electron affinities of oxygen and fluorine.

10. *(a)* (i) Define the term *molar first ionisation energy*. **3**
(ii) State and explain the trend in the value of the molar first ionisation energy that occurs down a group in the Periodic Table. **3**
(b) (i) State how the acid/base character of the oxides of the elements of Period 3 (Na to Cl) varies across the period. **1**
(ii) Copy and complete the table by writing the formula of an oxide of sodium, phosphorus and sulphur respectively. For each of the oxides, write an equation to show its reaction with water. **6**

Element	Sodium	Phosphorus	Sulphur
Formula of the oxide			

(c) For each of the molecules Cl_2O_7 and HClO, deduce the oxidation number shown by the chlorine atom. **2**

15 *AEB (AS/AL) 1998*

11. The elements of the third period are as follows.

Na Mg Al Si P S Cl Ar

All of your answers below should relate to these elements.

(a) Which elements can exist
(i) as diatomic molecules at room temperature, **2**
(ii) as macromolecular structures? **2**
(b) Which pairs of elements combine to produce compounds with formulae of the type *XY*? **2**
(c) Two elements form chlorides with formulae of the type XCl_3. Draw displayed formulae for these two chlorides, and suggest values for the bond angles. **4**
(d) (i) One element combines with oxygen to form an oxide which reacts with water to give a strongly alkaline solution. Name the element, and write a balanced equation for the oxide reacting with water.
(ii) One element combines with oxygen to form an oxide of the type XO_2, which reacts with water to given an acidic solution. Name the element and write a balanced equation for the oxide reacting with water. **4**

12 *C (AS/AL)*

12. *(a)* (i) **Three** factors are dominant in determining whether a given combination of elements will show ionic or covalent bonding. State these factors which favour **ionic** bonding as they relate to:
I. The sizes of the atoms involved.
II. The magnitudes of the formal charges on the ions.
III. The electronic structures of the resulting ions. **3**
(ii) State whether each of the following statements about **ionic** oxides is true **or** false.
(More than one statement may be correct.)
I. Such oxides can always react as a base.
II. Such oxides readily sublime on heating.
III. Such oxides when molten, conduct electricity.
IV. Such oxides are always unreactive to water.
V. Such oxides are always solids at room temperature.
VI. Such oxides react with both acids and bases. **3**
(b) (i) In each of the following species state whether the metal to non-metal bond is ionic or covalent: Na_2O; Al_2O_3; NaCl; Al_2Cl_6. **2**
(ii) An inorganic oxide is a white solid of melting temperature 1710 °C and is unreactive towards water.
I. State and explain whether or not this is sufficient information to deduce whether the bonding is ionic or covalent.
II. **Outline one** additional test which could be applied to substantiate your conclusion. **3**

11 *WJEC (AS/AL)*

13. The table below shows electronegativity values for some atoms.

H	N	O	F	Cl	Cs
2.1	3.0	3.5	4.0	3.0	0.7

(a) What do you understand by the term **electronegativity**? **1**

(b) The nature of the bonding in substances depends partly on the electronegativities of the atoms concerned. *Use the data* in the table above to suggest the nature of the bonding in each of the following substances.
(i) caesium fluoride; (ii) water; (iii) chlorine **1**

(c) Ammonia, NH_3 is a polar covalent molecule.
(i) State the general rules which determine the shape of a covalent molecule. **3**
(ii) Draw the shape of the ammonia molecule. **1**
(iii) Why is the bond angle in ammonia 107° rather than 109°28′? **1**
(iv) Explain why the molecule of ammonia is polar. **1**

(d) Each of the elements sodium to chlorine in Period 3 will react with oxygen given suitable conditions.
(i) Choose an element from this period which gives a basic oxide, and write equations both for its reaction with oxygen and to illustrate the basic nature of the oxide. **2**
(ii) Choose an element which forms an amphoteric oxide and write equations which illustrate this amphoteric nature. **2**
(iii) Carbon dioxide reacts readily with dilute aqueous sodium hydroxide whereas silicon dioxide does not. Explain this difference and suggest conditions under which silicon dioxide would react. **2**

16 *L (AS/AL)*

14. *(a)* Explain what is meant by the term *first ionisation energy*. **2**

(b) The graph below shows how the first ionisation energy varies with proton number across the period sodium to argon.

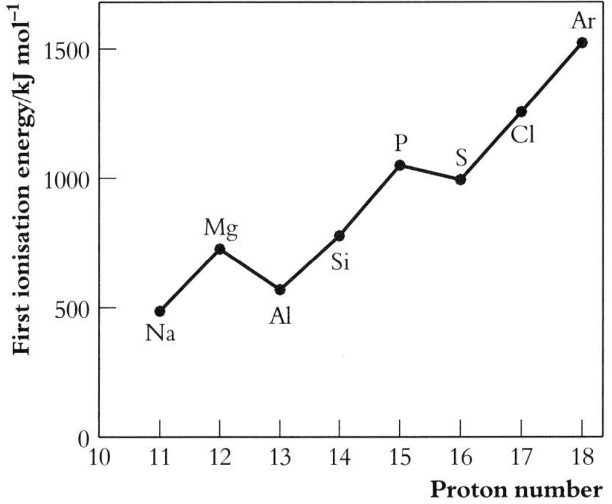

Explain
(i) why the first ionisation energy increases overall across the period from sodium to argon **2**
(ii) why the first ionisation energy of aluminium is less than that of magnesium **2**
(iii) why the first ionisation energy of sulphur is less than that of phosphorus. **2**

(c) Use the graph in part *(b)* to explain why sodium and magnesium are metals whilst sulphur and chlorine are non-metals. **3**

(d) Describe the bonding found in
(i) magnesium metal **2**
(ii) solid magnesium chloride. **2**

15 *O&C (AS/AL)*

15. *(a)* Explain the meaning of the term *periodic trend* when applied to trends in the Periodic Table. **2**

(b) Explain why atomic radius decreases across Period 2 from lithium to fluorine. **2**

(c) The table below shows the melting temperatures, T_m of the Period 3 elements.

Element	Na	Mg	Al	Si	P	S	Cl	Ar
T_m **/K**	371	923	933	1680	317	392	172	84

Explain the following in terms of structure and bonding.
(i) Magnesium has a higher melting temperature than sodium.
(ii) Silicon has a very high melting temperature.
(iii) Sulphur has a higher melting temperature than phosphorus.
(iv) Argon has the lowest melting temperature in Period 3. **8**

12 *NEAB (AS/AL)*

16

GROUP 0: THE NOBLE GASES

16.1 THE 'NEW GAS'

The relative atomic masses of the elements were found by the accurate, painstaking work of scientists in the nineteenth century. This may not sound a very exciting research topic, but a startling discovery came out of it. It began with the experiments of Lord Rayleigh, a Cambridge professor, to find the relative atomic mass of nitrogen. He noticed that nitrogen obtained from the air (by removing oxygen, carbon dioxide and water vapour) was 0.5% denser than nitrogen obtained from compounds (e.g. by the oxidation of ammonia). He was puzzled by the discrepancy. He tackled the problem by taking a measured volume of air and removing all the nitrogen from it by combining nitrogen with oxygen in an electric discharge and dissolving the nitrogen dioxide formed in a solution of alkali. Rayleigh found that 0.6% of the original volume of air remained; this gas was denser than nitrogen and even less reactive. What was it?

Group	1	2	3	4	5	6	7	
Element A_r	Li 7	Be 9	B 11	C 12	N 14	O 16	F 19	
Element A_r	Na 23	Mg 24	Al 27	Si 28	P 31	S 32	Cl 35.5	? 40
Element A_r	K 39	Ca 40						

Figure 16.1A
Part of the Periodic Table

William Ramsey, a London professor, was fascinated by Rayleigh's discovery. He tried passing a sample of nitrogen obtained from air over heated magnesium. After many passes to allow the nitrogen to combine with the magnesium, 1.25% of the original volume of nitrogen remained. Ramsey called this very unreactive gas the 'new gas'. He asked Sir William Crookes to look at its emission spectrum [§ 2.2]. To Crookes' amazement the emission spectrum was different from those of all the known elements. The 'new gas' was a new element! Rayleigh and Ramsey had discovered a new element. They examined the Periodic Table to see where it fitted in. They had found a relative atomic mass of 40, but this position was already occupied by calcium (see Figure 16.1). It seemed best to put the new element between A_r(Cl) = 35.5 and A_r(K) = 39. This position was later shown to be correct when elements were arranged in order of atomic number [§§ 1.7, 2.6]. The two scientists announced their discovery in 1894 and called their element **argon** (Greek *argos* = lazy).

Ramsey went on to discover helium, neon, krypton and xenon. Their relative atomic masses and their lack of chemical reactivity placed them in a group with argon, and they formed a new Group 0 of the Periodic Table. They were called the **inert gases** but are now known as the **noble gases**. Their discovery was a spectacular success for Mendeleev's classification.

16.2 MEMBERS OF THE GROUP

The members of Group 0 are helium, neon, argon, krypton, xenon and radon. The outermost shell of electrons is full with two electrons in the case of helium and eight for the other elements [see Table 16.2]. There is stability associated with a full valence shell, and the noble gases are very unreactive.

16.2.1 HELIUM

Uses of helium

Helium is obtained from some natural gas wells where the gas may contain up to 5% of helium. It is used to inflate airships, weather balloons and aeroplane tyres, being safer than hydrogen for these purposes. A mixture of oxygen and helium is used by divers instead of air. If they use air, nitrogen dissolves in the blood under pressure and then comes out of the blood when they surface, causing painful spasms which they call 'the bends'.

16.2.2 ARGON

Uses of argon

Argon forms 1% of air and is obtained during the fractional distillation of liquid air. It is used to provide an inert atmosphere for gas–liquid chromatography, for risky welding jobs and for some chemical reactions.

16.2.3 NEON, KRYPTON AND XENON

Neon and krypton lights

Neon, krypton and xenon are also obtained from air. Neon gives out a red glow when an electrical discharge is passed through the gas at low pressure, and finds widespread use in neon lights. Krypton also is used in discharge tubes.

16.2.4 RADON

Radon is a radioactive gas which is formed when radium decays.

16.3 COMPOUNDS OF THE NOBLE GASES

Compounds of xenon ...

... XeF_2, XeF_4, XeF_6 ...

The first noble gas compound ever made was the ionic solid xenon hexafluoroplatinate, $XePtF_6$. This discovery, in 1962, was followed by the synthesis of xenon(II) fluoride, XeF_2; xenon(IV) fluoride, XeF_4; and xenon(VI) fluoride, XeF_6. The bonding is covalent, with one, two and three of xenon's 5p electrons being promoted to 5d orbitals. The energy required for promotion is great and can be compensated for only by reaction with the most electronegative of elements, fluorine.

Compounds of krypton ...

... KrF_2, KrF_4

The krypton compounds krypton(II) fluoride, KrF_2, and krypton(IV) fluoride, KrF_4, have been synthesised.

QUESTIONS ON CHAPTER 16

1. The electron configurations of the noble gases were, until 1962, thought to be perfectly stable. Explain how this belief acted as a starting point in explanations of the nature of the chemical bond.

2. How did the discovery of the noble gases confirm people's acceptance of the Periodic Table?

3. Mention three uses of noble gases and say what property of each gas fits it for the use you mention.

4. Why is it more difficult for krypton to form compounds than for xenon? Would you expect argon to be more or less reactive than krypton?

17 HYDROGEN

17.1 OCCURRENCE

Hydrogen is the most abundant element in the universe. The Sun and the stars derive their energy from the nuclear fusion reaction

$$4\,{}^{1}_{1}\mathrm{H} \longrightarrow {}^{4}_{2}\mathrm{He} + 2\,{}^{0}_{1}\mathrm{e}\ (\textit{positron}) + \gamma(\textit{radiation})$$

There is very little free hydrogen in the Earth's atmosphere, but there is plenty of hydrogen on the Earth, combined as water and organic compounds.

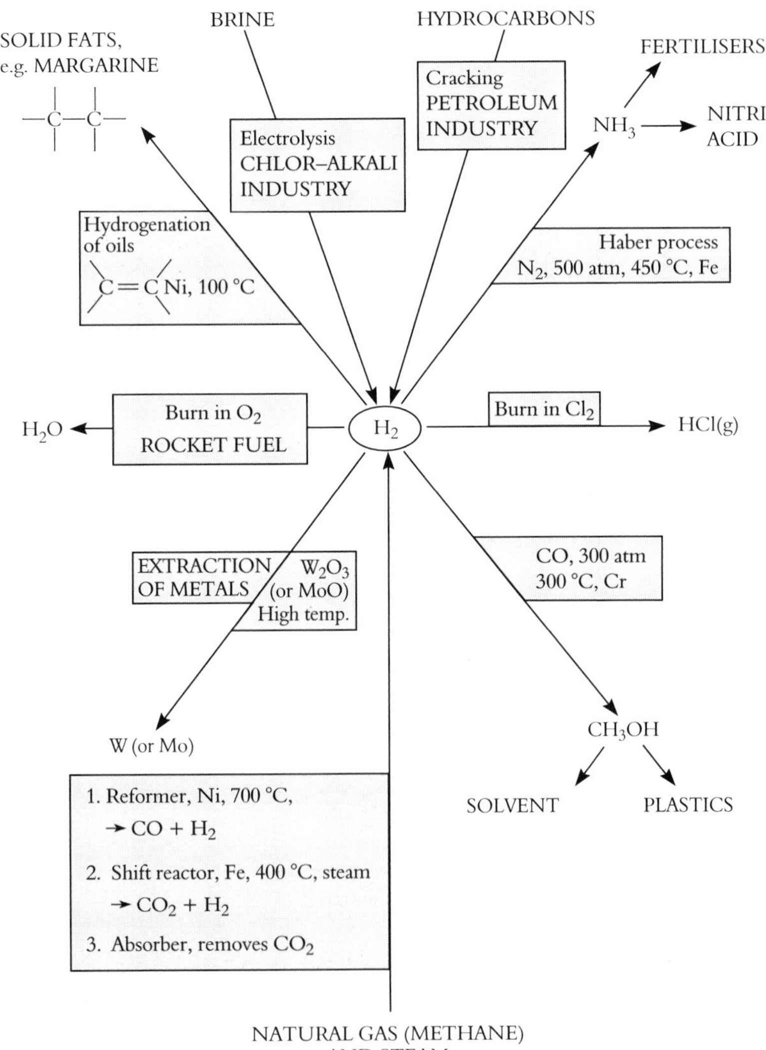

Figure 17.2A Some industrial uses of hydrogen

17.2 MANUFACTURE AND USES

Figure 17.2A shows the manufacture and main industrial uses of hydrogen.

17.3 LABORATORY PREPARATION AND REACTIONS

Hydrogen is prepared in the laboratory from a metal + HCl(aq) or H_2SO_4(aq)

The most convenient laboratory preparation is the action of a dilute acid (not nitric acid) on zinc or another metal which is fairly high in the electrochemical series. Made in this way, hydrogen contains some hydrogen sulphide, which comes from the zinc sulphide present as an impurity in zinc.

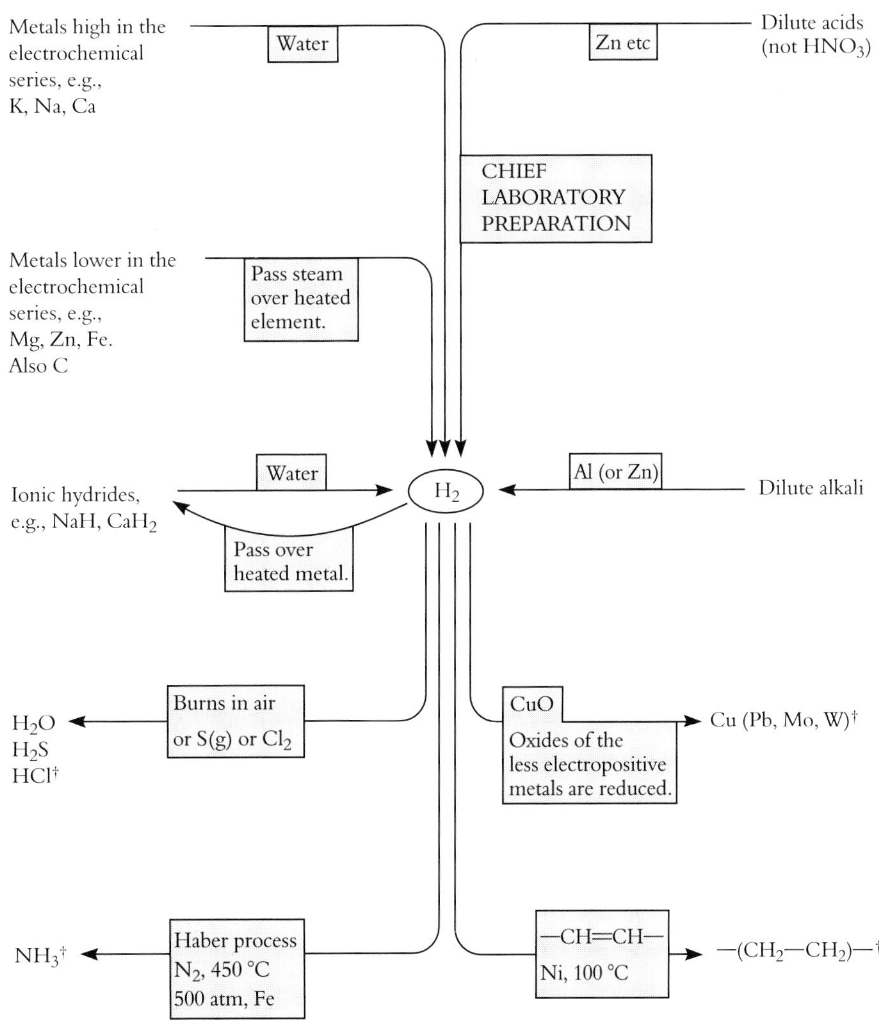

Figure 17.3A Some reactions of hydrogen

17.4 WATER

The physical properties of water

Some of the physical properties of water have been mentioned. These are: bond angle in H_2O [§ 5.2.3 and Figure 5.2F], hydrogen bonding [§ 4.8.3], melting and boiling temperatures [§ 4.8.3], ice [§ 4.8.3] and the dissolution of organic solutes [§ 4.8.3].

The polar nature of water molecules

$$\overset{\delta+}{H}-\overset{\delta-}{O}-\overset{\delta+}{H}$$

explains why water is a good solvent for ionic compounds. Figure 10.14A shows how a salt dissolves in water. More detail is given in § 10.14.

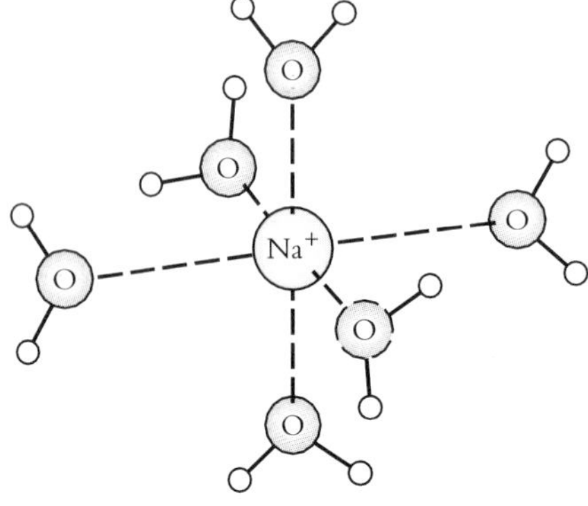

Figure 17.4A The aquasodium ion, $[Na(H_2O)_6]^+$ or $Na^+(aq)$

Hydrated ions (ions surrounded by water molecules) are called **aqua ions**. Figure 17.4A shows the aquasodium ion, $[Na(H_2O)_6]^+$ or $Na^+(aq)$ for short.

Aqua ions

The acid–base behaviour of water has been described in § 12.6.1. When salts dissolve in water, their solutions are often neutral. Some salts, however, react with water to form acidic or alkaline solutions [see §§ 12.9, 19.4.1, 24.14.2].

Water hydrolyses some salts, covalent inorganic compounds and organic compounds

Water hydrolyses many inorganic compounds (e.g., $SiCl_4$, PCl_3) and organic compounds (e.g., acid chlorides, amides, anhydrides and esters). Unlike the hydrolysis of salts, many of these reactions go to completion, sometimes slowly and sometimes rapidly.

CHECKPOINT 17.4: WATER

1. Explain the following statements:

   (a) Ponds freeze from the top downwards.
   (b) Oil and water do not mix.
   (c) Whisky and water do mix.
   (d) Water is a good solvent for ionic compounds.
   (e) Many classes of organic compounds dissolve in water.
   (f) A solution of sodium sulphide has an unpleasant smell.
   (g) It is dangerous to dissolve potassium cyanide in water.
   (h) Aluminium sulphate solution is used in fire-extinguishers to generate carbon dioxide.

*17.5 FLUORIDATION OF WATER

Tooth enamel consists of calcium hydroxide phosphate, $Ca_5(PO_4)_3OH$, also called calcium hydroxyapatite. In this ionic structure, an equilibrium exists:

$$Ca_5(PO_4)_3OH(s) + aq \underset{\text{Remineralisation}}{\overset{\text{Demineralisation}}{\rightleftharpoons}} 5Ca^{2+}(aq) + 3PO_4{}^{3-}(aq) + OH^-(aq)$$

Calcium hydroxide phosphate (calcium hydroxyapatite) in tooth enamel is attacked by acids

In the environment of the mouth, calcium hydroxyapatite can be dissolved and reformed. The equilibrium lies to the left. If acids are present in the mouth, however, they react with hydroxide ions and favour demineralisation. Normally the pH of the mouth is 6.8. Within the plaque (a gelatinous mass of microorganisms) which coats teeth, the pH may be much lower. Sugar is the chief culprit because bacteria in the plaque convert sugars into acids.

Fluoride ion can replace some of the hydroxide ion in hydroxyapatite. The fluoridated compound is less susceptible to attack by acids

It has been observed that in cities where the concentration of natural fluoride in the water is high the incidence of tooth decay is lower than average. Fluoride ion substitutes for some of the hydroxide ion in calcium hydroxyapatite to form $Ca_5(PO_4)_3(OH)_{1-x}F_x$. The fluoridated hydroxyapatite is less easily attacked by acidic solutions than is hydroxyapatite.

Many people have objected to fluoridation of drinking water as a kind of compulsory mass medication. Very high levels of fluoride cause damage to teeth. A study by the Royal College of Physicians has concluded, however, that there is no risk to the individual or the environment from levels of fluoride up to 1 ppm. Many water authorities now add sodium fluoride to bring the level of fluoride in drinking water up to 1 ppm. The alternative of taking fluoride tablets is less effective because a large dose of fluoride is rapidly excreted.

QUESTIONS ON CHAPTER 17

1. *(a)* Give one example each of the importance of hydrogen in (i) the fertiliser industry (ii) the food industry (iii) the extraction of metals from their ores.
(b) State three industrial sources of hydrogen.

2. Name two elements which fit each of the following descriptions

(a) gives hydrogen when heated in steam
(b) gives hydrogen in a reaction with cold water
(c) gives hydrogen in a reaction with dilute alkali
(d) added to dilute sulphuric acid, gives hydrogen at a moderate rate
(e) can be obtained by reduction of the oxide with hydrogen.

3. Write equations for the reactions of hydrogen with

(a) An alkene RCH=CHR′
(b) tungsten(III) oxide
(c) oxygen
(d) sulphur
(e) chlorine
(f) nitrogen.

Which of these reactions are important in industry? What conditions are employed to give good yields?

4. In 1781, the British chemist Henry Cavendish prepared water in the laboratory by exploding hydrogen in air.

(a) In a modern simulation of Cavendish's experiment, a chemist burnt 0.50 g of hydrogen in oxygen.
(i) Write an equation for the reaction of hydrogen with oxygen.
(ii) How many moles of hydrogen, H_2, did the chemist react?
(iii) Calculate the mass of water formed. **4**

(b) Water is an example of a compound that possesses hydrogen bonding.
(i) Draw a clear diagram to show how hydrogen bonding occurs in water. Your diagram should include bond angles and relevant dipoles.
(ii) State **two** properties of water that result from hydrogen bonding. **5**

(c) The chemist added water to sodium oxide, Na_2O, forming an aqueous solution.
(i) Write an equation for the reaction that took place.
(ii) Suggest a value for the pH of this solution. **2**

11 *C (AS/AL)*

18

THE s BLOCK METALS: GROUPS 1 AND 2

18.1 TWO INDUSTRIES BASED ON SALT

The best-known compound of the s-block metals is sodium chloride, common salt. Two important industries are based on common salt. One of these is the Solvay process, the ammonia–soda process, for making sodium hydrogencarbonate. The overall reaction is:

$$NaCl(aq) + NH_3(aq) + CO_2(aq) + H_2O(l) \longrightarrow NaHCO_3(s) + NH_4Cl(aq)$$

Sodium hydrogencarbonate is heated to form sodium carbonate. The importance of sodium carbonate lies in its use in the glass industry and in the paper industry. The other salt-based industry is the chlor–alkali industry, the electrolysis of brine to form sodium hydroxide, chlorine and hydrogen.

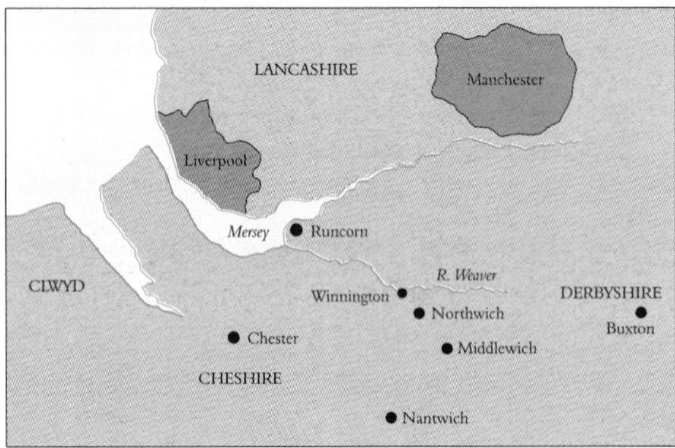

Figure 18.1A
Winnington, Runcorn and environment

The ammonia–soda process [§ 18.6.7] was established in the UK in 1872. John Brunner and Ludwig Mond acquired the right to make sodium carbonate by the process which had been patented by Alfred and Ernest Solvay in Belgium in 1861. The Brunner–Mond plant, built at Winnington in Cheshire [see Figure 18.1A], is now a part of ICI. At nearby Northwich, Middlewich and Nantwich, there are vast salt deposits underground. Salt was mined to such an extent that Northwich was suffering from subsidence by 1890. Another method of extracting salt is employed now. A hole is drilled in the ground, and water is pumped in. After the water has been left underground long enough to become saturated with salt, brine is pumped out. Pillars of salt are left intact at intervals. The holes are left full of saturated brine so that the ground above will not subside.

The ammonia–soda process discharges no pollutant gases into the air. The only by-product is calcium chloride. Some of this is sold for use as a drying agent. Most of it is waste. It is discharged into a short stretch of river which flows into the sea. The amount of chloride ion dumped in the sea annually is negligible compared with that present in sea water. Calcium ion is constantly added to sea water by the action of rainwater on limestone. Marine creatures remove calcium ion from the sea to build their shells. The contribution which the ammonia–soda process makes to the calcium content of the sea is negligible.

18.2 THE MEMBERS OF THE GROUPS

s block: Group 1; (core)ns Group 2; (core)ns^2

The s block metals are the metals in Group 1 and Group 2 of the Periodic Table. They are called the s block elements because they occupy an area of the Periodic Table following the noble gases, an area in which the s orbitals are being filled. The s block metals have much in common, and aluminium in Group 3 shares many of their properties. The s block metals are listed in Table 18.2.

Table 18.2 Physical properties of s block metals

Group 1				T_m/°C	T_b/°C	AR/nm	IR/nm	IE/kJ mol^{-1}	$E^{\ominus}$/V
Lithium	Li	3	$1s^22s$	180	1330	0.15	0.06	519	−3.05
Sodium	Na	11	(Ne)3s	98	892	0.19	0.10	494	−2.71
Potassium	K	19	(Ar)4s	64	760	0.23	0.13	418	−2.93
Rubidium	Rb	37	(Kr)5s	39	688	0.24	0.15	402	−2.92
Caesium	Cs	55	(Xe)6s	39	690	0.26	0.17	376	−2.92
Francium	Fr	87	(Rn)7s	A radioactive element which is artificially made					
Group 2									
Beryllium	Be	4	$1s^22s^2$	1280	2770	0.11	0.03	2660	−1.85
Magnesium	Mg	12	(Ne)$3s^2$	650	1110	0.16	0.07	2186	−2.37
Calcium	Ca	20	(Ar)$4s^2$	840	1440	0.20	0.10	1740	−2.87
Strontium	Sr	38	(Kr)$5s^2$	768	1380	0.21	0.11	1608	−2.89
Barium	Ba	56	(Xe)$6s^2$	714	1640	0.22	0.13	1468	−2.91
Radium	Ra	88	(Rn)$7s^2$	A radioactive element					

(Where T_m = melting temperature, AR = atomic radius, IR = ionic radius, IE = ionisation energy; for Group 2, the sum of the first and second ionisation energies, $E^{\ominus}$ = standard electrode potential.)

In these metals, the metallic bond is relatively weak

The members of Groups 1 and 2 are all metals. They are silvery coloured and tarnish rapidly in air. They show relatively weak metallic bonding because they have only one or two valence electrons. They differ in a number of ways from metals later in the Periodic Table:

1. They are soft: they can be cut with a knife.
2. Their melting and boiling temperatures are low.
3. They have low standard enthalpies (heats) of melting and vaporisation.
4. They have low densities. (Li, Na, K are less dense than water.) Group 2, with two valence electrons, show stronger metallic bonding, which is reflected in their physical properties.

The flame colours

The outer electron or electrons can be excited to a higher energy level. When they fall to a lower energy level, energy is emitted. For these metals, the energy is sufficiently low to have a wavelength in the visible spectrum [§ 2.2]. These elements therefore colour flames: Li — red, Na — yellow, K — lilac, Rb — red, Cs — blue, (Be — colourless), Mg — brilliant white, Ca — brick red, Sr — crimson, Ba — apple green.

Ionisation energies are low

The elements show constant oxidation numbers of +1 in Group 1 and +2 in Group 2. The ionisation energy required for the process

$$\mathbf{M}(g) \longrightarrow \mathbf{M}^{n+}(g) + ne^-$$

is low. The s electrons are shielded from the attraction of the nucleus by the noble gas core and are easily removed. As the size of the atoms increases down the groups, the electrons to be removed become more distant from the nuclear charge, and the ionisation energy decreases.

Metals are reducing agents. The highly negative $E^{\ominus}$ values show that s block metals are powerful reducing agents

Metals are reducing agents. The power of s block metals as reducing agents is shown by the vigour with which they reduce water to hydrogen. The standard electrode potential, $E^{\ominus}$ [§ 13.2.1], measures the tendency for the reduction process

$$\mathbf{M}^{n+}(\text{aq}) + ne^{-} \longrightarrow \mathbf{M}(\text{s})$$

to occur. A highly negative value for $E^{\ominus}$ indicates that the metal atoms will form ions and electrons.

Figure 18.2A Trends in ionisation energy and standard electrode potential

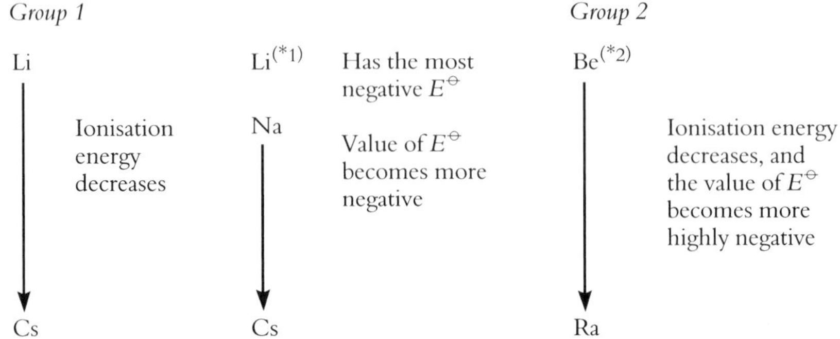

(***1**) The small size of Li^+ gives it a high standard enthalpy of hydration. This is why Li^+ has the most negative value of $E^{\ominus}$ in the group. The small size of Li^+ enables it to polarise anions, and its compounds have some covalent character.

(***2**) Be^{2+} is small and highly charged. It polarises anions, and beryllium compounds are mainly covalent.

18.3 USES

Sodium

Sodium is used in some nuclear reactors ... in electrical circuits ... in lamps ... as a reducing agent ...

1. Molten sodium is used as a coolant in some types of nuclear reactor. Its high thermal conductivity and low melting temperature and the fact that its boiling temperature is much higher than that of water make sodium suitable for this purpose.
2. Sodium wire is used in electrical circuits for special applications. It is very flexible and has a high electrical conductivity. The wire is coated with plastics to exclude moisture.
3. Sodium vapour lamps are used for street lighting.
4. Sodium amalgam and sodium tetrahydridoborate, $NaBH_4$ [§ 36.3], are used as reducing agents.
5. Sodium cyanide is used in the extraction of silver and gold.

Magnesium

Magnesium is used in alloys and in flares

1. Magnesium is alloyed with aluminium to make Duralumin® [§ 19.3.1].
2. Magnesium is used as a sacrificial anode to prevent iron from rusting [§ 13.2.6].
3. The intense white light of burning magnesium is used in flares and distress signals.
4. Magnesium is used in the extraction of titanium [§ 24.12.2].
5. Magnesium oxide is used in refractory (heat-resisting) linings of furnaces.

18.4 OCCURRENCE AND EXTRACTION

Extraction is by electrolysis

The s block metals are too reactive to occur uncombined. Group 1 are found as chlorides. Group 2 are found as chlorides, carbonates and sulphates. The metals are obtained by electrolysis of the molten chlorides.

Chemical reducing agents cannot reduce the oxides of s block metals

Electrolysis is an expensive way of obtaining metals. When possible, a reducing agent, such as carbon monoxide or carbon, is employed to reduce a metal oxide to the metal. The metals of Groups 1 and 2 are themselves such powerful reducing agents that their oxides cannot be reduced by chemical reducing agents. The metallurgist must resort to electrolysis.

CHECKPOINT 18.4: THE METALS

1. Refer to Table 18.2, § 18.2.

(a) Considering Group 1, explain why the radius of each ion is smaller than that of the atom.

(b) Explain why the ionic radius increases from Li^+ to Cs^+.

(c) Explain why, although Na^+ and Mg^{2+} have the same electron configuration, the radius of Mg^{2+} is only 0.07 nm.

(d) Which of the ions listed in Table 18.2 is likely to have the highest enthalpy (energy) of hydration? Explain your choice.

2. Explain why Group 2 metals have higher boiling temperatures and melting temperatures than Group 1 metals. What other physical properties differ?

3. For the process

$\mathbf{M}(s) \longrightarrow \mathbf{M}^{n+}(aq) + ne^-$

list the individual steps by which this process can be considered to occur [see Born–Haber cycle, § 10.12]. Name the enthalpy changes associated with each step.

*18.5 REACTIONS OF GROUP 1

Typical reactions of sodium are shown in Figure 18.5A. The reactions of the alkali metals are summarised in Table 18.5.

Table 18.5
Some reactions of Group 1

Reactant	Reaction
Hydrogen	All form hydrides, containing the H^- ion. The hydrides react with water to give hydrogen and the metal hydroxide.
Water	All react with cold water to give hydrogen + metal hydroxide. The metals are kept under oil to protect them from water vapour. The vigour of the reaction increases down the group. *Lithium* reacts slowly with water, violently with acids. *Sodium* reacts vigorously with water; violently with acids. $2Na(s) + 2H_2O(l) \longrightarrow 2NaOH(aq) + H_2(g)$ *Potassium* reacts so vigorously with water that the hydrogen formed catches fire. *Rubidium and caesium* react even more violently
Oxygen	They all burn readily in air. Caesium and rubidium inflame spontaneously. The rest react with air to form a surface film of oxide.
Halogens	All react on heating to form halides, $\mathbf{M}^+X^-$

Figure 18.5A
Some reactions of sodium and its compounds

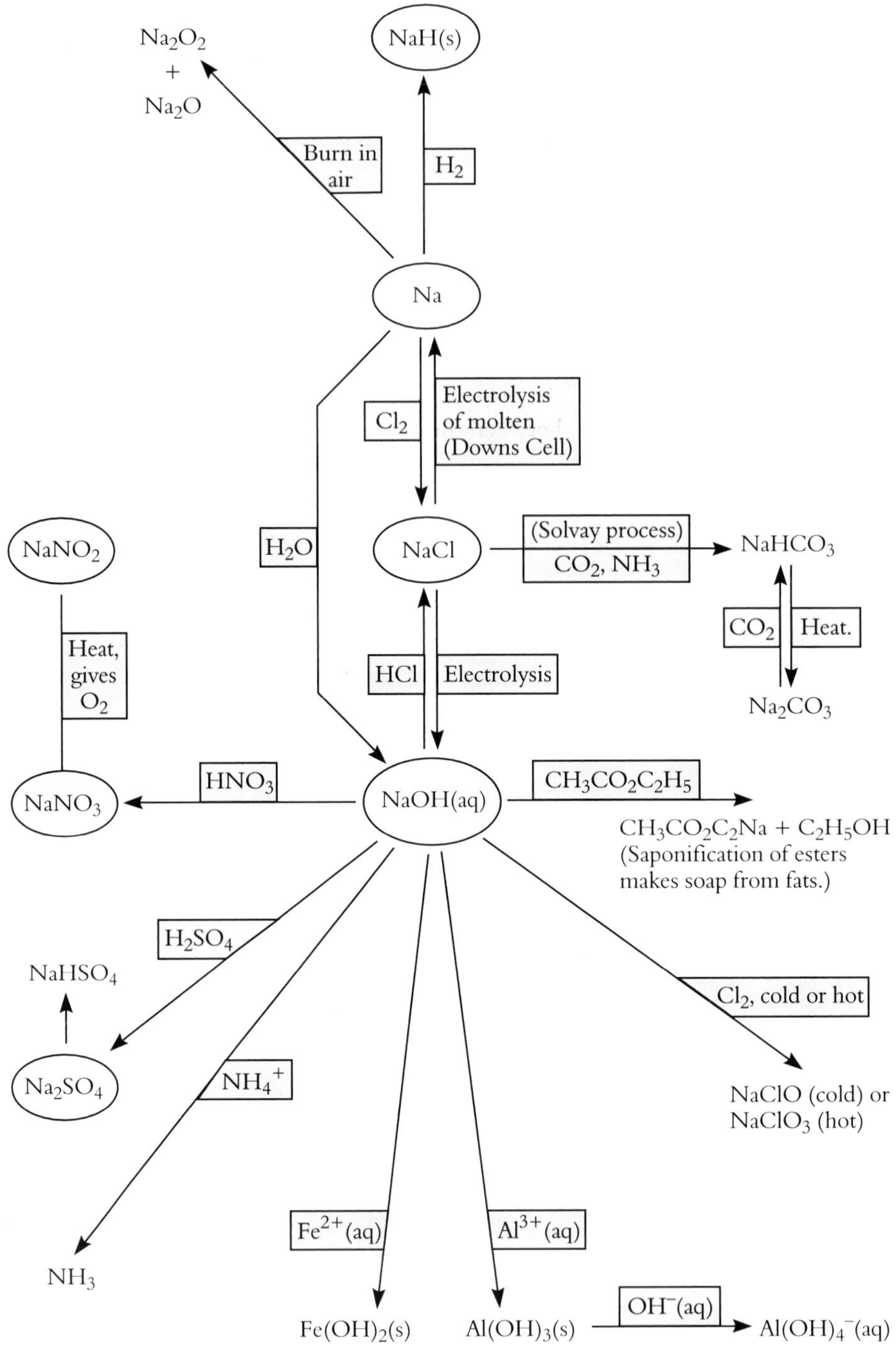

*18.6 COMPOUNDS OF GROUP 1

18.6.1 IONIC CHARACTER

The salts of Group 1 have the highest amount of electrovalent character of any salts. The metal ions are hydrated in solution [§ 17.4]. Only the salts of weak acids, e.g. sodium carbonate, are hydrolysed in solution [§ 12.9]. The ions of Group 1 form very few complex compounds.

18.6.2 SOLUBILITY

The salts and hydroxides of Group 1 are soluble because:

- the amounts of energy required to separate the ions (the lattice dissociation enthalpies) are not highly endothermic because the cations are singly charged
- the amounts of energy released (enthalpies of hydration of the small cations) are highly exothermic.

18.6.3 THERMAL STABILITY OF COMPOUNDS

The thermal stability of an ionic solid depends on:

- the charges on the ions. The greater the charges, the greater is the attraction between them and the higher is the energy needed to separate them: the lattice dissociation enthalpy is highly endothermic
- the sizes of the ions. Small ions can approach closely and much energy is needed to separate them: the lattice dissociation enthalpy is highly endothermic.

Nitrates

Nitrates of Group 1 decompose when heated to form the nitrites (nitrates(III)) and oxygen.

$$2KNO_3(s) \longrightarrow 2KNO_2(s) + O_2(g)$$

Since the NO_2^- ion is smaller than the NO_3^- ion, the $NaNO_2$ lattice is more stable than the $NaNO_3$ lattice. The nitrite lattice is sufficiently stable to avoid further decomposition. This is not the case in Group 2 where the nitrates decompose to form oxides, $\mathbf{M}^{2+}O^{2-}$ which have very high lattice enthalpies [§ 18.9.3].

Carbonates, hydrogencarbonates and hydroxides

The carbonates and hydroxides of Group 1 (except Li) are thermally stable. The hydrogencarbonates decompose at 100 °C, forming the carbonate and carbon dioxide. This is why sodium hydrogencarbonate is used in baking powder.

$$2NaHCO_3(s) \longrightarrow Na_2CO_3(s) + CO_2(g) + H_2O(g)$$

18.6.4 OXIDES

The s block metals form **normal oxides**, containing the O^{2-} ion and **peroxides** containing the O_2^{2-} ion. The oxides react with water to give solutions of hydroxide ions. Peroxides give hydrogen peroxide also.

$$O^{2-}(s) + H_2O(l) \longrightarrow 2OH^-(aq)$$
$$O_2^{2-}(s) + 2H_2O(l) \longrightarrow 2OH^-(aq) + H_2O_2(aq)$$

18.6.5 HYDROXIDES

All the hydroxides are soluble; they are strong alkalis. These strong alkalis give Group 1 their name of the **alkali metals**. All (except LiOH) are **deliquescent**, that is, they absorb water vapour from the air and dissolve in it. The reactions of alkalis are:

Alkalis react with acids ...

1. They neutralise acids to form salts:

$$Na^+(aq)OH^-(aq) + H^+(aq)Cl^-(aq) \longrightarrow Na^+(aq)Cl^-(aq) + H_2O(l)$$

... metal cations ...

2. They precipitate insoluble metal hydroxides from solution:

$$Cu^{2+}(aq) + 2OH^-(aq) \longrightarrow Cu(OH)_2(s)$$

3. They precipitate and then dissolve amphoteric hydroxides:

$$Al^{3+}(aq) + 3OH^-(aq) \longrightarrow Al(OH)_3(s) \xrightarrow{OH^-(aq)} Al(OH)_4^-(aq)$$

Aluminium hydroxide   Aluminate ion

... with salts of weak bases ...

4. They displace weak bases (ammonia and amines) from their salts:

$$NH_4^+(s) + OH^-(aq) \longrightarrow NH_3(g) + H_2O(l)$$

... organic compounds, e.g., esters ...

5. They hydrolyse esters and other organic compounds:

$$CH_3CO_2C_2H_5(aq) + OH^-(aq) \longrightarrow CH_3CO_2^-(aq) + C_2H_5OH(aq)$$

Ethyl ethanoate   Ethanoate ion   Ethanol

... and with halogens

6. Reactions with halogens are covered in § 20.8;

The manufacture of sodium hydroxide by the electrolysis of brine is illustrated in Figure 18.6A.

Figure 18.6A
A diaphragm cell

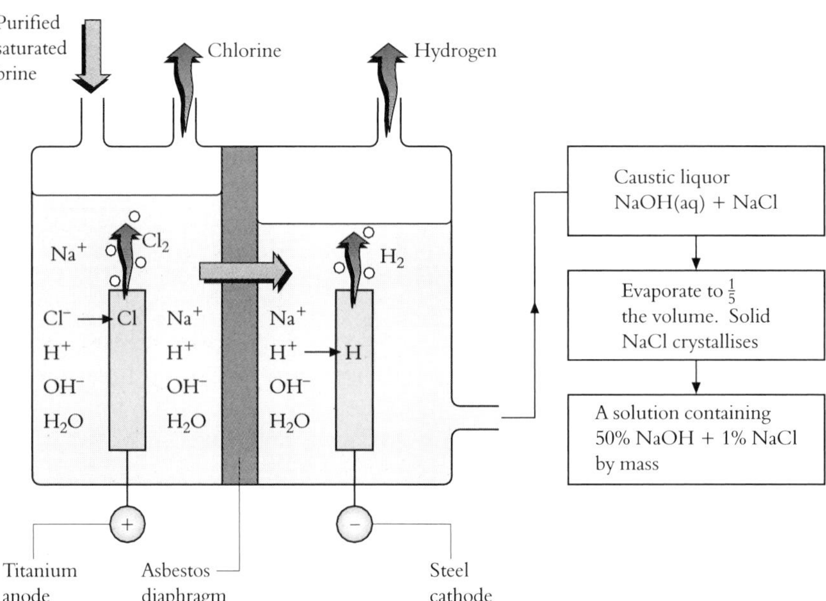

In the **diaphragm cell**, a porous asbestos diaphragm separates the cathode and anode. Purified saturated brine is fed into the anode compartment. Purification is necessary to remove calcium and magnesium ions which would precipitate as insoluble hydroxides and block the pores of the diaphragm. The level of liquid in the anode compartment is higher than that in the cathode compartment so that brine will seep through the diaphragm. The cell reactions are shown in Figure 18.6A.

Installations contain 50 to 100 single cells joined in series. The electrolysis of brine is the basis of the **chlor–alkali industry**. Originally sodium hydroxide was the product of prime importance, but with the development of plastics containing chlorine, e.g. PVC, chlorine has become an equally important product. The cell can be adjusted to allow chlorine and sodium hydroxide to mix and react to form sodium chlorate(I), NaClO. At a higher temperature, sodium chlorate(V), $NaClO_3$ is formed.

$$Cl_2(g) + NaOH(aq) \longrightarrow NaClO(aq) + HCl(aq)$$
$$3NaClO(aq) \longrightarrow NaClO_3(aq) + 2NaCl(aq)$$

Table 18.6A summarises the uses of the products.

Table 18.6A Products of the chlor-alkali industry

Product	Uses
Sodium hydroxide	1. Titrimetric analysis of acids [§ 3.14] 2. Hydrolysis of organic compounds, e.g. esters [§ 33.13.1], amides [§ 33.15] and nitriles [§ 33.16] and **saponification**, the hydrolysis of esters of glycerol to give soaps [§ 33.14].
Hydrogen	see §§ 17.2, 17.3
Chlorine	see § 20.11
Sodium chlorate(I)	antiseptic, e.g. Milton®
Sodium chlorate(V)	weedkiller, e.g. Tandol®

18.6.6 HALIDES, NITRATES, SULPHATES, CARBONATES

Table 18.6B lists some important compounds

Table 18.6B Some important compounds of Group 1

Compound	Uses
Sodium chloride	Common salt. Mined e.g. in Cheshire. Used in the manufacture of sodium [§ 18.4], sodium carbonate [§§ 18.1, 18.6.7], sodium hydroxide [§ 18.6.5], sodium chlorate(I) and sodium chlorate(V) [§§ 18.6.5, 20.7]
Potassium bromide	Used as a sedative and in the production of silver bromide for photographic films.
Sodium nitrate	Mined, used as a fertiliser
Potassium nitrate	Mined, used as a fertiliser
Sodium sulphate	Made from sodium hydroxide and sulphuric acid, crystallises as $Na_2SO_4.10H_2O$
Sodium hydrogensulphate	Made from sodium hydroxide and sulphuric acid. An acidic salt; dissociating partially to give $H^+(aq)$ and $SO_4^{2-}(aq)$ ions.
Sodium carbonate	Washing soda, used to soften hard water
$Na_2CO_3.10H_2O$	Effloresces to form $Na_2CO_3.H_2O$. On heating forms anhydrous Na_2CO_3.
Anhydrous, Na_2CO_3	Used as a primary standard in titrimetric analysis (K_2CO_3 is deliquescent). Used in the glass industry, the paper industry and in the manufacture of soaps and detergents.

18.6.7 THE SOLVAY PROCESS

The manufacture of sodium carbonate by the Solvay process (or ammonia–soda process) is illustrated in Figure 18.6B. The process uses the raw materials coke, limestone, sodium chloride and ammonia. It employs a clever recycling of materials so that the only by-product is calcium chloride. Although some of this is used as a drying agent, it is largely a waste product.

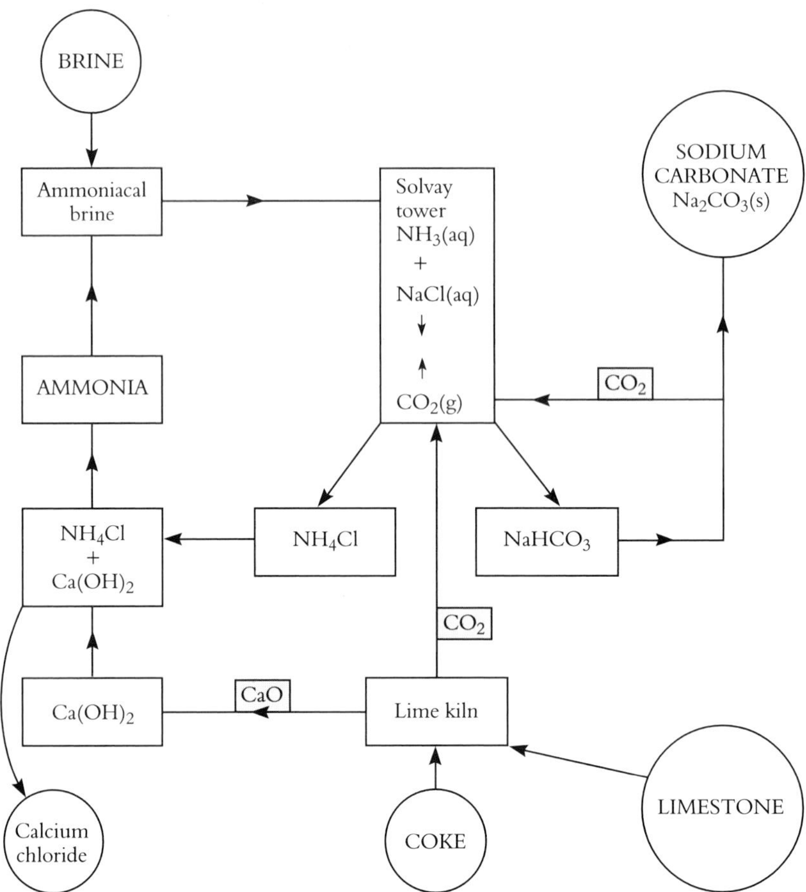

Figure 18.6B
The Solvay process for the manufacture of sodium carbonate

1. The Solvay process begins by heating limestone in a kiln.

   $$CaCO_3(s) \longrightarrow CaO(s) + CO_2(g)$$

2. In the Solvay tower, the carbon dioxide produced in (1) reacts with ammoniacal brine:

   $$CO_2(g) + NaCl(aq) + NH_3(aq) + H_2O(l) \longrightarrow NaHCO_3(s) + NH_4Cl(s)$$

3. The sodium hydrogencarbonate which is precipitated is washed, dried and heated.

   $$2NaHCO_3(s) \longrightarrow Na_2CO_3(s) + CO_2(g) + H_2O(l)$$

   The carbon dioxide produced is sent to the Solvay tower.

4. The calcium oxide formed in **1** is slaked.

   $$CaO(s) + H_2O(l) \longrightarrow Ca(OH)_2(s)$$

5. The calcium hydroxide produced in **4** and the ammonium chloride produced in **2** react to form ammonia, which is recycled, and calcium chloride, the waste product.

   $$2NH_4Cl(s) + Ca(OH)_2(s) \longrightarrow 2NH_3(g) + CaCl_2(s) + 2H_2O(l)$$

   Potassium carbonate is not made in the same way because potassium hydrogencarbonate is too soluble to crystallise at the bottom of the Solvay tower.

Figure 18.6C Winnington ICI, including the Solvay plant

18.7 LITHIUM

The ions Li^+ and Mg^{2+} are similar in size, and their compounds are therefore similar in having some covalent character. Features in which lithium resembles magnesium and differs from other alkali metals are:

1. the action of heat on the hydroxide, carbonate and nitrate
2. the formation of a nitride
3. the absence of a peroxide.

CHECKPOINT 18.7: GROUP 1

1. Explain why the ionisation $M \longrightarrow M^+ + e^-$ takes place more readily as you pass down Group 1. Illustrate the gradation in reactivity by referring to the reactions of the alkali metals with water.

2. Explain why
   (a) sodium salts have a large degree of ionic character
   (b) the salt NaCl is neutral in solution
   (c) sodium carbonate is alkaline in solution
   (d) sodium forms two sulphates.

3. Explain why
   (a) sodium nitrate decomposes on heating to form sodium nitrite
   (b) calcium nitrate decomposes on heating to form calcium oxide
   (c) Group 1 are called the alkali metals.

4. Write equations for the reactions of sodium hydroxide with *(a)* HCl(aq) *(b)* H_2SO_4(aq) *(c)* $MgSO_4$(aq) *(d)* NH_4Cl(s).

5. *(a)* What is the purpose of the diaphragm in the diaphragm cell for the electrolysis of aqueous sodium chloride?
   (b) Name four products of the diaphragm cell, other than sodium hydroxide.

6. A pellet of sodium hydroxide weighing 0.254 g was left to stand in the air. It changed to a colourless liquid and then to colourless, transparent crystals. After some days it formed a white solid, weighing 0.394 g. Explain the changes that have occurred.

7. *(a)* What is manufactured in the Solvay process? What uses are made of this product?
   (b) Comment on the cost and availability of the raw materials used. What techniques are used to keep running costs to a minimum?
   (c) What is the by-product? What use is made of it?

18.8 REACTIONS OF GROUP 2

The reactions of Group 2 are summarised in Table 18.8. The vigour of the reactions increases down the group. Figure 18.8A summarises the reactions of calcium and magnesium.

Figure 18.8A
Reactions of calcium and magnesium

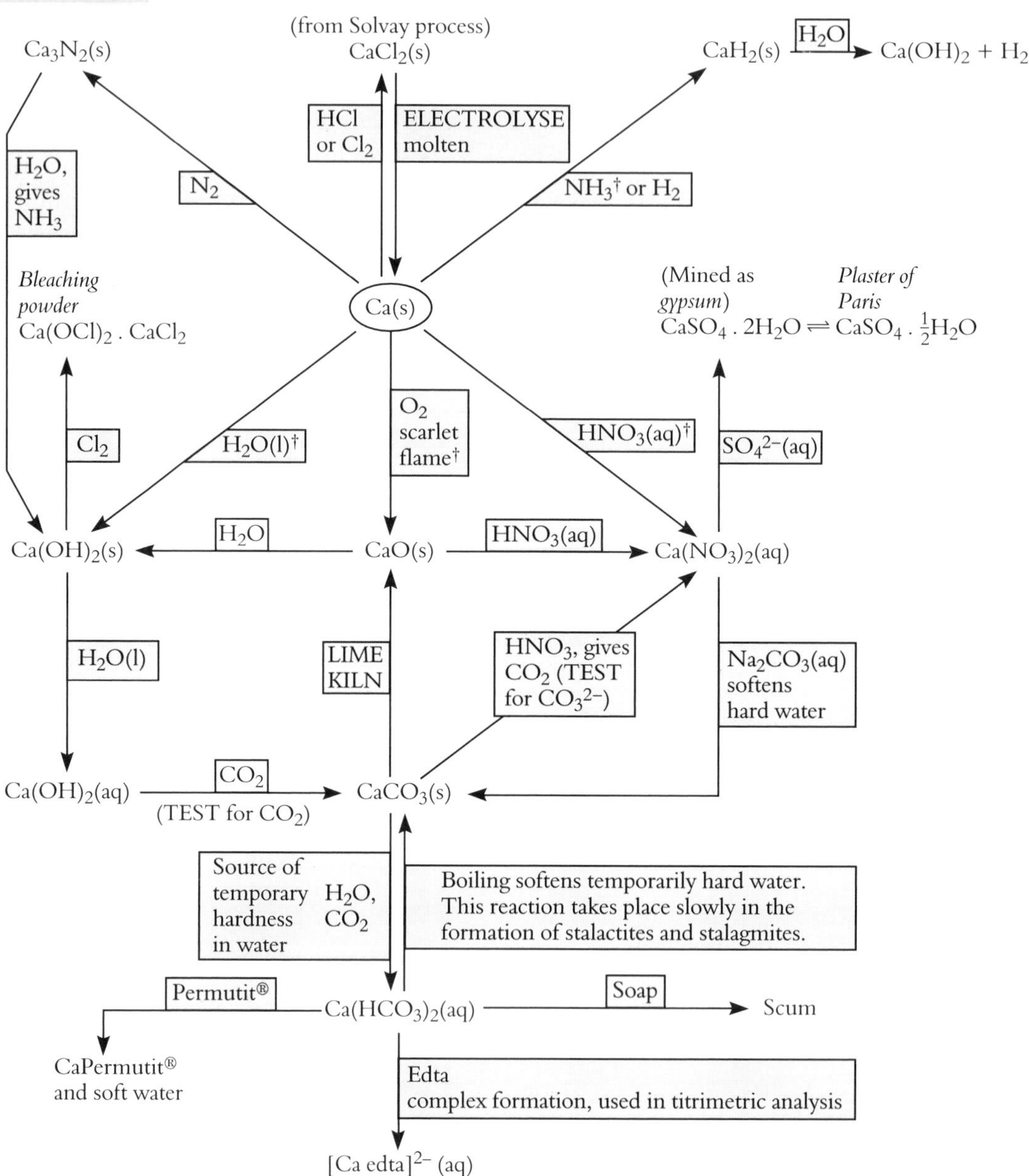

The reaction of magnesium are similar, but
† $Mg + NH_3 \longrightarrow Mg_3N_2$
† Mg burns with a bright white flame.
† Mg reacts with steam to form MgO, but only slowly with water.
† Mg + cold, dilute $HNO_3(aq)$ gives $H_2(g)$, not $NO_2(g)$.
Magnesium forms Grignard reagents [§ 29.11] e.g. C_2H_5MgBr

Table 18.8
The reactions of Group 2

Reactant	Reaction
Water	*Beryllium* does not react with water. *Magnesium* reacts slowly with cold water, rapidly with steam: $Mg(s) + H_2O(g) \longrightarrow MgO(s) + H_2(g)$ *Calcium* reacts steadily with water: $Ca(s) + 2H_2O(l) \longrightarrow Ca(OH)_2(aq) + H_2(g)$ *Strontium and barium* are more reactive than calcium.
Dilute acids	*Beryllium and magnesium* react steadily. Calcium reacts vigorously. *Strontium and barium* are more reactive than calcium.
Oxygen	All tarnish in air as a surface film of oxide forms. All burn readily in air.
Halogens	All react on heating to form halides, $M^{2+}2X^-$.
Nitrogen	All (except Be) react on heating to form nitrides. On addition of water, nitrides give ammonia: $Mg_3N_2(s) + 6H_2O(l) \longrightarrow 3Mg(OH)_2(aq) + 2NH_3(g)$

18.9 COMPOUNDS OF GROUP 2

18.9.1 IONIC CHARACTER

The metal ions are hydrated in solution. Most of them are not hydrolysed. However, when magnesium chloride solutions are evaporated, they do not give anhydrous magnesium chloride. Hydrolysis occurs with the formation of the basic chloride:

$$MgCl_2(aq) + H_2O(l) \rightleftharpoons Mg(OH)Cl(s) + HCl(aq)$$

If evaporation is carried out in a stream of hydrogen chloride, the equilibrium is reversed, and magnesium chloride can be obtained.

18.9.2 SOLUBILITY

Solubilities are determined by two factors

Solubilities are not easy to explain because they are determined by two factors. Differences in ionic size have opposing effects on the two factors:

Small ions {

Factor 1 Crystal lattices are hard to break up. (The lattice dissociation enthalpy is highly endothermic.)

Factor 2 Much energy is released when the ions are hydrated. (The enthalpy of hydration is highly exothermic.)

Table 18.9A Solubility of Group 2 compounds

... *in Group 2*

Sparingly soluble compounds				Soluble salts	
MSO_4 and MCO_3		$M(OH)_2$		$M(NO_3)_2$	
Mg Ca Sr Ba	↓ *Decrease* in solubility is due to the decrease in the enthalpy of hydration *(Note 1).*	↓	*Increase* in solubility is due to the decrease in the lattice dissociation enthalpy *(Note 2).*	Mg Ca Sr Ba	↓ *Increase* ↓ *Decrease*

Note 1 The lattice enthalpies do not vary much. Since in size
Anion ≫ Cation
the differences in cation size do not greatly affect the lattice enthalpy.
Note 2 This outweighs the change in the enthalpy of hydration.

18.9.3 THERMAL STABILITY OF COMPOUNDS

The thermal stability of a compound is measured by its standard lattice enthalpy [§§ 10.12, 10.13]. This depends on two factors:

- The greater the charges on the ions, the greater is the attraction between them and the higher is the standard lattice enthalpy.
- The smaller the ions, the more closely they can approach in the lattice, and the higher is the standard lattice enthalpy.

Nitrates

The nitrates decompose on heating to form oxides. The O^{2-} ion is smaller and more highly charged than the NO_3^- ion. The oxides $\mathbf{M}^{2+}O^{2-}$ have high standard lattice enthalpies, and nitrates of Group 2 metals therefore decompose to form oxides, in contrast to Group 1 [§ 18.6.3].

$$2Mg(NO_3)_2(s) \longrightarrow 2MgO(s) + 4NO_2(g) + O_2(g)$$

Carbonates and hydrogencarbonates

The carbonates decompose on heating to form oxides:

$$CaCO_3(s) \longrightarrow CaO(s) + CO_2(g)$$

The hydrogencarbonates exist only in solution. When the solutions are boiled, they decompose to form the carbonates:

$$Ca(HCO_3)_2(aq) \longrightarrow CaCO_3(s) + CO_2(g) + H_2O(l)$$

This is why hard water that owes its hardness to calcium hydrogencarbonate is softened by boiling.

Hydroxides

The hydroxides are decomposed by heat to form oxides:

$$Mg(OH)_2(s) \longrightarrow MgO(s) + H_2O(l)$$

18.9.4 OXIDES

Group 2 (like Group 1) form **normal oxides**, containing the O^{2-} ion and **peroxides** containing the O_2^{2-} ion. The oxides react with water to give solutions of hydroxide ions. Peroxides give hydrogen peroxide also

$$O^{2-}(s) + H_2O(l) \longrightarrow 2OH^-(aq)$$

$$O_2^{2-}(s) + 2H_2O(l) \longrightarrow 2OH^-(aq) + H_2O_2(aq)$$

18.9.5 HYDROXIDES

The hydroxides of Ca, Sr and Ba dissolve in water, but their low solubilities make them weak alkalis. The hydroxides of Be and Mg have very low solubilities. Beryllium hydroxide is amphoteric. For the reactions of alkalis, see § 18.6.5. The uses of calcium hydroxide are listed below.

Table 18.9B
Uses of calcium hydroxide

Calcium hydroxide	Uses
Manufacture: Limestone is heated in a lime kiln to give calcium oxide, called *quicklime*, which is 'slaked' with water to give *slaked lime*, $Ca(OH)_2$.	Called *slaked lime*. The aqueous solution is called *limewater*. 1. Treatment of fields which are too acidic for healthly plant growth 2. Mortar = slaked lime + sand + water 3. Manufacture of calcium hydrogensulphite The paper industry needs this to remove lignin from wood and leave cellulose, ready to be made into paper. 4. Reaction with chlorine to form bleaching powder, $Ca(OCl)_2.CaCl_2$. This is a useful source of chlorine, which it liberates readily when an acid is added.

18.9.6 SALTS

Some important salts are mentioned in Table 18.9C.

Table 18.9C
Salts of Group 2 metals

Calcium carbonate	Mined as calcite, marble, limestone, chalk, aragonite. Used in the Solvay process, in the iron and steel industry, in the glass industry and in making cement.
Magnesium carbonate	Mined as magnesite and as dolomite, $CaCO_3.MgCO_3$.
Calcium chloride	The by-product of the Solvay process [§ 18.6.7]. Deliquescent; used as a drying agent, except for ammonia and ethanol, with which it forms complexes.
Nitrates	Made from nitric acid + metal oxide or carbonate.
Magnesium sulphate	Mined as $MgSO_4.7H_2O$. Present in tap water, causing permanent hardness. Used as the laxative Epsom salts.
Calcium sulphate	Mined as anhydrite $CaSO_4$, and as gypsum, $CaSO_4.2H_2O$. Gypsum gives plaster of Paris when heated to 100 °C: $2CaSO_4.2H_2O(s) \rightleftharpoons (CaSO_4)_2.H_2O(s) + 3H_2O(l)$ calcium sulphate-2-water plaster of Paris. When water is added, plaster of Paris expands slightly and sets to form calcium sulphate-2-water. Used to make the fertiliser ammonium sulphate. Present in tap water, it is a cause of permanent hardness.
Barium sulphate	Used in 'X ray meals'

GROUP 1 AND GROUP 2

Group 1:
Li, Na, K, Rb, Cs
Group 2:
Be, Mg, Ca, Sr, Ba

s block elements

Properties
Group 1, the **alkali metals**, are soft, with low density, low melting and boiling temperatures. Group 2, the **alkaline earths**, are harder and denser with higher melting and boiling temperatures. Both groups form ionic compounds; Group 1 form ions $\mathbf{M}^+$ and Group 2 form ions $\mathbf{M}^{2+}$. The elements **colour flames**. § 18.2

Reaction with oxygen
All react with oxygen, tarnish in air and burn readily to form oxides. § 18.5, § 18.8
Reaction with halogens
All react with halogens on heating to form halides.
Reaction with water
All, except Be and Mg, react with cold water to form **M**OH or $\mathbf{M}(OH)_2 + H_2$. Group 1 metals are kept under oil.
§ 18.5, § 18.8

Reaction of oxides with water
The oxides react with water to form solutions of hydroxides. Group 1 **M**OH are strong alkalis. Group 2 $\mathbf{M}(OH)_2$ are less soluble. § 18.6.4, § 18.9.4

Trends down Groups 1 and 2
1. First **ionisation energy** decreases down each group. § 18.2
2. Passing down Group 2.

- **Solubility of $\mathbf{M}SO_4$** decreases as $\Delta H^\ominus$ of hydration of $\mathbf{M}^{2+}$ decreases with increase in size of $\mathbf{M}^{2+}$.
- **Solubility of $\mathbf{M}(OH)_2$** increases down the group as lattice enthalpy decreases with increase in size of $\mathbf{M}^{2+}$. § 18.9.2

3. **Nitrates** decompose on heating:
Group 1 (except Li) → $\mathbf{M}NO_2 + O_2$,
Group 2 and Li → $\mathbf{M}O + NO_2 + O_2$.
Group 1 $Na^+NO_2^-$ etc have a high lattice enthalpy and do not decompose further.
Group 2 nitrites decompose because $\mathbf{M}^{2+}O^{2-}$ have very high lattice enthalpies.
4. **Carbonates** of Group 1 (except Li) are thermally stable, of Group 2 and Li form $\mathbf{M}O + CO_2$ on heating.
5. **Hydrogencarbonates** of Group 1 decompose readily, of Group 2 exist only in solution.
6. Hydroxides of Group 1 (except Li) are thermally stable, of Group 2 and Li form oxides on heating.
§§ 18.6.3, 18.9.3

Chapter 18 deals with Group 1 and Group 2. OCR candidates and NEAB candidates are required to know about sodium and Group 2.

Sources of compounds of s block elements § 18.1, § 18.4, § 18.6

GROUP 2

Group 2 is the **alkaline earth metals**, Be, Mg, Ca, Sr, Ba. They have strongly basic oxides. The chlorides are crystalline solids. They form salts including sulphates, nitrates, carbonates and ionic hydrides. §18.2

Trends down Group 2

1. First ionisation energy decreases down the group. §18.2

2. Passing down Group 2.

- solubility of **M**SO_4 decreases as $\Delta H^{\ominus}$ of hydration of $\mathbf{M}^{2+}$ decreases with increase in size of $\mathbf{M}^{2+}$ §18.9.2
- solubility of **M**$(OH)_2$ increases down the group as lattice enthalpy decreases with increase in size of $\mathbf{M}^{2+}$. §18.9.2

3. Nitrates M$(NO_3)_2$ decompose on heating to give **M**O + NO_2 + O_2. §18.9.3

4. Carbonates MCO_3 decompose on heating to give **M**O + CO_2. §18.9.3

5. Hydrogencarbonates M$(HCO_3)_2$ exist only in solution. §18.9.3

6. Hydroxides M$(OH)_2$ form oxides **M**O + H_2O on heating. §18.9.3

Properties

Group 2 are harder and denser than Group 1 with higher melting and boiling temperatures. They are softer than transition metals with lower melting temperatures. They form ionic compounds in which the metals have oxidation number = +2. § 18.2

Important uses of Group 2 compounds include

- MgO as a refractory lining in furnaces § 18.9.5
- CaO and $Ca(OH)_2$ in agriculture to neutralise excessive acidity. § 18.9.5

Reactions

From the first member to the last, there is an increase in reactivity down the group.

Reaction with oxygen

All react with oxygen, tarnish in air and burn readily to form oxides. § 18.8

Reaction with halogens

All react with halogens on heating to form halides. § 18.8

Reaction with water

All, except Be and Mg, react with cold water to form **M**OH or **M**$(OH)_2$ + H_2. § 18.8

Reaction of oxides with water

The oxides react with water to form solutions of hydroxides. Group 2 hydroxides are less soluble than Group 1 hydroxides, e.g. NaOH. § 18.9.4

Note: Chapter 18 deals with Group 1 and Group 2. OCR candidates and NEAB candidates are required to know about sodium and Group 2.

CHECKPOINT 18.9: GROUP 2

1. Explain why the reactivity of the metals of Group 2 increases down the group. Illustrate the trend in reactivity by reference to the reactions with water.

2. *(a)* What two factors determine the solubility of a compound?
   (b) Explain why the solubility of $\mathbf{M}(OH)_2$ increases down Group 2.
   (c) Explain why the solubility of $\mathbf{M}CO_3$ decreases down Group 2.

3. Explain why the nitrate $M(NO_3)_2$ decomposes on heating to form MO.

4. Why is calcium hydroxide regarded as an important compound?

5. Explain the following statements
   (a) Calcium sulphate is a compound with (i) medical and (ii) agricultural importance.
   (b) Beryllium chloride is a covalent substance in the vapour state.
   (c) CaO has a higher standard lattice enthalpy than $CaCO_3$.
   (d) $MgCO_3$ is more easily decomposed by heat than $BaCO_3$.

18.10 EXTRACTING METALS FROM THEIR ORES

The methods used in extracting metals from their ores will be found under each group of the Periodic Table. References are given here.

Groups 1 and 2: The metals of Groups 1 and 2 are very reactive, and their compounds are difficult to reduce. The method of extracting the metal is electrolysis of the molten anhydrous chloride; see § 18.4.

Aluminium: Aluminium, in Group 3, is also obtained by electrolysis. In this case the molten anhydrous oxide is used. The oxide has a very high melting temperature and to lower the temperature at which electrolysis can be carried out the oxide is dissolved in molten sodium aluminium fluoride; see §§ 19.1, 19.3.2.

Iron: Less reactive metals are obtained from their compounds by chemical reducing agents. Transition metals are mined as sulphides and oxides; see § 24.5. The sulphide ores are roasted in air to form oxides. The method of reducing an oxide is exemplified by the extraction of iron in the blast furnace. Carbon monoxide reduces iron oxide to iron; see § 24.17.1.

Titanium: The ore rutile is titanium(IV) oxide. This is converted into the chloride, and magnesium is employed as reducing agent. Being higher in the electrochemical series, magnesium displaces titanium from its chloride; see § 24.12.2.

Copper: Copper is mined as sulphide ores. Roasting in air gives impure copper. The electrochemical method of purifying copper is described in § 24.18.

Zinc: Zinc is mined as the sulphide and as the carbonate. These ores are converted into zinc oxide, which is reduced by heating it with coke; see § 24.19.

EXTRACTION OF METALS I

Aluminium is a p block metal and is high in the electrochemical series. It is mined as the oxide, **bauxite**, which cannot be reduced by chemical reducing agents. It is purified, dissolved in a molten salt, $NaAlF_3$, and electrolysed. The process is costly because of the electricity consumed in melting and electrolysing the electrolyte. § 19.3.2

Transition metals are of moderate reactivity. Many are extracted by reduction of the oxides by carbon or carbon monoxide, e.g. iron in a blast furnace. §§ 24.5, 24.17.1

Steel is made by oxidising the carbon content of cast iron in a converter. § 24.17.3

The **choice of reducing agent** depends on

- the cost (carbon is the cheapest)
- the cost of the energy required (as in the high temperature required for reduction of $TiCl_4$ and the electricity required in the extraction of Al)
- whether a continuous process is employed, e.g. the blast furnace for iron, or a batch process, e.g. the Kroll process for titanium
- whether the metal is required in a high state of purity.

Al § 19.3.2, Ti § 24.12.2, Fe § 24.17.1, Cu § 24.18, Zn § 24.19

Many transition metals are mined as **sulphide ores**. These are roasted in air to form oxides. Then the oxides are reduced. The sulphur dioxide produced in the roasting is a **pollutant**. § 21.10, § 24.15

Groups 1 and 2
Electrolysis of the molten anhydrous chloride. § 18.4

Some transition metals form **carbides**; therefore carbon cannot be used as the reducing agent. Chromium and tungsten are obtained by reduction of the oxide by aluminium. § 24.13.1

Titanium is extracted by converting the oxide into $TiCl_4$ and reducing the chloride with a more reactive metal, Na or Mg. § 24.12.2

QUESTIONS ON CHAPTER 18

*Questions on Group 1 and Group 2

1. Francium, the last member of Group 1, is a short-lived radioactive element. From what you know of the chemistry of Group 1, deduce what the properties of francium are likely to be, with respect to

   (a) the nature of its hydride and the reaction of the hydride with water
   (b) combination with halogens
   (c) combination with oxygen
   (d) the action of heat on the carbonate, hydrogencarbonate and nitrate
   (e) the solubility of its salts in water and in organic solvents.

2. Comment on the statement, 'The metals of Group 1 are a similar set of elements, yet some gradation in properties can be observed from top to bottom of the group.'

3. The element **A** is a member of Group 1 or Group 2. Some reactions of **A** are shown below.

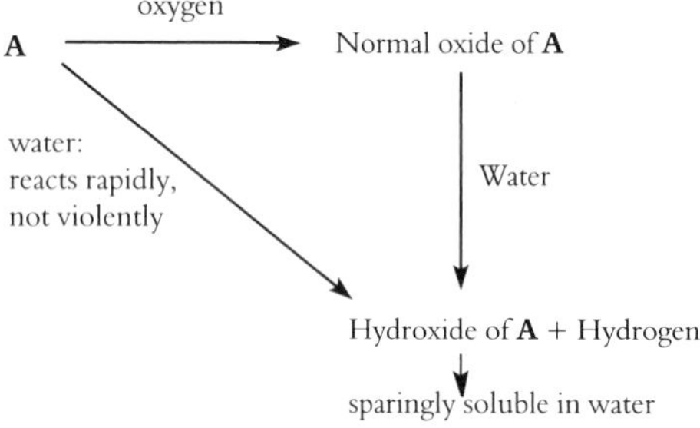

   (a) On the basis of this information, what can you deduce about the identity of **A**?
   (b) In its reaction with oxygen, x g of **A** forms $1.3992x$ g of the normal oxide. Calculate the relative atomic mass of **A**. Identify **A**.
   (c) What colour would the salts of **A** give in a flame test?
   (d) Write equations for the reactions of **A** with oxygen and with water and for the reaction of the oxide of **A** with water.

4. An element **Y** has standard first, second and third ionisation enthalpies 738, 1449 and 7728 kJ mol^{-1} respectively. The halides of **Y** do not colour a flame. A solution of the nitrate of **Y**

   (1) gave no precipitate with dilute sulphuric acid
   (2) gave a white precipitate with sodium carbonate solution
   (3) gave a white precipitate with sodium hydroxide solution, which did not dissolve in excess sodium hydroxide.

   (a) Suggest the identity of **Y**, explaining your reasoning.
   (b) Give equations for reactions (2) and (3).
   (c) Predict the thermal stabilities of the carbonate and hydroxide of **Y**, giving reasons.

5. *(a)* The table gives information about the fluorides of some Group (1) metals.

Salt	Lattice enthalpy /kJ mol^{-1}	The sum of the hydration enthalpies of the two ions /kJ mol^{-1}	Solubility in water /g kg^{-1}
Lithium fluoride	+1022	−1025	2.7
Sodium fluoride	+902	−912	40
Potassium fluoride	+801	−828	950
Rubidium fluoride	+767	−807	1310

   (i) Why is energy required to break up an ionic lattice? **2**
   (ii) Why is energy released when ions are hydrated? **2**
   (iii) Account for the trend in the solubilities of the four fluorides. **3**

   (b) Explain why it is not possible to determine the solubilities of the **oxides** of Group (1) metals in water. **2**
   (c) Why do the elements in Group (1) mainly form ionic compounds? **2**

   11 *O&C (AS/AL)*

6. *(a)* Describe, with the aid of diagrams. the structure of, and bonding in, sodium chloride, iodine, diamond and graphite. How do the properties of these different types of crystal enable you to distinguish between them? **15**
   (b) Write equations for the reactions of sodium oxide and phosphorus(V) oxide with water and explain, in terms of the bonding present in the oxides, why the resulting solutions have different pH values. **6**
   (c) (i) Explain the trends in first ionisation energy and electronegativity down Group I.
   (ii) Discuss two ways in which lithium is an atypical member of Group I. **9**

   30 *NEAB (AS/AL)*

7. *(a)* (i) Give **balanced** equations describing **two** reactions of calcium metal in which Ca^{2+} ions are formed. Compare the reactions of zinc metal with the same reagents. **4**

(ii) State how the thermal stability of hydroxides and carbonates changes as Group II is descended. Describe **briefly** a simple laboratory experiment which would confirm whether or not a carbonate is thermally stable or unstable. **3**

(b) 14.78 g of a pure sample of carbonate of an element **Z**, which belongs to **either** Group I **or** Group II, was completely decomposed upon heating, producing exactly 4.48 dm^3 of carbon dioxide at 0 °C and 1 atm pressure (101 kPa). Calculate the number of moles of CO_2 produced. Deduce the relative atomic mass of **Z**. Hence, using the Periodic Table, identify **Z**. **5**
[The molar volume of an ideal gas at 0 °C and 1 atm pressure (101 kPa) is 22.4 dm^3.]

(c) Tests were carried out on a colourless solution which contains a soluble salt $\mathbf{MX_2}$ at a concentration of about 0.05 mol dm^{-3}. The following table shows the reagents used and the observations made.

Reactions of MX_2	
Reagent added	**Observation**
Dilute sodium hydroxide	No precipitate
Dilute sulphuric acid	Dense white precipitate
Copper(II) sulphate	Off-white precipitate in a brown solution

Identify the cation **M** and the anion **X**, carefully explaining your reasoning. Write an ionic equation for the reaction that occurs when copper(II) ions are added to the solution. **4**

(d) Use the standard enthalpy change of formation data given below to calculate the enthalpy change of reaction for the reduction of **0.10 mol** of iron(III) oxide by carbon monoxide.

$$Fe_2O_3(s) + 3CO(g) \longrightarrow 2Fe(s) + 3CO_2(g)$$

$\Delta H_f^{\ominus}$ (298 K)/kJ mol^{-1} $Fe_2O_3(s)$ −824;
$CO(g)$ −110; $Fe(s)$ 0; $CO_2(g)$ −394. **4**

20 *WJEC (AL)*

8. Sodium and sodium hydroxide are both manufactured by electrolytic processes.

(a) Name the electrolyte used in the manufacture of: (i) sodium, (ii) sodium hydroxide. **2**

(b) (i) What is produced at the anode during the manufacture of sodium hydroxide? Write an equation for its formation.
(ii) What other gaseous product might be given off at the anode under other conditions? Write an equation for its formation. **5**

(c) Suggest a reason why the product in *(b)*(i) is formed in the industrial process rather than that in *(b)*(ii). **3**

(d) Describe what you would observe when dilute sodium hydroxide solution is added dropwise to a solution of aluminium sulphate until in excess. Give the formulae of the aluminium-containing species present in the original solution, and responsible for the observations you have described. **6**

See Chapter 19 for part *(d)*.

16 *L (AL)*

9. *(a)* (i) Define the term *standard enthalpy of formation* ($\Delta H_f^{\ominus}$) of a compound.
(ii) Write an equation, including state symbols, for the reaction which illustrates the enthalpy of formation of lithium carbonate. **5**

(b) The following table gives some values of standard enthalpies of formation.

Compound	Li_2CO_3	Na_2CO_3	Li_2O	Na_2O	CO_2
$\Delta H_f^{\ominus}$/**kJ mol**$^{-1}$	−1216	−1131	−596	−416	−394

(i) Use the data from the table to calculate the enthalpy changes for the decomposition of lithium carbonate and of sodium carbonate into the metal oxide and carbon dioxide.
(ii) Comment on the thermal stability of Group I carbonates using your answers to part *(b)*(i). **8**

(c) Give the meaning of the term *first ionisation energy* of an element. State and explain the trends in first ionisation energy and in melting point of the Group I elements from lithium to caesium. **8**

(d) A solid sample of hydrated sodium carbonate ($Na_2CO_3.xH_2O$) contained an unknown amount of water of crystallisation. The sample (2.995 g) was dissolved in water and made up to exactly 250 cm^3 of solution. When 25.0 cm^3 of this solution were titrated against 0.113 M hydrochloric acid, 21.2 cm^3 of the acid were required for complete neutralisation.

$$Na_2CO_3 + 2HCl \longrightarrow 2NaCl + CO_2 + H_2O$$

Calculate the mass of sodium carbonate in the original sample and hence the value of x, where x is the number of moles of water combined with one mole of sodium carbonate in the solid. **9**

30 *NEAB (AS/AL)*

Questions on Group 2

1. From your knowledge of the chemistry of Group 2, predict what you can of the properties of radium. In particular, comment on

(a) the reaction of radium with water
(b) the solubility of its hydroxide and a likely value for the pH of its solution

(c) the solubility of the sulphate, chloride and carbonate
(d) the action of heat on the nitrate and carbonate.

2. Comment on the statement. 'In some ways, the members of Group 2 are a very similar set of elements; yet one can also look at them as a pair of similar elements (Be, Mg) and a trio of elements (Ca, Sr, Ba).'

3. *(a)* (i) State the conditions under which magnesium and calcium will react with water, and write balanced equations for the reactions.
(ii) Explain any differences between the two reactions in terms of the atomic properties of the two metals.
(b) Compare the chemistries of magnesium and calcium with reference to the following:
(i) the solubilities of their sulphates in water
(ii) the thermal stabilities of their carbonates
(iii) the reaction of their oxides with water.

4. Refer to the data on Group 2.

Element	**Mg**	**Ca**	**Sr**	**Ba**
Standard enthalpy of hydration/kJ mol^{-1} $M^{2+}(g) + aq \longrightarrow M^{2+}(aq)$	−1920	−1650	−1480	−1360
Standard electrode potential $E^\ominus$/V	−2.37	−2.87	−2.89	−2.91

(a) Which element is the most powerful reducing agent? Explain your answer.
(b) Write an equation for the reduction of hydrogen gas by magnesium.
(c) Deduce from $E^\ominus$ values how the elements of Group 2 could be manufactured from their chlorides.
(d) Suggest a reason for the trend in enthalpy of hydration.
(e) Which of these elements forms (i) the sulphate which is least soluble in water (ii) the carbonate which is most stable to heat?

5. *(a)* Select **three** different general methods for the extraction of metals. For **each** method you select, state the starting materials, the conditions used and give one example of a metal extracted by this method. **9**
(b) (i) Indicate the essential chemistry involved in the removal of carbon from impure iron in the manufacture of steel.
(ii) Give **two** reasons why steel is less expensive to produce than titanium.
(iii) Give **one** reason why titanium is used for certain applications despite the extra cost of this metal as compared to steel. **5**

14 *NEAB (AL)*

6. Calcium fluoride and calcium chloride are typical ionic compounds which can be made from their respective elements.

$$Ca(s) + F_2(g) \longrightarrow CaF_2(s)$$
$$Ca(s) + Cl_2(g) \longrightarrow CaCl_2(s)$$

(a) (i) Explain why the two compounds have the general formula CaX_2. **1**
(ii) Describe, using trends in electronegativities, why both compounds consist of structures containing ions. **1**
(iii) Explain, in terms of the ions present, why calcium chloride has a melting temperature of 772 °C and calcium fluoride has a melting temperature of 1423 °C. **1**
(b) Calculate, using the data given below, the enthalpy of lattice **formation** at 298 K of calcium fluoride, CaF_2, in kJ mol^{-1}. **4**

Process		*Enthalpy/*kJ mol^{-1}
$\Delta H^\ominus$ atomisation of calcium	$Ca(s) \longrightarrow Ca(g)$	193
$\Delta H^\ominus$ atomisation of fluorine	$\frac{1}{2}F_2(g) \longrightarrow F(g)$	79
$\Delta H^\ominus$ adding electron to fluorine	$F(g) + e^- \longrightarrow F^-(g)$	−348
$\Delta H^\ominus$ formation of $CaF_2(s)$	$Ca(s) + F_2(g) \longrightarrow CaF_2(s)$	−1214

Process		*Ionisation energy/*kJ mol^{-1}
Ionisation energies of calcium		
First	$Ca(g) \longrightarrow Ca^+(g) + e^-$	590
Second	$Ca^+(g) \longrightarrow Ca^{2+}(g) + e^-$	1150

(c) (i) A common method for making metal halides is to react the metal with the appropriate acid. State why it is undesirable to attempt to make a solution of calcium chloride by reacting calcium with concentrated hydrochloric acid. **1**
(ii) A scheme for making an aqueous solution of calcium chloride is shown below.

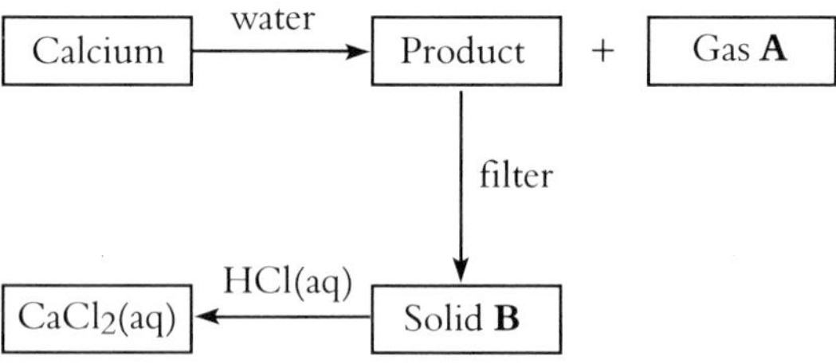

I. Identify gas **A** and the solid product **B** and give a balanced equation for the reaction of calcium with water. **3**
II. Give the balanced equation for the reaction of solid **B** with dilute hydrochloric acid and state how you would obtain crystals of hydrated calcium chloride from the resulting solution. **2**

(d) Rocks which principally contain calcium compounds give a characteristic flame colour. State the flame colour and explain why many rocks containing calcium carbonate have a biological origin. **2**

15 *WJEC (AL)*

7. *(a)* (i) Write an equation for the reaction of barium with water. **1**

(ii) Would the reaction in *(a)*(i) occur more vigorously or less vigorously than the reaction of calcium with water? Identify one contributory factor and use it to justify your answer. **2**

(iii) Write an equation for the action of heat on solid barium carbonate. **1**

(iv) At a given high temperature which of the two carbonates, barium carbonate or calcium carbonate, would decompose more easily? **1**

(v) How would you distinguish between solutions of barium chloride and calcium chloride? State in each case what you would see as a result of the test on each solution. **2**

(b) 1.71 g of barium reacts with oxygen to form 2.11 g of an oxide **X**.

(i) Calculate the formula of **X**. **2**

(ii) Give the formula of the anion present in **X**. **1**

(iii) What is the oxidation number of oxygen in this anion? **1**

(iv) Sodium forms an oxide, **Y**, which contains this same anion. Give the formula of **Y**. **1**

(c) Treatment of either **X** or **Y** with dilute sulphuric acid leads to the formation of the sulphate of the metal, together with an aqueous solution of hydrogen peroxide, H_2O_2.

(i) Write an equation for the reaction of **Y** with dilute sulphuric acid. **1**

(ii) The hydrogen peroxide solution produced may be separated from the other reaction product. Explain briefly why this is easier to achieve if **X** is used as the initial reagent rather than **Y**. **2**

15 *L (AS/AL)*

8. This question concerns the chemistry of calcium compounds.

(a) Calcium oxide can be prepared in several ways. One way is to heat the metal in oxygen, and another is to decompose calcium carbonate.

(i) Write down the equations for each of these reactions.

(ii) Calcium carbonate decomposes at about 900 °C, but magnesium carbonate decomposes at about 550 °C. Explain why these two carbonates decompose at different temperatures. **4**

(b) When water is added to solid calcium oxide, a white powder is produced which can be dissolved in water.

(i) What is the formula of the white powder?

(ii) Suggest a value for the pH of the resulting solution. **2**

(c) During the production of iron, calcium oxide reacts with silicon dioxide:

$$CaO(s) + SiO_2(s) \longrightarrow CaSiO_3(l)$$

Suggest a reason why these two oxides react together. **1**

(d) Calcium carbide, CaC_2, reacts with water to produce ethyne, C_2H_2.

$$CaC_2(s) + 2H_2O(l) \longrightarrow Ca(OH)_2(s) + C_2H_2(g)$$

(i) What volume of ethyne, C_2H_2, is produced at 101 kPa and 20 °C when 6.41 g of calcium carbide reacts as shown?

(ii) The calculation in *(d)*(i) assumes that ethyne behaves as an ideal gas. However, real gases do **not** behave like ideal gases at very high pressures. Give **one** reason why this is so. **5**

12 *C (AL)*

9. *(a)* Magnesium occurs naturally as the mineral *carnallite*, $KCl.MgCl_2.6H_2O$.

(i) State what is **observed** and give a balanced equation for the reaction which occurs when a solution of carnallite is treated with sodium hydroxide solution. **2**

(ii) State how to test for the presence of chloride ions in the carnallite solution, giving details of the reagents added, **observation** and an **ionic** equation for any precipitation reaction which may occur. **3**

(b) Both magnesium sulphate and barium sulphate occur naturally as minerals but only magnesium sulphate is soluble in water.

(i) Explain, in terms of hydration and lattice enthalpies, the reason why only one of the compounds is soluble in water. **2**

(ii) Name **two** features of magnesium sulphate which identify it as a **typical ionic compound**. **1**

(c) State why magnesium is an essential element for plant growth. **1**

9 *WJEC (AL)*

19 GROUP 3

19.1 THE ALUMINIUM PROBLEM

Aluminium is the most abundant metal in the Earth's crust, yet the metal was extracted from its ore only 150 years ago and remained a rarity for another 60 years. The Tsar of Russia gave his baby son an aluminium rattle to play with as it was more expensive than gold and he wanted the little fellow to have only the best.

A compound of aluminium which has been known for hundreds of years is a basic substance which chemists of old called *alumina*. They deduced that it was the oxide of a metal, the metal aluminum, which no-one had ever seen. Sir Humphry Davy tried in 1807 to obtain aluminium by electrolysing alumina, but he failed. In 1825 the Danish scientist H C Oersted succeeded in obtaining aluminium from a reaction between potassium and aluminium chloride. The German chemist Friedrich Wöhler improved on the method, and in 1827 his experiments yielded enough aluminium powder for him to melt down to a lump of metal. Next, a French chemist, Henri Sainte-Claire Déville, made sodium hexachloroaluminate, Na_3AlCl_6, and heated it with sodium to obtain molten aluminium and sodium chloride. Since molten aluminium was formed, he was able to run it off and cast it into ingots. This enabled him to turn his preparation into a commercial process. By the time that aluminium was obtained in quantity, in 1860, iron had been known for 5000 years.

Now that there was a commercial process for making aluminium on a large scale, people began to discover what a useful metal aluminium is and to invent new uses for it. It was still an expensive metal because of the high cost of the large quantities of sodium used in its extraction. Since the reactive metals (Na, K, Ca, Mg) are extracted from their ores by electrolysis, it was natural that people should return to the possibility of using an electrolytic method for the extraction of aluminium. The big problem was that the oxide could not be melted to give a conducting liquid.

The problem was solved by a young American student called Charles Martin Hall. Hall's professor had worked under Wöhler, and his accounts of the early research work had made Hall impatient to contribute his chapter to the aluminium story. Hall bought some batteries and found an outhouse where he could experiment. He discovered that he could melt the ore **cryolite**, sodium hexafluoroaluminate, Na_3AlF_6, at 1000 °C, and then dissolve aluminium oxide in the molten cryolite to form a conducting solution. Hall electrolysed the melt, using graphite electrodes and, to his joy, he obtained aluminium at the cathode. The year was 1886, and Hall was just 21 years old! The Aluminium Company of America was founded to develop the process he had discovered.

Across the Atlantic, another young man was obsessed with the same problem. In France, Paul Héroult (23 years old) was working away in another makeshift laboratory. He arrived at the same conclusion as Hall. The electrolytic cell used in aluminium plants is called the Hall–Héroult cell [see Figure 19.3C, § 19.3.2].

19.2 THE MEMBERS OF THE GROUP

The elements of Group 3 of the Periodic Table are boron, aluminium, gallium, indium and thallium. Aluminium is by far the most important of them.

19.3 ALUMINIUM

19.3.1 USES OF ALUMINIUM

Aluminium has a low density and it is not corroded

It is a good thermal conductor and also a reflector of heat

It is used in headlights ...

... in electrical cables ...

Every month new uses are being found for this metal which resists corrosion and which has a low density. Being completely resistant to corrosion, it is ideal for packaging food. Aluminium is amazing in being a good thermal conductor which can also be used as a thermal insulator. As a thermal conductor, it is used for the manufacture of saucepans and cooking foil. The insulating property of aluminium arises from its ability to reflect radiant heat (i.e., infrared rays). Prematurely born babies are sometimes wrapped in aluminium foil, which keeps them warm by reflecting heat lost from the body. Firefighters in the USA wear suits which are coated with aluminium to reflect the heat from the fire and keep them cool. The polished surface of aluminium finds it a use in the reflectors of car headlights. Aluminium is a good electrical conductor and is replacing copper in overhead cables: to support aluminium cables, which are lighter, the pylons can be spaced at longer intervals.

... and for the construction of boats and planes.

Some parts of cars are made of aluminium

Since it has a low density and is not corroded, aluminium has obvious advantages over iron as a manufacturing material. Pure aluminium is too soft for construction purposes, but alloys (e.g., Al/Mg and Duralumin®, Al/Mg/Cu) have a higher tensile strength and are used for the construction of aeroplanes and small boats [see Figure 19.3A]. More and more parts of cars are being made of aluminium: engine blocks can be cast from aluminium; piston heads are made of aluminium and encircled by steel rings, and rocker covers are made of aluminium. Some vehicles have an aluminium body, but the chassis needs the strength of steel, and the engine is of cast iron. When a car is scrapped because the iron in it has rusted, the aluminium parts are as good as new and can be recycled. Another advantage of incorporating aluminium parts is that the vehicle becomes lighter and consumes less petrol.

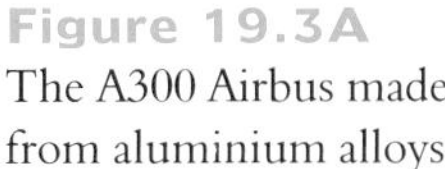

Figure 19.3A
The A300 Airbus made from aluminium alloys

The metal is coated with a film of aluminium oxide, which is unreactive

The use of aluminium alloys in construction is possible because the metal is coated with a thin film of aluminium oxide, which resists attack by corrosive reagents.

Anodised aluminium can be dyed, and is used for construction purposes

When aluminium is **anodised**, that is, made the anode in an electrolytic cell of sulphuric acid or chromic acid, the layer of oxide is thickened. When formed in this way, the oxide is hydrated and can absorb dyes. Dyed anodised aluminium is used for door frames and window frames, which are decorative as well as weatherproof.

19.3.2 EXTRACTION OF ALUMINIUM

Purificationof Al_2O_3 requires separation from Fe_2O_3 and SiO_2

Aluminium is mined as the ore bauxite, aluminium oxide-2-water, $Al_2O_3{\cdot}2H_2O$, which contains silicon(IV) oxide and iron(III) oxide as impurities. Pure aluminium oxide is obtained from the ore by utilising the fact that it is amphoteric, whereas, of the impurities, silicon(IV) oxide is acidic and iron(III) oxide is basic. After being ground, the ore is treated as shown in Figure 19.3B.

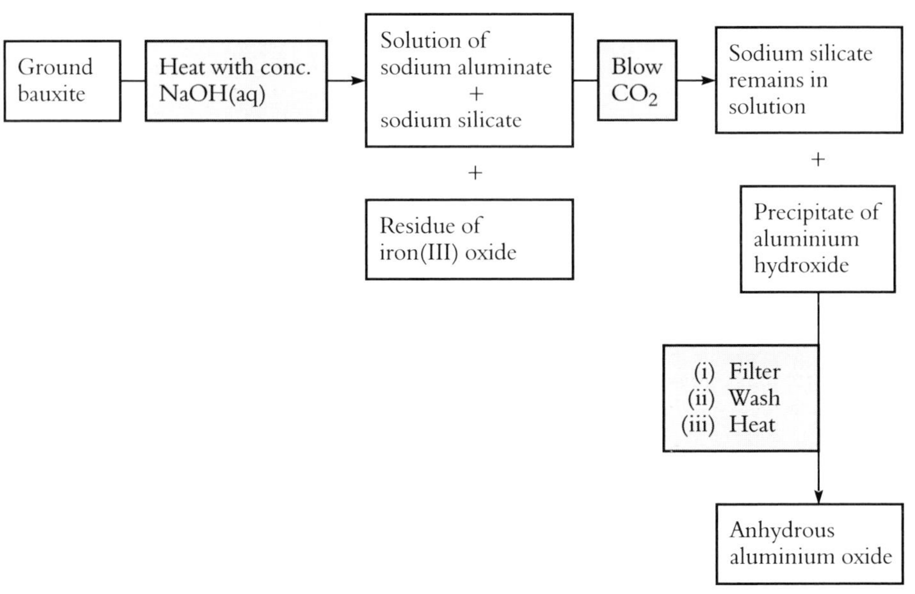

Figure 19.3B Purification of aluminium oxide

Electrolysis of Al_2O_3 in molten Na_3AlF_6 is used to extract aluminium

The next step is electrolysis. Since the melting temperature of aluminium oxide (2050 °C) is so high that electrolysis of the molten oxide cannot be accomplished, a solvent must be used. The cell shown in Figure 19.3C contains a molten mixture of the ore cryolite, Na_3AlF_6, with calcium fluoride and aluminium fluoride added to lower its melting temperature. Aluminium oxide is dissolved in this melt, and electrolysed at 850 °C to give aluminium and oxygen.

The electrode processes

It is postulated that the equilibrium

$$Al_2O_3 \rightleftharpoons Al^{3+} + AlO_3^{3-}$$

gives rise to the electrode processes

Cathode: $Al^{3+} + 3e^- \longrightarrow Al$
Anode: $4AlO_3^{3-} \longrightarrow 2Al_2O_3 + 3O_2 + 12e^-$

Figure 19.3C
A Hall–Héroult cell (5 m × 3 m × 1 m, 30 000 A, 5 V)

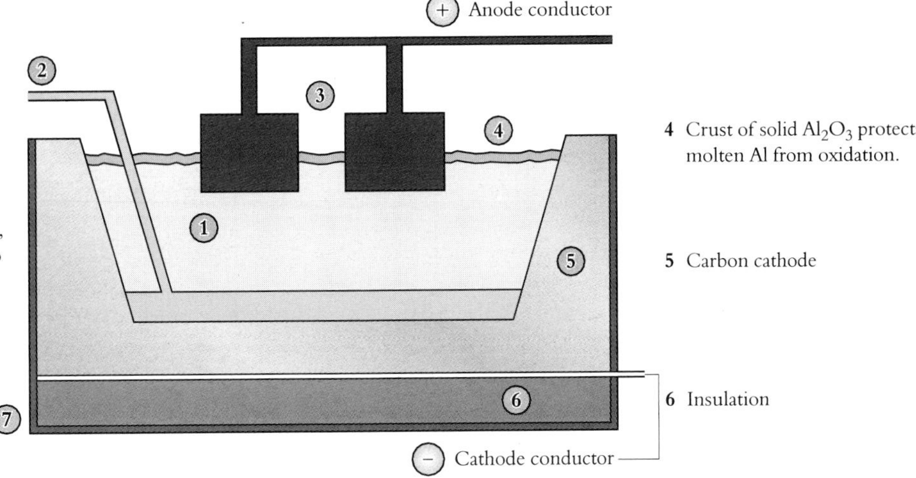
3 Carbon anode blocks replaced often because of oxidation to CO_2 by the O_2 evolved.
2 Molten Al is siphoned off.
1 Electrolyte: molten cryolite, Na_3AlF_6 (+ CaF_2 + AlF_3 to lower T_m) + Al_2O_3. More Al_2O_3 is added periodically.
7 Steel case
+ Anode conductor
4 Crust of solid Al_2O_3 protects molten Al from oxidation.
5 Carbon cathode
6 Insulation
– Cathode conductor

The process consumes much electricity

There may be a conflict between industry and conservation

The production of aluminium uses a great deal of electricity. It takes 15 000 kilowatt hours to make 1 tonne (10^3 kg) of aluminium. Hydroelectric power is the most economical form of electricity, and aluminium plants are built close to a waterfall or dam which can be used as a source of power. Hydroelectric power stations are often in regions of great natural beauty; aluminium plants are unsightly. There can be a conflict of interest between conservationists, who want to preserve the landscape, and industrialists, who want to provide us with more and more of this amazingly useful metal.

19.3.3 THE METAL

A fresh surface of aluminium reacts rapidly with water vapour in the air

Aluminium is high in the electrochemical series [see Table 12.4, §12.4.3], but its true reactivity is masked by the presence of a layer of aluminium oxide on its surface. This can be removed by reaction with mercury or mercury(II) chloride. When a fresh aluminium surface is exposed, it reacts immediately with water vapour in the air to form strands of aluminium hydroxide.

19.4 ALUMINIUM COMPOUNDS

19.4.1 THE Al^{3+} ION

The small size of the Al^{3+} ion gives its compounds a high degree of covalent character

The charge/radius ratio of the Al^{3+} ion is high and it therefore forms bonds with a high degree of covalent character. Aluminium fluoride is ionic, the oxide is largely ionic, with some covalent character, and the anhydrous chloride, bromide and iodide have polar covalent bonds.

In aqueous solution, H_2O molecules coordinate to the Al^{3+} ion

In aqueous solution, the Al^{3+} ion is stabilised by the coordination of water molecules to form the complex ion, $[Al(H_2O)_6]^{3+}$ [see Figure 19.4A(a)]. The coordination of water molecules occurs through the donation by the oxygen atom of a lone pair of electrons [see Figure 19.4A(b)].

Coordinated water molecules tend to donate protons to free water molecules

In consequence, the electrons in the O—H bond move closer to the oxygen atom, and the hydrogen atoms have a greater degree of positive charge than in free water molecules. The partial positive charges on the hydrogen atoms attract

Figure 19.4A
(a) $[Al(H_2O)_6]^{3+}$
(b) Polarity of O—H bonds

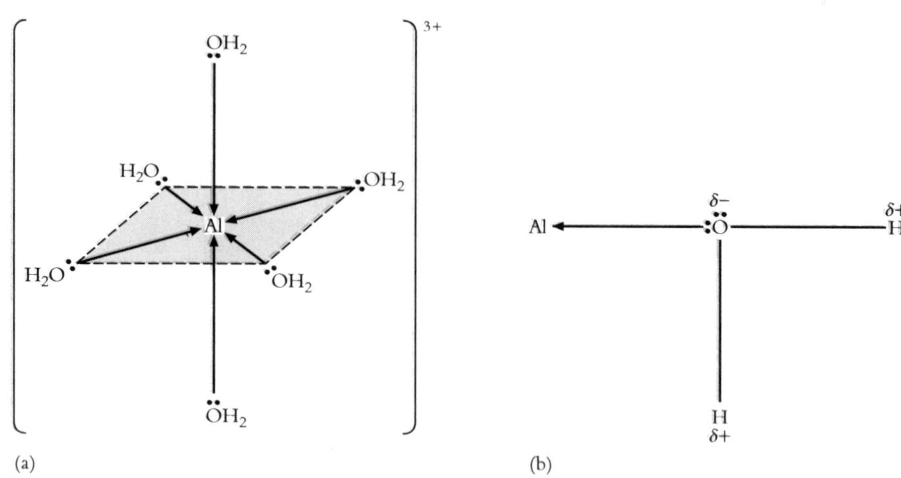

water molecules which may take protons from the coordinated water molecules.

This makes aluminium salts acidic in solution

$$[Al(H_2O)_6]^{3+} + H_2O \rightleftharpoons [Al(OH)(H_2O)_5]^{2+} + H_3O^+$$

$$[Al(OH)(H_2O)_5]^{2+} + H_2O \rightleftharpoons [Al(OH)_2(H_2O)_4]^{+} + H_3O^+$$

As they do so, oxonium ions are formed, and the solution becomes acidic.

The process is called salt hydrolysis

These equilibria are examples of salt hydrolysis [§12.9]. If stronger bases are present (e.g., OH^-, CO_3^{2-}, S^{2-}), H_3O^+ ions are removed, and the hydrolysis equilibria move to the right. The stronger base can also remove the third proton:

$$[Al(OH)_2(H_2O)_4]^{+} + OH^- \rightleftharpoons [Al(OH)_3(H_2O)_3](s) + H_2O$$

Addition of OH^-(aq) or CO_3^{2-}(aq) to a solution containing Al^{3+}(aq) precipitates $Al(OH)_3(H_2O)_3$

A precipitate of hydrated aluminium hydroxide appears. If an excess of hydroxide ions is added, protons are removed from the precipitate:

$$[Al(OH)_3(H_2O)_3](s) + OH^- \rightleftharpoons [Al(OH)_4(H_2O)_2]^-(aq) + H_2O$$

The solution formed contains diaquatetrahydroxoaluminate(III) ions:

$$[Al(OH)_4(H_2O)_2]^-$$

usually written $Al(OH)_4^-$(aq) and called simply aluminate ions. If an acid is added to the solution, the above equilibrium is reversed, and hydrated aluminium hydroxide is precipitated.

Al^{3+}(aq) ions are used as coagulating agents

The high charge/radius ratio of aluminium ions makes them useful coagulating agents. Aluminium ions are adsorbed on to the surface of negatively charged colloidal particles. The charges on the surfaces of the colloidal particles are reduced, and they are able to join to form a solid precipitate. Aluminium sulphate is used in water treatment plants to remove colloidal organic material from water.

19.4.2 ALUMINIUM HALIDES

Aluminium chloride, bromide and iodide dissolve in covalent solvents such as benzene. They are catalysts in the Friedel–Crafts reactions [§§ 28.8.4–28.8.5]. In the gaseous state they exist as dimers, e.g. Al_2Cl_6. These react readily with water to give e.g. Al^{3+}(aq) and Cl^-(aq).

Dimerisation of $AlCl_3$

Cl Cl Cl
Al Al
Cl Cl Cl

19.4.3 ALUMINIUM OXIDE AND HYDROXIDE

Aluminium oxide and hydroxide are amphoteric. With acids they react to form salts of the $Al^{3+}(aq)$ ion, and with alkalis they form salts of the aluminate ion, $Al(OH)_4^-(aq)$.

QUESTIONS ON CHAPTER 19

1. Describe the manufacture of aluminium, explaining the reasons for the conditions employed.

2. *(a)* What are the advantages and disadvantages of steel and aluminium for use in car engines and car bodies?
   (b) Why was iron discovered so many centuries earlier than aluminium?
   (c) Explain why aluminium finds use in (i) mirrors (ii) overhead electric cables (iii) milk-bottle tops and (iv) window frames.

3. In the Hall–Héroult cell, cryolite, Na_3AlF_3, is present. Why is sodium not formed at the cathode? Why does aluminium not form at the cathode when molten cryolite is electrolysed without added alumina?

4. Calculate how many coulombs of electricity are required to produce 1 tonne of aluminium (1 tonne = 10^3 kg, $A_r(Al) = 27$, Faraday constant = 96 500 C mol^{-1}).

5. Why is aluminium sulphate used in water treatment?

6. Why would it be dangerous to allow a solution of aluminium sulphate to come into contact with sodium cyanide?

7. *(a)* (i) Describe what you would observe when anhydrous aluminium chloride is added to an excess of water. Write an equation for the reaction.
   (ii) Write an equation to show why an aqueous solution containing aluminium ions is acidic.
   (iii) Describe what you would observe when solid sodium carbonate is added to a solution containing aluminium ions. Write an equation (or equations) for the reaction which occurs. **8**
   (b) (i) Describe what is observed when dilute ammonia solution is added, dropwise until in excess, to a solution containing aluminium ions. Give the formula of the final aluminium-containing species.
   (ii) How would your observations differ if dilute sodium hydroxide solution was used instead of dilute ammonia solution?
   Give the formula of any different aluminium-containing species formed. **4**

   12 *NEAB (AL)*

8. Aluminium is produced commercially by the electrolysis of a 5% solution of aluminium oxide in molten cryolite. The cathode and anode can be made of carbon and the temperature of the electrolyte is maintained at around 1200 K.
   (a) (i) Explain, with reference to economic considerations, why pure molten aluminium oxide is not used as the electrolyte. **2**
   (ii) Write an equation for the reaction at the cathode and indicate the physical state of the aluminium as it is formed. **2**
   (iii) Write an equation for the reaction at the anode and explain why the regular replacement of anodes is necessary. **3**
   (b) The electrolysis uses large quantities of electricity. Identify the **two** main processes that have high electrical energy requirements. **2**
   (c) Aluminium is readily recycled.
   (i) Give **two** benefits of recycling rather than extracting aluminium from its ore. **2**
   (ii) Identify **one** cost, other than electrical energy, in the process of recycling aluminium. **1**

   12 *AEB(AL)*

9. *(a)* Describe in outline how aluminium is manufactured from purified aluminium oxide, including in your account the electrode reactions. **5**
   (b) Aluminium fluoride and aluminium chloride each sublime when heated (i.e. they vaporise directly from the solid). The former has a sublimation point of 1270 °C whereas the latter's sublimation point is 178 °C.
   (i) What do these figures suggest about the nature of the bonding in these two compounds?
   (ii) Suggest reasons for any differences you suggest in their bonding. **3**
   (c) A 0.500 g sample of aluminium chloride was heated to 200 °C at a pressure of 1.00×10^5 Pa. The volume of its vapour was found to be 73.6 cm^3.
   (i) Calculate the M_r, of the vapour at this temperature.
   (ii) Draw a displayed formula to show the types of bonding in the molecules of the vapour. **4**

   C (AS/AL)

20

GROUP 7: THE HALOGENS

20.1 DDT: A LIFE-SAVING COMPOUND OF CHLORINE

A research student called Othmar Geidler made DDT, in 1874. Sixty years later, another chemist called Paul Mueller repeated the synthesis so that he could try out DDT in his work on insecticides. He found that it was extremely poisonous to houseflies and other insects.

$$\text{(Cl-C}_6\text{H}_4)_2\text{CH-CCl}_3$$

1,1,1-Trichloro-2,2-bis(4-chlorophenyl)ethane (The letters DDT come from its former name, dichlorodiphenyltrichloroethane.)

This work was done in the 1930s, and when the Second World War started in 1939, chemists had found more uses for DDT. During a war, in addition to those killed in action, many people die of disease through lack of medical supplies, shortage of water, overcrowding and poor sanitation. The use of DDT in the Second World War helped to alleviate some of this misery. When the Allies landed in islands in the Pacific, they faced the danger of malaria as well as enemy forces. By spraying DDT from aeroplanes, they were able to wipe out the mosquito population and remove the source of malaria. Dusting with DDT kept the troops free from the body lice which had plagued soldiers in earlier wars. After the Allies landed in Italy and occupied Naples, an epidemic of typhus broke out. To kill the lice which carry the disease, the whole area was sprayed with DDT and the population dusted themselves with DDT. Within days, the epidemic was over. In 1948, Mueller was awarded the Nobel prize, for discovering the life-saving properties of DDT.

After the war, farmers welcomed DDT-related compounds to replace the non-selective insecticides and herbicides which they had been using. DDT killed insects but not farm animals.

As early as 1946, however, it began to appear that the new chloro-compounds were not a perfect solution to the insect problem. Some species of housefly soon became resistant to DDT. In the USA, the cotton farmers had been delighted with the way DDT had attacked the cotton boll weevil, but, by 1960, they were having to spray more and more frequently with higher and higher doses as the weevil became resistant to the insecticide.

In 1962, Rachel Carson, in her book *The Silent Spring*, called for a halt to the widespread and indiscriminate spraying of insecticides and herbicides. These compounds are so stable that they persist for a long time in areas where they have been sprayed. If they are eaten by birds and animals, they cannot be excreted because they are insoluble in water, but they can be stored in the body because they are soluble in fat. Scientists began to investigate the spreading of insecticides and herbicides. They found DDT in penguins in the Antarctic, where the spray had never been used. They reasoned that if DDT is used in a malarial region, it will be taken up by small organisms. If these are eaten by a fish and the fish is eaten by a bird, the bird may carry DDT for hundreds of miles. DDT can concentrate up a food chain. It is possible that fish and birds which have high DDT contents may be eaten by human beings. If DDT is so toxic to insects, can it be completely harmless to human beings? In 1964, the Advisory Committee on Poisonous Substances used in Agriculture and Food Storage placed restrictions on the use of DDT, and consumption of DDT has fallen to about half of what it was in 1964.

20.2 THE MEMBERS OF THE GROUP

The appearance and physical state of the halogens

The elements of Group 7 are fluorine, chlorine, bromine, iodine and astatine. Fluorine is a poisonous pale yellow gas, chlorine is a poisonous dense green gas (Greek: *chloros*, green), bromine is a caustic and toxic brown volatile liquid (Greek: *bromos*, stench) and iodine is a shiny black solid which sublimes to form a violet vapour on gentle heating. Astatine is radioactive and does not occur naturally. Some properties of the elements are shown in Table 20.2.

Table 20.2 Physical properties of the halogens

Property	Fluorine F	Chlorine Cl	Bromine Br	Iodine I
Proton number (Atomic number)	9	17	35	53
Outer electron configuration	$2s^2 2p^5$	$3s^2 3p^5$	$3d^{10} 4s^2 4p^5$	$4d^{10} 5s^2 5p^5$
Atomic radius/nm	0.072	0.099	0.114	0.133
Boiling temperature/°C	−187	−35	59	183
Standard enthalpy of dissociation/kJ mol^{-1} of X	79.1	122	111	106
Standard electrode potential/V	+2.87	+1.36	+1.07	+0.54
Electronegativity	4.00	2.85	2.75	2.20

Van der Waals forces between halogen molecules

The halogens are a very similar set of non-metallic elements. Their name 'halogens' is derived from the Greek for 'salt formers'. They exist as diatomic molecules, X_2. The strength of the van der Waals forces between X_2 molecules increases as the number of electrons in the molecule of X_2 increases, i.e.

$$F_2 < Cl_2 < Br_2 < I_2$$

This explains the order of melting and boiling temperatures. In iodine, the van der Waals forces are strong enough to sustain a solid structure of iodine molecules [see Figure 6.5B, § 6.5].

20.3 BOND FORMATION

20.3.1 IONIC BONDS

Halogens react to form X^- ions ...

... in the formation of metal halides

The halogens form salts by accepting one electron to complete an octet of valence electrons, with the formation of a halide ion X^-. The steps involved in the formation of ionic halides are described in the Born–Haber cycle in § 10.12. The relative ease with which the halogens form ionic halides is determined by three factors. They are

Factor (a) The standard bond dissociation enthalpy of $X_2(g)$:

Energy is absorbed in $X_2(g) \longrightarrow 2X(g)$

Factor (b) The first electron affinity of X:

Energy is released in $X(g) + e^- \longrightarrow X^-(g)$

Factor (c) The standard lattice enthalpy of the halide formed:

Energy is released in $\mathbf{M}^+(g) + X^-(g) \longrightarrow \mathbf{M}X(s)$

Factor (a) The X—X bond decreases in strength $Cl_2 > Br_2 > I_2$ as the size of X increases [see Table 20.2 and Figure 20.3A].

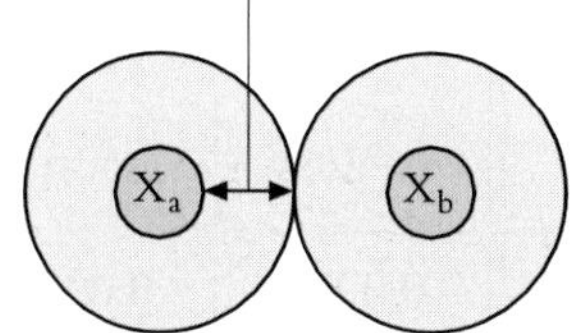
Attraction of nucleus of X_a for electrons of X_b decreases with increasing bond length.
X_a
X_b

Figure 20.3A The bonding in a halogen molecule

The F—F bond is, however, weaker than the other X—X bonds. This may be because the small size of the fluorine atom brings the lone pairs of electrons closer together than in other halogen molecules. As a result repulsion between lone pairs of electrons weakens the bond in :F̤—F̤:

Factor (b) Values of first electron affinity are similar for all the halogens.

Factor (c) The standard lattice enthalpies of halides have high negative values. For the sodium halides, the order from most exothermic to least exothermic is

NaF > NaCl > NaBr > NaI

In the formation of a metal halide, energy is required to ionise the metal. Since factors (b) and (c) are highly exothermic they compensate for the ionisation enthalpy of the metal. The halogens therefore readily form ionic compounds with metals. Fluorine is extremely reactive because factor(a) is less endothermic than for the other halogens, and factors (b) and (c) are highly exothermic. The standard enthalpies of formation of NaX have high negative values. The order from most exothermic to least exothermic is

NaF > NaCl > NaBr > NaI

The order of reactivity of the halogens in the formation of ionic bonds is

$F_2 > Cl_2 > Br_2 > I_2$

20.3.2 COVALENT BONDS

Halogens form covalent bonds in the diatomic molecules X_2 and in compounds with other non-metallic elements. Examples are HCl, CCl_4, $SiCl_4$, Cl_2O_7, SCl_4. Fluorine is restricted to the $n = 2$ shell and has only one oxidation state, −1. The other elements have empty d orbitals which allow promotion of electrons from p orbitals to d orbitals [see Figure 20.3B]. The other halogens have oxidation states +1, −1, −3, −5 and −7.

Figure 20.3B
Electron configurations in F, Cl and Cl*

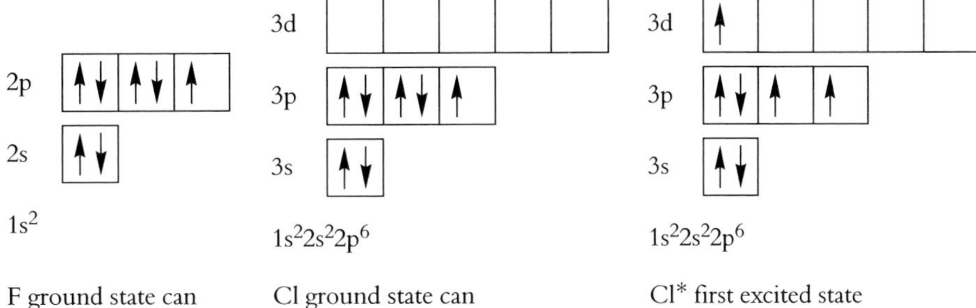

F ground state can form one covalent bond. | Cl ground state can form one covalent bond. | Cl* first excited state can form three covalent bonds.

The relative reactivities of the halogens in covalent bond formation are shown in their reactions with hydrogen [see Table 20.3].

Table 20.3
Reactions of hydrogen with halogens

Element	Fluorine	Chlorine	Bromine	Iodine
H_2	Reacts explosively, even in the dark at −200 °C	Reacts explosively in sunlight; slowly in the dark below 200 °C	Reacts above 200 °C and at lower temperatures with Pt catalyst	Reacts to form an equilibrium mixture of H_2, I_2, HI

Elements employ their highest oxidation states in combination with fluorine

Fluorine brings out the highest oxidation states of elements with which it combines. The reason is the high standard bond enthalpies of covalent bonds between fluorine and other elements. Much energy is given out when these bonds are formed. Atoms can promote electrons from shared orbitals to unoccupied orbitals because the energy required for promotion will be repaid when covalent bonds are formed.

CHECKPOINT 20.3: BONDING

1. Why is fluorine more reactive than the other halogens?

2. *(a)* Why do metals show their highest oxidation states in combination with fluorine?

(b) Explain why sulphur combines with fluorine to form SF_6 but with chlorine forms SCl_4.

3. For bromine (Ar)$4s^2 3d^{10} 4p^5$, draw electrons-in-boxes diagrams for the ground state and the first and second excited states, formed by the promotion of one p electron and a second p electron. Say how many covalent bonds can be formed in each state. Explain why electron promotion occurs more readily in bromine than in chlorine.

4. For the halogens, referring to Table 20.2, plot:

(a) atomic radius against the proton (atomic) number of the halogen,

(b) standard enthalpy of dissociation against proton number.

20.4 OXIDISING REACTIONS

The oxidising power of the halogens is measured by the value of the standard electrode potential, $E^\ominus$. You can see from Table 20.2 that the value of the standard electrode potential decreases down the group; that is the oxidising power decreases down the group.

$$\text{For chlorine,}\quad Cl_2(aq) + 2e^- \rightleftharpoons 2Cl^-(aq);\ E^\ominus = +1.36\ V$$
$$\text{For bromine,}\quad Br_2(aq) + 2e^- \rightleftharpoons 2Br^-(aq);\ E^\ominus = +1.07\ V$$

In a solution containing chlorine, chloride ions, bromine and bromide ions, the reaction that happens is the one for which $E^\ominus$ is positive; that is

$$Cl_2(aq) + 2Br^-(aq) \longrightarrow Br_2(aq) + 2Cl^-(aq);\ E^\ominus = +0.29\ V$$

When chlorine is passed through a solution of a bromide, bromine and chloride ions are formed: chlorine has oxidised bromide ions to bromine. Combining $E^\ominus$ values shows

- chlorine will oxidise bromide ions and iodide ions
- bromine will oxidise iodide ions
- fluorine will oxidise chloride, bromide and iodide ions.

Fluorine is too reactive to be a useful oxidising agent. Chlorine and its aqueous solution, 'chlorine water' are often used as oxidising agents. Chlorine water contains the oxidising agent chloric(I) acid, HClO, in addition to chlorine [§ 20.7].

Table 20.4A Some oxidising reactions of the halogens

Oxidant	Reaction
All halogens	Sulphite, $SO_3^{2-} \longrightarrow$ Sulphate, SO_4^{2-}
All halogens	Hydrogen sulphide, $H_2S \longrightarrow$ Sulphur, S
Cl_2, Br_2	Thiosulphate, $S_2O_3^{2-} \longrightarrow$ Sulphate, SO_4^{2-}
I_2	Thiosulphate, $S_2O_3^{2-} \longrightarrow$ Tetrathionate, $S_4O_6^{2-}$ (This reaction is used for titrimetric analysis of iodine [§ 3.15.1].)
Cl_2, Br_2	Organic compounds are oxidised, e.g., methane, CH_4,
F_2	Reacts explosively with organic compounds
I_2	Does not oxidise organic compounds

20.4.1 TESTS FOR HALOGENS

Table 20.4B Tests for halogens

Halogen	State at room temperature	Odour	Action on damp blue litmus paper	Colour in organic solvents
Chlorine	Pale yellow-green gas, poisonous	Pungent	Turns red and is later bleached	Colourless
Bromine	Reddish brown liquid, poisonous, and its vapour burns the skin	Pungent	Less powerful bleach than chlorine	Orange-red
Iodine	Shiny black solid, sublimes to form a purple vapour. Iodine turns starch blue	Pungent	No bleaching action	Purple

CHECKPOINT 20.4: REACTIVITY

1. Explain why bromine oxidises iodides, but bromides are oxidised by chlorine.

2. Some half-reaction equations are listed below:

   (a) $SO_3^{2-}(aq) + H_2O(l) \longrightarrow SO_4^{2-}(aq) + 2H^+(aq) + 2e^-$

   (b) $H_2S(aq) \longrightarrow S(s) + 2H^+(aq) + 2e^-$

   (c) $X_2(aq) + 2e^- \longrightarrow 2X^-(aq)$

   (d) $S_2O_3^{2-}(aq) + 5H_2O(l) \longrightarrow 2SO_4^{2-}(aq) + 10H^+(aq) + 8e^-$

   (e) $2S_2O_3^{2-}(aq) \longrightarrow S_4O_6^{2-}(aq) + 2e^-$

   By combining half-reaction equations, obtain equations for the reactions of chlorine with (i) sulphites, (ii) hydrogen sulphide, (iii) a thiosulphate and (iv) obtain an equation for the reaction of iodine with a thiosulphate.

3. Choose two reactions which illustrate the gradation in reactivity down Group 7.

4. Plot a graph of $E^\ominus$ against atomic number for the halogens (see values in Table 20.2).

20.5 COMMERCIAL EXTRACTION

The halogens are too reactive to occur free. The methods used for the commercial extraction of the halogens and the laboratory preparation of the halogens are related to their oxidising power. There is no oxidant that will oxidise fluorides to fluorine, and electrolysis is used. Chlorine also is made by electrolysis industrially. Sufficiently powerful oxidants will oxidise bromides and iodides to the halogens.

Chlorine is obtained by the electrolysis of molten sodium chloride [§ 18.4] and the electrolysis of brine [§ 18.6.5].

Bromine is obtained from seawater as shown in Figure 20.5A.

Iodine is mined as sodium iodate(V), $NaIO_3$, which is present in Chile saltpetre, $NaNO_3$. Sodium hydrogensulphite is employed to reduce iodate(V) ions to iodide ions. A reaction between iodide ions and iodate(V) ions produces iodine:

$$IO_3^-(aq) + 3HSO_3^-(aq) \longrightarrow I^-(aq) + 3HSO_4^-(aq)$$
$$IO_3^-(aq) + 5I^-(aq) + 6H^+(aq) \longrightarrow 3I_2(s) + 3H_2O(l)$$

Figure 20.5A
Extraction of bromine from sea water

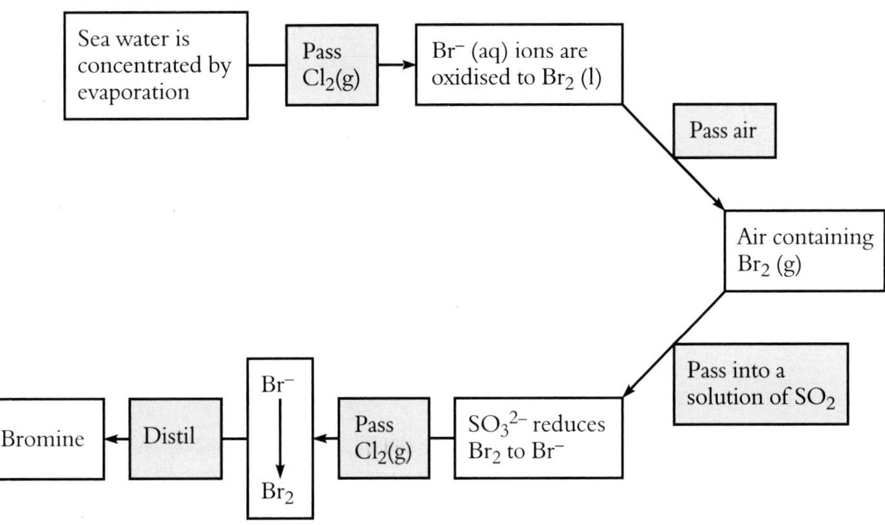

20.5.1 LABORATORY PREPARATION

In the laboratory, chlorine, bromine and iodine are obtained by oxidation of the halides:

$$2X^- \rightleftharpoons X_2 + 2e^-$$

Figure 20.5B shows the preparation of chlorine by the action of concentrated sulphuric acid and manganese(IV) oxide on sodium chloride.

Figure 20.5B Laboratory preparation of dry chlorine (in fume cupboard)

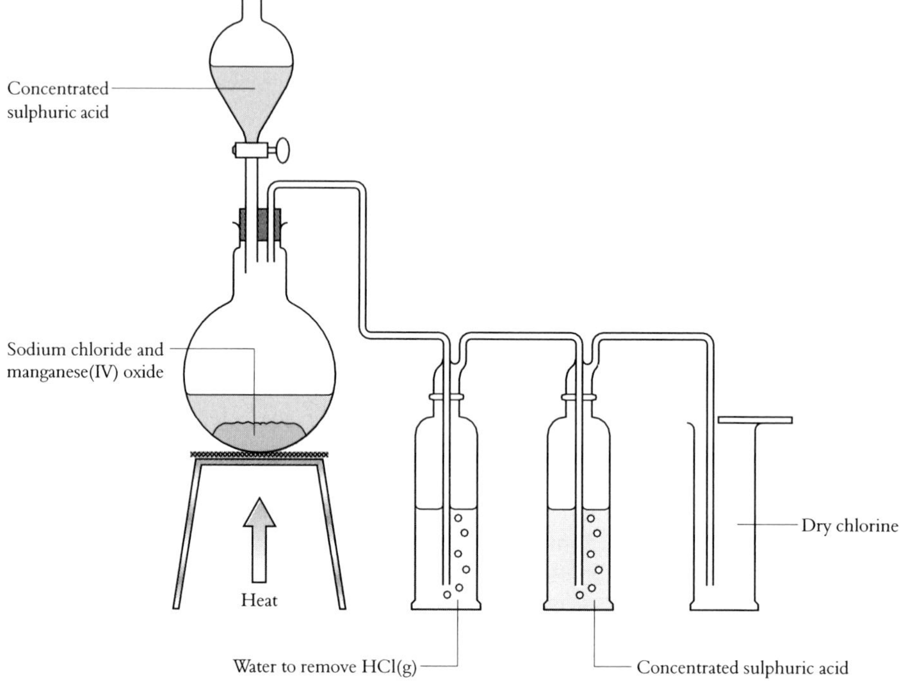

Figure 20.5C Laboratory preparation of bromine

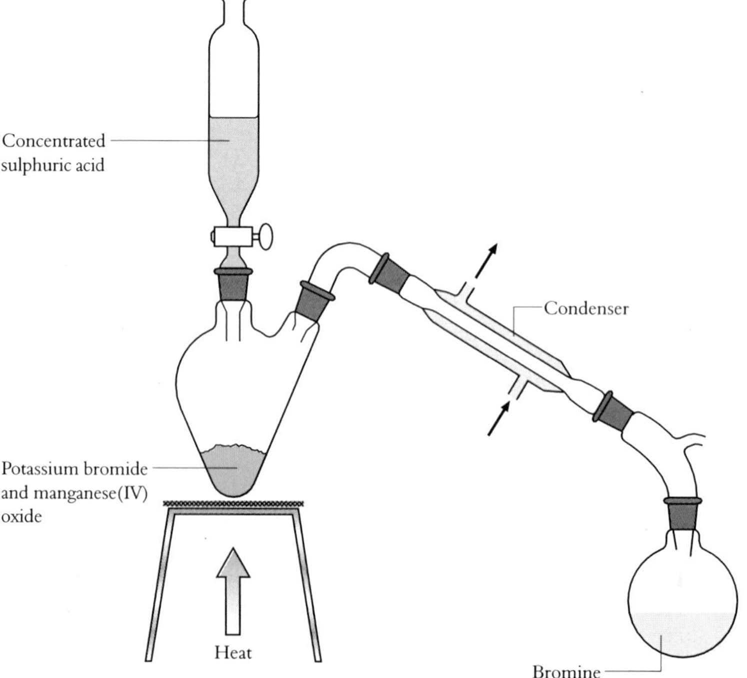

Alternatively, concentrated hydrochloric acid can be used as a source of chlorine. It can be run from a tap funnel on to potassium manganate(VII) or warmed with manganese(IV) oxide or lead(IV) oxide. If the chlorine is not required dry, it can be collected over water, in which it is only slightly soluble.

The laboratory preparation of bromine is shown in Figure 20.5C and that of iodine in Figure 20.5D. Concentrated sulphuric acid displaces the hydrogen halide from its salt, and manganese(IV) oxide oxidises it to the halogen [see Question 3, Checkpoint 20.5].

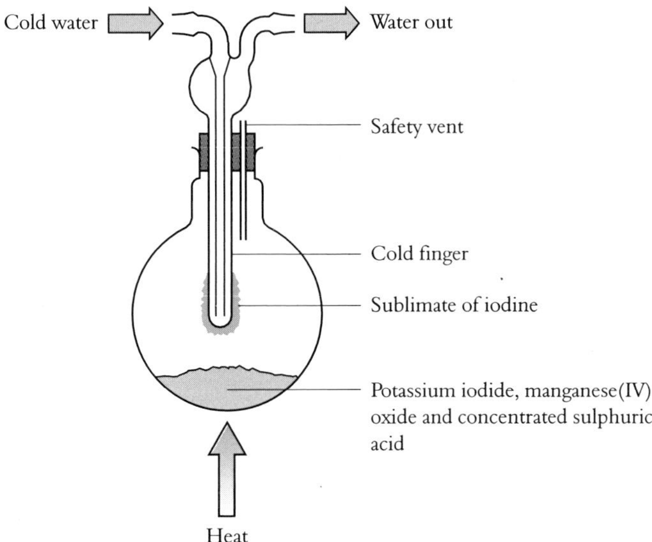

Figure 20.5D Laboratory preparation of iodine

A solution of chlorine in water is readily made by dissolving bleaching powder, $CaCl_2 . Ca(ClO)_2$, in water and adding dilute hydrochloric acid:

$$Ca(ClO)_2 + 4H^+(aq) + 2Cl^-(aq) \longrightarrow Ca^{2+}(aq) + 2H_2O(l) + 2Cl_2(aq)$$

CHECKPOINT 20.5: PREPARATIONS

*1. Draw an apparatus in which concentrated hydrochloric acid is run from a tap funnel on to potassium manganate(VII) in a flask fitted with a delivery tube. Illustrate how chlorine can be collected over water. What are the advantages of collecting chlorine *(a)* over water and *(b)* downwards?

*2. What do the laboratory preparations of chlorine, bromine and iodine have in common? Why cannot fluorine be prepared in this way?

3. Concentrated sulphuric acid displaces HX from KX, where X = Cl, Br, I. Write equations for the reactions of concentrated H_2SO_4 with KCl, KBr and KI. One product of the reaction is $KHSO_4$.
Write a half-reaction equation for the oxidation of X^- to X_2.
Write a half-reaction equation for the reduction of acidified MnO_2 to Mn^{2+}.
Combine the half-reaction equations to give the equation for the oxidation of X^- by MnO_2 and acid.

4. Fluorine has an oxidation state of −1 only. Iodine has oxidation states of −1, +1, +3, +5 and +7 in I^-, ICl, IC_3, IF_5 and IF_7. What is the reason for this difference between the two halogens?

5. Refer to Table 20.4A, § 20.4. Write equations for the reactions between *(a)* Cl_2 and Br^-, *(b)* SO_3^{2-} and Br_2. (This may be obtained by combining the half-reaction equations for $Br_2 \longrightarrow Br^-$ and $SO_3^{2-} \longrightarrow SO_4^{2-}$.)

6. Little is known of the chemistry of astatine.
(a) Describe the physical characteristics which you would expect for this halogen. *(b)* What do you think would be formed in a reaction between sodium astatide and concentrated sulphuric acid?

20.6 REACTION WITH WATER

The value of $E^{\ominus}$ measures the strength of an oxidising agent

The standard electrode potentials of the halogens are shown in Table 20.2, § 20.2. The standard electrode potential (reduction potential) for oxygen is +1.23 V.

$$O_2(g) + 4H^+(aq) + 4e^- \rightleftharpoons 2H_2O(l);\ E^{\ominus} = +1.23\ V$$

Fluorine and chlorine are thus capable of oxidising water, while bromine and iodine are not.

Chlorine reacts slowly with water to form hydrochloric acid and chloric(I) acid, HClO:

$$Cl_2(g) + H_2O(l) \longrightarrow HClO(aq) + H^+(aq) + Cl^-(aq)$$

Chloric(I) acid decomposes to give oxygen slowly, unless sunlight is present to accelerate the decomposition:

Chlorine slowly oxidises water, with the formation of HClO

$$2HClO(aq) \longrightarrow 2H^+(aq) + 2Cl^-(aq) + O_2(g)$$

In the presence of a reducing agent, chloric(I) acid acts as an oxidising agent:

$$HClO(aq) + H^+(aq) + 2e^- \rightleftharpoons Cl^-(aq) + H_2O(l)$$

20.7 REACTION WITH ALKALIS

Chlorine reacts with alkalis, to form ClO^- and ClO_3^- ...

Chlorine reacts faster with dilute alkalis than with water:

$$Cl_2(g) + 2OH^-(aq) \longrightarrow Cl^-(aq) + ClO^-(aq) + H_2O(l)$$

The chlorate(I) that is formed may decompose to form a chloride and a chlorate(V):

$$3ClO^-(aq) \longrightarrow 2Cl^-(aq) + ClO_3^-(aq)$$

... The reaction has commercial use

The decomposition is slow at room temperature but fast at 70 °C. A reaction like this, in which a species is simultaneously oxidised and reduced, is called a **disproportionation** reaction. Part of the ClO^- is oxidised to ClO_3^-, while the rest is reduced to Cl^-. These reactions are used commercially in the manufacture of sodium chlorate(I), NaClO, a widely used mild antiseptic (Milton®), and sodium chlorate(V), $NaClO_3$, a powerful weedkiller (Tandar®). Both chlorine and sodium hydroxide are products of the electrolysis of brine. If they are allowed to come into contact, sodium chlorate(I) is produced; if the temperature is raised, they can be made to form sodium chlorate(V). Chlorine reacts with calcium hydroxide to form bleaching powder, $Ca(OCl)_2 . CaCl_2$. This is a useful source of chlorine, which it yields readily on treatment with a dilute acid.

Br_2 and I_2 react with alkalis to form the halate(V) ions

Bromine and iodine react with dilute alkalis, either cold or warm, to give a mixture of halide and halate(V):

$$3Br_2(aq) + 6OH^-(aq) \longrightarrow BrO_3^-(aq) + 5Br^-(aq) + 3H_2O(l)$$

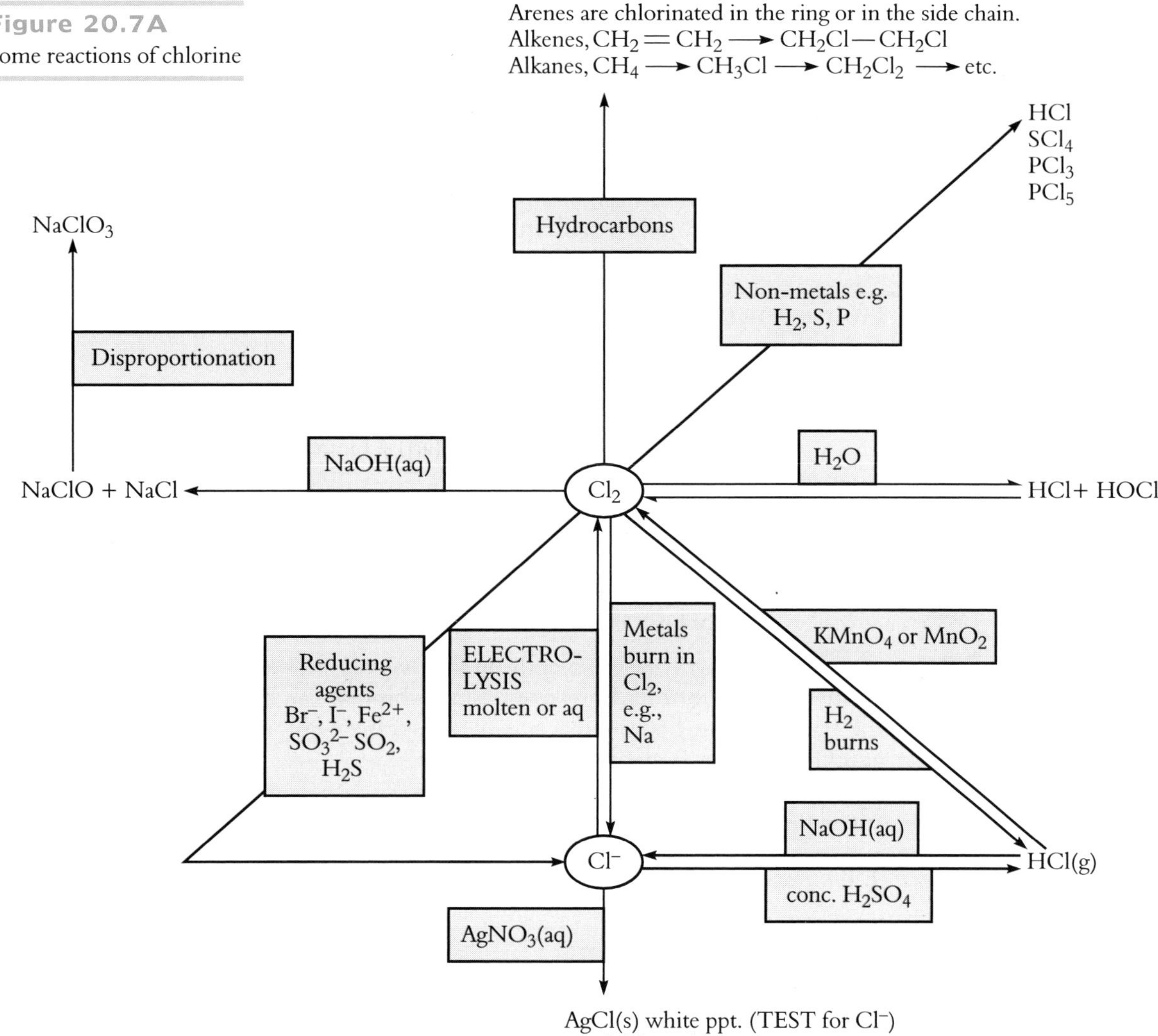

Figure 20.7A
Some reactions of chlorine

CHECKPOINT 20.7: REACTIONS

1. *(a)* Write the equation for the reaction of chlorine with water. Explain how the addition of an alkali affects the reaction. *(b)* What is the quickest way of preparing chlorine water?

2. Write the oxidation number of chlorine in each of the species

$$3\underline{Cl}O^-(aq) \longrightarrow 2\underline{Cl}^-(aq) + \underline{Cl}O_3^-(aq)$$

Why is this reaction described as *disproportionation*?

3. Write the oxidation numbers of oxygen and fluorine in each of the species

(a) $2\underline{F}_2(g) + 2\underline{O}H^-(aq) \longrightarrow \underline{O}\underline{F}_2(g) + 2\underline{F}^-(aq) + H_2\underline{O}(l)$

(b) $2\underline{F}_2(g) + 4\underline{O}H^-(aq) \longrightarrow \underline{O}_2(g) + 4\underline{F}^-(aq) + 2H_2\underline{O}(l)$

4. How do the standard electrode potentials of the X_2/X^- half-cells determine *(a)* the reactions of the halogens with water and *(b)* the methods of commercial extraction of the elements?

20.8 METAL HALIDES

Metal + X_2 or HX(g) gives metal halide

Anhydrous metal halides are prepared by heating the metal in a stream of dry halogen or hydrogen halide. Metals with variable oxidation states usually give the halide of a higher oxidation state with the halogen and the halide of a lower oxidation state with the hydrogen halide. Iron gives iron(III) chloride, $FeCl_3$, with chlorine, and iron(II) chloride, $FeCl_2$, with hydrogen chloride. Exceptions are the reaction of iodine with iron to form iron(II) iodide, FeI_2, and with copper to form copper(I) iodide, CuI.

Metal + HX(aq) gives hydrated metal halide

Hydrated metal halides are prepared by the reaction of a hydrohalic acid with a metal or its oxide, hydroxide or carbonate:

$$Zn(s) + 2HCl(aq) \longrightarrow ZnCl_2(aq) + H_2(g)$$
$$CuO(s) + 2HCl(aq) \longrightarrow CuCl_2(aq) + H_2O(l)$$

Many halides crystallise with water of crystallisation, e.g., $MgCl_2 . 6H_2O$, $AlCl_3 . 6H_2O$. When the hydrates are heated in an attempt to obtain the anhydrous salt, they are often hydrolysed to give a basic chloride or hydroxide:

$$MgCl_2 . 6H_2O(s) \rightleftharpoons MgCl(OH)(s) + HCl(g) + 5H_2O(g)$$
$$AlCl_3 . 6H_2O(s) \rightleftharpoons Al(OH)_3(s) + 3HCl(g) + 3H_2O(g)$$

Some hydrates are hydrolysed on heating

The degree of hydrolysis depends on the degree of covalent character in the bonds [§ 15.5.2]. Anhydrous magnesium chloride can be obtained by heating the crystals of the hydrate in a stream of hydrogen chloride gas. This moves the equilibrium over to the left-hand side.

Most metal halides are soluble

Metal halides are soluble, except for the lead halides and the chloride, bromide and iodide of mercury(I) and silver. They can be distinguished as shown in Table 20.8.

Table 20.8
Reactions of metal halides

Test reactions of solid halides ...

Reactions of solid halides			
Reagent	Chloride	Bromide	Iodide
Conc. H_2SO_4 Phosphoric(V) acid H_3PO_4	HCl(g) HCl(g)	HBr(g) + Br_2(g) HBr(g)	I_2(g) HI(g)
Conc. H_2SO_4 + MnO_2	Cl_2(g)	Br_2(g)	I_2(g)

... and of solutions

Reactions of halide ions in aqueous solution			
Reagent	Chloride	Bromide	Iodide
$Pb(NO_3)_2$(aq)	$PbCl_2$(s) white	$PbBr_2$(s) cream	PbI_2(s) yellow
$AgNO_3$(aq) + HNO_3(aq)	AgCl(s) white soluble in dil. NH_3(aq)	AgBr(s) pale yellow soluble in conc. NH_3(aq)	AgI(s) yellow insoluble in NH_3(aq)
Effect of light on AgX(s)	Turns black	Turns yellow	No change

20.9 NON-METAL HALIDES

Combination with non-metallic elements

Halogens combine with many non-metallic elements. The halides of non-metallic elements are covalent and are hydrolysed by water, with the exception of tetrachloromethane, CCl_4 [§ 23.7.2]:

Halides of non-metallic elements are hydrolysed with the formation of HX(g)

$$\underset{\text{Silicon(IV) chloride}}{SiCl_4(l) + 2H_2O(l)} \longrightarrow \underset{\text{Silicon(IV) oxide}}{SiO_2(aq) + 4HCl(aq)}$$

$$\underset{\text{Phosphorus(III) chloride}}{PCl_3(l) + 3H_2O(l)} \longrightarrow \underset{\text{Phosphonic acid}}{H_3PO_3(aq) + 3HCl(g)}$$

For chlorofluoroalkanes (CFCs) see § 23.6.

20.9.1 HYDROGEN HALIDES

The hydrogen halides, hydrogen fluoride, hydrogen chloride, hydrogen bromide and hydrogen iodide, can be made by direct synthesis. The method works well for hydrogen chloride, which is manufactured industrially by burning a stream of hydrogen in chlorine:

$$H_2(g) + Cl_2(g) \longrightarrow 2HCl(g)$$

Industrial manufacture of HCl(g) ...

The reaction between hydrogen and fluorine is dangerously fast, and the reactions between hydrogen and bromine or iodine do not give a good yield.

Laboratory preparation of hydrogen chloride ...

A displacement reaction between sodium chloride and concentrated sulphuric acid can be used for the laboratory preparation of hydrogen chloride [see Figure 20.9A].

Figure 20.9A
Laboratory preparation of hydrogen chloride

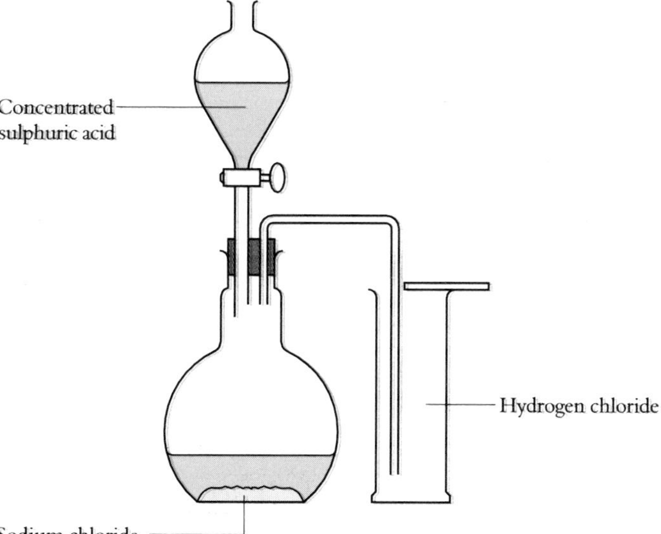

... hydrogen bromide and hydrogen iodide

Hydrogen bromide and iodide cannot be made in this way because concentrated sulphuric acid oxidises them to the halogens. They are made by hydrolysis of phosphorus tribromide or phosphorus triiodide.

From Table 20.9, you can see that the standard bond dissociation enthalpy decreases in the order

$$HF > HCl > HBr > HI$$

Table 20.9
The physical properties of the hydrogen halides

	HF	HCl	HBr	HI
$\Delta H^{\ominus}_{\text{Formation}}$/kJ mol^{-1}	−270	−92	−36	+26
$\Delta H^{\ominus}_{\text{Bond dissociation}}$/kJ mol^{-1}	+560	+430	+370	+300
Boiling temperature/°C	20	−85	−67	−35
pK_a	3.25	−7.4	−9.5	−10

Thermal stability of HX
Acid strength of HX

This is why the thermal stability of the compounds decreases in the same order and why the acidic strength increases in the order

$$HF \ll HCl < HBr < HI$$

Hydrogen fluoride is a much weaker acid than the rest.

HX(aq) show typical acid reactions

The hydrohalic acids react with metals above hydrogen in the electrochemical series and with metal oxides, hydroxides and carbonates.

The characteristics of chlorides are summarised in Figure 20.9B.

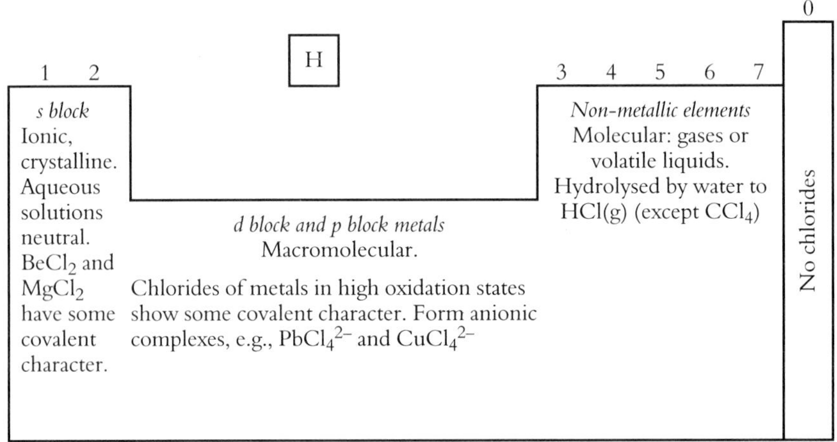

0
H
1 2
3 4 5 6 7
s block
Ionic, crystalline. Aqueous solutions neutral. $BeCl_2$ and $MgCl_2$ have some covalent character.
Non-metallic elements
Molecular: gases or volatile liquids. Hydrolysed by water to HCl(g) (except CCl_4)
d block and p block metals
Macromolecular.
Chlorides of metals in high oxidation states show some covalent character. Form anionic complexes, e.g., $PbCl_4^{2-}$ and $CuCl_4^{2-}$
No chlorides

Figure 20.9B
The nature of chlorides

CHECKPOINT 20.9: HALIDES

1. Which of the halogens has *(a)* the smallest electron affinity, *(b)* the greatest oxidising power, *(c)* the greatest ability to form hydrogen bonds and *(d)* the weakest acid HX?

2. What is meant by describing the hydrohalic acids as 'fuming' gases? Why do they behave in this way?

†3. The boiling temperatures of the hydrogen halides depend on intermolecular forces.

   (a) Explain the order of T_b: HCl < HBr < HI
   (b) Explain why T_b of HF is so much greater than those of the other hydrogen halides [see § 4.8.3].

4. Would you expect a solution of HAt to be a strong acid or a weak acid? What would you expect to see when a solution of silver nitrate was added to a solution of HAt(aq)?

5. Tetrachloromethane, CCl_4, does not react with water, but silicon(IV) chloride, $SiCl_4$, is rapidly hydrolysed. Explain this difference.

6. *(a)* In standard bond dissociation enthalpy

   HF > HCl > HBr > HI

   (b) In thermal stability

   HF > HCl > HBr > HI

   (c) In acid strength

   $HF \ll HCl < HBr < HI$

   Explain how statements *(b)* and *(c)* follow from statement *(a)*.

Figure 20.9C
Chlorine and some of its compounds

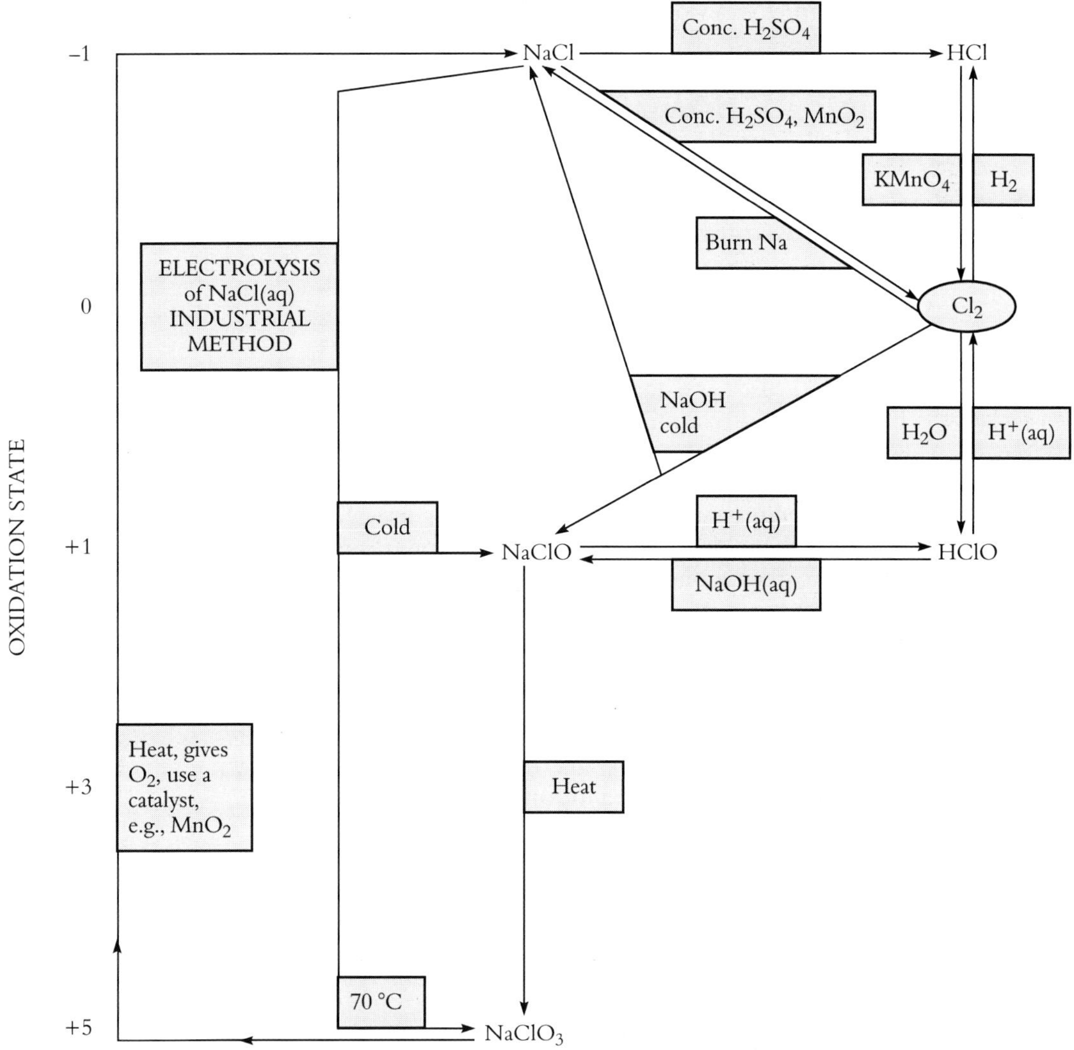

20.10 SUMMARY OF GROUP 7

Summary of Group 7

1. The ions X^- are readily formed. Electron affinity decreases down the group. All the halogens react vigorously with metals. With the s block metals, they form ionic compounds.
2. The ions XO^- and XO_3^- are formed, except by fluorine.
3. The halogens form molecular compounds with non-metallic elements.
4. The halogens are all oxidising agents. Oxidising power decreases in the order

   $$F_2 > Cl_2 > Br_2 > I_2$$

   Reducing power of X^- decreases in the order

   $$I^- > Br^-$$

   Cl^- and F^- are not reductants.

5. Intermolecular forces increase down the group, and lead to a transition in physical state from fluorine (pale yellow gas) and chlorine (green gas), to bromine (brown volatile liquid) and iodine (shiny black solid which sublimes to a violet vapour).

20.11 USES OF HALOGENS

Uses of fluorine include the manufacture of 'freon', PTFE

Fluorine is used in the manufacture of fluorohydrocarbons. The commercial refrigerant gas *freon* contains CCl_2F_2, $CClF_3$ and other chlorofluoromethanes. It is extremely unreactive. Tetrafluoroethene, $CF_2{=}CF_2$, can be polymerised to give poly(tetrafluoroethene), $\text{+}CF_2{-}CF_2\text{+}_n$, PTFE [§ 34.4.2].

For fluoridation of water see § 17.5.

Uses of chlorine include the manufacture of disinfectants ...

Chlorine is used as a domestic bleach, and as a disinfectant in swimming baths. It is used in the manufacture of the gentle antiseptic, sodium chlorate(I), and the powerful weedkiller sodium chlorate(V).

... solvents ...

... antiseptics ...

... and insecticides

Chlorinated organic compounds find many uses. Solvents such as trichloroethene, $CHCl{=}CCl_2$, are used for grease removal. Many antiseptics are chloro-compounds. TCP contains 2,4,6-trichlorophenol. Chlorocompounds, for example, DDT, have proved to be valuable insecticides [§ 20.1].

Bromine compounds are used as petrol additives

Bromine compounds, e.g., $C_2H_4Br_2$, are used as petrol additives in leaded petrol. They convert lead into volatile compounds, which are discharged through the vehicle exhaust into the air. Since 1990, all new motor vehicles in the UK have been designed to run on unleaded petrol.

20.11.1 THE CHEMISTRY OF THE SWIMMING POOL

Swimming pools are filled with mains water. The warmth of the water helps to wash dirt, sweat, oil and other substances into the water. Micro-organisms will breed if the water is not disinfected. Chlorine and chlorine compounds are employed as powerful and inexpensive disinfectants. At one time chlorine from a cylinder of the gas was used to treat swimming pool water. The disadvantage is that there is a danger of some of the deadly poisonous gas escaping. It is safer to use sodium chlorate(I), NaClO, or calcium chlorate(I), $Ca(ClO)_2$, as a source of chlorate(I) ions. Although chlorate(I) ions are not oxidising agents, they set up an equilibrium in solution:

$$ClO^-(aq) + H_2O(l) \rightleftharpoons OH^-(aq) + HClO(aq)$$

As a result, the powerful oxidant chloric(I) acid, HClO, is formed.

The pH affects the position of equilibrium in the dissociation of HClO

$$HClO(aq) + H_2O(l) \rightleftharpoons H_3O^+(aq) + ClO^-(aq)$$

Chloric(I) acid is a powerful oxidising agent, but chlorate(I) ion is very weak.

At low pH, the equilibrium lies over to the left-hand side. It is advantageous to keep the pH of the pool below 6.0; this ensures that HClO is largely undissociated. There is, however, a disadvantage of keeping the pool water at this low pH. The acidic water will react with the mortar, concrete, tiles and grouting of the pool.

A compromise pH of 7.3–7.4 is chosen to be comfortable to the skin of the bather, to protect the fabric of the pool, and to keep a large proportion of HClO undissociated. A solution of NaClO or $Ca(ClO)_2$ has a pH which is higher than this, and the pH must be lowered by adding some acid. The pool water must be buffered against changes in pH due to the addition of alkaline sweat, acidic urine and other substances. Often there is enough hydrogencarbonate ion in the water to act as a buffer.

THE HALOGENS

THE HALOGENS
F, Cl, Br, I

States at room temperature are:
F_2 yellow-green gas,
Cl_2 green gas,
Br_2 volatile brown liquid, forms orange vapour,
I_2 volatile black solid, forms violet vapour.
Van der Waals forces between molecules increase from F_2 to I_2. § 20.2

Trends down the group:
1. Electronegativity decreases down the group. § 20.2
2. Boiling temperature increases down the group. § 20.2
3. Power as oxidising agents decreases down the group, in accordance with the standard values of electrode potential. §§ 20.2, 20.4
- Cl_2 displaces Br_2 from bromides in sea water. § 20.5
- Reaction with thiosulphate

$S_2O_3^{2-} + Cl_2$ or $Br_2 \rightarrow SO_4^{2-}$
$S_2O_3^{2-} + I_2 \rightarrow S_4O_6^{2-}$ § 20.4

Hydrogen halides
The **stability** of the hydrogen halides decreases in the order in which the standard bond enthalpy decreases
HF > HCl > HBr > HI § 20.19.1
All except HF are strong acids in solution. The strength of the H—F bond makes HF a weak acid. § 20.19.1

Reaction with hydrogen
The vigour of the reaction decreases
$F_2 > Cl_2 > Br_2 > I_2$ § 20.3.2

Uses
Halogens are used in the **manufacture** of: bleaches, § 20.11
solvents and refrigerants, § 20.11
aerosols, §§ 21.5, 29.6
pesticides, § 20.1
polymers § 27.8.10
Chlorates(I) are disinfectants and bleaches are used in **water treatment**. §§ 20.11, 20.11.1
The stability of the C—Cl bond makes organic compounds of chlorine non-biodegradable and therefore **pollutant**. § 29.6

Chlorine
Made by the electrolysis of brine in e.g. a diaphragm cell. Other products of the **chlor–alkali industry** are:
$NaOH + H_2$ § 18.6.5
NaClO and $NaClO_3$. § 20.7
1. Reacts with water to form hydrochloric acid and chloric(I) acid, HClO.
2. Reacts with cold dilute NaOH(aq) to form sodium chloride and sodium chlorate(I), NaClO.
3. Reacts with hot dilute NaOH(aq) to form sodium chloride + sodium chlorate(V), $NaClO_3$. § 20.7

Halides
1. The reducing power of the halide ion, X^-, increases down the group. §§ 20.2, 20.9
2. Reaction of NaX(s) with concentrated sulphuric acid gives
$NaF(s) \rightarrow HF(g)$
$NaCl(s) \rightarrow HCl(g)$
$NaBr(s) \rightarrow HBr(g) + Br_2(g)$
$NaI(s) \rightarrow I_2(g)$ § 20.8
3. Reaction of $X^-(aq)$ with $AgNO_3(aq) + HNO_3(aq)$ gives:
AgCl(s) white, soluble in $NH_3(aq)$,
AgBr(s) pale yellow, sparingly soluble in $NH_3(aq)$,
AgI(s) deep yellow, insoluble in $NH_3(aq)$. § 20.8

QUESTIONS ON CHAPTER 20

1. Compare the compounds HF, HCl, HBr and HI with respect to *(a)* the strengths of the acids formed in aqueous solution and *(b)* thermal stability.

2. *(a)* How does the oxidising power of the halogens change with increasing atomic number?
(b) How does the trend in oxidising power show itself in the reactions of the halogens with (i) hydrogen (ii) metals (iii) water?

3. Explain how you would distinguish between sodium bromide and sodium iodide by means of a chemical test.

4. Describe the reactions of chlorine with *(a)* water *(b)* aqueous sodium hydroxide *(c)* aqueous sodium bromide. State the products of the reactions.

5. State the products of the reactions of concentrated sulphuric acid with *(a)* sodium chloride *(b)* sodium bromide *(c)* sodium iodide.

6. Predict the chemistry of astatine, the last member of Group 7. How would you expect it to react with *(a)* hydrogen *(b)* metals *(c)* water *(d)* alkalis?

7. *(a)* What is the oxidation number of chlorine in (i) ClO_3^- (ii) ClO^-?
(b) Write an equation for the formation of ClO^- by the reaction of chlorine and aqueous alkali.
(c) When $KClO_3$ is heated to just above its melting temperature it forms KCl and $KClO_4$. Write an equation for this reaction. Why is it described as disproportionation?
(d) Describe the reaction that occurs when iodine is added to cold dilute sodium hydroxide solution. State the type of reaction that takes place, and write an equation for it.
(e) Describe the reaction that occurs when iodine is heated with excess concentrated sodium hydroxide solution. State the type of reaction that takes place, and write an equation for it.
(f) A quantity of iodine was treated as in *(e)* and the resulting colourless solution was acidified with dilute sulphuric acid. A brown mixture containing iodine was formed. Find the relationship between the amount of iodine originally used and the amount liberated on acidification.

8. Values of the acid dissociation constants are hydrofluoric acid 5.6×10^{-4} mol dm^{-3} and hydrochloric acid 1×10^{7} mol dm^{-3}.

(a) What difference in properties do these values indicate?
(b) Give two factors that are responsible for the difference in values.
(c) What is meant by the term 'polar covalent bond' (e.g. the bond C—Cl)?
(d) Explain why the H—F bond is more polar than the H—Cl bond.

9. *(a)* Describe the industrial production of sodium hydroxide and chlorine by electrolysis of aqueous sodium chloride. Give a labelled sketch of the cell, and explain the steps in the process.
(b) One major use of chlorine is in the manufacture of sodium chlorate(I), which is used in solution as a household bleach. State how you could make sodium chlorate(I) in the laboratory.
(c) Give two other large-scale uses of chlorine.
(d) Mention two uses of sodium hydroxide.

10. Explain the following.

(a) When aqueous sodium chloride is added to silver nitrate, a white precipitate forms. This precipitate dissolves in an excess of dilute ammonia. When aqueous sodium bromide is added to this solution, a cream-coloured solid precipitates.
(b) The boiling temperature of hydrogen fluoride is much higher than the boiling temperatures of other hydrogen halides.
(c) Liquid hydrogen fluoride is used in biochemical research for dissolving proteins which are insoluble in water. (Hint: Protein structures contain many hydrogen bonds.)

11. Laminaria seaweed contains iodine derived from sea water. The amount of iodine in the seaweed can be estimated as follows:
The seaweed is heated in air to form an ash. The ash is boiled with water, filtered and the filtrate treated with an excess of acidified sodium nitrite solution.

$$2NO_2^- + 2I^- + 4H^+ \longrightarrow 2NO + 2H_2O + I_2$$

The liberated iodine is extracted using 1,1,1-trichloroethane (TCE). and then titrated against 0.01 mol dm^{-3} sodium thiosulphate solution

(a) (i) Write a balanced equation for the reaction of sodium thiosulphate with iodine.
(ii) Name the indicator which is used in the titration of iodine with sodium thiosulphate.
(iii) Why is the indicator named in part (ii) added towards the end of the titration?
(iv) If 250 g of seaweed produces sufficient iodine to react with 20 cm^3 of 0.01 mol dm^{-3} thiosulphate solution calculate the percentage of iodine, by mass, in the seaweed. **7**
(b) Iodide ions can be oxidised by iodate ions, IO_3^-, in the presence of acid to form iodine and water. Write a balanced equation for the reaction. **2**
(c) Iodide, $^{131}I^-$, is used as a radioactive tracer in the treatment of the thyroid gland. It can be taken as the sodium salt.
(i) The half-life of $^{131}I^-$ is 8 days. Explain the term **half-life**.

(ii) $^{131}I^-$ decays by beta emission ($^{0}_{-1}e$). Write a balanced nuclear equation for the reaction taking place.

(iii) Explain whether you would expect the solubility of $Na^{131}I$ to differ from that of $Na^{129}I$. **6**

(d) Iodine may be used to test for a $CH_3CO—$ group in the 'iodoform reaction'. Organic compounds containing this group give a yellow precipitate of iodoform when warmed with iodine in the presence of alkali.

(i) Name a compound which contains the $CH_3CO—$ group.

(ii) Calculate the empirical formula of iodoform from the following data. **3**

	g in 100 g of iodoform
Iodine	96.700
Carbon	3.050
Hydrogen	0.254

For (*d*) see § 31.2.

CCEA (AS/AL)

12. *(a)* State and explain the trend in the boiling points of the halogens chlorine, bromine and iodine. **3**

(b) State what is meant by the term *electronegativity*. Explain the trend in electronegativity for the halogens chlorine, bromine and iodine. **5**

(c) State and explain the trend in the reducing properties of the halide ions Cl^-, Br^- and I^-. Describe what you would observe when an aqueous solution containing chlorine is added to separate aqueous solutions containing bromide and iodide ions. Write equations for any reactions which occur. **8**

16 *NEAB (AS/AL)*

13. *(a)* Describe how the addition of aqueous silver nitrate followed by aqueous ammonia can be used to distinguish between aqueous solutions containing chloride, bromide or iodide ions. **6**

(b) A 1.20 g sample of a solid mixture of sodium carbonate and sodium iodide was treated with an excess of dilute nitric acid. When effervescence stopped an excess of silver nitrate solution was added. A precipitate was formed which weighed 0.850 g.

State the role of nitric acid in this analysis, write equations for the reactions which occurred and calculate the percentage by mass of sodium iodide in the mixture. **7**

(c) Define the terms *electronegativity* and *reducing agent*. Explain why fluorine has the highest electronegativity of the halogen elements and the fluoride ion has the lowest reducing power of the halide ions. **7**

(d) Describe what you see when concentrated sulphuric acid is added to separate solid samples of sodium chloride and sodium bromide and the mixtures are gently warmed. In each case, identify any gaseous products formed and write equations for the reactions in which they are formed. **10**

30 *NEAB (AS/AL)*

14. Select appropriate data from the table given below to answer the questions which follow.

Element X	Atomic number	Atomic radius/nm	Radius of X^-/nm	Boiling point/K	Electro-negativity
F	9	0.064	0.133	86	4.0
Cl	17	0.099	0.181	238	3.0
Br	35	0.111	0.196	332	2.8
I	53	0.130	0.219	456	2.5

(a) Explain why

(i) iodide ions can be oxidised more easily than bromide ions

(ii) the electronegativity of fluorine is much larger than that of chlorine

(iii) the boiling point of bromine is higher than that of chlorine. **9**

(b) The halogen below iodine in Group VII is astatine(At). Predict, giving an explanation, whether or not

(i) hydrogen astatide will be evolved when concentrated sulphuric acid is added to a solid sample of sodium astatide, NaAt

(ii) astatine will be precipitated as a solid when chlorine is bubbled through an aqueous solution of sodium astatide

(iii) silver astatide will dissolve in concentrated aqueous ammonia. **10**

19 *NEAB (AS/AL)*

21 GROUP 6

21.1 OXYGEN – THE BREATH OF LIFE

The first element in Group 6 is oxygen. Most of us can obtain enough oxygen by our normal breathing. Some people, however, have difficulty in breathing and need to inhale pure oxygen. Hospitals need oxygen for accident victims, such as survivors of fires and survivors of near-drowning, for premature babies and for patients with complaints such as pneumonia and asthma. Patients who are having operations are given an anaesthetic mixed with oxygen [see Figure 21.1A].

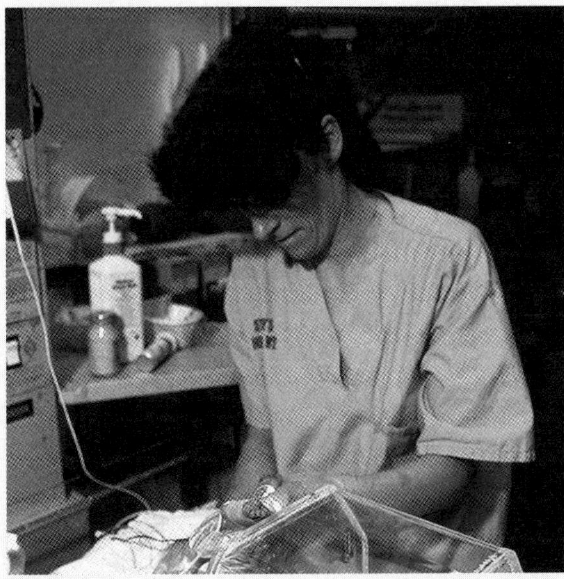

Figure 21.1A
Hospitals use oxygen

Space flights consume an enormous amount of oxygen. The Saturn rockets which launched American astronauts on their journey to the Moon carried 2200 tonnes of liquid oxygen. The first stage burned 450 tonnes of kerosene in 1800 tonnes of oxygen, and stages two and three were powered by hydrogen burning in oxygen. In addition, there was oxygen for the astronauts to breathe.

Figure 21.1B
Launch of space shuttle

The steel industry makes cast iron into steel. The carbon which is present as impurity in cast iron makes the iron brittle [§ 24.17.3]. Oxygen is used to burn off the carbon and other impurities, e.g. sulphur and phosphorus. Since so much oxygen is needed (one tonne of oxygen for every tonne of steel), steel works have their own oxygen plants on site.

Since fuels burn in oxygen more rapidly than in air, higher temperatures are reached. The oxy-acetylene torch burns ethyne (formerly called acetylene) in oxygen at a temperature of about 4000 °C. It is used in cutting and welding metals.

21.2 THE MEMBERS OF THE GROUP

The members of Group 6 are oxygen, sulphur, selenium, tellurium and polonium. There is a gradual transition down the group from non-metallic to metallic properties.

Oxygen

Oxygen constitutes 21% by volume of air. It is an essential part of the air for respiration. Oxygen is sufficiently soluble in water to be able to support the life of marine animals. The Earth's crust is 89% by mass oxygen.

Oxygen boils at −183 °C and freezes at −219 °C. It reacts directly with most other elements and forms compounds with all elements except the noble gases helium, neon, argon and krypton.

Oxygen is obtained industrially by the fractional distillation of liquid air. It is stored under pressure in cylinders. A laboratory method of preparation is the decomposition of hydrogen peroxide, H_2O_2, with a catalyst, e.g. manganese(IV) oxide. Oxygen can be collected over water.

21.3 REACTIONS OF OXYGEN AND SULPHUR

Oxygen combines directly with most elements [see Figure 21.3A]. Sulphur combines with most metals when heated and with the non-metallic elements fluorine, chlorine, oxygen and carbon. It is oxidised by concentrated nitric acid and by concentrated sulphuric acid to sulphur dioxide.

Figure 21.3A
Some reactions of oxygen

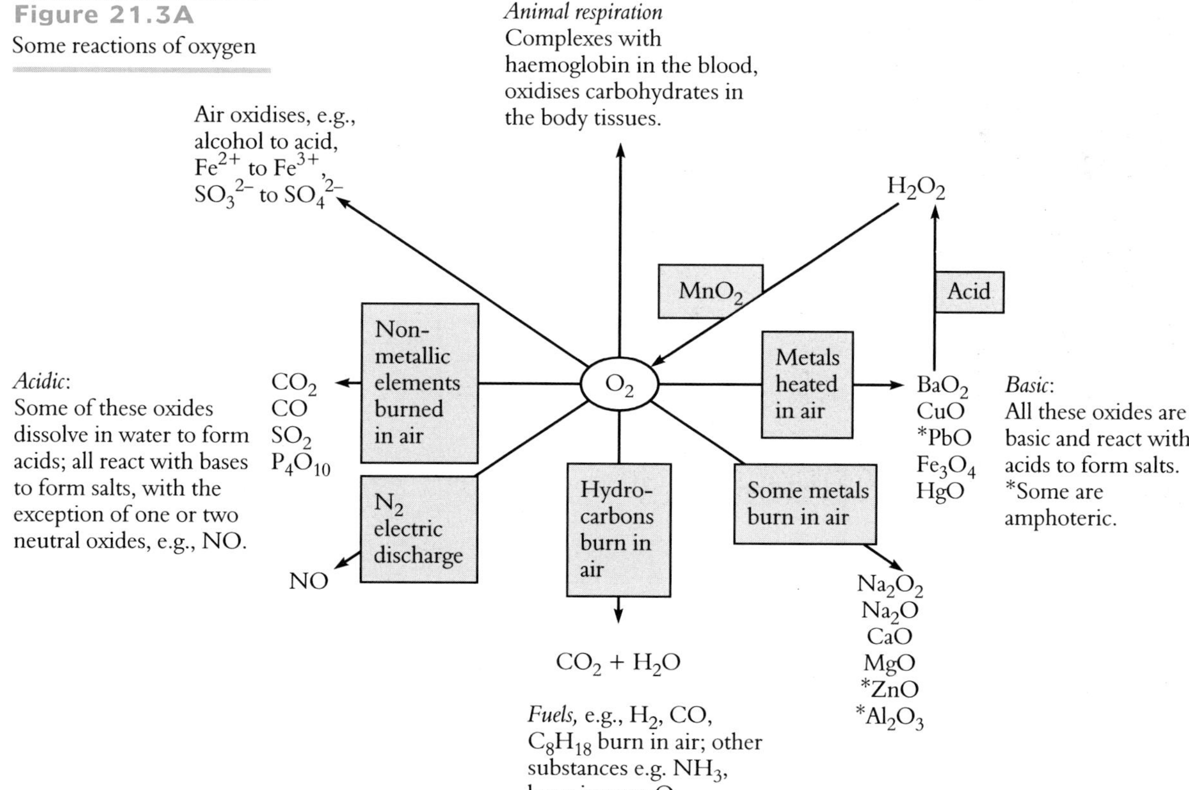

CHECKPOINT 21.3: REACTIONS OF OXYGEN AND SULPHUR

1. *(a)* Mention two uses of pure oxygen in (i) medicine (ii) industry (iii) exploration.
   (b) How is oxygen obtained industrially?
   (c) Outline a convenient laboratory preparation of oxygen. Write an equation for the reaction.
2. Give two examples of oxidation by oxygen of (i) metals (ii) non-metallic elements (iii) compounds. Give equations for one reaction of each type.

21.4 ALLOTROPES OF OXYGEN

The element oxygen exists in two forms, dioxygen, O_2, and trioxygen, O_3. The existence of an element in two forms is called **allotropy**. Trioxygen, ozone, is the less stable allotrope:

Allotropy of oxygen ...

... O_2 and O_3

$$3O_2(g) \longrightarrow 2O_3(g); \quad \Delta H^{\ominus} = 142 \text{ kJ mol}^{-1} \text{ of } O_3$$

Dioxygen must absorb energy in order to form trioxygen, but trioxygen decomposes spontaneously to form dioxygen. Since the direction of spontaneous change is always the same the allotropy is described as **monotropic** (moving in one direction).

Ozone has a characteristic smell. In concentrations above 1000 ppm, it is damaging to health.

The oxygen molecule $:\ddot{O}=\ddot{O}:$ has a σ bond and a π bond between the two atoms. In the ozone molecule, there is an angle of 117° between the bonds. The structure can be represented by the delocalised electron structure:

The structure of ozone

O⋯O⋯O (delocalised)

21.5 THE OZONE LAYER

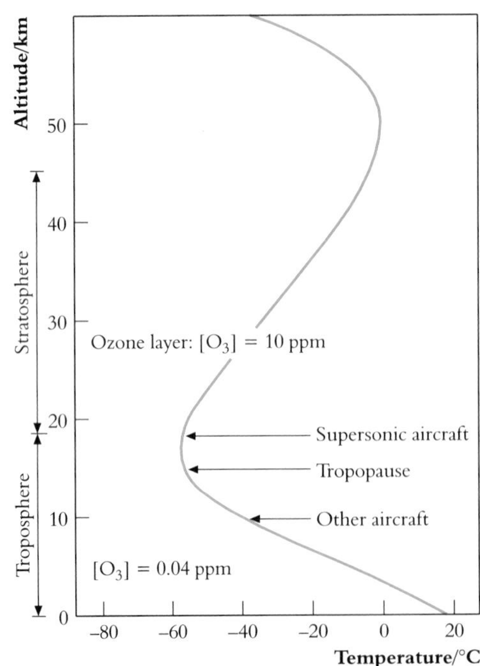

Figure 21.5A The structure of the atmosphere

There is a layer of ozone in the stratosphere (upper atmosphere). The first step in the formation of ozone in the stratosphere is the dissociation of oxygen molecules by UV light. The ozone layer protects oxygen in the troposphere (lower atmosphere) from UV light

The sunlight that falls on the upper layers of the atmosphere (the stratosphere; see Figure 21.5A) contains much more ultraviolet (UV) light than the radiation which reaches the surface of the Earth. Ultraviolet light has enough energy to bring about photochemical reactions that convert dioxygen, O_2 into ozone, O_3. Ozone can absorb UV light. In doing so, ozone protects oxygen in the lower atmosphere from being dissociated and keeps most of these harmful rays from penetrating to the Earth's surface. The maximum concentration of ozone, about 10 ppm, occurs 25–50 km from the surface of the Earth.

Some of the reactions involving dioxygen, O_2, and ozone (trioxygen, O_3) are:

(a) Dioxygen is dissociated by solar UV rays:

$$O_2 + h\nu(\text{UV}) \longrightarrow O\cdot + O\cdot$$

(b) Some of the oxygen atoms formed combine with dioxygen molecules to form ozone:

$$O\cdot + O_2 \longrightarrow O_3$$

(c) Ozone absorbs UV light and dissociates:

$$O_3 + h\nu(\text{UV}) \longrightarrow O_2 + O\cdot$$

(d) Ozone reacts with oxygen atoms to form dioxygen

$$O_3 + O\cdot \longrightarrow 2O_2$$

The rates of formation and destruction result in a steady state concentration of ozone in the stratosphere.

The ozone layer is attacked by natural and man-made chemicals

Ozone is destroyed by chemical reactions with a number of substances that occur naturally in the stratosphere, e.g. nitrogen oxides (from microbes and the combustion of fossil fuels) and methane (produced by microbes in swamps, rice paddies and the intestines of cattle and sheep). Increasing quantities of man-made substances are now attacking the ozone layer.

Scientists have discovered a decrease in the ozone layer over Antarctica. They relate it to the attack by CFCs on the ozone layer

A decrease in the ozone layer was discovered in 1983, when scientists with the British Antarctic Survey observed that the concentration of ozone in the stratosphere dropped rapidly over Antarctica each spring, to be replenished by the end of November. In 1986 the USA Antarctic National Ozone Expedition confirmed the decrease in the ozone layer and linked it to the presence in the stratosphere of chlorofluoroalkanes. These compounds, e.g. CCl_2F_2, are referred to as CFCs, short for chlorofluorocarbons [see §29.6]. They were at the time widely used as refrigerants, thermal insulation and aerosols. Being chemically unreactive, they pass through the troposphere to the stratosphere, where the ultraviolet light causes them to dissociate. As a result, atomic chlorine and chlorine oxide, ClO, are formed. Information gathered by satellites of the USA's National Aeronautics and Space Administration (NASA) showed that 50% of the ozone had disappeared during the polar thaw, and demonstrated a connection with CFCs. The satellites detected concentrations of chlorine oxide a hundred times as great as concentrations in latitudes where there is less change in the ozone layer. NASA predicted that, if the present rate of CFC production were to continue the ozone layer would decrease by 3% by the year 2010. The US Government Environmental Protection Agency, on the other hand, estimated a possible destruction of 50% of the ozone layer by the year 2050.

Reactions which take place after the photolysis of CFCs are:

(e) $O_3 + Cl\cdot \longrightarrow O_2 + ClO$
(f) $ClO + O\cdot \longrightarrow Cl\cdot + O_2$
(g) $O_3 + ClO \longrightarrow O_2 + ClO_2$

Reactions with atomic chlorine, chlorine oxide and nitrogen oxide destroy ozone. Chain reactions are set up

Notice that the reactions *(e)* and *(f)* constitute a chain reaction. As a result, one Cl atom can destroy thousands of molecules of ozone. Nitrogen oxide from the exhausts of the supersonic aircraft which fly in the stratosphere takes part in the reactions:

(h) $O_3 + NO \longrightarrow O_2 + NO_2$
(i) $NO_2 + O\cdot \longrightarrow NO + O_2$

Reactions *(h)* and *(i)* form a chain reaction. The effect of reactions *(e)*–*(i)* is to lower the steady state concentration of ozone.

In 1989 an international team of scientists working in northern Canada detected a decrease in the ozone layer over the Arctic. This is even more serious than the 'hole' over the Antarctic as it is closer to densely populated countries.

The result that is forecast for a decrease in the ozone layer is an increase in the amount of UV radiation reaching the Earth, with a consequent increase in the incidence of skin cancer

The results that are forecast for a decrease in the ozone layer are alarming. There is a strong connection between UV radiation and the incidence of non-melanoma skin cancer in humans. This is generally a non-fatal cancer, but there is some evidence to support a connection between UV radiation and melanoma skin cancer which is a more frequently fatal form of the disease. The US National Academy of Sciences report in 1986 concluded that there would be an increase in malignant melanoma, a serious form of cancer which frequently causes death. The report predicted that a 10% reduction in stratospheric ozone would result in a 20% increase in both forms of skin cancer. This would mean 160 000 extra cases of non-melanoma skin cancers per year in the US and 8000 in the UK. Excessive UV radiation has also been linked to cataracts and lowered immunity to disease.

Figure 21.5B
The ozone 'hole' over the Antarctic

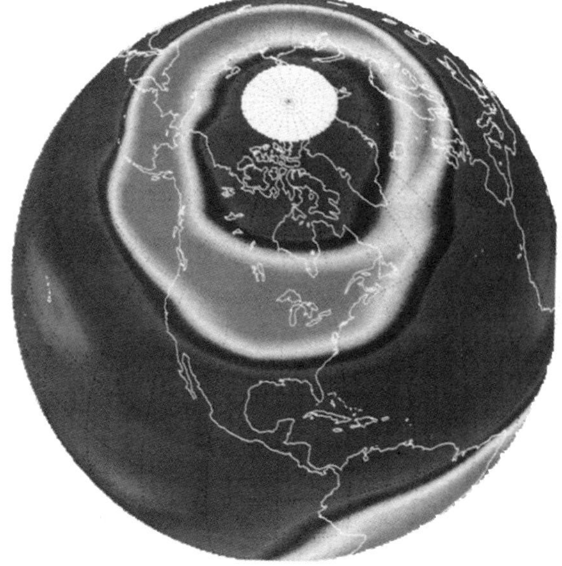

Ozone levels in the troposphere are rising. Often the levels exceed that recommended by the WHO

In the stratosphere ozone is beneficial, shielding the earth from UV radiation. In the troposphere, ozone is an undesirable chemical that causes damage to humans, vegetation and many materials, e.g. rubber and textiles. Ozone is suspected of triggering asthma attacks and bronchitis. In unpolluted air over Britain and the rest of Europe, the ozone concentration is 20–50 ppb. On still, sunny days, it can rise to over 60 ppb. World Health Organisation guidelines for human health state that the ozone concentration should not exceed a mean of 60 ppb over an 8 hour interval. This is often exceeded in southern England.

In the troposphere ozone is a dangerous substance, which damages plants, animals and materials. It causes respiratory diseases in humans, serious damage to trees and contributes to the greenhouse effect

Ozone in the troposphere has been linked to dying forests. Originally acid rain [see § 21.10] was thought to be the decisive factor in damage to forests. Now data for North America and Europe suggest that ozone may be of primary importance, with acid rain ranking second in Europe and perhaps third in North America.

Ozone also contributes to the greenhouse effect [see § 23.6].

Ozone is formed in the troposphere by photochemical reactions between oxygen, hydrocarbons and oxides of nitrogen. A reduction in the emission of oxides of nitrogen by motor vehicles would solve the problem

In the troposphere ozone is formed from oxygen by photochemical reactions with oxides of nitrogen and hydrocarbons. Both of these are emitted by motor vehicles [see § 26.5]. Power stations are another source of nitrogen oxides. Hydrocarbons come from many sources: ruminating cattle, marshes, rice paddies and leaking North Sea gas as well as vehicle exhausts. The concentration of ozone close to the ground has doubled in Europe over the past 30 years.

The production of ozone in the troposphere must be stemmed. It is difficult to remove hydrocarbons when they come from so many natural sources. The best hope is to reduce the emission of nitrogen oxides from vehicle exhausts [see § 26.5].

CHECKPOINT 21.5: THE OZONE LAYER

1. (*a*) What is meant by a *chain reaction?*
   (*b*) What chain reaction results in the formation of ozone from oxygen?
   (*c*) What is meant by *steady state concentration*?
   (*d*) Why are oxygen molecules dissociated in the stratosphere and not in the troposphere?
   (*e*) If the rate of destruction of ozone increases, will the ozone concentration decrease to a lower level or fall to zero? Explain your answer.

2. What results are predicted for the decrease in the ozone layer? Mention effects on the climate and on human health.

3. Ozone is something of an enigma. In the stratosphere it is beneficial to human health; in the troposphere it is detrimental to human health. Discuss this statement.

21.6 HYDRIDES OF OXYGEN AND SULPHUR

21.6.1 WATER

See § 17.4.

21.6.2 HYDROGEN SULPHIDE

Hydrogen sulphide ...

Hydrogen sulphide, H_2S, is an unpleasant-smelling gas which is poisonous at a level of 1000 ppm.

... a weak acid and a reducing agent

Hydrogen sulphide is a weak acid and a reducing agent, being oxidised to sulphur.

21.6.3 HYDROGEN PEROXIDE

The arrangement of bonds in H_2O_2

Hydrogen peroxide is a liquid of formula H_2O_2. Its structure is illustrated in Figure 21.6A. The molecules are associated by hydrogen bonding, and the liquid is viscous.

Hydrogen peroxide is used as a source of oxygen

Hydrogen peroxide readily decomposes to give oxygen:

$$2H_2O_2(aq) \longrightarrow 2H_2O(l) + O_2(g)$$

Figure 21.6A
The structure of H_2O_2

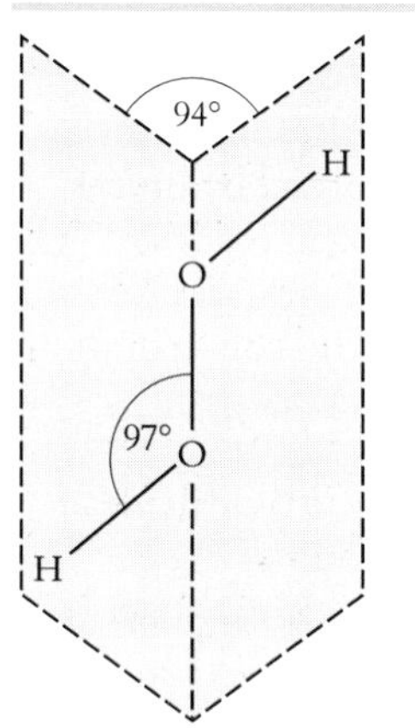

The decomposition is catalysed by a number of metals and metal oxides, the most popular being manganese(IV) oxide, MnO_2. In the laboratory, hydrogen peroxide is used as aqueous solutions which are designated as, for example, '20 volume', meaning that 1 volume of the solution will provide 20 volumes of oxygen at stp.

Oxidising agent/reducing agent

Hydrogen peroxide acts as an oxidising agent:

$$H_2O_2(aq) + 2H^+(aq) + 2e^- \longrightarrow 2H_2O(l); \quad E^\ominus = +1.77\text{ V}$$

and as a reducing agent:

$$H_2O_2(aq) \longrightarrow O_2(g) + 2H^+(aq) + 2e^-; \quad E^\ominus = -0.68\text{ V}$$

Hydrogen peroxide is an oxidising agent in acidic conditions ...

Acidic solutions favour the oxidising action e.g., Fe^{2+} to Fe^{3+}; PbS to $PbSO_4$; I^- to I_2. Alkaline solutions favour the reducing reaction, e.g. ClO^- to Cl^-. Even in acid solution, hydrogen peroxide reduces manganate(VII), MnO_4^-, to Mn^{2+}, with the formation of oxygen [see Question 2, Checkpoint 21.6].

Use

... and a reducing agent in alkaline conditions. It is used as a bleach

Hydrogen peroxide is used as a bleach for textiles, wood pulp and human hair. The bleaching action depends on its oxidising property. A convenient feature is that the only by-product is water.

CHECKPOINT 21.6: HYDRIDES

1. *(a)* Write the half-reaction equation for (i) hydrogen peroxide acting as an oxidant, (ii) hydrogen peroxide acting as a reductant.
   (b) Referring to Chapter 3 if you need to, write half-reaction equations for the reduction of
   (i) acidic manganate(VII), MnO_4^-,
   (ii) acidic dichromate(VI), $Cr_2O_7^{2-}$, (iii) iron(III) ions, Fe^{3+}, (iv) Cl_2 (v) I_2.

2. By combining half-reaction equations, construct equations for the following oxidation reactions.
   (a) I^- to I_2 by acidified H_2O_2
   (b) PbS to $PbSO_4$ by H_2O_2
   (c) H_2O_2 to O_2 by Cl_2
   (d) H_2O_2 to O_2 by acidified $KMnO_4$
   (e) Fe^{2+} to Fe^{3+} by acidified H_2O_2.

3. *(a)* What is the concentration/mol dm^{-3} of '20 volume' hydrogen peroxide?
   (b) Suggest how you could find the concentration of a hydrogen peroxide solution by titrimetric analysis.

21.7 OXIDES

Ionic oxides containing the ion O^{2-} are formed

The formation of O^{2-} is an endothermic process. When it forms an ionic structure, however, the small size and double charge on the anion result in a highly negative (exothermic) value for the lattice enthalpy. When the lattice enthalpy compensates for the ionisation enthalpies of the cation and anion, an ionic compound may be formed [see MgO, § 10.12]. Many ionic oxides are formed.

Oxides that dissolve in water give alkaline solutions due to the basic nature of the O^{2-} ion:

$$O^{2-}(aq) + H_2O(l) \longrightarrow 2OH^-(aq)$$

Table 21.7A Classification of oxides according to formula

Oxide	Types of bonds in the oxides E_xO_y
Normal oxides	Bonds are between **E** and O only. Some are ionic, e.g., $Ca^{2+}O^{2-}$; others are covalent, e.g., CO_2, $(SiO_2)_n$
Peroxides	Bonds are between **E** and O and also between O atoms. Some are ionic, e.g., $2Na^+$ $O—O^{2-}$; others are covalent, e.g., H—O—O—H
Mixed oxides	e.g., Pb_3O_4, which reacts as a mixture $2PbO . PbO_2$ and Fe_3O_4, which reacts as a mixture $FeO . Fe_2O_3$
Non-stoichiometric oxides	Transition metals form oxides of formula $\mathbf{M}_{0-1}O$, e.g., $Fe_{0.9}O$

Table 21.7B Oxides of metallic and non-metallic elements

	Oxides of metallic elements	Oxides of non-metallic elements
(a)	Oxides of metals in lower oxidation states are basic. Some react with water to give $OH^-(aq)$, e.g., CaO, MgO.	Most are acidic. Some dissolve in water to form solutions with a high concentration of hydrogen ions, e.g., SO_3.
(b)	Others are insoluble in water, but react with acids and acidic oxides, e.g., Fe_2O_3, CuO.	Macromolecular oxides, e.g., $(SiO_2)_n$, $(B_2O_3)_n$, do not dissolve, but react with basic oxides and amphoteric oxides to give salts.
(c)	Strongly basic oxides, e.g., K_2O, CaO react with amphoteric oxides.	A few are neutral, e.g., N_2O, NO.
(d)	Some metal oxides are amphoteric, reacting with both basic oxides and acidic oxides, e.g., ZnO, SnO, SnO_2, PbO, PbO_2, Cr_2O_3, Al_2O_3.	

21.7.1 CLASSIFICATION OF OXIDES

Oxides may be classified according to their formulae [see Table 21.7A] or structure or acid–base properties [see Table 21.7B]. In Chapter 15 we looked at oxides in relation to the Periodic Table. Figure 15.6C shows a classification of oxides according to structure and Figure 15.6D shows an analysis of the acid–base character of oxides. You will find it helpful to repeat Checkpoint 15.7.

21.8 SULPHUR DIOXIDE

Sulphur dioxide, SO_2, is a dense gas with a choking smell. It fumes in air and dissolves in water to form a solution which contains sulphurous acid.

Sulphur dioxide is the anhydride of sulphurous acid, which is a weak acid ...

$$SO_2(aq) + H_2O(l) \rightleftharpoons H_2SO_3(aq)$$

The acid is a weak acid, ionising to give hydrogensulphite ions, HSO_3^-, and sulphite ions, SO_3^{2-}. Attempts to concentrate the solution result in the decomposition of sulphurous acid and the evolution of sulphur dioxide gas.

... It reacts with alkalis to form sulphites or hydrogensulphites

Sulphur dioxide reacts with aqueous sodium hydroxide to form sodium sulphite solution. If more sulphur dioxide is passed into this solution, sodium hydrogensulphite is formed:

$$SO_2(g) + 2NaOH(aq) \longrightarrow Na_2SO_3(aq) + H_2O(l)$$
$$Na_2SO_3(aq) + SO_2(g) + H_2O(l) \longrightarrow 2NaHSO_3(aq)$$

Moist sulphur dioxide, or sulphurous acid, and sulphites behave as reducing agents:

SO_2 and SO_3^{2-} are reducing agents

$$SO_3^{2-}(aq) + H_2O(l) \rightleftharpoons SO_4^{2-}(aq) + 2H^+(aq) + 2e^-; \quad E^\ominus = -0.17\text{ V}$$

Sulphite ions are oxidised by air, chlorine, iron(III) ions, dichromate(VI) ions and manganate(VII) ions. The reaction with dichromate(VI), resulting in a change from orange to blue, and the decolourisation of manganate(VII) are used as tests for sulphur dioxide. Hydrogen sulphide is a stronger reducing agent than sulphur dioxide, and it reduces sulphur dioxide to sulphur.

Sulphur dioxide and sulphites are used as food additives. They act as reducing agents to inhibit the oxidation of food components, e.g. unsaturated fats, and also as acids to restrict the growth of moulds and aerobic bacteria.

CHECKPOINT 21.8: SULPHUR DIOXIDE

1. Write equations for the ionisation of sulphurous acid.

2. *(a)* Write half-reaction equations for the oxidants mentioned in § 21.8.

(b) Combine these with the half-reaction equation for $SO_3^{2-}(aq)$ to obtain equations for the reactions of $SO_3^{2-}(aq)$ with O_2, Cl_2, $Fe^{3+}(aq)$, $Cr_2O_7^{2-}(aq)$, $MnO_4^-(aq)$.

21.9 SULPHURIC ACID

Manufacture of sulphuric acid needs SO_2

Sulphuric acid is made by the Contact process. The anhydride of sulphuric acid is sulphur(VI) oxide, SO_3. It is made by the oxidation of sulphur dioxide, which is obtained from the following sources:

Sources of sulphur dioxide

1. Sulphur is burned in air.
2. Hydrogen sulphide from crude oil is burned in air.
3. During the extraction of metals from sulphide ores, the ores are roasted in air, giving sulphur dioxide.
4. When anhydrite, $CaSO_4$, is heated with coke and sand in the manufacture of cement, $CaSiO_3$, sulphur dioxide is a by-product.
5. Flue-gas desulphurisation in power stations supplies sulphur dioxide [see § 21.10].

Sulphur dioxide is passed through an electrostatic precipitator to remove dust and impurities.

$SO_2 + O_2$ combine to form SO_3 in a reaction which reaches equilibrium

When sulphur dioxide and oxygen combine

$$2SO_2(g) + O_2(g) \rightleftharpoons 2SO_3(g); \quad \Delta H^{\ominus} = -197\ \text{kJ mol}^{-1}$$

the reactants and product reach a state of equilibrium. According to Le Chatelier's Principle [§ 11.5.1] the reaction will be favoured by a low temperature and a high pressure. At high pressures, however, sulphur dioxide liquefies, and at low temperatures the rate of attainment of equilibrium is slow. The catalyst vanadium(V) oxide is so effective that a 95% conversion is achieved at 450 °C and 2 atm. The catalyst is easily 'poisoned' by absorbing impurities in the sulphur dioxide. These take up the surface of the catalyst and impair its efficiency. The temperature of the catalyst is maintained at 450 °C in spite of the exothermicity of the reaction by using the heat generated in the catalyst chamber to heat the incoming gases [see Figure 21.9A].

The principles which govern the position of equilibrium are discussed in the light of Le Chatelier's Principle

The sulphur(VI) oxide formed is not absorbed in water because the reaction

$$SO_3(g) + H_2O(l) \longrightarrow H_2SO_4(l)$$

SO_3 is absorbed in H_2SO_4

is so intensely exothermic as to vaporise the sulphuric acid formed. It can be safely absorbed in 98% sulphuric acid to form fuming sulphuric acid, *oleum*, $H_2S_2O_7$, which reacts with water to form sulphuric acid:

$$SO_3(g) + H_2SO_4(l) \longrightarrow H_2S_2O_7(l)$$
$$H_2S_2O_7(l) + H_2O(l) \longrightarrow 2H_2SO_4(l)$$

Figure 21.9A
The Contact process

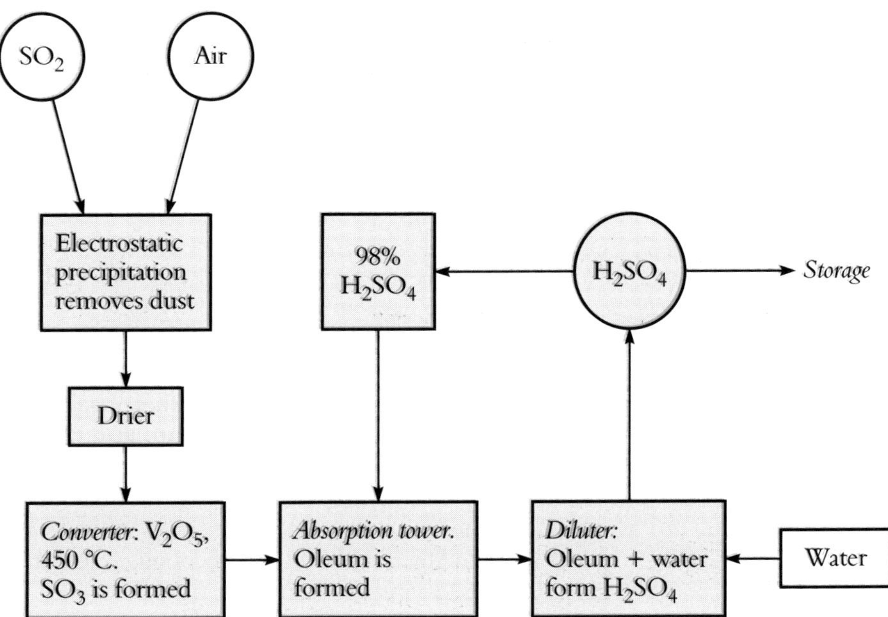

21.9.1 PHYSICAL PROPERTIES

Physical properties of H_2SO_4

Sulphuric acid is a viscous liquid. It is a covalent substance with the structure shown in Figure 21.9B(a). The viscosity and the high boiling temperature (338 °C) are due to hydrogen bonding [see Figure 21.9B(b)].

Figure 21.9B Sulphuric acid (a) the structure (b) the intermolecular hydrogen bonding

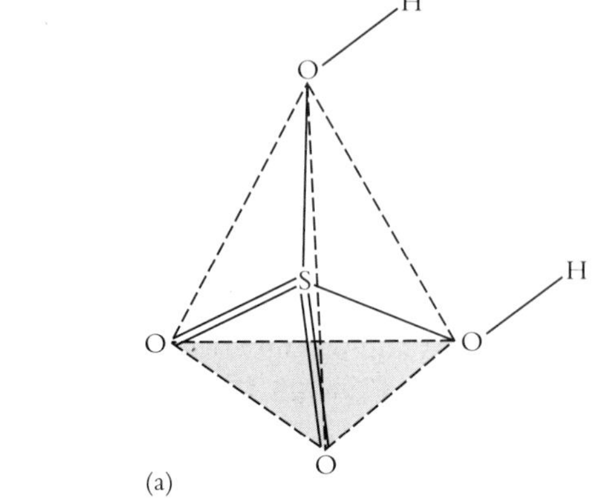

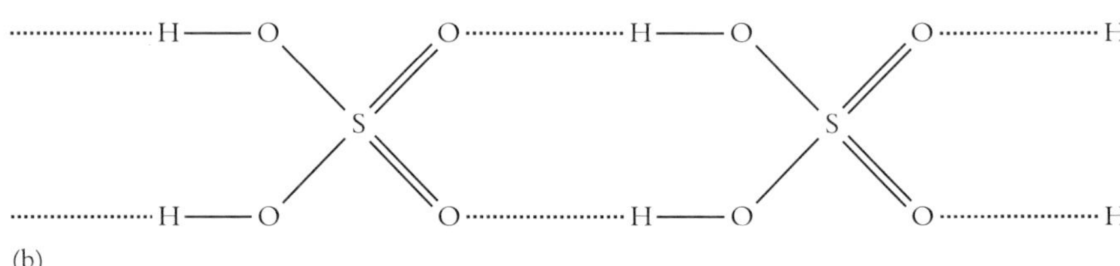

With water, sulphuric acid forms a constant-boiling [§ 8.5.4] mixture containing 98.3% acid.

21.9.2 CHEMICAL PROPERTIES

Acid properties

Sulphuric acid is a diprotic acid, the second ionisation being incomplete ...

In aqueous solution, sulphuric acid functions as an acid. The ionisation

$$H_2SO_4(aq) + H_2O(l) \longrightarrow H_3O^+(aq) + HSO_4^-(aq)$$

is virtually complete. The ionisation

$$HSO_4^-(aq) + H_2O(l) \rightleftharpoons H_3O^+(aq) + SO_4^{2-}(aq)$$

is about 10% complete.

... It reacts as a typical acid

Sulphuric acid in solution reacts with metals and their oxides, hydroxides and carbonates in typical acid fashion.

Concentrated sulphuric acid as a drying agent

Concentrated sulphuric acid must be diluted with care

Concentrated sulphuric acid reacts exothermically with water. When a solution is made it is essential to pour the acid into water, stirring to disperse the heat evolved. It is dangerous to add water to concentrated sulphuric acid as small pockets of water are likely to boil.

It is used to dry gases ...

Gases are dried by bubbling them through concentrated sulphuric acid. For basic gases, another drying agent must be used.

Concentrated sulphuric acid as a dehydrating agent

... and as a dehydrating agent

When concentrated sulphuric acid removes water from a mixture, it is termed a **drying agent**. When it removes the elements of water from a compound, with the formation of a new compound, it is described as a **dehydrating agent**. Some dehydrating reactions of concentrated sulphuric acid are:

$$\text{Glucose, } C_6H_{12}O_6(s) \longrightarrow \text{Sugar charcoal, } C(s) + H_2O(l)$$
$$\text{Methanoic acid, } HCO_2H(l) \longrightarrow CO(g) + H_2O(l)$$
$$\text{Ethanedioic acid, } HO_2CCO_2H(l) \longrightarrow CO(g) + CO_2(g) + H_2O(l)$$
$$\text{Ethanol, } C_2H_5OH(l) \longrightarrow \text{Ethene, } CH_2{=}CH_2(g) + H_2O(l)$$

Concentrated sulphuric acid as an oxidising agent

As an oxidising agent, concentrated sulphuric acid oxidises metals ...

Hot concentrated sulphuric acid is an oxidising agent. It may be reduced to sulphurous acid, which dissociates to form sulphur dioxide and water:

$$SO_4^{2-}(aq) + 4H^+(aq) + 2e^- \rightleftharpoons SO_2(aq) + 2H_2O(l); \quad E^{\ominus} = +0.17\text{ V}$$

Some oxidising reactions are:

1. Metals are oxidised to sulphates. Equations represent these reactions only approximately as a mixture of products, SO_2, H_2S and S, is formed:

$$Cu(s) + 2H_2SO_4(l) \longrightarrow CuSO_4(aq) + SO_2(g) + 2H_2O(l)$$

... non-metallic elements ...

2. Non-metallic elements, e.g., carbon and sulphur, are converted into oxides:

$$C(s) + 2H_2SO_4(l) \longrightarrow CO_2(g) + 2SO_2(g) + 2H_2O(l)$$

Figure 21.9C
Reactions of concentrated sulphuric acid

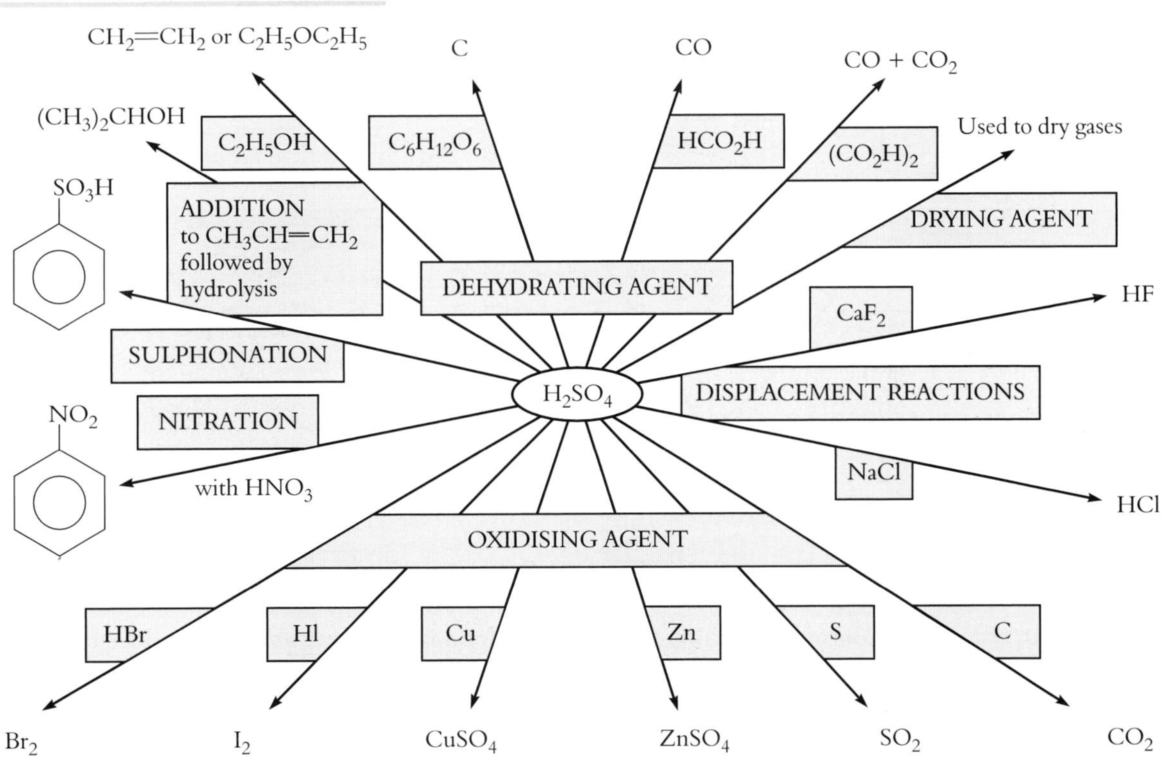

... and compounds. e.g., HBr, HI, H_2S ...

3. Compounds are oxidised, with the formation of sulphur dioxide or hydrogen sulphide. Hydrogen bromide, iodide and sulphide are oxidised:

$$2HBr(g) + H_2SO_4(l) \longrightarrow Br_2(l) + SO_2(g) + 2H_2O(g)$$

$$3H_2S(g) + H_2SO_4(l) \longrightarrow 4S(s) + 4H_2O(l)$$

Displacement reactions of concentrated sulphuric acid

Concentrated sulphuric acid displaces other acids from their salts

When concentrated sulphuric acid is heated with a nitrate, it displaces the other acid from its salt, and nitric acid is formed. In the cold, an equilibrium is set up:

$$NO_3^-(s) + H_2SO_4(l) \rightleftharpoons HNO_3(l) + HSO_4^-(s)$$

If the mixture is heated, nitric acid distils at 120 °C, while sulphuric acid, with T_b 270 °C remains. The removal of nitric acid moves the equilibrium over to the right-hand side. Similarly, the more volatile acids, hydrogen fluoride, hydrogen chloride and phosphoric(V) acid, are displaced from fluorides, chlorides and phosphates(V).

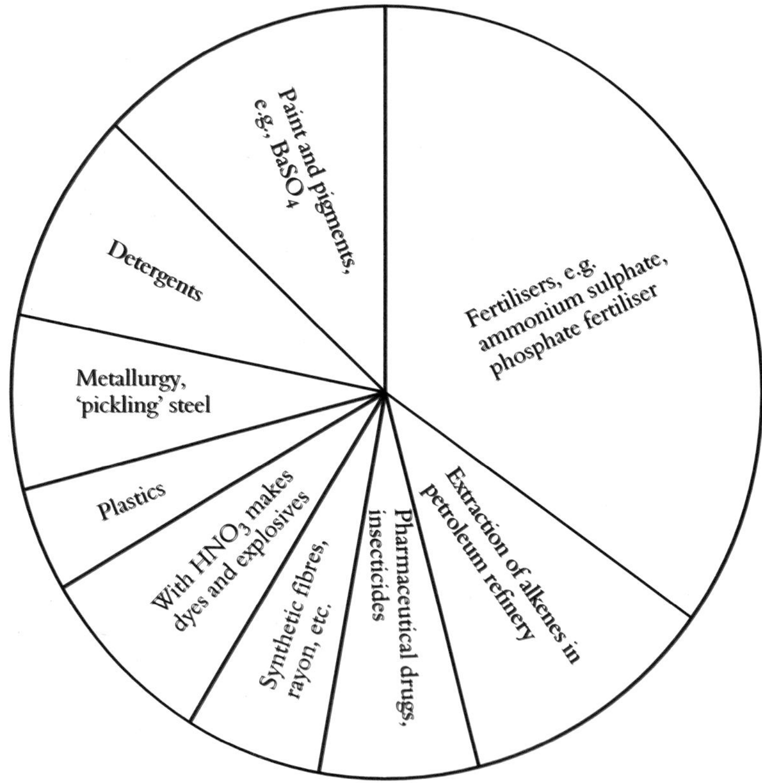

Figure 21.9D Some of the industrial uses of sulphuric acid (The world consumption totals 130 million tonnes per annum.)

Reactions with organic compounds

Conc. H_2SO_4 in organic chemistry

Concentrated sulphuric acid is a sulphonating agent, able to introduce the $—SO_3H$ group into an aromatic ring [§ 28.8.3]. It is used in nitration [§ 28.8.1] and in addition to alkenes [§ 27.8.7].

CHECKPOINT 21.9: SULPHURIC ACID

1. Concentrated sulphuric acid reacts with sodium iodide in a two-stage reaction to give iodine. Explain how sulphuric acid is reacting, and write equations for the two steps.

2. Concentrated sulphuric acid reacts with potassium bromide to form a solid and three gases. Name these products, and explain how they come to be formed.

3. Explain what would you expect to see when concentrated sulphuric acid is added to copper(II) sulphate crystals.

4. Phosphorus(V) oxide, P_2O_5, is a more powerful dehydrating agent than sulphuric acid. What is the fuming gas that is formed in a reaction between the two chemicals?

5. Give examples, with equations, of reactions in which sulphuric acid acts as *(a)* an acid, *(b)* a dehydrating agent, *(c)* an oxidising agent, *(d)* a sulphonating agent.

6. The Contact process achieves a 95% conversion of sulphur dioxide. What happens to the rest?

21.10 ACID RAIN

All rain water is slightly acidic. Acid rain has a pH below 5.0

Rain is naturally slightly acidic because it reacts with carbon dioxide in the air to form carbonic acid. Natural rain-water has a pH of about 5.6. In central Europe, rain-water is much more acidic, with a pH of about 4.1, and on the fringes of Europe, e.g. Ireland and Portugal, rain-water has a pH of about 4.9. Rain from individual storms can have a pH below 3. Rain with pH below 5 is described as **acid rain**.

Acid rain is now thought to be the cause of the extensive damage to Europe's trees and to the death of fish in the lakes of Canada, Norway, Sweden, Wales, Scotland and other countries. Europe shows the worst signs of damage by acid rain, but acid rain is becoming a global phenomenon.

Lakes in many countries of Europe and North America have become so acidic that their stocks of fish have died

Lakes in Scandinavia, Scotland, Canada and the USA have become much more acidic. Thousands of lakes which once stocked fish are now dead. The death of fish is usually attributed to poisoning by aluminium. Sulphate ions in acid rain can combine with aluminium in complex compounds to form soluble aluminium sulphate, which washes into streams. There it interferes with the operation of fish gills, so that they become clogged with mucus. The fish die from lack of oxygen.

Many of Europe's trees have lost much of their foliage. In some countries large areas of forest have died. The cause is thought to be acid rain

Forests have declined. In the mid-1970s Alpine forests started to lose their fir trees. Then in Germany, including the famous Black Forest, spruce began to thin and their needles turned brown. The Netherlands, Czechoslovakia, Switzerland and Britain recorded that 20–30% of their trees had lost much of their foliage. Acidic rain-water draining from soils washes out nutrients. Without essential nutrients, e.g. calcium and magnesium, trees starve to death.

Building materials which are basic ... and iron and steel ... are attacked by acid rain

Building stone may be limestone or marble (both calcium carbonate) or a sandstone in which quartz grains are held together by a coating of calcium carbonate or iron oxide. These materials (apart from quartz) are attacked by acid rain to form soluble substances or solids which flake off the surface. Metallic structures, e.g. bridges, ships and motor vehicles, are also attacked by acid rain.

Sulphur dioxide is one of the causes of acid rain. It is formed in the combustion of fossil fuels.

One of the chief culprits in the formation of acid rain is sulphur dioxide. Natural sources, such as volcanos, sea spray, rotting vegetation and plankton send sulphur dioxide into the atmosphere. Half the sulphur dioxide in the

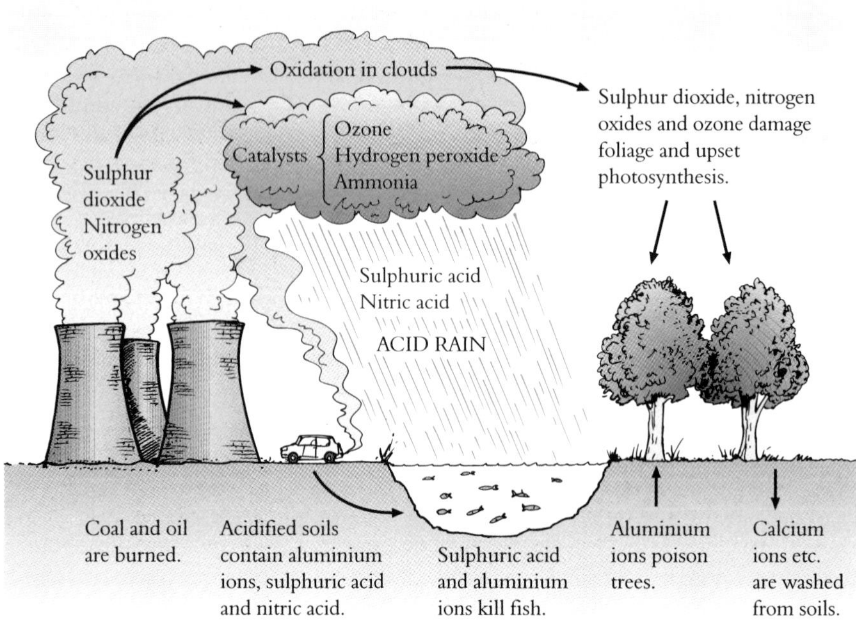

Figure 21.10A
The formation of acid rain

In the atmosphere, sulphur dioxide reacts with oxygen and water vapour to form sulphuric acid

atmosphere, however, comes from the combustion of fossil fuels. Over Europe the proportion of sulphur dioxide in the air that comes from fuels is 85%. When sulphur dioxide reaches the atmosphere, it reacts with moisture and oxygen to form sulphuric acid.

The 'Thirty Per Cent Club' of nations have agreed to reduce their emission of sulphur dioxide

Acid rain has become an important political issue, souring relationships between countries which export pollution and those which receive it; for example, between Britain and Norway and between the USA and Canada. In 1984 the UK joined the 'Thirty Percent Club', a group of nations committed to reducing sulphur dioxide emission by 30% by the year 2000. Figure 21.10B shows that the biggest source of sulphur dioxide is power stations. Several measures can be taken to reduce this pollution.

Low sulphur fuels

Some of the sulphur in oil and coal can be removed before the fuels are burnt

Oil power stations emit on average about 10% more sulphur dioxide than coal power stations. Modern oil refineries can, however, produce low-sulphur oil. The coal used in British power stations contains an average of 1.6% sulphur. About half the sulphur is combined as 'iron pyrites', FeS_2. Of this, 80% can be removed by grinding the coal and using various separation techniques. Removal of organically

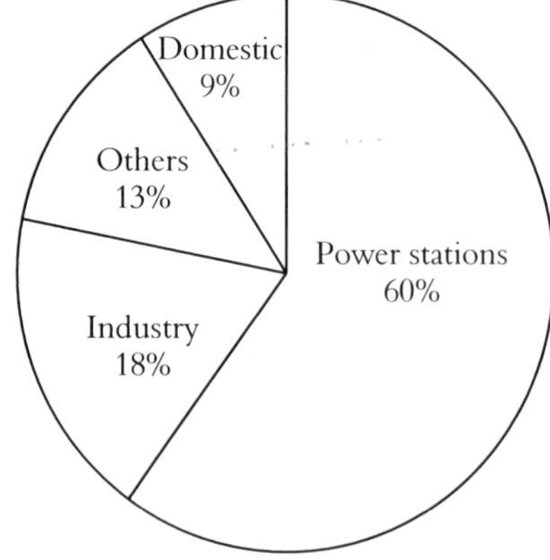

Figure 21.10B
Sources of sulphur dioxide

bound sulphur is more difficult. If the coal is converted into a gaseous fuel by one of the modern methods of *coal gasification*, sulphur can be removed as the gas is formed. Coal gasification has another advantage in that it may supplement our diminishing reserves of natural gas. Of the sulphur in bituminous coal, 25% can be removed by crushing and cleaning. A number of countries in Eastern Europe use lignite which cannot be cleaned by present technology.

New burners

Pulverised fluidised bed combustion (PFBC) is a technique for removing sulphur from coal as it burns. PFBC can be built into new power stations

Pulverised fluidised bed combustion (PFBC) offers a new, more efficient way of burning coal in a bed of limestone, which removes sulphur as the coal burns. PFBC can cut the emission of nitrogen oxides as well as sulphur dioxide. Pilot tests suggest that such burners could remove up to 80% of sulphur dioxide. These systems may soon be widely used in pollution-conscious countries.

PFBC has the advantage over other systems (for FGD, see later) that it is cheaper and more energy-efficient and creates fewer problems in the disposal of waste sludges. British Coal is developing the technology but for the present the alternative method (flue-gas desulphurisation; see below) is being adopted.

Removal of sulphur from exhaust gases

A large power plant emits about 10^8 m^3 of gases daily – roughly equal to the volume of air in 100 000 houses. Tall chimneys take the exhaust gases, including sulphur dioxide and other pollutants, high up into the air before they are discharged. The work force and the local community are protected. At one time, the managers of power stations and factories thought that was the end of the problem. Now that we know acidic pollutants are to blame for the acid rain that falls perhaps thousands of kilometres away, we realise that tall chimneys do not solve the problem; they merely transfer it to another region.

Flue gas desulphurisation (FGD) is a technique for removing sulphur dioxide from the exhaust gases in the chimney stacks of power stations. FGD plants can be fitted on to existing power stations. The UK is investing in FGD

Sulphur dioxide can be removed from the exhaust gases before they leave the chimney stack of the power station. Systems that carry out such **flue gas desulphurisation** (FGD) have been developed and are increasingly used in power stations. The exhaust gases are washed by an alkaline solution, which converts the sulphur dioxide into a waste sludge. The systems can remove up to 95% of the sulphur in the flue gases. The UK is fitting FGD plants to power stations. Sulphur dioxide is converted into calcium sulphate, which is used in the manufacture of plaster and cement and for land-filling.

Other solutions

Nuclear power stations and renewable energy sources are other options for reducing acid rain

Other solutions to the problem of pollution from power stations are a switch to nuclear power stations and the development of technology for harnessing 'renewable energy sourees', e.g. solar energy, wind, waves, hydropower and geothermal energy.

CHECKPOINT 21.10: ACID RAIN

1. *(a)* All rain water is weakly acidic. Explain why this is so.
   (b) What is meant by 'acid rain'?
2. The UK imports 750 000 tonnes of sulphur annually. The emission of sulphur from one power station is 1.3×10^6 tonnes of sulphur per year. Comment on these figures.
3. Explain how acid rain attacks (*a*) trees (*b*) fish (*c*) building materials.
4. *(a)* Briefly state how the amount of sulphur in fossil fuels can be reduced.
   (b) What do the letters PFBC and FGD stand for? Briefly explain the difference between the two processes.

QUESTIONS ON CHAPTER 21

1. Describe the Contact process for the manufacture of sulphuric acid. Explain the choice of conditions used for the process. Mention three uses for sulphuric acid.

2. How and under what conditions does sulphuric acid react with *(a)* potassium bromide, *(b)* zinc, *(c)* copper, *(d)* sucrose?

3. Illustrate the use of sulphuric acid *(a)* as an acid, *(b)* as an oxidising agent, *(c)* as a dehydrating agent, *(d)* in the displacement of other acids from their salts and *(e)* as a sulphonating agent.

4. Sulphur dioxide and chlorine are both used as bleaches. They act on different dyes. Why can sulphur dioxide bleach materials that are not bleached by chlorine?

5. Many metals are mined as sulphides. The process of extraction involves roasting the metal sulphide in air, with the production of an exhaust gas containing a high concentration of sulphur dioxide. This pollutant can be removed by converting it into sulphuric acid.

(a) What mass of sulphuric acid could be produced from one tonne of pyrites, FeS_2, if all the sulphur were converted into sulphuric acid?
(b) Suggest two advantages of removing sulphur dioxide from the exhaust gas.
(c) Suggest an advantage of converting sulphur dioxide into sulphuric acid rather than into calcium sulphate, as happens in other processes.
(d) Briefly state how sulphur dioxide can be converted into sulphuric acid.

6. *(a)* Give equations and essential conditions for the chemical reactions which take place during the manufacture of sulphuric acid from sulphur dioxide.
(b) Give one reaction, with an equation, to illustrate the reaction of sulphuric acid as (i) an oxidising agent (ii) a dehydrating agent (iii) a catalyst (iv) a dibasic acid.

7. Outline the chemical reactions which lead to

(a) The formation of ozone in the upper atmosphere.
(b) The destruction of ozone in the upper atmosphere by chlorofluorocarbons (CFCs).
(c) The mechanisms of the chemical reactions in *(a)* and *(b)* have features in common. Identify three of these features.
(d) Give two uses of CFCs, and say what properties of CFCs have led to these uses.
(e) What property of CFCs has caused environmental problems?

8. The liquid dichlorodifluoromethane CF_2Cl_2, has the trade name 'freon 12'. It is used as a refrigerant and was formerly used also as an aerosol propellant. 'Freon 12' is very unreactive and, when released, drifts slowly into the stratosphere, 20–40 km above the surface of the Earth. There it reacts with the ozone layer. The reactions which take place include:

(i) $CF_2Cl_2 + hf \longrightarrow CF_2Cl\cdot + Cl\cdot$
(ii) $Cl\cdot + O_3 \longrightarrow ClO\cdot + O_2$

(where hf represents radiation of frequency f)

(a) Suggest why the drift of 'freon 12' into the stratosphere is fairly slow.
(b) Why is the inertness of 'freon 12' an advantage as a refrigerant and an aerosol propellant?
(c) How does this very inertness lead to damage of the ozone layer?
(d) What term is used to describe reactions such as (i)?
(e) What term is used to describe particles such as $CF_2Cl\cdot + ClO\cdot$?
(f) Nitrogen monoxide is present in the stratosphere. Write an equation for the destruction of ozone by nitrogen monoxide.

9. Sulphur dioxide occurs naturally in volcanic springs where it reacts with water.

(a) Write an equation for this reaction. **2**
(b) State the colour changes observed when sulphur dioxide gas is bubbled through solutions containing: Fe^{3+}, MnO_4^-. **4**
(c) Sulphur dioxide may be converted to sodium thiosulphate via the following route:

$$SO_2 \xrightarrow{A} Na_2SO_3 \xrightarrow{B} Na_2S_2O_3$$

Identify substances A and B and write chemical equations for the changes. **4**
(d) Calculate the maximum mass of anhydrous sodium thiosulphate which could be produced from $10^6\ dm^3$ of sulphur dioxide. All measurements are at 20 °C and a pressure of 1 atmosphere. **2**
(e) State **two** uses of sodium thiosulphate. **2**
(f) Sulphur dioxide is used to make sulphuric acid, an important chemical. The acid is used in the manufacture of phosphate fertilisers and ammonium sulphate fertilisers; it is an oxidising agent and a catalyst in many reactions.
(i) Describe, using balanced equations, how sulphuric acid converts phosphate rock to a phosphate fertiliser. **3**
(ii) Concentrated sulphuric acid oxidises potassium iodide. Write an equation for the reaction indicating the oxidation states of iodine and sulphur at the beginning and end of the reaction. **4**

For *(f)* (i) see § 22.7.

18 *CCEA (AL)*

10. In this country almost all the sulphuric acid manufactured uses elemental sulphur as a starting material, although in some countries metal ores such as zinc sulphide, ZnS, are roasted to give sulphur dioxide. Catalytic oxidation of the sulphur dioxide takes place in a four-stage converter. The reaction is reversible.

$$2SO_2(g) + O_2(g) \rightleftharpoons 2SO_3(g) \quad \Delta H = -196\ \text{kJ mol}^{-1}$$

The converter operates at a temperature of about 500 °C and at normal atmospheric pressure. At each stage in the converter heat is removed. A 99.5% conversion is achieved by removing the sulphur trioxide between the third and fourth stages.

Water cannot be used for sulphur trioxide absorption. Instead the absorbent is 98% sulphuric acid. During the absorption of sulphur trioxide, water is added to maintain the concentration of the sulphuric acid.

(a) Explain why the conversion of sulphur dioxide and oxygen into sulphur trioxide is favoured by high pressure but, in practice, normal pressure is used. **3**

(b) The table gives values of the equilibrium constant, K_p, at different temperatures for the conversion of sulphur dioxide and oxygen into sulphur trioxide.

(i) Write an expression for the equilibrium constant, K_p and, by referring to the data in the table, explain why K_p varies with temperature. **5**

Temperature/°C	**K_p/Pa^{-1}**
25	4.0×10^{19}
200	2.5×10^{5}
400	3.0×10^{-1}
800	1.3×10^{-6}

(ii) Explain, in terms of the equilibrium, why heat is removed at each stage of the conversion. **2**

(c) Briefly explain how the catalyst speeds up the rate of the conversion of sulphur dioxide into sulphur trioxide and give the name of the catalyst used. **4**

(d) Suggest a reason for removing the sulphur trioxide between the third and fourth stages of the converter. Explain your answer. **2**

(e) Why is sulphur trioxide absorbed in 98% sulphuric acid rather than in water? **2**

(f) Write an equation to show the reaction occurring when zinc sulphide is roasted in air. **2**

(g) Discuss the factors that would influence the location of a sulphuric acid plant. **6**

(h) Give **two** variable costs in the production of sulphuric acid. **2**

AEB (AL) 1998

22 GROUP 5

22.1 TWO CHEMICAL MESSENGERS

The vital job of the heart is to pump blood through the blood vessels. Blood is pumped out of the heart through the arteries and returns to it in the veins. When the body's demand for oxygen increases, as during exertion, the heart pumps more blood round the body. The blood vessels dilate so that more blood can flow through them. This dilation happens through relaxation of the muscular part of the blood vessel [see Figure 22.1A].

How does the muscle know what to do? When an increased flow of blood is required, a **chemical messenger** passes from the blood to the muscle and tells the muscle to relax and allow the blood vessel to dilate. The chemical messenger is a nitrogen compound, **acetylcholine** [see § 32.9.2]. In some people the blood vessels do not dilate with ease, and these people suffer from high blood pressure. When the exertion and the need for increased blood flow cease, the muscles contract and the diameter of the blood vessel returns to normal.

The endothelium is a layer of cells lining the cavity of the blood vessel. Research workers found that removing the endothelium prevented a vessel from dilating. Their explanation was that a second chemical messenger is involved. They called this the endothelium-derived relaxing factor, EDRF. They suggested the sequence:

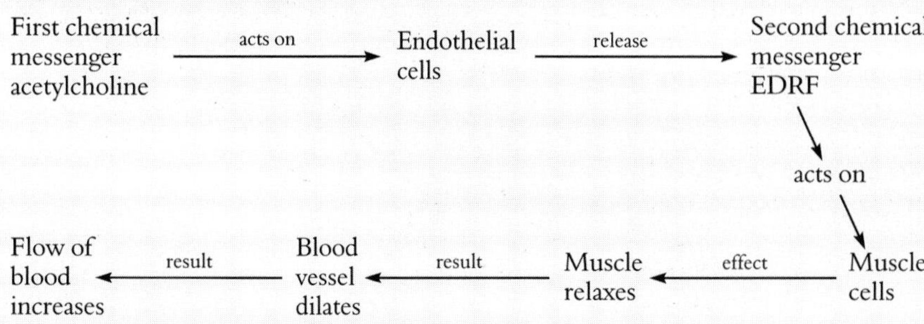

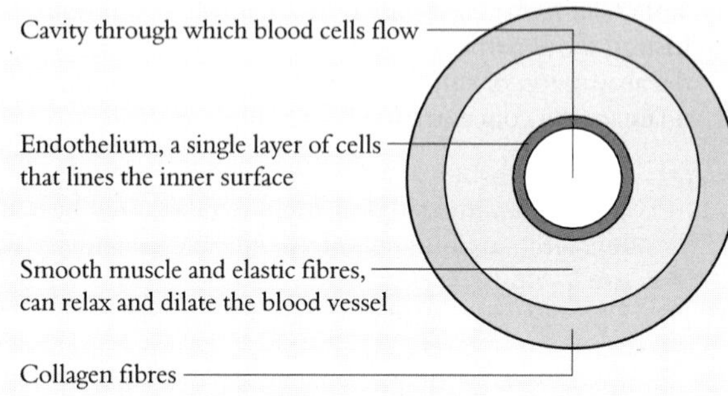

Figure 22.1A
Cross-section of an artery

A number of scientists joined in the search for the mysterious second messenger. A clue was that muscles could be induced to relax and dilate the blood vessels by another important nitrogen compound glyceryl trinitrate, GTN [see § 30.8]. In people who suffer from angina the arteries supplying the heart cannot dilate enough to allow enough blood to flow during exertion. The result is a severe pain in the chest which is relieved by taking GTN. It was shown that GTN releases nitrogen monoxide through the action of an enzyme in the body. Eventually the second messenger EDRF was identified as nitrogen monoxide, NO. This was a surprise as people had been expecting a larger and more complicated molecule than NO to be responsible for this vital function. In a healthy person, nitrogen monoxide is made from another important nitrogen compound, the amino acid arginine.

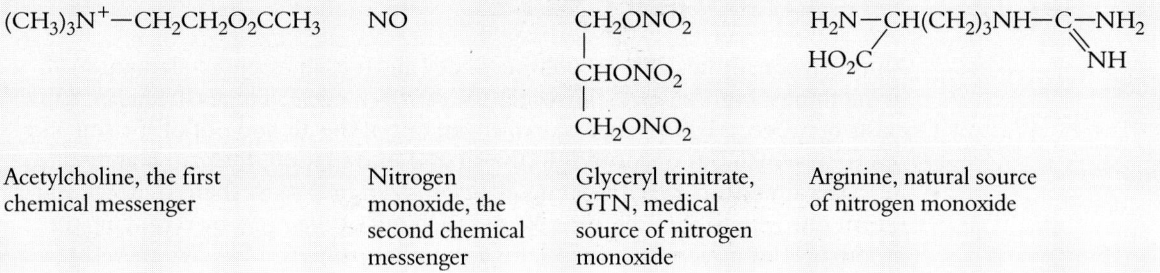

Acetylcholine, the first chemical messenger

Nitrogen monoxide, the second chemical messenger

Glyceryl trinitrate, GTN, medical source of nitrogen monoxide

Arginine, natural source of nitrogen monoxide

Figure 22.2A
The nitrogen cycle

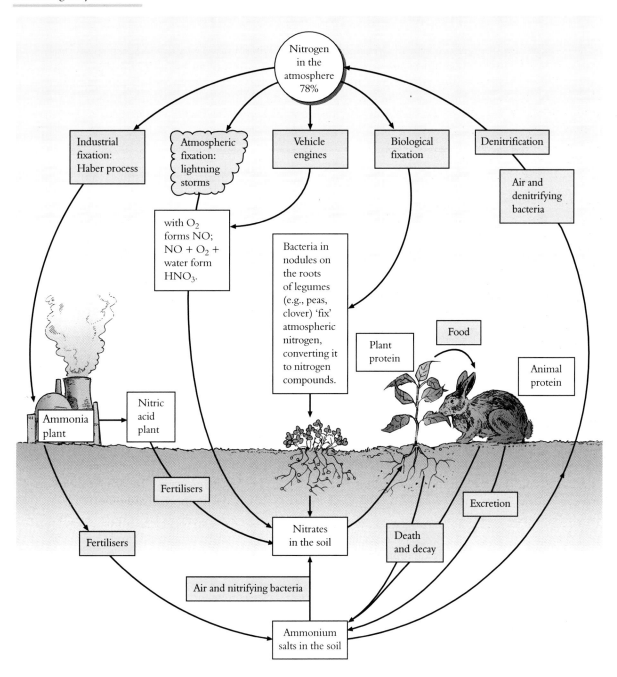

Nitrogen in the atmosphere 78%

Industrial fixation: Haber process

Atmospheric fixation: lightning storms

Vehicle engines

Biological fixation

Denitrification

Air and denitrifying bacteria

with O_2 forms NO; NO + O_2 + water form HNO_3.

Bacteria in nodules on the roots of legumes (e.g., peas, clover) 'fix' atmospheric nitrogen, converting it to nitrogen compounds.

Plant protein

Food

Animal protein

Ammonia plant

Nitric acid plant

Fertilisers

Fertilisers

Excretion

Nitrates in the soil

Death and decay

Air and nitrifying bacteria

Ammonium salts in the soil

22.2 THE NITROGEN CYCLE

Nitrogen constitutes 78% by volume of dry air. It is an essential element in all living things. Figure 22.2A illustrates the nitrogen cycle. This is the balance that exists between reactions that take nitrogen out of the air and out of the soil and reactions which put nitrogen into the air and into the soil. Human beings take nitrogen out of the cycle by eating plants, instead of leaving them to decay and return nitrogen to the soil. We take nitrogen out of the cycle by removing the products of excretion instead of letting them return to the soil. We therefore need to make a contribution to the nitrogen that returns to the soil. We do this through the manufacture of ammonia in the Haber process [§ 22.5] and the manufacture of ammonium salts and nitrates to be used as fertilisers.

22.3 THE MEMBERS OF THE GROUP

The elements in Group 5 are nitrogen, phosphorus, arsenic, antimony and bismuth. Nitrogen is the major component of air [see § 22.2]. Phosphorus is a very reactive non-metallic element. It is mined as phosphate ores, e.g. calcium phosphate which is important in the manufacture of NPK fertilisers [§ 22.8]. Arsenic, antimony and bismuth are somewhat metallic in nature.

22.3.1 NITROGEN

For industrial use, nitrogen is obtained by the fractional distillation of liquid air. Nitrogen distils at −196 °C, while the liquid becomes richer in oxygen, which has a boiling temperature of −183 °C.

Gaseous nitrogen is used to provide an inert atmosphere for reactions which cannot be carried out in the presence of oxygen. It is used as a carrier in gas–liquid chromatography.

22.3.2 BOND FORMATION

Nitrogen consists of molecules $N \equiv N$. The low reactivity of nitrogen arises from the strength of the triple bond:

$$N_2(g) \longrightarrow 2N(g); \quad \Delta H^\ominus = +940 \text{ kJ mol}^{-1}.$$

Nitrogen forms compounds in several ways.

1. Nitrogen can accept three electrons to form N^{3-} ions (oxidation state −3). The process is highly endothermic, and ions are formed only if the lattice enthalpies are high, i.e. by Groups 1 and 2.
2. Nitrogen forms covalent bonds by sharing electrons. In NH_3 it shows an oxidation state of −3.
3. Nitrogen can use its lone pair of electrons to form a coordinate bond to an acceptor atom, e.g. H^+, as in NH_4^+ (oxidation state −3) and O in NO_2^- (oxidation state +3) and NO_3^- (oxidation state +5).

Nitrogen does not react with halogens, alkalis, concentrated or dilute acids. It combines with oxygen at high temperature, for instance in lightning storms [Figure 22.2A] and in vehicle engines [§ 26.5]. Nitrogen combines with hydrogen in the Haber process [§ 22.5].

22.4 THE HABER PROCESS

The Haber process for ammonia

Ammonia is made by the Haber process from nitrogen and hydrogen:

$$N_2(g) + 3H_2(g) \rightleftharpoons 2NH_3(g); \quad \Delta H^{\ominus} = -92 \text{ kJ mol}^{-1}$$

The reaction is exothermic, and involves a decrease in the number of moles of gas. An application of Le Chatelier's Principle [§ 11.5.1] shows that the forward reaction should be assisted by a low temperature and a high pressure. At low temperature, the rate of attainment of equilibrium is low. At high temperature, the position of equilibrium is over to the left. A compromise temperature is adopted, and a catalyst is employed to speed up the attainment of equilibrium concentrations. The conditions employed in industrial plants are 200–1000 atm, 500 °C and, as catalyst, iron with aluminium oxide as a promoter. The yield is about 10%, and unreacted gases are recycled [see Figure 22.4A, B].

Figure 22.4A
A flow diagram of the Haber process

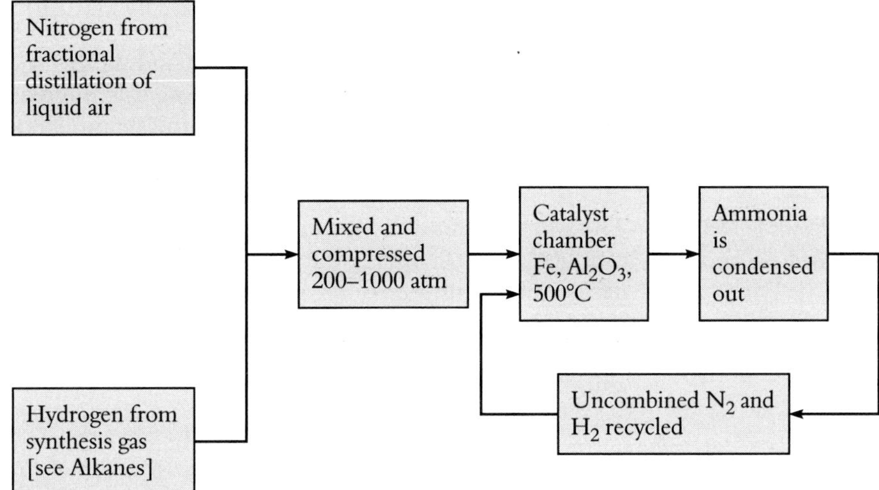

Figure 22.4B
ICI Billingham. The Haber process

CHECKPOINT 22.4: NITROGEN

1. *(a)* State three processes by which nitrogen in the atmosphere is converted into nitrogen compounds in the soil.
   (b) How is nitrogen returned to the atmosphere to keep the level of atmospheric nitrogen constant?
   (c) Why is it necessary to add to the nitrogen content of the soil?

2. Give the oxidation state of nitrogen in *(a)* nitric acid *(b)* nitrous acid *(c)* nitrogen dioxide *(d)* nitrogen monoxide *(e)* ammonia.

3. *(a)* State how nitrogen is obtained industrially.
   (b) Mention one industrial use of nitrogen (apart from the manufacture of ammonia).

22.5 AMMONIA

The laboratory preparation of ammonia from an ammonium salt and a base

In the laboratory, ammonia is made by the action of a strong base on an ammonium salt:

$$NH_4^+(s) + OH^-(s) \longrightarrow NH_3(g) + H_2O(g)$$

The reaction is accelerated by warming, and the gas is collected upwards after passing through a drying tower containing calcium oxide. Acidic drying agents cannot be used, and ammonia reacts with calcium chloride to form complex ions.

The bonds in NH_3 are tetrahedrally arranged

The arrangement of bonds in ammonia is tetrahedral, with the lone pair on the nitrogen atom occupying the fourth apex [see Figure 5.2F, § 5.2.3]. The lone pair enables ammonia to act as a base, accepting a proton to form the ammonium ion, NH_4^+. This is what happens when ammonia and hydrogen chloride come into contact and form ammonium chloride. This reaction is used as a test for ammonia:

The reaction with hydrogen chloride is a test for ammonia

$$H_3N\!: (g) + H{-}Cl(g) \longrightarrow [H_4N]^+\ Cl^-(s)$$

Ammonia is a Lewis base, able to use its lone pair to form a coordinate bond to another molecule [§ 12.6.1].

Ammonia forms hydrogen bonds with water

Ammonia is extremely soluble in water (1300 volumes per unit volume of water) because it can form hydrogen bonds with water. An aqueous solution of ammonia is ionised to a small extent, to form NH_4^+ ions and OH^- ions. The basic dissociation constant, K_b, is 1.8×10^{-5} mol dm^{-3} at 298 K [§ 12.6.4].

Reactions of aqueous ammonia

Basic reactions of NH_3 ...

1. Ammonia neutralises acids to form ammonium salts.

... precipitation reactions ...

2. It precipitates insoluble metal hydroxides from solutions of metal salts. Ammonia is useful for precipitating amphoteric metal hydroxides which dissolve in an excess of a strong alkali (e.g., $Al(OH)_3$, $Pb(OH)_2$).

... complex ion formation

3. Some metal hydroxides dissolve in an excess of ammonia to form soluble ammine complex ions, e.g., $[Ag(NH_3)_2]^+(aq)$ and $[Cu(NH_3)_4]^{2+}(aq)$.

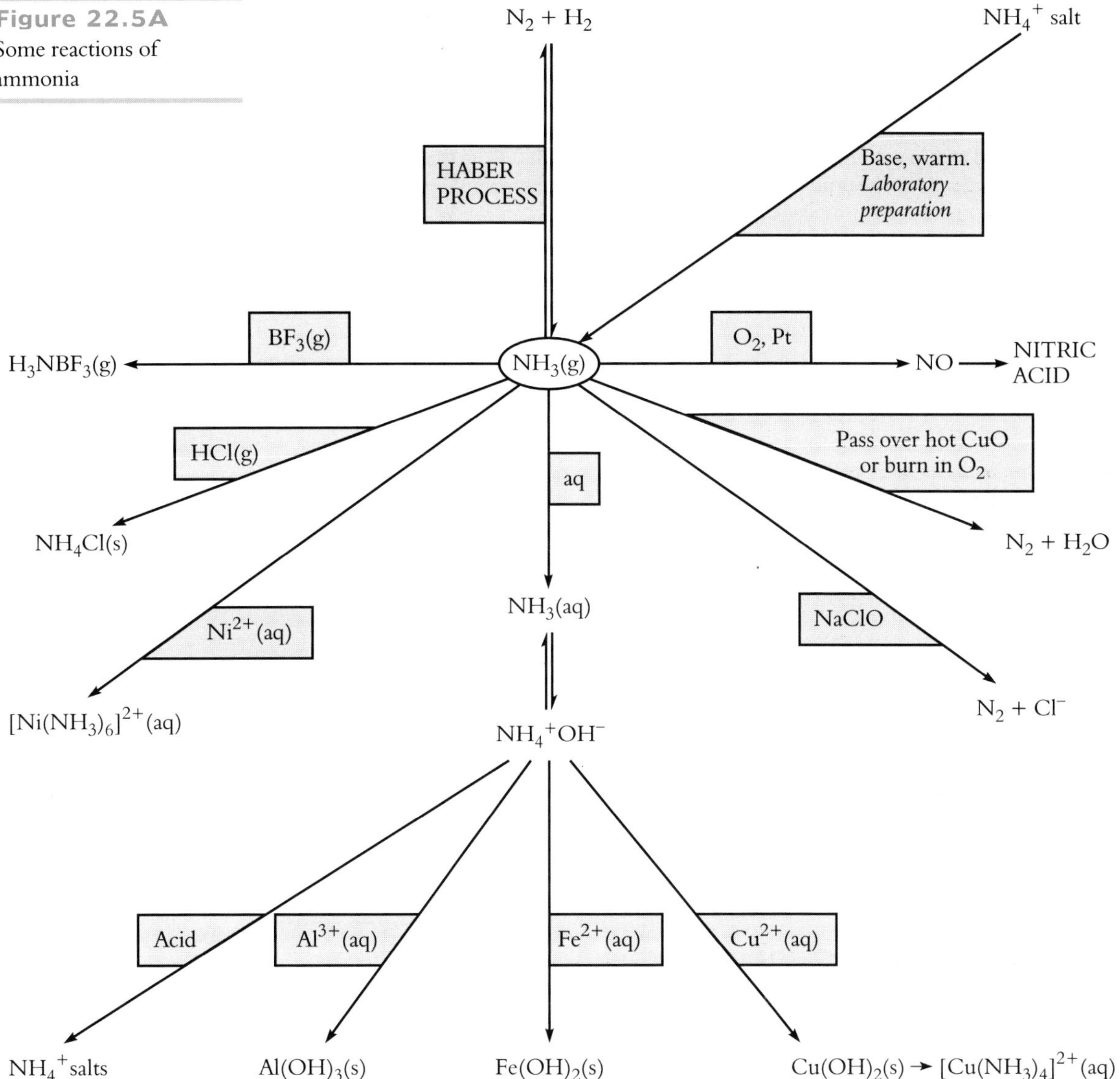

Figure 22.5A Some reactions of ammonia

Ammonium salts

Ammonium salts are important fertilisers [Figure 22.2A]. They are easily decomposed by heat. Ammonium chloride dissociates

$$NH_4Cl(s) \rightleftharpoons NH_3(g) + HCl(g)$$

When the ammonium salts of oxidising acids, e.g. nitric and chromic acid, are heated, the ammonia formed is oxidised to nitrogen. The decomposition can be explosive, e.g. ammonium nitrate

$$NH_4NO_3(s) \xrightarrow{\text{explosive decomposition}} N_2O(g) + 2H_2O(g)$$

A **test for ammonium salts** is to warm the salt with a strong base and test for the evolution of ammonia. If the gas turns red litmus blue and turns aqueous copper(II) sulphate from pale blue to deep blue, it is ammonia. The reaction with a base can be used in analysis to find the amount of ammonium salt present [§ 3.14].

CHECKPOINT 22.5: AMMONIA

1. Describe the industrial manufacture of ammonia. Explain the chemical principles involved.

2. Describe and explain what happens when aqueous ammonia is added slowly, until it is in excess, to solutions of *(a)* copper(II) sulphate, *(b)* copper(I) chloride and *(c)* silver nitrate. What use is made of the final products of *(a)* and *(c)* in organic chemistry?

3. Give examples of a reaction in which ammonia acts as *(a)* a Brönsted–Lowry base *(b)* a Lewis base.

4. Why is the solubility of ammonia in water extremely high?

5. Why is ammonia used as a household cleaner? Why should it not be used together with a bleach?

6. Sketch the shapes of the molecule NH_3 and the ion NH_4^+ and comment on their shapes.

22.6 NITRIC ACID

The industrial method of making nitric acid is the catalytic oxidation of ammonia. A flow diagram for the process, which was invented by Ostwald, is shown in Figure 22.6A. The reactions which take place are

$$4NH_3(g) + 5O_2(g) \xrightarrow[900\,°C]{Pt,\,Rh} 4NO(g) + 6H_2O(l) \quad [1]$$

$$2NO(g) + O_2(g) \longrightarrow 2NO_2(g) \quad [2]$$

$$4NO_2(g) + O_2(g) + 2H_2O(l) \longrightarrow 4HNO_3(l) \quad [3]$$

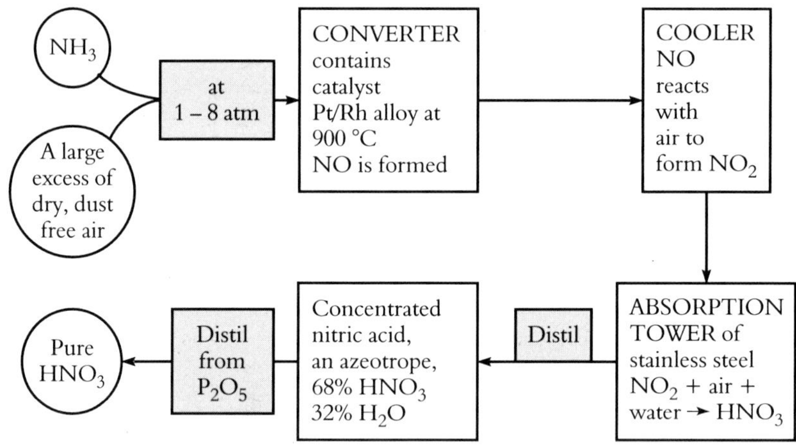

Figure 22.6A A flow diagram for the manufacture of nitric acid

Properties of nitric acid

Pure nitric acid is a colourless liquid which boils at 86 °C. The vapour consists of molecules which have the structure

The structure of HNO_3

H–O–N(=O)=O (delocalised N–O bonds)

Ionisation in the presence of water

In aqueous solution nitric acid is a strong acid:

$$HNO_3(l) + H_2O(l) \rightleftharpoons H_3O^+(aq) + NO_3^-(aq)$$

Reactions of nitric acid

Aqueous nitric acid gives the reactions typical of mineral acids [§ 12.6.1]. Metals react to form nitrates. Since nitric acid is an oxidising agent, hydrogen is rarely formed. Only magnesium and calcium react with cold, dilute nitric acid to give hydrogen:

$$Mg(s) + 2HNO_3(aq) \xrightarrow[\text{dilute}]{\text{cold}} Mg(NO_3)_2(aq) + H_2(g)$$

Nitric acid is an oxidising agent, being reduced to a number of different products

Other metals reduce the nitrate ion in preference to the hydrogen ion. The reduction products formed are water and NO_2, NO, N_2O, H_3N^+OH, NH_4^+, depending on the position of the metal in the electrochemical series, the concentration and the temperature.

The reaction of conc. HNO_3 with metal cations ...

Cations of metals with variable oxidation states (e.g., Fe^{2+}, Sn^{2+}) are oxidised by concentrated nitric acid.

... non-metals ...

Non-metallic elements (e.g., S, P) are oxidised by concentrated nitric acid to the acids corresponding to their highest oxidation states (e.g., H_2SO_4, H_3PO_4).

and anions

Anions oxidised by concentrated nitric acid include Cl^-, Br^-, I^-, S^{2-}.

Uses of nitric acid

Nitric acid is used in the manufacture of some explosives and dyes

Nitric acid is used in the manufacture of organic nitro-compounds [§ 28.8.1]. Some of these are useful explosives (e.g., dynamite and TNT). Others are useful intermediates in the dye industry. A large amount of nitric acid is used in the manufacture of nitrate fertilisers [see § 22.7].

Nitrates

Nitrates are soluble and are decomposed by heat

Nitrates are prepared by the action of nitric acid on metals, metal oxides, hydroxides and carbonates. They all dissolve in water and decompose when heated [see Table 22.6].

Table 22.6 Thermal decomposition of nitrates

Nitrate	Products of thermal decomposition
Group 1 (except Li)	Nitrite + Oxygen: $\mathbf{M}NO_2 + O_2$
Most metals	Oxide + Oxygen + Nitrogen dioxide: $\mathbf{M}O + O_2 + NO_2$
Unreactive metals (Ag, Hg)	Metal + Oxygen + Nitrogen dioxide: $\mathbf{M} + O_2 + NO_2$
Ammonium	Explosive ⟶ Dinitrogen oxide, $N_2O + H_2O$

The brown ring test for a nitrate or a nitrite

One test for a nitrate, the **brown ring test**, requires the addition to a solution of the suspected nitrate of iron(II) sulphate solution and a few drops of concentrated sulphuric acid. The formation of a ring of the brown complex, $FeSO_4 \cdot NO$, proves the presence of a nitrate or a nitrite.

The structure of NO_3^-

The nitrate ion has the planar structure represented:

$$\left[\begin{array}{ccc} O & & O \\ & N & \\ & O & \end{array} \right]^-$$

CHECKPOINT 22.6: NITRIC ACID

1. In the industrial manufacture of nitric acid, name *(a)* the two starting materials *(b)* the two intermediate nitrogen compounds in the process and *(c)* state the catalyst used.
2. State three important classes of compounds made from nitric acid.
3. What reagents must be added to a nitrate to produce a brown ring?

22.7 NPK FERTILISERS

22.7.1 THE NEED FOR FERTILISERS

The importance of nitrogen in the growth of plants was introduced in § 22.2.1 and the nitrogen cycle is shown in Figure 22.2A. Crops take nutrients from the soil, and the supply of natural fertilisers, manure and compost, is insufficient to replenish the soil. To feed the world's growing population, farming has to be intensive, growing large crops on the available agricultural land. Synthetic fertilisers are an essential part of the picture.

Figure 22.7A The manufacture of NPK fertilisers

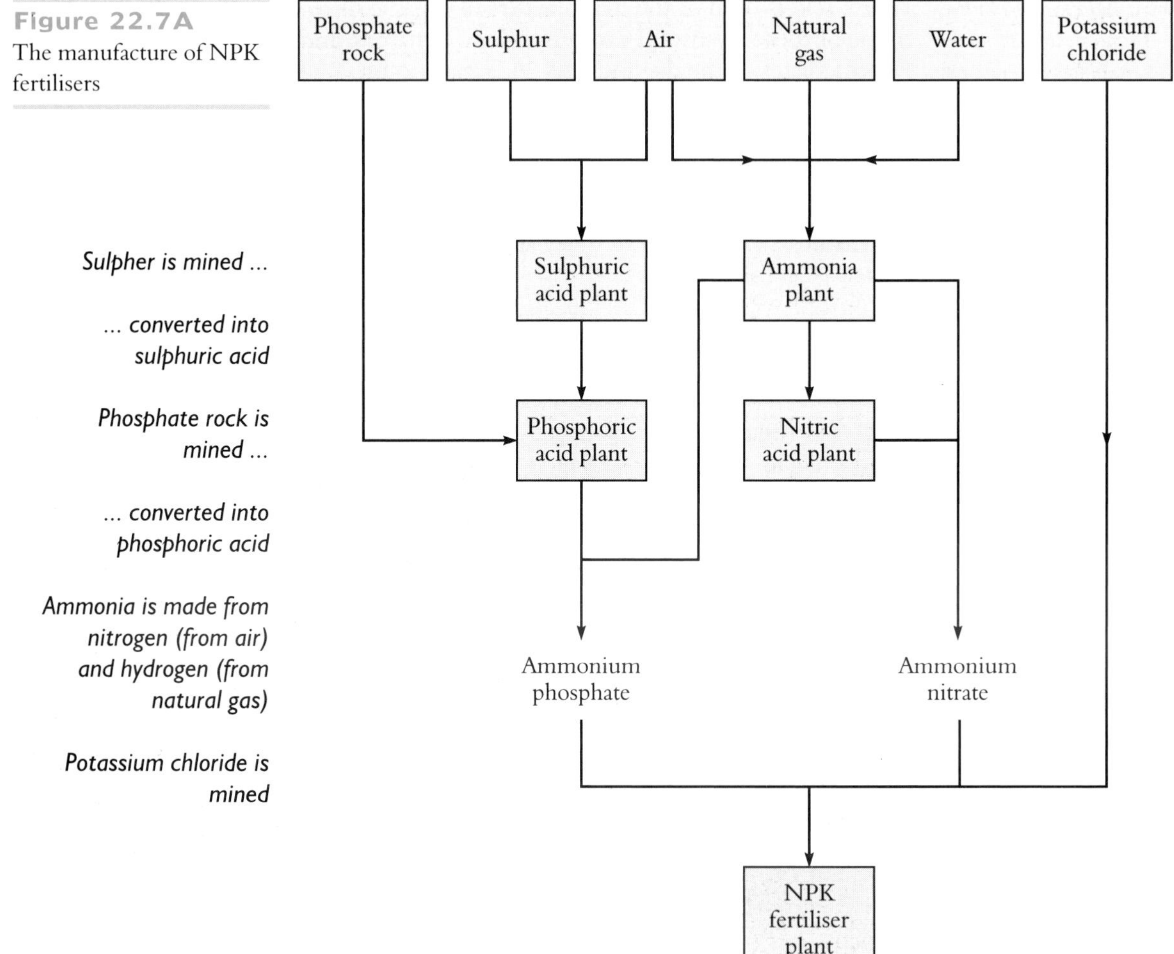

Sulpher is mined ...

... converted into sulphuric acid

Phosphate rock is mined ...

... converted into phosphoric acid

Ammonia is made from nitrogen (from air) and hydrogen (from natural gas)

Potassium chloride is mined

NPK fertiliser: ammonium sulphate or nitrate + ammonium phosphate + potassium chloride

The most popular commercial fertilisers contain nitrogen, phosphorus and potassium and are known as **NPK fertilisers**. Nitrogen is present as ammonium sulphate or nitrate, phosphorus is present as calcium phosphate and potassium as potassium chloride. Big manufacturers make ammonia, nitric acid, sulphuric acid and phosphoric acid on the same site. They mix the products, ammonium phosphate and ammonium nitrate, with potassium chloride which is mined. A range of NPK fertilisers is made by mixing different proportions of the components to suit different soils and crops.

22.7.2 EUTROPHIC LAKES

Lake water is poor in nutrients ...

... supports little plant life ...

... and is rich in oxygen ...
... so fish can thrive

Lakes age slowly through receiving plant nutrients

Natural lake water is poor in nutrients and supports little plant life. With no decaying vegetation to use up dissolved oxygen, the lake water is saturated with oxygen, and fish can thrive. Under natural conditions, streams flow into the lake bringing nutrients which are formed from the decay of plant and animal refuse on land. The lake water is gradually enriched with nutrients, a process called **eutrophication**. Water plants grow more vigorously, animal populations increase, then as plants and animals die deposits of organic matter on the bottom of the lake build up. Over a long period of time, the lake becomes gradually more shallow until eventually it is completely filled in. This natural process of **ageing** takes thousands of years.

Human activities may accelerate ageing ...

... accidentally enriching lake water with nutrients ...

... through leaching of fertilisers from the surrounding land ...

... in a process called eutrophication ...

The time scale is very different when plant nutrients are fed into a lake by human activities. When lake water is enriched with nitrates and phosphates, algae flourish, and produce an **algal bloom**, a green scum with an unpleasant smell. When algae die they are decomposed by **aerobic bacteria**. When the oxygen content falls too low to support aerobic bacteria, **anaerobic bacteria** take over. They convert the dead matter into unpleasant-smelling decay products and debris which falls to the bottom. Gradually a layer of dead plant material builds up on the bottom of the lake. The lowering of the oxygen concentration leads to the death of fish, which is a disaster for fishermen. The mats of algae ruin the lake for recreational use by swimmers, sail-boarders, dinghy-sailors, rowing boats and yachts.

Figure 22.7B
Algal bloom

A eutrophic lake supports plant growth ...

... algal bloom ...

... the buildup of dead matter on the bottom of the lake ...

... and the death of fish

The sources of the nitrates and phosphates are sewage and fertilisers. Intensively cultivated land receives generous applications of fertilisers containing nitrates and phosphates. Plants can absorb only a limited quantity of nitrate through their roots. The rest is leached out of the soil by rain. Nitrates are very soluble, phosphates are sparingly soluble and are leached from the soil more slowly than nitrates. Another source of phosphates is household detergents.

22.7.3 GROUNDWATER

The level of nitrates in groundwater is rising ...

Excessive fertiliser can also find its way into **groundwater**. This is the water held underground in porous layers of rock. One third of UK drinking water comes from groundwater. The World Health Organisation recommends that the level of nitrogen in the form of nitrates should not exceed 50 ppm. The average level in the UK is far short of this value (about 11 ppm), but it is increasing, and in parts of the country where the level is highest there is concern. There are two reasons for this concern.

... causing concern over the possibility of nitrite poisoning and carcinogenic nitrosamines

1. Increased nitrate levels increase the risk of 'blue baby syndrome'. This rare form of anaemia affects babies below 6 months of age. The cause is the oxidation by nitrite ions of Fe^{2+} in haemoglobin to Fe^{3+}. The oxidised haemoglobin cannot bind oxygen [§ 24.16.5], and the baby turns blue from lack of oxygen. Conditions in the digestive tracts of very young children are more favourable to the bacteria which reduce nitrates to nitrites than those in adults.
2. Another potential hazard is that carcinogenic nitrosoamines may be formed in the human digestive tract by the conversion of nitrate into nitrite and the reaction of nitrite with amino acids.

22.7.4 TACKLING THE PROBLEM

The use of the correct amount of fertiliser in the growing season will reduce pollution

Nitrates and ammonium compounds are very soluble and easily washed off fields into groundwater. As a result, the availability of fertiliser decreases with time and may be insufficient when the weather is warm and the plants are growing fast. Fertiliser manufacturers are trying to solve the problem by adding a coating to granules of fertiliser to allow the nutrients to be released slowly over the growing season. Farmers can seek advice from agricultural chemists on how much and what kind of fertiliser to use on their ground for the crop they want to grow. The UK uses about 7 million tonnes of NPK fertilisers a year, at a cost of about £80/tonne. Applying the correct amount of fertiliser in the right season will save farmers money as well as keeping the groundwater safe.

CHECKPOINT 22.7: WATER POLLUTION

1. What factor is it that most frequently limits the rate of plant growth?
2. Why is a lake a poor environment for plants and a good environment for fish?
3. Describe the conditions under which a lake becomes more able to support plant growth. Explain why this does not benefit the environment.
4. Why are nitrates used as fertilisers? What measures can be taken to prevent nitrates finding their way into lakes and groundwater?

QUESTIONS ON CHAPTER 22

1. Describe the Haber process for the manufacture of ammonia. State the sources of the raw materials used. Explain the reasons for the conditions employed by using the principles of kinetics and equilibrium.

2. Explain why nitrogen compounds are added to farmland. Find, by calculation, which is the best source of nitrogen: sodium nitrate, ammonium nitrate, ammonium sulphate or urea, CH_4N_2O.

3. What is the importance of nitrogen monoxide in relation to *(a)* the ozone layer [§ 21.5] and *(b)* photochemical smog [§ 27.11]?

4. *(a)* What part do oxides of nitrogen play in the formation of acid rain?
(b) How can the emission of oxides of nitrogen be reduced? [see §§ 21.5 and 26.5].

5. A laboratory method of finding the percentage of ammonia in a fertiliser involves boiling the fertiliser with an excess of aqueous sodium hydroxide until no more ammonia is given off. The excess of sodium hydroxide is found by titration with standard hydrochloric acid.

(a) Write an equation for the reaction between ammonium ions and hydroxide ions to give ammonia.
(b) Say how you could test for ammonia gas.
(c) In an experiment 10.0 g of fertiliser need 0.0500 mol of sodium hydroxide. Calculate the percentage of ammonium ion in the fertiliser.

6. *(a)* Explain how nitrogen is 'fixed' on an industrial scale.
(b) Name three nitrogen compounds and one phosphorus compound that are used as fertilisers.
(c) Mention two environmental problems that are associated with their large-scale use.

7. The fertiliser 'Cropinc' contains the compounds potassium chloride, ammonium nitrate and ammonium phosphate.

(a) Describe how you could test the fertiliser for (i) potassium ions (ii) chloride ions (iii) ammonium ions (iv) nitrate ions.
(b) Suggest two reasons why a farmer might use 'Cropinc' on his fields.
(c) Sometimes excess acidity in fields needs treatment by 'liming'. What measurement would a farmer make to decide whether liming is needed? Write an equation to show how lime (calcium hydroxide) neutralises excess acidity.
(d) What is the disadvantage of 'liming' at the same time as applying 'Cropinc'?

8. *(a)* Ammonia is manufactured by direct synthesis in the Haber process:

$$N_2(g) + 3H_2(g) \rightleftharpoons 2NH_3(g) \qquad \Delta H = -92 \text{ kJ mol}^{-1}$$

(i) Write an expression for the equilibrium constant, K_c, for this reaction and give its units.
(ii) When 3 mol of hydrogen and 1 mol of nitrogen were allowed to reach equilibrium in a vessel of 1 dm^3 capacity at 500 °C and 1000 atm pressure, the equilibrium mixture contained 0.27 mol of N_2, 0.81 mol of H_2 and 1.46 mol of NH_3. Calculate K_c at this temperature. **4**

(b) Predict and explain the effect of an increase in temperature on:
(i) the value of K_c
(ii) the rate of the forward reaction. **6**

(c) When ammonium salts are dissolved in water, the following reaction occurs.

$$NH_4^+(aq) + H_2O(l) \rightleftharpoons NH_3(aq) + H_3O^+(aq)$$

(i) Identify the acid/base conjugate pairs in this reaction.
(ii) Write an expression for the dissociation constant, K_a, for $NH_4^+(aq)$.
(iii) Calculate the pH of a solution of ammonium chloride of concentration 0.100 mol dm^{-3} at 298 K, the K_a value for NH_4^+ being 5.62×10^{-10} mol dm^{-3} at this temperature. **5**

(d) How would you modify the solution of ammonium chloride in *(c)*(iii) to make it more resistant to changes in pH when small amounts of acid or base are added to it? **1**

16 *L (AS/AL)*

9. One of the major ammonia-producing plants in this country is located at Billingham on the North Sea coast. The source of the hydrogen needed is methane, CH_4, although water has been used in the past and could become important again in the future.

The synthesis gas from which ammonia is produced has a typical composition of 74.2% hydrogen, 24.7% nitrogen, 0.8% methane, 0.3% argon.

The gas is compressed to about 200 atmospheres pressure and passed over a catalyst at a temperature of 450 °C. Under these conditions about 15% of the synthesis gas is converted into ammonia.

$$N_2(g) + 3H_2(g) \rightleftharpoons 2NH_3(g) \qquad \Delta H = -92.0 \text{ kJ mol}^{-1}$$

On leaving the converter, ammonia is removed from the unreacted gases.

Some ammonia is used to produce nitric acid. A further reaction between ammonia and nitric acid gives ammonium nitrate, most of which is used as fertiliser. Since ammonium nitrate is potentially explosive, great care has to be taken in its manufacture.

(a) (i) Suggest **one** reason why an ammonia plant is located at Billingham. **1**

(ii) Under what circumstances could water *become important again in the future* as a source of hydrogen? **1**

(iii) Explain the presence of argon in the synthesis gas. **1**

(b) Explain **one** advantage and **two** disadvantages of using high pressure for the production of ammonia. **3**

(c)

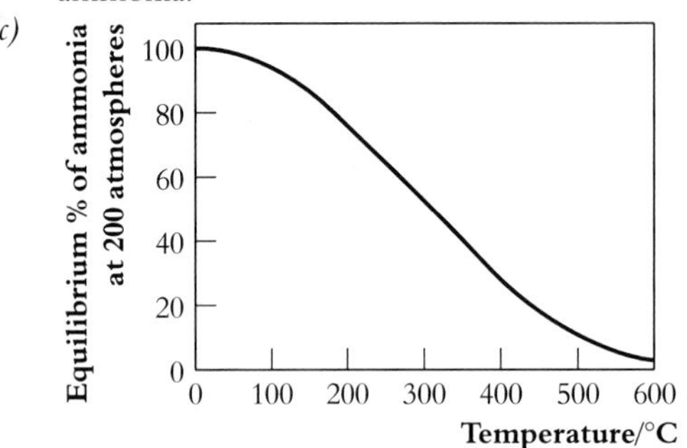

(i) By referring to the figure, discuss the choice of 450 °C as a temperature and the use of a catalyst for the conversion. **4**

(ii) Give the name of the catalyst used and the type of catalysis occurring. **2**

(d) Describe:

(i) how ammonia is removed from the unreacted gases **2**

(ii) what happens to the unreacted gases. **1**

(e) Write equations for the reactions in which:

(i) ammonia is oxidised to give nitrogen monoxide (NO) and water as the first stage of nitric acid production **2**

(ii) ammonia reacts with nitric acid. **1**

(f) (i) Explain why an ammonium nitrate plant is located on the same site as the ammonia production unit in Billingham. **3**

(ii) State a safety precaution which would be taken in and around an ammonium nitrate plant. **1**

(g) Ammonia is used to manufacture 1,6-diaminohexane, which reacts with hexanedioic acid to form a polymer.

$H_2N(CH_2)_6NH_2$	$HOOC(CH_2)_4COOH$
1,6-Diaminohexane	Hexanedioic acid

Give the name of the polymer formed, draw the graphical formula of the repeating unit and classify the type of polymerisation reaction. Give **two** uses of the polymer. **6**

For *(g)* see § 34.5.4.

AEB (AL) 1998

10. One of the elements needed by plants for growth is nitrogen. Farmers add nitrogen-containing compounds, such as ammonium salts, to replace nitrogen in the soil.

Reaction of an ammonium salt with excess aqueous sodium hydroxide produces ammonia which is evolved on boiling.

$$NH_4^+(aq) + OH^-(aq) \longrightarrow NH_3(g) + H_2O(l)$$

To determine the percentage of nitrogen in an ammonium salt, 0.500 g of the salt was dissolved in 50.0 cm^3 of sodium hydroxide solution containing 0.300 mol dm^{-3}. After expelling all the ammonia gas by boiling the mixture, the excess sodium hydroxide solution required 25.0 cm^3 of hydrochloric acid, containing 0.100 mol dm^{-3}, for neutralisation. [A_r(N) 14.0; A_r(H) 1.0; A_r(O) 16.0; A_r(S) 32.0; A_r(C) 12.0]

(a) Suggest why it is advisable to boil the ammonium salt with alkali in a fume cupboard. **1**

(b) Write the ionic equation for the reaction between sodium hydroxide and hydrochloric acid. **1**

(c) (i) Calculate the number of moles of hydroxide ions in the 50.0 cm^3 of solution taken. **1**

(ii) Calculate the number of moles of excess hydroxide ions. **1**

(iii) Calculate the number of moles of hydroxide ions which have reacted with the ammonium salt. **1**

(d) Determine the percentage of nitrogen, by mass, in the ammonium salt. **3**

(e) The following table contains information about two nitrogen-containing compounds. Copy and complete the table to determine which of these nitrogen-containing compounds would provide the best value for money for a farmer. **4**

Formula	Percentage by mass of nitrogen	Cost per tonne of compound	Cost to provide one tonne of nitrogen
$(NH_4)_2SO_4$		£3800	
$CO(NH_2)_2$		£5500	

12 *O&C*

23
GROUP 4

23.1 SILICON THE SEMICONDUCTOR

Everyone has heard of the silicon chip. Before the invention of the silicon chip, a computer was the size of a large room. Now a computer is small enough to occupy a briefcase or even less space. The development was made possible by the properties of silicon. As well as being a member of Group 4, silicon is a semiconductor.

Semiconductors are a special class of materials which have an electrical resistivity between those of electrical conductors and electrical insulators. They all conduct electricity better as the temperature rises; that is, the resistivity falls as the temperature rises.

Energy bands are used to explain the conduction of electricity by semiconductors. The electrons in the shell of an isolated atom have discrete energy levels [see § 2.3.1]. Those in the highest energy level are the valence electrons, which are used in bond formation. When atoms are present in a solid structure, the vibration of the atoms about their mean positions causes the energy levels to spread out into narrow bands. The band containing the valence electrons is called the **valence band**. In metals, electrons in the valence band can move easily to an unfilled energy level in a band of higher energy. This band is called the **conduction band**. Electrons in the conduction band are delocalised; they have broken free from individual atoms and are able to move through the material when a potential difference is applied across it.

In metals there is no energy gap between the valence band and the conduction band. In insulators the energy gap is too great (about 5 eV) to allow electrons to move between the bands. In semiconductors, the energy gap is smaller (about 1 eV), and the chance of electrons jumping the gap increases as the temperature rises [see Figure 23.1A].

Semiconductors include both elements and compounds. The semiconductors which are pure elements and compounds are described as **intrinsic semiconductors**. Silicon is the intrinsic semiconductor which is used for the production of

Figure 23.1A Valence bands and conduction bands

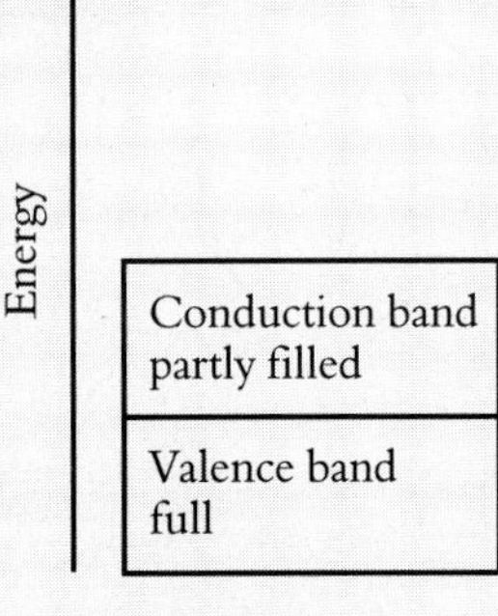

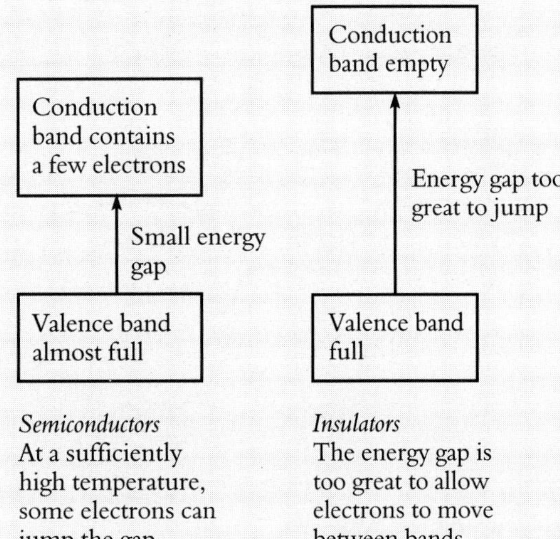

Metals
There is no energy gap between valence and conduction bands.

Semiconductors
At a sufficiently high temperature, some electrons can jump the gap.

Insulators
The energy gap is too great to allow electrons to move between bands.

the integrated circuits known as **silicon chips**. The electron configuration of silicon is (2.8.4). As each atom forms four tetrahedrally directed covalent bonds to other silicon atoms, a macromolecular crystal structure in three dimensions is built up.

At absolute zero, all the electrons are localised in the valence band, and the material is an electrical insulator. When heat is supplied to the piece of silicon, some of the electrons gain enough energy to jump from the valence band into the conduction band. These delocalised electrons can carry charge through the material when a potential difference is applied across it. As the temperature rises, more electrons jump into the conduction band, and the resistance of the material falls.

Extrinsic semiconductors are made by adding certain substances to intrinsic semiconductors. This can be done by exposing the semiconductor to the vapour of the substance to be added in a furnace. The chosen substance is added in carefully controlled amounts to bring its content up to only a few parts per million. The process is called **doping**, and the added substances are called **dopants**. The dopants are chosen to produce the required change in the electrical properties.

There are two types of extrinsic semiconductors: **n-type** and **p-type**. In n-type semiconductors, the dopants are Group 5 elements, such as phosphorus and antimony, which have five electrons in the outer shell. When a dopant atom replaces an atom of silicon in the structure, it uses four of its five valence electrons to form covalent bonds with silicon atoms. The fifth electron is supplied to the material [see Figure 23.1B], creating a negative charge. This is why such semiconductors are called n-type (n for negative). The added atoms are called **donor atoms** because they donate electrons to the material. The possession of delocalised electrons makes the extrinsic semiconductor a much better electrical conductor than the parent intrinsic conductor. The addition of donor atoms leaves the crystal uncharged overall because the additional electrons are associated with additional positive charges on the donor nuclei.

In p-type semiconductors, the dopants are Group 3 elements, such as boron and aluminium, which have three electrons in the outer shell. When a dopant atom replaces a silicon atom, it forms three electron-pair bonds with three silicon atoms, but the fourth bond is incomplete: it has only one electron. The vacancy is called an **electron hole** and is positively charged. This is why this type of semiconductor is called p-type (p for positive). The added atoms are called **acceptor atoms** because they can accept electrons to fill the 'holes' in the bonds.

In p-type semiconductors, conduction takes place by means of electrons jumping across from bonds into holes [see Figure 23.1C]. The electron which jumps from a bond into a hole creates a hole in the bond which it left. Another electron jumps from a bond into this hole, and so it goes on across the structure.

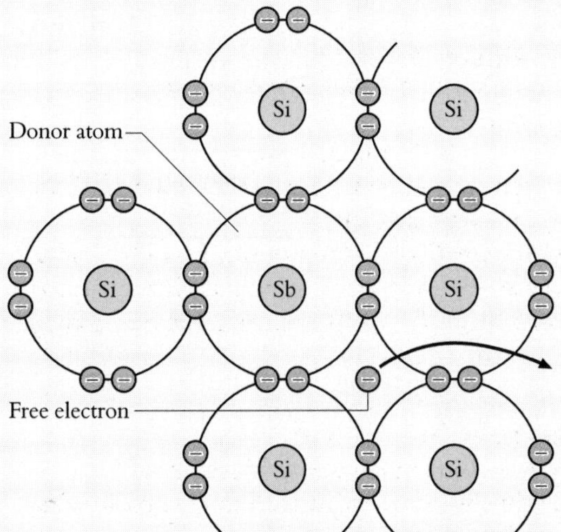

Donor atom
Free electron
Si
Sb

Figure 23.1B
n-type extrinsic semiconductor

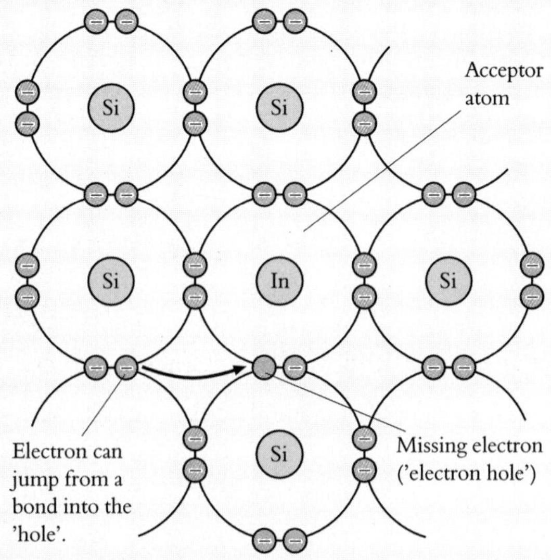

Figure 23.1C
p-type extrinsic semiconductor

23.2 A COMPARATIVE LOOK AT GROUP 4

There is a non-metal to metal transition down Group 4 ...

... This is illustrated by the reactions with dilute acid and with conc. nitric acid

In Group 4, the differences between the first member of the group and the last have reached a maximum. There is a transition from carbon and silicon, which are non-metallic elements, through germanium, which is intermediate in character, a **metalloid**, to tin and lead which are metals. Carbon and silicon have covalent, macromolecular structures. Carbon (except for graphite), is a non-conductor. Silicon and germanium are semiconductors. Tin and lead have metallic structures and are conductors. The reactions with dilute acids and with concentrated nitric acid illustrate the non-metal to metal transition:

Dilute acid + C, Si, Ge – no reaction
Dilute acid + Sn – reacts
Dilute acid + Pb – too low in the electrochemical series to react
Concentrated nitric acid + C, Si, Ge, Sn $\longrightarrow$ hydrated oxides
Concentrated nitric acid + Pb $\longrightarrow$ metal nitrate

Valencies of 2 and 4 are met in the group

The members of the group show valencies of 4 or 2. Passing down the group, there is an increase in the tendency to use a valency of 2 and an increase in the electrovalent character of the bonds.

There is a transition from covalency to electrovalency down the group

Sn and Pb show a valency of 2

C, Si, Ge:	covalent compounds, almost exclusively 4-valent
Sn:	covalent +4 and ionic +2 states are formed with almost equal ease, but Sn^{2+} is a reducing agent
Pb:	mainly ionic +2 state; also covalent +4 state. Pb(IV) is an oxidising state.

The 'inert pair effect' favours a valency of 2

The behaviour of later members of a group, showing a valency of 2 less than the group valency through a failure to use their s electrons, is called the '**inert pair effect**'. Tin(II) compounds are reducing agents because tin(IV) is the more stable oxidation state of tin, but lead(IV) compounds are oxidising agents because lead(II) is the stable oxidation state for lead.

Oxides increase in basicity down the group

The oxides increase in basicity down the group, and the +2 oxides are more basic than the +4 oxides. The oxides of carbon and silicon are acidic; those of germanium, tin and lead are amphoteric:

CO_2	dissolves in water to form a weak acid
CO	reacts with molten NaOH $\longrightarrow$ sodium methanoate
SiO_2	reacts with a molten base $\longrightarrow$ a silicate
SnO_2, SnO	react with dilute acid $\longrightarrow$ Sn^{4+} and Sn^{2+} salts and react with a molten base or concentrated aqueous alkali to form a stannate(IV) or a stannate(II).
PbO_2, PbO	react with dilute acids $\longrightarrow$ Pb(IV) compounds and Pb^{2+} salts and with a molten base or concentrated aqueous alkali to form a plumbate(IV) or a plumbate(II). PbO is more basic than PbO_2.

Tetrahalides are covalent

The tetrahalides $\mathbf{EX_4}$ are

CX_4	covalent and stable to hydrolysis
SiX_4	covalent and readily hydrolysed to silicates
SnX_4 and PbX_4	predominantly covalent, hydrolysed to give basic salts or stannates(IV) or plumbates(IV)

Dihalides are ionic

The dihalides $\mathbf{EX_2}$ are

SnX_2 and PbX_2 formed by tin and lead, predominantly ionic.

23.3 SPECIAL FEATURES OF CARBON CHEMISTRY

1. Carbon is restricted to the $n = 2$ shell of electrons. Carbon, with electron configuration $1s^2 2s^2 2p^2$ can form 4 covalent bonds by promoting an electron from a 2s orbital to a vacant p orbital.

	1s	2s	2p
C (ground state)	↑↓	↑↓	↑ ↑ _
C* (excited state)	↑↓	↑	↑ ↑ ↑

The absence of empty d orbitals explains why carbon halides are stable to hydrolysis whereas silicon halides are readily hydrolysed [23.5.2].

2. Carbon can **catenate**. **Catenation** is the ability of an element to form bonds between atoms of the same element. Consider an element **E**, exposed to air and water and the possibility of forming **E**—O bonds. If the **E**—O bond is stronger than the **E**—**E** bond, then energy considerations will make the element form compounds containing **E**—O bonds, rather than **E**—**E**—**E** chains. For silicon, energy considerations favour a chain

   —Si—O—Si—O—Si—

   as in silica and silicates. For carbon energy considerations favour a chain

   —C—C—C—C—

   as in hydrocarbons. Its ability to catenate enables carbon to form a huge number of hydrocarbons. Silicon forms only a few silanes, e.g. SiH_4, which are spontaneously flammable in air.
3. Carbon is the only member of the group which forms π bonds [§ 5.3.4]. This is why carbon can form multiple bonds. There are >C=C< double bonds in alkenes and arenes, >C=O bonds in aldehydes, ketones and carboxylic acids, —C≡C— triple bonds in alkynes and —C≡N bonds in nitriles. These compounds involve π bonds.
4. Carbon forms gaseous oxides, CO and CO_2, in contrast to the other members. For carbon energy considerations favour the formation of two double bonds in O=C=O rather than four C—O bonds. In silicon, the formation of four Si—O bonds is preferred on energy grounds and silicon forms polymeric $(SiO_2)_n$.

23.4 SOURCES AND USES OF GROUP 4 ELEMENTS

23.4.1 CARBON

Diamond

Carbon has two well-known allotropes, diamond and graphite, and also the allotropes known as 'bucky balls' [§ 35.1]. The structures of diamond and graphite have been described in §§ 6.6 and 6.7 and shown in Figures 6.6A and 6.7A.

Diamonds are mined in Brazil and South Africa. They are prized as jewellery because of their high reflectivity and the high refractive index which enables them to split white light into its different colours. Diamond is the hardest naturally occurring substance, and industry finds many applications for diamond-edged tools in grinding, cutting, and sharpening. The bits of oil-rig drills are studded with small diamonds, which enable them to drill through up to 7000 metres of hard rock. Industrial diamonds are made by subjecting graphite to high temperatures and pressures. e.g. 2000 °C, 10^5 atm, with a catalyst.

Graphite

Some countries have natural supplies of graphite. Others manufacture it by heating coke and sand in an electric furnace.

Graphite is used for making electrodes and in the construction of nuclear reactors. It is used as a lubricant and as a mixture with clay in 'lead' pencils.

Impure forms of carbon include charcoal, coal and coke.

23.4.2 SILICON

Silicon occurs naturally as silicon(IV) oxide (silica), SiO_2, in sand and quartz and also as a number of silicates. The element is extracted by reduction of the oxide with carbon (coke). The importance of silicon as a semiconductor was described in § 23.1.

23.4.3 GERMANIUM

Germanium is a semiconductor and is used in transistors.

23.4.4 TIN

Tin is mined as tinstone, SnO_2. The metal is obtained by reducing the oxide with coke.

Tin is used to plate iron to prevent it from rusting [§ 24.17.6] and as alloys such as bronze (tin and copper) and solder (tin and lead). Both tin and lead have low melting temperatures, and solders have melting temperatures lower than either of the pure metals.

23.4.5 LEAD

Lead is mined as galena, PbS. The ore is roasted to form the oxide PbO, which is reduced to the metal by coke.

Lead was for centuries used in plumbing, but now that it is known that lead slowly reacts with water to form poisonous lead compounds, copper pipes and plastic pipes are used instead. It is used to make solder (tin and lead) and lead–acid accumulators [§ 13.5.3]. Lead is used as a shield from radioactivity because of its high density [§ 1.10.5]. Antiknock, tetraethyllead, has been used for many years to improve petrol, but its use is being phased out because it leads to pollution of the air with lead compounds [§ 26.5]. Some lead compounds are used as pigments, e.g. lead chromate(VI) (yellow), lead carbonate hydroxide (white) and trilead tetraoxide Pb_3O_4 (orange).

CHECKPOINT 23.4: STRUCTURE AND REACTIONS

1. *(a)* How does the structure of diamond explain its hardness and brilliance?
(b) What industrial use is made of diamond?

2. How does the structure of graphite explain *(a)* its ability to mark paper *(b)* its use as a lubricant and *(c)* its use as an electrical conductor?

3. *(a)* How does silicon with the electron configuration (Ne) $3s^2 3p^2$ form four covalent bonds?
(b) Why is silicon(IV) oxide a solid of high melting point?
(c) Why are high pressure and high temperature employed to convert graphite into diamond? Note that diamond has a higher density than graphite.
(d) Why does lead frequently employ a valency of 2, rather than the group valency of 4?
(e) Which compound has more covalent character $PbCl_2$ or $SnCl_2$?

23.5 THE COMPOUNDS OF GROUP 4

23.5.1 THE HYDRIDES

Carbon forms alkanes, alkenes, alkynes and arenes [§§ 25.2–6]. Silicon forms silanes, hydrides of formula Si_nH_{2n+2} ($n = 1–10$), which are so unstable that they inflame spontaneously in air. The hydrides of tin and lead are very unstable.

23.5.2 THE HALIDES

Stability of $\mathbf{E}X_4$ decreases down the group and decreases from F to I

All the elements of Group 4 form tetrahalides of formula $\mathbf{E}X_4$, where X = F, Cl, Br or I, except for $PbBr_4$ and PbI_4. All the tetrahalides are volatile covalent compounds, except for SnF_4 and PbF_4. Except for tetrachloromethane and tetrafluoromethane, the tetrahalides are hydrolysed in solution:

$$SiCl_4(l) + 2H_2O(l) \longrightarrow SiO_2(s) + 4HCl(g)$$

Table 23.5A summarises the preparations of the halides.

Hydrolysis

CCl_4 is stable to hydrolysis

The hydrolysis of $SiCl_4$ employs 3d orbitals

Although silicon tetrachloride is hydrolysed with ease, tetrachloromethane is inert to hydrolysis. The reason for the difference in reactivity is that a different mechanism of hydrolysis operates in the two cases. It is believed that, in the hydrolysis of silicon tetrachloride, the first step is the attack by the negatively charged oxygen atom in a water molecule or hydroxide ion on the positively charged silicon atom. Silicon can use an unoccupied 3d orbital to accommodate the lone pair from the oxygen atom, forming a short-lived intermediate which dissociates to form hydrogen chloride and silicon trichloride hydroxide,

Table 23.5A
Halides of Group 4

Halide	Preparation	Properties
CCl_4	$Cl_2 + CS_2$, Fe catalyst $\longrightarrow CCl_4 + S_2Cl_2$	Uses [§ 29.5] No reaction with water.
$\mathbf{E}Cl_4$ (except CCl_4)	$\mathbf{E} + Cl_2$, heat $\mathbf{E}O_2 + HCl(g)$ $\mathbf{E}O_2 + HCl(aq, conc.)$	Hydrolysed: $SiCl_4 \longrightarrow SiO_2 + HCl(g)$ $\mathbf{E}Cl_4 \longrightarrow \mathbf{E}(OH)_6^{2-}$ ($\mathbf{E}$ = Ge, Sn, Pb) $PbCl_4$ must be kept below 5 °C or it dissociates $\longrightarrow PbCl_2 + Cl_2$
$SnCl_2$	Sn + HCl(g), heat Sn + HCl(aq, conc.)	Covalent when anhydrous; in solution, $Sn^{2+}(aq)$ is formed; partially hydrolysed to Sn(OH)Cl(s). Reducing agent [§ 32.6.2]
$PbCl_2$	$Pb + Cl_2(g)$, heat Pb + HCl(aq, conc.)	Ionic The reaction is possible because the insoluble $PbCl_2$ is converted by conc. HCl into soluble $PbCl_4^{2-}(aq)$.
PbX_2	Precipitation	Ionic

$SiCl_3OH$. Repetition of these steps gives hydrated silicon(IV) oxide:

$$\overset{\delta-}{Cl}\!-\!\overset{\delta+}{Si}Cl_3 \;+\; \overset{\delta-}{O}(\overset{\delta+}{H})_2 \longrightarrow \left[Cl_3Si\!-\!OH_2\right] \quad (\text{i.e. } [SiCl_4(OH_2)])$$

$$\left[SiCl_4(OH_2)\right] \longrightarrow Si(OH)Cl_3 + HCl(g)$$

$$Si(OH)Cl_3 \longrightarrow Si(OH)_4 \text{ or } SiO_2\,.\,xH_2O$$

Since carbon cannot expand its octet because its empty 3d orbitals are too different in energy from the 2p orbitals to come into play, a similar mechanism cannot operate for the hydrolysis of tetrachloromethane. This must proceed by a mechanism with a much higher activation energy and takes place extremely slowly.

CHECKPOINT 23.5A: HALIDES

1. *(a)* Why is an aqueous solution of tin(II) chloride cloudy? How can a clear solution be made?
   (b) What happens when this solution reacts with NaOH(aq)?

2. *(a)* Outline the methods used in the preparation of Group 4 chlorides from the elements.
   (b) Discuss the trends in valency and bond type shown by the chlorides.

3. *(a)* Suggest three methods for the preparation of $PbCl_2$.
   (b) Why is $PbCl_2$ very soluble in concentrated hydrochloric acid?

4. How does CCl_4 differ from $SiCl_4$ in its hydrolysis?

23.5.3 THE OXIDES

The basic character of the oxides increases as the group is descended. The +4 oxides are more acidic than the +2 oxides.

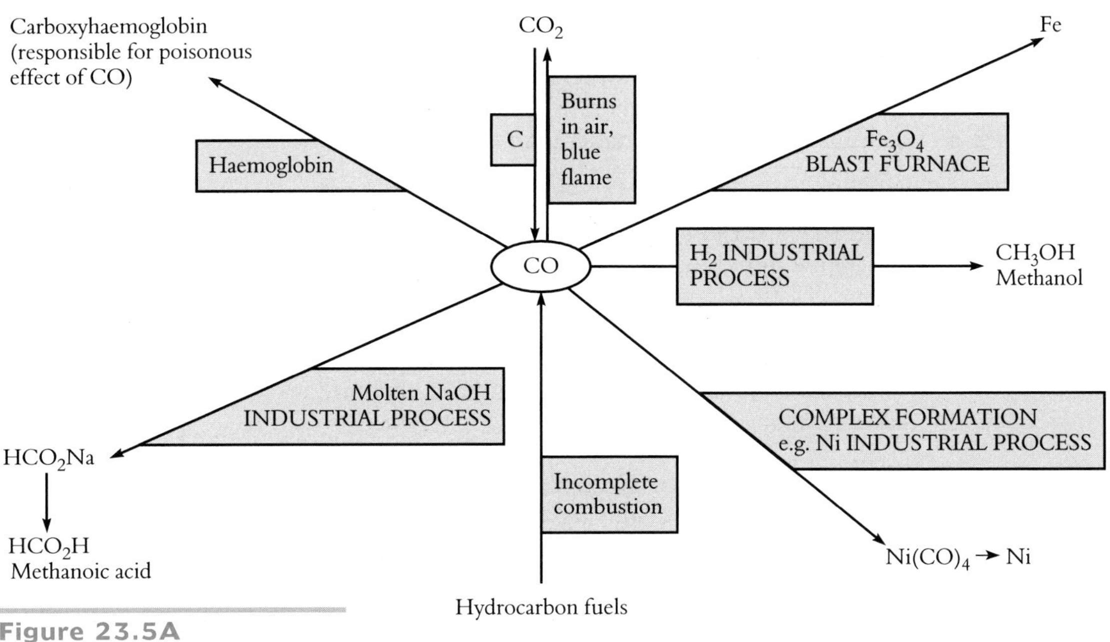

Figure 23.5A
Some reactions of carbon monoxide

Carbon monoxide, CO

CO is produced by petrol engines

It is a poisonous gas …

Carbon monoxide is formed when carbon and hydrocarbons burn incompletely. It is present in the exhaust fumes of petrol-driven vehicles [§ 26.4.1]. At concentrations above 0.1% carbon monoxide is poisonous [§ 24.16.5]. It is the more dangerous for being colourless and odourless.

… and a reducing agent

Carbon monoxide is an important reducing agent. It is used for the reduction of iron ores [§ 24.17.1].

… electronic structure …

… and complex formation

The electronic structure of carbon monoxide can be represented as

$$:C \equiv O:$$

The lone pair of electrons on the carbon atom enable it to act as a ligand in the formation of carbonyl complexes, e.g., $[Cr(CO)_6]^{3+}$ and $Ni(CO)_4$ [§ 24.10].

Carbon dioxide, CO_2

Photosynthesis

Carbon dioxide is present in air at a level of 0.03% by volume. Plants use carbon dioxide in the process of photosynthesis and both plants and animals evolve carbon dioxide in respiration. The balance of processes which give out carbon dioxide and those which use carbon dioxide is called the carbon cycle and is illustrated in Figure 23.5B.

CO_2 is used in soft drinks

Carbon dioxide is colourless and odourless, and is slightly soluble in water. Flavoured solutions of carbon dioxide under pressure are the basis of the soft drinks industry.

CO_2 is used in fire-extinguishers

Carbon dioxide is non-poisonous, denser than air and does not support combustion. These three factors find carbon dioxide a use in fire-extinguishers. It is stored in cylinders under pressure and released by the opening of a valve [see Figure 23.5C]. There are substances which burn with such a hot flame that they can decompose carbon dioxide and continue to burn in the oxygen formed. A magnesium fire could not be extinguished by carbon dioxide.

CO_2 is easily liquefied

Solid CO_2 is called 'dry ice'

The gas can be liquefied at room temperature by a pressure of 60 atm. If the compressed gas is allowed to expand rapidly, solid carbon dioxide is obtained. It is a white solid, with the structure described in § 6.5. Its high standard enthalpy of vaporisation makes it a useful refrigerant, and, since it sublimes to form gaseous carbon dioxide, rather than melting, it is known as 'dry ice' or 'Drikold' [see Figure 23.5D].

The laboratory preparation of CO_2

Industrially, carbon dioxide is formed as a by-product during the manufacture of quicklime from limestone [§ 18.9.5] and in the fermentation of sugars to ethanol [§ 30.4.3]. In the laboratory, it can be made by the action of dilute hydrochloric acid or dilute nitric acid on marble chips. The gas can be collected over water. If required pure and dry, it is bubbled through water (to remove any hydrogen chloride), through concentrated sulphuric acid and then collected downwards.

Carbonic acid and carbonates

The solution of CO_2 forms carbonic acid

Carbon dioxide is an acidic gas. It is the anhydride of carbonic acid, H_2CO_3, but only about 0.4% of dissolved carbon dioxide is converted into carbonic acid, and when a solution is boiled nearly all the dissolved carbon dioxide is expelled:

$$CO_2(g) + H_2O(l) \rightleftharpoons CO_2(aq) + H_2O(l) \rightleftharpoons H_2CO_3(aq)$$

Carbonic acid is a weak diprotic acid.

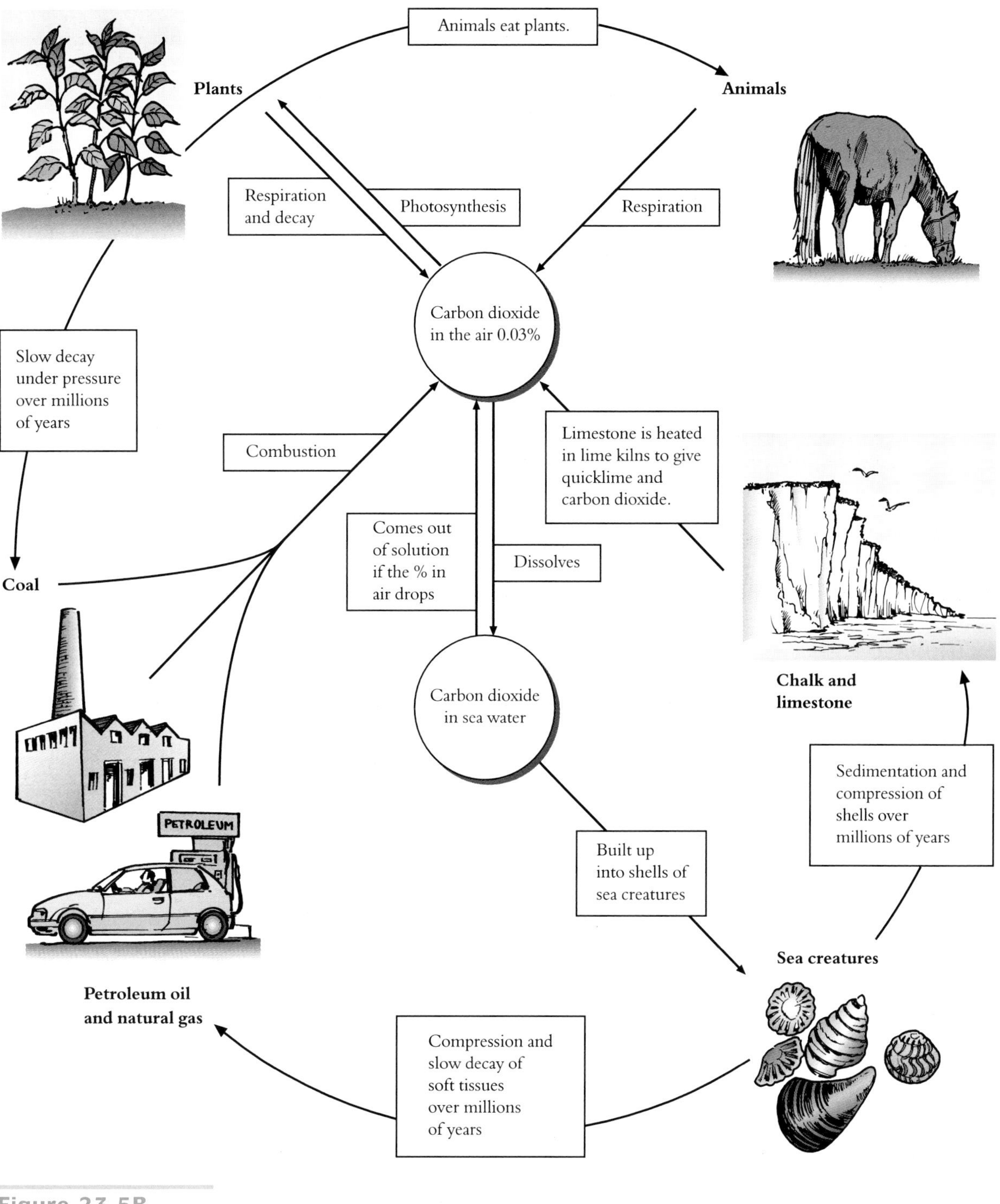

Figure 23.5B
The carbon cycle

Carbon dioxide reacts with bases. When passed into sodium hydroxide solution, it forms first sodium carbonate and then sodium hydrogencarbonate:

$$CO_2(g) + 2NaOH(aq) \longrightarrow Na_2CO_3(aq) + H_2O(l) \xrightarrow{CO_2(g)} 2NaHCO_3(aq)$$

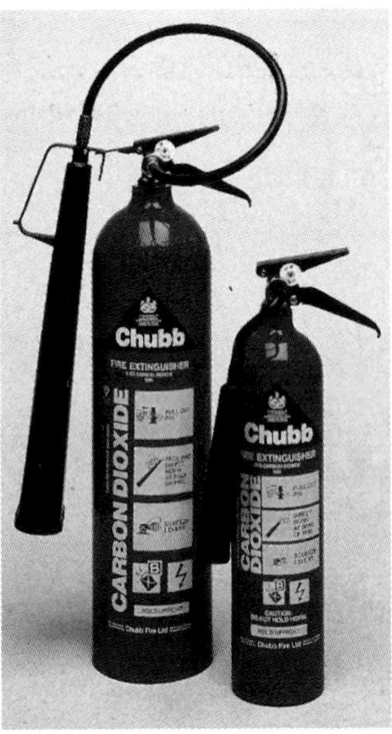

Figure 23.5C
A carbon dioxide fire-extinguisher

Figure 23.5D
Drikold, ICI's solid CO_2

Carbonates and hydrogencarbonates

A similar reaction with calcium hydroxide solution (limewater) is used as a test for carbon dioxide. A precipitate of calcium carbonate appears and then dissolves to form soluble calcium hydrogencarbonate:

The limewater test for CO_2

$$CO_2(g) + Ca(OH)_2(aq) \longrightarrow CaCO_3(s) + H_2O(l) \xrightarrow{CO_2(g)} Ca(HCO_3)_2(aq)$$

The structure of CO_3^{2-}

The structure of the planar carbonate ion can be represented by a molecular orbital structure:

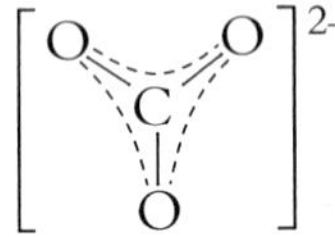

Silicon(IV) oxide, silica, SiO_2

Silica glass is chemically inert and transmits infrared and ultraviolet light

Several forms of silicon(IV) oxide or *silica*, SiO_2, are known. The structure of quartz and other crystalline forms of silica is shown in Figure 6.6D, § 6.6. Silica melts at 1710 °C to form a viscous liquid. When this liquid is cooled, it forms a glass. Silica glass is used for the manufacture of specialised laboratory glassware. Silica glass transmits infrared and ultraviolet light. It is chemically inert, being attacked only by hydrogen fluoride, damp fluorine and molten bases.

Uses for silica

Silica is used in the manufacture of cement and mortar and the manufacture of the abrasive, silicon carbide, SiC.

Quartz watches

Quartz is used in the manufacture of electronic equipment and timing devices. Quartz watches have recently taken a large slice of the market away from watches with moving parts. The timing mechanism is controlled by a quartz crystal which is induced to vibrate at 32 768 oscillations per second by the application of a small electric field.

Silicates

When silicates are formed by the reaction of silicon(IV) oxide and a molten base, the silicate ion that is formed depends on the amounts of base and silica present. Some silicate ions are shown:

Figure 23.5E
Silicate ions

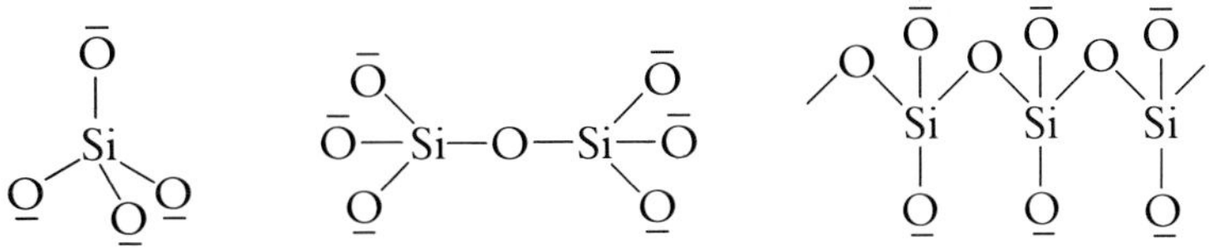

(a) SiO_4^{4-} (b) $Si_2O_7^{6-}$ (c) $(SiO_3)_n^{2n-}$

Figure 23.5F
A continuous sheet of glass leaving the Pilkington float glass plant

Soda glass is made from silica ...

... as are cobalt glass ...

... and water glass

Silica is used in the manufacture of glass. *Soda glass* is a mixture of sodium silicate and calcium silicate, which is made by melting the carbonates with silica at 1500 °C. Coloured glasses are made by adding metal oxides to the melt, e.g., cobalt(II) oxide for *blue glass*. Pyrex® is made by adding boron oxide to the melt. It can withstand higher temperatures than soda glass. The addition of lead oxides gives *lead glass*, which is dense and is very suitable for making cut glass objects.

Oxides of tin and lead

The oxides of tin and lead are amphoteric. Their preparation and properties are summarised in Table 23.5B.

23.5.4 SALTS

Tin(II) and lead(II) compounds are ionic. Tin(IV) is the preferred oxidation state to tin(II), so tin(II) compounds are reducing agents. Tin(II) salts are oxidised to tin(IV) salts by acidified potassium manganate(VII), acidified potassium dichromate(VI) and iron(III) salts.

Table 23.5B
Oxides of tin and lead

Oxide	Preparation	Properties
Tin(IV) oxide, SnO_2, the more stable oxide	(a) Heat Sn in air. (b) Sn + conc. HNO_3	Amphoteric; with conc. $H_2SO_4 \longrightarrow Sn(SO_4)_2$; with conc. alkali or molten base $\longrightarrow$ stannate(IV), $Sn(OH)_6^{2-}$
Tin(II) oxide, SnO	Heat SnC_2O_4, tin(II) ethanedioate. (CO and CO_2 are formed and prevent oxidation of SnO.)	Amphoteric, but more basic than SnO_2; with dilute acid $\longrightarrow Sn^{2+}$ salt; with alkali $\longrightarrow$ stannate(II), $Sn(OH)_4^{2-}$
Lead(IV) oxide, PbO_2, brown	Warm Pb^{2+} salt with oxidising agent, e.g., $ClO^-(aq)$.	Lead–acid accumulator [§ 13.5.3] Powerful oxidising agent; Heat $\longrightarrow PbO + O_2$ with $SO_2 \longrightarrow PbSO_4$ with warm HCl $\longrightarrow Cl_2$ Amphoteric; with HCl(aq) below 20 °C $\longrightarrow PbCl_4$(l); with alkali or molten base $\longrightarrow$ plumbate(IV), $Pb(OH)_6^{2-}$
Lead(II) oxide, PbO, yellow and red forms	(a) Heat $Pb(NO_3)_2$ (b) Heat $PbCO_3$.	Amphoteric; with HCl(aq) $\longrightarrow PbCl_2$; with alkali or molten base $\longrightarrow$ plumbate(II), $Pb(OH)_4^{2-}$
Dilead(II) lead(IV) oxide, Pb_3O_4, 'red lead'	Heat PbO at 400 °C in air.	Behaves as a mixed oxide, $2PbO \cdot PbO_2$ With $HNO_3(aq) \longrightarrow Pb(NO_3)_2 + PbO_2$ With conc. HCl(aq) $\longrightarrow PbCl_2 + Cl_2$

Lead(II) is the preferred oxidation state to lead(IV) so lead(IV) compounds are oxidising agents. The oxidation of concentrated hydrochloric acid to chlorine by lead(IV) oxide illustrates this:

$$PbO_2(s) + 4HCl(aq) \longrightarrow Cl_2(g) + PbCl_2(s) + 2H_2O(l)$$

Lead(IV) oxide; Concentrated hydrochloric acid; Chlorine; Lead(II) chloride

Tests for lead(II) salts

Apart from the nitrate and ethanoate, lead salts are insoluble. They can be made by adding a solution of the required anion to aqueous lead(II) nitrate, e.g.

$$Pb^{2+}(aq) + SO_4^{2-}(aq) \longrightarrow PbSO_4(s)$$

Lead(II) ion; Sulphate ion; Lead(II) sulphate

Lead(II) salts in solution can be recognised by precipitation of the insoluble compounds lead(II) chloride $PbCl_2$ (white), lead(II) iodide PbI_2 (yellow) and lead(II) hydroxide (white), which dissolves in an excess of alkali to form aqueous plumbate(II) ion.

$$Pb^{2+}(aq) + 2OH^-(aq) \longrightarrow Pb(OH)_2(s)$$

$$Pb(OH)_2(s) + 4OH^-(aq) \longrightarrow Pb(OH)_6^{4-}(aq)$$

Lead(II) hydroxide　　　　　　Plumbate(II)

CHECKPOINT 23.5B: OXIDES AND SALTS

1. Explain why silicon(IV) oxide has a different type of structure from carbon dioxide.

2. Discuss the trends in valency and bond type shown by the oxides of Group 4.

3. *(a)* State two characteristics of (i) metal oxides (ii) non-metal oxides.
   (b) Illustrate the gradation from non-metallic properties to metallic properties in Group 4 by considering the oxides.

4. Explain why carbon monoxide is present in vehicle exhaust gases and say what ill effects it has. Suggest how the emission of carbon monoxide can be reduced [see § 26.5].

5. *(a)* Outline a method of preparing carbon dioxide in the laboratory. Say how the gas can be collected.
   (b) Why is no preparation of silicon(IV) oxide mentioned in the text?

6. *(a)* Which is the more stable oxide (i) SnO or SnO_2 (ii) PbO or PbO_2?
   (b) Suggest a method of making lead(II) oxide PbO in the laboratory. Why is this method not suitable for tin(II) oxide, SnO?

7. *(a)* Write ionic equations for *(a)* the precipitation of lead(II) iodide when solutions of lead(II) ethanoate and potassium iodide are added *(b)* the reduction of iron(III) chloride by tin(II) sulphate.

23.6 THE GREENHOUSE EFFECT

The mean temperature of our planet is fixed by a steady state balance between the energy received from the Sun and an equal quantity of heat energy radiated back into space by the Earth. If disturbances in either incoming or outgoing energy upset this balance, the average temperature of the Earth's surface will drift to a different steady state value. The resulting changes in the Earth's climate could upset food production, create deserts, raise the level of oceans or start a new ice age. One mechanism for regulating the Earth's temperature is the **greenhouse effect** [see Figure 23.6A].

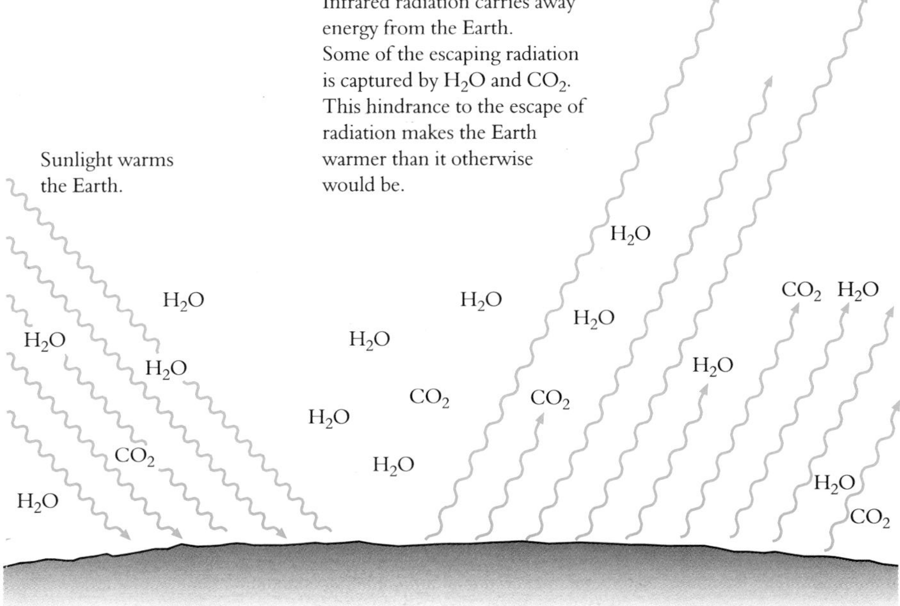

Infrared radiation carries away energy from the Earth.
Some of the escaping radiation is captured by H_2O and CO_2.
This hindrance to the escape of radiation makes the Earth warmer than it otherwise would be.
Sunlight warms the Earth.
H_2O
CO_2

Figure 23.6A The atmospheric greenhouse effect

The Earth receives energy from the Sun ...

The Sun emits radiation in a band of wavelengths from ultraviolet to infrared (UV to IR, 200 nm to 3000 nm) with a maximum in the visible spectrum at 500 nm. This radiation passes through the atmosphere of the Earth with very little absorption. When the radiation reaches the Earth, it warms the ground or sea. The warm surface of the Earth radiates energy outwards at the longer infrared wavelengths. Unlike sunlight, infrared radiation cannot travel freely through air. Infrared radiation is absorbed by water vapour, carbon dioxide, ozone and other gases in the lower atmosphere, warming up the lower layers of the atmosphere, which radiate some heat back to the ground and some out into space.

... and radiates energy back into space at the IR wavelengths

Gases in the lower atmosphere, e.g. water vapour and carbon dioxide, absorb IR radiation from the Earth and radiate energy back to Earth. This warming effect is called the greenhouse effect

The warming effect of carbon dioxide and water vapour has been named the **greenhouse effect**. The effect is compared with the glass of a greenhouse, which lets sunlight enter and prevents infrared radiation from leaving. (Actually, in addition to reflecting infrared radiation, real greenhouses trap heat mainly by preventing warm air from escaping by convection.)

There is more water in the atmosphere than carbon dioxide, so most of the greenhouse heating of the Earth's surface is due to water vapour. However, there is a gap in the absorption by water, which is partly plugged by carbon dioxide [see Figure 23.6B].

Figure 23.6B
Carbon dioxide partially plugs the gap in the water cover

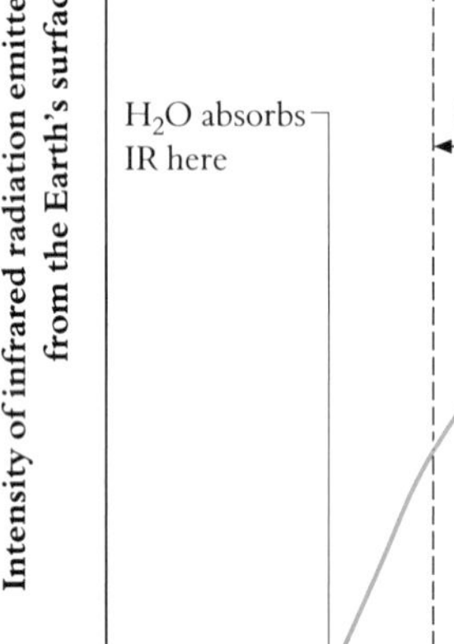

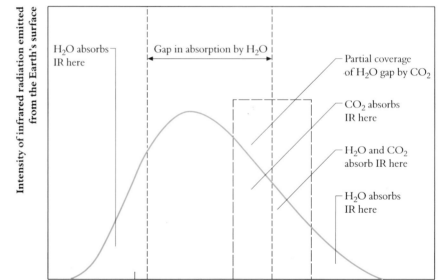

Carbon dioxide partially fills a gap in the absorption of IR radiation by water vapour

Without the warming greenhouse effect, the earth would be uninhabitable

The greenhouse effect is a natural phenomenon. Without it, the Earth would be uninhabitable. It is what keeps us from being a frozen planet. If gases such as carbon dioxide did not trap the Sun's energy, the Earth's mean temperature would be about −20 °C, rather than its current 15 °C.

The concentration of carbon dioxide in the atmosphere is increasing

Before the Industrial Revolution, the carbon dioxide concentration was 280–90 ppm; now it is 340 ppm. By 2100, if the burning of fossil fuels continues at its present level, the concentration of carbon dioxide will have doubled. Climatologists estimate that doubling the carbon dioxide concentration would increase mean temperatures over the globe by 2–3 °C. They also predict that the increase would be greatest at high latitudes (near the poles) with Russia warming

It is forecast that by the year 2100 the level of carbon dioxide will be twice what it was before industrialisation. The predicted result is an increase in global temperature of 2–3 °C

up by, say 3 °C, North America by 1–2 °C and some countries, e.g. Japan, India and Spain showing little change in temperature.

The combustion of fossil fuels is the source of most of the carbon dioxide

Plants take carbon dioxide from the atmosphere to use in photosynthesis. Unfortunately, the world's forest cover is being drastically cut. In Brazil, the Amazon rain forest once covered 3 million square miles. As the region has been developed for agriculture and mining, 20% to 25% of the forest has been destroyed and a further 20% has been seriously disturbed. When the felled trees are burned or left to rot, carbon dioxide and other greenhouse gases are released. The same kind of deforestation is going on in some African countries, Indonesia, India and the Philippines. The loss of trees may already be making the Earth warmer.

As a result of extensive felling of the world's forests, there are fewer trees to take in carbon dioxide in photosynthesis

Methane is another greenhouse gas. Methane enters the atmosphere from a number of sources. A single termite mound can emit 5 litres of methane per minute. Swamps, rice fields, leaking North Sea gas pipes and herds of cows all send methane into the atmosphere. Chlorofluoroalkanes, CFCs [§ 29.6] are powerful greenhouse gases.

Methane is another greenhouse gas

A temperature increase of 0.5 °C has been observed since the beginning of the century. This may not sound very alarming, but scientists expect any greenhouse warming to be masked for quite a time by the enormous capacity of the oceans to absorb heat.

Chlorofluorocarbons, CFCs, are powerful greenhouse gases

The Intergovernmental Panel on Climate Change asked in 1990 for an immediate reduction in the emission of greenhouse gases. The results of global warming would include melting of polar ice caps and flooding of low-lying countries

The Intergovernmental Panel on Climate Change was set up in 1988, and in 1990 asked for an immediate reduction in the emission of greenhouse gases. The Panel predicted that if emissions continue at the present levels the temperature of the Earth will rise by 3 °C in the next century and the sea level will rise by 65 cm. The predictions led to a great deal of alarm, with forecasts of the sea rising by up to 9 metres. The predictions calculated that as the temperature rose more water would vaporise from the oceans and the increase in water vapour would enhance global warming, increasing the temperature still further, leading to further vaporisation of water from the oceans and so on.

In 1992 the Panel issued a second report in which they gave a lower estimate of global warming and asked for more research work to be done

In 1992 the panel published a second report in which they admitted that further computer modelling had not confirmed the predictions of the 1990 report. The Panel gave a revised estimate of global warming between 1 °C and 4.5 °C. It stated that the increase in temperature over the past 100 years fell within the normal variation in the climate and might even be due to causes other than carbon dioxide. The 1992 report stated that the estimates of global warming over the past 100 years due to greenhouse gases made in the 1990 report were too high. The report stated that better temperature measurements were needed as well as better computer simulations.

At the 1992 Earth Summit, 200 nations agreed to stabilise carbon dioxide emissions at the 1990 levels. This would not be enough if the worst warnings of global warming were true. At a conference of industrial nations in Berlin in 1995, the Berlin Mandate agreed to continue to reduce emissions of all greenhouse gas after the year 2000.

The Earth Summit of 1992 and the Berlin Mandate of 1995 agreed to reduce emissions of greenhouse gases

One move to reduce the greenhouse effect is to cut back on the production of 'chlorofluorocarbons' [CFCs, see § 29.6]. The most far-reaching step would be to cut back on the use of fossil fuels. This would be hard to do in industrialised countries without a tremendous effort in energy conservation and the development of alternative energy sources. In developing countries, reductions in the burning of fossil fuels would be difficult to impose because they would delay industrialisation of these countries.

To avoid an increase in global temperature, it is essential to stop using CFCs ...

... to decrease the combustion of fossil fuels ...

... and to put a stop to deforestation

A very effective step would be to protect the tropical rain forests. There is a scheme for achieving this aim. It is called **debt-for-nature exchange** or **debt-swap**. The scheme recognises that many of the countries which are cutting down forests are struggling to repay foreign debts and are desperately in need of the revenue they will get for timber. In debt-swap, a rich country agrees to cancel a debt from a poor country if that country will agree to conserve a certain area of forest in exchange. Cameroon and Madagascar in Africa and Costa Rica and Bolivia in South America are some of the countries which have benefited from the scheme. The whole world will benefit if the scheme expands.

CHECKPOINT 23.6: THE GREENHOUSE EFFECT

1. *(a)* About 6×10^9 tonnes of carbon in fossil fuels are burned annually. If this mass of carbon were burned completely, what mass of carbon dioxide would it produce?
   (b) The present mass of carbon dioxide in the atmosphere is 2.5×10^{12} tonnes. It is assumed that half of the additional carbon dioxide will enter the sea and other 'sinks'. In how many years will the carbon dioxide content double?

2. An unknown number of 'multiplier effects' could make the greenhouse effect worse than expected.

   (a) If the polar ice caps began to melt, the land beneath them would be exposed. How would the exposure of this land add to global warming?
   (b) An increase in temperature would lead to evaporation of water from the oceans. How would this add to global warming?
   (c) An increase in temperature would decrease the solubility of carbon dioxide. How would this add to global warming?
   (d) An increase in temperature might lead to increased volcanic activity. How would this affect the global temperature?

3. How does deforestation add to the greenhouse problem? Which countries are cutting down forests to make arable land? Which countries are cutting down trees for use as fuel? If it were up to you to persuade such countries to sacrifice their short-term ends for the long-term future of the planet, what would you say? What resistance do you think you might meet?

   This is a tough problem, and a group discussion might be the best way to approach it.

4. Suppose experts were to state that, beyond reasonable doubt, the present rate of carbon dioxide emission would have catastrophic repercussions (melting of the polar ice caps etc.) by the year 2100. Draw up a plan for world response to the threat. Discuss the political aspects of international cooperation versus national goals, the desire of underdeveloped nations to industrialise, the standard of living in developed countries, the role of electrical energy and nuclear power and other factors.

   This is a problem which is baffling nations: you may want to form a group to discuss it and pool ideas!

GROUP 4

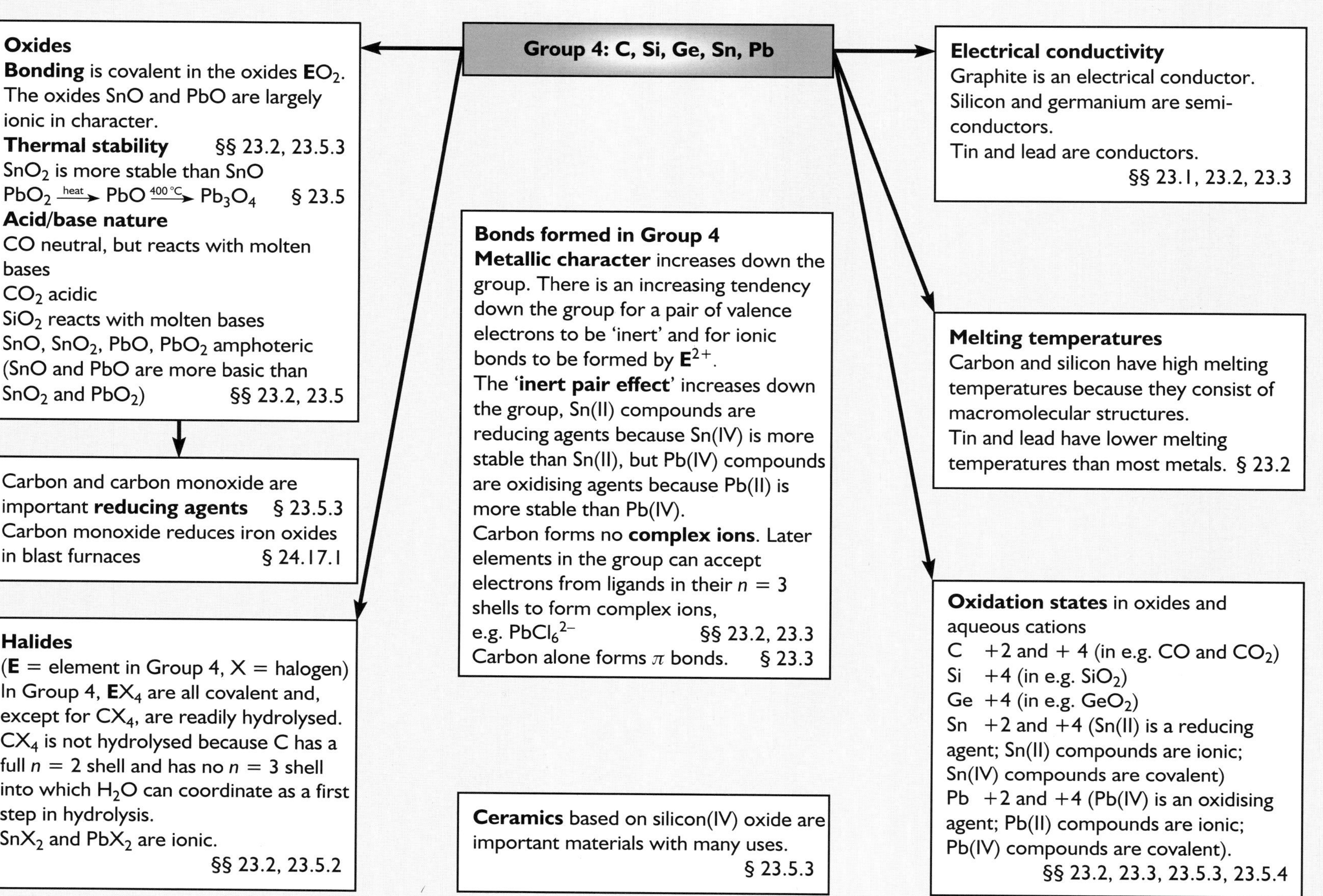
Group 4: C, Si, Ge, Sn, Pb
Oxides
Bonding is covalent in the oxides EO_2. The oxides SnO and PbO are largely ionic in character.
Thermal stability §§ 23.2, 23.5.3
SnO_2 is more stable than SnO
$PbO_2 \xrightarrow{heat} PbO \xrightarrow{400\,^{\circ}C} Pb_3O_4$ § 23.5
Acid/base nature
CO neutral, but reacts with molten bases
CO_2 acidic
SiO_2 reacts with molten bases
SnO, SnO_2, PbO, PbO_2 amphoteric (SnO and PbO are more basic than SnO_2 and PbO_2) §§ 23.2, 23.5
Carbon and carbon monoxide are important reducing agents § 23.5.3
Carbon monoxide reduces iron oxides in blast furnaces § 24.17.1
Halides
(E = element in Group 4, X = halogen)
In Group 4, EX_4 are all covalent and, except for CX_4, are readily hydrolysed. CX_4 is not hydrolysed because C has a full $n = 2$ shell and has no $n = 3$ shell into which H_2O can coordinate as a first step in hydrolysis.
SnX_2 and PbX_2 are ionic.
§§ 23.2, 23.5.2
Bonds formed in Group 4
Metallic character increases down the group. There is an increasing tendency down the group for a pair of valence electrons to be 'inert' and for ionic bonds to be formed by E^{2+}.
The 'inert pair effect' increases down the group, Sn(II) compounds are reducing agents because Sn(IV) is more stable than Sn(II), but Pb(IV) compounds are oxidising agents because Pb(II) is more stable than Pb(IV).
Carbon forms no complex ions. Later elements in the group can accept electrons from ligands in their $n = 3$ shells to form complex ions, e.g. $PbCl_6^{2-}$ §§ 23.2, 23.3
Carbon alone forms π bonds. § 23.3
Ceramics based on silicon(IV) oxide are important materials with many uses.
§ 23.5.3
Electrical conductivity
Graphite is an electrical conductor.
Silicon and germanium are semi-conductors.
Tin and lead are conductors.
§§ 23.1, 23.2, 23.3
Melting temperatures
Carbon and silicon have high melting temperatures because they consist of macromolecular structures.
Tin and lead have lower melting temperatures than most metals. § 23.2
Oxidation states in oxides and aqueous cations
C +2 and + 4 (in e.g. CO and CO_2)
Si +4 (in e.g. SiO_2)
Ge +4 (in e.g. GeO_2)
Sn +2 and +4 (Sn(II) is a reducing agent; Sn(II) compounds are ionic; Sn(IV) compounds are covalent)
Pb +2 and +4 (Pb(IV) is an oxidising agent; Pb(II) compounds are ionic; Pb(IV) compounds are covalent).
§§ 23.2, 23.3, 23.5.3, 23.5.4

QUESTIONS ON CHAPTER 23

1. Give two properties of *(a)* metallic oxides *(b)* non-metallic oxides *(c)* metallic chlorides *(d)* non-metallic chlorides. Illustrate the transition in properties down Group 4 by considering the properties of (i) their oxides and (ii) their chlorides.

2. Compare the elements of Group 4 with respect to

   (a) the crystal structures of the elements
   (b) the thermal stability of the hydrides
   (c) the stability to hydrolysis of the halides
   (d) the oxidation states of the elements
   (e) the basicity of the oxides $\mathbf{EO_2}$ and **EO**.

3. Explain why the chemistry of carbon is dominated by the ability of the carbon atom to bond to other carbon atoms and to form multiple bonds to atoms of carbon and other elements.

4. What are allotropes?

   Sketch the structures of the two chief allotropes of carbon. How does the structure explain the physical properties of each allotrope? How could you demonstrate that the allotropes are both pure carbon?

5. Outline how you could prepare *(a)* Na_2CO_3 from C *(b)* PbO from Pb *(c)* SnO_2 from Sn *(d)* PbI_2 from Pb. More than one step may be needed.

6. In what ways does carbon differ from the other members of Group 4?

7. Metals are good electrical conductors, and non-metallic elements, other than graphite, are insulators. Silicon is a semiconductor.

   (a) Explain metallic conduction in terms of delocalised bonding.
   (b) Explain why graphite is an electrical conductor.
   (c) Name an element that is added to silicon to convert it into a semiconductor (i) of the p-type (ii) of the n-type.

8. *(a)* Copy and compete the table to show the reactions of the following chlorides with water.

Chloride	Reaction with water
Carbon	
Silicon	
Tin(IV)	
Lead(II)	
Lead(IV)	

   (b) Explain why two chlorides of lead are listed, whereas only one is given for silicon.
   (c) Choose two chlorides from *(a)* which differ in their reactions with water, and explain the difference.

9. Silica occurs mainly as quartz and sand. Glass is made by heating a mixture of sand, sodium carbonate and calcium carbonate.

   (a) Sketch the structural unit of quartz, showing two repeating units.
   (b) Write an equation for the reaction of molten sodium hydroxide and silica.

10. Bond enthalpies/kJ mol^{-1} are:

    Si—O 374, Si═O 638, C—O 360, C═O 743

    Use them to explain why silicon(IV) oxide has a giant molecular structure whereas carbon dioxide consists of individual CO_2 molecules.

11. *(a)* Silicon carbide is a very hard substance which is used as an abrasive. Suggest a structure for silicon carbide.
    (b) Suggest why, although silicon has a diamond structure, there is no structure for silicon corresponding to graphite.
    (c) The densities of diamond and graphite are 3.53 g cm^{-3} and 2.25 g cm^{-3} respectively. The standard enthalpy change for graphite $\longrightarrow$ diamond is +2.1 kJ mol^{-1}. Industrial diamonds are made by heating graphite to 2000 °C under 70 000 atm pressure with a catalyst. Explain the choice of conditions.

12. *(a)* (i) State the colour change that occurs when a solution of tin(II) chloride is added to acidified potassium dichromate(VI) solution. Name the coloured ion **produced** in this reaction. 2
    (ii) Sn(II) compounds slowly oxidise in air forming Sn(IV) compounds. State what this reaction shows about the relative stability of Sn(II) and Sn(IV). 1
    (iii) Lead(IV) oxide reacts with concentrated hydrochloric acid producing lead(II) chloride, chlorine and water. State what this reaction shows about the relative stability of Pb(IV) and Pb(II) compounds. 1
    (iv) I. Briefly describe the ***inert pair effect*** as applied to Group IV elements.
    II. Given that the outer electronic configuration of the lead atom is $6s^2\,6p^2$, state the outer electronic configuration of the lead(II) **ion**. 2
    (b) The first and second standard molar ionisation energies of lead are 716 and 1450 kJ mol^{-1} respectively. The standard enthalpy change of atomisation of Pb(s) is 196 kJ mol^{-1}. The sum of all three enthalpy terms is 2362 kJ mol^{-1}.
    (i) Give an equation showing the process to which the energy change of 716 kJ mol^{-1} applies. 1

(ii) Calculate the standard enthalpy change for the process represented by the equation

$$Pb^{2+}(g) + 2e^- \longrightarrow Pb(s)$$

when 0.10 mol of $Pb^{2+}(g)$ is converted to Pb(s). **2**

(c) (i) Write **balanced** equations representing reactions in which
I. carbon monoxide is formed as an undesirable product in domestic fossil fuel fires,
II. methane burns in an excess of air,
III. carbon monoxide burns in air forming carbon dioxide. **3**

(ii) State briefly how the conductivity of silicon may be enhanced using Group III and Group V elements, naming the type of semiconductors produced in each case. **3**

15 *WJEC (AL)*

13. This question is about the elements in Group IV of the Periodic Table.

(a) (i) Complete the electronic arrangements, in terms of *s*, *p*, and *d* energy levels, for the elements silicon and germanium:

Carbon $1s^2 2s^2 2p^2$
Silicon $1s^2$..........
Germanium $1s^2$.......... **1**

The atomic radii of carbon, silicon and germanium are 0.077, 0.117 and 0.122 nm respectively.

(ii) Explain why the atomic radii increase down the Group. **3**

(iii) Suggest why the difference in radius between silicon and germanium is much smaller than the difference in radius between carbon and silicon. **3**

(iv) Explain why the first ionisation energies of these three elements decrease in the following pattern: **2**

Carbon 1090 kJ mol^{-1}
Silicon 786 kJ mol^{-1}
Germanium 762 kJ mol^{-1}

(b) (i) Write the equation for the reaction between silicon(IV) chloride, $SiCl_4$, and cold water. Suggest the equation for the reaction between silane, SiH_4, and cold water. **2**

(ii) Draw a dot-and-cross diagram showing the bonding in silicon(IV) chloride and a diagram showing the three dimensional shape of a molecule of this compound. **3**

(c) Explain why methane, CH_4, has a lower boiling point than ammonia, NH_3. **4**

(d) Lead and tin are quite reactive when each is heated with chlorine, producing $PbCl_4$ and $SnCl_4$ respectively. Diamond does not react with chlorine. Why is diamond so unreactive? **2**

20 *O&C (AS/AL)*

14. A lead ore contains the mineral galena which has the formula PbS. Lead is extracted from the ore by the following steps.

1. *Ore separation* The ore is ground with water to a fine particle size. Detergent is added and air is blown through the suspension. This separates galena from the unwanted rock.
2. *Smelting* Galena is roasted in air to form lead oxide.

$$2PbS + 3O_2 \longrightarrow 2PbO + 2SO_2 \qquad [1]$$

3. The lead oxide is then reduced to lead by carbon monoxide in a blast furnace.
4. The lead is *refined* to remove impurities such as antimony.

(a) Give a name for the process which occurs when air is blown through the suspension in step 1. **1**

(b) (i) Which element is being oxidised in equation [1]?
Element: .... From oxidation state .... to **2**

(ii) Suggest an environmental hazard which may arise from a *named chemical* in equation [1]. **2**

(c) Write a balanced equation for a reaction which occurs when lead oxide, PbO, is reduced by carbon monoxide in a blast furnace in step 3. **1**

(d) The metal antimony is removed from the lead in the refining process by using its reaction with sodium hydroxide.

$$Sb(s) + NaOH(aq) + H_2O(l) \longrightarrow NaSbO_2(aq) + 1\tfrac{1}{2}H_2(g) \qquad [2]$$

Give the oxidation state of antimony in
(i) Sb (ii) $NaSbO_2$ **2**

(e) Classify the reaction in equation [2] by stating one of: *precipitation, acid–base, redox* **1**

9 *O&C (S) (AS/AL)*

24 THE TRANSITION METALS

24.1 METALS AND CIVILISATION

The metals of Groups 1 and 2 [see Chapter 18], are not metals that we see around us in everyday use. They are too reactive and too soft to be used as materials. Aluminium, in Group 3, and tin and lead in Group 4 are metals that we see in use in everyday objects, such as aluminium windowframes, tin-plated cans and lead–acid batteries. Most of the metals in the Periodic Table belong in the d block of transition metals. They are hard and strong, and many of them are very familiar to us. The transition metal copper and its alloy bronze played such an important part in the advance of civilisation that the Bronze Age is named after them. The transition metal iron and its steel alloys made such a difference to people's lives that the Iron Age was named after them.

The Bronze Age

In the Stone Age, about 5000 BC, the only tools and weapons which our ancestors had were made from stones. Among these stones, people began to find native metals – metals that occur uncombined. When the rocks of a camp fire became very hot, native metals might trickle out, metals such as gold, silver and copper. Our ancestors used the metals to make ornaments. In time they found that copper was useful as well as ornamental. It was harder than gold and silver but soft enough to be hammered into arrowheads, spears and knives. As well as occurring native, copper is found as compounds, and people discovered how to obtain copper by heating stones containing copper compounds with charcoal.

Some batches of copper were better than others; they were harder and therefore made better tools and weapons. The reason was that often tin compounds are present in the same ores as copper compounds and smelting produced an alloy of copper and tin, the alloy bronze. Bronze is harder than copper and can be ground to a sharper edge. Both these characteristics make bronze tools better than copper tools. Bronze weapons revolutionised hunting, and bronze tools speeded up every aspect of farming. The Bronze Age had begun. Hunting and farming no longer occupied all the time of every member of the community, and some people developed other skills as potters, painters and so on. By 3500 BC several parts of the world were civilised. It began with the smelting of copper ores! The chemistry of copper is described in § 24.18.

The Iron Age

The Bronze Age was followed by the Iron Age. In order to smelt iron ores, the air flow to the primitive furnaces had to be improved to make them operate at higher temperatures. Iron took a sharper cutting edge than bronze and made better tools and weapons. Over a long period of time, the early smelters learned how to improve iron by quenching (rapid cooling) to give a hard, brittle metal, annealing (slow cooling) to give a softer, more malleable metal and tempering (quenching followed by annealing). The tempered iron had the best characteristics; it was hard but it could be worked without breaking.

The quality of iron produced by the early smelters was very variable. Iron swords could be very unreliable, and the smelters never knew what had gone wrong if a batch of iron was not up to standard. The Gauls were defeated by the Romans at a battle near Milan in 223 BC. The long iron swords of the Gauls were easily bent, and after one mighty blow the edges might turn and the blade might bend. A warrior had to hold his blade against the ground and straighten it with his foot before he could deliver a second blow. While he was doing this he was likely to get a Roman legionnaire's sword through his ribs. A thousand years later the Vikings were still having the same trouble.

It was impossible to explain why some batches of iron turned out stronger than others. When by chance a good piece of metal was obtained and turned into a mighty sword, it was believed to have a magical origin. King Arthur led his Knights of the Round Table into battle with his magical sword Excalibur. Why did Arthur think that Excalibur was magical? A sword must be hard, and it must be flexible. These are two properties which are difficult to achieve at the same time. It can be done; the method is to heat iron and cool it rapidly so that layers of flexible pure iron and hard iron–carbon alloy are built up. Modern metallurgists have electron microscopes to help them to do this. Primitive metallurgists used the method of trial and error. When they succeeded in combining hardness and flexibility to make a sword that would take a sharp cutting edge and would not break or bend when it struck armour plate, it must have seemed like magic.

By the fourteenth century the Japanese had become very good at making swords. The smiths hammered out the metal into a thin layer, folded it, hammered it again and so on, repeating the process until they had a metal ten or twenty layers thick. Then they heated the metal and quenched it. The result was a metal that contained thousands of layers of flexible pure iron and thousands of layers of hard iron–carbon alloy. The Samurai were the military class, entitled to wear arms. They were willing to pay a large sum for a sword from a famous sword maker and also preferred to take the precaution of testing their swords before going into battle with them. A Samurai warrior might bribe the public executioner to use his new sword to execute a condemned criminal. Once he had seen his sword cut off a head, the warrior felt much happier about trusting his own life to it.

The Industrial Revolution

The Industrial Revolution took place in the UK between 1780 and 1860. Iron was an essential part of the revolution. It was needed to construct the machinery which produced rapidly and in quantity articles which had previously been made by hand. Iron was used to build the railway lines and the trains which revolutionised transport. Today iron and its alloys are the basis of our technology. Our cars, trucks, trains and ships are made of steel. Our buildings are built around a framework of steel girders. The extraction of iron is described in § 24.17.1 and the manufacture of steel in § 24.17.3.

24.2 THE FIRST TRANSITION SERIES

The first transition series in Period 4 are a very similar set of elements

Across the second and third periods of the Periodic Table, there is a gradation in properties, from the alkali metals to the halogens. The fourth period begins in the same way, with an alkali metal (potassium) and an alkaline earth (calcium). The next ten elements do not show the gradation in properties of previous periods: they are remarkably similar to one another in their properties and are all metals. They are called the **first transition series**. Periods five and six also contain transition series. The reason for the similarity of the first transition series is that, considering the series from left to right, while each additional electron is entering the 3d shell, the chemistry of the elements continues to be determined largely by the 4s electrons. From one transition element to the next, the nuclear charge increases by 1 unit, and the number of electrons also increases by 1. Since each additional electron enters the 3d shell, it helps to shield the 4s electrons from the increased nuclear charge, with the result that the effective nuclear charge remains fairly constant across the series of transition elements. The sizes of the atoms and the magnitudes of the first ionisation energies are therefore very similar and the elements have comparable electropositivities. The electron configurations of the first transition series are shown in Table 24.2, in which (Ar) = $1s^2 2s^2 2p^6 3s^2 3p^6$.

The difference between transition metals is the number of d electrons. This affects their chemistry less than a difference in s or p electrons

You will notice that chromium completes occupying its d orbitals with unpaired electrons at the expense of its 4s electrons, and copper completes its full d^{10} shell at the expense of its 4s electrons. There appears to be a certain measure of stability associated with a full d^{10} shell and with a half-filled d^5 shell.

Table 24.2
Electron configurations of the first transition series

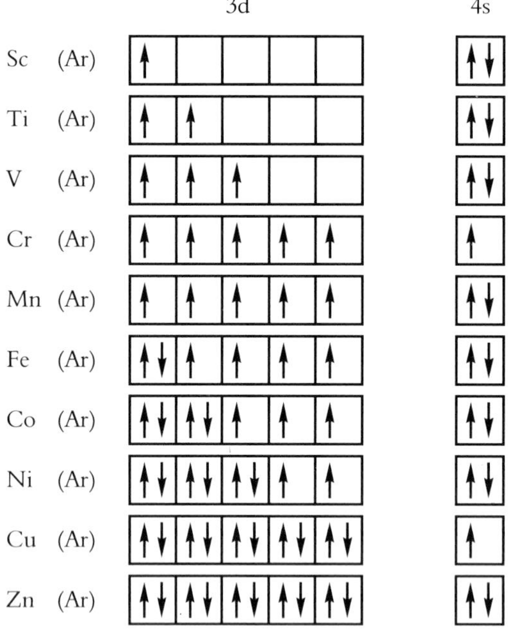

Transition metals are called d block metals

Scandium, zinc and copper

Transition metals are often referred to as d block metals. They are defined as elements which form some compounds in which there is an incomplete subshell of d electrons. Scandium ($3d^0$ in compounds) and zinc ($3d^{10}$ in compounds) are excluded by this definition, and copper is included only in copper(II) ($3d^9$) compounds. It is convenient to include these metals with transition metals, however, on account of the chemical resemblance of their compounds to transition metal compounds.

24.3 PHYSICAL PROPERTIES OF TRANSITION METALS

The metallic bond is stronger than in s block metals

The first transition metals are all hard and dense, good conductors of heat and electricity and possessing useful mechanical properties. Their melting and boiling temperatures and standard enthalpies of melting are higher than those of s block metals. All these properties are a measure of the strength of the metallic bond. With d electrons as well as s electrons available to take part in delocalisation, the metallic bond is strong in transition metals. The d block metals are, in general, denser than the s block metals.

Ionisation energy

Figure 24.3A shows how the first and second ionisation energies increase only slightly from scandium to zinc. The reason is that, as discussed in § 24.2, the effective nuclear charge increases only slightly across the series. The increases in the third and fourth ionisation energies across the series are more rapid as d electrons are being removed, and the effective nuclear charge therefore increases by a significant amount from one element to the next.

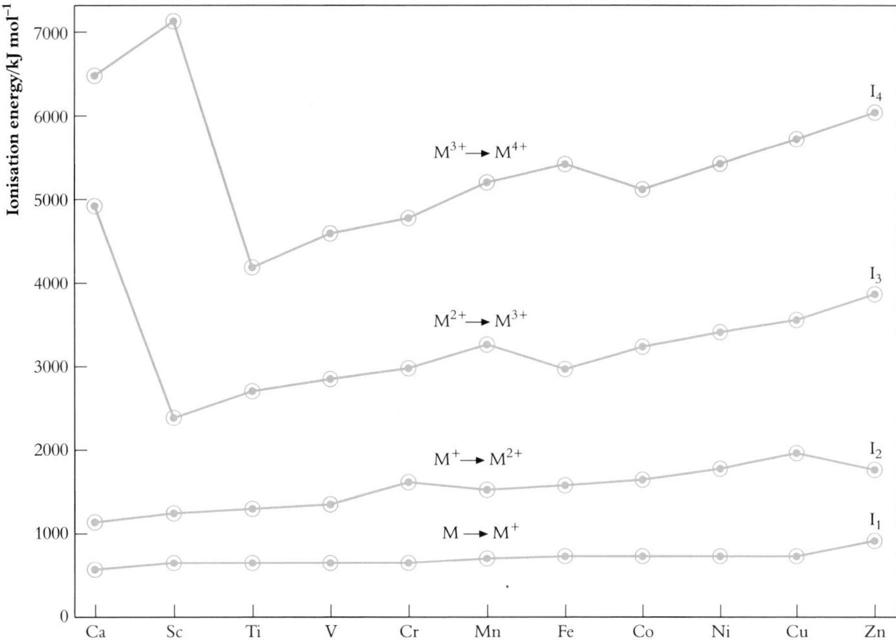

Figure 24.3A Ionisation energies of transition metals

Ionisation energies increase across the period

Each of the curves shows a maximum; these are at Cr^{2+}, at Mn^{3+} and at Fe^{4+}. In each case, the peak in ionisation energy occurs when the removal of an electron disturbs a half-full d^5 shell.

24.4 CHEMICAL PROPERTIES

Features of transition metals

A summary of the differences between d block metals and s block and p block metals is given in Table 24.3. Outstanding features of transition metals are:

1. The variable oxidation state [§ 24.7].
2. Their use as catalysts [§ 24.8].
3. Paramagnetism [§ 24.9].
4. The formation of complex ions [§ 24.10].
5. The formation of coloured ions [§ 24.10.2].

24.5 METHODS OF EXTRACTION

Extraction by reduction of the oxide

Transition metals occur mainly as sulphides and oxides. The least reactive (e.g., Cu, Au, Pt) are also found **native** (i.e., uncombined). Reduction of the oxide by carbon or carbon monoxide is the usual method of extraction. The following steps may be involved in the extraction:

1. The ore is concentrated. Sometimes flotation is employed: a stream of water carries away debris and leaves the denser ore behind.
2. Sulphide ores are then roasted to convert them to oxides.
3. Heating with coke reduces the oxide to the metal.
4. Carbon is present as a major impurity in the metal. It is removed by heating the metal in a stream of air. Further purification may be achieved by electrolysis (e.g., Cu, Ag, Cr).

Table 24.3
A comparison of metals

	s block	p block	d block
Physical properties	Soft, low melting temperature	Harder, with higher T_m than s block	Harder still, with higher T_m than p block
Reaction with water	React, often vigorously	React only slowly with cold water	
Reaction with non-metallic elements	React vigorously	React less vigorously than s block metals	
Reaction with hydrogen	Form ionic hydrides	Form no hydrides	Some form interstitial hydrides
Bonding	Usually ionic	Usually covalent or complex ions	
Properties of ions	Form simple ions with noble gas configuration	Simple ions have a completed d shell. Easily form complex ions	Some simple ions are formed. Many complex ions are formed
Complex ions	Simple ions can be loosely hydrated to form colourless complex ions	Colourless complex ions are formed, rather than simple ions	Complex ions are formed readily. Usually coloured
Oxidation numbers	Oxidation No. = Group No.	Employ Ox. No. = Gp. No. and also Ox. No. = Gp. No. − 2	Ox. No. varies usually by 1 unit at a time, +2, +3 are common.

24.6 USES OF TRANSITION METALS

Uses of iron, steel ...

Iron is our most important metal. Steels are made by alloying iron with carbon and with other transition metals, such as vanadium, manganese, cobalt and nickel. Titanium is a metal with the same kind of mechanical strength as steel and two big advantages: it is less dense than steel, and it does not corrode. It is stronger than aluminium. The high cost of titanium has limited its use to applications where no expense is spared. It is used in the construction of space capsules. It is better able than steel to withstand the high temperatures that are experienced when a space capsule re-enters the Earth's atmosphere.

... and titanium

24.7 OXIDATION STATES

The electron configuration is (Ar) $4s^2 3d^n$. Once the 4s electrons have been removed, the 3d electrons may also be removed. The difference in energy between the 3d and the 4s electrons is much smaller than the difference between the 3s and the 3p electrons. The chief oxidation states employed by the elements are Sc +3, Ti +3, +4, V +4, +5, Cr +2, +3, +6, Mn +2, +4, +7, Fe +2, +3, Co +2, +3, Ni +2, Cu +2, Zn +2.

The stability of the +2 oxidation state relative to +3 and higher oxidation states increases from left to right across the series. It reflects the increasing difficulty of removing a 3d electron as the nuclear charge increases.

24.8 CATALYSIS BY TRANSITION METALS

Many important reactions are catalysed by transition metals ...

Transition metals and their compounds are important catalysts. Some industrial reactions which are catalysed by transition metals are: the Contact process (vanadium(V) oxide), the Haber process (iron or iron(III) oxide), the hydrogenation of unsaturated oils (finely divided nickel) and the oxidation of ammonia (platinum or platinum–rhodium alloy). These examples all involve heterogeneous catalysis [§ 14.6], in which the reactant molecules are adsorbed on the surface of the catalyst. It is likely that the 3d electrons enable the transition metal catalyst to form temporary bonds with reactant molecules. In homogeneous catalysis [§ 14.5], which is usually found in reactions in solution, the variable oxidation number of the transition metal may enable it to take part in a sequence of reaction stages and emerge unchanged at the end. An example is the oxidation of iodide ions by peroxodisulphate ions, $S_2O_8^{2-}$, according to the equation

$$S_2O_8^{2-}(aq) + 2I^-(aq) \longrightarrow 2SO_4^{2-}(aq) + I_2(aq) \qquad [1]$$

... The oxidation of iodide ions is an example of homogeneous catalysis ...

... due to the variable oxidation state of the transition metal

Iron(II) ions catalyse the reaction, and it is thought that they may provide an alternative route for the reaction via steps [2] and [3]:

$$Fe^{2+}(aq) + \tfrac{1}{2}S_2O_8^{2-}(aq) \longrightarrow Fe^{3+}(aq) + SO_4^{2-}(aq) \qquad [2]$$

$$Fe^{3+}(aq) + I^-(aq) \longrightarrow Fe^{2+}(aq) + \tfrac{1}{2}I_2(aq) \qquad [3]$$

The alternative route involves two reactions between oppositely charged ions and therefore has a lower activation energy [§ 14.3] than reaction [1] between ions of the same charge. The Fe^{2+} catalyst is regenerated in step [3].

24.9 PARAMAGNETISM

Transition metal ions with unpaired electron spins are paramagnetic

Paramagnetic substances are weakly attracted by a magnetic field. Any species with an unpaired electron is paramagnetic because there is a magnetic moment associated with the spinning electron. Transition metal ions that have unpaired d electrons are paramagnetic. The greater the number of unpaired electrons, the more paramagnetic is the ion. The metals, iron, cobalt and nickel are **ferromagnetic**, that is, they are strongly attracted to a magnetic field.

24.10 COMPLEX COMPOUNDS

Transition metal ions form complex ions by coordination

Transition metals form complexes or coordination compounds see § 12.10. Such complexes are formed by the coordination of lone pairs of electrons from a donor (called a **ligand**) to an atom or cation (called an **acceptor**) which has empty orbitals to accommodate them. A cation may form a complex with a neutral molecule, e.g.

$$[Cu(NH_3)_4]^{2+}$$

or with an oppositely charged ion, e.g.

$$[CuCl_4]^{2-}$$

An atom may form a complex, e.g., $Ni(CO)_4$. The charge remaining on the central atom or ion when the ligands are removed together with their lone pairs is the **oxidation number** of the metal in the complex. The **coordination number** is the number of atoms forming coordinate bonds with the central atom or ion: 2, 4 and 6 are common.

A ligand shares a lone pair of electrons with a transition metal ion

Ligands must possess one or more unshared pairs of electrons. A ligand which can only form one bond to a central atom or ion is called a **monodentate** (literally, 'one tooth') ligand, e.g., NH_3, CN^-. A **polydentate** ('many teeth') ligand can form more than one bond. Examples are the ethanedioate ion

Polydentate ligands form chelates ...

$$\begin{array}{c} CO_2^- \\ | \\ CO_2^- \end{array}$$

and ethane-1,2-diamine $H_2\ddot{N}CH_2CH_2\ddot{N}H_2$

in which the lone pairs on the two nitrogen atoms can form coordinate bonds. The two coordinate bonds formed by each ligand are thought to resemble the claws of a crab (Greek: *chele*), and such compounds are named **chelate compounds** or **chelates** [see Figure 24.16C, § 24.16.4]. In the polydentate ligand,

... edta is an important ligand

$$\begin{array}{ccc} {}^-O_2C{-}H_2C & & CH_2{-}CO_2^- \\ & \diagdown\; N{-}CH_2{-}CH_2{-}N \;\diagup & \\ & \diagup \qquad\qquad\qquad \diagdown & \\ {}^-O_2C{-}H_2C & & CH_2{-}CO_2^- \end{array}$$

which is called **edta** (short for its old name), there are unshared electron pairs on four oxygen atoms and on the two nitrogen atoms. This ligand forms six coordinate bonds, and its complex ions are very stable. Zinc ions and most other metal ions can be estimated by complexometric titration against a solution of edta [§ 12.10].

24.10.1 NAMING

Formula gives central atom followed by ligands

In the formula of a complex ion, the symbol for the central atom appears first and is followed by the anionic ligands and then by neutral ligands, e.g.

$[CoCl_2(NH_3)_4]^+$

The formula for the complex ion may be enclosed in square brackets.

The name of the complex gives the name and oxidation state of the central metal cation, e.g., cobalt(III), preceded by the name and number of ligands attached to it, e.g., hexaamminecobalt(III) ion

$[Co(NH_3)_6]^{3+}$

The prefixes

di tri tetra penta hexa

are used to show the number of ligands. If several ligands are present, they are listed in alphabetical order, and the prefixes, di, tri, etc., are not allowed to alter this order, e.g.

$[CrCl_2(H_2O)_4]^+$

Tetraaquadichlorochromium(III) ion

The system for naming complex ions

If the complex is an anion, the suffix -ate follows the name of the metal, e.g., zincate and chromate. If the metal has a Latin name, then in the complex anion the Latin name of the metal is used, followed by the suffix -ate e.g.

$[Fe(CN)_6]^{4-}$

Hexacyanoferrate(II)

Aqua ions can be present in the solid state, e.g. $FeSO_4 . 7H_2O$ and $CoCl_2 . 6H_2O$.

24.10.2 COLOUR

Transition metal ions are often coloured because electrons move between non-degenerate d orbitals

Transition metal ions are often coloured. In an isolated transition metal atom, the five d orbitals are **degenerate**, that is, they are all at the same energy level. In a complex ion, the d orbitals differ slightly in energy as a result of overlapping differently with the ligands: they are **non-degenerate**. Electrons can jump from one d orbital to another if they absorb energy. For most transition metal complexes, the frequency of light absorbed in these energy transitions is in the visible region of the spectrum, and the ion appears coloured. The colour of the ion is complementary to the colours absorbed. The Sc^{3+}(aq) ion has no d electrons, and is colourless. In the ions Cu^+ and Zn^{2+}, with a d^{10} configuration, no d–d transition is possible, and these ions are colourless. Different ligands affect the energy levels of the d orbitals: $[Cu(H_2O)_4]^{2+}$ is blue, whereas $[Cu(NH_3)_4]^{2+}$ is a very intense deep blue.

CHECKPOINT 24.10: PROPERTIES OF TRANSITION METALS

1. Explain why the first transition series (Sc to Zn) is a similar set of elements in *(a)* size *(b)* first ionisation energy *(c)* chemical properties.

2. *(a)* Why is iron a stronger metal than magnesium?
   (b) Why does iron form two ions whereas magnesium forms only one?

3. *(a)* Mention three applications of transition metals as catalysts in industry.
   (b) List three transition metals which are ferromagnetic.
   (c) What is the condition for an atom or an ion to be paramagnetic?

4. In each of the complex ions *(a)* $[Cu(NH_3)_4]^{2+}$ and *(b)* $[CuCl_4]^{2-}$ name the acceptor and the ligand. Give sketches to show how the ligand bonds to the acceptor [see § 4.7 if necessary]. Why do the complexes have the charges shown?

FURTHER STUDY ▼

24.11 TRANSITION METALS IN GEMSTONES

Gemstones are valued for their ability to reflect light. They appear coloured when part of the light travelling through them is absorbed within the crystal structure. The cause of the absorption is usually small traces of impurities; for example, diamonds often contain nitrogen, and can be yellow, brown or green, depending on the amount of nitrogen present.
The commonest metals in gemstones are transition metals:

- chromium, which gives the red colour of ruby and the green of emeralds
- iron, which gives the red of garnet, the green of spinel and peridot, the blue of sapphire, the greenish-blue of aquamarine and the green and brown of jades
- manganese which gives the pink and orange of garnets
- titanium which, together with iron, gives the bluest sapphires
- copper which gives the sky-blue of turquoise and the green of malachite.

TRANSITION METAL I

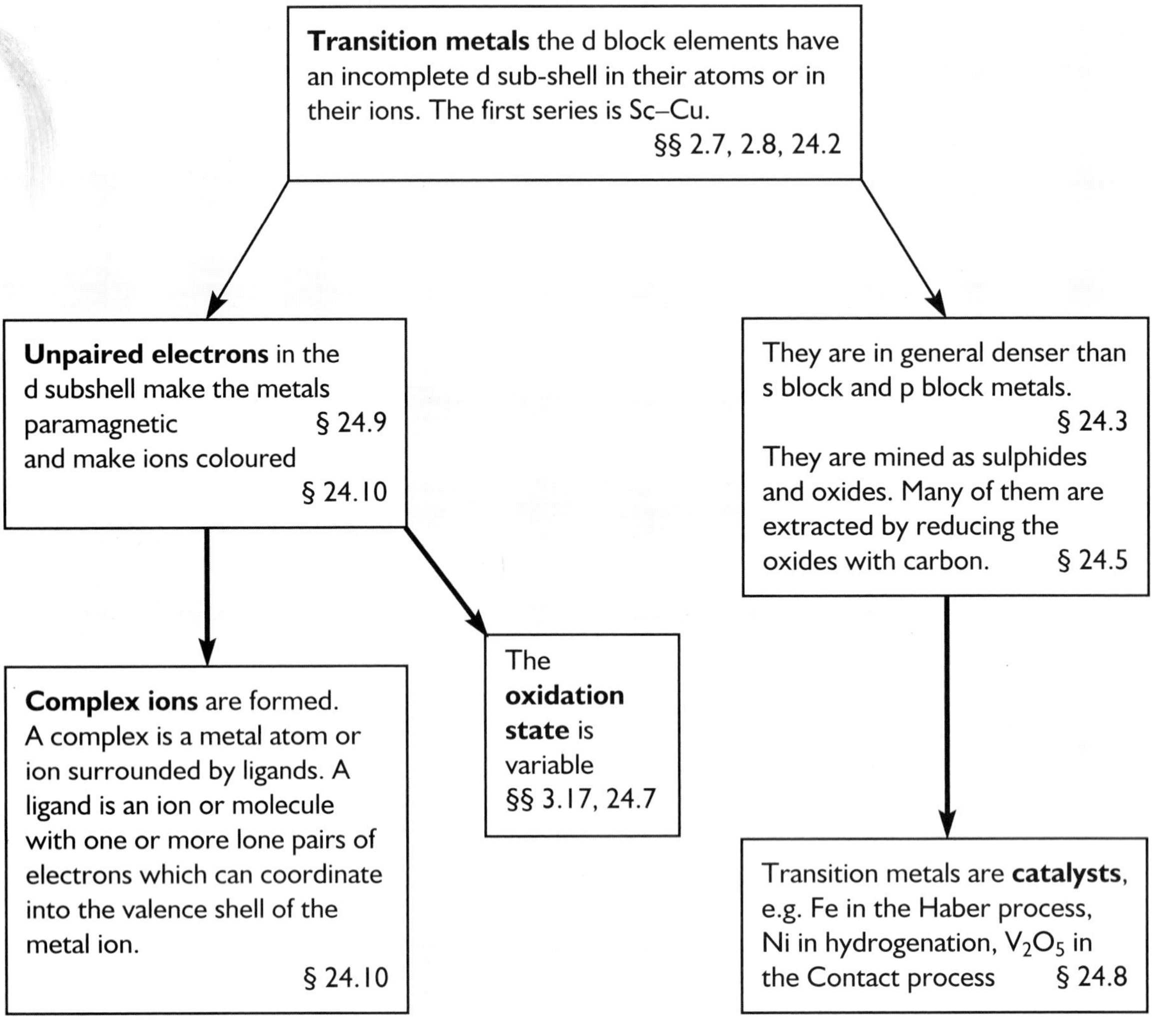

Transition metals the d block elements have an incomplete d sub-shell in their atoms or in their ions. The first series is Sc–Cu. §§ 2.7, 2.8, 24.2

Unpaired electrons in the d subshell make the metals paramagnetic § 24.9 and make ions coloured § 24.10

Complex ions are formed. A complex is a metal atom or ion surrounded by ligands. A ligand is an ion or molecule with one or more lone pairs of electrons which can coordinate into the valence shell of the metal ion. § 24.10

The **oxidation state** is variable §§ 3.17, 24.7

They are in general denser than s block and p block metals. § 24.3 They are mined as sulphides and oxides. Many of them are extracted by reducing the oxides with carbon. § 24.5

Transition metals are **catalysts**, e.g. Fe in the Haber process, Ni in hydrogenation, V_2O_5 in the Contact process § 24.8

24.12 OXIDES AND HYDROXIDES OF TRANSITION METALS

The oxides are insoluble, black or coloured, with covalent character ...

Transition metals react with oxygen to form oxides, with the exception of those (e.g., Ag, Au) that are low in the electrochemical series. The oxides are nearly all insoluble in water and either black or coloured. The covalent character of the bonds is appreciable.

24.12.1 BASICITY

... some are acidic, some basic, some amphoteric

The basicity of the oxides of the transition metals in an oxidation state of +2 decreases from left to right across the series. For any one metal, the basicity of the oxides decreases as the oxidation state of the metal increases.

24.12.2 REDUCTION OF OXIDES

The metals can be extracted from oxides

Iron

Transition metal oxides can be reduced to the metal. For the less electropositive metals (excluding Ti and V), carbon and carbon monoxide are often used as the reducing agents. In the blast furnace for the extraction of iron, carbon monoxide is the reducing agent [§ 24.17.1]. The ore *chromite*, $FeO.Cr_2O_3$, is reduced to an alloy of iron and chromium by heating it with carbon:

$$FeO \cdot Cr_2O_3(s) + 4C(s) \longrightarrow Fe(s) + 2Cr(s) + 4CO(g)$$

Ferrochrome alloy

The ferrochrome alloy produced is used in the production of stainless steel.

Titanium

Titanium is mined as rutile, TiO_2 and other oxide ores which contain iron. Titanium is not extracted from titanium(IV) oxide because of the high value of the enthalpy of this reaction, 800 kJ mol^{-1}. Titanium(IV) oxide is converted into titanium(IV) chloride and this is reduced by magnesium to titanium. For the conversion,

$$TiO_2(s) + 2Cl_2(g) + 2C(s) \longrightarrow TiCl_4(g) + 2CO(g)$$

the values of both ΔH and ΔS are favourable. The steps in the conversion are illustrated in Figure 24.12A.

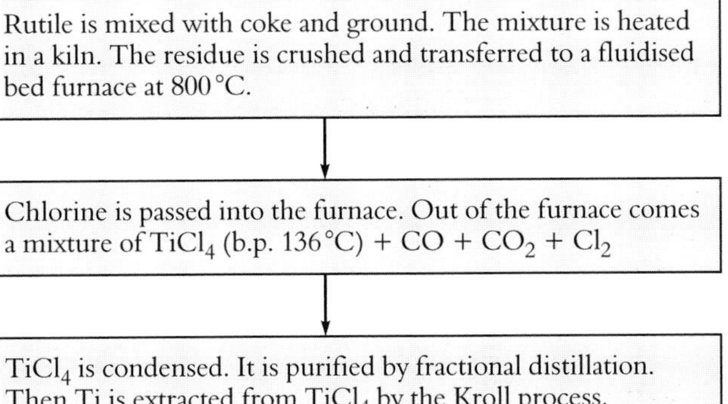

Rutile is mixed with coke and ground. The mixture is heated in a kiln. The residue is crushed and transferred to a fluidised bed furnace at 800 °C.
Chlorine is passed into the furnace. Out of the furnace comes a mixture of $TiCl_4$ (b.p. 136 °C) + CO + CO_2 + Cl_2
$TiCl_4$ is condensed. It is purified by fractional distillation. Then Ti is extracted from $TiCl_4$ by the Kroll process.

Figure 24.12A The conversion of rutile, TiO_2, into titanium(IV) chloride, $TiCl_4$

The Kroll process (named after its inventor) for the reduction of titanium(IV) chloride is illustrated in Figure 24.12B. The process is a **batch process**. Titanium forms inside the reactor, which must be allowed to cool so that the 'sponge' of titanium can be scraped out. The reactor is cleaned after each batch. The titanium sponge from the reactor is purified either by leaching with acid to remove magnesium and magnesium chloride (some titanium reacts with the acid) or by distillation under reduced pressure. Then the metal is melted in an electric arc furnace and converted into ingots.

Nickel

Nickel is obtained by reducing nickel(II) oxide with carbon.

Thermit reactions use Al to reduce metal oxides

The more electropositive elements cannot be obtained from their oxides by reduction with carbon. Aluminium is used. It reduces metal oxides with the evolution of heat in a set of reactions called **thermit reactions**. Chromium(III) oxide, Cr_2O_3, vanadium(V) oxide, V_2O_5 and cobalt(II) dicobalt(III) oxide, Co_3O_4, are reduced in this way:

$$Cr_2O_3(s) + 2Al(s) \longrightarrow 2Cr(s) + Al_2O_3(s)$$

Figure 24.12B The Kroll process for the reduction of titanium(IV) chloride to titanium

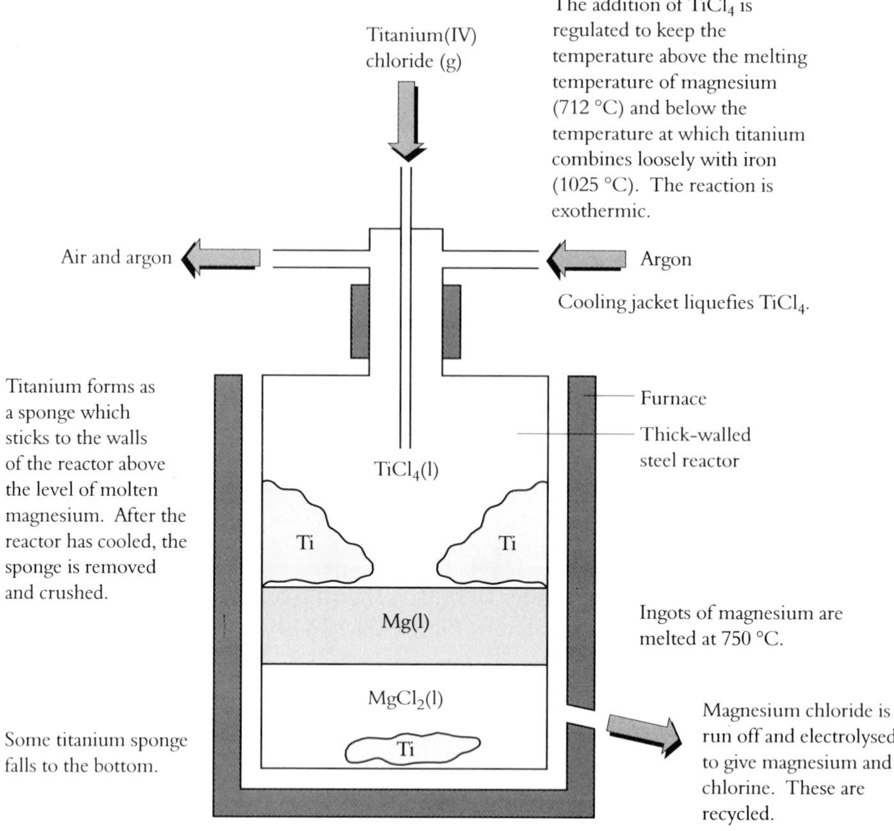
Titanium(IV) chloride (g)
The addition of $TiCl_4$ is regulated to keep the temperature above the melting temperature of magnesium (712 °C) and below the temperature at which titanium combines loosely with iron (1025 °C). The reaction is exothermic.
Air and argon
Argon
Cooling jacket liquefies $TiCl_4$.
Titanium forms as a sponge which sticks to the walls of the reactor above the level of molten magnesium. After the reactor has cooled, the sponge is removed and crushed.
Furnace
Thick-walled steel reactor
$TiCl_4$(l)
Ti
Ti
Mg(l)
Ingots of magnesium are melted at 750 °C.
$MgCl_2$(l)
Ti
Some titanium sponge falls to the bottom.
Magnesium chloride is run off and electrolysed to give magnesium and chlorine. These are recycled.

24.12.3 HYDROXIDES

The hydroxides can be made by precipitation

The hydroxides of transition metals are precipitated from solutions of the metal ions by the addition of hydroxide ions. The colour of the precipitate can often be used to identify the metal present. All the precipitates are gelatinous, owing to hydration, and all are basic. Some are amphoteric, and some form soluble complex ions with ammonia [see Table 24.12].

Table 24.12 Transition metal hydroxides

The precipitates are gelatinous and often coloured ...

... Some dissolve in ammonia

Cation	Precipitate	Colour	Reaction with NaOH(aq)	Reaction with NH_3(aq)
Cr^{3+}(aq)	$Cr(OH)_3$	Green	Chromate(III) ion, CrO_3^{3-}(aq)	–
Mn^{2+}(aq)	$Mn(OH)_2$	Beige	–	–
Fe^{2+}(aq)	$Fe(OH)_2$	Green	–	–
Fe^{3+}(aq)	$Fe(OH)_3$	Rust	–	–
Co^{2+}(aq)	$Co(OH)_2$	Pink	Cobaltate(II) ion, $Co(OH)_4^{2-}$(aq)	$Co(NH_3)_6^{2+}$(aq)
Ni^{2+}(aq)	$Ni(OH)_2$	Green	–	$Ni(NH_3)_6^{2+}$(aq)
Cu^{2+}(aq)	$Cu(OH)_2$	Blue	–	$Cu(NH_3)_4^{2+}$(aq)
Zn^{2+}(aq)	$Zn(OH)_2$	White	Zincate ion, $Zn(OH)_4^{2-}$(aq)	$Zn(NH_3)_4^{2+}$(aq)

24.13 OXO-IONS OF TRANSITION METALS

In their higher oxidation states, the transition metals occur, not as cations $\mathbf{M}^{2+}$, but combined with oxygen as anions $\mathbf{MO}_4^{3-}$ and $\mathbf{MO}_4^{2-}$. The most important of these are the vanadate(V) ion, $V_3O_9^{3-}$; chromate(VI), CrO_4^{2-}; dichromate(VI),

Dichromate(VI) and manganate(VII) are powerful oxidising agents

$Cr_2O_7^{2-}$; manganate(VI), MnO_4^{2-}; and manganate(VII) MnO_4^-. The sodium and potassium salts of these anions are soluble in water. Their power as oxidising agents enables them to be used in titrimetric analysis.

24.13.1 CHROMATES

When a chromium(III) salt is heated in alkaline solution with a peroxide, a chromate(VI) ion is formed:

$$2Cr^{3+}(aq) + 4OH^-(aq) + 3O_2^{2-}(aq) \longrightarrow 2CrO_4^{2-}(aq) + 2H_2O(l)$$

The preparation of chromate(VI) and of dichromate(VI)

The Cr^{3+}(aq) ion is blue-violet; the CrO_4^{2-}(aq) ion is yellow. In acid solution, yellow chromate(VI) ions condense to form orange dichromate(VI) ions:

$$\underset{\text{Yellow chromate(VI) ion}}{2CrO_4^{2-}(aq) + 2H^+(aq)} \rightleftharpoons \underset{\text{Orange dichromate(VI) ion}}{Cr_2O_7^{2-}(aq) + H_2O(l)}$$

The oxidation state of chromium is +6 in both ions, as can be seen from the structures:

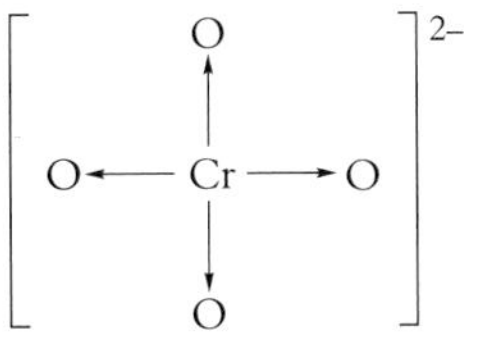

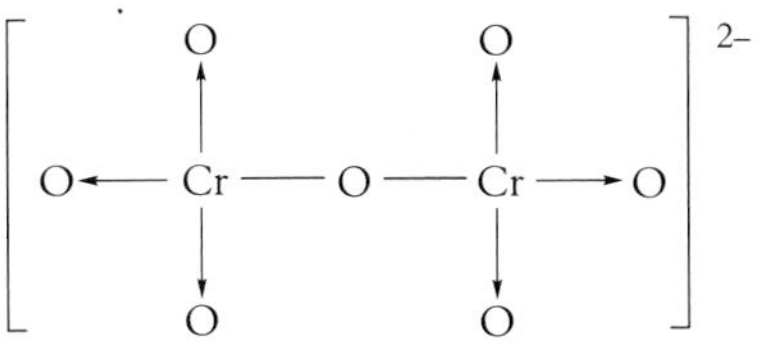

Potassium dichromate(VI) is an oxidising agent used in titrimetric analysis

For complexes of chromium see § 24.16.4

Potassium dichromate(VI) is used as an oxidising agent in titrimetric analysis. Since it can be obtained in a high degree of purity and is not deliquescent, it can be used as a primary standard. In the half-reaction [see § 3.15.1 for half-reactions, and § 13.2.1 for $E^\ominus$]

$$Cr_2O_7^{2-}(aq) + 14H^+(aq) + 6e^- \longrightarrow 2Cr^{3+}(aq) + 7H_2O(l); \quad E^\ominus = +1.33\ V$$

there is a colour change from orange to blue-violet. To give a sharper end-point, a redox indicator such as barium diphenylamine sulphonate is added. Potassium dichromate(VI) in acid solution can be used to estimate iron(II) salts, ethanedioates, iodides and other reducing agents [§ 3.15.1]. It can be used in the presence of chloride ions since the standard reduction potential for the system

$$Cl_2(g) + 2e^- \rightleftharpoons 2Cl^-(aq)$$

is +1.36 V, and chloride ions are not oxidised by potassium dichromate(VI).

24.13.2 MANGANATES

Potassium manganate(VII) is a powerful oxidising agent used in titrimetric analysis ...

Potassium manganate(VII), often called potassium permanganate, is widely used in acid solution as an oxidising agent in titrimetric analysis:

$$MnO_4^-(aq) + 8H^+(aq) + 5e^- \rightleftharpoons Mn^{2+}(aq) + 4H_2O(l); \quad E^\ominus = +1.51\ V$$

... It oxidises Cl^- ions

With a standard reduction potential of +1.51 V, it is a more powerful oxidising agent than potassium dichromate(VI), for which $E^\ominus = +1.33$ V. Potassium manganate(VII) cannot be used in solutions containing chloride ions, as it oxidises them to chlorine. Solutions of potassium manganate(VII) are kept in brown bottles because in the presence of light they slowly oxidise water to oxygen. Substances which can be estimated by titration against acidified potassium manganate(VII) are iron(II) salts, hydrogen peroxide and ethanedioates [§ 3.15.1].

CHECKPOINT 24.13: OXIDES AND OXO-IONS

1. Which of the oxides is the most strongly basic?

   (a) CrO or FeO or NiO
   (b) MnO_2 or Mn_2O_3 or MnO

2. *(a)* What reducing agent is employed to extract titanium from its compounds?
   (b) Why is the ore titanium(IV) oxide converted into the chloride before reduction?
   (c) What reducing agent is used to convert nickel(II) oxide into nickel?
   (d) What is the thermit reaction? Name three metals which are obtained by this method.
   (e) Why is the thermit reaction not used in the extraction of iron?

3. Aqueous sodium hydroxide is added separately to solutions of salts of the transition metals **A**, **B** and **C**. Identify **A**, **B** and **C** from the following observations.
   A: The white precipitate which appears is soluble in an excess of aqueous sodium hydroxide and also in aqueous ammonia.
   B: The blue precipitate which appears is insoluble in an excess of aqueous sodium hydroxide but dissolves in aqueous ammonia to form a deep blue solution.
   C: The green precipitate which appears is insoluble in an excess of aqueous sodium hydroxide and also in aqueous ammonia.

4. *(a)* Write the equation for converting chromate(VI) ions into dichromate(VI) ions. State the colour change that occurs.
   (b) Give reasons for the use of potassium dichromate(VI) in titrimetric analysis.
   (c) In which type of oxidation is potassium dichromate(VI) preferred to potassium manganate(VII)?

5. When manganese(IV) oxide is melted with potassium hydroxide and an oxidising agent potassium manganate(VI) is formed. In acid solution, potassium manganate(VI) undergoes a reaction for which the unbalanced equation is:

$$a\mathrm{MnO_4}^{2-}(\mathrm{aq}) + b\mathrm{H}^+(\mathrm{aq}) \longrightarrow c\mathrm{MnO_4}^-(\mathrm{aq}) + d\mathrm{MnO_2}(\mathrm{s}) + e\mathrm{H_2O}(\mathrm{l})$$

   (a) Balance the equation.
   (b) Explain why it is described as disproportionation.
   (c) State the colours and names of MnO_2(s) and MnO_4^-(aq).
   (d) Write the half-reaction equation for the action of MnO_4^- as an oxidising agent.
   (e) Name three substances which can be oxidised by MnO_4^-.

24.14 CHLORIDES

The reactions of transition metals with halogens

Chlorine attacks all the metals in the first transition series. Bromine and iodine also react with metals of the first transition series, although the metals may need to be heated to speed up their reactions with iodine.

The anhydrous chloride is made by heating crystals of the hydrate in a stream of dry HCl(g)

Hydrated chlorides can be made by reacting the metal or the metal oxide with hydrochloric acid and allowing the solution to crystallise. An attempt to obtain an anhydrous chloride by heating a hydrate results in hydrolysis and the formation of a basic chloride. If iron(II) chloride crystals are heated, iron(II) chloride hydroxide is formed:

$$\mathrm{FeCl_2.6H_2O(s)} \rightleftharpoons \mathrm{Fe(OH)Cl(s) + HCl(g) + 5H_2O(g)}$$

Anhydrous chlorides are made by synthesis

Anhydrous iron(II) chloride can be obtained, however, if a stream of dry hydrogen chloride is passed over the heated crystals. Equilibrium is driven over to the left. Anhydrous chlorides are usually made by reacting the metal with a stream of dry hydrogen chloride or chlorine. The apparatus shown in Figure 24.14A can be used for the preparation of anhydrous iron(III) chloride. The product sublimes over as molecules of Fe_2Cl_6. If water is added, Fe^{3+}(aq) and Cl^-(aq) ions are formed. If dry hydrogen chloride is passed over heated iron, anhydrous iron(II) chloride is formed.

Most transition metal chlorides are macromolecular. Ions are formed when the macromolecular chlorides are either melted or dissolved in water.

Many chlorides of the transition metals are soluble

Most transition metal chlorides are soluble; exceptions are copper(I) chloride, silver chloride and mercury(I) chloride. The ions are stabilised in solution by hydration with the formation of complex aqua ions, e.g., $Fe(H_2O)_6^{3+}$. In the presence of chloride ions, many transition metal chlorides form soluble chloride complex ions, e.g., $CuCl_4^{2-}$.

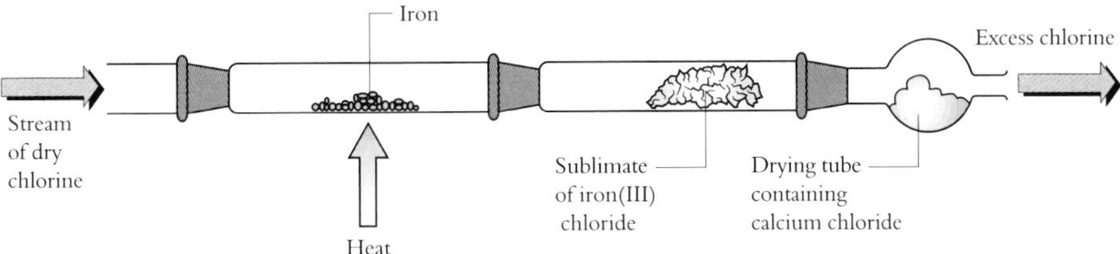

Figure 24.14A Preparation of iron(III) chloride

24.14.1 SALT HYDROLYSIS

The charge/radius ratio of the Fe^{3+} ion is high compared with ions such as Na^+ and Mg^{2+}. Being small and highly charged, the Fe^{3+} ion exerts a powerful attraction for water molecules [see Figure 24.14B(a)]. The coordination of a water molecule occurs through the donation of a lone pair of electrons by the oxygen atom. As a result the electrons of the O—H bond move closer to the oxygen atom and the hydrogen atoms have more positive charge than in free H_2O molecules [see Figure 24.14B(b)].

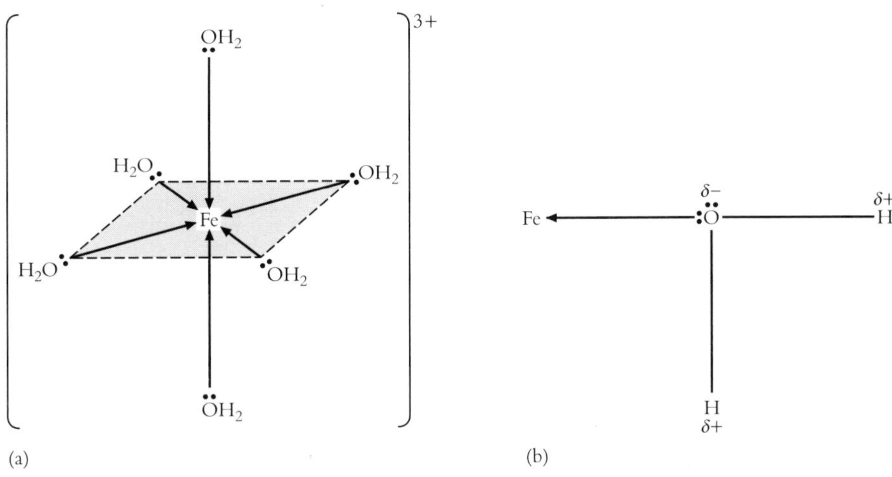

Figure 24.14B (a) $[Fe(H_2O)_6]^{3+}$ (b) Polarity of O—H bonds

The partial positive charges on the hydrogen atoms attract bases, for example the water molecules of the solvent, which may abstract protons from the coordinated water molecules

$$[Fe(H_2O)_6]^{3+}(aq) + H_2O(l) \rightleftharpoons [Fe(OH)(H_2O)_5]^{2+}(aq) + H_3O^+(aq)$$

As they do so hydrogen ions are formed and the solution becomes acidic. This is an example of **salt hydrolysis**.

Why is iron(III) carbonate, $Fe_2(CO_3)_3$, not formed?

When aqueous sodium carbonate Na_2CO_3 is added to an aqueous iron(III) salt, the product is not iron(III) carbonate, $Fe_2(CO_3)_3$. Instead a precipitate of hydrated iron(III) hydroxide, $Fe(OH)_3$ is formed. In the equilibrium shown above, if stronger bases, e.g. CO_3^{2-} are present, H_3O^+ ions are removed and the equilibrium moves to the right-hand side

$$[Fe(OH)(H_2O)_5]^{2+}(aq) + CO_3^{2-}(aq) \rightleftharpoons [Fe(OH)_2(H_2O)_4]^+(aq) + HCO_3^-(aq)$$

$$[Fe(OH)_2(H_2O)_4]^+ + CO_3^{2-} \rightleftharpoons Fe(OH)_3(H_2O)_3(s) + HCO_3^-(aq)$$

A precipitate of hydrated iron(III) hydroxide is formed. Similar reactions take place with $[Al(H_2O)_6]^{3+}$ salts and $[Cr(H_2O)_6]^{3+}$ salts.

24.15 SULPHIDES

Sulphides are black or coloured and macromolecular

Many transition metals are found as sulphide ores, for example, FeS_2, $CuFeS_2$, CuS, MnS, NiS, Ag_2S and HgS. The sulphides are black or coloured and are macromolecular. When sulphide ores are roasted in air, the metal oxide and sulphur dioxide are formed. The oxide can be reduced to yield the metal.

CHECKPOINT 24.15: SALTS

1. Explain why Fe^{3+} forms bonds with more covalent character than does Na^+.
2. Explain why aqueous $FeCl_3$ is ionic whereas anhydrous $FeCl_3$ is molecular.
3. State how you could prepare *(a)* $FeCl_2 . 6H_2O$ crystals, *(b)* anhydrous $FeCl_2$, *(c)* anhydrous $FeCl_3$, *(d)* $FeCl_3(aq)$.
4. When aqueous sodium carbonate is added to aqueous iron(II) sulphate, a precipitate of $FeCO_3$ forms, but when it is added to aqueous iron(III) sulphate the precipitate is $Fe(OH)_3(s)$. Why is $Fe_2(CO_3)_3$ not formed?
5. What method is generally employed to obtain a metal from a sulphide ore?

24.16 COMPLEX IONS

24.16.1 STEREOCHEMISTRY

The coordination number of the central metal ion or atom in a complex is the number of ligands bonded to it. This depends on the number of vacant orbitals of the metal ion or atom available for the formation of coordinate bonds. In general – but not always – the coordination number of an ion is the same for different ligands. The ion Ag^+ has a coordination number of 2 and a linear distribution of bonds. The arrangement of bonds in some complexes with coordination number 4 is tetrahedral, e.g. $Zn(NH_3)_4^{2+}$ and $Ni(CO)_4$. In others it is square planar, e.g. $CuCl_4^{2-}$ and $Ni(CN)_4^{2-}$. Chromium Cr^{2+} and Cr^{3+} and iron Fe^{2+} and Fe^{3+} and many other transition metal ions have coordination number 6 and an octahedral arrangement of bonds. The shapes of some complex ions are shown in Figure 24.16A.

Table 24.16

Coordination number	Shape of complex	Examples
2	linear	$[Ag(NH_3)_2]^+$
4	tetrahedral	$[Zn(NH_3)_4]^{2+}$, $[NiCl_4]^{2-}$
4	square planar	$[CuCl_4]^{2-}$, $[Ni(CN)_4]^{2-}$
6	octahedral	$[Co(NH_3)_6]^{3+}$, $[Fe(CN)_6]^{3-}$

Figure 24.16A The shapes of some complex ions

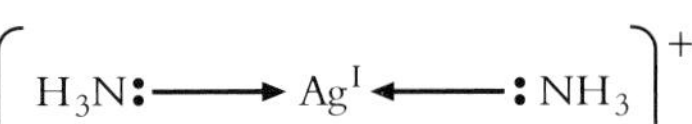

Diamminesilver(I) ion, coordination number 2, linear

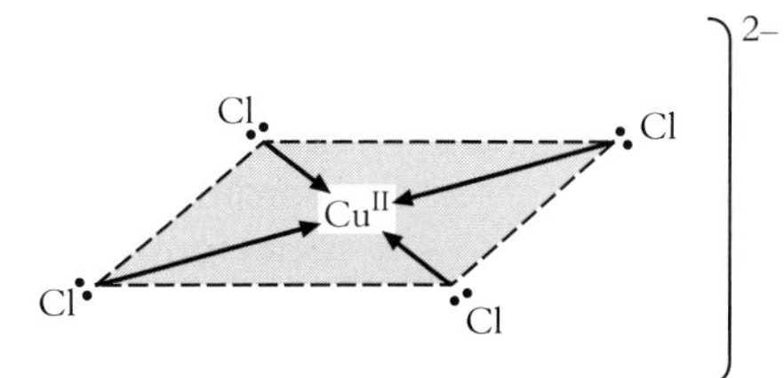

Tetrachlorocuprate(II) ion, $CuCl_4^{2-}$, coordination number 4, square planar

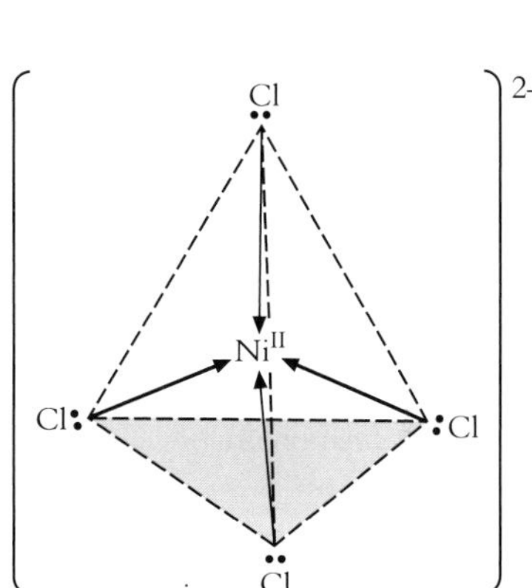

Tetrachloronickelate(II) ion, coordination number 4, tetrahedral

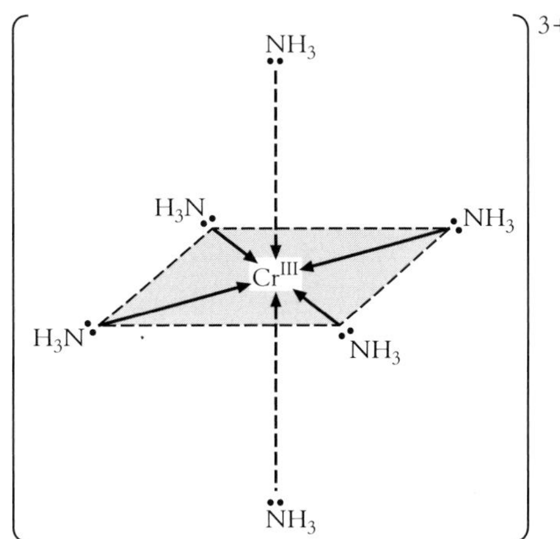

Hexaamminechromium(III) ion, coordination number 6, octahedral

24.16.2 PLATINUM COMPLEXES FIGHT CANCER

Platinum is a transition metal. In 1969, B. Rosenberg discovered that a number of platinum complex compounds have anti-tumour activity. The compound *cis*-$PtCl_2(NH_3)_2$, which is known as 'cisplatin', is especially potent. Together with the compound $PtCl_2(H_2NCH_2CH_2NH_2)$, which has a similar action, cisplatin provides one of the best current treatments for some types of cancer. The arrangement of bonds about the platinum atom is square planar.

```
Cl    NH3
  \  /
   Pt
  /  \
Cl    NH3
```

cis-diamminedichloroplatinum (cisplatin)

```
Cl    NH2
  \  /    \
   Pt      (CH2)2
  /  \    /
Cl    NH2
```

dichloroethane-1,2-diamineplatinum

The mode of action is not understood, but since only *cis*-isomers are effective it suggests that the two chlorine ligands may form a chelate complex. It is thought that DNA is involved. The two Cl ligands in cisplatin might be replaced by the nitrogen atoms of two bases in neighbouring positions of one strand of DNA. The formation of such a complex would distort the structure of DNA and prevent its replication (the synthesis of a new strand of DNA).

24.16.3 LIGAND EXCHANGE

A ligand contains a lone pair (or pairs) of electrons which form coordinate bonds to an acceptor atom or ion. The ligand is acting as a Lewis base [§ 12.6.1]. The strength of ligands as Lewis bases varies from ligand to ligand. A ligand which is a strong Lewis base can displace a weaker Lewis base from a complex ion. The ligand NH_3 is a stronger Lewis base than H_2O. These two ligands are similar in size and one can be exchanged for the other without a change in the coordination number of the acceptor atom. The ion Cl^- is larger and many metal ions can accommodate fewer Cl^- ions than H_2O or NH_3 ligands. Consequently there is a change in coordination number when chloride ion replaces a smaller ligand. The following ligand exchanges illustrate this.

1. Copper(II) complexes

$$[Cu(H_2O)_6]^{2+}(aq) \underset{H_2O}{\overset{NH_3}{\rightleftharpoons}} [Cu(NH_3)_4(H_2O)_2]^{2+}(aq) \underset{NH_3}{\overset{Cl^-}{\rightleftharpoons}} [CuCl_4]^{2-}(aq)$$

Blue	*Deep blue*	*Yellow*
Hexaaquacopper(II) ion	Hexaamminediaquacopper(II) ion	Tetrachlorocuprate(II) ion
coordination no. 6	coordination no. 6	coordination no. 4
octahedral	octahedral	square planar

2. Cobalt(II) complexes

$$[Co(H_2O)_6]^{2+}(aq) \underset{H_2O}{\overset{NH_3}{\rightleftharpoons}} [Co(NH_3)_6]^{2+}(aq) \underset{NH_3}{\overset{Cl^-}{\rightleftharpoons}} [CoCl_4]^{2-}(aq)$$

Pink	*Brown*	*Blue*
Hexaaquacobalt(II) ion	Hexaamminecobalt(II) ion	Tetrachlorocobaltate(II) ion
coordination no. 6	coordination no. 6	coordination no. 4
octahedral	octahedral	square planar

H_2O ⇅ OH^-

$[Co(OH)_4]^{2-}(aq)$

Blue

Tetrahydroxycobaltate(II) ion

coordination number 4, square planar

3. Iron(II) complexes

$$[Fe(H_2O)_6]^{2+}(aq) + 6CN^-(aq) \longrightarrow [Fe(CN)_6]^{4-}(aq) + 6H_2O(l)$$

Hexaaquairon(II) ion → Hexacyanoferrate(II) ion

4. Copper(II) complexes

Aqueous copper(II) sulphate is blue and contains $[Cu(H_2O)_6]^{2+}$ ions. When concentrated hydrochloric acid is added dropwise to the solution the solution changes to green and then to yellow.

$$[Cu(H_2O)_6]^{2+}(aq) + 4Cl^-(aq) \rightleftharpoons [CuCl_4(H_2O)_2]^{2-}(aq) + 4H_2O(l)$$

Hexaaquacopper(II) ion → Diaquatetrachlorocopper(II) ion

If concentrated ammonia is added dropwise to the yellow solution, it turns a deep blue.

$$[CuCl_4(H_2O)_2]^{2-}(aq) + 4NH_3(aq) \rightleftharpoons [Cu(NH_3)_4(H_2O)_2]^{2+}(aq) + 4Cl^-(aq)$$

Diaquatetrachlorocopper(II) ion → Tetraamminediaquacopper(II) ion

The order of ligand strengths as Lewis bases is $CN^- > NH_3 > Cl^- > H_2O$. The stability of a complex ion is measured by its **stability constant**. For example the tetraamminediaquacopper(II) ion:

$$[Cu(H_2O)_6]^{2+}(aq) + 4NH_3(aq) \rightleftharpoons [Cu(NH_3)_4(H_2O)_2]^{2+}(aq) + 4H_2O(l)$$

has a stability constant K_{st} given by

$$K_{st} = \frac{[[Cu(NH_3)_4(H_2O)_2]^{2+}(aq)]}{[[Cu(H_2O)_6]^{2+}(aq)][NH_3(aq)]^4}$$

*24.16.4 STEREOISOMERISM

If the six ligands of the octahedron are not identical, isomerism occurs

In a complex with six ligands, if the six ligands are not identical, *cis-trans* geometrical isomerism will occur. In the tetraamminedichlorochromium(III) ions, shown in Figure 24.16B, the *cis*-form (a) has two chlorine ligands adjacent, whereas in the *trans*-form (b), the chlorine atoms are diagonally opposite.

Stereoisomerism

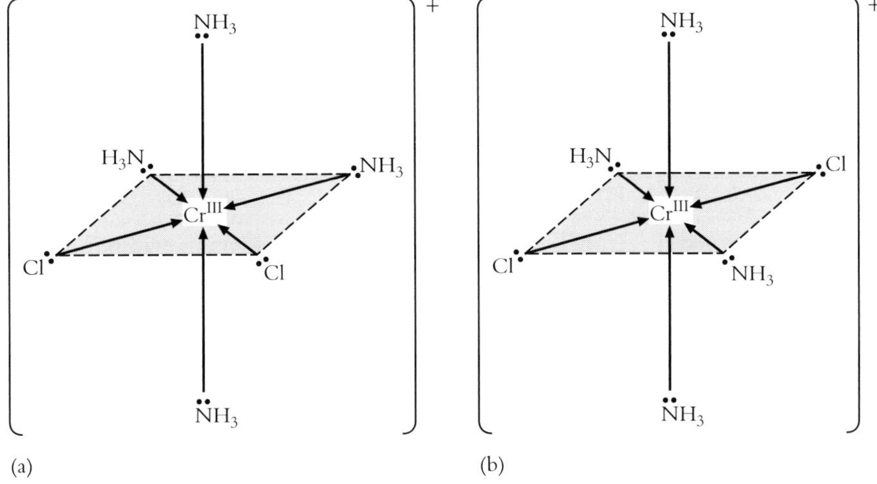

Figure 24.16B *Cis*- and *trans*-tetraammine-dichlorochromium(III) ions

Bidentate ligands, such as the ethanedioate ion

$$^{-}O_2C{-}CO_2^{-}$$

and ethane-1,2-diamine

$$H_2\ddot{N}CH_2CH_2\ddot{N}H_2$$

Optical isomerism occurs in complexes with bidentate ligands

form complexes which show optical isomerism. Figure 24.16C shows how the two chelate complexes, exist as mirror image forms or **enantiomers**.

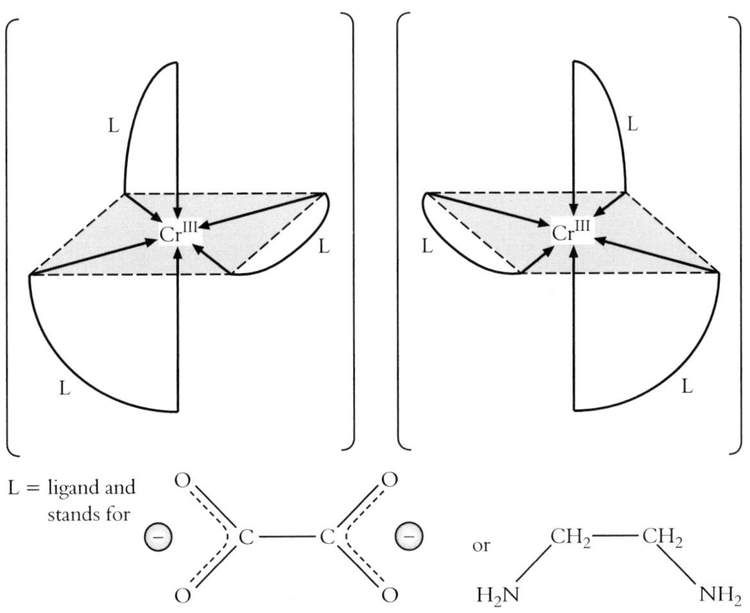

Figure 24.16C Enantiomers of $[Cr(C_2O_4)_3]^{3-}$ and $[Cr(C_2N_2H_8)_3]^{3+}$

The complex $Cr(H_2O)_6Cl_3$ crystallises as four isomers. In one isomer three Cl^- ions are free; in the others one or two are ligands bonded to the central Cr atom. The isomers are

$$[Cr(H_2O)_6]^{3+}3Cl^-$$
$$[Cr(H_2O)_5Cl]^{2+}2Cl^- . H_2O$$
cis- and *trans*- $[Cr(H_2O)_4Cl_2]^+Cl^- . 2H_2O$

A method of distinguishing between the isomers is to find the mass of silver chloride precipitated from a certain mass of the compound. The free chloride ions precipitate with aqueous silver nitrate; the ligands do not. This method does not distinguish between the *cis*- and *trans*- geometrical isomers.

24.16.5 IRON COMPLEXES

Complexes are formed by Fe^{2+} and Fe^{3+} with CN^- and H_2O

Iron(II) and iron(III) ions employ six hybrid bonds to form the complexes hexacyanoferrate(II), $[Fe(CN)_6]^{4-}$, and hexacyanoferrate(III), $[Fe(CN)_6]^{3-}$ [see Figure 24.16D]. So stable are these complexes that the hydroxides $Fe(OH)_2$ and $Fe(OH)_3$ are not precipitated when hydroxide ions are added to solutions of the complex ions. Both iron(II) and iron(III) ions are hydrated in solution as $[Fe(H_2O)_6]^{2+}$ and $[Fe(H_2O)_6]^{3+}$ [see Figure 24.16D].

Thiocyanate ions are used in a test for $Fe^{3+}(aq)$

Iron(III) ions form a blood-red complex with thiocyanate ions, $[Fe(SCN)(H_2O)_5]^{2+}(aq)$. No such complex is formed by iron(II) ions. The formation of this blood-red complex is therefore used to distinguish iron(III) from iron(II).

Figure 24.16D
(a) Hexacyanoferrate(II) ion
(b) Hexaaquairon(III) ion

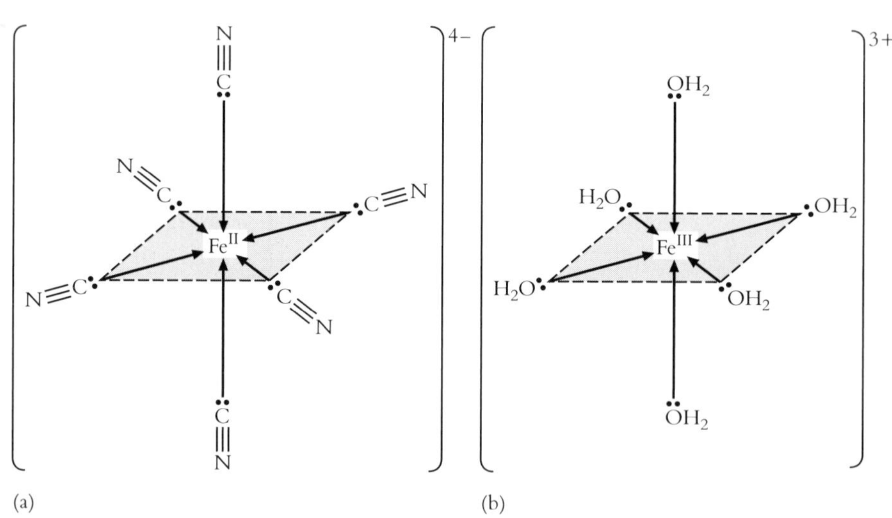

Complexes are formed by haemoglobin

Haemoglobin contains iron in the oxidation state +2, with coordination number 6. Figure 24.16E shows how oxygen molecules can form coordinate bonds to the iron atoms. The bonding is reversible, and enables haemoglobin to carry oxygen around the body and release it where it is needed. Carbon monoxide and cyanide ions coordinate more strongly than oxygen to form very stable complexes. They prevent haemoglobin from taking up oxygen, thus acting as poisons.

Figure 24.16E
Part of the haemoglobin molecule

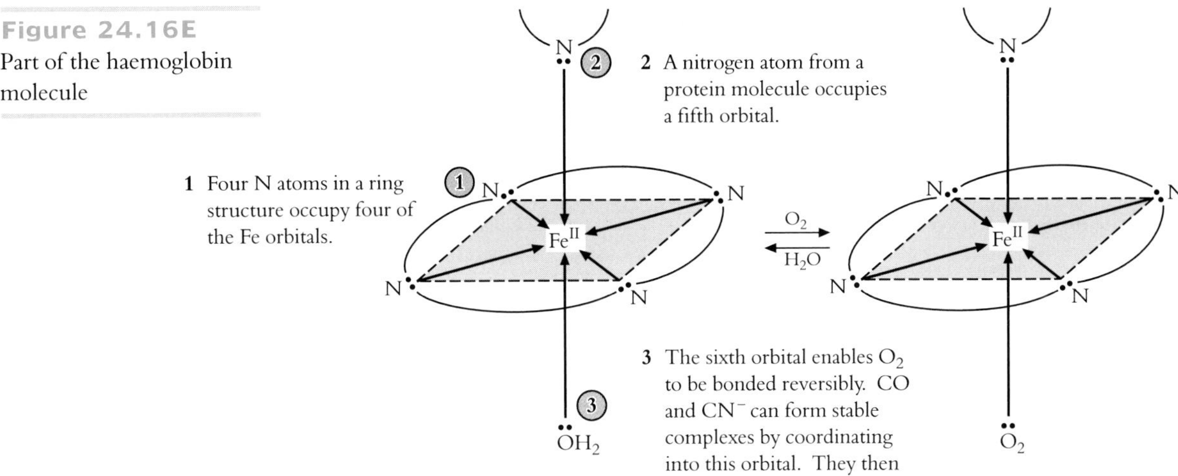
1 Four N atoms in a ring structure occupy four of the Fe orbitals.
2 A nitrogen atom from a protein molecule occupies a fifth orbital.
3 The sixth orbital enables O_2 to be bonded reversibly. CO and CN^- can form stable complexes by coordinating into this orbital. They then prevent the uptake of O_2.
N
Fe^{II}
OH_2
O_2
H_2O

24.16.6 COMPLEXES OF COPPER(II) IONS

Cu^{2+} uses four orbitals for complex formation

Copper(II) ions use four bonds in complex formation. The spatial arrangement of the bonds in the hydrated $[Cu(H_2O)_4]^{2+}$ ion is square planar. In solution, two water molecules are loosely coordinated at right angles in a distorted octahedral configuration [similar to Figure 24.16D(b)]. Ammonia displaces water molecules from the pale blue tetraaquacopper(II) ions, converting them into the deep blue tetraamminecopper(II) ions, $[Cu(NH_3)_4]^{2+}$.

The six ligands take up an octahedral arrangement

In concentrated hydrochloric acid, water molecules in $[Cu(H_2O)_4]^{2+}$ are replaced by chloride ions, with the formation of yellow tetrachlorocuprate(II) ions, $[CuCl_4]^{2-}$

Cl^- and edta complex with Cu^{2+}

$$\underset{Blue}{[Cu(H_2O)_4]^{2+}(aq)} + 4Cl^-(aq) \rightleftharpoons \underset{Yellow}{[CuCl_4]^{2-}(aq)} + 4H_2O(l)$$

In more dilute hydrochloric acid, the solution is green due to the presence of both blue $[Cu(H_2O)_4]^{2+}$ ions and yellow $[CuCl_4]^{2-}$ ions. Copper(II) ions also form complexes with ethane-1,2-diamine and with edta. Copper(II) is the most strongly complexing of all the first long period dipositive cations.

24.16.7 COBALT IN ANAEMIA AND IN ALLOYS

There is a wasting disease in sheep and cattle known as 'pine' in the UK, 'bush sickness' in New Zealand, 'coast disease' in Australia and 'salt sick' in the USA. It has been known for 100 years and was diagnosed as an anaemic condition and treated with iron salts. The treatment was sometimes successful and sometimes disappointing. In the 1930s it was found that the iron treatment worked only because of the cobalt impurity in the iron, but the role of cobalt was not understood. It became clearer when vitamin B_{12} was extracted from the liver and shown to be an effective treatment for anaemia and also to contain cobalt. It is now known that vitamin B_{12} is a coenzyme in the formation of red blood cells. It contains complexed cobalt(III) ions. Cobalt is obtained from dairy products and meat in the diet.

CHECKPOINT 24.16: COMPLEXES

1. *(a)* Define the term complex ion.
   (b) Why are the transition metals able to form complexes?
   (c) Why are these complexes coloured?

2. Name the following complex ions.

   $[Cr(H_2O)_6]^{3+}$, $[Al(OH)_4]^-$,
   $[Cu(NH_3)_4]^{2+}$, $[CuCl_4]^{2-}$

3. State the oxidation number and the coordination number of the metal in each of these complexes:

   (a) $Ag(NH_3)_2NO_3$
   (b) $Ni(CO)_4$
   (c) $K_3Fe(CN)_6$
   (d) $K_4Fe(CN)_6$
   (e) $[CuCl_4]^{2-}$
   (f) $[CuCl_4]^{3-}$
   (g) $Zn(NH_3)_4SO_4$

*4. Explain, with the aid of a sketch, why these ions show stereoisomerism,

   $[CrCl_2(H_2NCH_2CH_2NH_2)_2]^+$ and
   $[Co(H_2NCH_2CH_2NH_2)_3]^{3+}$

5. Sketch the arrangement of bonds in the complexes *(a)* hexaaquacobalt(III) ion *(b)* hexacyanoiron(III) ion *(c)* diamminesilver(I) ion and *(d)* the complex compound tetracarbonylnickel.

6. The addition of aqueous ammonia to aqueous copper(II) sulphate resulted in the formation of a pale blue precipitate. This dissolved on the addition of more aqueous ammonia to give a deep blue solution. The precipitate also dissolved in dilute hydrochloric acid to give a pale blue solution and in concentrated hydrochloric acid to give a green solution. Explain these colour changes, giving the name and formula of each coloured species formed.

7. The ligand Cl^- often displaces the ligand NH_3 in complex ions. What name is given to this behaviour? Why is Cl^- able to do this? How can the strength of Cl^- and NH_3 as ligands be compared quantitatively?

24.17 IRON

Iron is the most important of metals. The whole of our twenty first century way of life is based on the use of iron machinery. We shall look in more detail at iron than the rest of the transition metals.

24.17.1 EXTRACTION

Iron is mined as its oxides and sulphides

Iron is mined as its oxides, Fe_2O_3, *haematite*, and Fe_3O_4, *magnetite*, and FeS_2, *iron pyrites*. It is obtained by the reduction of the oxides by carbon monoxide

$$Fe_2O_3(s) + 3CO(g) \longrightarrow 2Fe(s) + 3CO_2(g); \quad \Delta H^\ominus = -27 \text{ kJ mol}^{-1}$$

The **blast furnace**, in which this process takes place, is illustrated in Figure 24.17A. At the bottom of the furnace, coke is oxidised exothermically to carbon dioxide

$$C(s) + O_2(g) \longrightarrow CO_2(g); \quad \Delta H^\ominus = -392 \text{ kJ mol}^{-1}$$

and the temperature of the furnace is about 1900 °C in this region. Higher up the furnace, carbon dioxide reacts with coke to form carbon monoxide. This reaction is endothermic, and the furnace in this region has a temperature of 1100 °C

$$CO_2(g) + C(s) \longrightarrow 2CO(g); \quad \Delta H^\ominus = 172 \text{ kJ mol}^{-1}$$

Carbon monoxide reduces iron oxides in the blast furnace

Iron oxides are reduced exothermically, and the iron produced falls to the bottom of the furnace, where the temperature is high enough to melt it, and a layer of molten iron lies on the bottom of the furnace. At the same time, the limestone in the charge dissociates to form calcium oxide and carbon dioxide

$$CaCO_3(s) \rightleftharpoons CaO(s) + CO_2(g); \quad \Delta H^\ominus = 178 \text{ kJ mol}^{-1}$$

Iron oxide + coke + limestone produce iron + 'slag'

Calcium oxide combines with silicon(IV) oxide and aluminium oxide, the impurities in the ore, to form a molten 'slag' of calcium silicate(IV) and calcium aluminate(III), which trickles down the stack

$$CaO(s) + SiO_2(s) \longrightarrow CaSiO_3(l)$$
$$CaO(s) + Al_2O_3(s) \longrightarrow CaAl_2O_4(l)$$

The blast furnace runs continuously

At the bottom of the furnace, iron and slag are tapped off every few hours. A modern furnace makes 3000 tonnes of iron in a day, using 3000 tonnes of coke and 4000 tonnes of air. If natural gas is injected with the hot air, the consumption of coke can be halved. The quantity of slag produced is about 1 tonne for every tonne of iron. It is used for road making and in the manufacture of cement. A furnace can operate continuously for several years before it needs relining.

24.17.2 CAST IRON

Cast iron or pig iron contains carbon, which decreases its ductility

Wrought iron is the purest form of iron

The iron leaving the blast furnace is run into moulds, where it forms solid blocks called 'pigs'. This *pig iron* or *cast iron* contains about 4% carbon. The carbon present lowers the melting temperature of the iron, increases its hardness and decreases its ductility. Strength increases up to 1% carbon and then decreases. If the carbon is removed, a malleable iron is produced, which can be readily worked by ironsmiths. It is called *wrought iron*.

Figure 24.17A
The blast furnace

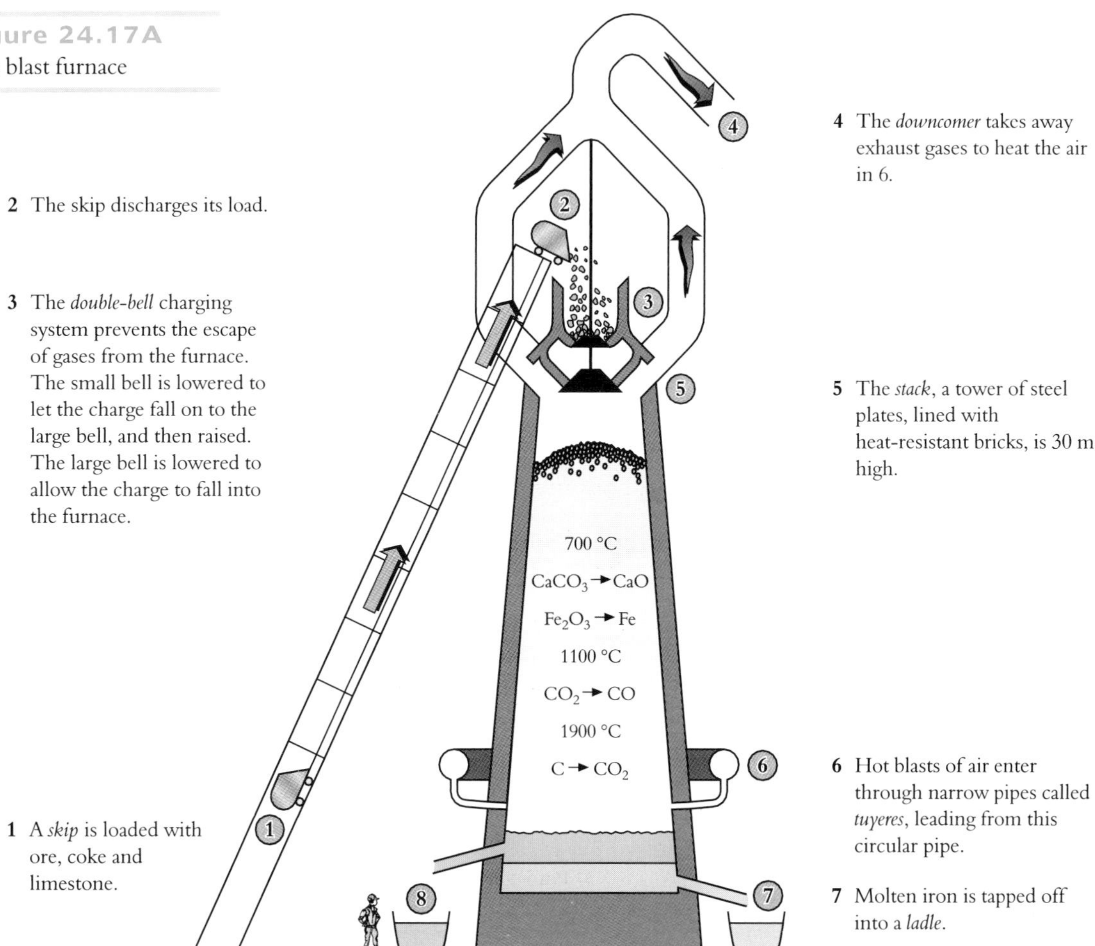

4 The *downcomer* takes away exhaust gases to heat the air in 6.
2 The skip discharges its load.
3 The *double-bell* charging system prevents the escape of gases from the furnace. The small bell is lowered to let the charge fall on to the large bell, and then raised. The large bell is lowered to allow the charge to fall into the furnace.
5 The *stack*, a tower of steel plates, lined with heat-resistant bricks, is 30 m high.
700 °C
$CaCO_3 \rightarrow CaO$
$Fe_2O_3 \rightarrow Fe$
1100 °C
$CO_2 \rightarrow CO$
1900 °C
$C \rightarrow CO_2$
6 Hot blasts of air enter through narrow pipes called *tuyeres*, leading from this circular pipe.
1 A *skip* is loaded with ore, coke and limestone.
7 Molten iron is tapped off into a *ladle*.
8 Slag is run off.

24.17.3 STEEL

Steels contain carbon and other metals alloyed with iron

Some of the iron leaving the blast furnace is not cast; it is run into giant crucibles and transported to the steel-making section of the iron and steel works. Steels contain less than 1.5% carbon, and contain added metals. Many thousands of different steels are made, with different properties suited to different uses.

Making steel requires the removal of C, S, P

In making steel from iron, carbon and other impurities such as sulphur and phosphorus are converted into their oxides. Gaseous oxides remove themselves; other oxides are removed by combination with a base such as calcium oxide to form a slag.

The Bessemer process

Oxygen-blown converters are used in modern steelworks

The LD process ...

... and the Kaldo process

In the Bessemer process, invented in Britain 125 years ago, molten iron was poured into a large tub, the *converter*, and air was blown on to it to oxidise the impurities. Modern steelworks use oxygen-blown converters. The steel is better because it contains no nitrogen, which makes steel brittle. Shown in Figure 24.17B is the LD converter, named after the Austrian towns of Linz and Donawitz, where it was invented. One converter makes about 500 tonnes of steel in an hour. The Kaldo process used in Sweden is similar, but employs a converter which can be rotated. Oxygen is blown in at a pressure of only 2 atm; the rotation assists oxidation.

Figure 24.17B
The LD process of steel making

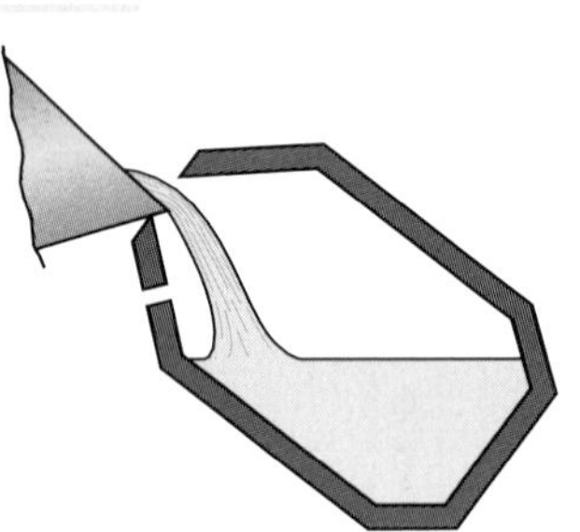

1 **Charging.** The *converter* tips to receive a charge of 300 tonnes of molten iron.

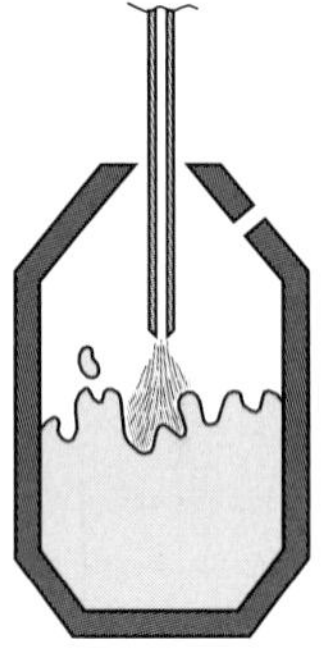

2 **The first blow.** The water-cooled *lance* directs oxygen and powdered calcium oxide on to the surface at a pressure of 10–15 atm

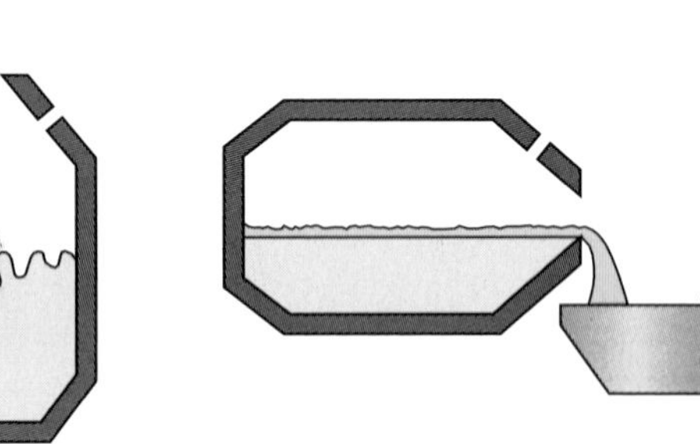

3 **Slagging.** The converter tips backwards to pour out the primary slag.

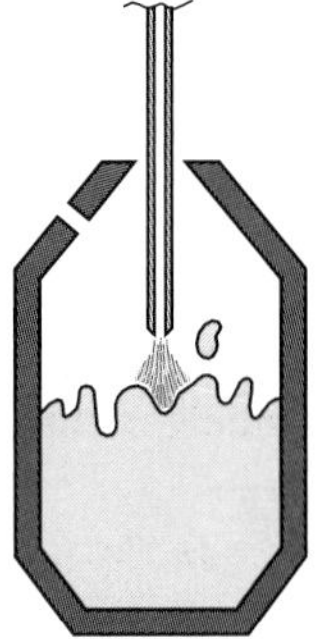

4 **The second blow.** The *lance* directs oxygen and lime into the converter.

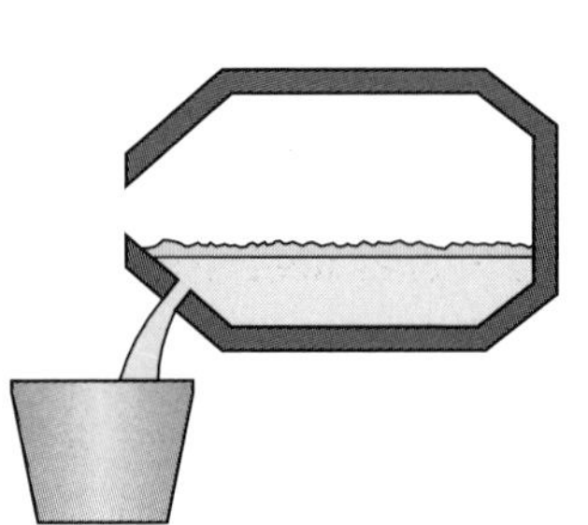

5 **Pouring.** The converter tilts forward to pour the steel into a *ladle*, while the slag remains on top.

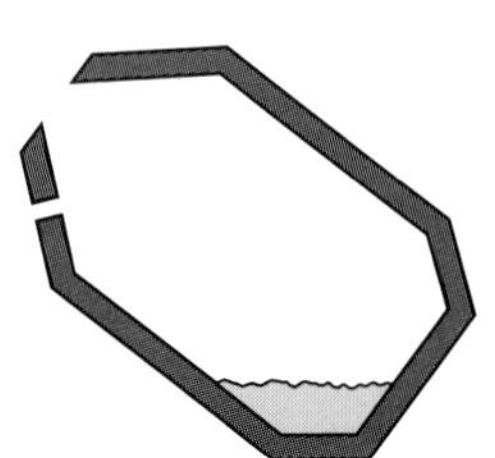

6 Slag remains in the converter, which tilts to receive the next charge.

Figure 24.17C
Charging the converter

24.17.4 A LIFE-SAVING ALLOY

Primo Levi was an Italian chemist born in 1919. During the Second World War he joined a partisan group fighting against the fascist goverment of Italy. He was captured in 1943 and sent to the concentration camp at Auschwitz. Fortunately for Levi, he and a fellow partisan called Alberto were taken from the camp each day to work in a chemical plant, making buna rubber for the German war effort. In his book *The Periodic Table*, Levi describes how his knowledge of chemistry saved his life [§ 15.1].

In his laboratory Primo Levi found a jar containing 40 metallic rods. He and Alberto tested the rods and found that they sparked when scratched with a knife. They were able to identify the rods as iron–cerium alloy. Alberto worked out a strategy for turning the rods into lighter flints for a secret industry which some prisoners had, making cigarette lighters which they sold to civilian workers in the chemical plant. For 120 flints the two men were able to buy 120 rations of bread – 60 days of life for the two of them.

*24.17.5 NICKEL ALLOYS

The most important element for alloying with iron to make engineering steels is nickel. Nickel steels have greater strength and toughness, especially at low temperatures, down to −200 °C. Nickel steels include the following:

- 'Superalloy steels' are used in aircraft, gas turbine engines, nuclear reactors, power generators and space vehicles.
- Heat-resisting alloys, e.g. nickel–chromium, are used in e.g. electric cooker hobs and industrial furnaces.
- The ferromagnetic alloy *Alnico* (Al, Ni, Co) is used in generators, electric motors, mass spectrometers and other equipment.

The production of pure nickel

Nickel is mined as the sulphide. This is roasted in air to form nickel oxide, which is reduced to nickel by heating with carbon (or hydrogen). The metal is obtained in an impure form. The Mond process for purifying nickel depends on the formation of the complex tetracarbonyl nickel(0), $Ni(CO)_4$, which is volatile. Impure nickel is fed into a kiln where it meets a counter flow of carbon monoxide at 60 °C. The volatile $Ni(CO)_4$ is distilled off, leaving impurities behind. On heating to 200 °C the complex decomposes to give 99.99% pure nickel and carbon monoxide.

$$Ni(s) + 4CO(g) \underset{\text{heat at } 200^\circ C}{\overset{\text{CO at } 60^\circ C}{\rightleftharpoons}} Ni(CO)_4(g)$$

24.17.6 PREVENTION OF RUSTING

Rusting is a serious problem. The electrochemical reactions that take place have been discussed in § 13.4. Methods of preventing or delaying rusting include the following.

Coating

Rusting can be prevented by ...

Various methods are used to provide a protective coat to exclude water and oxygen.

... paint ...

(*a*) Paint is used for many large objects, e.g., ships and bridges. A paint containing phosphoric(V) acid is effective as it forms a layer of insoluble iron(III) phosphate(V) with any rust present on the surface.

... oil, grease ...

(*b*) A coat of grease or oil is used for moving parts of machinery.

... zinc coating ...
... tin plating ...

(*c*) A coat of another metal may be used; zinc coating, i.e., *galvanising*, and tin plating are used. Since zinc is higher than iron in the electrochemical series, even if the coating of zinc is scratched, it will continue to protect the iron underneath from rusting. Tin cans rust if they are scratched because, being higher in the electrochemical series, iron is corroded in preference to tin.

... chromium plating ...

(*d*) Chromium plating is used for many car accessories because it is decorative as well as protective. An electrolytic method is used for plating [see Figure 24.17D].

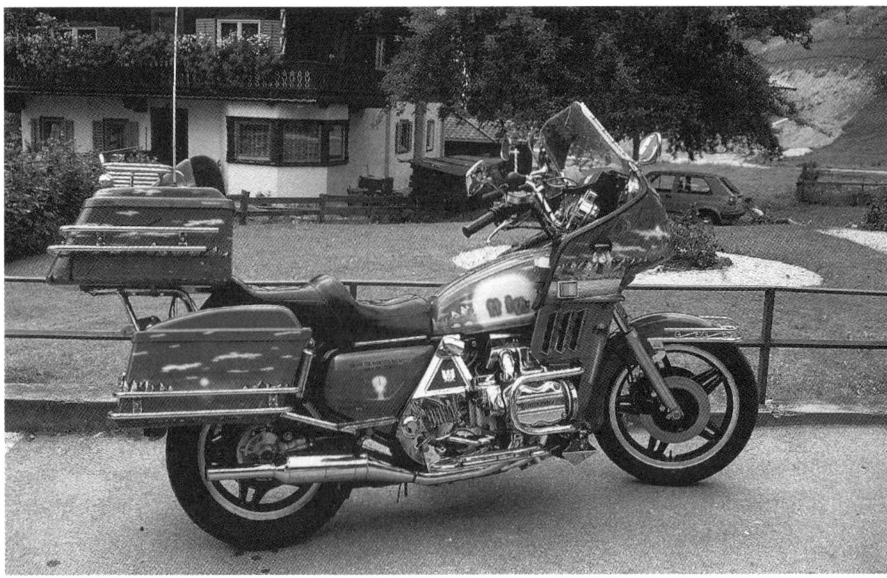

Figure 24.17D
Decorative chromium plating

Alloying

Stainless steel is an alloy of iron with nickel and chromium (e g. 18% Ni, 8% Cr). The added metals produce a surface film of metal oxides which is impervious to water.

Cathodic protection

The method of cathodic protection has been discussed in § 13.4.

Figure 24.17E
Many coats of paint are used to protect a car body from rusting

24.17.7 THE CHEMISTRY OF IRON

Iron combines with O_2, S, N_2, C and the halogens

Iron combines on heating with the non-metallic elements oxygen, nitrogen, the halogens, sulphur and carbon. It reacts with water and air to form rust, $Fe_2O_3 . xH_2O$, and with steam to form iron(II) iron(III) oxide, Fe_3O_4, which is magnetic

$$3Fe(s) + 4H_2O(g) \longrightarrow 4H_2(g) + Fe_3O_4(s)$$

This oxide is also formed when iron is heated in air.

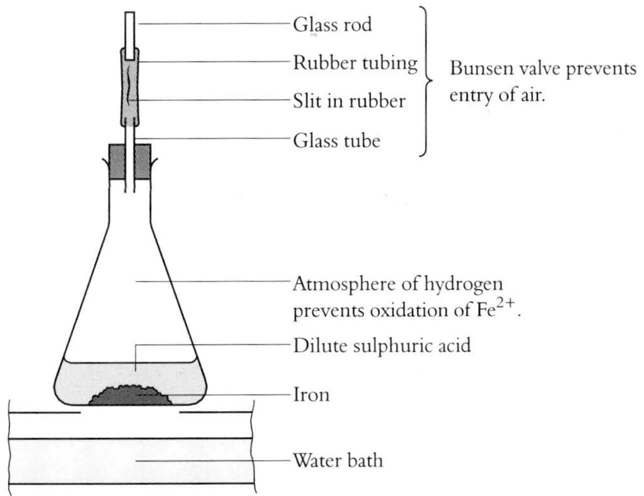

Figure 24.17F
Preparation of iron(II) sulphate-7-water, $FeSO_4 \cdot 7H_2O$
(Hydrochloric acid gives $FeCl_2 \cdot 6H_2O$.)

It reacts with dilute acids

Iron reacts with dilute sulphuric acid and hydrochloric acid to form iron(II) salts and hydrogen. It is attacked by dilute nitric acid to form iron(III) nitrate and is rendered 'passive' by concentrated nitric acid.

Table 24.17A
Iron(II) compounds

Compound	Preparation and Comments
Iron(II) sulphide, FeS, black	(a) Heat iron with sulphur. (b) Add $S^{2-}(aq)$ to a solution of either an Fe^{3+} salt or an Fe^{2+} salt: $2Fe^{3+}(aq) + S^{2-}(aq) \longrightarrow 2Fe^{2+}(aq) + S(s)$ $Fe^{2+}(aq) + S^{2-}(aq) \longrightarrow FeS(s)$
Iron(II) hydroxide, $Fe(OH)_2$, green	Add $OH^-(aq)$ to $Fe^{2+}(aq)$.
Iron(II) chloride $FeCl_2$ $FeCl_2 . 6H_2O$	Pass HCl(g) over heated iron. See Figure 24.14A.
Iron(II) sulphate $FeSO_4 . 7H_2O$, green $FeSO_4 . NO$	[See Figure 24.17F.] When heated, gives the anhydrous salt. On further heating, forms iron(III) oxide: $2FeSO_4(s) \longrightarrow Fe_2O_3(s) + SO_2(g) + SO_3(g)$ The brown ring test for nitrates involves the formation of the brown complex ion, $[Fe(NO)(H_2O)_5]^{2+}$

Table 24.17B
Iron(III) compounds

Fe^{3+} salts are acidic in solution due to hydrolysis

... see § 24.14.2

Compound	Preparation and Comments
Iron(III) oxide, Fe_2O_3, rust	(a) Mined as *haematite*. (b) Add $OH^-(aq)$ to $Fe^{3+}(aq)$. No $Fe(OH)_3$ exists; hydrated Fe_2O_3 is precipitated.
Iron(II) iron(III) oxide, Fe_3O_4 blue-black	(a) Mined as *magnetite*. (b) Heat iron in air or steam: $3Fe(s) + 4H_2O(g) \longrightarrow Fe_3O_4(s) + 4H_2(g)$ It reacts as a mixture of $FeO . Fe_2O_3$
Iron(III) halides $FeCl_3$, $FeBr_3$	Pass Cl_2 or $Br_2(g)$ over heated iron. [See Figure 24.14A.] (The reaction between Fe and I_2 gives FeI_2 because Fe^{3+} oxidises I^- to I_2.)

Table 24.17C
Tests for Fe^{2+} and Fe^{3+} ions

Reagent	Reaction of $Fe^{2+}(aq)$	Reaction of $Fe^{3+}(aq)$
$OH^-(aq)$	Gelatinous green ppt. of $Fe(OH)_2(s)$	Gelatinous rust ppt. of Fe_2O_3
Potassium hexacyanoferrate(II), $K_4Fe(CN)_6(aq)$	Green or brown colour	Prussian-blue $KFe[Fe(CN)_6](s)$
Potassium hexacyanoferrate(III), $K_3Fe(CN)_6(aq)$	Turnbull's blue, $KFe[Fe(CN)_6](s)$	Green or brown colour
Potassium thiocyanate, KCNS	No reaction	Blood-red colour, $Fe(CNS)^{2+}(aq)$

Turnbull's blue and Prussian blue are both $K^+Fe^{3+}[Fe^{II}(CN)_6]^{4-}$.

CHECKPOINT 24.17: IRON

1. Name *(a)* the four substances that are fed into a blast furnace and *(b)* the two substances removed from the blast furnace.

2. *(a)* Why is the iron produced in the blast furnace brittle?
   (b) What further treatment does iron receive after leaving the blast furnace?

3. Discuss the methods employed to prevent or delay the rusting of iron and steel.

4. Give the names and formulae of the compounds formed
   (a) when iron reacts with (i) S (ii) HCl (iii) Cl_2 (iv) I_2 (v) water and air (vi) steam and air (vii) $H_2SO_4(aq)$
   (b) when iron(II) ions react with (i) aqueous sodium hydroxide (ii) aqueous sodium cyanide.

5. Describe tests for *(a)* iron(II) ions and *(b)* iron(III) ions.

24.18 COPPER

Copper is low in the electrochemical series

Extraction and purification

Copper is low in the electrochemical series and is found 'native', i.e., uncombined. The chief ores are *copper pyrites*, $CuFeS_2$, and *copper glance*, CuS. Copper is obtained by roasting copper pyrites with silica and air in a furnace or by roasting copper glance with air. Impure copper is formed. The electrolytic method used to purify copper is illustrated in Figure 24.18A

Figure 24.18A
Purification of copper

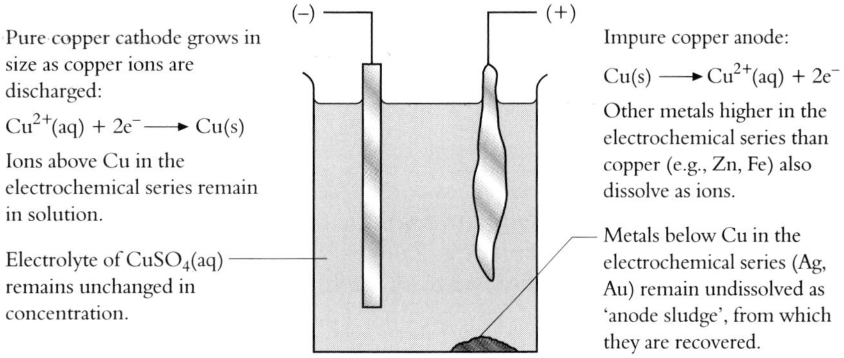

(–)

(+)

Pure copper cathode grows in size as copper ions are discharged:

$Cu^{2+}(aq) + 2e^- \longrightarrow Cu(s)$

Ions above Cu in the electrochemical series remain in solution.

Electrolyte of $CuSO_4(aq)$ remains unchanged in concentration.

Impure copper anode:

$Cu(s) \longrightarrow Cu^{2+}(aq) + 2e^-$

Other metals higher in the electrochemical series than copper (e.g., Zn, Fe) also dissolve as ions.

Metals below Cu in the electrochemical series (Ag, Au) remain undissolved as 'anode sludge', from which they are recovered.

24.18.1 USES OF COPPER

The uses of copper include cooking ware ...

... electrical cables ...

... and roofing ...

The high thermal conductivity of copper leads to its use for cooking ware. The high electrical conductivity makes copper wire admirably suitable for electrical circuits and cables. The resistance to corrosion makes copper useful for water pipes. Copper is used as a roofing material because it weathers to acquire a coating of green basic copper carbonate, $CuCO_3 . Cu(OH)_2 . nH_2O$, which lends a colourful touch to a building. Alloys of copper are *coinage metal* (Cu, Ni), *brass* (Cu, Zn) and *bronze* (Cu, Sn).

24.18.2 REACTIONS OF COPPER

Copper reacts with O_2, S, the halogens, and oxidising acids

Copper is low in the electrochemical series. It reacts with the oxidising acids, dilute nitric acid, concentrated nitric acid and concentrated sulphuric acid to form copper(II) salts. It combines directly with oxygen, sulphur and the halogens to form copper(II) compounds, except in the case of iodine, with which it forms copper(I) iodide. At temperatures of 800–1000 °C, it combines with oxygen to form copper(I) oxide.

24.18.3 COPPER COMPOUNDS

Table 24.18
Copper compounds

Compound	Preparation and Comments
Copper(II) oxide, CuO, black	Heat copper(II) nitrate, carbonate or hydroxide. Basic
Copper(II) hydroxide, $Cu(OH)_2$, blue	Add $OH^-(aq)$ to $Cu^{2+}(aq)$. A gelatinous blue precipitate forms. It dissolves in $NH_3(aq)$ to form $Cu(NH_3)_4{}^{2+}(aq)$. Basic
Copper(I) oxide, Cu_2O, reddish solid	Reduce $Cu^{2+}(aq)$ in alkaline solution with an aldehyde or by SO_2: $2Cu^{2+}(aq) + 2OH^-(aq) + 2e^- \longrightarrow Cu_2O(s) + H_2O(l)$ The formation of Cu_2O is a test for an aldehyde [§ 31.7.3]. With dilute acids, Cu(s) and a Cu^{2+} salt are formed.
Copper(I) chloride, CuCl, white solid	(a) Reduce $CuCl_2(aq)$ by Cu(s) or by $SO_2(g)$ to $[CuCl_2]^-(aq)$. (b) Add oxygen-free water. White CuCl(s) precipitates: $Cu^{2+}(aq) + 4Cl^-(aq) + Cu(s) \longrightarrow 2[CuCl_2]^-(aq)$ $[CuCl_2]^-(aq) \rightleftharpoons CuCl(s) + Cl^-(aq)$

24.19 ZINC

Zinc is extracted from zinc blende or zinc carbonate

Zinc is mined as *zinc blende*, ZnS, and as zinc carbonate, $ZnCO_3$. The sulphide ore is roasted to form the oxide and sulphur dioxide, which is used in the Contact process. Zinc oxide is reduced by coke, and, being a volatile metal, zinc can be distilled from the furnace, leaving less volatile impurities behind.

24.19.1 USES

Its uses include galvanising steel

Zinc is used in galvanising steel and in the production of *brass* (Cu, Zn).

24.19.2 REACTIONS

Zn and Zn^{2+} both have a full d subshell

Zinc resembles transition metals

Zinc is not a transition metal. It has a full d subshell of electrons, and its ions, Zn^{2+}, which also have a full d subshell, are colourless. The inability of the d electrons to take part in the metallic bonding gives zinc a lower melting temperature and boiling temperature than the transition metals. Zinc resembles transition metals in forming some complex ions and does not resemble the metals of Group 2.

It reacts with acids and alkalis

Zinc reacts with acids to form salts of the hydrated zinc(II) ion, $Zn(H_2O)_6{}^{2+}$. It also reacts with alkalis to form salts of the zincate ion, $Zn(OH)_4{}^{2-}$, with the evolution of hydrogen

$$Zn(s) + 2OH^-(aq) + 2H_2O(l) \longrightarrow [Zn(OH)_4]^{2-}(aq) + H_2(g)$$

CHECKPOINT 24.19: COPPER AND ZINC

1. Explain what happens when impure copper is purified electrolytically. Explain why the process is not disturbed by the presence of *(a)* metals higher than copper in the electrochemical series and *(b)* metals lower than copper in the electrochemical series.

2. List three uses of copper and say what properties of the metal enable it to be used in these ways.

3. Name and give the formulae of the products formed when copper(II) ions react with

   (a) $OH^-(aq)$
   (b) $Cl^-(aq, conc)$
   (c) $S^{2-}(aq)$
   (d) $CO_3{}^{2-}(aq)$
   (e) $NH_3(aq)$

4. *(a)* Name three compounds which yield copper(II) oxide on heating.
   (b) Give a reaction which produces copper(I) oxide.
   (c) Give a reaction which produces copper(I) chloride.

5. *(a)* Why is zinc, with electron configuration $(Ar)4s^2 3d^{10}$, included with the transition metals?
   (b) What uses are made of zinc?

FURTHER STUDY ▲

TRANSITION METALS II

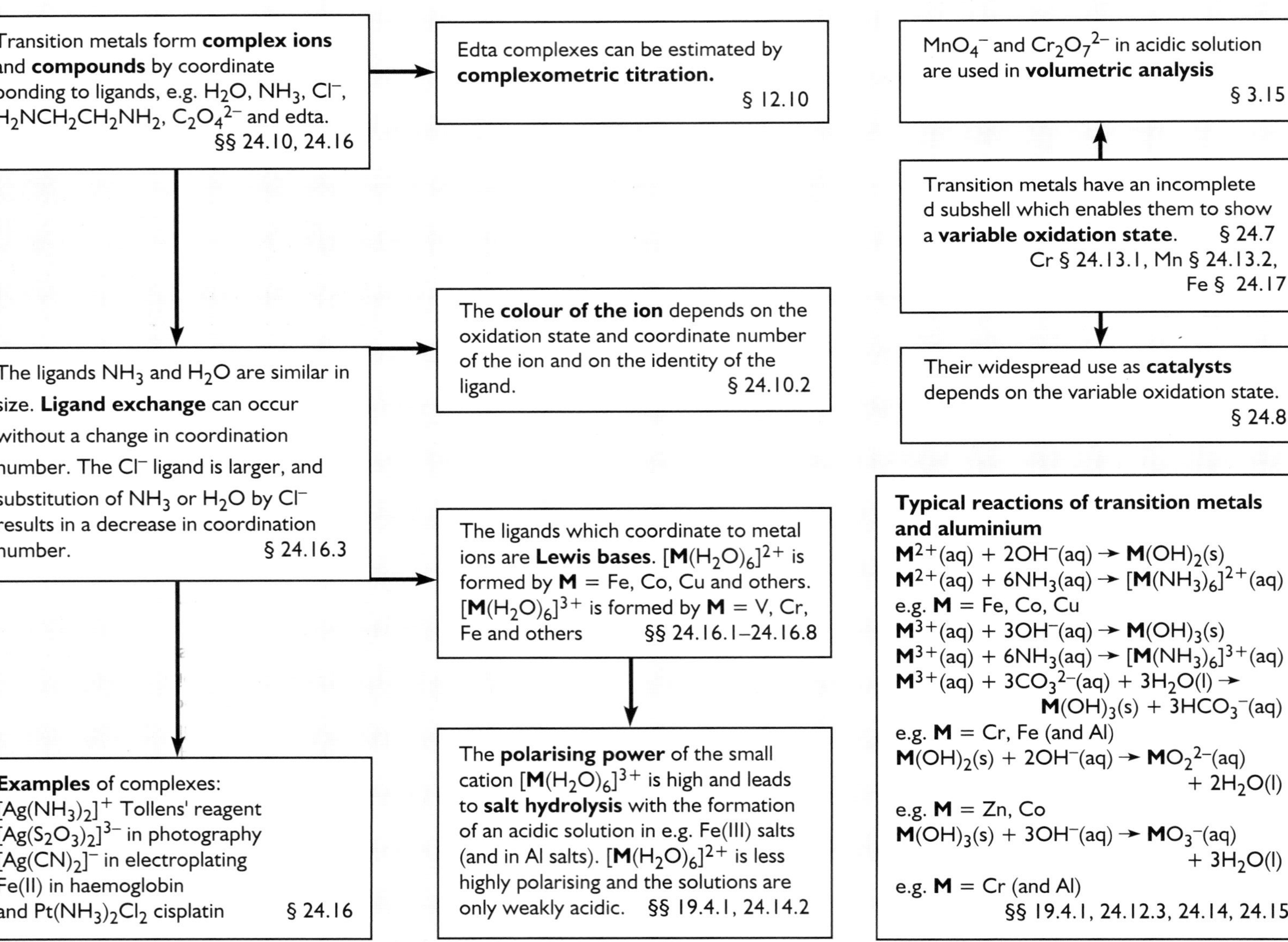

Transition metals form **complex ions** and **compounds** by coordinate bonding to ligands, e.g. H_2O, NH_3, Cl^-, $H_2NCH_2CH_2NH_2$, $C_2O_4^{2-}$ and edta. §§ 24.10, 24.16
Edta complexes can be estimated by **complexometric titration.** § 12.10
MnO_4^- and $Cr_2O_7^{2-}$ in acidic solution are used in **volumetric analysis** § 3.15
Transition metals have an incomplete d subshell which enables them to show a **variable oxidation state**. § 24.7 Cr § 24.13.1, Mn § 24.13.2, Fe § 24.17
Their widespread use as **catalysts** depends on the variable oxidation state. § 24.8
The **colour of the ion** depends on the oxidation state and coordinate number of the ion and on the identity of the ligand. § 24.10.2
The ligands NH_3 and H_2O are similar in size. **Ligand exchange** can occur without a change in coordination number. The Cl^- ligand is larger, and substitution of NH_3 or H_2O by Cl^- results in a decrease in coordination number. § 24.16.3
The ligands which coordinate to metal ions are **Lewis bases**. $[\mathbf{M}(H_2O)_6]^{2+}$ is formed by **M** = Fe, Co, Cu and others. $[\mathbf{M}(H_2O)_6]^{3+}$ is formed by **M** = V, Cr, Fe and others §§ 24.16.1–24.16.8
Typical reactions of transition metals and aluminium
$\mathbf{M}^{2+}(aq) + 2OH^-(aq) \rightarrow \mathbf{M}(OH)_2(s)$
$\mathbf{M}^{2+}(aq) + 6NH_3(aq) \rightarrow [\mathbf{M}(NH_3)_6]^{2+}(aq)$
e.g. **M** = Fe, Co, Cu
$\mathbf{M}^{3+}(aq) + 3OH^-(aq) \rightarrow \mathbf{M}(OH)_3(s)$
$\mathbf{M}^{3+}(aq) + 6NH_3(aq) \rightarrow [\mathbf{M}(NH_3)_6]^{3+}(aq)$
$\mathbf{M}^{3+}(aq) + 3CO_3^{2-}(aq) + 3H_2O(l) \rightarrow \mathbf{M}(OH)_3(s) + 3HCO_3^-(aq)$
e.g. **M** = Cr, Fe (and Al)
$\mathbf{M}(OH)_2(s) + 2OH^-(aq) \rightarrow \mathbf{M}O_2^{2-}(aq) + 2H_2O(l)$
e.g. **M** = Zn, Co
$\mathbf{M}(OH)_3(s) + 3OH^-(aq) \rightarrow \mathbf{M}O_3^-(aq) + 3H_2O(l)$
e.g. **M** = Cr (and Al)
§§ 19.4.1, 24.12.3, 24.14, 24.15
Examples of complexes:
$[Ag(NH_3)_2]^+$ Tollens' reagent
$[Ag(S_2O_3)_2]^{3-}$ in photography
$[Ag(CN)_2]^-$ in electroplating
Fe(II) in haemoglobin
and $Pt(NH_3)_2Cl_2$ cisplatin § 24.16
The **polarising power** of the small cation $[\mathbf{M}(H_2O)_6]^{3+}$ is high and leads to **salt hydrolysis** with the formation of an acidic solution in e.g. Fe(III) salts (and in Al salts). $[\mathbf{M}(H_2O)_6]^{2+}$ is less highly polarising and the solutions are only weakly acidic. §§ 19.4.1, 24.14.2

EXTRACTION OF METALS II

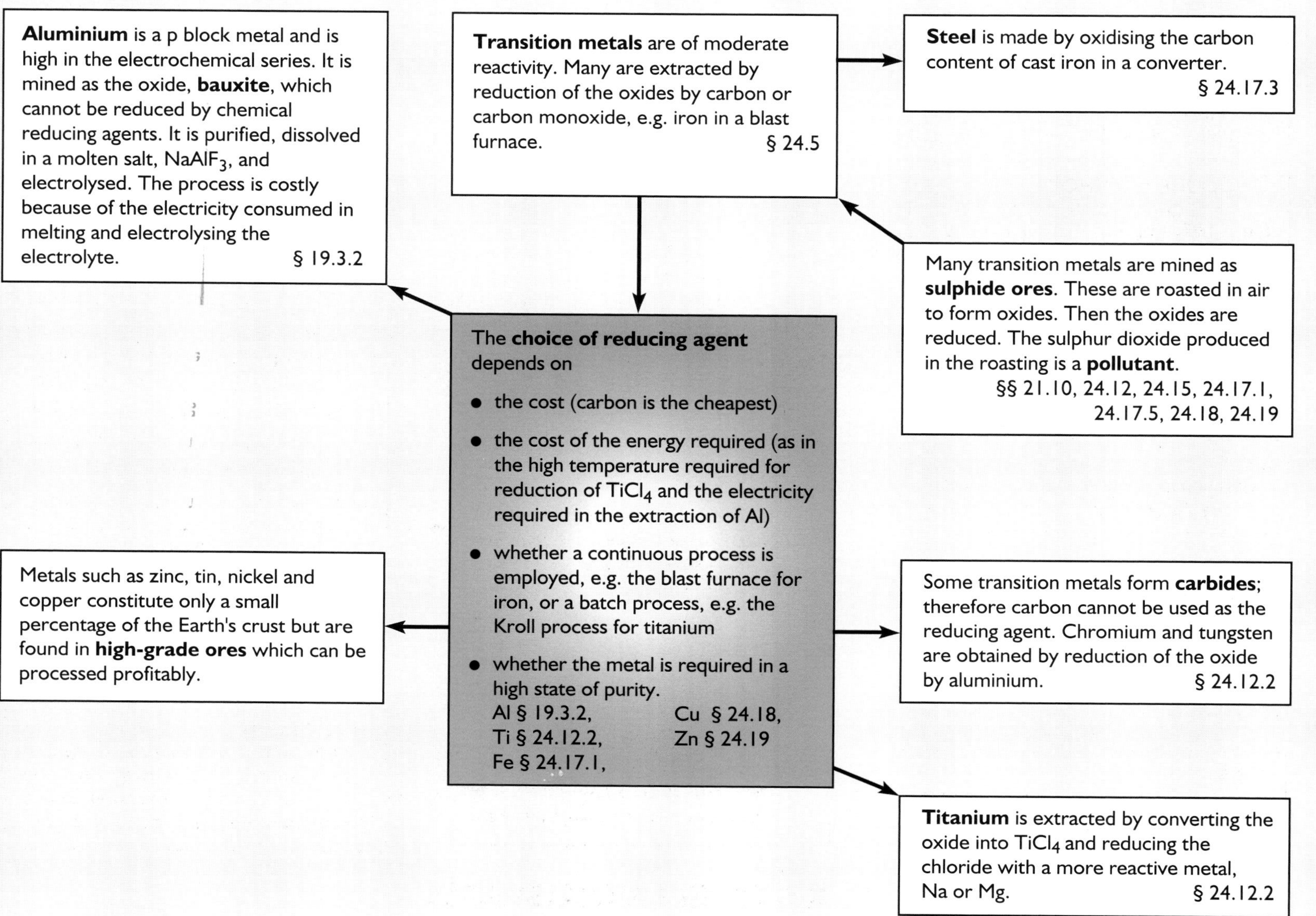
Aluminium is a p block metal and is high in the electrochemical series. It is mined as the oxide, bauxite, which cannot be reduced by chemical reducing agents. It is purified, dissolved in a molten salt, $NaAlF_3$, and electrolysed. The process is costly because of the electricity consumed in melting and electrolysing the electrolyte. § 19.3.2
Transition metals are of moderate reactivity. Many are extracted by reduction of the oxides by carbon or carbon monoxide, e.g. iron in a blast furnace. § 24.5
Steel is made by oxidising the carbon content of cast iron in a converter. § 24.17.3
Many transition metals are mined as sulphide ores. These are roasted in air to form oxides. Then the oxides are reduced. The sulphur dioxide produced in the roasting is a pollutant. §§ 21.10, 24.12, 24.15, 24.17.1, 24.17.5, 24.18, 24.19
The choice of reducing agent depends on
• the cost (carbon is the cheapest)
• the cost of the energy required (as in the high temperature required for reduction of $TiCl_4$ and the electricity required in the extraction of Al)
• whether a continuous process is employed, e.g. the blast furnace for iron, or a batch process, e.g. the Kroll process for titanium
• whether the metal is required in a high state of purity.
Al § 19.3.2, Cu § 24.18,
Ti § 24.12.2, Zn § 24.19
Fe § 24.17.1,
Metals such as zinc, tin, nickel and copper constitute only a small percentage of the Earth's crust but are found in high-grade ores which can be processed profitably.
Some transition metals form carbides; therefore carbon cannot be used as the reducing agent. Chromium and tungsten are obtained by reduction of the oxide by aluminium. § 24.12.2
Titanium is extracted by converting the oxide into $TiCl_4$ and reducing the chloride with a more reactive metal, Na or Mg. § 24.12.2

QUESTIONS ON CHAPTER 24

1. Explain the meaning of each of the following terms.

(a) transition element *(b)* oxidation number
(c) complex ion *(d)* coordination number

2. State three characteristics of a transition element, illustrating them by reference to the chemistry of iron. In your answer refer to the electron configurations of Fe, Fe^{2+} and Fe^{3+}.

3. List the chemical reactions that take place in the extraction of iron in the blast furnace.

4. Discuss the electrochemical methods used to prevent the rusting of iron and steel.

5. *(a)* Outline a laboratory preparation of copper(I) chloride.
(b) Explain the reactions that occur when aqueous ammonia is added to aqueous copper(II) sulphate until no further change occurs. What happens when aqueous potassium cyanide is added to this solution?

6. *(a)* Explain why potassium manganate(VII) is a useful reagent in titrimetric analysis. Why is it not used as a primary standard? When is potassium dichromate preferred?
(b) Construct equations for the oxidation by acidified potassium manganate(VII) of

(i) $Fe^{2+}(aq)$ to $Fe^{3+}(aq)$
(ii) $H_2O_2(aq)$ to $O_2(g) + 2H^+(aq)$

7. Describe the manufacture of steel. What are the advantages and disadvantages of steel compared with aluminium? Why is the production of steel greater than that of all other metals?

8. Sketch the spatial arrangement of bonds in the ions *(a)* $[Cu(NH_3)_4]^{2+}$ *(b)* $[Ag(NH_3)_2]^+$ *(c)* $[Fe(CN)_6]^{3-}$ *(d)* $[NiCl_4]^{2-}$

9. *(a)* Explain the terms complex ion and ligand.
(b) Draw the structure of NH_3 and Cl^-. Explain why NH_3 and Cl^- are able to act as ligands. Say what type of bonds they use in complex formation.
(c) Give examples of complex ions containing (i) NH_3 and (ii) Cl.

10. Explain the following statements:

(a) The compound $CrCl_3 \cdot 6H_2O$ exists in three forms. All produce a precipitate with silver nitrate solution, but in different molar proportions.
(b) Addition of alkali to potassium dichromate(VI) solution produces a yellow solution.
(c) The addition of barium chloride solution to a solution of potassium dichromate(VI) gives a yellow precipitate.
(d) When an excess of potassium iodide solution is added to an aqueous solution of a copper(II) salt, a white precipitate and a brown solution are formed.
(e) When dilute sulphuric acid is added to copper(I) oxide, a reddish brown solid and a blue solution are formed.

11. A 25.0 cm^3 portion of iron(II) chloride solution required, after acidification, 15.0 cm^3 of a 0.0100 mol dm^{-3} solution of potassium dichromate(VI). Calculate the concentration of the solution. Explain why potassium manganate(VII) would give an inaccurate result if it were used for this estimation.

12. Explain the following observations:

(a) When a solution of iron(II) sulphate is added to a solution of a nitrate in concentrated sulphuric acid, a brown ring forms at the junction of the two layers.
(b) When copper(I) sulphate is dissolved in water, a solution of copper(II) sulphate is obtained.
(c) Addition of water to copper(II) chloride gives first a green solution and, on dilution, a blue solution.

13. Chromium (atomic number 24), manganese (25) and iron (26) are in the d block of the Periodic Table.

(a) What is a d block element?
(b) Explain why the atomic radii of the three elements are similar.
(c) Give three properties characteristic of d block elements, with examples from the three elements mentioned above.
(d) Write the electronic configurations of Cr^{2+}, Mn^{3+} and Mn^{7+}.

14. Identify the metal ions present in solution A. Give the formulae of all the metal-containing species involved in B to I.

A: aqueous solution containing three metal ions
↓ Add NaOH(aq)
B: Coloured precipitate
↓ Add excess NaOH(aq). Filter
C: Coloured precipitate and D: Colourless solution
↓ Add excess $NH_3(aq)$. Filter ↓ Add $H_2SO_4(aq)$
E: Green solid + G: Dark blue solution H: White precipitate
↓ Air ↓ Add $H_2SO_4(aq)$
F: Brown solid I: Colourless solution

15. When concentrated ammonia solution is added dropwise to a solution of chromium(III) sulphate, a green precipitate **A** forms. **A** dissolves slowly in an excess of ammonia to give a purple solution, **B**. **A** also dissolves in aqueous sodium hydroxide to give a green solution, **C**. Addition of hydrogen peroxide to **C** gives a yellow solution, **D**. When **D** is treated with dilute sulphuric acid, it gives an orange solution **E**.

Identify **A**, **B**, **C**, **D** and **E**.

16. Aqueous copper(II) sulphate contains the copper species **W**. On treatment with an excess of concentrated hydrochloric acid, it turns into a yellow-green solution of copper-containing species **X**. When copper metal is added to this solution and boiled, a new copper species **Y** is produced. When solution of **Y** is poured into distilled water, a white precipitate of **Z** is obtained. When **Z** was filtered off and allowed to stand in the air, it slowly turned green.

Indentify the copper species **W**–**Z**. Explain the action of copper in the conversion of **X** into **Y**. Write equations for the conversions **W** into **X** and **X** into **Y**. Explain why **Z** turns green on standing in air.

17. Mild steel is an alloy of iron and carbon. The percentage of iron in steel can be found by dissolving steel in dilute sulphuric acid and titrating the iron(II) produced against potassium manganate(VII).

(a) Write half-reaction equations for (i) the oxidation of iron(II) to iron(III) (ii) the reduction of manganate(VII) to manganese(II).
(b) Combine the two half-reaction equations to give the equation for the reaction between iron(II) and acidified manganate(VII).
(c) What colour change would be observed during the titration?
(d) How could you test to confirm that (i) iron(II) was present (ii) iron(III) had been formed?
(e) A sample of steel, 0.1646 g, was dissolved in aqueous sulphuric acid. It required 27.4 cm^3 of potassium manganate(VII) of concentration 0.0200 mol dm^{-3} in a titration. Calculate the percentage by mass of iron in the steel.

18. Edta is the ligand whose structure in alkaline solution is:

$$\begin{array}{ccc} {}^{-}O_2CCH_2 & & CH_2CO_2^{-} \\ \quad\backslash & & / \\ & N-CH_2-CH_2-N & \\ \quad / & & \backslash \\ {}^{-}O_2CCH_2 & & CH_2CO_2^{-} \end{array}$$

It can be used to titrate $Zn^{2+}(aq)$ in solution.

(a) Explain how edta can form coordinate bonds to $Zn^{2+}(aq)$.
(b) Why is edta used in alkaline solution?
(c) A sample 5.3000 g of $ZnSO_4 \cdot 7H_2O$ was dissolved in water and made up to 250 cm^3 of solution. A 25.0 cm^3 portion of the solution required 18.45 cm^3 of 0.1000 mol dm^{-3} edta for reaction. In what molar ratio do $Zn^{2+}(aq)$ and edta react?

19. Brass is an alloy of copper and zinc. It reacts with nitric acid to give a solution containing $Cu^{2+}(aq)$ and $Zn^{2+}(aq)$ ions. The amount of $Cu^{2+}(aq)$ in the solution can be found by adding an excess of potassium iodide solution and titrating the iodine formed against standard thiosulphate solution.

$$2Cu^{2+}(aq) + 4I^{-}(aq) \longrightarrow 2CuI(s) + I_2(aq)$$
$$I_2(aq) + 2S_2O_3^{2-}(aq) \longrightarrow 2I^{-}(aq) + S_4O_6^{2-}(aq)$$

Zinc ions do not take part in the reactions.

In an analysis, 0.2685 g of brass was used and 27.50 cm^3 of 0.1000 mol dm^{-3} sodium thiosulphate were required in the titration. Calculate the percentage by mass of copper in the specimen of brass.

20. *(a)* The stability constants for the reaction of ammonia with the hexaaquanickel(II) ion, at 27 °C, are given below:

	K_1	K_2	K_3	K_4	K_5	K_6
$\log_{10}$	2.67	2.12	1.61	1.07	0.63	−0.09

(i) Write an expression for the stability constant K_1. 2
(ii) Deduce the units for the stability constant K_3. 2
(iii) Calculate the value of the overall stability constant for the reaction. 1

(b) Ligand replacement reactions of the hexaaqua cations of copper(II) and cobalt(II) with chloride ions produce $CuCl_4^{2-}$ and $CoCl_4^{2-}$ respectively. Copy and complete the table below. 4

Ion	Colour
$[Cu(H_2O)_6]^{2+}$	
$[Co(H_2O)_6]^{2+}$	
$[CuCl_4]^{2-}$	
$[CoCl_4]^{2-}$	

(c) The stability constants for two silver complexes are shown below:

$[Ag(CN)_2]^-$ 1.0×10^{21} mol^{-2} dm^6
$[Ag(NH_3)_2]^+$ 1.7×10^{7} mol^{-2} dm^6

(i) Both ammonia and cyanide ions are monodentate ligands. Explain the term **monodentate**. 2
(ii) Deduce the coordination number of silver in either of the complexes. 1
(iii) Deduce the result of adding a solution of ammonia to the dicyanoargentate(I) complex, $[Ag(CN)_2]^-$. 2

CCEA (AL)

21. *(a)* What is meant by the terms *ligand* and *complex ion*? 2
(b) Give the **full** electronic configuration of the copper(II) ion. 1
(c) When anhydrous $CuCl_2$ is dissolved in water a blue solution is formed.
Identify the species responsible for the blue colour and state the shape of this species. 2
(d) When anhydrous $CuCl_2$ is dissolved in concentrated hydrochloric acid, a yellow-green solution is formed due to the presence of the copper species **X**. If sulphur dioxide is bubbled through this yellow-green solution in the presence of an excess of hydrochloric acid, the colourless species $[CuCl_2]^-$ is formed together with SO_4^{2-} ions.

(i) Identify the yellow-green copper species, **X**, state its shape and give the oxidation state of copper in this species.
(ii) State the role of sulphur dioxide in the conversion of species **X** into $[CuCl_2]^-$
(iii) Explain, in terms of electronic configuration, why $[CuCl_2]^-$ is colourless.
(iv) When the solution containing the yellow-green copper species **X** is added to water, a blue solution is obtained. Write an ionic equation for this reaction. **6**

11 *NEAB (AS/AL)*

22. A solution of cobalt(II) chloride was reacted with ammonia and ammonium chloride while a current of air was blown through the mixture. A red compound, **X**, was produced which contained a complex ion of cobalt. The compound had the following composition:

	% by mass
Co	23.6
N	27.9
H	6.0
Cl	42.5

(a) Use the data to confirm that the empirical formula of **X** is $CoN_5H_{15}Cl_3$. **2**

(b) A sample of **X** with mass 1.00 g reacted with silver nitrate solution and 1.15 g of silver chloride was formed.

Calculate the number of moles of free chloride ions per mole of **X** using the information that the molecular and empirical formulae of **X** are the same. Hence deduce the formula of the complex ion in **X**.

Predict the three dimensional structure of the ion and show this in a sketch indicating the positions of the different ligands. **6**

(c) Explain the role of the air, the ammonia and the ammonium chloride in the formation of **X**, giving equations where appropriate. An overall equation is not required. **4**

(d) A solution of cobalt(II) chloride reacts with concentrated hydrochloric acid forming a complex ion which is tetrahedral. With solutions containing the bidentate ethanedioate ions, $C_2O_4^{2-}$, cobalt forms an octahedral complex ion.

Deduce the formula for these two complexes and suggest why different coordination numbers occur in these complexes. **3**

15 *L(N) (AL)*

23. *(a)* (i) Titanium can be prepared in the laboratory by passing titanium(IV) chloride vapour and hydrogen through a furnace at 1000 °C. Construct an equation for the reaction which occurs.
(ii) Industrially, a different reducing agent is used with titanium(IV) chloride. Give the reducing agent, state two essential conditions used and write an equation for the process.
(iii) Suggest two reasons why the method described in part (i) is not the preferred route on an industrial scale.
(iv) Suggest why titanium is manufactured by the method you have given in part (ii) rather than by the reduction of titanium(IV) oxide with carbon.
(v) Suggest two reasons why there has been much research into finding cheaper routes for the manufacture of titanium. **11**

(b) Nickel is refined by a process in which the impure metal is treated with carbon monoxide at 60 °C to form nickel tetracarbonyl, $Ni(CO)_4$, b.p. 43 °C. This compound is then decomposed at 160 °C to give pure nickel and carbon monoxide.
(i) What is the oxidation state of nickel in $Ni(CO)_4$?
(ii) What property of $Ni(CO)_4$ enables it to be separated from impurities such as carbon in the nickel?
(iii) Suggest two reasons why this purification route for nickel is economically viable. **4**

15 *NEAB (AL)*

24. *(a)* (i) Explain, with the help of an equation, what is meant by the *acidity* or *hydrolysis* reaction of a metal-aqua ion. State how the charge on the metal ion affects this acidity.
(ii) Describe what you would observe if sodium carbonate was added to a solution of a metal(III)-aqua ion.
(iii) When pale violet crystals of iron(III) nitrate, $Fe(NO_3)_3 \cdot 9H_2O$, were added to water, a brown solution was formed. The addition of concentrated nitric acid to this solution removed the brown colour. Suggest an interpretation of these observations with the help of your answer to part *(a)*(i). **10**

(b) (i) Given a solution of ammonium vanadate(V), NH_4VO_3, describe an experiment which demonstrates that vanadium has several oxidation states.
(ii) Potassium manganate(VII) solution oxidises all vanadium species to the vanadium(V) state.

When 0.234 g of ammonium vanadate(V) was dissolved in dilute sulphuric acid and the solution treated with an excess of sulphur dioxide, a blue solution was formed. After the excess of sulphur dioxide was removed by boiling, the blue solution decolourised 20.0 cm^3 of 0.0200 M $KMnO_4$ solution.

Use the data to calculate the oxidation state of vanadium in the blue solution. **11**

(c) Explain why transition metals and their compounds often prove effective as catalysts. Give **one** example of heterogeneous and **one** of homogeneous catalysis by a transition metal or one of its compounds. For **one** of your examples, outline the mode of action of the catalyst. **9**

30 *NEAB (AL)*

Part 4

Organic Chemistry

25 ORGANIC CHEMISTRY

25.1 CARBON COMPOUNDS

Carbon is an extraordinary element. All living things depend on carbon compounds. We depend on carbon compounds for the food we eat, the clothes we wear, the medicines we take and the fuels that warm our homes. Materials based on carbon compounds have properties which range from the softness of silk to the toughness of the carbon-fibre-reinforced plastics, as strong as mild steel, which are used in the construction of boats and aircraft.

Carbon compounds play a major role in all life processes. All the intricate reactions that are needed to keep us alive involve carbon compounds. A living cell is an amazing chemical factory. It maintains the concentrations of many dissolved substances, interacts with its environment and synthesises the substances which it needs for survival and growth. It can even reproduce itself. This impressive chemical machine consists largely of organic substances: amino acids, proteins, sugars, lipids, nucleic acids and others. There are, in addition, water and a few inorganic salts.

The reason why carbon is uniquely important is that carbon atoms link together by covalent bonds to form a great variety of chains and rings. The properties and reactions of organic compounds in all their variety can be understood in terms of the reactions of small groups of atoms, e.g. —OH the hydroxyl group and $—NH_2$ the amine group. The complex subject of organic chemistry is simplified when it is studied as the chemistry of families of compounds, e.g. alcohols and amines, and the reactions of the groups of atoms – the functional groups – which these compounds contain.

Organic chemistry was originally described as the chemistry of compounds found in living things; in plants and animals. All such naturally occurring compounds contain carbon, and it was thought that some 'vital force' was needed for their formation. Today, the term organic chemistry refers to the chemistry of millions of carbon compounds. Some of them have been extracted from plant or animal sources, but many more have been made by organic chemists in their laboratories.

Carbon forms more compounds than any other element. With the ground state electron configuration $1s^22s^22p^2$, it has very little tendency to form positive or negative ions. In order to achieve a stable outer octet of electrons, it forms four covalent bonds. The electron pairs in these bonds occupy sp^3 hybrid orbitals. They have axes directed to the corners of a regular tetrahedron with the carbon nucleus at the centre.

When a carbon atom combines with four hydrogen atoms, it forms a molecule of methane, CH_4 [see Figure 25.1A(a)]. If two carbon atoms join, each can still combine with three hydrogen atoms to form a molecule of ethane, C_2H_6 [see Figure 25.1A(b)]. A model of a molecule of propane, C_3H_8, is shown in Figure 25.1A(c).

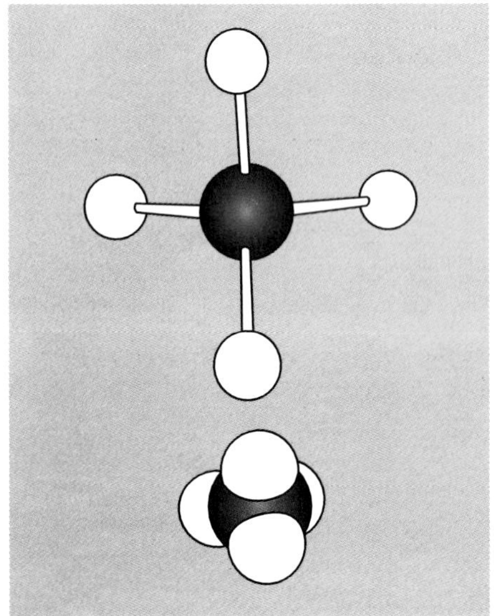

Figure 25.1A(a)
Models of methane, CH_4

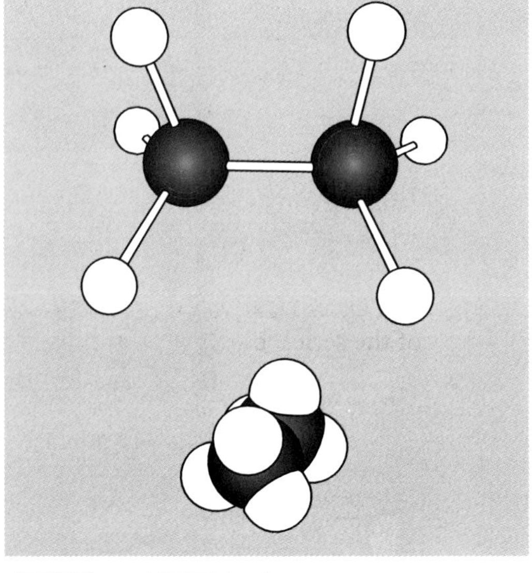

Figure 25.1A(b)
Models of ethane, C_2H_6

Figure 25.1A(c)
Models of propane, C_3H_8

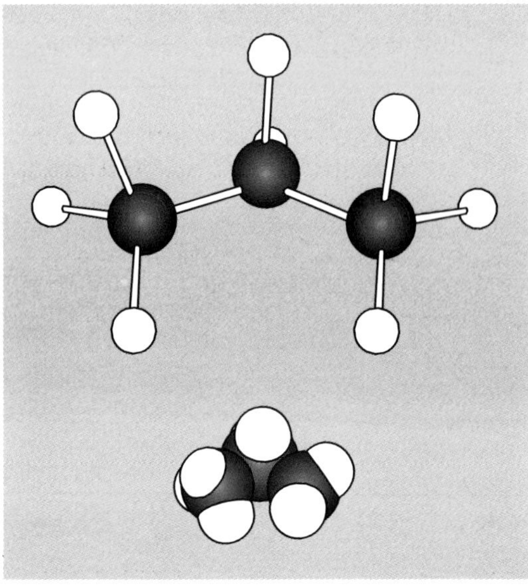

25.2 HYDROCARBONS

Compounds of carbon and hydrogen are called **hydrocarbons**. The compounds shown above, methane, ethane and propane, are **alkanes**. They possess only single bonds and have the general formula

$$C_nH_{2n+2}$$

The alkanes are a homologous series

A series of compounds with similar chemical properties, in which members differ from one another by the possession of an additional CH_2 group, is called a **homologous series**. The first ten members of the unbranched-chain alkane series are:

The names of alkanes and alkyl groups

CH_4	Methane	C_6H_{14}	Hexane
C_2H_6	Ethane	C_7H_{16}	Heptane

C_3H_8	Propane	C_8H_{18}	Octane
C_4H_{10}	Butane	C_9H_{20}	Nonane
C_5H_{12}	Pentane	$C_{10}H_{22}$	Decane

The groups of atoms CH_3—, C_2H_5— and C_nH_{2n+1}— are called methyl, ethyl and alkyl groups.

25.3 ISOMERISM AMONG ALKANES

Figure 25.1A shows models of methane, ethane and propane. The next member of the series, butane, C_4H_{10}, can be modelled in two ways:

(a)

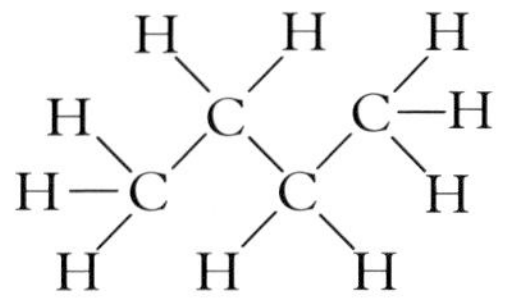

```
       H     H     H
        \   /     /
   H     C     C—H
    \   / \   / \
  H—C      C     H
    /     / \
   H     H   H
```

or (b)

```
  H  H       H  H
   \ |   H   | /
     C   |   C
    / \  |  / \
   H    C      H
        |
        C
       /|\
      H H H
```

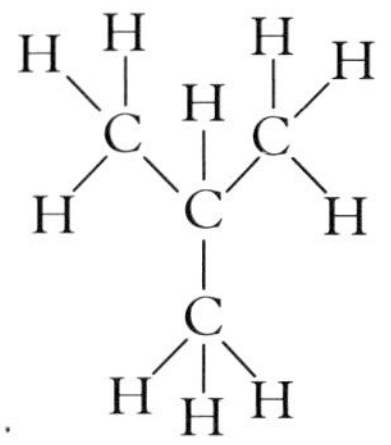

Isomeric compounds have the some molecular formulae but different structural formulae

Molecule (a) is a continuous-chain or unbranched-chain molecule, and (b) is a branched-chain molecule. The two formulae correspond to different compounds. Compound (a) is called butane (or sometimes, *normal*-butane or *n*-butane) and boils at −0.5 °C. Compound (b) is called 2-methylpropane, and boils at −12 °C.

The existence of different compounds with the same molecular formulae but different structural formulae is called **isomerism**. There are five isomeric hexanes of formula C_6H_{14}. They are:

The isomers of hexane

```
CH3CH2CH2CH2CH2CH3
```
Hexane (sometimes referred to as *normal*-hexane or *n*-hexane)

```
CH3CH2CH2CHCH3
            |
           CH3
```
2-Methylpentane

```
CH3CH2CHCH2CH3
       |
      CH3
```
3-Methylpentane

```
      CH3
       |
CH3CH2C—CH3
       |
      CH3
```
2,2-Dimethylbutane

```
CH3CH—CHCH3
    |   |
   CH3 CH3
```
2,3-Dimethylbutane

25.4 SYSTEM OF NAMING HYDROCARBONS

The IUPAC system of nomenclature

The names of these isomers are given in accordance with the International Union of Pure and Applied Chemistry (IUPAC) system of nomenclature. The procedure followed is:

1. Name the longest unbranched carbon chain.
2. Name the substituent groups.
3. Give the positions of the substituent groups.

Name the longest chain ...

... name the substituents ...

... in alphabetical order ...

... and number them

$$\begin{array}{l} CH_3{-}CH_2{-}CH_2{-}\underset{\displaystyle CH_3}{\underset{|}{CH}}{-}CH_3 \end{array}$$

2-Methylpentane
One could count from the other end and call it 4-methylpentane, but the IUPAC system is to count from the end which will give the lower locant (number) for the position of a substituent group.

$$CH_3{-}CH_2{-}\underset{\displaystyle H_3C}{\underset{|}{CH}}{-}\underset{\displaystyle CH_3}{\underset{|}{CH}}{-}CH_3$$

2,3-Dimethylpentane

$$CH_3{-}CH_2{-}CH_2{-}\underset{\displaystyle H_3C}{\underset{|}{CH}}{-}\underset{\displaystyle C_2H_5}{\underset{|}{CH}}{-}CH_3$$

3,4-Dimethylheptane
In spite of the way it is written, you should be able to see that the longest unbranched chain has seven carbon atoms.

$$CH_3{-}\underset{\displaystyle Cl}{\underset{|}{CH}}{-}CH_2{-}\underset{\displaystyle Br}{\underset{|}{CH}}{-}CH_3$$

2-Bromo-4-chloropentane
The substituent groups are named in alphabetical order.

$$CH_3{-}CH_2{-}CH_2{-}\underset{\displaystyle Br}{\underset{|}{CH}}{-}\underset{\displaystyle Cl}{\underset{|}{CH}}{-}CH_3$$

3-Bromo-2-chlorohexane
The substituents are named in alphabetical order, not in the numerical order of the locants.

To construct the formula of a compound from its name, e.g., 2,3-dichloro-4-methylhexane, first write the carbon atoms of the hexane part of the molecule:

$$C{-}C{-}C{-}C{-}C{-}C$$

Then put in Cl atoms on carbons 2 and 3, and a CH_3 group on carbon 4:

$$C{-}\underset{\displaystyle Cl}{\underset{|}{C}}{-}\underset{\displaystyle Cl}{\underset{|}{C}}{-}\underset{\displaystyle CH_3}{\underset{|}{C}}{-}C{-}C$$

Fill in the hydrogen atoms to give each carbon atom a valency of four:

$$CH_3{-}\underset{\displaystyle Cl}{\underset{|}{CH}}{-}\underset{\displaystyle Cl}{\underset{|}{CH}}{-}\underset{\displaystyle CH_3}{\underset{|}{CH}}{-}CH_2{-}CH_3$$

25.4.1 UNSATURATED HYDROCARBONS

Alkanes are saturated hydrocarbons

Alkenes and alkynes are unsaturated hydrocarbons: they contain multiple bonds

Alkanes, alkenes and alkynes are aliphatic hydrocarbons

The alkanes are not the only hydrocarbons. There are also alkenes and alkynes. Alkanes are said to be **saturated** hydrocarbons as they contain only single bonds between carbon atoms. Alkenes and alkynes are **unsaturated** hydrocarbons: they contain multiple bonds between carbon atoms. The simplest alkene is ethene

$$H_2C{=}CH_2$$

formerly called ethylene. It is the first member of the homologous series of alkenes, which have the general formula C_nH_{2n}. Alkynes contain one or more carbon–carbon triple bonds. Ethyne

$$HC{\equiv}CH$$

is the first member of the homologous series of alkynes, which have the general formula C_nH_{2n-2}.

Table 25.4
Names of aliphatic hydrocarbons

No. of C atoms	Alkane		Alkene		Alkyne		Alkyl group	
1	CH_4	Methane					$CH_3—$	Methyl
2	C_2H_6	Ethane	C_2H_4	Ethene	C_2H_2	Ethyne	$C_2H_5—$	Ethyl
3	C_3H_8	Propane	C_3H_6	Propene	C_3H_4	Propyne	$C_3H_7—$	Propyl
4	C_4H_{10}	Butane	C_4H_8	Butene	C_4H_6	Butyne	$C_4H_9—$	Butyl
5	C_5H_{12}	Pentane	C_5H_{10}	Pentene	C_5H_8	Pentyne	$C_5H_{11}—$	Pentyl
n	C_nH_{2n+2}		C_nH_{2n}		C_nH_{2n-2}		$C_nH_{2n+1}—$	

The names of the other members of the series are given in Table 25.4. All these hydrocarbons are classified as **aliphatic** hydrocarbons. Aliphatic means 'fatty' in Greek, the connection being that fats contain large alkyl groups, e.g. $C_{15}H_{31}—$.

In naming alkenes and alkynes, the positions of the multiple bonds must be stated:

How to state the position of the double bond in an alkene

$CH_2{=}CH—CH_2—CH_3$

But-1-ene
The *but*-part of the name shows that there are 4 carbon atoms. The *-ene* suffix shows that there is a C=C double bond. The number 1 indicates that the double bond is between carbon atoms 1 and 2. Count from the end that will give the lowest numbers, not 3 and 4.

$CH_3—CH{=}CH—CH_3$ But-2-ene

$CH_2{=}CH—CH{=}CH_2$ Buta-1,3-diene

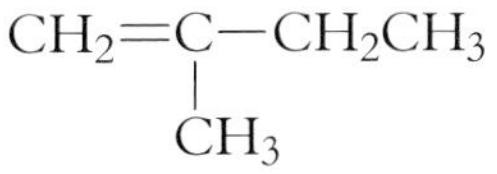

$CH_2{=}C(CH_3)—CH_2CH_3$ 2-Methylbut-1-ene

$CH_3—CH{=}CH—CH(CH_3)—CH_3$

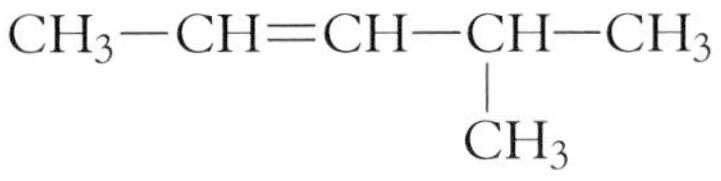

4-Methylpent-2-ene
The double bond is numbered first and then the methyl group.

25.5 ALICYCLIC HYDROCARBONS

A second set of hydrocarbons is the alicyclic hydrocarbons. They contain rings of carbon atoms. Examples are

Alicyclic hydrocarbons

```
H2C—CH2          CH2—CH2            CH2
  \ /            |    |         H2C/   \CH
  CH2            CH2—CH2         |      ||
                                H2C     CH
                                   \CH2/
```

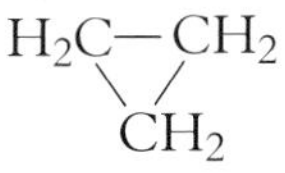

Cyclopropane 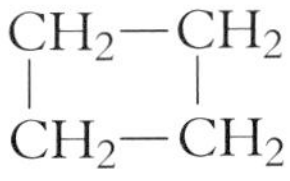 Cyclobutane 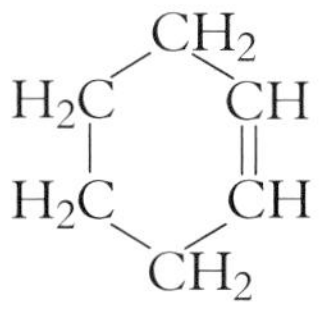 Cyclohexene

[See Figure 25.6A.]

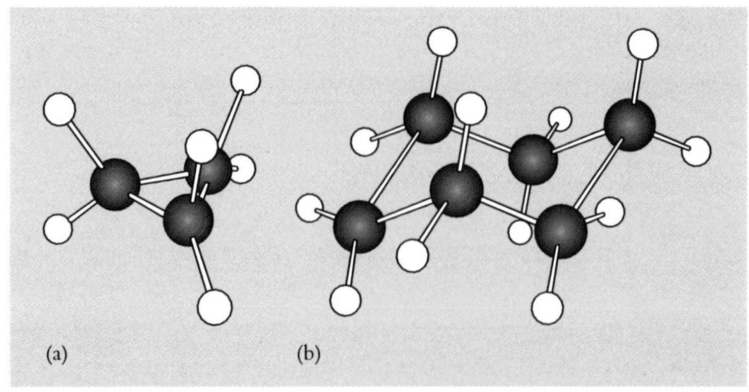

Figure 25.5A
(a) Cyclopropane
(b) Cyclohexane

25.6 AROMATIC HYDROCARBONS

Aromatic hydrocarbons are related to benzene; they are called arenes

A third group of hydrocarbons is the **aromatic** hydrocarbons. They are related to benzene. The first benzene compounds to be isolated had pleasant aromas, and gave this group of hydrocarbons their name. Members of the group of aromatic hydrocarbons, the **arenes**, are: benzene, C_6H_6; methylbenzene, $C_6H_5CH_3$; and naphthalene, $C_{10}H_8$. The structure of these compounds is discussed in § 5.4 and in Chapter 28. The group C_6H_5— is called a **phenyl** group, and the group $C_{10}H_{17}$— is called a **naphthyl** group. Both are **aryl** groups.

The phenyl and naphthyl groups are aryl groups

The subsets of hydrocarbons are summarised in Figure 25.6A.

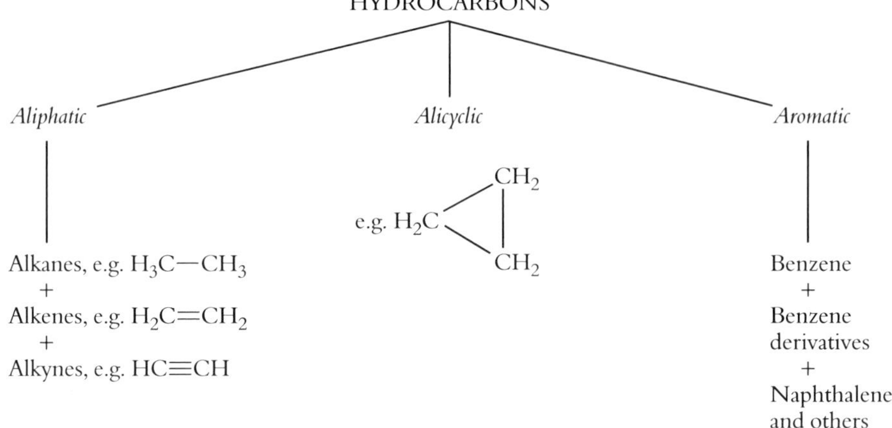

Figure 25.6A
Classes of hydrocarbons

25.7 FUNCTIONAL GROUPS

The double bond in the alkenes is responsible for most of the chemical reactions of these compounds. The group of atoms

$$>C=C<$$

is called the **functional group** of the alkenes.

Similarly, alkynes possess the functional group

$$—C{\equiv}C—$$

The reactions of a homologous series depend on the functional group

In your study of the reactions of the hydrocarbons, you will come across some other classes of compounds with different functional groups, and it will help to deal with their names in advance.

The halogenoalkanes (or haloalkanes) have the formula RX where R is an alkyl group and X is a halogen:

CH_3Br	Bromomethane
C_2H_5Cl	Chloroethane
$C_2H_4Cl_2$	Dichloroethane

How to name halogenoalkanes

There are two isomers. Figure 25.7A(a) shows 1,2-dichloroethane, $ClCH_2CH_2Cl$, with the chlorine atoms attached to different carbon atoms. Figure 25.7A(b) shows 1,1-dichloroethane, CH_3CHCl_2, with the chlorine atoms on the same carbon atom.

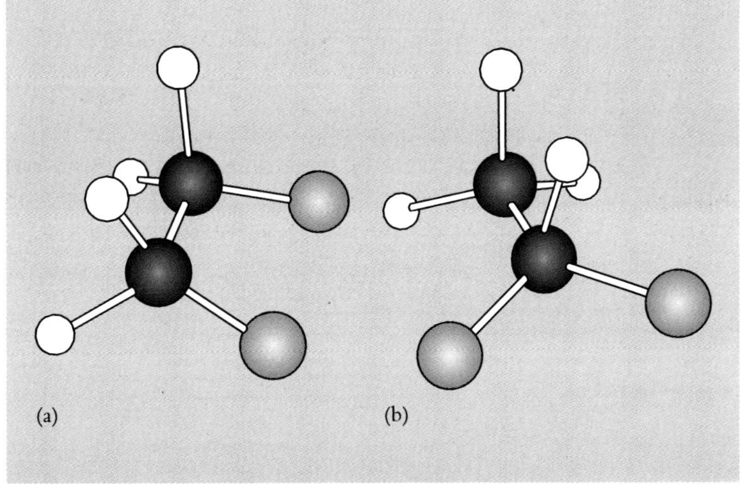

Figure 25.7A
(a) 1,2-Dichloroethane,
(b) 1,1-Dichloroethane

How to name alcohols ...

The **alcohols** or **alkanols** are compounds with the functional group —OH, a hydroxyl group. They are named by taking the name of the alkane with the same number of carbon atoms and changing the ending from *-ane* to *-anol*:

CH_3OH	Methanol	
C_2H_5OH	Ethanol	
C_3H_7OH	Propanol. There are two isomers	
	$CH_3CH_2CH_2OH$	Propan-1-ol
	$CH_3CH(CH_3)OH$	Propan-2-ol

A number must be used to show the position of the hydroxyl group in the carbon chain.

Carboxylic acids or **alkanoic acids** have the functional group

$$-C\begin{matrix} \diagup\!\!\!\diagup O \\ \diagdown O-H \end{matrix}$$

... and carboxylic acids They are named by taking the alkane with the same number of carbon atoms and changing the ending from *-ane* to *-anoic acid*.

CH_3CO_2H	Ethanoic acid (There are *two* carbon atoms, that in CH_3 and that in CO_2H.)
$CH_3CH_2CH{=}CHCO_2H$	Pent-2-enoic acid (The C of the CO_2 group is C-1, and the double bond lies between C-2 and C-3.)

The functional groups of other series are:

halogenoalkanes:	 $>C{-}$Halogen (C with three bonds)	(Chapter 29)
aldehydes:	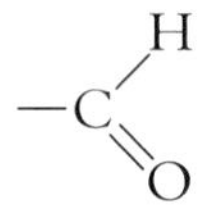 $-C(=O)H$	(Chapter 31)
ketones:	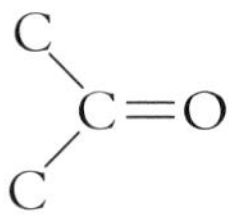 $C{-}C({=}O){-}C$	(Chapter 31)
amines:	$-NH_2$	(Chapter 32)
amides:	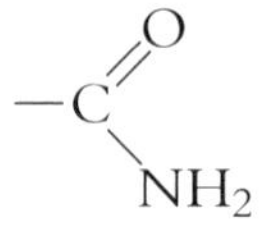 $-C(=O)NH_2$	(Chapter 33)
nitriles:	$-C{\equiv}N$	(Chapter 33)

SUMMARY

Hydrocarbons of general formula C_nH_{2n+2} are members of the homologous series called alkanes. They are saturated; that is, they have single C—C bonds.
Unsaturated hydrocarbon series are:
alkenes C_nH_{2n}, e.g. $H_2C{=}CH_2$ and alkynes C_nH_{2n-2}, e.g. $HC{\equiv}CH$.
The reactions of a homologous series of compounds depend on the functional group, e.g. $>C{=}C<$ in alkenes, —OH in alcohols, $-CO_2H$ in carboxylic acids.

CHECKPOINT 25.7: NOMENCLATURE

1. Name the following compounds:

(*a*) 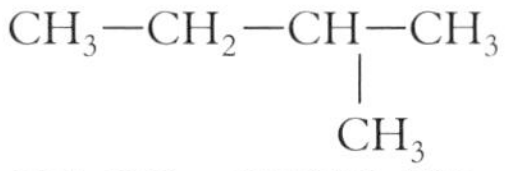 $CH_3{-}CH_2{-}CH(CH_3){-}CH_3$ (CH$_3$ branch on the third carbon)

(*b*) $CH_3CH{=}CHCH_2CH_3$

(*c*) 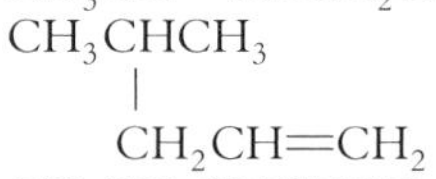 CH_3CHCH_3 with $CH_2CH{=}CH_2$ on the middle carbon

(*d*) $CH_3CH_2CHCHCH_3$ with Cl on C-3 and Br on C-4

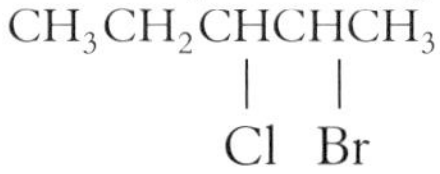

(*e*) $CH_3CH_2CH_2OH$

(*f*) $CH_3CH_2CH_2CH_2CO_2H$

2. Write structural formulae for:

(*a*) Heptane
(*b*) 2-Chloro-3-methylhexane
(*c*) 3-Bromo-2-chloroheptane
(*d*) Pentan-2-ol
(*e*) Hex-2-ene
(*f*) Butanoic acid

25.8 REACTIONS OF ORGANIC COMPOUNDS

25.8.1 TYPES OF REACTIONS

The reactions of organic compounds fall into four classes. These are listed below.

Substitution

There are four types of organic reactions

An atom or group of atoms replaces another, e.g. in hydrolysis:

$$C_2H_5Cl + OH^- \longrightarrow C_2H_5OH + Cl^-$$

Chloroethane + Hydroxide ion ⟶ Ethanol + Chloride ion

Addition

Two molecules react to form one:

$$Br_2 + CH_2{=}CH_2 \longrightarrow BrCH_2CH_2Br$$

Bromine + Ethene ⟶ 1,2-Dibromoethane

Elimination

One molecule reacts to form more than one:

$$C_2H_5OH \longrightarrow C_2H_4 + H_2O$$

Ethanol ⟶ Ethene + Water

Rearrangement

One molecule reacts to give a different molecule:

$$CH_3{-}\underset{\displaystyle Cl}{\underset{|}{CH}}{-}CH{=}CH_2 \rightleftharpoons CH_3{-}CH{=}CH{-}CH_2{-}Cl$$

3-Chlorobut-1-ene ⇌ 1-Chlorobut-2-ene

25.8.2 TYPES OF BOND FISSION

There are two types of bond fission ...

When organic compounds react, their bonds can split in either of two ways, by **homolytic** or **heterolytic** fission.

Homolytic fission

... homolysis ...

When the bond breaks, each of the bonded atoms takes one of the pair of electrons. **Free radicals** are formed. These are atoms or groups of atoms with unpaired electrons:

... each atom takes one of the bonding pair of electrons ...

$$CH_3CH_2CH_3 \xrightarrow{\text{heat}} CH_3CH_2\cdot + \cdot CH_3$$

Propane ⟶ Ethyl radical + Methyl radical

$$Cl_2 \xrightarrow{\text{sunlight}} 2Cl\cdot$$

Chlorine molecule ⟶ Chlorine atoms or free radicals

Energy must be supplied, either as heat or light, to break the bond. The free radicals formed possess this energy, and are very reactive.

The steps in a free radical reaction are initiation, propagation and termination. They are illustrated in the chlorination of methane, § 26.4.7.

Heterolytic fission

... and heterolysis

When the bond breaks, one of the bonded atoms takes both of the bonding electrons to form an anion. The rest of the molecule becomes a cation. An ion with a positively charged carbon atom is called a **carbocation**. An ion with a negatively charged carbon atom is called a **carbanion**:

... one of the atoms takes both of the bonding electrons

$$(CH_3)_3C—Cl \longrightarrow (CH_3)_3C^+ + Cl^-$$

2-Chloro–2-methylpropane ⟶ A carbocation + Chloride ion

$$CH_3COCH_2CO_2C_2H_5 + OH^- \rightleftharpoons CH_3COC^-HCO_2C_2H_5 + H_2O$$

Ethyl 3-oxobutanoate ⇌ A carbanion + Water

25.8.3 TYPES OF REAGENT

In a covalent bond between **A** and **B**, if **A** is more **electronegative** [§ 4.6.3] than **B**, the distribution of bonding electrons can be represented as

$$\mathbf{A}^{\delta-}—\mathbf{B}^{\delta+}$$

Covalent bonds can be polar

The bond is described as **polar**. The reagents which attack organic compounds seek out either the slightly positive ($\delta+$) end of the bond or the slightly negative ($\delta-$) end of the bond. There are two main classes of reagent:

Nucleophilic reagents

Negative ions, e.g., OH^-, CN^-, and compounds in which an atom has an unshared pair of electrons, e.g., NH_3, are **nucleophilic** (nucleus-seeking). They attack the electron-deficient end of a polar bond, the $\delta+$ end.

Nucleophiles attack centres of positive charge

Examples are the hydrolysis of a primary halogenoalkane [§ 29.9.1]

$$R^{\delta+}—X^{\delta-} + OH^- \longrightarrow ROH + X^-$$

and the substitution of a cyanide group for a halogen atom [§ 29.8.1]

$$R^{\delta+}—X^{\delta-} + CN^- \longrightarrow RCN + X^-$$

Electrophilic reagents

A reagent which attacks a region where the electron density is high is called an **electrophile**. Examples of electrophilic reagents are the nitryl cation, NO_2^+, and sulphur(VI) oxide, SO_3. The nitryl cation is involved in the nitration of benzene [§ 28.8.2]. Sulphur(VI) oxide is involved in introducing a sulphonic acid group – SO_3H into benzene [§ 28.8.3].

Electrophiles attack centres of negative charge

A reaction may involve

- substitution by a nucleophile – called an S_N reaction
- substitution by an electrophile – called an S_E reaction
- addition by an electrophile – called an A_E reaction
- addition by a nucleophile – called an A_N reaction.

25.8.4 REACTION MECHANISM

The mechanism of a reaction is the sequence of steps from start to finish

The stoichiometric equation for an organic reaction does not tell you how the reaction takes place. There may be a series of reactions in between the mixing of the reactants and the formation of the products. The sequence of steps by which the reaction takes place is called the **reaction mechanism**. The mechanism is worked out from a study of the kinetics of the reaction [see Chapter 14]. Other techniques, such as spectroscopy [see Chapter 35] and the incorporation of radioisotopes into a reactant, are also used.

SUMMARY

The reactions of organic compounds are:
substitution, addition, elimination, rearrangement.
Bonds can break

- by homolytic fission, A—B ⟶ A· ·B
- or by heterolytic fission, A—B ⟶ A: B

Reagents are nucleophilic, e.g. CN^-, or electrophilic, e.g. NO_2^+.

CHECKPOINT 25.8: SOME REACTIONS

1. Explain the difference between
   (a) homolytic fission and heterolytic fission, giving an example of each
   (b) a substitution reaction and an addition reaction, giving an example of each
   (c) an ion and a free radical, giving an example of each.

2. *(a)* State the difference between
   (i) a carbocation and a carbanion, giving an example of each
   (ii) a nucleophilic reagent and an electrophilic reagent, giving an example of each.
   (b) Is a nucleophilic reagent more likely to attack a carbocation or a carbanion? Explain your answer.

3. *(a)* What type of reaction is the following: addition, substitution, elimination or rearrangement?

$$C_2H_5Br + OH^- \longrightarrow C_2H_5OH + Br^-$$

   (b) What type of reagent is OH^-: a nucleophile or an electrophile?
   (c) Name the organic compounds in the equation.

25.9 ISOMERISM

Isomerism is the existence of different compounds with the same molecular formulae but different structural formulae. There are various types of isomerism.

25.9.1 STRUCTURAL ISOMERISM

A structural formula shows the order in which atoms are bonded together

The structural formula shows the sequence in which the atoms in a molecule are bonded. A structural formula can be written in full, with every bond drawn (a displayed formula), or it can be written by joining groups of atoms in sequence, provided the formula for each of the groups is unambiguous. The structural formula for 2-methylpropane is shown below.

```
H H     H  H
 \|     | /
H—C  H  C—H
   \ | /
     C
     |
   H—C—H
     |
     H
```
(a)

CH_3CHCH_3 with CH_3 on the central C

(b)

$CH_3CH(CH_3)C$

(c)

Formula (b) is unambiguous because there is only one way of writing the bonds in a $—CH_3$ group. Formula (c) is a condensed way of writing formula (b).

Chain isomerism

In chain isomers, the carbon 'skeletons' differ

The isomers have different carbon chains. They possess the same functional group, and belong to the same homologous series, e.g.

$CH_3CH_2CH_2CH_3$ — Butane

$$\begin{array}{c} CH_3CHCH_3 \\ | \\ CH_3 \end{array}$$

2-Methylpropane

Positional isomerism

The position of a functional group in the carbon skeleton differs between positional isomers

These isomers have a substituent group in different positions in the same carbon 'skeleton'. The isomers are chemically similar because they possess the same functional group, e.g.

(a) Propan-1-ol, $CH_3CH_2CH_2OH$ and propan-2-ol, $CH_3CH(OH)CH_3$.
(b) Pent-1-ene, $CH_3CH_2CH_2CH{=}CH_2$ and pent-2-ene, $CH_3CH_2CH{=}CHCH_3$.
(c) This type of isomerism is found in benzene derivatives. The structural formula of benzene, C_6H_6, was discussed in § 5.4.1. If two chlorine atoms replace two hydrogen atoms to form $C_6H_4Cl_2$, three different compounds can be formed. Their names are given below:

1,2-Dichlorobenzene (*ortho*-dichlorobenzene)

1,3-Dichlorobenzene (*meta*-dichlorobenzene)

1,4-Dichlorobenzene (*para*-dichlorobenzene)

Functional group isomerism

In another type of isomerism, the functional group is different in the isomers

These isomers have different fuctional groups and belong to different homologous series. Some examples follow:

(a) An alcohol and an ether, e.g.,
ethanol, C_2H_5OH, and methoxymethane (dimethyl ether) CH_3OCH_3.
(b) An aldehyde and a ketone, e.g.,
propanal, CH_3CH_2CHO, and propanone, CH_3COCH_3.
(c) A carboxylic acid and one or more esters, e.g.,
butanoic acid, $CH_3CH_2CH_2CO_2H$ and methyl propanoate, $C_2H_5CO_2CH_3$, ethyl ethanoate, $CH_3CO_2C_2H_5$, propyl methanoate, $HCO_2CH_2CH_2CH_3$, and methylethyl methanoate, $HCO_2CH(CH_3)_2$.

CHECKPOINT 25.9A: STRUCTURAL ISOMERISM

1. Write structural formulae for three compounds of formula C_5H_{12}.

2. Write the formulae of the three structural isomers, of formula C_3H_8O.

3. There are five linear molecules with the formula $C_4H_8Cl_2$. Give their structural formulae.

4. The molecular formula $C_5H_{10}O$ corresponds to a number of carbonyl compounds, which contain the group

$$>C=O$$

Write their structural formulae.

5. Five unbranched molecules have the formula $C_5H_{10}O_2$. Write their structural formulae.

25.9.2 STEREOISOMERISM

There are various types of stereoisomerism

Stereoisomers have the same molecular formula and also the same structural formula. The difference between them is the arrangement of the bonds in space. Stereoisomerism can be (1) *cis-trans* isomerism, or (2) optical isomerism.

Cis-trans isomerism

Restriction of rotation about a C=C double bond gives rise to cis-trans *isomerism*

The planar arrangement of bonds in $R_2C{=}CR_2$ was discussed in § 5.3.4. The CR_2 groups are not free to rotate about the double bond. In a compound

$$R_1R_2C{=}CR_3R_4$$

R_1 and R_3 may be on the same side of the double bond (the *cis*-isomer) or on opposite sides (the *trans*-isomer). *Cis*- and *trans*-butenedioic acid are shown in Figure 25.9A.

Figure 25.9A Models of butenedioic acids

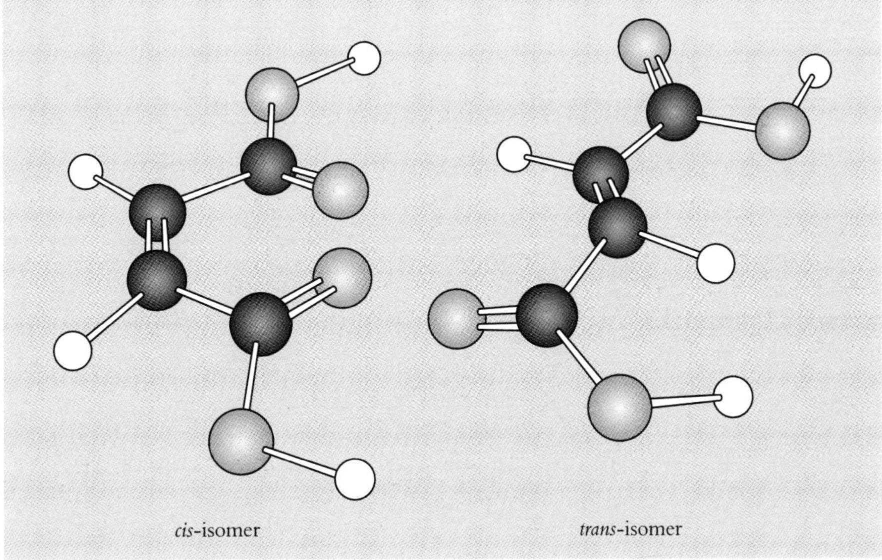

$HO_2C(H)C{=}C(H)CO_2H$ — *cis*-Butenedioic acid (both CO_2H groups on the same side)

$HO_2C(H)C{=}C(H)CO_2H$ — *trans*-Butenedioic acid (CO_2H groups on opposite sides)

The physical and chemical properties of cis-trans isomers differ

They do not have the same physical and chemical properties. They differ in melting temperature (*cis* 135 °C, *trans* 287 °C), in solubility (*cis* being 100 times more soluble than *trans*) and in dipole moments. The *cis* acid (formerly called maleic acid) forms an anhydride on gentle heating, but the *trans* isomer (fumaric acid) does not:

$$\text{cis-HO}_2\text{C–CH=CH–CO}_2\text{H} \xrightarrow{160\ ^\circ\text{C}} \text{(CH=CH)(C=O)}_2\text{O} + H_2O$$

cis-Butenedioic acid → Butenedioic anhydride

Examples are butenedioic acid ...

On strong heating, *trans*-butenedioic acid forms the anhydride of *cis*-butenedioic acid, showing that rotation about the double bond is possible at higher temperatures.

... and some inorganic compounds ...

For *cis-trans* isomerism in inorganic compounds, see § 24.6.4.

SUMMARY

Isomerism, the existence of different compounds with the same molecular formula may be structural isomerism, arising from:

- different carbon chains
- different positions of the functional group
- different functional groups

or stereoisomerism, e.g.

- *cis-trans* isomerism in $R_1R_2C{=}CR_3R_4$

CHECKPOINT 25.9B: STEREOISOMERISM

1. Write displayed formulae for *(a)* $CH_3CH{=}CHCl$ *(b)* $CH_3CH{=}CHC_2H_5$, *(c)* $ClCH{=}C(CH_3)CO_2H$.
2. Explain how the nature of the C=C bond gives rise to *cis-trans* isomerism.

FURTHER STUDY ▼

25.9.3 OPTICAL ISOMERISM

If a beam of light is passed through a Nicol prism (of *calcite*, $CaCO_3$) or a piece of polaroid, the emergent light vibrates in a single plane. It is said to be **plane-polarised** [see Figure 25.9B].

Optical activity is the ability to rotate the plane of polarisation of plane-polarised light

Certain substances, either in crystalline form or in solution, have the ability to rotate the plane of polarisation of plane-polarised light. They are said to be **optically active**. The effect is measured in an instrument called a **polarimeter**, which sends a beam of plane-polarised light through a solution of the substance. If the plane of polarisation rotates in a clockwise direction, viewed from the direction of the emergent beam, the rotation is designated (+); an anticlockwise rotation is designated (–).

Figure 25.9B
Plane-polarisation of light

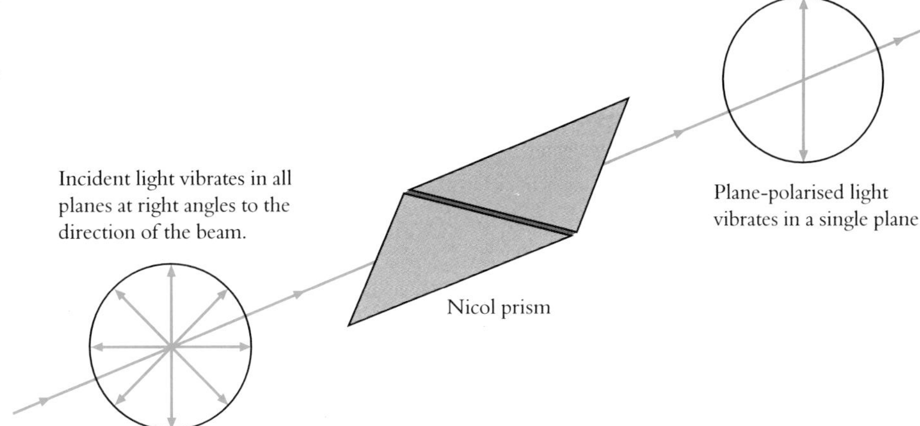
Incident light vibrates in all planes at right angles to the direction of the beam.
Nicol prism
Plane-polarised light vibrates in a single plane.

Optically active compounds are chiral ...

... Their molecules have no plane of symmetry ...

... They exist in (+)- and (−)-forms and as optically inactive mixtures called racemates ...

Compounds which are optically active are **chiral**. Their molecules have no plane or axis or centre of symmetry [§ 5.2.4]. Chiral compounds have two different types of molecules. They are **enantiomers**, i.e., mirror images of one another [see Figure 5.2K, § 5.2.4 and Question 1, Checkpoint 25.9C]. The (+)-enantiomer rotates the plane of polarisation of plane-polarised light clockwise. The (−)-enantiomer produces an anticlockwise rotation. An equimolar mixture of (+) and (−)-enantiomers is optically inactive and is called a **racemic mixture** or **racemate**.

There are three forms of 2-hydroxypropanoic acid (lactic acid), $CH_3CH(OH)CO_2H$. The racemic mixture is found in sour milk; the (+)-enantiomer is present in muscle. The (−)-enantiomer does not occur naturally and has to be obtained from the racemic mixture. All three forms have the same chemical properties. The physical properties of the (+)- and (−)-enantiomers are the same. They differ from those of the racemate because it is a mixture, not a pure compound.

... (+)- and (−)-forms are called enantiomers

They may differ in biochemical reactions

Enantiomers react in the same way in chemical reactions with achiral reagents. They may differ in biochemical reactions. Enzymes are catalysts found in plants and animals. An enzyme and its substrate (the substance which requires the enzyme in order to react) fit together 'like a key in a lock'. The geometry of the substrate is important, and enzymes can distinguish between enantiomers. *Penicillium glaucum* (a mould) feeds on (+)-2-hydroxypropanoic acid (lactic acid) but not on its (−)-enantiomer.

Inorganic compounds

Optical isomerism among inorganic compounds is met in § 24.16.4.

CHECKPOINT 25.9C: OPTICAL ACTIVITY

1. Alanine, an amino acid, has the formula $H_2NCH(CH_3)CO_2H$. Construct a model of the molecule. Do you think that alanine should be optically active? Replace $—CH_3$ by —H. This is the formula of the amino acid, glycine. Do you think that glycine should be optically active?

2. What is *optical activity*? How could you demonstrate that a compound is optically active?
Draw structural formulae for the optical isomers of molecular formula $C_3H_6O_3$.

3. Discuss the isomerism shown by the following compounds. State the differences (if any) in physical and chemical properties that exist between the isomers of each formula.

   (a) C_4H_{10}
   (b) C_4H_8
   (c) $CH_3CH(OH)CO_2H$
   (d) $C_6H_5CH{=}CH{-}CO_2H$
   (e) $C_6H_5{-}CH(OH){-}CN$
   ```
   C6H5—CH—CN
         |
         OH
   ```
   (f) $C_3H_6Br_2$
   (g) C_3H_7Cl
   (h) $C_6H_5CH_2CH(NH_2)CO_2H$

FURTHER STUDY ▲

QUESTIONS ON CHAPTER 25

1. *(a)* Say what the terms S_N, S_E, A_E, A_N mean.
   (b) Classify the following reactions as S_N, S_E, A_E or A_N
   (i) $C_6H_6 + NO_2^+ \longrightarrow C_6H_5NO_2 + H^+$
   (ii) $C_2H_5Br + OH^- \longrightarrow C_2H_5OH + Br^-$

2. Explain the difference between homolytic fission and heterolytic fission. Say which type of fission occurs in each of these reactions

   (a) $CH_3(CH_2)_2CH_2Br \longrightarrow CH_3(CH_2)_2CH_2\cdot + Br\cdot$
   (b) $(CH_3)_3CBr \longrightarrow (CH_3)_3C^+Br^-$

3. What is meant by the term '*cis-trans* isomerism'? What requirement must a compound have to show '*cis-trans* isomerism'?

*4. *(a)* Explain the term 'chiral molecule'.
   (b) The compound $CH_3CHBrCO_2H$ exists as two optical isomers. Draw the structures of the two isomers.
   (c) How do optical isomers compare in their chemical reactions?

*5. Say which of the following structures may show stereoisomerism. State what kind of stereisomerism is involved, and draw the stereoisomers.

   (a) $CH_3CH_2CH_2CHOHCH_3$
   (b) $CH_3CH_2CHOHCH_2CH_3$
   (c) $CH_3CH{=}CHCO_2H$

6. *(a)* Explain why *cis-trans* isomerism
   (i) does exist for 1,2-dibromoethene,
   (ii) does **not** exist for 1,2-dibromoethane,
   (iii) does **not** exist for 1,1-dibromoethene. 3

   (b) (i) Draw displayed formulae for the following compounds: hex-3-ene, butenedioic acid.
   (ii) State and explain which, if either, of hex-3-ene and butenedioic acid can display *cis-trans* isomerism. 3

   (c) (i) State what is meant by a chiral centre.
   (ii) Explain how such a centre gives rise to optical isomerism.
   (iii) Which of the compounds below has optical isomers?

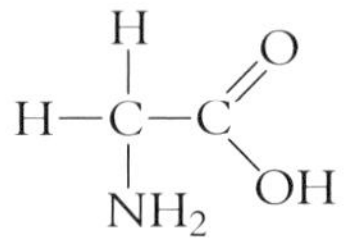

```
     H      O              H   H      O
     |    //               |   |    //
  H—C—C               H—C—C—C
     |    \               |   |    \
    NH2    OH             HO  H     OH

     L                        M

     H   H      O
     |   |    //
  H—C—C—C
     |   |    \
     H   OH    OH

        N
```

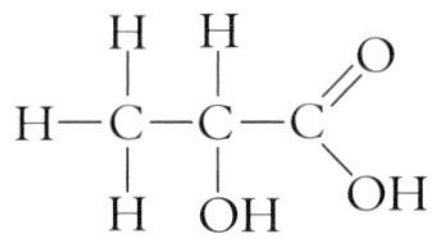

   (iv) Epinephrine (adrenalin), shown below, can exist as optical isomers. Draw these two isomers and indicate with an asterisk (*) the chiral carbon atom.

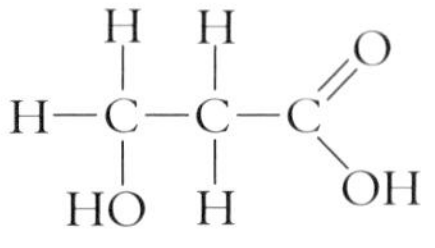

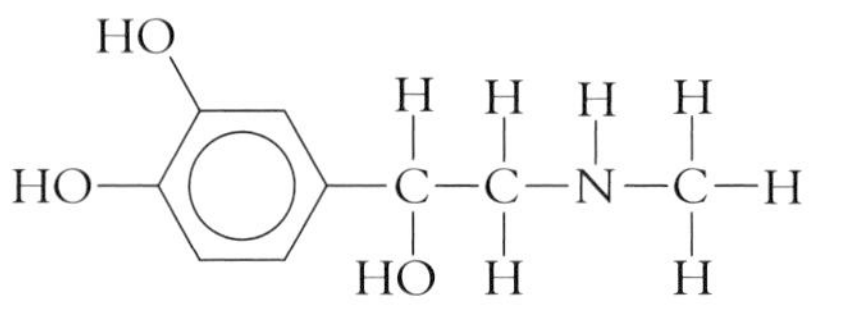

```
     HO
       \          H   H   H   H
        ⌒         |   |   |   |
  HO—( ○ )—C—C—N—C—H
                   |   |       |
                   HO  H       H
```

5

11 *C (AS/AL)*

26 THE ALKANES

26.1 METHANE

Methane, CH_4, is the simplest member of the alkane series. It is produced in large quantities by living organisms. In the cells of living organisms aerobic (with air) respiration takes place: glucose is oxidised to carbon dioxide and water

$$C_6H_{12}O_6(aq) + 6O_2(g) \longrightarrow 6CO_2(g) + 6H_2O(l)$$

Some organisms do not obtain enough oxygen for aerobic respiration, and respire anaerobically (without air). There is a group of bacteria called **methanogenic bacteria** which respire anaerobically to convert glucose into methane and other compounds. The reaction is not simple, but the equation for one methanogenic process is

$$C_6H_{12}O_6(aq) \longrightarrow 3CH_4(g) + 3CO_2(g)$$

Methanogenic bacteria are very common, and methane is produced whenever material containing carbohydrates is left in anaerobic conditions. Some examples are:

- In compost heaps, waste tips and marshes, vegetation rots in the absence of air, and methane is formed. Sometimes the methane ignites. This is the origin of the 'Will o' the wisp' flames that dance over marshes.
- Biogas digesters are containers in which waste material is left to decay in the absence of air in order to generate methane for use as a fuel.
- In the guts of animals, partly digested food is acted on by bacteria to form methane. One cow releases about 500 dm^3 of methane a day!
- Natural gas is mainly methane. It was formed geological ages ago by the anaerobic decay of vegetation.
- Rice paddies containing rotting vegetation covered by water are a major source of methane.

The level of methane in the atmosphere is rising. When water freezes, bubbles of air become trapped in the ice. When ice forms at the North and South Poles, bubbles of air become trapped in polar ice and remain there for centuries. By analysing bubbles of air in polar ice of different ages, a picture of the methane content of the air at different times in the past can be obtained. The increase in methane concentration over the last century has been much greater than in the preceding 3000 years.

Methane is, like carbon dioxide, a greenhouse gas [§ 23.6] and it also destroys ozone in the upper atmosphere [§ 21.5].

26.2 PETROLEUM OIL

Alkanes are fuels ...

The most important feature of the alkanes is their use as fuels. A huge fraction of the energy we use comes from the combustion of alkanes. The gas in our cookers, the petrol in our cars, aviation fuel and diesel oil for powering ships and electric generators – all these fuels are mixtures of alkanes. The source of these fuels is either crude petroleum oil or natural gas. Deposits of crude oil and natural gas usually occur together as they are formed by the same slow decay of marine animals and plants. Crude oil is found in many parts of the world. Figure 26.2A shows an oil rig in the North Sea.

Figure 26.2A
An oil rig

26.2.1 FRACTIONAL DISTILLATION OF CRUDE OIL

... They are obtained from crude oil by fractional distillation

Crude oil is a mixture of about 150 compounds. It is difficult to ignite. To yield volatile substances which can be used as fuels, crude oil is fractionally distilled. The distillation theory is covered in § 8.5.2. Figure 26.2B represents a fractionating column for use in separating crude oil into fractions. Each fraction is a mixture of hydrocarbons which boil over a limited range of temperature [see Table 26.2].

26.2.2 PETROCHEMICALS

Besides being used as fuels, all these fractions have another, very important use. They are the foundation of the petrochemicals industry. From them are manufactured thousands of compounds: plastics, paints, solvents, rubbers, detergents and many medicines are petrochemicals.

Figure 26.2B
An industrial fractionating column

Table 26.2
The fractions obtained from crude oil

Fraction	Boiling temperature/°C	Length of carbon chain	Use
Refinery gas	20	C_1–C_4	Fuel: domestic heating, gas cookers
Light petroleum	20–60	C_5–C_6	Solvent
Light naphtha	60–100	C_6–C_7	Solvent
Gasoline (petrol)	40–205	C_5–C_{12}	Fuel for the internal combustion engine (cars etc.)
Kerosene (paraffin)	175–325	C_{12}–C_{18}	Fuel for jet engines
Gas oil	275–400	C_{18}–C_{25}	Does not vaporise easily. Used in diesel engines, where it is injected into compressed air to make it ignite. Used in industrial furnaces, being introduced as a fine mist to help the oil to burn.
Lubricating oil	Non-volatile	C_{20}–C_{34}	Lubrication
Paraffin wax	Solidifies from lubricating oil fraction	C_{25}–C_{40}	Polishing waxes, petroleum jelly
Bitumen (asphalt)	Residue	$>C_{30}$	Road surfacing, roofing

26.3 PHYSICAL PROPERTIES

The C—H bond has a weak dipole ...

Since the electronegativities of carbon and hydrogen are 2.5 and 2.1 respectively

$$\overset{\delta-}{C}—\overset{\delta+}{H} \text{ bonds}$$

... and the intermolecular forces are weak

have only very weak dipole moments. Weak attractive forces exist between dipoles in neighbouring molecules [§ 4.8.1], and van der Waals forces [§ 4.8.2] also come into play. The attractive forces are so weak that the lower alkanes, from methane to butane, are gases at room temperature and pressure. Linear molecules of higher homologues can align themselves in a parallel arrangement so that dipole–dipole interactions and van der Waals forces can operate along the whole length of the molecule. The alkanes from C_5 to C_{17} are liquids, while

Branched-chain alkanes are more volatile than the unbranched-chain isomers

those with larger molecules are solids. Since branched-chain molecules are more spherical in shape than unbranched-chain hydrocarbons, the attractive forces between molecules are more restricted. The boiling temperatures of branched alkanes are therefore lower than those of their straight-chain isomers. The boiling temperatures of unbranched-chain alkanes are plotted against molar mass in Figure 26.3A.

Figure 26.3A Boiling temperatures of *n*-alkanes

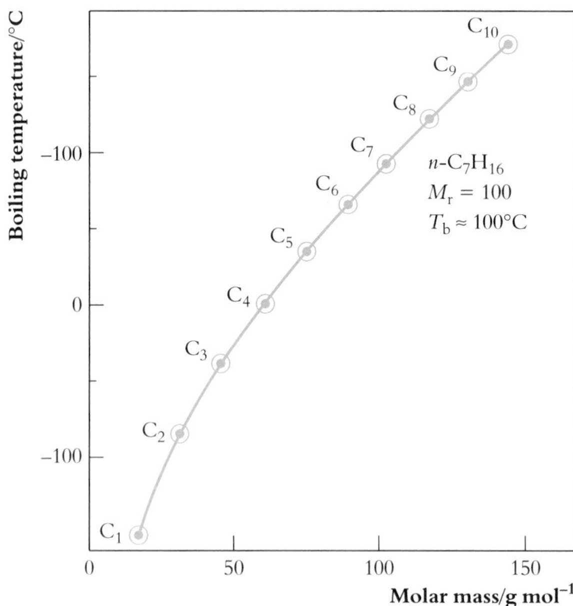

The difference in boiling temperatures between the C_1 and C_2 alkanes is 73 °C, while the C_9 and C_{10} alkanes differ by only 25 °C. It is therefore more difficult to separate the higher members by fractional distillation.

Liquid alkanes float on water ...

The liquid alkanes are less dense than water: oil floats on water. The higher members are viscous liquids, the viscosity increasing with increasing molecular mass as the attractive forces between molecules increase.

Alkanes are only slightly soluble in water. Water molecules interact because of the strong dipoles in the

... Alkanes cannot form hydrogen bonds with water and are therefore insoluble

$$\overset{\delta-}{O}—\overset{\delta+}{H} \text{ bonds}$$

[See Figure 4.8G, § 4.8.3.] The hydrogen bonds formed are stronger than any interaction which can occur between water molecules and the non-polar alkane molecules. Dissolution is therefore not favoured by energy considerations.

26.4 REACTIONS OF ALKANES

26.4.1 COMBUSTION

Alkanes burn with the release of energy

The most important reaction of the alkanes is combustion. They burn to form the harmless products, carbon dioxide and water, in an exothermic reaction [for $\Delta H^{\ominus}$, see § 10.6]:

$$CH_4(g) + 2O_2(g) \longrightarrow CO_2(g) + 2H_2O(l); \quad \Delta H^{\ominus} = -890 \text{ kJ mol}^{-1}$$

Liquid alkanes, such as octane, must be vaporised before they will burn:

$$C_8H_{18}(g) + 12\tfrac{1}{2}O_2(g) \longrightarrow 8CO_2(g) + 9H_2O(l);\ \Delta H^{\ominus} = -5510\ \text{kJ mol}^{-1}$$

Incomplete combustion gives the poisonous gas, carbon monoxide

This is the reaction which takes place in the internal combustion engine. The hydrocarbons in gasoline have boiling temperatures of about 150 °C and will vaporise in the internal combustion engine. If the supply of oxygen is insufficient, incomplete combustion will take place, and the poisonous gas carbon monoxide will be formed. There is always a certain percentage of carbon monoxide, about 5%, in the exhaust gases of motor vehicles.

In the internal combustion engine, compression of petrol vapour and air can lead to auto-ignition ...

... which causes 'knocking'

When combustion of petrol vapour occurs inside the cylinders in a car engine, a large volume of hot gases is formed. The gases force the piston down the cylinder, and the power generated is transmitted to the wheels. For smooth running it is essential that the ignition of gasoline vapour and air takes place when the piston is at the right point in the cylinder. To obtain the maximum energy from the fuel, the engine design must ensure that all the petrol vapour is burnt. For this reason the mixture of petrol vapour and air is compressed, and this compression can lead to **auto-ignition**. The ignition of gases takes place before the spark. It results in a sudden rise in pressure, which delivers a blow to the piston. The engine makes a metallic sound called **knocking**. The best fuels for resistance to knocking are branched-chain hydrocarbons, such as 2,2,4-trimethylpentane:

```
                          CH3
                          |
H3C—CH—CH2—C—CH3
        |                 |
       CH3               CH3
```

This compound, which used to be called *iso*-octane, has been assigned an *octane number* of 100. Heptane, $CH_3CH_2CH_2CH_2CH_2CH_2CH_3$, has very bad knocking properties and has been assigned an *octane number* of 0. The **octane number** of a petrol is found by comparing its performance with a mixture of heptane and 2,2,4-trimethylpentane. If it has the same performance as a mixture of 25% heptane and 75% 2,2,4-trimethylpentane, then its octane number is 75.

The octane number of a fuel measures its resistance to knocking

The addition of lead tetraethyl, $Pb(C_2H_5)_4$, TEL, reduces knocking. However, it also releases lead compounds into the air in the exhaust gases, and concern over the increasing levels of these toxic compounds has lead to a phasing-out of TEL. Most vehicles now use unleaded petrol. Another solution to the problem of knocking has to be employed. Fuels with poor knocking properties can be converted into branched-chain hydrocarbons by cracking, which is discussed below.

26.4.2 CRACKING

'Cracking' is used to obtain more of lower molecular mass alkanes, which are more volatile

The petroleum fractions with 1 to 12 carbon atoms in the molecule are in demand in larger quantities than the fractions with bigger molecules. The petroleum industry uses **pyrolysis** (splitting by heat) of high molar mass alkanes to give hydrocarbons with smaller molecules, which are more easily vaporised and are therefore more useful fuels:

$$\text{Alkane with large molecules} \xrightarrow[\text{at 450 °C over catalyst of } Al_2O_3/SiO_2]{\text{Vapour passed}} \text{Alkane with smaller molecules} + \text{Alkene} + \text{Hydrogen}$$

e.g.

$$2CH_3CH_2CH_3(g) \longrightarrow CH_4(g) + CH_3CH{=}CH_2(g) + CH_2{=}CH_2(g) + H_2(g)$$

Propane — Methane — Propene — Ethene — Hydrogen

The industry calls this type of reaction **cracking**.

26.4.3 ALKYLATION

'Alkylation' is used to make branched-chain alkanes

Since branched-chain compounds have higher octane numbers than straight-chain compounds, a good deal of research has gone into the synthesis of branched-chain compounds. They are used in unleaded fuels. They are made by the **alkylation** of alkenes by alkanes in the presence of a catalyst.

In alkylation reactions

$$\text{Tertiary alkane} + \text{Alkene} \xrightarrow[\text{conc. } H_2SO_4 \text{ as catalyst}]{20\,^\circ\text{C}} \text{Branched-chain alkane}$$

26.4.4 REFORMING

'Reforming' converts alkanes into aromatic compounds, e.g., benzene

A huge number of important chemicals are derived from benzene. One source of benzene is petroleum. Unbranched-chain alkanes are converted into aromatic compounds by the process of **reforming**. For example

$$\underset{\text{Hexane}}{C_6H_{14}(l)} \xrightarrow[\text{compressed to 40 atm, passed over } Al_2O_3 \text{ as catalyst (or Pt), 10 atm}]{\text{Vaporised at } 500\,^\circ\text{C}} \underset{\text{Benzene}}{C_6H_6(l)} + \underset{\text{Hydrogen}}{4H_2(g)}$$

'Platforming' uses a platinum catalyst

Much work has gone into the effectiveness of different catalysts. If a platinum catalyst is used, the process is called **platforming**.

26.4.5 OTHER REACTIONS OF ALKANES

Alkanes are not reactive

They undergo some substitution reactions

The alkanes used to be called the **paraffins**, a name derived from the Latin for *little liking*, implying that this class of compounds had little liking for the usual chemical reagents. Alkanes do not react with dilute acids or alkalis or with oxidising agents. At high temperatures they react with nitric acid vapour. All the reactions of alkanes, apart from combustion, are **substitution** reactions, of the form

$$RH + \mathbf{XY} \longrightarrow R\mathbf{X} + H\mathbf{Y}$$

26.4.6 HALOGENATION

Substitution by halogens occurs in sunlight

Under certain conditions, alkanes react with halogens. If a stoppered test-tube containing hexane and a drop of liquid bromine is left to stand at room temperature in the dark, nothing happens. The colour of the bromine is still as intense after three or four days. If the solution is exposed to sunlight, the colour fades in a few minutes, and the acidic, fuming gas hydrogen bromide can be detected. The reaction that has occurred is

$$C_6H_{14}(l) + Br_2(l) \longrightarrow C_6H_{13}Br(l) + HBr(g)$$

with the formation of bromohexanes. Since it takes place in the presence of light, it is called a **photochemical reaction**. Alkanes can be chlorinated and brominated photochemically.

Produced by chlorination of CH_4 are CH_3Cl, CH_2Cl_2, $CHCl_3$, CCl_4

When methane reacts with chlorine in sunlight, one or more chlorine atoms may replace hydrogen atoms, depending on the amounts of halogen and alkane present. The formation of chloromethane

$$CH_4(g) + Cl_2(g) \longrightarrow CH_3Cl(g) + HCl(g)$$

may be followed by the formation of dichloromethane, CH_2Cl_2, trichloromethane (chloroform), $CHCl_3$, and tetrachloromethane, CCl_4. Chloroalkanes are useful solvents, and the mixture of products formed in the chlorination of an alkane may find use as a solvent, without the need for isolation of individual compounds.

... They are useful solvents

Much work has been done on the subject of how this reaction takes place, when methane is so unreactive towards other reagents. The mechanism proposed for the chlorination of methane takes into account the following experimental observations:

Experimental observations on the reaction, $CH_4 + Cl_2$

1. Reaction takes place rapidly in sunlight or above 300 °C but not in the dark at room temperature.

Figure 26.4A Reactions of ethane

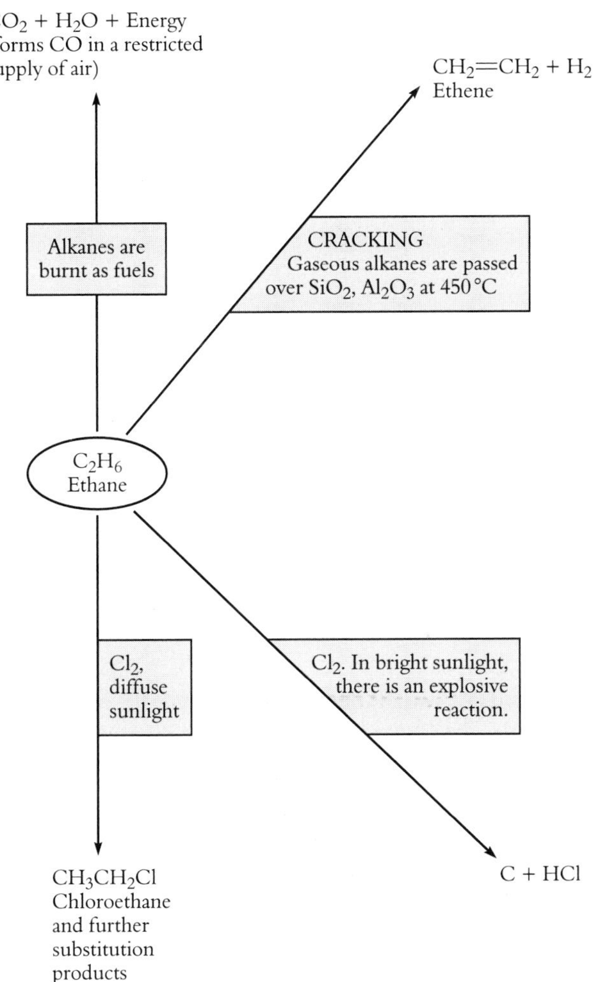

2. Thousands of molecules of chloromethane are formed for each photon of light absorbed.
3. A little ethane is formed.
4. If a trace of tetramethyllead, $Pb(CH_3)_4$, is added, the reaction will take place in the dark or at room temperature. This substance is known to dissociate into methyl radicals, $\cdot CH_3$.

A mechanism which fits these observations is as follows.

26.4.7 THE MECHANISM OF THE CHLORINATION OF METHANE

Step 1

Photochemical homolysis of Cl_2

Homolysis of the Cl—Cl bond. The necessary energy comes from the light absorbed or the heat supplied. It is easier to split the Cl—Cl bond than the C—H bond. (Bond energy terms are Cl—Cl, 242 kJ mol^{-1}; C—H, 435 kJ mol^{-1} [§ 10.11].)

$$Cl_2 \xrightarrow[\text{or heat}]{\text{light}} 2Cl\cdot$$

Step 2

There are two possible reactions of $Cl\cdot$ with CH_4

The chlorine atoms formed are very reactive. Since they are surrounded by methane molecules, there are two possible reactions:

$$Cl\cdot + CH_4 \longrightarrow CH_3Cl + H\cdot$$
$$Cl\cdot + CH_4 \longrightarrow HCl + \cdot CH_3$$

The second possibility is more likely because the formation of an H—Cl bond is more exothermic than the formation of a C—Cl bond. (Bond energy terms are H—Cl, 431 kJ mol^{-1}; C—Cl, 350 kJ mol^{-1}.)

Step 3

The reaction of $CH_3\cdot$ with Cl_2 results in a chain reaction

The methyl radicals formed collide with methane molecules and chlorine molecules. The reaction

$$\cdot CH_3 + CH_4 \longrightarrow CH_4 + \cdot CH_3$$

results in no net change.

The reaction

$$\cdot CH_3 + Cl_2 \longrightarrow CH_3Cl + \cdot Cl$$

leads to a chain reaction because the chlorine atoms formed react as in Step 2.

Step 4

Thousands of molecules of chloromethane are formed for every photon of light absorbed. The high yield is due to the chain reaction – Steps 2 and 3. The reason why the yield is not higher is that radicals can combine with each other and bring the chain to an end. The reactions

There are three chain-terminating reactions

$$2Cl\cdot \longrightarrow Cl_2$$
$$2CH_3\cdot \longrightarrow C_2H_6$$
$$Cl\cdot + \cdot CH_3 \longrightarrow CH_3Cl$$

bring the chain reaction to an end. Some ethane can be detected in the product.

SUMMARY

To summarise, the steps in the chlorination of methane to chloromethane are:

Summary of the mechanism

Chain initiation $Cl_2 \xrightarrow[\text{or heat}]{\text{light}} 2Cl\cdot$

Chain propagation $Cl\cdot + CH_4 \longrightarrow HCl + \cdot CH_3$

$\cdot CH_3 + Cl_2 \longrightarrow CH_3Cl + Cl\cdot$

Chain termination $2Cl\cdot \longrightarrow Cl_2$

$2\cdot CH_3 \longrightarrow C_2H_6$

$Cl\cdot + \cdot CH_3 \longrightarrow CH_3Cl$

Formation of CH_2Cl_2

Further Cl atoms can be introduced

Step 3 can give rise to the chain:

$$CH_3Cl + Cl\cdot \longrightarrow HCl + \cdot CH_2Cl$$
$$\cdot CH_2Cl + Cl_2 \longrightarrow CH_2Cl_2 + Cl\cdot$$

CH_2Cl_2 can undergo further chlorination to $CHCl_3$ and CCl_4.

26.4.8 OTHER HALOGENS

Methane reacts with the other halogens

The yield per photon is less for bromination than for chlorination because the step

$$Br\cdot + CH_4 \longrightarrow HBr + \cdot CH_3$$

is more endothermic than the corresponding step in chlorination. Iodination is slow and reversible. Fluorination is dangerously exothermic.

26.5 THE CAR

Two hundred million motor vehicles are travelling the roads of the world. They are covering 2 million million miles a year and burning 70 billion litres of petrol. From their exhausts come 40 million tonnes of carbon monoxide, 4 million tonnes of nitrogen oxide, 4 million tonnes of hydrocarbons and 0.2 million tonnes of lead. These pollutants are discharged at street level where people cannot avoid inhaling them.

The petrol engine powers most of our motor vehicles. § 26.4.1 gave a brief description of the combustion of petrol in a vehicle engine. The burning of the fuel in the cylinder after ignition by the spark is sudden and intense. The temperature soars to about 2800 °C, and some of the nitrogen and oxygen in the cylinder combine to form nitrogen oxide. As the piston is pushed out of the cylinder, the combustion gases expand and cool in less than a hundredth of a second. The heating–cooling cycle occurs so rapidly that much of the fuel is not completely oxidised to carbon dioxide and water. Some carbon monoxide is formed, and some hydrocarbons remain unburned. The simultaneous production of overoxidised pollutants, nitrogen oxide, and underoxidised pollutants, hydrocarbons and carbon monoxide, makes it difficult to clean up the exhaust with a single chemical treatment. The equation for the complete combustion of octane is

$$2C_8H_{18} + 25O_2 \longrightarrow 16CO_2(g) + 18H_2O(g)$$

From the equation the ratio (mass of air/mass of fuel) can be worked out. This ratio, called the **air/fuel ratio**, should be 14–15. The production of each major pollutant depends on the air/fuel ratio. A rich mixture (with less air than the ideal ratio) leads to the formation of carbon monoxide and hydrocarbons. A lean mixture (with an excess of air) leads to the formation of nitrogen oxide. Altering the air/fuel ratio merely trades one set of pollutants for another.

Nitrogen oxide is a pollutant which attacks the ozone layer [see § 21.5]. It also contributes to the formation of photochemical smog [see § 27.11] and to the formation of nitric acid in acid rain [see § 21.10].

Carbon monoxide is a poisonous gas [see § 24.16.5]. Hydrocarbons by themselves cause little damage, but in the presence of sunlight react with oxygen, ozone and oxides of nitrogen to form photochemical smog [see § 27.11]. One solution to the problem of pollution by vehicle exhausts is the catalytic converter [see § 26.6].

26.6 CATALYTIC CONVERTERS

Carbon monoxide and hydrocarbons are present in vehicle exhaust gases as a result of the incomplete combustion of hydrocarbons in the engine. Oxides of nitrogen are present as a result of the combination of oxygen and nitrogen in the air at the temperature of the car engine. Carbon monoxide and hydrocarbons can be rendered harmless by oxidation. The exhaust gases are passed through a **catalytic converter**, a pipe containing a solid catalyst, platinum and rhodium. This catalyses the oxidations

$$2C_8H_{18}(g) + 25O_2(g) \longrightarrow 16CO_2(g) + 18H_2O(g)$$
$$2CO(g) + O_2(g) \longrightarrow 2CO_2(g)$$

Figure 26.6A
A catalytic converter

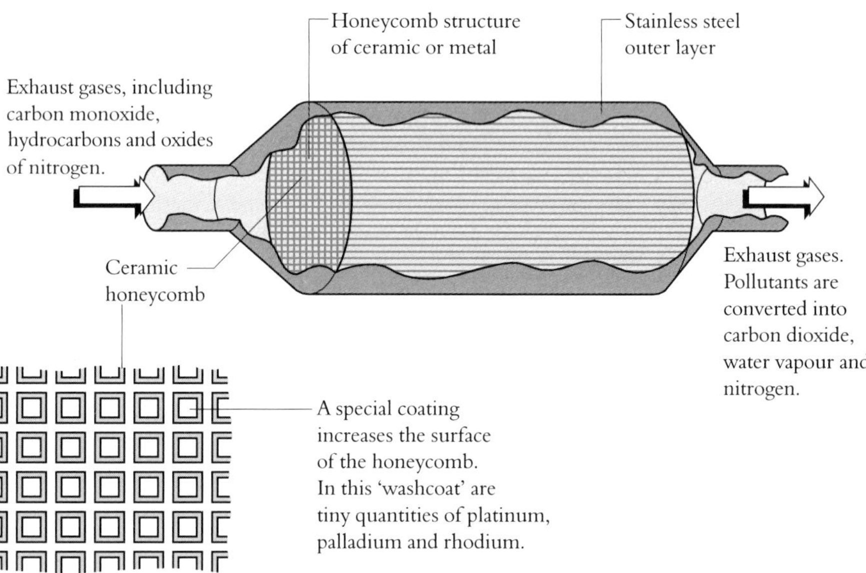

A catalytic converter is a pipe containing a catalyst of platinum and rhodium which catalyses the oxidation of carbon monoxide and hydrocarbons to carbon dioxide and water

It cannot remove oxides of nitrogen by oxidation. Catalytic converters have been fitted to all new UK vehicles since 1990 and can be retrofitted to older vehicles.

An engine can be fitted with a **three-way catalytic system**, which can both oxidise hydrocarbons and carbon monoxide and also reduce oxides of nitrogen. The exhaust gas is passed over a reduction catalyst of e.g. platinum, palladium, cobalt or nickel, which catalyses the reaction

$$2NO(g) + 2CO(g) \longrightarrow N_2(g) + 2CO_2(g)$$

It is not possible to fit a three-way catalytic system to an existing car.

Lead 'poisons' the catalysts in catalytic converters (destroys their catalytic action). This is why unleaded petrol must be used in vehicles with catalytic converters. For more about the choice of catalyst see § 14.7.

QUESTIONS ON CHAPTER 26

1. Write structural formulae for the following:

(a) 2,3-dimethylpentane
(b) 2,4,5-trimethylheptane
(c) 3-bromo-2-chloropentane
(d) cyclohexane

2. Which member of the following pairs of alkanes has the higher boiling temperature?

(a) (i) butane and (ii) heptane
(b) (i) 2-methylbutane and (ii) pentane
(c) (i) hexane and (ii) 2,3-dimethylbutane
(d) (i) hexane and (ii) cyclohexane

3. Name all the products formed in the following reactions:

(a) the combustion of octane
(b) the chlorination of methane
(c) the reactions between $CH_3\cdot$ and $H\cdot$

4. Write structural formulae for the compounds with the formulae *(a)* C_4H_{10} *(b)* $C_2H_4Cl_2$ *(c)* C_5H_{12}

5. Write equations for all the steps that occur in the reaction

$$CH_3Cl + Cl_2 \longrightarrow CH_2Cl_2 + HCl$$

What names are given to these steps?

6. Explain briefly what is meant by the following terms *(a)* fractional distillation, *(b)* pyrolysis, *(c)* catalytic cracking, *(d)* reforming, *(e)* platforming, *(f)* knocking, *(g)* octane number.

7. *(a)* The major processes used in the refining of crude petroleum oil are fractional distillation and cracking. Briefly state the principles of each process and explain why each process is necessary.
(b) Some of the compounds in crude oil contain sulphur. Give two reasons why sulphur compounds are removed from petrol during its manufacture.

8. The three-way catalytic systems which are fitted to the exhaust systems of cars remove nitrogen oxides, carbon monoxide and unburnt hydrocarbons. Oxidation and reduction are both involved.

(a) State whether the catalysis is homogeneous or heterogeneous.
(b) What reaction or reactions are catalysed by the reduction catalyst?
(c) What reaction or reactions are catalysed by the oxidation catalyst?
(d) Name a substance which poisons the catalyst and suggest a remedy for this poisoning.

9. In diffused light methane combines readily with chlorine in a photochemical substitution reaction forming chloromethane as the initial organic product. Further chlorination produces a mixture of chlorinated hydrocarbons.

(a) Explain the term **photochemical substitution.** 2
(b) Write equations to represent the following steps in the reaction:
(i) The initiation step 1
(ii) Two propagation steps 2
(iii) A termination step. 1
(c) Including an equation, define **standard enthalpy change of formation** of chloromethane. 4
(d) Given the standard enthalpy changes of formation below, calculate the enthalpy change for the overall reaction.

$$CH_4(g) + Cl_2(g) \longrightarrow CH_3Cl(g) + HCl(g)$$

3

Compound	$\Delta H_f^\ominus$/kJ mol^{-1}
HCl	−72
CH_4	−75
CH_3Cl	−82

(e) An organic compound has the composition by mass.

Element	Percentage
Hydrogen	4.1
Carbon	24.2
Chlorine	71.7

(i) Calculate the empirical formula of the compound. 3
(ii) The mass spectrum of the compound showed the molecular ion peak at a mass/charge ratio of 98. Suggest a structural formula for the compound. 1
(iii) Suggest why smaller peaks may be found at 100 and 102. 2

18 *CCEA (AL)*

10. *(a)* When excess methane reacts with chlorine, the main product is chloromethane.
(i) Under what conditions does this reaction occur? **1**
(ii) Give the mechanism for this reaction. **4**

(b) Heptane and octane are liquids at room temperature with boiling points of 98 °C and 126 °C respectively.
(i) What type of intermolecular forces of attraction are present in such liquids? **1**
(ii) A mixture of these two liquids is almost ideal. Explain, in terms of the intermolecular forces present, why this mixture approximates to ideal behaviour. **2**
(iii) On a copy of the axes below, sketch a graph of how the boiling points of various mixtures of these two liquids vary with composition. Show clearly the composition of the liquid and vapour phases. **2**

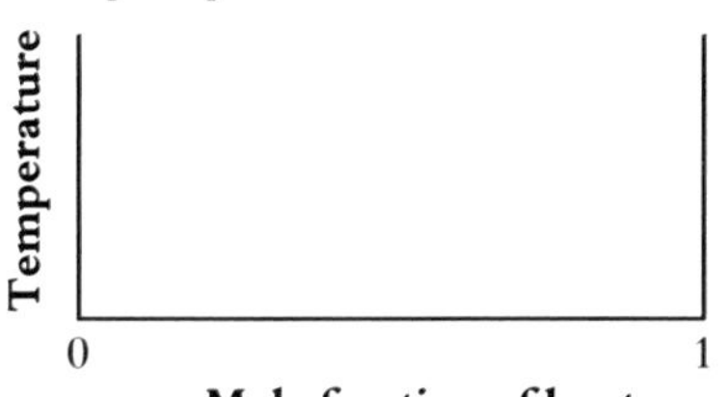

(iv) By what process would a mixture containing a mole fraction of 0.5 heptane be separated into pure heptane and pure octane? **1**

(c) Liquid alkanes such as heptane and octane occur in petrol used as fuel to drive cars.
(i) Give two reasons why liquid fuels are generally preferred to gaseous ones. **2**
(ii) The combustion characteristics of fuels for internal combustion engines in cars can be considerably improved by adding branched chain alkanes, cycloalkanes, aromatic hydrocarbons or tetraethyllead(IV) (lead tetraethyl). Two of these are now considered to be hazardous to health. Select these two and identify the health hazard with which each is associated. **2**

15 *L (AL)*

11. *(a)* (i) Octane, C_8H_{18}, is a member of the homologous series of alkanes. Explain the term *homologous series*.
(ii) The combustion of octane in car engines is often incomplete. Write an equation for the combustion of octane to form carbon monoxide and water only. **2**

(b) Write an equation for the reaction between carbon monoxide and nitrogen monoxide which occurs in a catalytic converter. State the role of carbon monoxide in this reaction. **2**

(c) A catalytic converter contains a coating of metal catalyst on a ceramic base.
(i) Give one metal used in such a catalyst.
(ii) Name the type of catalysis performed by a catalytic converter. **2**

(d) The free radical $\cdot CH_2Cl$ is produced during the reaction of methane with an excess of chlorine. Write equations for
(i) the propagation step in which $\cdot CH_2Cl$ is formed
(ii) the propagation step in which $\cdot CH_2Cl$ reacts
(iii) a termination step involving $\cdot CH_2Cl$. **3**

(e) Name another type of organic reaction which involves free radicals. **1**

(f) When the reaction of methane with a large excess of chlorine was complete, the mixture of products formed was investigated by mass spectrometry. Calculate the highest *m/z* value peak in the spectrum of the mixture and show your working. (Chlorine exists as a mixture of the isotopes, ^{35}Cl and ^{37}Cl.) **3**

13 *NEAB (AS/AL)*

27

ALKENES AND ALKYNES

27.1 ETHENE

Ethene, $CH_2{=}CH_2$, is the simplest member of the alkenes. Ethene has a number of effects on plant growth. It inhibits stem growth, especially during periods of physiological stress, such as drought and flooding. It encourages buds to open, and it promotes the ripening of fruit. In 1934, it was shown that ripening apples emit ethene, and subsequently ethene was detected emanating from a wide variety of ripening fruits and other plant organisms. When the plant starts to ripen it produces ethene, and the ethene produced promotes further ripening.

Fruits such as bananas, citrus fruits and tomatoes are picked green and transported to be stored in a warehouse before sale. They are often prevented from ripening by being stored in an atmosphere which lacks oxygen. When they are needed for sale, ripening is accelerated by applying ethene and oxygen. An 'ethylene gas generator' in the warehouse consists of a box packed with catalyst and containing a heating element. Ethanol is poured into the generator and the heated catalyst causes its dehydration to ethene. If the concentration of ethene rises to too high a level, the excess is absorbed by acidified potassium manganate(VII).

Pears, peaches and avocados are also ripened in this way because it is easier to transport the hard green fruits than soft ripe fruits. The unripe fruits are stored until required. The ethene concentration needed to start ripening is low – only about one part per million. Perhaps your family grows tomatoes and you find that you have to pick your tomatoes while they are still green. Here is a tip from chemistry: you can use a few ripe tomatoes to ripen the green ones. If you put the green tomatoes in a box or drawer or cupboard with a few ripe tomatoes, the ethene given off by the ripe tomatoes will start the ripening process in the rest of the batch.

27.2 ALKENES

The names of alkenes

Alkenes are a homologous series of aliphatic hydrocarbons with the general formula C_nH_{2n}. Typical members of the series are listed below:

$CH_2{=}CH_2$	Ethene
$CH_3CH{=}CH_2$	Propene
$CH_3CH_2CH{=}CH_2$	But-1-ene
$CH_3CH{=}CHCH_3$	But-2-ene

They are called **unsaturated** hydrocarbons because the double bond can open to allow them to take up more hydrogen atoms or other species.

27.3 SOURCE OF ALKENES

Alkenes are obtained in the petroleum industry by the process of cracking [§ 26.4.2]. The laboratory preparation is usually the dehydration of an alcohol. Figure 27.3A summarises the methods employed.

Figure 27.3A Preparations of alkenes

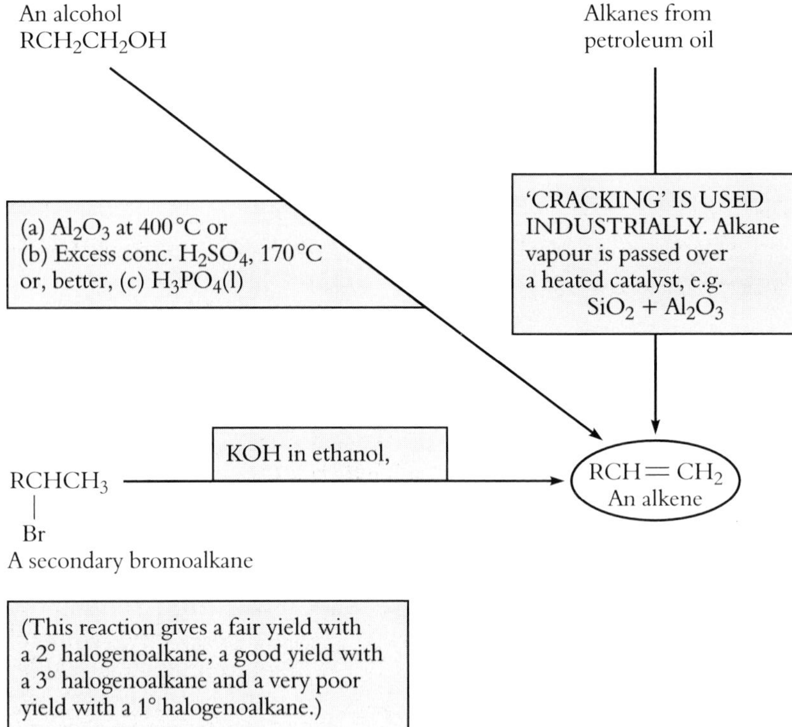

27.4 PHYSICAL PROPERTIES OF ALKENES

The volatility and solubility of alkenes are similar to those of the corresponding alkanes.

27.5 STRUCTURAL ISOMERISM

27.5.1 CARBON CHAIN

Both branched-chain and unbranched-chain isomers of alkenes exist, as with alkanes.

27.5.2 POSITIONAL ISOMERISM

Isomerism can involve the carbon chain or the position of the double bond

The position of the double bond must be given in the name of an alkene. Butene has two isomers, but-1-ene, with the double bond between carbon atoms 1 and 2; and but-2-ene, with the double bond between carbon atoms 2 and 3.

27.5.3 *CIS-TRANS* ISOMERISM OF ALKENES

Inhibition of rotation about the C=C bond gives rise to cis-trans *isomerism*

In ethene [see Figure 27.5A(a)], the four hydrogen atoms all lie in the same plane. Rotation of the CH_2 groups about the C=C bond is inhibited by the requirements for the formation of a strong π bond [§ 5.3.4]. The restriction of rotation is the reason why but-2-ene has two isomers, which are shown in Figure 27.5A(b) and (c). The geometry of the molecules is different. The isomer with both CH_3— groups on the same side of the double bond is called the *cis*-isomer, and the isomer with the CH_3— groups on opposite sides of the double bond is called the *trans*-isomer.

The cis-trans *isomers of but-2-ene*

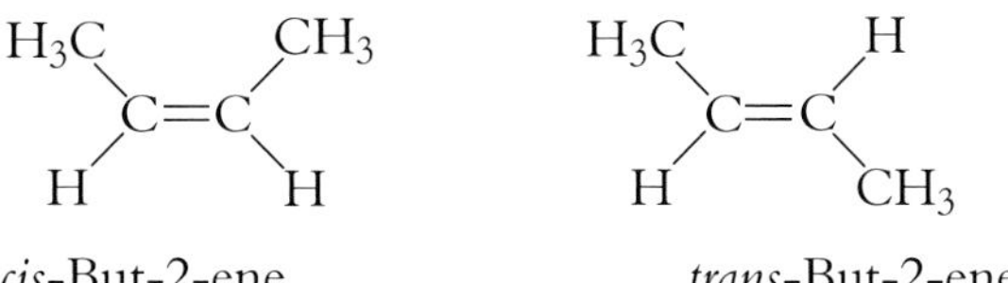

cis-But-2-ene *trans*-But-2-ene

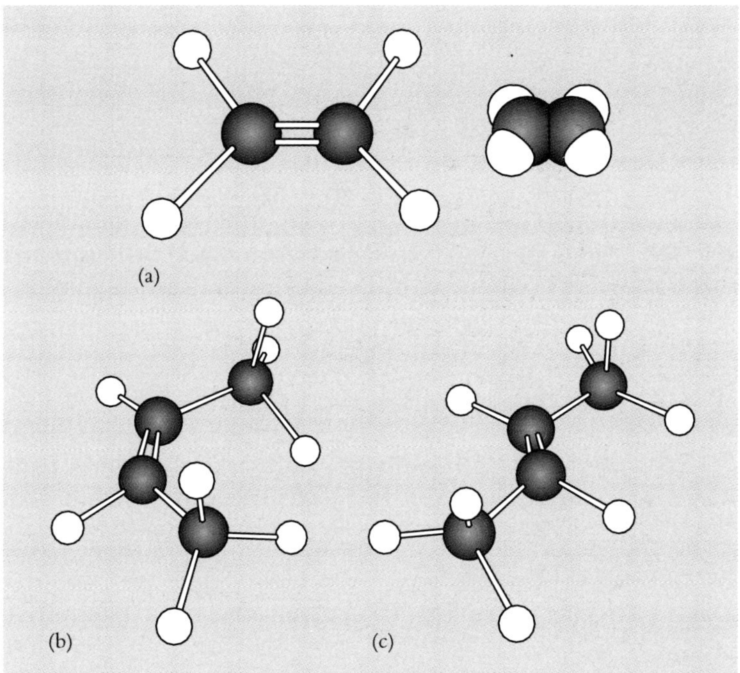

Figure 27.5A
(a) Ethene
(b) *cis*-but-2-ene
(c) *trans*-but-2-ene

27.6 REACTIVITY OF ALKENES

The two unsaturated carbon atoms in the

$$>C=C< \text{ bond}$$

are joined by a σ bond and also by a π bond [§ 5.3.4]. The cloud of electrons which forms the π bond lies above and below the plane of the three sp^2 hybrid bonds formed by each of the unsaturated carbon atoms. In this position, the π electrons are more susceptible than the σ electrons to attack by an electrophilic reagent [see Figure 27.6A]. This is why alkenes are so much more reactive than alkanes.

Alkenes are more reactive than alkanes

Figure 27.6A
The reactivity of ethene

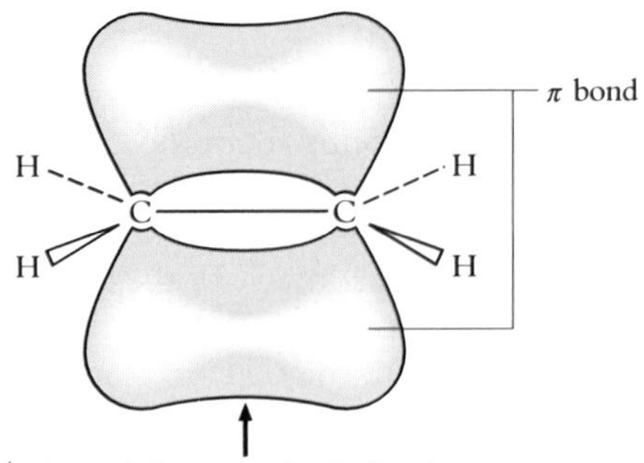

X$^+$ electrophile attacks cloud of π electrons.

SUMMARY

Alkenes, C_nH_{2n}, show positional isomerism and *cis-trans* isomerism.

The double bond of alkenes consists of a σ-bond and a π-bond. The π-bond is above and below the line joining the nuclei of the bonded carbon atoms and therefore accessible to electrophiles. This is why a C═C bond is reactive.

27.7 COMBUSTION

They are not used as fuels, because they can be used instead for the manufacture of important chemicals

Alkenes burn to form carbon dioxide and water. They show a greater tendency than alkanes to undergo incomplete combustion to form carbon monoxide and carbon. In any case, alkenes are not used as fuels. The double bond makes them reactive compounds, and they are the starting point in the manufacture of many important chemicals.

27.8 ADDITION REACTIONS

Although alkenes can form substitution products, with chlorine in sunlight for example, they more frequently take part in addition reactions.

27.8.1 HYDROGEN

Hydrogen adds to alkenes with Ni powder as catalyst

In the presence of a catalyst, hydrogen adds across a

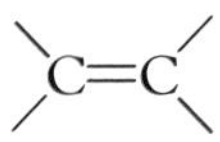

$$\gt C{=}C\lt \quad \text{bond}$$

to form a saturated compound. Palladium is an extremely effective catalyst, but nickel is used industrially because it is less expensive:

$$R_2C{=}CR'_2(g) + H_2(g) \xrightarrow[100\,°C,\ 4\ atm]{Ni\ powder} R_2CH{-}CHR'_2$$

Hydrogenation converts unsaturated edible oils into edible fats ...

... e.g., margarine

Catalytic hydrogenation, using nickel powder, is an important process in the food industry. Plant oils, such as sunflower seed oil and peanut oil, are 'polyunsaturates': they are esters of carboxylic acids which contain more than one carbon–carbon double bond. Although they are useful for cooking, these oils are less valuable commercially than expensive animal fats such as butter. Animal fats are esters of saturated carboxylic acids. Hydrogenation is used to convert unsaturated edible oils into edible fats. In soft margarine some of the double bonds remain: the degree of softness can be controlled by regulating the amount of hydrogenation.

Hydrogenation of alkenes; choice of a catalyst

The choice of catalyst in many reactions, including the hydrogenation of alkenes, is a transition metal. These metals are able to act as catalysts because they have a variable oxidation state. The reactants must be adsorbed on the catalyst and, after the reaction, the products must be desorbed. Some transition metals, e.g. tungsten, adsorb too strongly and are therefore not effective. Others, e.g. silver, adsorb too weakly to catalyse the reaction. Platinum and nickel and other transition metals have the right ability to adsorb the reactants and release the products.

27.8.2 ALKANES

Tertiary alkanes add to alkenes

Tertiary alkanes will add to alkenes [§ 26.4.3].

27.8.3 HALOGENS

Chlorine adds across a double bond …

Chlorine gas will react with ethene at room temperature, without catalysis by light or peroxides, to form liquid 1,2-dichloroethane:

$$H_2C{=}CH_2(g) + Cl_2(g) \longrightarrow ClCH_2CH_2Cl(l)$$

… as do bromine and iodine

A test for a multiple bond is the decolourisation of a bromine solution without gas evolution

Bromine and iodine will also add across the double bond. When an alkene is bubbled through bromine in an organic solvent, e.g. trichloroethane, the bromine colour disappears as bromine adds across the double bond:

$$CH_2{=}CH_2(g) + Br_2(C_2H_3Cl_3) \longrightarrow \underset{\text{1,2-Dibromoethane}}{BrCH_2CH_2Br(C_2H_3Cl_3)}$$

The decolourisation of bromine solutions without the evolution of hydrogen bromide (a fuming, pungent gas) is a test for a multiple bond. Bromine water is also decolourised by alkenes to give a different product, a **bromoalcohol**:

$$CH_2{=}CH_2(g) + Br_2(aq) + H_2O(l) \longrightarrow \underset{\text{2-Bromoethanol}}{BrCH_2CH_2OH(aq)} + HBr(aq)$$

27.8.4 HYDROGEN HALIDES

Hydrogen halides add across the double bond

The hydrogen halides, HCl, HBr and HI, all add across the double bond. When hydrogen bromide adds to ethene, bromoethane is formed:

$$CH_2{=}CH_2(g) + HBr(g) \longrightarrow CH_3CH_2Br(g)$$

Addition of hydrogen halide

When hydrogen halides add to an unsymmetrical alkene, two products are possible.

The addition of hydrogen bromide to propene could give either

$$CH_3CH{=}CH_2(g) + HBr(g) \longrightarrow \underset{\text{2-Bromopropane}}{CH_3\underset{\substack{|\\ Br}}{C}HCH_3(l)}$$

$$\textit{or}\ \underset{\text{1-Bromopropane}}{CH_3CH_2CH_2Br(l)}$$

In fact, the product is almost entirely 2-bromopropane. The Russian chemist, Markovnikov, formulated a rule for predicting which addition product would be

Markovnikov's Rule predicts the product of the addition of HX to an alkene

formed. **Markovnikov's Rule** can be stated in this form: *in addition of a compound HX to an unsaturated compound, hydrogen becomes attached to the unsaturated carbon atom which carries the larger number of hydrogen atoms*. In this case, hydrogen adds to the CH_2 group, not to the CH group.

27.8.5 MECHANISM OF ADDITION OF BROMINE TO ALKENES

At first sight it might seem that the reaction could take place in one step if the bromine molecule approaches the alkene from the right direction:

A simple one step mechanism for the addition of bromine to alkenes does not fit the facts ...

$$H_2C{=}CH_2 + Br{-}Br \longrightarrow H_2C(Br){-}C(Br)H_2$$

A number of workers have found evidence that this is not the course of the reaction. Only two of these pieces of evidence are given here.

Evidence

... For example, these two experimental observations on the addition of bromine to alkenes

(*a*) Ethene reacts with bromine in an organic solvent, e.g. trichloroethane, to give 1,2-dibromoethane. With bromine water, it gives a mixture of 2-bromoethanol, 1,2-dibromoethane and hydrogen bromide:

$$CH_2{=}CH_2 + Br_2 \xrightarrow{\text{organic solvent}} BrCH_2CH_2Br$$
$$CH_2{=}CH_2 + Br_2 \xrightarrow{\text{in } H_2O} BrCH_2CH_2OH + BrCH_2CH_2Br + HBr$$

(The yield of $BrCH_2CH_2OH$ is too great to be explained by hydrolysis of $BrCH_2CH_2Br$.)

(*b*) Ethene and bromine water containing sodium chloride give three products, 2-bromoethanol, 1-bromo-2-chloroethane and 1,2-dibromoethane:

$$CH_2{=}CH_2 + Br_2 + H_2O + NaCl \longrightarrow HOCH_2CH_2Br \text{ (major product)} + ClCH_2CH_2Br \text{ (smaller amount)} + BrCH_2CH_2Br \text{ (a trace)}$$

The proposed mechanism

The formation of the chloro-compound in (b) needs some explaining. Since sodium chloride does not react with ethene, there must be formed from bromine and ethene a positively charged species which will react with chloride ions:

$$CH_2{=}CH_2 \xrightarrow{Br_2} H_2C{-}CH_2 \text{ (bridged by } Br^+\text{)} \xrightarrow{Cl^-} ClCH_2CH_2Br$$

The following mechanism is proposed for the formation of the positive ion.

1. Ethene has both a σ bond and a π bond between the two carbon atoms. Bromine is an electrophile. When a bromine molecule approaches an ethene molecule, the π electron cloud interacts with the approaching bromine

molecule, causing a polarisation of the Br—Br bond:

When the π electron cloud of ethene reacts with bromine ...

$$H_2C{=}CH_2 + Br{-}Br \longrightarrow H_2C{=}CH_2 \triangleright \overset{\delta+}{Br}{-}\overset{\delta-}{Br}$$

2. The π electrons become gradually more attached to the δ+ Br atom, and the electrons of the Br—Br bond become gradually more polarised until the association of ethene and bromine is transformed into a positive ion, called a *bromonium ion*, and a bromide ion:

... the bromonium ion intermediate which is formed ...

$$H_2C{=}CH_2 \triangleright \overset{\delta+}{Br}{-}\overset{\delta-}{Br} \longrightarrow H_2C{-}CH_2\,(\overset{+}{Br}\text{ bridged}) + Br^-$$

The positive charge is stabilised by delocalisation. The negative charge resides on Br^-, which is a good leaving group.

3. The positive bromonium ion is immediately attacked by a bromide ion to form the product, 1,2-dibromoethane:

... quickly reacts to form the product

$$Br^- + H_2C{-}CH_2\,(\overset{+}{Br}\text{ bridged}) \longrightarrow H(Br)HC{-}CH(Br)H$$

This reaction step is very fast compared with the other steps.

4. In reaction (b), i.e., with $Br_2(aq) + NaCl(aq)$, the cation reacts in three ways:

$$H_2C{-}CH_2\,(\overset{+}{Br}) + Br^- \longrightarrow BrCH_2CH_2Br \quad [1]$$

$$H_2C{-}CH_2\,(\overset{+}{Br}) + H_2O \longrightarrow H_2\overset{+}{O}{-}CH_2CH_2Br \xrightarrow{H_2O} HOCH_2CH_2Br + H_3O^+ \quad [2]$$

$$H_2C{-}CH_2\,(\overset{+}{Br}) + Cl^- \longrightarrow ClCH_2CH_2Br \quad [3]$$

SUMMARY

A summary

The addition of halogens to alkenes can be formulated as

$$CH_2{=}CH_2 \;\; X{-}X \longrightarrow CH_2{-}CH_2\,(\overset{+}{X}) + X^- \longrightarrow X{-}CH_2CH_2{-}X$$

A curly arrow represents the movement of a pair of electrons from the tail of the arrow to its tip. The addition of an electrophile to an alkene results in the formation of a cation which can accept electrons from a nucleophile such as Br^-, Cl^-, NO_3^-, HSO_3^- or H_2O.

27.8.6 MECHANISM OF ADDITION OF HYDROGEN HALIDE TO ALKENES

Addition of HX to an alkene involves interaction of the electrophile with the double bond

A mechanism similar to that for the addition of bromine is proposed. A molecule of hydrogen halide, H—X is permanently polarised as $\overset{\delta+}{H}-\overset{\delta-}{X}$. The π electrons of the alkene bond to the electrophilic H atom:

$$CH_2{=}CH_2 \; \overset{\delta+}{H}-\overset{\delta-}{X} \longrightarrow \overset{+}{C}H_2-CH_3 + X^-$$

This is the slow step in the reaction. Once formed, the carbocation reacts rapidly with halide ions:

$$\overset{+}{C}H_2CH_3 + X^- \longrightarrow XCH_2CH_3$$

Evidence

The rate of addition increases with increasing acid strength of HX

(a) The rate of addition increases in the order

$$HF < HCl < HBr < HI$$

This is the order of increasing acid strengths, i.e., the order of readiness to release a proton. The reaction of the carbocation with X^- is rapid, and its speed is much the same for all X^- ions.

Markovnikov's Rule is explained by the carbocation theory

(b) Addition follows Markovnikov's Rule:

$$CH_3CH{=}CH_2 + HCl \longrightarrow CH_3CHClCH_3\,(90\%) + CH_3CH_2CH_2Cl\,(10\%)$$

Carbocations are produced during the formation of the two products:

$$CH_3-CH{=}CH_2 + HCl \longrightarrow CH_3-\overset{+}{C}H-CH_3 + Cl^- \quad [1]$$

$$CH_3-CH{=}CH_2 + HCl \longrightarrow CH_3-CH_2-\overset{+}{C}H_2 + Cl^- \quad [2]$$

Since alkyl groups are electron-releasing, they tend to decrease the charge on a cation and thus stabilise it. The carbocation $CH_3 \longrightarrow \overset{+}{C}H \longleftarrow CH_3$ [1] is therefore more stable than $CH_3 \longrightarrow CH_2 \longrightarrow \overset{+}{C}H_2$ [2] and reaction 1 predominates.

SUMMARY

Reactions of alkenes:

- **combustion**
- **addition of hydrogen, alkanes, halogens, hydrogen halides.**

The addition of bromine to an alkene involves an attack on the π-electrons of the double bond by the electrophile Br_2. A bromonium ion is formed and quickly reacts to form the product.

The addition of hydrogen halides to alkenes is similar: $H^{\delta+}$—$X^{\delta-}$ attacks the π-electrons of the double bond.

27.8.7 CONCENTRATED SULPHURIC ACID

Addition of conc. H_2SO_4 yields the hydrogensulphate ...

The addition of sulphuric acid across a double bond is similar to that of a hydrogen halide. Markovnikov's Rule is followed. When ethene is bubbled into concentrated sulphuric acid at room temperature, ethyl hydrogensulphate is formed:

$$CH_2{=}CH_2(g) + H_2SO_4(l) \xrightarrow{\text{cold}} \underset{\substack{| \\ H}}{H_2C}{-}\underset{\substack{| \\ OSO_2OH}}{CH_2(l)}$$

Ethyl hydrogensulphate

... which can be hydrolysed to the alcohol

The product, ethyl hydrogensulphate, when added to water and warmed, is hydrolysed to ethanol:

$$CH_2{=}CH_2 + H_2SO_4 \longrightarrow C_2H_5SO_4H \xrightarrow[\text{warm}]{H_2O} C_2H_5OH + H_2SO_4$$

The net result is the addition of H·OH across the double bond. The industrial method of accomplishing this is the catalytic hydration of ethene. Ethene and steam are passed at 300 °C and 60 atm over phosphoric acid absorbed on silica pellets:

$$\underset{\text{Ethene}}{CH_2{=}CH_2(g)} + H_2O(g) \xrightarrow[H_3PO_4\ \text{catalyst}]{300\ °C,\ 60\ \text{atm}} \underset{\text{Ethanol}}{C_2H_5OH(g)}$$

The reaction makes possible the manufacture of the important solvent ethanol from ethene, which is a product of the petroleum industry.

27.8.8 ALKALINE POTASSIUM MANGANATE(VII)

Oxidation of C=C by alkaline $KMnO_4$ ...

Potassium manganate(VII) in alkaline solution is a weak oxidising agent. When ethene is bubbled into alkaline potassium manganate(VII) solution, the purple colour fades as ethene is oxidised to ethane-1,2-diol:

... forming a diol ...

$$\underset{\text{Ethene}}{H_2C{=}CH_2} \xrightarrow{KMnO_4,\ OH^-} \underset{\text{Ethane-1,2-diol}}{HOCH_2{-}CH_2OH}$$

... with decolourisation is a test for C=C

A diol is a compound containing two hydroxyl groups [§ 30.2.3]. A brown suspension of manganese(IV) oxide, MnO_2, appears. The disappearance of the purple colour of ice-cold, dilute alkaline manganate(VII) solution is a test for a carbon–carbon multiple bond as few organic compounds are oxidised by this weak oxidising agent.

Ethane-1,2-diol (formerly called ethylene glycol) is used as 'antifreeze' in vehicle radiators to lower the freezing temperature of water.

27.8.9 OZONE

Ozone adds to alkenes to form ozonides

Ozone, or trioxygen, O_3, adds across the double bond to form an **ozonide**. Since the ozonide formed is an explosive compound, it is prepared below 20 °C in a non-aqueous solvent:

$$RHC{=}CR'R'' + O_3 \xrightarrow[\text{organic solvent}]{<20\ °C} \text{ozonide (RHC–O–CR'R'' ring with O–O bridge)}$$

On hydrolysis, it splits up into two carbonyl compounds [see Chapter 31], and hydrogen peroxide, which may oxidise the other products:

$$\text{RHC(O}_3\text{)CR}'\text{R}'' \text{ (ozonide)} + H_2O \longrightarrow \text{RHC}{=}O + O{=}C\text{R}'\text{R}'' + H_2O_2$$

When zinc and ethanoic acid are used as a combined hydrolysing and reducing agent, the possibility of oxidation is avoided. The carbonyl compounds formed in the hydrolysis of the ozonide can be identified. The usefulness of **reductive ozonolysis**, i.e., the formation of an ozonide followed by its hydrolysis under mildly reducing conditions, is that it can be used to locate the position of the double bond in an alkene. If the ozonolysis of hexene gives the carbonyl compounds

Ozonolysis can be used to locate the position of the C=C bond ...

$$CH_3{-}C(H){=}O \quad \text{and} \quad CH_3{-}CH_2{-}C(CH_3){=}O$$

Hot, concentrated $KMnO_4$ ruptures the C=C bond in the same way

then putting these compounds together gives the formula of the original hexene:

$$CH_3{-}\overset{H}{\overset{|}{C}}{=}O \quad \text{and} \quad O{=}\overset{CH_3}{\overset{|}{C}}{-}CH_2CH_3 \quad \text{come from} \quad CH_3{-}\overset{H}{\overset{|}{C}}{=}\overset{CH_3}{\overset{|}{C}}{-}CH_2CH_3$$

Hot concentrated potassium manganate(VII) reacts in a similar manner with alkenes.

$$\text{R(H)C}{=}C\text{R}'\text{R}'' + KMnO_4(aq) \longrightarrow H{-}\text{C(R)(OH)}{-}\text{C(R}')(OH){-}\text{R}''$$

On further oxidation, it splits into two carbonyl compounds and water:

$$H{-}\text{C(R)(HO)}{-}\text{C(R}')(OH){-}\text{R}'' + [O] \longrightarrow \text{R(H)C}{=}O + O{=}C\text{R}'\text{R}'' + H_2O$$

Example How many double bonds are present in a molecule of C_6H_{10}, given that 0.082 g of the compound absorbs 48 cm^3 of ozone at room temperature?

Method Gas molar volume = 24 dm^3 at rtp

Moles of ozone = 48/24000 = 2.0×10^{-3} mol

Moles of hydrocarbon = 0.082/82 = 1.0×10^{-3} mol

2 moles of ozone react with 1 mole of hydrocarbon: C_6H_{10} must contain 2 double bonds per molecule.

27.8.10 POLYMERISATION

Monomers and polymers

Another addition reaction of alkenes is **polymerisation**. Many molecules of **monomer** add together to form one large molecule of **polymer**. The polymer may have a molecular mass of several thousand. In the case of ethene

$$nCH_2{=}CH_2(g) \longrightarrow {+}CH_2{-}CH_2{+}_n(s)$$

The polymer formed is named poly(ethene), often called *polythene* for short.

Ethene polymerises to form polyethene ...

... which is an important plastics material

The first polymerisation of ethene was accomplished in 1933 by the use of very high pressure (1000 atm) and oxygen as the catalyst. A free radical mechanism operates. Research has developed the use of powerful catalysts, which enable the addition to take place at atmospheric pressure. The poly(ethene) formed under high pressure is a low density, extremely pliable material, while the polymer formed at low pressure is of a higher density and is tougher. Poly(ethene)is a **plastic**: it can be moulded easily into a multitude of shapes. Some polyalkenes are listed in Table 27.8. Plastics are discussed in § 34.1.

The disposal of plastics waste, including biodegradable plastics, is discussed in § 34.7

Table 27.8 Poly(alkenes)

Name	Monomer	Polymer	Uses
Poly(ethene) (polythene)	$CH_2{=}CH_2$	$\text{+}CH_2{-}CH_2\text{+}_n$	Polyethene has 40% of the polyalkene market. Low density poly(ethene) is made at high pressure (1000–2000 atm) at 100–300 °C. It has a low melting temperature. It is used in packaging and for plastic bags and toy making. High density poly(ethene) is made at low pressure (5–25 atm) at 20–50 °C, with a catalyst. It has a higher T_m and is used for kitchenware, food boxes, bowls, buckets, etc.
Poly(propene)	$CH_3CH{=}CH_2$	$\left(CH_2{-}CH(CH_3) \right)_n$	Poly(propene) is tougher than poly(ethene). Used to make ropes and for packaging. Its high T_m enables it to resist boiling water.
Poly(chloroethene) (polyvinylchloride, PVC)	$CH_2{=}CHCl$	$\text{+}CH_2{-}CHCl\text{+}_n$	PVC is more rigid than polyethene. Used as a building material, e.g., guttering and electrical insulation. With added plasticisers, PVC is used for macintoshes, wellingtons, etc.
Poly(tetrafluoro-ethene) (PTFE, Teflon®)	$CF_2{=}CF_2$	$\text{+}CF_2{-}CF_2\text{+}_n$	Used for coating surfaces to reduce friction, e.g., non-stick frying pans
Poly(phenylethene) (polystyrene)	$CH_2{=}CH(C_6H_5)$	$\text{+}CH_2{-}CH(C_6H_5)\text{+}_n$	Made into polystyrene foam by dissolving the polymer in a solvent and then vaporising the solvent. Used as insulation and for packaging. Polystyrene is flammable.

continued

Name	Monomer	Polymer	Uses
Poly(2-chloro-butadiene) (Neoprene)	$CH_2{=}CH{-}ClC{=}CH_2$	$\left(\!-CH_2{-}CH(ClC{=}CH_2)-\!\right)_n$	Used in the manufacture of synthetic rubber.
Poly(ethenyl-ethanoate) (polyvinylacetate, PVA)	$CH_3CO_2{-}CH{=}CH_2$	$\left(\!-CH(CH_3CO_2){-}CH_2-\!\right)_n$	Records. More flexible than PVC. Used in emulsion paints.
Poly(methyl 2-methyl-propenoate) (Perspex®, Plexiglass®)	$CH_2{=}C(CH_3)CO_2CH_3$	$\left(\!-C(CH_3)(CO_2CH_3){-}CH_2-\!\right)_n$	Transparent. Used as a substitute for glass.
Poly(propenenitrile) (polyacrylonitrile) (Acrilan®, Orlon®, Courtelle®)	$CH_2{=}CHCN$	$\left(\!-CH_2{-}CH(CN)-\!\right)_n$	Used as *acrylic* fibres for making clothing, blankets and carpets.

SUMMARY

The addition of water to ethene gives ethanol. Catalytic hydration employs steam and concentrated sulphuric acid (in the laboratory) or phosphoric(V) acid (in industry). Addition of potassium manganate(VII) to an alkene

- **cold, dilute ⟶ a diol**
- **hot, concentrated ⟶ a diol ⟶ carbonyl compounds.**

Addition of ozone ⟶ ozonide ⟶ hydrolysed to carbonyl compounds. Alkenes form polymers which are important plastics; see Table 27.8 and Chapter 34.

27.8.11 MECHANISM OF ADDITION POLYMERISATION OF ALKENES

The mechanism of addition polymerisation involves four steps:

1. *Generation of free radicals*: Chain reactions require free radicals and are initiated by ultraviolet light and by the decomposition of compounds such as benzoyl peroxide.

$$\underset{\text{Benzoyl peroxide}}{C_6H_5C{-}CO{-}O{-}O{-}CO{-}C_6H_5} \longrightarrow \underset{\text{Benzoyl radical}}{2C_6H_5COO\cdot} \longrightarrow \underset{\text{Phenyl radical}}{2C_6H_5\cdot} + 2CO_2$$

The decomposition can be initiated by heat or light. Thermal decomposition cannot be controlled very accurately because the system does not cool instantly when heating is stopped. The rate of photochemical decomposition can be controlled by altering the intensity of the initiating light.

2. *Initiation*: Free radicals Ra· add to C═C bonds. A new free radical is generated in each reaction.

$$Ra\cdot + CH_2{=}CH_2 \longrightarrow RaCH_2CH_2\cdot$$

3. *Chain propagation*: The free radical adds to another C═C bond. Chain propagation continues to form $Ra(CH_2)_n\cdot$

$$Ra\cdot + CH_2{=}CH_2 \longrightarrow RaCH_2CH_2\cdot \xrightarrow{CH_2=CH_2} Ra(CH_2)_4\cdot \xrightarrow{CH_2=CH_2} Ra(CH_2)_n\cdot$$

4. *Termination*: Chain propagation may continue until there is no monomer remaining. It is more usual, however, for the chain to be terminated before this by either (a) the combination of two radicals or (b) disproportionation. Disproportionation of the polymer chain leads to branching.

(a) $Ra(CH_2)_n\cdot + \cdot Ra \longrightarrow Ra(CH_2)_nRa$
(b) $Ra(CH_2)_n\cdot + \cdot(CH_2)_nRa \longrightarrow Ra(CH_2)_nH + CH_2{=}CH(CH_2)_mRa$

Alkenes can absorb oxygen from the air to form peroxides which lead to polymerisation. A small quantity of an inhibitor, e.g. quinone, is added to absorb oxygen and stabilise the monomer during storage.

SUMMARY

The mechanism of addition polymerisation involves four steps:

1. **Generation of free radicals.**
2. **Initiation: the addition of a free radical to a molecule of alkene to form another free radical.**
3. **Chain propagation, the formation of larger and larger molecules of alkene.**
4. **Chain termination by the addition of two free radicals or by disproportionation.**

CHECKPOINT 27.8: ALKENES

1. Name the following compounds:

(*a*) $CH_3CH{=}CHCH_2CH_2CH_3$
(*b*) $CH_3C(CH_3){=}CHCH_2CH(CH_3)CH_3$ (printed with CH_3 branches below the second and fifth carbons)
(*c*) $CH_3CH_2(H)C{=}C(H)CH_2CH_2C$ (drawn with CH_3CH_2 and H on one carbon, H and CH_2CH_2C on the other; CH_3CH_2 and H above, H and CH_2CH_2C below)
(*d*) $CH_3CHClCH{=}CH_2$
(*e*) $H_3C(H)C{=}C(H)CH_2CH_2CH_3$ (drawn with H_3C and $CH_2CH_2CH_3$ above, H and H below)

2. Write structural formulae for the following:

(*a*) Pent-1-ene
(*b*) 3-Chlorohexa-2,4-diene
(*c*) Buta-1,3-diene
(*d*) 4,4-Dimethylpent-2-ene

3. A compound C_6H_{12}, after ozonolysis, gives two products, one of which is propanone, $(CH_3)_2CO$. Which of the following is its formula?

(*a*) $CH_3CH{=}CHCH(CH_3)_2$
(*b*) $CH_3CH_2CH{=}C(CH_3)_2$
(*c*) $CH_3C(C_2H_5){=}CHCH_3$ (printed with C_2H_5 below the second carbon)
(*d*) $(CH_3)_2C{=}C(CH_3)_2$
(*e*) $CH_3CH{=}C(CH_3)C_2H_5$

4. Give the names and formulae of the products formed when the following reagents add to propene:

(*a*) chlorine in tetrachloromethane
(*b*) chlorine water.

Write the mechanism for each reaction.

5. Complete the following equations. Indicate the conditions needed for reaction.

(*a*) $CH_2{=}CH_2 + \quad\longrightarrow CH_3CH_2OH$
(*b*) $CH_3CH{=}CH_2 + HBr \longrightarrow$
(*c*) $(CH_3)_2C{=}CH_2 + Br_2 + H_2O \longrightarrow$
(*d*) $CH_3CH{=}CH_2 + H_2O(H_3PO_4\ \text{catalyst}) \longrightarrow$
(*e*) $CH_3CH_2CH{=}CH_2 + \quad\longrightarrow CH_3CH_2CHOHCH_2OH$
(*f*) $(CH_3)_2CHCH{=}CHCH_3 + \quad\longrightarrow (CH_3)_2CHCHO + CH_3CHO$

6. When propene is bubbled through chlorine water containing nitrate ions, three products are formed. Give the names and formulae of the three products, and explain how they come to be formed. What product do you think would be formed in the reaction between propene and nitrogen chloride oxide, NOCl, which reacts as NO^+Cl^-?

7. Why do alkenes show geometrical isomerism, whereas alkanes do not? Draw and name the isomers of $CH_3CH{=}CHCl$, $CH_3CH{=}CHC_2H_5$, $ClCH{=}CHBr$.

Figure 27.8A
Reactions of alkenes

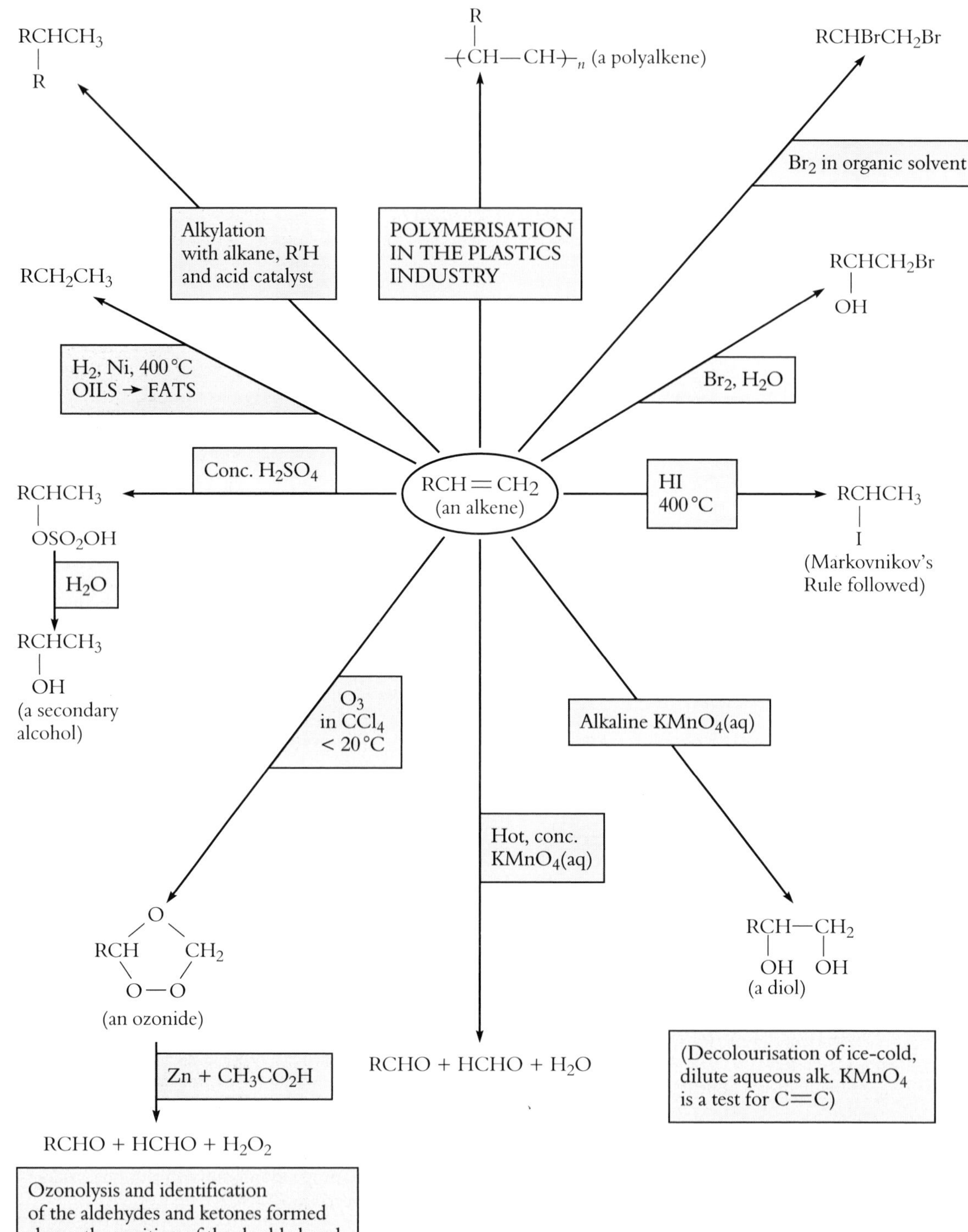

27.9 ALKYNES

Alkynes are aliphatic hydrocarbons with a C≡C triple bond.
The general formula of this homologous series is C_nH_{2n-2}. They are named by changing the name of the corresponding alk*ane* to end in *-yne*. Some members are:

The names of alkynes

HC≡CH	Ethyne
CH_3—C≡C—CH_3	But-2-yne

The model of ethyne in Figure 27.9A shows that the four atoms lie in a straight line. The triple bond —C≡C— makes alkynes very reactive.

Figure 27.9A
Models of ethyne

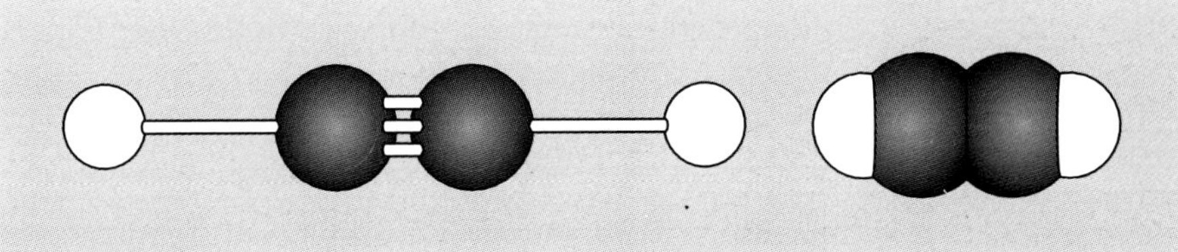

27.10 ETHYNE

The combustion of ethyne ...

Ethyne (formerly called acetylene) is a valuable compound. It burns in oxygen in an extremely exothermic reaction.

$$CH{\equiv}CH(g) + 2\tfrac{1}{2}\,O_2(g) \longrightarrow 2CO_2(g) + H_2O(l);\ \Delta H^{\ominus} = -1300\ \text{kJ mol}^{-1}$$

... is used in welding and cutting

Oxy-acetylene torches are used for cutting and welding metals. Ethyne is also valued because its unsaturated nature makes it a starting-point for the manufacture of a variety of organic compounds.

27.11 PHOTOCHEMICAL SMOG

Hydrocarbons enter the atmosphere from natural sources, such as the anaerobic decomposition of dead plant material, and from man-made sources. The atmospheric hydrocarbons that come from human activity account for only 15% of the total, but their effect on humans is great because they are released mostly in urban air. The outstanding emitter of hydrocarbons is the petrol engine [see § 26.5]. Incineration of rubbish is another source. By themselves hydrocarbons cause little damage, but in the presence of light, *photochemical oxidants* (oxidising agents produced by photochemical reactions) are formed. These oxidising agents react with many of the unsaturated hydrocarbons in the air. Ultimately, all atmospheric hydrocarbons are oxidised to carbon dioxide and water, but some of the intermediate oxidation products are irritating and toxic.

Hydrocarbons enter the atmosphere from natural sources and from petrol engines. Unsaturated hydrocarbons react with photochemical oxidants

A key step in the formation of photochemical oxidants is the photochemical decomposition of nitrogen dioxide:

$$NO_2 \xrightarrow{h\nu} NO + O$$

Oxygen atoms are extremely reactive. Many react with dioxygen to form ozone:

$$O + O_2 \longrightarrow O_3$$

Photochemical oxidants, e.g. atomic oxygen, ozone and peroxides, are formed in sunlight. Nitrogen oxides play an important part in their formation. They oxidise unsaturated hydrocarbons to irritating and toxic compounds

The photochemical oxidants, atomic oxygen, ozone and peroxides, attack alkenes in a large number of complex reactions. Research workers have outlined a scheme of 81 different reactions that may occur in the photochemical oxidation of propene! Among the varied products of the photochemical oxidation of alkenes are the irritating, lachrymatory (tear-producing) compounds, methanal, HCHO, propenal, 'acrolein', $H_2C{=}CHCHO$ and the notorious 'peroxyacetyl nitrate' (PAN),

$$H_3C{-}\underset{\underset{\displaystyle O}{\|}}{C}{-}O{-}O{-}NO_2$$

PAN is both toxic and irritating. At concentrations as low as a few parts per billion it causes eye irritation. Plants are very sensitive: a fraction of 1 ppm causes extensive damage to vegetation. Los Angeles is a city where conditions favour photochemical reactions, and levels of PAN are high enough to damage crops in the neighbouring countryside.

Figure 27.11A The formation of photochemical smog

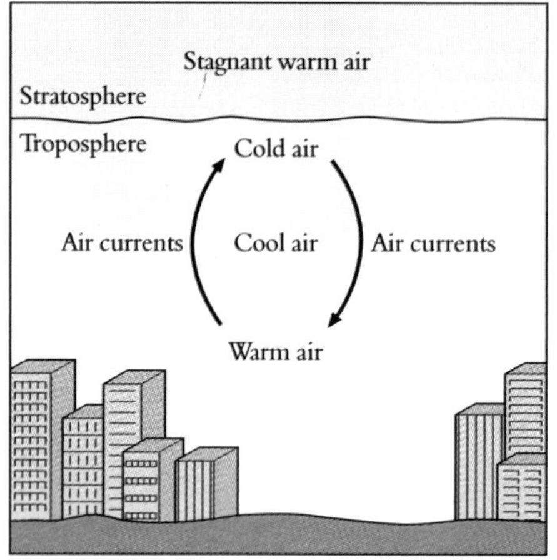

Air near the ground is warmed by the Earth. As warm air rises and cold air descends, vertical air currents carry pollutants into the upper layers of the atmosphere.

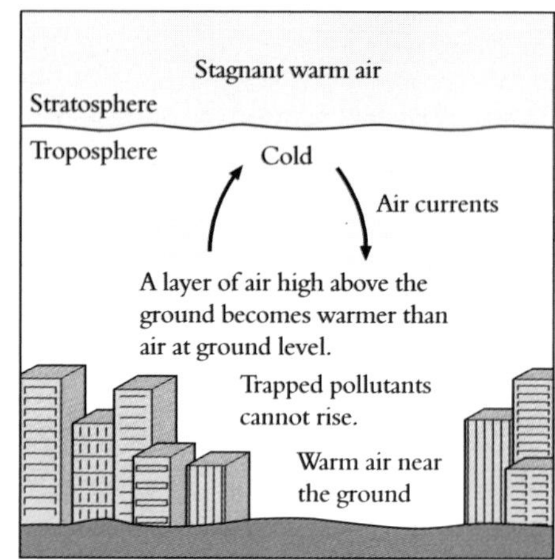

Sometimes a temperature inversion occurs. Brilliant sunshine makes a layer of air above the ground warmer than air at ground level. The stagnant air traps pollutants near the ground. These conditions favour the formation of photochemical smog.

A photochemical smog of oxidants and oxidation products is formed. It includes irritating and toxic compounds

As chemical reactions continue in the atmosphere, more ozone is produced and ozone levels soar [see § 21.5]. Small molecules combine to form larger molecules and eventually tiny particles. These particles and the brown colour of nitrogen dioxide give the air a dirty, 'smoggy' appearance. Eventually, the intermediate oxidation products are converted into carbon dioxide and water. Until that happens, they are irritating and toxic to the population. This type of pollution is called **photochemical smog**.

The control of photochemical smog can be achieved by reducing emission of hydrocarbons and oxides of nitrogen. This is discussed in § 26.5.

QUESTIONS ON CHAPTER 27

1. When ethene is bubbled through a solution containing bromine and potassium chloride, the products are CH_2BrCH_2Br and CH_2BrCH_2Cl, but no CH_2ClCH_2Cl is formed. Explain this behaviour.

2. What is the major product of the reaction between hydrogen bromide and but-1-ene in solution in an organic solvent at room temperature? What other product is formed in small amount? Explain, by referring to the mechanism of the reaction, why the first product is the major one.

3. What is formed when propene reacts with *(a)* bromine water and *(b)* concentrated sulphuric acid? What is the product of hydrolysis of the compound formed in reaction *(b)*? Discuss the mechanisms of these two reactions, pointing out the similarity between them.

4. Measurements on an alkene showed that 100 cm^3 of the gas weighed 0.250 g at 0 °C and 1 atm. 25.0 cm^3 of the alkene reacted with 25.0 cm^3 of hydrogen.

Find the molar mass of the alkene, and give its molecular formula.

Give the names and structural formulae of alkenes with this formula.

5. *(a)* What kind of weather favours the formation of photochemical smog?
(b) What kind of geographical locations are prone to photochemical smog?
(c) Explain why the conditions you mention in *(a)* and *(b)* favour the formation of photochemical smog.

6. *(a)* Describe the nature of the bonding in alkenes. Explain how this determines their shape and chemical reactivity.
(b) What is the most common type of reaction of alkenes? State the type of reagents with which they frequently react, giving two examples.

7. A $CH_3-CH_2-CH(CH_3)-CH(CH_3)-CH_3$

B $CH_3-CH(CH_3)-CH(CH_3)-CH_3$

C $CH_3-C(CH_3)_2-CH_3$

D $CH_3-CH<$ (cyclobutyl ring: $CH-CH_2-CH_2-CH_2$)

E $H_3C-CH<$ ring: $H_2C-C(H)(CH_3)-CH_2-H_2C$ (methylcyclobutane: ring $H_2C-CH-CH_2-CH_2$ with CH_3 on the CH)

(a) Give the letter of the structure
(i) which is isomeric with pentane
(ii) which has a systematic name ending in -propane
(iii) which forms two and only two different monochloro-compounds when hydrogen atoms are replaced by chlorine
(iv) which is under the greatest strain
(v) which could not be made by the addition of hydrogen to an alkene.

(b) Which two pairs of structures are structural isomers?

8. *(a)* What is the industrial source of propene?
(b) Explain why propene is a more useful starting point for the manufacture of organic compounds than propane.
(c) State the conditions under which propene is converted into poly(propene). Draw the structural formula of poly(propene).
(d) Poly(propene) can be made into fibres. The fibres can be made into ropes, but are unsuitable for the manufacture of clothing. Explain what properties of the fibres make them unsuitable for clothing.
(e) For what uses are the following polymers suited?
(i) poly(2-chlorobutadiene)
(ii) poly(phenylethene)
(iii) poly(tetrafluoroethene)
(iv) poly(propenenitrile)
(v) poly(chloroethene)?

9. *(a)* (i) Describe, giving outline conditions and equations, the laboratory preparation of but-1-ene from **two** different classes of organic compound. **4**
(ii) State the type of reaction taking place in **each** case. **1**

(b) (i) Describe the reaction of ethene with,
I. water, and
II. hydrogen.

State the reaction conditions in **each** case. **3**
(ii) Comment on the industrial importance of the ethene-water reaction. **1**

(c) The **mechanism** of the ethene-water reaction is believed to include the following steps;

$$CH_2{=}CH_2 + H^+ \xrightarrow{\text{SLOW}} CH_3CH_2^+ \xrightarrow[\text{FAST}]{H_2O} C_2H_5OH + H^+$$

(i) State the name given to this type of mechanism. **1**

(ii) State the name given to species such as the intermediate $CH_3CH_2^+$. ½

(iii) State another reaction of ethene which follows this type of mechanism. **1**

(iv) **State** and **explain** the effect which increasing the pressure of steam would be expected to have on,

I. the rate of the reaction in the industrial process, and

II. the equilibrium yield of ethanol. **2**

(v) State, giving a **reason**, what would be the main product of the reaction of water with propene. **1½**

(d) Use the bond energy data below to calculate the enthalpy change in the reaction of ethene with water above. **Explain** the basis of your calculation.

bond	C=C	O—H	C—C	C—O	C—H
bond energy/ kJ mol^{-1}	612	463	348	360	412

5

20 *WJEC (AL)*

10. *(a)* The reaction between ethene and concentrated sulphuric acid is an example of the first stage of a general method used to form alcohols from alkenes. Write an equation for the reaction and name the type of mechanism involved. **2**

(b) Suggest a reagent which could be used to convert the product obtained in part *(a)* into ethanol. Name the type of reaction taking place. **2**

(c) Write an equation for the formation of epoxyethane from ethene and explain briefly why the product is highly reactive. **4**

(d) Predict the major organic product of the reaction between but-1-ene and hydrogen bromide. Explain the basis of your prediction. **4**

12 *NEAB (AL)*

11. *(a)* Alkane **C** has a relative molecular mass of 170 and occurs in the kerosene fraction obtained by the fractional distillation of petroleum.

(i) Write the general formula for the homologous series of alkanes.

(ii) Deduce the molecular formula of alkane **C**.

(iii) Give one use for the kerosene fraction.

(iv) Name one fraction which is obtained higher up the fractioning column than kerosene and explain why it is obtained higher up. **6**

(b) Three hydrocarbons, **D**, **E** and **F**, all have the molecular formula C_6H_{12}.
D decolourises an aqueous solution of bromine and shows geometric isomerism.
E also decolourises an aqueous solution of bromine but does not show geometric isomerism.
F does not decolourise an aqueous solution of bromine.
Draw one possible structure each for **D**, **E** and **F**. **3**

9 *NEAB (AS/AL)*

28 AROMATIC COMPOUNDS

28.1 PROFILE: KATHLEEN LONSDALE (1903–1971)

Benzene is the characteristic part of all aromatic compounds. Identifying its structure was therefore a matter of some importance. The German chemist Kekulé had suggested a ring structure for benzene, a hexagonal ring with alternate single and double bonds. Benzene, being a liquid, was not suitable for crystallographic work. Professor Ingold of London University prepared some crystals of hexamethylbenzene and sent them to the crystallographers in the University of Leeds for analysis. The scientist who was entrusted with the job was a newly married young woman called Kathleen Lonsdale. To everyone's surprise, her mathematical analysis of her X ray diffraction patterns showed that the benzene molecule was planar.

H_3C CH_3 H_3C CH_3 H_3C CH_3

Hexamethylbenzene

Kathleen became a professor in University College, London. In recognition of her crystallographic work on the structure of diamond, one form of diamond is named lonsdaleite after her. She was one of the first two women to be elected Fellow of the Royal Society.

Kathleen's early life was hard. Kathleen Yardley was the youngest of ten children in a poor family. Kathleen was a great success at school and went on to read physics in the University of London. While working for her PhD, Kathleen married an engineering student called Thomas Lonsdale. Thomas never failed to support his wife in her work, recognising the problems which she had to cope with as wife, mother and scientist and sharing them with her. The Lonsdales were active members of the Society of Friends. During the Second World War, they were pacifists. Kathleen was sent to prison for a month for refusing to register for Civil Defence. After the war, she was made a Dame Commander of the Order of the British Empire. She travelled widely in support of peace and better East–West relations.

28.2 BENZENE

Benzene is an arene

Benzene, C_6H_6, is the simplest member of the class of hydrocarbons called **aromatic hydrocarbons** or **arenes**. It is a colourless liquid with a characteristic smell.

For many years there was speculation over the structure of benzene. After 1834, when the molecular formula was established as C_6H_6, people put forward unsaturated formulae, such as

From the formula, C_6H_6, one might expect it to be an unsaturated compound ...

$$CH_2{=}C{=}CH{-}CH{=}C{=}CH_2$$

In fact, benzene appeared to be strangely unreactive in comparison with the alkenes. In 1865, A Kekulé suggested the cyclic hexatriene structure, written in full in (a) and schematically in (b).

... but in fact it does not show much resemblance to the alkenes

(a) (b)

The Kekulé formula was not accepted by all chemists because it implies that benzene should show the same addition reactions as alkenes.

The Kekulé formula has now been superseded by the delocalised formula

Years later, X ray work showed that all the C—C bonds in benzene have the same length, 0.139 nm. This is intermediate between the C—C single bond length of 0.154 nm and the C=C double bond length of 0.133 nm. The Kekulé structure was superseded, by the delocalised structure, which has been described and shown in Figure 5.4A, § 5.4.1. The delocalised formulae for benzene and other arenes are shown below:

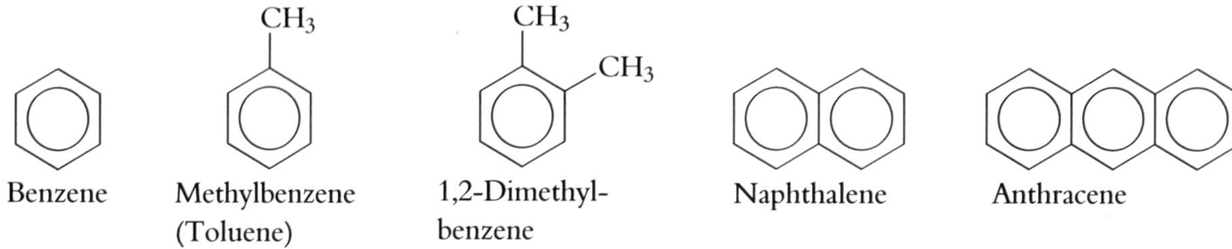

Benzene Methylbenzene (Toluene) 1,2-Dimethylbenzene Naphthalene Anthracene

Other arenes

The group C_6H_5- is called a *phenyl* group. Phenyl and substituted phenyl groups are called *aryl* groups. Some derivatives of benzene are:

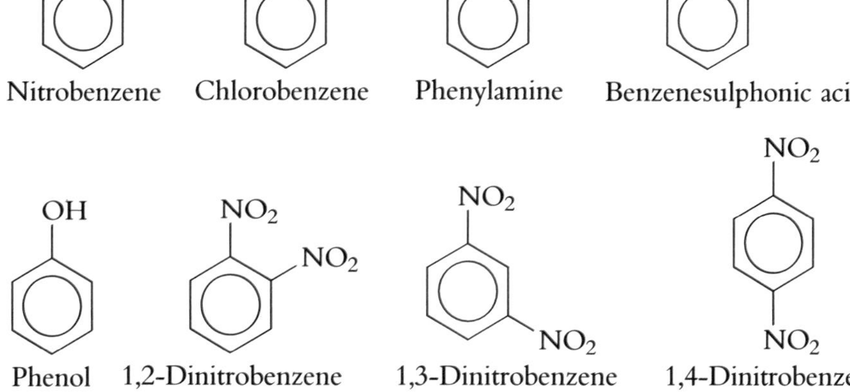

Nitrobenzene Chlorobenzene Phenylamine Benzenesulphonic acid

Phenol 1,2-Dinitrobenzene 1,3-Dinitrobenzene 1,4-Dinitrobenzene

Naming substituted benzenes

Isomers with two substituents in the 1,2-, 1,3-, and 1,4-positions were formerly called *ortho*-, *meta*- and *para*-isomers or *o*-, *m*- and *p*- for short.

28.3 MORE NAMES OF AROMATIC COMPOUNDS

Derivatives of benzene are:

$C_6H_5NH_3^+Cl^-$	Phenylammonium chloride	(Anilinium chloride)
$C_6H_5CO_2H$	Benzenecarboxylic acid	(Benzoic acid)
$C_6H_5CO_2C_2H_5$	Ethyl benzenecarboxylate	(Ethyl benzoate)
C_6H_5COCl	Benzenecarbonyl chloride	(Benzoyl chloride)
$C_6H_5CONH_2$	Benzenecarboxamide	(Benzamide)
C_6H_5CN	Benzenecarbonitrile	(Benzonitrile)
C_6H_5CHO	Benzenecarbaldehyde	(Benzaldehyde)
$C_6H_5COCH_3$	Phenylethanone	
C_6H_5OH	Phenol	
$C_6H_5NH_2$	Phenylamine	(Aniline)
$C_6H_5OCH_3$	Methoxybenzene	

The names given are the IUPAC names. The names in brackets are traditional names which are still in widespread use. The naming of benzene derivatives containing two or more substituent groups is done by giving the group which is nearest to the top of the list the number 1 position in the benzene ring. Other groups are numbered by counting from position 1 in the manner which gives them the lowest **locants** (numbers). Some examples are given below:

CO_2H 1 2 3 4 5 6 OH

3-Hydroxybenzenecarboxylic acid
(3-Hydroxybenzoic acid)

The $-CO_2H$ group is higher up the list than $-OH$, and is regarded as the principal group, occupying position 1. The $-OH$ group is in position 3, not 5, as the lower locant (number) is used.

NH_2 CHO 1 2 3 4 5 6

2-Aminobenzenecarbaldehyde
(2-Aminobenzaldehyde)

The $-CHO$ group, being higher up the list, is regarded as the principal group and given position 1.

The $-NH_2$ group is in position 2, not 6.

28.4 PHYSICAL PROPERTIES OF BENZENE

Benzene burns, and has a toxic vapour

Benzene is a colourless liquid of boiling temperature 80 °C and melting temperature 5.5 °C. It is immiscible with water. It dissolves many substances and is soluble in other organic solvents. When benzene and other arenes burn, particles of carbon are produced in addition to carbon dioxide and water, and these particles make the flame smoky and luminous. Benzene is toxic; inhalation over a period of time leads to anaemia and leukaemia. Other arenes, e.g., methylbenzene, are much less toxic.

28.5 SOURCES OF BENZENE

Benzene is obtained from petroleum oil by fractional distillation followed by reforming, and also by destructive distillation of coal.

28.6 REACTIVITY OF BENZENE

The benzene ring is a planar hexagon with a cloud of delocalised π electrons lying above and below the ring [Figure 28.6A]. The reactions of benzene usually involve the attack of an electrophile on the cloud of π electrons.

Figure 28.6A
A model of the benzene molecule

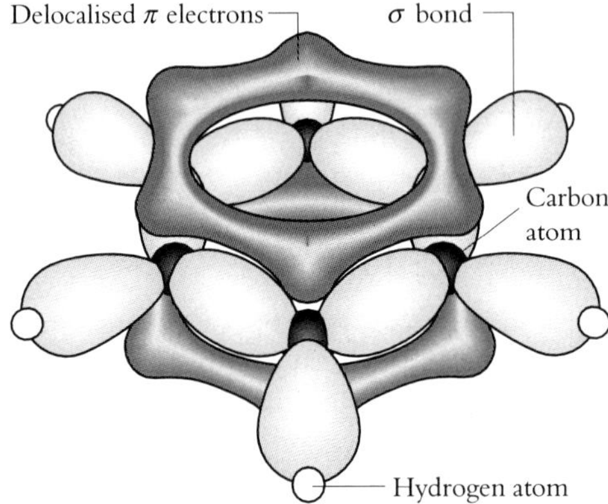

28.7 ADDITION REACTIONS OF BENZENE

The unsaturated character of benzene allows it to undergo addition reactions. Figure 28.8C shows the addition of hydrogen, ozone and halogen.

28.8 SUBSTITUTION REACTIONS

28.8.1 NITRATION

Benzene reacts with a 'nitrating mixture' to give nitrobenzene ...

The substitution of a —H atom by a $—NO_2$ group is called **nitration**. To obtain nitrobenzene, $C_6H_5NO_2$, benzene is refluxed on a water bath at 60 °C with a 'nitrating mixture', as shown in Figure 28.8A. Although concentrated nitric acid will slowly nitrate benzene, a mixture of concentrated nitric and sulphuric acids is much more effective:

$$\underset{\text{Benzene}}{C_6H_6(l)} + HNO_3(l) \xrightarrow[\text{conc. } HNO_3 + \text{ conc. } H_2SO_4]{\text{Reflux, 60°C}} \underset{\text{Nitrobenzene}}{C_6H_5NO_2(l)} + H_2O(l)$$

... which is an intermediate in the preparation of many derivatives of benzene

Nitrobenzene is a pale yellow liquid which can be separated from benzene by distillation under reduced pressure. Its preparation from benzene is an important reaction as it is the first step in the introduction of many different substituent groups into the benzene ring. The $—NO_2$ group can be reduced to $—NH_2$, to

Figure 28.8A
Apparatus for refluxing benzene and nitrating mixture

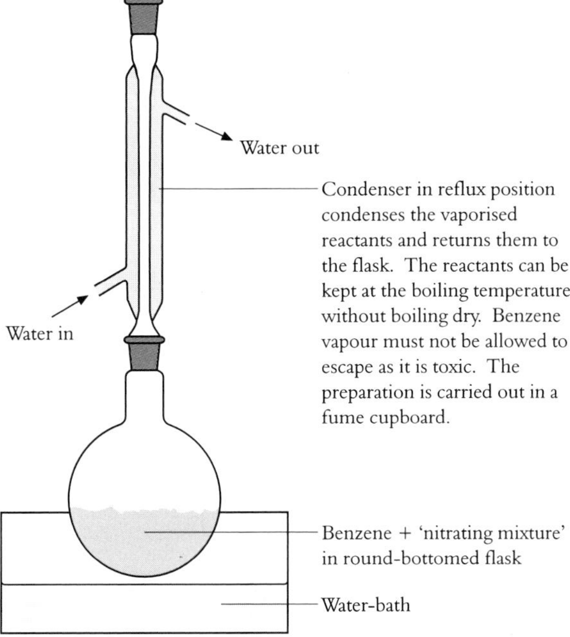

Water out
Condenser in reflux position condenses the vaporised reactants and returns them to the flask. The reactants can be kept at the boiling temperature without boiling dry. Benzene vapour must not be allowed to escape as it is toxic. The preparation is carried out in a fume cupboard.
Water in
Benzene + 'nitrating mixture' in round-bottomed flask
Water-bath

form phenylamine, $C_6H_5NH_2$, which gives rise to all the compounds described in Figure 32.8B, § 32.8.4.

If the temperature is raised to 95 °C, and fuming nitric acid is used, 1,3-dinitrobenzene is formed.

$$C_6H_5NO_2(l) + HNO_3(l)\text{ (fuming)} \xrightarrow[\text{nitrating mixture}]{95\,°C} C_6H_4(NO_2)_2(s) + H_2O(l)$$

1,3-Dinitrobenzene

Dinitrobenzene can also be formed by nitrating benzene

Nitrobenzene is less reactive than benzene, and 1,3-dinitrobenzene is obtained together with nitrobenzene and benzene. Since it is a solid, 1,3-dinitrobenzene crystallises out when the mixture is poured into cold water. No trinitrobenzene is formed.

Nitration of methylbenzene produces TNT

When methylbenzene (toluene), $C_6H_5CH_3$, is nitrated, 2,4,6-trinitromethylbenzene (2,4,6-trinitrotoluene) is formed.

$$CH_3C_6H_2(NO_2)_3$$

This shows that the methyl group makes methylbenzene more reactive than benzene. The product, which is called TNT for short, is a powerful explosive.

28.8.2 MECHANISM OF NITRATION

What is the nitrating agent?

Chemists were intrigued by the puzzle that, although concentrated nitric acid alone is a poor nitrating agent, a mixture of concentrated nitric and sulphuric acids is a powerful reagent. It seemed that the two acids must react to produce a species which can attack benzene. Much research work has gone into the unravelling of this mystery, and only a glimpse of this exciting area of study can be given here.

It is suggested that sulphuric acid, being a stronger acid than nitric acid, can to some extent donate a proton to nitric acid:

$$H_2SO_4 + HNO_3 \rightleftharpoons HSO_4^- + H_2\overset{+}{O}{-}NO_2$$

Nitric acid, in accepting a proton, is acting as a base. The protonated nitric acid splits up to form a molecule of water and the cation, NO_2^+, which is called a *nitryl cation*.

Formation of the nitryl cation ...

$$H_2\overset{+}{O}{-}NO_2 \rightleftharpoons H_2O + NO_2^+$$

The molecule of water is protonated by a second molecule of sulphuric acid:

$$H_2O + H_2SO_4 \rightleftharpoons H_3O^+ + HSO_4^-$$

Adding up the three steps gives

$$HNO_3 + 2H_2SO_4 \rightleftharpoons NO_2^+ + H_3O^+ + 2HSO_4^-$$

NO_2^+ is an electrophile ...

... which adds to benzene to form a reactive intermediate ...

This intermediate rapidly loses a proton to form the product

Once formed, the nitryl cation is attracted by the cloud of delocalised π electrons which lies above and below the plane of the benzene ring [see Figure 28.6A, § 28.6]. The NO_2^+ cation adds to the ring to form a short-lived intermediate, in which both the entering —NO_2 group and the leaving —H atom are bonded to the ring. The symmetry of the annular cloud of delocalised π electrons is disrupted. The positive charge contributed by the NO_2^+ ion is distributed over the ring. The intermediate rapidly loses a proton, restoring the symmetry and stability of the benzene ring. The proton is immediately picked up by a hydrogensulphate ion:

$$C_6H_6 + NO_2^+ \xrightarrow{\text{slow}} [C_6H_6NO_2]^+ \xrightarrow{\text{fast}} C_6H_5NO_2 + H^+$$

Reactive intermediate

$$H^+ + HSO_4^- \xrightarrow{\text{fast}} H_2SO_4$$

The overall rate of the reaction is determined by the slowest step in the sequence. This is the rate at which the C—NO_2 bond is formed in the intermediate.

28.8.3 SULPHONATION

Benzene can be sulphonated by fuming H_2SO_4

Sulphonation is the substitution of an —H atom by an —SO_3H group. Benzenesulphonic acid, $C_6H_5SO_3H$, is obtained by refluxing benzene with concentrated sulphuric acid for many hours, or warming with fuming sulphuric acid (which contains SO_3) at 40 °C for 20–30 minutes:

$$\text{C}_6\text{H}_6(\text{l}) + \text{H}_2\text{SO}_4(\text{l}) \xrightarrow[\text{warm with fuming acid}]{\text{reflux with conc. acid}} \text{C}_6\text{H}_5\text{SO}_3\text{H}(\text{l}) + \text{H}_2\text{O}(\text{l})$$

Benzenesulphonic acid

Sulphonic acids are important in the preparation of phenols and in the manufacture of detergents

The importance of this reaction is that the $—SO_3H$ group can be replaced by hydrolysis to give an —OH group. Benzenesulphonic acid is converted into phenol, C_6H_5OH. This is the easiest way of preparing phenol in the laboratory. Sulphonation is also important in the manufacture of detergents. The $—SO_3H$ group confers solubility in water. It is because detergents contain $—SO_3H$ groups that they lather even in hard water, as the calcium and magnesium salts of sulphonic acids are soluble. Sulphonic acids are important in the manufacture of dyes and drugs.

28.8.4 ALKYLATION

An alkyl group can be introduced into the benzene ring by the reaction of a halogenoalkane with benzene:

$$\text{C}_6\text{H}_6(\text{l}) + \text{C}_2\text{H}_5\text{Br}(\text{l}) \xrightarrow[\text{warm}]{\text{FeBr}_3} \text{C}_6\text{H}_5\text{C}_2\text{H}_5(\text{l}) + \text{HBr}(\text{g}) \longrightarrow \text{Further substitution}$$

Benzene Bromoethane Ethylbenzene

Alkylation needs a catalyst + RX

The reaction takes place under the influence of a catalyst, such as aluminium chloride or iron(III) bromide. More than one alkyl group enters the ring as the alkylbenzene formed is more reactive than benzene. The effectiveness of the catalysts was discovered by C Friedel and J M Crafts, and reactions of this type are called **Friedel–Crafts reactions**. The mechanism is discussed in § 28.8.7.

Alkenes are alkylating agents ...

Alkylation can also be effected by the use of an alkene and a Friedel–Crafts catalyst together with an acid such as HCl or H_3PO_4:

$$\text{C}_6\text{H}_6(\text{l}) + \text{CH}_2{=}\text{CH}_2(\text{g}) \xrightarrow[\text{HCl or H}_3\text{PO}_4]{\text{AlCl}_3\text{ catalyst}} \text{C}_6\text{H}_5\text{C}_2\text{H}_5(\text{l})$$

Benzene Ethene Ethylbenzene

... used to make styrene and cumene

This is the method used industrially to make ethylbenzene. The product is dehydrogenated to give phenylethene (styrene), from which polystyrene is made [see Table 27.8, § 27.8.10].

28.8.5 ACYLATION

Acylation, the introduction of an acyl group

$$-\mathrm{C}\begin{matrix}{/\!/}\mathrm{O}\\ \diagdown\mathrm{R}\end{matrix}$$

is another Friedel–Crafts reaction:

Acylation needs RCOCl + catalyst

$$C_6H_6(l) + CH_3COCl(l) \xrightarrow[60°C]{AlCl_3} C_6H_5COCH_3(l) + HCl(g)$$

Ethanoyl chloride — Phenylethanone

An acid chloride or acid anhydride, in the presence of a Friedel–Crafts catalyst, will acylate benzene and its derivatives [see Figure 28.8B]. Only one acyl group enters the ring as the acyl compound formed is less reactive than benzene. This gives acylation an advantage over alkylation, which yields polyalkyl derivatives. A method of preparing pure ethylbenzene is to introduce the group $-COCH_3$ by acylation and then reduce it to $-CH_2CH_3$:

Figure 28.8B
Friedel–Crafts acylation

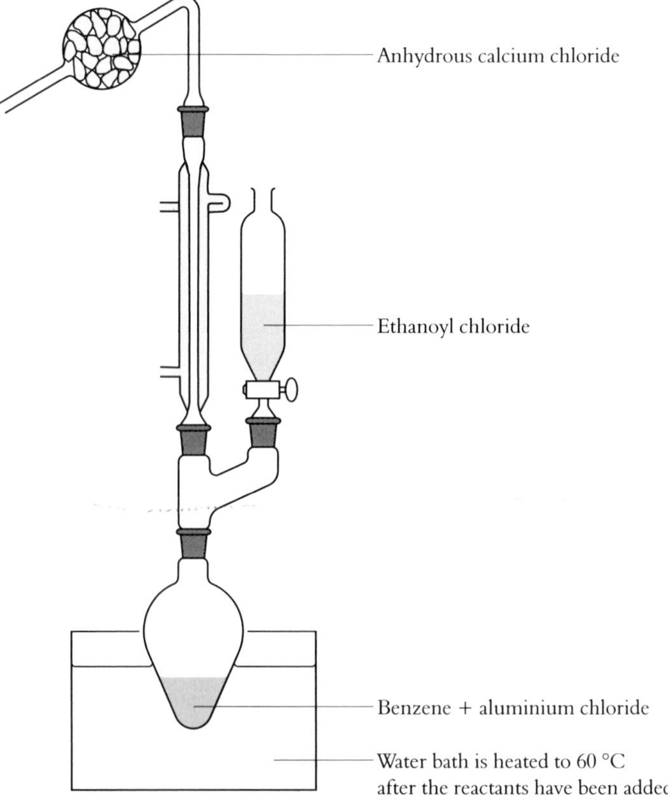

The reduction of an acyl group gives an alkyl group

$$C_6H_6 + CH_3COCl \xrightarrow{AlCl_3} C_6H_5COCH_3 \xrightarrow{\text{reduce}} C_6H_5CH_2CH_3$$

$$C_6H_6 \xrightarrow{C_2H_5Cl,\ AlCl_3} \text{Ethylbenzene}$$

Gives some polyethylbenzenes in addition.

28.8.6 HALOGENATION

The addition of halogens to the benzene ring [Figure 28.8C] takes place under conditions which favour the formation of free radicals, i.e., sunlight or high temperature. The first step in the addition is the homolysis of the halogen molecule, e.g.

$$Cl_2 \xrightarrow{h\nu} 2Cl\cdot$$

Chlorine and bromine can be substituted in the ring in the presence of a Friedel-Crafts catalyst ...

A different type of reaction takes place at room temperature in the presence of a Friedel–Crafts catalyst. Substitution then occurs. In recognition of the way in which they ease the path of halogens into the benzene ring the Friedel–Crafts catalysts are referred to as **halogen-carriers**. Aluminium and iron act as catalysts because during the reaction they are converted into their chlorides or bromides, which act as halogen-carriers:

$$C_6H_6(l) + Cl_2(g) \xrightarrow[\text{room temperature}]{\text{Al or } AlCl_3} C_6H_5Cl(l) + HCl(g)$$

Chlorobenzene

$$C_6H_6(l) + Br_2(l) \xrightarrow[\text{room temperature}]{\text{Fe or } FeBr_3} C_6H_5Br(l) + HBr(g)$$

Bromobenzene

Iodine cannot be introduced in this way.

... which is called a halogen-carrier

Why does the presence of a halogen-carrier result in a completely different reaction – substitution instead of addition? Much work has been done on the mode of action of Friedel–Crafts catalysts, and only an outline can be given here.

28.8.7 MECHANISMS OF ALKYLATION, ACYLATION AND HALOGENATION

The mechanism: Friedel–Crafts catalysts act as Lewis acids ...

Friedel–Crafts catalysts are halides like $AlCl_3$, $AlBr_3$, BF_3 and $FeBr_3$. They function as Lewis acids [§ 12.6.1] since the central atom can accept a pair of electrons. Aluminium chloride can accept a pair of electrons from a chloride ion.

In the complex ion formed, $AlCl_4^-$, aluminium has eight electrons in its outer shell:

$$\text{Cl—}\underset{\underset{\text{Cl}}{|}}{\text{Al}}\text{—Cl} + \text{Cl}^- \longrightarrow \left[\text{Cl—}\overset{\overset{\text{Cl}}{|}}{\underset{\underset{\text{Cl}}{|}}{\text{Al}}}\text{—Cl}\right]^-$$

Let us see how the ability to form complex ions by acting as Lewis acids can explain the catalytic effect of these halides.

Alkylation

... They form complexes with halogenoalkanes

When a Friedel–Crafts catalyst is dissolved in a halogenoalkane, the solution formed is a weak electrical conductor. This could be explained by the formation of an ionic complex. In the case of bromoethane and aluminium bromide

$$C_2H_5Br + AlBr_3 \rightleftharpoons C_2H_5^+AlBr_4^-$$

A carbocation complex attacks benzene to form an intermediate ...

The attack on the π electron cloud of the benzene ring could be by the electrophilic carbocation, $C_2H_5^+$ or by the complex, $C_2H_5^+\,AlBr_4^-$. In the case of primary and secondary carbocations, it is thought that the complex is the attacking electrophile. Tertiary carbocations are more stable, and in the case of $(CH_3)_3C^+$, for example, it may be the carbocation that attacks benzene:

$$C_6H_6 + C_2H_5^+AlBr_4^- \xrightarrow{\text{slow}} [C_6H_6(C_2H_5)]^+\,AlBr_4^-$$

... which loses a proton to form the product

The reactive intermediate that is formed quickly loses a proton to form the product, ethylbenzene. The catalyst, aluminium bromide, is regenerated, and hydrogen bromide is evolved:

$$[C_6H_6(C_2H_5)]^+\,AlBr_4^- \xrightarrow{\text{fast}} C_6H_5C_2H_5 + HBr + AlBr_3$$

Reactive intermediate   Ethylbenzene

When alkenes are used for alkylation a proton acid must be present, in addition to a Lewis acid. The first step in the reaction is protonation of the alkene to form a carbocation which takes part in an electrophilic substitution, S_E, of benzene.

Acylation

The mechanism of acylation: the F–C catalyst + RCOCl form a complex ...

Again, the Friedel–Crafts catalyst acts as a Lewis acid. In the reaction with an acyl chloride, a complex containing an acyl cation, $R\overset{+}{C}O$, is formed:

$$RCOCl + AlCl_3 \rightleftharpoons R\overset{+}{C}O\,AlCl_4^-$$

The electrophile which attacks the benzene ring could be either the acyl cation, $R\overset{+}{C}O$, or the complex, $R\overset{+}{C}O\ AlCl_4^-$. There is evidence that the attacking reagent is the whole complex:

... the electrophilic complex attacks benzene ...

$$C_6H_6 + R\overset{+}{C}O\,AlCl_4^- \xrightarrow{\text{slow}} [C_6H_6(H)(COR)]^+ AlCl_4^- \xrightarrow{\text{fast}} C_6H_5COR + HCl + AlCl_3$$

Reactive intermediate — Phenylketone

... to form a reactive intermediate which is rapidly converted into the product

The rate-determining step is the formation of the RCO—ring bond. The intermediate that is formed rapidly loses a proton to yield the product.

Halogenation with the aid of a halogen-carrier

The role of a halogen-carrier is explained in terms of polarising the X—X bond

In the bromination of benzene a bromine molecule approaches a benzene molecule and meets the annular cloud of delocalised π electrons above and below the plane of the ring [see Figure 28.6A, § 28.6]. As the π electron cloud interacts with the bromine molecule, the Br—Br bond becomes polarised:

$$C_6H_6 + Br—Br \longrightarrow C_6H_6 \cdots \overset{\delta+}{Br}—\overset{\delta-}{Br}$$

This will remind you of the interaction between Br_2 and the π electrons of the C=C bond in alkenes [§ 27.8.5]. Benzene is less reactive than an alkene, and the reaction proceeds only if there is a Lewis acid (e.g., $FeBr_3$) present. This accepts a pair of electrons from the $\delta-$ Br atom in the polarised Br_2 molecule, and enables the Br—Br bond to split. A bromonium ion and a $FeBr_4^-$ complex ion are formed.

The bromonium ion rapidly loses a proton to form bromobenzene with the regeneration of the catalyst, $FeBr_3$:

$$C_6H_6 + Br—Br\ FeBr_3 \xrightarrow{\text{slow}} [C_6H_6(H)(Br)]^+ FeBr_4^- \xrightarrow{\text{fast}} C_6H_5Br + HBr + FeBr_3$$

Iodination of benzene does not occur, but some benzene derivatives, containing activating substituents, are iodinated.

A summary

Alkylation, first step

$$RX + AlX_3 \rightleftharpoons R^+AlX_4^-$$

followed by

Summary of the role of Friedel–Crafts catalysts

$$C_6H_6 + R^+\,AlX_4^- \xrightarrow{\text{slow}} [C_6H_6(H)(R)]^+ AlX_4^- \xrightarrow{\text{fast}} C_6H_5R + HX + AlX_3$$

Acylation, first step

$$RCOX + AlX_3 \rightleftharpoons R\overset{+}{C}O\ AlX_4^-$$

followed by

All these reactions involve substitution by an electrophile ...

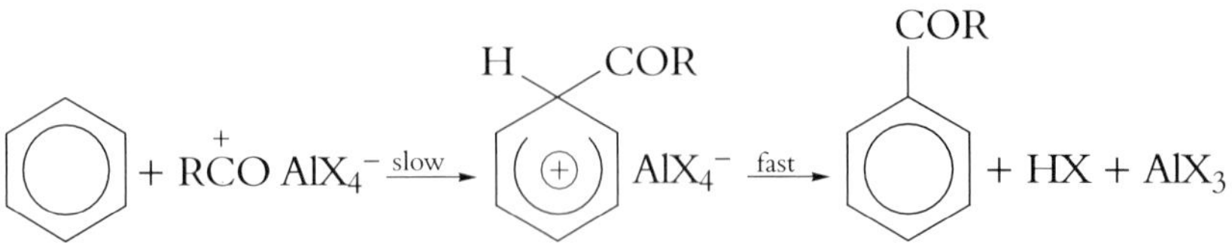

$$C_6H_6 + R\overset{+}{C}O\ AlX_4^- \xrightarrow{slow} [C_6H_6(COR)]^+ AlX_4^- \xrightarrow{fast} C_6H_5COR + HX + AlX_3$$

Halogenation, first step

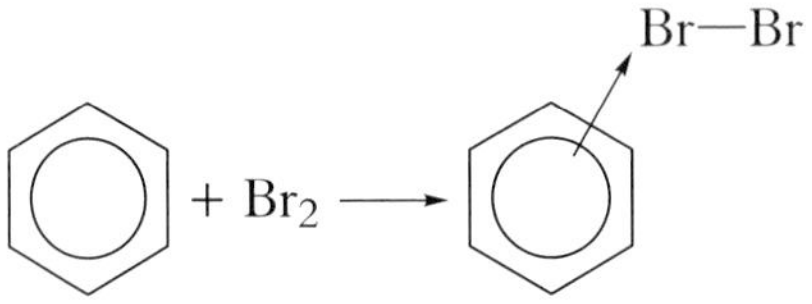

$$C_6H_6 + Br_2 \longrightarrow C_6H_6 \cdots Br{-}Br$$

followed by

... they are S_E reactions

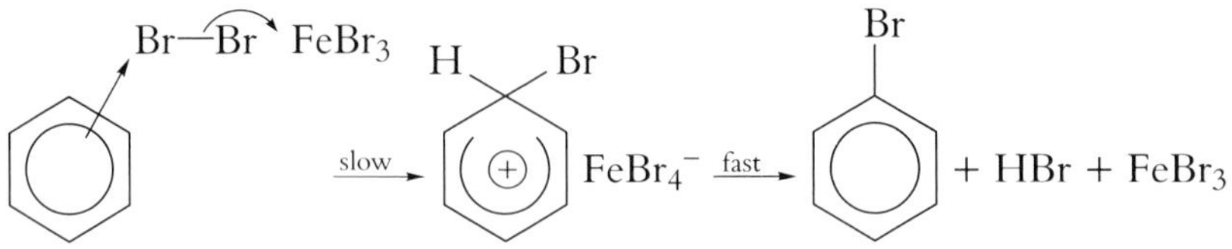

$$C_6H_6 \cdots Br{-}Br\ FeBr_3 \xrightarrow{slow} [C_6H_6Br]^+ FeBr_4^- \xrightarrow{fast} C_6H_5Br + HBr + FeBr_3$$

Table 28.8
A comparison of an arene and a cycloalkene

Reagent	Benzene 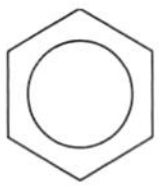	Cyclohexene 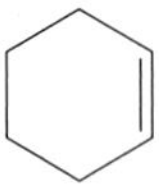
Br_2(organic solvent)	No reaction	Rapid decolourisation
$KMnO_4$, alkaline, cold	No reaction	Rapid decolourisation
HBr(g)	No reaction	Rapid addition
Air, catalyst, 150 °C	No reaction	Oxidised to hexane-1,6-dioic acid
H_2, catalyst, 200 °C	Slowly adds to give cyclohexane	Adds rapidly to give cyclohexane
Halogen, sunlight, boil	Slow addition gives $C_6H_6X_6$	Rapid addition gives $C_6H_{10}X_2$
Conc. HNO_3, H_2SO_4	Substitution gives nitrobenzene	Oxidation gives hexane-1,6-dioic acid
Conc. H_2SO_4	No reaction when cold. Slow sulphonation when heated	Absorbed by H_2SO_4
Friedel–Crafts reagents	Alkylation or acylation	No reaction

Figure 28.8C
Some reactions of benzene

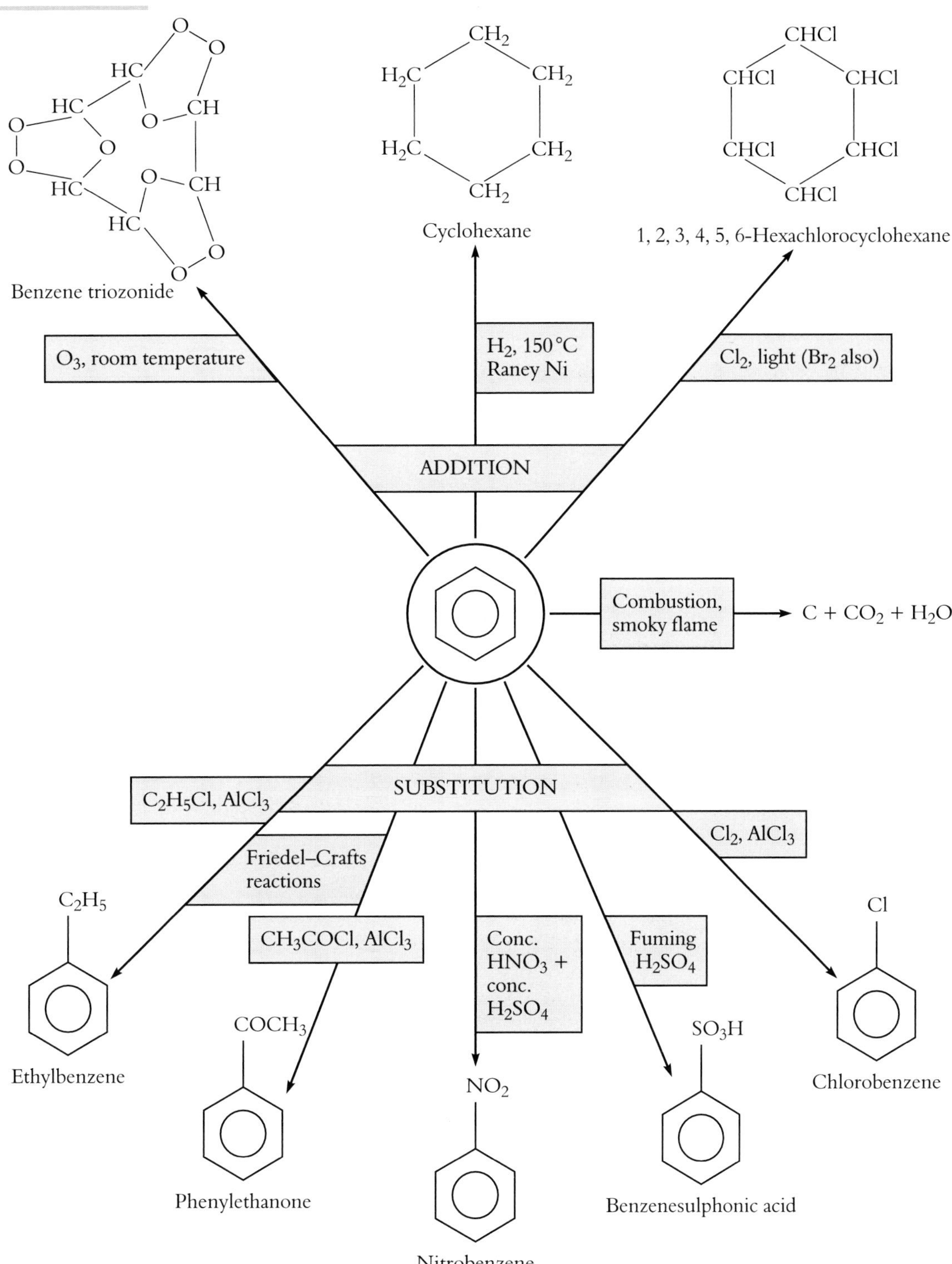

Benzene triozonide
Cyclohexane
1, 2, 3, 4, 5, 6-Hexachlorocyclohexane
O$_3$, room temperature
H$_2$, 150 °C Raney Ni
Cl$_2$, light (Br$_2$ also)
ADDITION
Combustion, smoky flame
C + CO$_2$ + H$_2$O
SUBSTITUTION
C$_2$H$_5$Cl, AlCl$_3$
Friedel–Crafts reactions
CH$_3$COCl, AlCl$_3$
Conc. HNO$_3$ + conc. H$_2$SO$_4$
Fuming H$_2$SO$_4$
Cl$_2$, AlCl$_3$
Ethylbenzene
Phenylethanone
Nitrobenzene
Benzenesulphonic acid
Chlorobenzene

CHECKPOINT 28.8: BENZENE

1. Explain why:

   (a) 1 mole of benzene reacts with 3 moles of chlorine in the presence of ultraviolet light, without the formation of hydrogen chloride, and
   (b) 1 mole of benzene reacts with 1 mole of chlorine in the presence of iron(III) chloride with the formation of hydrogen chloride.

2. What type of reagent attacks the benzene ring to form substitution compounds? State the attacking reagents which are involved in *(a)* the nitration of benzene, *(b)* the alkylation of benzene, and *(c)* the acylation of benzene.

3. What are the components of the 'nitrating mixture' used to make nitrobenzene from benzene? What is the nitrating agent that they produce? Give an equation for the formation of this nitrating agent from the nitrating mixture, and explain how benzene reacts with it.

4. Name two reactions in which benzene differs from ethene. Point out the structural resemblance between benzene and ethene and the reason for the difference in reactivity towards electrophiles.

5. *(a)* Briefly describe how you could carry out the reaction between ethanoyl choride and benzene.
   (b) Explain the mechanism of this reaction.

6. Give three methods which would enable you to distinguish between benzene and cyclohexene.

28.9 METHYLBENZENE (TOLUENE)

Methylbenzene is used as a solvent and a petrol additive

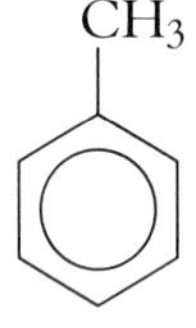

Methylbenzene, or toluene, $C_6H_5CH_3$, resembles benzene. Like benzene, methylbenzene is obtained from petroleum oil and from coal tar [see Figure 28.11A]. Methylbenzene is used as a solvent, as a source of the explosive trinitrotoluene (TNT), and as an additive to petrol, in which it improves the antiknock quality. The disadvantage of adding aromatic hydrocarbons to petrol is that they are carcinogenic.

28.10 REACTIONS OF THE RING

There are two sets of reactions, one set involving the aromatic ring in substitution or addition and the other involving reactions of the methyl group or 'side chain', as it is called.

28.10.1 SUBSTITUTION

Methylbenzene is more reactive than benzene towards electrophiles ...

... Conditions for substitution are milder ...

Methylbenzene is more reactive than benzene towards the electrophilic reagents which substitute in the benzene ring. This is because the methyl group pushes electrons into the ring [§ 28.12]. Milder conditions are employed than in the reactions of benzene. They are:

Chlorination/bromination: Cl_2/Br_2 with halogen carrier at 20 °C
Nitration: conc. HNO_3 + conc. H_2SO_4 at 30 °C
Sulphonation: fuming H_2SO_4 (containing SO_3) at 0 °C
Alkylation: RCl, $AlCl_3$ at 20 °C
Acylation: RCOCl or $(RCO)_2O$, $AlCl_3$, warm

A mixture of 1,2- and 1,4-substituted methylbenzenes is obtained in each case, e.g.

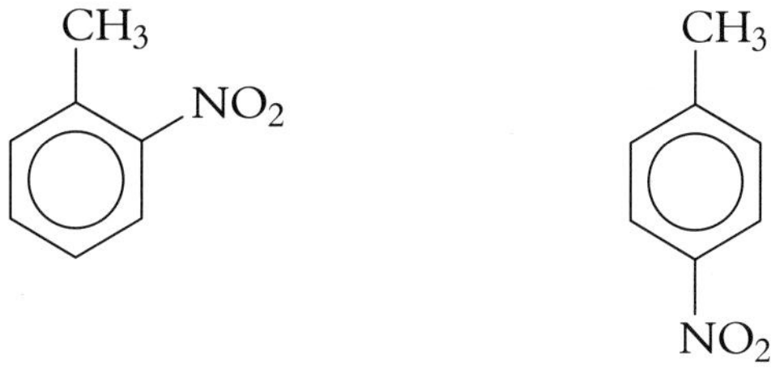

CH_3 NO_2 CH_3 NO_2

1-Methyl-2-nitrobenzene and 1-Methyl-4-nitrobenzene

... More than one group is substituted

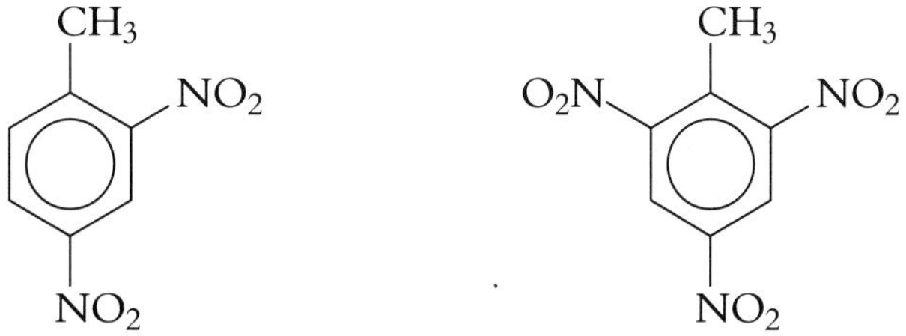

If the temperature is raised, two groups or three groups are introduced, e.g.

CH_3 NO_2 NO_2 CH_3 O_2N NO_2 NO_2

1-Methyl-2,4-dinitrobenzene and 1-Methyl-2,4,6-trinitrobenzene

28.10.2 ADDITION

Addition reactions are the same as those of benzene [see Figure 28.8C].

28.11 REACTIONS OF THE SIDE CHAIN

28.11.1 HALOGENATION

In sunlight, Cl_2 and Br_2 substitute the CH_3— group, not the ring

When chlorine is bubbled through boiling methylbenzene in strong sunlight or ultraviolet light, substitution occurs in the side chain:

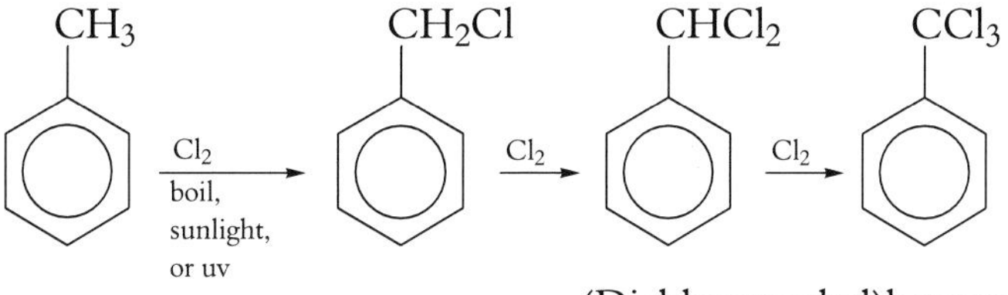

CH_3 —(Cl_2, boil, sunlight, or uv)→ CH_2Cl —(Cl_2)→ $CHCl_2$ —(Cl_2)→ CCl_3

(Dichloromethyl)benzene

(Chloromethyl)benzene (Trichloromethyl)benzene

Reaction involves free radicals, as does the halogenation of methane

Bromination occurs under the same conditions to give similar products. The conditions employed here favour the formation of free radicals, and the reaction proceeds in a similar way to the halogenation of alkanes [§ 26.4.7]. Note the difference between the product of free radical halogenation in the side chain and halogenation in the presence of a halogen carrier, when the halogen substitutes in the aromatic ring.

28.11.2 OXIDATION

Oxidation of —CH_3 gives —CHO or —CO_2H

The powerful oxidising agents acidified potassium manganate(VII) and acidified potassium dichromate(VI) will oxidise the side chain, —CH_3, to the carboxylic acid group, —CO_2H. The reaction mixture must be refluxed for several hours.

A milder oxidising agent, manganese(IV) oxide, MnO_2, or chromium dichloride dioxide, CrO_2Cl_2, oxidises $-CH_3$ to the aldehyde group, $-CHO$

CH3
CHO
CO2H
MnO2 or CrO2Cl2
Reflux, KMnO4, acid or K2Cr2O7, acid

Benzenecarbaldehyde (benzaldehyde)

Benzenecarboxylic acid (benzoic acid)

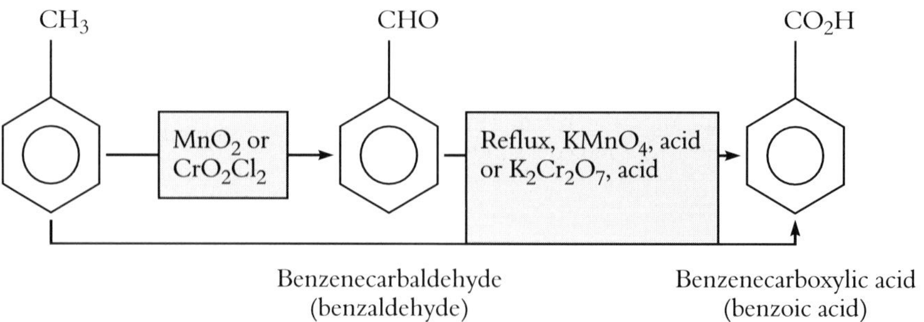

Figure 28.11A
The reactions of methylbenzene (toluene)

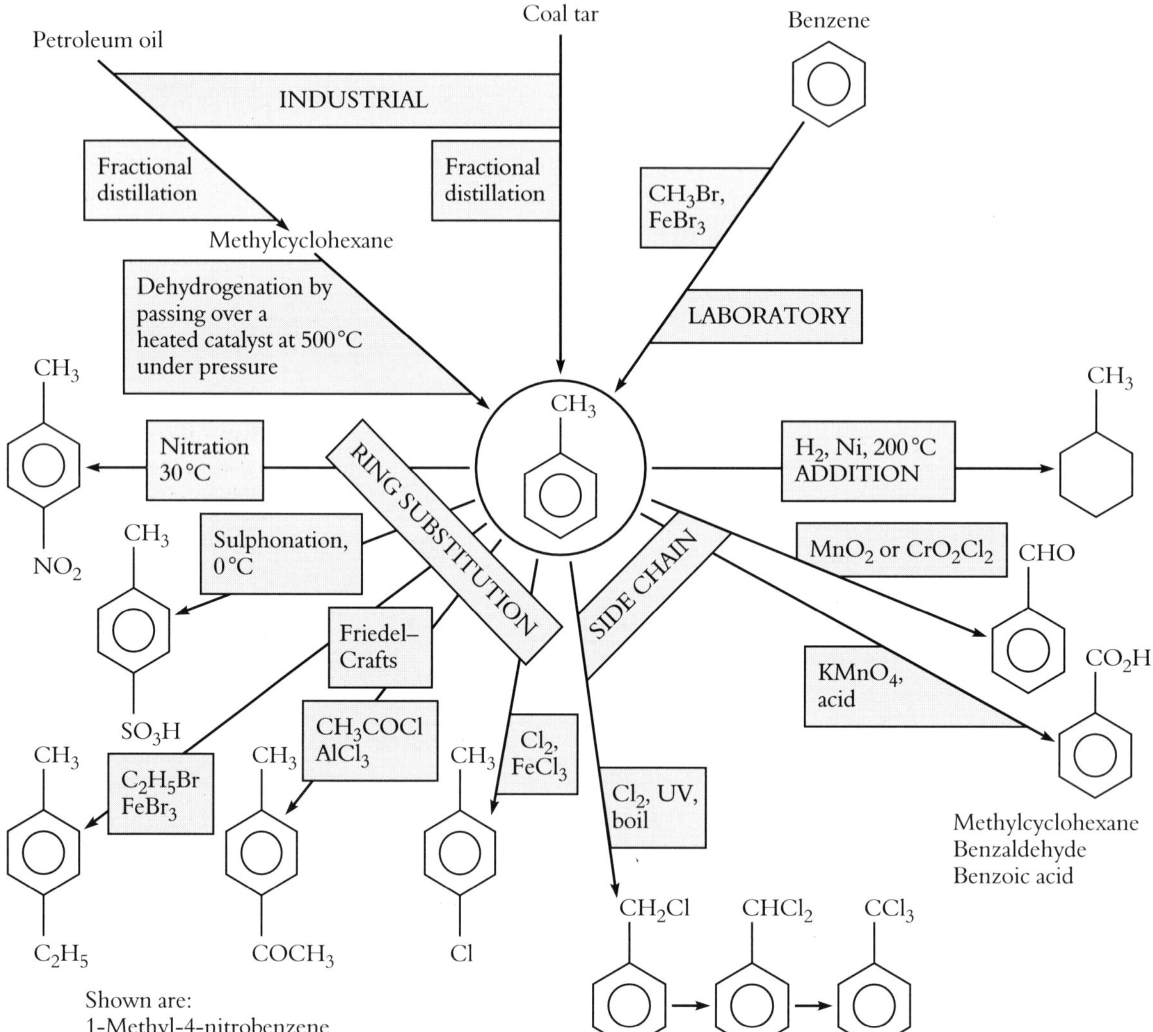

Coal tar
Benzene
Petroleum oil
INDUSTRIAL
Fractional distillation
Fractional distillation
CH3Br, FeBr3
Methylcyclohexane
Dehydrogenation by passing over a heated catalyst at 500 °C under pressure
LABORATORY
CH3
CH3
CH3
Nitration 30 °C
RING SUBSTITUTION
H2, Ni, 200 °C ADDITION
NO2
CH3
Sulphonation, 0 °C
SIDE CHAIN
MnO2 or CrO2Cl2
CHO
Friedel–Crafts
CO2H
KMnO4, acid
SO3H
CH3COCl AlCl3
CH3
CH3
CH3
Cl2, FeCl3
C2H5Br FeBr3
Cl2, UV, boil
C2H5
COCH3
Cl
CH2Cl
CHCl2
CCl3

Methylcyclohexane
Benzaldehyde
Benzoic acid

Shown are:
1-Methyl-4-nitrobenzene
4-Methylbenzenesulphonic acid
1-Ethyl-4-methylbenzene
(4-Methylphenyl)ethanone
1-Chloro-4-methylbenzene
The 2-product is also formed in each case.

(Chloromethyl)benzene
(Dichloromethyl)benzene
(Trichloromethyl)benzene

CHECKPOINT 28.11: METHYLBENZENE

1. State the products of the reactions between (i) benzene (ii) methylbenzene with
   (a) chloromethane and aluminium chloride
   (b) ethanoyl chloride and iron(III) chloride
   (c) concentrated nitric acid and concentrated sulphuric acid
   (d) manganese(IV) oxide
   (e) chlorine in diffuse sunlight
   (f) chlorine and aluminium chloride

2. Write structural formulae for 1-chloro-2-methylbenzene, 1-chloro-4-methylbenzene and (chloromethyl)benzene. How can each of these compounds be made from methylbenzene?

3. How is methylbenzene obtained *(a)* industrially and *(b)* in the laboratory, starting from benzene?

28.12 THE EFFECT OF SUBSTITUENT GROUPS ON THE BENZENE RING

When the benzene ring contains a substituent X as in 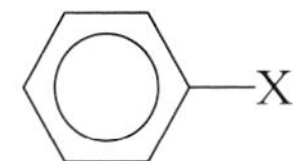

the reactions of the compound C_6H_5X may be faster or slower than those of benzene. If substitution takes place faster in C_6H_5X, then X is said to **activate** the ring. If substitution is slower in C_6H_5X, then X is said to **deactivate** the ring. The reagents which attack the benzene ring are mainly electrophilic reagents (such as NO_2^+). A substituent such as —CH_3 increases the electron density in the benzene ring and therefore activates the ring. Methylbenzene reacts faster than benzene. A substituent such as —Cl, which withdraws electrons from the ring, deactivates the ring. Chlorobenzene reacts more slowly than benzene in substitution reactions.

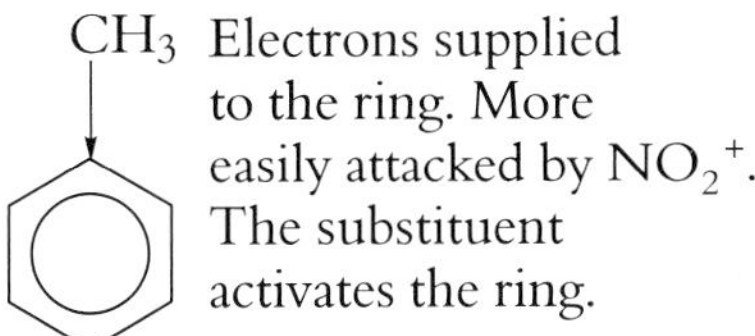

CH3
Electrons supplied to the ring. More easily attacked by NO2+. The substituent activates the ring.

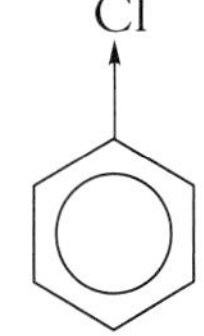

Electrons withdrawn from the ring. Less easily attacked by NO_2^+. The substituent deactivates the ring.

The reason why a —Cl substituent deactivates the benzene ring is that chlorine is more electronegative than carbon. The bond —$C^{\delta+}$—$Cl^{\delta-}$ is polarised and withdraws electrons from the ring towards the Cl atom.

The compounds phenol and phenylamine

—OH

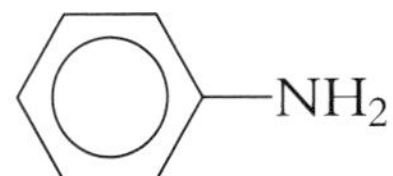

react a great deal faster than benzene in substitution reactions. The reason is the lone pairs of electrons on the oxygen and nitrogen atoms. These are fed into the π orbitals of the benzene ring. The movement of a pair of electrons is shown by a curved arrow.

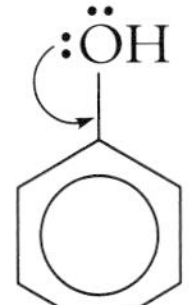

Lone pairs of electrons on the oxygen atom or the nitrogen atom are fed into the π orbitals of the benzene ring. The ring is activated: it is more easily attacked by NO_2^+.

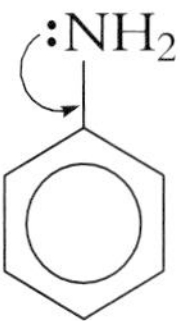

In a group with a multiple bond, the π electrons shift towards the more electronegative atom in the bond. In the carboxyl group this shift can be shown as

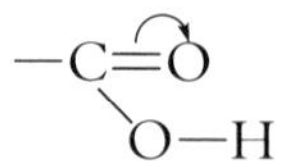

$$-\overset{}{C}{=}O \quad (\text{with } O{-}H \text{ on } C)$$

The carboxyl group therefore withdraws electrons: it is a deactivating group. The same is true of groups such as CO_2R (ester), —CHO (aldehyde) and —COR (ketone).

Activating substituents, e.g. $—CH_3$, direct groups entering the benzene ring into the 2- and 4- positions. Thus methylbenzene yields on nitration a mixture of 1-methyl-2-nitrobenzene and 1-methyl-4-nitrobenzene. Deactivating substituents, e.g. $—CO_2H$, direct entering groups into the 3-position in the ring. Thus benzoic acid is nitrated to 3-nitrobenzoic acid.

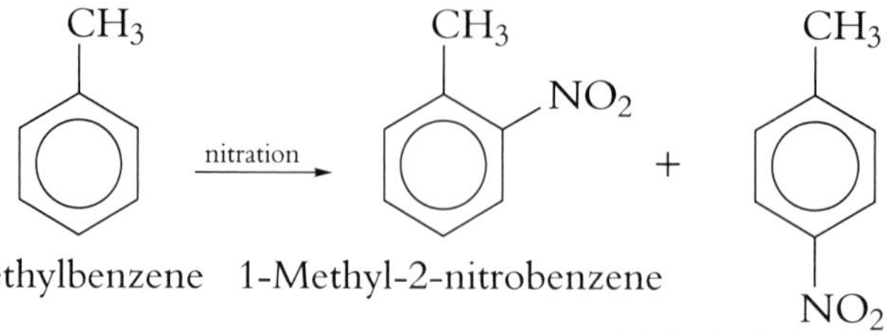

CH_3 CH_3 NO_2 CH_3

nitration → +

Methylbenzene 1-Methyl-2-nitrobenzene

NO_2

1-Methyl-4-nitrobenzene

CO_2H CO_2H

nitration →

NO_2

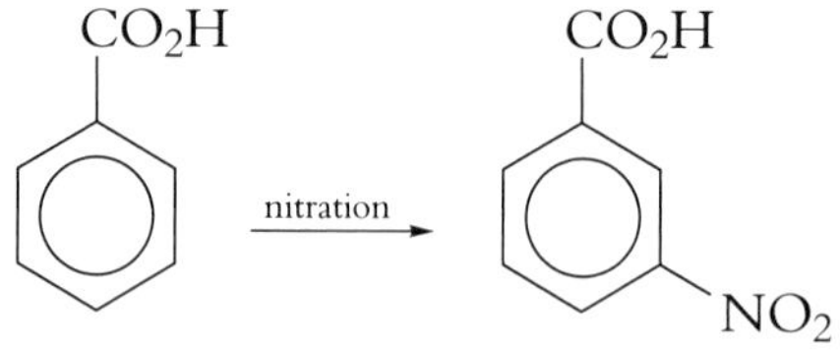

Benzoic acid 3-Nitrobenzoic acid

QUESTIONS ON CHAPTER 28

1. Describe a laboratory method for the preparation of nitrobenzene from benzene. Include: the reagents, the necessary conditions, the equation for the reaction and the method of purifying the product.

2. Under what conditions does benzene react with *(a)* chlorine, *(b)* chloroethane, *(c)* ethanoyl chloride and *(d)* sulphuric acid?

3. List the reactions of methylbenzene which are *(a)* typical of benzene and *(b)* not shared by benzene.

4. *(a)* Contrast the reactions of propene and benzene with (i) bromine (ii) sulphuric acid.
(b) Contrast the reactions of cyclohexene and benzene with (i) potassium manganate(VII) (ii) ethanoyl chloride + $AlCl_3$

5. What type of reaction is most typical of benzene? Give three examples of this type of reaction, with equations. What type of catalyst is employed in Friedel–Crafts reactions? Outline the mechanism of one such catalysed reaction, showing the importance of the catalyst.

6. Nitrobenzene is obtained from benzene and a mixture of concentrated nitric and sulphuric acids.

(a) What type of mechanism is involved?
(b) Name the nitrating species, and show how it is formed from the mixture of acids.
(c) Show how the nitrating species attacks benzene.

7. 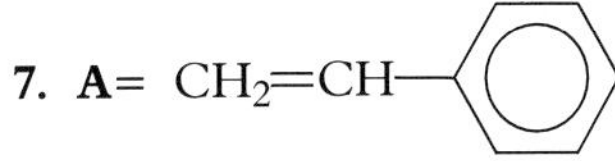

(a) Give the systematic name and the trade name of compound **A** shown above.

(b) Treatment of **A** with hydrogen gives compound **B** at room temperature and compound **C** at high temperature and pressure. Name **B** and **C**, and give their formulae.

(c) Compound **A** polymerises to give **D**. Name the polymer and draw its structure.

8. The preparation of bromobenzene can be carried out in the laboratory as follows:

In a fume cupboard, place 30 cm^3 of benzene (density 0.87 g cm^{-3}) and about 1 g of iron filings in a flask fitted with a reflux condenser. Add 35 cm^3 of bromine (density 3.2 g cm^{-3}) and heat at 70 °C for 30 minutes. Cool the flask and pour the contents into a beaker of cold water. Decant off the aqueous layer. The organic layer has a red-brown colour. Place the organic layer in a separating funnel and wash it several times with aqueous sodium hydroxide solution until almost colourless. Finally shake the organic layer with water, separate it and treat with anhydrous calcium chloride.

(a) Write an equation for the conversion of benzene to bromobenzene. **1**

(b) Explain why the experiment was conducted in a fume cupboard. **2**

(c) (i) Describe, giving practical details, the use of a separating funnel in washing the organic layer with sodium hydroxide solution. **3**

(ii) Explain, using an equation, why the organic layer changed colour during this washing. **3**

(d) Suggest why the organic layer is treated with anhydrous calcium chloride. **1**

(e) Calculate the percentage yield if 25.9 g of purified product was obtained. **4**

14 *CCEA (AS/AL)*

9. Three reactions of methylbenzene can be summarised on a flow sheet.

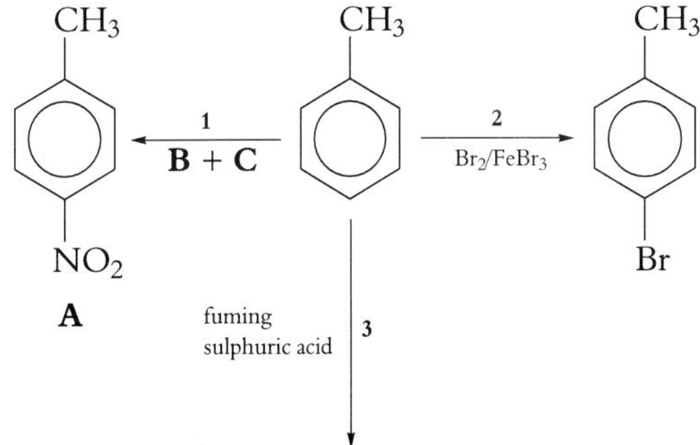

(a) Name substance **A** and the reagents **B** and **C**. **3**

(b) (i) Draw the structural formula for 4-methylbenzenesulphonic acid. **1**

(ii) State the formula of the molecule which attacks the benzene ring in reaction **3**. **1**

(iii) Name an important group of industrial products that is made from benzenesulphonic acids. **1**

(c) (i) Write an equation to show how the catalyst iron(III) bromide induces polarization of a bromine molecule in reaction **2**. **1**

(ii) Classify reaction **2** as fully as possible. **2**

(d) A student tried to make 4-methylcyclohexanol by the following route

CH_3 / Br (benzene ring) —Step 1→ CH_3 / Br (cyclohexane ring) —Step 2→ CH_3 / OH (cyclohexane ring)

(i) Suggest reagents and conditions for each step. **4**

(ii) The final yield of 4-methylcyclohexanol was rather low. This was thought to be due to the formation of some 4-methylcyclohexene. Suggest a chemical test and the colour change you would expect to observe for an alkene such as 4-methylcyclohexene. **2**

15 *L(N) (AS/AL)*

10. *(a)* (i) The enthalpy of hydrogenation of cyclohexene is -119.6 kJ mol^{-1}.

Use this information to calculate the enthalpy of hydrogenation of the hypothetical molecule cyclohexa-1,3,5-triene.

(ii) The enthalpy of hydrogenation of benzene is -208.4 kJ mol^{-1}. Explain why this value differs from the one you have calculated for cyclohexa-1,3,5-triene. **4**

(b) Name the electrophile involved in the nitration of benzene. Write an equation showing how this species is formed in a mixture of concentrated nitric acid and concentrated sulphuric acid. Outline a mechanism for the nitration of benzene. **5**

(c) Give **one** use for polynitro-compounds such as 2,4,6-trinitromethylbenzene. **1**

10 *NEAB (AL)*

11. Some important reactions of benzene can be summarised in a flow sheet:

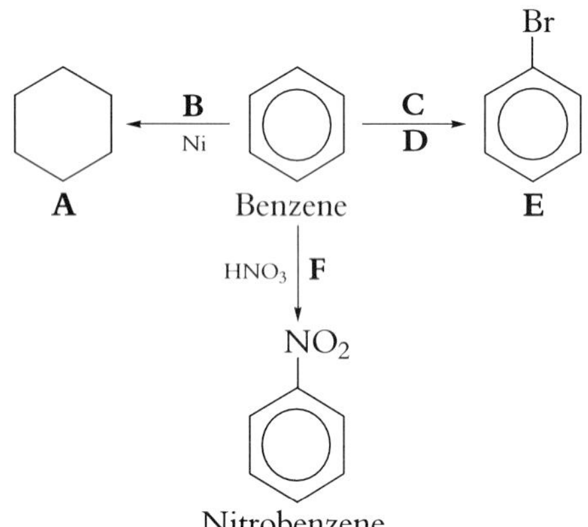

(a) Name substances **A** to **F**. 6

(b) (i) The reactions to form **E** and nitrobenzene are electrophilic substitutions. Give the formula and charge of the electrophile in each reaction. 2

(ii) In the reaction to form nitrobenzene it is important to keep the temperature below 55 °C. Suggest a reason for this. 1

(iii) Suggest **ONE** use for compounds formed by the nitration of arenes. 1

(c) (i) State the conditions for the reaction to form **A** from benzene. 2

(ii) Benzene reacts with chlorine to form 1,2,3,4,5,6-hexachlorocyclohexane, an important insecticide. Write a **balanced** equation for this reaction. 1

13 *L(N) (AS/AL)*

12. Ethylbenzene is obtained by the reaction of benzene, in the presence of aluminium chloride, either with chloroethane or with ethene and hydrogen chloride.

(a) (i) Write an equation showing how a reactive species is generated from chloroethane and aluminium chloride.

(ii) Write an equation showing how the same reactive species is formed from ethene, hydrogen chloride and aluminium chloride.

(iii) Name the type of reaction between the reactive species formed and benzene. Outline a mechanism for this reaction. 7

(b) Suggest two reasons why the ethene route is preferred for the industrial manufacture of ethylbenzene. 2

†*(c)* Give the structure of the monomer obtained by dehydrogenation of ethylbenzene. Name the polymer obtained from this monomer. 2

11 *NEAB (AL)*

† For polymers see Chapter 27.

13. Cyclohexene and benzene form a mixture which is approximately ideal.

(a) Explain the meaning of the term **ideal mixture**. 2

(b) (i) Calculate the number of moles of cyclohexene and benzene in a mixture which contains 25 g of cyclohexene and 40 g of benzene.

(ii) The vapour pressures of cyclohexene and benzene at 25 °C are 11 830 and 9975 Pa respectively.
Calculate the total vapour pressure above a mixture of 25 g of cyclohexene and 40 g of benzene. 3

(c) The enthalpies of combustion of cyclohexene and benzene are −3752 and −3267 kJ mol^{-1} respectively.

Calculate the enthalpy change produced by the complete combustion of the mixture of cyclohexene and benzene described in part *(b)*. 2

(d) The mixture can be catalytically hydrogenated in a similar way to alkenes to form a single product.

(i) Name a suitable catalyst.

(ii) Name the product after the mixture has been completely hydrogenated. 2

(e) Hydrogen bromide reacts with cyclohexene and ethene in a similar way. Draw a flow scheme to illustrate the mechanism of the reaction with cyclohexene. 3

†*(f)* Cyclohexene may be prepared by the following route:

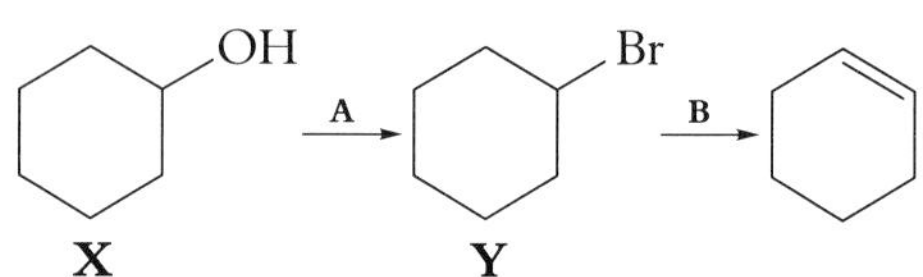

(i) Name, or give the formulae of, the reagents **A** and **B**. 2

(ii) Suggest the names of the substances **X** and **Y**. 2

(g) State **one** industrial source of benzene. 1

† See §§ 29.4, 29.7

CCEA (AS/AL)

29

HALOGENOALKANES AND HALOGENOARENES

29.1 ANAESTHETICS

Trichloromethane (chloroform), $CHCl_3$, is a clear dense liquid with a pleasant smell and a boiling temperature of 61 °C. Its first use in medicine was as an inhalant for people suffering from asthma. In 1847, chloroform was used as a total anaesthetic by Dr Simpson in Edinburgh. Ethoxyethane (ether), $C_2H_5OC_2H_5$, was first used as an anaesthetic in dentistry in the same year. Dinitrogen oxide (laughing gas), N_2O, was another pioneering anaesthetic. These anaesthetics revolutionised surgery. Before total anaesthesia, a surgeon adopted the fastest possible procedure so as to minimise pain and shock to the patient. With the patient unconscious, a surgeon could undertake more complicated operations and explore new methods of surgery. The risk of the patients dying of shock was much reduced.

The night before surgery, the patient is usually given a tranquilliser to ensure a good night's sleep. During surgery, a combination of drugs is given to achieve anaesthesia, analgesia (relief of pain) and muscle relaxation. Following surgery, a drug may be given for the relief of severe pain. All these drugs are part of the chemical industry's contribution to medicine. Life is now longer, safer and freer from pain than in any previous century.

Ethoxyethane (ether), $C_2H_5OC_2H_5$, is a good anaesthetic and there is a big difference between the amount that causes unconsciousness and the lethal dose. It is given in a stream of oxygen containing 10–30% ether. The disadvantage of ether is that it is extremely flammable and precautions have to be taken over its use and storage, and surgeons cannot use electrical equipment which may cause a spark and ignite it.

Trichloromethane (chloroform), $CHCl_3$, is non-flammable, but it causes liver damage and there is a small difference between the dose which produces anaesthesia and the lethal dose. It is seldom used now.

Dinitrogen oxide (laughing gas), N_2O, is an excellent anaesthetic.

The majority of operations use 2-bromo-2-chloro-1,1,1-trifluoroethane, $CF_3CHBrCl$, which has the trade names Fluothane® and Halothane®. It was introduced by ICI in 1957. ICI was looking for a new anaesthetic which would satisfy these requirements:

1. It should work rapidly and smoothly.
2. It should not be unpleasant or irritating to breathe.
3. There should be a good safety margin between the anaesthetic dose and the lethal dose.
4. It should not affect the heart or other organs.
5. Recovery should not have unpleasant effects such as nausea.
6. It should be non-flammable and non-explosive.

The chemists realised that fluoroalkanes satisfy these requirements. Selected fluoroalkanes were tested on animals and the one which emerged as the best was 2-bromo-2-chloro-1,1,1-trifluoroethane. Clinical trials on humans showed that the new anaesthetic was excellent and it was marketed as Fluothane®.

For minor surgery, a local anaesthetic is sometimes used. Chloroethane, C_2H_5Cl, is a gas at room temperature. With boiling temperature 12 °C, it can be liquefied under pressure in a sealed container. When the liquid is sprayed on to the surface of the skin, it evaporates so quickly and cools the surface so rapidly that it freezes tissues near the surface, making them insensitive to pain. It is used in minor surgery and in sporting injuries.

Most of the anaesthetics referred to are halogenoalkanes.

29.2 HALOGENOALKANES

The homologous series called **halogenoalkanes** (or haloalkanes) have the functional group

$$—C—X$$

where X = F, Cl, Br or I. Some members of the series are listed:

Names of some halogenoalkanes ...

C_2H_5Cl — Chloroethane. The name is derived from ethane, as the compound can be thought of as ethane, C_2H_6, with a Cl atom substituted for an H atom.

$CH_3CH_2CHCH_3$ (Cl on C-2) — 2-Chlorobutane. The name is taken from that of the longest unbranched chain. The **locant**, 2, gives the position of the chlorine atom, counting from the end which gives the lower number for the locant.

... and halogenoalkenes

$CH_3CH{=}CHCHCH_3$ (Cl on C-4) — 4-Chloropent-2-ene. The double bond is numbered first; then the chlorine atom.

$CH_3CH_2CHCH_2CHCH_3$ (CH_3 on C-4, Br on C-2) — 2-Bromo-4-methylhexane. The longest chain is 6C: the compound is a derivative of hexane. The substituents are named in alphabetical order and then numbered.

CH_3CHCl_2 — 1,1-Dichloroethane

CH_2ClCH_2Cl — 1,2-Dichloroethane

$CHCl_3$ — Trichloromethane (chloroform)

CCl_4 — Tetrachloromethane (carbon tetrachloride)

Primary, secondary and tertiary

There are three types of halogenoalkanes:

Primary (1°)	Secondary (2°)	Tertiary (3°)
C bonded to R, H, H, X	C bonded to R, H, R′, X	C bonded to R, R″, R′, X

29.3 PHYSICAL PROPERTIES

29.3.1 VOLATILITY

Electronegativities are F = 4.0, Cl = 3.0, Br = 2.8, I = 2.5, C = 2.5. The strong

Fluoroalkanes and chloroalkanes have polar molecules ...

polarity of C—F and C—Cl bonds gives rise to attraction between the dipoles in neighbouring molecules:

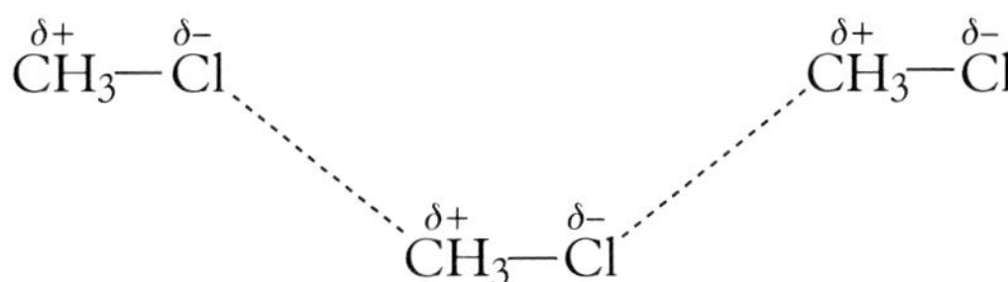

... As a result, their boiling temperatures are higher than those of the corresponding alkanes ...

Energy must be supplied to separate the molecules, and the boiling temperatures of fluoroalkanes and chloroalkanes are therefore higher than those of alkanes of similar molecular mass. This is not the case for bromoalkanes and iodoalkanes.

29.3.2 SOLUBILITY

Halogenoalkanes have a low solubility in water

The polar molecules can interact with water molecules, but the attractive forces set up are not as strong as the hydrogen bonds present in water. Halogenoalkanes therefore, although they dissolve more than alkanes, are only slightly soluble in water.

29.3.3 SMELL

These compounds have a sweet, slightly sickly smell.

29.4 LABORATORY METHODS OF PREPARING HALOGENOALKANES

Preparative methods are summarised in Figures 29.4A and 29.4B.

Figure 29.4A Laboratory preparations of halogenoalkanes

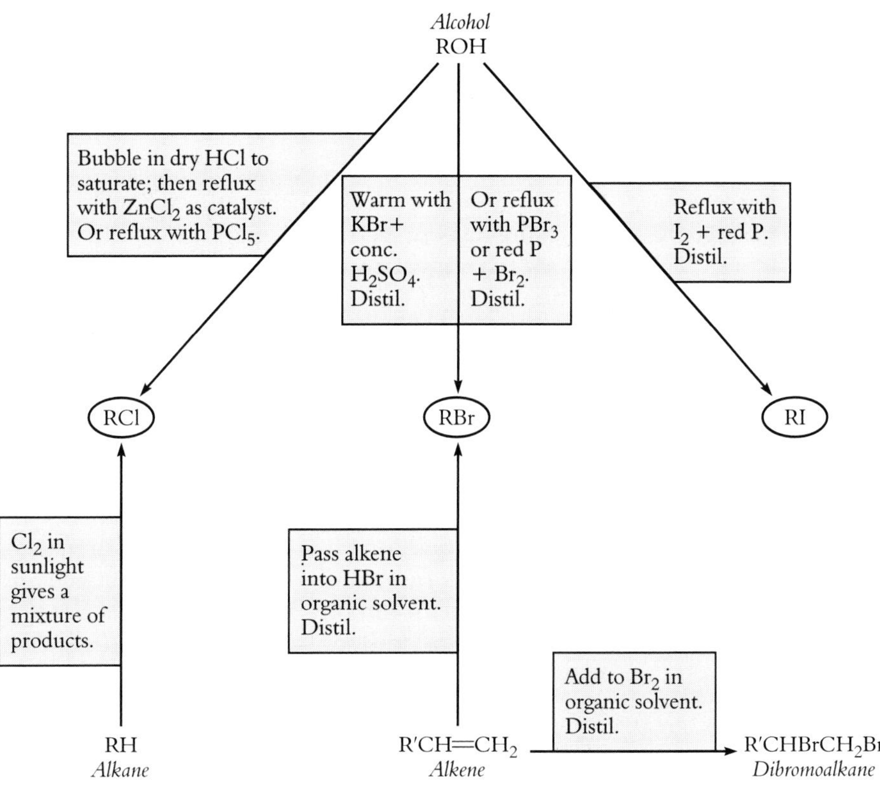

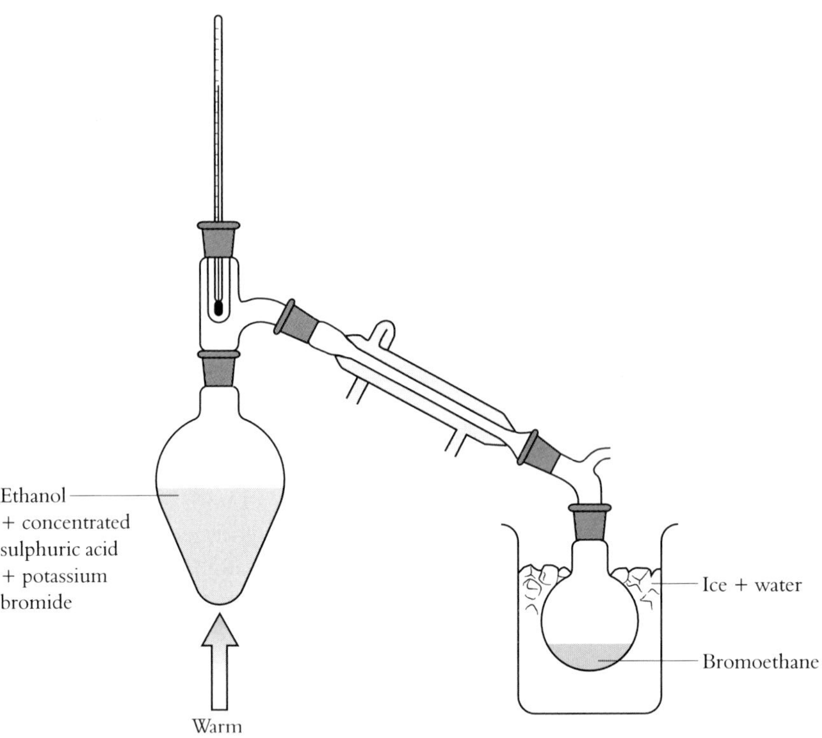

Figure 29.4B Preparation of bromoethane from ethanol

29.5 USES OF HALOGENOALKANES

Halogenoalkanes are used as grease solvents …

… CCl_4 is toxic …

Halogenoalkanes dissolve oil and grease. They are used in the dry-cleaning industry and also for cleaning articles which carry a film of oil or grease from the machinery used in their manufacture. Tetrachloromethane was once the prime halogenoalkane in this area. It is toxic as, when inhaled in quantity, it dissolves fat from the liver and kidneys. Less toxic are dichloromethane, CH_2Cl_2, trichloroethene, $CCl_2{=}CHCl$, called 'trichlor' in industry, and tetrachloroethene, $CCl_2{=}CCl_2$.

Halogenoalkanes are also used as fire-extinguishers

Fully halogenated alkanes, being non-flammable, volatile and dense, are used in fire-extinguishers. Tetrachloromethane, CCl_4, was used as a fire-extinguisher, but, at high temperatures, there is a danger of its being oxidised to the poisonous gas phosgene, $COCl_2$. It also has a toxic effect on the liver and kidneys and has been replaced by the safer gas dibromochlorofluoromethane, CBr_2ClF, called 'BCF'.

29.6 CHLOROFLUOROCARBONS, CFCs

Halogenoalkanes which have boiling temperatures just below room temperature can easily be liquefied by a slight increase in pressure. Halogenoalkanes containing chlorine and fluorine and no hydrogen are chlorofluorohydrocarbons. Examples are $CFCl_3$, CCl_2F_2 and $C_2Cl_2F_4$. They are usually called chlorofluorocarbons or CFCs (trade name Freons®). In addition to having low boiling temperatures, they are non-flammable, odourless, stable, non-toxic and good solvents. They were developed in the 1920s as what appeared to be ideal replacements for liquid ammonia and liquid sulphur

CFCs appeared to be ideal for use as fluids in refrigerators and as solvents in aerosol sprays ...

dioxide, which were formerly used as fluids in refrigerators and air-conditioning units. Being good solvents, they were also ideal as the solvents in aerosol sprays. Aerosols were used to dispense insecticides, hairsprays, perfumes and deodorants, window-cleaning liquids, polishes, waxes and laundry products. As more and more uses were found for these remarkable compounds, CFCs became big business, with hundreds of thousands of tonnes being produced yearly. Now they are being phased out. These stable, non-toxic compounds are dangerous!

... but their very stability has turned out to be a problem

During all the time that the use of CFCs was increasing, no-one thought about what would happen to the gases in the atmosphere. Because of their lack of reactivity and insolubility in water, there is no natural process for removing CFCs. In fact they drift up into the stratosphere (the upper atmosphere; see Figure 21.5A) where ultraviolet light causes photolysis. The chlorine radicals formed in photolysis take part in a chain reaction which converts ozone into oxygen.

CFCs destroy ozone in the stratosphere ...

(a) $Cl\cdot + O_3 \longrightarrow ClO\cdot + O_2$

(b) $ClO\cdot + O \longrightarrow Cl\cdot + O_2$

(c) $ClO\cdot + O_3 \longrightarrow ClO_2 + O_2$

... reduce the thickness of the ozone layer ...

Reactions (a) and (b) form a chain. This is why one chlorine radical from one CFC molecule can destroy thousands of ozone molecules. These reactions and the importance of the ozone layer in absorbing excess ultraviolet light are discussed in § 21.5.

... increasing the threat of global warming

Infrared radiation is strongly absorbed by CFCs. They therefore act as 'greenhouse gases', increasing the threat of global warming [see § 23.6]. CFCs are 10 000 times more effective than carbon dioxide in absorbing infrared radiation.

Because of concern over the decrease in the ozone layer, many nations have agreed to cut down their use of CFCs [see § 21.5]. Alternative compounds are already in production. Hydrohalocarbons contain at least one hydrogen atom per molecule. The C—H bond can be attacked by $HO\cdot$ radicals in the lower

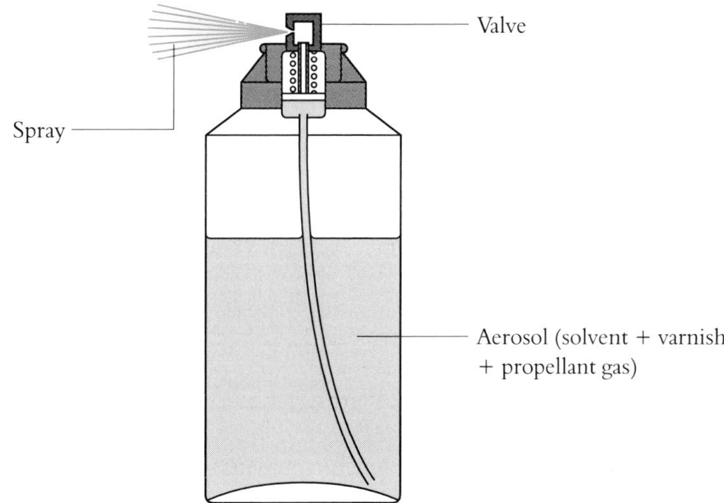

Figure 29.6A
An aerosol can

Replacements for CFCs have been found

atmosphere and the compounds do not reach the upper atmosphere. Hydrohalocarbons include

- hydrochlorofluorocarbons, HCFCs, e.g. $CHCl_2CF_3$, used in blowing plastics foam and $CHClF_2$, used in air-conditioners
- hydrofluorocarbons, HFCs, e.g. CH_2FCF_3, used in air-conditioners and refrigerators. HFCs cause no damage to the ozone layer, although they are greenhouse gases.

CHECKPOINT 29.6: CFCs

1. The substance CF_2Cl_2 is a Freon®. It is used as a refrigerant and in the recent past was used as an aerosol propellant. The Freon passes through the lower atmosphere and reaches the stratosphere. There it reacts with the ozone layer in reactions which include the following:

   (i) $CF_2Cl_2 \xrightarrow{uv} CF_2Cl\cdot + Cl\cdot$
   (ii) $Cl\cdot + O_3 \longrightarrow ClO\cdot + O_2$

   (a) Why is the lack of reactivity of the Freon an advantage in its use as an aerosol propellant?
   (b) How does it compare with the substances previously used as refrigerant liquids?
   (c) What is the connection between the Freon's lack of reactivity and the damage which it does to the ozone layer?
   (d) Name the type of reaction in (i).

2. What solutions have been proposed to the problem of CFCs in the upper atmosphere?

Figure 29.7A
Reactions of halogenoalkanes

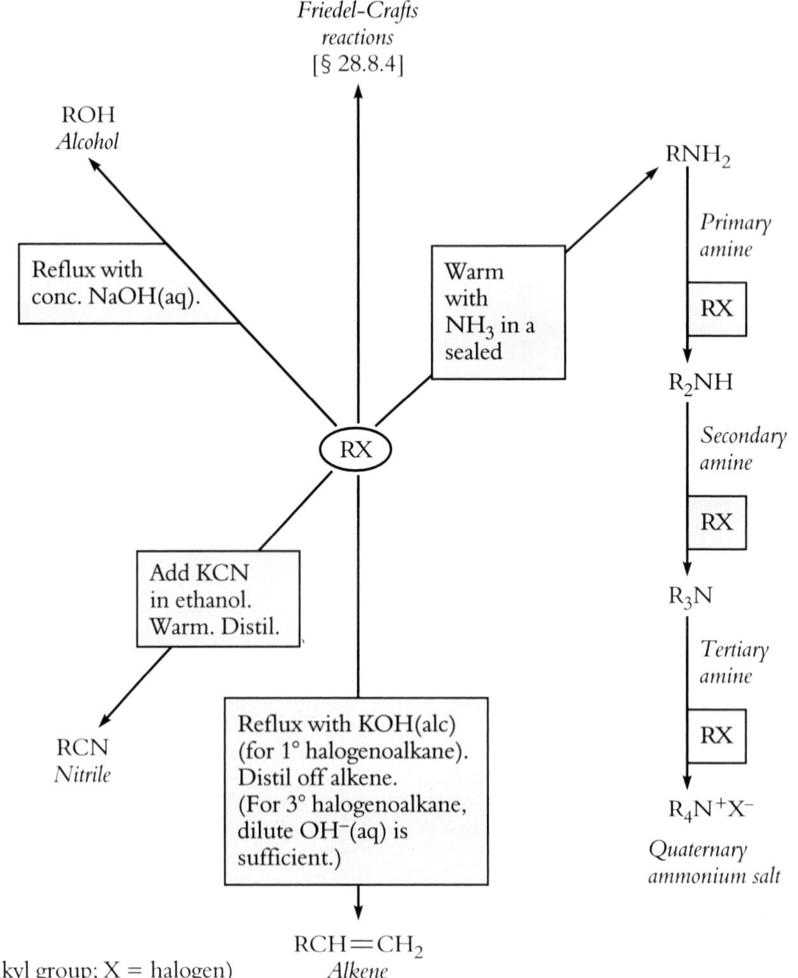

29.7 REACTIONS

The reactions of halogenoalkanes are shown in Figure 29.7A.

CHECKPOINT 29.7: REACTIONS

1. Name the following:

   (a) $CH_3CH_2CH(I)CH_3$ *(b)* $(CH_3)_3CBr$

2. Explain the differences between the boiling temperatures of the following compounds:

Compound	Molar mass/ g mol^{-1}	Boiling temperature/°C
C_5H_{12}	72	36
C_3H_7Cl	79	46
C_3H_8	44	−42

3. State four industrial uses of the halogenoalkanes. Why do fluoroalkanes find special uses?

4. Which of the following would you choose for hydrolysing bromoethane to ethanol?

   A water
   B aqueous potassium hydroxide
   C alcoholic potassium hydroxide
   D aqueous sulphuric acid

5. *(a)* Draw the structural formula of 1-bromo-2-methylpropane.
   (b) Give the name and the structural formula of the alcohol obtained when this compound is treated with aqueous potassium hydroxide.
   (c) When 1-bromo-2-methylpropane is treated with alcoholic potassium hydroxide, a different product is formed. Give the name and structural formula of this compound.
   (d) When the product in *(c)* is treated with hydrogen bromide, an isomer of 1-bromo-2-methylpropane is formed. Give its name and structural formula.

6. Say how you could distinguish between these pairs by means of simple chemical tests.

   (a) **A** $CH_3CH_2CH_2Cl$ and **B** CH_3CH_2OH
   (b) **C** $CH_3CH_2CH_2Cl$ and **D** $CH_3CH_2CH_2Br$
   (c) **E** $CH_3CH_2CH_2Cl$ and **F** CH_3CH_2COCl

7. The conversion $C_3H_8 \longrightarrow C_3H_7OH$ cannot be achieved directly. How can it be done?

8. Name four series of compounds that can be made from halogenoalkanes. Show your answer in the form of a reaction scheme.

29.8 REACTIVITY

29.8.1 SUBSTITUTION REACTIONS

Why are halogenoalkanes reactive?

Halogenoalkanes undergo substitution reactions. The electronegativity values C = 2.5, I = 2.5, Br = 2.8, Cl = 3.0, F = 4.0 show that, except for C—I, the C—X bond is polarised:

The C—X bond is polarised ...

$$\overset{\delta+}{C}-\overset{\delta-}{X}$$

The positively charged carbon atom is susceptible to attack by nucleophiles, anions such as OH^- and CN^- and compounds with lone pairs of electrons such as $:NH_3$ and $H_2\ddot{O}$. The halide ion, X^-, makes a stable leaving group, and substitution reactions of the type

... Nucleophiles attack the carbon in C—X

$$Y: + \overset{\delta+}{C}-\overset{\delta-}{X} \longrightarrow Y-C + X^-$$

therefore occur.

These substitution reactions are not very fast. To substitute —X by $—NH_2$, the halogenoalkane must be heated with 0.880 ammonia in a 'bomb', a sealed metal container:

Conditions needed for reaction

$$C_2H_5Cl(g) + NH_3(aq) \xrightarrow[\text{pressure develops}]{\text{heat in a 'bomb'}} C_2H_5NH_2(g) + HCl(g)$$

Chloroethane — Ethylamine or Aminoethane

To substitute —X by —OH, a primary halogenoalkane must be refluxed with aqueous alkali for about an hour:

$$C_2H_5Cl(g) + OH^-(aq) \xrightarrow{\text{reflux}} C_2H_5OH(g) + Cl^-(aq)$$

Chloroethane — Ethanol

A suitable apparatus is shown in Figure 28.8A.

The ease of reaction depends on the ease of breaking the C—X bond. Average standard bond enthalpies in kJ mol^{-1} are C—F = 484, C—Cl = 338, C—Br = 276, and C—I = 238. Thus, the ease of bond-breaking is

$$C—I > C—Br > C—Cl > C—F$$

The ease of reaction also depends on the stability of the *leaving group*. The ease of formation of $X^-(aq)$ is

$$F^-(aq) > Cl^-(aq) > Br^-(aq) > I^-(aq)$$

The two factors are in opposition. The C—F bond is so strong that fluoroalkanes are extremely unreactive. The order of reactivity of the other halogenoalkanes towards nucleophiles is usually

$$RI > RBr > RCl$$

29.8.2 ELIMINATION REACTIONS

Elimination of HX with the formation of an alkene takes place in reactions with strong bases

Halogenoalkanes also undergo reactions in which hydrogen halide is eliminated and an alkene is formed. The nucleophiles which take part in substitution reactions (e.g., OH^-, CN^-) have basic properties. They are able to abstract a hydrogen atom from a halogenoalkane, while the halogen is expelled as a halide ion. Thus, two reactions, substitution and elimination, are in competition:

$$RCH_2CH_2X + OH^-(aq) \longrightarrow RCH_2CH_2OH + X^-(aq); \textit{ Substitution}$$

$$RCH_2CH_2X + OH^-(aq) \longrightarrow RCH{=}CH_2 + H_2O + X^-(aq); \textit{ Elimination}$$

Substitution reactions are favoured by weakly basic nucleophiles, e.g., CN^-; elimination by strongly basic reagents, e.g., OH^-. A high concentration of base in a non-aqueous solvent (e.g., potassium hydroxide in ethanol) at a reflux temperature favours elimination.

The relative extents to which substitution and elimination take place depend on the structure of the molecule, as well as on the basic strength of the nucleophile. Elimination becomes progressively more important in the order

Primary < Secondary < Tertiary halogenoalkane

When a tertiary halogenoalkane reacts with a base, the main product is an alkene:

$$(CH_3)_3CCl(l) + OH^-(aq) \longrightarrow (CH_3)_2C{=}CH_2(l) + H_2O(l) + Cl^-(aq)$$

Tertiary halogenoalkanes tend to undergo elimination reactions

To replace —X by —OH in a tertiary halogenoalkane, no base is needed:

$$(CH_3)_3CCl(l) + H_2O(l) \xrightarrow[\text{at 25 °C}]{\text{80\% aqueous ethanol}} (CH_3)_3COH(l) + HCl(aq)$$

For a given aqueous base, the reactions which take place are

Primary halogenoalkane: Substitution
Secondary halogenoalkane: Substitution/Elimination
Tertiary halogenoalkane: Elimination.

29.8.3 OUTLINE MECHANISM OF HYDROLYSIS

The hydrolysis of a halogenoalkane is a **substitution reaction**; —OH is substituted for —Halogen. The attacking species is a **nucleophile**; the OH^- ion has a pair of electrons to donate in bond-making. Substitutions by nucleophiles are described as **S_N reactions**.

The rate of hydrolysis of a primary halogenoalkane, e.g. CH_3CH_2Br, is proportional to the concentrations of both halogenoalkane and hydroxide ions. The mechanism proposed is an attack by OH^- on the $\delta+$ carbon atom in $>C^{\delta+}—Br^{\delta-}$ to form a reactive intermediate. As a bond forms between the $\delta+$ carbon atom and the OH^- ion, the bond from the carbon atom to the bromine atom weakens, until eventually the bromine atom is displaced as a bromide ion.

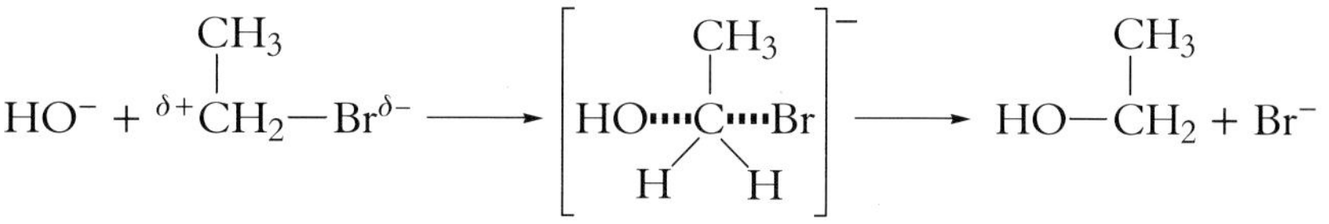

$$HO^- + {}^{\delta+}\underset{}{\overset{CH_3}{|}}{CH_2}—Br^{\delta-} \longrightarrow \left[HO\cdots\overset{CH_3}{\underset{H\quad H}{C}}\cdots Br \right]^- \longrightarrow HO—\overset{CH_3}{CH_2} + Br^-$$

Hydroxide ion + Halogenoalkane	In the intermediate negative ion, both HO and Br are partially bonded to the carbon atom.	Alcohol + Bromide ion

Two species are needed to form the reactive intermediate so this type of reaction is described as an **S_N2 reaction**.

The rate of hydrolysis of a tertiary halogenoalkane, e.g. $(CH_3)_3CBr$, is proportional to the concentration of the halogenoalkane but does not depend on the concentration of hydroxide ions. To fit in with this experimental observation, it has been proposed that the reaction takes place in two steps. The first is a slow step in which hydroxide ions play no part. The carbon—bromine bond breaks to form a bromide ion and a positive ion called a carbocation. The second step is a fast reaction between the carbocation and a hydroxide ion.

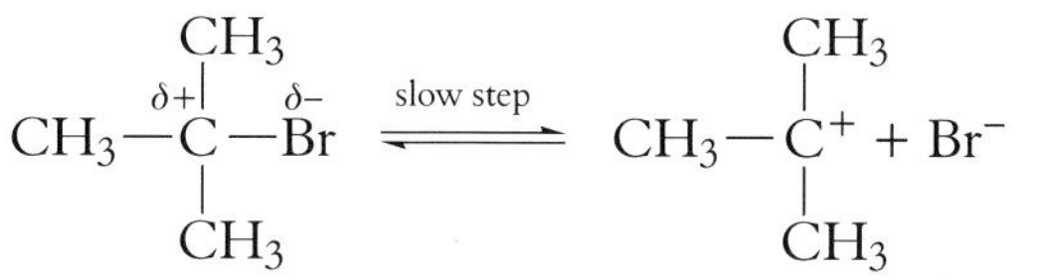

$$CH_3—\overset{CH_3}{\underset{CH_3}{C^{\delta+}}}—Br^{\delta-} \xrightleftharpoons{\text{slow step}} CH_3—\overset{CH_3}{\underset{CH_3}{C^+}} + Br^- \qquad [1]$$

Reactive intermediate: a carbocation

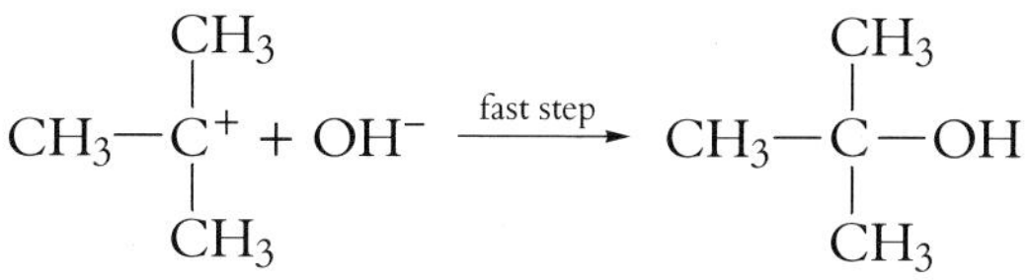

$$CH_3—\overset{CH_3}{\underset{CH_3}{C^+}} + OH^- \xrightarrow{\text{fast step}} CH_3—\overset{CH_3}{\underset{CH_3}{C}}—OH \qquad [2]$$

One molecule is involved in forming the reactive intermediate, and the reaction is therefore described as an **S_N1 reaction**.

The study of these mechanisms continues in § 29.9.

29.8.4 PATHWAYS BETWEEN HALOGENOALKANES AND OTHER SERIES OF COMPOUNDS

Note Alkanes and alkenes are obtained from the petroleum industry. They can be converted into halogenoalkanes, from which a variety of different compounds can be made. The halogenation of alkanes gives a mixture of halogenoalkanes. If the appropriate alcohol is available, it is a more convenient source of a monohalogenoalkane.

Figure 29.8A
Relating halogenoalkanes to other series of compounds

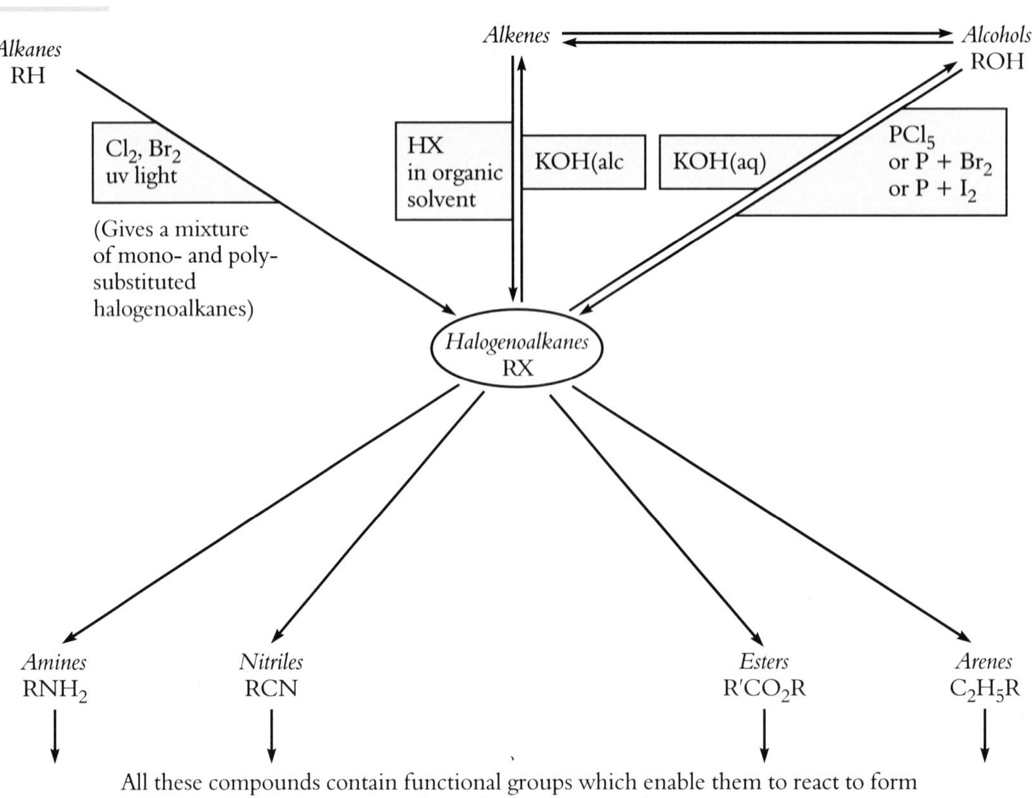

CHECKPOINT 29.8: REACTIVITY

1. Give the reagents and conditions needed to make the following compounds from 1-bromopropane.

   (a) propan-1-ol *(b)* propene
   (c) propylamine, $CH_3CH_2CH_2NH_2$,
   (d) butanenitrile, $CH_3CH_2CH_2CN$

2. Explain why halogenoalkanes undergo substitution reactions *(a)* more readily than alkanes *(b)* fairly slowly.

3. *(a)* What is a nucleophile? Give two examples.
   (b) Why do nucleophiles attack halogenoalkanes?
   (c) What two types of reaction can follow the attack of a nucleophile on a halogenoalkane? Name two products which can be formed from 1-bromopropane by these reactions.

4. This question is about the hydrolysis of halogenoalkanes by aqueous alkali. When OH^- attacks the primary halogenoalkane RCH_2X, a reactive intermediate is formed. Once formed, this intermediate is rapidly converted into the products.

   (a) Sketch the intermediate.
   (b) Explain why it reacts rapidly.

   In the hydrolysis of the tertiary halogenoalkane R_3CX, a different reactive intermediate is formed.

   (c) Sketch the intermediate.
   (d) Explain why it reacts rapidly.

FURTHER STUDY ▼

29.9 THE MECHANISMS OF HYDROLYSIS AND ELIMINATION REACTIONS OF HALOGENOALKANES

Hydrolyses of halogenoalkanes are S_N reactions

The hydrolysis of a halogenoalkane is a **substitution** reaction. The attacking species is a **nucleophile**. Hydrolyses are therefore described as **S_N reactions**. We shall now extend the study of the mechanisms of S_N reactions which we began in § 29.8.3.

29.9.1 HYDROLYSIS OF PRIMARY HALOGENOALKANES

Experimental evidence

The alkaline hydrolysis of a primary halogenoalkane

$$RX(l) + OH^-(aq) \longrightarrow ROH(aq) + X^-(aq)$$

is second-order, following the rate expression

$$Rate = k[RX][OH^-]$$

where [RX] = concentration of RX and k is a constant.

Experimental evidence on the hydrolysis of primary halogenoalkanes

One can deduce from this evidence that a molecule of RX and an OH^- ion must collide before reaction will occur. Calculation shows that the rate of reaction is much less than the rate at which RX and OH^- collide. Only a small fraction of the collisions result in reaction. It must be necessary for the two species not only to collide but also to collide with enough energy to overcome the repulsion between the hydroxide ion and the halogenoalkane molecule.

Mechanism

The mechanism ...

It is known that, in order to minimise repulsion between RX and OH^-, OH^- approaches the carbon atom attached to X on the opposite side of the molecule

∴ The making of a new bond eases the breaking of the old bond …

… A transition state is formed

from X:

$$HO^- \text{ approaches } H-\underset{CH_3}{\overset{H}{\underset{|}{\overset{|}{C}}}}-X \text{ to form } \left[H-O\cdots\underset{CH_3}{\overset{H\;\;H}{C}}\cdots X \right]^-$$

As a bond forms between O and C, the C—X bond weakens, and a transition state is reached in which C is partially bonded both to O and X. Once the transition state has been formed, it is rapidly converted into the products. A transition state has a momentary existence [§ 14.17.2]. The reactive intermediates that have been postulated in other reactions are longer-lived.

A summary of the mechanism

The proposed mechanism for the alkaline hydrolysis of primary halogenoalkanes is

$$HO^- + H-\underset{CH_3}{\overset{H}{\underset{|}{\overset{|}{C}}}}-X \xrightarrow[\text{step}]{\text{rate-determining}} \left[H-O\cdots\underset{CH_3}{\overset{H\;\;H}{C}}\cdots X \right]^- \xrightarrow{\text{fast}} H-O-\underset{CH_3}{\overset{H}{\underset{|}{\overset{|}{C}}}}-H + X^-$$

Transition state

The movement of the bonding electrons is often shown by a curved arrow

$$HO^- \quad \underset{CH_3}{\underset{|}{CH_2}}-I \longrightarrow HO-\underset{CH_3}{\underset{|}{CH_2}} + I^-$$

29.9.2 HYDROLYSIS OF TERTIARY HALOGENOALKANES

The hydrolysis of tertiary halogenoalkanes is first-order

An example of the hydrolysis of a tertiary halogenoalkane is the hydrolysis of 2-chloro-2-methylpropane in aqueous ethanol:

$$(CH_3)_3CCl(l) + H_2O(l) \longrightarrow (CH_3)_3COH(l) + HCl(aq)$$

Experimental results show that the reaction is first order with respect to the halogenoalkane:

$$\text{Rate} = k[(CH_3)_3CCl]$$

This shows that the slow, rate-determining step must involve the halogenoalkane alone. It is suggested that this step is the dissociation of the halogenoalkane into a carbocation and a halide ion:

A carbocation intermediate is postulated

$$(CH_3)_3CCl \xrightarrow{\text{slow, rate-determining step}} (CH_3)_3C^+ + Cl^-$$

The carbocation formed reacts rapidly with water molecules:

$$(CH_3)_3C^+ + H_2O \xrightarrow{\text{fast step}} (CH_3)_3COH + H^+(aq)$$

This step is fast, and the rate of the overall reaction is determined by the rate at which the carbocations are formed.

A summary of the mechanism

The proposed mechanism for the hydrolysis of tertiary halogenoalkanes is

$$R_3CX \xrightarrow{\text{rate-determining step}} R_3C^+ + X^-$$

followed by $R_3C^+ + H_2O \xrightarrow{\text{fast}} R_3COH + H^+(aq)$

29.9.3 HYDROLYSIS OF SECONDARY HALOGENOALKANES

Secondary halogenoalkanes

The behaviour of secondary halogenoalkanes is intermediate between primary and tertiary halogenoalkanes, depending on the nature of the halogenoalkane and the solvent.

29.9.4 *DESCRIPTION OF MECHANISM

Hydrolysis ...

... by S_N1 and S_N2 reactions

The hydrolyses of halogenoalkanes are S_N reactions, nucleophilic substitution reactions. In the hydrolysis of primary (and some secondary) halogenoalkanes, two species, RX and OH^-, are involved in the formation of the transition state: the rate-determining step is bimolecular. The reaction is described as an **S_N2 reaction**. In the hydrolysis of tertiary (and some secondary) halogenoalkanes, only one species is required for the formation of the transition state: the rate-determining step is unimolecular. Such reactions are designated as **S_N1 reactions**. [See § 29.8.3.]

29.9.5 *MECHANISMS OF ELIMINATION REACTIONS

Attack by a nucleophile can eliminate HX to form an alkene

Elimination is favoured by the same conditions as substitution: the attack by a nucleophile such as a hydroxide ion, OH^-. The OH^- ion is attracted to a hydrogen atom in the molecule of halogenoalkane. This causes a movement of electron pairs. The OH^- ion removes the hydrogen atom to form H_2O, while the C—X bond weakens and breaks to release X^-.

$$HO^-: + H{-}CH_2{-}CH_2{-}X \longrightarrow HO^-: \cdots H{-}CH_2{-}CH_2{-}X \longrightarrow H_2C{=}CH_2 + X^- + H_2O$$

This mechanism for the reaction of primary halogenoalkanes is termed an **E_N2 reaction** (corresponding to the S_N2 reaction in hydrolysis).

Elimination...

... by E_N1 and E_N2 reactions

Tertiary halogenoalkanes employ a different mechanism. The halogenoalkane ionises to give a halide ion and a carbonium ion, which is unstable and loses a hydrogen ion.

$$H{-}CH_2{-}CR_2{-}X \longrightarrow H{-}CH_2{-}CR_2^+ + X^- \longrightarrow H_2C{=}CR_2 + H^+ + X^-$$

This mechanism is termed an **E_N1 reaction** (corresponding to the S_N1 reaction in hydrolysis).

CHECKPOINT 29.9: REACTION MECHANISMS

1. *(a)* Describe the reaction of aqueous potassium hydroxide with $CH_3CH_2CH_2Br$.
   (b) Sketch the mechanism for this reaction.

2. The rate equation for the alkaline hydrolysis of the halogenoalkane RBr is

   $$\text{Rate} = k[\text{RBr}][\text{OH}^-]$$

   Which of the following statements does *not* fit in with this observation?

   A The reaction is first order with respect to RBr.
   B The reaction is second order overall.
   C The final step in the reaction is the attack of OH^- on R^+.
   D The rate-determining step is bimolecular.
   E OH^- attacks RBr before the R—Br bond is broken.

3. Explain the difference between
   (a) a substitution reaction and an elimination reaction
   (b) a nucleophile and an electrophile
   (c) a first order reaction and a second order reaction
   (d) a unimolecular reaction and a bimolecular reaction.

4. Sketch the transition state which is thought to be formed in the alkaline hydrolysis of 1-bromopropane. What leads a molecule of bromopropane to form this transition state?

   Why do hydroxide ions not attack alkanes?

   What is the difference between an S_N1 reaction and an S_N2 reaction? Explain how the hydrolysis of 2-bromo-2-methylpropane differs from that of 1-bromopropane in the manner in which the rate of the reaction depends on the concentrations of the reactants.

   What reason has been suggested to account for the difference?

29.10 HALOGENOARENES

Halogenoarenes

Compounds in which halogens are substituents in the benzene ring are called halogenoarenes. An example is chlorobenzene:

The halogen atom is difficult to displace

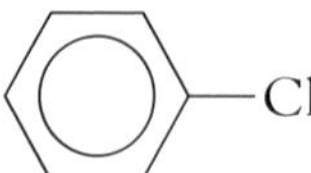

The halogen atom in a halogenoarene is very much less reactive than that in a halogenoalkane. The p orbitals of the halogen atom interact with the p orbitals of the six carbon atoms in the benzene ring to form a delocalised cloud of π electrons [see Figure 29.10A]. This π bond adds to the strength of the σ bond between the halogen atom and the ring, and makes the halogen atom very difficult to displace.

Figure 29.10A Electron distribution in halogenoarenes

The C—X bond interacts with the ring ...

... π bonding adds to the strength of the σ bond ...

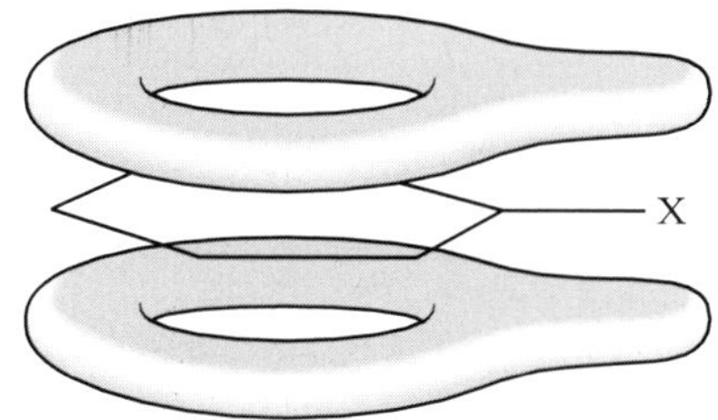

29.10.1 REPLACEMENT OF THE HALOGEN ATOM

Stringent conditions are needed for hydrolysis of halogenoarenes

The conditions needed for the replacement of a halogen atom are exemplified in the manufacture of phenol, C_6H_5OH. One industrial method is to heat chlorobenzene and aqueous sodium hydroxide together at 150 atm and 350 °C. The sodium phenoxide formed is converted into phenol by the action of dilute acid:

$$\underset{\text{Chlorobenzene}}{C_6H_5Cl(l)} + 2NaOH(aq) \xrightarrow[350\,°C]{150\text{ atm}} \underset{\text{Sodium phenoxide}}{C_6H_5ONa(aq)} + NaCl(aq) + H_2O(l)$$

$$\underset{\text{Sodium phenoxide}}{C_6H_5ONa(aq)} + HCl(aq) \longrightarrow \underset{\text{Phenol}}{C_6H_5OH(l)} + NaCl(aq)$$

29.10.2 SUBSTITUTION IN THE BENZENE RING

Halogenoarenes are less reactive than benzene

The benzene ring undergoes the usual reactions. Halogen substituents withdraw electrons, thus deactivating the benzene ring [§ 28.12] and are 2/4-directing [§ 28.12].

Figure 29.11A
Grignard reagents

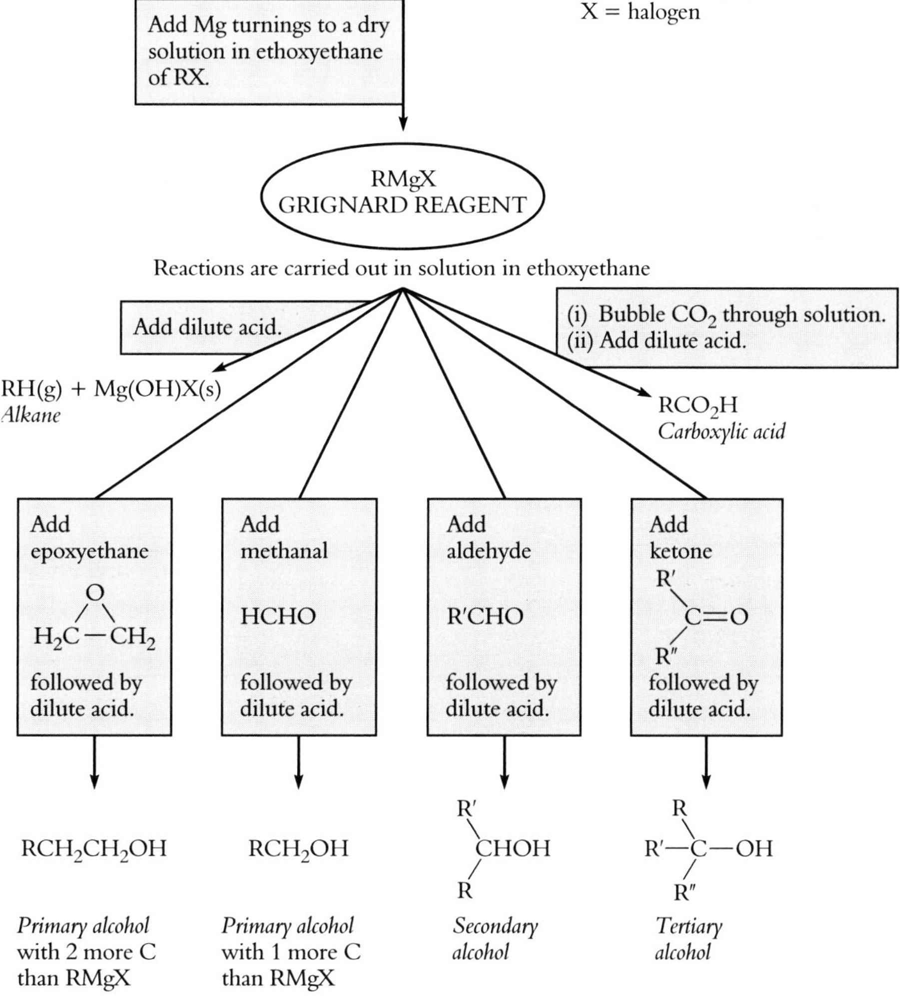

HALOGENOALKANES AND HALOGENOARENES

Halogenoalkanes
Reactions include:
OH^-(aq) → alcohols, ROH § 29.8.1
OH^-(ethanolic) → alkenes §§ 29.7, 29.8.2
CN^- → nitriles, RCN §§ 29.7, 29.8.1
NH_3 → amines, RNH_2 §§ 29.7, 29.8.1
Mg → Grignard reagent RMgX § 29.11

Halogenoarenes, e.g. chlorobenzene are much less reactive than halogenoalkanes. § 29.10

In these reactions, the polar $C^{\delta+}$–Halogen$^{\delta-}$ bond is attacked by nucleophiles, e.g. OH^-, CN^-, NH_3, The result may be
a **substitution reaction** § 29.8.1 or
an **elimination raction** § 29.8.2

Uses of halogenoalkanes include:
anaesthetics, e.g. Fluothane® $CF_3CHBrCl$, local anaesthetics, e.g. chloroethane. § 29.1
solvents, § 29.5
fire-extinguishers, § 29.5
coolants in refrigerators and air conditioners. § 29.5

Chlorofluoroalkanes, **CFCs**, attack the **ozone layer**. Their use is now restricted. § 29.6

The **mechanism** of substitution reactions is discussed in § 29.8.3.

***S_N1 and S_N2 reactions** are taken further in § 29.9.

* **Grignard reagents**, e.g. C_2H_5MgBr, are intermediates in the synthesis of other compounds, e.g. alcohols. § 29.11

*29.11 GRIGNARD REAGENTS

Grignard reagents, R Mg X

When a dry ethereal solution of a halogenoalkane is added to magnesium turnings, an exothermic reaction occurs. The cloudy liquid formed is a solution of a Grignard reagent. These compounds have the formula RMgX, where R is an alkyl or an aryl group and X is a halogen. They undergo many reactions which enable them to be used as the starting point in the synthesis of a variety of compounds [see Figure 29.11A].

*CHECKPOINT 29.11: GRIGNARD REAGENTS

1. Chloroethane is hydrolysed to ethanol by aqueous alkali at 60 °C whereas to convert chlorobenzene into phenol the conditions needed are 360 °C and 150 atm. Explain the reason for the difference.

*__2.__ Give the names and formulae of all the different alcohols that can be made from the Grignard reagent C_2H_5MgBr, provided that the following reagents are also available: epoxyethane, methanal, ethanal, propanone.

*__3.__ Outline two methods of making butanoic acid from propyl magnesium bromide, $CH_3CH_2CH_2MgBr$.

*__4.__ Outline methods of making *(a)* propanoic acid and *(b)* butanoic acid from ethyl magnesium bromide, C_2H_5MgBr.

FURTHER STUDY ▲

QUESTIONS ON CHAPTER 29

1. State the conditions needed for bromine to react with

(a) ethane to form bromoethane
(b) benzene to form bromobenzene
(c) benzene to form hexabromocyclohexane
(d) methylbenzene to form (bromomethyl)benzene
(e) methylbenzene to form 2-bromomethylbenzene

Compare the reactions of (i) silver nitrate solution and (ii) aqueous alkali with the products in *(a)*, *(b)* and *(d)*.

2. Describe a laboratory preparation for bromoethane. Write an equation for the reaction. State what impurities may be present, and how you would purify a sample of bromoethane made in this way.

3. *(a)* Describe how you could make 2-bromobutane from a named alcohol.
(b) Give the structures and names of the isomers of 2-bromobutane. Compare the rates at which they react with aqueous sodium hydroxide.
(c) State the conditions under which 2-bromobutane reacts with the following reagents. Name the products, and write equations for the reactions.

(i) $OH^-(aq)$ (iii) KCN
(ii) KOH(ethanol) (iv) NH_3

4. *(a)* Name four series of compounds which can be made from halogenoalkanes.
(b) Draw a reaction scheme to illustrate the reactions.
*(c) How is the use of halogenoalkanes in synthesis extended by Grignard reagents?

5. *(a)* (i) State, giving reagents and conditions, how benzene is converted into bromobenzene. **1**
(ii) Write an equation for the reaction between ethene and bromine, giving full structural (graphic) formulae for the organic substances. State the type of reaction taking place. **3**
(iii) Explain why ethene and benzene may both react with bromine but by different types of reaction. **2**

(b) CFCs (chlorofluorocarbons) such as 1,2-dichloro-1,1,2,2-tetrafluoroethane, which are used for example in aerosols, destroy ozone in the stratosphere. It is believed that CFCs produce chlorine radicals Cl• which then change ozone into oxygen.

Hydrofluoroalkanes (HFAs) are being developed to replace CFCs. An example of an HFA is 1,1,1,2-tetrafluoroethane.
(i) Give the full structural (graphic) formula of 1,2-dichloro-1,1,2,2-tetrafluoroethane, showing all the covalent bonds. **1**
(ii) Suggest why the HFA 1,1,1,2-tetrafluoroethane is believed not to destroy ozone. **1**
(iii) The HFA in *(b)*(ii), 1,1,1,2-tetrafluoroethane, can be made by a process involving two reactions which are carried out under different conditions. Write a balanced equation to represent each reaction.

I. The reaction of trichloroethene with HF giving 2-chloro-1,1,1-trifluoroethane and HCl.

II. The reaction of 2-chloro-1,1,1-trifluoroethane with HF to form the required HFA and HCl. **2**

10 *WJEC (AS/AL)*

6. *(a)* This part of the question is about the hydrolysis of halogenoalkanes.

$2\ cm^3$ of ethanol is added to each of three test-tubes.

Three drops of 1-chlorobutane are added to the first, three drops of 1-bromobutane to the second, and three drops of 1-iodobutane to the third test-tube.

$2\ cm^3$ portions of hot aqueous silver nitrate solution are added to each test-tube. A precipitate forms immediately in the third test-tube, slowly in the second test-tube and extremely slowly in the first test-tube. In each reaction the precipitate is formed by silver ions, $Ag^+(aq)$, reacting with the halide ions formed by hydrolysis of the halogenoalkane.

(i) Why was ethanol added to each test-tube? **1**

(ii) The same organic product forms in each reaction. Name this organic product. **1**

(iii) Complete the equation for the hydrolysis of 1-bromobutane.

$$C_4H_9Br + H_2O \longrightarrow$$ **2**

(iv) What is the colour of the precipitate in the **third** test-tube? **1**

(v) Name the precipitate which forms extremely slowly in the **first** test-tube and write the **ionic** equation, including state symbols, for its formation. **3**

(vi) Ammonia solution is added to the precipitate formed in the **first** test-tube. Describe and explain what you would observe. **2**

(vii) Explain why the rates of hydrolysis of the three halogenoalkanes are different. **2**

(b) 1-bromobutane reacts with an alcoholic solution of potassium hydroxide at high temperature to form but-1-ene.

(i) Draw a fully labelled diagram to show an apparatus for carrying out this reaction in the laboratory and collecting the gaseous but-1-ene. **3**

(ii) Suggest a chemical test for an alkene such as but-1-ene. State the colour change you would observe. **2**

17 *L(N) (AL)*

7. *(a)* Bromobenzene can be prepared from benzene by reaction with bromine and a suitable catalyst.

(i) Write the balanced equation for the reaction. **2**

(ii) Name a suitable catalyst for this reaction. **1**

(b) This part of the question is about three ways of preparing bromoethane, C_2H_5Br. Each method has a different mechanism or reaction type.

(i) State the reagent and the condition needed to prepare bromoethane from **ethane**. State the mechanism and type of this reaction. **4**

(ii) Bromoethane can also be prepared by an **electrophilic addition** reaction. Write the balanced equation for this reaction. **2**

(iii) In a third route bromoethane can also be prepared by adding potassium bromide to concentrated sulphuric acid and ethanol.

The potassium bromide reacts with the concentrated sulphuric acid to produce hydrogen bromide, which then reacts with the ethanol. Name the mechanism and type of the reaction between hydrogen bromide and ethanol. **2**

(c) When samples of pure hydrogen bromide gas are needed, concentrated sulphuric acid is not an appropriate acid to react with potassium bromide.

(i) Which acid should be used? **1**

(ii) Why is concentrated sulphuric acid **not** suitable for preparing a sample of pure hydrogen bromide? **1**

(iii) Suggest a reason why concentrated sulphuric acid is suitable for making hydrogen bromide in the preparation of bromoethane from ethanol in *(b)*(iii). **1**

(iv) Describe and explain what you would observe when a test tube containing hydrogen bromide gas is inverted in water. **2**

16 *L(N) (AL)*

8. *(a)* There are four **structural** isomers of molecular formula C_4H_9Br. The formulae of two of these isomers are given.

```
    H  H  H  H            H  CH3 H
    |  |  |  |            |  |   |
 H—C—C—C—C—H         H—C—C——C—H
    |  |  |  |            |  |   |
    H  H  Br H            H  Br  H
```

Isomer 1 **Isomer 2**

(i) Draw the remaining **two** structural isomers. **2**

(ii) Give the name of **Isomer 2**. **1**

(b) All four structural isomers of C_4H_9Br undergo similar reactions with ammonia.

(i) Give the name of the mechanism involved in these reactions. **1**

(ii) Draw the structural formula of the product formed by the reaction of **Isomer 1** with ammonia. **1**

(iii) Select the isomer of molecular formula C_4H_9Br that would be the most reactive with ammonia. State the structural feature of your chosen isomer that makes it the most reactive of the four isomers. **2**

(c) The elimination of HBr from **Isomer 1** produces two structural isomers, compounds **A** and **B**.

(i) Give the reagent and conditions required for this elimination reaction. **3**

(ii) Give the structural formulae of the two isomers, **A** and **B**, formed by elimination of HBr from **Isomer 1**. **2**

(d) Ethene, C_2H_4, reacts with bromine to give 1,2-dibromoethane.

(i) Give the name of the mechanism involved. **1**

(ii) Show the mechanism for this reaction. **3**

16 *AEB (AL)* 1998

9. *(a)* (i) Give the reaction mechanism for the reaction of 1-bromobutane with aqueous sodium hydroxide. **2**

(ii) Name this type of mechanism. **1**

(iii) State the evidence which can be used to support the mechanism you have given. **2**

(b) A few drops of 1-chlorobutane are added to a test-tube containing aqueous sodium hydroxide and the tube is shaken and the mixture gently warmed. The mixture is then acidified with dilute nitric acid before adding aqueous silver nitrate. State what will be observed.

State the difference in behaviour if chlorobenzene replaced 1-chlorobutane and explain the difference in terms of the structures of the compounds. **3**

(c) 1-Bromopropane can be converted into butanoic acid in a two stage process. Name suitable reagents for **each** stage of the conversion and calculate the percentage yield if 24.60 g of the halogen compound produces 8.80 g of the acid. **3**

(d) (i) Under **each** of the following categories, name a compound of your own choice which is used, or has been used, on a large scale as

I. a pesticide;

II. an aerosol propellant;

III. a chlorinated solvent.

(ii) For **one** of the compounds chosen in *(d)*(i), describe an adverse effect of its long term use **or** ready availability. **4**

15 *WJEC (AL)*

10. *(a)* The free radical reaction between methane and chlorine in ultraviolet light has a mechanism whose initiation step is

$$Cl_2 \longrightarrow 2Cl\cdot$$

This is followed by one of two possible propagation steps

$$CH_4 + Cl\cdot \longrightarrow \cdot CH_3 + HCl \qquad I$$

$$CH_4 + Cl\cdot \longrightarrow CH_3Cl + H\cdot \qquad II$$

(i) Use the data given below to predict which is the more likely propagation step.
bond enthalpies/kJ mol^{-1}: Cl—Cl 243; C—H 435; H—Cl 432; C—Cl 346. **4**

(ii) Identify one product you might expect from a termination step following each reaction that would not be formed in the alternative reaction scheme. **2**

(b) Iodomethane reacts with cyanide ions from sodium cyanide solution as follows:

$$CH_3I + CN^- \longrightarrow CH_3CN + I^-$$

The rate equation for this reaction is

$$\text{rate} = k[CH_3I][CN^-]$$

Give the mechanism for the reaction and justify it on the basis of the rate equation. **3**

(c) Suggest the reagents and conditions for a series of reactions by means of which iodomethane may be converted into methanoic acid. **3**

12 *(AS)*

11. *(a)* Give the structural formulae, showing all covalent bonds, for all the isomers of C_4H_9Br, and name them. **3**

(b) The compounds in *(a)* all react on heating with aqueous sodium hydroxide by a nucleophilic substitution reaction. For one of these isomers, the reaction at a given temperature was found to be first order with respect to the organic molecule and first order overall.

(i) Explain what is meant by the term *nucleophile*.

(ii) Identify the nucleophile in this reaction.

(iii) Write the rate expression for this reaction.

15 *L (AS/AL)*

30

ALCOHOLS AND PHENOLS

30.1 ALCOHOLIC DRINKS

Ethanol, C_2H_5OH, commonly referred to as 'alcohol', is the most commonly used drug used legally by adults and illegally by young people. It is a depressant of the central nervous system. Addiction to ethanol is the greatest medical problem resulting from the use of drugs: there are 30 000 alcoholics in the UK alone. Alcohol addiction develops slowly and is a disease of middle age. Addiction to heroin and other drugs develops quickly and is a disease of young people.

Ethanol does not have to be digested: it can be absorbed into the blood stream through the walls of the stomach and intestine. Carbon dioxide relaxes the pyloric valve, between the stomach and the small intestine, and so sparkling wines have a faster effect than still wines.

For convenience, the amount of ethanol in drinks is measured in 'units'. The number of units in some common drinks is shown in Figure 30.1B. The effects of a given amount of ethanol depend on the rate of drinking, the size of the drinker and other factors. Some people are visibly affected after 2 units of alcohol; others can drink 5 units over a period of 2–3 hours before they appear drunk. In an average man, at blood alcohol levels of up to 0.05% by volume, ethanol produces a sense of well being and some impairment of coordination. A level of 0.08% is the legal limit for driving. A man of average weight can drink 5 units of alcohol before he reaches this level. A lighter man could drink less, and a woman of the same weight could drink less before reaching this level. Above 0.08% accidents can result from faulty judgement and coordination and increased aggressiveness. At 0.15% drowsiness and vomiting occur, and at

Figure 30.1A
Social drinking

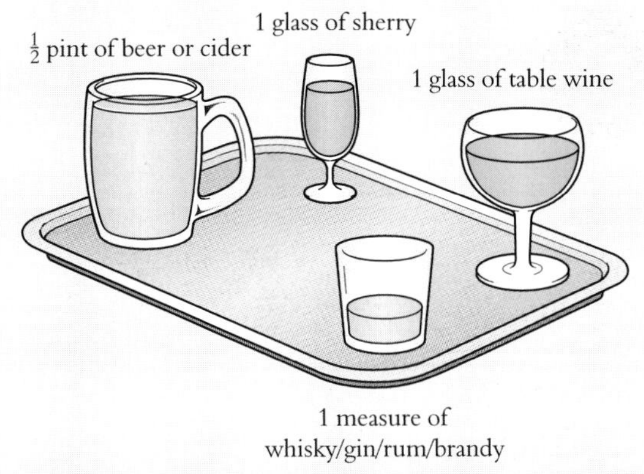

Figure 30.1B
One 'unit' of alcohol

0.30% breathing difficulties may cause death. People usually become unconscious before they can take a lethal dose.

Alcoholism is physical and psychological dependence on alcohol. Withdrawal symptoms can be very severe. The heavy drinker tends to behave in antisocial ways and deteriorates socially and personally. Irreversible damage is caused to the heart, brain, liver and kidneys. Even moderate drinking during pregnancy may result in the birth of a deformed baby.

One method of treatment for alcoholism is group and individual psychotherapy for hospital inpatients and outpatients. Alcoholics Anonymous is a self-help organisation founded in 1935 which has probably had more success than any other treatment. Another treatment is aversion therapy. Ethanol is oxidised in the liver to ethanal, which is then oxidised further. A drug called Antabuse® inhibits the oxidation of ethanal, a build-up of which causes great discomfort, with sweating and nausea, and discourages the individual from drinking more alcohol.

30.2 ALCOHOLS

30.2.1 NOMENCLATURE

Aliphatic alcohols have the general formula $C_nH_{2n+1}OH$

Aliphatic alcohols are a homologous series. They all possess the same functional group, a hydroxyl group attached to a saturated carbon atom

$$-\overset{\diagdown}{\underset{\diagup}{C}}-OH$$

They all have the formula ROH, where R is an alkyl group C_nH_{2n+1}—

They can be regarded as alkanes in which one H atom is replaced by a —OH group, and are often called the **alkanols**. The names are arrived at by adapting the name of the parent alkane by changing the terminal *-ane* to *-anol*. The first members of the series are:

Names

CH_3OH	Methanol	C_4H_9OH	Butanol
C_2H_5OH	Ethanol	$C_5H_{11}OH$	Pentanol
C_3H_7OH	Propanol	$C_6H_{13}OH$	Hexanol

Figure 30.2A
Ethanol

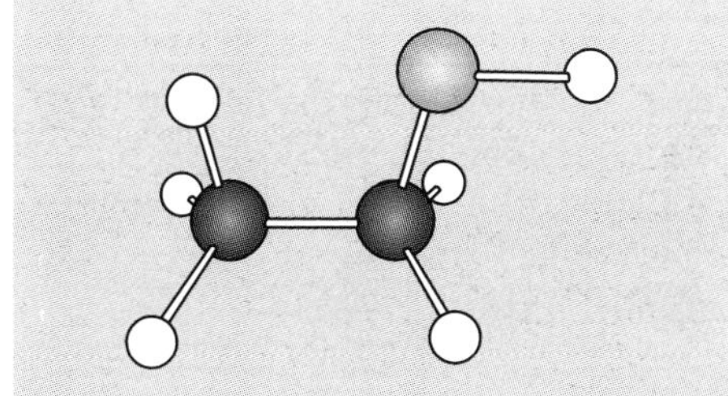

'Alcohol' Ethanol is by far the most important member of the series [see Figure 30.2A], and is often referred to simply as 'alcohol'.

30.2.2 ISOMERISM

Primary, secondary and tertiary alcohols Isomerism occurs. Primary (1°), secondary (2°) and tertiary (3°) alcohols exist. Their formulae are:

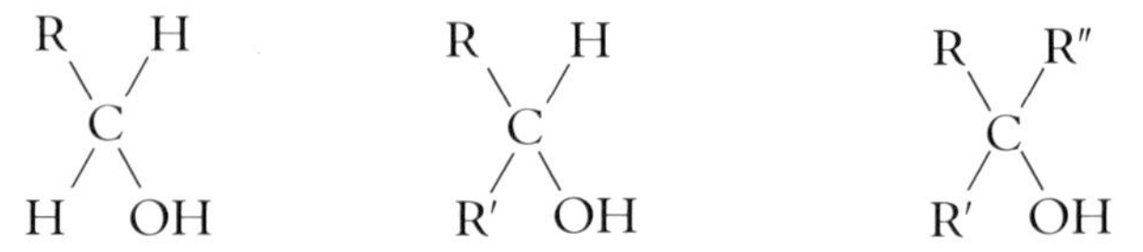

```
R   H          R   H          R   R″
 \ /            \ /            \ /
  C              C              C
 / \            / \            / \
H   OH         R′  OH         R′  OH
```

Primary (1°) Secondary (2°) Tertiary (3°) alcohols

The three isomeric butanols, of formula C_4H_9OH, are:

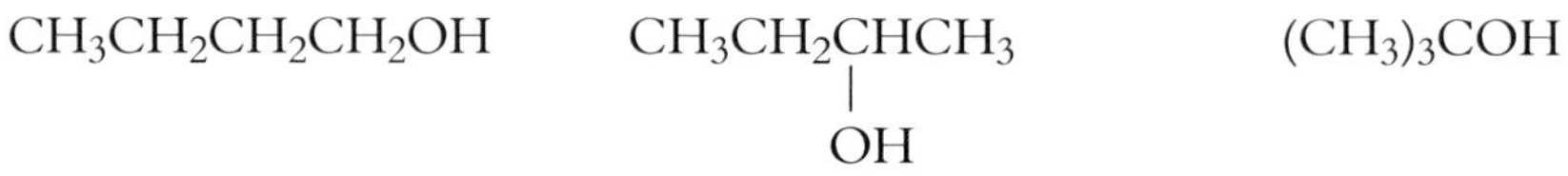

$CH_3CH_2CH_2CH_2OH$ — Butan-1-ol (1°)

$CH_3CH_2CH(OH)CH_3$ — Butan-2-ol (2°)

$(CH_3)_3COH$ — 2-Methylpropan-2-ol (3°)

30.2.3 POLYHYDRIC ALCOHOLS

Diols and triols Polyhydric alcohols contain more than one —OH group. Diols contain two, and triols contain three hydroxyl groups:

```
CH2—CH2            CH2—CH—CH2
|    |             |   |   |
OH   OH            OH  OH  OH
```

Ethane-1,2-diol (ethylene glycol, used as antifreeze) Propane-1,2,3-triol (glycerol)

30.2.4 ARYL ALCOHOLS

Aromatic alcohols or aryl alcohols contain a benzene ring. The simplest are:

$C_6H_5CH_2OH$ — Phenylmethanol (benzyl alcohol)

$C_6H_5CH_2CH_2OH$ — 2-Phenylethanol

Aryl alcohols differ from phenols The hydroxyl group is separated from the benzene ring by a saturated carbon atom and behaves as it does in aliphatic alcohols. If the hydroxyl group is in the ring, as in C_6H_5OH, the compound is not an alcohol: it is a **phenol**, and its reactions are quite different.

30.3 PHYSICAL PROPERTIES

30.3.1 VOLATILITY

Hydrogen bonding raises the boiling temperatures of alcohols ...

Aliphatic alcohols with less than 12 carbon atoms and the lower aryl alcohols are liquids at room temperature. The boiling temperatures are higher than those of alkanes of comparable molecular mass (e.g., values of T_b/°C are: $C_2H_5OH = 78°$ and $C_3H_8 = -42°$; $C_7H_{15}OH = 180°$ and $C_8H_{18} = 126°$). The reason is that the highly polar nature of the $-\overset{\delta-}{O}-\overset{\delta+}{H}$ bond leads to hydrogen bonding between molecules of alcohols [see Figure 30.3A].

Figure 30.3A Hydrogen bonding between alcohol molecules

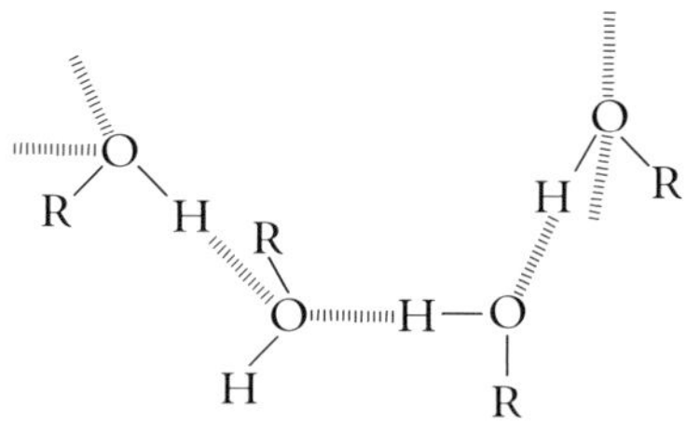

When the liquid is vaporised, energy must be supplied to break the hydrogen bonds and to convert the association of molecules into monomers in the vapour phase.

... as do an increasing molecular mass and reduced branching

The boiling temperatures increase with increasing molecular mass, and branched-chain isomers have lower boiling temperatures than unbranched-chain isomers.

30.3.2 SOLUBILITY

Lower alcohols are soluble in water. Molecules of alcohol can displace molecules of water in the hydrogen-bonded association of water molecules.

Figure 30.3B Hydrogen bonding in aqueous ethanol

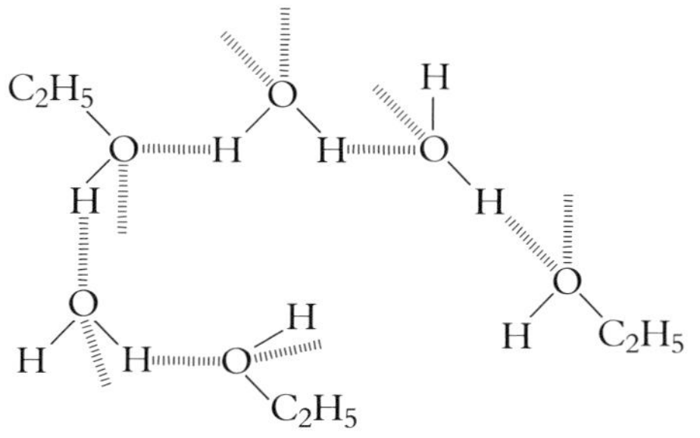

Alcohols of high molecular mass form fewer hydrogen bonds than the numerous water molecules they displace. Energy considerations therefore make the higher alcohols less soluble than the lower members.

An azeotrope is formed by ethanol and water

Ethanol forms a constant-boiling azeotropic mixture with water [§ 8.5.4]. It contains 95.6% ethanol and boils at 78.1 °C. Pure ethanol can be obtained by distilling the azeotrope from a drying agent such as calcium oxide. Other alcohols also form azeotropes with water.

Alcohols dissolve polar solutes and non-polar solutes

Alcohols are good solvents. The polar —OH group enables them to dissolve sodium hydroxide and potassium hydroxide. The non-polar hydrocarbon part of the molecule enables alcohols to dissolve substances like hexane.

30.4 INDUSTRIAL SOURCES OF ALCOHOLS

30.4.1 HYDRATION OF ALKENES

Catalytic hydration of alkenes gives alcohols

Ethanol is manufactured by the hydration of ethene [§ 27.8.7]. Hydration of alkenes in the presence of a catalyst is used for the preparation of other alcohols too.

30.4.2 NATURAL GAS: A SOURCE OF METHANOL

Methanol is obtained from synthesis gas

Methanol can be manufactured from natural gas. First, methane from natural gas is passed with steam over a catalyst to convert it into a mixture of carbon monoxide and hydrogen known as *synthesis gas*. Then synthesis gas is passed over another catalyst at high pressure and moderate temperature to give methanol.

30.4.3 FERMENTATION: A METHOD USED FOR ETHANOL

Ethanol is made by the fermentation of sugars

The fermentation method is used to make alcoholic drinks. Fruit juices such as grape juice contain the sugar glucose, $C_6H_{12}O_6$. When yeast is added, the sugar 'ferments' to form wine (a solution of ethanol) and carbon dioxide:

$$\underset{\text{Glucose}}{C_6H_{12}O_6(aq)} \xrightarrow{\text{yeast}} \underset{\text{Ethanol}}{2C_2H_5OH(aq)} + 2CO_2(g)$$

Wines contain about 12% ethanol

Yeast is a living plant, containing the enzyme zymase, which catalyses the reaction. Juices of other fruits (plums, apples, pears, etc.) can be fermented to give wine. Wines contain about 12% ethanol. When the ethanol content reaches this level it kills the yeast, and fermentation stops.

Fractional distillation of fermented liquors gives 'spirits'

The starch present in potatoes and grain is also used as a source of ethanol. It is first hydrolysed enzymically to give glucose. Ethanol produced from fermentation has a concentration of 7–14% of ethanol by volume. Beer and cider are 7–8% ethanol, and wines are about 12% ethanol. Some people like drinks with a higher ethanol content. Whisky, gin and brandy have 40% ethanol. These 'spirits' are made by fractional distillation of fermented liquors [see Figure 8.5D, § 8.5.2 for fractional distillation].

SUMMARY

Sources of alcohols:
Industrial:

- ethanol from ethene by catalytic hydration
- ethanol from carbohydrates by fermentation
- methanol from natural gas.

Laboratory:

- from halogenoalkanes by hydrolysis
- from aldehydes and ketones by reduction
- from carboxylic acids, acid chlorides and esters by reduction.

Figure 30.4A Methods of preparing ethanol, a typical alcohol

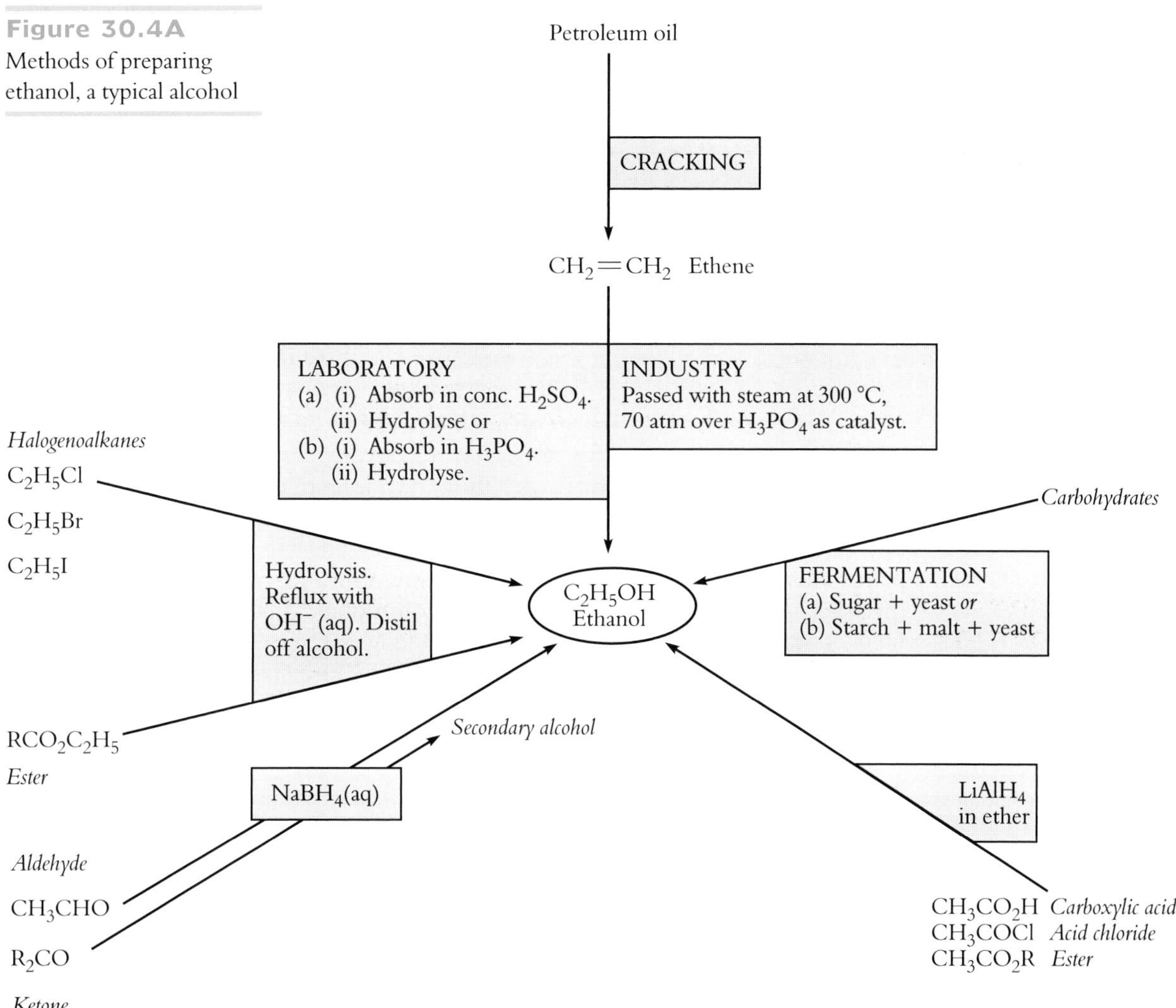

CHECKPOINT 30.4: PROPERTIES

1. *(a)* What is the functional group of the alcohol series? Give the name and formula.
   (b) Draw the structural formulae of ethanol and propan-2-ol. Circle the functional group in each formula.

2. There is hydrogen bonding between molecules of ethanol. Explain why: *(a)* hydrogen bonding is possible, *(b)* ethanol dissolves in water, *(c)* the boiling temperature of ethanol is higher than that of ethane.

3. What are the two main methods used for the large-scale manufacture of ethanol? Compare the two processes with respect to *(a)* whether they can be made to run continuously or are operated as batch processes *(b)* whether they use up non-renewable resources *(c)* the energy consumed *(d)* the time required.

30.5 USES OF ETHANOL AND METHANOL

30.5.1 ETHANOL IN ALCOHOLIC DRINKS

Ethanol is the only alcohol which people drink, and it is referred to as 'alcohol'. It is a toxic substance if taken in large quantities. Methanol is much more toxic than ethanol: drinking methanol leads to blindness and then to death. Methanol

'Meths' is a mixture of methanol and ethanol

is added to ethanol to give 'industrial methylated spirit', which is unfit to drink and therefore carries tax at a lower rate than ethanol. For domestic sale, industrial methylated spirit has an unpleasant-tasting substance added to it and also a purple dye as a warning. This liquid is referred to as 'meths'.

30.5.2 USES AS SOLVENTS

Ethanol is widely used as a solvent. It is used in the manufacture of esters, e.g. ethyl ethanoate, of cosmetics and as a fuel. Methanol is used as a solvent, in the manufacture of methanal and as a fuel.

30.5.3 USES AS MOTOR FUELS

Countries which have to import oil are interested in finding alternatives. The major use of petroleum oil fractions is in vehicle engines. The use of alternative fuels would extend the life of Earth's finite resources of oil and also allow crude oil to be used in the petrochemicals industry. Petrol engines are designed to operate over a temperature range at which petrol will vaporise. Any fuel added to petrol must vaporise at the engine temperature, must dissolve in petrol and must form harmless combustion products. The commonest substances which are blended with petrol are ethanol, methanol and MTBE (see later). The substances are referred to as **oxygenates** in the oil industry. They increase the octane number of petrol and cause less pollution, especially in the emission of carbon monoxide.

Fuels used in the internal combustion engine must vaporise at the engine temperature, must dissolve in petrol, and must form harmless combustion products

Ethanol dissolves in petrol and it boils at the same temperature as heptane, 79 °C. Ethanol burns well in petrol engines, producing about 70% as much energy per litre as petrol. A petrol engine will take 10% ethanol in the fuel without any adjustments to the carburettor (which controls the ratio of air to fuel in the cylinders). Ethanol burns to form carbon dioxide and water and a little ethanal; there is little atmospheric pollution. A mixture of ethanol and petrol is described as **gasohol**.

Figure 30.5A Petrol pumps in Brazil selling 'Alcool' gasohol

Ethanol burns well in petrol engines as mixtures containing up to 10% ethanol

Brazil makes ethanol from sugar cane, ...

... and clears forests to obtain more agricultural land

It is an attractive proposition for a country with no oil but plenty of arable land to grow crops for fermentation. Brazil has plenty of land on which to grow sugar cane and has the sunshine needed to ripen the crop. Most of the petrol sold in Brazil contains 10% ethanol, and this reduces the cost of oil imports. Brazil hopes to increase the content of ethanol in petrol in the future. To obtain more agricultural land, Brazil is clearing large areas of rain forest. You will be aware of the environmental cost: forest species lose their habitats, trees which absorb carbon dioxide and maintain a steady level of carbon dioxide in the carbon cycle are lost, and the burning of trees emits carbon dioxide, which contributes to the greenhouse effect.

Methanol and MTBE

Methanol also is used as a fuel. It burns cleanly in a vehicle engine to form the harmless combustion products carbon dioxide and water, with little carbon monoxide emission. Methanol produces 60% of the energy of an equal volume of petrol so tanks have to be larger. The octane number of methanol is high (114), and the carburettor requires only a small adjustment. Racing drivers use methanol as fuel because, being less volatile than petrol, it is less likely to explode in a collision.

Methanol is used as a mixture with petrol in some countries. Methanol and hydrocarbons do not mix well unless a co-solvent is added, such as $(CH_3)_3COH$, 2-methylpropan-2-ol (called TBA after its old name of tertiary butyl alcohol). The mixture is hygroscopic (absorbing water from the air). The water content makes the fuel corrosive so fuel tanks must be made out of stainless steel. The fuel conducts electricity so electrical components of engines need to be well insulated. Methanol is a toxic substance. There is concern that petrol pump attendants and mechanics who are exposed to the vapour for many hours could suffer damage. The exhaust fumes of cars driven on methanol contain methanal, which is lachrymatory and carcinogenic.

Another contribution of methanol to alternative fuels is as a source of 2-methoxy-2-methylpropane, $(CH_3)_3COCH_3$, known as MTBE after its old name of methyl tertiary butyl ether. MTBE is the fastest growing chemical in the world. It is manufactured from methanol and 2-methylpropene:

$$CH_3OH(g) + (CH_3)_2C{=}CH_2(g) \longrightarrow (CH_3)_3COCH_3(g)$$

MTBE is another oxygenate. With octane number 118, MTBE is blended with petrol up to 15% by volume to increase the octane number and reduce the emission of carbon monoxide. More than 15% by volume of MTBE (b.p. 55 °C) would make the mixture too volatile. In practice, the addition is limited to 7% in unleaded super plus. MTBE costs more than gasoline. The cost of using a high percentage of MTBE has to be weighed against the cost of increasing the octane number of gasoline by isomerising, cracking and reforming the alkanes in it.

CHECKPOINT 30.5: USES

1. 'Oxygenates' are used in vehicle engines as alternative fuels to hydrocarbons. Say how oxygenates differ from fuels derived from petroleum oil *(a)* in structure *(b)* in the way in which they burn *(c)* in conserving Earth's resources.
2. *(a)* List three characteristics which a fuel that is to be used in petrol engines must possess.
   (b) Review the advantages of using ethanol as a fuel for cars. In which type of country is there special interest in this fuel?

30.6 REACTIVITY OF ALCOHOLS

Alcohols are weak acids with very low values of K_a

In alcohols, the —O—H bond is polarised as

$$-\overset{\delta-}{O}-\overset{\delta+}{H}$$

There is a tendency for H^+ to dissociate in the presence of a base. An example of ethanol acting as an acid is its slight dissociation in water to form ethoxide ions and hydrogen ions:

$$C_2H_5OH + H_2O \rightleftharpoons C_2H_5O^- + H_3O^+$$

It is an even weaker acid than water, with $K_a = 10^{-16}$ mol dm^{-3} at 25 °C [§ 12.6.4].

RO—H bond fission is involved in some reactions

Reactions of ethanol which involve fission of the RO—H bond are the reactions with sodium and with carboxylic acids [§ 30.7.1].

R—OH bond fission is involved in other reactions ...

Other reactions involve fission of the R—OH bond, e.g. in the reaction of hydrogen iodide with ethanol:

$$\underset{\text{Ethanol}}{C_2H_5OH} + HI \longrightarrow \underset{\text{Iodoethane}}{C_2H_5I} + H_2O$$

30.7 REACTIONS OF ALCOHOLS

30.7.1 FISSION OF RO—H

Reaction with sodium

Alcohols react with sodium

Alcohols react with sodium to give hydrogen and a sodium alkoxide. The reaction is much slower than the reaction of water and sodium:

$$\underset{\text{Ethanol}}{2C_2H_5OH(l)} + 2Na(s) \longrightarrow \underset{\text{Sodium ethoxide}}{2C_2H_5O^-Na^+(s)} + H_2(g)$$

Esterification

They react with carboxylic acids to give esters

Alcohols and carboxylic acids react to give esters. The functional groups of acids and esters are (where R is an alkyl group)

$$-C\begin{matrix}{\diagup\!\!\diagup O}\\{\diagdown O-H}\end{matrix} \qquad\qquad -C\begin{matrix}{\diagup\!\!\diagup O}\\{\diagdown O-R}\end{matrix}$$

Carboxylic acid   Carboxylic ester

Esterification takes place much faster in the presence of a catalyst, such as hydrogen chloride or concentrated sulphuric acid. Acid anhydrides and acid chlorides also react with alcohols to give esters:

$$\underset{\text{Ethanoic acid}}{CH_3CO_2H(l)} + \underset{\text{Ethanol}}{C_2H_5OH(l)} \xrightleftharpoons[\text{conc. } H_2SO_4]{HCl(g) \text{ or}} \underset{\text{Ethyl ethanoate}}{CH_3CO_2C_2H_5(l)} + H_2O(l)$$

30.7.2 FISSION OF R—OH BOND

Halogenation

(a) Chlorination

Dry HCl(g) or conc. HCl replaces OH by Cl

Dry hydrogen chloride is bubbled through the anhydrous alcohol in the presence of anhydrous zinc chloride as catalyst. When the solution is saturated, it is refluxed on a water bath. For secondary and tertiary alcohols concentrated hydrochloric acid can be used. For tertiary alcohols no zinc chloride catalyst is needed:

$$C_2H_5OH(l) + HCl(g) \xrightarrow[\text{ZnCl}_2\text{ catalyst}]{\text{reflux}} C_2H_5Cl(l) + H_2O(l)$$

The order of rates of reaction is

tertiary > secondary > primary alcohol

(b) Bromination

HBr replaces —OH by —Br

Sodium bromide and concentrated sulphuric acid are used as a source of hydrogen bromide.

$$\underset{\text{Ethanol}}{C_2H_5OH(l)} + HBr(g) \xrightarrow{\text{distil}} \underset{\text{Bromoethane}}{C_2H_5Br(l)} + H_2O(l)$$

(c) Iodination

Red P + I_2 replace OH by I

Red phosphorus and iodine are used as a source of hydrogen iodide (concentrated sulphuric acid cannot be used because it oxidises HI to I_2):

$$\underset{\text{Ethanol}}{C_2H_5OH(l)} + HI(g) \xrightarrow{\text{distil}} \underset{\text{Iodoethane}}{C_2H_5I(l)} + H_2O(l)$$

(d) Halogenation using phosphorus halides

RBr and RI are formed by reaction with PBr_3 and PI_3

If it is available, phosphorus(III) bromide is used. Alternatively, red phosphorus and bromine are used to generate PBr_3 in the reaction mixture. Phosphorus(III) iodide is generated by the addition of iodine to red phosphorus and the alcohol. The mixture is refluxed on a water bath, and then the halogenoalkane is distilled over:

$$\underset{\text{Ethanol}}{3C_2H_5OH(l)} + PBr_3(l) \xrightarrow[\text{distil}]{\text{reflux}} \underset{\text{Bromoethane}}{3C_2H_5Br(l)} + \underset{\text{Phosphoric(III) acid}}{H_3PO_3(l)}$$

PCl_5 replaces OH by Cl

Phosphorus(V) chloride reacts in the cold with alcohols. Phosphorus(III) chloride cannot be used:

$$\underset{\text{Ethanol}}{C_2H_5OH(l)} + PCl_5(l) \xrightarrow{\text{room temperature}} \underset{\text{Chloroethane}}{C_2H_5Cl(l)} + \underset{\text{Phosphorus trichloride oxide}}{POCl_3(l)} + HCl(g)$$

The formation of hydrogen chloride on reaction with phosphorus(V) chloride is a test for the presence of a hydroxyl group in a compound.

*Mechanisms of reactions of alcohols with hydrogen halides

A test for the —OH group

The alcohol reacts with the hydrogen halide, e.g. hydrogen bromide. The oxygen atom of the alcohol uses one of its lone pairs of electrons to accept a proton from the hydrogen halide:

$$RCH_2OH + HX \rightleftharpoons RCH_2{-}\overset{\overset{\displaystyle H}{|}}{O^+}{-}H + X^-$$

The halide ion then attacks the positive ion to displace a molecule of water.

$$X^- \quad \underset{\underset{\displaystyle R}{/}}{CH_2}{-}O^+\!\begin{smallmatrix}/H\\ \backslash H\end{smallmatrix} \longrightarrow R{-}CH_2{-}X + H_2O$$

This is an S_N2 reaction.

With tertiary alcohols, the second step is dissociation of the R—O⁺—H ion:

$$R_3C{-}OH_2^+ \longrightarrow R_3C^+ + H_2O$$

This is an S_N1 reaction. The carbonium ion immediately reacts with a halide ion:

$$R_3C^+ + X^- \longrightarrow R_3CX$$

Dehydration

Reaction with conc. H_2SO_4 ...

A primary alcohol reacts with cold concentrated sulphuric acid to form an alkyl hydrogensulphate:

$$C_2H_5OH(l) + HO{-}SO_2{-}OH(l) \longrightarrow C_2H_5O{-}SO_2{-}OH(l) + H_2O(l)$$

Ethanol Sulphuric acid Ethyl hydrogensulphate

... results in dehydration to give an ether ... (excess alcohol, < 170 °C)

If an excess of alcohol is used and the reaction mixture is warmed to 140 °C, an ether is formed:

$$C_2H_5OH(l) + C_2H_5HSO_4 \xrightarrow[C_2H_5OH\text{ (in excess)}]{140\,°C} C_2H_5OC_2H_5(g) + H_2SO_4(l)$$

Ethoxyethane (or diethyl ether)

Ethers possess the group

$$>C{-}O{-}C<$$

... or to give an alkene (excess conc. H_2SO_4, > 170 °C)

If an excess of concentrated sulphuric acid is used, and the temperature is raised to 170 °C, water is eliminated, with the formation of an alkene:

$$C_2H_5HSO_4(l) \xrightarrow[\text{conc. } H_2SO_4\text{ (in excess)}]{170\,°C} C_2H_4(g) + H_2SO_4(l)$$

Ethyl hydrogensulphate Ethene

Other dehydrating agents are Al_2O_3 and H_3PO_4

Phosphoric(V) acid and aluminium oxide at 300 °C [see Figure 30.7A] also act as dehydrating agents.

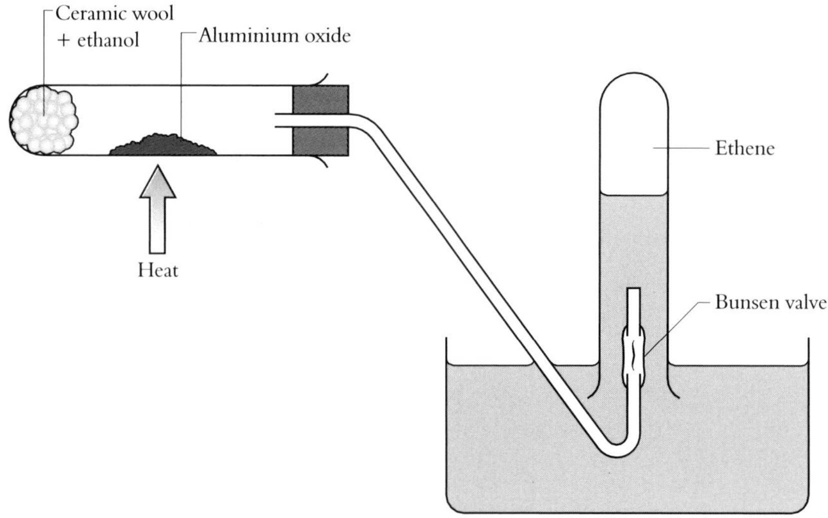

Figure 30.7A Dehydration of ethanol to ethene

Mechanism of dehydration

The mechanism of dehydration involves the addition of a proton from the sulphuric acid or phosphoric acid to a lone pair of electrons on the oxygen atom of the alcohol. The intermediate formed has a positive charge on the oxygen atom, which leads to the movement of electron pairs and the elimination of H_2O and a proton.

$$\mathrm{H_3C{-}CH_2{-}\ddot{O}{-}H} \xrightarrow{H^+} \mathrm{H_3C{-}CH_2{-}O^+H_2} \longrightarrow \mathrm{H_2C{=}CH_2} + H^+ + H_2O$$

Oxidation

Primary alcohols are oxidised to aldehydes and to acids ...

Primary alcohols are oxidised to **aldehydes**, a series of compounds which have the functional group

$$\mathrm{R{-}C(H){=}O}$$

They can be further oxidised to carboxylic acids:

$$\underset{\text{Primary alcohol}}{\mathrm{R{-}CH_2OH}} \xrightarrow{[-2H]} \underset{\text{Aldehyde}}{\mathrm{R{-}C(H){=}O}} \xrightarrow{[+O]} \underset{\text{Carboxylic acid}}{\mathrm{R{-}C(O{-}H){=}O}}$$

... Secondary alcohols are oxidised to ketones ...

Secondary alcohols are oxidised to ketones, a series of compounds possessing the functional group

$$\mathrm{RR'C{=}O}$$

$$\underset{\text{Secondary alcohol}}{\mathrm{R_2C(H)OH}} \xrightarrow{[-2H]} \underset{\text{Ketone}}{\mathrm{RR'C{=}O}}$$

... Tertiary alcohols resist oxidation

Tertiary alcohols are resistant to oxidation. A powerful acidic oxidising agent converts them into a mixture of carboxylic acids.

A number of oxidising agents can be used. Acidified sodium dichromate solution at room temperature will oxidise primary alcohols to aldehydes and secondary alcohols to ketones. At higher temperatures primary alcohols are oxidised further to acids:

$$\mathrm{RCH_2OH} \xrightarrow[\text{room temperature}]{Na_2Cr_2O_7,\ H^+(aq)} \mathrm{RC(H){=}O} \xrightarrow[50\ ^\circ C]{Na_2Cr_2O_7,\ H^+(aq)} \mathrm{RC(OH){=}O}$$

$$\mathrm{R_2CHOH} \xrightarrow{Na_2Cr_2O_7,\ H^+(aq)} \mathrm{R_2C{=}O}$$

Acid + dichromate were used in a breathalyser test

The dichromate solution turns from the orange colour of $Cr_2O_7^{2-}(aq)$ to the blue colour of $Cr^{3+}(aq)$. This colour change was the basis for the original 'breathalyser test'. The police could ask a motorist to exhale through a tube containing some orange crystals. If the crystals turned blue, it showed that the breath contained a considerable amount of ethanol vapour.

Acid + $KMnO_4$ ...

Acidified potassium manganate(VII) solution. This is too powerful an oxidising agent to stop at the aldehyde: it oxidises primary alcohols to acids. It oxidises secondary alcohols to ketones.

All secondary alcohols and one primary alcohol, ethanol, give a positive result in the haloform reaction [§ 31.7.7].

The reactions of alcohols are summarised in Figure 30.7B and Table 30.7.

Figure 30.7B
Reactions of ethanol, a typical aliphatic alcohol

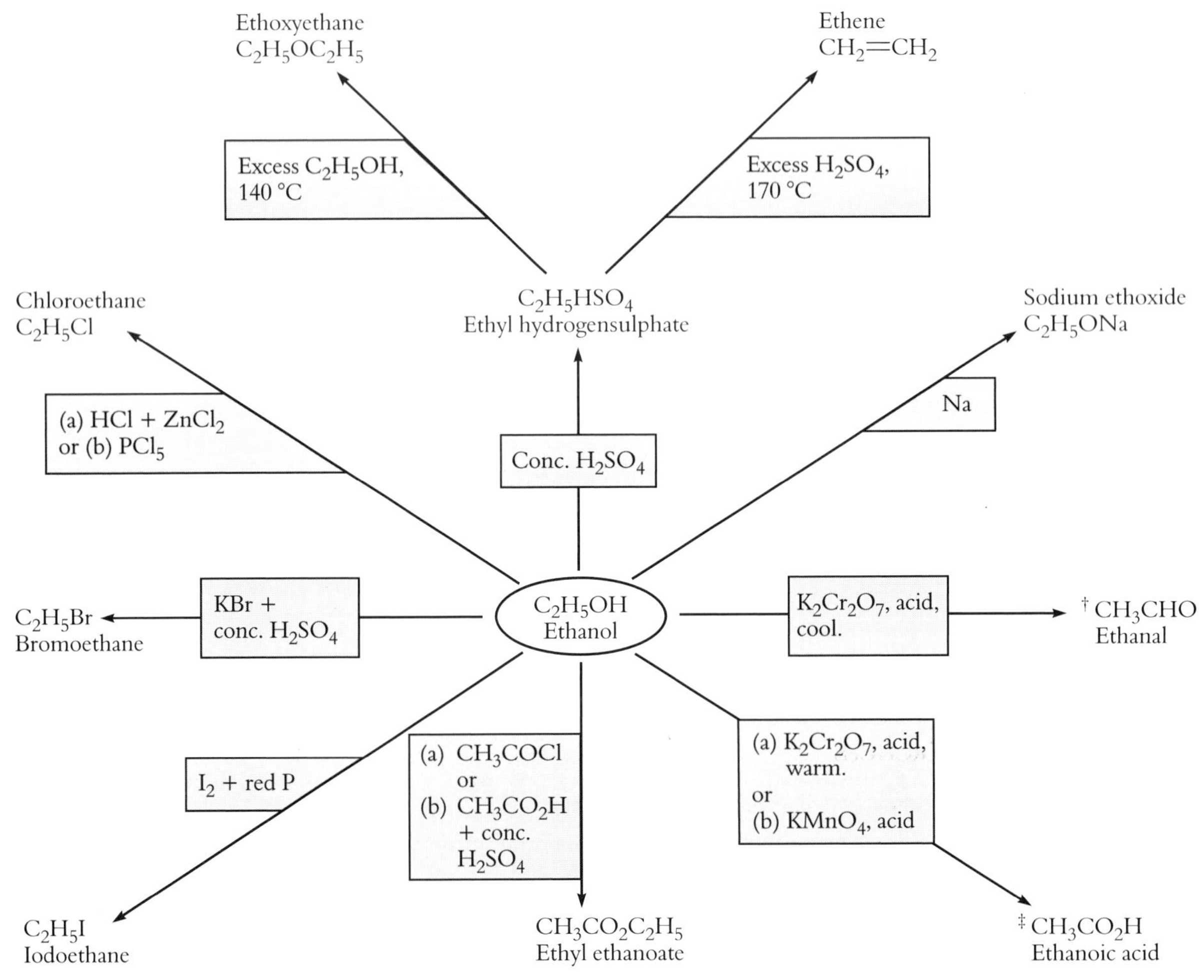

† Secondary alcohols are oxidised to ketones.

‡ Tertiary alcohols are more difficult to oxidise. With powerful, acidic oxidising agents, they form a mixture of acids.

Table 30.7 Methods of distinguishing between primary, secondary and tertiary alcohols

Reagent	Primary alcohol	Secondary alcohol	Tertiary alcohol
Acidified $K_2Cr_2O_7$ (orange)	Aldehyde, RCHO formed (blue $Cr^{3+}(aq)$ formed)	Ketone, R_2CO formed (blue $Cr^{3+}(aq)$ formed)	Resists oxidation
Conc. H_2SO_4	Alkene formed slowly	Intermediate in speed	Alkene formed fast
Conc. HCl + $ZnCl_2$. Add to alcohol and place in boiling water bath. Lucas test	Cloudiness due to formation of RCl is slow to appear. (Anhydrous conditions are needed for primary alcohols.)	Cloudiness appears in 5 minutes.	Cloudiness appears in 1 minute owing to the formation of RCl, which is insoluble in water.

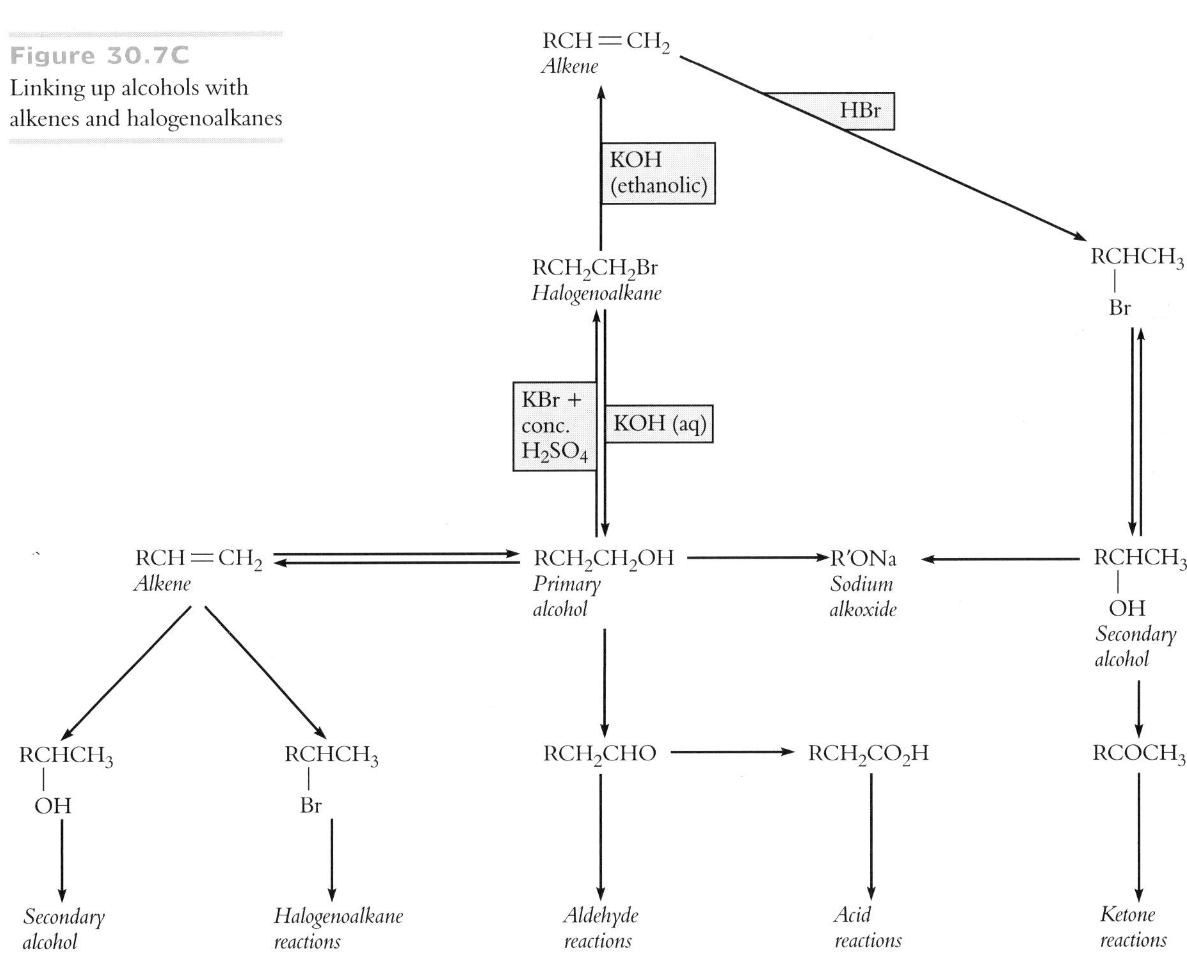

Figure 30.7C Linking up alcohols with alkenes and halogenoalkanes

Note that the group $—CH_2CH_2OH$ can be converted into a number of different functional groups. All these groups have characteristic reactions, to be studied in later chapters. By means of these reactions it is possible to make a large number of compounds in two stages starting from alcohols.

SUMMARY

Reactions of alcohols:
Weakly acidic: H of —OH group replaced by sodium
Esterification: reaction with carboxylic acids
Replacement of —OH group by —Cl or —Br or —I
Dehydration to form alkenes or ethers
Oxidation: primary ⟶ aldehydes ⟶ carboxylic acids
secondary ⟶ ketones

CHECKPOINT 30.7: REACTIONS

1. *(a)* Give two examples of reactions of ethanol which involve fission of the RO—H bond.
(b) Give two examples of reactions of ethanol which involve fission of the R—OH bond.

2. Name the reagents and describe the conditions needed to convert C_2H_5OH into

(a) C_2H_4 *(d)* C_2H_5Cl
(b) C_2H_5I *(e)* $C_2H_5OCOCH_3$
(c) C_2H_5Br *(f)* C_2H_5CN

3. Name and give the structural formulae of the products formed when propan-1-ol reacts with

(a) Na *(c)* conc. $H_2SO_4 > 170\,°C$
(b) PCl_5 *(d)* CH_3CO_2H

4. Explain how you could synthesise from propan-1-ol *(a)* propan-2-ol *(b)* 1,2-dichloropropane *(c)* propane-1,2-diol.

5. Give the names and structural formulae of a primary, a secondary and a tertiary alcohol, and explain how you could distinguish between them.

6. Identify the intermediate compound in each of the following syntheses. State the reagents and the conditions which are needed for each step in the synthesis:

$$C_2H_5OH \xrightarrow{(a)} A \xrightarrow{(b)} C_2H_5CN$$

$$CH_3CH(Br)CH_3 \xrightarrow{(c)} B \xrightarrow{(d)} CH_3COCH_3$$

7. *(a)* A compound **X** has the composition by mass C 35.0%, H 6.6%, Br 58.4%. Calculate the empirical formula of **X**.
(b) When **X** was boiled with dilute aqueous sodium hydroxide solution, 2-methylpropan-1-ol was formed. Draw the structural formula of this alcohol and deduce the structural formula of **X**.
(c) When **X** was heated with alcoholic potassium hydroxide, the alkene **Y** was formed. Give the name and structural formula of **Y**.
(d) When **Y** was treated with hydrogen bromide, the product was **Z**, an isomer of **X**. Give the name and structural formula of **Z**.

FURTHER STUDY ▼

30.8 POLYHYDRIC ALCOHOLS

Ethane-1,2-diol is used as antifreeze

Ethane-1,2-diol, $HOCH_2CH_2OH$, is used in car radiators as antifreeze and as a de-icing fluid on aeroplane wings. It is often called by its old name of ethylene glycol. Its reactions are similar to those of monohydric alcohols.

Glycerol is a triol

Propane-1,2,3-triol is usually known by its old name of glycerol. It is a by-product in the manufacture of soap [§ 33.14]. Glycerol is used for the manufacture of the ester which it forms with nitric acid, propane-1,2,3-triyl trinitrate or glyceryl trinitrate:

$$\begin{array}{l} CH_2OH \\ | \\ CHOH \\ | \\ CH_2OH \end{array} + 3HNO_3 \longrightarrow \begin{array}{l} CH_2ONO_2 \\ | \\ CHONO_2 \\ | \\ CH_2ONO_2 \end{array} + 3H_2O$$

Propane-1,2,3-triol (*Glycerol*) Propane-1,2,3-triyl trinitrate (*Glyceryl trinitrate*)

Nitroglycerine is the major component of dynamite

This ester is also called *nitroglycerine*. It is an explosive which is detonated by shock. When nitroglycerine is absorbed on kieselguhr (a type of clay), it is called *dynamite*. It is still a powerful explosive, but it is less sensitive to shock and can be handled with safety. Dynamite was invented by the Swedish chemist, A B Nobel, who invested some of the rewards for his invention to endow the Nobel prizes for great achievements in science and the arts.

30.9 PHENOLS

30.9.1 FUNCTIONAL GROUP

Phenols

Phenols are compounds containing a hydroxyl group attached to an aromatic ring. Some members of the group are listed below:

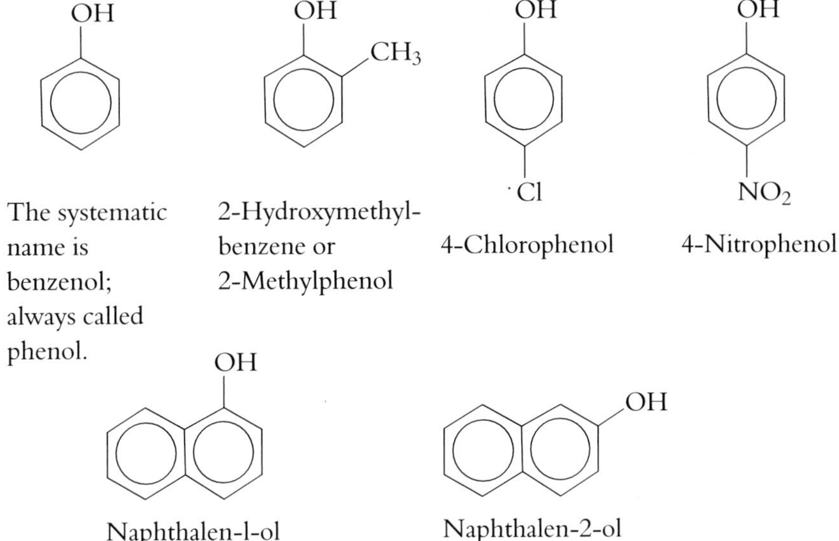

The systematic name is benzenol; always called phenol.

2-Hydroxymethylbenzene or 2-Methylphenol

4-Chlorophenol

4-Nitrophenol

Naphthalen-1-ol

Naphthalen-2-ol

30.9.2 PHYSICAL PROPERTIES

Phenol + water = the antiseptic, 'carbolic acid'

The vapour is toxic. Solid or liquid phenol burns the skin

Most phenols are colourless solids. Phenol itself melts at 42 °C. The presence of water lowers its melting temperature, and a mixture of phenol and water is a liquid at room temperature. It is called 'carbolic acid', and is a powerful antiseptic. A century ago, a surgeon called J Lister began to wonder why patients were dying some time after operations when the initial surgery appeared to have been satisfactory. He realised that their wounds were becoming infected both during the operation and afterwards in the ward. By spraying phenol mist around the operating theatre and the wards, he was able to reduce the death rate among his patients. Phenol is not a convenient antiseptic as the vapour is toxic, and if the solid or liquid comes into contact with the skin, it causes burns. Research workers have come up with a number of substituted phenols which are better at killing bacteria and less caustic to the skin. One of these is 2,4,6-trichlorophenol, which is present in TCP® and Dettol®.

TCP®

OH

Cl Cl

Cl

2,4,6-Trichlorophenol

Hydrogen bonding raises T_m

The presence of hydrogen bonds makes the melting temperatures of phenols higher than those of hydrocarbons of comparable molecular mass.

30.10 SOURCES OF PHENOL

30.10.1 SODIUM BENZENESULPHONATE

Phenol from sodium benzenesulphonate ...

Both in industry and in the laboratory, fusion of sodium benzenesulphonate with sodium hydroxide is used to make sodium phenoxide. This is dissolved in water and acidified to give phenol:

$$\underset{\text{Sodium benzenesulphonate}}{C_6H_5SO_3Na} + 2NaOH(s) \xrightarrow{\text{fuse}} \underset{\text{Sodium phenoxide}}{C_6H_5ONa(s)} + Na_2SO_3(s) + H_2O(g)$$

$$\underset{\text{Sodium phenoxide}}{2C_6H_5ONa(aq)} + CO_2(g) + H_2O(l) \longrightarrow \underset{\text{Phenol}}{2C_6H_5OH(l)} + Na_2CO_3(aq)$$

30.10.2 DIAZONIUM COMPOUNDS

The laboratory preparation of phenol

The easiest method to use in the laboratory is the hydrolysis of diazonium compounds [§ 32.8.1].

30.11 REACTIONS OF PHENOL

The reactions of phenol are of two kinds: (a) the reactions of the hydroxyl group and (b) substitution in the aromatic ring.

30.11.1 REACTIONS OF THE —OH GROUP

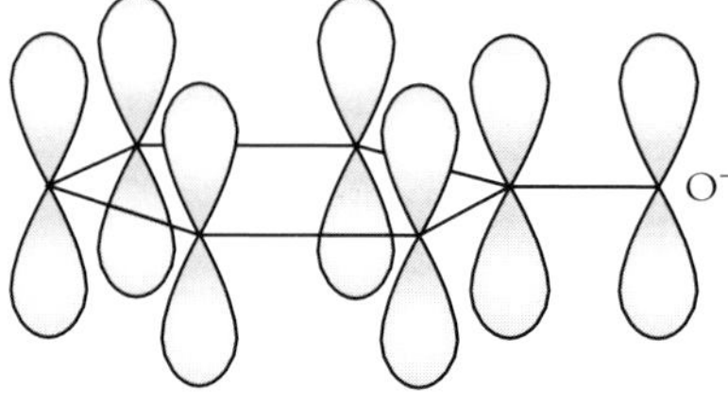

Figure 30.11A Interaction between the charge on the oxygen atom and the ring in the phenoxide ion

The phenoxide ion is stabilised by charge delocalisation

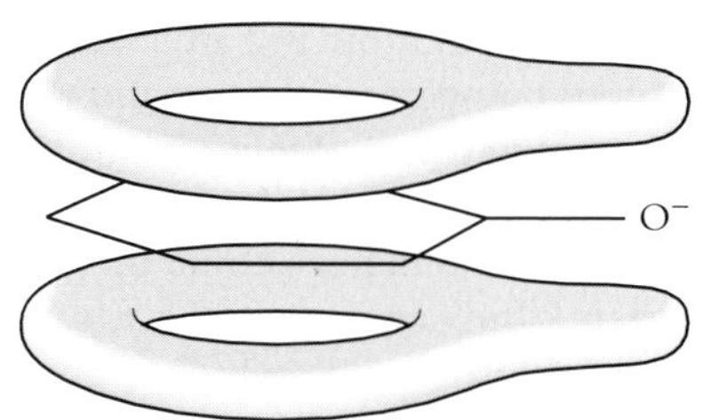

Dissociation

Phenol dissociates in water:

$$C_6H_5OH(aq) + H_2O(l) \rightleftharpoons C_6H_5O^-(aq) + H_3O^+(aq)$$

Phenol is a weak acid ...

Dissociation occurs to only a slight extent: phenol is a very weak acid, with $pK_a = 10.0$. The fact that it is an acid makes it dissolve more readily in sodium hydroxide solution than in water. It forms a solution of sodium phenoxide, $C_6H_5O^- Na^+$:

$$C_6H_5OH(l) + NaOH(aq) \longrightarrow C_6H_5ONa(aq) + H_2O(l)$$

... weaker than H_2CO_3 ...

... but stronger than C_2H_5OH

Since phenol is a weaker acid than carbonic acid, it will not displace carbon dioxide from carbonates as carboxylic acids do. Phenols are stronger acids than alcohols. The ethoxide ion, $C_2H_5O^-$, readily accepts a proton to form C_2H_5OH: ethanol is a very weak acid, with $pK_a = 16.0$. In the phenoxide ion, a p orbital of the oxygen atom overlaps with the π orbital of the ring carbon atoms [see Figure 30.11A].

The negative charge is to some extent spread around the ring, and the —O^- group is less likely to accept a proton to form C_6H_5OH.

Esterification

Before esterification, a phenol, ArOH, must be converted into the ion ArO^-

Esterification occurs less readily than with alcohols. First, the phenol must be converted into the phenoxide ion; then it will react with acid chlorides and anhydrides:

$$\underset{\text{Phenol}}{C_6H_5OH} \xrightarrow[\text{room temperature}]{NaOH(aq)} \underset{\text{Sodium phenoxide}}{C_6H_5O^- Na^+} \xrightarrow{(CH_3CO)_2O} \underset{\text{Phenyl ethanoate}}{C_6H_5OCOCH_3} + CH_3CO_2Na$$

Replacement by halogens

Reaction with PCl_5 is slow

The replacement of —OH by a halogen takes place much less readily than with alcohols. Hydrogen halides do not react, and phosphorus(V) halides react slowly to give a poor yield of the halogenoarene:

$$\underset{\text{Phenol}}{C_6H_5OH(s)} + \underset{\text{Phosphorus(V) chloride}}{PCl_5(s)} \longrightarrow \underset{\text{Chlorobenzene}}{C_6H_5Cl(l)} + \underset{\text{Phosphorus trichloride oxide}}{POCl_3(s)} + HCl(g)$$

30.11.2 SUBSTITUTION IN THE RING

Phenol is more reactive than benzene towards electrophiles ...

Phenol is more reactive than benzene towards electrophilic reagents. The reason is that the interaction between the lone pairs on the oxygen atom in —OH or —O^- and the ring increases the availability of electrons in the aromatic ring [§ 28.12]. Milder conditions are employed in substitution reactions than are needed for benzene. These are shown in Figure 30.11A.

SUMMARY

Phenols are aromatic compounds with an —OH group attached to the benzene ring.
Some phenols are used as antiseptics.
Phenols are weakly acidic but more strongly acidic than alcohols.
Phenols form esters with carboxylic acids.
The —OH group can be replaced by a halogen atom.
Phenol is more reactive than benzene toward electrophilic reagents, e.g. NO_2^+.

30.11.3 TEST FOR PHENOL

A test for phenol In solution, phenol reacts with iron(III) chloride to form a violet complex. Other phenols give different colours, and this reaction is used as a test for phenols.

Table 30.11
Comparison of ethanol and phenol

	Reagent	Ethanol, C_2H_5OH	Phenol, C_6H_5OH
		Liquid at room temperature Characteristic smell	Solid at room temperature Characteristic smell
(1)	Water	Miscible Neutral solution	Partially miscible Weakly acidic solution
(2)	Sodium	$C_2H_5ONa + H_2(g)$ formed	$C_6H_5ONa + H_2(g)$ formed
(3)	PCl_5	Vigorous reaction $\longrightarrow$ HCl(g)	Slow reaction. Poor yield
(4)	CH_3COCl, $(CH_3CO)_2O$	Readily $\longrightarrow$ ester	Readily $\longrightarrow$ ester
(5)	$C_6H_5COCl + NaOH$	Readily $\longrightarrow$ ester	Readily $\longrightarrow$ ester
(6)	Organic acid	Ethyl ester formed	No reaction
(7)	HCl, HBr, HI	Halogenoethanes formed	No reaction
(8)	H_2, Ni catalyst	No addition	Addition $\longrightarrow$ $C_6H_{11}OH$
(9)	$HNO_3(aq)$	No reaction	$O_2NC_6H_4OH$ formed
(10)	Conc. H_2SO_4	Dehydration $\longrightarrow$ ethene or ether	Sulphonation $\longrightarrow$ $HOC_6H_4SO_3H$
(11)	Br_2, H_2O	No reaction	White ppt. of $HOC_6H_2Br_3$
(12)	I_2, OH(aq)	CHI_3, yellow ppt. formed	No reaction
(13)	Distil with Zn	No reaction	Benzene formed
(14)	Acid $KMnO_4$	Aldehyde, CH_3CHO formed	Mixture of oxidation products
(15)	Diazonium salt	No reaction	Yellow dye produced
(16)	$FeCl_3$	No colour produced	Violet colour produced

Figure 30.11B
Reactions of phenol

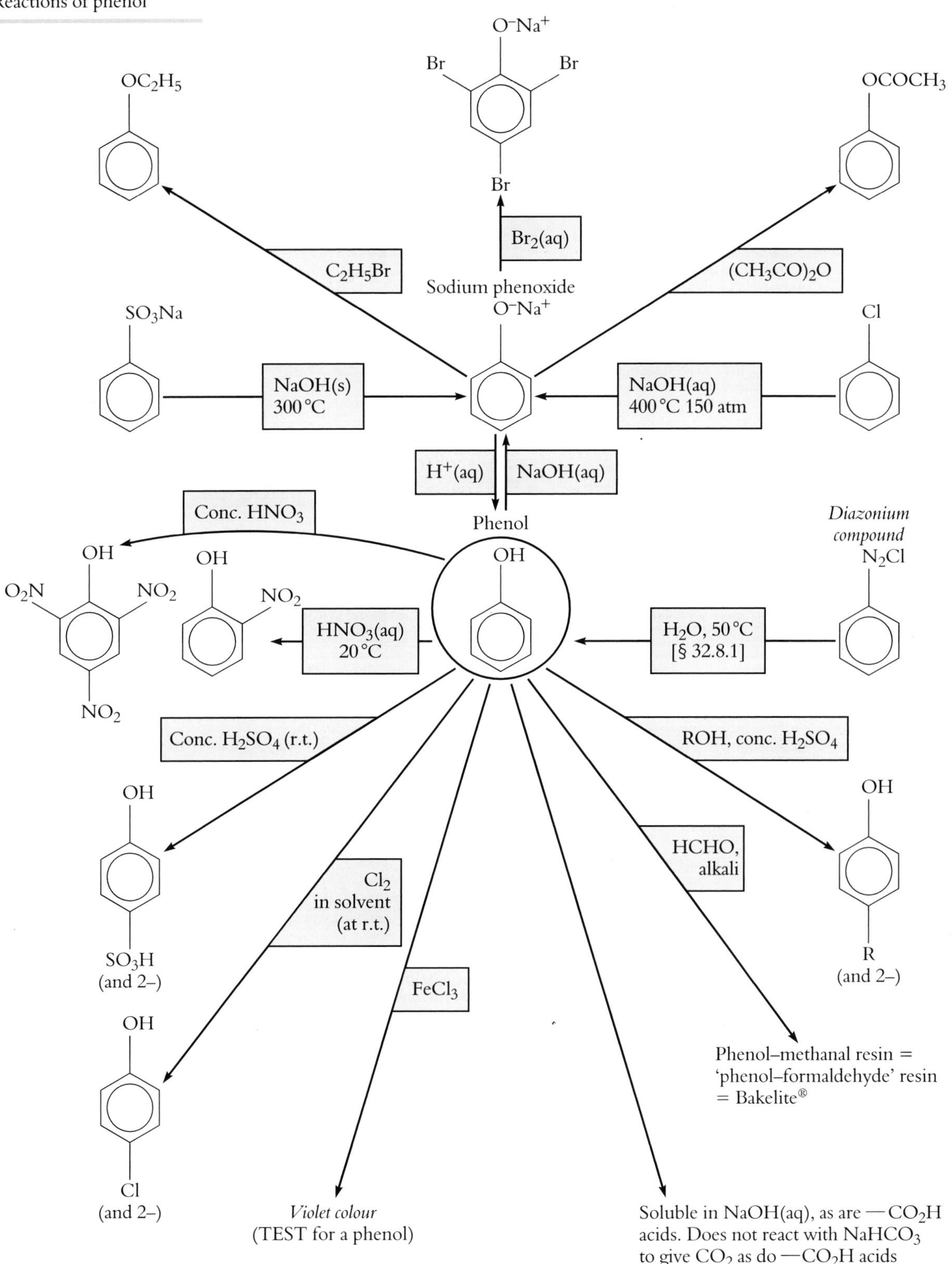

Figure 30.11C Some of the routes by which aromatic compounds are made from phenol

The OH group directs other substituents into the ring.

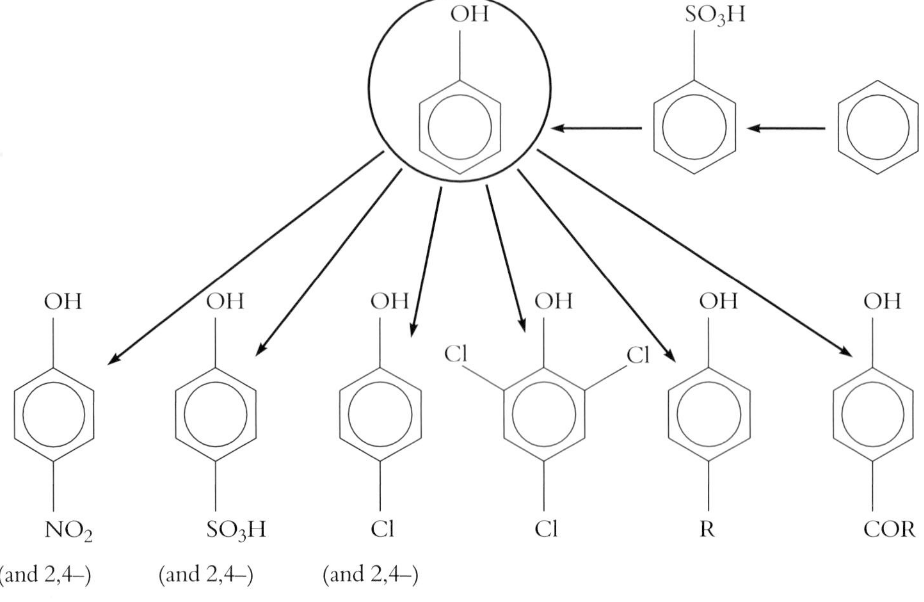

In each of the products, —OH can be replaced by —OR or —OCOR.

CHECKPOINT 30.11: PHENOLS

1. Copy the scheme shown below and fill in the necessary reagents and conditions. After reading Chapter 32, come back and fill in **X** and **Y**.

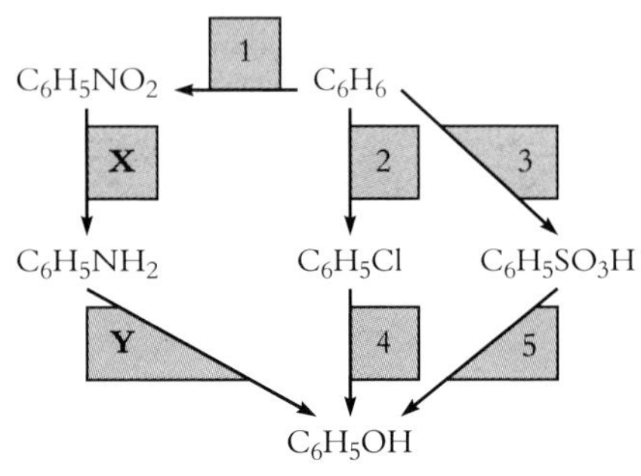

2. Starting from phenol, how could you make the following compounds?

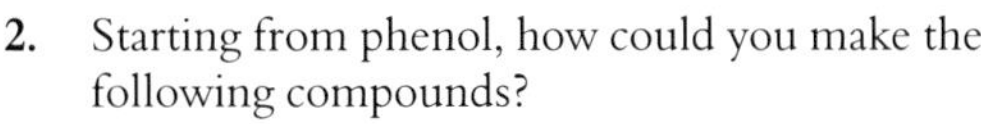

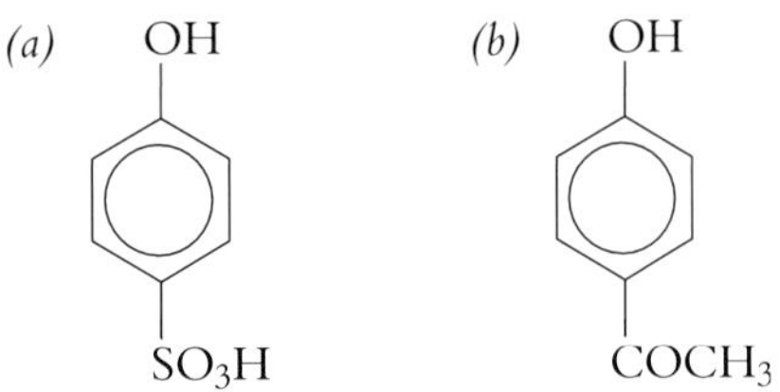

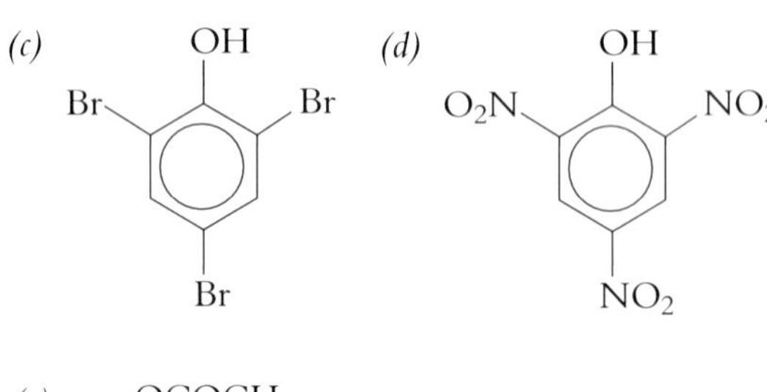

(e) $OCOCH_3$

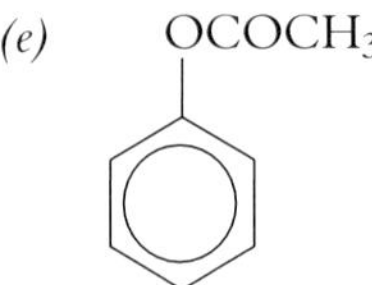

3. Explain why phenol is a stronger acid than ethanol.

FURTHER STUDY ▲

30.12 THE EXPLOSION AT SEVESO

The Seveso incident focused world attention on the dangers connected with the chemical industry. A chemical plant in Seveso, Italy, employed 200 people in the production of 2,4,5-trichlorophenol for the Swiss Givaudan Corporation, which is a subsidiary of Hoffman–LaRoche. On 10 July 1976 a build-up of pressure in a reaction vessel caused a safety valve to rupture, and a cloud of 2,4,5-trichlorophenol and other chemicals was released into the atmosphere.

One of the chemicals released was dioxin. It is an accidental by-product of the manufacture of 2,4,5-trichlorophenol, an important intermediate in the manufacture of 2,4,5-T, an agricultural herbicide used to control brushwood, and 'hexachlorophene', a bactericide which is used for the treatment of acne, for sterilising wounds and in skin cleansers.

Dioxin is stable to heat, acids and alkalis, is almost insoluble in water but soluble in some organic solvents. It is 500 times more poisonous than strychnine and 10 000 times more poisonous than cyanide ion. When a person is exposed to dioxin over a long time, dioxin residues accumulate in the liver and fat cells. Symptoms are cirrhosis of the liver, damage to the heart, kidney, spleen, central nervous system, lungs and pancreas, memory and concentration disturbances and depression. The skin disease, *chloracne*, is caused by the body's attempt to get rid of the poison through the skin. Dioxin also has a *teratogenic* effect, an effect on the genes, which results in birth defects [see below].

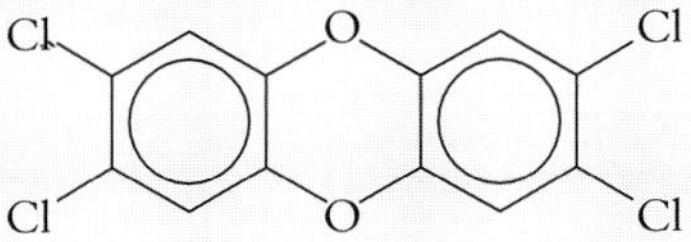

Dioxin

As a result of the accident, thousands of sheep and cows died, and people became ill, particularly with the terrible skin sores of chloracne. Nine days after the accident, people were told that the dust that had settled all over the town contained dioxin, and the town was evacuated. Among the children of Seveso, there were 134 confirmed cases of chloracne and 600 suspected cases. Of the 730 pregnant women in the town, 250 applied for abortions, and the Italian Government changed the law to allow the women to end their pregnancies.

The contaminated area of Seveso was sealed off, and experts debated how to tackle the pollution. Incineration of the contaminated soil and bacterial degradation of the dioxin in the soil were suggested. One expert recommended dismantling buildings and planting forests, rather than moving earth and washing buildings. Over a period of many years, layers of topsoil were removed and buried 10 m down beneath plastic and cement. Hoffman-LaRoche agreed to pay for all material damage and set up a fund to pay compensation to individuals.

The trichlorophenol made at Seveso was used in the manufacture of hexachlorophene, which is used for the treatment of acne. Ironically, the most common symptom of dioxin poisoning in humans is a form of acne. In 1972, in France, talcum powder accidentally contaminated with 6% hexachlorophene caused the death of 35 babies. Since then, hexachlorophene has been used only in prescription drugs. A replacement bactericide is chlorhexidine.

The Seveso accident moved the European Community (EC) to bring out a set of guidelines aimed at preventing similar accidents. This legislation, known as the *Seveso Directive*, laid down regulations for the control of hazardous industrial activities in member countries of the EC.

ALCOHOLS AND PHENOLS

Alcohols: include:
primary alcohols, I°, RCH_2OH
secondary alcohols, 2°, R_2CHOH
tertiary alcohols, 3°, R_3COH
§§ 30.2.1, 30.2.2
They can be distinguished by the ease of oxidation and Lucas' test.
Table 30.7

Ethanol, C_2H_5OH
The fermentation method is used to make alcoholic drinks. § 30.4.3
The hydration of ethene is used to make ethanol for industrial use. § 27.8.7

The **iodoform test** for $CH_3CH(OH)$– compounds is the formation of tri-iodomethane with I_2 + NaOH(aq).
§ 31.7.7

Reactions of alcohols

1. Oxidation by e.g. $Cr_2O_7^{2-}/H^+$
I°, RCH_2OH oxidised to aldehydes RCHO and then to carboxylic acids RCO_2H
2°, R_2CHOH oxidised to ketones, R_2CO
3°, R_3COH oxidised with difficulty.
§ 30.7.2

2. Combustion $\rightarrow CO_2 + H_2O$ § 30.5.3

3. Reaction with Na $\rightarrow$ RONa + H_2
§ 30.7.1

4. Esterification by acids or acid chlorides to esters RCO_2R'. § 30.7.1

5. Reaction with HCl or P + I_2 or PBr_3 or PCl_5 or $SOCl_2$ $\rightarrow$ halogenoalkanes § 30.7.2

6. Dehydration by conc H_2SO_4 to alkenes.
§ 30.7.2

Phenol, C_6H_5OH, is a weak acid, stronger than C_2H_5OH and weaker than CH_3CO_2H. § 30.11.1

- The –OH group reacts with NaOH or $NaHCO_3$ to give C_6H_5ONa. § 30.11.1
- Phenol is more reactive than benzene to electrophiles, e.g. Br_2, NO_2^+, which substitute in the ring to give $C_6H_2Br_3OH$ and $C_6H_2(NO_2)_3OH$.
§ 30.11.2

RELATIONSHIPS BETWEEN ALCOHOLS AND OTHER SERIES

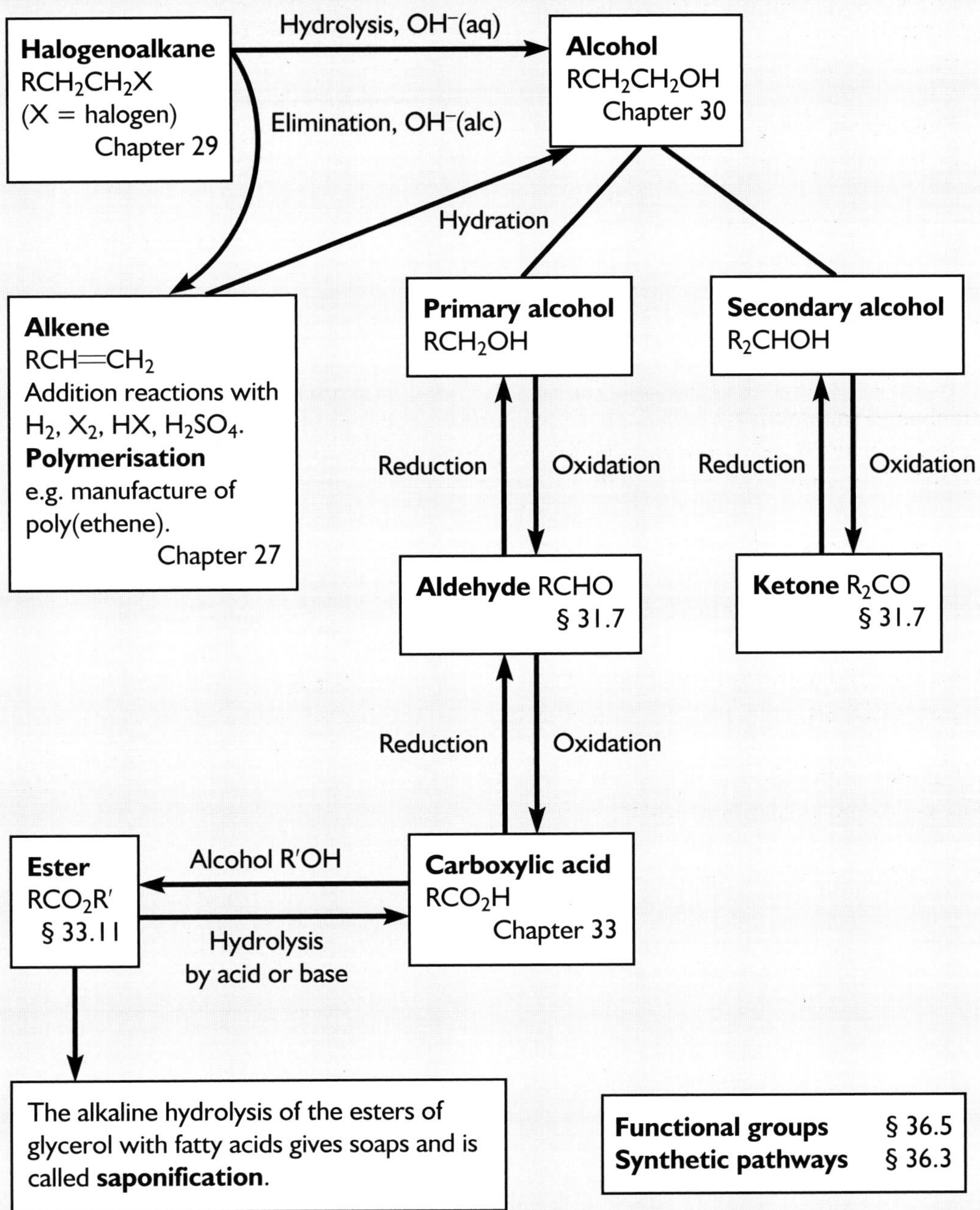

Functional groups § 36.5
Synthetic pathways § 36.3

QUESTIONS ON CHAPTER 30

1. Explain why *(a)* ethanol dissolves in water but ethane does not *(b)* ethanol has a higher boiling temperature than ethane.

2. Ethanol can be manufactured by the direct hydration of ethene.

(a) What is the source of ethene for this process?
(b) Write a reaction scheme for preparing ethanol from ethene by a method different from direct hydration.
(c) Ethanol blended with petrol is used as a motor fuel. Give two advantages of this fuel over normal petrol.
(d) Write an equation for the complete combustion of ethanol.
(e) In the first 'breathalyser' test for ethanol, drivers were asked to expel air from their lungs into acidified potassium dichromate(VI) crystals. State what was observed when the test was positive and explain what reaction had happened.

3. A phenol *(a)* and an aryl alcohol *(b)* have the same molecular formula. Their structural formulae are:

(a) $H_3C-\bigcirc-OH$

(b) $\bigcirc-CH_2OH$

How could you use the reagents sodium hydroxide, iron(III) chloride and phosphorus(V) chloride to distinguish between them?

4. *(a)* Outline an industrial method for the manufacture of ethanol.
(b) Give the names and formulae of the products formed when ethanol reacts with (i) sodium (ii) ethanoic acid (iii) sodium bromide + conc. sulphuric acid. State the conditions under which the reactions occur.

5. *(a)* Outline a method of preparing phenol, starting from benzene. Give the necessary reagents and conditions.
(b) Explain the following observations:
(i) When aqueous bromine is added to a solution of phenol a white precipitate forms.
(ii) When a solution of phenol in aqueous sodium hydroxide is shaken with benzoyl chloride a white precipitate forms.
(c) Give the formulae of the organic products formed when phenol reacts with (i) concentrated sulphuric acid (ii) ethanoic anhydride.

6. Prop-2-en-1-ol, $CH_2{=}CHCH_2OH$, is an unsaturated primary alcohol.

(a) Give the formula of the product that is formed when prop-2-en-1-ol is treated with a mild oxidising agent. State what happens on further oxidation.
(b) Give the name and formula of the product formed when bromine in an organic solvent reacts with prop-2-en-1-ol. State whether the reaction is substitution or elimination or addition and whether it involves attack by a nucleophile or an electrophile.
(c) Give the name and formula of the product formed when prop-2-en-1-ol is warmed with hydrogen in the presence of nickel as catalyst.
(d) Give the formula of the organic product formed when prop-2-en-1-ol is heated with ethanoic acid and sulphuric acid as catalyst. What type of reaction is this?

7. Complete the following reactions. State the necessary conditions for reaction to occur:

(a) $CH_3CH(OH)CH_3 + H_2SO_4 \longrightarrow$
(b) $C_6H_5SO_3Na + NaOH \longrightarrow$.
(c) $CH_3CH_2OH + Na \longrightarrow$
(d) $CH_3CH_2C(CH_3){=}CH_2 + H_2SO_4 \longrightarrow$
(e) $(CH_3)_3COH + HCl \longrightarrow$

8. How can ethanol be made from *(a)* ethene and *(b)* bromoethane? How and under what conditions does ethanol react with (i) ethanoic acid and (ii) sulphuric acid?

9. How would you prepare the following compounds from phenol? *(a)* sodium phenoxide, *(b)* 2,4,6-trinitrophenol and *(c)* 2,4,6-tribromophenol.

10. Both the compounds **A** and **B** are soluble in ethoxyethane (ether). Neither is soluble in water. Describe how you could separate **A** and **B** from a mixture of the two compounds, using only sodium hydroxide, hydrochloric acid and ethoxyethane.

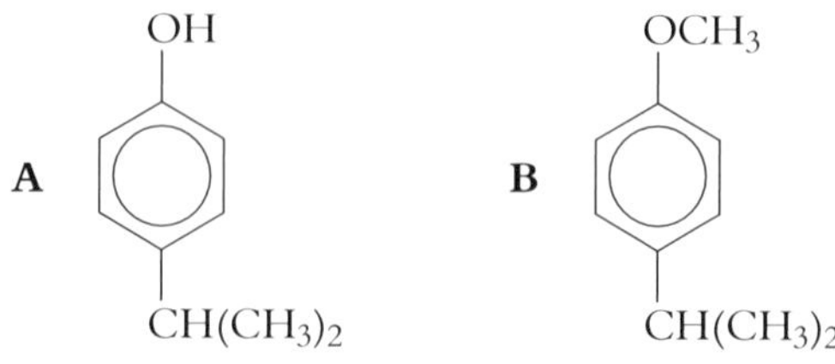

11. Compare the reactions of ethanol and phenol with *(a)* sodium, *(b)* sodium hydroxide, *(c)* sulphuric acid and *(d)* ethanoic acid.

12. *(a)* Explain the relative acidities of water, phenol and ethanol. **4**

(b) Compare and contrast the reactions, if any, of (i) ethanol, (ii) phenol with

- sodium,
- ethanoyl chloride,
- ethanoic acid,
- bromine.

Write equations for each reaction that occurs and name any organic products formed. **12**

(c) Ethane-1,2-diol, phenylmethanol and propane-1,2,3-triol each contain at least one —OH group in their molecules and react in a similar way to ethanol. Suggest equations for the following reactions:

(i) ethane-1,2-diol + sodium,

(ii) phenylmethanol + ethanoic acid,

(iii) propane-1,2,3-triol + ethanoyl chloride. **5**

25 *C (AS/AL)*

13. There are four alcohols of molecular formula $C_4H_{10}O$ which are structural isomers. Three of these alcohols are given below.

$CH_3CH_2CH_2CH_2OH$	$CH_3CH(OH)CH_2CH_3$
Butan-1-ol	Butan-2-ol

$(CH_3)_3COH$
2-Methylpropan-2-ol

(a) Give the structural formula of the fourth alcohol that is isomeric with those above. **1**

(b) On heating with concentrated sulphuric acid, butan-2-ol is converted into a mixture of alkenes.

(i) Give the name of the type of reaction taking place. **1**

(ii) Give the structural formula of one of the alkenes formed. **1**

(c) (i) Give the name or structural formula of the organic compound produced when butan-1-ol is heated with acidified potassium dichromate(VI) and the product is removed by distillation as it forms. **1**

(ii) Give the name or structural formula of the organic compound produced when butan-1-ol is heated under reflux for 20 minutes with acidified potassium dichromate(VI). **1**

(iii) Give the structural formula of the organic product formed when butan-2-ol is heated under reflux for 20 minutes with acidified potassium dichromate(VI). **1**

(iv) State the type of reaction occurring in *(c)*(iii). **1**

(d) When 2-methylpropan-2-ol is heated with a carboxylic acid in the presence of a catalyst, an ester, $C_6H_{12}O_2$, is formed.

(i) Give the structural formula of this ester of molecular formula $C_6H_{12}O_2$. **1**

(ii) Give the name of the carboxylic acid needed to form the ester in *(d)*(i). **1**

(iii) Suggest a suitable catalyst for the reaction. **1**

(e) Lucas' test may be used to distinguish between the three alcohols butan-1-ol, butan-2-ol and 2-methylpropan-2-ol. A mixture of anhydrous zinc chloride and hydrochloric acid is warmed with each of the three alcohols under identical conditions.

(i) Describe how the results of Lucas' test lead to the identification of the three alcohols. **2**

(ii) Give the name and the structural formula of the organic compound formed by butan-2-ol in Lucas' test. **2**

14 *AEB (AL)* 1998

14. *(a)* (i) Ethanol can be produced industrially either from ethene or from sugars such as glucose, $C_6H_{12}O_6$. For each route, name the method used, write an equation for the reaction and give one necessary condition. Suggest, with a reason, which route is likely to become the major method of production in the future.

(ii) Ethanol can also be produced in a reaction involving reduction. Give a suitable reducing agent and write an equation for the reaction. You may use [H] to indicate the reductant in your equation. **10**

(b) Ethene can be converted into two different saturated hydrocarbons one of which is a gas and the other a solid of high relative molecular mass. Give the structure of each product, state the types of reaction involved and, for the gaseous product only, give the conditions necessary for its formation. **5**

15 *NEAB (AS/AL)*

15. Under appropriate reaction conditions, 2-bromo-3-methylbutane, $(CH_3)_2CHCHBrCH_3$, can be converted into an alcohol or into two isomeric alkenes.

(a) Name the type of reaction taking place and give the role of the reagent when 2-bromo-3-methylbutane reacts with **aqueous** potassium hydroxide. Give the structure of the alcohol and outline a mechanism for this reaction. **5**

(b) (i) Name the type of reaction taking place and give the role of the reagent when 2-bromo-3-methylbutane reacts with **ethanolic** potassium hydroxide.

(ii) One of the reaction products is 2-methylbut-2-ene. Outline a mechanism for the formation of this compound and give the structure of the second alkene which is also formed. **7**

12 *NEAB (AL)*

31

ALDEHYDES AND KETONES

31.1 POISONS, FLAVOURS AND PERFUMES

A solution called formalin is used for preserving biological specimens. If you have come across it, you will remember how it irritates your eyes, nose and throat. It is a solution of methanal, HCHO, (previously called formaldehyde) the first member of the aldehyde series. Methanal is able to preserve tissues because it combines readily with proteins, reacting with —NH— groups and $—NH_2$ groups. The results are that it kills microorganisms and hardens tissues. A similar reaction between methanal and urea produces urea-formaldehyde resins, as described in § 34.5.6.

In § 30.1, the danger of drinking methanol was mentioned. When the body attempts to get rid of methanol it oxidises methanol to methanal, under the influence of the enzyme catalase. The reaction takes place in the liver and also in the retina because catalase is involved in the chemistry of vision. In the retina, methanal produces cross-links between the proteins of the retina, preventing them from playing their part in vision. A dose of more than 2 grams results in blindness.

Other aldehydes are more pleasant substances. Benzaldehyde, C_6H_5CHO, is used to give the flavour of almonds to food. Phenylethanal, $C_6H_5CH_2CHO$ is similar in structure to benzaldehyde but has a very different use: it smells of hyacinths and is used in perfumes. Vanillin is one of the most widely used flavour compounds. It is added to chocolate, and it is present in wine because it leaches out of the oak barrels which are used to store wine.

Vanillin

OCH_3

HO—⟨benzene ring⟩—CHO

The aldehyde 2-methylundecanal, $CH_3(CH_2)_8CHCH_3CHO$, has a special place in perfume chemistry. When Marilyn Monroe was asked what she wore in bed, she replied '*Chanel No 5*'. The perfume she referred to was created in 1921 by Ernest Beaux, a 'grand nez', that is an expert in blending perfumes, for the clothes designer 'Coco' Chanel. In addition to natural oils derived from flowers, Beaux included a synthetic substance, 2-methylundecanal, to provide the most volatile part of the perfume. This was the first time a synthetic substance had been included in a prestigious perfume. *Chanel No 5* has been a market-leader since 1921 because the components are blended to vaporise at the same rate so that the scent does not change over the course of the day.

The first member of the ketone series is propanone, CH_3COCH_3 (formerly called acetone). It can be produced in the body. In people suffering from diabetes mellitus, abnormal metabolic reactions lead to the production of propanone. It builds up in the tissues and appears in the breath and urine of diabetics, unless they are successfully treated. The next member of the ketone series is butanone. It is the pleasant-smelling solvent used in nail varnish remover. Many higher members of the series have pleasant smells and are used in perfumes.

31.2 THE FUNCTIONAL GROUP

In aldehydes and ketones the functional group is the carbonyl group

$$>C=O$$

The formulae of aldehydes and ketones

Aldehydes have a hydrogen atom attached to the carbonyl carbon atom; ketones have two alkyl or aryl groups:

$R(H)C=O$	$R'(R)C=O$
Aldehyde	Ketone
(In the first member of the series, methanal, R = H.)	(R and R′ may be aliphatic or aromatic.)

The general formula of saturated aliphatic aldehydes and ketones is $C_nH_{2n}O$. In addition, there are cyclic ketones of formula $C_nH_{2n-2}O$, for example, cyclohexanone. The simplest aromatic aldehyde and ketone are benzenecarbaldehyde and phenylethanone:

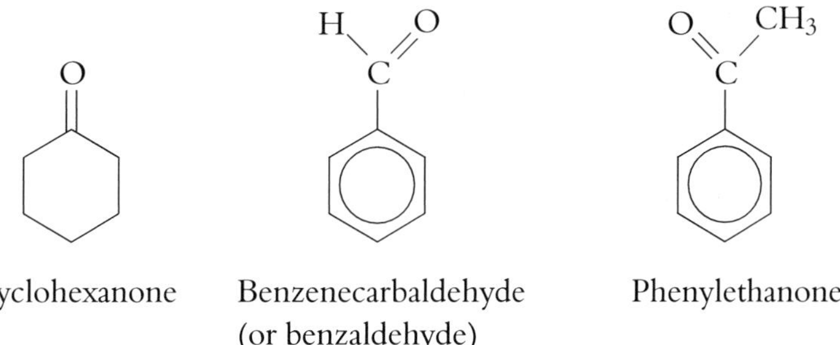

Cyclohexanone — Benzenecarbaldehyde (or benzaldehyde) — Phenylethanone

31.3 NOMENCLATURE FOR ALDEHYDES AND KETONES

The system of naming carbonyl compounds

Aldehydes and ketones are named after the hydrocarbon with the same number of carbon atoms. The name of the parent alkane is changed to end in *-al* for aldehydes and *-one* for ketones. The position of the carbonyl group in ketones is indicated by a locant preceding the suffix, *-one* [see Table 31.3(b)].

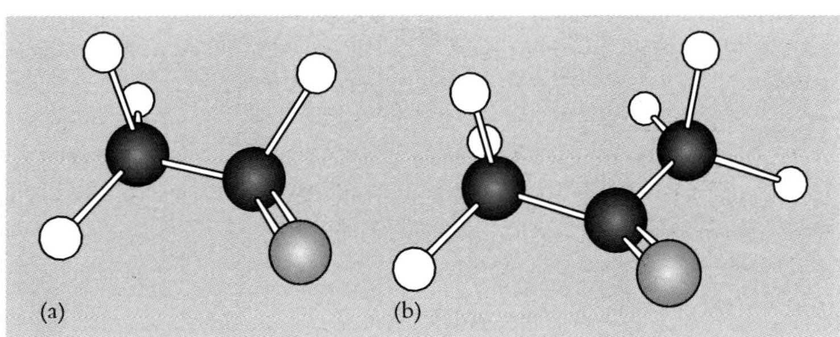

Figure 31.3A
(a) Ethanal, CH_3CHO and (b) propanone, $(CH_3)_2CO$

Table 31.3
The names of some aldehydes and ketones

(a) *Aldehydes*

Formula	Name
$HCHO$	Methanal (formerly formaldehyde)
CH_3CHO	Ethanal (formerly acetaldehyde)
$CH_3(CH_2)_3CHO$	Pentanal
$C_6H_5CH_2CHO$	Phenylethanal
C_6H_5CHO	Benzenecarbaldehyde (or benzaldehyde)

(b) *Ketones*

Formula	Name
CH_3COCH_3	Propanone (formerly acetone)
$CH_3COCH_2CH_3$	Butanone (formerly ethyl methyl ketone)
$CH_3CO(CH_2)_2CH_3$	Pentan-2-one
$CH_3CH_2COCH_2CH_3$	Pentan-3-one
$C_6H_5COCH_3$	Phenylethanone

31.4 THE CARBONYL GROUP

The π electrons in the $\rangle C=O$ bond are extensively polarised

In the group $R_2C=O$, the carbon atom forms three single σ bonds, which are coplanar. The unused p orbital of the carbon atom overlaps with one of the unused p orbitals of the oxygen atom to form a π orbital [see Figure 31.4A].

Figure 31.4A
The carbonyl group
$\rangle C=O$

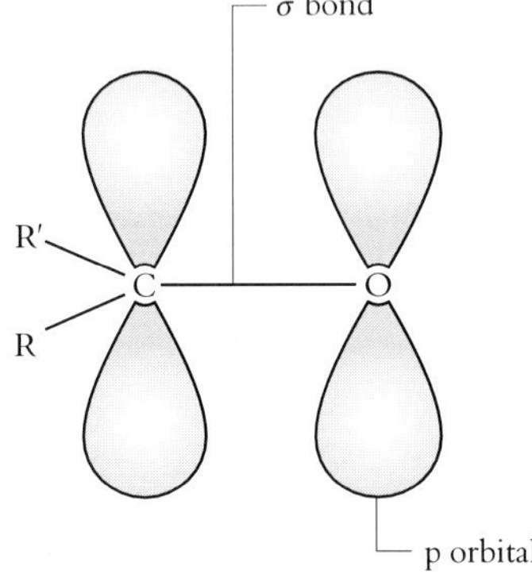

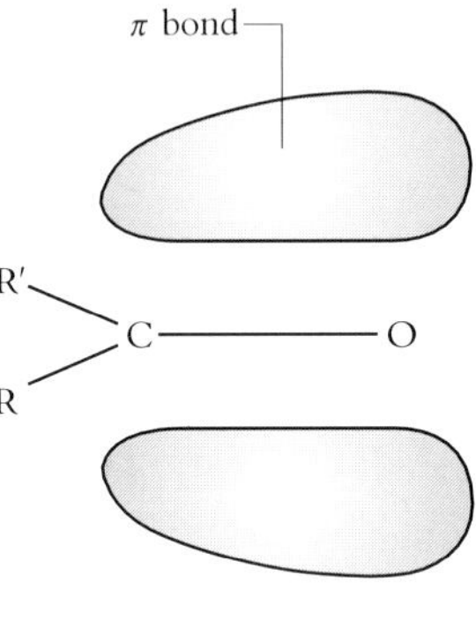

Aldehydes and ketones owe their reactivity to the polarised $\rangle \overset{\delta+}{C}=\overset{\delta-}{O}$ group

The electrons in the σ and π are drawn towards the more electronegative oxygen atom. To represent this polarisation the carbonyl group is written as

$$\rangle C\overset{\frown}{=}O \quad \text{or} \quad \rangle \overset{\delta+}{C}=\overset{\delta-}{O}$$

Since π electrons are polarised much more extensively than σ electrons, the dipole in a

$$\rangle C\overset{\frown}{=}O \text{ bond}$$

is greater than that in a

$$\geq C \longrightarrow Cl \text{ Bond}$$

The electronic structure of the carbonyl group is responsible for the reactions of aldehydes and ketones.

31.4.1 ATTACK BY NUCLEOPHILES

Nucleophiles attack $>\overset{\delta+}{C}=\overset{\delta-}{O}$ to form an intermediate ...

Nucleophilic reagents are attracted by the partial positive charge on the carbon atom of the

$$>C=\ddot{O} \text{ group}$$

At the approach of a nucleophile (e.g., CN^-), the π electrons are repelled away from the carbon atom towards the oxygen atom. In the reactive intermediate which is formed, oxygen bears a negative charge, and the carbon atom is surrounded by four electron pairs.

$$N\equiv C:^- \; + \; R'RC=\ddot{O}: \longrightarrow \left[N\equiv C-C(R')(R)-\ddot{O}:^- \right]$$

Nucleophile attacks $\overset{\delta+}{C}$ C is 4-valent; O is negative

... which acts as a base

The intermediate acts as a base:

$$\left[N\equiv C-C(R')(R)-\ddot{O}:^- \right] + HA \longrightarrow N\equiv C-C(R')(R)-\ddot{O}-H + A^-$$

The complete reaction is the addition of HNu across the double bond

The overall reaction is the addition of HCN across the double bond. In general, if Nu: is the nucleophile

$$R_2C=O + HNu \longrightarrow R_2C(OH)(Nu)$$

The ease with which the reaction occurs is in contrast with the attack by a nucleophile on a saturated carbon atom [see the hydrolysis of halogenoalkanes, § 29.9].

31.4.2 COMPARISON WITH THE ALKENE GROUP

$>C=O$ does not react with electrophiles as does $>C=C<$

The carbonyl group resembles the alkene group in that both groups contain a σ bond and a π bond between the bonded atoms. One might expect that the π electrons in the

$>C=O$ bond like those in $>C=C<$

would react with electrophiles, such as Br_2 and HBr. This does not happen. The reason is that the oxygen atom in the carbonyl group, being electronegative, is able to keep control of the π electrons and does not make them available for bonding to an electrophile. To summarise:

$\overset{\delta+}{>C}=\overset{\delta-}{O}$	$>C=C<$
polar bond	non-polar bond
The electron-deficient carbon atom is attacked by nucleophiles.	The double bond is attacked by electrophiles.

31.4.3 COMPARISON OF ALDEHYDES AND KETONES

Aldehydes are more reactive than ketones

Aldehydes are more reactive than ketones. The reason is chiefly that the presence of two alkyl groups in ketones hinders the approach of attacking reagents to the carbonyl group. Another factor is that alkyl groups are electron-donating and reduce the partial positive charge on the carbonyl atom [see Figure 31.4B].

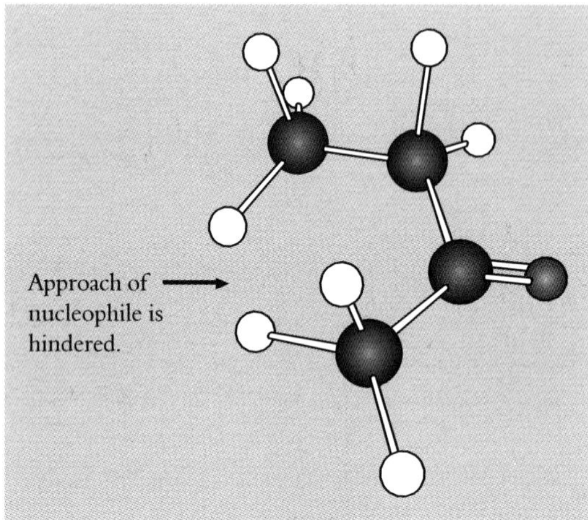
Approach of nucleophile is hindered.

Figure 31.4B Hindrance by alkyl groups to nucleophilic attack

CHECKPOINT 31.4: CARBONYL COMPOUNDS

1. Write structural formulae for
   (*a*) butanal
   (*b*) pentan-2-one
   (*c*) pentan-3-one
   (*d*) cyclohexanone
   (*e*) 3-methylhexanal
   (*f*) pentane-2,4-dione
   (*g*) 4-hydroxybenzaldehyde
   (*h*) 2-chlorobenzaldehyde

2. Name the following:
   (*a*) $CH_3CH_2CH_2CHO$
   (*b*) $CH_3CH_2COCH_3$
   (*c*) $C_2H_5COC_2H_5$
   (*d*) $(CH_3)_2CHCHO$
   (*e*) $(CH_3)_2CHCOCH_3$
   (*f*) $CH_3CH{=}CHCHO$
   (*g*) C_6H_5–$COCH_3$ (benzene ring–$COCH_3$)
   (*h*) O_2N–C_6H_4–CHO (benzene ring with O_2N– and –CHO)

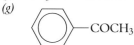

3. What feature of the carbonyl group is chiefly responsible for the reactive nature of aldehydes and ketones?

4. Why does bromine attack $>C=C<$ but not $>C=O$?

5. Explain why there is intermolecular hydrogen bonding in $C_2H_5CH_2OH$ but not in C_2H_5CHO.

31.5 SOME MEMBERS OF THE SERIES

Methanal

Methanal is used as formalin and in the manufacture of plastics, antiseptics and explosives

Methanal (formaldehyde) is a gas (T_b = −21 °C). It is used as a 40% aqueous solution, called *formalin*, for the preservation of biological specimens. Methanal is used in the manufacture of the tough plastics, Bakelite® [§ 34.5.5], Formica® [§ 34.5.6] and the explosive *cyclonite*.

Ethanal

Ethanal is made from ethanol ...

Ethanal (acetaldehyde) is made from ethanol by aerial oxidation in the presence of silver as catalyst.

... It is used to make chloral and DDT

Ethanal is used as a starting point in the manufacture of certain organic compounds, e.g. ethanoic acid, DDT and chloral ('knockout drops').

Propanone

Propanone is made from propan-2-ol by dehydrogenation and is used as a solvent

Propanone (acetone) is made by the dehydrogenation of propan-2-ol using a copper catalyst. It is a widely used, inexpensive industrial solvent, with T_b = 56 °C. If a solvent with a higher boiling temperature is needed, butanone is used (T_b = 80 °C). Propanone is formed in the fermentation of some sugars and starches and is found in the breath and urine of diabetics.

31.6 LABORATORY PREPARATIONS

31.6.1 FROM ALCOHOLS

Oxidation of alcohols

A laboratory method of preparation is the oxidation of alcohols to aldehydes and ketones ...

Oxidation of primary alcohols gives aldehydes, provided that conditions are controlled to avoid further oxidation to a carboxylic acid [§ 30.7.2]. Oxidation of secondary alcohols gives ketones. Acidified sodium dichromate(VI) is often employed for oxidation in the laboratory. The oxidising agent is added slowly to the alcohol [see Figure 31.6A]. The temperature is kept below the boiling

Figure 31.6A Preparation of aldehydes and ketones

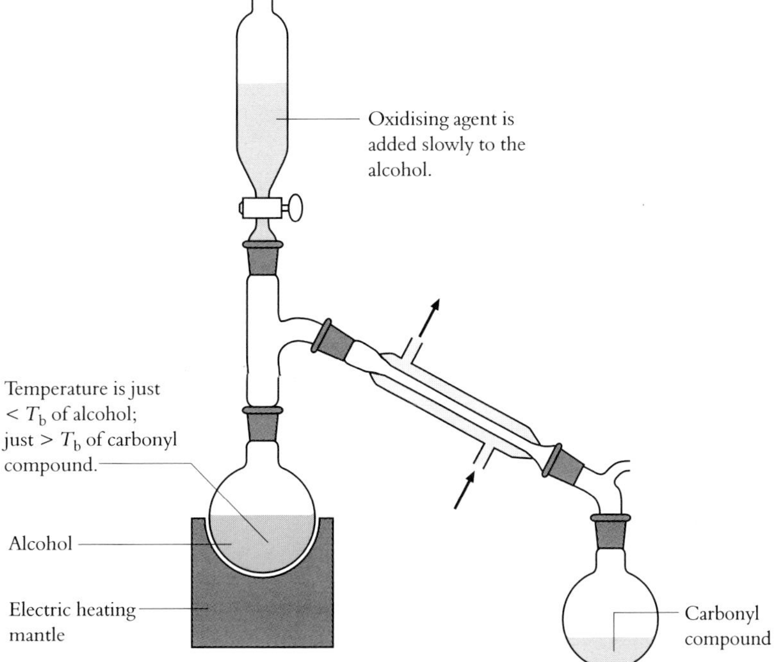
Oxidising agent is added slowly to the alcohol.
Temperature is just < T_b of alcohol; just > T_b of carbonyl compound.
Alcohol
Electric heating mantle
Carbonyl compound

temperature of the alcohol and above that of the carbonyl compound. (Carbonyl compounds are more volatile than the corresponding alcohols.) Arranging for an excess of alcohol over oxidant and distilling off the aldehyde as it is formed avoid further oxidation. Ketones are in little danger of further oxidation:

$$\underset{\text{Ethanol}}{C_2H_5OH(l)} + [O] \xrightarrow[\text{acid, warm}]{Na_2Cr_2O_7} \underset{\text{Ethanal}}{CH_3CHO(l)} + H_2O(l)$$

$$\underset{\text{Propan-2-ol}}{(CH_3)_2CHOH(l)} + [O] \xrightarrow[\text{acid, warm}]{Na_2Cr_2O_7} \underset{\text{Propanone}}{(CH_3)_2CO(l)} + H_2O(l)$$

Dehydrogenation of alcohols

Carbonyl compounds are made from alcohols ...

Catalytic dehydrogenation of alcohols can be used in the laboratory as well as in industry [§ 31.5].

31.6.2 AROMATIC KETONES BY FRIEDEL–CRAFTS ACYLATION

Friedel–Crafts acylation gives aromatic ketones

Aromatic ketones are obtained by the Friedel–Crafts acylation reactions [§ 28.8.5].

Figure 31.6B Methods of preparing aldehydes and ketones

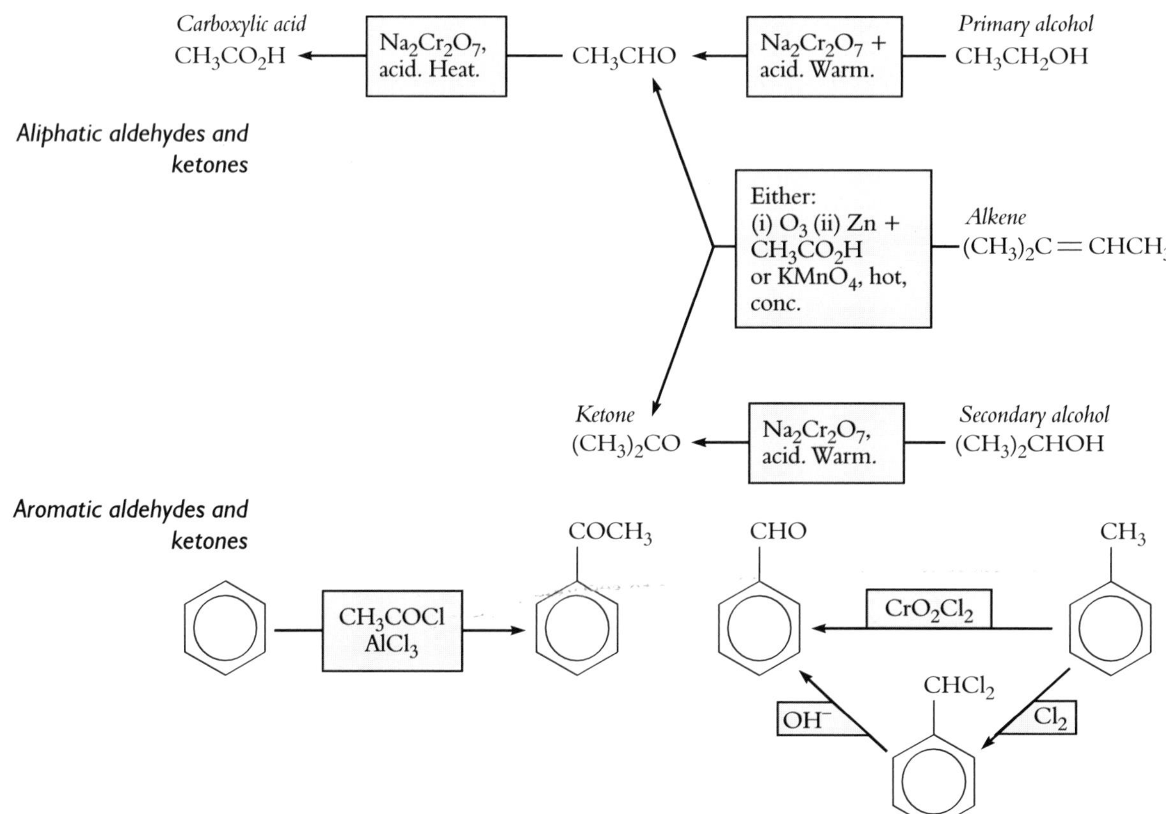

31.6.3 AROMATIC ALDEHYDES FROM METHYLARENES

Oxidation of —CH_3 attached to an aromatic ring gives an aromatic aldehyde

Aromatic aldehydes can be made by the oxidation of a methyl group attached to the benzene ring. Chromium dichloride dioxide is often employed as the oxidising agent:

$$\underset{\text{Methylbenzene}}{C_6H_5CH_3} \xrightarrow{CrO_2Cl_2} \underset{\text{Benzenecarbaldehyde (benzaldehyde)}}{C_6H_5CHO}$$

Another route is the alkaline hydrolysis of 1,1-dihalogenocompounds:

$$C_6H_5CH_3 \xrightarrow{Cl_2} C_6H_5CHCl_2 \xrightarrow{OH^-(aq)} C_6H_5CHO$$

Aliphatic aldehydes can be made in this way, but the dihalogenocompounds are usually made from the aldehydes.

SUMMARY

Sources of carbonyl compounds are:

Aldehydes: oxidation of primary alcohols.
Ketones: oxidation of secondary alcohols.
Aromatic aldehydes: oxidation of —CH_3 attached to the benzene ring.
Aromatic ketones: Friedel–Crafts reaction.

CHECKPOINT 31.6: PREPARATIONS

1. Aldehydes are often prepared from alcohols. What difficulty arises in this method of preparation? How is it overcome?

2. Draw the structural formulae of the alkenes that yield the following products on ozonolysis:

   (*a*) ethanal and propanone
   (*b*) propanal, propanone and butanedial ($OHCCH_2CH_2CHO$)
   (*c*) methanal and pentane-2,4-dione ($CH_3COCH_2COCH_3$)
   (*d*) methanal and propanedial ($OHCCH_2CHO$)

31.7 REACTIONS OF CARBONYL COMPOUNDS

31.7.1 REDUCTION

Aldehydes are reduced to primary alcohols by H_2 or by $LiAlH_4$ in ether or by $NaBH_4(aq)$

Hydrogen adds across the carbonyl group, reducing aldehydes to primary alcohols and ketones to secondary alcohols. The reaction is more difficult than addition to a carbon–carbon double bond, and usually requires heat, pressure and a metal catalyst (Pt or Ni):

$$RR'C{=}O + H_2 \xrightarrow[\text{Pt or Ni}]{H_2\text{, pressure}} RR'C(H)OH$$

(R′ can be H)

In the laboratory, reduction is usually effected by the use of a metal hydride. Lithium tetrahydridoaluminate, $LiAlH_4$, is used in solution in ethoxyethane, which must be dry because the hydride reacts with water. It releases hydride ions which attack the carbonyl carbon atom. The reaction with aldehydes is very vigorous, and the less powerful reducing agent, sodium tetrahydridoborate, $NaBH_4$, can be used. This is a more convenient reagent as it can be used in aqueous or methanolic solution:

Ketones are reduced by H_2 or by $LiAlH_4$ in ether to give secondary alcohols

$$\mathrm{R(H)C{=}O} \xrightarrow[\text{or NaBH}_4\text{ (methanol/water)}]{\text{LiAlH}_4\text{ (ethoxyethane)}} \mathrm{R(OH)C(H)H}$$

$$\mathrm{R'(R)C{=}O} \xrightarrow{\text{LiAlH}_4\text{ (ethoxyethane)}} \mathrm{R'(OH)C(R)H}$$

31.7.2 OXIDATION

Aldehydes are oxidised to carboxylic acids ...

Aldehydes are readily oxidised to carboxylic acids by acidified potassium dichromate(VI) or acidified potassium manganate(VII). Alternatively, the aldehyde vapour can be passed with air over a heated catalyst

$$\mathrm{R{-}C(H){=}O(g)} + \mathrm{O_2(g)} \xrightarrow[\text{Pt or Cu as catalyst}]{\text{heat}} \mathrm{R{-}C({=}O)OH}\ (g)$$

Aromatic aldehydes are less easily oxidised than aliphatic aldehydes.

... Ketones are more difficult to oxidise

Ketones are more difficult to oxidise than aldehydes. With a powerful oxidising agent, in acid conditions, they are oxidised to a mixture of carboxylic acids.

31.7.3 DISTINGUISHING BETWEEN ALDEHYDES AND KETONES

The difference in the ease of oxidation provides a means of distinguishing between aldehydes and ketones

The relative ease with which aldehydes are oxidised gives a means of distinguishing them from ketones. The gentle oxidising agents, Benedict's reagent and Fehling's solution, contain complex copper(II) ions. They are reduced to a reddish precipitate of copper(I) oxide, Cu_2O, by aldehydes but not by ketones. Tollens' reagent, a solution of complex silver ions, $[Ag(NH_3)_2]^+$, is reduced to a silver mirror when warmed with aldehydes but not by ketones.

31.7.4 ADDITION REACTIONS

Hydrogen cyanide

The carbonyl group is attacked by nucleophiles. As mentioned in § 31.4.1, the cyanide ion, CN^-, is such a nucleophile. When hydrogen cyanide is added to a cold (10–20 °C) solution of a carbonyl compound, it adds across the double bond. The reaction occurs rapidly in the presence of a base and is

HCN adds across >C=O to give a hydroxynitrile (a cyanhydrin) ...

inhibited by acids:

$$CH_3(C_2H_5)C{=}O(l) + HCN(g) \xrightarrow{\text{base}} CH_3(C_2H_5)C(OH)(CN)(s)$$

Butanone → 2-Hydroxy-2-methylbutanenitrile (the cyanhydrin of butanone)

The product that is formed is described as the **cyanhydrin** of the carbonyl compound. It is a **hydroxynitrile**. The **nitrile** group

$$-C{\equiv}N$$

can be hydrolysed to the carboxyl group

... which can be hydrolysed to a hydroxyacid

$$-CO_2H$$

The hydrolysis of hydroxynitriles is a useful way of making hydroxycarboxylic acids:

$$CH_3-CHO \xrightarrow{HCN} CH_3-CH(OH)-CN \xrightarrow[\text{reflux}]{\text{conc. HCl(aq)}} H_3C-CH(OH)-CO_2H$$

Ethanal → 2- Hydroxypropanenitrile → 2-Hydroxypropanoic acid (lactic acid)

Note that the reaction with hydrogen cyanide is a method of adding one carbon atom to the chain.

To avoid the danger of working with the poisonous gas hydrogen cyanide, it is generated in the reaction mixture by the action of dilute sulphuric acid on potassium cyanide.

* Sodium hydrogensulphite

$NaHSO_3$ adds across the C=O bond to give a crystalline hydrogensulphite compound, from which the carbonyl compound can be regenerated

Another reagent which adds across the carbonyl group is sodium hydrogensulphite:

$$(C_2H_5)_2C{=}O + NaHSO_3 \underset{\text{acid or alkali}}{\overset{\text{excess of } NaHSO_3}{\rightleftharpoons}} (C_2H_5)_2C(OH)(SO_3Na)$$

Pentan-3-one → Sodium hydrogensulphite compound of pentan-3-one

These derivatives are used to purify carbonyl compounds ...

Aldehydes react more readily than ketones, possibly on account of the steric hindrance to the approach of the attacking reagent offered by two alkyl groups. The reaction is carried out by shaking together the carbonyl compound and an excess of a saturated sodium hydrogensulphite solution at room temperature. The hydrogensulphite derivative crystallises out of the solution. On treatment with acid or alkali, it liberates the free aldehyde or ketone. The pair of reactions are used to purify carbonyl compounds. The hydrogensulphite compound is precipitated, recrystallised from water and then decomposed.

... and for conversion into hydroxynitriles

With sodium cyanide, hydrogensulphite compounds react to give hydroxynitriles (cyanhydrins):

$$C(C_2H_5)(OH)(CH_3)(SO_3^-\,Na^+) + Na^+\,CN^- \longrightarrow C(C_2H_5)(OH)(CH_3)(CN) + 2Na^+ + SO_3^{2-}$$

Sodium hydrogensulphite compound of butanone — 2-Hydroxy-2-methylbutanenitrile

Hydroxynitriles are often made from carbonyl compounds by these two steps in order to avoid the use of hydrogen cyanide.

31.7.5 ADDITION–ELIMINATION (OR CONDENSATION) REACTIONS

Compounds XNH_2 give products as the result of addition followed by elimination

Carbonyl compounds form adducts with many derivatives of ammonia, $\mathbf{X}NH_2$, adducts which immediately eliminate a molecule of water to form a product $R_2C{=}N\mathbf{X}$. In the reactions with hydrazine, the products are **hydrazones**:

$$RR'C{=}O + H_2NNH_2 \rightleftharpoons [RR'C(OH)(NHNH_2)] \longrightarrow RR'C{=}N{-}NH_2 + H_2O$$

Hydrazine — Intermediate — A hydrazone

Since H_2O is eliminated, these **addition–elimination** reactions are described as **condensation** reactions [see Table 31.7A].

Table 31.7A Addition–elimination reactions of carbonyl compounds (Reactions of RCOR′, where R can be H (in aldehydes) or an alkyl group (in ketones).)

Reaction with a weak base	Product
Hydrazine H_2NNH_2	A hydrazone $RR'C{=}N{-}NH_2 + H_2O$
Phenylhydrazine $H_2N{-}NH{-}C_6H_5$	A phenylhydrazone $RR'C{=}N{-}NH{-}C_6H_5 + H_2O$
2,4-Dinitrophenylhydrazine $H_2N{-}NH{-}C_6H_3(NO_2)_2$	A 2,4-dinitrophenylhydrazone $RR'C{=}N{-}NH{-}C_6H_3(NO_2)_2 + H_2O$

... Brady's reagent is used to detect carbonyl compounds ...

A solution of 2,4-dinitrophenylhydrazine in methanol and sulphuric acid is called Brady's reagent. When it is added to a solution of a carbonyl compound, the solid 2,4-dinitrophenylhydrazone separates rapidly. Aliphatic carbonyl compounds give orange or yellow derivatives, while those of aromatic carbonyl compounds are darker in colour. Brady's reagent can therefore be used to test for the presence of carbonyl compounds. The reaction is quantitative, and the precipitate can be dried and weighed to reveal the amount of carbonyl compound present.

... and to identify them from the melting temperatures of their 2,4-dinitrophenylhydrazones

The 2,4-dinitrophenylhydrazone derivatives can also be used to identify carbonyl compounds. They are solids of limited solubility in ethanol so they can be recrystallised and obtained pure. The melting temperatures can be taken and compared with the melting temperatures of a list of 2,4-dinitrophenylhydrazones of known aldehydes and ketones.

A number of useful plastics are formed by addition–elimination reactions. Methanal is used to make Formica® and Bakelite®; see § 34.5.5.

31.7.6 CHLORINATION

Chlorine replaces the hydrogens on the 2-carbon atoms ...

... e.g., in the formation of CCl_3CHO, chloral

Chlorine reacts with carbonyl compounds. The hydrogen atoms on the 2-carbon atoms (those adjacent to the carbonyl atoms) are easily substituted by chlorine atoms. This is because the electron-withdrawing effect of the carbonyl group weakens the C—H bonds. When chlorine is bubbled through ethanal, trichloroethanal, CCl_3CHO, is formed. Propanone forms a mixture of products, and benzaldehyde forms benzoyl chloride:

$$CH_3C(H){=}O + 3Cl_2 \longrightarrow CCl_3C(H){=}O + 3HCl$$

Ethanal → 2,2,2-Trichloroethanal

$$CH_3COCH_3 \xrightarrow{Cl_2} CH_3COCH_2Cl \xrightarrow{Cl_2} CH_3COCHCl_2 \longrightarrow \text{etc.}$$

Propanone → Chloropropanone → 1,1-Dichloropropanone

$$C_6H_5C(H){=}O + Cl_2 \longrightarrow C_6H_5C(Cl){=}O + HCl$$

Benzaldehyde → Benzoyl chloride

31.7.7 THE HALOFORM REACTION

Chloroform from CH_3COR, chlorine and alkali

Methyl carbonyl compounds, i.e., compounds containing a

$$CH_3{-}C({=}O){-}C{<} \quad \text{group and} \quad CH_3{-}C({=}O){-}H$$

give chloroform, $CHCl_3$, when warmed with an aqueous solution of sodium chlorate(I), NaClO (or with chlorine and alkali):

$$CH_3(R)C{=}O + 3Cl_2 \longrightarrow CCl_3(R)C{=}O + 3HCl$$

$$\begin{matrix} CCl_3 \\ \quad\backslash \\ \quad C{=}O + OH^- \longrightarrow CHCl_3 + RCO_2^- \\ / \\ R \end{matrix}$$

Iodoform from CH_3COR and I_2 and alkali or NaIO ...

If potassium bromide or potassium iodide is dissolved in the reacting mixture, bromoform, $CHBr_3$, or iodoform, CHI_3, is formed. Iodoform, tri-iodomethane, is precipitated as fine yellow crystals with a characteristic smell. The reaction is used as a test for the group

$$CH_3{-}\underset{\underset{O}{\|}}{C}{-}\overset{|}{\underset{|}{C}}{-}$$

... and from $CH_3CH(OH)R$

Alcohols of formula $CH_3CH(OH)R$ are oxidised by sodium iodate(I) to CH_3COR and therefore give a positive iodoform test:

$$\underset{\text{Ethanol}}{CH_3CH_2OH} \xrightarrow{NaIO} \underset{\text{Ethanal}}{CH_3CHO} \xrightarrow{NaIO} \underset{\text{Tri-iodo-ethanal}}{CI_3CHO} \xrightarrow{OH^-} \underset{\text{Tri-iodomethane}}{CHI_3}$$

Figure 31.7A
A summary of the reactions of aldehydes

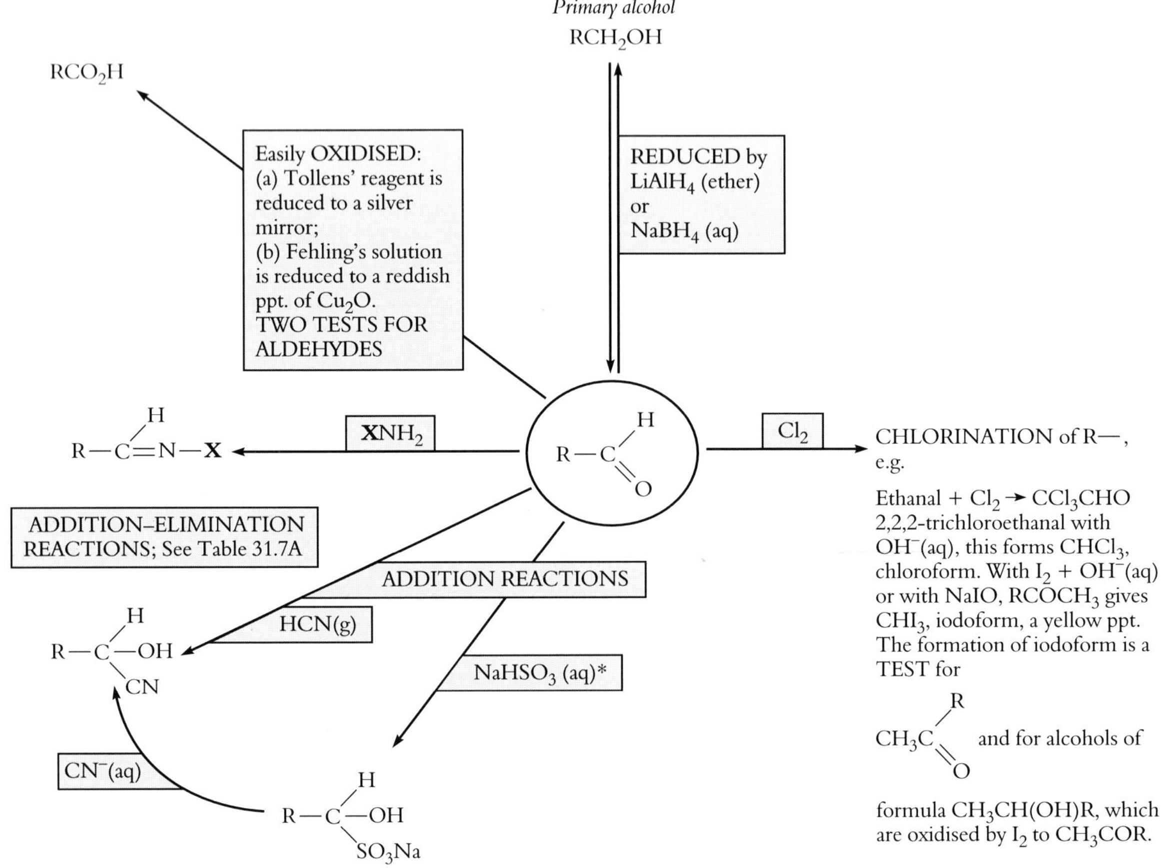

Table 31.7B A Comparison of aliphatic and aromatic aldehydes

	Ethanal, CH_3CHO	Benzaldehyde, C_6H_5CHO
Reagent	Product	Product
$KMnO_4$, acid	CH_3CO_2H	$C_6H_5CO_2H$
$LiAlH_4$ (ethoxyethane)	CH_3CH_2OH	$C_6H_5CH_2OH$
HCN	$CH_3CH(OH)CN$	$C_6H_5CH(OH)CN$
* $NaHSO_3$	$CH_3CH(OH)SO_3Na$	$C_6H_5CH(OH)SO_3Na$
$C_6H_5NHNH_2$	$CH_3CH{=}NNHC_6H_5$	$C_6H_5CH{=}NNHC_6H_5$
Cl_2	CCl_3CHO	C_6H_5COCl
Cl_2 + Halogen-carrier	—	3-ClC_6H_4CHO
Conc. HNO_3 + conc. H_2SO_4	—	3-$O_2NC_6H_4CHO$
Hot conc. H_2SO_4	—	3-$HO_3SC_6H_4CHO$
Fehling's solution	Red Cu_2O	—
Tollens' reagent	Ag mirror	—
KI, NaClO	CHI_3, iodoform	—

Figure 31.7B A summary of the reactions of ketones

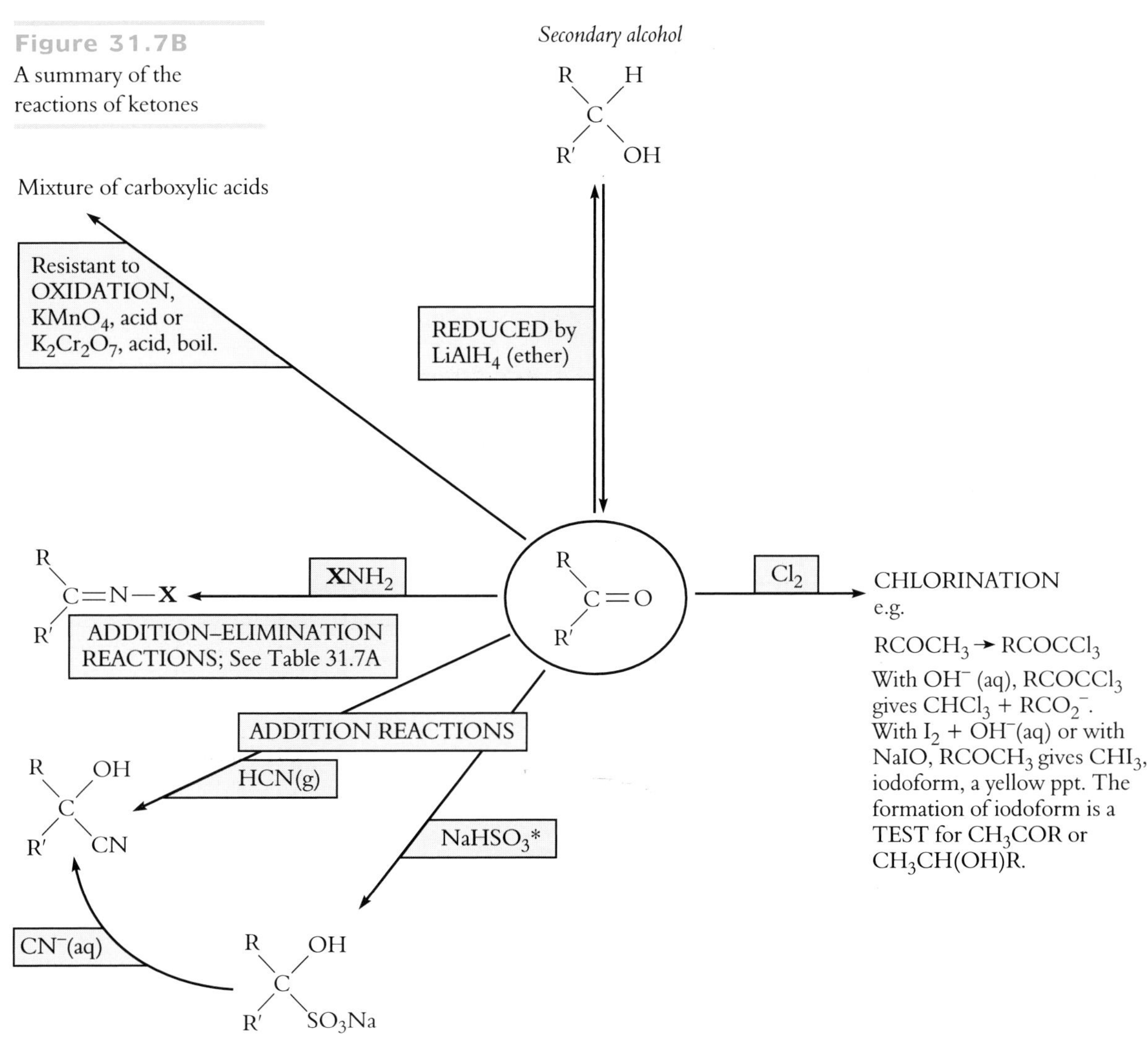

CHECKPOINT 31.7: REACTIONS

1. Draw structural formulae for
   (a) propanone hydrazone
   (b) propanone phenylhydrazone
   (c) pentan-2-one phenylhydrazone
   (d) butanone 2,4-dinitrophenylhydrazone
   (e) * the hydrogensulphite compound of benzaldehyde.

2. Write the formulae for the organic products formed in these reactions:
   (a) $C_2H_5COCH_3 + C_6H_5NHNH_2 \longrightarrow$
   (b) $C_6H_5CHO + LiAlH_4 \longrightarrow$
   (c) $C_2H_5CHO + [Ag(NH_3)_2]^+(aq) \longrightarrow$
   (d) $C_6H_5CH{=}CHCHO + H_2 \xrightarrow{Pd}$
   (e) $C_6H_5CH{=}CHCHO + NaBH_4 \longrightarrow$
   (f) $C_2H_5COCH_3 + LiAlH_4 \longrightarrow$

3. State two reactions of aldehydes which are also characteristic of ketones and two which are not shared by ketones. What is the reason for the difference?

4. Give the structural formulae of the products of the reactions of butanone with
   (a) HCN
   (b) $C_6H_5NHNH_2$
   (c) Br_2
   (d) $I_2 + NaOH(aq)$
   (e) $LiAlH_4$

5. State the reagents and conditions needed for the preparation of the following from ethanal:
   (a) chloroform, (b) ethene, (c) ethyl ethanoate.

6. Of the compounds **A** to **H**, which compounds (a) give a positive iodoform test, (b) are reduced to a secondary alcohol?
   A $CH_3CH_2COCH_2CH_3$
   B $CH_3CH_2CH_2CH_2CHO$
   C $C_6H_5COCH_3$
   D C_6H_5CHO
   E $CH_3COC_2H_5$
   F $CH_3CH_2CH(OH)CH_3$
   G CH_3CH_2CHO
   H ICH_2CHO

7. Which of the following compounds reacts faster with sodium iodate(I): $CH_3CH(OH)CH_2CH_2OH$ or $CH_3CH_2COCH_3$? Explain your choice and state the products of the reactions.

*31.8 THE MECHANISMS OF THE REACTIONS OF ALDEHYDES AND KETONES

In the $\overset{\delta+}{C}{=}\overset{\delta-}{O}$ *group, there are two reactive sites ...*

Aldehydes and ketones possess the polar carbonyl group

$$\overset{\delta+}{C}{=}\overset{\delta-}{O}$$

Carbonyl compounds undergo addition reactions and condensation reactions. There are two possibilities to be considered: the δ– oxygen atom of the carbonyl group may react with an electrophile, or the δ+ carbon atom of the carbonyl group may react with a nucleophile.

31.8.1 THE ADDITION OF HYDROGEN CYANIDE TO FORM A HYDROXYNITRILE

... In the addition of HCN, it is the $\overset{\delta+}{C}$ *that is attacked by* CN^-*...*

$$RR'C{=}O + HCN \longrightarrow RR'C(OH)CN$$

The rate of reaction is increased by the presence of a base and decreased by hydrogen ions. This is what would happen if the concentration of cyanide ions were a crucial factor. Hydrogen cyanide dissociates:

$$HCN(aq) + H_2O(l) \rightleftharpoons H_3O^+(aq) + CN^-(aq)$$

... The evidence comes from kinetic studies

Addition of hydrogen ions suppresses dissociation, while addition of hydroxide ions removes hydrogen ions and increases the degree of dissociation and the concentration of CN^- ions. It is likely that the rate-determining step (R.D.S.) involves CN^- ions. There are two possibilities:

$$>C{=}O + H^+(aq) \underset{\text{fast}}{\rightleftharpoons} >\overset{+}{C}{-}O{-}H \xrightarrow[\text{R.D.S.}]{CN^-} >C(CN)(OH) \qquad [1]$$

$$>C{=}O + CN^-(aq) \xrightarrow[\text{R.D.S.}]{} >C(CN)(O^-) \xrightarrow[\text{fast}]{H^+(aq)} >C(CN)(OH) \qquad [2]$$

In [1], the step involving CN^- is the attack on a positive ion. This type of reaction is fast, and is unlikely to be the rate-determining step. In [2], CN^- is involved in an attack on the polar carbonyl group. The second step is a fast reaction between $H^+(aq)$ and a negatively charged intermediate. Mechanism [2] involves CN^- in the rate-determining step, and is more likely to be correct.

A summary of the mechanism of hydroxynitrile formation

The proposed mechanism for the addition reactions of aldehydes and ketones is illustrated by the formation of a hydroxynitrile:

$$HCN(aq) \rightleftharpoons H^+(aq) + CN^-(aq)$$

$$RR'C{=}O + CN^- \xrightarrow[\text{R.D.S.}]{} RR'C(CN)O^- \xrightarrow[\text{fast}]{H^+(aq)} RR'C(CN)OH$$

31.8.2 A COMPARISON OF ALKENES AND CARBONYL COMPOUNDS

The addition reactions of carbonyl compounds contrast with those of alkenes.

Addition to C=O starts with attack by a nucleophile; addition to C=C starts with attack by an electrophile

The first step in addition to $>C{=}O$ is the addition of a nucleophile, Nu: (e.g. CN^-) to form $>C(Nu){-}O^-$

In the addition to $>C{=}C<$ the first step is the addition of an electrophile (e.g. $Br^{\delta+}{-}Br^{\delta-}$). A nucleophile does not add to a carbon–carbon double bond because it is repelled by the unpolarised π electrons of the C=C bond.

The hypothetical adduct $>C(Nu){-}\overset{-}{C}<$ is not formed.

Carbon is less electronegative than oxygen, and this type of species is much less stable than

$>C(Nu){-}O^-$ where the negative charge is carried by an oxygen atom.

Figure 31.8A Aldehydes in the preparation of other classes of compounds

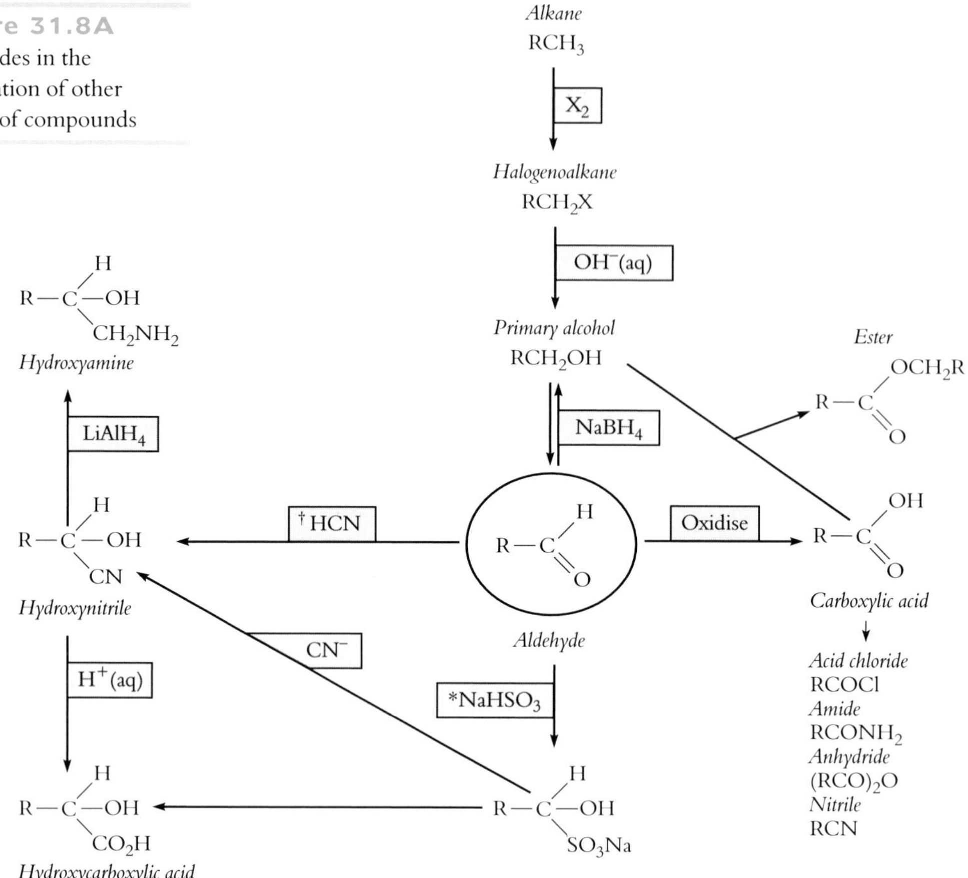

† The addition of HCN increases the length of the chain by 1 carbon atom.

Figure 31.8B Some of the ways of making other classes of compounds from ketones

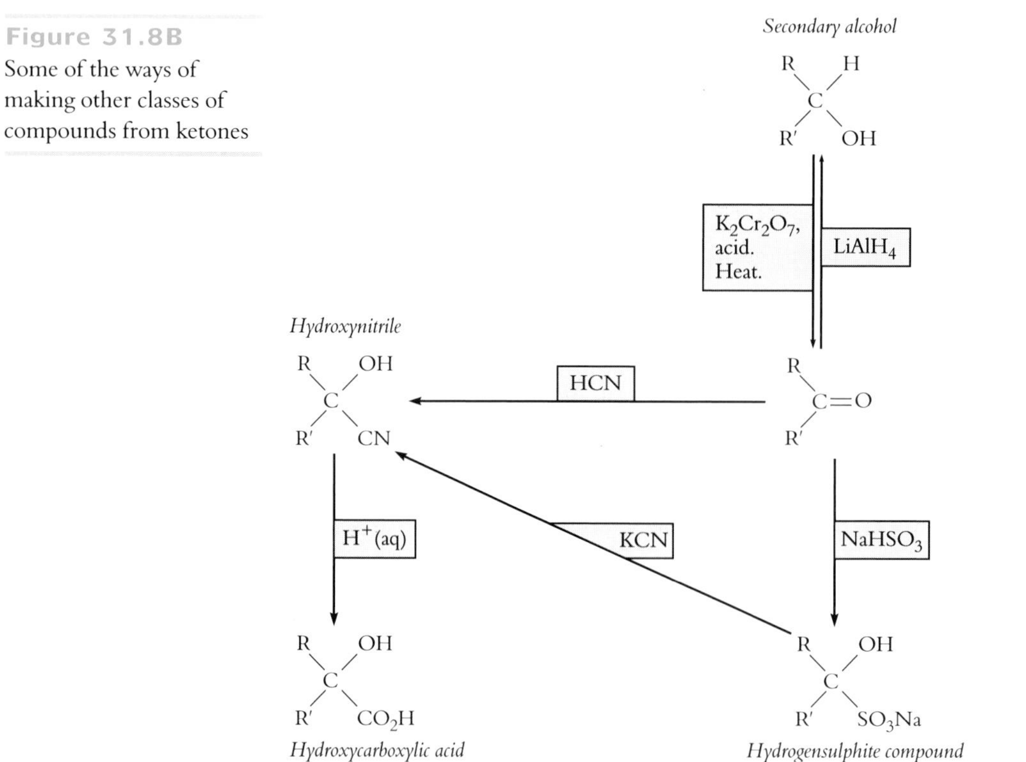

CHECKPOINT 31.8: REACTIVITY

1. The alkene group reacts with compounds of formula HX. What can **X** be?
The carbonyl group reacts with compounds of formula HY. What can **Y** be? Compare the reactivity of the alkene group and the carbonyl group in reactions of this type.

2. The formation of butanone hydroxynitrile is speeded up by the addition of cyanide ions and retarded by the addition of hydrogen ions. Explain these observations.

3. Why does HBr attack $\rangle C{=}C\langle$ but not $\rangle C{=}O$?

4. *(a)* Sketch the activated complex that is formed in the reaction

$$OH^- + CH_3CH_2Br \longrightarrow CH_3CH_2OH + Br^-$$

How many pairs of bonding electrons surround the carbon atom attached to the halogen? Which atoms carry negative charge?

(b) Sketch the intermediate in the reaction

$$HCN + CH_3CHO \longrightarrow CH_3CH(OH)CN$$

How many pairs of bonding electrons surround the central carbon atom? Which atoms carry negative charge?

(c) Which of the two species, the activated complex in *(a)* or the intermediate in *(b)*, appears to you to be the more stable? Explain your answer.

*31.9 CARBOHYDRATES

The general formula of carbohydrates is $C_m(H_2O)_n$

Carbohydrates are an important, naturally occurring group of compounds which contain carbonyl groups. This group includes sugars and starch, cellulose and a number of antibiotics. The name carbohydrate is derived from the molecular formulae of these compounds, which in many cases can be written $C_m(H_2O)_n$, e.g., glucose, $C_6H_{12}O_6$.

Sugars contain one or more carbonyl groups and other functional groups

Carbohydrates are polyfunctional. Glucose contains five hydroxyl groups as well as an aldehyde group. Fructose contains a ketone group. Both glucose and fructose have the formula $C_6H_{12}O_2$ and are **monosaccharides**:

$$\begin{array}{c} CHO \\ | \\ H-C-OH \\ | \\ HO-C-H \\ | \\ H-C-OH \\ | \\ H-C-OH \\ | \\ CH_2OH \end{array} \qquad\qquad \begin{array}{c} CH_2OH \\ | \\ C=O \\ | \\ HO-C-H \\ | \\ H-C-OH \\ | \\ H-C-OH \\ | \\ CH_2OH \end{array}$$

(+)-Glucose (an **aldose**) (−)-Fructose (an **ketose**)

31.9.1 DISACCHARIDES

Sucrose, maltose and lactose are disaccharides

Disaccharides have the formula $C_{12}H_{22}O_{11}$. They are hydrolysed by dilute acids and by enzymes to monosaccharides. Sucrose (cane or beet sugar), maltose (in malt) and lactose (in milk) are the commonest disaccharides.

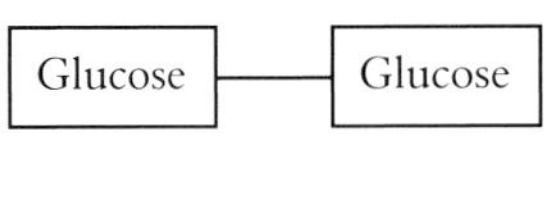

Maltose: 2 glucose ring molecules linked by the elimination of H_2O between 2OH groups.

Glucose — Fructose

Sucrose: a glucose molecule and a fructose molecule linked as in maltose.

Figure 31.9A The disaccharides maltose and sucrose

31.9.2 POLYSACCHARIDES

Starch is made by photosynthesis ...

Starch and cellulose are **polysaccharides**. They are polymers of glucose. Starch is formed as a result of photosynthesis in green plants:

$$6CO_2(g) + 6H_2O(l) \xrightarrow[\text{sunlight and chlorophyll in green plants}]{\text{photosynthesis}} \underset{\text{Glucose}}{C_6H_{12}O_6(aq)} + 6O_2(g)$$

$$\underset{\text{Glucose}}{nC_6H_{12}O_6(aq)} \longrightarrow \underset{\text{Starch}}{(C_6H_{10}O_5)_n(s)} + nH_2O(l)$$

... A test for starch is the formation of a blue colour with iodine

Starch is found in potatoes, cereals and rice. With iodine, it gives an intense blue colour, which is used as a test for iodine or for starch. Starch does not reduce Fehling's solution or Tollens' reagent.

Cellulose is another polymer of glucose

Cellulose is the main structural component of cell walls, and wood contains large quantities of cellulose. It cannot be digested by animals: they do not produce the enzymes which will hydrolyse it. Cows have living in their alimentary canals bacteria which hydrolyse cellulose.

CARBONYL COMPOUNDS

Aldehydes, **R**CHO, can be made by oxidation of primary alcohols, **R**CH_2OH. **Ketones** can be made by the oxidation of secondary alcohols, **R**$_2$CHOH, by e.g. acidified dichromate. §§ 31.6.1–3

→

Common to **aldehydes and ketones** are:

- **addition reactions** with nucleophiles, e.g. +HCN → hydroxynitrile, **R**$_2$COHCN §§ 31.4, 31.4.1, 31.4.2, 31.8
  mechanism. § 31.8
- addition–elimination reactions with e.g. 2,4-dinitrophenylhydrazine (Brady's reagent) give coloured crystalline solids used for identification. § 31.7.5

↓

Aldehydes, **R**CHO, are reduced to primary alcohols by e.g. $NaBH_4$. § 31.7.1

Ketones, **R**$_2$CO, are reduced to secondary alcohols by e.g. $NaBH_4$. § 31.7.1

↓

Oxidation
Aldehydes are oxidised to carboxylic acids by e.g. acidified dichromate; ketones are difficult to oxidise. § 31.7.2
Aldehydes (not ketones) **reduce**

- Tollens' reagent, $[Ag(NH_3)_2]^+(aq)$ to Ag,
- Benedict's solution and Fehling's solution, Cu^{2+} complex ions, to Cu_2O. § 31.7.3

→

The **iodoform test** for $CH_3\overset{|}{C}{=}O$ or CH_3CHOH—: Add I_2 + OH^-(aq) or KIO; tri-iodomethane (iodoform), CHI_3, is formed. § 31.7.7

QUESTIONS ON CHAPTER 31

1. Pentan-2-ol yields a ketone when heated with potassium dichromate solution under acid conditions.
(a) Give the structural formula and name of the ketone.
(b) What is the change in the oxidation number of chromium during the reaction?
(c) What other reagents will oxidise the alcohol?
(d) How can the product be identified as a ketone?

2. State the reactions of propanone with *(a)* acidified potassium manganate(VII), *(b)* lithium tetrahydridoaluminate, *(c)* iodine and alkali, *(d)* hydrogen cyanide. Why is reaction *(d)* so slow in the absence of a catalyst? What catalyst is used to speed up the reaction? How does it function?

3. Outline a method for the preparation of propanone from propene. What structural isomer of propanone exists? How would you distinguish between the two isomers?

4. A compound, **P**, of formula C_6H_{10}, was subjected to reductive ozonolysis. **Q**, $C_6H_{10}O_2$, was formed. **Q** gave a silver mirror with ammoniacal silver nitrate solution and reacted with 2,4-dinitrophenylhydrazine. One mole of **Q** reacted with 2 moles of DNP to form a yellow, crystalline solid. Deduce the identity of **P** and **Q**.

5. **A** has the formula $C_5H_{12}O$. On oxidation it gives **B**, of formula $C_5H_{10}O$. **B** reacts with phenylhydrazine and gives a positive result in an iodoform test. **A** is dehydrated by concentrated sulphuric acid to **C**, C_5H_{10}. Reductive ozonolysis of **C** gives butanal. What is **A**?

6. **P** has the formula $C_5H_8O_2$. It forms a compound by reaction with hydrogen cyanide which has the formula $C_7H_{10}O_2N_2$. **P** gives a positive iodoform test, a silver mirror with Tollens' reagent and can be reduced to pentane. What is **P**?

7. Describe chemical tests which you could do to distinguish between the following compounds:
(a) C_2H_5CHO, CH_3COCH_3, $C_2H_5CH_2OH$ and $(CH_3)_2CHOH$
(b) C_6H_5CHO, $C_6H_5CH_2OH$, $C_6H_5COCH_3$ and $C_6H_5CH{=}CH_2$

8. How can the following conversions be brought about?
(a) $CH_3COCH_3 \longrightarrow CH_3COCH_2OH$
(b) $CH_3COCH_3 \longrightarrow CH_3CH{=}CH_2$
(c) $C_6H_5CHO \longrightarrow C_6H_5CH_2Br$

9. *(a)* Compare the reactions of ethanal and propanone with (i) aqueous ammoniacal silver nitrate, (ii) acidified potassium dichromate (VI), (iii) hydrogen cyanide, (iv) 2,4-dinitrophenylhydrazine.
(b) How does the presence of a catalyst assist reaction (iii)?

10. This question is about ethanal, CH_3CHO
(a) Give a reagent or set of reagents that will convert ethanal into (i) CH_3CO_2H (ii) $CHI_3 + HCO_2Na$ (iii) $CH_3CHOHCO_2H$
(b) Give balanced equations for reactions (ii) and (iii).
(c) Give the structural formulae of the products obtained when ethanal reacts with 2,4-dinitrophenylhydrazine.

11. Some reactions of benzaldehyde, C_6H_5CHO, are represented below.

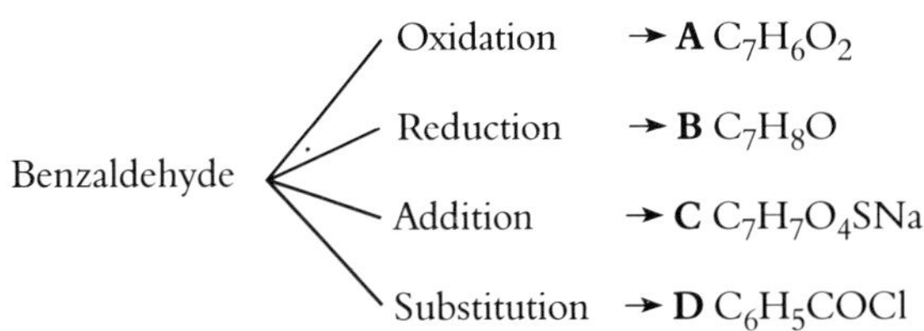

(a) Give the structural formula of each compound **A–D**
(b) For each reaction, give the reagents and conditions needed.
(c) Explain how the addition reactions of benzaldehyde differ from those of alkenes.
(d) Give two chemical tests that could be used to distinguish between benzaldehyde and phenylethanone, $C_6H_5COCH_3$.

12. An aliphatic aldehyde **A** has the formula RCHO.
(a) A reacts with 2,4-dinitrophenylhydrazine. Explain what happens and name the type of reaction. Say how the product of the reaction could be used to identify **A**.
(b) When **A** is treated with warm, acidified potassium dichromate solution, **B** is formed. Give the structural formula of **B**.
(c) When **A** reacts with lithium tetrahydridoaluminate(III) in ethoxyethane solution, **C** is formed. Give the structural formula of **C**.
(d) **A** is warmed gently with ammoniacal silver nitrate. Explain what happens, and say what is observed.
(e) **B** and **C** react to form **D**. Write the structural formula of **D**.
(f) Of the compounds **A**, **B**, **C** and **D**, which would you expect to have (i) the highest boiling temperature (ii) the lowest boiling temperature? Explain your answers.

13. *(a)* Propene and propanal undergo addition reactions. Compare the mechanisms of the addition of hydrogen bromide to propene and the addition of hydrogen cyanide to propanal.

(b) Propanal and propanone differ in their behaviour towards oxidising agents. Describe a reaction with an oxidising agent that can be used to distinguish propanal from propanone and that can be carried out in a test tube. Say what would happen with each compound.

14. *(a)* Ethanal can be prepared in the laboratory from ethanol. Acidified potassium dichromate(VI) is slowly added to ethanol, while the mixture is heated to distil off the product.

(i) Draw a labelled diagram to show the apparatus used.

(ii) Explain the advantage of distilling off the product continuously.

(b) Hydrogen cyanide adds to ethanal in the presence of a small amount of potassium cyanide. Write the equation for the reaction. Sketch the mechanism of the addition, and explain why potassium cyanide helps the reaction.

15. A carbonyl compound **X** undergoes the following reactions:

X gives an orange precipitate with 2,4-dinitrophenylhydrazine.

X gives a pale yellow precipitate with sodium iodate(I).

X does not react with warm, acidified potassium dichromate(VI).

X does not react with aqueous bromine.

X is reduced by hydrogen in the presence of a catalyst to a mixture of isomers **Y** and **Z**, of formula $C_4H_{10}O$.

Identify **X**, and give the structural formulae of **X**, **Y** and **Z**.

16. *(a)* Three compounds **E**, **F** and **G** all have the molecular formula C_3H_6O. **E** is an alcohol, **F** is a ketone, and **G** is an aldehyde.

(i) Draw possible structural formulae for **E**, **F** and **G**.

(ii) Describe tests (reagents, conditions and observations with **each** compound) that would allow you to show that:

1. **E** is an alcohol, whereas **F** and **G** are not.
2. **F** and **G** are carbonyl compounds, whereas **E** is not.
3. **G** is an aldehyde, whereas **E** and **F** are not.

Write balanced equations for all reactions that occur. **9**

(b) One of the compounds responsible for the flavour of butter is butane-2,3-dione:

$$CH_3-\underset{\underset{O}{\|}}{C}-\underset{\underset{O}{\|}}{C}-CH_3$$

Give the structural formulae of the organic products formed when butane-2,3-dione reacts completely with (i) H_2/Ni, (ii) $I_2/OH^-(aq)$. **3**

C (AS/AL)

17. Cinnamaldehyde, one of the stereoisomers of which has the structure

(benzene ring)–CH=CH–CHO, drawn with H and H on opposite sides of C=C:

H (upper right), C=C, H (lower left), CHO (lower right)

is used in fragrances for its jasmine-like odour. It contains two functional groups, other than the benzene ring, which may be assumed to have their normal reactions and to behave independently of each other.

(a) Name the two functional groups present in the molecule. **2**

(b) The presence of unsaturation in this molecule can be shown by the reaction with bromine dissolved in hexane.

(i) What is the observable result of this test? **1**

(ii) Give the equation for the reaction occurring. **1**

(iii) Give the mechanism for this reaction. You may use —CH=CH— to represent the molecule of cinnamaldehyde if you wish. **3**

(iv) What type of mechanism is this? **1**

(c) How would you test for the presence of a >C=O group in this molecule? **2**

(d) Give the structure of another stereoisomer of cinnamaldehyde and explain how it arises. **3**

(e) (i) Give the structure of a molecule produced by reduction of cinnamaldehyde. **1**

(ii) Suggest a reagent you would use to achieve this reduction. **1**

15 *L (AS/AL)*

18. Vanillin is a constituent ot the vanilla bean and has the structure

(benzene ring with OH at top, OCH_3 adjacent, and CHO at the bottom)

(a) In each of the following only **one** group of the vanillin molecule reacts. In each case state which group reacts and what you would observe, when

(i) an ethanolic solution of vanillin is treated with 2,4-dinitrophenylhydrazine reagent **1**

(ii) a few drops of an ethanolic solution of vanillin are added to Tollens' reagent and the mixture warmed gently. **1**

(b) Explain your observations for the reaction in (a)(ii). **1**

(c) (i) Write down the structure of the compound, **X**, which would be formed if vanillin were treated with sodium tetrahydridoborate(III), $NaBH_4$. **1**

(ii) Both vanillin and **X** react with dilute aqueous sodium hydroxide.

I. State the number of moles of sodium hydroxide which reacts with one mole of vanillin. **1**

II. State whether one mole of **X** would need the same number of moles of sodium hydroxide as reacts with one mole of vanillin. Give a reason to support your answer. **1**

(d) (i) Give the systematic name for iodoform. **1**

(ii) State the reagents required to carry out the iodoform reaction. **2**

(iii) State, giving a reason, whether vanillin would give a positive iodoform reaction. **1**

10 *WJEC (AL)*

19. (a) Outline a mechanism for the reaction of but-2-ene with bromine. **4**

(b) Give the structure of the organic product of the reaction between $CH_3COCH_2CH_3$ and aqueous $NaBH_4$. **1**

(c)

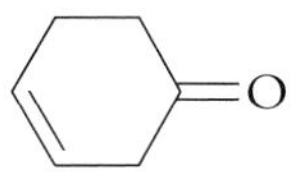

=O

X

Compound **X** reacts separately with HBr, with HCN and with an excess of hydrogen in the presence of a Ni catalyst.

(i) Name the type of reaction and give the structure of the product formed in the reaction of compound **X** with HBr.

(ii) Name the type of reaction and give the structure of the product formed in the reaction of compound **X** with HCN.

(iii) Suggest the structure of the fully saturated product formed when compound **X** reacts with an excess of hydrogen in the presence of a Ni catalyst. **5**

10 *NEAB (AS/AL)*

20. Compound **X** has the molecular formula C_5H_8O.

(a) Calculate the percentage composition of **X**. **2**

(b) Give the reagents you would use to show the presence of each of the following groups which are present in **X**. State what you would observe as the result of each test.

(i) $\gt C{=}C\lt$ (ii) $\gt C{=}O$ **4**

(c) Given that **X** also forms a silver mirror when warmed with ammoniacal silver nitrate solution and shows no geometrical isomerism, write TWO structures for **X**, neither of which is chiral. **2**

(d) Select ONE of your structures for **X**. This reacts with HBr to form two products, one of which is chiral.

(i) Give the structures of your two products, indicating which one is chiral.

(ii) Write the mechanism for the formation of either product.

(iii) Indicate which of your two products is likely to be in the greater yield. Give a reason for your choice. **7**

15 *L (AL)*

21. (a) Warfarin is a compound which has been used widely to kill rats and other rodents. Its molecule has the structure given below.

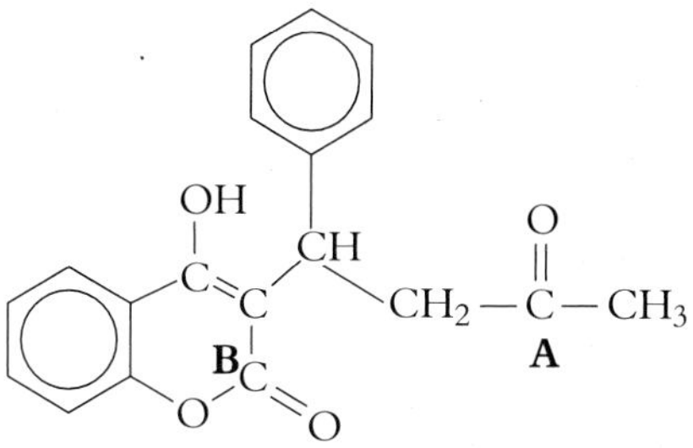

OH CH O

C=C CH₂—C—CH₃

B C A

O O

Name any **three** different functional groups present in the warfarin molecule. **3**

(b) State what **would be expected to be observed** when warfarin is treated under appropriate conditions with **each** of the following reagents. In **each** case give a reason for your prediction.

(i) Bromine in tetrachloromethane in the absence of a halogen carrier. **2**

(ii) Iodine and aqueous sodium hydroxide. **2**

(c) (i) The structure shows two $\gt C{=}O$ groups the carbon atoms of which are labelled **A** and **B**.
State how many molecules of 2,4-dinitrophenylhydrazine would react with one molecule of warfarin. Give a reason for your answer. **2**

(ii) State, giving a reason, whether you would expect warfarin to reduce Tollens' reagent. **1**

10 *WJEC (AL)*

32

AMINES

32.1 WILLIAM PERKIN AND MAUVE

In the early nineteenth century chemists did not know what to do with coal tar. The destructive distillation of coal (in the absence of air) gave coke, which they needed for smelting iron ores, coal gas, ammonia and coal tar. Some of the coal tar was used to make creosote, which is a wood preservative, but most of it was dumped in rivers and on waste ground. From 1830 onwards, chemists began to separate coal tar into fractions. They obtained benzene, methylbenzene, naphthalene, anthracene, phenol and phenylamine. In the early nineteenth century, the textile industry was the main source of employment for chemists. They extracted dyes from plants, e.g. indigo, madder and French purple, but were dissatisfied with the way many of the dyes faded rapidly. Chemists started to try to make dyes from their new coal tar derivatives.

The big breakthrough was made by William H Perkin. He was an 18 year old pupil at the City of London School whose enthusiasm for chemistry had led him to set up a small laboratory in the family home. During his Easter holidays in 1856, Perkin decided to try to make quinine by oxidising phenylamine (then called aniline), a coal tar derivative. Phenylamine is a primary aromatic amine [see § 32.2] with the formula $C_6H_5NH_2$. The result was not what he expected: he obtained a black solid from which he extracted a beautiful intense purple compound. He found that a solution of the compound would dye silk.

Inspite of his youth, Perkin decided to explore the possibility of marketing his discovery. He sent his new dye to a manufacturer who tested it and found that it was more colour-fast and more resistant to fading in sunlight than any similar dye available. Perkin patented his discovery. Then he went into business with his father and brother to manufacture the dye.

Perkin started preparing phenylamine from benzene. He was unaware that benzene from coal tar contained methylbenzene and his product therefore contained methylphenylamines. There were chemical problems to be solved in the dyeing process. The new dye took to silk and wool but not cotton. Existing mordants (substances which help dyes to stick to cloth) were only suitable for the acidic plant dyes. Perkin had to discover new mordants which were suited to his new dye, which he called *aniline purple*, and which was basic.

For some time, the firm of Perkin & Sons was unable to attract large orders. In 1857, William Perkin went on a promotion campaign, travelling through Britain, demonstrating how 'aniline purple' could be used on different fabrics. Fortunately, purple was a very fashionable colour. By 1858, the superiority of Perkin's dye over those of competitors had been demonstrated, and orders poured in. The apparatus had to be scaled up from bench scale to commercial scale. Perkin's 'aniline purple' became known as 'Mauve'. The fashion for Mauve lasted until 1863, and Perkin made a fortune.

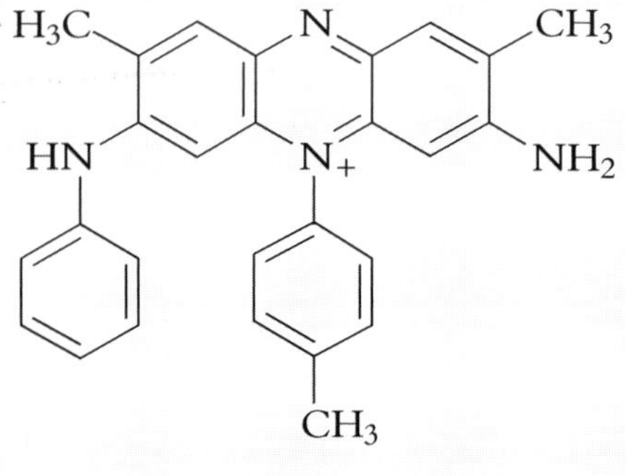

Mauveine, the chief component of Mauve

32.2 NOMENCLATURE FOR AMINES

Primary, secondary and tertiary (1°, 2°, 3°) amines ...

Amines are organic derivatives of ammonia. They are classified as primary, secondary or tertiary amines according to the number of alkyl or aryl groups attached to the nitrogen atom. Examples are:

... some examples ...

Aromatic amines ...

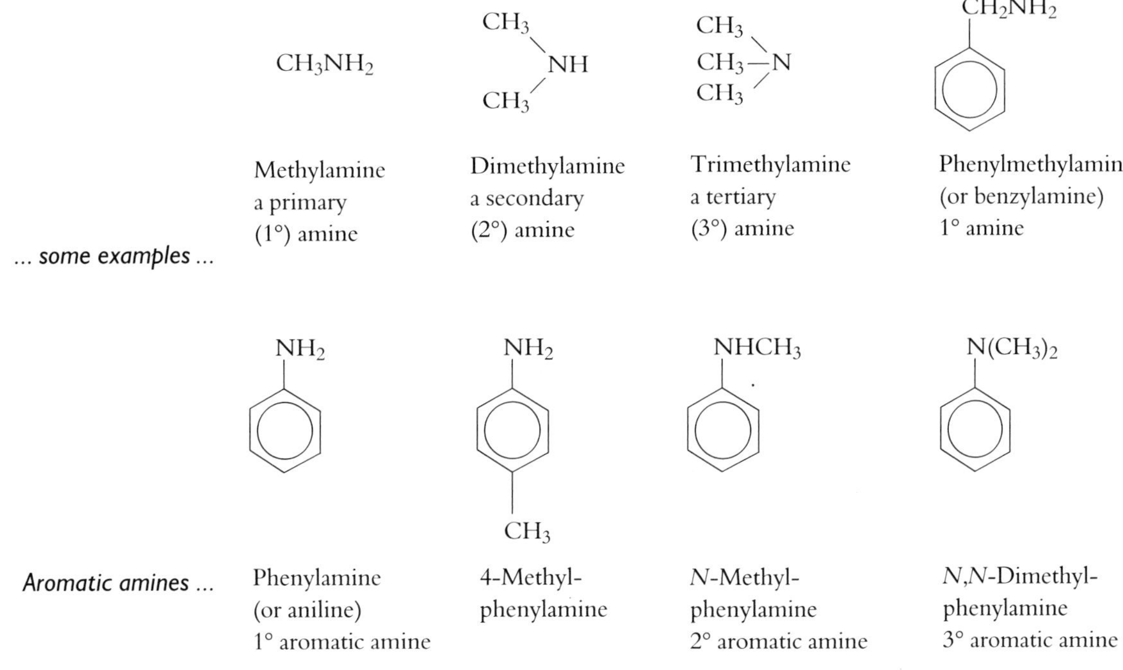

An aromatic amine has the nitrogen atom directly attached to the aromatic ring. In phenylmethylamine the nitrogen atom is not directly attached to the ring: this is not an aromatic amine; it is a phenyl-substituted alkylamine.

The system of naming is illustrated by the examples given above. The alkyl groups or aryl groups attached to the nitrogen atom are named, and the ending *-amine* is added. Some amines are named as amino-substituted compounds:

... Amino-compounds

$H_2NCH_2CO_2H$ — 2-Aminoethanoic acid

HO—C_6H_4—NH_2 — 4-Aminophenol

Related to the amines are the **quaternary** (4°) ammonium compounds, in which nitrogen is tetravalent. They are named as alkyl-substituted ammonium salts:

Quaternary ammonium compounds

$(CH_3)_4N^+I^-$ — Tetramethylammonium iodide

$CH_3—N^+(C_2H_5)(CH_3)—C_3H_7\ Br^-$ — Ethyldimethylpropylammonium bromide

Other nitrogen compounds will be covered in Chapter 33. They are the amides and nitriles. They are named as derivatives of acids:

Amides, nitriles

Carboxylic acid	R—C(=O)—OH	e.g. $CH_3CH_2C(=O)—OH$ Propanoic acid
Amide	R—C(=O)—NH_2	e.g. $CH_3CH_2C(=O)—NH_2$ Propanamide
Nitrile	R—C≡N	e.g. $CH_3CH_2C≡N$ Propanenitrile

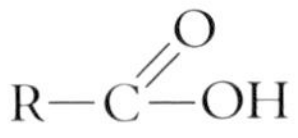

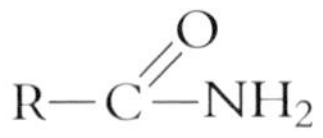

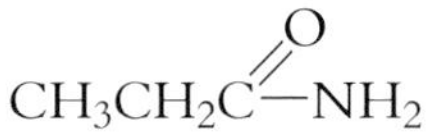

Other derivatives of acids which will be met in this chapter are:

Acid chlorides and anhydrides

Acid chloride	R—C(=O)—Cl	e.g. $CH_3CH_2C(=O)—Cl$ Propanoyl chloride
Acid anhydride	R—C(=O)—O—C(=O)—R	e.g. $CH_3CH_2C(=O)—O—C(=O)CH_2CH_3$ Propanoic anhydride

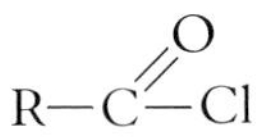

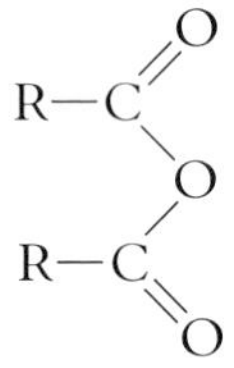

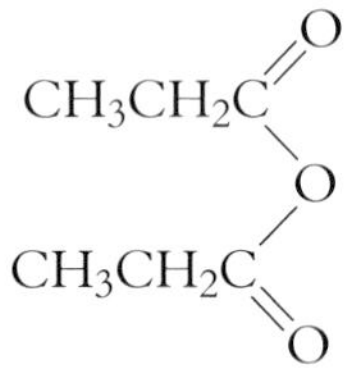

32.3 NATURAL OCCURRENCE

Proteins ... ... amino acids ... ... DNA ...

Compounds with amino groups are widely distributed in nature. Amino acids, of formula $H_2NCH(R)CO_2H$, are the building blocks from which proteins are made. Their chemistry is described in § 33.17. The way in which hydrogen bonding between amino groups and other groups maintains the three-dimensional configuration of proteins is illustrated in Figure 4.8K, § 4.8.3. The bases in DNA (deoxyribonucleic acid) are amines. Hydrogen bonding between the bases maintains the double helical structure of DNA, as shown in Figures 4.8L, M, § 4.8.3.

... drugs ... dyes

Many powerful drugs and medicines are amines, e.g., morphine (a powerful painkiller), valium (a tranquilliser), LSD (a hallucinogen), amphetamines (stimulants) and barbiturates (depressants) [§ 36.1]. Many dyes are amines [§ 32.1].

32.4 PHYSICAL PROPERTIES

Intermolecular hydrogen bonding is weak

The N—H bond is polar, more polar than C—H but less polar than O—H. Intermolecular hydrogen bonding in amines is therefore weaker than in alcohols, and the lower molecular mass amines (up to C_3) are gases at room temperature. The lower molecular mass amines dissolve in water as they can

Solubility *Toxicity*

form hydrogen bonds with water molecules. Phenylamine and other amines with a large hydrocarbon part of the molecule are only sparingly soluble in water but are soluble in organic solvents. Since phenylamine is soluble in fatty tissues, it can be absorbed through the skin. The ease of absorption, combined with its toxicity, makes phenylamine a somewhat dangerous substance.

Odour

Amines are formed when proteins decay

Amines have a characteristic smell. That of the lower members of the series resembles ammonia. Higher members have an odour described as 'fishy'. This is because amines are formed when protein material decomposes; dimethylamine and trimethylamine are found in rotting fish. Higher amines are found in decaying animal flesh.

32.5 BASICITY OF AMINES

Amines are weak bases, resembling ammonia

The nitrogen atom in ammonia has a lone pair of electrons, which enable it to act as a base ($pK_b = 4.74$) [§ 12.6.4]. In the same way, amines are weak bases. Their solutions are alkaline, and they react with acids to form salts:

$$CH_3NH_2{:} + H_2O \underset{pK_b = 3.36}{\rightleftharpoons} CH_3\overset{+}{N}H_3 + OH^-$$

$$(CH_3)_3N{:} + H_2O \underset{pK_b = 4.20}{\rightleftharpoons} (CH_3)_3\overset{+}{N}{-}H + OH^-$$

$$C_6H_5NH_2{:} + H_2O \underset{pK_b = 9.4}{\rightleftharpoons} C_6H_5\overset{+}{N}H_3 + OH^-$$

The gas methylamine reacts with hydrogen chloride to form a white crystalline solid, methylammonium chloride, $CH_3NH_3^+Cl^-$. The salt is involatile, odourless and soluble in water. The addition of a strong base, such as sodium hydroxide, liberates the weak base methylamine from its salt:

$$CH_3NH_3^+Cl^-(s) + OH^-(aq) \longrightarrow CH_3NH_2(g) + H_2O(l) + Cl^-(aq)$$

Trimethylamine forms trimethylammonium salts, e.g., $(CH_3)_3NH^+\ Br^-$, and the salts of phenylamine are called phenylammonium salts, e.g., $C_6H_5NH_3^+I^-$.

Aliphatic amines are more basic than ammonia. Aromatic amines are weaker bases because of the delocalisation of the N lone pair

Aliphatic amines are stronger bases than ammonia in aqueous solution because the positive charge in the alkylammonium ion can be shared between the nitrogen atom and a carbon atom. This distribution of charge stabilises the cation. Aromatic amines are weaker bases. The electron pair on the nitrogen atom is partially delocalised by interaction with the π electron cloud of the benzene ring [see Figure 32.5A]. As a result of delocalisation, the lone pair is less available for coordination to a proton.

Substituents affect basicity

Substituents in the benzene ring affect the basicity of phenylamine. Electron-donating groups, e.g. $CH_3O—$ and $HO—$, especially in the 4-position, increase the basicity; electron-withdrawing groups, e.g. $—NO_2$ and $—Cl$, decrease the basicity.

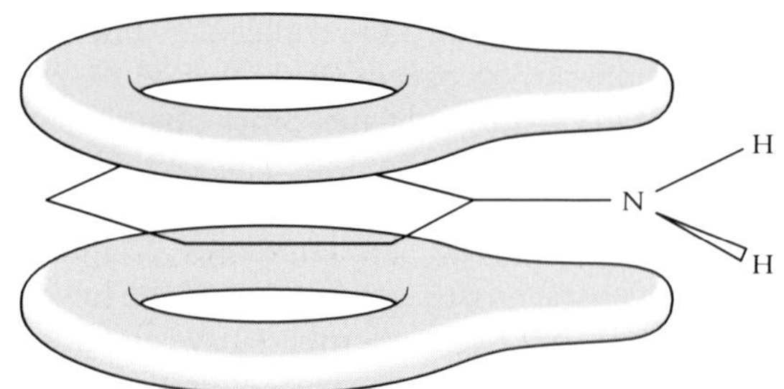

Figure 32.5A Phenylamine: the interaction between the lone pair of electrons on N and the ring

32.5.1 NOTE ON AMIDES

Amides

$$R-C(=O)-NH_2$$

Amides, $RCONH_2$, are not bases

are not basic. The electron-withdrawing character of the carbonyl group reduces the electron density on the nitrogen atom:

$$-C(=O)-\ddot{N}H_2$$

SUMMARY

Amines: Primary 1° RNH_2, Secondary: 2°, RR′NH, Tertiary: 3°, RR′R″N
Quaternary ammonium salts: $RR'R''R'''N^+ X^-$
Lower molar mass aliphatic amines are soluble in water. They have a characteristic fishy smell.
Amines are basic and form salts. The order of basicity is:
$NH_3 < 3° < 1° < 2°$
Aromatic amines are weaker bases than aliphatic amines.

CHECKPOINT 32.5: FORMULAE

1. Draw structural formulae for
   (a) ethylmethylpropylamine
   (b) (1-methylpropyl)amine
   (c) (2-chloropropyl) amine
   (d) diethyldimethylammonium chloride
   (e) *N,N*-dimethylphenylamine
   (f) 4-nitrophenylamine.

2. Name the following compounds, and say whether they are 1°, 2°, 3° or 4° amines:
   (a) $CH_3-N(CH_3)-H$
   (b) $C_2H_5-N(CH_3)-C_2H_5$
   (c) $C_6H_5-N(H)-C_6H_5$
   (d) $I-C_6H_4-NHC_2H_5$
   (e) $(CH_3)_3N^+-C_2H_5\ Br^-$
   (f) $H_3C-C_6H_4-NHCH_3$

3. Which of the following has the highest boiling temperature?
   $(C_2H_5)_3N$, $(CH_3CH_2CH_2)_2NH$, $CH_3(CH_2)_5NH_2$

32.6 LABORATORY PREPARATIONS

32.6.1 REACTION OF AMMONIA AND A HALOGENOALKANE

RX + NH_3 form a mixture of 1°, 2°, 3° and 4° ammonium salts

Ammonia and a halogenoalkane (in solution in ethanol) are heated together in a **bomb** (a metal container which will withstand high pressure):

$$RX(alc) + NH_3(alc) \xrightarrow{\text{heat in a bomb}} R\overset{+}{N}H_3\,X^-(s)$$

A mixture of primary, secondary, tertiary and quaternary ammonium salts is formed:

$$NH_3 \xrightarrow{CH_3I} CH_3\overset{+}{N}H_3I^- \xrightarrow{CH_3I} (CH_3)_2\overset{+}{N}H_2I^- \xrightarrow{CH_3I} (CH_3)_3\overset{+}{N}H\,I^- \xrightarrow{CH_3I} (CH_3)_4\overset{+}{N}\,I^-$$

$$+HI \qquad\qquad +HI \qquad\qquad +HI$$

The products are separated by the addition of alkali to liberate the free amines, followed by fractional distillation.

32.6.2 REDUCTION OF NITROGEN COMPOUNDS

(a) Reduction of a Nitrile, R—C≡N

(b) Reduction of an Amide, $RCONH_2$

Lithium tetrahydridoaluminate in ethoxyethane solution is used as the reducing agent for (**a**) and (**b**). The use of an acidic reagent which would hydrolyse the starting material must be avoided.

(c) Reduction of a Nitro-compound

$ArNH_2$ is made by the reduction of $ArNO_2$

Phenylamine, $C_6H_5NH_2$, is made by the reduction of nitrobenzene, $C_6H_5NO_2$. Tin(II) chloride is made in the reaction mixture from tin and hydrochloric acid [see Figure 32.6A]. In reducing the nitro-compound, tin(II)

Figure 32.6A Reduction of nitrobenzene to phenylamine

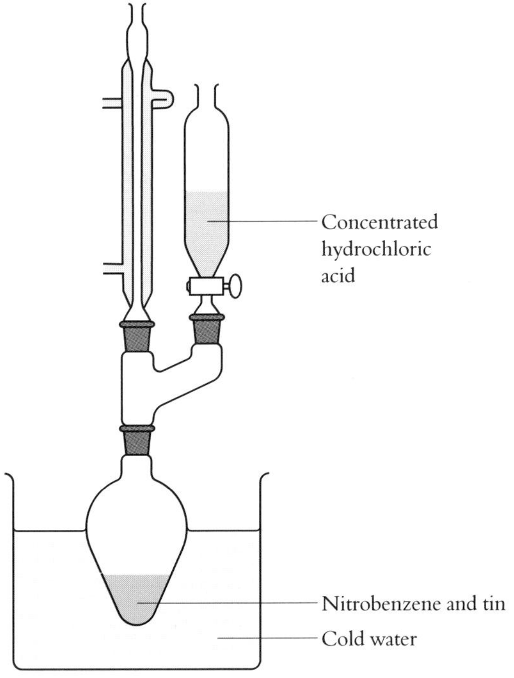

Practical details for the preparation of phenylamine

chloride is oxidised to tin(IV) chloride, which reacts with hydrochloric acid to form the complex ion, $[SnCl_6]^{2-}$. The reaction product is therefore $(C_6H_5NH_3^+)_2\,[SnCl_6]^{2-}$. The addition of concentrated alkali liberates phenylamine, and the addition of sodium chloride to the mixture reduces the solubility of phenylamine in water. Steam distillation [§ 8.5.5] is employed to separate phenylamine. Extraction with ethoxyethane separates phenylamine from the water in the distillate. Distillation removes ethoxyethane from the dried extract. The product can be purified by distillation under reduced pressure. Phenylamine decomposes when heated to its normal boiling temperature.

Figure 32.6B
Methods of preparing amines

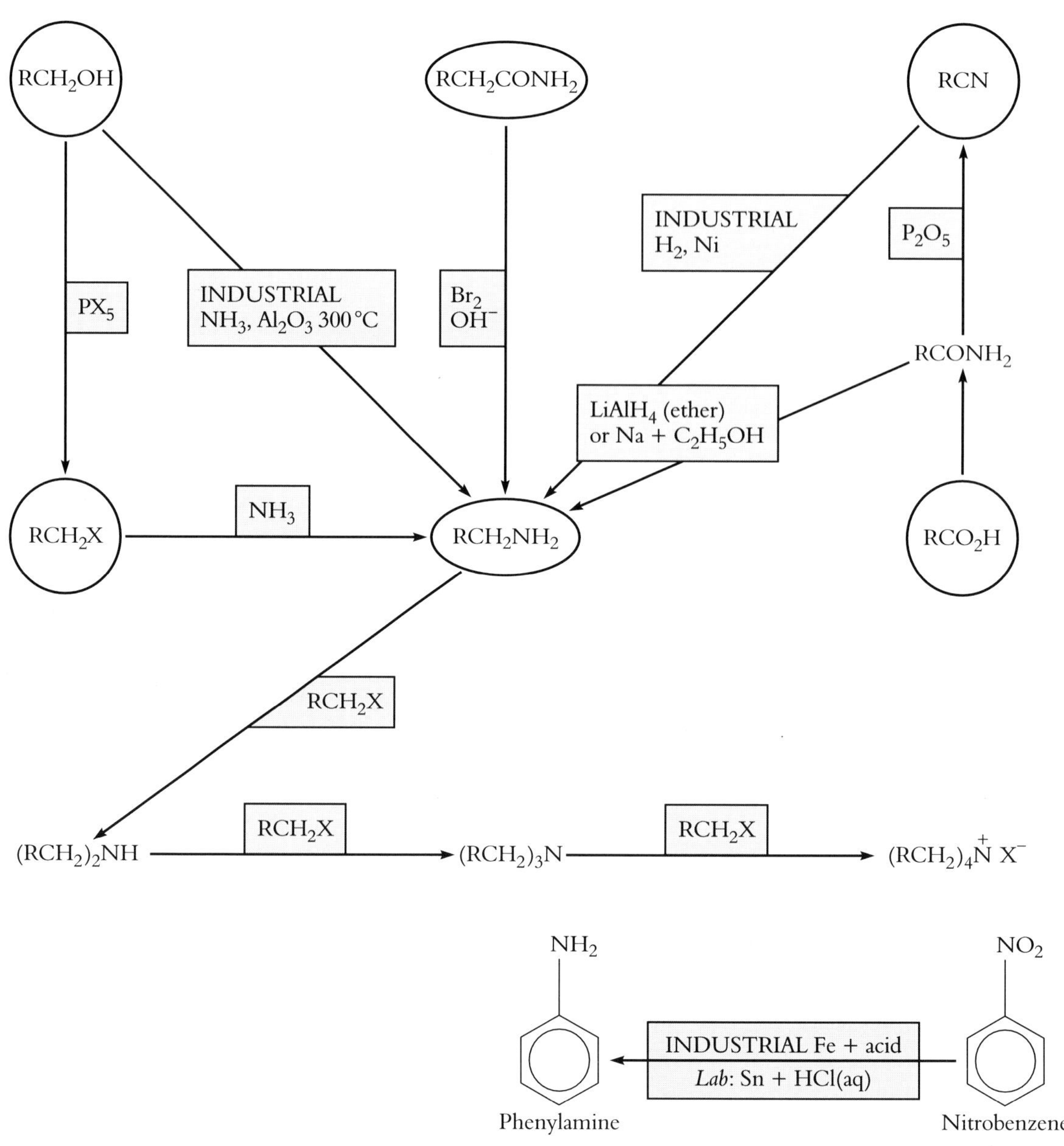

32.6.3 FROM AN AMIDE BY THE HOFMANN DEGRADATION

$RCONH_2$ can be converted into RNH_2 by the Hofmann degradation

The action of bromine and concentrated alkali is to convert an amide of formula $RCONH_2$ into an amine of formula RNH_2:

$$\underset{\text{Ethanamide}}{CH_3CONH_2} + Br_2 + 4NaOH \longrightarrow \underset{\text{Methylamine}}{CH_3NH_2} + 2NaBr + Na_2CO_3 + 2H_2O$$

This reaction, the Hofmann degradation, reduces the number of atoms in the carbon chain [§ 36.2].

SUMMARY

Laboratory preparations of amines:

- reaction of ammonia + halogenoalkane
- reduction of nitrile, amide, aromatic nitro-compound
- Hofmann degradation (Br_2 + conc alkali) of an amide, e.g. $RCONH_2 \longrightarrow RNH_2$

CHECKPOINT 32.6: PREPARATIONS OF AMINES

1. Explain how you would carry out the following conversions. Some of them involve more than one step:
   (a) $CH_3CH_2Cl \longrightarrow CH_3CH_2NH_2$
   (b) $CH_3CH_2Cl \longrightarrow CH_3CH_2CH_2NH_2$
   (c) 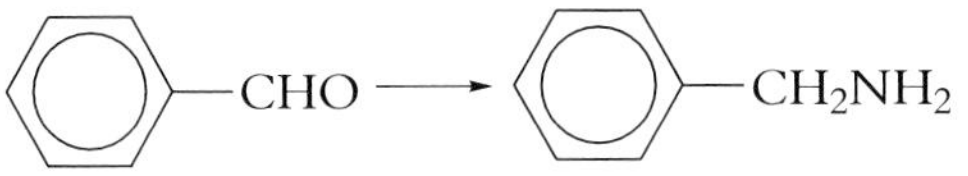 C_6H_5—CHO $\longrightarrow$ C_6H_5—CH_2NH_2
   (d) 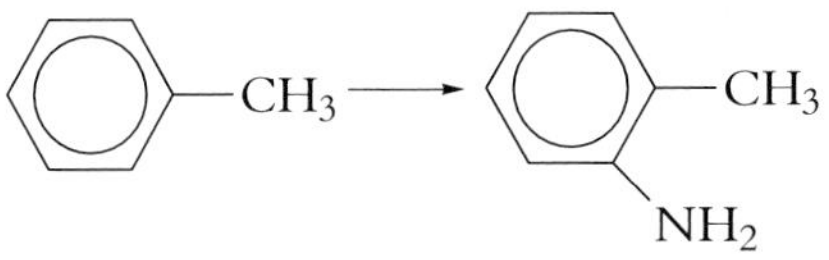 C_6H_5—CH_3 $\longrightarrow$ C_6H_4(—CH_3)(—NH_2) (NH₂ ortho to CH₃)
   (e) $CH_3CH{=}CH_2 \longrightarrow (CH_3)_2CHNH_2$
   (f) 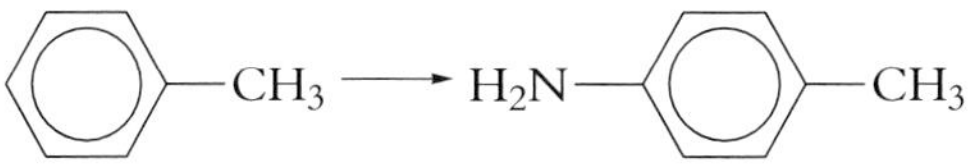 C_6H_5—CH_3 $\longrightarrow$ H_2N—C_6H_4—CH_3 (para)

2. Phenylamine can be made from nitrobenzene. Describe how the reduction is carried out. Explain why the product of the reaction is made alkaline before being steam-distilled. Describe how the product is extracted from the steam-distillate, and how it is finally purified.

3. Complete the following equations, indicating reagents and conditions:
   (a) $(CH_3)_3N + C_2H_5I \xrightarrow{\text{conditions}}$
   (b) 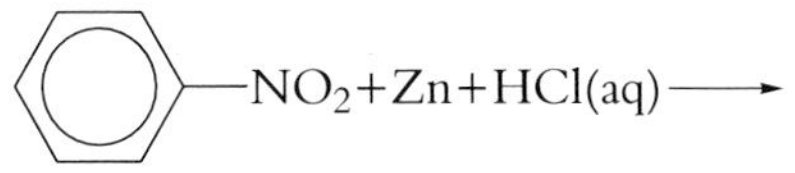 C_6H_5—$NO_2 + Zn + HCl(aq) \longrightarrow$
   (c) $C_3H_7OH + NH_3 \xrightarrow[\text{conditions}]{\text{catalyst}}$
   (d) $C_2H_5CONH_2 \xrightarrow[\text{solvent}]{\text{reducing agent}} C_2H_5CH_2NH_2$
   (e) $C_2H_5Br \longrightarrow \mathbf{X} \longrightarrow (C_2H_5)_4NBr$

4. **A** is an amine which is insoluble in water. **P** is a phenol which is insoluble in water. **E** is a solution of **A** and **P** in ethoxyethane. How could you separate **A** and **P** from the solution **E**, using only acid and alkali and ethoxyethane?

32.7 THE REACTIONS OF AMINES

32.7.1 ACYLATION

—NH_2 can be acylated to form —NHCOR ...

Acylation is the conversion of the groups

—N(H)(H) into —N(H)—C(=O)—R

and

$$-N(R')(H) \quad \text{into} \quad -N(R')(C(=O)R)$$

Tertiary amines, having no replaceable hydrogen atoms, are not acylated. The compounds formed are **amides**. The acylating agent can be an acid chloride, RCOCl or, preferably, an acid anhydride, $(RCO)_2O$:

... by RCOCl or by $(RCO)_2O$ ...

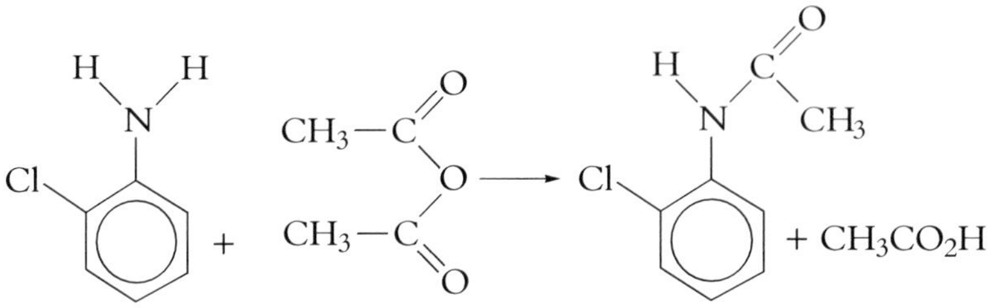

$$\text{2-ClC}_6\text{H}_4\text{NH}_2 + (CH_3CO)_2O \longrightarrow \text{2-ClC}_6\text{H}_4\text{NHCOCH}_3 + CH_3CO_2H$$

2-Chlorophenylamine   Ethanoic anhydride   *N*-(2-Chlorophenyl)ethanamide

Benzoylation is accomplished by the use of benzoyl chloride and an excess of alkali:

and by $C_6H_5COCl + NaOH$

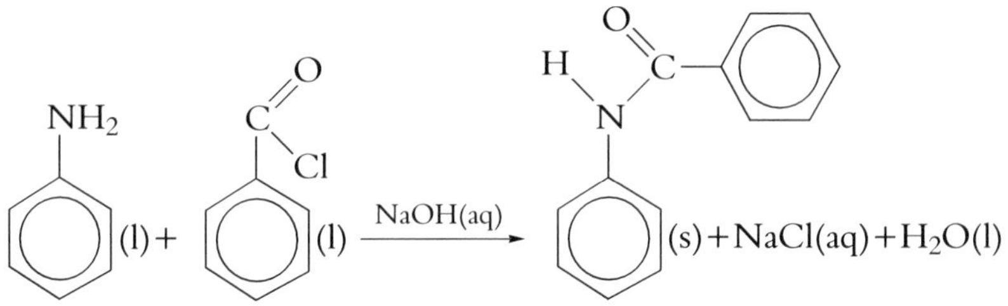

$$C_6H_5NH_2(l) + C_6H_5COCl(l) \xrightarrow{NaOH(aq)} C_6H_5NHCOC_6H_5(s) + NaCl(aq) + H_2O(l)$$

When chemists want to identify an amine their first choice is infrared spectrometry [see § 35.8]. If this is not available, they can make an amide derivative. Amides are solids, so the melting temperature of the amide can be found and compared with a table of the melting temperatures of different amides. This identifies the amide and the parent amine.

32.7.2 ALKYLATION

Alkylation of RNH_2 gives 1°, 2°, 3° amines and quaternary ammonium salts

Halogenoalkanes react with ammonia and with amines, replacing the hydrogen atoms by alkyl groups:

$$NH_3 \xrightarrow{RX} RNH_2 \xrightarrow{RX} R_2NH \xrightarrow{RX} R_3N \xrightarrow{RX} R_4N^+X^-$$

Halogenoarenes do not react in this way.

Primary amines are converted into secondary amines and secondary amines into tertiary amines. The final product is a quaternary ammonium salt, e.g., $(CH_3)_4N^+I^-$, tetramethylammonium iodide.

32.7.3 REACTIONS WITH NITROUS ACID

Nitrous acid HNO_2 reacts with aliphatic 1° amines to form unstable $R-\overset{+}{N}\equiv N$ ions which decompose to give N_2 ...

Amines undergo a number of different reactions with nitrous acid. Since it is an unstable compound, nitrous acid is generated in the reaction mixture by the action of a mineral acid on sodium nitrite at 5 °C [Figure 32.7A].

Aliphatic primary amines and aromatic primary amines react to form a cation

$$R-\overset{+}{N}\equiv N$$

called a **diazonium ion**. The diazonium compounds from primary aromatic amines are stable in solution at 5 °C. Their reactions are important, and are described in § 32.8. The diazonium compounds from primary aliphatic amines decompose to form carbocations, with the elimination of nitrogen:

$$R\overset{+}{N}H_3(aq)+HNO_2(aq) \longrightarrow R-\overset{+}{N}\equiv N(aq)+2H_2O(l)$$

$$R-\overset{+}{N}\equiv N(aq) \longrightarrow R^+(aq)+N_2(g)$$

... with aromatic 1° amines to form more stable diazonium compounds ...

The reaction is quantitative: the volume of nitrogen evolved is a measure of the amount of primary aliphatic amine present. The carbocation, R^+, can react in a number of ways, to form an alkene, an alcohol, ROH, an ether, ROR, and also, by reaction with the $NaNO_2$ + HCl(aq) present, RCl and RNO_2.

Secondary amines, both aliphatic and aromatic, are nitrosated (i.e. a —NO group is introduced) by nitrous acid to form *N*-nitrosoamines, which are yellow oils of a highly carcinogenic nature.

Aromatic tertiary amines are nitrosated in the ring.

Figure 32.7A
Reactions of nitrous acid with amines

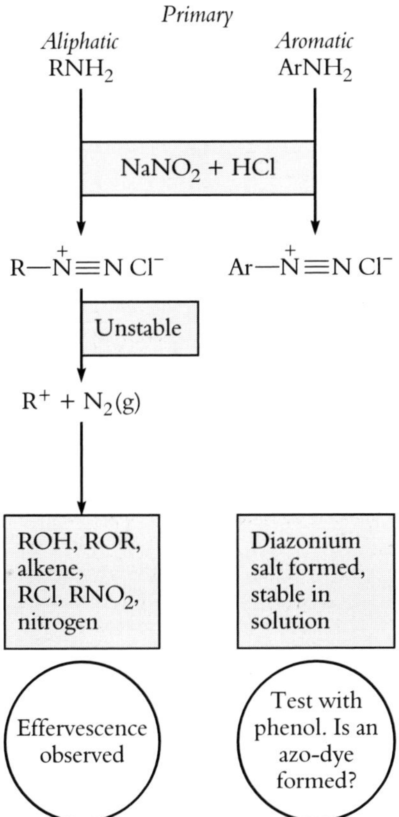

32.7.4 SUBSTITUTION IN THE AROMATIC RING

The —NH_2 group activates the ring

Aromatic amines undergo substitution in the ring. The —NH_2 group activates the ring, and is 2/4-directing (like —OH [§ 28.12]). In many cases, it is difficult to obtain a monosubstituted compound.

Halogenation

Halogenation gives a tri-halogeno-derivative ...

Phenylamine reacts quantitatively with bromine water to give a white precipitate of 2,4,6-tribromophenylamine:

$$C_6H_5NH_2(l) + 3Br_2(aq) \longrightarrow C_6H_2Br_3NH_2(s) + 3HBr(aq)$$

2,4,6-Tribromophenylamine

... unless the —NHCOR derivative is halogenated

If a monobromo-derivative is required, the —NH_2 group is ethanoylated to convert it into the less powerfully activating —$NHCOCH_3$ group. Then the acyl derivative is brominated to give a mixture of the 2- and 4-bromo-derivatives. Hydrolysis restores the —NH_2 group:

$$C_6H_5NH_2 \xrightarrow{(CH_3CO)_2O} C_6H_5NHCOCH_3 \xrightarrow{Br_2,\ H_2O} BrC_6H_4NHCOCH_3 \xrightarrow[\text{boil}]{H_2SO_4(aq)} BrC_6H_4NH_2$$

Phenylamine — *N*-Ethanoylphenylamine — *N*-Ethanoyl-4-bromophenylamine — 4-Bromophenylamine (and 2-)

CHECKPOINT 32.7: REACTIONS OF AMINES

1. Explain why *(a)* methylamine is a stronger base than ammonia *(b)* phenylamine is a weaker base than ammonia.

2. Give the names and formulae of the products formed when phenylamine reacts with
   (a) hydrochloric acid
   (b) aqueous bromine
   (c) ethanoic anhydride
   (d) ethanoic anhydride followed by bromine water
   (e) sodium nitrite and hydrochloric acid (i) above 10 °C (ii) below 10 °C.

3. State how you could prepare from phenylamine
   (a) C_6H_5—$NHCOCH_3$
   (b) C_6H_5—$N_2^+Cl^-$
   (c) O_2N—C_6H_4—NH_2

4. State the conditions under which butylamine reacts with *(a)* concentrated hydrobromic acid *(b)* ethanoyl chloride *(c)* nitrous acid.
   In each case give the names and formulae of the products of the reaction, and write an equation.

SUMMARY

Reactions of amines:

Acylation: $RNH_2 + R'COCl$ (or $(R'CO)_2O$) $\longrightarrow$ $RNHCOR'$

Alkylation with halogenoalkanes, not with halogenoarenes:

$RNH_2 \xrightarrow{R'X} RR'NH \xrightarrow{R'X} RR'_2N \xrightarrow{R'X} RR'_3N^+X^-$

With nitrous acid:

1° aliphatic amines $\longrightarrow$ nitrogen;
1° aromatic amines $\longrightarrow$ diazonium compound

With aqueous bromine, substitution in the ring, e.g.

Phenylamine + Bromine $\longrightarrow$ 2,4,6-Tribromophenylamine

32.8 DIAZONIUM COMPOUNDS

$ArNH_2 + HNO_2$ form a diazonium compound

Diazonium compounds result from the reaction between primary aromatic amines and nitrous acid. The reaction is carried out below 10 °C to avoid the decomposition of both nitrous acid and the product:

$$C_6H_5NH_2(l) + NaNO_2(aq) + 2HCl(aq) \xrightarrow{<10°C} C_6H_5\overset{+}{N}{\equiv}N\ Cl^- + NaCl(aq) + 2H_2O$$

Phenylamine — Benzenediazonium chloride

Solid diazonium compounds are explosive

Diazonium salts can be isolated, but they are unstable and explosive in the solid state. They are used in solution in a number of important preparations. The $-\overset{+}{N}_2$ group can be replaced by a number of different groups.

32.8.1 REPLACEMENT OF $-\overset{+}{N}{\equiv}N$ BY $-OH$

They are used in solution

When warmed in acidic solution, diazonium compounds form phenols, with the evolution of nitrogen:

$$C_6H_5-\overset{+}{N}{\equiv}N + H_2O \xrightarrow[H_2SO_4\text{(aq, dilute)}]{\text{warm} > 10°C \text{ with}} C_6H_5-OH + N_2(g) + H^+(aq)$$

On warming, they give phenols

To avoid a reaction between the phenol formed and the diazonium compound, [§ 32.8.3] the diazonium compound is added slowly to a large excess of boiling dilute sulphuric acid. The volatile phenol distils over.

32.8.2 REPLACEMENT OF $-\overset{+}{N}{\equiv}N$ BY $-$HALOGEN OR $-CN$

$-\overset{+}{N}{\equiv}N$ can be replaced by —Cl, —Br, —I or —CN …

In a set of reactions named after T Sandmeyer, $-\overset{+}{N}{\equiv}N$ is replaced by —Cl, —Br or —CN. The diazonium compound is warmed to 100 °C with the appropriate reagent and catalyst. These are shown in Figure 32.8A. If potassium iodide solution is added, $-\overset{+}{N}{\equiv}N$ is replaced by —I.

Figure 32.8A

Replacement of $-\overset{+}{N}_2$ by $-Cl$, $-Br$, $-I$, $-CN$

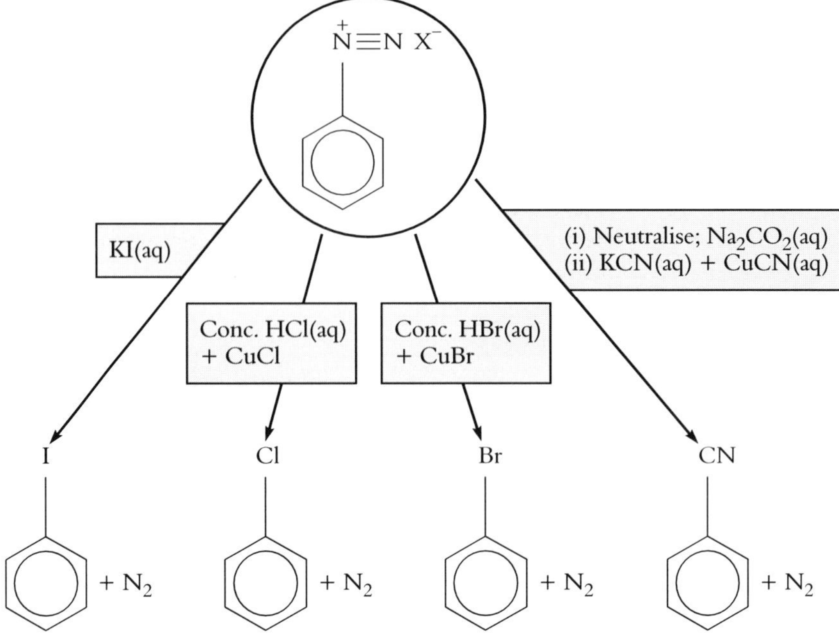

32.8.3 COUPLING REACTIONS

Diazonium compounds will react with phenols and aromatic amines to form azo-dyes

Diazonium ions are electrophiles. The diazonium cation will attack reactive nucleophilic sites, such as the 4-positions in phenols and aromatic amines. The reaction results in the formation of a —N=N— bond between two aromatic rings, and is called **azo-coupling**. The compounds formed are highly coloured, and many are used as dyes. The group —N=N— is responsible for the colour of azo-dyes and is termed a **chromophore**.

$$C_6H_5-\overset{+}{N}{\equiv}N\,X^- + C_6H_5-NH_2 \longrightarrow C_6H_5-N{=}N-C_6H_4-NH_2 + HX$$

Phenylamine (aniline)

4-(Phenylazo)phenylamine (aniline yellow, the first azo-dye to be made)

$$Na^+\ {}^-O_3S-C_6H_4-\overset{+}{N}{\equiv}N\,X^- \quad + \quad C_{10}H_7O^-$$

Sodium 4-sulphonatobenzenediazonium halide

Naphthalen-2-ol

$$\xrightarrow[\text{pH}]{\text{controlled}} Na^+\ {}^-O_3S-C_6H_4-N{=}N-C_{10}H_6OH + HX(aq)$$

'Orange 11'

(The $-SO_3^-\,Na^+$ substituent makes this dye soluble in water.)

32.8.4 THE IMPORTANCE OF DIAZONIUM COMPOUNDS

Diazonium compounds are used in synthesis

Diazonium compounds open up the possibility of making a number of benzene derivatives which cannot be made directly from benzene [See Figure 32.8B.] You saw how halogenoalkanes opened up the route

Alkane, RH ⟶ *Halogenoalkane*, RX ⟶ RY

where Y can be any of a number of groups. Halogenoarenes are not very reactive, and it is often diazonium compounds that open up routes to aromatic compounds:

Arene ⟶ *Nitroarene* ⟶ *Arylamine* ⟶ *Diazonium compound* ⟶ *Substituted arene*

$ArH \longrightarrow ArNO_2 \longrightarrow ArNH_2 \longrightarrow ArN_2^+X^- \longrightarrow ArY$

Figure 32.8B
Reactions of phenylamine and benzene diazonium chloride

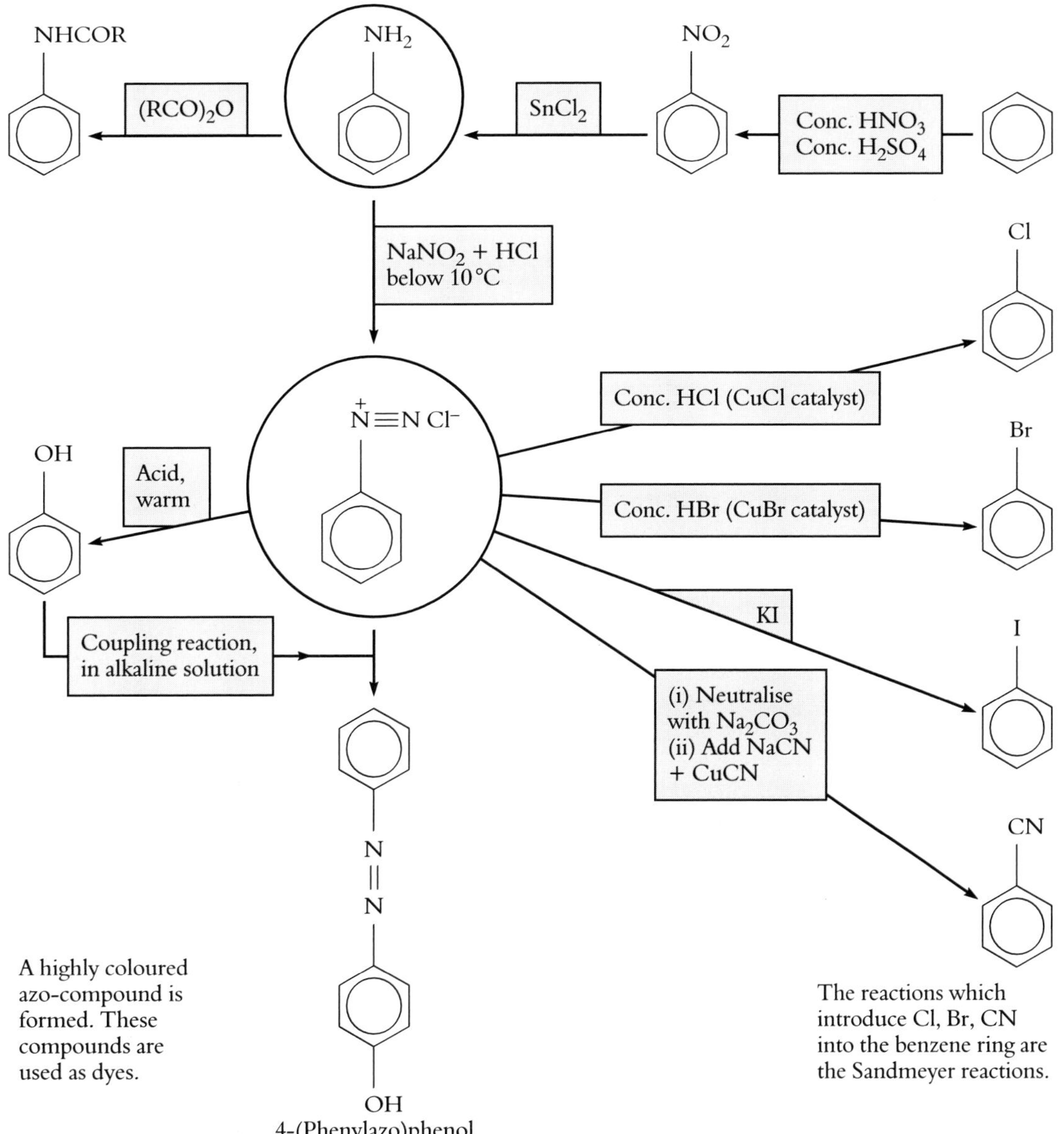

CHECKPOINT 32.8: DIAZONIUM COMPOUNDS

1. (*a*) Describe how you could prepare a solution of benzene diazonium chloride. How can the compound be converted into phenol? How can phenol be separated from the solution?
   (*b*) Under what conditions will a phenol react with a diazonium salt? What is the electrophile in this reaction? Why does benzene not react with benzenediazonium chloride?
   (*c*) How could you use the solution of benzenediazonium chloride to obtain (i) chlorobenzene (ii) iodobenzene (iii) benzonitrile?

2. Outline how you could convert nitrobenzene into (*a*) phenylamine (*b*) phenol (*c*) chlorobenzene (*d*) 1,3-dinitrobenzene (*e*) 1,3-dichlorobenzene (*f*) *N*-benzoylphenylamine

32.9 QUATERNARY AMMONIUM COMPOUNDS

32.9.1 CATIONIC DETERGENTS

Quaternary ammonium compounds which contain a long non-polar group in addition to the polar group can mix with water and also with water-repelling substances. This property enables them to act as detergents, and they are described as cationic detergents. An example is $C_{15}H_{31}CH_2N^+(CH_3)_3Br^-$.

Quaternary ammonium compounds are cationic detergents used in hair conditioners and fabric conditioners

Quaternary ammonium compounds are not very effective detergents. They have germicidal properties, however, and are often incorporated into detergents used for washing nappies. Quaternary ammonium compounds are also used in hair conditioners and fabric conditioners. Both wet hair and wet fabrics pick up negative charges on their surfaces. These attract positively charged cationic detergent ions, which coat the surface of the hair or fabric by preventing the build-up of static electricity and lubricate the surfaces.

32.9.2 THE TRANSMISSION OF NERVE IMPULSES

A quaternary ammonium compound which occurs naturally is acetylcholine chloride. (*Acetyl* = ethanoyl; *choline* is shown below.) When a voluntary nerve cell is stimulated, acetylcholine is released at the nerve endings. After it has been active in transmitting a nerve impulse, acetylcholine is removed. The removal is achieved by the enzyme acetylcholinesterase which catalyses the hydrolysis of acetylcholine to choline and ethanoic (acetic) acid:

Only in the presence of the specific enzyme does the reaction take place rapidly. An enzyme and its **substrate** (e.g., acetylcholinesterase and acetylcholine) are described as fitting together 'like a lock and key'. There are substances which attach themselves to the site on the enzyme that is needed by the substrate. The quaternary ammonium compound, '*decamethonium*' will do this:

$$(CH_3)_3\overset{+}{N}(CH_2)_{10}\overset{+}{N}(CH_3)_3$$

decamethonium

$$CH_3CO_2CH_2CH_2{-}\overset{+}{N}(CH_3)_3\ Cl^- + H_2O \xrightarrow[\text{acetylcholinesterase}]{\text{catalysed by}} CH_3CO_2H + HOCH_2CH_2{-}\overset{+}{N}(CH_3)_3\ Cl^-$$

Acetylcholine chloride → Ethanoic (acetic) acid + Choline chloride

You can see the resemblance between 'decamethonium' and acetylcholine. Once decamethonium is attached to the enzyme, acetylcholinesterase is prevented or **inhibited** from doing its job, and nerve impulses are not transmitted. Decamethonium was used as a muscle relaxant. The naturally occurring base, *curare*, which acts in a similar way, was also used medicinally. Curare has other uses too: some South American Indians use it to tip their poison arrows. The quantities which they use cause paralysis and death in their victims.

Other compounds which do not resemble acetylcholine in structure also inhibit the enzyme acetylcholinesterase. Nerve gases (e.g., DFP) work in this way. Insecticides like Parathion® have the same effect on insects:

$$[(CH_3)_2CHO]_2P(=O)-F$$

DFP

$$(C_2H_5O)_2P(=S)-O-C_6H_4-NO_2$$

Parathion®

Parathion® and similar insecticides can affect human beings. Agricultural workers who use the compounds regularly and who do not wear protective clothing can absorb sufficient insecticide to damage their health.

AMINES

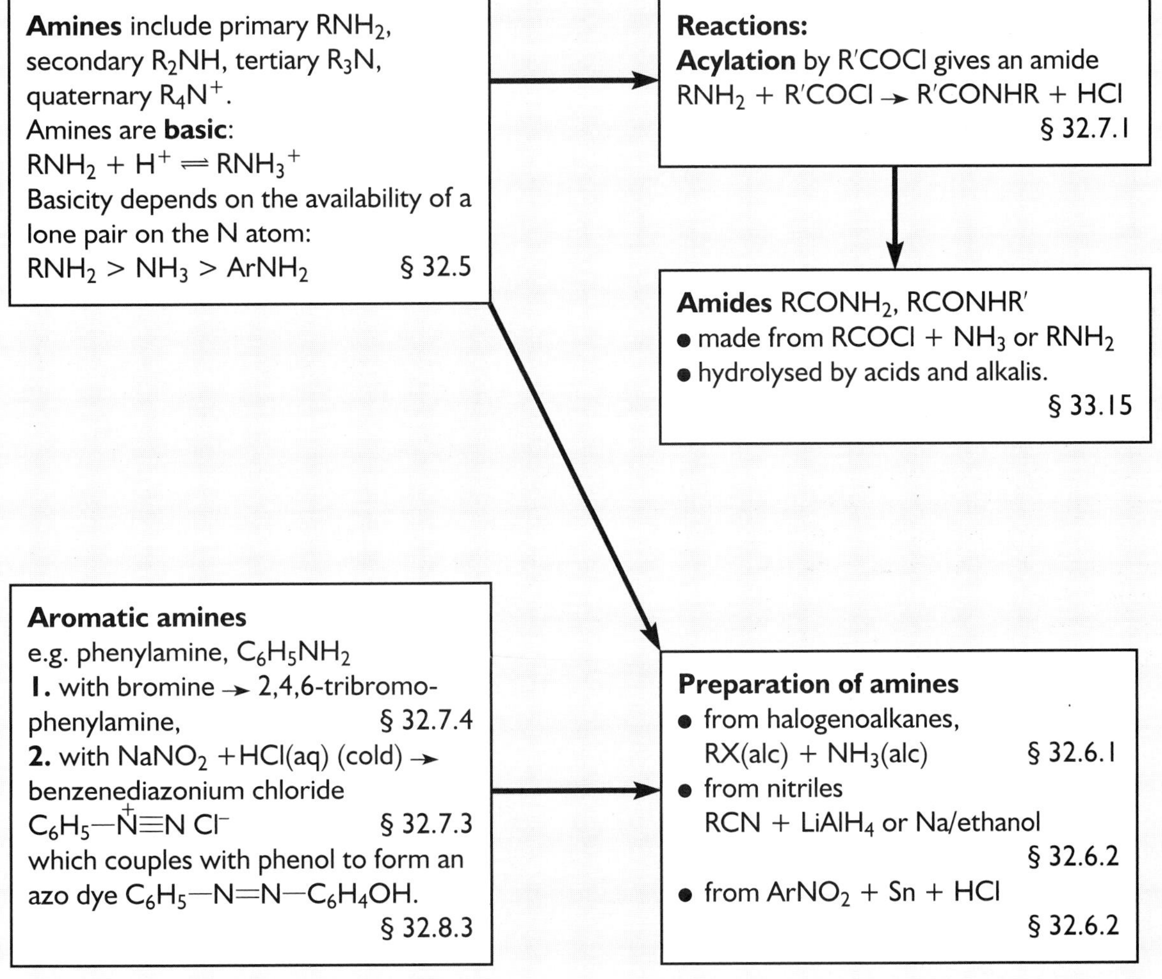

Amines include primary RNH_2, secondary R_2NH, tertiary R_3N, quaternary R_4N^+.
Amines are **basic**:
$RNH_2 + H^+ \rightleftharpoons RNH_3^+$
Basicity depends on the availability of a lone pair on the N atom:
$RNH_2 > NH_3 > ArNH_2$ § 32.5
Reactions:
Acylation by R′COCl gives an amide
$RNH_2 + R'COCl \rightarrow R'CONHR + HCl$
§ 32.7.1
Amides $RCONH_2$, $RCONHR'$
• made from $RCOCl + NH_3$ or RNH_2
• hydrolysed by acids and alkalis.
§ 33.15
Aromatic amines
e.g. phenylamine, $C_6H_5NH_2$
1. with bromine → 2,4,6-tribromophenylamine, § 32.7.4
2. with $NaNO_2$ + HCl(aq) (cold) → benzenediazonium chloride
$C_6H_5-\overset{+}{N}\equiv N\ Cl^-$ § 32.7.3
which couples with phenol to form an azo dye $C_6H_5-N=N-C_6H_4OH$.
§ 32.8.3
Preparation of amines
• from halogenoalkanes,
RX(alc) + NH_3(alc) § 32.6.1
• from nitriles
RCN + $LiAlH_4$ or Na/ethanol
§ 32.6.2
• from $ArNO_2$ + Sn + HCl
§ 32.6.2

QUESTIONS ON CHAPTER 32

1. Outline the preparation, starting from benzene, of *(a)* nitrobenzene *(b)* phenylamine *(c)* bromobenzene *(d)* iodobenzene *(e)* a named azo-compound.

2.

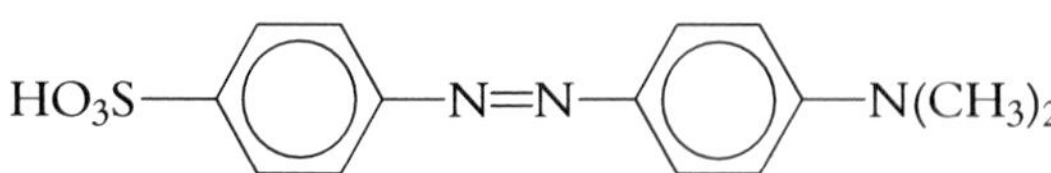

Shown above is the formula for methyl orange. Suggest a method of making the compound from 4-aminobenzenesulphonic acid, 4-$HO_3SC_6H_4NH_2$ and phenylamine.

3. Outline the preparation from propane of *(a)* 1-aminopropane *(b)* 2-aminopropane. For each step, state the type of reaction involved (S_N, A_E etc.)

4. *(a)* Phenylamine can be obtained from benzene by a two-stage process.
(i) Give equations for the reactions involved, and state the conditions required.
(ii) Say how phenylamine is isolated from the reaction mixture.
(b) Give an example of a reaction which shows phenylamine acting as (i) a Brönsted–Lowry base (ii) a Lewis base.
(c) Give the names and structural formulae of the organic products formed when phenylamine reacts with (i) ethanoic anhydride (ii) nitrous acid below 10 °C, (iii) nitrous acid above 10 °C, (iv) aqueous bromine.

5. *(a)* Suggest a synthesis for each of the following compounds, starting from benzene.

A 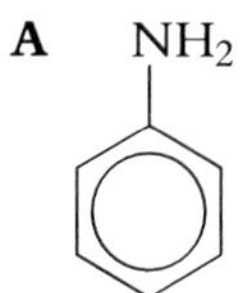

B 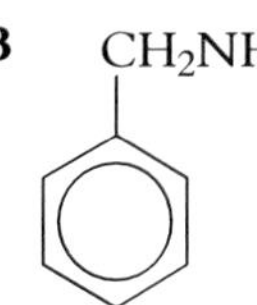

C 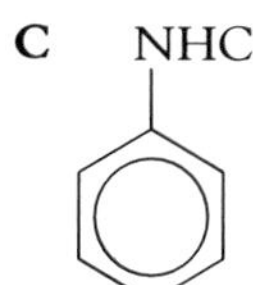

D 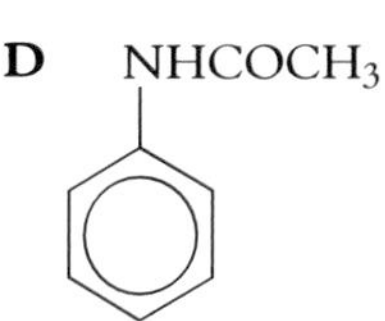

(b) Suggest a method of distinguishing between **A** and **B**.

6. Compare the reactions of ethylamine and phenylamine with reference to
(a) their basicities, explaining why they differ
(b) their reactions with bromine
(c) their reactions with ethanoyl chloride.

7. *(a)* Briefly describe how an aqueous solution of benzenediazonium chloride can be (i) made from phenylamine (ii) converted into iodobenzene (iii) converted into phenylhydrazine, $C_6H_5NHNH_2$.
(b) Draw the structural formula of the compound formed when benzenediazonium chloride is coupled with phenol.
(c) How does phenylhydrazine react with (i) hydrochloric acid (ii) ethanal?
(d) Explain how 2,4-dinitrophenylhydrazine is used for identifying aldehydes and ketones. Say why it is preferred for this purpose to phenylhydrazine.
(e) What other method could be used for identifying these compounds?

8. Compound **A**, C_2H_7N, a primary amine, is obtained by treatment of an iodoalkane, **B**, with ammonia. It is also obtained, together with compound **C**, by boiling compound **D**, C_4H_9NO, with aqueous sodium hydroxide. Boiling the iodoalkane, **B**, with aqueous sodium hydroxide yields compound **E**, C_2H_6O.
(a) Deduce structures for compounds **A** to **E**, giving your reasoning and accounting for all of the reactions involved. **9**
(b) (i) Name the type of reaction which occurs when the iodoalkane, **B**, reacts with aqueous sodium hydroxide. **1**
(ii) In this reaction, what is the reactive species provided by aqueous sodium hydroxide? **1**
(iii) To which mechanistic class does this reactive species belong, electrophile, nucleophile, free radical or oxidant? **1**

12 *O&C (AL)*

9. The reaction scheme below shows a route by which the drug benzedrine may be prepared from benzene.

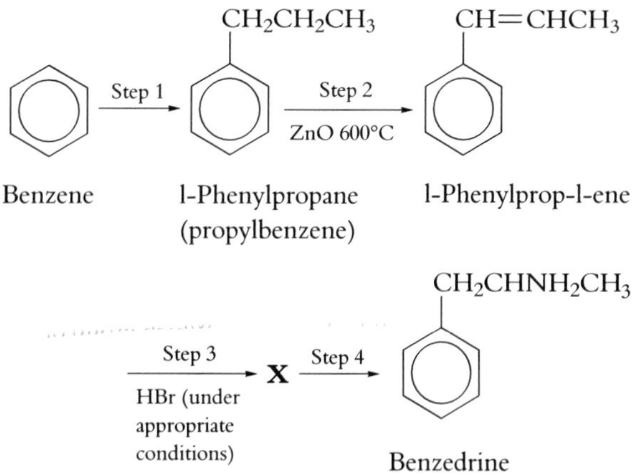

Benzene —Step 1→ 1-Phenylpropane (propylbenzene) ($CH_2CH_2CH_3$) —Step 2, ZnO 600°C→ 1-Phenylprop-1-ene ($CH{=}CHCH_3$)

—Step 3, HBr (under appropriate conditions)→ **X** —Step 4→ Benzedrine ($CH_2CHNH_2CH_3$)

(a) (i) Give the reagents and conditions for the conversion of benzene to 1-phenylpropane in Step 1. Write an equation for the reaction.
(ii) Identify the compound **X** and state the reagents and conditions for its conversion to benzedrine. **3**

(b) The yield of benzedrine produced at the end of this route is low as a number of other unwanted reactions occur. Suggest TWO organic products which may be formed as well as benzedrine and show how they might be produced. **3**

(c) What would you predict for the physical properties and solubility of benzedrine? Give reasons for your suggestions. **3**

(d) Benzedrine has an isomer, 4-amino-propylbenzene.

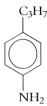

C_3H_7

NH_2

Describe a chemical reaction you could use to distinguish between samples of benzedrine and 4-amino-propylbenzene. You should include the observations you would expect to make with each compound and equations for the reactions involved. **2**

(e) The identity of benzedrine can be confirmed by instrumental methods. Predict the results which benzedrine would give in investigations using TWO different methods, making your reasoning clear. **4**

15 *L(N) (AL)*

10. This question concerns the reaction sequence shown below:

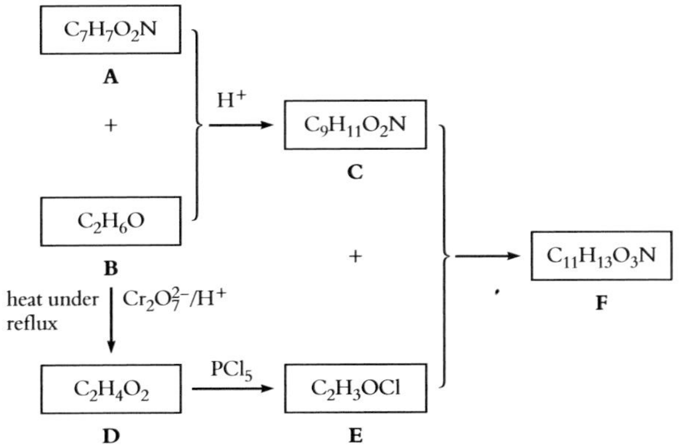

(a) (i) Substances **A**, **B** and **D** all react with sodium to give off hydrogen gas. Which group of atoms does this suggest is present in these three substances? **1**

(ii) Three pieces of information are given below about substance **A**. What does each piece of evidence suggest about the structure of **A**?
A reacts with sodium carbonate solution to give off carbon dioxide gas.
A forms a coloured complex ion with copper(II) cations.
A is unable to undergo addition reactions easily. **3**

(iii) Draw a possible displayed formula for **A**. **2**

(iv) Explain why there are two other possible isomers of **A** which have a similar structure. **1**

(b) (i) What type of reaction is occurring when **B** is changed into **D**? **1**

(ii) Draw the displayed formulae for **B** and **D**. **2**

(c) (i) Suggest what other molecule forms when **A** and **B** react to form **C**, by considering the number of atoms present in each substance. **1**

(ii) What type of reaction is this? **1**

(iii) Draw a structural formula for **C**. **1**

(d) Substance **E** reacts vigorously with water to form an acidic solution.

(i) Give the name of substance **E**. **1**

(ii) Write a balanced equation for the reaction of **E** with water. **2**

(e) Using your answers to *(c)* and *(d)*, suggest the structural formula for substance **F**. **2**

18 *L(N) (AL)*

11. *(a)* Explain why phenylamine is a weaker base than (phenylmethyl)amine, $C_6H_5CH_2NH_2$. **2**

(b) Write an equation for the formation of (phenylmethyl)amine from benzenecarbonitrile, C_6H_5CN, and name the type of reaction involved. **2**

(c) (Phenylmethyl)amine can also be obtained from the reaction between (bromomethyl)benzene. $C_6H_5CH_2Br$, and ammonia. Name the type of reaction involved and explain why this method of synthesis is not as effective as that in part *(b)*. **3**

(d) The secondary amine $(C_6H_5CH_2)NH(CH_2)_{11}CH_3$ can be converted into a cationic fabric-softening product by reaction with an excess of chloromethane. Name the type of product formed and give the structural formula of this compound. **3**

10 *NEAB (AL)*

12. *(a)* Explain why propylamine is classed as a Brönsted–Lowry base. **2**

(b) Explain why propylamine is a stronger base than ammonia. **2**

(c) Propylamine is obtained from the reaction between 1-bromopropane and an excess of ammonia. Name the type of reaction taking place and outline a mechanism. **5**

(d) Propylamine can also be prepared from propanenitrile, CH_3CH_2CN. Name the type of reaction involved and write an equation for this reaction. What advantage does this method of synthesis have over that in part *(c)*? **3**

12 *NEAB (AL)*

13. 3-Aminophenylethanone can be obtained from benzene in three steps:

$$C_6H_6 \xrightarrow{\text{Step 1}} C_6H_5COCH_3 \xrightarrow{\text{Step 2}} 3\text{-}NO_2C_6H_4COCH_3 \xrightarrow{\text{Step 3}} 3\text{-}NH_2C_6H_4COCH_3$$

(a) For Step 1 and Step 3 name the type of reaction taking place and suggest a suitable reagent or a combination of reagents. **5**

(b) Write an equation showing how the electrophile is formed from the reagent(s) in Step 1. Outline a mechanism for the subsequent reaction between benzene and this electrophile. **5**

10 *NEAB (AL)*

14. *(a)* Give the reagents and conditions necessary to convert nitrobenzene to phenylamine. **3**

(b) In order to prepare a diazonium salt, phenylamine is dissolved in excess hydrochloric acid. A solution of sodium nitrite is added in small quantities to this solution until in excess, maintaining the temperature between 0 and 10 °C, preferably around 5 °C. The diazonium salt is formed in solution according to the equation:

$$C_6H_5NH_2 + 2HCl + NaNO_2 \longrightarrow C_6H_5\overset{+}{N}{\equiv}NCl^- + NaCl + 2H_2O$$

(i) What structural feature of the phenylamine molecule allows it to dissolve readily in hydrochloric acid? Give the formulae of the species formed and hence explain, in energetic terms, why phenylamine is soluble in hydrochloric acid but not very soluble in water, despite its ability to form hydrogen bonds.

(ii) The function of the sodium nitrite is to react with excess hydrochloric acid to form nitrous acid. Write an ionic equation for this reaction. Identify the acid/base conjugate pairs.

(iii) Suggest why the sodium nitrite solution is added in small quantities. **10**

(c) If the solution of the diazonium salt is warmed to about 50 °C, it decomposes as follows:

$$C_6H_5\overset{+}{N}{\equiv}NCl^- + H_2O \longrightarrow C_6H_5OH + N_2 + HCl$$

Describe in detail an experiment you would perform in order to measure the rate of this reaction. Include in your account how the measurements taken could be used to determine the rate at a particular time during the decomposition. **12**

25 *L (AL)*

33 ORGANIC ACIDS AND THEIR DERIVATIVES

33.1 THE SULPHONAMIDE ANTIBIOTICS

The simplest aromatic carboxylic acid is benzoic acid, C_6H_5—CO_2H. Bacteria need the related compound 4-aminobenzoic acid (Figure 33.1A) to keep them alive. A compound which can prevent bacteria from using 4-aminobenzoic acid will kill bacteria. Such a compound is 4-aminobenzenesulphonamide (Figure 33.1A), the first of the sulphonamide antibiotics.

In 1932, a new drug, called Prontosil®, was patented. It had been developed for use as a red dye by a German manufacturer and then found to have an antibacterial action. Gerhard Domagh showed that Prontosil® was good at fighting streptococcal infections. Soon afterwards, it was found that Prontosil® is converted into 4-aminobenzenesulphonamide (sulphanilamide) in the body and that sulphanilamide has the same antibacterial activity as Prontosil®. The first person to receive the new antibiotic was Domagh's daughter. As she lay close to death with 'child bed fever', Domagh decided to risk treating her with his discovery. Regulations for testing new drugs were not as strict then as they are now. Fortunately, the antibiotic cured her infection, and she recovered. In 1939, Domagh received the Nobel Prize for his work on antibiotics. A number of derivatives of sulphanilamide, called the *sulphonamides* or *sulpha-drugs* were synthesised.

Figure 33.1A The action of sulphonamide drugs

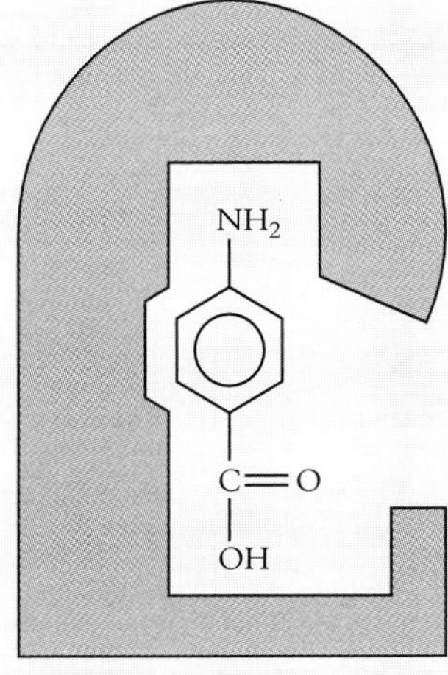

4-Aminobenzoic acid

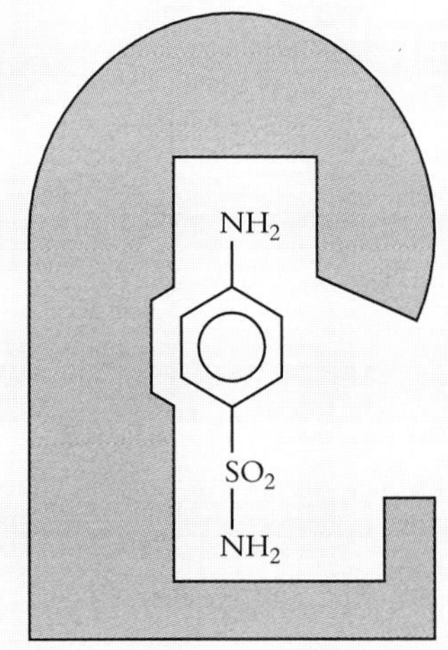

4-Aminobenzenesulphonamide

Bacteria use 4-aminobenzoic acid in the synthesis of folic acid, which is a vital substance for them. 4-Aminobenzenesulphonamide, 'sulphanilamide', bears a close structural resemblance to 4-aminobenzoic acid. Molecules of this and other similar sulpha-drugs compete with 4-aminobenzoic acid for the active site on the bacterial enzyme, inhibit the synthesis of folic acid [Figure 33.1A] and kill the bacteria.

Sulpha-drugs were widely used in the Second World War. They were sprinkled on open wounds to prevent infection. They produce a number of side effects with prolonged use, including kidney damage, and have now given way to penicillin and other antibiotics.

33.2 THE CARBOXYL GROUP

Most organic acids contain a carboxyl group or a sulphonic acid group:

$$\mathrm{R{-}C({=}O){-}O{-}H}$$

Carboxylic acid

$$\mathrm{R{-}S({=}O)_2{-}O{-}H}$$

Sulphonic acid

This chapter is devoted to carboxylic acids and their derivatives. The carboxylic acids derived from alkanes are sometimes called alkanoic acids.

Carboxylic acids are weakly ionised ...

Carboxylic acids ionise to some extent to give hydrogen ions, and are neutralised by bases to form salts:

$$\mathrm{RCO_2H(aq) + H_2O(l) \rightleftharpoons RCO_2^-(aq) + H_3O^+(aq)}$$

$$\mathrm{RCO_2H(aq) + OH^-(aq) \longrightarrow RCO_2^-(aq) + H_2O(l)}$$

... They form salts, many of which are soluble

The salts are usually soluble in water. This property makes carboxylic acids easy to extract from natural sources, and is the reason why they were among the first organic compounds to be isolated. Ethanoic (acetic) acid was obtained from sour wine; butanoic (butyric) acid from butter; 2-hydroxypropanoic (lactic) acid from sour milk and benzoic acid from gum benzoin. Aliphatic carboxylic acids were called 'fatty acids' because esters of several of the higher members are fats.

33.3 NOMENCLATURE FOR ORGANIC ACIDS AND THEIR DERIVATIVES

Aliphatic carboxylic acids are named after the alkane with the same number of carbon atoms. The ending *-ane* is changed to *-anoic acid*. Examples are:

The IUPAC names of some acids

HCO_2H	Methanoic acid (formerly called formic acid)
CH_3CO_2H	Ethanoic acid (formerly acetic acid)
$CH_3CH_2CO_2H$	Propanoic acid (formerly propionic acid)

The C in the $-CO_2H$ group is always given the number 1, and substituents are given locants:

$CH_3-CH(OH)-CO_2H$	2-Hydroxypropanoic acid
HO_2C-CO_2H	Ethanedioic acid (formerly oxalic acid)
$CH_2CO_2H-CH_2CO_2H$	Butanedioic acid (formerly succinic acid)

Aromatic carboxylic acids are named by adding the suffix *-carboxylic acid* to the name of the parent hydrocarbon. Alternatively, the suffix *-oic acid* can be used:

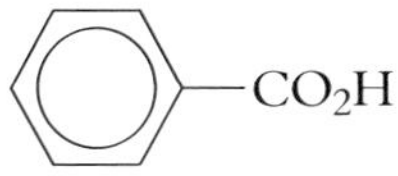

$C_6H_5-CO_2H$	Benzenecarboxylic acid (or benzoic acid)
$O_2N-C_6H_4-CO_2H$	4-Nitrobenzenecarboxylic acid (or 4-nitrobenzoic acid)

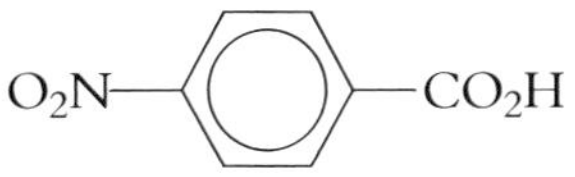

The derivatives of carboxylic acids covered in this chapter contain the **acyl** group

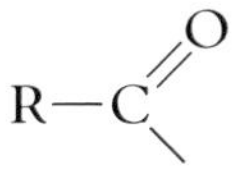

$$R-C(=O)-$$

and are listed in Table 33.4. The nitriles are included because their reactions link up with those of acids.

33.4 PHYSICAL PROPERTIES OF ACIDS AND THEIR DERIVATIVES

33.4.1 ACIDS

The lower aliphatic acids are liquids ...

The aliphatic acids C_1–C_{10} are liquids. Anhydrous ethanoic acid freezes at 17 °C, and is often called *glacial* ethanoic acid. The boiling temperatures increase with increasing molecular mass. Aromatic acids are crystalline solids with melting temperatures above those of aliphatic acids of comparable molecular mass.

The lower members of the aliphatic carboxylic acids have penetrating odours. Vinegar is a 3% solution of ethanoic acid. Butanoic acid is the substance you smell in rancid butter.

Hydrogen bonding takes place between molecules of carboxylic acids. In the vapour phase and in solution in organic solvents, dimerisation occurs:

... associated by hydrogen bonds

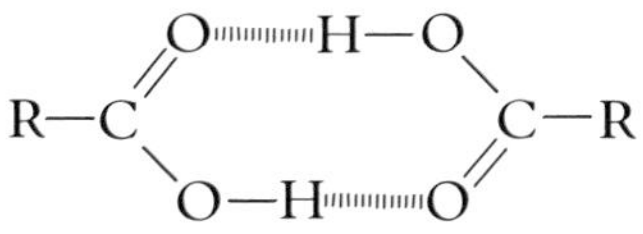

$$R-C(=O\cdots H-O)(-O-H\cdots O=)C-R$$

The facility which acids have for forming two hydrogen bonds per molecule makes their boiling temperatures higher than those of corresponding alcohols.

Table 33.4
The names and formulae of some derivatives of acids

Formula	Name	Example	
$R-C(=O)-$	Acyl group	CH_3CO-	Ethanoyl group
$\{R-CO_2\}^-$	Carboxylate ion	$CH_3CO_2^-$	Ethanoate ion
$R-C(=O)Cl$	Acid chloride	CH_3COCl	Ethanoyl chloride
$R-C(=O)-O-C(=O)-R$	Acid anhydride	$(CH_3CO)_2O$	Ethanoic anhydride
$R-C(=O)NH_2$	Amide	CH_3CONH_2	Ethanamide
$R-C(=O)OR'$	Carboxylic ester	$CH_3CO_2C_2H_5$	Ethyl ethanoate
$R-C\equiv N$	Nitrile	CH_3CN	Ethanenitrile

Carboxylic acids dissolve in water as they form hydrogen bonds to water molecules

Carboxylic acids of fairly low molecular mass dissolve in water. The dimers dissociate to form monomers in order to form hydrogen bonds to water molecules:

33.4.2 AMIDES

Amides are solids ...

... which dissolve in water

Amides, with two hydrogen atoms in each $—CONH_2$ group, can form more hydrogen bonds than acids. Even the lowest members of the series (except $HCONH_2$) are solids, have higher boiling temperatures than the corresponding acids, and have little odour. The ability to form hydrogen bonds with water molecules makes amides more soluble in water than other acid derivatives.

33.4.3 ESTERS, CHLORIDES, ANHYDRIDES AND NITRILES

Other derivatives are volatile liquids ...

Esters, chlorides, anhydrides and nitriles form no hydrogen bonds. Except for anhydrides, their boiling temperatures are lower than those of the corresponding acids. These derivatives are volatile and odorous. There are, however, strong dipole–dipole interactions between the molecules, arising from the polarity of the

$$\rangle C{=}O \text{ group}$$

Even the lowest members of the series are liquids at room temperature.

... Chlorides are fuming liquids ...

Acid chlorides are colourless liquids with a pungent smell and a lachrymatory action. Anhydrides also have a pungent smell. Aromatic anhydrides are solids.

... Esters have a fruity smell

Esters are colourless liquids with pleasant fruity odours. Examples are:

$CH_3CH_2CH_2CO_2C_2H_5$	Ethyl butanoate	*apple odour*
$CH_3CO_2(CH_2)_7CH_3$	Octyl ethanoate	*orange odour*

33.4.4 SALTS

The salts of carboxylic acids are ionic solids

The salts of carboxylic acids are electrovalent compounds and are therefore involatile crystalline solids. They are often soluble as, on dissolution, the ions are hydrated as the $—CO_2^-$ group forms hydrogen bonds with water molecules. The lower members of the series are soluble, but in higher members the large hydrocarbon part of the molecule makes the compounds insoluble.

33.5 REACTIVITY OF CARBOXYLIC ACIDS

The $\rangle C{=}O$ and $—OH$ groups in $—CO_2H$ do not react as they do in carbonyl compounds and alcohols

The carboxyl group is so named because it contains a *carb*onyl group and a hydr*oxyl* group. The two groups influence each other to such an extent that the reactions of carboxylic acids bear little resemblance to those of either carbonyl compounds or alcohols

The polar C=O group makes it easier for H to ionise as H^+

The carboxylic acid group is

$$—C\begin{matrix}{=}O \\ —O—H\end{matrix}$$

The OH group reduces the δ+ charge on the C of C=O so it is not attacked by nucleophiles ...

The polar carbonyl group in

attracts electrons away from the —O—H bond, and makes it easier for the hydrogen atom to ionise than is the case in the —O—H bond in an alcohol. The flow of electrons from the —OH group towards the carbonyl carbon atom reduces the δ+ charge on the carbonyl carbon atom, with the result that it is not attacked by the nucleophiles that attack carbonyl compounds.

The charge in RCO_2^- is delocalised. RCO_2^- is a weaker base than RO^-...

... and RCO_2H is a stronger acid than ROH

When a carboxylate ion, RCO_2^-, is formed, the negative charge on the ion is shared equally between two oxygen atoms. The structure of RCO_2^- can be represented by a molecular orbital picture:

The delocalisation of the charge makes the carboxylate ion less ready to accept a proton than is an alkoxide ion, $R—O^-$. The carboxylate ion, RCO_2^- is therefore a weaker base than the alkoxide ion, RO^-, and RCO_2H is a stronger acid than ROH.

Carboxylic acids are weak acids ...

... Substituents affect the strength of acids ...

Carboxylic acids are much weaker than the common mineral acids. The dissociation constant, K_a, for ethanoic acid is 1.8×10^{-5} mol dm^{-3} [§ 12.6.4]. This means that in a 1 mol dm^{-3} solution of the acid, 3 molecules in a thousand are ionised. Substituents affect the strength of acids [see § 12.6.6].

... Benzoic acid is stronger than aliphatic acids ...

Benzoic acid is stronger than ethanoic acid because the negative charge on the $—CO_2^-$ group, being delocalised by interaction with the π electron cloud of the benzene ring, is less available for attaching a proton to form $—CO_2H$ [see Figure 33.5A].

... Carboxylic acids are stronger than carbonic acid

Carboxylic acids are stronger than carbonic acid. The evolution of carbon dioxide that occurs when a carboxylic acid reacts with sodium hydrogencarbonate is used to distinguish carboxylic acids from weaker acids, such as phenols. Some substituted phenols also give a positive result.

Figure 33.5A
The benzoate ion

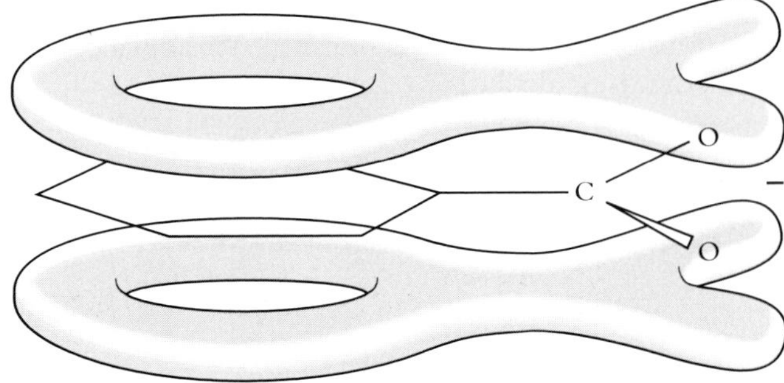

SUMMARY

The carboxyl group is polarised: $—C^{\delta+}(=O^{\delta-})—O—H$

This makes it easier for H to ionise as H^+. Electrons from the OH group reduce the $\delta+$ charge on the C atom, and the C=O group is not attacked by nucleophiles.

RCO_2H is a stronger acid than ROH because the negative charge in RCO_2^- is delocalised. Benzoic acid is stronger than ethanoic acid because there is greater delocalisation of charge in $C_6H_5CO_2^-$ over the benzene ring.

33.6 LABORATORY PREPARATIONS OF CARBOXYLIC ACIDS

33.6.1 OXIDATION

Primary alcohols and aldehydes

Oxidation of alcohols, aldehydes ...

Potassium dichromate(VI) and potassium manganate(VII), both in acid solution, are often used as oxidising agents. Primary alcohols are oxidised via aldehydes to carboxylic acids:

$$RCH_2OH \longrightarrow RCHO \longrightarrow RCO_2H$$

Alkenes

... and alkenes gives acids

Alkenes are oxidised by acidified potassium manganate(VII):

$$\text{Cyclohexene} \xrightarrow[\text{acid}]{KMnO_4} \text{Hexane-1,6-dioic acid } (HO_2C(CH_2)_4CO_2H)$$

Cyclohexene Hexane-1,6-dioic acid

Methylbenzene

Aromatic side chains are oxidised to CO_2H

Benzoic acid is made by oxidising methylbenzene. Industrially, air is used as the oxidising agent. Even if the group attached to the aromatic ring is larger than $—CH_3$, benzoic acid is the oxidation product:

$$C_6H_5—CH_3 \xrightarrow[\text{acid, heat}]{KMnO_4} C_6H_5—CO_2H \xleftarrow[\text{acid, heat}]{KMnO_4} C_6H_5—CH_2CH_2OH$$

33.6.2 HYDROLYSIS

Nitriles

Hydrolysis of RCN gives RCO_2H

Nitriles are hydrolysed to carboxylic acids by being boiled under reflux with aqueous alkali or mineral acid. The amide, $RCONH_2$, is an intermediate:

$$RCN(l) \xrightarrow[\textit{or } H^+(aq)]{NaOH(aq)} RCONH_2(aq) \xrightarrow{H_2O} RCO_2Na(aq) + NH_3(g)$$
$$\textit{or } RCO_2NH_4(aq)$$

Esters

Esters are hydrolysed to acids

Esters

$$R-\overset{\overset{\displaystyle O}{\|}}{C}-OR'$$

are hydrolysed to alcohols and carboxylic acid salts when they are boiled under reflux with aqueous alkali:

$$RCO_2R'(l) + NaOH(aq) \longrightarrow RCO_2Na(aq) + R'OH(aq)$$

The carboxylic acid is obtained by acidifying the salt with a mineral acid:

$$RCO_2Na(aq) + HCl(aq) \longrightarrow RCO_2H(aq) + NaCl(aq)$$

Esters are also hydrolysed by mineral acids.

SUMMARY

Carboxylic acids are made by

- oxidation of primary alcohols, aldehydes and alkenes
- hydrolysis of esters and nitriles
- oxidation of $C_6H_5CH_3$ and other compounds gives benzoic acid.

CHECKPOINT 33.6: ACIDS

1. Name the following acids:

   (a) $CH_3(CH_2)_5CO_2H$

   (b) $(CH_3)_2CHCO_2H$ *(c)* CF_3CO_2H

   (d) $CO_2H-CH_2-CO_2H$ *(e)*

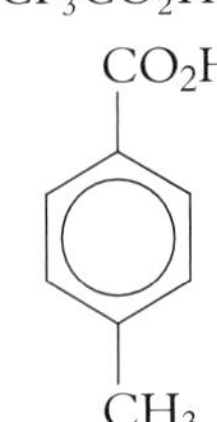

   (f) *(g)*

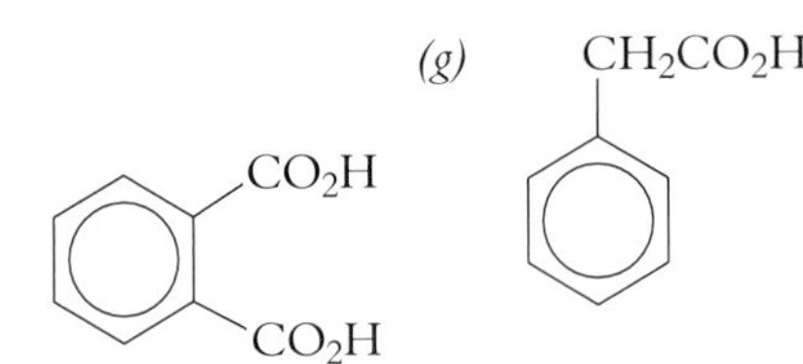

2. Arrange in order of increasing boiling temperature the compounds: *(a)* $CH_3CH_2CO_2H$, *(b)* CH_3CH_2CHO, *(c)* $CH_3CH_2CH_2OH$, *(d)* $CH_3CH_2CH_3$. Explain the reasons for the order you give.

3. Arrange in order of increasing acid strength: *(a)* phenol, *(b)* benzoic acid, *(c)* carbonic acid, *(d)* ethanoic acid.

4. Complete the following equations:

   (a) $RCH_2Cl + \qquad \longrightarrow RCH_2CN$

   (b) $H_2C{=}CHCN + \qquad \longrightarrow H_2C{=}CHCO_2H$

   Indicate the conditions needed for each reaction. What is the IUPAC name for $H_2C{=}CHCO_2H$? The trivial name for this compound is acrylic acid. What is its industrial importance?

5. *(a)* Outline a method for the conversion

   $$CH_3CHO \longrightarrow CH_3CH(OH)CO_2H$$

   (b) Name the product.

   (c) How could you show that the product is (i) a secondary alcohol and (ii) an acid?

   (d) What effect would you expect the acid to have on plane-polarised light? Explain your answer.

33.7 VITAMIN C: ASCORBIC ACID

Vitamin C is the most unstable of the vitamins: it is easily destroyed by oxidation, which is speeded up by heat or by a base.

Vitamin C is needed for the formation of tissues, teeth and bone. It plays a role in the metabolism of carbohydrates and proteins and in the formation of adrenalin, serotonin, haemoglobin and collagen. Lack of collagen results in the symptoms of scurvy.

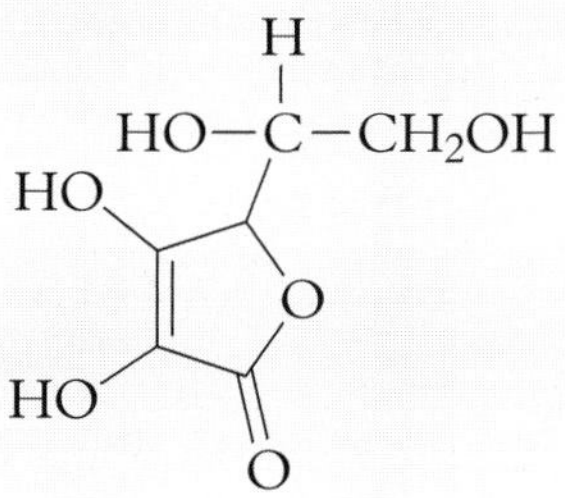

H
HO—C—CH_2OH
HO, HO, O, O

Ascorbic acid

Many people believe that large doses of Vitamin C prevent colds. (The late Linus Pauling [see § 4.6.3] was one of these.) It has been proved that it helps wounds to heal.

33.8 REACTIONS OF CARBOXYLIC ACIDS

33.8.1 SALT FORMATION

Carboxylic acids form salts by reaction with metals, carbonates, hydrogencarbonates and alkalis:

A test for $—CO_2H$ is the evolution of CO_2 from $NaHCO_3$

$$2RCO_2H(aq) + Mg(s) \longrightarrow (RCO_2)_2Mg(aq) + H_2(g)$$
$$2RCO_2H(aq) + Na_2CO_3(s) \longrightarrow 2RCO_2Na(aq) + CO_2(g) + H_2O(l)$$
$$RCO_2H(aq) + NaOH(aq) \longrightarrow RCO_2Na(aq) + H_2O(l)$$

The evolution of carbon dioxide from sodium hydrogencarbonate is used as a test to distinguish carboxylic acids from weaker acids, such as phenols. The reaction with alkali can be carried out by titration against a standard alkali to find the concentration of a solution of the acid. A suitable indicator must be used [§ 12.7.1].

Stronger acids displace carboxylic acids from their salts

The salts are crystalline solids, most of which are soluble in water. Organic acids can be extracted from a mixture of products by dissolving them in aqueous sodium hydroxide. Acidification with a mineral acid may precipitate the free acid from its salt:

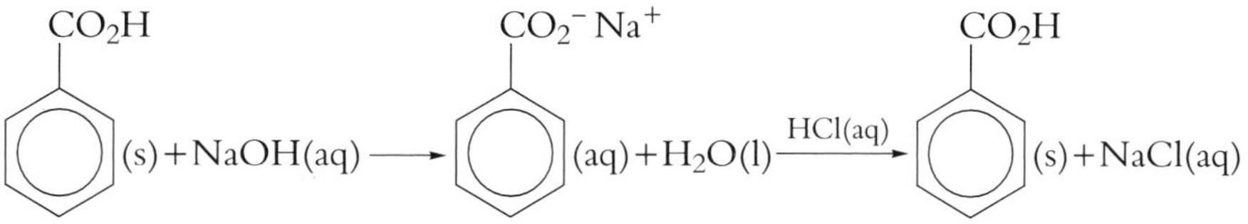

$$C_6H_5CO_2H(s) + NaOH(aq) \longrightarrow C_6H_5CO_2^-Na^+(aq) + H_2O(l) \xrightarrow{HCl(aq)} C_6H_5CO_2H(s) + NaCl(aq)$$

33.8.2 ESTERIFICATION

Reaction with an alcohol gives an ester

Carboxylic acids react with alcohols in the presence of concentrated sulphuric acid to form esters. The acid and alcohol are boiled under reflux with the concentrated acid or, alternatively, hydrogen chloride can be bubbled through the mixture. If the alcohol is labelled with ^{18}O, analysis of the products in a mass spectrometer shows that all the ^{18}O is present in the ester and none is in the water. This proves that the bonds that are broken in the reaction are:

$$R{-}C(=O){-}OH + R'{-}^{18}O{-}H \longrightarrow R{-}C(=O){-}^{18}O{-}R' + H_2O$$

The reverse of esterification is ester hydrolysis. Experiments on the hydrolysis of esters of the type $R{-}C(=O){-}^{18}O{-}R'$ show that the RCO—OR′ bond is broken in hydrolysis.

$$R{-}C(=O){-}^{18}O{-}R' + OH^- \longrightarrow R{-}C(=O){-}O{-}H + {}^{18}OR'^-$$

33.8.3 CONVERSION INTO ACID CHLORIDES

RCOCl is formed by reaction with PCl_5 or $SOCl_2$

The conversion

$$R{-}C(=O){-}OH \longrightarrow R{-}C(=O){-}Cl$$

is accomplished by the action of phosphorus(V) chloride, PCl_5, or by sulphur dichloride oxide, $SOCl_2$. The latter is more convenient as the by-products are gaseous:

$$C_2H_5CO_2H(l) + PCl_5(l) \longrightarrow C_2H_5COCl(l) + POCl_3(l) + HCl(g)$$

Propanoic acid — Propanoyl chloride

$$C_6H_5CO_2H(s) + SOCl_2(l) \longrightarrow C_6H_5COCl(l) + SO_2(g) + HCl(g)$$

Benzoic acid — Benzoyl chloride

The reverse of esterification is ester hydrolysis. Experiments on the hydrolysis of esters of the type $R{-}C(=O){-}^{18}O{-}R'$ show that the RCO—OR′ bond is broken in hydrolysis.

$$R{-}C(=O){-}^{18}O{-}R' + OH^- \longrightarrow R{-}C(=O){-}O{-}H + {}^{18}OR'^-$$

33.8.4 CONVERSION INTO AMIDES

The conversion

$$R{-}C(=O){-}OH \longrightarrow R{-}C(=O){-}NH_2$$

Amides are made from ammonium salts by distillation with acid

is accomplished via the ammonium salt of the acid. The ammonium salt is made by the action of the acid on ammonium carbonate (not ammonia solution because of salt hydrolysis). When heated to 100–200 °C, with an excess of the parent acid, the ammonium salt is dehydrated to give the amide:

$$R{-}CO_2^-\ NH_4^+ \xrightarrow[100\text{–}200\ °C]{RCO_2H} R{-}C(=O)NH_2 + H_2O$$

The reason for the presence of an excess of the free acid is that ammonium salts dissociate on being heated:

$$RCO_2^-NH_4^+(s) \rightleftharpoons RCO_2H(g) + NH_3(g)$$

The addition of acid displaces the equilibrium to the left.

33.8.5 REDUCTION

Acids can be reduced to alcohols ...

To convert a carboxylic acid into an alcohol, a powerful reducing agent is needed. Lithium tetrahydridoaluminate dissolved in ethoxyethane will effect reduction:

$$RCO_2H \xrightarrow[\text{in ethoxyethane}]{LiAlH_4} RCH_2OH$$

33.8.6 DECARBOXYLATION OF A SODIUM SALT

... decarboxylated to give a hydrocarbon

When the sodium salt of a carboxylic acid is heated with soda lime, the carboxyl group is removed as a carbonate and a hydrocarbon is formed:

$$C_6H_5{-}CO_2Na(s) + NaOH(s) \longrightarrow C_6H_6 + Na_2CO_3$$

SUMMARY

Carboxylic acids react to form

- salts by reaction with metals, bases and carbonates
- esters by reaction with alcohols
- acid chlorides by reaction with PCl_5 or $SOCl_2$
- amides by the action of heat on the acid + its ammonium salt
- alcohols by reduction with $LiAlH_4$.

CHECKPOINT 33.8: REACTIONS OF ACIDS

1. A student has read that rancid butter contains butanoic acid. She has supplies of ethoxyethane and laboratory acids and alkalis. How can she obtain a sample of butanoic acid? How can she *(a)* purify her sample and *(b)* test its purity? Butanoic acid is a liquid at room temperature.

2. How could propanoic acid be converted into *(a)* a stronger acid, *(b)* a buffer solution? [§ 12.8]

3. Suggest a method for the conversion

   $C_2H_5OH \longrightarrow CH_3CH(OH)CO_2H$

4. Predict the results of the reactions of the compound shown below with *(a)* bromine water, *(b)* ethanoic anhydride, *(c)* hydrogen and nickel, *(d)* lithium tetrahydridoaluminate, *(e)* hot, concentrated potassium manganate(VII).

   $H_2N{-}C_6H_4{-}CH{=}CHCO_2H$

5. Explain why the carbonyl group in a carboxylic acid is less reactive than that in an aldehyde or a ketone.

6. Compare the —OH group in ethanol, ethanoic acid and phenol.

Figure 33.8A
Reactions of carboxylic acids

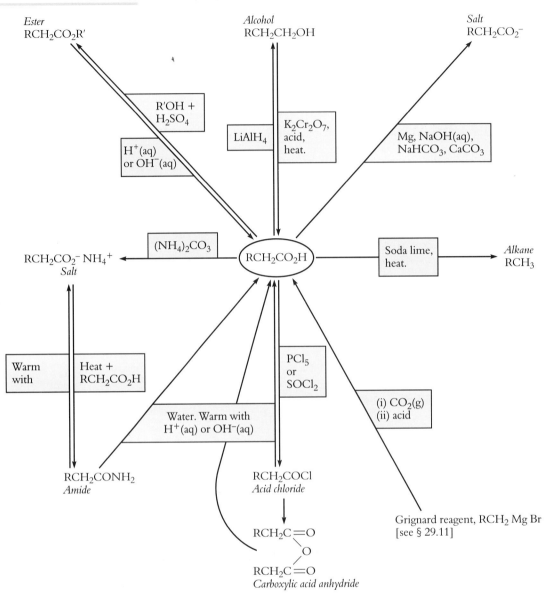

Ester
RCH_2CO_2R'

Alcohol
RCH_2CH_2OH

Salt
$RCH_2CO_2^-$

$R'OH + H_2SO_4$

$H^+(aq)$ or $OH^-(aq)$

$LiAlH_4$

$K_2Cr_2O_7$, acid, heat.

Mg, NaOH(aq), $NaHCO_3$, $CaCO_3$

$(NH_4)_2CO_3$

$RCH_2CO_2^-\ NH_4^+$
Salt

RCH_2CO_2H

Soda lime, heat.

Alkane
RCH_3

Warm with

Heat + RCH_2CO_2H

PCl_5 or $SOCl_2$

Water. Warm with $H^+(aq)$ or $OH^-(aq)$

(i) $CO_2(g)$
(ii) acid

RCH_2CONH_2
Amide

RCH_2COCl
Acid chloride

Grignard reagent, $RCH_2\ Mg\ Br$
[see § 29.11]

$RCH_2C{=}O$
O
$RCH_2C{=}O$
Carboxylic acid anhydride

Figure 33.9A
Acid derivatives in order of decreasing reactivity

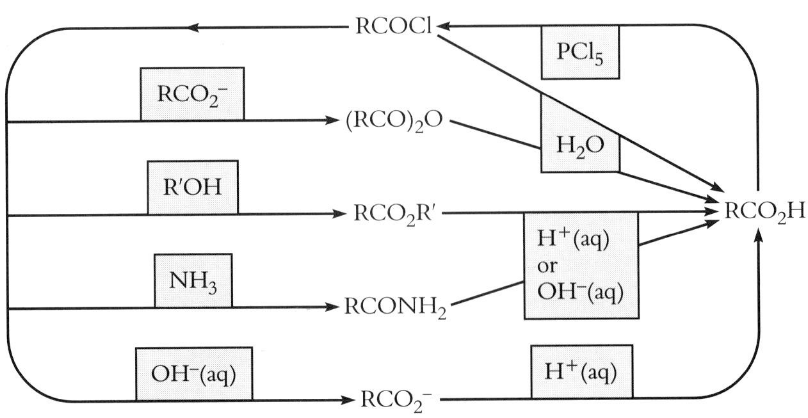

33.9 DERIVATIVES OF CARBOXYLIC ACIDS

The derivatives of carboxylic acids that contain an acyl group are shown in Figure 33.9A in order of decreasing reactivity.

33.10 ACID CHLORIDES

33.10.1 REACTIVITY

With the functional group

$$-\overset{\delta+}{C}(=O^{\delta-})-Cl$$

Nucleophiles attack the carbonyl C atom in RCOCl

the reactions of acid chlorides depend upon the attack by nucleophiles on the δ+ carbon atoms. This is why acid chlorides are more readily hydrolysed than chloroalkanes or chloroarenes. In benzoyl chloride and other aromatic acid chlorides, the δ+ charge is spread over the ring, and the carbon atom is less susceptible to nucleophilic attack than it is in aliphatic acid chlorides.

33.10.2 REACTIONS OF ACID CHLORIDES

Hydrolysis

Hydrolysis by H_2O is an example

Acid chlorides are insoluble in water but are hydrolysed quickly to a carboxylic acid and hydrochloric acid. Aliphatic acid chlorides fume in moist air to give hydrogen chloride:

$$CH_3COCl(l) + H_2O(l) \longrightarrow CH_3CO_2H(l) + HCl(g)$$

Ester formation

Alcohols, ROH and phenols in alkaline solution, ArO^- react to form esters

Acid chlorides, RCOCl, and aroyl chlorides, ArCOCl, react readily with phenols in alkaline solution and with alcohols in the presence of pyridine (a base which removes the HCl produced) to form esters:

$$\underset{\text{Ethanoyl chloride}}{CH_3COCl(l)} + \underset{\text{Ethanol}}{C_2H_5OH(l)} \xrightarrow{\text{pyridine}} \underset{\text{Ethyl ethanoate}}{CH_3CO_2C_2H_5(l)} + HCl$$

$$CH_3COCl(l) + \underset{\text{Sodium phenoxide}}{C_6H_5-O^-\,Na^+(s)} \longrightarrow \underset{\text{Phenyl ethanoate}}{C_6H_5-O-\overset{\overset{O}{\|}}{C}-CH_3(l)} + NaCl(s)$$

Amide formation

Ammonia and amines give amides

Amides are formed by reactions between acid chlorides and ammonia or a primary or secondary amine:

$$RCOCl + R'_2NH \longrightarrow RCONR'_2 + HCl$$

The reaction takes place readily at room temperature.

Anhydride formation

With RCO_2^- an anhydride is formed

When an acid chloride is heated with the sodium salt of the carboxylic acid, the acid anhydride distils over:

$$\underset{\text{Sodium propanoate}}{C_2H_5CO_2Na(s)} + \underset{\text{Propanoyl chloride}}{C_2H_5COCl(l)} \xrightarrow{\text{distil}} \underset{\text{Propanoic anhydride}}{(C_2H_5CO)_2O(l)} + NaCl(s)$$

Ketone formation

See Friedel–Crafts acylation, § 28.8.5 and Figure 33.11A.

*Mechanisms of reactions of acid chlorides

The reactivity of the functional group has been described in § 33.10.1.

Reaction of an acid chloride with water

The lone pair of electrons on the oxygen atom in H_2O makes H_2O a nucleophile. It attacks the C atom of the $C^{\delta+}{=}O^{\delta-}$ group on account of its $\delta+$ charge. The intermediate formed loses a proton followed by a chloride ion.

$$H_2\ddot{O}: + R{-}C^{\delta+}({=}O^{\delta-}){-}Cl \rightleftharpoons H_2O^{+}{-}C(R)(O^-){-}Cl \rightleftharpoons H{-}O{-}C(R)(O^-){-}Cl + H^+ \longrightarrow H{-}O{-}C(R){=}O + H^+ + Cl^-$$

Reactions of acid chlorides with alcohols and phenols

The mechanism is similar to the reaction with water, with $R{-}\ddot{O}{-}H$ acting as a nucleophile.

Reaction of an acid chloride with ammonia

The lone pair on the nitrogen atom in NH_3 makes NH_3 a nucleophile. It attacks the $\delta+$ carbon atom of the C=O group. The ion formed loses a proton and a chloride ion as in the reaction with water.

$$H_3N: + R{-}C^{\delta+}({=}O^{\delta-}){-}Cl \longrightarrow H_3N^{+}{-}C(R)(O^-){-}Cl \longrightarrow H_2N{-}C(R)(O^-){-}Cl + H^+ \longrightarrow H_2N{-}C(R){=}O + H^+ + Cl^-$$

33.11 ACID ANHYDRIDES

The reactions of acid anhydrides are similar to those of acid chlorides

Anhydrides are insoluble in water but react with it to form the parent acid:

$$(RCO)_2O(l) + H_2O(l) \longrightarrow 2RCO_2H(aq)$$

Anhydrides react in a similar way to acid chlorides. They are often preferred to acid chlorides because no hydrogen chloride is formed in the reaction. They form esters with alcohols and phenols [see Figure 33.11A]; they form amides with ammonia and primary and secondary amines and are Friedel–Crafts acylating agents:

$$(CH_3CO)_2O + C_6H_5O^- Na^+ \longrightarrow C_6H_5O_2CCH_3 + CH_3CO_2^- Na^+$$

Ethanoic anhydride   Sodium phenoxide   Phenyl ethanoate

$$(CH_3CO)_2O + C_2H_5NHCH_3 \longrightarrow C_2H_5N(COCH_3)CH_3 + CH_3CO_2H$$

N-Methylethylamine   *N*-Ethyl-*N*-methylethanamide

The industrial importance of ethanoic anhydride is featured in § 33.12.

Methanoic acid does not form an anhydride. Dehydration of this acid yields carbon monoxide.

33.12 ASPIRIN

Ethanoic anhydride is used in the manufacture of aspirin. The method of synthesis is

$$C_6H_5OH + CO_2 \xrightarrow[\text{(ii) HCl(aq)}]{\text{(i) NaOH, heat under pressure}} C_6H_4(OH)CO_2H \xrightarrow[\text{alkali}]{(CH_3CO)_2O} C_6H_4(OCOCH_3)CO_2H$$

Phenol   2-Hydroxybenzoic acid (*Salicylic acid*)   2-Ethanoyloxybenzoic acid (*Aspirin*)

Ethanoic anhydride is used for the manufacture of aspirin ...

... which is an analgesic ...

The analgesic (pain-killing) effects of the esters of 'salicylic acid' have been known for centuries. Physicians in ancient times made up remedies for fever and pain from leaves which are now known to contain these compounds. 2-Hydroxybenzoic acid (salicylic acid) is the active analgesic. While it is a sufficiently strong acid to irritate the stomach, its ester, aspirin, is less acidic and less irritating. Aspirin passes through the acidic stomach contents unchanged, to be hydrolysed in the alkaline conditions of the intestines to sodium 2-hydroxybenzoate (sodium salicylate), which dissolves in the blood stream.

SUMMARY

Acid chlorides contain the group $-\overset{\delta+}{C}(=O^{\delta-})Cl$

Nucleophiles attack the $\delta+$ carbon atom as in the reactions:

- water ⟶ carboxylic acid
- alcohols and phenols ⟶ esters
- ammonia and amines ⟶ amides
- sodium salts of carboxylic acid ⟶ anhydride

Anhydrides have similar reactions.

Figure 33.11A
Reactions of acid chlorides and anhydrides

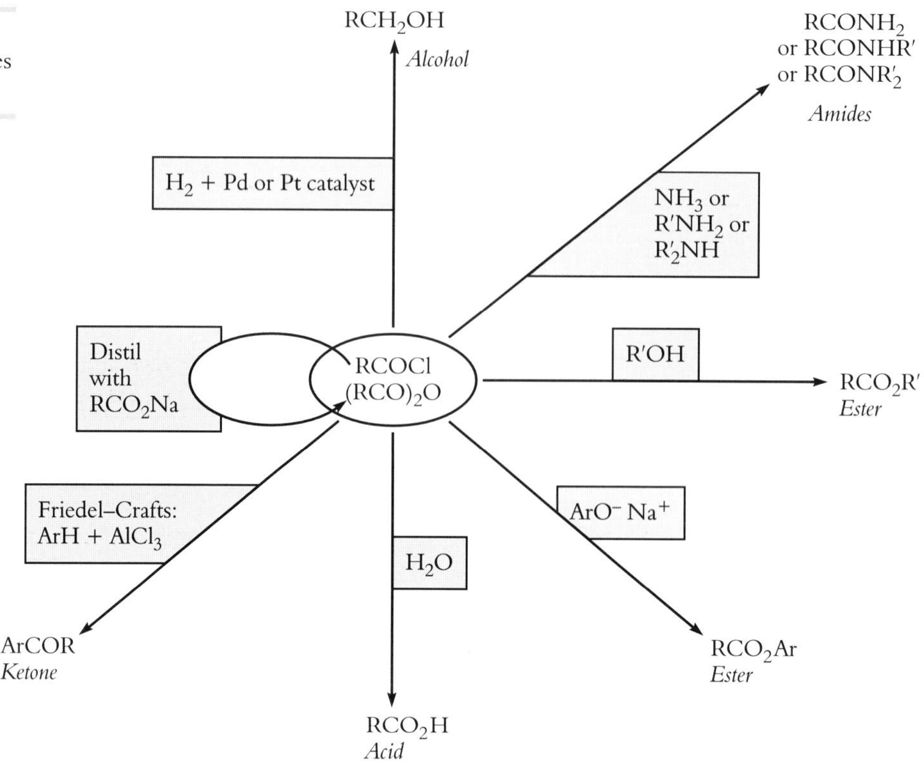

CHECKPOINT 33.12: ACID CHLORIDES AND ANHYDRIDES

1. Complete the following equations:

   (a) $C_2H_5CO_2H + PCl_5 \longrightarrow$
   (b) $CH_3CH_2CO_2Na + CH_3CH_2COCl \longrightarrow$
   (c) $(C_6H_5CO)_2O + NH_3 \longrightarrow$
   (d) $(CH_3)_2CHCOCl + C_2H_5OH \longrightarrow$

2. State how you could prepare *(a)* ethanoyl chloride and *(b)* ethanoic anhydride. Compare the reactions of the two compounds with (i) water, (ii) ethanol, (iii) phenol, (iv) phenylamine.

3. Compare the reactions of chloroethane and ethanoyl chloride with *(a)* water, *(b)* ethanol, *(c)* ammonia, *(d)* KCN. Explain the reason for the differences in behaviour.

4. Into a large excess of water were put *m* grams of ethanoyl chloride. The products remained in solution and were neutralised by 50.0 cm^3 of a solution of sodium hydroxide of concentration 0.100 mol dm^{-3}. Find *m*.

5. *(a)* Describe how you could use benzoyl chloride, C_6H_5COCl, to distinguish between phenylamine and phenylmethylamine.
   (b) Describe how you could distinguish ethanoyl chloride from benzoyl chloride. [Remember Chapter 32.]

33.13 ESTERS

Esters contain the functional group

$$R-C\begin{matrix}\nearrow O \\ \searrow O-R'\end{matrix}$$

where R and R′ may be alkyl or aryl groups and may be the same or different. The reactions of esters are summarised in Figure 33.13A. For polyesters, see § 34.5.1.

Esters are used as solvents in the laboratory and in industry.

Many esters are volatile liquids with fruity smells and fruity tastes. In the food industry, esters used as food flavourings include ethyl methanoate (rum flavour), propyl pentanoate (pineapple flavour), ethyl butanoate (apple odour) and octyl ethanoate (orange odour). Esters used as emulsifiers include mixed esters of glycerol with fatty acids and another acid, e.g. ethanoic acid, citric acid or tartaric acid. Antioxidants include the esters 2-butyl-4-methoxyphenol (known as BHA, with E number E 320) and 2,6-dibutyl-4-methylphenol (known as BHT, with E number E 321).

Esters are used in the manufacture of plastics; see polyesters, § 34.5.1. Esters are added as plasticisers in the polymerisation of other polymers, e.g. PVC, to improve the flexibility of those plastics.

Esters of glycerol are used in soap manufacture; see § 33.14.

Esters used in medicine include methyl 2-hydroxybenzoate, 'oil of wintergreen', 2-ethanoyloxybenzoic acid, aspirin, § 33.12, soluble aspirin and paracetamol. Benzocaine and novocaine are used as local anaesthetics in dentistry.

Figure 33.13A
Reactions of esters

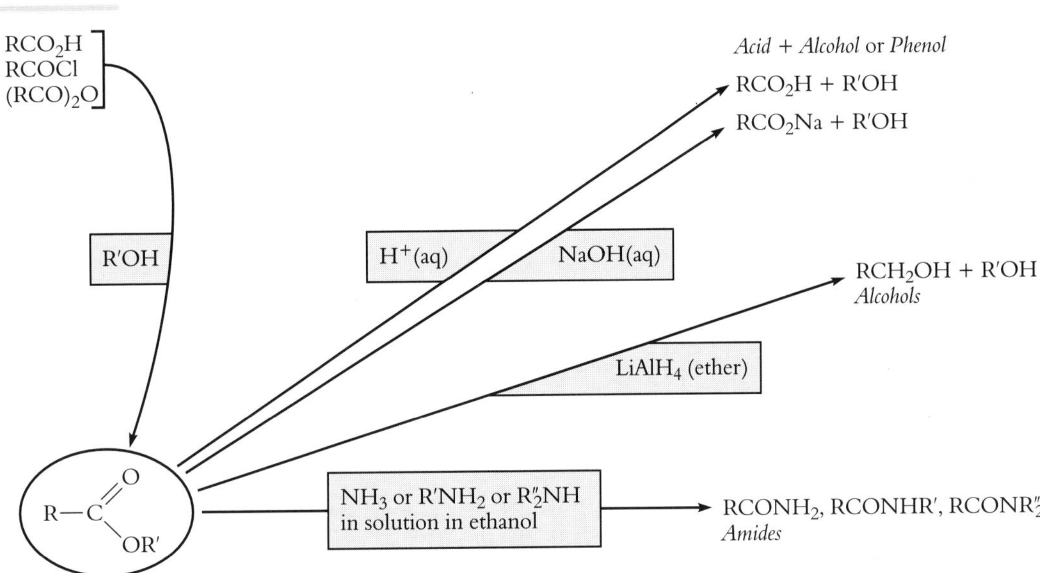

33.13.1 ESTER HYDROLYSIS

The hydrolysis of esters to acid + alcohol is catalysed by H⁺(aq)

The hydrolysis of an ester gives an equilibrium mixture of carboxylic acid and alcohol or phenol. The attainment of equilibrium is catalysed by the presence of mineral acids:

$$CH_3CO_2C_2H_5(l) + H_2O(l) \xrightleftharpoons{H^+(aq)} CH_3CO_2H(aq) + C_2H_5OH(aq)$$

OH^- ions catalyse the R.D.S. and also move the position of equilibrium in favour of the products

The reaction also proceeds faster in alkaline conditions. Hydroxide ions catalyse the rate-determining step in the hydrolysis:

$$RCO_2R' + H_2O \xrightleftharpoons{OH^-} RCO_2H + R'OH$$

When some RCO_2H molecules have been formed, OH^- ions react with them to form a salt:

$$RCO_2H + OH^- \longrightarrow RCO_2^- + H_2O$$

Salt formation removes RCO_2H from the equilibrium mixture. The hydrolysis reaction therefore goes further towards completion in the presence of a base. The base is used up in the reaction:

$$RCO_2R' + NaOH \longrightarrow RCO_2Na + R'OH$$

For the alkaline hydrolysis of oils see § 33.14.

*Mechanism of hydrolysis of esters

It has been established that the bond broken in hydrolysis is the RCO—OR′ bond of the ester group (§ 33.8.2). The mechanism of base-catalysed hydrolysis is similar to the reactions of acid chlorides (§ 33.10.2). The nucleophile OH^- attacks the $\delta+$ carbon atom of the C═O group. The negatively charged ion produced may revert to the ester or lose an alkoxide ion $R'O^-$. The carboxylic acid formed reacts with hydroxide ion to form a carboxylate ion, and the alkoxide ion reacts with water to form an alcohol.

$$H\ddot{O}^- \quad R{-}C^{\delta+}(=O^{\delta-}){-}OR' \rightleftharpoons HO{-}C(R)(O^-){-}OR' \longrightarrow HO{-}C(R){=}O + R'O^-$$

$$RCO_2H + OH^- \rightleftharpoons RCO_2^- + H_2O$$
$$R'O^- + H_2O \rightleftharpoons R'OH + OH^-$$

33.14 FATS AND OILS; SOAPS AND DETERGENTS

Fats and oils are esters of propane-1,2,3-triol and acids

Fats and oils are together classified as **lipids**. They are esters of propane-1,2,3-triol (*glycerol*) and a long chain carboxylic acid. Two examples are:

$$\begin{array}{l} CH_2OCOC_{15}H_{31} \\ | \\ CHOCOC_{15}H_{31} \\ | \\ CH_2OCOC_{15}H_{31} \end{array}$$

Propane-1,2,3-triyl trihexadecanoate (*glyceryl palmitate*) is a component of animal fats.

$$\begin{array}{l} CH_2OCO(CH_2)_7CH{=}CH(CH_2)_7CH_3 \\ | \\ CHOCO(CH_2)_7CH{=}CH(CH_2)_7CH_3 \\ | \\ CH_2OCO(CH_2)_7CH{=}CH(CH_2)_7CH_3 \end{array}$$

Propane-1,2,3-triyl trioctadec-9-enoate (*glyceryl oleate*) occurs in olive oil.

Fats are saturated; oils are unsaturated ...

... Hydrogenation converts oils into fats

Fats occur both in plants and animals, and are a source of energy in our diet. If the carboxylic acid groups are largely saturated, the ester is a solid fat; if the hydrocarbon chains are highly unsaturated, the ester is an oil. The difference in melting temperatures arises from the fact that saturated hydrocarbon chains can pack more closely together than unsaturated chains. In these esters, the configuration at the double bonds is always *cis*, so that the molecules are bent, and cannot pack closely. There is a better sale for solid fats than for oils, and the conversion of oils to fats is a profitable business. It is accomplished by hydrogenation in the presence of nickel as a catalyst. Margarine is made in this way from corn oil and soya bean oil. Recently, it has been suggested that unsaturated esters are metabolised more easily than saturated fats, and lead to less cholesterol in the blood. Sales of unsaturated fats have increased, but the matter is far from settled, and debate continues.

Saponification is the alkaline hydrolysis of fats to give soaps

An important use of fats is soap-making. The alkaline hydrolysis or **saponification** (Latin: soap-making) of fats gives glycerol and the sodium or potassium salt of the carboxylic acid. This is a soap. On addition of salt, the soap separates from solution, and can be skimmed off the surface:

$$\begin{array}{lcl} CH_2OCOC_{17}H_{35} & & CH_2OH \\ | & & | \\ CHOCOC_{17}H_{35}(s) + 3NaOH(aq) & \longrightarrow & CHOH(l) + 3C_{17}H_{35}CO_2Na(s) \\ | & & | \\ CH_2OCOC_{17}H_{35} & & CH_2OH \end{array}$$

Propane-1,2,3-triyl trioctadecanoate (*glyceryl tristearate*) — Sodium octadecanoate (*stearate*), Propane-1,2,3-triol (*glycerol*)

This process has been carried out since the Iron Age. Animal fats were boiled with water and the ashes of a wood fire, which contained potassium carbonate.

Soaps cannot work in hard water

Sodium soaps have limited solubility in water and can be obtained as solid cakes. Potassium soaps are more soluble and are used as gels in shampoos and shaving creams. The calcium and magnesium salts of soaps are insoluble. When a soap, such as sodium hexadecanoate, is added to hard water, an insoluble 'scum' of calcium and magnesium hexadecanoates is formed. To allow soap to lather readily, hard water must be softened by the removal of Ca^{2+} and Mg^{2+} ions.

Soaps can emulsify fats and oils because they contain both hydrophilic and lipophilic groups

Soap cleans because it can **emulsify** fats and oils, i.e., convert them into a suspension of tiny droplets in water. Dirt is held to fabrics by a thin film of oil or grease, and this film must be removed before the dirt can be rinsed away. Soap owes its emulsifying action to the combination of polar and non-polar groups in its structure. At one end is a highly polar carboxylate ion, which is **hydrophilic** (attracted to water) and **lipophobic** (repelled by oils and fats). At the other end is a long hydrocarbon chain which is **hydrophobic** and **lipophilic**. When soap is added to water, the hydrophilic carboxylate ions dissolve in the water, while the hydrophobic hydrocarbon ends do not. The result is a surface layer one molecule thick [see Figure 33.14A] which reduces the surface tension of water.

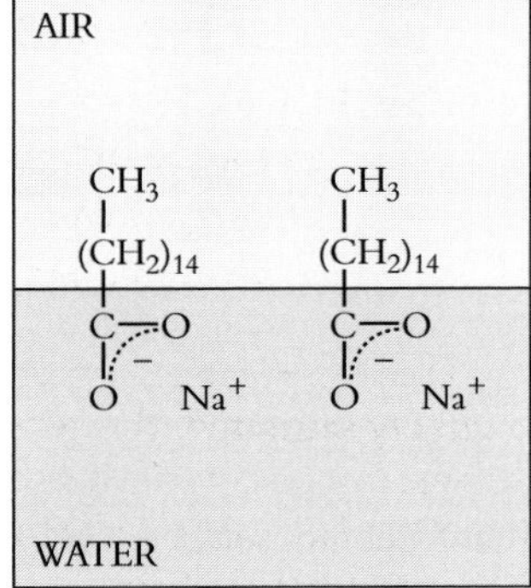

Figure 33.14A A monomolecular layer of soap on the surface of water

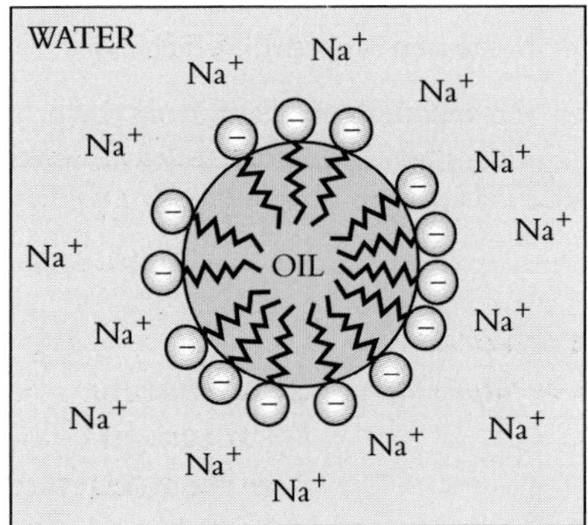

Figure 33.14B A drop of oil surrounded by soap ions

Soap interacts with an interface between oil and water when clothes are washed. The carboxylate end dissolves in the water, and the hydrocarbon end dissolves in the oil. Soap ions arrange themselves round each droplet of oil [see Figure 33.14B]. If the water is agitated, the emulsified oil droplet can float free of the fabric. As the surface of each droplet is negatively charged, the drops repel one another and do not coalesce.

Synthetic detergents are the sodium salts of sulphonic acids ...

The sodium salts of certain alkyl hydrogensulphates and benzenesulphonic acids also have detergent (cleaning) properties:

$$CH_3(CH_2)_{10}CH_2O-\overset{O}{\underset{O}{\overset{\|}{\underset{\|}{S}}}}-O^- Na^+ \qquad C_{12}H_{25}-C_6H_4-\overset{O}{\underset{O}{\overset{\|}{\underset{\|}{S}}}}-O^- Na^+$$

Sodium dodecanyl sulphate — Sodium 4-dodecanylbenzenesulphonate

... They can work in hard water

They have an advantage over soaps in that their calcium and magnesium salts are soluble, allowing them to work in hard water. A disadvantage of the first soapless detergents is that they were not biodegradable. When these detergents were discharged into rivers, the microorganisms present in the water could not destroy them. This problem has been solved by modifying the structure of the detergents.

CHECKPOINT 33.14: ESTERS

1. Write the names and structural formulae of nine isomeric esters with the formula $C_5H_{10}O_2$.

2. What products are formed in the alkaline hydrolysis of this ester?

$$C_3H_7\underset{\underset{O}{\|}}{C}-{}^{18}OC_2H_5$$

3. From which alcohol and acid are the following esters made?

   (*a*) propyl propanoate
   (*b*) methyl benzoate
   (*c*) $HCO_2C_2H_5$
   (*d*) $CH_3(CH_2)_3CO_2CH_3$
   (*e*) $(CH_3)_2CHO_2CCH_3$
   (*f*) $C_6H_5CO_2CH(CH_3)_2$

4. Outline how you would carry out the following conversions:

   (*a*) $C_6H_5NH_2 \longrightarrow C_6H_5OH$
   (*b*) $C_6H_5NH_2 \longrightarrow C_6H_5COCl$

   More than one step is needed in each case.

   (i) How could you obtain from the products of (*a*) and (*b*) a sample of phenyl benzoate, $C_6H_5CO_2C_6H_5$?

   (ii) Starting from 10.0 g of phenylamine, and using no other organic compound, what is the maximum yield of the ester?

5. The complete hydrolysis of 1.76 g of an ester of a monocarboxylic acid and a monohydric alcohol required 2.0×10^{-2} mol of sodium hydroxide. Find the molar mass of the ester. Deduce its molecular formula, and write the names and structural formulae of all the esters with this molecular formula.

6. $CH_3CH{=}CHCH_2CO_2C_2H_5$

   What are the products of the reactions of this ester with the following reagents?

   (*a*) NaOH, heat, (*b*) H_2, Ni, (*c*) Br_2,
   (*d*) O_3 followed by Zn + CH_3CO_2H, (*e*) $LiAlH_4$.

7. Naturally occurring fats and oils are the esters of acids with an even number of carbon atoms. Acids with an odd number of carbon atoms are rare. Suggest a method of increasing the length of an aliphatic acid chain by one carbon atom:

   $$RCH_2CO_2H \longrightarrow RCH_2CH_2CO_2H$$

   More than one step will be needed.

33.15 AMIDES

The methods of preparation have been described, and are summarised in Figure 33.16A.

Amides are hydrolysed to ammonium salts or acids ...

Amides are hydrolysed in a similar way to esters:

$$RCONH_2(aq) + H_2O(l) \xrightarrow{H^+(aq),\ heat} RCO_2NH_4(aq)$$

$$RCONH_2(aq) + H_2O(l) \xrightarrow{OH^-(aq),\ heat} RCO_2^-(aq) + NH_3(aq)$$

Dehydration of an amide by distillation with phosphorus(V) oxide gives a nitrile:

$$RCONH_2(s) \xrightarrow{P_2O_5\ \text{distil}} RCN(l) + H_2O(l)$$

... and converted into nitriles, acids, amines ...

The action of nitrous acid ($NaNO_2 + HCl$) is to replace $—NH_2$ by $—OH$, giving a carboxylic acid and nitrogen:

$$RCONH_2(aq) + HNO_2(aq) \longrightarrow RCO_2H(aq) + N_2(g) + H_2O(l)$$

Reduction of an amide to an amine can be accomplished by using hydrogen with nickel as a catalyst or an ethoxyethane solution of lithium tetrahydridoaluminate:

$$RCONH_2(s) \xrightarrow[\text{or } LiAlH_4(\text{ether})]{H_2,\ \text{Ni, heat}} RCH_2NH_2(l)$$

... and into the amines which have one carbon atom less than the amides

In the Hofmann degradation [see Figure 33.16A], reaction with bromine and concentrated alkali gives an amine with one carbon atom less than the parent amide:

$$RCONH_2(aq) + Br_2(aq) + 4OH^-(aq) \longrightarrow RNH_2(l) + 2Br^-(aq) + CO_3^{2-}(aq) + 2H_2O(l)$$

For polyamides see § 34.5.4.

Figure 33.16A
Reactions of amides and nitriles

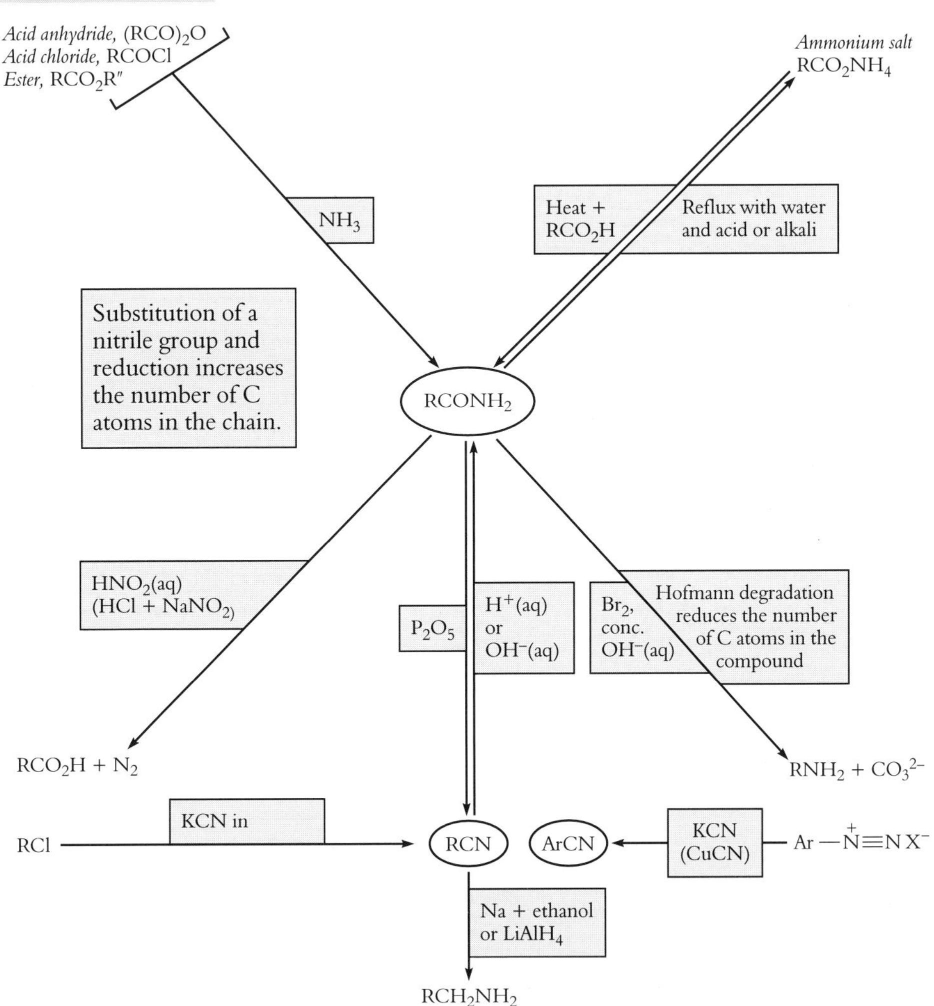

33.16 NITRILES

Aliphatic nitriles, RCN are made from halogenoalkanes, RX ...

... Aromatic nitriles, ArCN, are made from diazonium compounds

Nitriles

$$R—C\equiv N$$

differ from the other acid derivatives in that they do not possess a carbonyl group. They are readily prepared from acids and readily yield acids on hydrolysis. Aliphatic nitriles are prepared from halogenoalkanes; aromatic nitriles from diazonium compounds [see Figure 32.8A, § 32.8.2 and Figure 33.16A].

Figure 33.16B Relationships between carboxylic acids and other series

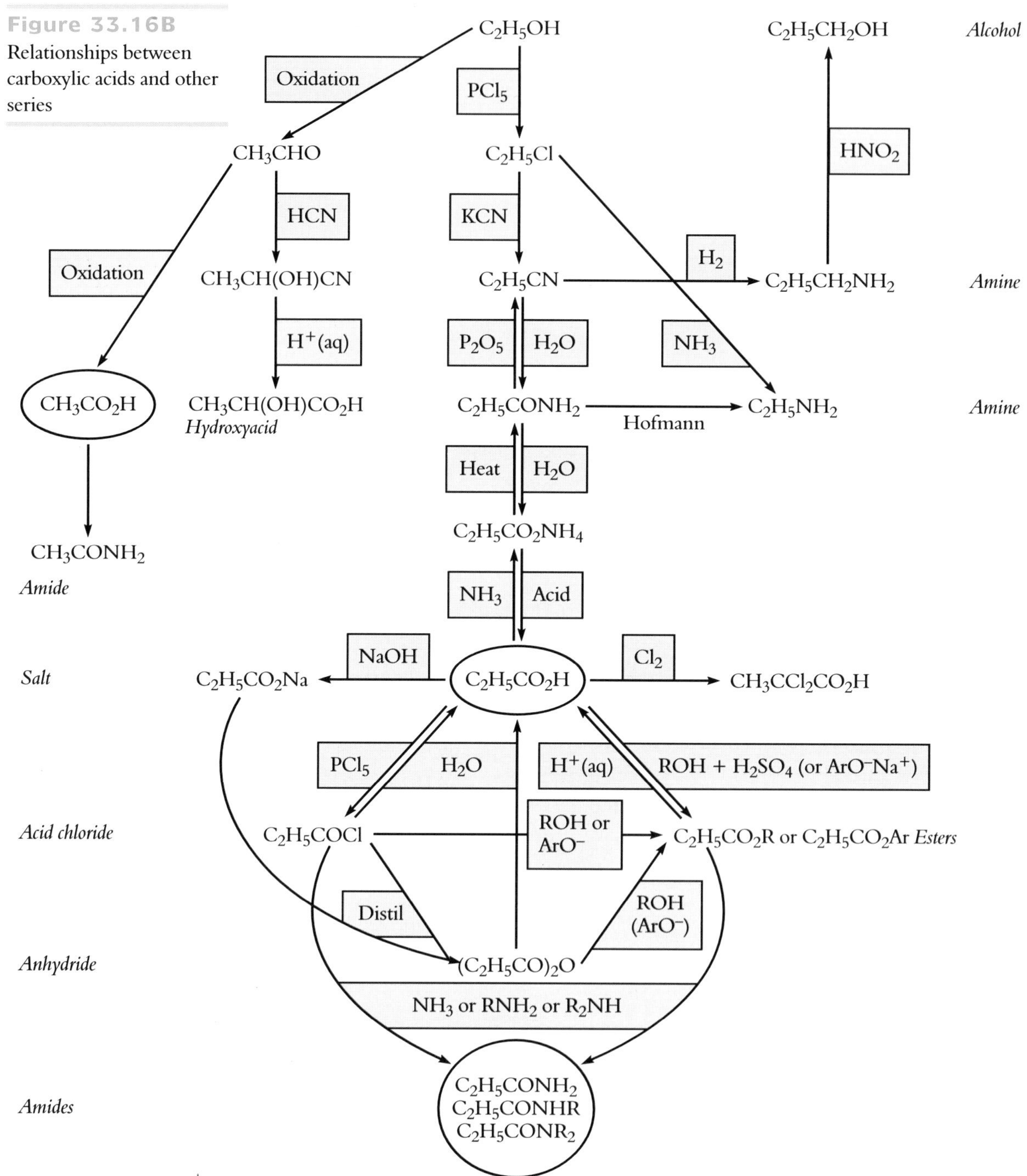

SUMMARY

Amides, $RCONH_2$

- hydrolysed to carboxylic acids RCO_2H
- dehydrated to nitriles RCN
- reduced by $LiAlH_4$ to amines RCH_2NH_2
- with bromine + conc. alkali ⟶ amine with one less carbon atom RNH_2.

Nitriles, RCN

- hydrolysed to carboxylic acids RCO_2H
- reduced by $LiAlH_4$ to amines RCH_2NH_2

To increase the number of C atoms in an alkyl group:

$$RCl \xrightarrow{\text{KCN (ethanol)}} RCN \xrightarrow{\text{LiAlH}_4\text{ (ether)}} RCH_2NH_2$$

CHECKPOINT 33.16: AMIDES AND NITRILES

1. Say how the following conversions could be brought about:

   (a) $RCONH_2 \longrightarrow RCO_2H$
   (b) $RCONH_2 \longrightarrow RCH_2NH_2$
   (c) $RCONH_2 \longrightarrow RNH_2$
   (d) $RCO_2R' \longrightarrow RCONH_2 + R'OH$
   (e) $(RCO)_2O \longrightarrow RCONH_2$
   (f) $RCOCl \longrightarrow RCH_2OH$

2. *(a)* Write the name and structural formula of (i) a carboxylic acid amide and (ii) a primary aliphatic amine.
   (b) Describe two ways in which the properties of amides differ from those of amines and two ways in which they show resemblance.
   (c) How can amide (i) be converted into an amine?
   (d) How can amine (ii) be converted into an amide?

3. State how you could prepare the following compounds, using the organic compound cited as the only organic substance and any inorganic reagents you need:

   (a) $CH_3CH_2CH_2CO_2H$ from $CH_3CH_2CH_2OH$
   (b) $CH_3CO_2C_2H_5$ from CH_3CH_2OH
   (c)

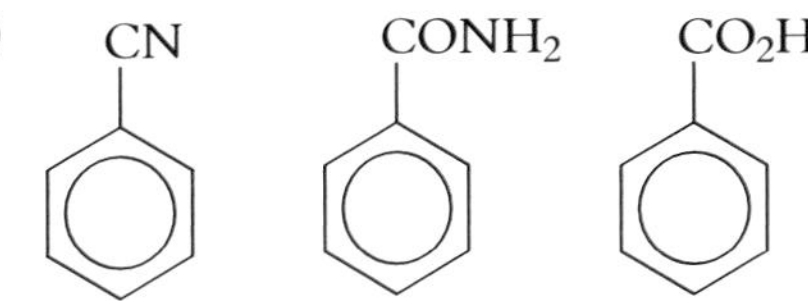

all from benzene.

4. Identify compounds **A**, **B**, **C**, **D** and **E**, and reagents *x*, *y* and *z* in the following scheme of reactions. Write equations for the reactions involved:

$$C_2H_5CO_2H \xrightarrow{PCl_5} \mathbf{A} \xrightarrow{NH_3} \mathbf{B} \xrightarrow{x} C_2H_5NH_2$$

$C_2H_5CO_2H \xrightarrow{z} \mathbf{E}(C_4H_8O_2)$; $C_2H_5NH_2 \xrightarrow{y} \mathbf{C}$; $\mathbf{C} \xrightarrow[\text{(+ acid)}]{CH_3CO_2H} \mathbf{D}(C_4H_8O_2)$

5. Identify **A**, **B**, **C**, **D**, **E**, **F** and **G** in the scheme of reactions shown below. Write equations for the reactions portrayed:

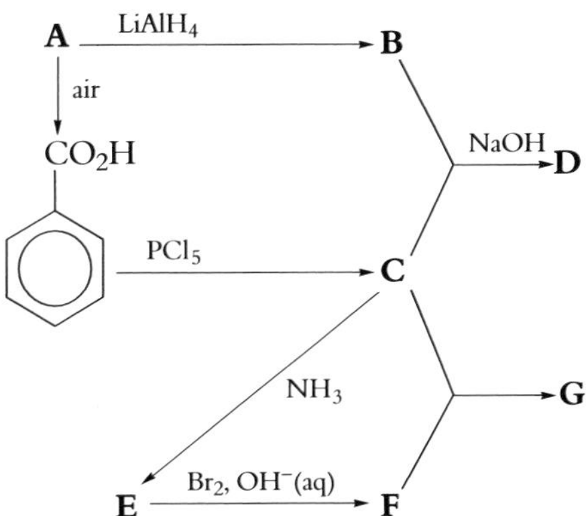

33.17 AMINO ACIDS AND PROTEINS

Proteins are one of the three large classes of natural polymers

There are three large classes of natural polymers: polysaccharides, nucleic acids and proteins. Proteins are polyamides. The units from which they are formed are amino acids, of formula

$$H_2NCH(R)CO_2H$$

Since the $—NH_2$ group is on the α carbon atom (the 2-carbon atom, that adjacent to the $—CO_2H$ group), they are correctly known as α amino acids. Table 33.17 lists some of the 20 naturally occurring α amino acids.

Table 33.17 Some naturally occurring α amino acids

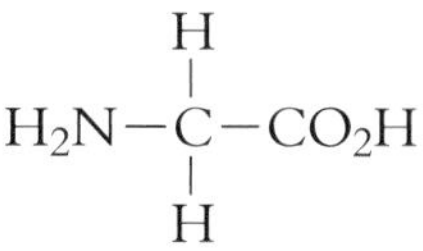 $H_2N—CH(H)—CO_2H$ 2-Aminoethanoic acid *(Glycine)*	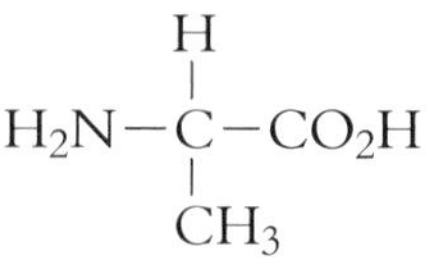 $H_2N—CH(CH_3)—CO_2H$ 2-Aminopropanoic acid *(Alanine)*
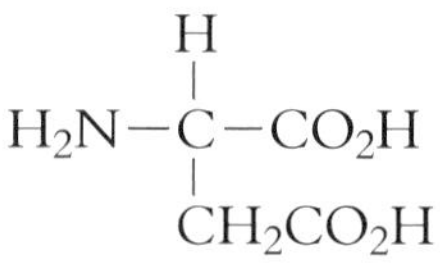 $H_2N—CH(CH_2CO_2H)—CO_2H$ 2-Aminobutane-1,4-dioic acid *(Aspartic acid)*	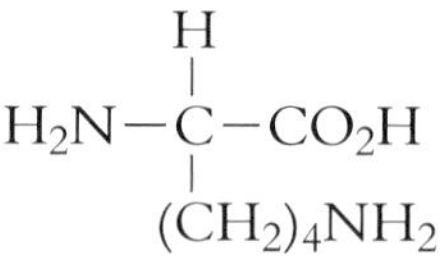 $H_2N—CH((CH_2)_4NH_2)—CO_2H$ 2,6-Diaminohexanoic acid *(Lysine)*
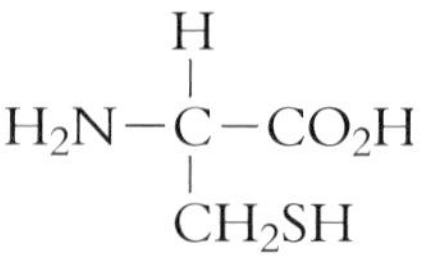 $H_2N—CH(CH_2SH)—CO_2H$ *Cysteine*	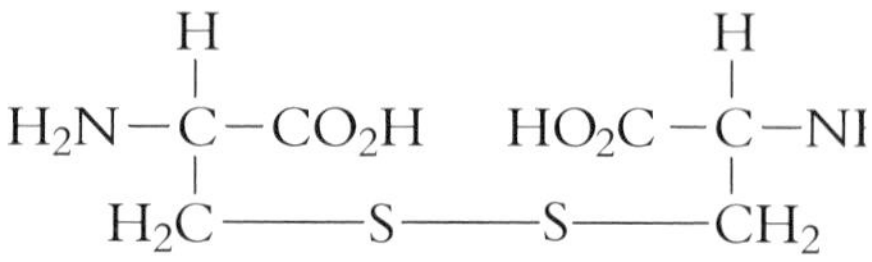 $H_2N—CH(CO_2H)—CH_2—S—S—CH_2—CH(NH_2)—CO_2H$ (printed: H₂N—C(H)—CO₂H, HO₂C—C(H)—NI; H₂C—S—S—CH₂) *Cystine*

Glycine can be made from ethanoic acid

All amino acids, except glycine, have an asymmetric carbon atom and show optical isomerism. Glycine is a white solid of T_m 232 °C. It can be made from ethanoic acid:

$$\underset{\text{Ethanoic acid}}{CH_3CO_2H} \xrightarrow[\text{pass } Cl_2]{\text{boil}} \underset{\text{Chloroethanoic acid}}{ClCH_2CO_2H} \xrightarrow[\text{large excess}]{\text{conc. } NH_3(aq)} \underset{\text{Glycine}}{H_2NCH_2CO_2H} + NH_4Cl$$

33.17.1 REACTIONS OF AMINO ACIDS

There are two functional groups, $—NH_2$ and $—CO_2H$, each of which is responsible for a typical set of reactions [see Figure 33.17A]. In acidic solution,

Amino acids form zwitterions, containing $—NH_3^+$ and $—CO_2^-$

the $—NH_2$ group ionises as $—\overset{+}{N}H_3$; in alkaline solution, the $—CO_2H$ group ionises as $—CO_2^-$. At intermediate values of pH, the **zwitterion**

$$H_3\overset{+}{N}CH(R)CO_2^-$$

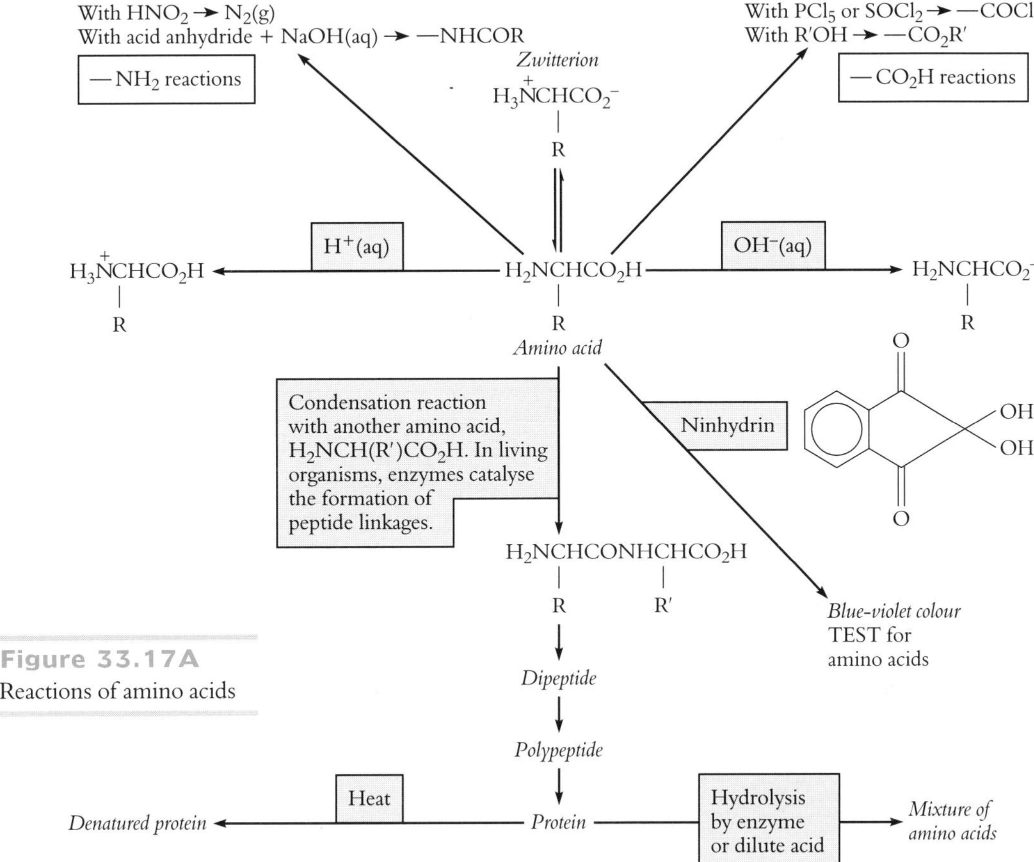

Figure 33.17A
Reactions of amino acids

They have a buffering action

exists (German: *Zwitter*, mongrel). The reason why amino acids are high-melting solids is that they crystallise as zwitterions. In solution, the pH at which the concentration of the zwitterion is maximal is called the **isoelectric point**. It differs for different amino acids. Since amino acids can accept both $H^+(aq)$ ions and $OH^-(aq)$ ions, they exert a buffering action; see § 12.8. In **electrophoresis**, an amino acid will move towards the cathode or the anode, depending on the pH. This behaviour is the basis of the electrophoretic method (Greek: *phoresis*, being carried) of separating amino acids according to their isoelectric points.

Amino acids can be separated by electrophoresis or by chromatography

Another method of separating amino acids is chromatography [§ 8.7]. The positions of the colourless amino acids on the chromatogram are revealed by spraying the paper with ninhydrin. This reagent gives a blue-violet colour with amino acids. Since amino acids can combine with $H^+(aq)$ ions and $OH^-(aq)$ ions, they exert a buffering action [see § 12.8].

33.17.2 PEPTIDES AND PROTEINS

The polyamides formed by amino acids are peptides and proteins

The most important reaction of amino acids is polymerisation. In living organisms, enzymes catalyse the polymerisation of amino acids to form peptides and proteins. Polymerisation occurs through the formation of amide groups by the reaction of an —NH_2 group in one molecule with a —CO_2H group in another molecule:

$$H_2N-\underset{H}{\overset{R}{C}}-\overset{O}{\overset{\|}{C}}-OH + H-\overset{H}{N}-\underset{H}{\overset{R}{C}}-\overset{O}{\overset{\|}{C}}-OH \longrightarrow \underset{\text{A dipeptide}}{H_2N-\underset{H}{\overset{R}{C}}-\overset{O}{\overset{\|}{C}}-\overset{H}{N}-\underset{H}{\overset{R}{C}}-\overset{O}{\overset{\|}{C}}-OH} + H_2O$$

The dipeptide formed can go on polymerising to form a long chain of amino acid residues linked through —CONH— groups. The —CONH— group is called a **peptide linkage**.

Fibrous proteins have linear molecules ...

When the number of amino acid residues is 100 or more, the polymer is called a **protein**. Proteins are a diverse group of polymers. Fibrous proteins have linear molecules, and are insoluble in water and resistant to acids and alkalis. Examples are *keratin* (in hair, nails, horn and feathers), *collagen* (in muscles and tendons), *elastin* (in arteries and tendons) and *fibroin* (in silk).

... Globular proteins have a complicated 3D structure ...

The molecules of globular proteins, e.g., *albumin* (in egg white) and *casein* (in milk), adopt a complicated three-dimensional structure. The polypeptide chain is held in position by attractions between polar groups, e.g., $—CO_2^-$ in aspartic acid and $—\overset{+}{N}H_3$ in lysine and also by cystine bridges [see Table 33.17]. The —SH groups in two cysteine amino acid residues can be oxidised to form an —S—S— bridge between two strands of polypeptide. Anything which interferes with the three-dimensional configuration, such as acids, alkalis and a rise in temperature, is said to **denature** the protein. Globular proteins are soluble in water and are easily denatured.

... e.g., enzymes

Enzymes are globular proteins, and their catalytic activity depends on their three-dimensional structure. If the enzyme is denatured by a rise in temperature or an extreme of pH, or even by violent agitation of the solution, it is no longer able to function as a catalyst.

Hair protein can be permanently waved

Hair protein owes some of its texture to —S—S— bonds in cystine. Permanent waving techniques utilise this fact. When hair is soaked in a gentle reducing agent, the —S—S— bridges are broken by reduction to —SH groups. While the hair proteins are no longer cross-linked, they are arranged into curls. Re-oxidation by soaking the hair in a mild oxidising agent reforms the disulphide bonds and holds the hair in its new configuration [see Figure 33.17B].

Figure 33.17B
Permanent waving

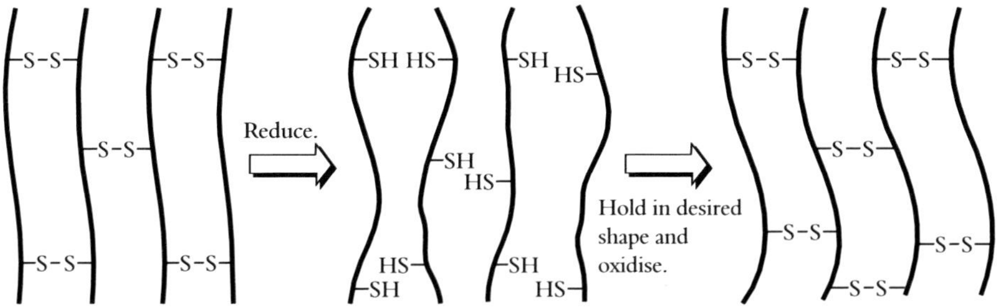

SUMMARY

Amino acids, with the general formula $H_2NCHRCO_2H$ (in which R can be about 20 groups)

- have the reactions of the $—NH_2$ group and those of the $—CO_2H$ group
- ionise as the zwitterion $H_3N^+CHRCO_2^-$
- exert a buffering action
- polymerise to form peptides and proteins.

CARBOXYLIC ACIDS AND DERIVATIVES

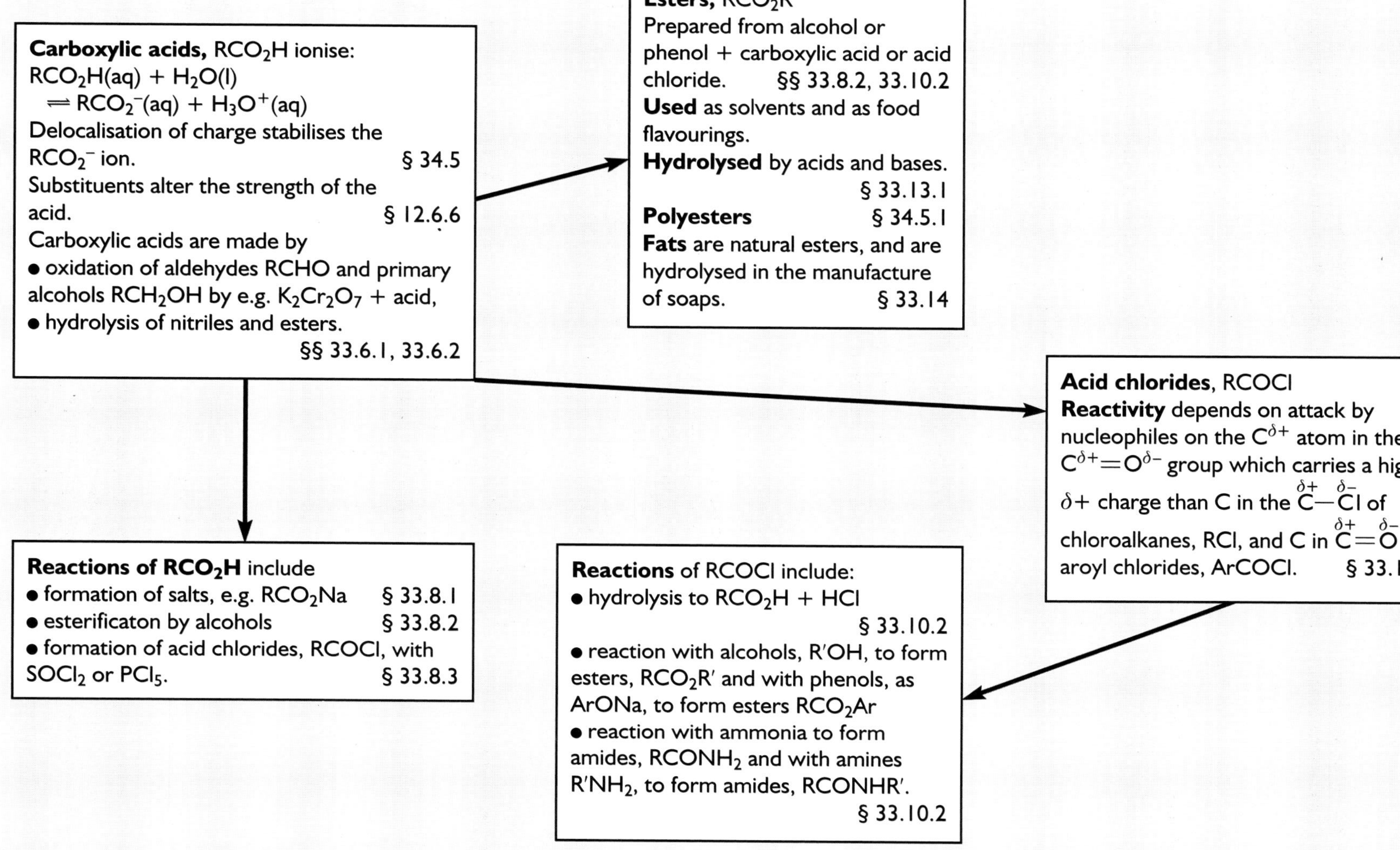

Carboxylic acids, RCO_2H ionise:
$RCO_2H(aq) + H_2O(l) \rightleftharpoons RCO_2^-(aq) + H_3O^+(aq)$
Delocalisation of charge stabilises the RCO_2^- ion. § 34.5
Substituents alter the strength of the acid. § 12.6.6
Carboxylic acids are made by
• oxidation of aldehydes RCHO and primary alcohols RCH_2OH by e.g. $K_2Cr_2O_7$ + acid,
• hydrolysis of nitriles and esters.
§§ 33.6.1, 33.6.2
Esters, RCO_2R'
Prepared from alcohol or phenol + carboxylic acid or acid chloride. §§ 33.8.2, 33.10.2
Used as solvents and as food flavourings.
Hydrolysed by acids and bases. § 33.13.1
Polyesters § 34.5.1
Fats are natural esters, and are hydrolysed in the manufacture of soaps. § 33.14
Acid chlorides, RCOCl
Reactivity depends on attack by nucleophiles on the $C^{\delta+}$ atom in the $C^{\delta+}{=}O^{\delta-}$ group which carries a higher $\delta+$ charge than C in the $\overset{\delta+}{C}{-}\overset{\delta-}{Cl}$ of chloroalkanes, RCl, and C in $\overset{\delta+}{C}{=}\overset{\delta-}{O}$ of aroyl chlorides, ArCOCl. § 33.10.1
Reactions of RCO_2H include
• formation of salts, e.g. RCO_2Na § 33.8.1
• esterificaton by alcohols § 33.8.2
• formation of acid chlorides, RCOCl, with $SOCl_2$ or PCl_5. § 33.8.3
Reactions of RCOCl include:
• hydrolysis to RCO_2H + HCl § 33.10.2
• reaction with alcohols, R'OH, to form esters, RCO_2R' and with phenols, as ArONa, to form esters RCO_2Ar
• reaction with ammonia to form amides, $RCONH_2$ and with amines $R'NH_2$, to form amides, RCONHR'. § 33.10.2

CHECKPOINT 33.17: AMINO ACIDS AND PROTEINS

1. Outline how you could make *alanine*, $H_2NCH(CH_3)CO_2H$, from propanoic acid. Give the IUPAC name for alanine. Predict two reactions which alanine would undergo.

2. *(a)* Outline the preparation of *glycine*, aminoethanoic acid, starting from ethanoic acid.
   (b) Which of these formulae is the better representation of the structure of glycine?

   $H_2NCH_2CO_2H$ or $H_3\overset{+}{N}CH_2CO_2^-$

   Explain your choice.
   (c) Write equations for two reactions of glycine.

3. What class of polymers are made from monomers such as glycine and alanine? Give the names of three of these polymers.
   What is the name and the structure of the functional group in these polymers? State one reaction which is typical of this functional group.

4. How could you distinguish between aminoethanoic acid and ethanamide?

5. A 0.1110 g specimen of an amino acid was dissolved in water and treated with nitrous acid. The volume of nitrogen produced was 16.0 cm^3 at 1 atm and 293 K. Calculate the molar mass of the amino acid.
   (GMV = 24.0 dm^3 at rtp.)

QUESTIONS ON CHAPTER 33

1. Identify the compounds **A**, **B**, **C**, **D** and **E** in the figure. Name the product. How would you expect this acid to react with *(a)* Cl_2, *(b)* PCl_5, *(c)* Na_2CO_3, *(d)* plane-polarised light?

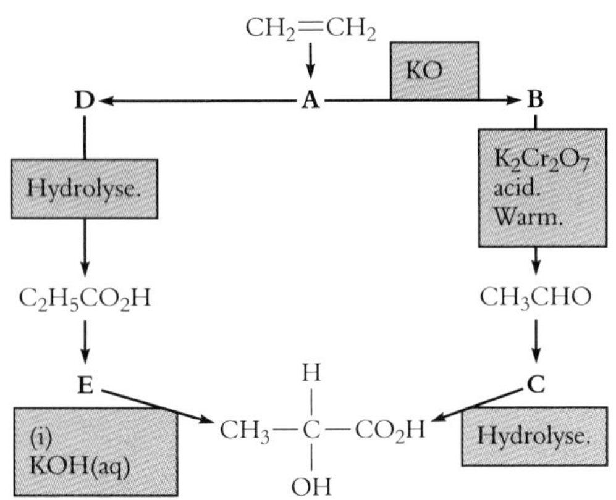

2. How could you carry out the following conversions? Use no organic compounds other than the one cited. Use any inorganic substances you need.
   (a) $CH_3CH_2CH_2OH \longrightarrow CH_3CH_2CO_2CH_2CH_2CH_3$
   (b) $(CH_3)_2CHOH \longrightarrow (CH_3)_2C(OH)CO_2H$
   (c) $CH_3CHO \longrightarrow CH_3CO_2CH_2CH_3$

   More than one step will be needed.

3. Starting from an aliphatic carboxylic acid, how could you make *(a)* an acid chloride, *(b)* an acid anhydride, *(c)* an amide, *(d)* an ester? Give the necessary conditions for reaction, and give equations for the reactions. State the conditions under which each of the products will react with water. Write equations, and name the products of the hydrolyses.

4.

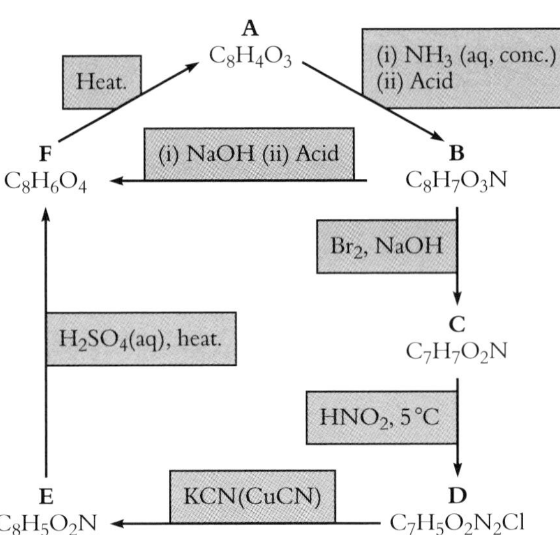

Write structural formulae for compounds **A**–**F**. Explain the reactions shown in the figure.

5. Explain why a carboxylic acid, although it possesses a $>C=O$ group does not react with the nucleophiles which attack aldehydes and ketones.

6. Two esters have the molecular formula $C_6H_{12}O_2$. Both show optical isomerism. When heated with aqueous sodium hydroxide, **A** gives sodium ethanoate and another product, and **B** gives methanol and another product. Write structural formulae for **A** and **B**.

7. An optically active ester of molecular formula $C_8H_{16}O_2$ was hydrolysed by aqueous alkali to give as one product a liquid of molecular formula C_3H_8O, which gave a positive iodoform test. Write a structural formula for the ester.

8. Explain these statements:

(a) Ethanoyl chloride is more reactive towards water than is chloroethane.
(b) Laboratory preparations of 2-hydroxypropanoic acid yield a racemic mixture.

9. You are asked to prepare a carboxylic acid from an aldehyde.

(a) Name the oxidising agent you would use. Give the conditions for the reaction.
(b) Write the equation for the reaction of a named aldehyde.
(c) For the carboxylic acid in *(b)* write the structural formulae of the anhydride and the amide.
(d) Give the conditions required to convert (i) the anhydride and (ii) the amide into the acid.

10. A liquid **A** of formula $C_5H_{10}O_2$ was reduced by $LiAlH_4$ to a mixture of two alcohols **B** and **C**. Both alcohols reacted with iodine in alkaline solution to give a pale yellow crystalline solid **D**. The liquid **A** is insoluble in cold, dilute aqueous sodium hydroxide but on boiling the mixture gradually becomes one layer.

Identify and draw structural formulae of **A**, **B**, **C** and **D**. Give your reasoning.

11. A hydrocarbon **A** does not react with chlorine in the dark. When a mixture of **A** and chlorine is irradiated with ultraviolet light two and only two monochlorinated products, **B** and **C**, are obtained. When **B** and **C** are treated separately with warm aqueous sodium hydroxide solution and then with potassium manganate(VII) solution, **B** gives an acid **D**, and **C** gives a ketone, **E**.

Identify **A**, **B**, **C**, **D** and **E**. Give your reasoning and draw their structural formulae.

12. One of the products from the vigorous hydrolysis of oil of bitter almonds is mandelic acid, **H**. It can be synthesised from benzaldehyde in two steps:

$$C_6H_5CHO \xrightarrow{\text{step I}} \mathbf{J} \xrightarrow{\text{step II}} \underset{\mathbf{H}}{C_6H_5CH(OH)CO_2H}$$

(a) Suggest reagents and conditions for steps I and II, and draw the structural formula of the intermediate **J**. **4**
(b) Suggest structural formulae for the products of the reaction of mandelic acid with the following reagents.
(i) HBr (iv) C_2H_5OH/H^+
(ii) PCl_5 (v) NaOH(aq)
(iii) CH_3COCl **6**
(c) A sample of mandelic acid isolated from bitter almonds was contaminated with a neutral impurity. A 0.100 g sample of the impure acid required 6.00 cm^3 of 0.100 mol dm^{-3} NaOH to neutralise it. Calculate the percentage purity of the mandelic acid. **2**

C (AS)

13. This question is about 2-hydroxybenzoic acid and some of its reactions.

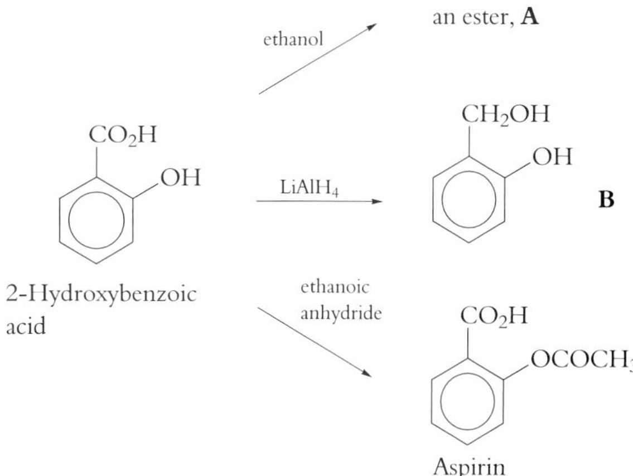

(a) How would you show that 2-hydroxybenzoic acid contains a phenolic —OH group? **2**
(b) When 2-hydroxybenzoic acid is reacted with ethanol under suitable conditions it forms an ester, **A**. Write a balanced equation for the reaction which occurs. **2**
(c) In the formation of compound **B**
(i) what is the attacking ion derived from $LiAlH_4$? **1**
(ii) explain why the 2-hydroxybenzoic acid molecule is susceptible to attack by this ion. **1**
(d) The reaction of 2-hydroxybenzoic acid with ethanoic anhydride can be described as a nucleophilic substitution.
(i) Draw the displayed formula of ethanoic anhydride. **2**
(ii) Suggest why ethanoic anhydride reacts more readily than ethanoic acid with the phenolic group. **2**
(e) Aspirin may be reacted with dilute aqueous sodium hydroxide to modify its structure before being used as a drug.
(i) Give the structural formula of the compound which forms when aspirin is reacted with dilute aqueous sodium hydroxide at room temperature. **1**
(ii) Suggest **TWO** advantages of using the compound in *(e)*(i) as a drug instead of aspirin. **2**

13 *L(N) (AL)*

14. Esters are often described as having 'fruity smells'. An ester can be prepared in the laboratory by the reaction of an alcohol and a carboxylic acid in the presence of an acid catalyst.

(a) Ethanol reacts with 2-methylbutanoic acid to produce an ester which is found in ripe apples.
(i) Draw the displayed formula of 2-methylbutanoic acid.
(ii) When ethanol reacts with 2-methylbutanoic acid, the ester produced has the formula $CH_3CH_2CH(CH_3)CO_2CH_2CH_3$. Write a balanced equation for the formation of this ester. **2**

(b) The following experiment was carried out by a student.

A 9.2 g sample of ethanol and 20.4 g of 2-methylbutanoic acid were mixed in a flask and 2.0 g of concentrated sulphuric acid was added. The mixture was refluxed for four hours and then fractionally distilled to give 17.4 g of the crude ester. The ester was washed repeatedly with aqueous sodium carbonate until there was no more effervescence. After further washing with distilled water and drying, 15.6 g of pure ester were obtained.

By referring to the experimental procedure above,
(i) explain the meaning of *refluxed*;
(ii) explain why the crude ester was *washed repeatedly* with aqueous sodium carbonate;
(iii) state which gas was responsible for the *effervescence*.
(iv) Calculate how many moles of each reactant were used: ethanol (M_r: 46); 2-methylbutanoic acid (M_r: 102)
(v) Use your answers to *(a)*(ii) and to *(b)*(iv) to calculate the percentage yield of pure ester obtained in the above experiment. **9**

(c) The ester produced in the above experiment smelt of apples. A similar ester, $CH_3CO_2CH_2CH_2CH(CH_3)_2$, contributes to the flavour of ripe pears.

Suggest structures of a carboxylic acid and an alcohol that could be used for the preparation of the ester $CH_3CO_2CH_2CH_2CH(CH_3)_2$ **2**

13 *C (AS/AL)*

15. *(a)* Compounds **A**, **B** and **C** are shown in order of **decreasing** acidity.

CH_3COOH (**A**) C_6H_5–OH (**B**) CH_3CH_2OH (**C**)

(i) State what is meant by a *Brönsted–Lowry acid*. **1**
(ii) With reference to the molecular structures of each compound, account for the variation in acidity of compounds **A**, **B** and **C**. **4**
(iii) Write an expression linking pK_a with K_a. **1**
(iv) The pK_a values of **A**, **B** and **C**, but not necessarily in that order, are 10.00, 4.76 and 16.00. Assign the appropriate pK_a value to each compound. **1**
(v) Two samples of each of the compounds **A**, **B** and **C** were taken. To the first sample of each compound was added aqueous sodium hydroxide. To the second sample of each compound was added aqueous sodium hydrogencarbonate. In a copy of the table below write, as appropriate, either 'reaction' or 'no reaction' in each box.

	A	B	C
Addition of aqueous sodium hydroxide			
Addition of aqueous sodium hydrogencarbonate			

2

(b) Compound **D** may be prepared from compound **B**.

C_6H_5–O–C(=O)–CH_3

D

Give the name of **D**, state a reagent you could use to prepare it from **B**, and write an equation for the reaction. **3**

(c) State **two** observations you would be able to make if bromine water were added to a solution of **B**. Draw the structure of the organic product formed. **4**

16 *AEB (AL)* 1998

16. *(a)* A compound **A**, $C_3H_5O_2Cl$, after boiling for half an hour with aqueous acid, yielded compounds **B**, $C_2H_3O_2Cl$, and **C**, CH_4O. Boiling **A** for half an hour with aqueous sodium hydroxide, yielded compounds **D**, $C_2H_3O_3Na$, and **C**.

When compound **B** was boiled with aqueous sodium hydroxide, and the mixture acidified, it gave **E**, $HOCH_2CO_2H$. Treatment of **E** with aqueous sodium carbonate resulted in a vigorous effervescence as **E** was converted into **D**.

Deduce the structures of compounds **A** to **D** inclusive, giving your reasoning and accounting for each of the given reactions. **13**

(b) What product would be obtained, and under what conditions, when compounds **E** and **C** react together? **3**

16 *O&C (AS/AL)*

17. 2-Aminopropanoic acid (alanine), $CH_3CH(NH_2)CO_2H$, has a chiral centre and hence can exist as two optical isomers.

(a) (i) State what is meant by a *chiral centre*.
(ii) Explain how a chiral centre gives rise to optical isomerism.
(iii) Draw diagrams to show the relationship between the two optical isomers. State the bond angle around the chiral centre. **5**

(b) In aqueous solution, 2-aminopropanoic acid exists in different forms at different pH values. The zwitterion predominates between pH values of 2.3 and 9.7.

Draw the displayed formula of the predominant form of 2-aminopropanoic acid at pH values of 2.0, 6.0 and 10.0. **3**

(c) 2-Aminopropanoic acid can react with an amino acid, **K**, to form the dipeptide below.

```
     H           H
     |           |          O
H—N—CH—C—N—CH—C//
        |   ||      |        \
       CH3  O      CH        OH
                  /  \
               CH3    CH3
```

(i) Identify the peptide linkage.
(ii) Draw the displayed formula of the amino acid, **K**. **3**
(iii) 2-Aminopropanoic acid can react with the amino acid, **K**, to form a different dipeptide from that shown above. Draw the structural formula of this other dipeptide. **2**

C (AS/AL)

18. An organic compound, **A**, has molecular formula $C_5H_{12}O$. On heating under reflux with acidified dichromate(VI), $Cr_2O_7^{2-}$, solution it is converted into **B**, which possesses a chiral centre.

Compound **B** is acidic and on heating with soda lime produces butane, $CH_3CH_2CH_2CH_3$.
On gently heating compound **A** with acidified $Cr_2O_7^{2-}$ solution, followed by *immediate* distillation of the initial product, compound **C** is obtained.

Compound **C** gives a silver mirror when heated with ammoniacal silver nitrate solution (Tollens' reagent).

(a) Name the reaction types for the conversions **A** ⟶ **B** and **A** ⟶ **C** **1**

(b) Give the systematic names and structural formulae for compounds **A**, **B** and **C**. **Briefly** give reasons for your identifications **4**

(c) State and explain how you would expect the boiling temperature of compound **C** to compare with those of **A** and **B**. **1½**

(d) Compound **B** may be directly converted back to compound **A**. State the reagent and reaction conditions required for this and name the reaction type. **1½**

(e) Explain what the term ***chiral centre*** means and mark that in your formula for compound **B** (in part *(b)* above) with an asterisk (*). **1½**

(f) Name a reagent which would yield a bright orange/red precipitate with compound **C** and state another functional group which would react similarly with this reagent. **1**

(g) Name the reaction type involved when compound **C** shows a positive test with Tollens' reagent. **1½**

11 *WJEC (AS/AL)*

19. *(a)* By varying the reaction conditions, the addition of HBr to propene can be made to produce either of two isomeric bromopropanes, **A** and **B**.

When reacted with aqueous NaOH, **A** and **B** give the isomeric alcohols, **C** and **D**, which are oxidised to **E** and **F** by acidified dichromate(VI).

E produces an orange crystalline derivative when reacted with 2,4-dinitrophenylhydrazine, and gives no reaction with Ag^+ ions.

F gives a sweet smelling liquid when reacted with ethanol and an acid catalyst.

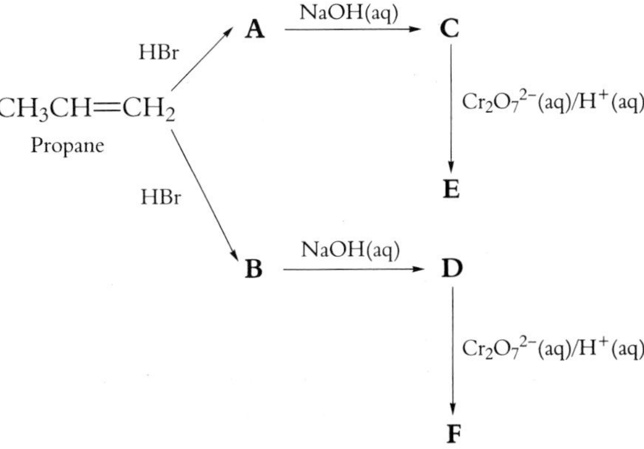

Deduce the **structural formulae** of **A**, **B**, **C**, **D**, **E** and **F**.
Give the systematic names for **A** and **B**, and give your reasoning for the identification of **C**, **D**, **E** and **F**.
(Hint: you may find that the best approach is to identify **E** and **F** first.) **6**

(b) (i) Explain what is meant by the expression ***Lowry–Brönsted acids and bases***. **1**
(ii) Write an expression for the dissociation constant, K_a, of an acid, HA, in water. **1**
(iii) This question refers to 0.1 mol dm^{-3} solutions of HCl and CH_3COOH.
I. Which of HCl and CH_3COOH is the stronger acid?
II. Which will have the larger K_a value?
III. Which will have the lower pH value? **2**

10 *WEJC*

34
POLYMERS

34.1 PLASTICS IN USE

Plastics are all around us. In the sitting room of fifty years ago, the settee and chairs would have been made of natural materials, such as wool, cotton and leather. Today the probability is that they are made of synthetic fibres such as nylons and polyesters. The carpet probably contains a high percentage of nylon or poly(propene), and the curtains contain polyester. The television set, the hi-fi, the video-recorder all have parts made of tough plastics.

In the kitchen, the worktop and table top may be made of thermosetting plastics which are heat-resistant as well as being easy to clean and hygienic. The insides of the refrigerator and freezer are made of plastics. The cupboards contain plastic bowls and buckets and possibly plastic microwave dishes and plates.

The bathroom is likely to contain an acrylic plastic bath with plumbing made of poly(propene) pipes.

On the outside of the house, the windowframes, the gutters, the drainpipes and sewage pipes may well be of PVC. The paint and varnish on the house contain plastics. The car which stands outside the door contains about 100 kg of plastics.

As well as replacing traditional materials, plastics find some completely new applications. The plastic poly(propenol) can absorb up to 600 times its own volume of water. An application that makes use of this remarkable property is to promote the growth of seedlings in desert conditions. A handful of plastic granules is placed at the bottom of a hole, and allowed to soak up water. A seedling is planted, the hole is filled, and the plastic releases water gradually at the demand of the seedling. The technique has been tried out successfully on acacia trees in the Sudan and eucalyptus trees in India.

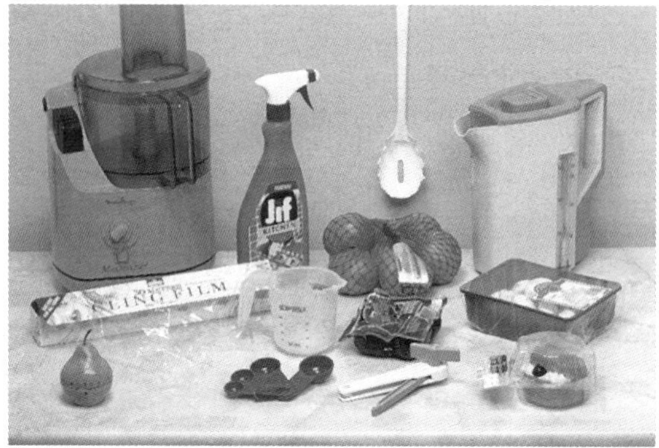

Figure 34.1A
Plastics in the kitchen

Plastics are **polymers**, and we shall now look into their chemical composition.

34.2 LONG MOLECULES

A polymer molecule is composed of a large number of repeating units ...

... each of which has two reacting groups – which enable it to form long chains ...

... or three reactive groups – which enable it to form a three-dimensional network of cross-linked chains

A **polymer** is a substance with molecules built up from many smaller repeating units, which are connected by covalent bonds: —P—P—P—P—P— or —$(P)_n$—. One molecule of polymer is formed by the combination of many molecules of **monomer**, for example, many molecules of the monomer ethene $CH_2{=}CH_2$ polymerise to form one molecule of the polymer poly(ethene), $(CH_2)_n$. The repeat unit, P, cannot exist on its own. P has a reactive group at each end, such as the —CH_2— group in poly(ethene). There may be an acidic group at one end and a basic group at the other end as in a polypeptide, $H_2N(CHRCONH)_nCHRCO_2H$. Linear polymers are not the only synthetic polymers. The monomers may have three functional groups so that polymer chains may be cross-linked to form three-dimensional networks, e.g. phenol–methanal resin [§ 34.5.5].

In many polymers, P_n, n is large and the relative molecular mass may be of the order of 1×10^6. A molecule of a polymer of relative molecular mass 1×10^7 would, if fully extended, be about 1 mm long with a diameter of about 0.5 nm. These dimensions would make it the same shape as a piece of spaghetti 2 km long. In reality, it is shaped more like a miniature ball of cooked spaghetti. A polymer chain is never fully extended; it coils in a random manner to occupy a sphere of about 200 nm in diameter.

34.3 STRUCTURE AND PROPERTIES

Polymer molecules contain a very large number of repeating units

... have a high relative molecular mass

... are solids ...

... able to be shaped by flow

When a piece of poly(ethene) is deformed, it stays in its new shape: the deformation is permanent. Substances like this are called **plastics**. *A plastic is a polymerised organic substance, a solid of high molar mass, which at some time in its manufacture can be shaped by flow*. Plastics find a large number of applications on account of their toughness, resistance to water and corrosive substances, ease of moulding and colour range. They are electrical and thermal insulators.

Elastomers return to their original shapes after being deformed, e.g. rubber

Polymers which are soft and springy and return to their original shape after being deformed are called **elastomers**. Rubber is an example.

Fibres are polymers which can be made into strong thin threads, e.g. nylon

Strong polymers which do not change shape easily are used in the textile industry. They are made into thin, strong threads which can be woven together. These polymers are called **fibres**. Nylon is an example.

Thermosoftening and thermosetting plastics

Thermosoftening plastics can be softened by heating, hardened and resoftened many times

Since plastics soften on heating and harden on cooling, objects can be moulded easily from plastics. There are two subsets of plastics: **thermosoftening plastics** (often called **thermoplastics**) and **thermosetting plastics** (often called **thermosets**).

Thermosoftening plastics (thermoplastics) can be softened by heating, allowed to cool and harden, and then resoftened many times.

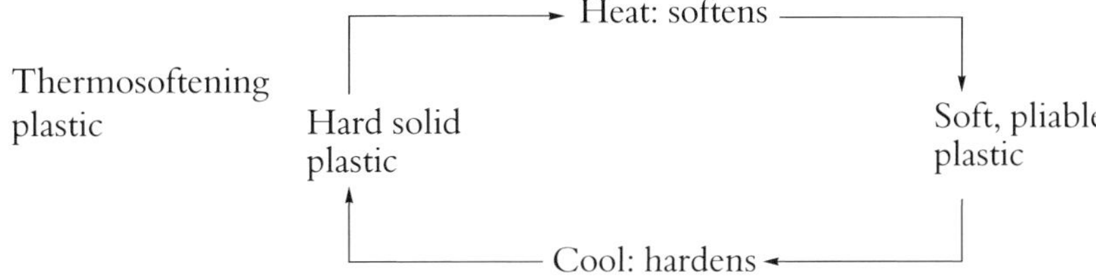

Thermosetting plastics once set, cannot be softened by heating. In thermoplastics the forces of attraction between chains are weak

Thermosetting plastics are plastic during the first stages of manufacture, but once moulded they set and cannot be resoftened by reheating.

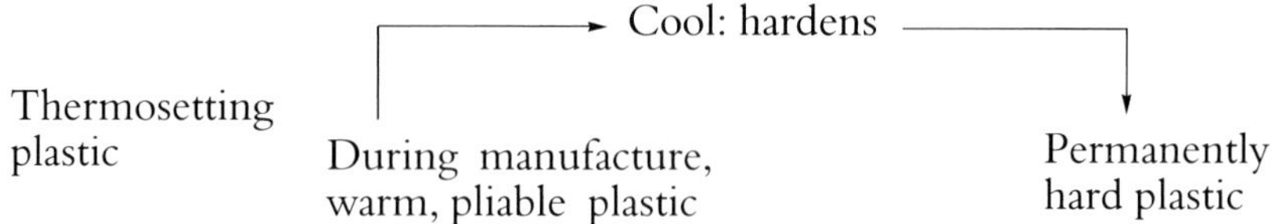

In thermosetting plastics, polymer chains are cross-linked to form a three-dimensional structure

The reason for the difference in behaviour is a difference in structure. Thermoplastics consist of long polymer chains. The forces of attraction between chains are weak [see Figure 34.3A]. Thermosetting plastics have a different structure. When a thermosetting plastic is moulded, covalent bonds form between the chains. Cross-links are formed, and a huge three-dimensional structure is built up. This is why thermosetting plastics can be formed only once.

Figure 34.3A
The structure of (a) a thermosoftening plastic (b) a thermosetting plastic

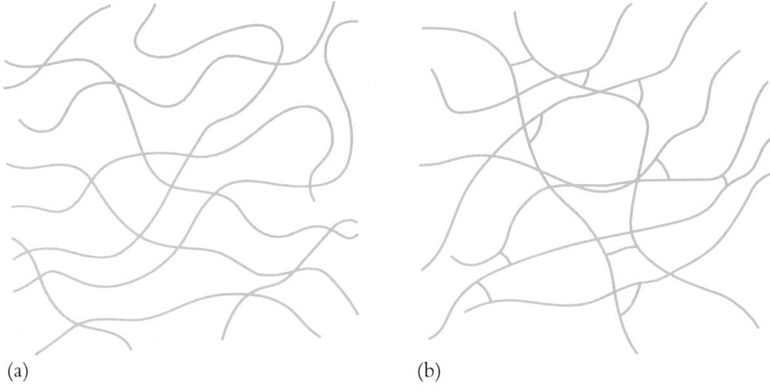

Thermoplastics are convenient to use in manufacturing as they are easily moulded

Thermoplastics are very convenient for manufacturers to use. Manufacturers can buy tonnes of thermoplastic in the form of granules, soften the material, and mould it into the shape of the object they want to make. Moulding can be a continuous process as the plastic granules are fed into one end of the moulding machine, the moulded plastic comes out of the other end in the shape of tubes, sheets or rods. Plastics of this sort can be moulded several times during the manufacture of an article. Coloured plastics can be made easily by adding a pigment, melting the plastic and mixing thoroughly. Moulded objects made from the plastic are coloured all through. This is a big advantage over a coat of paint which can become chipped.

Thermosets are polymerised and moulded at the same time

The moulding of thermosetting plastics must take place at the same time as polymerisation. The monomer is poured into the mould and heated to make it polymerise. As polymerisation occurs, a press forms the plastic into the required shape while it is setting. It is necessary to use a batch process. This is less efficient and more costly than a continuous process.

Within the polymer molecule covalent bonds operate. Between molecules and sections of the same molecule intermolecular forces of attraction operate

Both thermosoftening and thermosetting plastics can be strong, tough, rigid and stable towards chemical attack. Bonds within the molecule are covalent bonds. Between molecules and between segments of the same molecule intermolecular forces of attraction operate, e.g. van der Waals forces and forces of attraction between polar groups. When plastics melt or dissolve or flow, it is intermolecular bonds that are broken and made so that molecules and different segments of the same molecule can move past one another or away from one another. The strongest intermolecular bond, the hydrogen bond, operates in polyamides, e.g. nylon [see § 34.5.4], and plays a big part in determining the properties of the polymer.

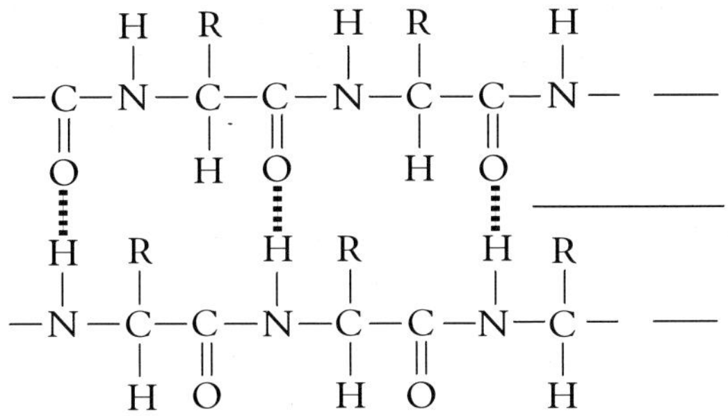

```
  H  R       H  R      H
  |  |       |  |      |
—C—N—C—C—N—C—C—N—        —— Polyamide chain in nylon
 ||   |  ||     |  ||
 O    H  O      H  O
 :       :         :      ———— Hydrogen bond between sections
 H  R    H  R      H  R        of the chain or between chains
 |  |    |  |      |  |
—N—C—C—N—C—C—N—C—        —— Polyamide chain
    |  ||     |  ||     |
    H  O      H  O      H
```

Dipole–dipole interactions are present in e.g. polyesters:

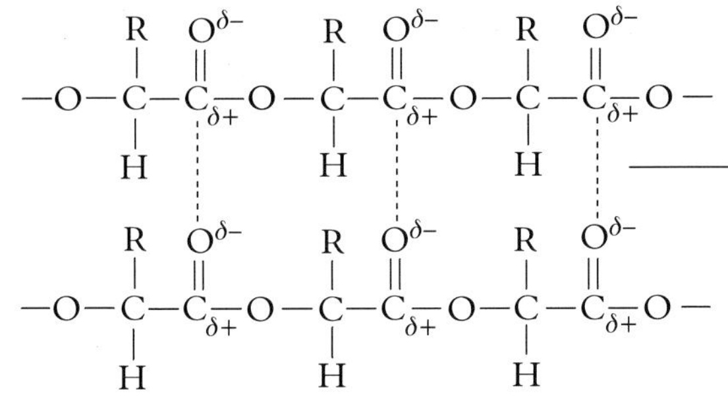

```
    R  Oδ−      R  Oδ−      R  Oδ−
    |  ||       |  ||       |  ||
—O—C—C—O—C—C—O—C—C—O—        —— Polyester chain
    |  δ+       |  δ+       |  δ+
    H  :        H  :        H  :        ———— Attraction between δ+ of
       :           :           :             Cδ+=Oδ− group and δ− of
    R  Oδ−      R  Oδ−      R  Oδ−           Cδ+=Oδ− group
    |  ||       |  ||       |  ||
—O—C—C—O—C—C—O—C—C—O—
    |  δ+       |  δ+       |  δ+
    H           H           H
```

34.4 ADDITION POLYMERS

Polymerisation may take place by **addition** or by **condensation** [see § 34.5]. The addition polymerisation of alkenes has been mentioned in § 27.8.10. Addition polymerisation results from chain reactions, which take place in the following stages.

Addition polymerisation involves the formation of free radicals ...

Initiation

Free radicals are formed by the decomposition of an initiator, e.g. benzoyl peroxide.

$$(C_6H_5CO_2)_2 \longrightarrow 2C_6H_5\cdot + 2CO_2$$

Benzoyl peroxide Phenyl radical

The decomposition can be initiated by heat or light. The phenyl radical, shown below as Ph•, reacts with a molecule of monomer, e.g. ethene, to form a new free radical.

$$Ph\cdot + CH_2{=}CH_2 \longrightarrow Ph{-}CH_2{-}CH_2\cdot$$

... which take part in a chain reaction

Propagation

The new free radical reacts with another molecule of monomer.

$$Ph{-}CH_2{-}CH_2\cdot + CH_2{=}CH_2 \longrightarrow Ph{-}CH_2{-}CH_2{-}CH_2{-}CH_2\cdot$$

The chain reaction continues until large polymer molecules have been formed.

Termination

There are a number of reactions which can bring the growth of a polymer chain P• to an end. One is the reaction

$$P\cdot + M \longrightarrow P + M\cdot$$

polymer free radical + monomer ⟶ polymer molecule + monomer free radical
The monomer free radical can start a new chain reaction.

34.4.1 THE DISCOVERY OF POLY(ETHENE)

In the 1930s, ICI was carrying out research into the synthesis of new dyes. In March 1933, Eric Fawcett and Reginald Gibson carried out a reaction between ethene and benzaldehyde under a pressure of 2000 atm. They hoped to achieve the formation of a ketone:

$$C_6H_5CHO + CH_2{=}CH_2 \longrightarrow C_6H_5COCH_2CH_3$$

The reaction was left to run over the weekend. When some ethene was lost by leakage, they added more ethene. When they opened the reaction vessel after the weekend, they found a white waxy solid. On analysis, it was shown to have the empirical formula CH_2. They called it polyethylene after the traditional name ethylene for the compound $H_2C{=}CH_2$. They repeated the synthesis many times, sometimes with success and at other times with an explosion. Because of the dangerous nature of the reaction, the work was terminated in July, 1933. Engineers were able to construct a reactor vessel which would withstand high pressure, and in December 1935 the work recommenced. By controlling the pressure, they could control the molar mass of the polymer. After one month, the researchers had produced enough material to show that it could be moulded, due to its low melting temperature, and that it was an electrical insulator.

Michael Perrin joined the project in 1935. He found that in the absence of oxygen no polymerisation occurred and that an excess of oxygen made the reaction run out of control. Fortunately the leak in the apparatus of Fawcett and Gibson had let in just the right amount of oxygen! Perrin showed that if benzaldehyde were omitted from the reaction mixture the product would still form.

The first poly(ethene) products appeared in the shops in 1948. They included washing-up bowls, carrier bags, squeezy bottles and sandwich bags.

34.4.2 USES OF SOME ADDITION POLYMERS

Low-density poly(ethene), ldpe

Poly(ethene) ...

... low density, made at high pressure ...

Low-density poly(ethene), ldpe, is made at high pressure (about 15 atm). It consists of branched molecules and has a density of 0.91–0.94 g cm^{-3}. Ldpe has many uses because the material is tough and durable and resists attack by water, weather, acids and alkalis. It is used in films to make plastic bags and to wrap produce, frozen foods and items of clothing. Electrical cables are insulated with poly(ethene) because it withstands bad weather conditions better than rubber which was used previously for this purpose.

High-density poly(ethene), hdpe

... and high density, made at atmospheric pressure

High-density poly(ethene), hdpe, is made at atmospheric pressure with the help of a Ziegler catalyst, e.g. titanium(IV) chloride + triethylaluminium. The molecules have little branching and pack more closely together to give a higher density, 0.95 – 0.97 g cm^{-3}, a higher melting temperature, 135 °C, and greater tensile strength. Hdpe is harder and stiffer than ldpe and is less easily softened by heating. It is used in the production of water tanks and pipes, crates, bottles, washing-up bowls and other housewares. Since it can be heat-sterilised, hdpe is used in the manufacture of hospital equipment such as trays and bed pans.

Poly(chloroethene) PVC

PVC, a versatile plastic

Poly(chloroethene), $-(CH_2CHCl)_n-$, PVC, is harder and stiffer than poly(ethene) because the chlorine atoms increase the forces of attraction between chains. It is used for windowframes, water pipes, drains and floor coverings. The addition of **plasticisers** opens up a wider range of uses. Plasticised PVC is used for footwear, bottles, food containers and cling film.

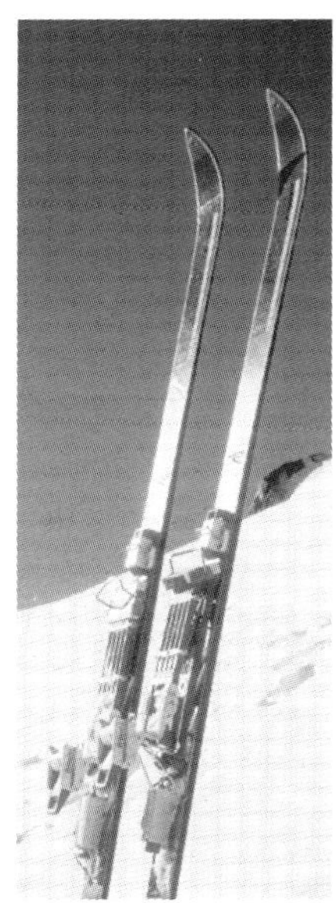

Figure 34.4A
Skis coated with PTFE

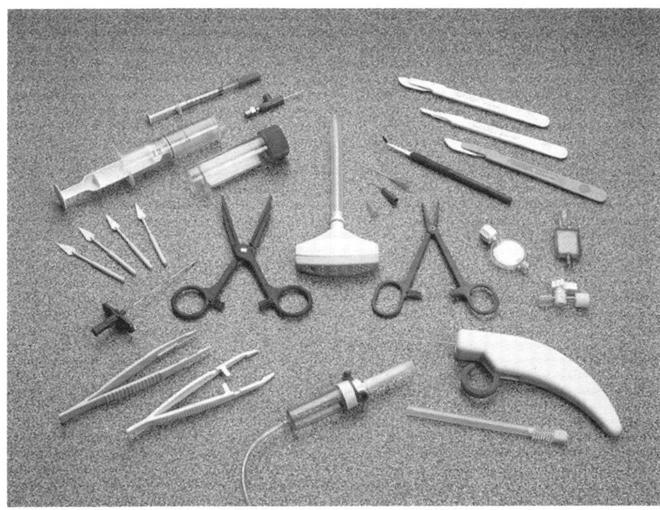

Figure 34.4B
Disposable hospital ware

Poly(phenylethene), i.e. poly(styrene)

Polystyrene and polystyrene foam

Poly(phenylethene) $\leftarrow CHC_6H_5—CH_2 \rightarrow_n$ is one of the easiest plastics to mould. Since it is transparent, it is usual to add a pigment to give an opaque material. Crystalline poly(styrene) is familiar as ball point pens. Disposable coffee cups and cartons for dairy produce are made of polystyrene with the white pigment titanium(IV) oxide. Polystyrene foam is made by adding hexane to the solid beads of polymer before they are heated and moulded. It is widely used for packaging breakable objects.

Other addition polymers

Table 27.8 of § 27.8.10 lists other poly(alkenes), including poly(tetrafluoroethene), (PTFE) and poly(methyl-2-methylpropenoate) (Perspex®).

34.4.3 RUBBERS

Natural rubber

Natural rubber is cis-poly(2-methylbuta-1,3-diene)

Natural rubber is a polymer of the monomer 2-methylbuta-1,3-diene (isoprene),

```
CH2     CH2
  \\    //
   C—C
  /    \
CH3     H
```

Rubber is an elastomer ...

Poly(2-methylbuta-1,3-diene) can exist in *cis*- and *trans*-isomeric forms. Natural rubber is the *cis*-form.

... with convoluted molecules ...

... which are straightened out by a force ...

... and contract when the force is removed ...

... but creep can occur

A rubber consists of a tangle of large numbers of convoluted molecules which can be straightened out if sufficient force is applied. If natural rubber is held in the extended state for some time, it **creeps**: the rubber does not contract to its original length; it remains permanently longer. Creep occurs because polymer chains become separated from one another. It would obviously be an advantage to eliminate creep, and this can be done by forming cross-links between molecules. In the process of **vulcanisation**, 1–3% by mass of sulphur is added to rubber and the mixture is heated. Short chains of sulphur atoms form between polymer chains.

Vulcanisation creates cross-links and reduces creep ...

```
                              H3C       H
                                 \     /
CH2             CH2               C=C
/   \          /    \      /      \   /
      C—CH2          CH2          CH2
     /|
H3C S
      |
H3C S
     \|
      C—CH2         CH2            CH2
\   /          \   /    \        /   \
CH2             CH2       C=C
                          /    \
                     CH3       H
```

Vulcanised *cis*-poly(2-methylbuta-1,3-diene)

Synthetic rubbers

Poly(buta-1,3-diene) is a synthetic rubber ...

... which can be vulcanised ...

neoprene is similar

Poly(buta-1,3-diene) has a high resistance to abrasion and retains its elasticity at low temperatures. It can be vulcanised. It is made by polymerising buta-1,3-diene, $CH_2{=}CH{-}CH{=}CH_2$. Poly(butadienes) are used in the manufacture of tyres, in coatings and adhesives and as additives to other polymers.

Buta-1,3-diene is often copolymerised with other alkenes to form polymers with a range of properties. These include styrene–butadiene rubber and neoprene rubber.

34.4.4 A PLASTIC THAT THINKS IT'S A METAL

When a stream of ethyne is directed onto the surface of a Ziegler catalyst [§ 34.4.2] at −78 °C, *cis*-poly(ethyne), $\left[CH{=}CH\right]_n$, is formed.

Cis- poly(ethyne) is red, unstable and on warming is converted into the *trans*- form.

Trans- poly(ethyne) is blue and is the stable form of the polymer.

The alternating double and single bonds mean that π-bonds are formed between adjacent carbon atoms, and a band of delocalised electrons stretches the full length of the molecule. Such a chain is an insulator. To make it into a conductor, some electrons must be removed from the band of delocalised electrons. This is done by doping – adding an oxidising agent, e.g. iodine, which will remove some electrons.

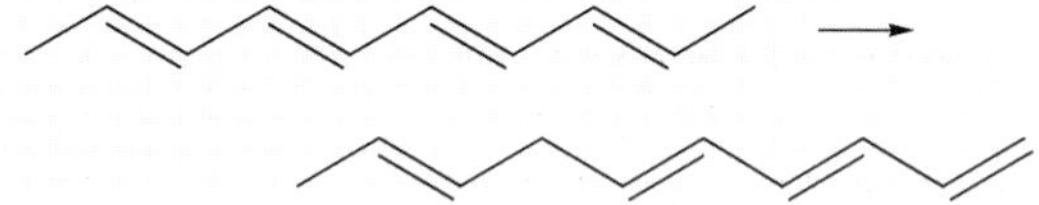

The conductivity increases to a value which is not far below that of a metal. Conducting plastics combine the conductivity of metals with the properties of plastics: flexibility, low density, freedom from corrosion and low cost. They could lead to plastic batteries and computer displays. Poly(ethyne) is an inconvenient material because it is oxidised by air. Other doped conducting polymers have better properties. Research continues on the design of lightweight rechargeable batteries using conducting polymer electrodes.

CHECKPOINT 34.4: ADDITION POLYMERS

1. *(a)* Outline the difference in manufacture of low-density poly(ethene), ldpe, and high-density poly(ethene), hdpe.
   (b) Describe the difference in structure between the two polymers and explain how this explains the difference in properties.
   (c) Give two examples of the uses of ldpe and two for hdpe.
   (d) How do the physical properties of poly(chloroethene), PVC, compare with those of poly(ethene)? How can the difference be explained on structural grounds?

2. *(a)* Write the structural formula of
   (i) poly(phenylethene), poly(styrene)
   (ii) poly(tetrafluoroethene), PTFE
   (iii) poly(methyl 2-methylpropenoate), perspex.
   [See § 27.8.10.]
   (b) Give two examples of the uses of each polymer. State what advantage the plastic has over the material that was used for the purpose before plastics came on the scene.
   (c) Give the formula of poly(buta-1,3-diene). Give an example of its use.

3. Suggest a use for the conducting polymer described in § 34.4.4. Say why it would be better than (i) a metal (ii) a plastic for the use you mention.

4. *(a)* Explain how the structure of rubber enables it to stretch.
   (b) What does vulcanisation do to the structure and properties of rubber?
   (c) How does vulcanisation of rubber improve it for use in car tyres?
   (d) Suggest why poly(ethene) stands up to bad weather conditions better than rubber does.

34.5 CONDENSATION POLYMERS

Condensation polymers are formed by addition–elimination reactions

Condensation polymers are formed by addition–elimination reactions. When molecules of monomer combine, molecules of water or another compound are eliminated. The elimination of water gives the name of **condensation polymerisation** to this type of reaction.

34.5.1 POLYESTERS

A **polyester** is formed by condensation – the elimination of water – between a diol and a dicarboxylic acid. A carboxylic acid reacts with an alcohol to form an ester and water:

$$R'CO_2H + R''OH \longrightarrow R'CO_2R'' + H_2O$$

A dicarboxylic acid and a diol react to form a diester.

Polyesters, e.g. Terylene®, are formed by condensation between a diol and a dicarboxylic acid

$$HO_2C{-}C_6H_4{-}CO_2H + 2HOCH_2CH_2OH \longrightarrow HOCH_2CH_2O_2C{-}C_6H_4{-}CO_2CH_2CH_2OH + 2H_2O$$

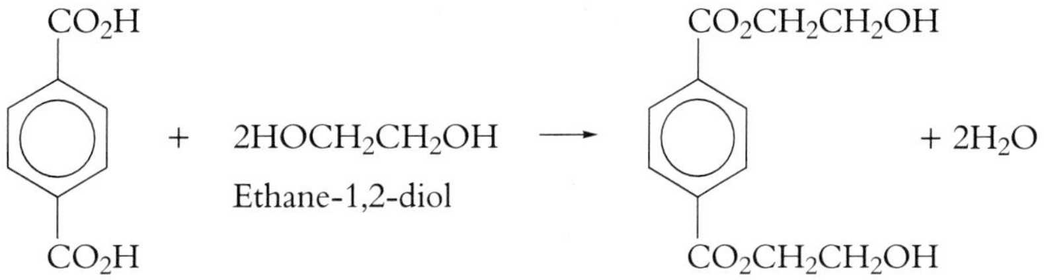

Benzene-1,4-dicarboxylic acid + Ethane-1,2-diol

Further condensation occurs to form a polyester,

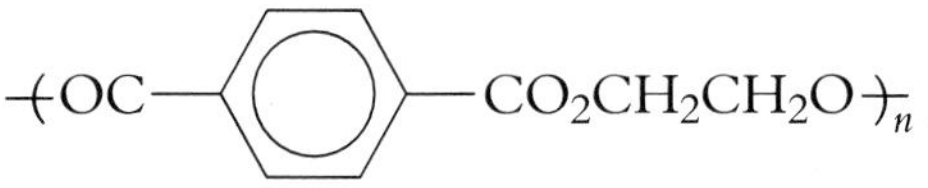

$$\text{-}\!\!\left(OC{-}C_6H_4{-}CO_2CH_2CH_2O\right)\!\!\text{-}_n$$

The product is the polyester with the ICI trade name Terylene® and the DuPont trade name Dacron®.

34.5.2 POLYESTER RESINS

Polyester resins are formed by a condensation reaction between a diol and a dicarboxylic acid or anhydride. If one of the reactants is unsaturated, an unsaturated polymer is formed.

Unsaturated polyesters can form thermosetting resins

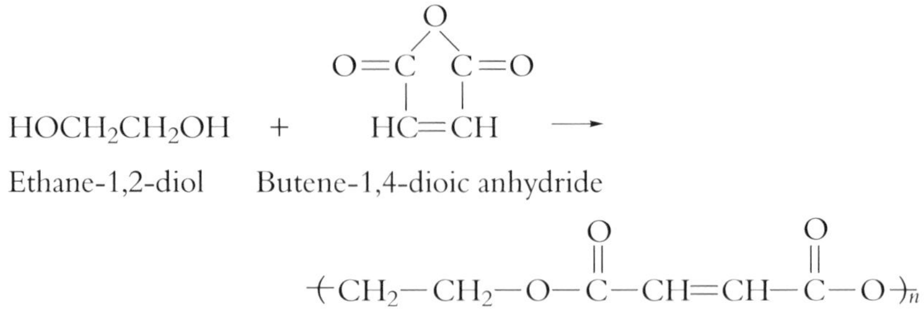

$$HOCH_2CH_2OH \quad + \quad \begin{matrix} & O & \\ O{=}C & & C{=}O \\ | & & | \\ HC & {=} & CH \end{matrix} \quad \longrightarrow$$

Ethane-1,2-diol Butene-1,4-dioic anhydride

$$-\!\!\left(CH_2{-}CH_2{-}O{-}\overset{\overset{O}{\|}}{C}{-}CH{=}CH{-}\overset{\overset{O}{\|}}{C}{-}O\right)\!\!-_n$$

Unsaturated polyester

An unsaturated polyester can form cross-links. For example, X—CH=CH—Y can add to the polyester shown above to form a **thermosetting polyester resin** with the cross-linked structure:

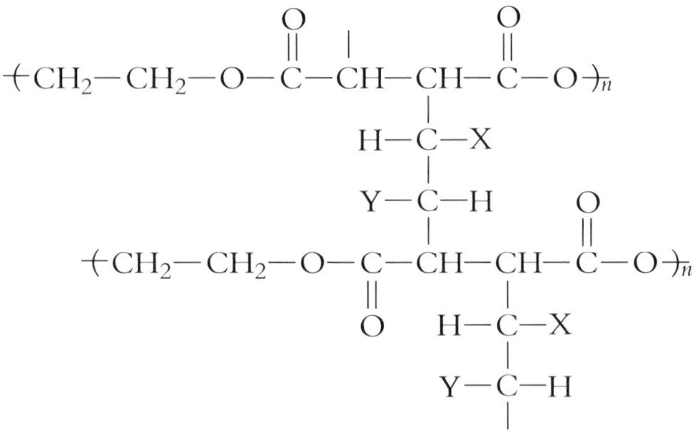

$$\begin{array}{l} -\!\!\left(CH_2{-}CH_2{-}O{-}\overset{\overset{O}{\|}}{C}{-}\overset{|}{C}H{-}CH{-}\overset{\overset{O}{\|}}{C}{-}O\right)\!\!-_n \\ \qquad\qquad\qquad\qquad\quad | \\ \qquad\qquad\qquad\quad H{-}C{-}X \\ \qquad\qquad\qquad\qquad\quad | \\ \qquad\qquad\qquad\quad Y{-}C{-}H \\ \qquad\qquad\qquad\qquad\quad | \\ -\!\!\left(CH_2{-}CH_2{-}O{-}\underset{\underset{O}{\|}}{C}{-}CH{-}CH{-}\overset{\overset{O}{\|}}{C}{-}O\right)\!\!-_n \\ \qquad\qquad\qquad\qquad\qquad\quad | \\ \qquad\qquad\qquad\qquad\quad H{-}C{-}X \\ \qquad\qquad\qquad\qquad\qquad\quad | \\ \qquad\qquad\qquad\qquad\quad Y{-}C{-}H \\ \qquad\qquad\qquad\qquad\qquad\quad | \end{array}$$

34.5.3 THE DISCOVERY OF NYLON

The chemical company of DuPont started a line of research in 1928 which was aimed at finding polymers that might be used in fibres. Wallace Carothers was part of the team. It was known from X-ray crystallography that some natural fibres, e.g. silk, are polyamides. Carothers decided to investigate polyamides and polyesters.

Carothers succeeded in obtaining polyesters by the reaction of a hexanedioic acid and a diol [§ 34.5.1]. The polyesters were gummy resins because the water formed in the condensation reaction was incorporated in the structure of the polymer. Carothers had the idea of carrying out the polymerisation in a heated still so that water would distil off. As a result, he obtained tough, solid polyesters. The new polymers did not at first seem suited for use as fibres because they had low melting temperatures and were slightly soluble in water. However, a method of overcoming these disadvantages was found. The polyester was heated, drawn into a long thin filament and then cooled. The drawn fibre was tough, flexible and elastic with the ability to be stretched to many times its original length. X-ray diffraction studies showed that the undrawn fibres were crystalline, and in the drawn fibres the crystals had been oriented along the length of the fibre, thus increasing its strength.

J R Whinfield and J T Dickson were members of the research group. They substituted benzene-1,4-dicarboxylic acid for hexanedioic acid and obtained a stronger polyester fibre. This fibre was manufactured by ICI and named Terylene®.

Carothers extended his work to making polyamides by condensation reactions between a diamine and a diacid chloride. When he used 1,6-diaminohexane and decanedioyl chloride, they reacted rapidly. The polyamide which he obtained could be drawn into strong fibres. This polyamide was named nylon 66 after the number of carbon atoms in the diamine and the diacid chloride used to make it.

The first nylon product, which appeared in 1938, was a toothbrush. Nylon stockings appeared in 1939, but with the outbreak of the Second World War nylon was devoted to the manufacture of parachutes. These had previously been made of silk, and supplies of silk would not have been enough to provide the huge number of parachutes needed by the Royal Air Force. Nylon stockings and other nylon goods became available after the end of the war.

34.5.4 POLYAMIDES: NYLONS

A dicarboxylic acid and a diamine react to form a **polyamide**. Polyamides are also called **nylons**. The reaction between hexanedioic acid and hexane-1,6-diamine begins with the step:

Condensation between a dicarboxylic acid and a diamide can give a polyamide – a nylon

$$H_2N(CH_2)_6NH_2 + HO_2C(CH_2)_4CO_2H \longrightarrow H_2N(CH_2)_6NHCO(CH_2)_4CO_2H + H_2O$$

Hexane-1,6-diamine   Hexane-1,6-dioic acid

The product, having an amino group and a carboxyl group can polymerise further to form $H_2N(CH_2)_6[NHCO(CH_2)_4CONH(CH_2)_6]_nNHCO(CH_2)_4CO_2H$. This is **nylon 66**. The name comes from the numbers of carbon atoms in the monomers. Nylon 6 is $\leftarrow\!\!(CH_2)_5CONH\!\!\rightarrow_n$, made from $H_2N(CH_2)_5CO_2H$.

A laboratory preparation of nylon 66 is shown in Figure 34.5A. Nylon forms at the interface between the aqueous solution of the diamine and the cyclohexane solution of the dicarboxylic acid. It is removed as it is formed to allow further reaction to take place.

Nylon is a tough, strong material

Nylon 66 is insoluble in common solvents, has melting temperature 263 °C, and the materials used to make it are readily available. Nylon is an **engineering plastic**. It is used in place of metals to make some machine parts. It is strong, tough, rigid and resistant to chemical attack. It is better as an engineering

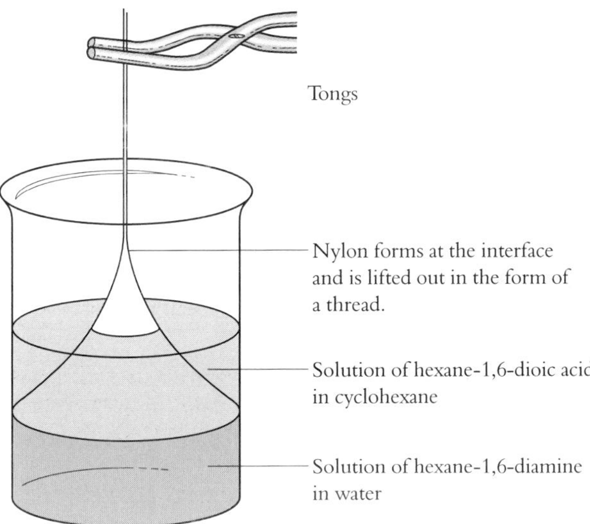

Figure 34.5A
Making nylon

material than poly(ethene) or poly(propene) because of the strong hydrogen bonds acting between chains [§ 34.3].

Polypeptides and proteins

Polypeptides are formed from amino acids by condensation polymerisation

Polypeptides are polyamides formed by condensation polymerisation between amino acids [see § 33.16]:

$$H_2NCHRCO_2H + H_2NCHRCO_2H \longrightarrow H_2NCHRCONHCHRCO_2H + H_2O$$

The group —CONH— is called the **peptide link**. This reaction is followed by further condensation of the $—NH_2$ group and $—CO_2H$ group with other amino acid molecules to form a polypeptide with molecules consisting of long chains of 100–300 amino acid residues. A polypeptide $H_2N(CHRCONH)_nCHRCO_2H$ can contain up to 20 different R groups corresponding to the different amino acids; see § 33.17. Further condensation of polypeptides produces proteins with still larger molecules.

SUMMARY

Polyesters and Polyamides

A **polyester** is formed by condensation – the elimination of water – between a diol and a dicarboxylic acid. An example is Terylene® (or Dacron®).

A **polyester resin** is formed when the diol or the dicarboxylic acid is unsaturated. The polymer forms cross-links which make it thermosetting.

Polyamides are formed by condensation polymerisation between a dicarboxylic acid and a diamine. Nylons, e.g. nylon 66, are polyamides.

Polypeptides and proteins are polyamides formed by condensation polymerisation of amino acids.

34.5.5 PHENOLIC RESINS

Another condensation reaction is that which occurs between phenol and methanal to form a phenol–methanal resin. The first step can be shown as:

Phenol and methanal condense to form phenol–methanal resin, a thermoset

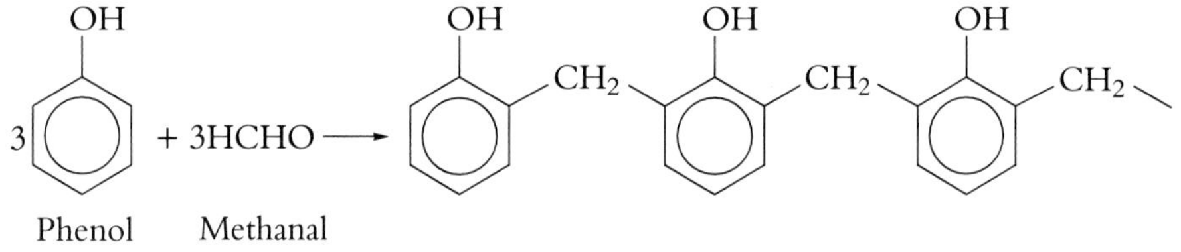

OH OH OH OH

3 (Phenol) + 3HCHO ⟶ chain of phenol rings (OH at 1-position) linked by CH_2 groups at the 2-positions: —CH_2—

Phenol Methanal

If the proportion of phenol in the reaction mixture is reduced, the 4-position is also utilised with the formation of the polymer

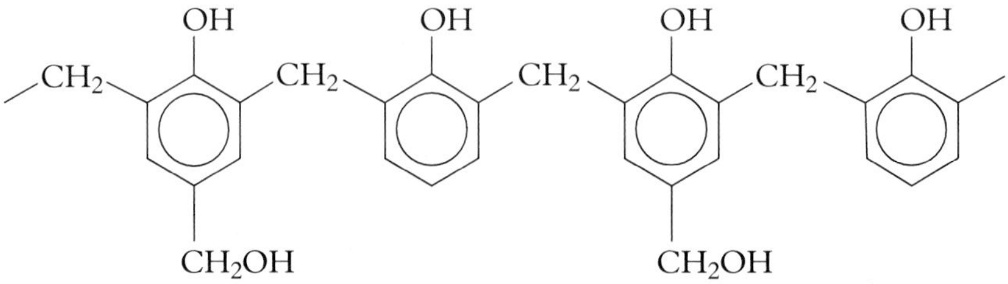

OH OH OH OH

CH_2 CH_2 CH_2 CH_2

CH_2OH CH_2OH

On further heating, cross-linking takes place with the formation of the phenol–methanal resin.

$$n\,C_6H_5OH + n\,HCHO \longrightarrow \text{(cross-linked phenol–methanal resin)} + n\,H_2O$$

Bakelite®

Phenol–methanal resins (formerly called phenol–formaldehyde resins) are thermosetting polymers. The first to be patented was named Bakelite® by its discoverer, Leo Baekland, in 1872. The uses of these resins include radio cases, electrical switches and sockets, fuse holders and door handles. Bakelite® is easily recognisable by its brown colour. It has to a large extent been replaced by urea–methanal resins.

34.5.6 EPOXY RESINS

The epoxide group is highly reactive

The epoxide group is $-CH-CH-$ (with the two carbons bridged by O)

Epoxyethane is formed by a reaction between ethene and the oxygen in the air.

$$2CH_2{=}CH_2 + O_2 \xrightarrow{\text{Ag catalyst, 300 °C, under CO}_2} 2CH_2-CH_2 \text{ (bridged by O)}$$

The angles between the bonds are only 60°, and this strained configuration makes the epoxide group very reactive.

Epoxy-3-chloropropane and a compound with two phenolic groups condense to form an epoxy resin ...

The epoxide of 3-chloropropane, $CH_2-CH-CH_2Cl$ (with the first two carbons bridged by O)

condenses with compounds with two phenolic groups, e.g.

$$HO-C_6H_4-C(CH_3)_2-C_6H_4-OH$$

The elimination of hydrogen chloride gives a polymer with the repeating unit

$$\left(O-C_6H_4-C(CH_3)_2-C_6H_4-O-CH_2-CH(OH)-CH_2\right)_n$$

... used in adhesives and coatings

If n is 25 or more, the resin is a hard tough solid. Any resin containing one or more epoxide groups is an **epoxy resin**. Epoxy resins are intermediates which must be cured, that is, **cross-linked**, to yield useful resins. Epoxy resins have uses which range from adhesives, e.g. Araldite®, to can and drum coatings. They have excellent resistance to chemicals, excellent adhesive properties and electrical insulating properties.

34.5.7 FORMICA AND MELAMINE

Urea, $CO(NH_2)_2$ and methanal CHO, condense to form a polymer, **urea-methanal**:

(Formica®) ...

$$n\text{HCHO} + n\text{H}_2\text{NCONH}_2 \longrightarrow -\!\!(\text{NHCONH})\!\!-_n + n\text{H}_2\text{O}$$

Methanal   Urea   Urea–methanal polymer

... methanal–melamine resins (Melamine®) ...

... have important uses ...

Methanal condenses with urea to form Formica® ...

... and with melamine to form the plastic Melamine®

Cross-linking produces a thermosetting resin which is used for trays, knobs, toilet seats etc. and also in laminates. It is often known by its older name of urea–formaldehyde resin and by its trade name of **Formica**®.

NH_2 / H_2N / NH_2 (triazine ring: C, N, C, N, C, N)

Melamine

Methanal also condenses with melamine to form a highly cross-linked polymer resin, **methanal–melamine**. It is thermosetting and, with its qualities of hardness, resistance to chemical attack and good finish, is used for tableware with the trade name **Melamine**®. It is also used in laminates.

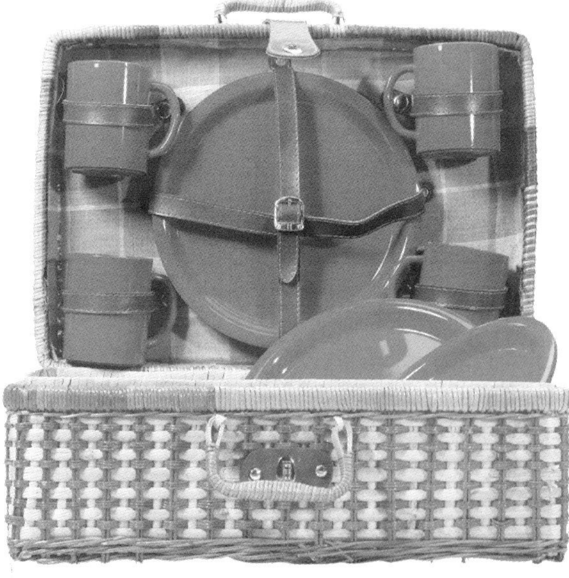

Figure 34.5B
Plastic tableware

34.5.8 LAMINATED MATERIALS

Thermosetting plastics are used in the manufacture of laminated materials

A **laminated material** is made from layers of material bonded together. Laminated plastics use the adhesives melamine, epoxy resins, unsaturated polyester resins and phenolic resins to bond layers of paper, wood or textiles together.

SUMMARY

Phenolic resins and epoxy resins
These polymers are cross-linked thermosets. They are made by condensation polymerisation.
A phenol and methanal give a **phenolic resin**.
A phenol and a compound possessing an epoxide group give an **epoxy resin**.
Methanal and urea give **urea–methanal**, trade name Formica®.
Methanal and melamine give **methanal–melamine**, trade name Melamine®.
These resins and others are used to make laminated materials.

CHECKPOINT 34.5: CONDENSATION POLYMERS

1. *(a)* Give an example of a pair of compounds that can react to form a polyester.
   (b) What structural feature is needed for the formation of a cross-linked polyester?
   (c) How do a straight-chain polyester and a cross-linked polyester differ in properties?

2. *(a)* Give an example of a pair of compounds that can react to form a polyamide.
   (b) By what trade name are polyamides known?

3. *(a)* What is the difference between a thermosoftening polymer and a thermosetting polymer?
   (b) What structural differences give rise to this difference in behaviour?
   (c) State two advantages to a manufacturer of working with thermosoftening plastics.
   (d) Suggest two uses for which a thermosetting plastic is chosen.

4. Phenol–methanal resins are thermosetting polymers.
   (a) What structural feature is responsible for the thermosetting property?
   (b) Give two uses of phenol–methanal resins.

5. *(a)* Name and give the formula of a compound which can condense with a phenol to form an epoxy resin.
   (b) Give two applications of epoxy resins.

34.6 SUMMARY

Table 34.6 lists some polymers and their uses.

Table 34.6
A summary of uses of polymers

The table gives a summary of some polymers and their uses

Polymer	Properties	Applications
Epoxy resins Thermosetting	Excellent chemical and thermal stability Strong and tough Electrical insulators Good adhesives	Adhesives Laminates Floorings Surface coatings Linings
Phenoxy resins Thermosetting	Easy to mould, with low shrinkage in mould Stable to heat	Surface coatings Adhesives and binders Electronic components
Poly(alkenes)	See § 27.8.10	See § 27.8.10

Poly(alkenes) are listed in Table 27.8 of § 27.8.10

Polymer	Properties	Applications
Polyamides, e.g. nylon Thermosoftening	Tough and strong Easily mouldable Chemically stable Can be drawn into fibres of high tensile strength	Unlubricated bearings Gears Sutures Fishing lines and nets Tyres. Bottles Ropes. Clothing
Polyesters Thermosoftening	Good thermal and chemical stability, though can be hydrolysed by alkali Low cost	Vehicle repair Laminates. Skis Fishing rods. Boats Aircraft parts Bottles
Polycarbonates Thermosoftening	High refractive index, transparent Good creep resistance Electrical insulators	Used as replacement for metals in e.g. safety helmets Lenses. Insulators Electrical components
Terylene® Thermosoftening	Electrical insulator Can be drawn into fibres	Fibres used to make textiles which are crease-resistant and easy to wash, dry and iron
Phenolic resins e.g. phenol–methanal, Bakelite® Thermosetting	Strong Chemically and thermally stable High T_m Machinable	Glues. Laminates Electrical components Switches and sockets Saucepan handles
Urea–methanal Thermosetting	Similar to Bakelite® but colourless and transparent Can be coloured	Electrical fittings

The table gives a summary of some polymers and their uses

34.7 DISPOSAL OF PLASTICS

34.7.1 DUMPING OF PLASTICS

A problem with most plastics is non-biodegradability ...

... and the disposal of plastic waste occupies more and more landfill sites

Plastic waste constitutes about 7% of household waste. Unlike some other wastes, e.g. kitchen waste and paper, plastics are **non-biodegradable**. They are synthetic materials, and soil and water do not contain micro-organisms with the enzymes needed to feed on plastics and degrade them. Plastic waste is buried in landfill sites, and there it remains unchanged for decades. Local authorities have to find more and more landfill sites.

34.7.2 INCINERATION OF PLASTICS

Incineration of plastic waste provides heat which can be utilised

An alternative to dumping is **incineration**, with the possibility of making use of the heat generated. In the UK only about 10% of plastic waste is incinerated, but in some other countries incinerators consume over 70% of domestic waste.

Some plastics, however, burn with the formation of toxic gases, e.g. hydrogen chloride, from PVC, and hydrogen cyanide, from poly(propenenitrile), and incinerators must be designed to remove these gases from the exhaust.

34.7.3 RECYCLING PLASTICS

Recycling is a very efficient way of dealing with waste plastics. A difficulty is that plastics waste consists of a mixture of plastics with different properties

Recycling is the most efficient use of resources. The UK produces annually about 1.8 million tonnes of plastics. A major difficulty in recycling is that, since different plastics have widely differing properties, mixed plastic waste is of limited use. While any two glass bottles can be recycled together, the same is not true of a PVC bottle and a poly(ethene) bottle. If these are melted down together, it may be possible to make some articles from the mixture, but mixtures of plastics are much weaker than individual plastics.

Individual plastics can be recycled with ease

The easiest plastics to recycle are off-cuts and substandard products, which consist of a known single plastic. When different plastics can be collected separately, e.g. poly(propene) car bumpers, recycling becomes more worth while.

34.7.4. PYROLYSIS

Pyrolysis is not cost-effective at present

When plastics are heated in the absence of air, they are **pyrolysed** (split up by heat). The products can be separated by fractional distillation and then used in the manufacture of other materials, including plastics. The process is expensive and at present costs about five times as much as dumping the plastics in a landfill site.

34.7.5 BIODEGRADABLE PLASTICS

Biodegradable plastics have been developed

They include ...

... biopolymers made from natural products ...

Biopolymers: In the UK, ICI markets the biopolymer poly(3-hydroxybutanoic acid), PHB, which has the trade name Biopol®. It is made by certain bacteria from glucose. Biopol® is made for special applications such as surgical stitches which dissolve in time inside the body. There is another potential use for Biopol®. When Biopol® is discarded, micro-organisms in the soil, in river water and in the body can break it down within 9 months. By incorporating copolymers, the properties of the polymer can be tailored to make it suitable for articles such as carrier bags. At present a Biopol® container is seven times the price of a poly(ethene) container. With increasing use, the price may well fall.

... photodegradable plastics which break down slowly in sunlight ...

Photodegradable plastics: A Canadian firm has produced a photodegradable polymer, which they incorporate in polystyrene cups. Exposed to sunlight for 60 days, the cups break down into dust particles.

... synthetic biodegradable plastics incorporating natural materials, e.g. starch ...

Synthetic biodegradable plastics: An Italian company, Feruzzi, has produced a material which consists of poly(ethene) and up to 50% starch. The poly(ethene) chains and starch chains interweave to form a material which is strong enough for shopping bags. When the material is buried, micro-organisms begin to feed on the starch, converting it into carbon dioxide and water, and in time the polymer chains dissolve in water. The cost at present is about twice that of a regular plastic bag. Europe uses 100 kilotonnes of degradable plastic a year, and the consumption is rising.

... soluble plastics which dissolve slowly

Soluble plastics: Plastics which dissolve in water include poly(ethenol), $\leftarrow CH_2—CHOH \rightarrow_n$, also called poly(vinyl alcohol), PVA. It is used as a packaging for swimming pool chemicals, descalers and seed strips.

POLYMERS

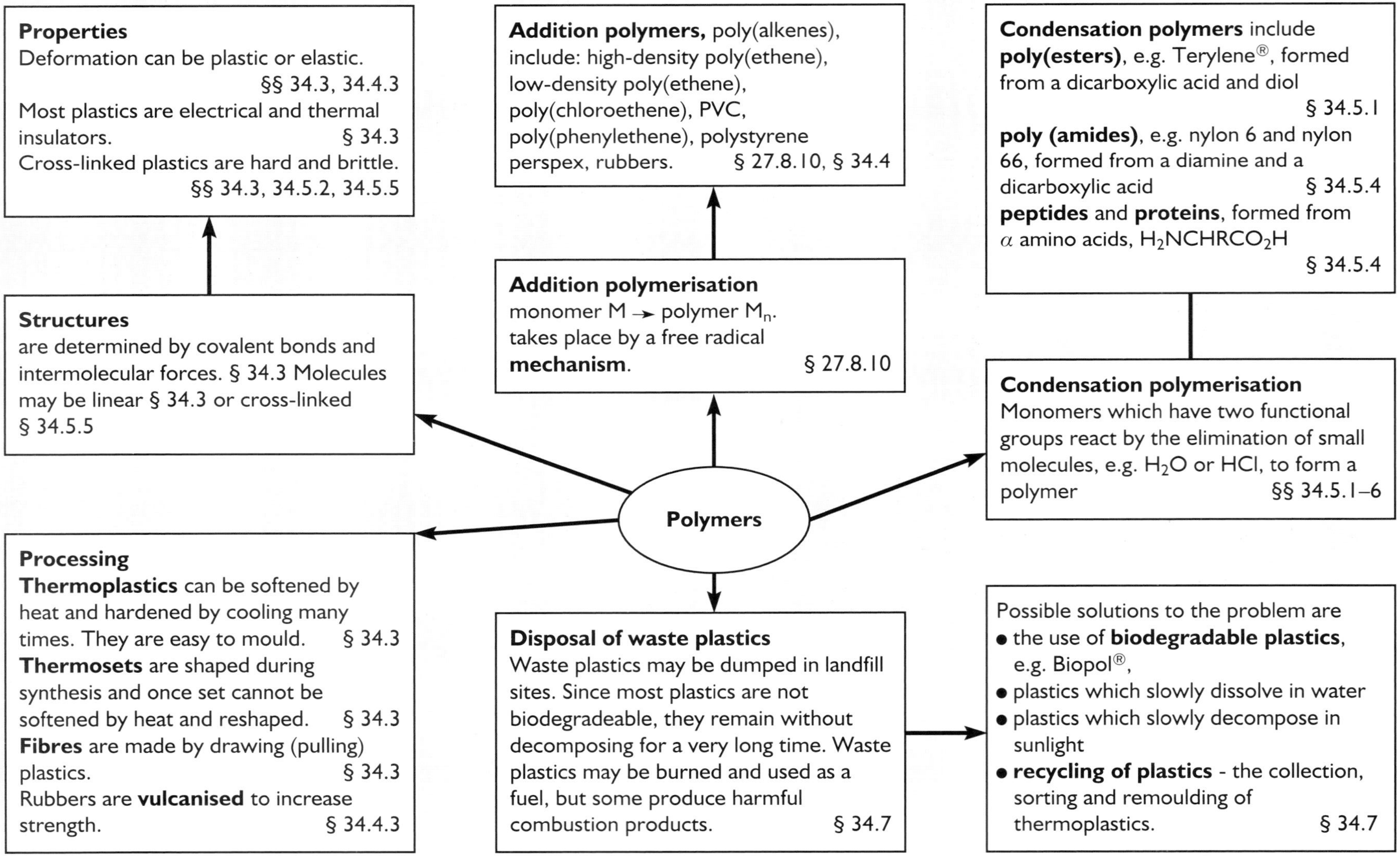

Polymers

Properties
Deformation can be plastic or elastic. §§ 34.3, 34.4.3
Most plastics are electrical and thermal insulators. § 34.3
Cross-linked plastics are hard and brittle. §§ 34.3, 34.5.2, 34.5.5

Structures
are determined by covalent bonds and intermolecular forces. § 34.3 Molecules may be linear § 34.3 or cross-linked § 34.5.5

Processing
Thermoplastics can be softened by heat and hardened by cooling many times. They are easy to mould. § 34.3
Thermosets are shaped during synthesis and once set cannot be softened by heat and reshaped. § 34.3
Fibres are made by drawing (pulling) plastics. § 34.3
Rubbers are **vulcanised** to increase strength. § 34.4.3

Addition polymers, poly(alkenes), include: high-density poly(ethene), low-density poly(ethene), poly(chloroethene), PVC, poly(phenylethene), polystyrene perspex, rubbers. § 27.8.10, § 34.4

Addition polymerisation
monomer M → polymer M_n.
takes place by a free radical **mechanism**. § 27.8.10

Disposal of waste plastics
Waste plastics may be dumped in landfill sites. Since most plastics are not biodegradeable, they remain without decomposing for a very long time. Waste plastics may be burned and used as a fuel, but some produce harmful combustion products. § 34.7

Condensation polymers include
poly(esters), e.g. Terylene®, formed from a dicarboxylic acid and diol § 34.5.1
poly (amides), e.g. nylon 6 and nylon 66, formed from a diamine and a dicarboxylic acid § 34.5.4
peptides and **proteins**, formed from α amino acids, $H_2NCHRCO_2H$ § 34.5.4

Condensation polymerisation
Monomers which have two functional groups react by the elimination of small molecules, e.g. H_2O or HCl, to form a polymer §§ 34.5.1–6

Possible solutions to the problem are
- the use of **biodegradable plastics**, e.g. Biopol®,
- plastics which slowly dissolve in water
- plastics which slowly decompose in sunlight
- **recycling of plastics** - the collection, sorting and remoulding of thermoplastics. § 34.7

34.7.6 RECYCLING VERSUS BIODEGRADABILITY

The governments of the USA, Sweden and Italy have passed laws making degradability compulsory for plastics in certain types of packaging. On the other hand, Friends of the Earth and British Plastics Federation now oppose degradable plastics on the grounds that it is better to recycle waste plastics.

Opinions differ. On the one hand, biodegradable plastics can be used to reduce plastic waste. On the other hand, recycling is a more efficient use of resources

Some plastic waste is never collected, e.g. plastics used for agricultural purposes and plastics thrown overboard from ships. There is much to be gained from using biodegradable plastics for such articles. Other plastic objects could be collected for recycling. Some plastics are easy to recyle, e.g. PET from which many bottles and jars are made. There are problems in recycling plastics, however, as described earlier.

There are two solutions to the problem of plastic waste. One is to make degradable plastics, and the other is to recycle plastics. The two solutions do not live well together. Although some biodegradable plastics, e.g. PHB, can be recycled with other plastics, photodegradable plastics cannot be included. Waste plastics can be turned into items such as sacks, park benches, roofing and drain pipes. You can imagine the accidents that could occur if such materials were to break up in sunlight.

QUESTIONS ON CHAPTER 34

1. Polymerisation reactions may be classified as **addition** or **condensation** reactions. Explain the meanings of these terms, and give two examples of each.

2. Explain the terms *crosslinking* and *thermosetting* with reference to condensation polymers. For what purposes are thermosetting polymers suitable?

3. How does the chemical inertness of poly(ethene)arise? How does it increase the usefulness of the material? How does it affect the disposal of waste polyethene?

4. *(a)* What type of functional group joins the repeating units in nylon?
   (b) In what way does the structure of nylon resemble that of a polypeptide?
   (c) What type of interaction takes place between polymer molecules which contain the functional group present in polypeptides?

5.

$$\mathbf{A} = \left(CH(C_6H_5)-CH_2 \right)_n$$

$$\mathbf{B} = \left(CO_2-C_6H_4-CO_2CH_2CH_2 \right)_n$$

Outline the preparation of the polymers **A** and **B** from $C_6H_5COCH_3$, (which must first be converted into phenylethene) and $H_3CC_6H_4COCH_3$ (which must first be converted into benzene-1,4-dicarboxylic acid).

If you wanted to manufacture plastics bottles, which would you choose **A** or **B**, for making bottles to contain *(a)* concentrated sodium hydroxide, *(b)* concentrated hydrochloric acid *(c)* concentrated sulphuric acid?

6. PHB stands for poly(hydroxybutanoate).
   (a) Write the structural formula of 4-hydroxybutanoic acid.
   (b) The $—CO_2H$ group of one molecule forms an ester with the —OH group of a second molecule. Write the structural formula of this ester.
   (c) Write the repeating unit in a long chain of poly(hydroxybutanoate).

7. *(a)* Write the structural formulae of (**A**) hexanedioic acid, (**B**) hexane-1,6-diamine.
   (b) Write the structural formula of the repeating unit in nylon 66, which is formed by condensation polymerisation of (**A**) and (**B**) in part *(a)*.
   (c) What is done to the nylon produced to enable it to be used as fibres?
   (d) Suggest why nylon has poor water absorbancy. How does this affect the use of nylon in the clothing industry?

8. Nylon-66 was first produced in 1935 by Wallace Carothers. Polyamides such as nylon are produced by condensation polymerisation from difunctional monomers.

   (a) Explain the following terms with reference to nylon-66.
   (i) difunctional monomer
   (ii) condensation polymerisation
   (iii) polyamide. 6

(*b*) Suggest why the monomers used must be free from traces of monofunctional compounds. **2**

(*c*) Explain, using equations and reaction conditions, how cyclohexane is converted to nylon-66 via adipic acid (hexanedioic acid) and hexamethylenediamine (1,6-diaminohexane). **10**

(*d*) Melamine is also used to make plastics. These are also produced by condensation polymerisation.

(i) Explain, giving a balanced equation and reaction conditions, how melamine is produced from urea. **4**

(ii) Give **two** properties of melamine plastics which make them particularly useful. **2**

(*e*) PVC is an addition polymer widely used in cables, bottles and pipes.

(i) Explain why burning plastics such as PVC is, particularly dangerous. **2**

(ii) Briefly discuss the use of flame retardants in PVC. **2**

CCEA (AL)

9. (*a*) (i) Draw structures for the products of the alkaline hydrolysis by sodium hydroxide of ethyl benzenecarboxylate (ethyl benzoate). **2**

(ii) Draw the structure of the organic product formed when phenylmethanol, $C_6H_5CH_2OH$, is treated with excess ethanoyl chloride. **1**

(*b*) A polyester is derived from ethane-1,2-diol and benzene-1,4-dicarboxylic acid, the structures of which are shown below.

$HOCH_2CH_2OH$

Ethane-1,2-diol

COOH

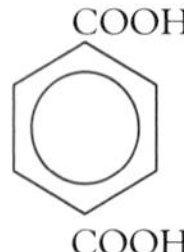

COOH

Benzene-1,4-dicarboxylic acid

(i) Draw a structure to show the repeating unit in the polymer. **1**

(ii) Give **one** large scale use for polyester polymers and state the property of polyesters on which the use depends. **2**

(iii) State **one** environmental problem caused by the widespread use of polymers such as polyesters and nylons, explaining the properties of the polymers which cause the problem. **2**

(*c*) (i) Write down the general formula of an amino acid derived from the hydrolysis of a protein. **1**

(ii) Explain why 2-aminoethanoic acid, NH_2CH_2COOH, has a high melting temperature and shows low solubility in hydrocarbon solvents. **2**

(iii) State how the optical properties of 2-aminoethanoic acid, NH_2CH_2COOH, differ from those of all the other amino acids derived from proteins. **2**

(iv) The hormone, insulin, has the chemical nature of a polypeptide. Explain why diabetics cannot take insulin by mouth but must inject the hormone directly into the blood stream. **2**

15 *WJEC (AL)*

10. (*a*) Explain the term *polymerisation*. **1**

(*b*) Polymers found in natural materials can be formed by the reaction between amino acids.

(i) Draw the graphical formula of the product formed when two molecules of alanine, $CH_3CH(NH_2)COOH$ react together. **1**

(ii) Give the name of the important linkage formed and draw a ring, round it on the formula drawn in (*b*)(i). **2**

(iii) Give the name of the type of naturally-occurring polymer containing this linkage. **1**

(*c*) Poly(ethene) is an example of a synthetic polymer. It is manufactured in two main forms, low-density poly(ethene) and high density poly(ethene).

(i) Write an equation to represent the polymerisation of ethene. **1**

(ii) What is the main structural difference between the polymer chains in the two main forms of poly(ethene)? Explain how this difference affects the densities of the polymers. **3**

(iii) Give **one** further physical property that is affected by the structural difference given in (*c*)(ii). **1**

(iv) Low-density poly(ethene) is manufactured via a free radical mechanism. Draw a graphical formula to represent the free radical formed between a free radical, **R·**, and a molecule of ethene in the reaction. **1**

(v) What type of catalyst is used in the manufacture of high-density poly(ethene)? **1**

(*d*) Poly(ethene) is a non-biodegradable plastic.

(i) Explain the term *non-biodegradable*. **1**

(ii) Give **one** environmental benefit of using biodegradable plastics. **1**

(iii) Developing biodegradable plastics involves compromise. Suggest **one** factor that requires careful consideration and explain your choice. **2**

16 *AEB (AL) 1998*

11. Polyesters are widely used as fibres, films and packaging materials. The most common polyester is made by condensing together compound **A** and compound **B**.

A HOOC—C_6H_4—COOH

B HO—CH_2—CH_2—OH

(a) (i) Name the functional group in compound **B**. **1**
(ii) Name compound **B**. **1**
(iii) Name the functional group in compound **A**. **1**

(b) In the laboratory, compound **B** can be made from 1,2-dichloroethane.
(i) Suggest a reagent to carry out this change. **1**
(ii) A sample of compound **B** made in this way is separated from the reaction mixture, washed and dried. Describe how a pure sample of the compound (a liquid boiling at 198 °C) can then be obtained from the impure liquid after drying. **2**

(c) In industry, compound **B** is made from ethene. Suggest a reason why the use of 1,2-dichloroethane would be more expensive. **2**

(d) Draw the formula of part of the polyester structure, showing how one molecule of compound **A** combines with one molecule of compound **B**. Show the ester link as a full structural formula. **2**

(e) State the strongest type of intermolecular force between polyester chains. **1**
Illustrate your answer with a diagram. **2**

(f) Compound **B** has also been polymerised with compound **C**.

C HOOC COOH (benzene-1,2-dicarboxylic acid structure)

The polyester produced using compound **C** is less flexible than that produced using compound **A**. Suggest reasons for this. **3**

16 *O&C (Salters) (AS/AL)*

12. *(a)* Consider the following reaction sequence, then answer the questions which follow.

$$C_2H_5Br \xrightarrow{\text{step 1}} C_2H_5CN \xrightarrow{\text{step 2}} C_2H_5CO_2H \xrightarrow[PCl_5]{\text{step 3}}$$

$$\mathbf{A} \xrightarrow[NH_3]{\text{step 4}} C_2H_5CONH_2 \xrightarrow{\text{step 5}} C_2H_5NH_2$$

(i) Give the reagents and conditions necessary for steps 1, 2 and 5. **7**
(ii) Identify **A**. **1**

(b) Nylon 6 : 6, a polyamide, has a structure containing the following repeat unit:

$$-\!\!\!+\!CO-(CH_2)_4-CONH-(CH_2)_6-NH\!+\!\!\!-$$

(i) Give the structures of the monomers from which this polymer could be made. **2**
(ii) What type of polymer is this? **1**

(c) (i) Show the structure of poly(tetrafluoroethene) **2**
(ii) State one use of this polymer. **1**

(d) Which of the two polymers in *(b)* and *(c)* is the easier to break down and hence constitutes the smaller environmental hazard? Give a reason for your answer. **2**

16 *L (AL)*

13. *(a)* Give the structure of (i) a carboxylic acid amide, (ii) a tertiary amine. **2**

(b) (i) Indicate, by means of an equation, a reaction which would lead to your amide, using **either** ammonia **or** an amine as the source of nitrogen. **2**
(ii) A tertiary amine cannot be used to prepare an amide in this way. Explain why. **1**

(c) (i) Show, by means of equations, how a polyamide may be obtained from ethane-1,2-diamine and benzene-1,4-dicarboxylic acid. You will need two steps for this synthesis. **3**
(ii) Give **one** way in which the essential chain structure of this polyamide differs from that of a protein, which is a naturally occurring polyamide. **2**

(d) (i) Compare the effect of aqueous sulphuric acid on your amide (*(a)*(i) above) with its effect upon your tertiary amine (*(a)*(ii) above). **2**
Contrast the **conditions** required for each reaction. **2**
(ii) Give **one** reason why amides do not react with acid in the same way as do amines. **3**

17 *(O&C) (AL)*

35 IDENTIFYING ORGANIC COMPOUNDS

35.1 'BUCKY BALLS'

The carbon allotropes, diamond and graphite, have been known for centuries. In 1985 scientists were excited by the discovery of a new family of carbon allotropes. They consist of molecules containing 30–70 carbon atoms and have been named **fullerenes**.

The discovery had its origins in space research. In 1980, Donald Huffmann at Arizona University, USA and Wolfgang Kratschmer at Heidelberg University, Germany, wanted to find out whether soot could form in outer space from heated graphite. They struck an electric arc between a graphite rod and a graphite disc in an inert atmosphere. Graphite vaporised, and soot was deposited on the container. When they studied the ultraviolet absorption spectrum of the soot, they were surprised to find some unexpected peaks, which corresponded to large molecules. They took the matter no further at that time.

In 1984, Harold Kroto and David Walton of Sussex University, UK were investigating their theory that molecules containing up to 33 carbon atoms might exist in the space surrounding red giant stars. In collaboration with Richard Smalley of Texas University, USA, they directed a laser at a graphite target and obtained the molecules they were looking for. They also detected by mass spectrometry some molecules of mass 720 u. They wondered whether these were molecules of formula C_{60}, but they did not have enough material to do a spectroscopic study. When Huffmann and Kratschmer learned about this discovery, they had another look at the soot which they had made in 1980. They measured the mass spectrum and the ultraviolet and infrared spectra and obtained results which agreed with the formula C_{60}. From the soot they separated by sublimation and dissolving in benzene a mixture of 90% C_{60} and 10% C_{70}. Using X ray and electron diffraction measurements, they showed that the C_{60} molecules are spherical and packed 1.04 nm apart.

Kroto discussed the possible structures of C_{60} with his team. Was a layer structure, like graphite, or a spherical structure more likely? Kroto recalled the geodesic domes created by the architect Buckminster Fuller. He suggested that C_{60} might be a perfect sphere of carbon atoms. The team tried to make a model of C_{60} from hexagons of carbon atoms, but they could not make a closed structure in this way. However, success came when they used a combination of 20 hexagons and 12 pentagons. The structure resembled a football! Electron microscope pictures show a mosaic of 20 hexagons and 12 pentagons on the surface of the molecule, which measures 1 nm in diameter.

The chemists decided to call the allotrope C_{60} **buckminsterfullerene**. The new molecule quickly became better known as a '**bucky ball**'. Huffmann and Kratschmer beat Kroto in the race to publish the structure by only a few days. However, Kroto and Roger Taylor were the first to separate C_{60} from C_{70}.

Kroto and Taylor measured the carbon-13 nuclear magnetic resonance (NMR) spectrum of a solution

Figure 35.1A
Fullerenes C_{28}, C_{32}, C_{50}, C_{60} and C_{70}

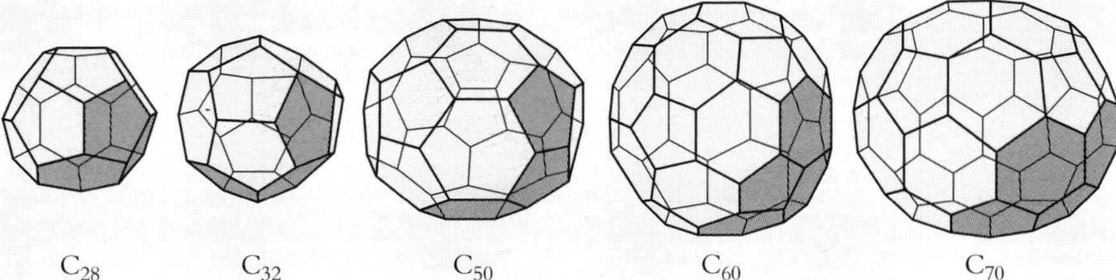

of C_{60}. Since the NMR spectrum depends on the extent to which a nucleus is screened from the magnetic field by electrons in neighbouring atoms, it therefore depends on the precise chemical environment of the nucleus. If buckminsterfullerene is a highly symmetrical structure as Kroto thought, then all 60 carbon atoms occupy identical positions in the structure. No matter where a carbon-13 nucleus is located in the C_{60} molecule, it will be in the same environment and will produce the same NMR frequency, and the spectrum should consist of a single line. What a triumph it was when the team discovered that the carbon-13 NMR spectrum consisted of a single line! Their structure was correct!

The story of the discovery of fullerenes illustrates the importance of modern instrumental techniques: mass spectrometry, ultraviolet and infrared spectrometry and nuclear magnetic resonance spectrometry in establishing the structures of organic compounds. These methods have not always been available and are not available in all laboratories today. This chapter begins with an account of the traditional methods by which organic chemists established the structures of most of the compounds we meet in this book.

35.2 METHODS OF PURIFICATION

The first step in identifying an organic compound is to obtain it in as pure a state as possible. Some of the methods of purifying organic compounds are listed below.

1. Recrystallisation and fractional recrystallisation [§ 9.3] are used to purify solids. The purity of the solid can be assessed by taking its melting temperature. A pure solid melts sharply at a certain temperature. The presence of impurity lowers the melting temperature and also results in a more gradual melting.
2. A liquid may be converted into a solid derivative so that the solid can be recrystallised (e.g. an aldehyde into a 2,4-dinitrophenylhydrazone [§ 31.7.5]).
3. Filtration and centrifugation remove suspended matter from liquids.
4. Distillation is used for liquids [§ 8.2] and fractional distillation for mixtures of liquids [§ 8.5.2]. A pure liquid distils at a certain temperature. Impurities raise the boiling temperature and also make the liquid boil over a range of temperature.
5. Solvent extraction is used for both liquids and solids [§ 8.6.1].
6. Various types of chromatography are widely used [§ 8.7].

*35.3 IDENTIFYING THE ELEMENTS PRESENT

Mass spectrometry [§§ 1.9 and 35.9] will detect the elements present in a compound. In the absence of this instrument, the Lassaigne test is used. A sample of the compound is heated with molten sodium. Nitrogen is converted into cyanide, sulphur into sulphide and halogens into halides. These ions are identified by chemical tests. It is assumed that carbon and hydrogen are present.

*35.4 EMPIRICAL, MOLECULAR AND STRUCTURAL FORMULAE

35.4.1 EMPIRICAL FORMULA

Quantitative analysis is used to find the percentage by mass of each element so that the empirical formula can be calculated [§ 3.8].

Carbon and hydrogen

A known mass of the compound is heated in a stream of pure dry oxygen. The carbon dioxide formed is absorbed in weighed bulbs of concentrated potassium hydroxide solution. The water vapour formed is absorbed in weighed anhydrous calcium chloride tubes. The increases in mass give the masses of carbon dioxide and water formed and from these the masses of carbon and hydrogen in the sample.

Nitrogen

A known mass of the compound is boiled with concentrated sulphuric acid and sodium sulphate. Ammonium sulphate is formed. The mass of the salt is found by adding a known excess of alkali to liberate ammonia, and then titrating the amount of alkali remaining against a standard acid.

Halogens

A known mass of the compound is heated in a sealed tube with fuming nitric acid and silver nitrate. Halogens are converted into silver halides, which can be filtered off, dried and weighed.

Sulphur

A known mass of compound is heated with fuming nitric acid. The sulphur content is converted into sulphate ion, which is precipitated as barium sulphate, dried and weighed.

Oxygen

The percentage by mass of oxygen is found by difference after the percentages of all the other elements have been found.

35.4.2 MOLECULAR FORMULA

To convert the empirical formula into the molecular formula [§ 3.9], the molar mass must be known. Methods of finding molar masses have been given for gases [§ 7.5], for liquids [§ 8.4], and by mass spectrometry [§§ 1.9, 35.9.1].

35.4.3 STRUCTURAL FORMULA

Chemical tests reveal the presence of functional groups, such as hydroxyl groups and carbonyl groups. The structural formula can often be worked out from the molecular formula and the functional groups present. Once the identity of the compound has been deduced it can be confirmed by matching the melting temperature or boiling temperature with that of a known compound. A better method for a liquid is to convert it into a solid derivative and find the melting temperature of the derivative, for example to convert a ketone into a solid 2,4-dinitrophenylhyrazone. Spectroscopic methods of identifying compounds are covered later in the chapter.

CHECKPOINT 35.4: ANALYSIS

1. *(a)* When 0.200 g of a compound **A**, which contains only carbon and hydrogen, was burned completely in a stream of dry oxygen, 0.629 g of carbon dioxide and 0.257 g of water were formed. Find the empirical formula of the compound.
   (b) When 0.200 g of **A** is vaporised, the volume which it occupies (corrected to stp) is 53.3 cm^3. The GMV = 22.4 dm^3 at stp. Find the molar mass of **A**.
   (c) Use the molar mass of **A** to convert the empirical formula, from *(a)*, into a molecular formula.

2. A Lassaigne test showed that a compound **X** contained sulphur and chlorine. From 0.1000 g of **X** was obtained 0.1322 g of $BaSO_4$. Another 0.1000 g sample of **X** gave 0.0813 g of AgCl. The combustion of 0.2000 g of **X** gave 0.2992 g of CO_2 and 0.0510 g of H_2O. Find the empirical formula of **X**. (Remember that, if the percentages of C, H, S and Cl do not total 100%, the difference is due to oxygen.)

3. The compound **Y** contains nitrogen. After conversion of 0.1850 g of **Y** into an ammonium salt, ammonia was expelled by warming the salt with 75.0 cm^3 of sodium hydroxide solution of concentration 0.100 mol dm^{-3}. Titration showed that 50.0 cm^3 of sodium hydroxide remained at the end of the reaction. Find the percentage by mass of nitrogen in **Y**.

   When 0.146 g of **Y** was burned completely in dry oxygen, 0.264 g of CO_2 and 0.126 g of H_2O were formed. Find the percentage by mass of C and H in the compound.

   Find the empirical formula of the compound.

35.5 INSTRUMENTAL METHODS

Instrumental methods have made the identification of organic compounds faster and more reliable. They require very small quantities of substances

Modern instrumental methods have made the identification of organic compounds much simpler. The spectrum of an unknown substance can be compared with those of known compounds. The spectra of a vast number of compounds can be kept on a computer for easy reference. The instrumental methods which will be covered here are:

- mass spectrometry, which has been introduced in § 1.9
- visible–ultraviolet spectrophotometry, infrared spectrophotometry and nuclear magnetic resonance spectrometry, which absorb energy in different regions of the electromagnetic spectrum.

Instrumental methods take little time and little material. A visible–ultraviolet spectrum can be obtained with 0.1–0.01 g of substance, an infrared spectrum or a nuclear resonance spectrum with as little as 10^{-9} g and a mass spectrum with only about 10^{-12} g.

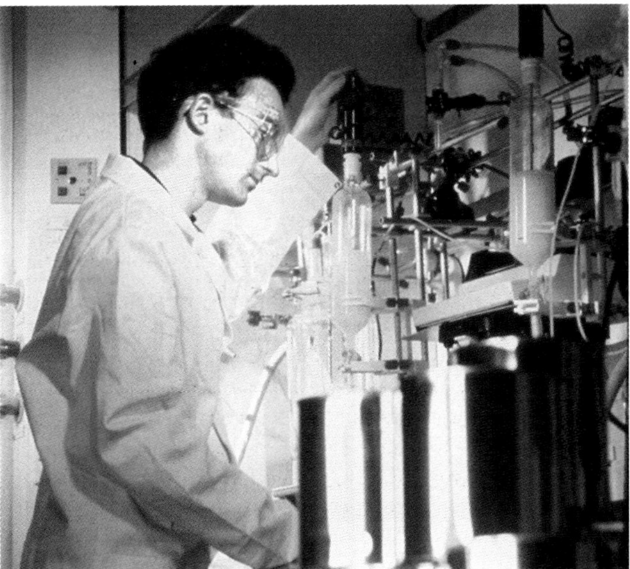

Figure 35.5A
An analytical chemist at work

35.6 THE ENERGY LEVELS OF A MOLECULE

When a molecule absorbs a photon of energy, it may increase ...

... the electronic energy of the molecule (visible–ultraviolet radiation) ...

... or vibrational energy of the molecule (infrared radiation) ...

Molecules absorb photons of energy from radiation. The energy of the photon raises the molecule that has absorbed it to an **excited state**. Different types of radiation produce different kinds of excited states. Visible and ultraviolet light can excite a molecule to a higher electronic energy state by promoting an electron to an orbital of higher energy. Infrared radiation is of lower energy than UV radiation. It cannot promote electrons but it can increase the **vibrational energy** of a molecule. The ways in which the molecule H_2O can bend and vibrate are shown in Figure 35.6A. There are two conditions for absorption in the infrared. Firstly, the frequency of infrared radiation must match exactly the frequency of the bond vibration. Secondly, the molecule must have a **dipole moment**, and only vibrations which cause a change in the dipole moment can absorb infrared energy.

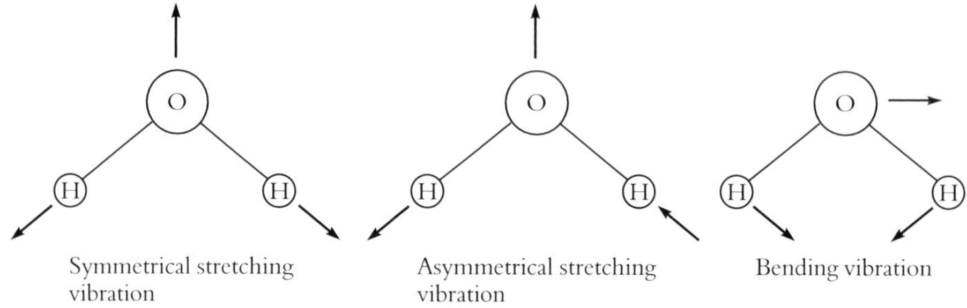

Figure 35.6A Modes of vibration of the H_2O molecule

... or the rotational energy of the molecule (microwave radiation)

In addition to vibrating and bending, a molecule may rotate. The absorption of **microwave radiation**, which is radiation of low photon energy, raises molecules to higher levels of rotational energy. The frequency of the radiation that can be absorbed by an atom or a molecule depends on the difference in energy ΔE between the ground state and the excited state. It is given by

The absorption of radiation is governed by the relationship ...

$$\Delta E = h\nu$$

...$\Delta E = h\nu$...

where h = Planck's constant, ν = frequency of radiation [see § 2.3.1].

...where ΔE = energy absorbed, h = Planck's constant, ν = frequency

The quantities of energy required for excitation are approximately: electronic energy 100 kJ mol^{-1}, vibrational energy 1 kJ mol^{-1} and rotational energy 0.01kJ mol^{-1}. Ultraviolet, visible, infrared and microwave radiation are all part of the **electromagnetic spectrum** (see any A-level Physics text for more information).

35.7 VISIBLE–ULTRAVIOLET SPECTRA

Visible–ultraviolet spectra arise from transitions between electronic energy levels

A transition between two electronic energy levels should give rise to a line in an absorption spectrum. In practice it may not be possible to distinguish individual lines. Large molecules have a large number of vibrational and rotational energy levels. Visible–ultraviolet radiation may raise molecules at many slightly different electronic energy levels to many slightly different higher levels. As a result, sharp lines are replaced by narrow bands – **absorption bands**.

*35.7.1 COLORIMETRY

The height of an absorption peak at a certain wavelength depends on the amount of absorbing substance present and can be used to measure its concentration.

The absorption of a solution can be measured and used to find the concentration of solute

A solvent which does not absorb at the same wavelength should be chosen if possible. The method compensates for any slight absorption due to the solvent. The height of an absorption peak due to an absorbing substance in a non-absorbing solvent is given by the Beer–Lambert Law:

$$A = \lg (I_0/I) = \varepsilon c l$$

where A = absorbance, I_0 = intensity of incident beam, I = intensity of transmitted beam, ε = the molar absorption coefficient, which is constant for a particular species (molecule or ion) at a certain wavelength, c = concentration, l = path length, that is, the thickness of the cell. The value of ε must be found for the species at the wavelength being used. Then the measured value of A for a solution of the species will give the concentration of the species.

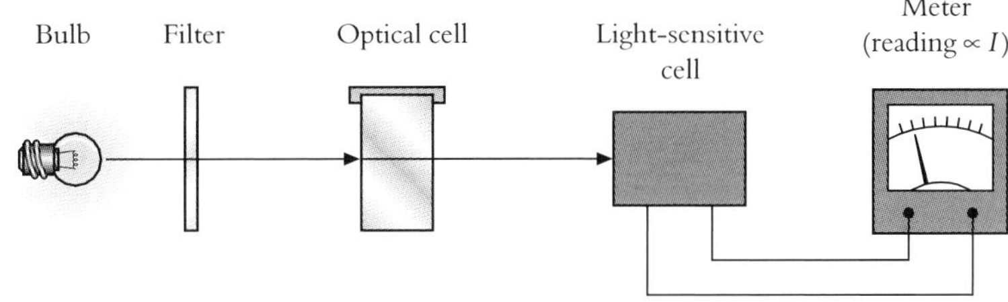

Figure 35.7A
A colorimeter

35.7.2 VISIBLE–ULTRAVIOLET SPECTROPHOTOMETRY

All substances have absorption spectra in the visible–ultraviolet region. Those with π electrons absorb light strongly

Since all matter contains electrons, all substances absorb light at some wavelength in the visible–ultraviolet (visible-UV) spectrum. Compounds with π electrons, since these are less strongly held than σ electrons, absorb light intensely. The ultraviolet spectrum of a compound is like a fingerprint. If it can be matched up with the spectrum of a known compound, it can be used as a means of identification. The spectrum is obtained by an instrument like that shown in Figure 35.7B.

Applications of visible–ultraviolet spectrophotometry

1. Identification. Some compounds can be identified by comparing their spectra with those of known compounds.
2. Analysis. The concentration of a solution can be found from the height of its absorption peak. A UV spectrophotometer can be used to analyse continuously the eluant from a chromatography column.
3. Reaction kinetics. The course of a reaction can be followed by measuring the absorption spectrum of a reactant or a product.
4. Detection and analysis. Low concentrations of transition metal ions can be measured by converting them into coloured complexes.
5. Monitoring pollution. A pollutant must be converted into a species which absorbs in the visible-UV region, e.g. nitrate ion into an azo dye.

Figure 35.7B
A UV spectrophotometer

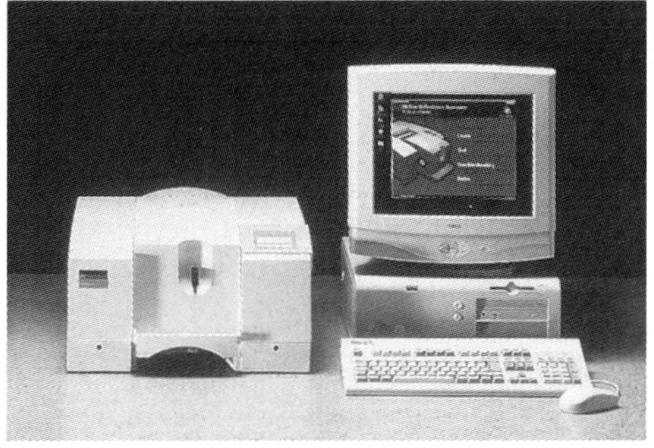

SUMMARY

Visible–ultraviolet spectrophotometry records the complete visible–ultraviolet spectrum of a compound. It is sometimes used to identify compounds. More often, it is used to measure concentrations, e.g. in following the course of a reaction and in detection and analysis of pollutants.

CHECKPOINT 35.7: VISIBLE–ULTRAVIOLET SPECTRA

1. Volumes of 0.100 mol dm^{-3} $\mathbf{M}^{2+}$(aq) and 0.100 ml dm^{-3} **L**(aq) are added as shown in the table. **M** is a transition metal, and **L** is a ligand which forms a complex with $\mathbf{M}^{2+}$.

   The absorbance of each solution is measured. Plot the absorbance against the number of the solution. Which solution gives the maximum colour intensity? This is the solution which has the composition of the complex. What is the formula of the complex ion?

Volume of M^{2+}(aq)/cm^3	Volume of L(aq)/cm^3	Absorbance
0	10.0	0
1.0	9.0	0.293
2.0	8.0	0.569
3.0	7.0	0.759
4.0	6.0	0.793
5.0	5.0	0.759
6.0	4.0	0.655
7.0	3.0	0.465
8.0	2.0	0.310
9.0	1.0	0.155
10.0	0	0

35.8 INFRARED SPECTROMETRY

All organic compounds absorb in the IR region ...

... and can be identified from their spectra

All organic compounds absorb infrared (IR) radiation. Books of recorded spectra are available for comparison. If the unknown compound is a new substance, its spectrum will not match with any recorded IR spectrum; yet it is still possible to infer a great deal about the structure of a compound. The C=O bond and the C—OH bond and others have characteristic absorption frequencies and can be identified. Functional groups can be identified from their IR spectra, and the IR spectrum is the most rapid method of detecting the series to which a compound belongs. Infrared spectrometry is valuable because of the great ease with which samples are prepared and spectra are taken. It requires about 1 mg of the compound.

35.8.1 THE INFRARED SPECTROMETER

In an IR spectrometer, one beam passes through the sample ...

... and meets the reference beam in the detector

An infrared spectrometer is a double beam instrument. Radiation from a hot wire is split into two beams. One beam passes through the sample, which absorbs radiation of the same frequency as the natural frequency with which the atoms vibrate. The other beam is a reference beam. Both beams meet in the detector, which compares the two beams and sends information to the chart recorder. This plots the transmittance (the percentage of radiation that passes through the sample) against either the frequency or the **wavenumber** (1/wavelength). The peaks extend from the top of the trace downwards [e.g. Figure 35.8B]. The instrument contains no glass or quartz because these absorb IR radiation; all the mirrors and prisms are cut from large sodium chloride crystals.

35.8.2 IDENTIFYING THE PEAKS

The absorption spectrum identifies functional groups ...

Apart from small molecules such as SO_2 and CO_2 it is not possible to identify each absorption. Each type of bond has a different natural vibration frequency, but the same type of bond in different molecules is in a different environment and its frequency changes; for example the C=O bond in —CHO has a different absorption from the C=O bond in $—CO_2H$. There are no two compounds with the same infrared spectrum. Organic compounds absorb infrared radiation over wavenumbers 4000 cm^{-1} to 650 cm^{-1}.

Figure 35.8A
The infrared absorption of a number of groups

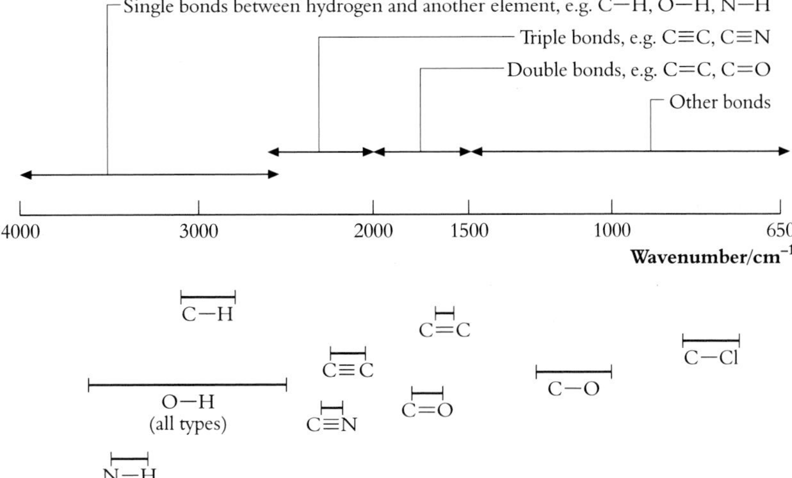

... the precise wavelength depending on the rest of the molecule

The regions in which a number of groups absorb are shown in Figure 35.8A. Table 35.8 shows the wavenumbers of various groups. You need not try to remember them! You will always be able to refer to a table of wave numbers when you need to identify a spectrum.

Table 35.8
Wavenumbers of functional groups

The wavenumbers of some functional groups are listed ...

... for reference, not for memorisation

Absorption bands above 1500 cm^{-1} identify functional groups

Functional group		Wavenumber/cm^{-1}
O—H	Aliphatic and aromatic	3600–3000
N—H	Primary, secondary, tertiary amines	3600–3100
N—H	Primary amines	3500–3350
C—H	Aromatic	3150–3000
C—H	Aliphatic	3000–2850
C≡N	Nitrile	2280–2200
C—H	of CHO	2830–2720
C—C	Arene	1600
C—O		1000–1300
O—H	Free	3580–3670
O—H	H-bonded in acids	2500–3300
O—H	H-bonded in alcohols and phenols	3230–3550
C≡C	Alkyne	2250–2070
CO_2R	Ester	1750–1700
CO_2H	Carboxylic acid	1740–1670
C=O	Aldehydes, ketones and esters	1750–1680
$CONH_2$	Amides	1720–1640
C=C	Alkene	1680–1610
C—O—R	Aromatic	1300–1180
R—O—R	Aliphatic	1160–1060
C—Cl		800–700

Hydrogen bonding

Hydrogen-bonded compounds show the O—H bond as a broad band

In hydrogen-bonded compounds, the O—H bond appears as broad bands instead of sharp peaks. This happens with alcohols and carboxylic acids.

35.8.3 FINGERPRINT REGIONS

Absorption bands below 1500 cm^{-1} give a fingerprint of the compound

In many compounds, vibrations of the molecule as a whole give rise to a series of absorption bands at low energy, below 1500 cm^{-1}, which are characteristic of that compound [see Figures 35.8B, C, D] and offer a **fingerprint** of the particular compound. Comparing the fingerprint region of a spectrum with that of an authentic sample of the compound is an extremely reliable method of identification.

35.8.4 INTERPRETING IR SPECTRA

Benzene has a molecule with only C—C and C—H bonds. The IR spectrum (Figure 35.8B) shows not only the C—C arene bond peak at 1480 cm^{-1} but also three C—H peaks. These are due to the C—H bend, the C—H stretch in the plane of the ring and the C—H stretch out of the plane of the ring. Small peaks can be overtones or harmonics of fundamental vibrations.

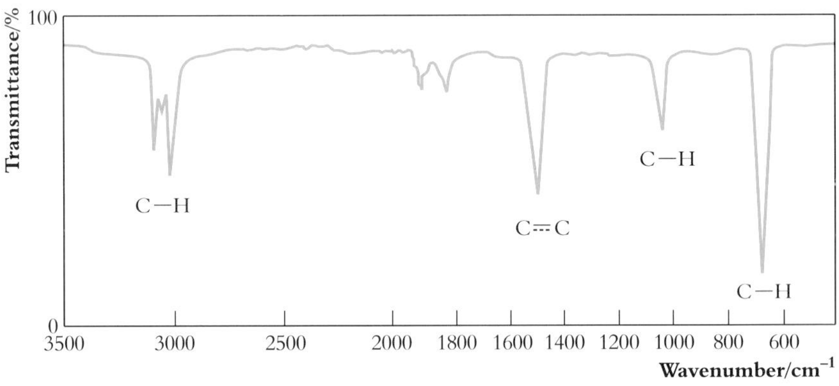

Figure 35.8B The IR spectrum of benzene

In organic compounds many absorption bands arise below 1500 cm^{-1}. The exact form of these bands gives a 'fingerprint' region which can be used to identify an organic compound

Many IR spectra are more complicated than that of benzene, and it is not possible to identify every peak. The best way to proceed is to identify important functional groups. The functional groups C=O, N—H, C—O, O—H, C=C, C≡C and C≡N have peaks which are easy to recognise [see Table 35.8 and Figure 35.8A].

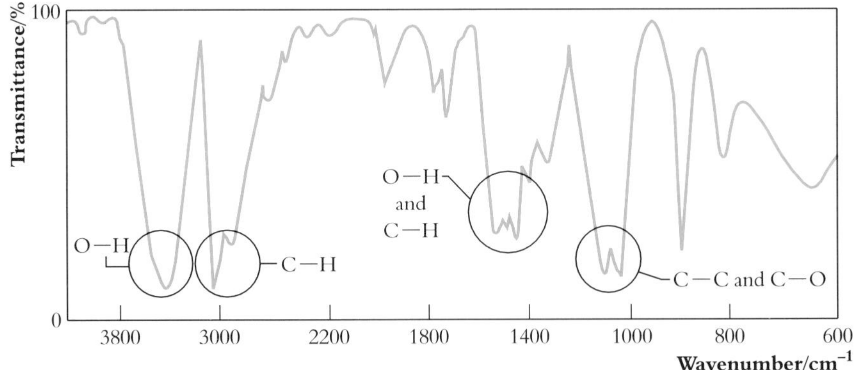

Figure 35.8C IR spectrum of ethanol, C_2H_5OH

The IR spectrum of ethanol is shown in Figure 35.8C. The O—H group absorbs in two regions: 3340 cm^{-1}, due to stretching, and 1450–1380 cm^{-1}, due to bending [see Figure 35.6A]. The broad absorption peak at 3550–3230 cm^{-1} characterises the hydrogen-bonded —O—H group in alcohols. The C—H bonds absorb at 2950 cm^{-1} due to stretching and 1450–1380 cm^{-1} due to bending. The C—O stretch and the C—C stretch both absorb at 1090–1050 cm^{-1}.

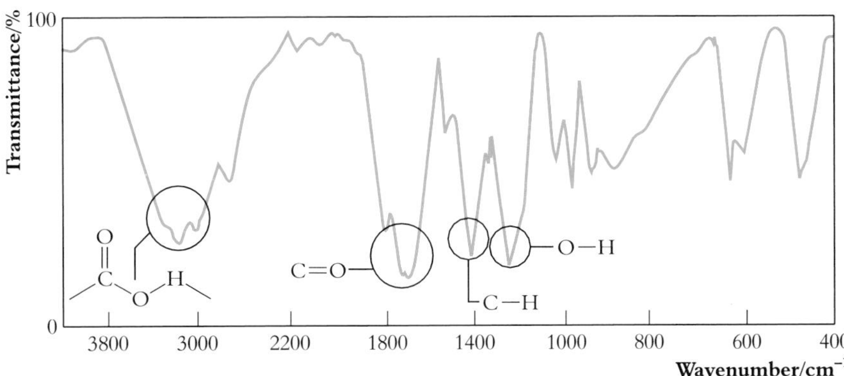

Figure 35.8D IR spectrum of ethanoic acid, CH_3CO_2H

The IR spectrum of ethanoic acid is shown in Figure 35.8D. The C═O stretch absorbs at 1800–1600 cm^{-1}, the C—H bending at 1450–1380 cm^{-1}. The free —O—H group absorbs at 1450–1380 cm^{-1}. The hydrogen-bonded —O—H group in carboxylic acids absorbs at 3300–2500 cm^{-1}.

35.8.5 A KEY TO INTERPRETING SPECTRA

Say that C═O is present in a compound:

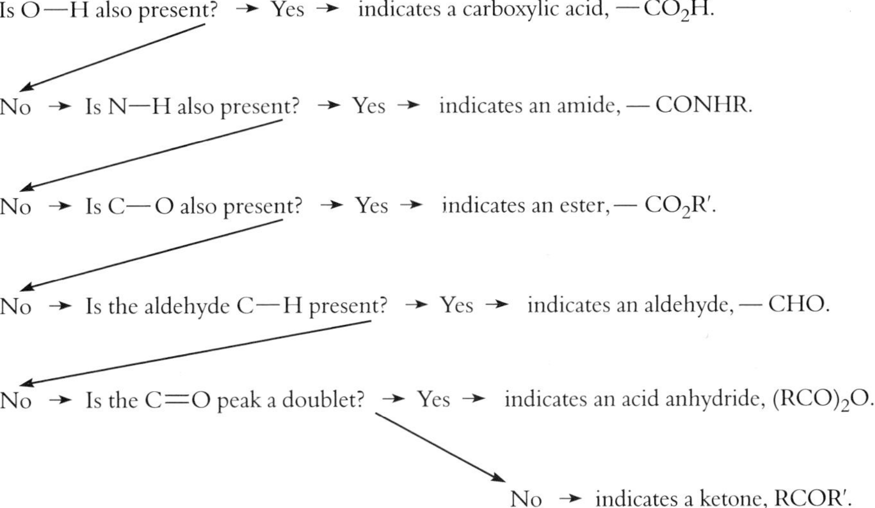

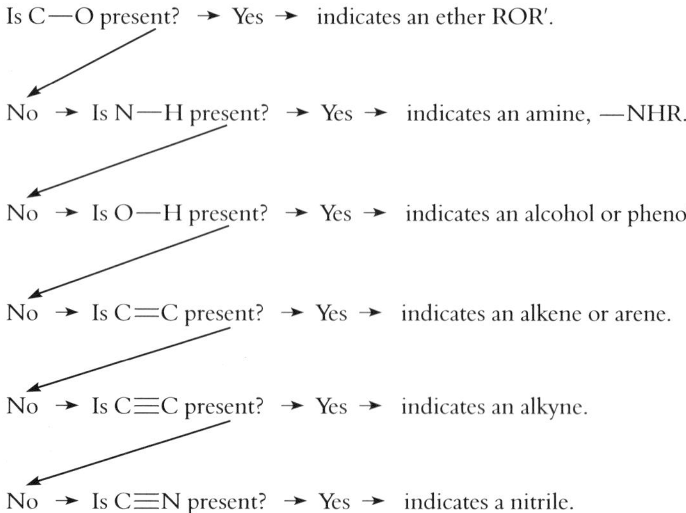
Say that C=O is absent in a compound:
Is C—O present? → Yes → indicates an ether ROR′.
No → Is N—H present? → Yes → indicates an amine, —NHR.
No → Is O—H present? → Yes → indicates an alcohol or phenol.
No → Is C=C present? → Yes → indicates an alkene or arene.
No → Is C≡C present? → Yes → indicates an alkyne.
No → Is C≡N present? → Yes → indicates a nitrile.

35.8.6 APPLICATIONS OF IR SPECTROMETRY

1. Identifying compounds. The functional groups can be detected. In order to identify the compound, it may be necessary to study the mass spectrum or the nuclear magnetic resonance spectrum also. The identity of a compound can be confirmed by comparing the whole of its IR spectrum with that of a known specimen of the compound.
2. Reaction kinetics. The course of a reaction may be followed by recording the replacement of an absorption band due to a reactant by an absorption band due to a product, e.g. a $>C{=}O$ band by a $\geqslant C{-}O{-}H$ band.
3. Forensic science. Part of the work of forensic scientists is to provide evidence that a suspect was present at the scene of a crime or that a tool found in his possession was used to force an entry. They frequently examine traces of paint because the resins in a paint can be identified by IR spectrometry.

Figure 35.8E
The Lion intoximeter

4. Breathalyser. The police frequently have to decide whether a person has been driving while 'under the influence' of alcohol. The police conduct a quick roadside test with a type of fuel cell. If the roadside test suggests that the driver is over the limit, he or she is taken to the police station where two breath samples are taken for IR spectrometry [see Figure 35.8C for the IR spectrum of ethanol].
5. Monitoring air pollution. Infrared spectrometry is used in the detection and analysis of air pollutants, in particular carbon monoxide.

INFRARED SPECTROMETRY

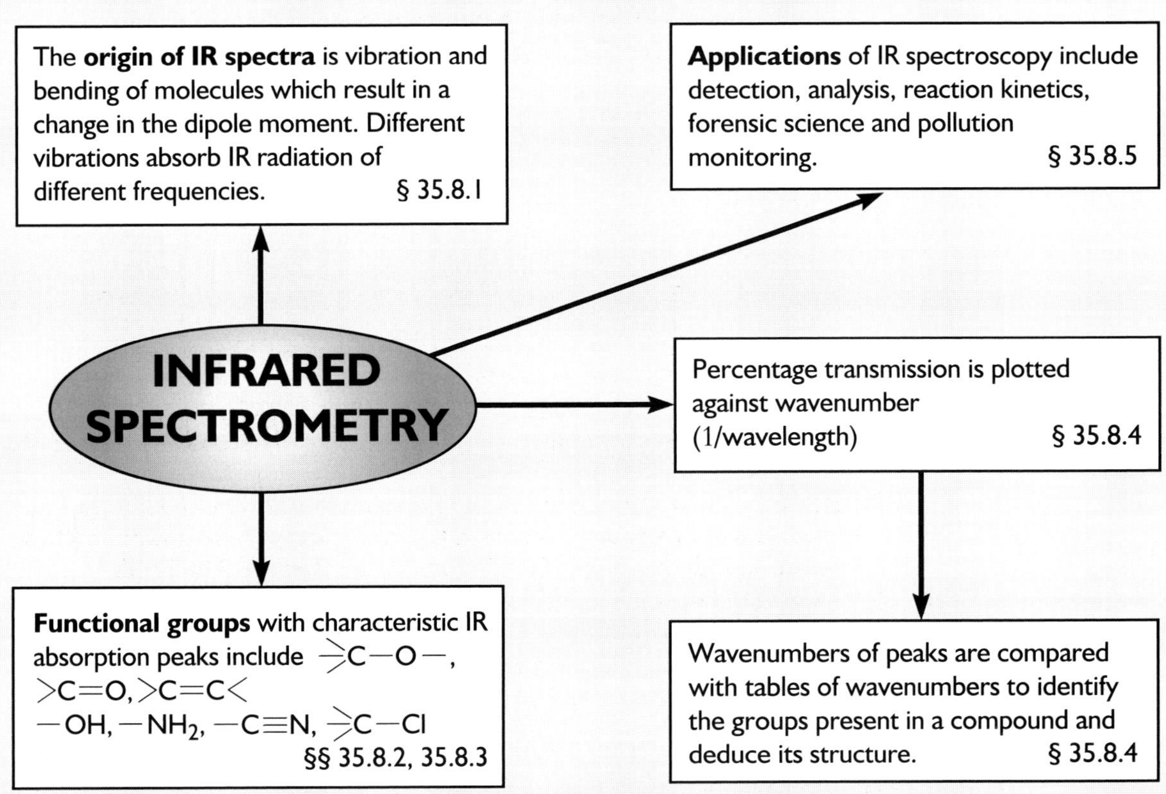

The **origin of IR spectra** is vibration and bending of molecules which result in a change in the dipole moment. Different vibrations absorb IR radiation of different frequencies. § 35.8.1
Applications of IR spectroscopy include detection, analysis, reaction kinetics, forensic science and pollution monitoring. § 35.8.5
INFRARED SPECTROMETRY
Percentage transmission is plotted against wavenumber (1/wavelength) § 35.8.4
Functional groups with characteristic IR absorption peaks include $\geq C-O-$, $>C=O$, $>C=C<$ $-OH$, $-NH_2$, $-C\equiv N$, $\geq C-Cl$ §§ 35.8.2, 35.8.3
Wavenumbers of peaks are compared with tables of wavenumbers to identify the groups present in a compound and deduce its structure. § 35.8.4

CHECKPOINT 35.8: INFRARED SPECTRA

1. What can you deduce from IR spectrum **A**?

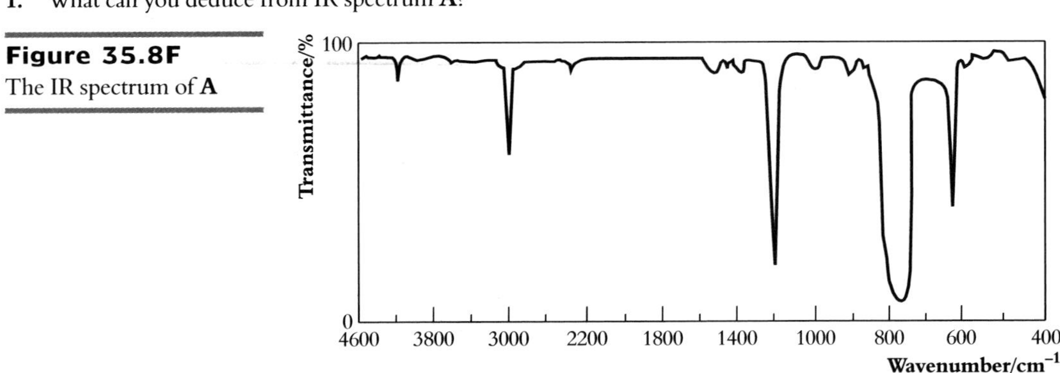

Figure 35.8F
The IR spectrum of **A**

2. What features can you identify in IR spectrum **B**?

Figure 35.8G
The IR spectrum of **B**

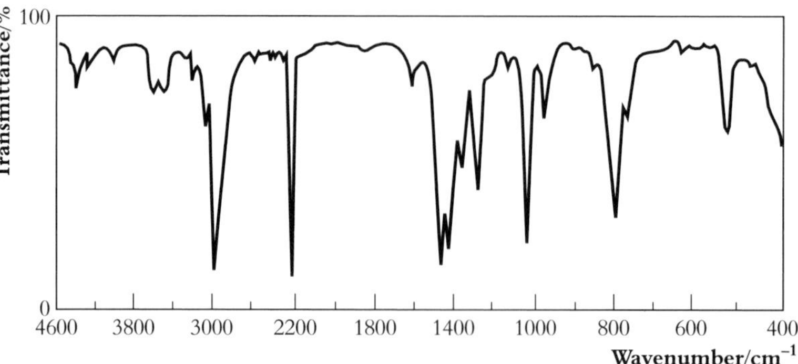

3. What features can you identify in IR spectrum **C**?

Figure 35.8H
The IR spectrum of **C**

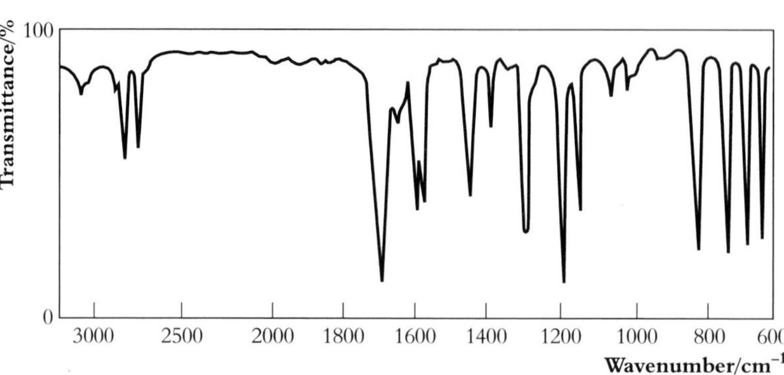

4. What can you say about IR spectrum **D**?

Figure 35.8I
The IR spectrum of **D**

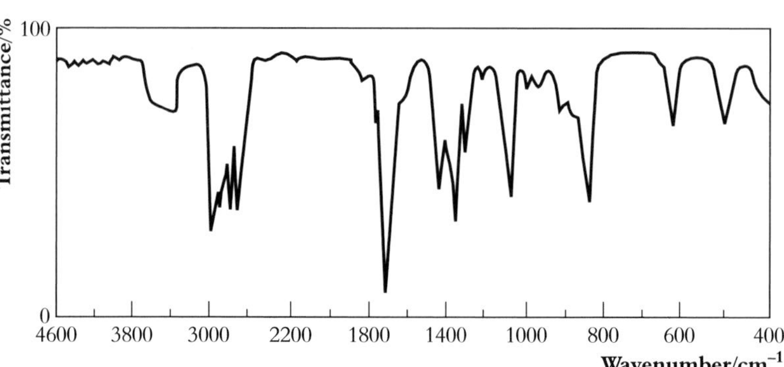

5. What type of compound gives rise to IR spectrum **E**? How could you identify the compound?

Figure 35.8J
The IR spectrum of **E**

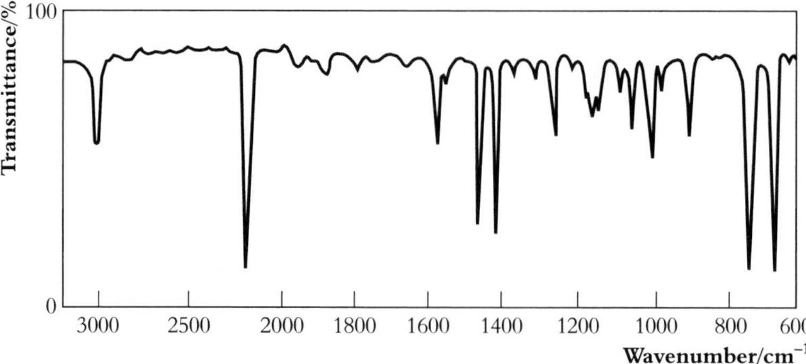

6. The IR spectrum of an organic liquid is shown below. What type of compound gave rise to spectrum **F**?

Figure 35.8K
The IR spectrum of **F**

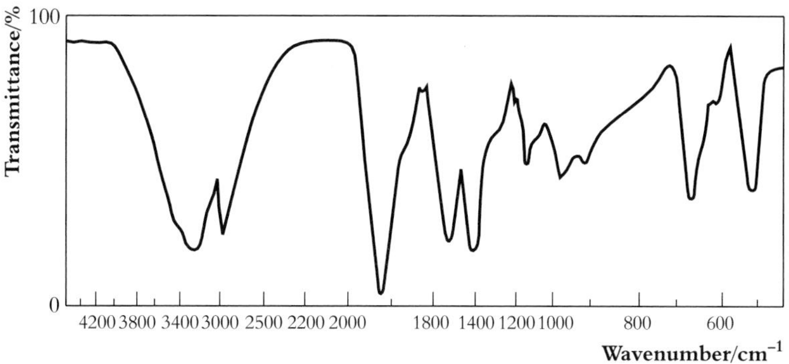

35.9 MASS SPECTROMETRY

35.9.1 MOLECULAR IONS

Mass spectrometry can analyse a sample of 10^{-12} g; it is a thousand times more sensitive than IR spectrophotometry or NMR. The use of a mass spectrometer to find relative atomic mass has been described in § 1.9 and Figure 1.9B.

When a compound is ionised in the mass spectrometer, a **molecular ion** is formed by the loss of one electron.

$$M \longrightarrow M^+ + e^-$$

The peak with the highest value of m/z is due to the molecular ion. Its mass gives the molecular mass of the compound. This is not the only peak because in addition to generating molecular ions the beam of electrons ruptures chemical bonds with the formation of positively charged fragments. The largest peak corresponds to the most stable fragment. It is called the **base peak** and allocated a relative abundance of 100%. Other peak heights are expressed as percentages of the base peak.

The mass spectrum of methanol is shown in Figure 35.9A. A peak for the molecular ion M is observed at $m/z = 32$. Peaks are obtained at $m/z = 31$, 29 and 28. The peak at $m/z = 31$ can be assigned to the ion CH_3CO^+ formed by the loss of one H atom. The loss of further H atoms gives the ions CHO^+ at $m/z = 29$ and CO^+ at $m/z = 28$.

Figure 35.9A
Mass spectrum of methanol

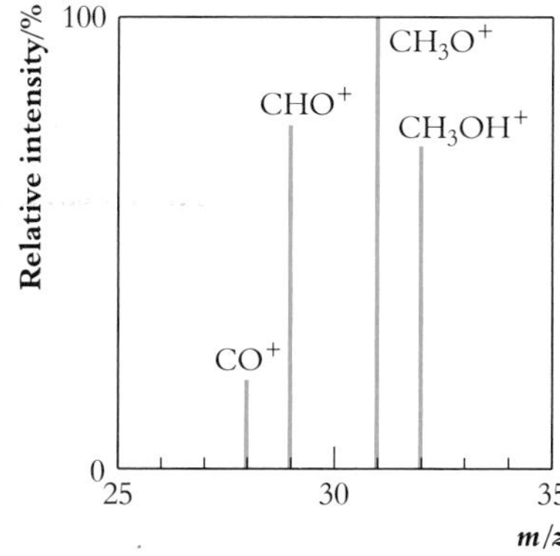

Other ions are fragments of molecules

If isotopes are present, more than one molecular ion is formed. Some large molecules, e.g. polymers, are easily fragmented and do not give molecular ions.

The accuracy of the molecular mass determination is such that it can be used to distinguish between compounds with the same mass number; see Table 35.9.

Table 35.9 Compounds with the same mass number

Compound	Mass number/u	Molecular mass/u
$C_{10}H_{20}NO$	170	170.1540
$C_9H_{18}N_2O$	170	170.1415
$C_9H_{16}NO_2$	170	170.1177

The accurate value of the molecular mass enables a distinction to be made between the compounds.

35.9.2 FRAGMENTATION PATTERNS

The molecular ion is formed by the process

$$M \longrightarrow M^+ + e^-$$

The energy of the bombarding electrons is so high compared with the bond energies that bonds in M^+ are broken, and fragments are formed. Only one fragment retains the positive charge. The other fragment is an electrically neutral species with an unpaired electron. Electrically neutral fragments such as $\cdot CH_3$ and CO do not appear as peaks in the spectrum. The ions may fragment further. The way in which a molecular ion fragments depends on the relative stabilities of the species that can be formed. The ion $\cdot CH_3{}^+$ is less stable than the ion $R_3C\cdot^+$ in which the positive charge is less localised.

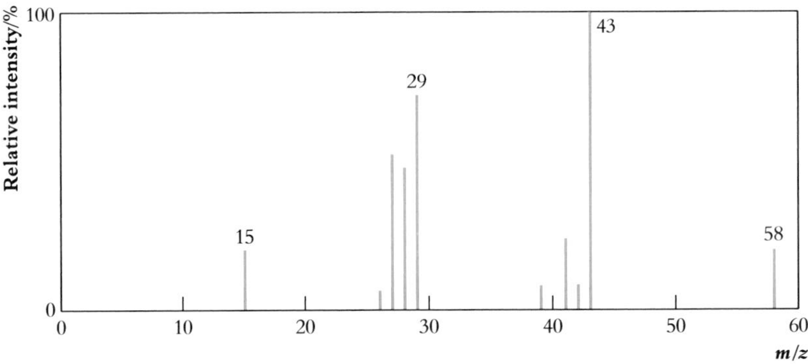

Figure 35.9B The mass spectrum of butane

Figure 35.9B shows the mass spectrum of butane. The peak at 58 is due to the molecular ion, $C_4H_{10}{}^+$. The peak at 15 corresponds to the $CH_3{}^+$ ion, that at 29 to the $C_2H_5{}^+$ ion and that at 43 to the $C_3H_7{}^+$ ion. The molecule has fragmented:

H_3C | CH_2 | CH_2 | CH_3

←15→

←29→

←43→

Figure 35.9C shows the mass spectrum of ethanal. The peak at 44 corresponds to the molecular ion. The peak at 15 corresponds to $CH_3{}^+$ and that at 29 to CHO^+. The molecule has fragmented:

H_3C | C(H)=O

←15→←29→

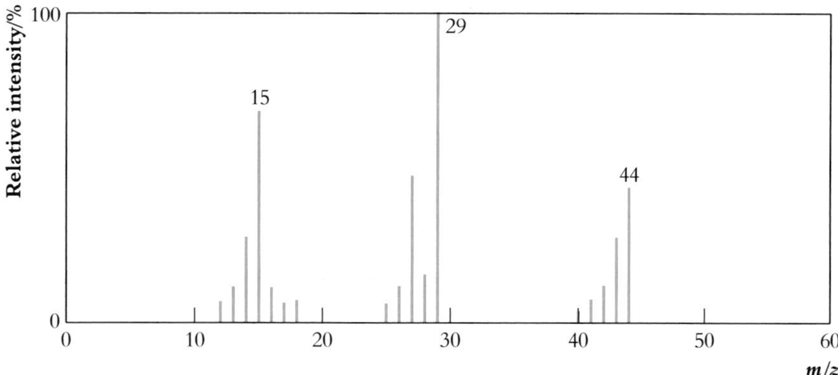

Figure 35.9C The mass spectrum of ethanal

Figure 35.9D shows the mass spectrum of a hydrocarbon. How much can we deduce about its identity? The peak at 77 is very common in aromatic compounds; it corresponds to $C_6H_5^+$. The peak at 106 is probably the molecular ion. Subtracting 77 from 106 leaves 29, and a peak at 29 can be seen. This is likely to be due to $C_2H_5^+$. The peak at 15 corresponds to CH_3^+ which could be formed by the loss of CH_2 from $C_2H_5^+$. Putting together the groups C_6H_5, C_2H_5 and CH_3 gives $C_6H_5CH_2CH_3$, ethyl benzene, of relative molecular mass 106.

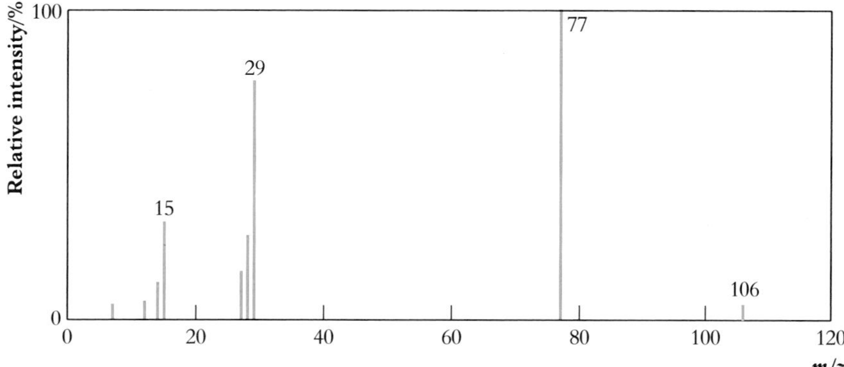

Figure 35.9D The mass spectrum of a hydrocarbon

The molecular ion and the fragmentation pattern can often identify an organic compound. Isomers have the same molecular ion but different fragmentation patterns. However when a spectrum is compared with a library of spectra of known compounds for identification, the operating conditions, including the energy of the ionising electrons, must be quoted.

35.9.3 APPLICATIONS OF MASS SPECTROMETRY

1. Determination of atomic masses and isotopic masses; see § 1.9.1.
2. Determination of molecular masses and identifying compounds from them; see § 35.9.1.
3. Identifying compounds from their molecular masses and fragmentation patterns; see § 35.9.1.
4. Forensic science. A mass spectrum can be obtained on as little as 10^{-12} g. A mass spectrometer is often used in conjunction with gas chromatography [§ 8.7.5]. Specimens found at the scene of a crime can be compared with drugs, fibres etc. found on a suspect.
5. Space exploration. A mass spectrometer on the Viking spacecraft was used with gas chromatography to analyse the atmosphere and soil of Mars.

6. Pollution monitoring. Gas chromatography and mass spectrometry can detect pollutants at low concentrations. An examination of an oil slick will identify the oil field from which it came and link it to the tanker responsible for the pollution.
7. Reaction mechanisms, e.g. esterification; see § 33.8.2.

CHECKPOINT 35.9: MASS SPECTROMETRY

1. Identify the compound which has the mass spectrum shown below.

Figure 35.9E

2. Interpret the mass spectrum shown below.

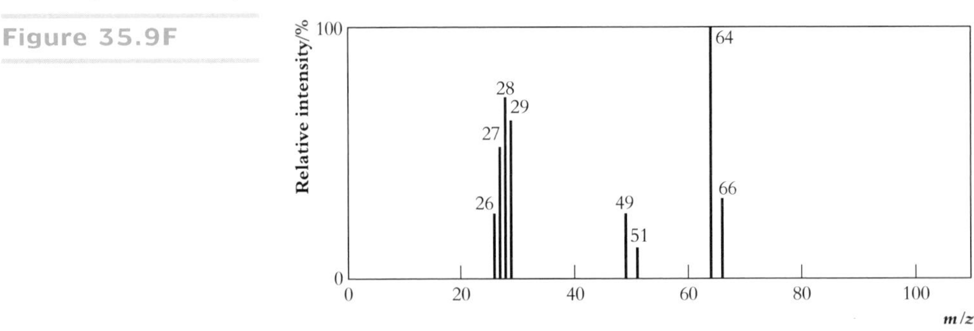

Figure 35.9F

MASS SPECTROMETRY

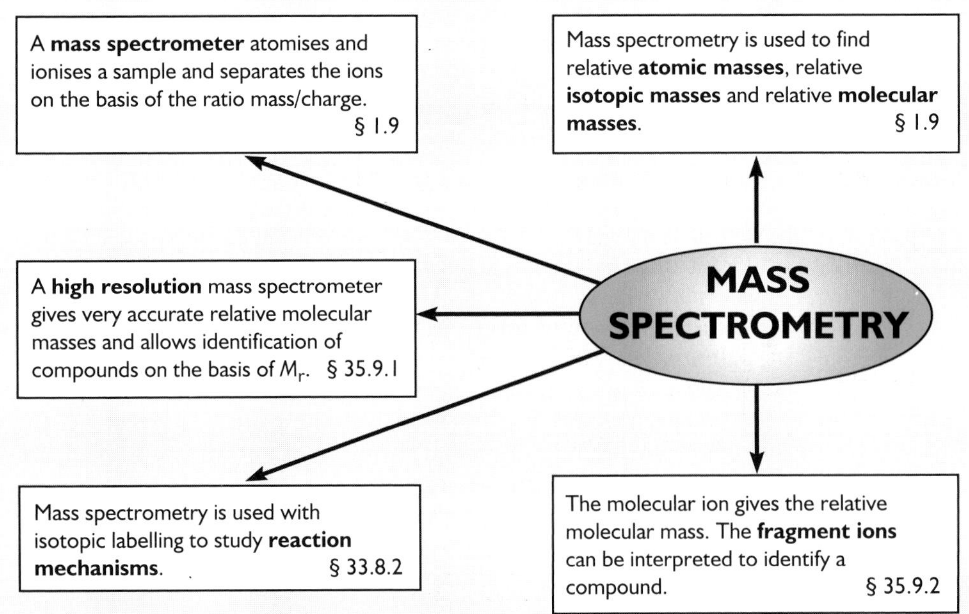

A **mass spectrometer** atomises and ionises a sample and separates the ions on the basis of the ratio mass/charge. § 1.9
Mass spectrometry is used to find relative **atomic masses**, relative **isotopic masses** and relative **molecular masses**. § 1.9
A **high resolution** mass spectrometer gives very accurate relative molecular masses and allows identification of compounds on the basis of M_r. § 35.9.1
MASS SPECTROMETRY
Mass spectrometry is used with isotopic labelling to study **reaction mechanisms**. § 33.8.2
The molecular ion gives the relative molecular mass. The **fragment ions** can be interpreted to identify a compound. § 35.9.2

35.10 NUCLEAR MAGNETIC RESONANCE SPECTROMETRY

35.10.1 NMR IN MEDICINE

Body scanners are NMR spectrometers ...

Many hospitals possess an instrument called a **body scanner**. This is an NMR spectrometer in which a patient can be placed so as to lie inside a large magnet. The patient has to remain still for about 20 minutes while scanning takes place. The machine obtains images called **magnetic resonance images** of soft tissue in any number of planes.

... used to give an image of body tissues ...

Protons in water, carbohydrates, proteins and lipids give different signals. Different parts of the body possess protons in different environments and therefore give different signals. These enable different organs in the body to be differentiated in the magnetic resonance image.

... which can assist in diagnosis

There is no damage to the tissues and no known side effects, so patients can be scanned regularly. The technique has been used to diagnose cancer, multiple sclerosis, hydrocephalus (water on the brain) and other diseases.

35.10.2 SPIN

Atomic nuclei are charged and spin about an axis ...

... giving rise to a magnetic moment

The nuclear spin is quantised ...

... therefore the magnetic moment of the nucleus is quantised

The basis of this powerful analytical technique, nuclear magnetic resonance, is that atomic nuclei can be thought of as tiny magnets. Atomic nuclei spin about an axis. The combination of charge and spin gives rise to a magnetic moment, that is, the nucleus behaves to some extent like a small bar magnet. The nuclear spin is quantised therefore the magnetic moment of the nucleus is quantised. For a proton, ^{1}H, the spin quantum number = $\frac{1}{2}$. Other nuclei which contain an odd number of protons or neutrons or both also have a spin quantum number = $\frac{1}{2}$. Examples are ^{13}C, ^{15}N, ^{19}F and ^{31}P. Two very common nuclei ^{12}C and ^{16}O have zero spin and zero magnetic moment and are invisible in NMR spectrometry. NMR is most widely used to identify ^{1}H nuclei.

Figure 35.10A
Nuclear spins

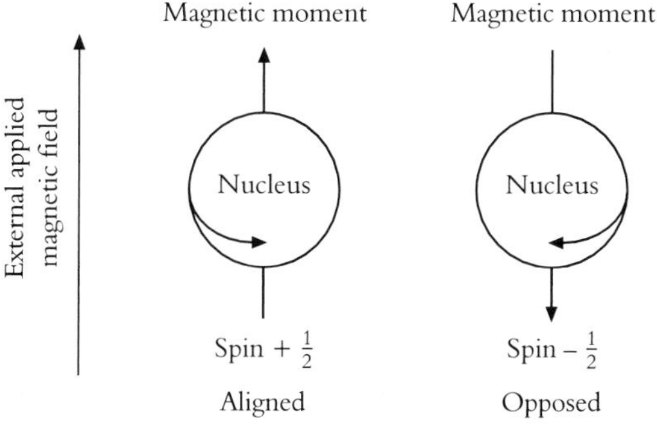

NMR is most widely used to identify ^{1}H nuclei. When a magnetic field is applied, a nuclear spin can align with the field: spin = $+\frac{1}{2}$ ...

... or oppose it: spin = $-\frac{1}{2}$

Figure 35.10B
The splitting of spin states in a magnetic field

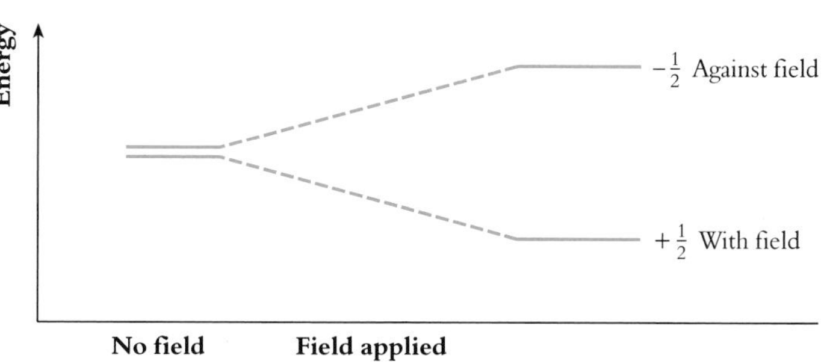

... therefore the spin state splits into two energy levels

A proton ^1H has spin quantum number = $\frac{1}{2}$

The spin states of a certain nucleus have equal energy in the absence of a magnetic field. If a magnetic field is applied, the spin states are no longer of equal energy. The magnetic moments of different nuclei may either align with or oppose the applied magnetic field [see Figure 35.10A]. The result is to split each spin state into two energy levels, $+\frac{1}{2}$ and $-\frac{1}{2}$ [see Figure 35.10B].

Nuclei can be 'brought into resonance' by absorbing energy to move from the lower-energy spin state to the higher-energy spin state. The absorption of energy is called nuclear magnetic resonance. Energy absorbed = hν ...

...ν is in the radiofrequency region

Nuclei with spins that are aligned with the applied field can absorb energy and change their orientation from with the field to against the field. They are described as **flipping**. The absorption of energy that makes the nuclear spin **flip** is called **nuclear magnetic resonance**. The quantity of energy absorbed depends on the identity of the nucleus and the magnitude of the applied field. It is given by

$$E_{\text{absorbed}} = (E_{-1/2} - E_{+1/2}) = h\nu \qquad (h = \text{Planck's constant; § 2.3.1})$$

For typical applied magnetic fields, ν is in the radiofrequency region of the electromagnetic spectrum: the absorbed energy is much smaller than the energy changes associated with electronic, vibrational or rotational changes.

The electromagnetic field required to bring the protons in a substance into resonance can be measured. Since the field depends on the characteristics of the brand of NMR spectrometer in use, a standard is needed. Tetramethylsilane, TMS, $Si(CH_3)_4$, is chosen as a standard. TMS gives a single NMR line because all the hydrogen atoms in it are equivalent. The magnetic field is increased until the exact condition for resonance is attained. Then the nucleus absorbs energy as the nuclear spin 'flips'. As the field is increased, protons in different groups, e.g. —OH and $—CH_3$, 'flip' (see Figure 35.10C). Signals can be detected from a few milligrams of substance.

The field required to bring protons into resonance is compared with that required for TMS, the standard ...

... and expressed as the **chemical shift**, δ

The difference ΔB between the field required to bring the protons in a substance into resonance and that required for TMS is measured. This difference in parts per million of the applied field B_0 is called the **chemical shift**, δ.

$$\text{Chemical shift,} \qquad \delta = \frac{\Delta B}{B_0} \times 10^6$$

where ΔB = difference in field, B_0 = applied field. For TMS, $\delta = 0$ by definition. The value of δ for the protons in most common organic molecules is between 0 and 10.

35.10.3 INTERPRETING SPECTRA

The NMR behaviour is a property of the nucleus. However, protons in different environments, that is different positions in a molecule, have different NMR spectra. The NMR spectrum of ethanol shows three absorption peaks at different values of applied field [Figure 35.10C].

The areas under the peaks are in the ratio 1 : 2 : 3. They correspond to the single —OH group, the two $—CH_2—$ protons and the three $—CH_3$ protons. In other compounds also, protons in different positions in a molecule give peaks at different field strengths. The ratio of the areas of the peaks is the ratio of the numbers of protons of each type.

Figure 35.10C
NMR spectrum of ethanol

The magnetic field needed to bring protons into resonance depends on the environment of the protons in the molecule ...

... e.g. in —CH_2— groups or —CH_3 groups or —OH groups

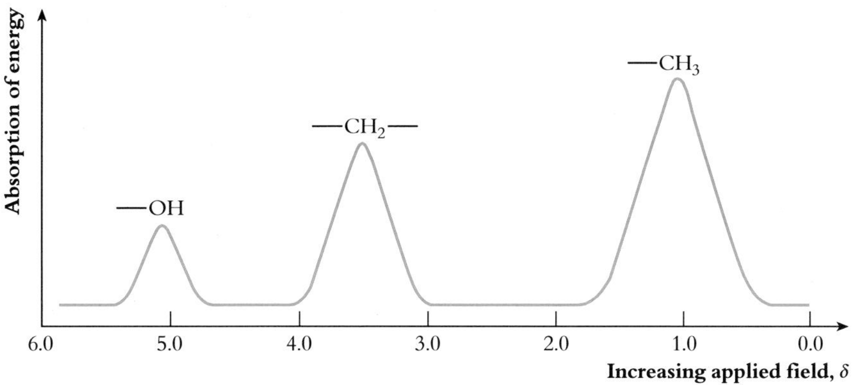

NMR spectrometers draw an integration trace over the spectrum. The height of a step on this trace is proportional to the area of the peak, and therefore to the number of protons responsible for that peak. In Figure 35.10D for ethoxyethane, $C_2H_5OC_2H_5$, the steps are in the ratio 2 : 3 and there are 10 protons so the protons form a set of four and a set of six. These are four protons in two —CH_2— groups and six protons in two —CH_3 groups.

Figure 35.10D
NMR spectrum of ethoxyethane

The area of each peak is a measure of the number of protons in a certain environment ...

... and is registered on the integration trace

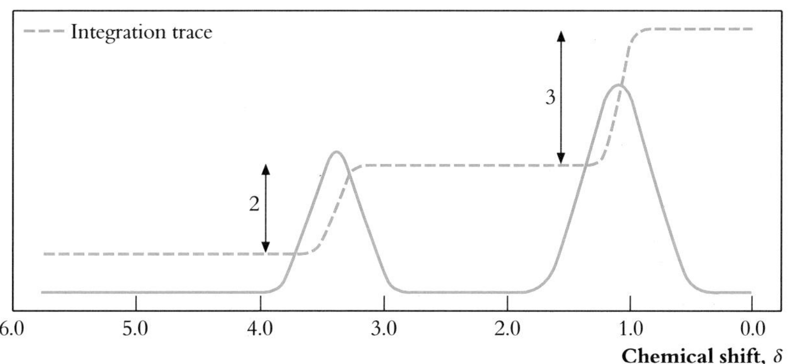

35.10.4 SPIN–SPIN COUPLING (OR SPLITTING)

The spectra shown so far are low-resolution spectra. In high-resolution spectra, some of the peaks seen at low resolution are split into a number of peaks close together. (High resolution means the ability to distinguish between objects or lines that are close together.) In 1,1,1-trichloroethane,

At high resolution, single peaks are resolved into multiple peaks

```
      Ha  Hb
      |   |
  Cl—C—C—Cl
      |   |
      Cl  Hb
```

there are protons of two types H_a and H_b. The low-resolution spectrum shows two peaks. The high-resolution spectrum is shown in Figure 33.10E. It contains five peaks: a triplet and a doublet. The three peaks of the triplet are in the intensity ratio 1 : 2 : 1. The two peaks in the doublet are equal in intensity.

The splitting of the peaks can be explained. The magnetic field which the protons H_b experience is due to the applied magnetic field and also the local field due to the magnetic moment of proton H_a. Some molecules have the local field of H_a aligned with the applied field and others have it in opposition. Thus two slightly different energy states arise, and the peak is split into a **doublet**. In the case of the proton H_a, the applied field is modified by local fields produced by

Figure 35.10E
High-resolution spectrum of 1,1,2-trichloroethane

The applied field which a certain proton experiences is modified depending on whether the field is aligned with, or opposed to, the local magnetic field of adjacent protons

both H_b protons. These local fields may both align with the applied field, or both may oppose it or one may align with the field and the other oppose it. In this way the peak becomes a **triplet** (with intensities in the ratio 1 : 2 : 1).

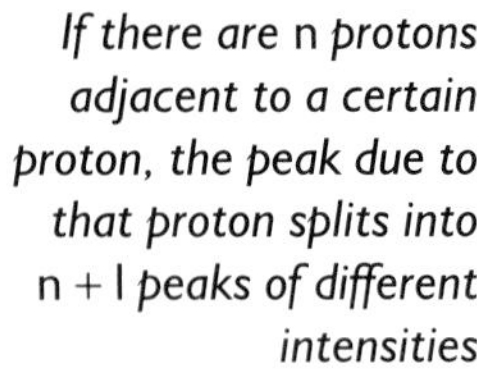

If there are n protons adjacent to a certain proton, the peak due to that proton splits into n + 1 peaks of different intensities

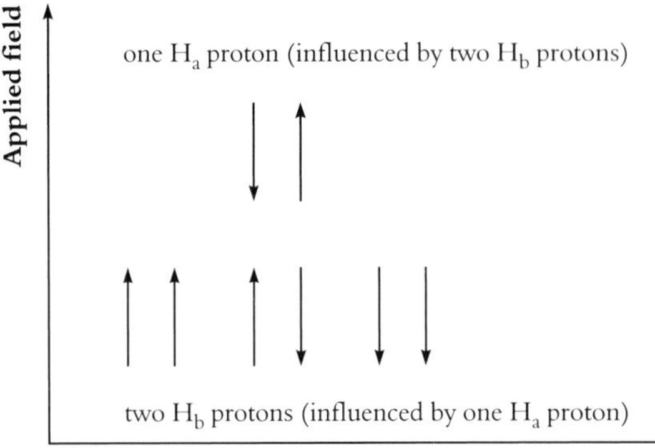

A peak due to n adjacent protons is split into $n + 1$ parts.

Figure 35.10F
High-resolution NMR spectrum of ethoxyethane

The area of each peak is a measure of the number of protons in a certain environment ...

... and is registered on the integration trace

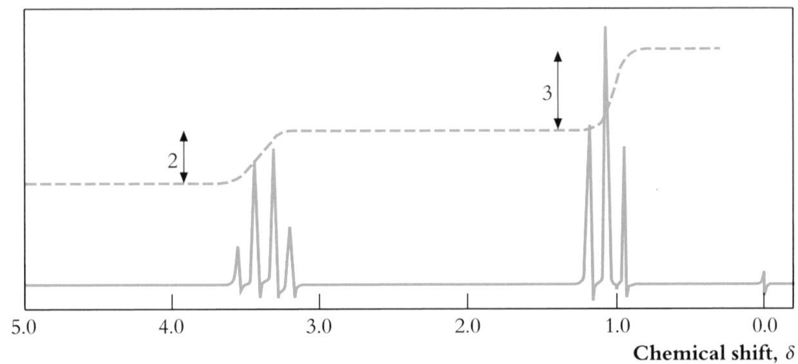

In the high-resolution spectrum of ethoxyethane, $CH_3CH_2OCH_2CH_3$ [Figure 35.10F], the $—CH_2—$ and $CH_3—$ groups produce different absorptions. The field experienced by the protons in each $CH_3—$ group is modified by the two protons in the $—CH_2—$ group adjacent to it. The result is a triplet of peaks. The integration trace shows that 3/5 of the protons produce this triplet, confirming that this absorption is produced by the $CH_3—$ protons. The $—CH_2—$ protons are influenced by the three protons on the adjacent $CH_3—$ group, giving a quartet (with relative intensities 1 : 3 : 3 : 1).

35.10.5 IDENTIFYING LABILE PROTONS

The NMR spectrum of ethanol [Figure 35.10C] shows three peaks. These are due to the —OH group, the $—CH_2—$ group and the $CH_3—$ group. When ^2H_2O (deuterium oxide) is added to ethanol, the peak due to —OH disappears. The reason is that the H of the —OH group is a 'labile' proton, and proton exchange takes place between the —OH group and ^2H_2O.

$$C_2H_5OH + {}^2HOH \rightleftharpoons C_2H_5O^2H + HOH$$

Since ^2H (deuterium) does not absorb in the NMR region of the spectrum, the peak due to the H of —OH disappears. The use of deuterium oxide to identify labile protons is an important tool in NMR spectroscopy [see Checkpoint 35.10, Question 1].

Some proton chemical shift, δ, values are tabulated. You do not need to memorise them as you will be able to refer to a table when you need to interpret spectra.

Table 35.10
Values of chemical shift, δ/ppm

When you need to interpret NMR spectra you will be able to refer to a table of δ values. You do not need to memorise them!

Type of proton	Chemical shift, δ/ppm
$R—CH_3$	0.85–0.95
$R—NH_2$	1.0
$R—CH_2—R$	1.3
$R_3C—H$	2.0
$CH_3—CO_2R$	2.0
$R—COCH_3$	2.1
$C_6H_5—CH_3$	2.3
$R—C{\equiv}C—H$	2.6
$R—CH_2—Hal$	3.2–3.7
$R—O—CH_3$	3.8
$R—O—CH_2R$	4.0
$R—O—H$	3.5–5.5
$RCH{=}CH_2$	4.9–5.9
$C_6H_5—OH$	7.0
$C_6H_5—H$	7.3
$R—CHO$	9.7
$R—CO_2H$	11.0–11.7

CHECKPOINT 35.10: NMR

1. Identify the peaks in the NMR spectrum below, and suggest the identity of the compound which gives rise to these peaks. Explain why the peak at $\delta = 5.2$ is absent when ^2H_2O is present.

Figure 35.10G

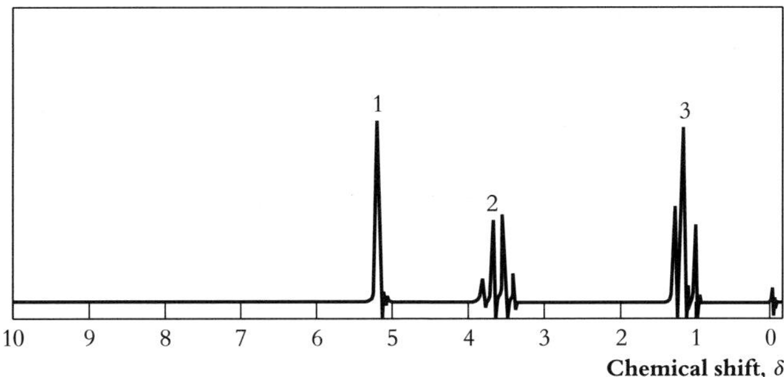

2. A compound contains the elements C, H, N. Suggest what groups may give rise to the peaks in its NMR spectrum. The peak at $\delta = 1.0$ is absent in ^2H_2O. Explain how this happens. Suggest the identity of the compound.

Figure 35.10H

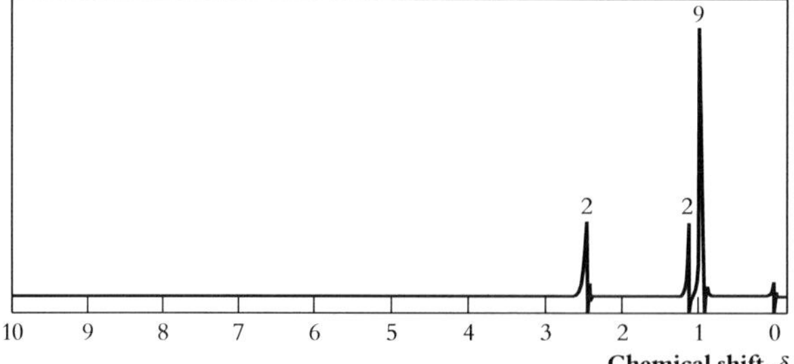

3. A compound contains the elements C, H, O. Identify the peaks in its NMR spectrum. Explain why the peak at $\delta = 11.7$ is absent when ^2H_2O is present.

Figure 35.10I

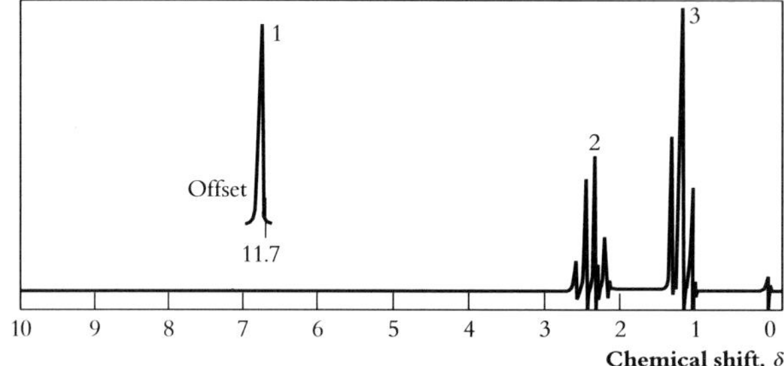

4. A compound is aromatic and contains only carbon and hydrogen. Identify the peaks in its NMR spectrum, and suggest what the compound may be.

Figure 35.10J

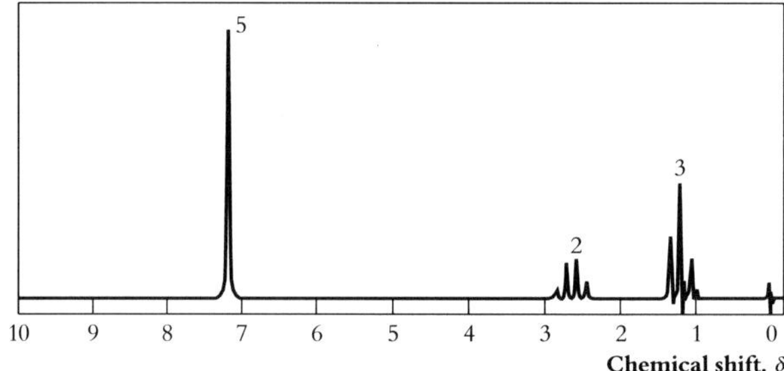

5. Identify the peaks in the NMR spectrum below, and suggest what the compound may be.

Figure 35.10K

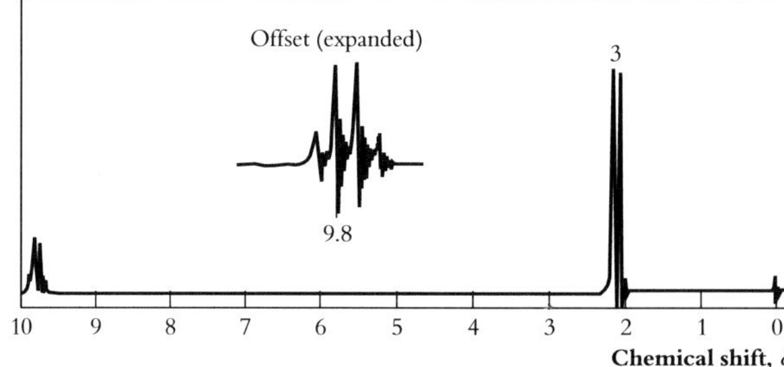

6. Identify the peaks in the NMR spectrum below and suggest what the compound may be.

Figure 35.10L

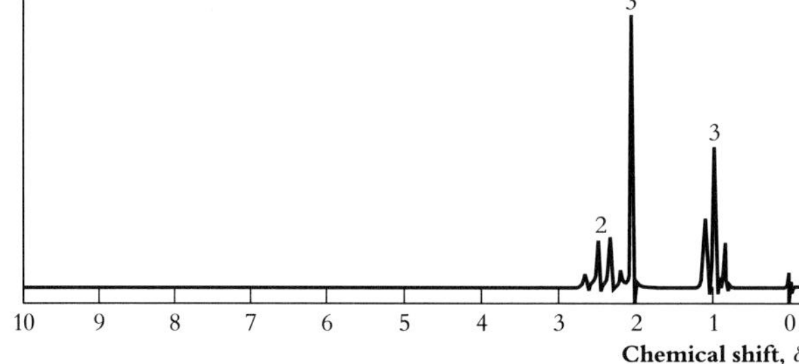

NUCLEAR MAGNETIC RESONANCE SPECTROMETRY

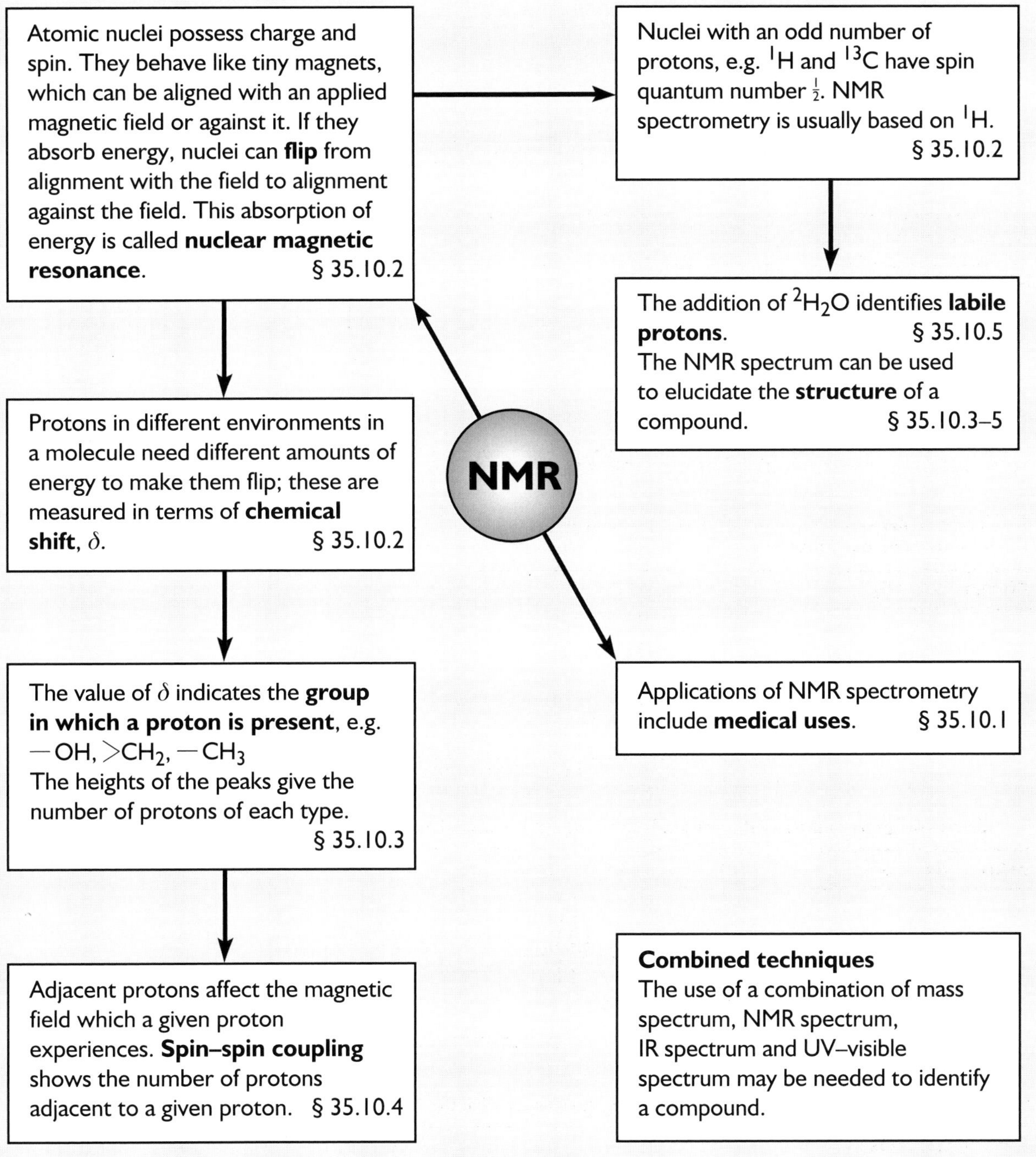

Atomic nuclei possess charge and spin. They behave like tiny magnets, which can be aligned with an applied magnetic field or against it. If they absorb energy, nuclei can **flip** from alignment with the field to alignment against the field. This absorption of energy is called **nuclear magnetic resonance**. § 35.10.2
Nuclei with an odd number of protons, e.g. ^{1}H and ^{13}C have spin quantum number $\frac{1}{2}$. NMR spectrometry is usually based on ^{1}H. § 35.10.2
The addition of $^{2}H_2O$ identifies **labile protons**. § 35.10.5
The NMR spectrum can be used to elucidate the **structure** of a compound. § 35.10.3–5
NMR
Protons in different environments in a molecule need different amounts of energy to make them flip; these are measured in terms of **chemical shift**, δ. § 35.10.2
The value of δ indicates the **group in which a proton is present**, e.g. $-OH$, $>CH_2$, $-CH_3$
The heights of the peaks give the number of protons of each type. § 35.10.3
Applications of NMR spectrometry include **medical uses**. § 35.10.1
Adjacent protons affect the magnetic field which a given proton experiences. **Spin–spin coupling** shows the number of protons adjacent to a given proton. § 35.10.4

Combined techniques

The use of a combination of mass spectrum, NMR spectrum, IR spectrum and UV–visible spectrum may be needed to identify a compound.

QUESTIONS ON CHAPTER 35

1. *(a)* Explain why compounds absorb energy in the IR region of the spectrum.

(b) The spectra shown below were obtained from two compounds **A** and **B** of formula $C_4H_{10}O$. Deduce the structures of **A** and **B**, explaining how you have used the information in the spectra in your deductions.

(c) What further spectroscopic evidence would enable a positive identification of **A** and **B** to be made?

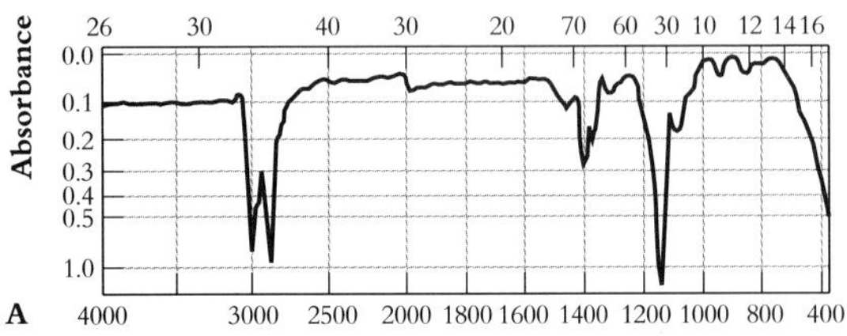

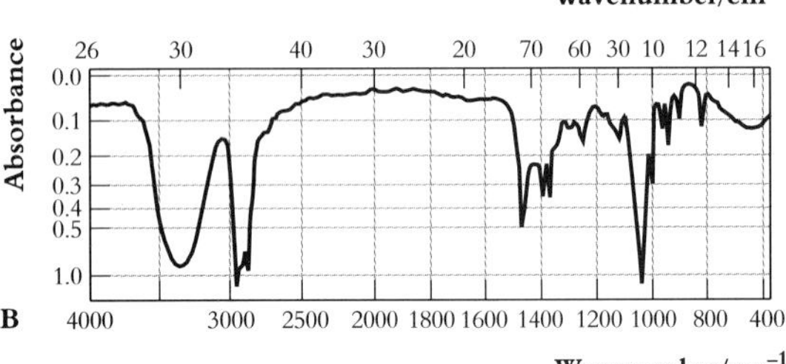

What can you say about the IR spectra of the organic compounds shown below? It is not possible to identify each compound, but you will detect functional groups and be able to say whether the compound is aliphatic or aromatic.

2.

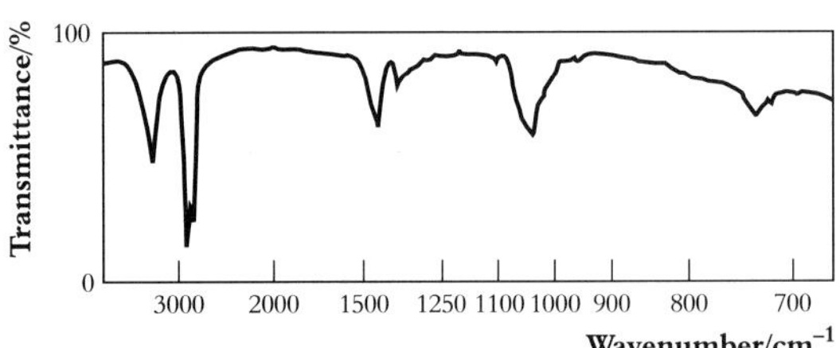

3. A liquid of M_r 98.

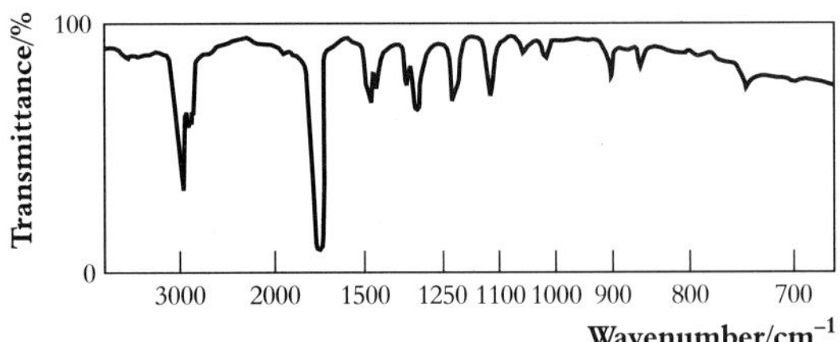

4.

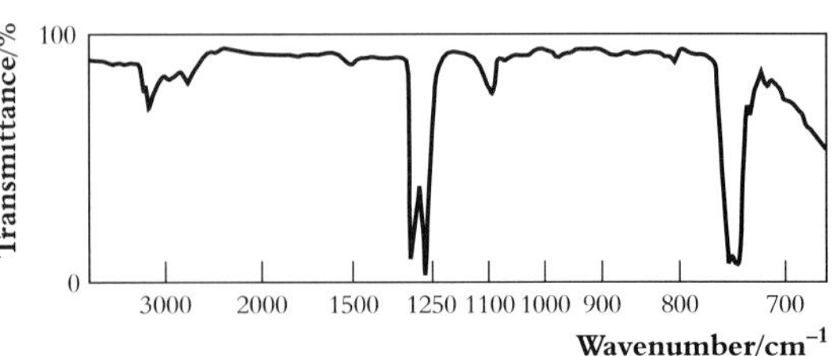

5.

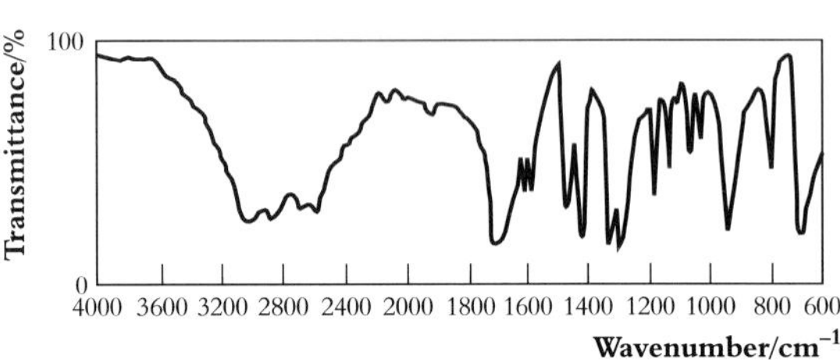

6.

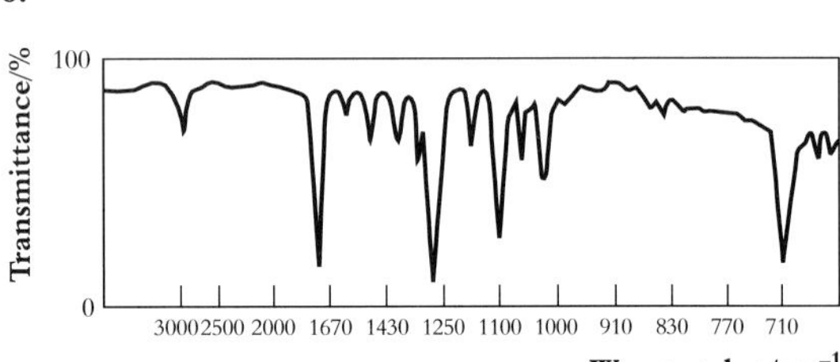

7. There are four esters of formula $C_4H_8O_2$.

(a) Write the formulae of the four compounds.

(b) Which of the four gives the mass spectrum below?

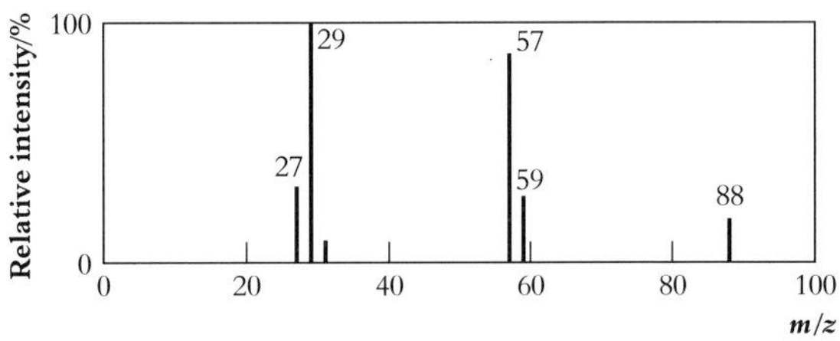

8. The figure shows a simplified version of the mass spectrum of ethanol, C_2H_5OH. Explain the origin of the six peaks.

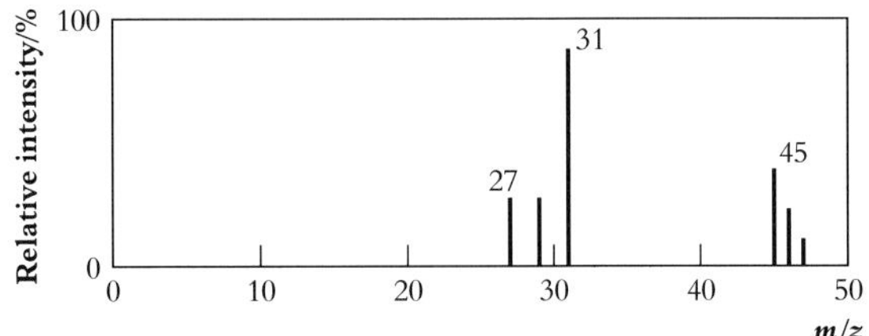

9. Identify the compound $C_8H_8O_2$ from its mass spectrum.

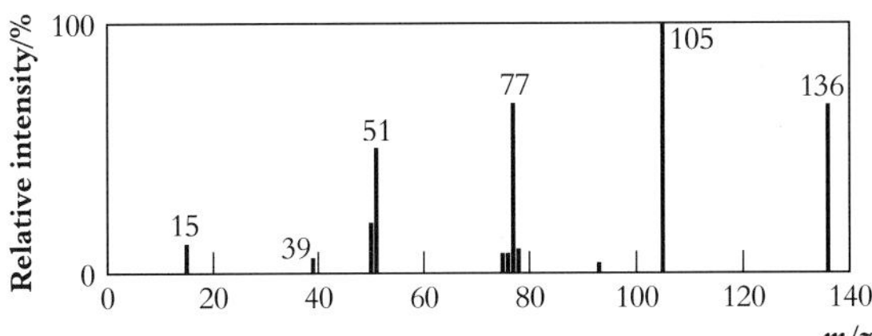

10. Identify the compound C_3H_6O from its mass spectrum.

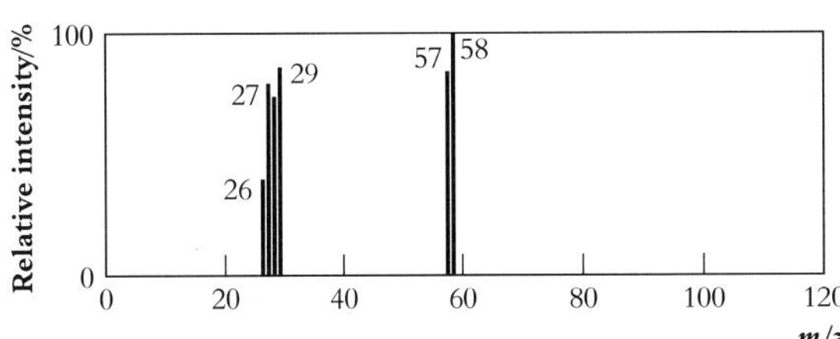

11. The mass spectrum of a compound $C_6H_{14}O$ is shown. From the smell and the high volatility the compound appears to be an ether. Identify the peaks in the spectrum and identify the compound as far as you can.

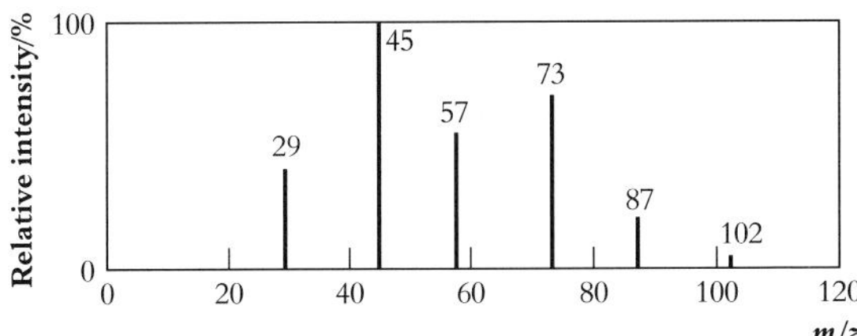

12. The NMR spectrum of an isomer of ethanol, is shown below.

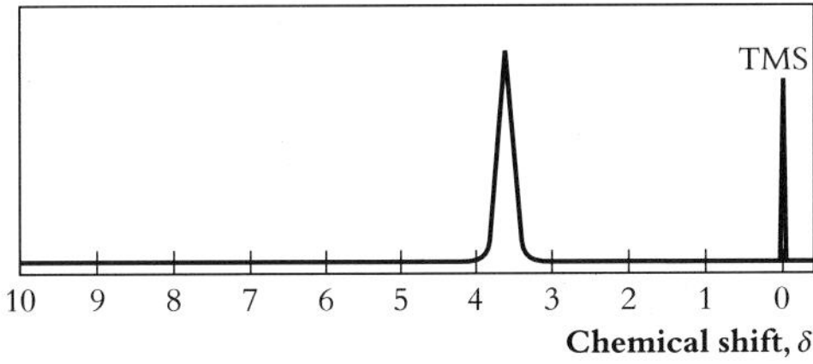

(*a*) Identify this compound and explain why its NMR spectrum is different from that of ethanol.
(*b*) Why is TMS used as a reference point?

13. Identify the ester with the following NMR spectrum.

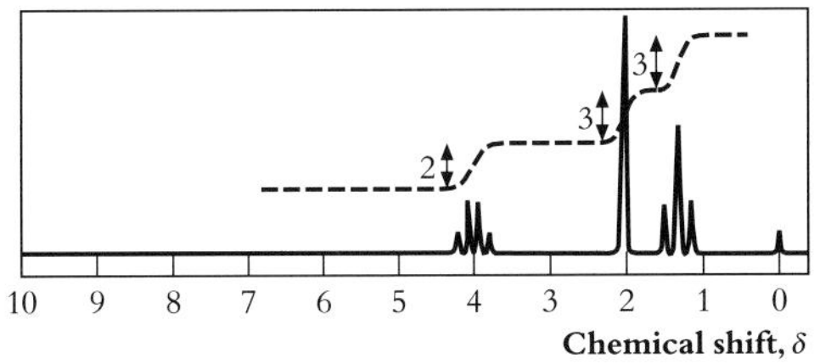

14. The proton n.m.r. spectrum of compound **A**, $C_3H_8O_2$, which can be made from epoxyethane and methanol, is shown below. The integration trace when measured gives the ratio 1.4 to 1.4 to 2.1 to 0.7

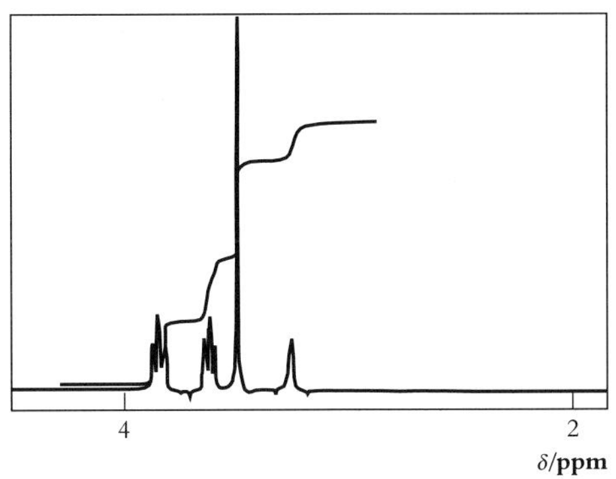

(*a*) How many different types of proton are present in compound **A**? **1**
(*b*) What is the actual ratio of the numbers of each type of proton? **1**
(*c*) By considering the splitting of the peaks in the above spectrum, deduce the structure of compound **A**. **5**

7 *NEAB (AL)*

15. (*a*) The infrared spectra of two cycloalkanes, cyclopentane and cyclohexane, and a table of i.r. absorption data are shown below.

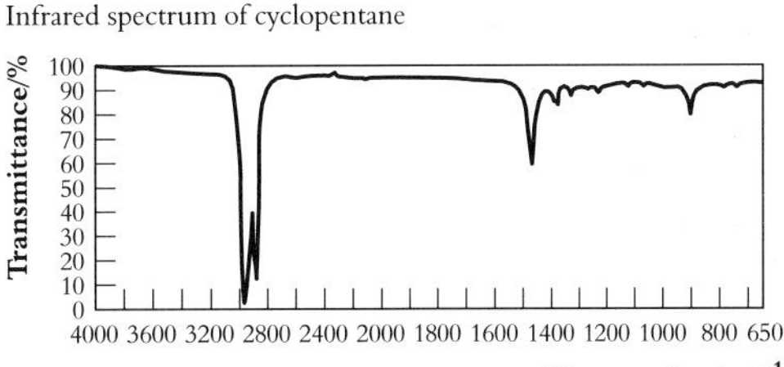

Infrared spectrum of cyclohexane

Transmittance/%

100 90 80 70 60 50 40 30 20 10 0

X

4000 3600 3200 2800 2400 2000 1800 1600 1400 1200 1000 800 650

Wavenumber/cm^{-1}

Bond	Wavenumber/cm^{-1}
C—H	2850–3300
C—C	750–1100
C=C	1620–1680
C=O	1680–1750
C—O	1000–1300
O—H (alcohols)	3230–3550
O—H (acids)	2500–3000

(i) Use the table of i.r. absorption data to identify the bond which results in the absorption labelled **X**.
(ii) Explain how 'fingerprinting' could be used to show that an unknown cycloalkane was cyclopentane and not cyclohexane. **3**

(b) (i) Explain why the proton (^1H) n.m.r. spectrum of cyclohexane, C_6H_{12}, and that of dichloromethane, CH_2Cl_2, each contain only one peak.
(ii) The proton n.m.r. spectrum of a mixture of equal numbers of moles of cyclohexane and dichloromethane shows two peaks. Explain why the areas under the two peaks are in the ratio 6 : 1.

(c) (i) Write an equation for the formation of chloromethane from methane and chlorine and give a necessary condition for the reaction to occur. **2**
(ii) Give equations for the two propagation steps in the mechanism for the reaction. **4**

(d) The mass spectrum of chloromethane shows two molecular ion peaks because chlorine exists as two isotopes ^{35}Cl and ^{37}Cl. Predict the number of molecular ion peaks for dichloromethane and deduce their m/z values. **3**

12 *NEAB (AS/AL)*

16. Keto-ethers contain both the —C(=O)— and

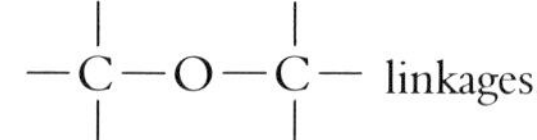

linkages.

The proton n.m.r. spectrum of keto-ether **A**, $C_6H_{12}O_3$, is shown below.

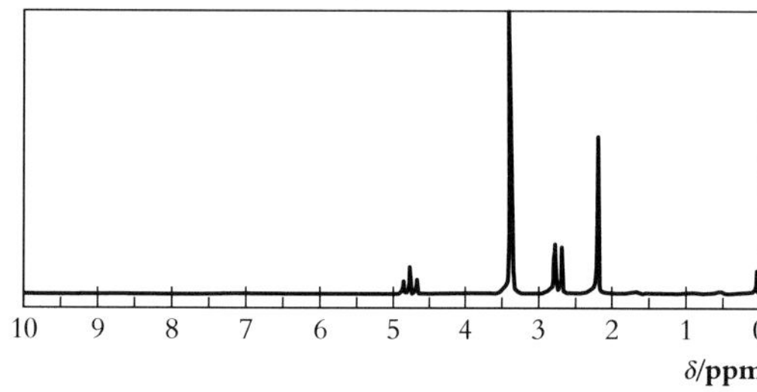

The measured integration trace gives the ratio 0.6 to 3.6 to 1.2 to 1.8 for the peaks at δ 4.8, 3.4, 2.7 and 2.2, respectively.

Refer to the spectrum above and to the information given in the table below in order to answer the following questions.

(a) How many different types of proton are present in compound **A**? **1**

(b) What is the actual ratio of the numbers of each type of proton? **1**

(c) What type of proton is responsible for the peak at δ 2.2? **1**

(d) What can be deduced from the splitting of the peaks at δ 4.8 and δ 2.7? **2**

(e) Suggest the structure of compound **A**. **2**

Type of proton	δ/ppm
RCH_3	0.7–1.2
R_2CH_2	1.2–1.4
R_3CH	1.4–1.6
$RCOCH_3$	2.1–2.6
$ROCH_3$	3.3–3.9
$RCH(OR)_2$	4.4–5.0

7 *NEAB (AL)*

17. *(a)* Explain what is meant by the term *fragmentation* as used in mass spectrometry. Include in your answer a general equation for the fragmentation of a molecular ion into two new species. **5**

(b) Fragmentation of the molecular ion of a ketone **B**, $C_5H_{10}O$, leads to a mass spectrum which has a major peak at $m/z = 57$. Fragmentation of an isomeric ketone **C**, $C_5H_{10}O$, results in a major peak at $m/z = 43$ and a minor peak at $m/z = 71$. Ketone **C** could have one of two alternative structures.

Deduce the structure of ketone **B**. Deduce the two possible structures for ketone **C** and explain how proton n.m.r. spectroscopy could be used to distinguish between these two isomeric structures. **10**

(c) An aromatic hydrocarbon **D**, C_7H_8, is converted into compound **E**, $C_9H_{10}O$, on treatment with ethanoyl chloride in the presence of $AlCl_3$. When **E** is treated with $NaBH_4$, it is converted into **F**, $C_9H_{12}O$. Compound **G** is formed when **F** is warmed with concentrated H_2SO_4. Note that compound **E** is formed as a mixture of three isomers.

Compound **E** has a strong absorption band in the infrared at 1685 cm^{-1}, compound **F** has a broad absorption at 3340 cm^{-1} and compound **G** has an absorption band close to 1630 cm^{-1}.

(i) Show how the information provided in the question and the data in the table in question 15 can be used to deduce structures for compounds **D**, **E**, **F** and **G**, respectively. Choose **one** of the isomers of compound **E** to show the formation of compounds **F** and **G**.
(ii) Name the types of reaction taking place and outline a mechanism for the formation of compound **G**. **15**

30 *NEAB (AL)*

18. The proton n.m.r. spectrum of a chloroalkyl ketone, **A**, C_5H_9ClO, is shown below.

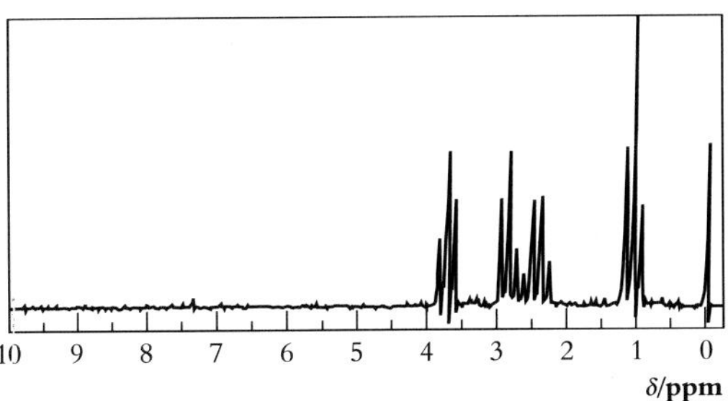

The measured integration trace gives the ratio 1.2 to 1.2 to 1.2 to 1.8 for the peaks at δ 3.8, 2.8, 2.4 and 1.1, respectively.
Refer to the spectrum above and to the information given in the data in question 15 as necessary, in order to answer the following questions.

(a) How many different types of proton are present in compound **A**? **1**

(b) What is the actual ratio of the numbers of each type of proton? **1**

(c) The peaks at δ 2.4 and δ 1.1 arise from the presence of an alkyl group. Identify the group and explain the splitting pattern. **3**

(d) What can be deduced from the splitting of the peaks at δ 3.8 and δ 2.8? **1**

(e) Deduce the structure of compound **A**. **1**

7 *NEAB (AL)*

19. *(a)* The low resolution ^{1}H NMR spectrum of compound **A**, with the molecular formula C_4H_8O, is shown in the figure.

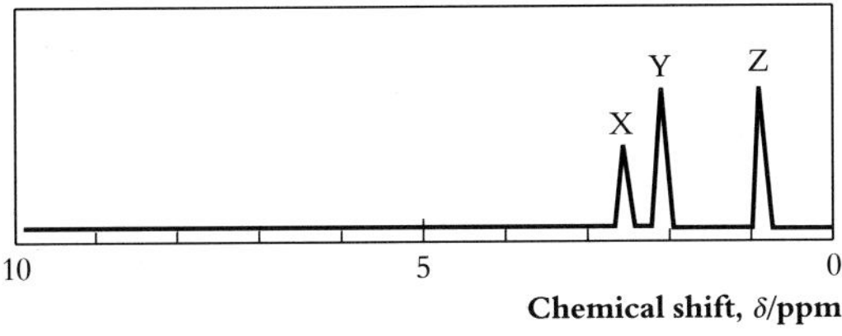

Type of proton	Chemical shift/ppm
R—CH_3	0.9
R—CH_2—R	1.4
R—CO—CH_3	2.1
R—CH_2—CO—	2.5
R—CHO	9.7

(i) Use the data given in the table to deduce the type of proton responsible for each peak in the figure. **3**

(ii) Draw the graphical formula of compound **A**. **1**

(iii) The ratio of the areas under the peaks **X**, **Y** and **Z** is 2 : 3 : 3. Suggest an explanation for this. **1**

(b) Butanal is a structural isomer of compound **A**.

(i) Explain the term *structural isomerism*. **2**

(ii) Draw the graphical formula of butanal. **1**

(iii) State **two** ways in which you would expect the ^{1}H NMR spectrum for butanal to differ from that of compound **A**. **2**

(c) Compound **A** can be obtained by the oxidation of butan-2-ol, $CH_3CH_2CH(OH)CH_3$, which exists as optical isomers.

(i) Explain the term *optical isomers*. **2**

(ii) Draw graphical formulae to show the spatial distribution of the groups in each optical isomer. **2**

(iii) Explain why butan-1-ol, $CH_3CH_2CH_2CH_2OH$, does not exhibit this type of isomerism. **1**

15 *AEB (AL) 1998*

36

SOME GENERAL TOPICS

This chapter deals with a number of topics which span different chapters. Students often complain of difficulty in getting from compound **A** to compound **D** via **B** and **C**; the pages on synthetic routes are an attempt to tackle this difficulty. The section 'What are these reagents used for?' is a summary of the plethora of different reagents and conditions which appear over the top of arrows, and which students find onerous to remember if tackled separately. There is a selection of questions on topics which span chapters, such as methods of distinguishing between compounds, reaction mechanisms, comparisons between functional groups and the identification of 'mystery' compounds from descriptions of their reactions.

36.1 DRUGS WHICH ALTER BEHAVIOUR

36.1.1 STIMULANTS

Amphetamines

Amphetamines ('uppers') stimulate the CNS. The once widely used drug benzedrine is addictive

Amphetamines (often called 'uppers') stimulate the central nervous system in a similar manner to adrenalin. They make the user feel less tired and more alert. Amphetamines were taken by some students when they wanted to stay awake all night to study for an exam and by some long distance lorry drivers. The drugs were found to have another effect: people became **addicted** to them. The results of using the drugs over a period of time are **dependence** and **tolerance** (see below). The drugs are now strictly controlled.

$$C_6H_5-CH_2-\underset{\large CH_3}{\overset{\large H}{C}}-\overset{\large H}{N}-H$$

Benzedrine, an amphetamine

Note the meanings of these terms:

Dependence on a drug may be physical, psychological or both

Dependence The user experiences withdrawal symptoms if he or she stops using the drug to which he or she has become addicted. Dependence may be physical or psychological or both. If the dependence is physical, withdrawal may lead to very severe symptoms. If the dependence is psychological, i.e. **habituation**, the user finds life unbearable unless he or she is experiencing the effects of the drug.

Tolerance is the user's need for increasing doses to produce the same 'high'

Tolerance The body becomes increasingly able to tolerate a drug without ill effects. The user requires increasing amounts of the drug in order to experience the effects.

The most serious drug problems arise with depressants of the central nervous system: the narcotics (heroin, morphine, etc.), the barbiturates, tranquillisers and alcohol.

Nicotine

Nicotine is a toxic substance in tobacco. Smokers build up a tolerance to the drug. The tar which condenses from cigarette smoke is carcinogenic

Nicotine is found in the leaves of tobacco. Nicotine was isolated in 1828 and named after a French ambassador, Jean Nicot, who was convinced that tobacco had medicinal uses. In fact, in its pure form, nicotine is an extremely toxic drug which acts as quickly as cyanide. Two to three drops of pure nicotine (about 60 mg) will kill if placed on the tongue. Nicotine has a complicated action on the human body. It stimulates the central nervous system, induces vomiting and diarrhoea, and first stimulates, then inhibits glandular secretions. Non-smokers can absorb only 4 mg of nicotine before nausea and vomiting begin. Smokers build up a tolerance: they can absorb twice as much without noticeable effects. The smoke from one cigarette may contain about 6 mg of nicotine, of which 0.2 mg is absorbed into the body. (How many cigarettes would it take to make a person ill if he or she had never smoked before?)

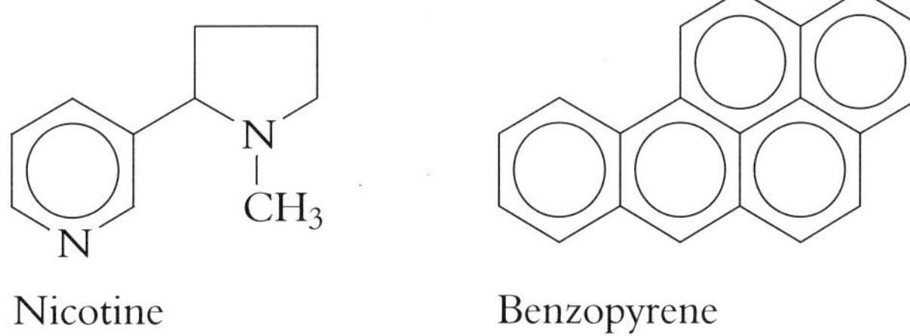

Nicotine Benzopyrene

Cigarette smoke condenses to form tar. The lungs of a smoker receive about 100 g of tar a year. This tar has been shown to be carcinogenic in all kinds of tissue. In laboratory tests on animals, it has induced cancer in epithelial tissues and in connective tissues, e.g. skin and bone. Known carcinogens in tobacco tar include benzopyrenes.

Cocaine

Cocaine stimulates the CNS. It has some use in medicine. People use it to reduce fatigue, but it gives rise to physical symptoms and psychological dependence

Cocaine, also called 'coke' and 'snow', is a fluffy white powder obtained from the leaves of the coca plant, which grows in South America, especially in Peru and Bolivia. It is used as a local anaesthetic, for example in the nose and throat. Cocaine stimulates the central nervous system even more strongly than amphetamines do. It quickens reflexes, reduces fatigue and makes the user feel exhilarated without appearing intoxicated. It has been used by some show business people to improve the sparkle of their performance. Although it is even more expensive than heroin, cocaine is popular among drug users because it

Cocaine

does not result in tolerance. However, it does cause nausea, weight loss, insomnia and psychological dependence. Chemists have tried to synthesise compounds with the local anaesthetic qualities of cocaine but without its other effects. Xylocaine® and Novocaine® have resulted from this research. They are used in dentistry [see § 29.1].

36.1.2 HALLUCINOGENS

LSD

LSD causes hallucinations. Dependence is psychological rather than physical

LSD, short for lysergic acid diethylamide, is a **hallucinogen**. It causes changes in perception, especially visual perception, with vivid, often brightly coloured hallucinations. No physical dependence develops, but bizarre mental effects and permanent personality changes sometimes occur. LSD works by disrupting the transmission of nerve impulses to the brain.

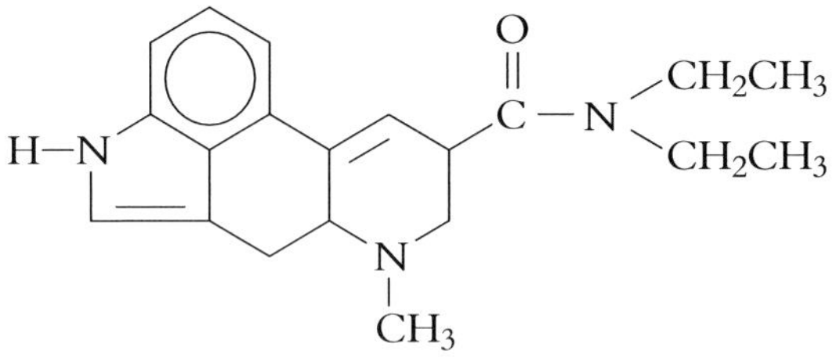

O
C—N
CH_2CH_3
CH_2CH_3
H—N
N
CH_3

LSD

Marijuana

Marijuana is the dried leaves, flowering parts, stems and seeds of the Indian hemp plant, *Cannabis*. The stems yield tough fibres that are used for making rope. Marijuana is often called 'cannabis', 'weed', 'grass', 'pot' or 'Mary Jane'. The dried resin from the flowering part of the plant is called **hashish**. It is more potent than marijuana.

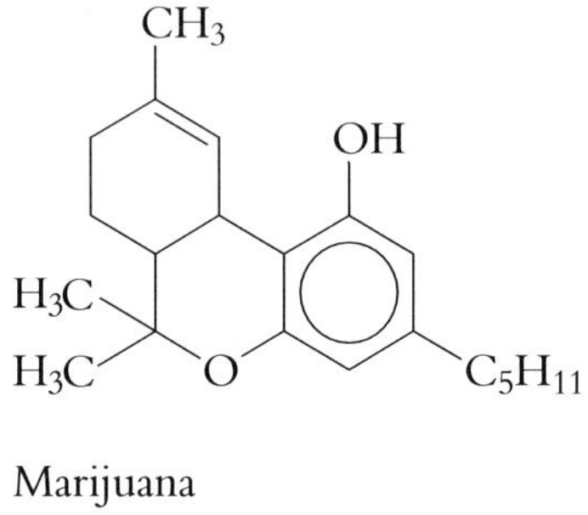

CH_3
OH
H_3C
H_3C
O
C_5H_{11}

Marijuana

Marijuana is a hallucinogen. Users develop psychological dependence. Tar from marijuana cigarettes is carcinogenic. Many marijuana users switch to heroin for a bigger 'high'

Marijuana is smoked in pipes and cigarettes. Its effects last two to three hours, producing a feeling of well-being, excitement, alterations in appreciation of time and space and hallucinations. There is controversy over its safety. It is not regarded as addictive but it can produce psychological dependence. There is some evidence that long-term users of marijuana are apathetic and sluggish. The psychological dependence shows as restlessness, anxiety, irritability and insomnia. Driving performance, work performance and relationships with other people are all affected by marijuana use. Regular use of marijuana also suppresses the body's immune response and makes the user more susceptible to disease. Marijuana cigarettes contain 50% more tar than ordinary cigarettes, and smoking them affects the bronchial tract and lungs. After a while, many marijuana users crave a better or a different sort of 'high', and turn to other drugs. Suppliers are keen to get their customers to switch from marijuana to heroin so that they will quickly become addicted to heroin.

36.1.3 DEPRESSANTS

The opiates

Heroin, morphine and codeine are analgesics (painkillers). Heroin and morphine are narcotics: they produce mental fogginess. Both are addictive drugs

Analgesics are substances that relieve pain; we call them 'painkillers'.

Narcotics are analgesics which also produce euphoria, a feeling of peace and tranquillity. They are addictive. Heroin and morphine are narcotics. Heroin is the more dangerous of the two because it causes rapid addiction. Heroin, morphine and codeine (which is not addictive) are called opiates because they are obtained from opium. Opium is the residue obtained by evaporating the juice of the opium poppy, which is grown in the Orient and the Middle East. It contains 10% morphine and 5% codeine. Heroin is made from morphine.

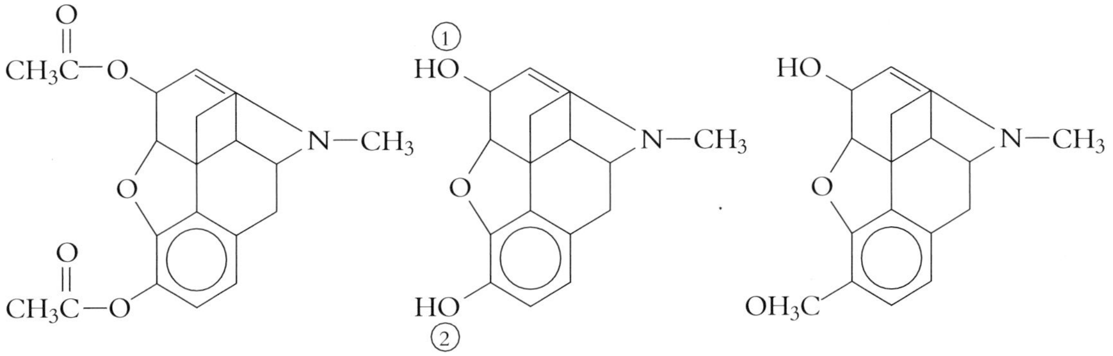

Heroin — Morphine — Codeine

Morphine has some use in medicine as an analgesic, but it is addictive. Codeine is widely used

Morphine is a wonderful pain reliever; it is used in cases of radical surgery and on the battlefield. Morphine is 50 times as potent as aspirin. It was widely used in the American Civil War. Sadly, it had another effect: 100 000 soldiers became addicted to it. People who take morphine suffer from changes in mood and mental fogginess.

Codeine is only one-sixth as effective as morphine, but it does not cause addiction. While morphine must be injected, codeine can be taken by mouth. Codeine is the least potent of the opiates; it is used as a cough suppressant.

Heroin is never used in medicine. It rapidly causes addiction. Users inject the drug, and infectious diseases spread through shared needles

Heroin is far more potent than morphine and much more addictive. As heroin depresses the central nervous system, it causes drowsiness, respiratory depression (which is why an overdose causes death) and decreased gastrointestinal movement which leads to constipation, nausea and vomiting. Heroin is rejected for use by the medical profession. It is very popular with 'drug pushers' because users very soon become addicted. The cost of supporting the heroin habit is about £100 a day. The high price encourages the drug user to administer the drug in the most effective manner: by intravenous injection 'mainlining'. The practice of giving injections under insanitary conditions and sharing needles leads to the spread of infectious hepatitis and AIDS. It is estimated that 60 000 people in Britain are hooked on addictive drugs such as heroin.

Treatments for heroin addiction include counselling, group therapy and transfer to methadone dependence

There are two ways of treating heroin addiction. One is a programme of withdrawal under medical supervision with the support of a counsellor and possibly support from a group of other users who are trying to stop the habit. The other is to substitute methadone. Like the opiates, methadone depresses the central nervous system, but it does not cause drowsiness or euphoria. It can be taken orally. An addict who switches to methadone is able to perform a job and return to a normal way of life. Methadone is addictive: addicts transfer their dependence from heroin to methadone. Since methadone is prescribed by a doctor, users do not have to steal to support their habit. Withdrawal from methadone is as difficult as withdrawal from heroin. Methadone does not treat the original motivation for drug abuse, but it allows addicts to live like normal members of society.

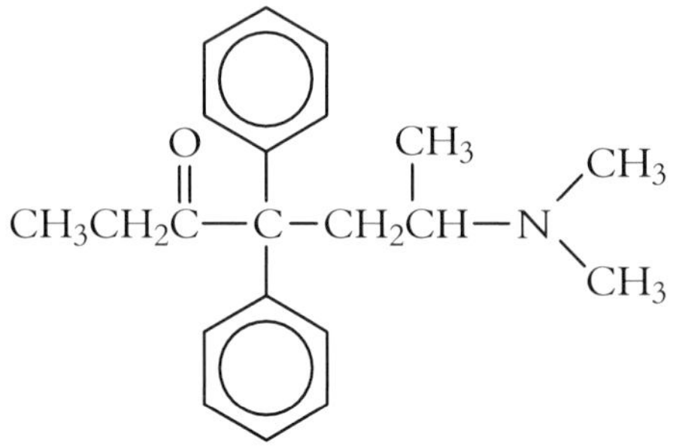

$$CH_3CH_2C(=O)-C(C_6H_5)_2-CH_2CH(CH_3)-N(CH_3)_2$$

Methadone

Barbiturates

Barbiturates depress the CNS. They are called 'downers' and are used as sleeping pills

Barbiturates are sometimes called 'downers' because they act in the opposite sense to 'uppers' (amphetamines, see earlier). They depress the central nervous system. They are prescribed as sleep-inducing drugs, but they can produce dependence. Seconal® (called 'red devils') and Amytal® (called 'blue heavens') are popular among users because they take effect within a few minutes and the effects last for only a few hours.

Valium

Tranquillisers such as Valium® are widely prescribed

Valium® (diazepam) is a depressant of the central nervous system, which is prescribed medicinally as a tranquilliser. It is prescribed for many complaints, such as muscular disorders and spasms and to promote sleep. A similar drug called Mogadon® (nitrazepam) is used in sleeping pills. Sometimes a physical dependence develops, but it is usually a psychological need that makes a person start using Valium®.

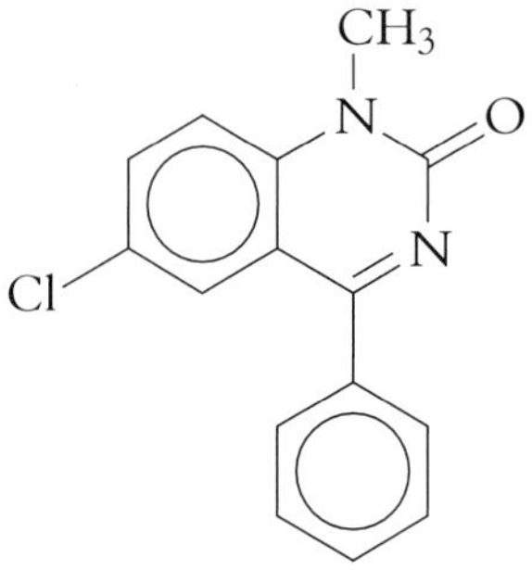

Valium®

Alcohol

Alcohol (ethanol) is the most widely used drug. Its effects as a depressant of the central nervous system were described in § 30.5.

36.1.4 DRUG ABUSE

The problem of drug abuse is not confined to a small group of drug addicts. A large percentage of the population takes pills of all kinds to deal with all sorts of problems, both real and imaginary.

Drugs never solve problems

A person can become addicted to a drug if he or she gets into the habit of turning to the drug for relief from every problem, mental, physical and social. If a person is nervous, worried, upset or depressed, drugs are not the cure. It is better to try to cope with the realities of life without the assistance of mind-altering and mood-altering chemicals. Once a person tries to solve a problem by escaping through a drug experience, he or she may want to repeat that drug experience every time a similar problem arises. Mood-altering drugs give us a holiday from problems; they do not solve our problems.

36.2 SYNTHETIC ROUTES

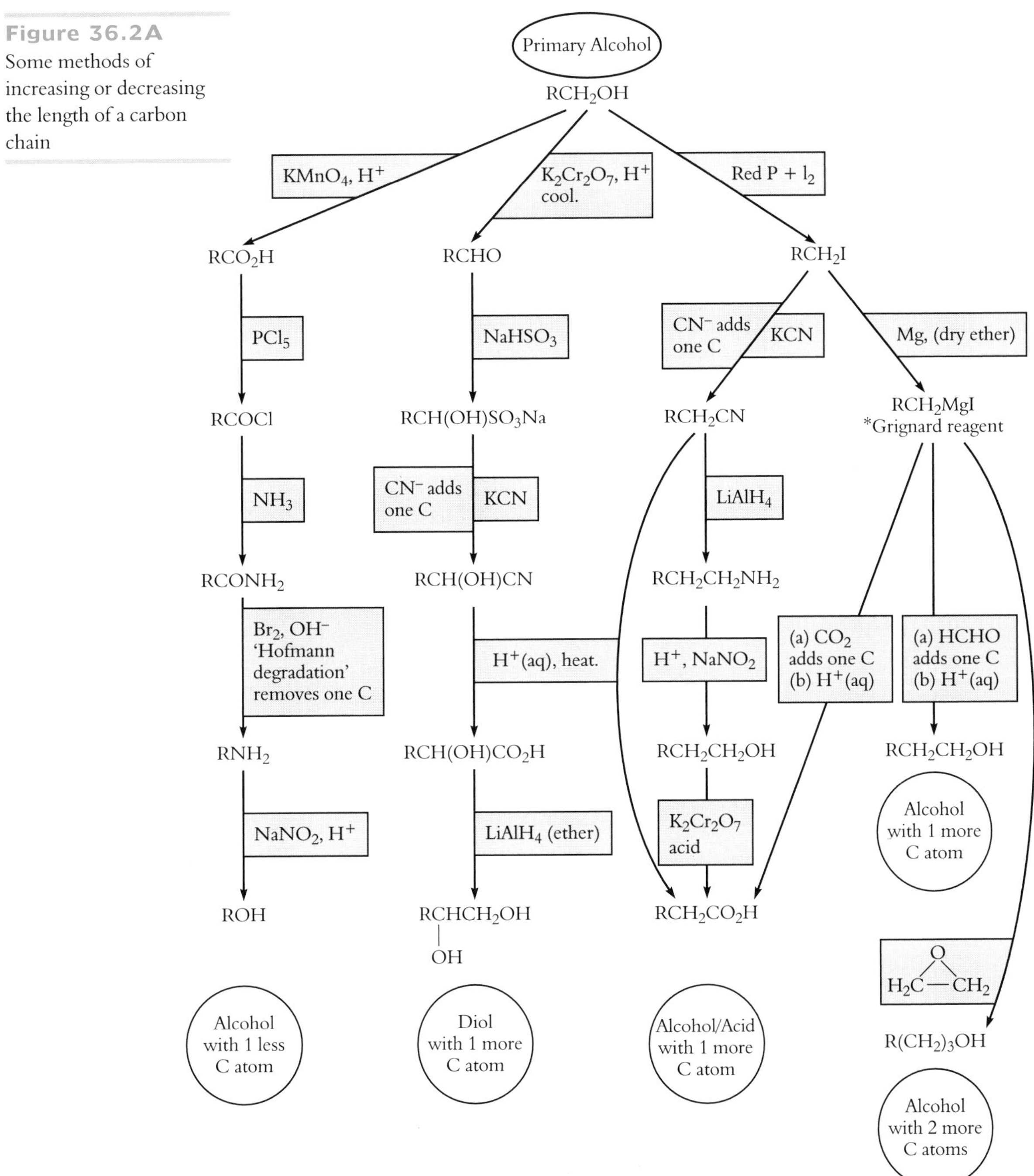

Figure 36.2A Some methods of increasing or decreasing the length of a carbon chain

Figure 36.2B Synthetic routes from a primary alcohol in two steps or more

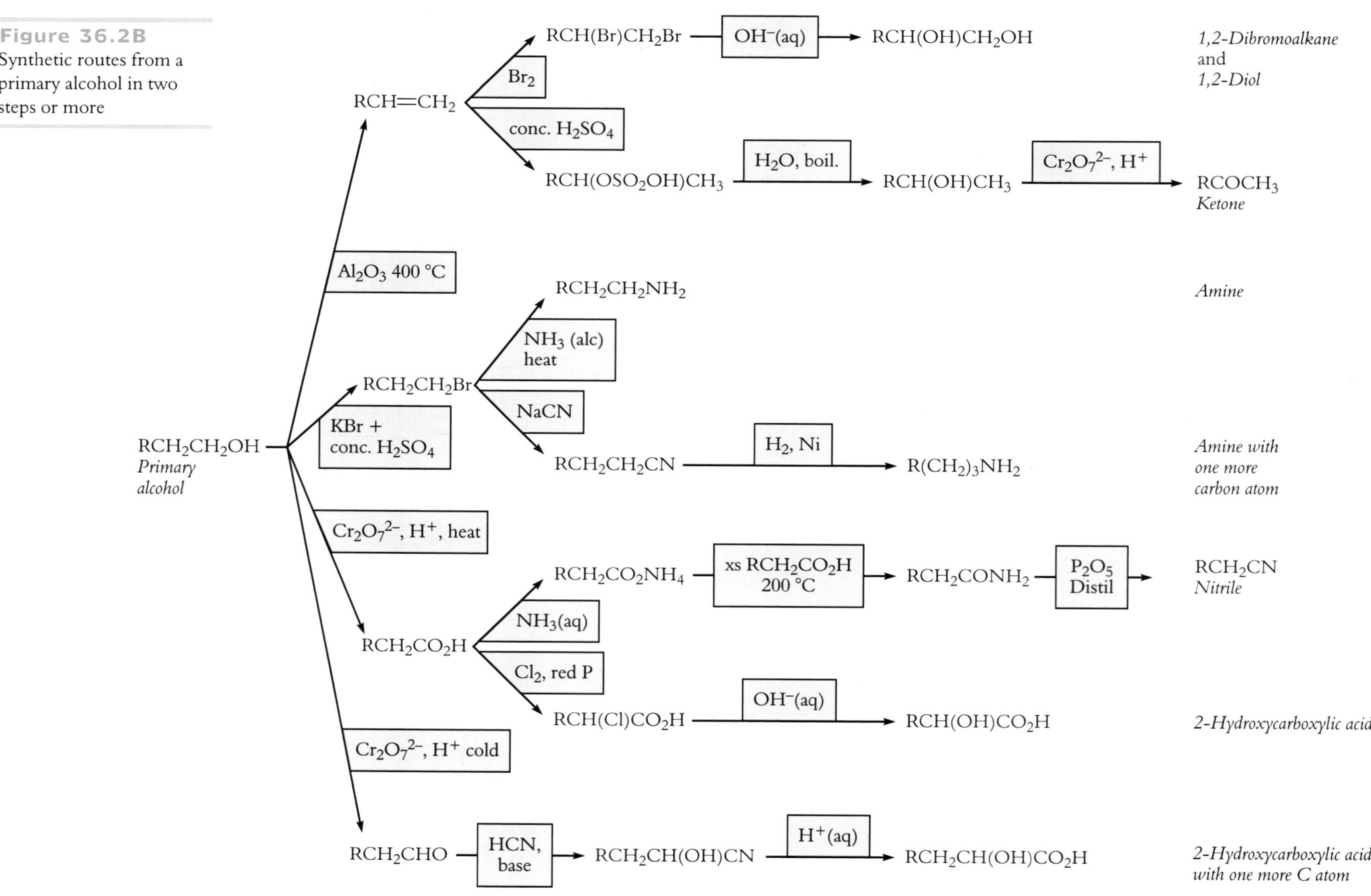

Figure 36.2C
Synthetic routes: schemes connecting aromatic compounds

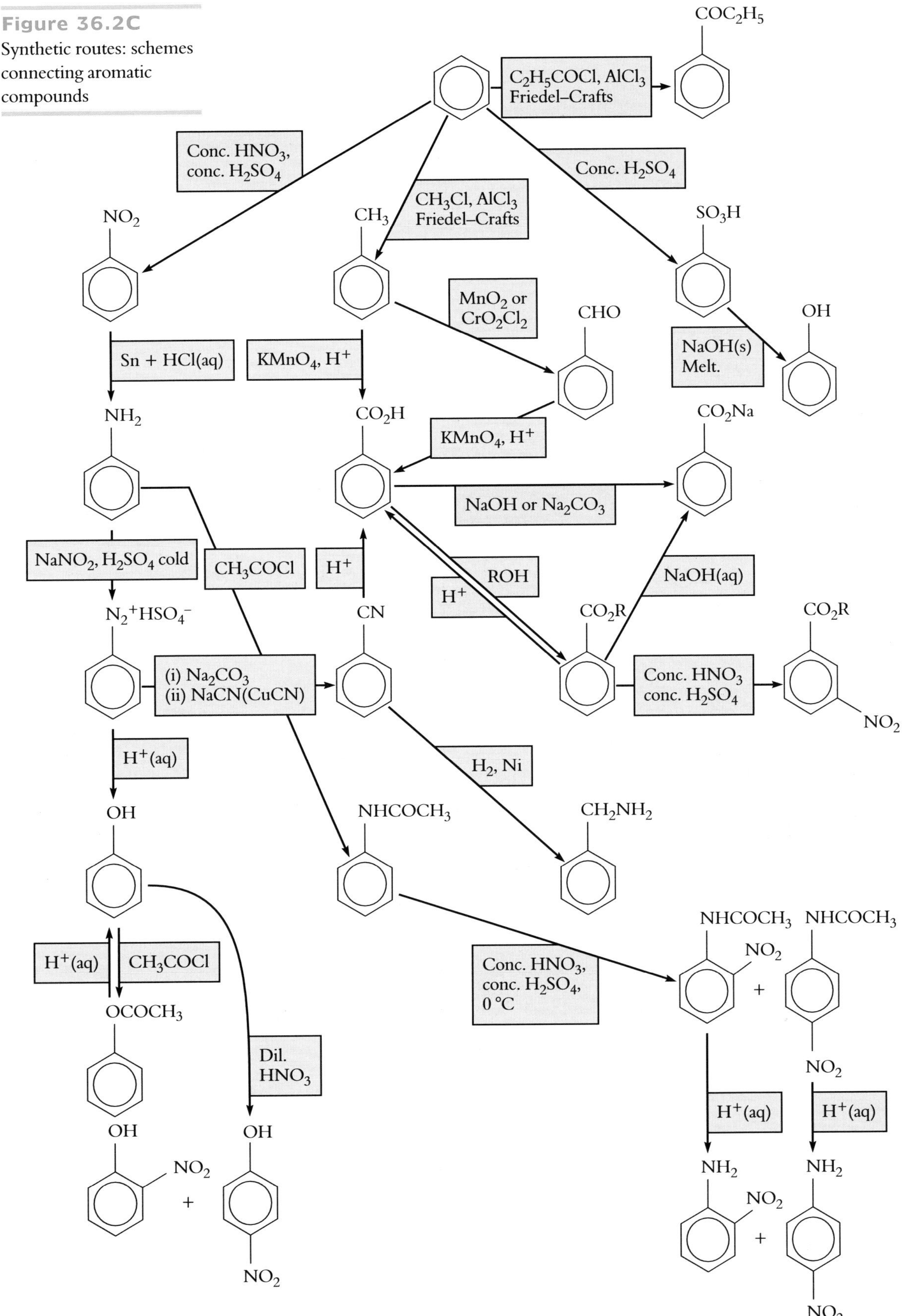

Figure 36.2D
Synthetic routes: products from methylbenzene (toluene) in two steps or more

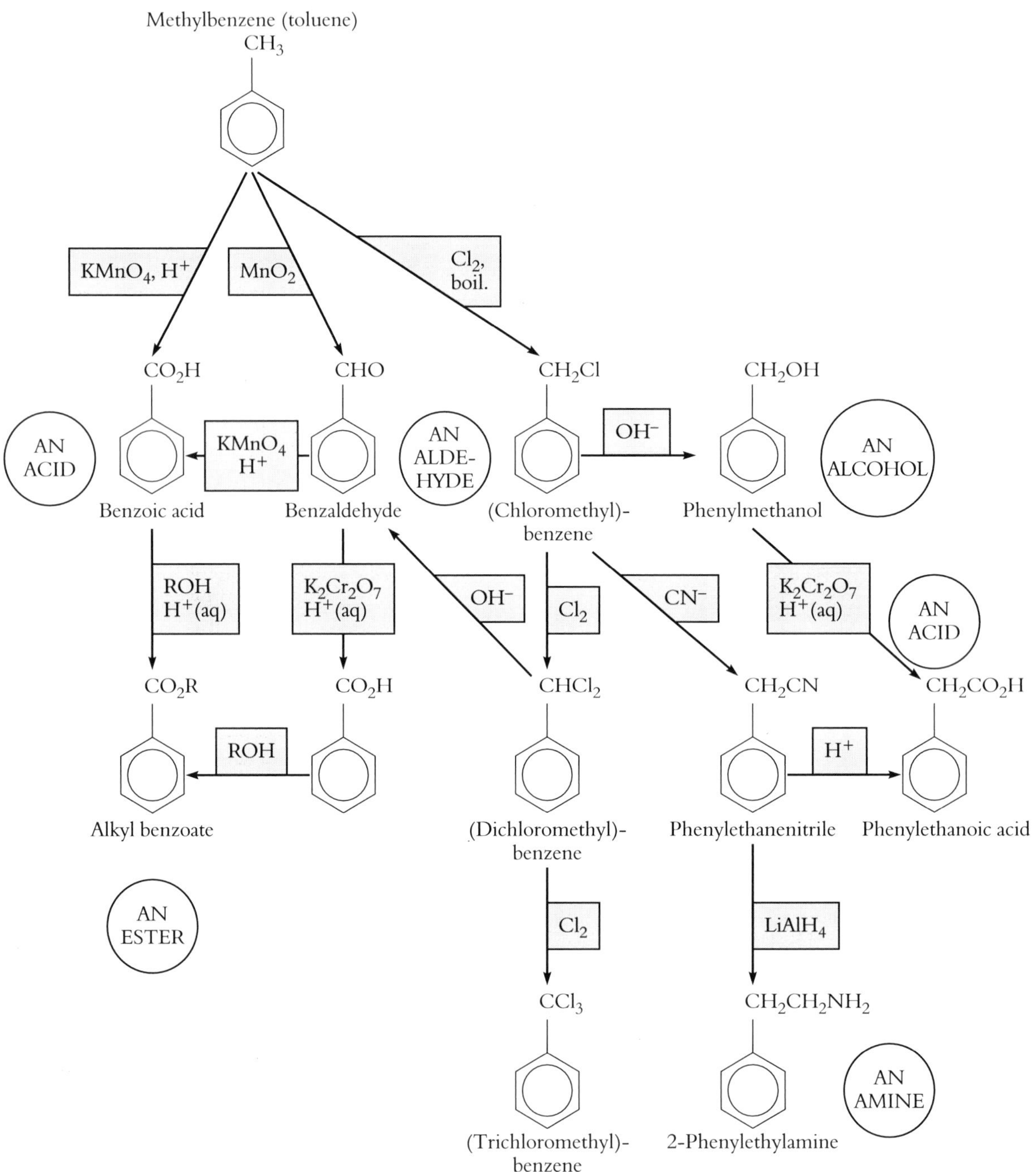

CHECKPOINT 36.2: SYNTHETIC ROUTES

1. State how you would carry out the following conversions:
   (a) $CH_3COCH_3 \longrightarrow CH_3CH(OH)CH_3$
   (b) $CH_3COCH_3 \longrightarrow CH_3CO_2H$
   (c) $CH_3CHO \longrightarrow CH_3CH(OH)CO_2H$
   (d) $CH_3CH_2CHO \longrightarrow CH_3CH_2CO_2CH_2CH_2CH_3$
   (e) $CH_3CH(OH)CH_3 \longrightarrow (CH_3)_2C(OH)(CN)$

2. How could you carry out the conversions listed below?
   (a) $CH_3COCH_3 \longrightarrow CH_3CHClCH_2Cl$
   (b) $(CH_3)_2CO \longrightarrow (CH_3)_2CHBr$
   (c) $CH_3CH_2CH_2OH \longrightarrow CH_3CH{=}CH_2$
   (d) $C_2H_6 \longrightarrow C_2H_5OH$
   (e) $C_2H_6 \longrightarrow C_2H_5NH_2$
   (f) $C_6H_6 \longrightarrow C_6H_5COCH_3$

3. With C_2H_5OH as your only organic starting-material, explain how you could make:
   (a) CH_3CHO, *(b)* $C_2H_5NH_2$,
   (c) $CH_3CH_2CH_2NH_2$, *(d)* CH_3CONH_2,
   (e) CH_3COCl, *(f)* $C_2H_5CO_2H$.

4. Starting from benzene or methylbenzene, say how you could make: *(a)* $C_6H_5CO_2H$,
   (b) $C_6H_5CH_2CO_2H$, *(c)* $C_6H_5CH_2CH_2NH_2$.

5. How can the following compounds be made from phenylamine? *(a)* C_6H_5OH,
   (b) $C_6H_5CO_2H$, *(c)* C_6H_5COCl, *(d)* C_6H_5I.

6. How could you make from propene:
   (a) CH_3COCH_3, *(b)* $(CH_3)_2CHCO_2H$,
   (c) $(CH_3)_2CHCH_2NH_2$,
   (d) $CH_3CH(OH)CH_2OH$?

7. Explain how you could make from chloroethane and any inorganic materials you need *(a)* CH_3CHO,
   (b) $(CH_3CO)_2O$, *(c)* $CH_3CO_2C_2H_5$,
   (d) $CH_3CH_2CO_2H$, *(e)* $CH_3CH_2CONH_2$.

8. Devise syntheses for the following compounds, using no organic compounds other than those stated. State the reagents and conditions needed for each step in the synthesis:
   (a) *N*-phenylbenzamide, $C_6H_5CONHC_6H_5$ from benzene and methylbenzene
   (b) phenylmethanol, $C_6H_5CH_2OH$ from benzene
   (c) benzenecarboxylic acid (benzoic acid) from benzene and methane.

36.3 WHAT ARE THESE REAGENTS USED FOR?

Table 36.3
Reagents and their uses

Reagent	Use	Example
$KMnO_4$, H^+ (aq)	Oxidation	$C_6H_5CH_3 \longrightarrow C_6H_5CO_2H$
MnO_2	Oxidation	$C_6H_5CH_3 \longrightarrow C_6H_5CHO$
CrO_3	Oxidation	$C_6H_5CH_3 \longrightarrow C_6H_5CHO$
$K_2Cr_2O_7$, H^+(aq) $Na_2Cr_2O_7$, H^+(aq)	Oxidation	$RCH_2OH \longrightarrow RCHO$ (Add oxidant slowly to hot alcohol, and distil off aldehyde as fast as it is formed.) $RCH_2OH \longrightarrow RCO_2H$ (Reflux the mixture until oxidation is complete.)
$KMnO_4$ (neutral)	Oxidation	Reflux: $C_6H_5{-}\overset{\vert}{\underset{\vert}{C}}{-}X \longrightarrow C_6H_5CO_2H$ (where X = $-CH_3$, $-CH_2OH$ etc.)
$KMnO_4$, conc, hot	Oxidation of alkenes to determine position of C=C bond	$RCH{=}CHR' \longrightarrow RCHO + R'CHO$

Reagent	Use	Example
H_2, Pt	Reduction	$C_6H_5NO_2 \longrightarrow C_6H_5NH_2$
$NaBH_4$(aq)	Reduction of C=O, not C=C	$RCHO \longrightarrow RCH_2OH$ $R_2C{=}O \longrightarrow R_2CHOH$
Fe + HCl(aq)	Reduction	$C_6H_5NO_2 \longrightarrow C_6H_5NH_2$
Sn + HCl(aq)	Reduction	$C_6H_5NO_2 \longrightarrow C_6H_5NH_2$
H_2, Ni	Reduction of C=C, not C=O	$>C{=}CH{-}C(R){=}O \longrightarrow >CH{-}CH_2{-}C(R){=}O$ $-C{\equiv}C- \longrightarrow -CH_2CH_2-$ $RCN \longrightarrow RCH_2NH_2$ Hydrogenation of oils to form fats
$LiAlH_4$ (ether)	Reduction of C=O, not C=C	$RCO_2H \longrightarrow RCH_2OH$ $RCO_2R' \longrightarrow RCH_2OH$ $RCONH_2 \longrightarrow RCH_2NH_2$
O_3 followed by Zn + CH_3CO_2H(aq)	Ozonolysis to determine position of C=C bond	$R{-}CH{=}CH{-}R' \longrightarrow RCHO + R'CHO$ $RCH{=}CH_2 \longrightarrow RCHO + HCHO$
Br_2 (organic solvent)	Test for unsaturation	$RCH{=}CHR' \longrightarrow RCH(Br){-}CHR'(Br)$ (no HBr evolved)
$KMnO_4$, OH^-(aq) dilute, 0 °C	Test for unsaturation	$RCH{=}CHR' \longrightarrow RCH(OH){-}CHR'(OH)$
NaOH, I_2	Iodoform test for CH_3CO- or $CH_3CH(OH)-$	$CH_3CHO + I_2 + NaOH \longrightarrow CHI_3$
$NaNO_2$, H^+(aq), <10 °C	Diazotisation	$C_6H_5NH_2 \longrightarrow C_6H_5\overset{+}{N}{\equiv}N$
Al_2O_3	Dehydration	$C_2H_5OH \longrightarrow CH_2{=}CH_2$
P_2O_5	Dehydration	$RCONH_2 \longrightarrow RCN$
Hot conc. H_2SO_4	Dehydration	$RCH(OH)CH_3 \longrightarrow RCH{=}CH_2$
Conc. H_2SO_4	Esterification	$ROH + R'CO_2H \longrightarrow R'CO_2R + H_2O$
Cold conc. H_2SO_4, then H_2O	Hydration	$RCH{=}CHR' \longrightarrow RCH(H){-}CHR'(OH)$

Reagent	Use	Example
H_3PO_4, then H_2O	Hydration	$RCH{=}CHR' \longrightarrow RCH(H){-}CHR'(OH)$
$FeCl_3$, $FeBr_3$	Halogen carriers	$C_6H_6 + Cl_2 \xrightarrow{FeCl_3} C_6H_5Cl + HCl$
$AlCl_3$	Friedel–Crafts catalyst	$C_6H_6 + RCl \xrightarrow{AlCl_3} C_6H_5R + HCl$ $C_6H_6 + RCOCl \xrightarrow{AlCl_3} C_6H_5COR + HCl$
Conc. HNO_3 + conc. H_2SO_4	Nitration	$C_6H_6 + NO_2^+ \longrightarrow C_6H_5NO_2$
SO_3 in conc. H_2SO_4	Sulphonation	$C_6H_6 \longrightarrow C_6H_5SO_3H$
Conc. H_2SO_4	Sulphonation of phenols and aromatic amines	$C_6H_5OH \longrightarrow HOC_6H_4SO_3H$ $C_6H_5NH_2 \longrightarrow H_2NC_6H_4SO_3H$
'Soda lime'	Decarboxylation	$RCO_2H \longrightarrow RH$
OH^- (ethanol)	Elimination of hydrogen halide	$RCH_2CH_2X \longrightarrow RCH{=}CH_2$
OH^- (aq)	Hydrolysis	$RCO_2R' \longrightarrow RCO_2^-(aq) + R'OH$ $RCN \longrightarrow RCO_2^-(aq)$ $RCONH_2 \longrightarrow RCO_2^-(aq) + NH_3(g)$
C_6H_5COCl, NaOH	Benzoylation	$C_6H_5OH \longrightarrow C_6H_5OCOC_6H_5$
$(C_6H_5CO_2)_2$ (Dibenzoyl peroxide)	Initiating chain reactions	Chlorination of alkanes Polymerisation of alkenes
$SOCl_2$ (Sulphur dichloride oxide)	Replacement of —OH by —Cl	$RCH_2OH \longrightarrow RCH_2Cl$ $RCO_2H \longrightarrow RCOCl$
PCl_5 (Phosphorus pentachloride)	Replacement of —OH by —Cl	$RCO_2H \longrightarrow RCOCl$ $RCH_2OH \longrightarrow RCH_2Cl$
$H^+(aq)$	Hydrolysis	RCO_2R', $RCONH_2$, $RCN \longrightarrow RCO_2H$
*C_2H_5MgBr (Ethylmagnesium bromide)	Grignard reactions	$HCHO \rightarrow$ Primary alcohol $R'CHO \rightarrow$ Secondary alcohol $R'R''C{=}O \longrightarrow$ Tertiary alcohol
2,4-Dinitrophenyl-hydrazine (DNP)	Test for a C=O group	Gives an orange ppt. with an aldehyde or ketone

Reagent	Use	Example
Fehling's solution Benedict's reagent (Cu^{2+} complex ions)	Test for an aliphatic aldehyde	Gives a reddish ppt. of Cu_2O with aliphatic aldehydes, but not with aromatic aldehydes or with ketones
Tollens' reagent ($Ag(NH_3)_2^+(aq)$)	Test for an aldehyde	Reduced to a silver mirror by aliphatic aldehydes and, when warmed, by aromatic aldehydes, but not by ketones

CHECKPOINT 36.3: REAGENTS

1. Compare the reactions of bromine with the following compounds. State the products of the reactions and the conditions necessary for reaction.
(a) methane, *(b)* benzene, *(c)* phenol,
(d) ethene, *(e)* ethanamide.

2. Explain the following terms. Give an example of each type of reaction, indicating the reactants and the conditions needed for reaction.
(a) reduction, *(b)* ozonolysis, *(c)* alkylation,
(d) acylation, *(e)* decarboxylation.

3. Give an example of each of the following types of reaction. Indicate the necessary conditions.
(a) nitration, *(b)* sulphonation, *(c)* oxidation,
(d) cracking, *(e)* the halogenation of an aromatic ring.

4. Give examples of the reduction of organic compounds by *(a)* hydrogen and a catalyst,
(b) zinc and hydrochloric acid, *(c)* lithium tetrahydridoaluminate, *(d)* sodium tetrahydridoborate.

5. Illustrate the use of the following reagents in organic chemistry. State the conditions necessary for reaction, and give equations.
(a) bromine, *(b)* aluminium chloride,
(c) sodium nitrite *(d)* hydrogen cyanide,
(e) nickel, *(f)* alkaline potassium manganate(VII),
(g) ozone, *(h)* hot, concentrated potassium manganate(VII).

6. State the products of the reaction between sodium hydroxide and each of the following compounds. For each reaction, state the necessary conditions, and write the equation.
(a) ethyl ethanoate, *(b)* 1-bromobutane,
(c) 1,1,1-trichloropropanone, CCl_3COCH_3.

7. The following are well-known reagents. State what use each finds in organic chemistry. Describe the bonding in each compound. Explain how the type of bonding enables the reagent to function as it does.
(a) $AlCl_3$, *(b)* HCN, *(c)* KOH(aq), *(d)* $LiAlH_4$.

36.4 HOW WOULD YOU DISTINGUISH BETWEEN THE MEMBERS OF THE FOLLOWING PAIRS OF COMPOUNDS BY CHEMICAL TESTS?

CHECKPOINT 36.4: PAIRS OF COMPOUNDS

1. **A** CH_3COCH_3 **B** C_2H_5OH
2. **A** CH_3COCH_3 **B** C_2H_5CHO
3. **A** $C_2H_5COCH_3$ **B** $C_2H_5CO_2H$
4.

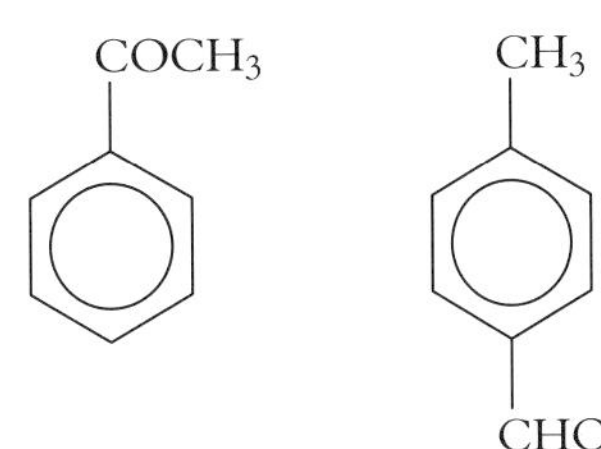

A **B**

5. **A** CH_3CHO **B** $CH_3CO_2CH_3$
6. **A** $CH_3CH_2CH_2OH$ **B** $CH_3CH(OH)CH_3$
7. **A** $(CH_3)_3COH$ **B** $(CH_3)_2CHCH_2OH$
8.

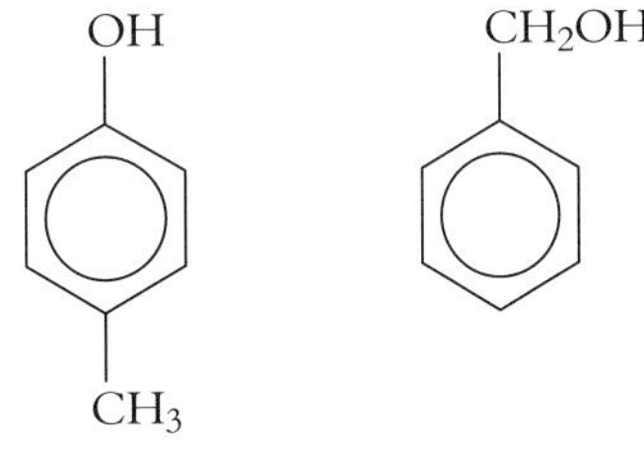

A **B**

9.

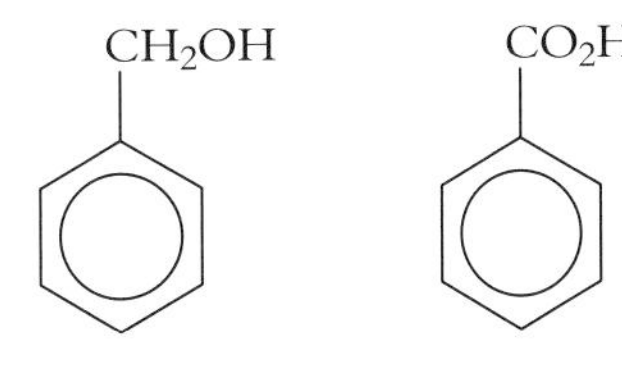

A **B**

10.

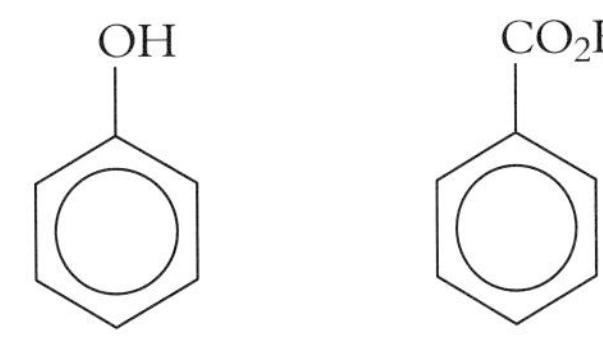

A **B**

11. **A** CH_3CHO **B** CH_3CO_2H
12. **A** CH_3COCH_3 **B** $C_2H_5CO_2H$
13. **A** CH_3CO_2H **B** HCO_2H
14. **A** Na_2SO_4 **B** CH_3CO_2Na
15. 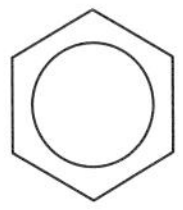 **A** (benzene) **B** $CH_3(CH_2)_3CH{=}CH_2$
16. **A** (benzene)

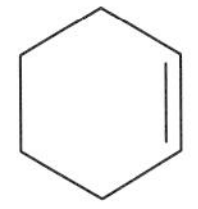

B

17. **A** $C_6H_5CH{=}CH_2$ **B** C_6H_5OH
18. **A** CH_4 **B** CH_3Cl
19. **A** CH_3COCl **B** C_6H_5COCl
20. **A** $ClCH_2CO_2H$ **B** CH_3COCl
21.

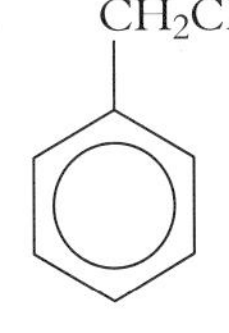

A

B (2-chlorotoluene: CH_3, Cl)

22. **A** CH_3CH_2I **B** CH_3CH_2Br
23. **A** $HCO_2C_2H_5$ **B** $CH_3CO_2CH_3$
24. **A** $CH_3CO_2C_6H_5$ **B** $C_6H_5CO_2CH_3$
25. **A** $ClCH_2CO_2H$ **B** CH_3CO_2H
26. **A** $CH_3CH_2NH_2$ **B** CH_3CN
27. **A** $CH_3CH_2NH_2$ **B** $H_2NCH(CH_3)CO_2H$
28. **A** CH_3CONH_2 **B** $CH_3CH_2NH_2$
29. **A** CH_3CONH_2 **B** $CO(NH_2)_2$
30. **A** $C_2H_5CONH_2$ **B** $C_6H_5NH_2$
31. **A** $C_3H_7NH_3^+Cl^-$ **B** $C_3H_7CO_2^-Na^+$
32. **A** $CH_3CO_2NH_4$ **B** CH_3CONH_2

33.

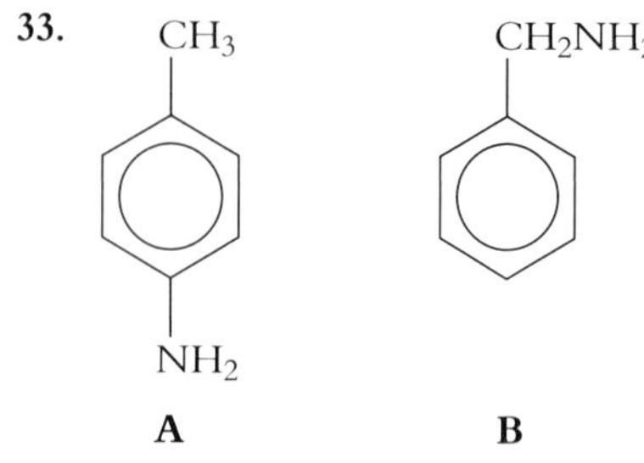

34. **A** C_2H_5OH
B $(CH_3)_3COH$

35. **A** CH_3CH_2COCl
B $CH_3CH_2CH_2Cl$

ANSWERS TO CHECKPOINT 36.4: PAIRS OF COMPOUNDS

1. **A**, a ketone, reacts with 2,4-dinitrophenylhydrazine, DNP, to give an orange-coloured precipitate of the 2,4-dinitrophenylhydrazone.

   B, an alcohol, reacts with $CH_3CO_2H(l)$ + conc. H_2SO_4 to give a fruity, ester smell of $CH_3CO_2C_2H_5$.

2. Both react with DNP. **A**, a ketone, is not easily oxidised. **B**, an aldehyde, reduces Tollens' reagent to silver and Fehling's solution to red Cu_2O.

3. **A** is a ketone and reacts with DNP. **B** is an acid, and liberates CO_2 from $NaHCO_3$.

4. **A** has no reducing action; **B** reduces Tollens' reagent but not Fehling's solution.

5. **A** + NaOH(aq, conc.) forms a brown resin, $(CH_3CHO)_n$, with a characteristic smell.

   B + NaOH(aq) when warmed give $CH_3CO_2Na + CH_3OH$. On addition of HCl(aq), CH_3CO_2H, with a characteristic smell, is formed.

6. **A**, a primary alcohol, is oxidised by acidified $Na_2Cr_2O_7$ in the cold to an aldehyde. (Test as in 2.)

   B, a secondary alcohol, is oxidised to a ketone. (Test as in 2.)

   Also Lucas test [Table 30.1].

7. **A**, a tertiary alcohol, is oxidised by acid $K_2Cr_2O_7$ to a mixture of products, including an acid.

   B, a primary alcohol, is oxidised to an aldehyde. (Test with DNP.)

8. **A** is a phenol. Its solution is acidic, reacting with NaOH(aq) to give the phenoxide, $H_3CC_6H_4O^-Na^+$. **A** gives a purple colour with $FeCl_3$(aq) and a white precipitate with Br_2(aq).

   B is an alcohol, neutral in solution. It does not react with NaOH(aq), but does react with Na(s) to give H_2(g) and the alkoxide, $C_6H_5CH_2O^-Na^+$.

9. **A** is an alcohol, neutral in solution.

   B is an acid: its solution is acidic, and it reacts with Mg to give H_2(g) and with $NaHCO_3$ to give CO_2(g).

10. **A** is weakly acidic: it dissolves in NaOH(aq), but does not give CO_2(g) with $NaHCO_3$. **B** is an acid, and liberates CO_2(g) from $NaHCO_3$.

    A gives a purple colour with $FeCl_3$(aq) and a white ppt. with Br_2(aq).

11. **A** + $AgNO_3$(aq) give a silver mirror when warmed. **B** smells of vinegar and liberates CO_2 from $NaHCO_3$.

12. **A** with DNP gives an orange crystalline ppt. **B** + C_2H_5OH + conc. H_2SO_4 gives, when warmed, an ester with a characteristic, fruity smell. **B** liberates CO_2(g) from $NaHCO_3$.

13. **A** smells of vinegar, **B** of formalin.

    When warmed with $AgNO_3$, NH_3(aq), **B** gives a silver mirror. Conc. H_2SO_4 + **B** gives CO(g), which burns with a blue flame.

    B + Fehling's solution gives a red ppt. of Cu_2O.

14. **A**(aq) + $BaCl_2$(aq) $\longrightarrow$ white ppt. of $BaSO_4$(s), **B**(aq) is weakly alkaline.

15. **A** is benzene. It does not show unsaturation, as does **B**, which decolourises Br_2 (organic solvent) and alkaline $KMnO_4$(aq).

16. **A** burns with a smoky flame, as aromatic compounds do. **B**, cyclohexene, shows unsaturation (see above).

17. **A** decolourises $Br_2(CCl_4)$, **B** decolourises $Br_2(CCl_4)$ with evolution of HBr and the formation of a precipitate.

18. **A** is unreactive. **B** is hydrolysed by alkali to form methanol (Test for alcohol) and Cl^- ions (Test with $AgNO_3$, HNO_3(aq)).

19. In water, **A** immediately forms CH_3CO_2H and HCl. A strongly acidic solution is formed. (Test with $NaHCO_3$). **B** is hydrolysed when warmed with NaOH(aq) to sodium benzoate. On acidification with HCl(aq), solid benzoic acid comes out of solution.

20. **A**, chloroethanoic acid, is a stronger acid than ethanoic acid. (Test: Mg; $NaHCO_3$).

B is not acidic; but it reacts with water to form a strongly acidic solution of CH_3CO_2H + HCl. Addition of $AgNO_3$, HNO_3(aq) will detect Cl^- ions, which are not present in a solution of **A**.

21. A —Cl in the side chain, as in **A**, is hydrolysed off by warming **A** with alkali, to give an alcohol (Test by oxidising it to an aldehyde), and Cl^- ions (Test with $AgNO_3$, HNO_3(aq)).

The —Cl in the ring in **B** cannot be removed in this way.

22. On hydrolysis with NaOH(aq), **A** gives I^- ions, and **B** gives Br^- ions. $Pb(NO_3)_2$(aq) gives a yellow ppt. of PbI_2 with the first, and a white ppt. of $PbBr_2$ with the second.

With **A**, CH_3CO_2Ag(aq) gives a yellow ppt. of AgI; with **B** it gives a pale yellow ppt. of AgBr.

23. **A** is hydrolysed when warmed with OH^- (aq) to $HCO_2H + C_2H_5OH$. Ethanol gives the iodoform test: with $I_2 + OH^-$(aq), a ppt. of CHI_3 is formed.

B is hydrolysed to CH_3CO_2H (which smells of vinegar) and CH_3OH. On oxidation, CH_3OH gives methanal, HCHO, with the smell characteristic of formalin.

24. **A** is an ester which is hydrolysed to CH_3CO_2H and C_6H_5OH. Phenol in alkaline solution couples with a diazonium salt to give a dye.

B is hydrolysed to $C_6H_5CO_2H + CH_3OH$. Benzoic acid liberates CO_2 from $NaHCO_3$.

25. **A** is a stronger acid than **B** (e.g. in reaction with $NaHCO_3$). **A** is hydrolysed by OH^-(aq) to give Cl^-(aq); tested with $AgNO_3$(aq).

26. **A** is an amine, with a characteristic smell. Its aqueous solution is alkaline. With acids, it forms salts, which are crystalline solids. With nitrous acid, **A** gives $C_2H_5OH + N_2$(g).

B is a nitrile, hydrolysed by H^+(aq) to CH_3CO_2H (Test for carboxylic acid).

27. **A** is a primary amine. With nitrous acid, it gives $C_2H_5OH + N_2$(g).

B is an amino acid, a crystalline solid.

28. **A** is an amide. It is hydrolysed by NaOH(aq) to give NH_3(g) (which is basic) and CH_3CO_2Na(aq).

B is an amine, and reacts with nitrous acid to give N_2(g).

29. **A**, an amide, is hydrolysed by OH^-(aq) to NH_3(g). **B**, urea, is hydrolysed by OH^-(aq) to NH_3(g) and by H^+(aq) to CO_2(g).

30. **A** + Br_2(aq) + OH^-(aq) give an amine, which is basic and has a 'fishy' smell.

B + Br_2(aq) gives a white ppt. of $C_6H_2Br_3NH_2$.

31. A solution of **A** in water gives a white ppt. of AgCl on treatment with $AgNO_3$, HNO_3(aq). Addition of alkali to **A** liberates the free amine, $C_3H_7NH_2$, which has the characteristic smell.

B gives a yellow flame test. Addition of HCl(aq) to **B** liberates butanoic acid, $C_3H_7CO_2H$, with a rancid smell.

32. With cold NaOH(aq), **A** gives NH_3(g). When strongly heated with NaOH(aq), **B** gives NH_3(g).

33. **A** is an aryl amine, which can be diazotised with cold $NaNO_2$ + HCl(aq). The diazonium salt formed gives, when warmed, N_2(g) and a smell of phenol. It gives a dye when coupled with phenol in alkaline solution. **B** reacts with $NaNO_2$ + HCl(aq) to give N_2(g).

34. **A** gives a yellow precipitate of iodoform with $I_2 + OH^-$(aq). **B** is not easily oxidised.

35. **A** is a fuming lachrimatory liquid easily hydrolysed to propanoic acid + HCl. **B** is a volatile liquid which is hydrolysed only by refluxing with alkali.

36.5 QUESTIONS ON SOME TOPICS WHICH SPAN CHAPTERS

CHECKPOINT 36.5A: FUNCTIONAL GROUPS AND REACTION MECHANISMS

1. Explain these statements:
   (a) Water and petrol do not mix.
   (b) Halogenoalkanes are more reactive than alkanes towards nucleophilic reagents.
   (c) Alkenes, unlike alkanes, react readily with bromine water.
   (d) Phenol, unlike benzene, reacts readily with bromine water.
   (e) Ethanoyl chloride is more reactive towards water than is chloroethane.

2. Compare and contrast the reactions of:
   (a) sulphuric acid with ethanol and phenol,
   (b) phosphorus(V) chloride with ethanol and phenol,
   (c) sodium hydroxide with 1-chlorohexane and chlorobenzene,
   (d) nitrous acid with ethylamine and phenylamine.

3. Describe the mechanisms of the reactions between the following substances:
   (a) ethene and bromine,
   (b) methane and chlorine in sunlight,
   (c) benzene and ethanoyl chloride in the presence of aluminium chloride,
   (d) 1-iodopropane and aqueous sodium hydroxide,
   (e) propanal and hydrogen cyanide.

4. Explain, by means of examples, what is meant by the following terms:
   (a) homolytic fission,
   (b) heterolytic fission,
   (c) nucleophilic substitution,
   (d) an addition–elimination reaction,
   (e) catalytic cracking.

5. Explain the following observations:
   (a) Ethene does not react with water, but, when ethene is passed into concentrated sulphuric acid and water is added to the solution, ethanol is formed.
   (b) Hydrogen cyanide adds to propanone in the presence of bases but not in the presence of acids.
   (c) Both propanone and ethanol give yellow precipitates when treated with iodine and alkali. Neither pentan-3-one nor pentan-3-ol gives a positive result in this test.
   (d) Carboxylic acids and their derivatives do not show the addition reactions of aldehydes and ketones with nucleophiles.

6. Adrenalin has the formula shown below. It is a water-soluble hormone:

   $(HO)_2C_6H_3-CH(OH)-CH_2-NH-CH_3$

   (a) Name the functional groups in the molecule.
   (b) Describe how you could test for the presence of two of these groups.
   (c) Where does the optical activity of adrenalin arise?

7. $H_2N-C_6H_4-CH{=}CH-CHBr-CO_2H$

   Imagine that you have synthesised the compound which has the formula shown above. Describe how you could test for each of the functional groups, and say what the results of these tests would be if the compound you have made is the correct substance.

   Say what further tests you could do to confirm that your product is indeed the substance you intended to make.

CHECKPOINT 36.5B: DRAW YOUR OWN CONCLUSIONS

1.

A and **B** give precipitates with DNP
No reaction
Tollens' reagent
Silver mirror
$KMnO_4$, acid
Tollens' reagent
A (C_3H_6O)
D (C_3H_8O)
C (C_3H_8O)
$KMnO_4$, acid
B (C_3H_6O)

Identify the compounds **A** to **D**. Explain the reactions depicted.

2.

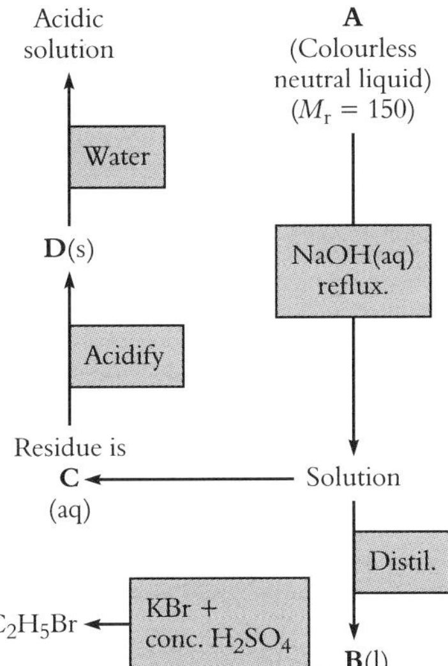

Suggest identities, giving reasons, for the compounds **A** to **D**.

3. The isomers **A** and **B** have the molecular formula $C_9H_8O_2$. They are insoluble in water but soluble in aqueous sodium carbonate. Both **A** and **B** are reduced by hydrogen in the presence of a platinum catalyst to **C**, $C_9H_{10}O_2$. Benzoic acid is formed when **A** or **B** or **C** is oxidised by alkaline potassium manganate(VII), and the solution is later acidified. **C** reacts with phosphorus(V) chloride to give **D**, and reacts with sodium carbonate solution to give carbon dioxide.

Identify the compounds **A** to **D**, giving reasons, and explaining the reactions mentioned.

4. **P** is a crystalline solid which melts at 200 °C. It dissolves in water to give a neutral solution. **P** forms salts with both acids and bases. When heated with soda lime, it gives **Q**, a pungent-smelling gas. **Q** burns in air and dissolves in water to give an alkaline solution. When treated with nitrous acid, **P** gives nitrogen and an acid, **R**, with a molar mass of 76 g mol^{-1}. Identify, with reasons, **P**, **Q** and **R**.

5. Identify compounds **A** to **G**. Explain the reactions mentioned.

A is a crystalline solid which is very soluble in water and gives a yellow flame test. When **A** is heated with ethanoyl chloride, **B** distils. **B** is a neutral liquid with a relative molar mass of 102. **B** reacts with water to give an acidic solution.

C is a colourless liquid which is sparingly soluble in water but dissolves readily in hydrochloric acid. Evaporation of the solution yields **D**, a crystalline solid. **D** reacts with alkali to give **C**. A solution of **C** in hydrochloric acid reacts with a cold solution of sodium nitrite, followed by an alkaline solution of phenol to give an orange dye. When **C** is treated with bromine water, it gives a white precipitate of **E**, which has a relative molar mass of 330.

F is a colourless solid with a distinctive smell. It dissolves in water to give a weakly acidic solution, which does not liberate carbon dioxide from a solution of sodium hydrogencarbonate. When an alkaline solution of **F** is shaken with benzoyl chloride, a solid, **G**, of molar mass 198 g mol^{-1}, is formed.

6. **W** is a colourless aliphatic liquid with a relative molar mass of 123. **W** is insoluble in water. When refluxed with aqueous sodium hydroxide, it gives a solution of **X** and **Y**.

A solution of **X** gives a creamy yellow precipitate with silver nitrate solution. **Y** can be distilled from the solution. It gives a positive iodoform test and is oxidised by chromic acid to **Z**. **Z** gives a positive iodoform test, and reacts with 2,4-dinitrophenylhydrazine but not with ammoniacal silver nitrate.

Identify **W**, **X**, **Y** and **Z**. Write equations for the reactions involved.

7. **A** ($C_7H_7NO_2$) is reduced by tin and concentrated hydrochloric acid, followed by alkali, to **B** (C_7H_9N).

B is converted by sodium nitrite and dilute hydrochloric acid into **C** (C_7H_8O).

C is converted by (i) sodium, (ii) iodomethane into **D** ($C_8H_{10}O$).

D is oxidised by acidified dichromate to **E** ($C_8H_8O_3$). When heated with soda lime and acidified, **E** gives **F** (C_7H_8O).

B dissolves in acid; **E**, dissolves in alkali; **C** gives a violet colour with iron(III) chloride.

Give the names and structural formulae for **A** to **F**. Explain what happens in each of the reactions mentioned.

8.

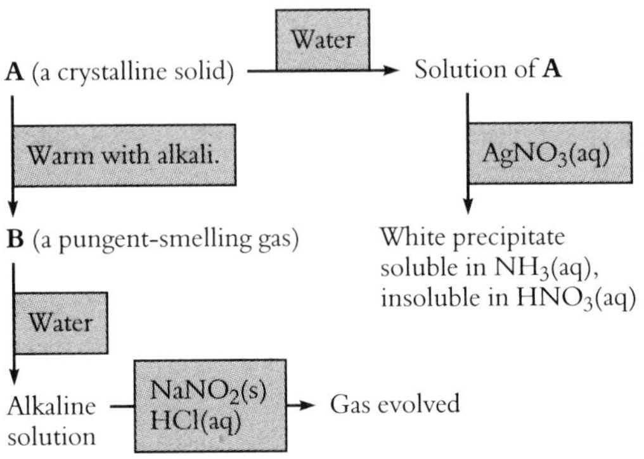

Name two compounds which could be **A** and **B**. Explain the reactions depicted above.

9. The formula of the hormone thyroxine is shown:

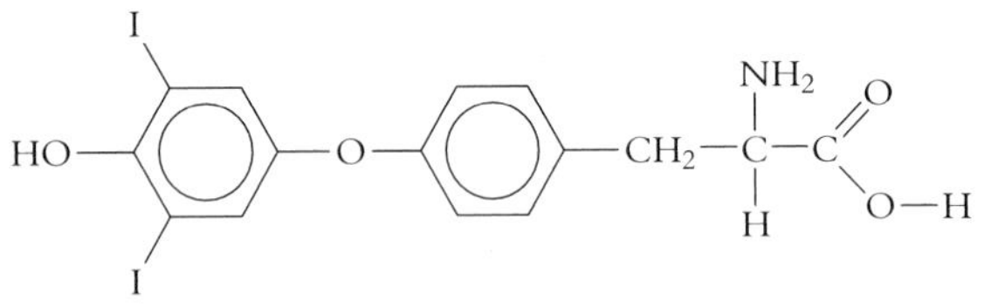

(a) How could you demonstrate the presence of an —OH group in thyroxine?
(b) How does thyroxine fit the formula $H_2NCHRCO_2H$? What name is given to compounds of this type? What properties would you expect a member of this group of compounds to have?
(c) What kind of reaction would you expect of the iodine atoms in thyroxine? Explain your answer.

10. An organic compound **H** ($C_2H_4O_3$) is oxidised to **I** ($C_2H_2O_3$), which can be oxidised to **J** ($C_2H_2O_4$). **H**, **I** and **J** dissolve in water to give acidic solutions which decolourise potassium manganate(VII) on warming. On treatment with phosphorus(V) chloride, all three react, **H** and **J** giving 2 moles of hydrogen chloride per mole, and **I** giving 1 mole of hydrogen chloride per mole. **I** gives a precipitate with 2,4-dinitrophenylhydrazine.

Deduce the identity of **H**, **I** and **J**. Explain the reactions described.

QUESTIONS ON CHAPTER 36

1. *(a)* For each of the following pairs of compounds describe a chemical test which would enable you to distinguish between the members of the pair.
(i) propanone and propanal
(ii) propanoyl chloride and 1-chloropropane
(iii) phenol and benzenecarboxylic acid (benzoic acid)
(iv) ethanamide and ethylamine

(b) Describe how you could obtain pure samples of benzenecarboxylic acid (benzoic acid) and phenol from a mixture of both.

2. Briefly state, giving the reagents and conditions, how you could carry out the following conversions in the laboratory.
(a) ethanal, CH_3CHO, into 2-hydroxypropanoic acid, $CH_3CH(OH)CO_2H$
(b) ethanoic acid, CH_3CO_2H, into *N*-phenylethanamide, $CH_3CONHC_6H_5$
(c) methylbenzene, $C_6H_5CH_3$, into phenylmethanol, $C_6H_5CH_2OH$
(d) phenylamine, $C_6H_5NH_2$, into iodobenzene, C_6H_5I
(e) benzene, C_6H_6, into benzenecarbonitrile (benzonitrile), C_6H_5CN
(f) phenylethanone, $C_6H_5COCH_3$, into methyl benzoate, $C_6H_5CO_2CH_3$
(g) propan-1-ol, $CH_3CH_2CH_2OH$, into propan-2-ol, $CH_3CHOHCH_3$.

3.

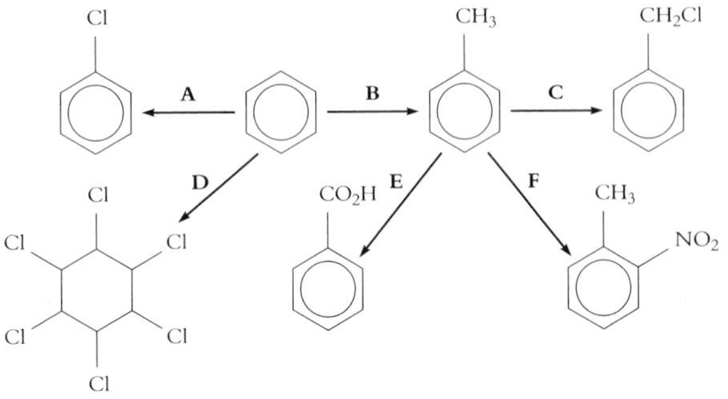

(a) Name the reagents represented by the letters **A**–**F**. State the conditions under which each reaction takes place.
(b) Give the name or structural formula of the organic product formed when (chloromethyl) benzene (benzyl chloride) reacts with
(i) aqueous sodium hydroxide
(ii) potassium cyanide
(iii) ammonia.

4.

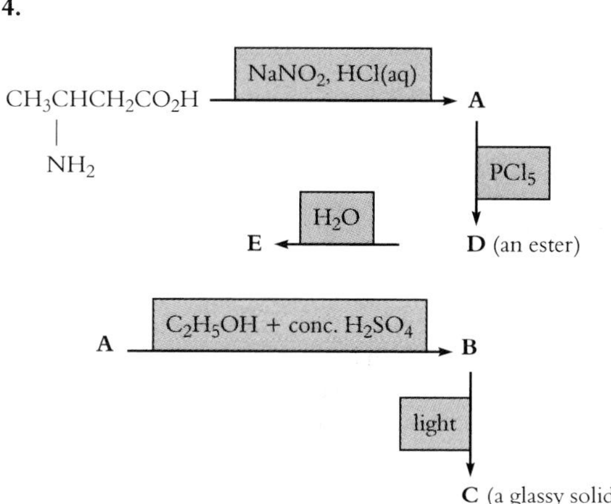

Some reactions starting from an amino acid are shown above. Write the structural formulae of substances **A** to **E**.

5. Say how you would distinguish between the following compounds:

A propanal CH_3CH_2CHO, **B** propanone CH_3COCH_3, **C** ethanoic acid CH_3CO_2H, **D** ethanamide CH_3CONH_2 and **E** 1-aminopropane, $CH_3CH_2CH_2NH_2$.

You are provided with 2,4-dinitrophenylhydrazine solution, ammoniacal silver nitrate solution, dilute aqueous sodium hydroxide, dilute hydrochloric acid and red litmus paper. Say what tests you would make and what results you would observe.

6. Give an example of an organic reaction which involves free radicals. Write a balanced equation. Outline a mechanism for the reaction.

7. Give an example of an organic reaction which involves the formation of carbocations. Write a balanced equation. Outline a mechanism for the reaction.

8. Give an example of an organic reaction which involves substitution by a nucleophile. Write a balanced equation. Outline a mechanism for the reaction.

9. *(a)* Study the reaction scheme given below. Compound **A** is a **straight** chain bromoalkane of molecular formula C_4H_9Br and compound **B** is a hydrocarbon: both can exist in **two** isomeric forms.

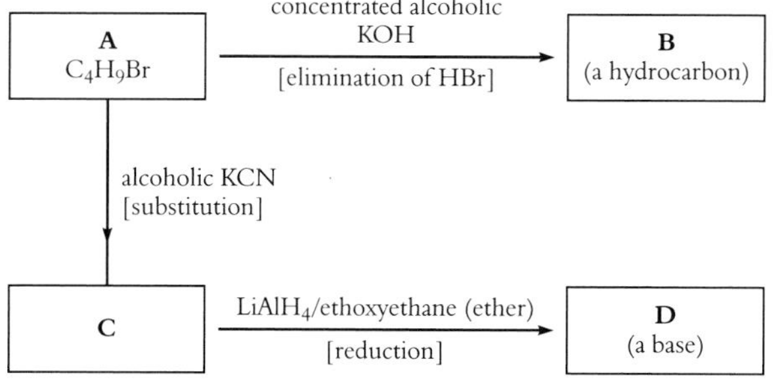

(i) State and explain the nature of the isomerism in **A** and draw diagrams illustrating the relationship of the two isomers of **A** to each other. **2**

(ii) State and explain the nature of the isomerism in **B** and draw diagrams illustrating the relationship of the two isomers of **B** to each other. **2**

(iii) Give the systematic names for **A** and **B**. **1**

(iv) Write down shortened structural formulae for **C** and **D**. **1**

(v) When **D** (1.0000 g) is titrated against 0.2000 mol dm^{-3} hydrochloric acid, 57.35 cm^3 of the latter is required for complete reaction. Calculate the relative molecular mass of **D**. **2**

(vi) Indicate and explain the approximate value of the pH which you would expect at the end-point of the titration in (v) above. **2**

(b) For each pair of compounds I to IV state a **chemical** test which will distinguish between the two compounds and describe the difference in what is observed in each case.

I. $CH_3COCH_2CH_2CH_3$ and $CH_3CH_2COCH_2CH_3$

II. $CH_3CH_2CH_2OH$ and $(CH_3)_2CHOH$

III. $CH_3CH_2CH_2CH_2COOH$ and $CH_3CH_2COCH_2CH_2OH$

IV. $CH_3CH_2NH_2$ and $C_6H_5NH_2$

WJEC (AS/AL)

10. *(a)* 2-Phenylethanol occurs naturally in rose oil and geranium oil; it is much used in the perfume industry. It can be synthesised from (chloromethyl)benzene by the following series of reactions.

$$C_6H_5CH_2Cl \xrightarrow{I} C_6H_5CH_2CN \xrightarrow{II}$$

(Chloromethyl)benzene

$$C_6H_5CH_2CO_2H \xrightarrow[LiAlH_4]{III} C_6H_5CH_2CH_2OH$$

2-Phenylethanol

(i) Suggest reagents and conditions for steps I and II.

(ii) What types of reacton are steps I and III? **4**

(b) Suggest simple one-step test-tube reactions by which the isomers in the following pairs can be distinguished from each other. You should state the reagents and conditions for each test, and describe how each of the isomers in the pair behaves.

(i) $CH_3CH(OH)CH_3$ and $CH_3CH_2CH_2OH$

(ii) CH_3COCH_3 and $CH_2{=}CHCH_2OH$

(iii) $CH_3CH_2COCH_2CH_3$ and $CH_3CH_2CH_2CH_2CHO$

(iv) $CH_3CH_2CO_2H$ and CH_3COCH_2OH **8**

C (AL)

11. *(a)* Each of the following conversions can be completed in not more than *three* steps. Use equations to show how you would carry out each of these conversions in the laboratory. For each conversion, give the reagent(s), conditions, and structure of the intermediate(s).

(i) $C_6H_5CH_2Br \rightarrow C_6H_5CH_2CH_2NH_2$

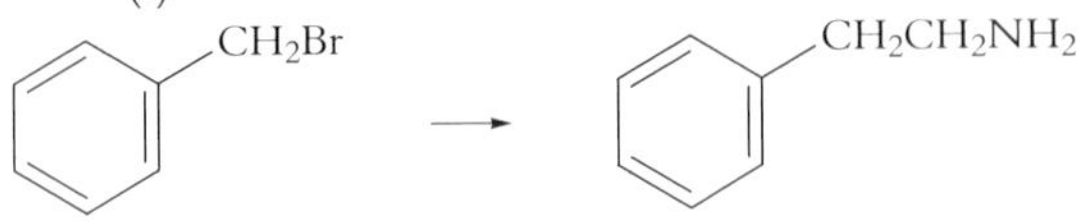

(ii) 1-bromo-1-$C(CH_3)_3$-cyclopentane $\rightarrow$ $H-\overset{O}{\overset{\|}{C}}-(CH_2)_3-\overset{O}{\overset{\|}{C}}-C(CH_3)_3$

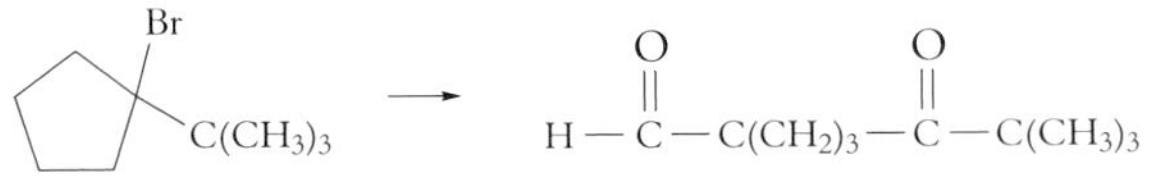

(iii) $C_6H_5COCl \rightarrow C_6H_5NH_2$

(iv) cyclohexanol (OH) $\rightarrow$ cyclohexane-1,2-diol (OH, OH)

12

(b) Give the structure of the major product in the following reaction and outline the mechanism of the reaction. (Movement of electron pairs should be indicated by curly arrows.)

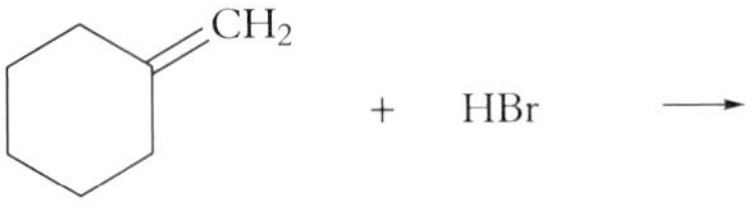

methylenecyclohexane ($=CH_2$) + HBr $\longrightarrow$

3

(c) Give a systematic name to each of the following compounds:

(i) $CH_3-\underset{|}{\overset{CH_3}{CH}}-CH_2-\overset{OH}{\underset{|}{CH}}-CH_3$

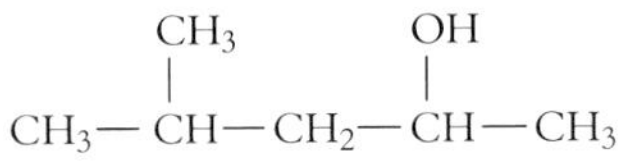

(ii) cyclohexenone ring bearing H_3C on the ring double bond carbon adjacent to C=O, and two CH_3 groups on one ring carbon

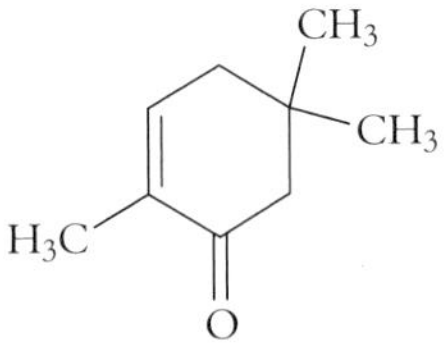

2

(HK)

12. *(a)* Suggest a chemical test to distinguish one compound from the other in each of the following pairs. Each test should include the reagent(s), the expected observation with each compound and the chemical equation(s).

(i) 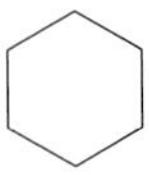 cyclohexane and cyclohexene

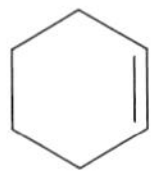

(ii) 1-ethylcyclopentanol (OH, CH_2CH_3) and cyclopentyl–$\overset{OH}{\overset{|}{C}H}-CH_3$

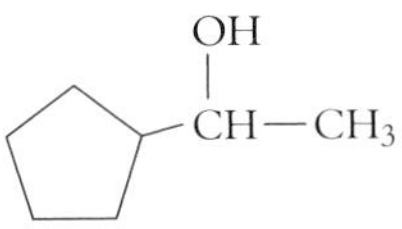

(iii) $C_2H_5-\overset{O}{\overset{\|}{C}}-O(CH_2)_2CH_3$ and $C_2H_5-\overset{O}{\overset{\|}{C}}-CH_2OC_2H_5$

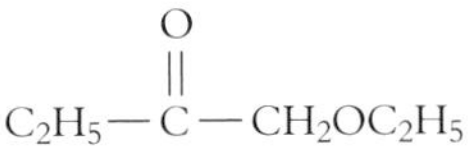

(iv) $C_2H_5-\overset{O}{\overset{\|}{C}}-C_2H_5$ and $CH_3(CH_2)_3-\overset{O}{\overset{\|}{C}}-H$

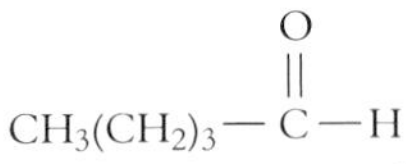

12

(b) Identify *J*, *K*, *L*, *M* and *N* in the following reactions.

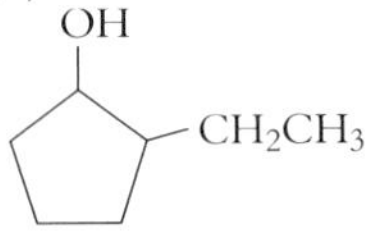

(i) $C_6H_5-\overset{O}{\overset{\|}{C}}-H \xrightarrow{J} C_6H_5CH=NOH$

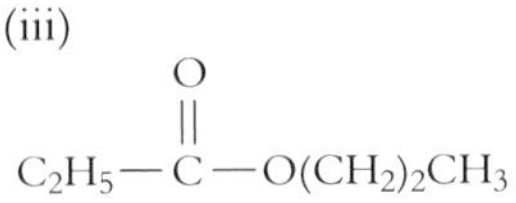

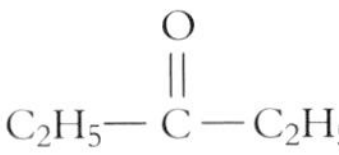

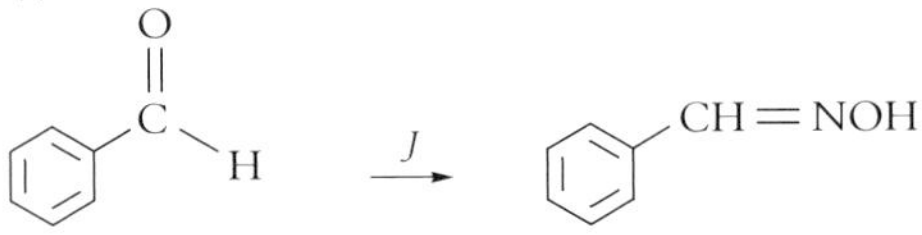

(ii) methylenecyclopentane ($=CH_2$) $\xrightarrow{K}$ 1-(OH)-1-(CH_2Br)cyclopentane

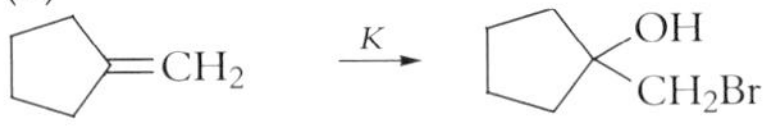

(iii) $(CH_3)_2C=CH\overset{O}{\overset{\|}{C}}-OCH_3 \xrightarrow{L} (CH_3)_2C=CHCH_2OH + CH_3OH$

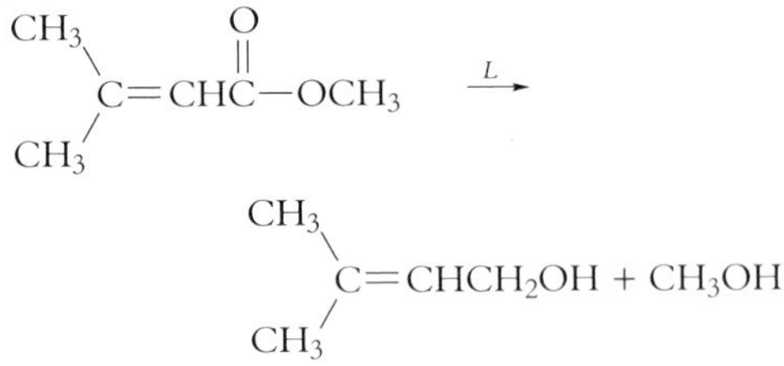

(iv) $CH_3CH_2CH_2-\overset{O}{\overset{\|}{C}}-NH_2 \xrightarrow[\text{heat}]{P_2O_5} M$

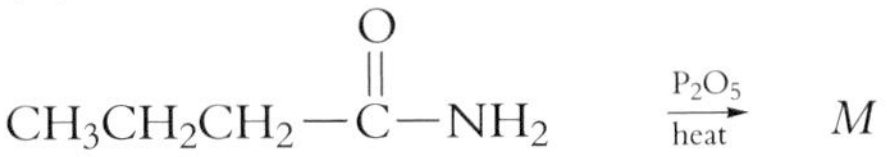

(v) benzene $\xrightarrow[\text{(2) NaOH(aq)}]{\text{(1) conc. } H_2SO_4} N$

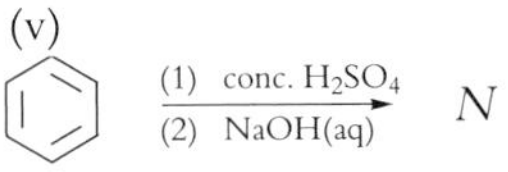

5

(HK)

13. The flow chart below summarises some of the organic reactions you have studied.

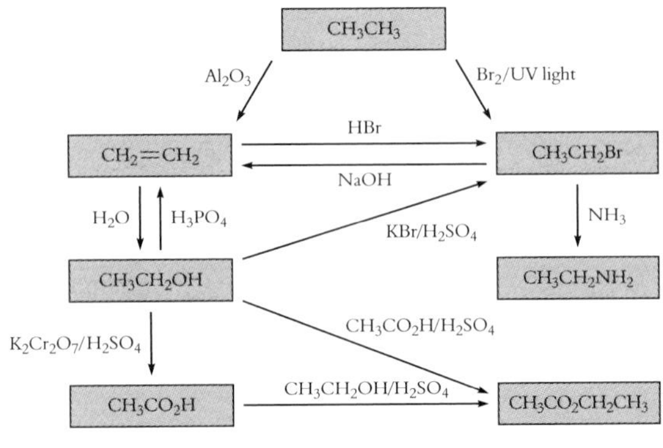

(a) Explain what is meant by the following reaction types. In each case select ONE example from the flow chart to illustrate your answer.
- free radical substitution
- nucleophilic substitution
- elimination
- oxidation
- esterification **10**

(b) How can ethene be converted to ethane? **1**

(c) An intermediate is sometimes isolated in the conversion of ethanol to ethanoic acid. What is its structural formula? **1**

(d) Using the chart, or otherwise, plan a two stage synthesis of propylamine, $CH_3CH_2CH_2NH_2$, starting from propan-1-ol. Give the name and formula of the product of the first stage of the synthesis and the reagents you would use for each stage. **2**

14 *L(N) (AS/AL)*

14. Consider the reaction sequence below.

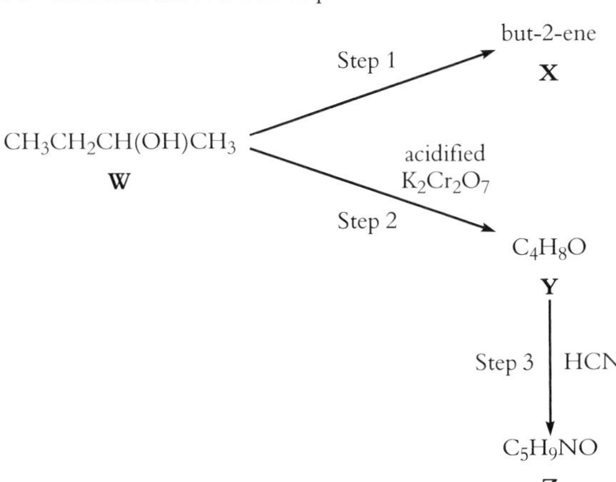

(a) Name **W**, deduce structures for **Y** and **Z** and state which of the four molecules **W**, **X**, **Y** and **Z** show optical isomerism. **5**

(b) (i) Give reagents and conditions and outline a mechanism for Step 1.

(ii) An alternative structural isomer of **X** is also formed in Step 1. Suggest a structure for this alternative product and, by considering the mechanism you have drawn in part *b*(i), explain why it is formed. **8**

(c) Give a reagent which could convert **Y** back into **W**. **1**

(d) In Step 3, **Y** undergoes nucleophilic addition. Explain the term *nucleophilic addition* and explain why compound **Y** undergoes this reaction. **4**

18 *NEAB (AS/AL)*

15. *(a)* A compound **H**, containing carbon, hydrogen and oxygen only, has the following composition by mass: C, 64.9%; H, 13.5%; O, 21.6%. **H** has two isomers **J** and **K**, which contain the same functional group. **H** does not react with acidified aqueous potassium dichromate(VI) solution, but both **J** and **K** are oxidised, **K** forming an acid.

Identify **H** and its isomers **J** and **K**, explaining how you arrive at your structures. **8**

(b) Isomer **J** also consists of two isomers and these react with concentrated sulphuric acid to form a hydrocarbon, which itself exists as two isomers, **L** and **M**.

(i) Identify the type of isomerism shown by **J**, and draw the structure of its two isomers.

(ii) Write an equation for the reaction of **J** with sulphuric acid, draw the structures of the isomers **L** and **M**, and identify the form of isomerism present in these two compounds. **8**

(c) Although ethanol and phenol both contain the —OH group, they sometimes react in different ways. Give **three** ways in which these two compounds differ chemically, giving equations where appropriate. **9**

25 *C (AS/AL)*

16. *(a)* Explain the meaning of the terms *empirical formula* and *molecular formula*. **3**

(b) Give the three molecular formulae for organic compounds which have the empirical formula CH_2O and relative molecular masses below 100.

In fact, four compounds, **H**, **J**, **K** and **L** fit this information.

H can be oxidised to a carboxylic acid, **M**, and also reduced to a primary alcohol, **N**.

M and **N** react together, when warmed in the presence of concentrated sulphuric acid, to form **J**.

L contains a carboxylic acid group and is a structural isomer of **J**.

K contains a carboxylic acid group and shows optical isomerism.

Draw the structures of the six compounds, **H**, **J**, **K**, **L**, **M** and **N**. **9**

(c) Glucose ($M_r = 180$) also has the empirical formula CH_2O. Using its molecular formula, write an equation for the fermentation of glucose to form ethanol and give a necessary condition for fermentation to occur. **3**

15 *NEAB (AS/AL)*

THE PERIODIC TABLE

Relative atomic mass — 1
Proton (atomic) number — 1
H Hydrogen

TRANSITION ELEMENTS

1	2											3	4	5	6	7	0
						1 H 1 Hydrogen											4 He 2 Helium
7 Li 3 Lithium	9 Be 4 Beryllium											11 B 5 Boron	12 C 6 Carbon	14 N 7 Nitrogen	16 O 8 Oxygen	19 F 9 Flourine	20 Ne 10 Neon
23 Na 11 Sodium	24 Mg 12 Magnesium											27 Al 13 Aluminium	28 Si 14 Silicon	31 P 15 Phosphorus	32 S 16 Sulphur	35.5 Cl 17 Chlorine	40 Ar 18 Argon
39 K 19 Potassium	40 Ca 20 Calcium	45 Sc 21 Scandium	48 Ti 22 Titanium	51 V 23 Vanadium	52 Cr 24 Chromium	55 Mn 25 Manganese	56 Fe 26 Iron	59 Co 27 Cobalt	59 Ni 28 Nickel	63·5 Cu 29 Copper	65 Zn 30 Zinc	70 Ga 31 Gallium	73 Ge 32 Germanium	75 As 33 Arsenic	79 Se 34 Selenium	80 Br 35 Bromine	84 Kr 36 Krypton
85 Rb 37 Rubidium	88 Sr 38 Strontium	89 Y 39 Yttrium	91 Zr 40 Zirconium	93 Nb 41 Niobium	96 Mo 42 Molybdenum	98 Tc 43 Technetium	101 Ru 44 Ruthenium	103 Rh 45 Rhodium	106 Pd 46 Palladium	108 Ag 47 Silver	112 Cd 48 Cadmium	115 In 49 Indium	119 Sn 50 Tin	122 Sb 51 Antimony	128 Te 52 Tellurium	127 I 53 Iodine	131 Xe 54 Xenon
133 Cs 55 Caesium	137 Ba 56 Barium	139 La 57 Lanthanum*	178.5 Hf 72 Hafnium	181 Ta 73 Tantalum	184 W 74 Tungsten	186 Re 75 Rhenium	190 Os 76 Osmium	192 Ir 77 Iridium	195 Pt 78 Platinum	197 Au 79 Gold	201 Hg 80 Mercury	204 Tl 81 Thallium	207 Pb 82 Lead	209 Bi 83 Bismuth	210 Po 84 Polonium	210 At 85 Astatine	222 Rn 86 Radon
223 Fr 87 Francium	226 Ra 88 Radium	227 Ac 89 Actinium†	Db 104 Dubnium	Jl 105 Joliotium	Rf 106 Rutherfordium	Bh 107 Bohrium	Hn 108 Hahnium	Mt 109 Meitnerium									

*58-71 Lanthanum series	140 Ce 58 Cerium	141 Pr 59 Praseodymium	144 Nd 60 Neodymium	147 Pm 61 Promethium	150 Sm 62 Samarium	152 Eu 63 Europium	157 Gd 64 Gadolinium	159 Tb 65 Terbium	162 Dy 66 Dysprosium	165 Ho 67 Holmium	167 Er 68 Erbium	169 Tm 69 Thulium	173 Yb 70 Ytterbium	175 Lu 71 Lutetium
†90-103 Actinium series	232 Th 90 Thorium	231 Pa 91 Protactinium	238 U 92 Uranium	237 Np 93 Neptunium	242 Pu 94 Plutonium	243 Am 95 Americium	247 Cm 96 Curium	245 Bk 97 Berkelium	251 Cf 98 Californium	254 Es 99 Einsteinium	253 Fm 100 Fermium	256 Md 101 Mendelevium	254 No 102 Nobelium	257 Lr 103 Lawrencium

UNITS

BASIC SI UNITS

Physical Quantity	*Name of unit*	*Symbol*
Length	metre	m
Mass	kilogram	kg
Time	second	s
Electric current	ampere	A
Temperature	kelvin	K
Amount of substance	mole	mol
Light intensity	candela	cd

DERIVED SI UNITS

Physical Quantity	*Name of unit*	*Symbol*	*Definition*
Energy	joule	J	$kg\ m^2\ s^{-2}$
Force	newton	N	$J\ m^{-1}$
Electric charge	coulomb	C	A s
Electric potential difference	volt	V	$J\ A^{-1}\ s^{-1}$
Electric resistance	ohm	Ω	$V\ A^{-1}$
Area	square metre		m^2
Volume	cubic metre		m^3
Density	kilogram per cubic metre		$kg\ m^{-3}$
Pressure	newton per square metre or pascal		$N\ m^{-2}$ or Pa
Molar mass	kilogram per mole		$kg\ mol^{-1}$

With all these units, the following prefixes (and others) may be used.

Prefix	*Symbol*	*Meaning*
deci	d	10^{-1}
centi	c	10^{-2}
milli	m	10^{-3}
micro	μ	10^{-6}
nano	n	10^{-9}
kilo	k	10^{3}
mega	M	10^{6}
giga	G	10^{9}
tera	T	10^{12}

INDEX OF SYMBOLS AND ABBREVIATIONS

A = area
A = mass (nucleon) number
A* = activated/excited A
[A] = concentration/mol dm^{-3} of A
$[A]_0$ = initial concentration of A
A_r = relative atomic mass
Ar = aryl group
AR = anionic radius
c = concentration/mol dm^{-3}
c = specific heat capacity
c = velocity of light
C = heat capacity
C = charge
C = coulomb
cpm = counts per minute
CR = cationic radius
d = distance
e = elementary charge
E = electromotive force (emf)
E = energy
$E^{\ominus}$ = standard electrode potential
E_a = activation energy
EA = electron affinity
f = force
F = Faraday constant
G = free energy
$G^{\ominus}$ = standard free energy
$\Delta G^{\ominus}$ = change in standard free energy
G.M.V. = gas molar volume
h = Planck constant
H = enthalpy
$\Delta H^{\ominus}$ = change in standard enthalpy
I = electric current
IE = ionisation energy
IR = ionic radius
IR = infrared
k = rate/velocity constant
K = equilibrium constant
K_a = acid dissociation constant
K_b = base dissociation constant
K_c = equilibrium constant in concentration terms
K_p = equilibrium constant in partial pressure terms
K_{sp} = solubility product
K_w = ionic product for water
l = length
l = second quantum number
L = Avogadro constant
m = mass
m_l = third quantum number
m_s = fourth (spin) quantum number
M = molar mass
M_r = relative molar/molecular mass
n = principal quantum number
n_A = amount/mol of A
N = number of molecules
Ox. No. = oxidation number
p_B = partial pressure of B
p_B^0 = vapour pressure of pure B
P = pressure
ppm = parts per million
ΔQ = heat absorbed
r = rate
R = electric resistance
R = universal gas constant
R = alkyl group
R_F value
R.D.S. = rate-determining step
rms = root mean square
rtp = room temperature and pressure
S = entropy
$\Delta S^{\ominus}$ = change in standard entropy
stp = standard temperature and pressure
svp = saturated vapour pressure
t = time
t = temperature/°C
$t_{1/2}$ = half-life

T = temperature/K
T_b = boiling temperature
T_f = freezing temperature
T_m = melting temperature
u = atomic mass unit
U = internal energy
ΔU = change in internal energy
UV = ultraviolet
v = velocity/rate
v_0 = initial velocity/rate
V = volume
V = potential difference
V = volts
V_m = molar volume
x_A = mole fraction of A
X^- = halide ion
X_2 = halogen
Z = proton (atomic) number

Greek letters:
α = position of group
λ = wavelength
ν = frequency
ρ = density

ANSWERS TO NUMERICAL PROBLEMS AND SELECTED QUESTIONS

PART 1 THE FOUNDATION

CHAPTER 1 THE ATOM

Checkpoint 1.8: Atoms

1. *(a)* 19p, 19e, 20n *(b)* 13p, 13e, 14n *(c)* 56p, 56e, 81n *(d)* 88p, 88e, 138n
2. *(a)* 6p, 6e, 6n *(b)* 6p, 6e, 8n *(c)* 1p, 1e *(d)* 1p, 1e, 1n *(e)* 1p, 1e, 2n *(f)* 38p, 38e, 49n *(g)* 38p 38e, 52n *(h)* 92p, 92e, 143n *(i)* 92p, 92e, 146n

Checkpoint 1.9: Mass spectrometry

2. Mean $A_r = 35.45$
3. The peak at 84 is due to $CH_2{}^{35}Cl_2$; that at 86 to $CH_2{}^{35}Cl^{37}Cl$; that at 88 to $CH_2{}^{37}Cl_2$. Ratios ${}^{35}Cl_2 : {}^{37}Cl_2 = 9 : 1$ and ${}^{35}Cl^{37}Cl : {}^{37}Cl_2 = 6 : 1$

Checkpoint 1.10A: Nuclear reactions 1

1. ${}^{35}Cl$, ${}^{37}Cl$
2. The half-life of the radioactive isotope is one year (the time taken for 8 g to decay to 4 g). In the same time, 6 g will decay to 3 g.
3. *(a)* ${}^{14}_{7}N$ *(b)* ${}^{17}_{8}O$ *(c)* ${}^{226}_{88}Ra$ ${}^{222}_{86}Rn$ *(d)* ${}^{73}_{33}As$ ${}^{73}_{32}Ge$ *(e)* ${}^{27}_{14}Si$ *(f)* ${}^{1}_{0}n$

Questions on Chapter 1

2. 64 and 73; ${}^{13}C_2{}^{1}H_5{}^{35}Cl$, ${}^{12}C_2{}^{1}H_5{}^{37}Cl$, ${}^{13}C_2{}^{1}H_3{}^{2}H_2{}^{35}Cl$, ${}^{12}C^{13}C^{1}H_4{}^{2}H^{35}Cl$
3. For peaks of mass = 2, 3, 4, 5, 6 Ratio of heights = 1 : 2 : 3 : 2 : 1
5. Peaks at 49 (${}^{14}N^{35}Cl$), 51 (${}^{14}N^{37}Cl$), 84 (${}^{14}N^{35}Cl_2$), 86 (${}^{14}N^{35}Cl^{37}Cl$), 88 (${}^{14}N^{37}Cl_2$), 119 (${}^{14}N^{35}Cl_3$), 121 (${}^{14}N^{35}Cl_2{}^{37}Cl$), 123 (${}^{14}N^{35}Cl^{37}Cl_2$), 125 (${}^{14}N^{37}Cl_3$)
6. ${}^{228}_{90}Z$
7. **X** is in Group 5, **Y** is in Group 3 and **Z** is in Group 4.
8. *(a)* $a = 35$, $b = 16$, **X** = S *(b)* $c = 4$, $d = 2$, **Y** = He
9. 80.0
10. Peaks: m/z is 46 molecular peak, 47 is ${}^{12}C^{13}CH_5OH$, 45 is $C_2H_5O^+$, 31 is CH_2OH^+, 29 is $C_2H_5{}^+$ and 27 is $C_2H_3{}^+$.
11. 44, 46, 48
12. Empirical and molecular formula C_4H_8O. Combining the fragments gives $CH_3CH_2CH_2CHO$.
13. *(b)* (ii) 24.320
14. *(d)* (i) ${}^{58}_{28}Ni^+$ (ii) 58.70
15. *(b)* 1.007 25
16. *(c)* Ratio ${}^{10}B : {}^{11}B = 1 : 4$
    (d) Ratio B : H = 1 : 2.5
    Empirical formula B_2H_5, molecular formula B_4H_{10}. Highest $m/z = {}^{11}B_4H_{10} = 54$
17. *(a)* (i) ${}^{222}_{86}Rn$ (ii) ${}^{17}_{7}N$ (iii) ${}^{60}_{28}Ni$
    (c) $${}^{241}_{95}Am \xrightarrow{{}^{0}_{-1}\beta} {}^{241}_{96}X \xrightarrow{{}^{0}_{-1}\beta} {}^{241}_{97}Y \xrightarrow{{}^{4}_{2}He} {}^{237}_{95}Am$$
    (d) (i) 24.32
    (ii) CH_3 has $m/e = 15$; CH_3CO has $m/e = 43$
    (iii) C_2H_5 has $m/e = 29$, suggesting propanal
18. *(c)* (i) ${}^{20}_{10}Ne$ (ii) ${}^{20}_{10}Ne^{2+}$ (iii) 20.2

CHAPTER 2 THE ATOM – THE ARRANGEMENT OF ELECTRONS

Checkpoint 2.8: Electronic configurations

1. Calcium has $Z = 20$; therefore each isotope has 20 protons, and Ca-40 has 20 neutrons, Ca-42 has 22n, Ca-43 has 23n, Ca-44 has 24n, Ca-46 has 26n and Ca-48 has 28n.

Questions on Chapter 2

6. *(b)* $^{7}_{3}X$ *(c)* (i) $(Ar)4s^2$
8. *(a)* 20.18 *(b)* (i) 3.67 dm^3 (ii) 0.734 dm^3
10. *(c)* (i) $1s^2 2s^2 2p^6 3p$

CHAPTER 3 EQUATIONS AND EQUILIBRIA

Checkpoint 3.2: Formulae

1. *(a)* 12 *(b)* 14 *(c)* 5 *(d)* 8 *(e)* 12 *(f)* 30 *(g)* 13 *(h)* 14 *(i)* 17 *(j)* 39

Checkpoint 3.5: Relative molecular mass

1. 40, 74.5, 40, 74, 63, 123.5, 80, 159.5, 146

Checkpoint 3.6: The Avogadro constant

1. £150 million million
2. *(a)* 1 mol = 39 g *(b)* 1300 g
3. 2.7×10^{-7} p

Checkpoint 3.7: The mole

1. *(a)* 72 g *(b)* 8 g *(c)* 16 g *(d)* 8 g *(e)* 64 g
2. *(a)* 0.33 mol *(b)* 0.25 mol *(c)* 2.0 mol *(d)* 0.010 mol *(e)* 0.33 mol
3. *(a)* 88 g *(b)* 980 g *(c)* 117 g *(d)* 37 g
4. *(a)* 482 g mol^{-1} *(b)* 342 g mol^{-1} *(c)* 368 g mol^{-1}
5. *(a)* 2.5×10^{-3} mol *(b)* 5.00×10^{-2} mol *(c)* 5.40×10^{-2} mol
6. *(a)* 3.0×10^{19} *(b)* 7.5×10^{16} *(c)* 3.0×10^{11}

Checkpoint 3.10: Formulae and percentage composition

1. *(a)* 72% *(b)* 39% *(c)* 80%
2. *(a)* CO_2 *(b)* C_3O_2 *(c)* $Na_2S_2O_3$ *(d)* Na_2SO_4
3. *(a)* MgO *(b)* $CaCl_2$ *(c)* $FeCl_3$
4. *(a)* $a = 7$ *(b)* $b = 2$ *(c)* $c = 1/2$

Checkpoint 3.11: Masses of reacting solids

1. 94 tonnes.
2. 25.4%
3. 21.0 g, 93%
4. 21.2 g, 96%

Checkpoint 3.12: Reacting volumes of gases

1. 2.80 dm^3 at stp
2. 26.8 g
3. 26.1 g NaCl, 43.8 g H_2SO_4

Checkpoint 3.13A: Concentration

1. *(a)* 0.20 M *(b)* 0.020 M *(c)* 0.20 M *(d)* 8 M
2. *(a)* 10 g *(b)* 0.365 g *(c)* 4.9 g *(d)* 0.14 g

Checkpoint 3.13B: Solutions

4. 140 cm^3
5. *(b)* 0.42 g

Checkpoint 3.14: Titration

1. 0.15 M
2. *(a)* 0.25 M *(b)* 10.0 cm^3
3. *(a)* 2 *(b)* 0.24 M
4. *(a)* Stopit
5. 1.0×10^{-5} mol dm^{-3}

6. 98.9%
7. $n = 10$
8. 58.2%

Checkpoint 3.15: Redox reactions

1. *(a)* $Sn^{2+}(aq) \longrightarrow Sn^{4+}(aq) + 2e^-$
   (b) $2Cl^-(aq) \longrightarrow Cl_2(aq) + 2e^-$
   (c) $H_2S(aq) \longrightarrow S(s) + 2H^+(aq) + 2e^-$
   (d) $SO_3^{2-}(aq) + H_2O(l) \longrightarrow SO_4^{2-}(aq) + 2H^+(aq) + 2e^-$
   (e) $H_2O_2(aq) \longrightarrow O_2(g) + 2H^+(aq) + 2e^-$

2. *(a)* $Br_2(aq) + 2e^- \longrightarrow 2Br^-(aq)$
   (b) $MnO_2(s) + 4H^+(aq) + 2e^- \longrightarrow Mn^{2+}(aq) + 2H_2O(l)$
   (c) $PbO_2(s) + 4H^+(aq) + 2e^- \longrightarrow Pb^{2+}(aq) + 2H_2O(l)$
   (d) $2IO^-(aq) + 4H^+(aq) + 2e^- \longrightarrow I_2(aq) + 2H_2O(l)$
   (e) $2ClO_3^-(aq) + 12H^+(aq) + 10e^- \longrightarrow Cl_2(aq) + 6H_2O(l)$

3. *(a)* $2Fe^{3+}(aq) + 2I^-(aq) \longrightarrow 2Fe^{2+}(aq) + Sn^{4+}(aq)$
   (b) $2Fe^{3+}(aq) + Sn^{2+}(aq) \longrightarrow 2Fe^{2+}(aq) + I_2(aq)$
   (c) $2MnO_4^-(aq) + 16H^+(aq) + 10Cl^-(aq) \longrightarrow 2Mn^{2+}(aq) + 8H_2O(l) + 5Cl_2(aq)$
   (d) $2MnO_4^-(aq) + 16H^+(aq) + 5H_2O_2(aq) \longrightarrow 2Mn^{2+}(aq) + 8H_2O(l) + 5O_2(aq)$
   (e) $2MnO_4^-(aq) + 16H^+(aq) + 10I^-(aq) \longrightarrow 2Mn^{2+}(aq) + 8H_2O(l) + 5I_2(aq)$
   (f) $Cr_2O_7^{2-}(aq) + 14H^+(aq) + 6I^-(aq) \longrightarrow 2Cr^{3+}(aq) + 7H_2O(l) + 3I_2(aq)$
   (g) $Cr_2O_7^{2-}(aq) + 14H^+(aq) + 6Fe^{2+}(aq) \longrightarrow 2Cr^{3+}(aq) + 7H_2O(l) + 6Fe^{3+}(aq)$
   (h) $2MnO_4^-(aq) + 16H^+(aq) + 5Sn^{2+}(aq) \longrightarrow 2Mn^{2+}(aq) + 8H_2O(l) + 5Sn^{4+}(aq)$
   (i) $Cr_2O_7^{2-}(aq) + 8H^+(aq) + 3SO_3^{2-}(aq) + 3H_2O(l) \longrightarrow 2Cr^{3+}(aq) + 7H_2O(l) + 3SO_4^{2-}(aq)$
   (j) $2MnO_4^-(aq) + 6H^+(aq) + 5SO_3^{2-}(aq) \longrightarrow 2Mn^{2+}(aq) + 3H_2O(l) + 5SO_4^{2-}(aq)$
   (k) $MnO_4^-(aq) + 8H^+(aq) + 5Fe^{2+}(aq) \longrightarrow Mn^{2+}(aq) + 4H_2O(l) + 5Fe^{3+}(aq)$
   (l) $Cr_2O_7^{2-}(aq) + 14H^+(aq) + 3Sn^{2+}(aq) \longrightarrow 2Cr^{3+}(aq) + 7H_2O(l) + 3Sn^{4+}(aq)$
   (m) $PbO_2(s) + 4H^+(aq) + 4Cl^-(aq) \longrightarrow PbCl_2(aq) + 2H_2O(l) + Cl_2(g)$
   (n) $ClO_3^-(aq) + 4H^+(aq) + 4I^-(aq) \longrightarrow ClO^-(aq) + 2H_2O(l) + 2I_2(aq)$
   (o) $Br_2(aq) + 2I^-(aq) \longrightarrow 2Br^-(aq) + I_2(aq)$
   (p) $2Cl_2(aq) + IO^-(aq) + 2H_2O(l) \longrightarrow 4Cl^-(aq) + IO_3^-(aq) + 4H^+(aq)$
   (q) $Br_2(aq) + H_2S(aq) \longrightarrow S(s) + 2Br^-(aq) + 2H^+(aq)$

Checkpoint 3.16A: Oxidation number

1. 0, +1, 0, +2, +1, 0, 0, 0, −3, −1, 0, +1, 0, −1
2. +2, +1, −2, +4, +6, +2, +4, +3, +6, +2, +4, +5
3. *(a)* +2, +4, +4, +1, +3, +5, +5 *(b)* +2, +3, +4, +7, +6
   (c) +3, +3, +5, +3 *(d)* +6, +6, +6
   (e) −1, +1, +5, 0, +3, +1

Checkpoint 3.16B: Equations and oxidation numbers

2. $ICl_3 + 3KI \longrightarrow 2I_2 + 3KCl$; +3
3. *(a)* −2, −1, −2, 0 *(b)* 0, +5, +2, +2
   (c) +2, +6, +3, +3 *(d)* +2, 0, +2.5, −1
   (e) +5, −1, −1, −1, +1, −1 *(f)* +6, +6

Checkpoint 3.18: Redox titrations

1. 96%
2. 5.50×10^{-2} mol dm^{-3}, 30.8 cm^3 O_2
3. 95.0%
4. 60.0 cm^3

Questions on Chapter 3

3. 30.0 cm^3
5. 20.0%
6. 20.1%
9. *(a)* 3.20×10^8 mol *(b)* 3.20×10^8 mol
   (c) 7.7×10^9 dm^3 SO_2
10. *(c)* (i) 3.0×10^{-3} mol (ii) 1.5×10^{-3} mol
    (iii) 1.905 g (3.0×10^{-2} mol) (iv) 95%
11. *(f)* 12.0%
12. *(c)* (i) −1 *(d)* 13.3 cm^3
13. *(b)* % Ag 87.900% and Cu 12.100%
    (c) (ii) % Ag = 91.83%
14. *(d)* (i) 38.0 g product
    (ii) mass of MgO = 28.0 g, mass of Mg_3N_2 = 10.0 g
    (iii) 2.3
15. *(a)* (ii) + 4 and + 6
    (b) (i) 1.0×10^{-3} mol $KMnO_4$ (ii) SO_2 0.32 g m^{-3}
16. *(b)* (ii) 69.3%
17. *(a)* (iii) 121.9
    (b) (i) + 3 (ii) 360 dm^3 CO_2
18. *(a)* (i) Ox. No. of N = + 5 (ii) Ox. No. of N = + 3
    (b) 1.304 g KNO_2
    (c) (i) MnO_4^- 4.335×10^{-4} mol (ii) 0.01084 mol
    (iii) 0.9214 g (iv) 70.7%
19. *(b)* (i) 3.448×10^{-2} mol HCl (ii) 345 cm^3
20. *(d)* $KMnO_4$ 3.62×10^{-4} mol, H_2O_2 9.05×10^{-4} mol,
    $[H_2O_2]$ 0.0362 mol dm^{-3}
21. *(a)* Thiosulphate 2.00×10^{-3} mol
    (c) I_2 1.00×10^{-3} mol
    (d) Cu^{2+} 2.00×10^{-3} mol
    (e) Changes in ox. no. are I^-, + 2 and Cu^{2+}, −2.

CHAPTER 4 THE CHEMICAL BOND

Questions on Chapter 4

7. *(b)* (ii) **D**: $(^{35}Cl^{37}Cl)^+$, **E**: $(^{37}Cl^{37}Cl)^+$
   (c) (iii) Li—H + 0.9, C—O + 1.0, B—C + 0.5
9. *(a)* (i) 19%
   (b) **X** = B, a = 5, b = 12

CHAPTER 5 THE SHAPES OF MOLECULES

Questions on Chapter 5

8. *(a)* Ne: 10p, 10n, 10e Na^+: 11p, 12n, 10e
   S^{2-}: 16p, 16n, 18e α particle : 2p, 2n

PART 2 PHYSICAL CHEMISTRY

CHAPTER 7 GASES

Checkpoint 7.6: Partial pressures

1. 2.5×10^4 N m^{-2}
2. *(a)* 3.03×10^4 N m^{-2} CO, 5.05×10^4 N m^{-2} O_2,
   2.02×10^4 N m^{-2} CO_2
   (b) 3.03×10^4 N m^{-2} CO, 5.05×10^4 N m^{-2} O_2
3. 5.33×10^5 N m^{-2}
4. *(a)* $p(NH_3) = 1.96 \times 10^4$ N m^{-2}, $p(H_2) = 5.39 \times 10^4$ N m^{-2};
   $p(N_2) = 2.45 \times 10^4$ N m^{-2}
   (b) No change
5. The volume is *(a)* halved, *(b)* increased by a factor of 3,
   (c) increased by a factor of 2.5, *(d)* decreased by a factor of 4.
6. The pressure *(a)* doubles, *(b)* increases by a factor 873/573,
   (c) increases by a factor of 4, *(d)* is halved.
7. V_2 = 819 dm^3
8. $O_2 = 5.88 \times 10^{-5}$ mol, Hb 1.47×10^{-5} mol, *M* of
   Hb = 6.80×10^{-4} g mol^{-1}
9. *(a)* (i) C_2H_4O (ii) M = 88 g mol^{-1} (iii) $C_4H_8O_2$
   (b) (ii) 84.0 dm^3 at rtp

Questions on Chapter 7

3. 24.9 dm^3 mol^{-1}

CHAPTER 8 LIQUIDS

Checkpoint 8.4: Molar mass of volatile liquid

1. 84 g mol^{-1}
2. M = 49.5 g mol^{-1}, formula is $C_2H_4Cl_2$ with M = 99 g mol^{-1}
3. M = 53 g mol^{-1}, formula is $(CH_2)_{14}$ so M = 56 g mol^{-1}
4. 154 g mol^{-1}

Checkpoint 8.5: Raoult's Law

1. 41 kPa
2. 30 kPa
4. svp is > calculated, and T_b is < calculated

Checkpoint 8.6: Partition

1. B
2. *(a)* 3.57 g *(b)* 4.01 g

Questions on Chapter 8

5. 3.55 g
11. *(a)* (i) a horizontal line, since PV = constant
    (c) $M = 120$
12. *(c)* (1) 2.92×10^3 Pa (2) $n = 3.85 \times 10^{-7}$ mol
13. *(a)* (ii) N_2 0.05 Pa, O_2 = 0.20 Pa, total pressure = 0.25 Pa
    (c) (i) $p_A = 8$ kPa, $p_B = 12$ kPa, 1. Total vp = 20 kPa, 2. n_A in vapour = 0.4
    (ii) 1. positive, 2. positive
14. *(b)* CH_3O
    (c) (iii) 62 (iv) $C_2H_6O_2$

CHAPTER 9 SOLUTIONS

Checkpoint 9.3: Solutions

2. *(a)* (i) 80 g per 100 g water
    (ii) 50 g (approximately) of solid crystallise
    (b) (i) 105 g per 100 g water (ii) 40 g per 100 g water
    (c) 2 g sodium chloride and 75 g potassium nitrate

CHAPTER 10 THERMOCHEMISTRY

Checkpoint 10.8: Combustion

1. 983 kJ; $\Delta H^{\ominus}_{C}$ at 37 °C, rather than at 25 °C
2. 12.0 g
3. *(c)* exothermic, $mc\Delta T = -836$ J *(d)* $mc\Delta T = 1.71$ kJ
4. Amount of Mg = 0.0833 mol, $\Delta H = -153$ kJ mol^{-1}
5. *(a)* $\Delta H = -1338$ J *(b)* $\Delta H = -53.5$ kJ mol^{-1}

Checkpoint 10.10: Enthalpy changes

1. *(a)* – *(b)* – *(c)* – *(d)* + *(e)* –

Checkpoint 10.11: Standard enthalpy changes

2. $\Delta H^{\ominus} = +1.9$ kJ mol^{-1}
3. $\Delta H^{\ominus} = +0.33$ kJ mol^{-1}
4. $\Delta H^{\ominus}_{r} = -142.9$ kJ mol^{-1}
5. *(a)* $\Delta H^{\ominus} = -1214$ kJ mol^{-1}
7. $\Delta H^{\ominus}_{F} = -457$ kJ mol^{-1}
8. *(a)* $\Delta H^{\ominus} = -126$ kJ mol^{-1}
    (b) $\Delta H^{\ominus} = -168$ kJ mol^{-1} of reaction
    (c) $\Delta H^{\ominus} = -92$ kJ mol^{-1}
    (d) prefer exothermic reaction *(b)*

Checkpoint 10.14: Standard enthalpy of reaction and average standard bond enthalpies

1. $\Delta H^{\ominus}_{6} = -604$ kJ mol^{-1}
2. *(a)* -1560 kJ mol^{-1}
    (b) -1370 kJ mol^{-1}
    (c) -286 kJ mol^{-1}
    (d) -5520 kJ mol^{-1}
3. **E** = -372 kJ mol^{-1}

Checkpoint 10.15: Entropy

1. *(a)* **B** *(b)* **B** *(c)* **A** *(d)* **B** *(e)* **A**
    (f) H-bonding in liquid phase
2. *(a)* decrease *(b)* increase *(c)* decrease *(d)* decrease
3. *(b)* < *(c)* < *(e)* < *(a)* < *(d)* < *(f)*
4. *(a)* – *(b)* + *(c)* – *(d)* – *(e)* +

Checkpoint 10.16: Free energy

2. *(a)* + *(b)* + *(c)* + *(d)* – *(e)* –
3. *(a)* Negative because both ΔH and $(-T\Delta S)$ are –ive.
    (b) Could be +ive or –ive because ΔH is +ive and $(-T\Delta S)$ is –ive.
    (c) Could be +ive or –ive because ΔH is –ive and $(-T\Delta S)$ is +ive.
    (d) Positive because both ΔH and $(-T\Delta S)$ are +ive.
    (e) In *(a)*, no: ΔH and $(-T\Delta S)$ are both –ive at all temperatures.
    In *(b)* yes: at high T $(-T\Delta S)$ has a higher –ive value and may outweigh +ive ΔH.
    In *(c)* yes, –ive ΔH could be outweighed by a +ive $(-T\Delta S)$ at high T.
    In *(d)*, no, ΔH and $(-T\Delta S)$ are both +ive at all temperatures.
4. $\Delta H^{\ominus} = 96$ kJ mol^{-1}; $\Delta S^{\ominus} = 0.138$ kJ mol^{-1}
    At 300 K, $\Delta G^{\ominus} = +54.6$ kJ mol^{-1}; reaction is not feasible.
    At 800 K, $\Delta G^{\ominus} = -14.4$ kJ mol^{-1}; reaction is feasible.

Questions on Chapter 10

2. $\Delta H^{\ominus} = -436$ kJ mol^{-1}
3. *(a)* (C—C) bond = 331 kJ mol^{-1} *(b)* 3990 kJ mol^{-1}
4. L.E. = -2346 kJ mol^{-1}
5. *(b)* $\Delta H^{\ominus} = -140$ kJ mol^{-1}
6. *(b)* both = -240 kJ mol^{-1}
7. $\Delta H_{solution} = -56$ kJ mol^{-1}
8. *(c)* The sum (+lattice enthalpy– hydration enthalpy) has values/kJ mol^{-1} of LiCl –40, LiF +6, NaCl –3, NaF +3. The salt with a negative value or the lower positive value of this sum is the more soluble.
9. *(b)* (i) decrease (ii) decrease
    (iii) increase (iv) no change
11. *(a)* -2250 kJ mol^{-1}
13. *(b)* coal -394 kJ mol^{-1}, gas -891 kJ mol^{-1}
14. *(a)* $\Delta H^{\ominus}_{L} = -717$ kJ mol^{-1} *(b)* $\Delta H^{\ominus}_{solution} = +32$ kJ mol^{-1}
    (c) $\Delta H^{\ominus} = -126$ kJ mol^{-1}
15. *(c)* $\Delta H^{\ominus}_{F} = -106$ kJ mol^{-1}
16. *(b)* (i) $\Delta G^{\ominus} = -543$ kJ mol^{-1}
    (ii) $\Delta S^{\ominus}$ is negative, therefore $\Delta G^{\ominus}$ becomes less negative with increasing T.
    (iii) Reactions **I** and **III** do not involve a change in the number of moles of gas. In **II**, 2.5 moles of gas form 3 moles of gas, and $\Delta S^{\ominus}$ is positive.
    (c) (i) $\Delta G^{\ominus}$ is negative, $K_c > 1$
    (ii) For all the reactions $\Delta G^{\ominus}$ is negative. Activation energy. At 3000 K reaction **I** is feasible.
    (d) (ii) 3200 K when K_c(**I**) = K_c(**II**), K_c(**IV**) = 1
17. *(b)* $\Delta S^{\ominus} = 119$ J K^{-1} mol^{-1}. With $\Delta H^{\ominus}$ negative and $\Delta S^{\ominus}$ positive $\Delta G^{\ominus}$ has a negative value. At equilibrium $\Delta G^{\ominus} = 0$. $\Delta H^{\ominus} = 44.4$ kJ mol^{-1}

(c) Since $\Delta S^\ominus$ is negative, $T\Delta S^\ominus$ is positive therefore $\Delta G^\ominus$ can become negative when the positive value of $\Delta H^\ominus$ is compensated by the negative value of $-T\Delta S^\ominus$.
(e) $T = 1085$ K

18. (b) $\Delta H^\ominus_{solution} = +969 - 991 = -22$ kJ mol^{-1}
(c) EA = −322 kJ mol^{-1}
19. (b) 30 kJ mol^{-1}

CHAPTER 11 EQUILIBRIA

Checkpoint 11.6: Equilibrium constants

3. $K_c = 2.5$
4. $K_c = 0.510$ dm^6 mol^{-2}
5. (a) $K_p = 126$ atm$^{-1/2}$ (b) $K_p = 1.60 \times 10^4$ atm^{-1}
6. $K_c = 3.79$

Questions on Chapter 11

2. (b) $K_p = 3.73$ atm^{-1}
3. (b) $K_p = 4.00$
4. $K_c = 2.89$ mol^{-1}dm^3
5. $K_c = 8.91$ mol dm^{-3}
6. (a) The unit of K_c is concentration^{-1}.
(b) The value of $[R]^2/([P]^2[Q]) = 0.33$, which is not the equilibrium value.
(c) $V = 18.0$ dm^3
7. (a) The value of $[SO_3]^2/([SO_2]^2[O_2]) = 19.6$ which is not the equilibrium value.
8. (c) $K_c = 3.98$
10. (b) (v) $K_p = 8.5$ kPa^{-1}
11. (c) **B**
(d) (i) x = **B**, y = **D** or **A**, z = **C**
12. (a) 0.12 mol NO_2 and 0.060 mol N_2O_4.
(b) (ii) $K_c = 0.05$ mol dm^{-3}
(c) (ii) 0.667 (iii) $K_p = 133$ kPa
(d) decrease
13. (c) (i) mole fractions: N_2 0.209 atm, H_2 0.655 atm, NH_3 0.137 atm
(ii) $K_p = 3.20 \times 10^{-3}$ atm^{-2}
(iii) K_p decreases with rising temperature
14. (a) Cl_2 0.156 mol, PCl_5 0.400 mol
(b) (ii) $K_c = 1.08 \times 10^{-2}$ mol dm^{-3}
(c) (ii) mole fraction of $Cl_2 = 0.281$
(iii) $p(PCl_5) = 109.7$ kPa
(iv) $K_p = 44.9$ kPa

CHAPTER 12 ELECTROCHEMISTRY

Checkpoint 12.5: Electrolysis

1. 1.8 g
2. (a) 2.40 g (b) 7.10 g (c) 6.35 g (d) 20.7 g (e) 0.200 g
3. (a) double (b) no change (c) no change (d) double
4. 1.23×10^{-3} mol metal, 2.46×10^{-3} mol electrons, +2

Checkpoint 12.6A: Acids and bases

5. (a) $pK_w = 15$ (b) pH = 13

Checkpoint 12.6B: pH and dissociation constants

1. pH < 7
5. (a) 3.0 (b) 1.6 (c) 4.4 (d) 12.4 (e) 11.8
6. (a) pH = 3.38 (b) pH = 2.37
7. (a) $K_a = 3.97 \times 10^{-10}$ mol dm^{-3}
(b) $K_a = 3.02 \times 10^{-11}$ mol dm^{-3}

Checkpoint 12.7: Titration

4. (a) pH = 3.00 (b) pH = 3.22 (c) pH = 3.70
(d) pH = 10.3 (e) pH = 11.0

Checkpoint 12.8: Buffers

4. pH = 4.76

Checkpoint 12.9: Salt hydrolysis

2. (a) 8.7 (b) 0.167 mol dm^{-3}

Questions on Chapter 12

3. time = 287 min
7. volume, at stp, of $H_2 = 55$ cm^3
9. (d) (i) pH = 2 (ii) pH = 3.37 (iii) pH = 4.74
10. (e) pH = 5.04
11. (a) (i) *A*: pH = 2.34, *B*: pH = 3.50
(b) $M = 29.9$, formula is CH_2O
12. (c) pH = 1
(d) As c increases, $[H^+(aq)]$ increases, and pH decreases. If c doubles, $[H^+(aq)]$ doubles and pH decreases by log 2, that is by 0.301.
13. (c) (ii) pH = 2.88
(d) (ii) The pH of methanoic acid solution is lower than that of ethanoic acid even though the concentration of methanoic acid is half that of ethanoic acid. The pH of ethanoic acid at 0.050 M would be 2.88 + log 2 = 3.18
14. (a) (ii) $pK_a = 4.91$
(c) (iii) pH = 4.91
15. (b) (i) pH = 1 (ii) pH = 12
(c) (i) pH = 2.95 (ii) See Figure 12.7A(b)
16. (b) (ii) 2.87 (iii) 1.0
(d) $[Na_2CO_3] = 6.25 \times 10^{-2}$ mol dm^{-3}
17. (c) $K_a = 5.01 \times 10^{-5}$ mol dm^{-3}, $c = 0.200$ mol dm^{-3}
18. (b) (ii) mass of $CH_3CO_2Na = 52.9$ g
19. (a) (ii) 1.00×10^{-14} mol^2 dm^{-6}
(b) (i) 0.126 mol Na_2CO_3
(ii) $[H^+(aq)] = 0.063$ mol dm^{-3}, pH = 1.20
(iii) pH = 12.3
20. (b) (i) 13.5 (ii) 12.5 (iii) 1.12

CHAPTER 13 OXIDATION–REDUCTION EQUILIBRIA

Checkpoint 13.3: Electrode potentials

3. *(a)* emf = −0.46 V
   (b) $Cu(s) + 2Ag^+(aq) \longrightarrow Cu^{2+}(aq) + 2Ag(s)$; $E^\ominus = +0.46$ V
   (c) from Cu to Ag
5. emf is positive for *(b)*, *(e)*, *(h)*

Questions on Chapter 13

5. Since emf is positive, chlorine is acting as an oxidising agent, with $E^\ominus = 1.36$ V.
   $E^\ominus$ for half cell = −0.27 V
6. *(a)* +0.40 V *(b)* +0.26 V *(c)* −0.29 V *(d)* +0.94 V
   (e) −0.46 V *(f)* +0.78 V *(g)* +0.63 V
7. *(c)* +0.36 V
8. *(b)* (i) +7 (ii) +1 (iii) +2
   (c) (ii) $KMnO_4$ 5.44×10^{-4} mol, $[Fe^{2+}] = 0.109$ mol dm^{-3}
   (iii) 51.6%
   (e) The reaction that happens is that with the higher value of $E^\ominus$; that is
   $2Fe^{3+} + Zn \longrightarrow 2Fe^{2+} + Zn^{2+}$
9. *(d)* (i) $E^\ominus = -0.30$ V
   (e) The reaction with the positive value of $E^\ominus$ takes place: Fe is oxidised to Fe^{2+}, and Sn^{2+} is reduced to Sn.
10. *(c)* (ii) $K_a = 1.76 \times 10^{-5}$ mol dm^{-3}
    $[H^+] = 1.76 \times 10^{-5}$ mol dm^{-3},
    $[CH_3CO_2H] = 0.20$ mol dm^{-3},
    $[CH_3CO_2^-] = 0.20$ mol dm^{-3}
11. *(c)* (i) $MnO_4^-(aq) + 8H^+(aq) + 5Fe^{2+}(aq) \longrightarrow Mn^{2+}(aq) + 4H_2O(l) + 5Fe^{3+}(aq)$
    (ii) Mass of Mn = 0.275 g and % = 11.9%
12. *(d)* $2Cu^+(aq) \longrightarrow Cu^{2+}(aq) + Cu(s)$; $E^\ominus = 0.37$ V
    (e) $2MnO_4^{2-} \longrightarrow MnO_4^- + 2MnO_2$; $E^\ominus = +0.95$ V

CHAPTER 14 REACTION KINETICS

Checkpoint 14.2: Reaction rates

3. *(a)* (i) $6a^2$ (ii) a^3 (iii) $6/a$
4. *(b)* 5.0×10^{-3} mol *(c)* 0.34 g *(d)* 1.75 min

Checkpoint 14.14A: Initial rates

3. *(a)* Rate ∝ $[\mathbf{A}]^2$, and rate is independent of $[\mathbf{B}]$. Rate = $k[\mathbf{A}]^2$.
   (b) $k = 5.0$ dm^3 mol^{-1} s^{-1}
   (c) 1.8 mol dm^{-3} s^{-1}
4. *(a)* The reaction is first order in $[\mathbf{A}]$ and second order in $[\mathbf{B}]$; overall order = 3.
   (b) $k = 1.5 \times 10^{-3}$ mol^{-2} dm^6 min^{-1}
   (c) 8.7×10^{-6} mol dm^{-3} min^{-1}

Checkpoint 14.14B: First-order reactions

1. $t_{1/2} = 1$ hour
2. 6.25%
3. *(a)* 1st order in $[\mathbf{A}]$ *(b)* 6.3×10^{-4} mol dm^{-3} s^{-1}
   (c) 7.9×10^{-4} s^{-1}
4. *(b)* 14.6 cm^3 *(c)* 10.3 min = 618 s
   (e) 1.06×10^{-6} mol dm^{-3} s^{-1} *(f)* 9.7×10^{-4} s^{-1}

Checkpoint 14.14C: Order of reaction

1. 1st order
2. 2nd order
3. Rate = $k[BrO_3^-][Br^-][H^+]^2$, 4th order, $k = 12$ mol^{-3} dm^9 s^{-1}
4. Rate = $[S_2O_8^{2-}(aq)][I^-(aq)]$; $k = 36$ dm^3 mol^{-1} s^{-1}
6. 1st order
7. *(a)* (i) 1 (ii) 1
   (b) Rate = $k[CH_3CO_2C_2H_5][OH^-]$
   (c) 0.195 dm^3 mol^{-1} s^{-1}
8. *(a)* 1.7×10^{-4} mol dm^{-3} s^{-1} *(b)* 5.1×10^{-4} mol dm^{-3} s^{-1}
   (c) 15.3×10^{-4} mol dm^{-3} s^{-1}
9. 1st order; $k = 1.75 \times 10^{-4}$ s^{-1}

Questions on Chapter 14

2. *(a)* 1
3. *(a)* concentration time^{-1} *(b)* time^{-1}
   (c) concentration^{-1} time^{-1}; see § 14.13.2
4. 30 hours
5. 9 days
6. *(a)* 1 *(b)* 1
   (c) 6.0×10^{-3} dm^3 mol^{-1} s^{-1}; 1.08×10^{-5} dm^3 mol^{-1} s^{-1}
8. *(d)* order = 1, $k = 6.0 \times 10^{-4}$ s^{-1}
9. *(b)* (i) $d[N_2O_5]/dt = k[N_2O_5]$ (ii) I
   (iii) $k = 8.6 \times 10^{-4}$ s^{-1}
10. *(a)* (ii) Rate = $k[CH_3COCH_3][H^+(aq)]$ (iii) **A**
11. *(c)* (ii) mol^{-n}dm^{3n}s^{-1}
12. *(a)* (i) mol^{-2}dm^{6}s^{-1} (ii) double
    (iii) decrease by a factor 4
    (b) (i) 3
13. *(d)* $[H_2SO_4] + [H_2C_2O_4] = 0.0500$ mol dm^{-3}
14. *(c)* (i) in mol dm^{-3} (2): 6.4×10^{-5}
    (3) 16.0×10^{-2} (4) 1.5×10^{-2}
    (ii) $k = 0.444$ mol^{-2} dm^6 s^{-1} (iii) increase
    (iv) no effect
15. *(a)* (i) 3 (ii) mol^{-2} dm^6 s^{-1} (iv) double (v) decrease
17. *(a)* (i) Compare expts. 1 and 2: rate doubles when $[(CH_3)_3CBr]$ doubles. Compare expts. 2 and 3: rate does not change when $[OH^-]$ doubles.
    (ii) $k = 3$ s^{-1} (iii) Rate = 1.2×10^{-2} mol dm^{-3} s^{-1}
    (b) S_N2; see § 29.9.1 *(c)* Lucas test; see Table 30.7
18. *(c)* (i) Rate = $k[CH_3CSNH_2][OH^-]$ (ii) 2
    (iii) $CH_3CONH_2 + OH^- \longrightarrow CH_3CO_2^{2-} + NH_3$

PART 3 INORGANIC CHEMISTRY

CHAPTER 17 HYDROGEN

Questions on Chapter 17

4. *(a)* (ii) 0.25 mol (iii) 4.5 g
   (b) (i) See Figure 4.8G

CHAPTER 18 THE s BLOCK METALS GROUPS 1 AND 2

Checkpoint 18.7: Group 1

6. 0.394 g $Na_2CO_3.H_2O$

Questions on Chapter 18

Questions on Group 1 and Group 2

3. *(b)* $A_r = 40$, **A** = Ca
5. (iii) The sums (lattice enthalpy + hydration enthalpy)/kJ mol^{-1} are LiF – 3, NaF – 10, KF – 27, RbF – 40. Thus, dissolution becomes more exothermic and solubility increases from LiF to RbF.
7. *(b)* 4.48 dm^3 CO_2 at stp = 0.200 mol therefore 0.200 mol of $\mathbf{Z}_2CO_3$ or $\mathbf{Z}CO_3$ was used.
Therefore 14.78 g = 0.200 mol, and M = 14.78/0.200 = 74.
M of CO_3 = 60, therefore either $2A_r(\mathbf{Z}) + 60 = 74$ and $A_r(\mathbf{Z}) = 7$ or Z + 60 = 74 and $A_r(\mathbf{Z}) = 14$. Values of A_r show that **Z** is Li.
(d) Putting values of $\Delta H_F^\ominus$ under each species,
$Fe_2O_3(s) + 3CO(g) \longrightarrow 2Fe(s) + 3CO_2(g)$
–824   3(–110)   0   3(–394)
$\Delta H_r^\ominus = 3(-394) + 824 + 330 = -28$ kJ mol^{-1}
For 0.1 mol, $\Delta H^\ominus = -2.8$ kJ
9. *(b)* (i) $Li_2CO_3(s) \longrightarrow Li_2O(s) + CO_2(g)$
–1216   –596   –394; $\Delta H = +226$ kJ mol^{-1}
$Na_2CO_3(s) \longrightarrow Na_2O(s) + CO_2(g)$
–1131   –416   –394; $\Delta H = +321$ kJ mol^{-1}
(d) Amount of HCl = 2.396×10^{-3} mol
Amount of Na_2CO_3 = 0.011 98 mol in 250 cm^3
Mass of Na_2CO_3 = 1.269 g. Mass of H_2O = 1.725 g
Molar ratio = 1 : 8, and $x = 8$

Questions on Group 2

6. *(b)* –1214 = 193 + 590 + 1150 + 158–796 + L.E.
L.E. = –2509 kJ mol^{-1}
7. *(b)* (i) Ratio Ba, 1.71/137 : O, 0.40/16 = 0.0125 : 0.025 = 1 : 2.
The formula is BaO_2.
(ii) O^{2-} (iii) –1 (iv) Na_2O_2
8. *(d)* (i) Amount of $CaCl_2$ = 0.100 mol
Volume of C_2H_2 = 2.40 dm^3

CHAPTER 19 GROUP 3

Questions on Chapter 19

4. Amount of Al = $1 \times 10^6/27$ mol
Coulombs needed = $3 \times 96\,500 \times 10^6/27 = 1.07 \times 10^{10}$ C
9. *(c)* (i) $pV = nRT$
1.00×10^5 Pa $\times 73.6 \times 10^{-6}$ m^3 = $(0.500/M) \times 8.314$ J K^{-1} $mol^{-1} \times 473$ K
$M_r = 267$

CHAPTER 20 GROUP 7 THE HALOGENS

Checkpoint 20.7: Reactions

2. +1, –1, +5
3. *(a)* F zero, O –2, F –1, F –1, O –2
(b) F zero, O –2, O zero, F –1, O –2

Questions on Chapter 20

11. *(a)* (iv) Amount of thio = $20.0 \times 10^{-3} \times 0.10$ mol
Amount of $I_2 = 1.00 \times 10^{-4}$ mol
Mass of $I_2 = 254 \times 10^{-4}$ g and % = 1.0×10^{-2}%
(b) $I^-(aq) + 5IO_3^-(aq) + 6H^+(aq) \longrightarrow 3I_2(aq) + 3H_2O(l)$
(c) (ii) ${}^{131}_{53}I^- \longrightarrow {}^{0}_{-1}e + {}^{131}_{54}Xe$
(d) (ii) I (96.70/127) : C (3.050/12) : H (0.254/1) gives CHI_3
13. *(b)* $HNO_3(aq)$ prevents precipitation of AgOH(s) in alkaline solution. 0.850 g AgI formed from 0.5425 g
NaI % of NaI = 45.2%

CHAPTER 21 GROUP 6

Checkpoint 21.6: Hydrides

3. *(a)* 1 dm^3 of $H_2O_2(aq) \longrightarrow$ 20 dm^3 of O_2 = 0.833 mol O_2
Since $2H_2O_2(aq) \longrightarrow O_2(g) + 2H_2O(l)$
1 dm^3 of solution contains 2×0.833 mol H_2O_2, and
$[H_2O_2] = 1.67$ mol dm^{-3}

Questions on Chapter 21

5. *(a)* FeS_2 (M_r = 120) $\longrightarrow$ $2H_2SO_4$ (M_r = 98)
Therefore 1 tonne $FeS_2 \longrightarrow$ 1.63 tonnes H_2SO_4
9. *(c)* A = NaOH(aq), B = S
(d) Mass of SO_2 = 6.58 tonnes
(f) (ii) S: +6 to +4 and I: –1 to 0

CHAPTER 22 GROUP 5

Checkpoint 22.4: Nitrogen

2. *(a)* +5 *(b)* +3 *(c)* +4 *(d)* +2 *(e)* –3

Questions on Chapter 22

2. Urea has the highest %N = 46%.
5. *(c)* 0.0500 mol NaOH, therefore 0.0500 mol NH_4^+ = 0.90 g NH_4^+, % = 9.0% NH_4^+
8. *(a)* (ii) $K_c = (1.46)^2/[0.27(0.81)^3] = 14.85$ mol^{-2} dm^6
(c) (ii) $K_a = [H_3O^+(aq)][NH_3(aq)]/[NH_4^+(aq)]$
$5.62 \times 10^{-10} = [H_3O^+(aq)]^2/0.100$, therefore pH =5.12

CHAPTER 23 GROUP 4

Checkpoint 23.6: The greenhouse effect

1. *(a)* C (12 g) ⟶ CO_2 (44 g)
   6×10^9 tonnes C ⟶ 2.2×10^{10} tonnes CO_2
   (b) If 1.1×10^{10} tonnes remain in the atmosphere p.a., to add 2.5×10^{12} tonnes will take 2.5×10^{12} tonnes/(1.1×10^{10} tonnes year^{-1}) = 227 years

Questions on Chapter 23

10. For polymerisation of $nSiO_2$ to $[-SiO_2-]_n$:
    bonds broken are 2(Si=O) = + 1276 kJ mol^{-1}
    bonds made are 4(Si—O) = −1496 kJ mol^{-1}
    Total = −220 kJ mol^{-1}, showing that energy considerations favour polymerisation.
    For polymerisation of nCO_2 to $[-CO_2-]_n$:
    bonds broken are 2(C=O) = + 1486 kJ mol^{-1}
    bonds made are 4(C—O) = −1440 kJ mol^{-1}
    Total = + 46 kJ mol^{-1}, therefore energy considerations favour CO_2 molecules.
12. *(b)* (ii) Total of steps from Pb(s) ⟶ $Pb^{2+} + 2e^-$
    = −2362 kJ mol^{-1}
    Total of steps for 0.1 mol Pb^{2+}(g) ⟶ Pb(s)
    = −236 kJ mol^{-1}

CHAPTER 24 THE TRANSITION METALS

Checkpoint 24.13: Oxides and oxo-ions

3. **A** = Zn, **B** = Cu, **C** = Fe
5. *(a)* $a = 3, b = 4, c = 2, d = 1, e = 2$

Checkpoint 24.16: Complexes

3. *(a)* + 1, 2 *(b)* 0, 4 *(c)* + 3, 6 *(d)* + 2, 6 *(e)* + 2, 4
   (f) + 1, 4 *(g)* + 2, 4

Questions on Chapter 24

11. $Cr_2O_7^{2-}(aq) + 14H^+(aq) + 6Fe^{2+}(aq) \longrightarrow$
    $2Cr^{3+}(aq) + 6Fe^{3+}(aq) + 7H_2O(l)$
    Amount of $Cr_2O_7^{2-} = 1.50 \times 10^{-4}$ mol
    Amount of $Fe^{2+} = 9.00 \times 10^{-4}$ mol, and
    $[Fe^{2+}] = 0.036$ mol dm^{-3}
14. C = $Fe(OH)_2(s) + Cu(OH)_2(s)$, D = $Na_2ZnO_2(aq)$,
    E = $Fe(OH)_2(s)$, F = Fe_2O_3, G = $Cu(NH_3)_4SO_4(aq)$,
    H = $Zn(OH)_2(s)$, I = $ZnSO_4(aq)$
    B = $Fe(OH)_2(s) + Cu(OH)_2(s) + Zn(OH)_2(s)$
    A contains $Fe^{2+}(aq)$, $Cu^{2+}(aq)$ and $Zn^{2+}(aq)$.
15. **A** = $Cr(OH)_3(s)$ chromium(III) hydroxide
    B = $[Cr(NH_3)_6]_2(SO_4)_3(aq)$, hexaamminechromium(III) sulphate
    C = $Na_3Cr(OH)_6(aq)$, sodium chromate(III)
    D = Na_2CrO_4, sodium chromate(VI), yellow
    E = $Na_2Cr_2O_7$, sodium dichromate(VI), orange
16. **W** = $Cu(H_2O)_4^{2+}(aq)$, **X** = $CuCl_4^{2-}(aq)$, **Y** = $CuCl_2^-(aq)$,
    Z = CuCl(s)
17. *(b)* $MnO_4^-(aq) + 8H^+(aq) + 5Fe^{2+}(aq) \longrightarrow$
    $Mn^{2+}(aq) + 4H_2O(l) + 5Fe^{3+}(aq)$
    (e) Mass of Fe = 0.1534 g, percentage = 93.2%
18. *(c)* M_r of $ZnSO_4.7H_2O$ = 287.4
    Amount of Zn = 0.0184 mol
    Amount of edta = $18.45 \times 0.1000 \times 10^{-3}$ mol
    Amount of edta for 250 cm^3 = 0.184 mol
    Molar ratio Zn : edta = 1 : 1
19. Amount of thio = 2.75×10^{-3} mol = amount of Cu
    Mass of Cu = $63.5 \times 2.75 \times 10^{-3}$ g, and percentage = 65.0%
20. *(a)* (i) $K_1 = \dfrac{[Ni(NH_3)(H_2O)_5{}^{2+}(aq)]}{[Ni(H_2O)_6{}^{2+}(aq)]\,[NH_3(aq)]}$
    (ii) Unit = mol^{-1} dm^3
    (iii) $K_1K_2K_3K_4K_5K_6$ =
    antilog −(2.67 + 2.12 + 1.61 + 1.07 + 0.63 −0.09) =
    9.77×10^{-9} mol^{-6} dm^{18}
22. *(a)* Ratio Co 23.6/59 : N 27.9/14 : H 6.0/1 : Cl 42.5/35.5 =
    Co 0.4 : N 5 : H 15 : Cl 3
    (b) 4.0×10^{-3} mol **X** contains 8.0×10^{-3} mol Cl^-
    $[CoN_5H_{15}Cl]^{2+}\ 2Cl^-$
    $[Co(NH_3)_5Cl]^{2+}\ 2Cl^-$, octahedral
24. *(b)* (ii) Amount of $NH_4VO_3 = 2.00 \times 10^{-3}$ mol
    Amount of $KMnO_4 = 0.40 \times 10^{-3}$ mol
    $MnO_4^-(aq) + 8H^+(aq) + 5e^- \longrightarrow Mn^{2+}(aq) + 4H_2O(l)$
    1 mol MnO_4^- requires 5 mol electrons; therefore 0.40×10^{-3} mol MnO_4^- requires 2.00×10^{-3} mol electrons; therefore 2.00×10^{-3} mol electrons is provided by 2.00×10^{-3} mol V in blue solution. This means that 1 mol electrons is provided by 1 mol V in blue solution, therefore V in blue solution changes in ox. no. by 1 unit.
    Ox. No. of V in NH_4VO_3. = 5, therefore ox. no. of V in blue solution = + 4.

PART 4 ORGANIC CHEMISTRY

CHAPTER 26 THE ALKANES

Questions on Chapter 26

9. *(d)* $CH_4(g) + Cl_2(g) \longrightarrow CH_3Cl(g) + HCl(g)$
   −75 0 −82 −72 kJ mol^{-1}
   $\Delta H_r^\ominus$ = −82−72−(−75) = −79 kJ mol^{-1}
   (e) (i) Ratio H 4.1/1 : C 24.2/12 : Cl 71.7/35.5 gives the empirical formula CH_2Cl
   (ii) M_r of CH_2Cl = 49.5 corresponding to $C_2H_4Cl_2$
   (iii) Peak at 100 is due to $C_2H_4{}^{35}Cl^{37}Cl$, and peak at 102 to $C_2H_4{}^{37}Cl_2$.
11. *(f)* The final mixture contains CCl_4 which contains ^{35}Cl and ^{37}Cl. The highest value of $M_r = C^{37}Cl_4$ with m/z = 160.

CHAPTER 27 ALKENES AND ALKYNES

Questions on Chapter 27

4. 100 cm^3 at stp is the volume of 0.250 g therefore 22.4 dm^3 (the gas molar volume) is the volume of (22400/100) × 0.250 g = 56 g. With molar mass = 56 g mol^{-1}, the formula is C_4H_8.

9. *(b)* (iv) I. Steam is involved in the fast step, so changing the pressure will not alter the rate.
II. Since $K_p = pC_2H_5OH(g)/(pC_2H_4(g) \times pH_2O(g))$ an increase in $pH_2O(g)$ increases $pC_2H_5OH(g)$, that is, increases the equilibrium yield.

(d) Bonds broken are (C═C) + (H—O) and bonds made are (C—C) + (C—O) + (C—H). The difference is −45 kJ mol^{-1}.

CHAPTER 28 AROMATIC COMPOUNDS

Questions on Chapter 28

8. *(e)* Mass of benzene = 26.1 g
M_r(benzene) = 78, M_r(bromobenzene) = 157
Mass of bromobenzene calculated = 26.1 × 157/78 = 52.53 g
% yield = 100 × 25.9/52.53 = 49.3%

10. *(a)* (i) 3 × (−119.6) = −358.8 kJ mol^{-1}
(ii) Benzene is more stable by 150.4 kJ mol^{-1} than expected from the Kekulé formula. The stabilisation is due to delocalisation of the π-electrons.

11. *(a)* **A** = cyclohexane, **B** = hydrogen, **C** = bromomethane, **D** = iron(III) bromide, **E** = bromobenzene, **F** = conc. sulphuric acid
(b) (i) **E**: $CH_3^+FeBr_4^-$; nitrobenzene: NO_2^+

13. *(b)* (i) Cyclohexane 0.305 mol, benzene 0.513 mol
(ii) 10.7 kPa
(c) −2820 kJ

CHAPTER 29 HALOGENOALKANES AND HALOGENARENES

Questions on Chapter 29

10. *(a)* (i) Values of ΔH for (bonds broken + bonds made) are I: +3 kJ mol^{-1} and II: +192 kJ mol^{-1} therefore I is more likely.

CHAPTER 30 ALCOHOLS AND PHENOLS

Checkpoint 30.7: Reactions

6. **A** = C_2H_5Br, **B** = $CH_3CH(OH)CH_3$

7. *(a)* **X** = C_4H_9Br
(b) $(CH_3)_2CHCH_2Br$
(c) **Y** = 2-methylpropene
(d) **Z** = 2-bromo-2-methylpropane

CHAPTER 31 ALDEHYDES AND KETONES

Questions on Chapter 31

4. **P** is an alkene, **Q** is an aldehyde with 2 CHO groups.

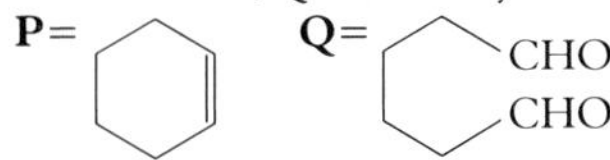

5. **A** is $CH_3CH_2CH_2CHOHCH_3$.

6. **P** is $CH_3COCH_2CH_2CHO$.

11. *(a)* **A**: $C_6H_5CO_2H$, **B**: $C_6H_5CH_2OH$, **C**: $C_6H_5CHOHSO_3Na$, **D**: C_6H_5CClO
(b) **A**: $K_2Cr_2O_7$, acid, warm, **B**: $NaBH_4$(aq), warm, **C**: $NaHSO_3$(aq), r.t., **D**: Cl_2, boil

12. *(b)* **A** must be an aldehyde RCHO, not a ketone, oxidised to **B**, a carboxylic acid, RCO_2H.
(c) **A** is reduced to an alcohol, **C**, RCH_2OH.
(d) **A** reduces Ag^+(aq) to Ag(s). **A** is oxidised to **B.**
(e) RCO_2CH_2R
(f) (i) RCO_2H: strong hydrogen bonds between molecules
(ii) RCO_2CH_2R: there is no $\delta+$ H atom to form hydrogen bonds.

15. The evidence shows: **X** is a carbonyl compound (adds DNP). **X** contains a CH_3CO– group (gives iodoform). It is not an aldehyde (not easily oxidised). It has no C═C bond (does not add Br_2), so hydrogen must reduce the >CO group to >CHOH. **X** = $CH_3COCH_2CH_3$, **Y** and **Z** = $CH_3C^*HOHCH_2CH_3$. Since C* is chiral, a mixture of stereoisomers is formed.

16. *(a)* (i) **E** = $CH_3CH═CHOH$ or $CH_2═CHCH_2OH$, **F** = CH_3COCH_3, **G** = CH_3CH_2CHO
(ii) **1.** With CH_3CO_2H + conc. H_2SO_4, **E** gives a sweet-smelling ester; **F** and **G** do not react.
2. With DNP, **F** and **G** give coloured ppts; **E** does not react.
3. *(a)* With Tollens' reagent, **G** gives a silver mirror; **E** and **F** do not react.
(b) (i) $CH_3CHOHCHOHCH_3$ (ii) $2CHI_3 + HO_2CCO_2H$

17. *(d)* *cis-* and *trans-* C_6H_5 CH═CHCHO

18. *(c)* (i)

HO–C_6H_3(–CH_3O)–CH_2OH

(ii) I: 1½ mol because 1 mol reacts with –OH, and ½ mol reacts with aromatic –CHO in the Cannizzaro reaction:

$$2ArCHO + NaOH \longrightarrow ArCH_2OH + ArCO_2Na$$

19. *(b)* $CH_3CHOHCH_2CH_3$
(c) (i) HBr adds across C═C by electrophilic addition.
(ii) HCN adds to C═O, by nucleophilic addition. H_2 adds to C═C.
(iii) H_2 adds to C═C, not C═O.

20. *(a)* C 71.4%, H 9.5%, O 19.1%
(c) **X** is $CH_2═CHCH_2CH_2CHO$ or $(CH_3)_2C═CHCHO$

CHAPTER 32 AMINES

Questions on Chapter 32

8. *(a)* **A** = $C_2H_5NH_2$, **B** = C_2H_5I, **C** = CH_3CO_2Na, **D** = $C_2H_5NHCOCH_3$ and **E** = C_2H_5OH

9. *(a)* (ii) **X** = $C_6H_5CH_2CHBrCH_3$

(b) $C_6H_5CHNH_2CH_2CH_3$ if HBr adds differently, 1,4-$CH_3NH_2CHCH_2C_6H_4CH_2CHNH_2CH_3$ if polyalkylation occurs at Step 1.

10. *(a)* (i) —OH

(iii) 1,4-$H_2NC_6H_4CO_2H$

(iv) the 1,2- and 1,3- isomers

(b) **B** is CH_3CH_2OH, **D** is CH_3CO_2H

(c) (i) H_2O

(iii) 1,4-$H_2NC_6H_4CO_2C_2H_5$

(d) (i) ethanoyl chloride

(e) **F** is 1,4-$CH_3CONHC_6H_5CO_2C_2H_5$

CHAPTER 33 ORGANIC ACIDS AND THEIR DERIVATIVES

Checkpoint 33.8: Reactions of acids

5. *(a)* $H_2NC_6H_4CHBrCHBrCO_2H$

(b) $CH_3COHNC_6H_4CH{=}CHCO_2H$

(c) $H_2NC_6H_4CH_2CH_2CO_2H$

(d) $H_2NC_6H_4CH{=}CHCH_2OH$

(e) $H_2NC_6H_4CHO + OHCCO_2H$

Checkpoint 33.12: Acid chlorides and anhydrides

4. $CH_3COCl(l) + H_2O(l) \longrightarrow CH_3CO_2H(aq) + HCl(aq)$
Amount of OH^- = 5.00×10^{-3} mol; amount of CH_3COCl = 2.5×10^{-3} mol; m = 0.196 g

Checkpoint 33.14: Esters

5. Ester RCO_2R'. Amount of ester = amount of NaOH = 2.0×10^{-2} mol = 1.76 g
M = 88 g mol^{-1}. Subtracting 44 for CO_2 leaves 44, which corresponds to C_3H_8. The formula is $C_4H_8O_2$.

Checkpoint 33.16: Amides and nitriles

4. **A** = C_2H_5COCl, **B** = $C_2H_5CONH_2$, **C** = C_2H_5OH, **D** = $CH_3CO_2C_2H_5$, **E** = $C_2H_5CO_2CH_3$
x = Br_2 + conc. NaOH(aq), y = $NaNO_2$ + HCl(aq), z = CH_3OH + conc. H_2SO_4

5. **A** = C_6H_5CHO, **B** = $C_6H_5CH_2OH$, **C** = C_6H_5COCl, **D** = $C_6H_5CO_2CH_2C_6H_5$, **E** = $C_6H_5CONH_2$, **F** = $C_6H_5NH_2$, **G** = $C_6H_5NHCOC_6H_5$

Checkpoint 33.17: Amino acids and proteins

5. 166 g mol^{+1}

Questions on Chapter 33

1. **A** = CH_3CH_2Br, **B** = CH_3CH_2OH, **C** = $CH_3CH(OH)CN$, **D** = CH_3CH_2CN, **E** = $CH_3CHClCO_2H$, product = 2-hydroxypropanoic acid

4. **D** is $HO_2CC_6H_4N_2^+\ Cl^-$
C is $HO_2CC_6H_4NH_2$
B is $HO_2CC_6H_4CONH_2$
F is $HO_2CC_6H_4CO_2H$
E is $HO_2CC_6H_4CN$
A, made by heating a dicarboxylic acid, must be an anhydride, and the carboxyl groups in **F** must be adjacent.

A is [benzene ring with C=O, O, C=O] **B** is [benzene ring with $CONH_2$, CO_2H]

C is [benzene ring with NH_2, CO_2H] **D** is [benzene ring with $N_2^+Cl^-$, CO_2H]

E is [benzene ring with CN, CO_2H] **F** is [benzene ring with CO_2H, CO_2H]

6. **A** = $H_3C{-}C({=}O){-}O{-}C(CH_3)(H){-}C_2H_5$

B = $C_2H_5{-}C(H_3C)(H){-}C({=}O){-}O{-}CH_3$

7. The ester is $C_2H_5CHCH_3CO_2CH(CH_3)_2$.

10. **A** is $CH_3CO_2CH(CH_3)_2$, **B** is $(CH_3)_2CHOH$, **C** is C_2H_5OH, **D** is CHI_3

11. **A** = $CH_3CH_2CH_3$, **B** = $CH_3CH_2CH_2Cl$, **C** = $CH_3CHClCH_3$, **D** = $CH_3CH_2CO_2H$, **E** = $(CH_3)_2CO$

13. *(c)* $C_6H_5CH(OH)CO_2H$ has M_r = 152
Amount of NaOH = 6.00×10^{-4} mol, so 0.100 g of sample contain 6.00×10^{-4} mol acid = 0.0912 g, therefore sample is 91.2% pure.

14. *(b)* (iv) Amount of ethanol = 0.20 mol, amount of acid = 0.20 mol

(v) M of ester $C_7H_{14}O_2$ = 130 g mol^{-1}
Mass expected = 130 g mol^{-1} × 0.20 mol = 26.0 g
% yield = 100 × 15.6/26.0 = 60%

(c) CH_3CO_2H and $(CH_3)_2CHCH_2CH_2OH$

16. **A** = $ClCH_2CO_2CH_3$, **B** = $ClCH_2CO_2H$, **C** = CH_3OH, **D** = $HOCH_2CO_2Na$, **E** = $HOCH_2CO_2H$
E + **C** $\longrightarrow$ $HOCH_2CO_2CH_3 + H_2O$

18. **A** is $C_2H_5CH_2CH(CH_3)OH$
B is $C_2H_5CH(CH_3)CO_2H$
C is $C_2H_5CH(CH_3)CHO$

CHAPTER 35 IDENTIFYING ORGANIC COMPOUNDS

Checkpoint 35.4: Analysis

1. *(a)* Mass of C = 0.1715 g, mass of H = 0.0285 g
   Ratio C 0.1715/12 : H 0.0285/1 gives formula CH_2.
   (b) Amount of **A** = 2.38×10^{-3} mol, therefore
   $M = 84.0 \text{ g mol}^{-1}$
   Molecular formula is C_6H_{12}.
2. Mass of S = 0.018 16 g, mass of Cl = 0.020 11 g
   Mass of C = 0.0408 g, mass of H = 0.002 885 g,
   mass of O = 0.018 055 g
   Ratio of amounts in moles gives the empirical formula $C_6H_5SO_2Cl$.
3. NaOH used = 25.0 cm^3 of 0.100 mol dm^{-3}
   Amount of $NH_4^+ = 2.5 \times 10^{-3}$ mol
   Mass of N = 3.5×10^{-2} g, % of N = 18.9%
   Mass of C = 0.072 g, % of C = 49.31%
   Mass of H = 0.014 g, % of H = 9.59%
   % of O = 22.2%
   Ratio of amounts ($\%/A_r$) gives the empirical formula C_3H_7NO.

Checkpoint 35.7: Visible–ultraviolet spectra

1. A plot of absorbance against volume of $\mathbf{M}^{2+}$(aq) or against volume of **L**(aq) gives a maximum at $\mathbf{M}^{2+}$(aq) = 3.3 cm^3, **L** = 6.6 cm^3, a ratio = 1 : 2, so formula is $\mathbf{ML_2}$.

Checkpoint 35.8: Infrared spectra

1. There is a C—H aliphatic absorption at 3000 cm^{-1} and a C—Cl absorption at 765 cm^{-1}. The compound is a chloroalkane. (In fact this is the IR spectrum of $CHCl_3$.)
2. C≡N at 2250 cm^{-1} and C—H aliphatic at 3000 cm^{-1} and C—H vibrations at 1460 and 1430 cm^{-1}. The presence of C≡N and the absence of aromatic C—H and aromatic C—C indicate an aliphatic nitrile. (In fact this is propanenitrile, CH_3CH_2CN.)
3. C—H aromatic above 3000 cm^{-1}, C—H of CHO at 2700, 2800 cm^{-1}, C=O at 1700 cm^{-1}, C—C aromatic at 1600 cm^{-1} and C—H at 1200 cm^{-1}. The spectrum is that of an aromatic aldehyde. (In fact it is that of benzaldehyde.)
4. C—H aliphatic at 3000 cm^{-1}, C—H of CHO at 2720 and 2830 cm^{-1}, C=O at 1730 cm^{-1}. The compound is an aliphatic aldehyde. (In fact it is propanal.)
5. C≡N at 2200 cm^{-1}, C—H aromatic at 3050 cm^{-1}, C—C aromatic at 1480 cm^{-1}. There are indications of an aromatic ring and a nitrile group. The simplest compound possessing these structures is benzonitrile or benzenecarbonitrile, C_6H_5CN. One could identify the compound by a molar mass determination. (In fact the spectrum is that of benzenecarbonitrile.)
6. C=O at 1700 cm^{-1} and O—H with hydrogen bonding at 2600–3400 cm^{-1} suggest a carboxylic acid. Since it is a liquid, it must be an aliphatic acid. (In fact the IR spectrum is that of ethanoic acid.)

Checkpoint 35.9: Mass spectra

1. E: The peak at 15 corresponds to CH_3. The peak at 94 is probably the molecular peak with a side-peak at 96 suggesting chlorine or bromine. Peaks at 79 and 81 of almost equal heights indicate bromine. CH_3 (15) + Br (79) = 94, and the compound is bromomethane, CH_3Br.
2. F: 64 is probably the molecular peak with a side-peak of 66 due to the presence of ^{37}Cl. There is a peak at 29 corresponding to the loss of ^{35}Cl from the molecular ion of $M = 64$. The peaks at 49 and 51 could correspond to the loss of CH_3 from $M = 64$ and from $M = 66$. The peaks at 29, 28 and 27 could be $C_2H_5^+$, $C_2H_4^+$ and $C_2H_3^+$. Combining C_2H_5 and Cl gives C_2H_5Cl, chloroethane.

Checkpoint 35.10: NMR

1. Ethanol, C_2H_5OH. The —CH_3 1 : 2 : 1 triplet at 1.0–1.3 δ and the —CH_2— 1 : 3 : 3 : 1 quartet at 3.5–3.9 δ are recognisable. The peak at 5.0–5.3 δ is due to —OH and is absent in ^2H_2O because the –OH proton is labile.
2. 1-Amino-2,2-dimethylpropane, $(CH_3)_3CCH_2NH_2$. The —CH_3 absorption at 0.9 δ is strong. The peak at 2.5 δ could be due to R—CH_2—R and that at 1.0 δ to RNH_2. The ratio 9 H in CH_3—groups : 2 H in —NH_2 groups : 2 H in —CH_2— groups makes the compound $(CH_3)_3CCH_2NH_2$ a possibility.
3. Propanoic acid, $CH_3CH_2CO_2H$. The —CH_3 and —CH_2— groups are recognisable – as in spectrum 1. The peak at 11.7 δ could be due to —CO_2H, which does not give a peak in ^2H_2O because the H is labile. The ratio 3 H in —CH_3 : 2 H in —CH_2— : 1 H in —CO_2H identifies the compound as $CH_3CH_2CO_2H$.
4. Ethylbenzene, $C_6H_5C_2H_5$. The CH_3CH_2— group is recognisable as in spectrum 1. The peak at 7.2 is due to C_6H_5—H. The ratio 5 H in aromatic ring : 2 H in —CH_2— : 3 H in —CH_3 gives $C_6H_5C_2H_5$ as the compound.
5. Ethanal, CH_3CHO. The peak at 9.7–9.8 δ is due to RCHO. The peak at 2.1 is probably due to R—$COCH_3$. The ratio 1 H in —CHO : 3 H in $RCOCH_3$ gives CH_3CHO as the formula.
6. Butanone, $CH_3COCH_2CH_3$. The —CH_3 peak at 0.9–1.1 δ and the —CH_2— peak at 2.2–2.5 δ are identifiable. The peak at 2.1 can be due to $RCOCH_3$. A compound with a —$COCH_3$ group and a C_2H_5— group is $CH_3COCH_2CH_3$.

Questions on Chapter 35

1. *(b)* The spectrum **A** has a peak at 2800 cm^{-1}, indicating a C—H bond and a peak at 1100 cm^{-1}, indicating a C—O bond in an alcohol, ether or ester. The spectrum **B** has a peak at 2800 cm^{-1} (C—H), a peak at 1100 cm^{-1} (C—O) and a peak at 3300 cm^{-1}, indicating the —OH group of an alcohol. **A** corresponds to an ether, e.g. $C_2H_5OC_2H_5$, $CH_3OCH(CH_3)_2$ or $CH_3OCH_2CH_2CH_3$.
   B corresponds to an alcohol, one of the isomers of C_4H_9OH.
   (c) Mass spectrometry would distinguish between accurate M_r values. NMR would identify the two compounds by showing which alkyl groups were present.
2. The O—H absorption is present at 3450 cm^{-1}, and the C—O absorption at 1050 cm^{-1}, the C—H aliphatic absorption at 2900 cm^{-1} suggesting an aliphatic chain. There is an absence of bands below 1000 cm^{-1} which aromatic compounds show. It looks like an aliphatic hydrocarbon chain with the bonds C—O and O—H, that is an alcohol. (In fact this is the spectrum of dodecanol, $CH_3(CH_2)_{10}CH_2OH$.)
3. The C=O absorption at 1715 cm^{-1} and CH_2 at 1460 cm^{-1} can be seen. The absence of a big absorption band below 1000 cm^{-1} indicates an aliphatic compound. The CO group is at the lower end of the absorption range and is probably a ketone. From $M_r = 98$, subtract 28 for CO, leaving 70. This could be C_5H_{10} and the compound could be $C_6H_{10}O$, cyclohexanone.
4. The C—Cl absorption is present at 750 cm^{-1}. There are aliphatic C—H vibrations at 2940 cm^{-1}. This seems to be a chloroalkane. (It is in fact dichloromethane.)
5. The C=O absorption is seen at 1700 cm^{-1}, C⩦C at 1400 cm^{-1}, C—H at 3000 cm^{-1}, O—H at 2500–2800 cm^{-1}. An aromatic compound with a C=O group and an O—H bond, e.g. an aromatic carboxylic acid. (In fact this is benzoic acid.)

6. The C=O absorption is seen at 1725 cm^{-1}, C—H aromatic at 3000 cm^{-1}, C—O—C aliphatic at 1100 cm^{-1}. An aromatic ester would fit the spectrum. (In fact this is ethyl benzoate.)

7. **A** = $HCO_2CH_2CH_2CH_3$, **B** = $HCO_2CH(CH_3)_2$,
C = $CH_3CO_2C_2H_5$, **D** = $C_2H_5CO_2CH_3$
All give a molecular peak at 88.
HCO_2^+ = 45, and this is absent, eliminating **A** and **B.**
$CH_3CO_2^+$ = 59, and C_2H_5 = 29, and these are both present so **C** is a possibility.$C_2H_5CO_2^+$ = 73, and CH_3 = 15, and both these are absent, so **D** is unlikely.The ester is **C**, $CH_3CO_2C_2H_5$.

8. The peak at m/z = 46 is the molecular peak. The peak at m/z = 47 corresponds to $^{12}C^{13}CH_5OH$. The peak at m/z = 45 is due to $C_2H_5O^+$, 31 to CH_2OH^+, 29 to $C_2H_5^+$ and 27 to $C_2H_3^+$.

9. The peak at 136 is the molecular peak. The peak at 77 corresponds to C_6H_5. The peak at 105 = 136–31 and could be due to the loss of OCH_3, leaving C_7H_5O. The peak at 51 could be $C_4H_3^+$. Combining the benzene ring and the ability to lose OCH_3 to form $C_6H_5CO^+$ indicates $C_6H_5CO_2CH_3$, methyl benzoate.

10. 58 = molecular peak, 57 = loss of one H = $C_3H_5O^+$, 29 could be $C_2H_5^+$ or CHO^+. 28 could be $C_2H_4^+$ or CO^+. Combining C_2H_5 and CO and H gives C_2H_5CHO, propanal.

11. 102 = molecular ion, peak at 87 = M—CH_3 = $C_5H_{11}O^+$, peak at 73 = 87–14 = 87–CH_2 = $C_4H_9O^+$, peak at 57 could be M—C_2H_5O = $C_4H_9^+$, peak at 45 could be $C_2H_5O^+$, 29 = $C_2H_5^+$. Combining C_2H_5O and C_4H_9O gives $C_4H_9OC_2H_5$. The isomer is in fact 2-ethoxybutane,

```
CH3CH2CHCH3
        |
       OC2H5
```

12. *(a)* There is only one type of H atom in **G**. It must be CH_3OCH_3, methoxymethane. In ethanol there are H atoms in three different environments, —OH, —CH_2— and —CH_3.
(b) Tetramethylsilane, has all its H atoms in identical environments and gives a single NMR line.

13. There are three types of H atom: two which are split into a quartet by 3 adjacent H atoms, could indicate CH_2CH_3; three which are not split; three which are split into a triplet by 2 adjacent H atoms, e.g. CH_2CH_3. The ester **J** could be $CH_3CO_2CH_2CH_3$.

14. *(a)* 4
(b) 2 : 2 : 3 : 1
(c) 3 H of one type suggests CH_3. 1 H of another type suggests OH. 2 H and 2 H suggest CH_2 and CH_2 in different environments. The formula $CH_3CH_2CH_2OH$ fits.

15. *(a)* (i) C—H
(ii) Compare the spectra with those of known compounds.
(b) (i) In cyclohexane the 12 H atoms are in identical environments; in dichloromethane the 2 H atoms are identical.
(ii) Equal amounts of cyclohexane and dichloromethane contain H atoms in a ratio 6 : 1.
(d) CH_2Cl_2 : $CH_2{}^{35}Cl^{35}Cl$ = 84, $CH_2{}^{35}Cl^{37}Cl$ = 86, $CH_2{}^{37}Cl^{37}Cl$ = 88. Three peaks at 84, 86, 88

16. *(a)* 4
(b) 1 : 6 : 2 : 3
(c) $RCOCH_3$
(d) Splitting patterns show that the peak at 4.8 δ is due to a proton with 2 adjacent protons and the peak at 2.7 δ is due to a proton with 1 adjacent proton.

Of 12 H atoms, 1 H of Type a has 2 adjacent protons, 6 H of Type b – value of δ suggests 2(—OCH_3), 2 H of Type c have 1 adjacent proton, are in a $COCH_3$ group (see part *(c)*). 3 H of Type d – value of δ suggests $RCOCH_3$.

```
      H  H                         H(c) H(a)
      |  |                          |    |
H3C—C—C—C—OCH3      H(d)3C—C—C—C—O—CH(b)3
    ‖  |  |                     ‖    |    |
    O  H  OCH3                  O  H(c) O—CH(b)3
```

The H atom *(a)* has 2 adjacent H atoms, the 2 H atoms *(c)* each have one adjacent H atom, the 6 H atoms *(b)* and the 3 H atoms *(d)* do not have hydrogen atoms bonded to adjacent atoms.

17. *(b)* $C_5H_{10}O$
B: Peak at 57 corresponds to loss of C_2H_5, suggesting $C_2H_5COC_2H_5$.
C: Peaks at 43 and 71 correspond to loss of C_3H_7 and CH_3 respectively.
C could be (1) $CH_{a3}COCH_{b2}CH_{c2}CH_{d3}$ or (2) $CH_{a3}COCH_b(CH_{c3})_3$
The NMR spectrum of (1) shows 4 peaks for 3 H_a, 2 H_b, 2 H_c, 3 H_d. The NMR spectrum of (2) shows 3 peaks for 3 H_a, 1 H_b, 9 H_c.
(c) **D** ⟶ **E**: substitution of CH_3CO— for H—. The IR peak at 1685 cm^{-1} shows the presence of a C=O group. **D** could be $C_6H_5CH_3$, **E** = $CH_3C_6H_4COCH_3$ with isomers 1, 2-, 1, 3- and 1, 4-. **E** ⟶ **F** adds 2 H; **F** is $CH_3C_6H_4CHOHCH_3$. The band at 3340 cm^{-1} shows the presence of —OH in an alcohol. Dehydration of **F** ⟶ **G. G** is $CH_3C_6H_4CH{=}CH_2$. The IR band at 1630 cm^{-1} indicates a C=C bond.

18. *(a)* 4
(b) 2 : 2 : 2 : 3
(c) $RCOCH_3$ (2.4 δ), RCH_3 (1.1 δ) 2.4 δ peak shows 3 adjacent protons, 1.1 δ peak shows 2 adjacent protons
(d) 3.8 δ peak: 2 adjacent protons, 2.8 δ peak: 2 adjacent protons
(e) C_5H_9ClO, $CH_3CH_2COCH_2CH_2Cl$

19. *(a)* (i) **X** RCH_2CO, **Y** $ROCH_3$, **Z** RCH_3
(ii)

```
   H  H  O  H
   |  |  ‖  |
H—C—C—C—C—H
   |  |     |
   H  H     H
```

(iii) $CH_{a3}CH_{b2}COCH_{c3}$
There are 3 H_a protons, 2 H_b protons, 3 H_c protons.
(b) (iii) There will be a band due to RCHO at 9.7 δ but no RCH_2CO at 2.5 δ.
(c) (ii)

```
      CH3              CH3
       |                |
       C                C
      / \ \           / / \
 C2H5  OH  H        H  HO  C2H5
```

(iii)

```
   C3H7
    |
    C
   / \ \
  H  OH  H
```

has a plane of symmetry.

CHAPTER 36 SOME GENERAL TOPICS

Checkpoint 36.2: Synthetic routes

4. *(a)* $C_6H_5CH_3$, oxidise with $KMnO_4$ + acid $\longrightarrow C_6H_5CO_2H$
   (b) Boil $C_6H_5CH_3$ and pass Cl_2 through in UV or sunlight $\longrightarrow C_6H_5CH_2Cl$. With KCN $\longrightarrow C_6H_5CH_2CN$. Reflux with HCl(aq) $\longrightarrow C_6H_5CH_2CO_2H$.
   (c) $C_6H_5CH_3$, oxidise with $MnO_2 \longrightarrow C_6H_5CHO$. Add $NaHSO_3$ followed by KCN $\longrightarrow C_6H_5CH(OH)CN$. With $LiAlH_4$ (ethoxyethane) $\longrightarrow C_6H_5CH_2CH_2NH_2$
7. *(a)* Reflux with NaOH(aq) $\longrightarrow C_2H_5OH$. Warm with $K_2Cr_2O_7$ + acid, and distil off CH_3CHO as it is formed.
   (b) Make C_2H_5OH as in *(a)*. Heat with $K_2Cr_2O_7$ + acid $\longrightarrow CH_3CO_2H$. Convert part of CH_3CO_2H into CH_3CO_2Na by reaction with NaOH(aq). Convert part into CH_3COCl, using PCl_5 or $SOCl_2$. Heat $CH_3CO_2Na + CH_3COCl \longrightarrow (CH_3CO)_2O$
   (c) Make C_2H_5OH as in *(a)*. Make CH_3CO_2H as in *(b)*. Esterify the acid and alcohol, using conc. sulphuric acid as catalyst.
   (d) Add KCN(ethanol) $\longrightarrow C_2H_5CN$. Reduce this with $LiAlH_4$(ethoxyethane) $\longrightarrow C_2H_5CH_2NH_2$. Then $NaNO_2$ + HCl(aq), warm $\longrightarrow C_2H_5CH_2OH$. Oxidise with $K_2Cr_2O_7$ + acid $\longrightarrow C_2H_5CO_2H$
   (e) Make $C_2H_5CO_2H$ as in *(d)*. With $(NH_4)_2CO_3 \longrightarrow C_2H_5CO_2NH_4$. Heat with excess $C_2H_5CO_2H \longrightarrow C_2H_5CONH_2$
8. *(a)* (i) C_6H_6 + conc. HNO_3 + conc. $H_2SO_4 \longrightarrow C_6H_5NO_2$. With Sn + HCl(aq) $\longrightarrow C_6H_5NH_2$
   (ii) $C_6H_5CH_3 + KMnO_4$ + acid, warm $\longrightarrow C_6H_5CO_2H$. With PCl_5 (or $SOCl_2$) $\longrightarrow C_6H_5COCl$
   (iii) $C_6H_5NH_2 + C_6H_5COCl \longrightarrow C_6H_5CONHC_6H_5$
   (b) $C_6H_6 + CH_3Cl + AlCl_3$ (catalyst) $\longrightarrow C_6H_5CH_3$. With Cl_2 + sunlight or UV $\longrightarrow C_6H_5CH_2Cl$. Reflux with NaOH(aq) $\longrightarrow C_6H_5CH_2OH$
   (c) (i) $CH_4 + Cl_2$ in sunlight $\longrightarrow CH_3Cl$
   (ii) $C_6H_6 + CH_3Cl + AlCl_3$ (catalyst) $\longrightarrow C_6H_5CH_3$ Warm with $KMnO_4$ + acid $\longrightarrow C_6H_5CO_2H$

Checkpoint 36.5B: Draw your own conclusions

1. **A** = propanone CH_3COCH_3, **B** = propanal CH_3CH_2CHO **C** = propan-1-ol $CH_3(CH_2)_2OH$, **D** = propan-2-ol $CH_3CH(OH)CH_3$,
2. The compounds could be **A** = ethyl benzoate $C_6H_5CO_2C_2H_5$, **B** = ethanol C_2H_5OH, **C** = sodium benzoate $C_6H_5CO_2Na$, **D** = benzoic acid $C_6H_5CO_2H$.
3. **A** and **B** are *cis-* and *trans*–3-phenylpropenoic acid $C_6H_5CH{=}CHCO_2H$, **C** is 3-phenylpropanoic acid, $C_6H_5CH_2CH_2CO_2H$, **D** is 3-phenylpropanoyl chloride, $C_6H_5CH_2CH_2COCl$.
4. **P** is 2-aminoethanoic acid (glycine), $H_2NCH_2CO_2H$, **Q** is aminomethane/methylamine CH_3NH_2, **R** is 2-hydroxyethanoic acid $HOCH_2CO_2H$.
5. **B** is ethanoic anhydride $(CH_3CO)_2O$ and **A** is sodium ethanoate CH_3CO_2Na.
   C is phenylamine, $C_6H_5NH_2$, **D** is phenylammonium chloride, $C_6H_5NH_3^+ Cl^-$.
   G is $C_6H_5OCOC_6H_5$. **F** is C_6H_5OH.
6. **W** can be hydrolysed by alkali to give a bromide **X** – as shown by the yellow ppt. of AgBr(s). **Y** gives a positive iodoform test, so it contains either CH_3CO— or CH_3CHOH— The fact that **Y** can be oxidised to **Z**, which gives a positive iodoform test, suggests that **Y** contains the group CH_3CHOH— and **Z** contains the group CH_3CO—. **Z** is an aldehyde, as shown by the DNP and $AgNO_3$ tests. If **Y** is a 2° alcohol, **W** = RR′CHBr. Subtracting 93 for CHBr from M_r = 123 leaves 30, which allows for $(CH_3)_2$; therefore **W** = $(CH_3)_2CHBr$, **X** = NaBr, **Y** = $(CH_3)_2CHOH$, **Z** = $(CH_3)_2CO$.
7. **A** seems to be an aromatic nitro compound, which is reduced to a 1° aromatic amine, e.g. $XC_6H_4NO_2 \longrightarrow XC_6H_4NH_2$. The formula $C_7H_7NO_2$ corresponds to $CH_3C_6H_4NO_2$. **B** would then be $CH_3C_6H_4NH_2$ (dissolves in acid). HNO_2(aq) converts —NH_2 into —OH, making **C** = $CH_3C_6H_4OH$ (violet colour with $FeCl_3$) and **D** = $CH_3C_6H_4OCH_3$, oxidised to **E** = $HO_2CC_6H_4OH + CH_3OH$. With soda lime, followed by acid **E** $\longrightarrow$ **F**, $C_6H_5OCH_3$.
8. **A** $\longrightarrow$ **B**, pungent-smelling gas, suggests ammonia or an amine. **A** is an ammonium salt or the salt of an amine. If **B** is an amine the reaction with HNO_2(aq) would give N_2(g) + alcohol. If **A** is the salt of an amine and it gives a white ppt. with $AgNO_3$, this is a chloride and the compounds could be **A** = $C_2H_5NH_3^+ Cl^-$, **B** = $C_2H_5NH_2$.
10. **H** can be oxidised to **I** and then further oxidised to **J**, suggesting alcohol $\longrightarrow$ aldehyde $\longrightarrow$ carboxylic acid. All **H**, **I** and **J** have acidic solutions and also contain a group which can be oxidised by MnO_4^-(aq); they could contain —CO_2H and also –OH. **I** is a carbonyl compound, and, since it can be oxidised, it is an aldehyde. The formulae of **H** and **I** contain 3 O atoms, therefore H could be HO_2CCH_2OH, **I** could be HO_2CCHO, **J** = HO_2CCO_2H.

Questions on Chapter 36

3. *(a)* **A**: $Cl_2 + AlCl_3$
   B: $CH_3Br + AlBr_3$ (or $CH_3Cl + AlCl_3$)
   C: Chlorine. Pass Cl_2 into methylbenzene under reflux.
   D: Chlorine. Sunlight or UV.
   E: $KMnO_4$ + acid. Reflux.
   F: conc. HNO_3 + conc. H_2SO_4, 30 °C, separate from the 4-nitro compound
4. **A** = $CH_3CHOHCH_2CO_2H$, **B** = $CH_3CH{=}CHCO_2C_2H_5$
   C = $(\text{—}\underset{\displaystyle CH_3}{CH}\text{—}\underset{\displaystyle CO_2C_2H_5}{CH}\text{—})_n$
   D = $CH_3CHClCH_2COCl$, **E** = $CH_3CHClCH_2CO_2H$
5. (i) Add DNP to each of **A** – **E**; **A** and **B** give coloured ppts. Add $AgNO_3$(ammoniacal) to **A** and **B**; **A** gives a silver mirror; **B** does not.
   (ii) To each of **C**–**E**, add NaOH(aq) and warm. **D** gives ammonia.
   (iii) To each of **C**–**E**, add HCl(aq), warm and cool. **E** gives a white crystalline solid.
   (iv) This leaves **C** to identify. Add red litmus paper to NaOH(aq). The paper turns blue. Add **C**. The paper turns purple and then red.
9. *(a)* (i) **A**: $CH_3CH_2CH_2CH_2Br$ or $CH_3CH_2CHBrCH_3$; positional isomerism
   (ii) **B**: $CH_3CH_2CH{=}CH_2$ or $CH_3CH{=}CHCH_3$; structural isomerism
   (iii) 1-bromobutane and 2-bromobutane; but-1-ene and but-2-ene
   (iv) **C**: $CH_3(CH_2)_3CN$, **D**: $CH_3(CH_2)_4NH_2$
   (v) Amount of HCl = 0.01147 mol, therefore M_r = 87.2
   (vi) pH ≈ 5

11. *(a)* (i) $C_6H_5CH_2Br \xrightarrow{1} C_6H_5CH_2CN \xrightarrow{2}$ $C_6H_5CH_2CH_2NH_2$
1. Reflux with KCN(in ethanol) 2. Reduce with $LiAlH_4$ (in ethoxyethane)
(ii) 1. Reflux with KOH(ethanolic) 2. either (i) ozone (ii) Zn + ethanoic acid or hot conc. $KMnO_4$
(iii) $C_6H_5COCl \xrightarrow{1} C_6H_5CONH_2 \xrightarrow{2} C_6H_5NH_2$
1. NH_3, room temperature 2. Br_2 + conc. NaOH(aq)
(iv) 1. conc. H_2SO_4 2. alkaline $KMnO_4$
(c) (i) 4-methylpentan-2-ol
(ii) 2,5,5-trimethylcyclohex-1-ene-3-one

12. *(a)* (ii) Add Br_2 (in organic solvent): no reaction with benzene; decolourised by cyclohexene.
(ii) Add $NaIO_3$(aq), warm. $C_5H_9CHOHCH_3$ is oxidised to $C_5H_9COCH_3$ which gives a yellow ppt. of iodoform CHI_3. The alcohol $C_4H_9CHOHCH_2CH_3$ does not react.
(iii) Add DNP. The ketone gives a yellow ppt. The ester does not react.
(iv) Add Tollens' reagent (ammoniacal $AgNO_3$), warm. The aldehyde gives a silver mirror. The ketone is not oxidised.
(b) $J = H_2NOH$ hydroxylamine
K = bromine water
$L = LiAlH_4$ (in ethoxyethane)
$M = CH_3CH_2CH_2CN$
$N = C_6H_5SO_3Na$

14. *(a)* **W** = butan-2-ol, **Y** = $CH_3CH_2COCH_3$,
Z = $CH_3CH_2C(CH_3)(CN)OH$
Z has a chiral C atom and shows optical isomerism.
(b) (ii) $CH_3CH_2CH{=}CH_2$, but-1-ene

15. *(a)* Formula = $C_4H_{10}O$. **K** is oxidised to an acid, therefore it is an aldehyde or an alcohol. The formula $C_4H_{10}O$ fits an alcohol, not an aldehyde. **K** could be $CH_3(CH_2)_2CH_2OH$, butan-1-ol, and **J** could be $CH_3CH_2CHOHCH_3$, butan-2-ol, which is oxidised to a ketone. Then **H**, which does not have the same functional group as **J** and **K**, must be an ether, $C_2H_5OC_2H_5$, ethoxyethane.
(b) (i) **J**, $CH_3CH_2C^*HOHCH_3$, has a chiral C atom and has optical isomers:
(ii) $CH_3CH_2CHOHCH_3 \longrightarrow$
$CH_3CH{=}CHCH_3 + H_2O$
cis–trans isomers, **L** and **M**

16. *(b)* M_r of $CH_2O = 30$, so formulae are CH_2O, $C_2H_4O_2$ and $C_3H_6O_3$.

H = H—C(H)=O, **J** = H—C(=O)—O—CH_3,

K = C(CO₂H)(H)(CH₃)(OH) [chiral centre: CO_2H above; H, CH_3, OH below]   **L** = CH_3—C(=O)—O—H

M = H—C(=O)—O—H, **N** = CH_3OH

APPENDIX: MATHEMATICS

This appendix is a reminder of some of the mathematics which you studied earlier in your school career. You may need to brush up on your mathematical skills as you tackle the calculations in the book.

1. USING EQUATIONS

Scientists measure quantities such as pressure, volume and electric potential difference. Sometimes they find that when one quantity changes another changes as a result. The related quantities are described as **variables**. The relationship between the variables can be written in the form of a mathematical equation.

Proportionality

For example the mass of a sample of a substance is proportional to its volume:

Mass $\propto$ Volume ($\propto$ means 'is proportional to')

Proportionality: $y \propto x$

This means that as the volume increases the mass increases in proportion. The relationship can be written as

Mass = k × Volume, where k is a constant

The constant k is the density of the substance:

Mass = Density × Volume

Inverse proportionality

Inverse proportionality: $y \propto 1/x$

The volume of a fixed mass of gas is inversely proportional to its pressure, that is, as the pressure increases the volume decreases in proportion:

Volume $\propto$ 1/Pressure

Volume = k/Pressure (where k is a constant)

Rearranging equations

The aim of rearranging an equation is to put the quantity you want
- *by itself on one side of the equation and*
- *positive*

Example (a) You might want to use the equation

Mass = Density × Volume

to find the density of an object from its mass and volume. Then you rearrange the equation to put density by itself on one side of the equation, that is, in the form

Density = ?

If you add to or subtract from or multiply or divide one side of the equation you must do the same to the other side

To obtain density by itself you must divide the right-hand side of the equation by volume. Naturally you must do the same to the left-hand side.

$$\frac{\text{Mass}}{\text{Volume}} = \frac{\text{Density} \times \cancel{\text{Volume}}}{\cancel{\text{Volume}}}$$

$$\text{Density} = \text{Mass/Volume}$$

Cross-multiplying

Once you have understood the idea behind rearranging equations you can try the method of cross-multiplying.

Cross-multiplying is a shortcut

Example (b) For one mole of a gas the relationship between pressure, temperature and volume is

$$\frac{P}{T} = \frac{R}{V}$$

If you cross-multiply:

$$\frac{P}{T} \times \frac{R}{V}$$

you obtain the equation

$$PV = RT$$

Cross-multiplying is putting into practice the method of multiplying or dividing both sides of the equation ($P/T = R/V$) by the same quantity. Check it:

The sign $\Rightarrow$ means 'it follows that'

- first multiply both sides of the equation by $T \Rightarrow P = RT/V$
- then multiply both sides of the equation by $V \Rightarrow PV = RT$

which is the equation you obtained by cross-multiplying.

Example (c) A zinc electrode and a copper electrode can be connected to form a chemical cell. The voltage of the cell under standard conditions is $E^{\ominus}$. It is given by

$$E^{\ominus}_{\text{cell}} = E^{\ominus}_{\text{Zn}} - E^{\ominus}_{\text{Cu}}$$

If $E^{\ominus}_{\text{cell}} = -1.10$ V and $E^{\ominus}_{\text{Zn}} = -0.76$ V, what is $E^{\ominus}_{\text{Cu}}$?

You need to get $E^{\ominus}_{\text{Cu}}$ by itself and positive. You must do the same to both sides of the equation. First, add $E^{\ominus}_{\text{Cu}}$ to both sides of the equation.

$$E^{\ominus}_{\text{Cu}} + E^{\ominus}_{\text{cell}} = E^{\ominus}_{\text{Zn}}$$

Now subtract $E^{\ominus}_{\text{cell}}$, from both sides:

$$E^{\ominus}_{\text{Cu}} = E^{\ominus}_{\text{Zn}} - E^{\ominus}_{\text{cell}}$$

$$= -0.76\text{ V} - (-1.10\text{ V}) = -0.76\text{ V} + 1.10\text{ V} = +0.34\text{ V}$$

Another shortcut is: 'Change the side; change the sign' plus $\longrightarrow$ minus; minus $\longrightarrow$ plus

Change the side; change the sign

You will have noticed that when a quantity changes from one side of the equation to the other side, it changes sign, from (+) to (−) or from (−) to (+).

EXERCISE 1: EQUATIONS

1. Rewrite each of the following equations
   (a) $a = b + c$ in the form $c = ?$
   (b) $x - y = z$ in the form (i) $x = ?$ (ii) $y = ?$
   (c) $p - q = r + s$ in the form (i) $p = ?$ (ii) $q = ?$

2. Rewrite each of the following equations.
   (a) speed = distance/time in the form
   (i) distance = ? (ii) time = ?
   (b) $K_c = \frac{[PCl_5]}{[PCl_3][Cl_2]}$ in the form

   (i) $[PCl_5] = ?$ (ii) $[Cl_2] = ?$

   (c) $K_c = \frac{[HI]^2}{[H_2][I_2]}$ in the form

   (i) $[H_2] = ?$ (ii) $[HI] = ?$

3. The pressure, volume and temperature of one mole of a gas are related to the gas constant R by the equation

   $$\frac{P}{T} = \frac{R}{V}$$

   Rearrange the equation to obtain equations for (a) T and (b) V.

4. The concentration of a solution can be expressed

   $$\text{Concentration} = \frac{\text{Amount of solute}}{\text{Volume of solution}}$$

   Rearrange the equation into the form
   (a) Amount of solute = ? and
   (b) Volume of solution = ?

5. The expression for the dissociation constant of a weak acid is

   $$K_a = \frac{[H^+]^2}{[HA]}$$

   Rearrange this expression to give an equation for $[H^+]$.

2. CALCULATIONS ON RATIO

Many of the calculations you meet involve ratios. You met this type of problem in your maths course. Do not forget how to solve them when you meet them in chemistry!

Multiplier method

Example (a) A cook makes a pie for 6 people. She uses 300 g flour, 150 g margarine and 450 g of fruit. What quantities should she use to make a pie for 8 people?

In ratio calculations use the multiplier or scale factor: (size you want / size you have)

Method You know that the quantities needed are proportional to the number of people.

The method is to multiply by a ratio called the **multiplier** (or **scale factor**).

$$\text{Multiplier} = \frac{\text{Size you want}}{\text{Size you have}}$$

Size cook wants = 8 people. Size she has = 6 people; therefore multiplier = 8/6 = 4/3.
Flour needed = 300 g × $\frac{4}{3}$ = 400 g
Margarine needed = 150 g × $\frac{4}{3}$ = 200 g
Fruit needed = 450 g × $\frac{4}{3}$ = 600 g

Example (b) If 24 g of magnesium give 40 g of magnesium oxide, what mass of magnesium is needed to give 60 g of magnesium oxide?

$$\text{Multiplier} = \frac{\text{Mass of MgO you want}}{\text{Mass of MgO you have}}$$

Multiplier = 60 g/40 g = 3/2
Multiply 24 g magnesium by 3/2 to get 36 g magnesium. The mass of magnesium needed = 36 g.

EXERCISE 2: RATIOS

1. The mass of copper deposited by electroplating is proportional to the time for which a constant current is passed. If the mass of copper deposited in 12.5 minutes is 27.5 mg, what mass of copper is deposited in 75 minutes?

2. Zinc reacts with dilute acids to give hydrogen. If 0.0400 g of hydrogen is formed when 1.30 g of zinc reacts with an excess of acid, what mass of zinc is needed to produce 6.00 g of hydrogen?

3. If 0.030 g of a gas has a volume of 150 cm^3 what is the volume of 45 g of the gas (at the same temperature and pressure)?

4. 88 g of iron(II) sulphide is the maximum quantity that can be obtained from the reaction of an excess of sulphur with 56 g of iron.
What is the maximum quantity of iron(II) sulphide that can be obtained from 7.00 g of iron?

5. A lime kiln obtains 56 tonnes of calcium oxide from 100 tonnes of limestone. What mass of limestone must be decomposed to yield 280 tonnes of calcium oxide?

3. NEGATIVE NUMBERS

You will often meet physical quantities with negative values. Electrode potentials can be positive or negative. Enthalpy changes can be positive (in endothermic reactions) or negative (in exothermic reactions).

In dealing with negative numbers,
$+(-) = -$
$-(-) = +$
$(+) \times (-) = -$
$(+) \div (-) = -$
$(-) \times (-) = +$
$(-) \div (-) = +$

Remember how to deal with negative numbers:

- adding a negative number, e.g. $7 + (-6) = 7 - 6 = 1$
- subtracting a negative number, e.g. $7 - (-6) = 7 + 6 = 13$
- multiplying by a negative number, e.g. $3 \times (-6) = -18$
- dividing by a negative number, e.g. $18 \div (-3) = -6$

Example In the reaction shown below the energy content of each species is shown beneath it in kJ mol^{-1}.

$$\begin{array}{ccccccc} \mathbf{A} & + & \mathbf{B} & \longrightarrow & \mathbf{C} & + & \mathbf{D} \\ (-100) & & (+50) & & (-200) & & (+60) \end{array}$$

To enter a negative number on a calculator, use the +/– change of sign key

(a) What is the total energy of the products? *(b)* What is the total energy of the reactants? *(c)* What is the energy change: (energy of products) – (energy of reactants)?

Method
(a) The total energy of **C** + **D** = (–200) + (+60) = –140 kJ mol^{-1}
(b) The total energy of **A** + **B** = (–100) + (+50) = –50 kJ mol^{-1}
(c) The difference (–140) – (–50) = –140 + 50 = –90 kJ mol^{-1}

When entering negative numbers on your calculator, enter the digits and then use the +/− key. For the number −6.7, enter 6.7 and then press the +/− key to see −6.7 on the display.

EXERCISE 3: NEGATIVE NUMBERS

1. A process consists of four steps. The heat changes in the four steps are +29.5 kJ mol^{-1}, −41.6 kJ mol^{-1}, −61.0 kJ mol^{-1}, +42.5 kJ mol^{-1}. Calculate the heat change for the complete process.

2. The voltage of a cell is given by (if RHS = right-hand side)

   $E_{cell} = E_{RHS\ electrode} - E_{LHS\ electrode}$

   For the following combinations of electrode, state E_{cell}.

	LHS electrode		RHS electrode	
(*a*)	Nickel	−0.25 V	Zinc	−0.76 V
(*b*)	Iron(II)	−0.44 V	Iron(III)	+0.77 V
(*c*)	Zinc	−0.76 V	Nickel	−0.25 V
(*d*)	Lead	−0.13 V	Silver	+0.80 V
(*e*)	Tin	−0.14 V	Zinc	−0.76 V

4. WORKING WITH NUMBERS IN STANDARD FORM

For working with large numbers and small numbers, it is convenient to write them in the form known as **scientific notation** or **standard form** or **standard index form**. Instead of writing two million as 2 000 000, we can write:

$$2\,000\,000 = 2 \times 10 \times 10 \times 10 \times 10 \times 10 \times 10 = 2 \times 10^6$$

In standard form, 2 000 000 = 2×10^6

Note $1 \times 10^2 = 100$ = 1 hundred
$1 \times 10^3 = 1000$ = 1 thousand
$1 \times 10^6 = 1\,000\,000$ = 1 million

Numbers less than one can also be written in standard form:

$$0.0002 = 2/10\,000 = 2/10^4 = 2 \times 10^{-4}$$

In standard form, 0.0002 = 2×10^{-4}

Note $1 \times 10^{-1} = 0.1$ = one tenth
$1 \times 10^{-2} = 0.01$ = one hundredth
$1 \times 10^{-3} = 0.001$ = one thousandth
$1 \times 10^{-6} = 0.000\,001$ = one millionth

You can see that in standard form, the number is written as a product of two factors. In the first factor the decimal point comes after the first digit. The second factor is a multiple of ten. For example,

$2123 = (2.123) \times (10^3)$

first factor: the decimal point comes after the first digit

second factor: a multiple of ten

$0.000\,167 = 1.67 \times 10^{-4}$

The number 3 or −4 is called the **power** or the **index** or the **exponent**, and the number 10 is the **base**. 10^3 is referred to as '10 to the power 3'.

If the power (exponent) is

- increased by 1, the decimal point must be moved one place to the left,
- decreased by 1, the decimal point must be moved one place to the right:

$$2.5 \times 10^3 = 0.25 \times 10^4 = 25 \times 10^2 = 250 \times 10^1 = 2500 \times 10^0.$$

Since $10^0 = 1$, this factor is omitted and 2500×10^0 is written as 2500.

To write a number in standard form: Put a decimal point after the first digit Count how many places the decimal point has moved. This number is the index. It is positive if the decimal point has moved to the left, negative if it has moved to the right. e.g. $2\,100 = 2.1 \times 10^3$ e.g. $0.0021 = 2.1 \times 10^{-3}$

To change a number into standard form you

- Put a decimal point after the first digit.
- Count how many places the decimal point has moved. This number gives the power of ten.
- If the decimal place has moved to the left the power is positive; if the decimal place has moved to the right the power is negative.

Example 1 $1234 = 1.234 \times 10^3$
The decimal point has moved 3 places to the left so the new number is multiplied by 10^3.

Example 2 $0.0037 = 3.7 \times 10^{-3}$
The decimal point has moved 3 places to the right so the new number is multiplied by 10^{-3}.

How to enter powers (indices) in a calculator

To enter 1.44×10^6, you enter 1.44; then press the EXP key, then the 6 key. The display reads 1.44 06 or 1.44 E 06.

To enter 4.50×10^{-2}, you enter 4.5; then press the EXP key, then the 2 key and lastly the +/− key. The display will read 4.50 −02 or 4.50 E −02.

To enter 10^{-3}, you enter 1; then press the EXP key, then the 3 key and finally the +/− key. The display will show 1. −03 or 1. E −03.

To multiply expressions involving powers of 10 (indices), add the powers. To divide expressions involving powers of 10, subtract the powers

When you are writing down an answer you have read from your calculator, write it as e.g. 1.44×10^6 and not as 1.44 06.

In multiplication, powers are added, e.g.

$$(1.40 \times 10^6) \times (1.20 \times 10^{-3}) = (1.40 \times 1.20) \times (10^6 \times 10^{-3}) = 1.68 \times 10^3$$

In division, powers are subtracted, e.g.

$$(2.8 \times 10^6) \div (1.4 \times 10^3) = (2.8 \div 1.4) \times (10^6 \div 10^3) = 2.0 \times 10^3$$

EXERCISE 4: STANDARD FORM

1. Express the following numbers in standard form
   (a) 600 000 *(b)* 4000 *(c)* 45 000 *(d)* 0.0123 *(e)* 0.006 *(f)* 0.000 49

5. DECIMAL PLACES

The number of decimal places = number of digits after the decimal point

Sometimes you are asked to give an answer to a certain number of decimal places. For example you are asked to quote your answer to four decimal places, and your calculator reads 0.0867318. You count four digits from the decimal point to obtain 0.0867.

However, if the display reads 0.0867618, the answer to four decimal places is 0.0868 because it has been rounded up. The first digit dropped after the 7 is a 6 so the fourth digit is rounded up from 7 to 8. If the first digit dropped is 5 or greater than 5, the last digit is rounded up. Note that 4.602 to 2 decimal places is 4.60.

6. SIGNIFICANT FIGURES

The number of significant figures = number of figures you are sure of

Rounding off: if the first insignificant figure is 5 or more, round up the last significant figure

Often your calculator will display an answer containing more digits than the numbers you fed into it. Suppose you are calculating the concentration of a sodium hydroxide solution. You have found that 18.6 cm^3 of aqueous sodium hydroxide neutralise 25.0 cm^3 of hydrochloric acid of concentration 0.100 mol dm^{-3}. On putting the numbers into your calculator you obtain a value of 0.134 408 6 mol dm^{-3}. If you take this as your answer, you are claiming an accuracy of one part in a million! Since you read the burette to three figures, you quote your answer to three figures, 0.134 mol dm^{-3}. The figures you are sure of are called the **significant figures**. The insignificant figures are dropped. This operation is called **rounding off**. If the answer had been 0.134 708 6, it would have been rounded off to 0.135. When the first of the figures to be dropped is 5 or greater, the last of the significant figures is rounded up to the next digit. The number 123 has 3 significant figures. The number 1.23×10^4 has 3 significant figures, but 12 300 has 5 significant figures because the final zeros mean that each of these digits is known to be zero and not some other digit.

EXERCISE 5: DECIMAL PLACES AND SIGNIFICANT FIGURES

1. State the number of significant figures in
   (a) 1.0×10^{-2} *(b)* 1.00×10^{-2} *(c)* 0.135
   (d) 1.2345 *(e)* 2500 *(f)* 25 *(g)* 2.5×10^{-3}
   (h) 2.500×10^{-3}
   (g) an enthalpy change of −108.6 kJ mol^{-1}
   (h) a rate constant of 2.048 dm^3 mol^{-1} s^{-1}
   (i) the ideal gas constant
   0.082 0578 dm^3 atm K^{-1} mol^{-1}

2. Write the following quantities to three significant figures.
   (a) a volume of 15.814 cm^3
   (b) a mass of 205.5 tonnes
   (c) a volume of 1120 cm^3
   (d) a percentage of 49.517%
   (e) an amount of 3.3333 mol
   (f) a concentration of 0.14535 mol dm^{-3}

3. Write the results of these calculations to the appropriate number of significant figures.
   (a) 2.00×41.72 *(b)* $3.00 \times 20.02 \times 40.1$
   (c) 20.5/32.0

4. Give each number correct to two decimal places.
   (a) 47.123 *(b)* 8.334 *(c)* 8.345 *(d)* 0.0643
   (e) 0.567

7. ESTIMATING YOUR ANSWER

One advantage of standard index form is that very large and very small numbers can easily be entered on a calculator. Another advantage is that you can easily estimate the answer to a calculation to the correct order of magnitude (the correct power of ten).

$$\text{For example, } \frac{2456 \times 0.0123}{5223 \times 60.7}$$

Putting the numbers into standard form gives

$$\frac{2.456 \times 10^3 \times 1.23 \times 10^{-2}}{5.223 \times 10^3 \times 6.07 \times 10}$$

This is approximately

$$\frac{2 \times 1}{5 \times 6} \times \frac{10^3 \times 10^{-2}}{10^3 \times 10} = \frac{2 \times 10^{-3}}{30} = 0.07 \times 10^{-3} = 7 \times 10^{-5}$$

By putting the numbers into standard form you can estimate the answer very quickly. A complete calculation gives the answer 9.53×10^{-5}. The rough estimate is sufficiently close to this to reassure you that you have not made any slips with powers of ten.

8. LOGARITHMS

The logarithm of a number (written as log or lg) is the power to which 10 must be raised to give the number. For example, the number $100 = 10^2$ therefore the logarithm of 100 to the base 10 is 2: lg 100 = 2.

There is also a set of logarithms to the base e. They are called natural logarithms as e is a significant quantity in mathematics. Natural logarithms are written as ln N and are related to logs to the base ten by

$$\ln N = 2.303 \lg N$$

To find the logarithm to the base 10 enter the number on your calculator and press the log key.
To find the natural logarithm enter the number on your calculator and press the ln key

With most calculators, to obtain the log of a number, enter the number on your calculator and press the log key. Some calculators, however, require you to press the log key first, then the number. The value of the log will appear in the display. This will happen whether you enter the number in standard form or another form. For example, lg 12 345 = 4.0915, whether you enter the number as 12 345 or as 1.2345×10^4. To obtain the natural logarithm of a number, enter the number in your calculator and press the ln key (unless your calculator requires the ln key first followed by the number).

Natural logarithms arise in problems in physical chemistry. Logarithms to the base 10 are used when a graph must be plotted of measurements over a wide range of values. You would find it difficult to fit the numbers 10, 1000 and 1 000 000 onto the same graph, but their logs, 1, 3 and 6 can easily fit on to the same scale.

9. GRAPHS

Here are some hints for drawing graphs.

(a) If the quantities x and y are related by the equation $y = ax + b$, then a plot of experimental values of y against corresponding values of x will give a straight line [see Figure A]. Values of y are plotted on the vertical axis, the y-axis, and values of x are plotted on the horizontal axis, the x-axis. From the graph,

gradient = (increase in y, δy)/(increase in x, δx) = a

intercept on the y-axis = b.

Figure A
Plotting a graph

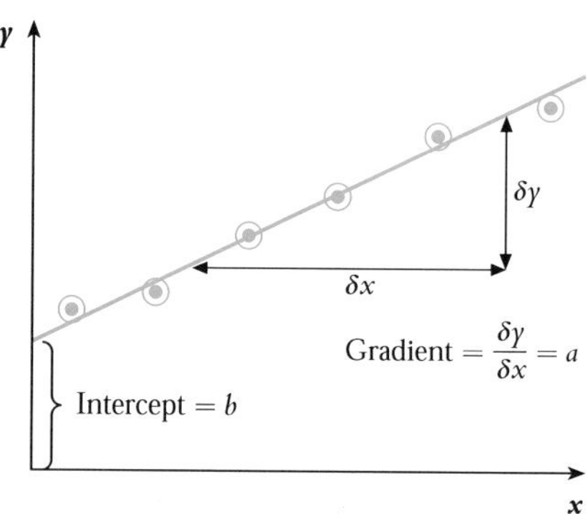

(b) Choose a scale which will which will allow the graph to cover as much of the piece of paper as possible. There is no need to start at zero. If the points lie between 80 and 100 to start at zero would cramp your graph into a small section at the top of the page [see Figure B].

Figure B
Don't cramp your graph!

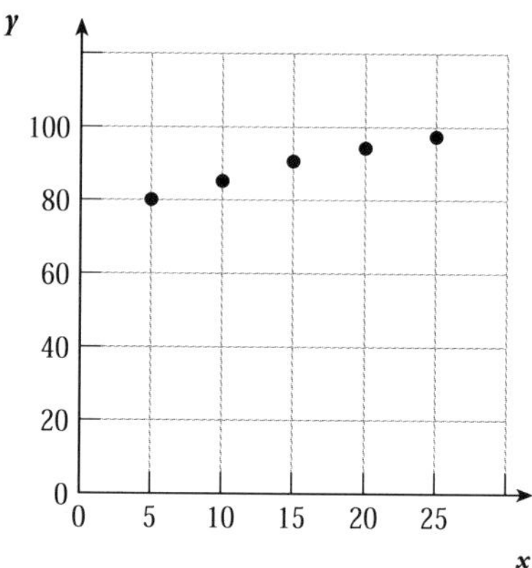

Figure C
Label the axes

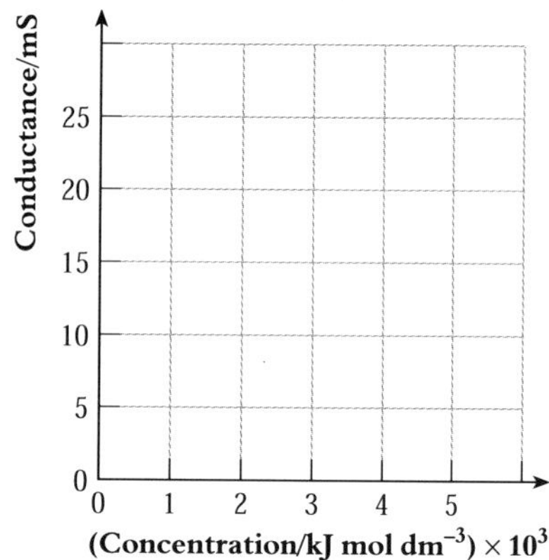

When plotting a graph,

- *spread the points over the graph paper*
- *label the axes*
- *draw the best straight line through the points ...*

... or a smooth curve

(c) Label the axes with the quantities and the units. Make the scale units as simple as possible. Instead of plotting 1×10^{-3} mol dm^{-3}, 2×10^{-3} mol dm^{-3}, 3×10^{-3} mol dm^{-3} etc. plot 1, 2, 3 etc. and label the axis as (Concentration/mol dm^{-3}) $\times 10^3$ [see Figure C]. The sign (/) means 'divided by'. The numbers 1, 2, 3 etc. are the values of the concentration divided by the unit, mol dm^{-3} and multiplied by 10^3.

(d) When you draw a straight line through the points, draw the best straight line you can, to pass through or close to as many points as possible. In Figure D, a line can be drawn close to four of the points, but one point does not fit in. It is better to assume that some experimental error was made in this point and draw the best fit to the other four points.

(e) If you are drawing a curve, draw a smooth curve [Figure E]. Do not join up the points with straight lines.

Figure D
Drawing the best line

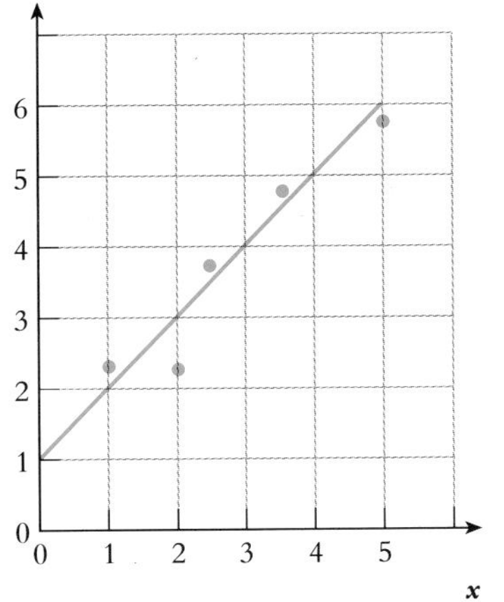

Figure E
Drawing a smooth curve

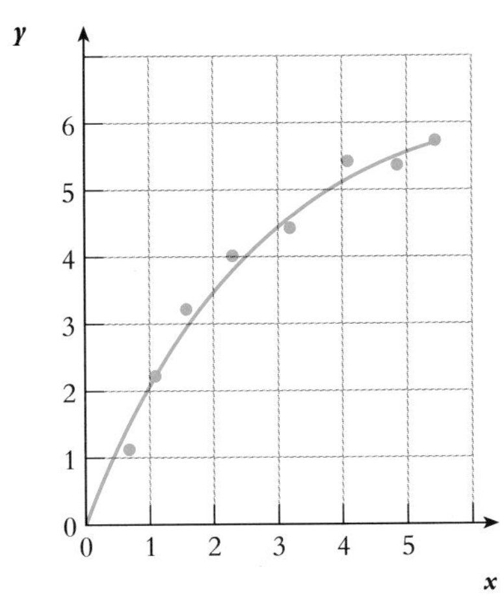

EXERCISE 6: GRAPHS

1. Plot the values of y against the corresponding values of x.

x	3	6	9	12
y	13	22	31	40

Deduce the equation of the straight line plot.

2. Plot a graph of values of y against x.

x	56	28	14	7
y	1	2	4	8

Deduce the equation of the straight line.

3. Hydrogen peroxide decomposes to give oxygen.

$$2H_2O_2(aq) \longrightarrow O_2(g) + 2H_2O(l)$$

The rate of decomposition is given by (change in concentration)/time and has the units mol dm^{-3} s^{-1}. The rate depends on the hydrogen peroxide concentration at the start of the reaction.

Rate/mol^{-3} s^{-1}	$[H_2O_2]$/mol dm^{-3}
3.64×10^{-5}	0.05
7.41×10^{-5}	0.10
1.51×10^{-4}	0.20
2.21×10^{-4}	0.30

Plot the values of rate (on the y-axis) against the concentration of hydrogen peroxide (on the x-axis). Write an equation relating rate and concentration of hydrogen peroxide.

4. Krypton is a radioactive element. The amount of radioactivity is measured by the count per second registered on a detector. Use the results shown to plot a graph of radioactivity (on the y-axis) against time (on the x-axis). From the graph, read off the time taken for the radioactivity to drop to half its original value.

Time/minutes	0	20	40	60	80	100	120	140	160	180
Radioactivity/counts per second	100	92	85	78	72	66	61	56	52	48

10. INCLUDE THE UNITS

When you are doing a calculation, include the units of each physical quantity as well as the numerical values. If the answer you obtain has the correct units, it is likely that you have the correct answer.

Include the units of measurement in your working as well as in your answer

Example Diana is calculating the concentration of a solution which contains 1.25 moles of solute in 250 cm^3 of solution. She obtains the answer 0.2 $dm^3\ mol^{-1}$. Immediately she realises that since concentration has the unit mol dm^{-3} her answer is wrong. This time she puts in the units.

$$\text{Concentration} = \frac{\text{Amount of solute}}{\text{Volume of solution}}$$

$$\text{Concentration} = 1.25\ \text{mol}/250\ \text{cm}^3 = 1.25\ \text{mol}/250 \times 10^{-3}\ \text{dm}^3$$
$$= 5.0\ \text{mol dm}^{-3}$$

Now the unit of concentration is correct, and Diana has more confidence in her answer.

11. A FINAL POINT

Does your answer look right? Does it have the right units? Does it have the right power of 10?

Always look critically at your answer. Ask yourself whether it is a reasonable answer. Does it have the right power of ten for the data? Be sure it is not a hundred times or a thousand times the correct answer. Is it in the right units? Many errors can be detected by a review of this kind.

ANSWERS: APPENDIX

Exercise 1: Equations

1. *(a)* $c = a - b$
   (b) (i) $x = y + z$
   (ii) $y = x - z$
   (c) (i) $p = r + s + q$
   (ii) $q = p - r - s$
2. *(a)* (i) distance = speed × time,
   (ii) time = distance / speed
   (b) (i) $[PCl_5] = K_c\ [PCl_3]\ [Cl_2]$
   (ii) $[Cl_2] = [PCl_5]/K_c\ [PCl_3]$
   (c) (i) $[H_2] = [HI]^2/K_c\ [I_2]$
   (ii) $[HI] = \sqrt{K_c[H_2][I_2]}$
3. *(a)* $T = PV/R$ *(b)* $V = RT/P$
4. *(a)* Amount of solute = Concentration × Volume
   (b) Volume = Amount of solute/Concentration
5. $[H^+] = \sqrt{K_a[HA]}$

Exercise 2: Ratios

1. 165 mg = 0.165 g
2. 195 g
3. 225 dm^3
4. 11 g
5. 500 tonnes

Exercise 3: Negative numbers

1. −30.6 kJ mol^{-1}
2. *(a)* −0.51 V *(b)* +1.21 V *(c)* +0.51 V *(d)* +0.93 V *(e)* −0.62 V

Exercise 4: Standard form

1. *(a)* 6×10^5
   (b) 4×10^3
   (c) 4.5×10^4
   (d) 1.23×10^{-2}
   (e) 6×10^{-3}
   (f) 4.9×10^{-4}

Exercise 5: Decimal places and significant figures

1. *(a)* 2 *(b)* 3 *(c)* 3 *(d)* 5 *(e)* 4 *(f)* 2 *(g)* 2 *(h)* 4
2. *(a)* 15.8 cm^3
   (b) 206 tonnes
   (c) 1120 cm^3
   (d) 49.5%
   (e) 3.33 mol
   (f) 0.145 mol dm^3
   (g) −109 kJ mol^{-1}
   (h) 2.05 $dm^3\ mol^{-1}\ s^{-1}$
   (i) 0.0821 dm^3 atm $K^{-1}\ mol^{-1}$
3. *(a)* 83.4
   (b) 2410
   (c) 0.641
4. *(a)* 47.12 *(b)* 8.33 *(c)* 8.35 *(d)* 0.06 *(e)* 0.57

Exercise 6: Graphs

1. $y = 3x + 4$
2. $x = 56/y$
3. On the vertical axis plot (Rate/mol $dm^{-3}\ s^{-1}$) × 10^4. The four values are 0.364, 0.741, 1.51, 2.21. On the horizontal axis plot $[H_2O_2]$. A straight line can be drawn through the origin and other four points, showing that Rate ∝ $[H_2O_2]$ or rate = $k[H_2O_2]$ where k is a constant.
4. Plot the count rate on the vertical axis against time on the horizontal axis. Spread the results from 100 cpm to 50 cpm over the whole of the graph paper; there is no need to go down to zero cpm. Draw a smooth curve through the points. The time at which the count rate has dropped to 50 cpm can be read from the curve. It is 170 minutes.

INDEX

Absolute temperature 153
Absorption spectrum 22, 298
Acceptor atom 106, 424
Accumulator 275
Acetylcholine 626
Acid
  Brönsted–Lowry 242–3
  carboxylic 631–61
  conjugate 243
  degree of dissociation 246–7
  dissociation constant 246–8, 636
  Lewis 243
  nomenclature 72, 632–3
  rain 405–7
  strength 248
Acid anhydrides 644–5, 642
Acid–base catalysis 291
Acid–base reactions 242–4
  titrations 250–2
Acid chlorides 635, 640, 642–4
Acidic oxides 329, 332
Acrilan® 518
Activation energy 287–90, 306–10
Acylation 532–5, 619–20
Acyl chloride *see* Acid chloride
Addition reactions 487, 510–19
Addition–elimination reactions 600–1
Aerosols 549
Air bags 151
Alanine 654
Alcoholism 564–5
Alcohols 564–89
  aryl 566
  concept maps 586, 587
  nomenclature 565–6
  polyhydric 566, 578
  preparation 569
  properties 567
  reactions 572–8, 582, 595, 715–16
  reactivity 571
  sources 568–9
  uses 570
Aldehydes and ketones 590–611
  carbonyl group 591–4
  nomenclature 591–2
  manufacture 595–7
  preparation 532, 595
  reactions 593, 597–608
Aldose 607
Alicyclic hydrocarbons 484
Aliphatic hydrocarbons 483
Alkali 330, 351
Alkali metals (Group 1) 34–5, 346–55
  compounds 350–5
  extraction 349
  uses 348
  properties 347–8
  reactions 349–50
Alkaline earths (Group 2) 35, 347, 356–61
  compounds 357–9
  extraction 349
  properties 347–8
  reactions 356–7
Alkanes 480–1, 495–506
  isomerism 481, 492
  nomenclature 481–2
  properties 497–8
  reactions 498–504
  source 495–7
Alkanols 565
Alkenes 507–22
  isomerism 508–9
  nomenclature 482–3, 507
  preparation 508–9, 574
  properties 508
  reactions 510–22, 569
    addition 510–20
    combustion 510
    mechanism 512–14
    polymerisation 516–19
    reactivity 509–10
  source 508
Alkylation 500, 531, 534, 620
Alkynes 521
Allotropes
  carbon 140–2
  oxygen 394
Alloy 465, 467
Alnico 465
Alpha particle 6–7, 14
Alumina 368
Aluminium 240, 368–73
  chloride 372, 531–6, 721
  extraction 240, 370
  halides 372, 531–6, 721
  hydroxide 372
  ion 371–2
  oxide 368, 373, 720
  reactivity 369–70
  uses 369
Amides 614, 650–3, 616
  preparation 640–1, 651
  reactions 650–1
Amines 612–23
  basicity 615–16
  concept map 627
  nomenclature 613
  occurrence 614
  preparation 617–19
  properties 614–15
  reactions 619–23
Amino acids 654–6
Aminoethanoic acid 654
Ammonia 414–16
  aqueous 242–3, 253, 414–15
  complexes 107
  hydrogen bonding 111–12
  manufacture 223–4, 413
  molecule 121–3
  reactions 243, 414–15
  –soda process 353–4
Ammonium
  ion 123
  salts 415
Amount 53
Amphetamines 710
Anaesthetics 545–6
Analgesics 713
Analysis 682–709
Anhydrides 642, 644–5
Aniline purple 612
Anion 85
  exchanger 177
  radius 96
Anode 85, 236
Anodising 370
Antimony 412
Aqua-ions 344
Arenes 484, 525–42
Argon 140, 340–1
Aromatic compounds 484, 525–42, 717–8
Aromatic hydrocarbons 484, 525–42
Aromatic substitution 528–42
Arrhenius theory 242
Arsenic 412
Aryl alcohols 566
Ascorbic acid 639
Aspirin 645
Astatine 375
Asymmetry 124–5
Atmosphere 394–7
Atom 4
  splitting 17
  wave theory of 27–9
Atomic
  bomb 17–18
  masses 6
  nucleus 6–8
  number 8–9
  orbitals 27–30
  radii 323–4
  spectra 22–5
  structure 6–8
    concept map 42
  theory 4
Average standard bond enthalpy 200
Avogadro constant 53, 235–6
Avogadro's hypothesis 154
Azeotrope 171
Azo dyes 624

Beta particle 14
Bakelite 672–3, 676
Balmer series 23–5

Barbiturates 714
Barium 347
Bases
  Brönsted–Lowry 242–3
  conjugate 243
  degree of ionisation 246–7
  Lewis 243
  strength 248
Basic oxides 329, 332
Basic salts 256
Batch process 451
Bauxite 370
Becquerel, A H 13
Benedict's reagent 598
Benzaldehyde 591, 597, 601, 603
Benzedrine 710
Benzene
  derivatives 525–7
  properties 527
  reactions 528–38
  reactivity 528
  sources 528
  structure 130–2, 528
Benzenecarbaldehyde *see* Benzaldehyde
Benzenecarboxylic acid *see* Benzoic acid
Benzenediazonium chloride 623
Benzenesulphonic acid 530
Benzoic acid 636
Benzoyl chloride 640
Benzyl alcohol 566
Beryllium
  chloride 121, 123
Bessemer process 464
Bimolecular reaction 309
Binding energy 17
Bismuth 412
  trichloride 223
Bitumen 497
Blast furnace 462–3
Bleaching powder 359, 382
Bodenstein, M 219
Body-centred cubic packing 143–4
Bohr model 23–5
Boiling temperature 111–12, 163
  –composition 168, 170, 683
Bond
  angle 120–30
  double 98, 128, 426, 510
  chemical, concept map 116
  coordinate 106–7
  covalent 97–9, 104–5
  electrovalent 89–93
  energy term 200
  hydrogen 111–15
  ionic 89–93, 104–5, 376
  metallic 137–8
  multiple 128–32
  π (pi) 129–32, 510
  polar 104–5
  σ (sigma) 129–32, 509
  triple 426, 521
  type 105–6
Bonding orbital 127–32
Born–Haber cycle 203–5
Boron 369
  trichloride 121–2
  trifluoride 243
Boyle, R 152
Boyle's Law 152
Brackett series 23, 25
Brady's reagent 600–1
Brass 470
Bromate(V) 382
Bromination 500–1, 533
Bromine 375
  extraction 379, 380
  reactions 376–8, 500–1, 511–12, 533
Brönsted–Lowry theory 242–3
Bronze 442, 470
Brown ring test 417
Buckminsterfullerene 682–3
'Bucky ball' 682–3
Buffer solutions 253–5
But-2-ene 122
Butenedioic acid 491

Caesium 347
Caesium chloride structure 145
Calcium
  carbonate 358–9
  hydroxide phosphate 344
  hydroxyapatite 344
  reactions 356–7
  sulphate 359
Calomel electrode 267–8
Calorimeter 194–5
Cancer 396, 457
Cannabis 712
Car 503–4
Carbanion 488
Carbocation 488
Carbohydrates 607–8
Carbon 425–31, 479
  allotropes 83, 101, 141–2, 426–7
  bonding 426, 479, 510, 528
  cycle 431
  dioxide 98, 121, 181, 229, 429–33, 435–8
  halides 428
  hydrides 428
  monoxide 429–30, 499, 503–4
  special features 426
  reactions 425–6
  structure 83, 101, 141–2
Carbon-14 dating 15
Carbonates 251, 358, 430–2
Carbonyl compounds 575, 590–608
  concept map 608
Carboxylic acids 485–6, 631–58
  derivatives 643–53
    concept map 657
  nomenclature 632–3
  preparation 637–8
  properties 633
  reactions 639–42
  reactivity 635–6
  salts 635, 639
Carothers, W 670
Catalysis
  by transition metals 292–3
  effect on equilibrium 77
  effect on reaction rate 290–4
  heterogeneous 292–3
  homogeneous 291–2
Catalytic
  converter 293, 504
Catenation 426
Cathode 85, 235–6
  rays 6
Cathodic protection 274, 467
Cation 85–8
  exchanger 177
  nomenclature 71
  polarising power 105–6, 330–1, 335
Cell
  chemical 266, 274–6
  Daniell 276
  diaphragm 352
  dry 275
  electrolytic 233
  electrochemical 274–6
  emf 266
  fuel 276
  galvanic 274–6
  membrane 352
  voltaic 274–6
Cellulose 608
Centrifugation 683
CFCs 395–7, 548–50
Chadwick, J 8
Chain reaction 17, 396, 502
Charles' law 152
Charge density 106
Charge on ion 88, 93–4
Chelate 448
Chemical bond 83–119, 137–8
Chemical cell 266, 274–6
Chemical equilibrium 76–7, 219–32
  concept map 230
Chirality 124–5, 493
Chlor–alkali industry 346, 352–5
Chlorate(I) 350, 352, 382, 388
Chlorate(V) 382, 388
Chlorides 330, 384–6
Chlorination 533, 535, 540, 572, 601
Chlorine 374–89
  manufacture 352
  preparation 380
  reactions 378, 382–3, 511, 533, 535, 540, 572, 601
Chlorofluorocarbons 395–7, 348–50
Chloroform 545, 601
Chromates 452–3
Chromatography 174–7
Chromium 444
  complexes 457, 459–60
  steels 466
Chromophore 624
Cisplatin 457
cis–trans Isomerism 122, 459, 491, 509
Close-packed structures 143–4
Coal gasification 407
Cobalt 444
  -60 15
  complexes 458
Cocaine 711
Codeine 713

Collision
theory 287–90, 307–8
Colour
spectra 22–3
indicators 252
transition metals 449
Combining volumes of gases 57–8, 154
Combustion 189, 191, 195–6
Comparison of
aldehydes and ketones 594
alkenes and carbonyl compounds 593, 605
arenes and cycloalkenes 536
fluorine and other halogens 376–7
Group 4 425
halogenoalkanes and halogenoarenes 558
lithium and magnesium 355
metals 446
Complex
chelate 448, 459
colour 449
ion 107, 258, 447–9
isomerism 60, 456–62
nomenclature 448
shapes 456–62
Complexometric titration 258, 448
Concentration 58–9
effect on equilibrium 220–2
effect on rate 284, 296–7, 299–305, 311–3
Concept maps
Acid-base equilibria 259
Alcohols and phenols 586
Amines 627
Atomic structure 42
Carbonyl compounds 608
Carboxylic acid derivatives 657
Chemical kinetics 314
Equilibrium 230
Extraction of metals I 363
Extraction of metals II 473
Gases 158
Group 1 and Group 2 360
Group 2 361
Group 4 439
Halogenoalkanes and halogenoarenes 560
Infrared spectrometry 693
Liquids 178
Mass spectrometry 698
Nuclear magnetic resonance spectrometry 705
Polymers 678
Quantitative chemistry 78
Redox equilibria 277
Relationships between alcohols and other series 587
Solutions 185
The chemical bond 116
The halogens 389
The Periodic Table 336
The shapes of molecules 133
The structures of solids 146
Thermochemistry 214
Transition metals I 450
Transition metals II 472
Condensation polymerisation 669–75
Conduction band 423
Conjugate acid–base pair 243
Conjugated double bonds 131
Conservation of energy 188
Constant-boiling mixture 170–1, 567
Constant
Avogadro 53
dissociation 246–8, 636
equilibrium 220–1
Faraday 234
Planck's 24
rate 299
Universal Gas 155
Contact process 401
Continuous spectrum 22
Convergence frequency 26
Coordinate bond 106–8
Coordination number 139, 448
Copper 469–70
complex ions 107, 461
compounds 470
purification 240, 469
reactions 470
Coupling reactions 624
Courtelle® 518
Covalent bond 97–9, 103–7
arrangement in space 120–132, 128–30
multiple 88–9, 426
polar 105, 325
radius 103
Covalent compounds 99–106
Covalent substances 99–102
Cracking 293, 499
Critical
mass 17
temperature 157
Crookes tube 6, 340
Crude oil 495
Cryolite 368, 371
Crystal structure 91–2, 137–9
Cubic close-packing 143–4
Curie, M and P 13
Cyanhydrins 599
Cyclohexene 484, 536

d block elements 326, 442–72
d orbitals 30
Dacron® 669
Dalton, J 3
Dalton's law of partial pressures 155–6
Daniell cell 275
Davy, H 87, 233
DDT 374
de Broglie, L 27
Debt for nature exchange 438
Decarboxylation 641
Decay, radioactive 13–16
Decreasing the length of a carbon chain 715
Deforestation 437
Degenerate orbitals 449
Dehydrating agent 403
Deioniser 177
Delocalisation energy 132
Delocalised electrons 130–2
Dental caries 344–5
Detection of elements 683
Detergents 648–9
Depressants 713
Diamond 83, 101, 141, 426
Diaphragm cell 352
Diazepam 714
Diazonium compounds 621, 623–6
Dichromate(VI) 66, 73, 453, 575, 595, 719
Diesel fuel 497
Diffusion 152, 157
Dilution 61
2,4-Dinitrophenylhydrazine 600–1, 721
Dioxin 585
Dipole–dipole interaction 108
Dipole 108
Disaccharide 607
Discharge of ions 85–7, 234–5, 238–9
Disorder 209–10
Disproportionation 70, 382, 470
Dissociation 246–8
Dissociation constant 246–7
Distillation 163–4, 683
fractional 168–9, 497
steam 171
under reduced pressure 164
Distribution 172–4
coefficient 173
of molecular energies 287–90
DNA 115
Domagh, G 631
Donor atom 106, 424
Dopant 424
Double bonds 98, 128–9, 426, 509
Double helix 115
Drug
abuse 714
addiction 710
dependence 710
tolerance 710
Dry cells 275
Ductility 137
Dyes 612, 624
Dynamic equilibrium 74–7, 219–32
Dynamite 579

edta 258, 461
Effective nuclear charge 324
Efflorescence 353
Ekasilicon 33
Electricity
conduction of 85–7, 137–8
Electrochemical
cell 266, 274–6
series 238–9
transport 264–5
Electrochemistry 233–63
Electrode 6, 85, 234
calomel 268–9
glass 271–2
hydrogen 267–8
potential 265
potential, standard 267

processes 85–6, 239
standard 267–8
Electrolysis 85–8, 233–41
Electrolyte 85, 233
strong 244–5
weak 88, 246–8
Electrolytic
cell 233
conduction 233
extraction 239–40
Electromotive force, emf 266
Electron
acceptor 106
affinity 26, 203–4, 376
charge 6, 9
cloud 103
delocalised 130–2, 138
density map 92, 103, 105
donor 106
hole 424
localised 130
spin 28
transitions 22–5
wave nature 27
Electronegativity 104–5, 325
Electronic
configuration 30–31, 40
shell 30
subshell 30
Electrons in boxes 30–31
Electrophile 488
Electrophilic
addition 512–4
reagent 488
substitution 533–6
Electroplating 240
Electrovalent bond 89–93
Elements, repeating pattern 32–7
Elimination reactions 487, 552, 557
Eluant 174
e/z 10–11, 14
Emission spectrum 22–3
Empirical formula 55, 684
Emulsion 649
Enantiomer 124, 459, 493
Endothermic
compound 199
reaction 190
End-point 63, 250–1
Energy 188
activation 213, 288–90, 306–10
distribution 289
free 211–13
ionisation 25–6, 203, 325
kinetic 188
lattice 203
profile 309–10
sources 186
potential 188
Enthalpy 188, 190–213
change 190–3
diagram 192, 193, 198
profile 309–10
standard: *see* Standard enthalpy
Entropy 209–11
Enzyme 291–2, 493, 626
Epoxy resins 673
Equation of state 153
Equations 48–51, 56–8, 65–7
Equilibrium 74–7, 219–232
acid–base 242–57
concept map 259
chemical 76–7, 219–30
concept map 230
complex ions 258
constant 220–2, 223–4
K_c 220–2
K_p 226–7
dynamic 75–6, 220
heterogeneous 221
homogeneous 221
Law 222
phase 228–9
position 222
redox 228, 264–80
concept map 277
Esters 646–50
preparation 639–40
reactions 646–8
Esterification 221, 225, 646–50
Ethanal 592, 595
Ethanamide 634
Ethane 480, 501–3
Ethanedioate 66–7
Ethanedioic acid 66, 633
Ethane-1,2-diol 578
Ethanoic acid 632, 636
Ethanoic anhydride 645
Ethanol 564–5, 568–71, 576
Ethanenitrile 634
Ethanoyl chloride 634, 643, 645
Ethene 122, 128–9, 482, 507, 509–17, 663
Ethers 574
Ether extraction 172–4, 683
Ethoxyethane 173, 574
Ethylamine 613
Ethyl ethanoate 221, 225–6, 572, 634
Ethyl hydrogensulphate 574
Ethyne 521
Exclusion Principle 30
Exothermic reaction 189
Extraction of metals 362
concept maps 363, 473

Face-centred cubic packing 143–4
Fajans' rules 105–6
Faraday 85, 87, 233–4
Faraday constant 234
Fats 648
Feasibility of reaction 187–8, 212
Fehling's solution 598, 722
Fermentation 568
Ferromagnetism 447
Fertilisers 411, 418–20
Fibres 663, 670–1, 676
Fireworks 21
First Law of Thermodynamics 188
First-order reaction 302–3
Fission
heterolytic 488
homolytic 306, 487
nuclear 17–18
Flavours 590
Flue gas desulphurisation 407
Fluoridation 344–5
Fluorides *see* Group 7
Fluorine *see* Group 7
Fluothane® 545
Forensic science 12, 692, 697, 705
Formica 674
Formula 55
empirical 55, 684
molecular 55, 684
structural 489, 684
unit 91
Formulae of ionic compounds 94–5
Fractional distillation 168–70, 496
Fragmentation pattern 696–7
Fossil fuels 186–7
Free
energy 211–2
radicals 17, 396, 487, 502
Freons 388, 549
Friedel–Crafts reactions 531–6, 596
Fructose 607
Fuels 186–7, 497, 498–500, 503–4
Fuel cells 276
Fullerene 682
Functional group 484–6
isomerism 485

Galvanic cell 266, 274–6
Galvanising 466, 471
Gamma rays 14
Gas
concept map 158
constant 155
ideal 153, 156–7
kinetic theory 156–7
laws 152–4
liquefaction 157
molar volume 57, 155
non-ideal/real 157
partial pressure 155
Gaseous diffusion 152, 157
Gasohol 570
Gasoline 497
Gay-Lussac, J L 152
Gemstones 449
Geometrical isomerism *see cis–trans* Isomerism
Germanium 425
Giant molecule 101, 140–2
Glass 433
Glass electrode 271–2
Glucose 607
Glycerol 578, 668–9
Glycine 654
Graphite 83, 142
Greenhouse effect 435–8, 549
Grignard reagents 559–61
Ground state 24
Group 0 340–1
Group 1 33–6, 346–55, 360–7
concept map 360
Group 2 33–6, 347–9, 356–7
concept maps 360–1

Group 3 33–6, 368–73
Group 4 33–6, 423–41
concept map 439
Group 5 33–6, 410–22
Group 6 33–6, 392–409
Group 7 33–6, 374–91

Haber, F 219
Haber process 219, 413
Haematite 462
Haemoglobin 120, 461
Hahn, O 17
Half-reaction equation 65–7
Half-life 15, 302–3
Halides 330, 332, 334, 384–6
Hall–Héroult cell 368, 371
Hallucinogens 712
Haloform reaction 601–2
Halogens 33–6, 374–91
bond formation 376–7
compounds 384–8
concept map 389
manufacture 352, 379
preparation 380–1
properties 375
reactions 378
uses 388
Halogenation of alcohols 572–3
carbonyl compounds 601
of alkanes 500–3
of alkenes 510
of arenes 533, 535, 540
Halogenoalkanes 486, 500–3, 545–57
concept map 560
nomenclature 546
preparation 548
properties 546–7
reactions 551–8
reactivity 551
uses 545, 548
Halogenoarenes 558–9
compared with halogenoalkanes 558
concept map 560
reactions 559
reactivity 551
Halothane® 545
Harcourt–Esson reaction 313
Hashish 712
Heat and energy 188–9
and temperature 194
capacity 194
of combustion 195
of reaction 191
Helium 341
Helix 114–15
Heptane 480, 499
Heroin 713
Hess's law 196–7
Heterogeneous catalysis 292–3
Heterolysis/heterolytic fission 488
Hexacyanoferrate ions 460
Hexagonal close-packing 143
Hexane 480
Hodgkin, D 136
Hofmann degradation 619, 651, 715
Homogeneous catalysis 291–2
Homologous series 480
Homolysis/homolytic fission 306, 487
Huffmann, D 682
Hund's multiplicity rule 38
Hybridisation 127–9
Hydration
of ions 107, 207–8, 344
Hydrazine 600
Hydrazones 600
Hydrides 331
Hydrocarbons 480
alicyclic 484
aliphatic 483
aromatic 484, 525–7
nomenclature 481–4
unsaturated 482
Hydrogen 342–5
atomic spectrum 23–5
bond 111–15
electrode 267–8
ion 238, 242–57
ion concentration 244–8
manufacture 240, 342
molecule 103
reactions 343, 510, 528, 720
uses 342
Hydrogen bromide 385–6, 511
Hydrogen chloride 97, 385–6, 511
Hydrogen cyanide 121, 598–600
Hydrogen fluoride 111, 385
Hydrogen iodide 219–20, 226, 511
Hydrogen peroxide 398
Hydrogen sulphide 397
Hydrogenation 510
Hydrolysis
of salts 256, 371–2
of organic compounds
acid chlorides 643
amides 650
anhydrides 644
esters 221, 647–8
fats 648–9
halogenoalkanes 552–7
halogenoarenes 558–9
nitriles 652
Hydrophilic group 649
Hydrophobic group 649
Hydroxide ions 244–5, 721
concentration 244–8
2-Hydroxybenzoic acid (salicylic acid) 645
2-Hydroxypropanoic acid (lactic acid) 493

Ice 113, 229
ICI 346, 355
Ideal gas 153
Ideal gas equation 153
Increase length of carbon chain 715
Indicators 249–52
Induced dipole 109–10
Inert gas 34, 340
Inert pair effect 425
Infrared
radiation 435, 688
spectrometry 688–95
concept map 693
Initial rate 296, 300
Initiation 502
Inorganic nomenclature 71–3
Insecticides 374, 627
Intermolecular forces 99–101, 108–115
Internal
combustion engine 498–9, 503–4
energy 190
Intramolecular bonds 97–9, 100
Iodides 384–6
Iodination 311–12
Iodine
extraction 379
reactions 67, 378, 720
structure 100, 140
structure of halides 125
Iodoform reaction 607, 720
Ion exchange 177
Ionic
bond 89–93, 105, 376
character of bonds 105
compounds 94–6, 101–2
product of water 244
radius 323–4
structures 145
Ionisation
enthalpy/energy 25–6, 31, 96, 325
Ions 84–97
complex 258, 447–9, 456–62
discharge of 238–9
formation of 88–93
hydration of 107, 108, 344, 371
Periodic Table 94, 732
Iron 442, 460–9
extraction from ore 462–3
complexes 460–1
compounds 467–9
reactions 227, 467–9
rusting 273–4, 466
Iron(II) 65–6
Isomerism 489–93
chain 490
cis–trans 122, 459, 491, 509
functional group 490
optical 125, 459, 492–3
positional 490, 508
stereo- 125, 456–60
structural 489, 491–4, 508
Isotope 9, 15–16
radioactive 15–16, 76
Isotopic tracer 15–16, 76
IUPAC nomenclature 71–3

K_a 246
K_b 246
K_c 225
K_p 226
K_w 244
Kelvin temperature 153
Kerosene 497
Ketones *see* Aldehydes and ketones
Ketose 607
Kinetic
control 213

energy 188
theory 156–7
Kinetics 281–318
concept map 314
Knocking 499
Kratschmer, W 682
Kroll process 452
Kroto, H 682
Krypton 341

Labelling 16, 76, 639–40
Lactic acid 493
Laminated materials 674
Lassaigne test 683
Lattice
crystal 138–9
enthalpy 203
Law
Avogadro's 154
Boyle's 152–3
Charles' 152–3
conservation of energy 188
Dalton's 155
equilibrium 222
Gay-Lussac's 154
Hess's 196–7
ideal gas 153
Le Chatelier's 222, 401, 413
Raoult's 166, 169
thermodynamics 188
Layer structure 142
LD converter 464
Lead 427, 433–4
–acid accumulator 275–6
compounds 433–4
in petrol 388
reactions 425
Lead(II) bromide 87
Le Chatelier's principle 222, 401, 413
Levi, P 321, 465
Lewis acid–base theory 243–4
Ligand 447–8, 457–8
Light
absorption of 22
and reaction rate 285
emission of 21, 22
plane-polarised 492
Limestone 359
Line spectrum 22–3
Liquefaction of gases 157
Liquid crystals 147
Liquids 160–80
concept map 178
miscible 166–70
immiscible 171–4
saturated vapour pressure 163
Lithium 347, 355
Lithium tetrahydridoaluminate 598, 618, 641, 651, 720
Litmus 249, 252
Lone pair 123, 126
Lonsdale, K 525
LSD 712
Lubricants 497
Lyman series 23, 25

Macromolecule 101, 140–2
Magnesium
compounds 357–9
extraction 240
properties 347
reactions 356
uses 348
Magnesium fluoride 92
Magnesium oxide 93
Malleability 137
Maltose 607
Manganate(VII) 66, 72, 73, 453, 515, 637, 719
Manganese 444
(IV) oxide 719
Margarine 510
Marijuana 712
Markovnikov's rule 512
Marsden, R 6
Mass
and energy 17
number 9
spectrometer and spectrometry 9–12, 695–8
concept map 698
Matter 3
Mauve 612
Maxwell–Boltzmann distribution 288–9
Mechanism 488
Melting temperature 111, 322, 683
Mendeleev classification 33
Metallic
bond 137–8
properties 137
radius 145
structure 143–4
Methadone 713
Methanal 592
Methane 98, 123, 128, 480, 495
Methanoic acid 632
Methanol 565, 569, 570–1
Methylamine 613
Methylated spirit 570
Methylbenzene (toluene) 526
reactions 538–40, 637, 718
Methyl orange 250, 252
Millikan, R A 6
Mirror image 125, 459, 493
Miscible liquids 166–70
Mixed oxide 399
Mogadon® 714
Molar
concentration 58–9
mass 53–4
mass of a liquid 165
volume of a gas 154
Mole 52–3
fraction 155
Molecular
crystal 140–2
energy 156–7, 288–9
formula 55, 684
ion 11, 695
mass 11
orbital 127–32
shapes 120–32, 133
solids 140–2
spectrometry 685–705
structures 140–2
Molecularity 309
Molecule
chiral 124
giant 140–2
linear 121
polar 104–6
shapes 120–32
concept map 133
Monosaccharide 607
Morphine 713
Moseley, H G 8
MTB 570–1
Multiple bonds 98–9, 129–32, 412, 426
Multiplicity rule 38

Naming inorganic compounds 71–3
Naming organic compounds 481–4, 525–7, 632–3
Naphthalene 526
Narcotics 713
Natural gas 495
Neon 341
Neoprene 518
Nerve gas 627
Nerve impulse 626
Neutralisation 250–2, 257
Neutron 8
Newlands, J 32
Nickel 444, 465–6
complexes 457
Nicotine 711
Nitrates 351, 358, 417
Nitration 528–30
Nitrazepam 714
Nitric acid 416–17
manufacture 416
reactions 417, 528–30, 721
Nitriles 634, 652
Nitrites 417
Nitrobenzene 528–9, 617–8
Nitrogen
compounds 413–17
cycle 411
molecule 99
source 412
reactions 219, 412–13
Nitroglycerine 579
Nitrous acid 417
Nitryl cation 530
Nobel, A B 579
Noble gases 33–4
Nomenclature (ASE) 71
Nomenclature (IUPAC) 71, 480–6, 525–7
Nomenclature (Stock) 71
Non-electrolyte 84, 88
Non-ideal gases 157
Non-ideal solutions 169
Novocaine 712
Nuclear
binding energy 17
decay 13–16
equations 14

fission 17
magnetic resonance 699–705
concept map 705
reactions 12–18
reactor 18
Nucleic acids 115
Nucleon 8–9
Nucleophile 488
Nucleophilic addition 604–5, 593
Nucleophilic substitution 553
Nucleus 6–7
Nuclide 9
Nylon 670–1

Octadecanoic acid (stearic acid) 648–9
Octane 498
number 499, 570–1
Oils (animal and vegetable) 648
Optical isomerism 124–5, 459, 492–3
Orbital
atomic 29–30
delocalised 130–2
hybrid 127–32
molecular 103
Order of reaction 299–305
Orlon® 518
Ostwald, W 242
Osmotic pressure 200
Ostwald dilution law 259
Oxidants
conc. sulphuric acid 403, 721
definition 65, 269
dichromate(VI) 66, 453, 575, 595, 719
hydrogen peroxide 398
list 719
manganate(VII) 66, 453, 515, 576, 637, 719
nitric acid 417
Oxidation number/state 67–72, 335
Oxides 351, 329, 332–3
classification 329, 332, 399
Oxidising agent *see* Oxidants
Oxygenates 570
Oxonium ion 107, 242
Oxygen 392–400
allotropy 394
source 393
reactions 393
uses 392
Ozone 394, 515, 720
layer 394–7
Ozonolysis 515–6, 720

Parathion® 627
p block elements 326
p orbital 29, 128–32
Paper chromatography 175–6
Paramagnetism 447
Partial pressure 155
Partition chromatography 174–6
Partition coefficient and law 173
Paschen series 23, 25
Pauli exclusion principle 30
Pauling, L 104, 325, 639
Pauling electronegativity values 325
Peptide link 656
Percentage composition 55–6
Percentage of ionic character 104–6
Perfumes 590
Period 3
compounds 329–32
elements 327–9
Periodic Table 32–43, 321–36, 732
concept map 236
Periodicity 322–34
Perkin, W H 612
Perspex® 518
Pesticides 374, 627
Petrochemicals 496
Petrol 497
Petroleum oil 495
Pfund series 23, 25
pH
calculations 244–8
curves 250
measurement 271–2
meter 271–2
of acids and bases 244–8
of buffers 253–4
of salts 256
scale 247
Phase 228–9
diagram 228–9
carbon dioxide 229
water 229
Phenolphthalein 251, 252
Phenolic resins 672
Phenols 579–84
concept map 586
nomenclature 519
sources 580
properties 579
reactions 580–4
Phenoxide ion 581
Phenylamine 171, 612–13, 615, 617, 619–22
Phosphoric(V) acid 514, 569
Phosphorus 412
halides 573, 581, 640, 721
oxide 574, 721
Photon 24
Photochemical
oxidants 521–2
reaction 501
smog 521–2
Photosynthesis 189
Pi (π) bond 129–30
Pi (π) orbital 129–30
pK_a 246
pK_b 246
pK_w 244
Planck's constant 23
Planck's equation 23
Plane-polarised light 492
Plastics 662–679
biodegradeable 677
disposal 676–7
Platforming 500
Plexiglass® 518
Plumbates 434
pOH 244
Polar
covalent bond 104–5
molecule 104–5
solvent 208
Polarimeter 492
Polarisation 104–5, 492
Polarisability 105–6
Polarising power 105–6
Pollution
acid rain 405–7, 504
air 430, 521–2
car 503–4
eutrophic lakes 419–20
greenhouse effect 435–8
ground water 419–20
ozone layer 394–7
plastics 676
Seveso accident 585
Polymer 662–79
concept map 678
polyalkene 516–19
polyamide 670–2
polyester 669–70
polyester resins 670
poly(ethene) 666
poly(ethyne) 668
polypeptide 672
Polymerisation
addition 516–9, 665–8
condensation 669–676
Polysaccharide 608
Position of equilibrium 222
Potassium 347–8
dichromate(VI) 66, 453
manganate(VII) 66, 453
Potential
difference 265
energy 186–9
standard electrode 267
Power stations 406–7
Pressure
effect on equilibrium 223
kinetic theory 157
partial 155, 157
Primary standard 60
Principal quantum number 27, 30
Probability 27–8
Prontosil® 631
Propane 481, 498
Propanol 565
Propanone 311, 592
Propane-1,2,3-triol 578
Propene 507
Proteins 655–6
structure of 114
Proton
acceptor 243
donor 242
in nucleus 8
transfer 243
PTFE 517
Pulverised fluidised bed combustion 406–7
Purification 683
PVA 518
PVC 517

Quantisation 23
Quantitative analysis 685–705
Quantitative chemistry
  concept map 78
Quantum
  mechanics 103
  number 24
  theory 23
Quaternary ammonium salts 626–7
Quartz 141–2

Racemate 493
Racemic mixture 493
Radiation 13–14
Radioactive
  decay 13–16, 303
  element 13–18
  tracer 16
Radioactivity 13–18
Radioisotope 15–16
Radius
  atomic 323–4
  ionic 324
Radon 341
Ramsey, W 340
Raoult's law 166
  deviations 170
Rate
  and catalysts 286, 290–3
  and collision theory 287–90
  and light 285
  and particle size 283
  and pressure 285
  concentration dependence 284–5, 296, 299–305
  constant 299
  determining step 296, 311
  equation 299
  initial 296
  of reaction 281–318
  temperature dependence 285, 306–7
Rayleigh, W 340
Reaction
  bimolecular 296, 309
  chain 17, 306
  coordinate 309–10
  endothermic 189
  exothermic 189
  first-order 302
  free radical 306
  kinetics 281–318
  molecularity 296, 309
  order 300–5
  photochemical 306
  profile 288
  pseudo-first order 303
  rate 281–318
  reversible 76–7, 219–228
  second-order 304–5
  unimolecular 296, 309
  zero-order 302
Real gas 157
Rearrangement 487
Recrystallisation 183–4, 683
Redox
  electrode potential 265–70
  equations 65–7
  equilibria concept map 277
  half-reaction equations 66–7, 270
  reactions 65–7
  titrations 65–7
Reducing agent *see* Reductants
Reductants
  carbon monoxide 430, 462
  definition 65, 269
  extraction of metals 451
  hydrogen 597–8
  hydrogen peroxide 398
  list 720
  lithium tetrahydridoaluminate 597–8, 617, 641, 720
  sodium tetrahydridoborate 597–8, 720
  sulphur dioxide 400
  thiosulphate 67
  tin(II) compounds 433, 617
Reduction *see* Reductants
Reference electrode 267–8
Refining of oil 495–7
Reforming 500
Refrigerants 548
Relative
  atomic mass 11, 32, 34, 51
  formula mass 52
  molecular mass 11, 52
Restricted rotation 491, 509
Reversible reaction 219–228, 74
Rubber 667–9
Rust 273–4, 466–7
Rutherford, E 6–8, 14
Rutile 451

s block 326, 346–63
s orbital 29
Sacrificial protection 274, 467
Salicylic acid 645
Salt bridge 268
Salt hydrolysis 256, 372
Salts
  Group 1 351, 353
  Group 2 357–9
  Group 7 384
  nitrates 417
  Periodic Table 331, 334
  tin and lead 433
  transition metals 453–6, 468–9
Sandmeyer reactions 623–5
Saponification 648–9
Saturated
  hydrocarbon 482
  solution 182
  vapour pressure 163
Screening 324, 443
Second-order reaction 304
Seveso 585
Shapes
  of molecules 120–32
    concept map 133
  of orbitals 29–30, 127–32
Shell 30–1
Shielding 324, 443
Sidgwick–Powell theory 121
Sigma (σ) bond 129–30
Silanes 428
Silica 141–2, 427, 432
Silicates 433
Silicon 423–5
  compounds 428–9
  extraction 427
  (IV) oxide 141–2, 425, 427
  semiconductor 423–4
Silver mirror test 598, 722
Slag 463–4
Smog, photochemical 521–2
S_N1 and S_N2 555–7
Soap 648–9
Soda lime 641, 722
Sodium
  carbonate 351, 353–4
  chlorate(I) 239
  chlorate(V) 239, 382
  chloride 90–2, 145, 352, 354
  chloride structure 145
  extraction 239, 349
  compounds 350–4
  hydrogensulphite 599
  hydroxide 239, 351–3
  nitrate 417, 419–20
  nitrite 621, 720
  reactions 349–50
  sulphate 353
  tetrahydridoborate 720
  thiosulphate 67
  uses 348
Solids *see* Structure
Solubility
  and bond type 102, 108
  and enthalpy of hydration 207–8
  and hydrogen bonding 113
  and lattice enthalpy 208
  curve 182
Solution
  concept map 185
  enthalpy of 192, 207–8
  ideal 167
  of liquid in liquid 166–71
  of solid in liquid 182
  real (non-ideal) 169
Solvation 108
  enthalpy 208
Solvay process 346, 353–5
Solvent
  extraction 173–4
  partition 172–7
Sommerfeld, A J W 27–8
Space 392
Specific heat capacity 194
Spectrum
  absorption 22–3, 298
  emission 21–3
  hydrogen 23
  infrared 688–95
    concept map 693
  line 23
  mass 9–12, 695–8
    concept map 698
  nuclear magnetic resonance 699–705
    concept map 705
  visible–ultraviolet 687–8

Spin 28
Spontaneous change 209
Stainless steels 467
Standard
calomel electrode 269
conditions 191
enthalpy change 191–3
of atomisation 192
of combustion 191
of condensation 193
of delocalisation 202
of dissolution 192, 207
of formation 191, 198
of fusion 192
of ionisation 203
of melting 192
of neutralisation 191, 257
of reaction 191, 198–9, 201
of sublimation 203
of vaporisation 193, 203
electrode potential 268
entropy change 209–10
free energy change 211–13
hydrogen electrode 267
solution 59, 60–2
states 191
temperature and pressure, stp 153
Stannates 434
Starch 608
States of matter 151
Steam distillation 171
Steel 392, 464
Stereoisomerism 555–6
Stimulants 710
Stoichiometry 56, 73
Strontium 347
Structural formula 489, 684
Structure
determination 682–705
of metals 143–4
of solids 136–50
concept map 146
Subshell 30–1
Substitution 487
Substituents in benzene ring 541–2
Sucrose 607
Sugars 607–8
Sulpha-drugs 651–2
Sulphates 353, 359
Sulphites 400
Sulphonamides 631–2
Sulphonation 404, 530–1
Sulphur 393
dichloride oxide 640, 721
dioxide 122, 400
(VI) oxide 401, 721
Sulphuric acid 400–4, 528–31, 720–1
Sulphurous acid 400
Sun 186, 342
Superheating 164
Surface charge density 106, 372, 455
Swimming pool 388
Synthetic routes 715–18
System 74

Teflon 517
Temperature
effect on equilibrium 223
effect on rate 285, 306–7
Terylene 669
Tetrachloromethane 503, 548
Tetraethyllead 499, 505
Tetrahydridoaluminate(III) 598, 641
Tetrahydridoborate(III) 598
Thermal conductivity 137
Thermal dissociation 220
Thermochemistry 186–218
concept map 214
Thermodynamic
control 213
stability 213
Thermodynamics 186–218
and feasibility of reaction 213
and kinetics 213
laws of 188, 210
Thermosetting plastics 663–4
Thermosoftening plastics 663–4
Thiocyanate 460
Thiosulphate 73
Thomson, J J 6
Tie-line 167
Tin 425
(II) chloride 122
compounds 433–4
plating 466
reactions 425, 617, 720
Titanic 240
Titanium 444, 451–2
Titration
acid–base 62–5, 73–4, 250–3
complexometric 258
curves 265
redox 250–3
Titrimetric analysis 62–5, 73–4, 250–3
Tollens' reagent 589, 722
Toluene *see* Methylbenzene
Tooth enamel 344–5
Tracer 16, 76
Transition metals 34–5, 442–473
catalysis 447
colour 449
comparison 446
complex ions 447–9, 456–61
concept maps 450, 472
extraction 445
halides 454–5
oxides and hydroxides 450
properties 444–5
in gemstones 449
sulphides 456
uses 446
Transition state 308–310, 556
Trichloroethanal 601
1,1,2-Trichloromethane 601
Trichloroethanoic acid 248
Tri-iodomethane 602
Trimethylamine 613
2,4,6-Trinitrophenol 583
TNT 529
Trioxygen (ozone) 394–7, 515–6, 720
Triple
bond 521
point 228
Types of reaction 487

Ultraviolet
radiation 394–7, 549
spectrophotometry 686–7
Unit cell 143–4
Universal indicator 252
Universal gas constant 155
Unsaturation 482–3, 507
Uranium 17–18
Urea–methanol 674

Valence band 423
Valence bond method 99
Valence shell electron pair repulsion theory 121
Valium® 714
van der Waals
forces 108–10
radius 110
Vanadium 444
Vapour 162–3
Vapour pressure
and boiling temperature 163
and composition 167–8
effect of temperature on 163
partial 166
Vitamin C 639
Voltaic cells 266, 274–6
Volumetric analysis 62–5, 73–4, 250–3
Vulcanisation 667

Water 343
acid–base properties 242–4
anomalous properties of 111
as a solvent 112–13
dissociation constant 244
fluoridation 344
hydrogen bonding in 112–13
ionic product 244
molecule 107, 123, 242
phase diagram 229
structure (ice) 113
Wave theory 27–8
Weak electrolytes 88, 246–8

X ray diffraction 137, 525
Xenon 341
Xylocaine 712

Zero-order reaction 302
Zinc 444, 470–1
Zwitterion 654